Basic Modern Theory of Linear Complex Analytic q-Difference Equations

Mathematical
Surveys
and
Monographs
Volume 287

Basic Modern Theory of Linear Complex Analytic q-Difference Equations

Jacques Sauloy

AMERICAN
MATHEMATICAL
SOCIETY
Providence, Rhode Island

2020 *Mathematics Subject Classification.* Primary 39A13; Secondary 33D15, 34M50, 34M55, 05A30, 12H10.

For additional information and updates on this book, visit
www.ams.org/bookpages/surv-287

Library of Congress Cataloging-in-Publication Data

Cataloging-in-Publication Data has been applied for by the AMS.
See http://www.loc.gov/publish/cip/.

Mathematical Surveys and Monographs ISSN: 0076-5376 (print); 2331-7159 (online)
DOI: https://doi.org/10.1090/surv/287

♾ The paper used in this book is acid-free and falls within the guidelines
established to ensure permanence and durability.
Visit the AMS home page at `https://www.ams.org/`

10 9 8 7 6 5 4 3 2 1   29 28 27 26 25 24

This book is dedicated to Iannis.

Contents

Foreword

There are, if we are restricted to functions in a single variable, three big families of functional equations:

- the differential equations connecting a function f to its successive derivatives f', f'', $\dots$;
- the difference equations connecting a function f to its successive translates $f(x+a)$, $f(x+2a)$, $\dots$;
- the q-difference equations connecting a function f to its successive "dilatations" $f(qx)$, $f(q^2x)$, $\dots$.

They are related to homographies of the Riemann sphere and to their infinitesimal counterpart. I shall only speak of linear equations with coefficients that are polynomial or, more generally, holomorphic. The subject of the book by Jacques Sauloy falls in the third case, analytic linear q-difference equations.

Discretizations of differential equations give difference and q-difference equations with constant coefficients. One is then interested in their discrete solutions. But it is more interesting to consider their meromorphic solutions. We may for instance consider the solution $(n-1)!$ of the discrete equation $f(n+1) = nf(n)$, but it is much more interesting to look at the solution $\Gamma(x)$ of $f(x+1) = xf(x)$, for x real, like Euler, or better for x complex. In his book, J. Sauloy studies meromorphic solutions[1] of q-difference equations.

The history of the subject begins with q-calculus, based on q-deformations of integers like: $n \mapsto [n]_q = \frac{q^n-1}{q-1}$, and of functions: $f(x) \mapsto f(qx)$. Later q-special functions appear. In an enthralling historical prelude, J. Sauloy shows its birth and development from Fermat and Euler to Watson. The formalized notion of q-difference equation and the separation of q-calculus and equations appear later, in the work of Jackson in 1910[2]. In the case of differential and difference equations, the notion of operator had been extracted from the coating of computations by the alsacian mathematician Louis François Antoine Arbogast more than a century earlier, in 1800.

It is interesting to note that, even recently, most experts in q-calculus, q-orthogonal polynomials, q-special functions or q-combinatorics did not relate these questions to q-difference equations. Ramanujan does not speak of q-difference equations, but neither do Gasper and Rahman in the first edition [3] of their "bible" on q-hypergeometric functions (although the link was clearly established by Jackson

[1] The "great elders", in particular G. D. Birkhoff, also considered ramified solutions, but I proposed, around 1990, a change of paradigm by limiting oneself to meromorphic solutions.

[2] After, however, works on functional equations in France at the end of nineteenth century, in particular by Picard and Poincaré, but with a different point of view.

[3] The expression "q-difference equations" appears marginally in two places in the second edition.

in his work on Heine's q-series ...). Richard Askey signalled the importance of the point of view of q-difference equations, but rather tardily. In this book, J. Sauloy shows how that point of view allows one to transform the jungle of q-calculus and q-special functions into a harmonious garden *à la française.*

When I was a student in the Sorbonne in Paris, at the beginning of the sixties, our masters thought that the history of analytic differential equations had come to an end in the late twenties[4]. But there was a vigorous revival at the beginning of the seventies, with, in particular, a return to Birkhoff, whose works on the subject had been completely forgotten. New concepts and techniques appeared (or reappeared): Riemann-Hilbert correspondence, Newton polygons, Gevrey asymptotic expansions, k-summability, index theorems, Stokes multipliers, sheaf cohomology, differential Galois theory, density theorems .... An essential point is the treatment of irregular equations. Later a modern parallel theory of q-difference equations followed[5]. The goal of this book is to present the foundations of that theory. The book contains, in particular, the classification of *regular singular* equations by crossing analytic and Galois theoretic points of view. There also is an introduction to the irregular case, which, as in the differential case, is much more difficult. Thus appear the notions of q-asymptotics, summability and q-Stokes phenomenon[6].

Recently some q-difference equations (possibly nonlinear, like the q-analogues of Painlevé equations) appeared in a large number of works in theoretical physics[7]. For instance, theoretical physicists think that a "q-deformation" is involved when going from "two-dimensional conformal field theory" to the theory in dimension 3[8]. Some q-difference equations also appear in many works by Andrei Okounkov, in relation with various subjects: "enumerative geometry, string theory, mirror symmetry, and quantum gravitation".

Given the important development of q-difference equations in various domains of mathematics and theoretical physics, the lack of a basic work crossing various approaches was a huge lacuna. Everything was scattered in various works or research articles, including the "classical" basics[9]. This book perfectly fills this gap. I am convinced that it will become an essential tool for a whole community of researchers verging on the q-world in their work, and a very good entry point for students. It should establish itself over the years as the reference book on the subject.

—Jean-Pierre Ramis, Member of the French Academy of Sciences

---

[4]It is clearly the point of view of J. Dieudonné in his *Abrégé d'histoire des mathématiques.*

[5]The roots of the modern theories of differential and q-difference equations lie in great part in a 1913 article by Birkhoff who tackles the three "sister theories" of differential, difference and q-difference equations from a point of view which J. Sauloy calls Riemann-Hilbert-Birkhoff theory.

[6]The author announces that those subjects (and others ...) will be deepened in a companion work in preparation.

[7]One even encounters the q-Stokes phenomenon in some works.

[8]According to Beem, Dimofte and Pasquetti "the natural 3D analogues of the differential equations whose solutions determine the partition function in two dimensions are q-difference equations".

[9]There did not even exist for q-differences the analogue of Batchelder's 1930 book for differences.

Preface

In 2003, Jean-Pierre Ramis (of whom I had been a student), Changgui Zhang (who also was a student of Ramis), Lucia Di Vizio (who was a student of Yves André) and myself together wrote a short survey article on q-difference equations for *La Gazette des Mathématiciens*, a publication of the Société Mathématique de France. The theory of q-difference equations had, in the preceding years, known a revival under (notably) the impulsion of such as Jean-Paul Bézivin, Hidetaka Sakai, Marius van der Put, Michael Singer, Yves André, and mostly Ramis himself.

Following our article, Yves André suggested that a team of valiant involved mathematicians should write a modern book on the subject. So the same group, under the wise direction of Ramis, began to prepare it. After (I believe) two years of false starts and procrastination, the project was abandoned. The life of "enseignants-chercheurs" in France does not favour long-term projects. Moreover, as Hardy said, "young mathematicians should prove theorems, old mathematicians should write books". (It is true that in the meantime I produced a book, but it was a more or less direct adaptation of courses I had given on very classical matters.)

Four years ago, the beginning of my retirement and the COVID pandemic compelled me to resume the project. But diminution of task force (to 25%) and difference in temperament (see below) make the result quite different from what it would have been. An obvious reason is my narrower domain of expertise. The scope of the resulting book is certainly narrower to the same extent. Another reason is that, as *enseignant-chercheur*, or research professor, I probably have more professor and less researcher in me than do my worthy colleagues. So the book is certainly more didactic and less systematic that it could have been. I favour examples, motivation, analogies, historical reminders, exercises in numbers, redundancy[10] ... Besides, all along, I wrote this book with the" Graduate Studies in Mathematics" series in mind, until the AMS assessed that rather it should belong to the "Mathematical Surveys and Monographs" series.

Also, I am more attracted to tools and techniques than to problems. In some sense, in Marx classification, this book belongs to sector I (where the instruments of production are produced). Over the past two decades, I have been (unhappily?) encouraged in this direction by my activity as referee for many interesting articles which tried to use the nascent modern theory of q-difference equations but revealed misunderstandings of its basics, although the attempt at using it was plainly justified and the rest of the articles was valuable. The reader will easily spot other

[10]Curious readers may notice that in many places my presentation of facts is longer than in the original articles they come from. Let me quote Johannes Huebschmann in the December 2023 issue of AMS *Notices*: "The diligent reader will notice that the proof is complete, i.e., no detail is left to the reader. It is an instance of the common observation to the effect that *mathematics consists in continuously improving notations and terminology.*" (Emphasis in the source.)

biases of mine (no Cauchy integral, function theory almost entirely through power series, etc).

For all the shortcomings of this book I can only propose a compensation: a kind of "companion volume" with broader scope and at a higher level of sophistication will be produced within a reasonable timeframe (say two years), tackling important omissions (q-asymptotics and summation, q-special functions, q-Painlevé equations, various aspects of Galois and of moduli theory, phenomena related to nonintegral slopes, effective methods, solutions of q-difference equations *vs* solutions of differential equations ...); this will be a collective work by Shaoshi Chen, Yousuke Ohyama, Jean-Pierre Ramis, Julien Roques, Changgui Zhang, and myself.

There is another shortcoming of the book[11] for which I can only hope that future work (presumably by others) will bring reparation. During the decades I devoted to q-difference equation theory, I stumbled upon a mass of facts for which an underlying order could be guessed, and here, I have tried to make that order explicit. There are quite a few places[12] where I think more order can be expected.

Paying dues. It is time to thank those who helped so much: above all, Jean-Pierre Ramis, of course; my colleagues, friends, and coauthors Lucia Di Vizio, Charlotte Hardouin, Yousuke Ohyama, Julien Roques, and Changgui Zhang; my colleagues, friends, and workmates Anne Duval, Michèle Loday, Claudine Mitschi, Vadim Schechtman, and Joseph Tapia; and a special word for Yannis with whom it all began.

I should add the six anonymous colleagues who read the manuscript and made helpful suggestions; and, last but not the least, the AMS team!

Dedication. I dedicate this book to Yannis Varouchas (1950–2003), also known as Jean Varouchas. Of greek origin, he came to study in France in 1969, when I met him in the "hypotaupe" of André Warusfel. When we started (under the guidance of Georges Maltsiniotis) to meet "true math", he expressed a strong dislike of mathematics based on arrows (as he said), but later was reconciled with algebraic geometry by the book of Griffiths and Harris. As a devoted teacher, he despised the formalist tendency and said that after three years in University, French students only knew three functions: exp, log, and f. In the last years before his untimely death, he caught the q-disease, which was a happy event for me since it allowed us to meet anew.

Obituary. Bernard Malgrange (1928–2024) passed away on January 5th, just after I finished writing this book. The few lines (in French, but it does not seem to matter too much) received along with the announcement of his passing may be found on the next page, but no doubt much more information will be published in the coming months.

Bernard Malgrange had a tremendous impact on the theory of complex differential equations, thus, indirectly, on the theory of q-difference equations. Quite a few essential results in this book were obtained trying to adapt work of Malgrange: this is of course true of the q-analogues of Birkhoff-Malgrange-Sibuya theorems, but also actually of many

[11] As for what the book is on and what it isn't, see more details in the Introduction.

[12] To name but a few: it seems to me that the mixing of *filtration by the slopes* and *graduation by the exponents* is not totally clarified by the use of the Grothendieck group in Chapter 10; the phenomenon of confluence (Chapter 11) could be proved more neatly using holomorphic functional calculus; and the determination of the fuchsian local Galois group in Chapter 13 could somehow be illuminated by the theory of decomposition of tensor products of irreducible representations of linear groups; last, the possibility to attach a vector bundle to each q-difference system (Chapter 7) is clearly underexploited. All this is beyond my forces !

things in Chapters 7 to 10. More recently, his contribution to nonlinear differential Galois theory received the beginning of an extension to q-differences in the work **[99]** by Anne Granier.

Malgrange never took a deep interest in q-difference equations, but, as a close friend of Ramis, he kept himself informed about them. He was the referee for my first articles and displayed a patience little deserved by them. His very professional attitude was a great help to me.

Né le 6 juillet 1928 à Paris, ancien élève de l'Ecole Normale Supérieure, Bernard Malgrange a été Attaché de Recherche au CNRS, Maître de Conférences et Professeur à l'Université de Strasbourg et Professeur à la Faculté des Sciences d'Orsay avant de rejoindre l'Université de Grenoble en 1969, d'abord en tant que Professeur et ensuite comme Directeur de Recherches au CNRS. Membre de l'Académie des Sciences depuis 1977, récipiendaire de nombreux prix pour ses travaux profonds sur les équations aux dérivées partielles, la géométrie différentielle, les singularités de fonctions analytiques et la théorie algébrique des équations différentielles, il a participé très largement au rayonnement de l'Institut Fourier dans le monde mathématique. Le théorème de préparation différentiable restera pour toujours associé à son nom.

Introduction

0.1. General orientation of this book

The greatest single impetus in the theory of q-difference equations was probably Birkhoff's article **[23]**. After some sleepy (though not totally void) decades, the theory experienced a revival at the very end of the previous millenium. This book revolves around one of the active threads of that revival during a quarter of a century.

0.1.1. Scope. There are many excellent books and survey articles about q-special functions (and also some about q-calculus and about difference algebra). This book is not one of them. It stands, in relation to the former, in more or less the same position as would a book on (linear, complex, analytic) differential equations with respect to a book on (classical[13]) special functions.

A theoretical approach is pursued. Functional equations are a powerful way to understand functions, and they in turn can be understood through underlying structures (more on this in 0.1.3). We shall try to detail and explain those structures as clearly and explicitly as possible, with many examples and exercises of application; however neither numerical nor symbolic algorithms will be displayed.

0.1.2. Origins and specificities of q-difference equation theory. While the Big Bang of q-difference equation theory is **[23]**, the history of q-calculus goes back at least to Euler, maybe even to Fermat (see Chapter 1). Apart from an untimely forerunner in Euler (the "fundamental trick", see 1.2), the first conscious use of q-difference equations seems to be due to Jackson **[126]** as a way to study Heine's basic hypergeometric series. Yet, according to Birkhoff's historical article **[24]** (see in particular p. 639), the Founding Fathers are Carmichael, Adams, Birkhoff himself, Trjitzinsky, ...

Although in **[23]** Birkhoff puts "the three sister theories" of differential, difference, and q-difference equations as far as possible on the same ground, he admits (p. 640 of **[24]**) that the latter is simpler. This is true and important and it can be understood at least in two ways:

- Any nontrivial analytic automorphism of the Riemann sphere is a Mobius transformation, thus conjugate either to the translation operator $x \mapsto x+1$ (if it has a single fixed point) or to some dilatation operator $x \mapsto qx$, $q \neq 0, 1$ (if it has two fixed points). The former case is clearly a degeneration of the latter (the fixed points merge).
- Now look at the action of said automorphism on $\mathbf{S} \setminus \{\text{fixed point(s)}\}$. The quotient of $\mathbf{C}$ by the action $x \mapsto x + 1$ is a cylinder, an open Riemann

[13]Throughout his book, I will generally use the word "classical" to qualify concepts and techniques related to ordinary differential equations, as opposed to q-difference equations.

surface, while the quotient of $\mathbf{C}^*$ by the action $x \mapsto qx$ is a torus, a compact Riemann surface. The latter is obviously an easier playground for complex analysis.

If we prefer a comparison with differential equations, we note that the relevant operator is an infinitesimal automorphism, the notion of fixed point is unclear; and the closest we get to a quotient is the "Malgrange circle of directions" S^1 (see 15.3.3 for an explanation), not even a Riemann surface, thus not such a nice playground for complex analysis.

Compared to q-difference theory, plain difference theory is much more complicated and we shall have nothing to say about it (look at **[23, 75, 122, 249]** to get a feeling of how much more complicated it can be!). However, if we rather compare q-difference theory to differential equation theory, it offers two striking properties which make life easy:

- The domain of definition of local solutions can usually be expanded through q-dilatations and most equations admit *uniform* solutions over the whole of $\mathbf{C}^*$, which is of course not to be dreamed of in classical theory; see for instance sections 5.3 and, around Praagman's theorem, 7.3.
- As for the *local* study, all analytic q-difference operators admit an *analytic* factorisation, while in the differential or difference case only a *formal* factorisation is possible. We call this fact Adams's lemma, see 5.3.2, although it seems to have been independently rediscovered by Birkhoff and Guenther, **[25]**, see Chapter 15.

Yet, classical theory of (linear, complex, analytic) differential equations is a reliable guide (we Frenchmen say "Fil d'Ariane", which translated into English can be rendered as "bread crumbs") in that new maze, while trying to stay conscious at all times of similarities and dissimilarities. A notable thread originating in classical theory is (under any of its thousand guises) Riemann-Hilbert correspondence (see Chapter 12), which we shall rather call Riemann-Hilbert-Birkhoff correspondence since Birkhoff has renewed it so much in **[23]**. In that respect, I must mention that the frequent reference to my book **[222]** just means that I presented there very classical results (none of them due to me!) but in a form that suits rather closely the points of view of the present book.

EXERCISE 0.1. One parallel is that differential equation theory is additive while q-difference theory is multiplicative. Try to find examples of that in the text (definition of basic functions, of exponents, properties of slopes, Fuchs relation ...).

0.1.3. Technical bias.

> Thus, after due respect has been paid to the algebraic verifications in their various shapes, I am still convinced against all Puritan doctrines that the analytic method is the least artificial, affording the deepest insight and best in keeping with our program: to solve concrete problems by means of general ideas which shed light upon a much wider range of mathematical facts that were needed for our immediate purpose.
>
> —Hermann Weyl, *The Classical Groups*, Chapter VII, §7.

As can be guessed from the previous considerations (and the above quotation), we favor a wholly transcendental approach. The course will involve only *complex analytic linear q-difference equations* and for them it seems that the strongest results have been obtained through function theoretic methods[14], appropriately seasoned with an adequate proportion of algebra.

A fundamental restriction is that $|q| \neq 1$. The case $|q| = 1$ is much more complicated and not so developed (see however for instance **[60]** for a significant advance and, for example, **[237]** for the usefulness of that case). We avoid the arithmetic theory **[58, 59]** and the more algebraic approach of **[249]** based on Picard-Vessiot theory (this, however, should be fixed later—see herebelow 0.1.4). To make our aforementioned restriction more precise, we shall require:

$$0 < |q| < 1.$$

The case $|q| > 1$ boils down to the previous one through easy procedures which will be commented on later in the course of the text.

Although the Founding Fathers used multivalued functions, we shall take advantage of the possibility explained above in 0.1.2 to totally avoid them and work with uniform functions. This idea originates in the programmatic text **[190]** by Ramis (itself related to a suggestion of Birkhoff at the end of **[23]**). (More will be said on that choice in 12.1.1.)

When tackling global problems, the base field will be $\mathbf{C}(x)$. When tackling local matters, it will be either $\mathbf{C}(\{x\})$ or $\mathbf{C}((x))$. It is possible to work over rings like $\mathbf{C}\{x\}$, $\mathbf{C}[[x]]$, or $\mathbf{C}[x, x^{-1}]$ but this leads to unnecessary algebraic complications (torsion modules, etc) in our setting—although it is plainly appropriate[15] when going beyond one variable as, for instance, Sabbah does in **[214]**. However, it is true that our choice to avoid q-difference rings or algebras sometimes leads to overelaborate explanations.

We have founded the algebraic part of the theory on q-analogues of differential modules (for us: q-difference modules), and, to a lesser extent, of $\mathcal{D}$-modules (for us: $\mathcal{D}_{K,q}$-modules): see Chapter 8. We did not find it necessary to introduce "discrete connections" of the kind used, for instance, by Tarasov and Varchenko in **[243]**.

0.1.4. A projected "Companion volume". Given all the important topics that I fail to address, we decided (along with a few colleagues working in the domain) to prepare a kind of "Companion volume" (in the sequel, to be referred to as [CV][16]) at a higher level of sophistication and tackling the following matters:

(1) Asymptotics and summation theory for irregular equations (here barely touched in Chapters 14 and 15).
(2) Important q-special functions, notably:
 - q-hypergeometric functions,
 - Rogers-Ramanujan continued fractions.
 - orthogonal polynomials,

 Note that the first two items do appear in the present volume at various places as illustration (and we initiate a study of the first in Chapter 4).
(3) q-Painlevé equations, from the analytic and the geometric points of view.

[14]There is a tradition originating, it seems, in physics which considers difference and q-difference equations as some kinds of recurrence relations; see e.g., **[184]** or **[92]**, and **[131]**. Here, we stick to the functional interpretation and enjoy the full power of function theory.

[15]A serious treatment of holonomy also requires base rings or algebras, so our baby study in 3.2.4 is really a low-level one.

[16]This book will be entitled: *Advanced Modern Theory of Linear Complex Analytic q-Difference Equations*

(4) Local classification and Galois theory for irregular equations with arbitrary slopes (in Chapters 14 and 15, I consider equations with integral slopes only; the general case is more complicated).
(5) Comparison of the present function theoretic approach to Galois theory with the more algebraic Picard-Vessiot approach.
(6) Effective methods, procedures and algorithms in resolution, classification and Galois theory.
(7) Simultaneous solutions of q-difference equations and of differential equations, hypertranscendence, simultaneous solutions of q-difference equations for more than one q.

0.2. Contents

Broadly speaking, the book has three parts. Chapters 1 to 4 are meant as an initiation to the q-difference world. Chapters 5 to 10 are the technical heart, they display the tools of the trade. Chapters 11 to 15 aim at substantial applications. (Of course, the distinction is not so clear-cut: the last five chapter also display important tools and Chapter 4 already tackles significant applications.) There are appendices (see Subsubsection 0.3.1.1). At the end of this introduction (from Section 0.3 on) we provide a practical guide to the book, including organization, prerequisites, notations, conventions, etc.

0.2.1. Initiation. Chapter 1 contains a series of historical examples predating the real period of q-difference equations; so, in some sense, they are prehistorical. They are mainly intended to get the reader accustomed to that "Brave New World" and its rich flavors; but they do come with explicit calculations and arguments.

Chapter 2 introduces the "basic bricks", i.e., the most elementary q-special functions needed to solve all q-difference equations. They rest on classical special functions, for which reminders are provided in Appendices A (on classical special functions) and B (on Riemann surfaces). We also begin there an initiation to the role of sheaves and vector bundles in q-difference equation theory (see the reason why further below 0.2.4).

Chapter 3 introduces elementary calculation techniques from difference algebra and from q-calculus.

Chapter 4 addresses equations of order 1 (in this way, it is but an extension of the previous chapter) and equations of order 2; and there, it already involves significant applications, like basic hypergeometric series. In order to emphasize the specificities of the latter, Appendix C summarizes the basics on classical hypergeometric functions. We continue with the initiation to the role of sheaves and vector bundles (the real thing will start in 7.3 and in 9.3). This chapter is also where the so-called "q-de Rham complex" appears for the first of quite a few times, although it will never receive a true general definition.

0.2.2. Tools of the trade. Chapter 5 adapts the classical methods inherited from the Founding Fathers to our framework based on uniform functions (in this way, it rests on Chapter 2). It also takes advantage of algebraic methods from differential and difference algebra (noncommutative polynomials) to display our first important theorem, Adams's lemma; and to introduce in his first avatar an essential tool, the Newton polygon.

Chapter 6 originates in the work of Bézivin and the work of Ramis. It is more "modern" than Chapter 5 in that it replaces resolution by the measure of obstructions to the possibility of resolution. It uses some functional analysis and very elementary homological methods, all recalled in Appendix D.

Chapter 7 displays the classical method of vectorization of scalar equations, well known in the study of differential equations. It is probably due (once again!) to Birkhoff. The complete understanding of relations between scalar equations and systems also requires some elementary homological methods, summarized in Appendix E. It is also the right place to start to systematize the use of sheaves of solutions and the attached vector bundles (which have, actually, already been touched upon in special cases).

Chapters 8, 9, and 10 are, in my opinion, the technical heart of our toolbox. First we start the heavy use of linear methods in Chapter 8 (modules over noncommutative rings), and this requires some linear and multilinear algebra; see Appendix F. We complete it with a more detailed algebraic study in Chapter 9. Then we apply it to elucidate the formal and analytic structure of (complex, analytic, linear) q-difference equations in Chapter 10.

0.2.3. Applications to "the real world". As in the classical case, there is a first division of the world between *regular singularities* (when solutions have moderate growth; they come here under the heading of *fuchsian equations and systems*) and *irregular singularities.*

In our framework, the only possible local studies are at 0 and at ∞ (which boils down to the previous one); this is explained in Chapters 11 (local study of fuchsian systems) and 12 (global study), where the original contribution of Birkhoff [**23**] is modernized, systematized, and extended. This is then completed by the Galois theory in Chapter 13, where we begin to require some more sophisticated tools (Tannakian duality) to be recalled in Appendix G. Chapter 12 also involves an opening toward a nonlinear theme, that of q-Painlevé equations, which will be more seriously adressed in [CV].

In Chapter 14 we treat irregular *scalar equations.* The scalar form is appropriate for the finer analytic theory, but sadly we can only touch upon it here. Hopefully this will be compensated by [CV]. Like the following one, this chapter uses some cohomology of sheaves, which is summarized in Appendices H (abelian cohomology) and I (nonabelian cohomology).

Chapter 15 adresses irregular *systems*, which are an appropriate framework for classification and Galois theory, so we again require information from Appendix G; but classification also requires sheaf-cohomological methods from Appendices H and I.

The role of the appendices will be explained in 0.3.1.1.

0.2.4. Why the fuss about vector bundles and sheaves. Theorems on factorisation of holomorphic matrices (by Hilbert, Plemelj, Birkhoff, Cartan ...) have been for a long time associated with the study of functional equations. Röhrl [**205**] formulated and solved a Riemann-Hilbert problem along those lines in terms of vector bundles (see also [**88**, Chapter 3 §31], [**182**], [**249**, 12.3.1], [**222**, Chapter 12, Theorem 12.8]). So the use of holomorphic vector bundles here seems unavoidable. On the other hand, it seems to me that their elementary study is easier in terms of the associated locally free sheaves, since the formalism of sheaves offers so much flexibility: whence Sections 7.3 and 9.3 and Appendices H and I.

0.3. Some practical tips

The numbering of chapters, sections, subsections, and subsubsections is digital: Subsubsection 5.1.2.3 appears in Subsection 5.1.2, which is part of Section 5.1 within Chapter 5. Moreover the reference system is absolute, not local: wherever you are in the book Subsection 5.1.2 is referred as 5.1.2. References beginning by a letter, like C.1.1, lead to an appendix.

0.3.1. Organisation and *mode d'emploi*.

0.3.1.1. *Prerequisites.* The main prerequisites are general algebra and analysis as they are taught in the first three years of University (or College if you prefer). However, specific higher-level math is needed at some places, most of the time at the level of fourth year. I expect that each particular reader has been confronted with some of them, but not all.

Therefore, each time it is so, I have tried to summarize it in the form of an appendix, as I usually do when delivering the corresponding graduate course. The appendix is meant to *avoid* the necessity to dwell on the subject, by providing a user-friendly access to terminology and results.

0.3.1.2. *Exercises.* Two kinds of exercises are presented: some serve as illustrations (e.g., examples, counterexamples, explicit calculations, etc). Some may propose deepening or extensions of the main text.

0.3.1.3. *Indexes.* I include a terminological index, a notation index, and an index of names. They take into account the main text only, not the appendices. The terminological index only mentions terms special to the book, terms of general mathematical use are listed further below in 0.4.2. The index of names mentions names only for relevant contributions, for instance expressions like "Riemann surface", "Galois groups", and "Hilbert theorem 90" do not give rise to items like "Galois", "Riemann", or "Hilbert". I found no definite rule for the notation index, which is maybe a mess (containing at times entire formulas ...): I only hope that a desperate reader may find it at times more convenient to browse through 12 pages of symbols than to browse through more than 650 pages of the book.

0.3.1.4. *Errata.* I cannot hope to have corrected all the typographical and more substantial errors that appeared in the long process of making this book. A list of errata will be maintained on the related AMS webpage.

0.4. General notations and conventions

0.4.1. Some general conventions. The end of a proof, of its absence, is marked by the symbol $\square$. Notation $A := B$ means that the term A is defined by formula B. New terminology is written in *emphatic style* when first defined. Example: "The q-difference module M is said to be *fuchsian (at* 0*)* if it is pure isoclinic of slope 0, i.e., if $S(M) = \{0\}$."
Note that a definition can appear in the course of a theorem, an example, an exercise, etc.

We use commutative diagrams. For instance, the commutativity of the diagram

$$\begin{array}{ccc} \mathscr{F}(U) & \xrightarrow{\phi_U} & \mathscr{F}'(U) \\ \rho_V^U \downarrow & & \downarrow \rho'^U_V \\ \mathscr{F}(V) & \xrightarrow[\phi_V]{} & \mathscr{F}'(V) \end{array}$$

means that $\phi_V \circ \rho_V^U = \rho'^U_V \circ \phi_U$.

0.4.2. Some general notations.

0.4.2.1. *Sets of numbers.*

- $\mathbf{C}, \mathbf{R}, \mathbf{Q}, \mathbf{Z}, \mathbf{N}$: complex, real, rational numbers; then rational, natural integers.
- $\mathbf{C}^*, \mathbf{R}_+, \mathbf{R}_+^*$: nonzero, nonnegative, strictly positive.
- $\mathbf{U}, \mu_n, \mu_n^*, \mu_\infty$: group of modulus one complex numbers, of nth roots of unity, of primitive nth roots, of arbitrary roots.
- $\Re, \Im$: real and imaginary part. (The image of a map is denoted $\operatorname{Im} f$.)

0.4.2.2. *Spaces.*

- $\mathbf{S}, \mathbf{C}_\infty$: Riemann sphere, the same deprived of 0.
- $\mathcal{H}$: Poincaré half-plane (defined by $\Im z > 0$).
- $\overline{\mathrm{D}}(a, R), \mathring{\mathrm{D}}(a, R), \dot{\mathrm{D}}(a, R), \partial \mathrm{D}(a, R)$: closed, open, punctured disk, circle.
- $\mathring{\mathcal{C}}(r, R), \overline{\mathcal{C}}(r, R), \mathcal{C}(r, R)$ open, closed, semi-open annulus $r < |z| \leq R$.
- $\Lambda, \Lambda_\tau, \mathbf{E}_\Lambda$: a lattice of $\mathbf{C}$, the lattice $\mathbf{Z} + \mathbf{Z}\tau$, the quotient $\mathbf{C}/\Lambda$.
- $\operatorname{div}, \operatorname{div}_V$: divisor of a function, of its restriction to V.
- $\deg, \operatorname{ev}$: degree of a divisor, evaluation (if defined on an abelian group).

0.4.2.3. *Sets of functions.*

- $\mathscr{O}(U), \mathscr{O}(U, a), \mathscr{M}(U), \mathscr{M}(U, a)$ holomorphic functions and germs, meromorphic functions, and germs.
- $\mathbf{C}[[x]], \mathbf{C}((x)), \mathbf{C}\{x\}, \mathbf{C}(\{x\})$ ring and field of formal, resp. convergent series.
- $\mathbf{C}[x], \mathbf{C}(x), \mathbf{C}[x]_d$ polynomials, rational fractions, polynomials of degree $\leq d$.
- $v_a(f), v(f) = v_0(f), \operatorname{Res}_{x=a}$ valuation or order of f at a, x-adic valuation, residue.
- f is *regular* at a if $v_a(f) = 0$.
- $\|f\|_K$: maximum of $|f(x)|$ on compact K.
- $D := d/dx, \delta := x d/dx$: plain and Euler differential operator.
- $T_n f, \overline{T}_n f$: truncatures of series, so that $\overline{T}_n f - T_n f$ is the x^n term.

0.4.2.4. *Matrices, linear algebra.*

- $\mathscr{L}_K(V, W), \mathscr{L}_K(V), \mathrm{GL}(V)$: K-linear maps, endomorphisms, automorphisms.
- $\mathrm{Mat}_{m,n}(K), \mathrm{Mat}_n(K), \mathrm{GL}_n(K)$: corresponding sets of matrices.
- $\mathrm{D}_n^*(K), \mathrm{Diag}(c_1, \ldots, c_n)$: invertible diagonal matrices (set, individual).
- $\mathrm{Sp}(A), [A, B]$: spectrum[17], commutator $AB - BA$.
- Sing $M(x)$: singular locus, i.e., where M is undefined or not invertible.
- $\begin{bmatrix} C_1 & \cdots & C_n \end{bmatrix}, \begin{bmatrix} L_1 \\ \vdots \\ L_m \end{bmatrix}$: denotations of an $m \times n$ matrix by columns, by lines.
- $|||A|||$: matricial norm associated to a norm $\|X\|$ on $\mathbf{C}^n$.

[17]Sometimes, the spectrum is seen as a plain set and, for instance, writing 0_n the null $n \times n$ matrix, $\mathrm{Sp}\, 0_n = \{0\}$. Sometimes, it is seen as a *multiset* (elements have multiplicities) and then $\mathrm{Sp}\, 0_n = \{\{0, \ldots, 0\}\}$ (counted n times).

0.4.2.5. *Categories, sheaves and vector bundles.*

• $\mathrm{Ob}(\mathscr{C}), \mathrm{Hom}_{\mathscr{C}}(X,Y), \mathrm{Id}_X$: class of objects, sets of morphisms, identity morphisms in a category.
• $\rightsquigarrow$: funny arrow reserved for functors acting on objects.
• $\mathcal{V}ect^f_{\mathbf{C}}, \mathcal{R}ep_{\mathbf{C}}(G), \mathcal{R}ep^f_{\mathbf{C}}(G), \mathcal{R}ep_{rat}(G)$: category of finite-dimensional complex vector spaces, of complex representations, of finite-dimensional complex representations, of rational representations (see Appendix G for the latter).
• $\varprojlim, \varinjlim$: inverse and direct limit.
• $\mathscr{F}, \mathfrak{F}$: locally free sheaf and associated bundle.
• $\rho^U_V, \mathscr{F}_a, \mathfrak{F}_a$: restriction map, stalk (or fiber) for a sheaf, for a vector bundle.

0.4.2.6. *Miscellany.*

• $\delta_{x,y}$: Kronecker index (1 if $x = y$, otherwise 0).
• $n >> -\infty, n >> 0, n << a, \ldots$: for n bounded below, for n big enough, etc.
• $\{\{c_1, \ldots, c_n\}\}$: multiset, or bag (of roots, of eigenvalues ...).
• Empty sum is 0; empty product is 1 or I_n, etc; $\min \emptyset, \max \emptyset$, respectively largest or smallest possible value.
• E^I, $E^{(I)}$: families of elements of E indexed by I, families with finite support (the latter presupposes that $0 \in E$).
• $\simeq, \sim, \mathrm{cl}, \approx$: isomorphism, equivalence relation, equivalence class, ad hoc relation of "similarity".
• $[\Gamma,\Gamma], \Gamma^{ab}, \mathrm{Hom}_{gr}, \mathrm{Hom}_{grtop}, \Gamma^{\vee}$: commutator subgroup, abelianized, set of morphisms of groups, of continuous morphisms, proalgebraic group of morphisms to $\mathbf{C}^*$ (see Appendix G for the latter).
• K^σ, R^σ: fixed subfield, subring under an automorphism.
• E^G fixed subset of E acted upon by the group G.

0.4.2.7. *The q-world.*

• $q \in \mathbf{C}^*$ s.t. $|q| < 1$ but for some explicit exceptions.
• $\equiv$: most of the time congruence modulo $q^{\mathbf{Z}}$ in $\mathbf{C}^*$.
• $\tau \in \mathcal{H}$ such that $q = e^{2i\pi\tau}$, $q^\alpha := e^{2i\pi\tau\alpha}$, $q^{\mathbf{R}} := e^{2i\pi\tau\mathbf{R}}$.
• $\sigma_q f(x) := f(qx)$, $D_q := \dfrac{\sigma_q - 1}{(q-1)x}$, $\delta_q := \dfrac{\sigma_q - 1}{q-1}$.
• $\mathscr{L}_{\sigma_q}(V,W)$: σ_q-linear maps, i.e., group morphisms s.t. $f(ax) = \sigma_q(a) f(x)$.
• $\mathbf{E}_q := \mathbf{C}^*/q^{\mathbf{Z}}$: group and Riemann surface, $\pi : \mathbf{C}^* \to \mathbf{E}_q, c \mapsto \overline{c} := c \pmod{q^{\mathbf{Z}}}$.
• $\mathcal{C}_q := \mathcal{C}(|q|, 1)$: fundamental annulus. 1
• $c = q^{\epsilon}(c)\overline{c}$, $\epsilon(c) \in \mathbf{Z}$, $\overline{c} \in \mathcal{C}_q$.
• $[a;q] := aq^{\mathbf{Z}}$, $[a;q]_0 := aq^{\mathbf{N}}$, $[a;q]_\infty := aq^{-\mathbf{N}}$: discrete q-spirals and half q-spirals.
• $(a;q)_n, (a;q)_\infty, [\alpha]_q, [\alpha]!_q, \begin{bmatrix} n \\ l \end{bmatrix}_q$: q-Pochhammer symbols (in two guises), q-numbers, q-factorials, q-binomial coefficients (Section 3.3).
• $\theta_q, e_{q,c}, \ell_q$: qh-Theta function, q-character, q-logarithm (Chapter 2).
• $A \underset{K}{\sim} F[A] := (\sigma_q F) A F^{-1}$, where $F, A \in \mathrm{GL}_n(K)$: gauge tranformation and equivalence over K.

0.4.3. A word about "coverings" and "etale coverings". I have chosen to call *etale covering* what is often called *covering space*, see in particular [**90**, **102**, **115**, **165**, **238**]. This is to distinguish this *fundamental tool* from the more standard but more banal notion of *open covering*.

On the other hand, the theory of Riemann surfaces makes heavy use of *ramified coverings*, which are a degenerated form of etale coverings (see Appendix B.3.3) and we also want to emphasize that distinction.

CHAPTER 1

Prelude

This chapter is not logically necessary[1] for the rest of the book. It is proposed for motivation, feeling, flavour. As Marx said, "The anatomy of man is a key to the anatomy of the ape", and so we consider the advanced theory as unveiling the true nature of the original objects. However, it seems that a historical perspective helps in understanding better conceptual constructions. Moreover this prelude will provide the readers with a wealth of pretty examples to exercise their newly acquired skills.

By historical necessity, this prelude is more oriented toward q-series, q-calculus, "q-disease" (as it is called by Richard Askey's in his foreword to the "bible" **[93]** of Gasper and Rahman) and less toward q-difference equations than the rest of the book. Besides, the "Founding Fathers" Adams, Carmichael, Birkhoff ..., do not appear here. Conversely, the themes of this chapter that will be pursued in the rest of the book pertain mainly to Sections 1.4 (on Cauchy) and 1.6 (on Heine), along with the intermediate Section 1.5.

We do give a hint in 1.12.1 on how, starting with Jackson and (separately) with Poincaré, a distinct thread appeared, centered on functional equations. For more scholarly historical information, the reader is warmy encouraged to look at Berndt **[18]** (for the early history of q-series) and at Adams **[2]** (for the early history of q-difference equations); both texts may well be complemented by Birkhoff [**24**, §8] (the section on linear difference equations).

1.1. Fermat (Pierre de Fermat; French; first decade of 17th century–1665)

What follows may be considered as the earliest occurrence of q-calculus. It features two characteristics of it: computations involve geometric series, which are rather easy to deal with; and degeneracy when $q \to 1$ yields results in infinitesimal calculus. The natural way to approximate an integral $\int_0^a f(x)\,dx$ (for instance an area or a volume in greek mathematics) is to divide the interval $[0; a]$ into N intervals of equal lengths $[ia/N; (i+1)a/N]$, $i = 0, \ldots, N-1$; and then to replace f on each such interval by one of its values, for instance $f(ia/N)$ or $f((i+1)a/N)$. Thus we could get (evaluating f at left extremities):

$$\int_0^a f(x)\,dx \approx \sum_{i=1}^{N} \frac{a}{N} f\left(\frac{ia}{N}\right),$$

actually a Riemann sum. Then, letting $N \to +\infty$, one may hope to get the exact value of the integral as a limit.

[1]Except for some formulas or isolated results which will be stated again further but not their proof.

For instance, to compute the area under a (generalized) parabola, a problem dating from antiquity, one approximates $\int_0^a x^n \, dx$ by $\frac{a^{n+1}}{N^{n+1}} \sum\limits_{i=1}^{N} i^n$. One then proves that there is a polynomial $P(X)$ with leading term $\frac{X^{n+1}}{n+1}$ such that $\sum\limits_{i=1}^{N} i^n = P(N)$ for all integers $N \geq 0$, and from that it is easy to conclude that $\int_0^a x^n \, dx = \frac{a^{n+1}}{n+1}$. What is not easy, in terms of ancient algebra, without proper notations, is the statement about the polynomial P. Fermat had the idea of using instead a "geometric" subdivision of the interval ("geometric" in the sense of geometric series). There are two ways to do it.

First, we may consider an infinite subdivision of $[0, a]$ by intervals $[q^{i+1}a; q^i a]$, $i \geq 0$, where $q \in]0; 1[$ is fixed (to begin with). We get for instance (evaluating f at right extremities):

$$\int_0^a f(x) \, dx \approx \sum_{i=0}^{\infty} (q^i a - q^{i+1} a) f(q^i a) = (1 - q) \sum_{i=0}^{\infty} q^i a \, f(q^i a),$$

thus an anticipation of Jackson's q-integral (see Section 1.9).

Second, we may rather consider integrals such as $\int_b^a f(x) \, dx$, where $0 < b < a$. For each $N > 0$, let q such $q^N = b/a$. Then the interval $[a; b]$ admits the finite subdivision by the $[q^{i+1}a; q^i a]$, $i = 0, \ldots, N - 1$ and:

$$\int_b^a f(x) \, dx \approx \sum_{i=0}^{N-1} (q^i a - q^{i+1} a) f(q^i a) = (1 - q) \sum_{i=0}^{N-1} q^i a \, f(q^i a).$$

Applying the first method to $f(x) := x^n$, we get the approximation:

$$\int_0^a x^n \, dx \approx (1 - q) \sum_{i=0}^{\infty} (q^i a)^{n+1} = a^{n+1} \frac{1 - q}{1 - q^{n+1}},$$

which tends to $\frac{a^{n+1}}{n+1}$ when $q \to 1$.

Applying the second method:

$$\begin{aligned}\int_b^a x^n \, dx \approx (1 - q) \sum_{i=0}^{N-1} (q^i a)^{n+1} &= a^{n+1} \frac{(1 - q)(1 - q^{N(n+1)})}{1 - q^{n+1}} \\ &= (a^{n+1} - b^{n+1}) \frac{1 - q}{1 - q^{n+1}},\end{aligned}$$

which tends to $\frac{a^{n+1} - b^{n+1}}{n+1}$ when $N \to +\infty$ (and so $q \to 1$).

In both cases we have to sum geometric series, which is easy and was known by antique mathematicians (who often solved them by geometrical methods). Of course, Fermat (one of the fathers of infinitesimal calculus!) mastered enough the calculus of limits to know that $\frac{1 - q}{1 - q^{n+1}}$ tends to $\frac{1}{n+1}$ when $q \to 1$.

REMARK 1.1. However it seems that he did not see it as a limit in the sense of analysis as we know it since Cauchy; but, under the term of *adaequatio* **[83]**, rather as a *specialisation* in the sense of algebraic geometry. That point of view in dealing with the $q \to 1$ degeneracy (so important in q-calculus; see for instance 1.4 and 1.6) found a modern echo in Yves André's paper [**4**].

1.2. Euler (Leonhard Euler, Swiss, 1707–1783)

1.2.1. Partitio numerorum. The basic reference here is [**112**].

Let $p(n)$ the number of *partitions* of a natural integer $n \in \mathbf{N}$, that is, the number of its decompositions in a sum of integers, duplications allowed and regardless of the order. For instance, since:

$$5 = 1+4 = 2+3 = 1+1+3 = 1+2+2 = 1+1+1+2 = 1+1+1+1+1,$$

$p(5) = 7$ (we consider 5 itself as a partition into one part). Euler found the generating series of the sequence $(p(n))_{n\geq 0}$:

$$\mathcal{P}(q) := \sum p(n)q^n = \frac{1}{(1-q)(1-q^2)(1-q^3)\cdots}.$$

Of course, we use here the letter q (Euler did not) to emphasize the connection to q-calculus. Euler then proved the formula:

$$\mathcal{P}(q) = \sum_{n\geq 0} \frac{q^n}{(1-q)(1-q^2)\cdots(1-q^n)}.$$

By the usual convention on empty products (see 0.4), the term with index $n = 0$ is 1. While a large part of what Euler proved about partitions pertains to combinatorics or to q-calculus but not to the world of q-difference equations, the proof of that particular formula may well be considered as the act of birth of q-difference equations. The *fundamental trick* is to introduce a "new parameter":

$$f_q(x) :== \frac{1}{(1-x)(1-qx)(1-q^2x)\cdots} = \frac{1}{(x;q)_\infty}, \text{ where } (x;q)_\infty := \prod_{n\geq 0}(1-q^n x).$$

The functional equation $f_q(qx) = (1-x)f_q(x)$ readily yields the power series expansion of f_q as $\sum\limits_{n\geq 0} f_n x^n$. Indeed, by identification of coefficients, it implies that $q^n f_n = f_n - f_{n-1}$ for all n (with the natural convention that $f_{-1} = 0$), whence the recurrence relation $f_n = \dfrac{f_{n-1}}{1-q^n}$ for $n \geq 1$. The initialisation comes from $f_0 = f_q(0) = 1$ (obviously) and then an easy induction yields:

$$f_n = \frac{1}{(q;q)_n}, \text{ where } \forall x \in \mathbf{C}\,,\ (x;q)_n := \prod_{0\leq i<n}(1-q^i x).$$

Quantities such as $(x;q)_\infty$ and $(x;q)_n$ will be studied in great detail in Chapter 3. Eventually, we find:

$$f_q(x) = \sum_{n\geq 0} \frac{1}{(q;q)_n} x^n,$$

whence (by taking $x := q$) an "eulerian" expansion:

$$\mathcal{P}(q) = \frac{1}{(q;q)_\infty} = f_q(q) = \sum_{n\geq 0} \frac{q^n}{(q;q)_n}.$$

Here "eulerian" refers to the denominator $(q;q)_n$.

EXERCISE 1.2. (i) Give a combinatorial interpretation of the coefficient of $q^n x^m$.

(ii) Give a power series expansion for $(x;q)_\infty$. (Calling it $g_q(x)$, find a functional equation involving $g_q(qx)$; or else, read Chapter 3.)

1.2.2. Around asymptotics. Generating series are a useful tool to study the asymptotic behaviour of their coefficients, see **[86, 112, 137, 257]**. Here is a simple example directly related to this section; a more important example will be featured in 1.10.

Let $p_m(n)$ the number of partitions of n into parts $\leq m$, i.e., the number of decompositions $n = n_1 + \cdots + n_k$, $0 \leq n_1 \leq \cdots \leq n_k \leq m$ (k is unspecified). Saying that 1 occurs x_1 times (among the n_i), that 2 occurs x_2 times, etc, we see that $p_m(n)$ is the number of solutions $(x_1, \ldots, x_m) \in \mathbf{N}^m$ of equation $x_1 + 2x_2 + \cdots + mx_m = n$. It is easy to check that:

$$\mathcal{P}_m(q) := \sum_{n \geq 0} p_m(n) q^n = \frac{1}{(1-q)\cdots(1-q^m)} = \frac{1}{(q;q)_m}, \text{ so for instance:}$$

$$\mathcal{P}_3(q) = \frac{1}{(1-q)(1-q^2)(1-q^3)} = \frac{1}{(1-q)^3(1+q)(1-jq)(1-j^2q)},$$
$$\text{where } j := e^{2i\pi/3}.$$

Expansion of that rational function into simple parts has the form:

$$\mathcal{P}_3(q) = \frac{a}{(1-q)} + \frac{b}{(1-q)^2} + \frac{c}{(1-q)^3} + \frac{d}{(1+q)} + \frac{e}{(1-jq)} + \frac{f}{(1-j^2q)},$$

where $a, b, c, d, e \in \mathbf{C}$ can be explicitly determined; for instance:

$$c = \lim_{q \to 1} (1-q)^3 \mathcal{P}_3(q) = \lim_{q \to 1} \frac{1}{(1+q)(1+q+q^2)} = \frac{1}{6}.$$

On the other hand, from the well-known power series expansions of $(1-x)^{-k}$, we get:

$$p_3(n) = a + b(n+1) + c(n+1)(n+2)/2 + d(-1)^n + ej^n + fj^{2n} \sim n^2/12.$$

While for general m the exact form of $p_m(n)$ will not be so easy to write down, it is clear that its dominant term will be provided by the dominant singularity $q = 1$ of the rational fraction $\dfrac{1}{(q;q)_m}$; and that the other singularities (roots of 1) will provided subsequent terms in an asymptotic expansion of $p_m(n)$ when $n \to +\infty$.

EXERCISE 1.3. (i) Show along the same lines that $p_m(n)$ admits (m being fixed and $n \to +\infty$) an asymptotic expansion with leading term $T_m(n) := \dfrac{n^{m-1}}{m!(m-1)!}$.

(ii) Check that $p(n) = p_n(n)$. Is $T_n(n)$ a reasonable asymptotic equivalent for $p(n)$?

(iii) Check that $p(n) = \max\limits_{m=1,\ldots,n} p_m(n)$. So let $T(n) := \max\limits_{m=1,\ldots,n} T_m(n)$. Give an asymptotic equivalent of $T(n)$, or at least a gross order. (Maximum should occur at $m \approx \sqrt{n}$, so first look for an equivalent of $T_m(m^2)$.) You should find something like $e^n/(2\pi n)$; compare it with the actual equivalent obtained by Hardy and Ramanujan (see 1.10 or **[109, 111]**).

Euler had already found an infinite product expansion for another generating series[2] arising naturally in number theory:

$$\zeta(s) := \sum_{n\geq 1} n^{-s} = \prod_{p\in P} \frac{1}{1-p^{-s}}, \text{ where } P := \{2,3,5,7,\ldots\}.$$

Although Euler saw s as a real variable, that formula was used along with complex function theory by Dirichlet, Riemann, Hadamard, de la Vallée-Poussin, Hardy-Littlewood, and so many more, ...to obtain information on the distribution of prime numbers and to asymptotic properties of related arithmetic functions. Euler himself deduced from that same formula a new proof of the infinity of prime numbers and other nice elementary analytic applications; see **[112]**. The deepest part of the whole story is related in some way to the singularity analysis.

However, it does not seem that Euler himself gave an analytic interpretation of the wealth of formulas he obtained in q-calculus: this would have to wait for Jacobi and Ramanujan (and possibly Gauß); we'll see that later. So the relevance of Euler's discoveries around *partitio numerorum* for the present book rather lie in the use of a functional equation in the proof of 1.2.1.

Obtaining the asymptotic behaviour of $p(n)$ by studying the singularities of $\mathcal{P}(q)$ is considerably more difficult than the little games above for $p_m(n)$; it was the work of Hardy and Ramanujan (and Rademacher); see 1.10.

1.2.3. Other kinds of expansions. Among so-called q-series involving a variable x, not all are power series. We just show a few examples (some, by the way, are instances of Heine's transformation formulas; see 1.6 below and 4.4 in Chapter 4. Note however that there is a wealth of formulas of various types in the q-world **[93, 130, 133, 185, 186]** and it is not the general task of this book to explore them thoroughly.

A symmetric formula. Here is a simple one:

$$x \sum_{n\geq 1} \frac{a^n}{(x;q)_n} = a \sum_{n\geq 1} \frac{x^n}{(a;q)_n}.$$

To prove it, let $g(x) := \sum\limits_{n\geq 0} \dfrac{a^n}{(x;q)_{n+1}}$, which, under appropriate conditions (easy to formulate - this is left to you!), is analytic at 0 and therefore admits a power series expansion $g(x) = \sum\limits_{n\geq 0} b_n x^n$. Noting that $(x;q)_{n+1} = (1-x)(qx;q)_n$, we compute:

$$g(x) = \frac{1}{1-x} \sum_{n\geq 0} \frac{a^n}{(qx;q)_n} = \frac{1}{1-x} + \frac{a}{1-x} \sum_{n\geq 1} \frac{a^{n-1}}{(qx;q)_n} = \frac{1}{1-x} + \frac{a}{1-x} g(qx),$$

whence $(1-x)g(x) = 1 + ag(qx)$, whence $b_n - b_{n-1} = aq^n b_n$, i.e., $b_n = \dfrac{b_{n-1}}{1-aq^n}$ for $n \geq 1$. Since $b_0 = g(0) = \sum\limits_{n\geq 0} a^n = \dfrac{1}{1-a}$, we find $b_n = \dfrac{1}{(a;q)_{n+1}}$ and:

$$\sum_{n\geq 0} \frac{a^n}{(x;q)_{n+1}} = \sum_{n>0} \frac{x^n}{(a;q)_{n+1}}.$$

---

[2]The words "eulerian expansion" are usually associated with such products indexed by the primes, or some fancy generalisations in algebraic and analytic number theory. The words "eulerian form" in relation to quantities such as $(q;q)_n$ or $(q;q)_\infty$ were those of Ramanujan; see **[253]**.

Multiplying by ax yields the stated formula.

A formula of Karpelevich. The second example is a formula atributed to Karpelevich in [177]:

$$\frac{(q;q)_\infty}{(x;q)_\infty} = \sum_{n\geq 0} \frac{(-1)^n q^{n(n+1)/2}}{(q;q)_n(1-q^n x)}.$$

Call $f(x)$ the left-hand side and $g(x) := \sum_{n\geq 0} \frac{c_n}{1-q^n x}$ the right-hand side, where we set $c_n := \frac{(-1)^n q^{n(n+1)/2}}{(q;q)_n}$. Actually those coefficients could be guessed as follows. The left-hand side has simple poles at all q^{-n}, $n \geq 0$, so one could hope for an expansion of the form $g(x)$, where c_n would be found as:

$$\lim_{x\to q^{-n}} (1-q^n x) f(x) = \lim_{x\to q^{-n}} (1-q^n x)\frac{(q;q)_\infty}{(x;q)_\infty} = \lim_{x\to q^{-n}} (1-q^n x)\frac{(q;q)_\infty}{(x;q)_n (q^n x;q)_\infty} =$$

$$\lim_{x\to q^{-n}} \frac{(q;q)_\infty}{(x;q)_n (q^{n+1}x;q)_\infty} = \frac{1}{(q^{-1};q)_n} = \frac{(-1)^n q^{n(n+1)/2}}{(q;q)_n}.$$

(We used the easily checked facts fact that $(x;q)_\infty = (x;q)_n(q^n x;q)_\infty$ and that $(q^{-1};q)_n = (-1)^n q^{-n(n+1)/2}(q;q)_n$.) So we could conclude the argument by saying that the left- and right-hand side differ by an entire function, and proving (that might be more complicated!) that both sides tend to 0 at infinity (so the said entire function is 0). We prefer to make the link with Euler's world as follows. Using the last exercise in 1.2.1 or anticipating 3.3.1 we have:

$$(x;q)_\infty = \sum_{n\geq 0} \frac{(-1)^n q^{n(n-1)/2}}{(q;q)_n} x^n = \sum_{n\geq 0} c_n (x/q)^n.$$

Also, $c_n/c_{n-1} = \dfrac{-q^n}{1-q^n} \Rightarrow c_{n-1} = (1-q^{-n})c_n$. Now:

$$g(qx) = \sum_{n\geq 0} \frac{c_n}{1-q^{n+1}x} = \sum_{n\geq 1} \frac{c_{n-1}}{1-q^n x} = \sum_{n\geq 1} \frac{(1-q^{-n})c_n}{1-q^n x} = \sum_{n\geq 0} \frac{(1-q^{-n})c_n}{1-q^n x} =$$

$$(1-x)g(x) + \sum_{n\geq 0} \frac{(x-q^{-n})c_n}{1-q^n x} = (1-x)g(x) + \sum_{n\geq 0} c_n q^{-n} = (1-x)g(x),$$

since $\sum_{n\geq 0} c_n q^{-n} = (1;q)_\infty = 0$. Now, using equality $g(0) = \sum_{n\geq 0} c_n = (q;q)_\infty$ and iterating the functional relation $g(x) = \dfrac{1}{1-x} g(qx)$, we get:

$$g(x) = \frac{1}{1-x} g(qx) = \frac{1}{(1-x)(1-qx)} g(q^2 x) = \cdots = \frac{1}{(x;q)_\infty} g(0) = \frac{(q;q)_\infty}{(x;q)_\infty}.$$

A formula due to Lucia Di Vizio. In [59, 60], the following is proved:

$$\sum_{n\geq 0} \frac{z^n}{(q\lambda;q)_n} = \frac{1-\lambda}{(z;q)_\infty} \sum_{n\geq 0} \frac{(-1)^n q^{n(n+1)/2} z^n}{(q;q)_n (1-q^n\lambda)}.$$

A proof is proposed using a second order q-difference equation (the variable being z). Here we choose a different path. We shall replace λ by x and see it as the variable of interest. Note that we have already been dealing with expansions of the form $g(x) := \sum_{n\geq 0} \frac{a_n}{1-q^n x}$ (see the formula of Karpelevich, and also the exercice below). Introducing the plain power series $f(x) := \sum_{n\geq 0} a_n x^n$, we easily obtain a relation:

$$g(x) = \sum_{n\geq 0} \frac{a_n}{1-q^n x} = \sum_{n\geq 0} a_n \sum_{p\geq 0} (q^n x)^p = \sum_{p\geq 0} \left(\sum_{n\geq 0} a_n q^{np} \right) x^p = \sum_{p\geq 0} f(q^p) x^p.$$

This is a purely formal computation, we leave it to the reader to state conditions of validity (since we have permuted two infinite summations). Admitting the result, we apply it with $a_n := \frac{(-1)^n q^{n(n+1)/2} z^n}{(q;q)_n}$, so that $f(x) = (qzx;q)_\infty$ (again Euler's formula) and:

$$\sum_{n\geq 0} \frac{(-1)^n q^{n(n+1)/2} z^n}{(q;q)_n (1-q^n x)} = \sum_{p\geq 0} (q^{p+1} z; q)_\infty x^p = \sum_{p\geq 0} \frac{(z;q)_\infty}{(z;q)_{p+1}} x^p = (z;q)_\infty \sum_{p\geq 0} \frac{x^p}{(z;q)_{p+1}}.$$

We used the easily checked fact that $(z;q)_\infty = (z;q)_{p+1}(q^{p+1}z;q)_\infty$. Now, since $\sum_{p\geq 0} \frac{x^p}{(z;q)_{p+1}} = \sum_{p\geq 0} \frac{z^p}{(x;q)_{p+1}}$ (our first example above) and since $(x;q)_{p+1} = (1-x)(qx;q)_p$, the above implies:

$$(1-x) \sum_{n\geq 0} \frac{(-1)^n q^{n(n+1)/2} z^n}{(q;q)_n (1-q^n x)} = (z;q)_\infty \sum_{p\geq 0} \frac{z^p}{(qx;q)_p},$$

in essence the desired formula.

EXERCISE 1.4. Prove the following more complicated formula **[110, 152]**:

$$\sum_{n\geq 0} \frac{a^n}{(x;q)_n} = 1 + \sum_{n\geq 0} \frac{(-1)^n q^{n(n+1)/2} a^{n+1}}{(q;q)_n (a;q)_\infty} \frac{1}{1-q^n x}.$$

1.3. Gauß(Carl Friedrich Gauß, German, 1777–1855)

It is likely that Gauß anticipated much of what will be recounted in this chapter, in particular Theta functions, including Jacobi's Triple Product Formula (1.7). However, because he published so scarcely, his name is associated to q-calculus mainly under the heading of *Gauß' q-binomial theorem.* Note that $(-x;q)_n = (1+x)(1+qx)\cdots(1+q^{n-1}x)$ tends to $(1+x)^n$ when $q \to 1$, so that we can consider the former as a "q-deformation" of the latter. Therefore we adopt the suggestive notation from **[133]**:

$$(1+x)_q^n := (-x;q)_n.$$

Then Gauß' q-binomial theorem consists in an explicit expansion:

$$(1+x)_q^n = \sum_{j=0}^{n} \begin{bmatrix} n \\ j \end{bmatrix}_q q^{j(j-1)/2} x^j.$$

The *q-binomial coefficients* $\begin{bmatrix} n \\ j \end{bmatrix}_q := \frac{[n]!_q}{[j]!_q [n-j]!_q}$ belong to the family of *q-numbers* $[n]_q := \frac{1-q^n}{1-q}$, *$q$-factorials* $[n]!_q := [1]_q \cdots [n]_q$, ..., which will be studied in more detail in Chapter 3. So, in usual language:

$$(1+x)_q^n = \sum_{j=0}^{n} q^{j(j-1)/2} \frac{(1-q^n)\cdots(1-q^{n-j+1})}{(1-q)\cdots(1-q^j)} x^j.$$

Let's prove the formula; maybe this is a good occasion to test the efficiency of q-difference equations in a baby case. The left-hand side is a priori a polynomial of degree n, say $f(x) = a_0 + \cdots + a_n x^n$. Of course $a_0 = 1$ and $a_n = 1 \times q \times \cdots \times q^{n-1} = q^{n(n-1)/2}$. We have:

$$(1+x)f(qx) = (1+x)(1+qx)\cdots(1+q^n x) = (1+q^n x)f(x),$$

so, identifying the coefficients:

$$\forall k \geq 1\,,\; q^k a_k + q^{k-1} a_{k-1} = a_k + q^n a_{k-1} \Longrightarrow a_k = \frac{q^{k-1}(1-q^{n-k+1})}{1-q^k} a_{k-1}.$$

Induction immediately gives as wanted $a_j = q^{j(j-1)/2} \dfrac{(1-q^n)\cdots(1-q^{n-j+1})}{(1-q)\cdots(1-q^j)}$.

The vague resemblance of the fraction with a usual binomial coefficient is strengthened by the fact that $1-q^k \sim k(1-q)$ when $q \to 1$, so that $\begin{bmatrix} n \\ j \end{bmatrix}_q \to \dbinom{n}{j}$ and Gauß' q-binomial theorem degenerates into Newton's binomial theorem (here in finite form, i.e., the exponent is a natural integer). We shall see in Chapter 3 another interpretation of the q-binomial coefficients for "noncommutative" variables: if $xy = qyx$, then the expansion of $(x+y)^n$ requires them.

EXERCISE 1.5. Writing more precisely $(1+x)_q^n = \sum\limits_{j=0}^{n} a_j^{(n)} x^j$, transform the recurrence relation $(1+x)_q^{n+1} = (1+q^n x)(1+x)_q^n$ into some relation between q-binomial coefficients.

EXERCISE 1.6. The q-difference equation $(1+x)f(qx) = (1+q^n x)f(x)$ may also be written in the form $\dfrac{f(x)-f(qx)}{(1-q)x} = A_q(x)f(x)$.

(i) Compute $A_q(x)$ and find its limit $A(x)$ when $q \to 1$.

(ii) Solve the differential equation $f' = Af$ with initial condition $f(0) = 1$.

That process of *degeneracy* (or *confluence*) will be further explored in 1.4 and 1.6, and then again, in great detail, in Sections 11.2 and 12.3.

1.4. Cauchy (Augustin Louis Cauchy, French, 1789–1857)

Here, around a basic example, we shall encounter many themes of this book. Recall the notations from 1.2.1; remember that we always assume that $|q| < 1$:

$$(a;q)_n := \prod_{0 \leq i < n} (1-aq^i) \quad \text{and} \quad (a;q)_\infty := \prod_{n \geq 0} (1-aq^n).$$

1.4.1. General q-binomial theorem. The *general q-binomial theorem*, due to Cauchy[3], is the following formula, valid for $|x| < 1$:

$$\frac{(ax;q)_\infty}{(x;q)_\infty} = \sum_{n \geq 0} \frac{(a;q)_n}{(q;q)_n} x^n.$$

[3] It is sometimes also called Cauchy-Heine's q-binomial theorem.

We shall prove it in two ways, reasoning either from the left-hand side $f(x)$, or from the right-hand side $g(x)$. Then, in 1.4.2, 1.4.3, and 1.4.4, we shall discuss why it is called a "q-binomial theorem", i.e., why it is a q-analogue of the binomial theorem; and, toward the end of section 1.4, we shall deal with the mystery of a uniform function degenerating to a multivalued one.

EXERCISE 1.7. (i) Which formulas of Euler and Gauß can be derived from the above one?

(ii) Deduce from it a power series expansion for $\dfrac{(ax;q)_\infty}{(bx;q)_\infty}$ and then produce nice new formulas by identification of the coefficients in the equality:

$$\frac{(ax;q)_\infty}{(bx;q)_\infty} \times \frac{(bx;q)_\infty}{(cx;q)_\infty} = \frac{(ax;q)_\infty}{(cx;q)_\infty}.$$

You could use the suggestive notations from **[133]**:

$$(a-b)_q^n := (a-b)(a-qb)\cdots(a-q^{n-1}b).$$

(iii) What happens if you respectively substitute q^α, q^β, q^γ, for a, b, c, in the general q-binomial theorem and then have $q \to 1$? (We return to that soon.)

First proof. From general theorems on infinite products (more precise justifications will be given in Chapter 3), $f(x)$ is holomorphic over $\overset{\circ}{\mathrm{D}}(0,1)$ and therefore admits a power series expansion $f(x) = \sum\limits_{n\geq 0} a_n x^n$. Since $f(0) = 1$, we have $a_0 = 1$.

Then, from equalities $(x;q)_\infty = (1-x)(qx;q)_\infty$ and $(ax;q)_\infty = (1-ax)(aqx;q)_\infty$, we deduce a functional relation (actually a q-difference equation):

$$f(x) = \frac{1-ax}{1-x} f(qx) \Longrightarrow (1-x)f(x) = (1-ax)f(qx).$$

Identification of the coefficients for x^n, $n \geq 1$, yields:

$$a_n - a_{n-1} = q^n a_n - aq^{n-1}a_{n-1} \Longrightarrow a_n = \frac{1-aq^{n-1}}{1-q^n} a_{n-1}.$$

Iterating (or using plain induction):

$$a_n = \frac{1-aq^{n-1}}{1-q^n} a_{n-1} = \frac{(1-aq^{n-1})(1-aq^{n-2})}{(1-q^n)(1-q^{n-1})} a_{n-2} = \cdots = \frac{(a;q)_n}{(q;q)_n},$$

since $a_0 = 1$. This concludes the first proof that $f(x) = g(x)$. Note that, since $\dfrac{a_n}{a_{n-1}} = \dfrac{1-aq^{n-1}}{1-q^n} \to 1$ as $n \to +\infty$, the radius of convergence of $g(x)$ is 1.

Second proof. As we just saw, the power series $g(x)$ has radius of convergence 1, so it defines a holomorphic function over $\overset{\circ}{\mathrm{D}}(0,1)$. Also, letting $b_n := \dfrac{(a;q)_n}{(q;q)_n}$ its coefficients, we see that $b_0 = 1$, so that $g(0) = 1$; and we deduce from the equalities $(q;q)_n = (1-q^n)(q;q)_{n-1}$ and $(a;q)_n = (1-aq^{n-1})(a;q)_{n-1}$ (valid for $n \geq 1$) that:

$$b_n = \frac{1-aq^{n-1}}{1-q^n} b_{n-1} \Longrightarrow b_n - q^n b_n - b_{n-1} - aq^{n-1} b_{n-1}.$$

The standard trick to transform such a recursive relation into a relation between generating series runs here as follows; we multiply both sides by x^n and sum all over $n \in \mathbf{N}$ (with the usual convention $b_{-1} := 0$, the equality above is valid for $n = 0$), getting:

$$\sum_{n \geq 0} b_n x^n - \sum_{n \geq 0} b_n (qx)^n = x \left(\sum_{n \geq 1} b_{n-1} x^{n-1} - a \sum_{n \geq 1} b_{n-1} (qx)^{n-1} \right) \Longrightarrow$$

$$g(x) - g(qx) = x\left(g(x) - ag(qx)\right) \Longrightarrow g(x) = \frac{1 - ax}{1 - x} g(qx).$$

We iterate that functional relation:

$$g(x) = \frac{1 - ax}{1 - x} g(qx) = \frac{(1 - ax)(1 - aqx)}{(1 - x)(1 - qx)} g(q^2 x) = \cdots = \frac{(ax; q)_\infty}{(x; q)_\infty},$$

since, by continuity, $\lim_{n \to +\infty} g(q^n x) = g(0) = 1$. This concludes the second proof that $f(x) = g(x)$.

REMARK 1.8. The second proof only works (i.e., provides an explicit solution) only in the case of a first order recursive relation on cooefficients, because it yields a first order q-difference equation $g(x) = u(x)g(qx)$ and the latter often admits an explicit solution in the form of an infinite product (see Chapters 3 and 4).

1.4.2. Why is Cauchy's formula a q-analogue of Newton's binomial theorem? We mean here the *generalized* binomial theorem $(1 + x)^\alpha = \sum \binom{\alpha}{n} x^n$, where $\alpha \in \mathbf{C}$ (all sums here are understood to be for $n \geq 0$ if not otherwise stated). If $\alpha \notin \mathbf{N}$, the right-hand side is an infinite sum. To give first hints of how both formulas are related, we shall rather use the following equivalent form of Newton's binomial theorem:

$$(1 - x)^{-\alpha} = \sum \frac{(\alpha)_n}{n!} x^n, \text{ where } (\alpha)_n := \alpha(\alpha + 1) \cdots (\alpha + n - 1),$$

so, by the usual convention, $(\alpha)_0 = 1$).

If you did the first question in the exercise of 1.4.1, you may have found that replacing a by q^{-n} (and rescaling x) in Cauchy's formula gave Gauß' binomial theorem; that one we already know to be a q-analogue. If you did the second question, you may have proved that:

$$\sum_{i+j=n} \frac{(b - a)_q^i}{(q; q)_i} \frac{(c - b)_q^j}{(q; q)_j} = \frac{(c - a)_q^n}{(q; q)_n},$$

which is formally similar to $\sum_{i+j=n} \frac{\alpha^i}{i!} \frac{\beta^j}{j!} = \frac{(\alpha + \beta)^n}{n!}$. We can make the resemblance less formal by setting $a := q^\alpha$, $b := q^\beta$ and $c := q^\gamma$ and using the q-number and q-factorial notations from q-calculus (already met in 1.3): $[\alpha]_q := \dfrac{1 - q^\alpha}{1 - q}$ (so that $[\alpha]_q \to \alpha$ when $q \to 1$) and $[n]!_q := [1]_q \cdots [n]_q = \dfrac{(q; q)_n}{(1 - q)^n}$. Then for instance:

$$\frac{(c - a)_q^n}{(q; q)_n} = q^{n\gamma} \frac{[\alpha - \gamma]_q [\alpha - \gamma + 1]_q \cdots [\alpha - \gamma + n - 1]_q}{[n]!_q} \to \frac{(\alpha)_n}{n!} \text{ when } q \to 1.$$

EXERCISE 1.9. Rewrite the formula from the exercise with those notations and explicit the degeneracy when $q \to 1$.

More formulas and degeneracies of this type will be looked at and precisely justified in Chapter 3. From now on, we just stress that the right-hand side of Cauchy's formula degenerates termwise, when $q \to 1$, to $\sum \frac{(\alpha)_n}{n!} x^n$. That the convergence does hold for the whole series is not tautological (it involves permuting a limit with an infinite sum) and will be properly proved in Chapter 3.

1.4.3. Is the left-hand side in Cauchy's formula a q-analogue? It seems more difficult to prove directly that $\frac{(q^\alpha x; q)_\infty}{(x; q)_\infty}$ degenerates to $(1-x)^{-\alpha}$ when $q \to 1$. Here is one possible approach (sketched). We introduce:

$$\phi(x) := \sum_{n \geq 1} \frac{x^n}{n(1-q^n)}, \text{ where } |q|, |x| < 1.$$

Then $\phi(0) = 0$ and $\phi(x) - \phi(qx) = \sum_{n \geq 1} \frac{x^n}{n} = -\log(1-x)$, so $\Phi := e^\phi$ satisfies $\Phi(0) = 1$ and $\Phi(x) = \frac{1}{1-x} \Phi(qx)$, so, by the usual iteration process (along with a continuity argument), $\Phi(x) = \frac{1}{(x; q)_\infty} \cdot$ Now:

$$\frac{(q^\alpha x; q)_\infty}{(x; q)_\infty} = \frac{\Phi(x)}{\Phi(q^\alpha x)} = \exp\left(\phi(x) - \phi(q^\alpha x)\right).$$

On the other hand:

$$\phi(x) - \phi(q^\alpha x) = \sum_{n \geq 1} \frac{1 - q^{\alpha n}}{n(1-q^n)} x^n,$$

which tends termwise to $\sum_{n \geq 1} \frac{\alpha}{n} x^n = -\alpha \log(1-x)$ when $q \to 1$. If we admit that the convergence holds for the functions, we conclude (taking exponentials) that indeed $\frac{(q^\alpha x; q)_\infty}{(x; q)_\infty} \to (1-x)^{-\alpha}$.

1.4.4. Yet another argument. All the methods above are ad hoc in that they rest on explicit power series or infinite product expansions, which are a rare gift. We now show an approach with a much broader scope of application. However, for the moment, we do it heuristically; complete proofs will come in 11.2.

We fix α but show explicitly dependency on q in notation $f_q(x) := \frac{(q^\alpha x; q)_\infty}{(x; q)_\infty}$. That function satisfies, as we already saw in the first proof in 1.4.1, the q-difference equation $f_q(qx) = \frac{1-x}{1-q^\alpha x} f_q(x)$, whence:

$$\frac{f_q(x) - f_q(qx)}{(1-q)x} = \frac{1-q^\alpha}{1-q} \frac{1}{1-q^\alpha x} f_q(x) = \frac{[\alpha]_q}{1-q^\alpha x} f_q(x).$$

(We have therefore generalized the notation $[\alpha]_q$ for q-numbers to the case of a complex α.) Using the "q-derivation operator" D_q, which sends $f(x)$ to $D_q f(x) := \frac{f(x) - f(qx)}{(1-q)x}$, i.e.,[4] $D_q = \frac{1-\sigma_q}{(1-q)x}$, where the "$q$-dilatation operator" σ_q sends $f(x)$

[4] We enter a noncommutative world; as we shall see in Chapter 3, σ_q and x do not commute, so $1 - \sigma_q$ and $\frac{1}{(1-q)x}$ don't either, and we should rather write $D_q := \frac{1}{(1-q)x}(1-\sigma_q)$.

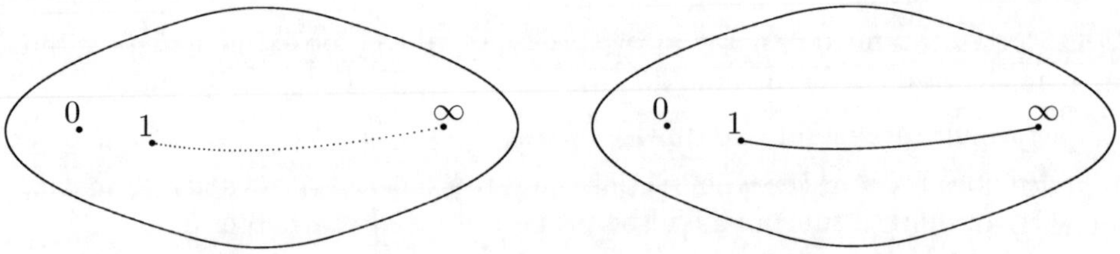

FIGURE 1.1. Condensation of a discrete half q-spiral

to $\sigma_q f(x) := f(qx)$, the q-difference equation has been rewritten:

$$D_q f_q = \frac{[\alpha]_q}{1 - q^\alpha x} f_q.$$

Now, D_q formally degenerates to the derivation operator $D : f \mapsto f'$, meaning that, if we fix f and let $q \to 1$, then $D_q f \to f'$. Moreover, there is an obvious initial condition $f_q(0) = 1$ that does not depend on q. So we could hope that f_q tends (when $q \to 1$) to some f such that $f(0) = 1$ and:

$$f' = \left(\lim_{q \to 1} \frac{[\alpha]_q}{1 - q^\alpha x} \right) f = \frac{\alpha}{1 - x} f.$$

In 11.2, we shall see that, under rather general conditions, this works; and then of course the solution to our differential equation with initial condition is $f(x) = (1 - x)^{-\alpha}$.

1.4.5. So, let's admit that we are able to prove that, if $|x| < 1$:

$$\lim_{q \to 1} \frac{(q^\alpha x; q)_\infty}{(x; q)_\infty} = (1 - x)^{-\alpha}.$$

However, for generic q, the left-hand side is a uniform meromorphic function on $\mathbf{C}$, with simple poles over the discrete logarithmic half-spiral $q^{-\mathbf{N}}$; while the right-hand side is a priori a multivalued function over $\mathbf{C} \setminus \{1\}$: analytic continuation along a simple loop turning once around 1 multiplies it by $e^{2i\pi\alpha}$ (see **[222]**). So first, we should examine the fate of the half q-spiral $q^{-\mathbf{N}}$ when $q \to 1$. If the limit process is arbitrary, poles will spread densely all around a wide portion of the plane.

EXERCISE 1.10. Let $a \in \mathbf{C}$, $|a| > 1$. Show that when $q \to 1$ with $|q| < 1$, some points of the set $q^{-\mathbf{N}}$ get infinitely often arbitrarily close to a.

So we impose a condition: the q-spiral should *condense*. One way to enforce this is to fix some $q_0 \in \mathbf{C}$, $0 < |q_0| < 1$, and set $q := q_0^\epsilon$ with $\epsilon \to 0$, $\epsilon > 0$, which we shall abbreviate into $\epsilon \to 0^+$. This means that $q_0 = e^{2i\pi\tau_0}$ for some τ_0 such that $\Im\, \tau_0 > 0$; and that $q = e^{2i\pi\tau}$, where $\tau := \tau_0 \epsilon$. Then the discrete logarithmic half-spiral $q^{-\mathbf{N}}$ condenses into the *continuous* logarithmic half-spiral $q^{-\mathbf{R}_+}$; see Figure 1.1.

Then we shall show in 11.2 that, when $q \to 1$ in the way stated above (the conditions can actually be slightly relaxed), $\dfrac{(q^\alpha x; q)_\infty}{(x; q)_\infty}$ tends to the principal determination of $(1 - x)^{-\alpha}$ (the one such that $0 \mapsto 1$) over the simply connected open set $\mathbf{C} \setminus q^{-\mathbf{R}_+}$: we have performed a *cut*.

1.4.6. Since we spoke of a cut from 1 to ∞, we should remember that 1 is not the only singularity of the multivalued function $(1-x)^{-\alpha}$ or of the differential equation $f' = \dfrac{\alpha}{1-x} f$. Indeed, setting $y := 1/x$, considered as a coordinate on the Riemann sphere $\mathbf{S}$ (so that the point at infinity has coordinate $y = 0$), and writing $g(y) := f(x)$, we see that g is solution of the differential equation $g'(y) = \dfrac{\alpha}{y(1-y)} g(y)$, which is singular (actually "regular singular"; see [**222**]) at $y = 0$. The basic solution is $g(y) = y^\alpha (1-y)^\alpha$, which is of course multivalued; but here we see that the cut needed to make it uniform must also prevent loops around $y = 0$. Connecting solutions "near 0" with solutions "near ∞" belongs to Riemann's theory of ordinary differential equations; can we do something similar here?

As a former president once said, "yes we can". Starting from q-difference equation $f(qx) = \dfrac{1-x}{1-ax} f(x)$, to look at its behaviour at ∞, we set $y := 1/x$ and $g(y) := f(x)$, so that:

$$g(qy) = f(1/qy) = f(x/q) = \frac{1-ax/q}{1-x/q} f(x) = \frac{1-a/qy}{1-1/qy} g(y) = a \frac{1-qy/a}{1-qy} g(y).$$

Because of the factor a, we cannot directly solve with a product expansion. Actually, for generic a, we cannot either solve with a power series.

EXERCISE 1.11. Prove that, if $a \notin q^{\mathbf{Z}}$, the equation admits no nontrivial power series solution (even allowing for negative exponents).

We shall suppose that some nontrivial analytic function e_a accounts for the factor a, that is: $e_a(qx) = e_a(x)$. We shall see in Chapter 2 that this can be done with e_a meromorphic over $\mathbf{C}^*$ (but, for generic a, with an essential singularity at 0). Then we see that the function $g(y) := e_a(y) \dfrac{(qy;q)_\infty}{(qy/a;q)_\infty}$ is a nontrivial solution, and that it is meromorphic over $\mathbf{C}^*$. We consider it as a solution "near ∞", which wants to be "connected" to our previous solution "near 0": $f(x) := \dfrac{(ax;q)_\infty}{(x;q)_\infty}$. The point is the following. Functions $f(x)$ and $g(1/x)$ behave in the same way when x is replaced by qx, they are multiplied by the same factor. So their quotient:

$$p(x) := g(1/x)/f(x) = e_a(1/x) \frac{(q/x;q)_\infty (x;q)_\infty}{(q/ax;q)_\infty (ax;q)_\infty}$$

is q-invariant, i.e., it satisfies $p(qx) = p(x)$. We shall see in Chapter 2 that q-invariant meromorphic functions over $\mathbf{C}^*$ are essentially the same thing as what is classically called "elliptic functions". The function p is a baby example of *Birkhoff's connection matrix* (see 12.1 in Chapter 12), an invertible matrix of elliptic functions connecting bases of solutions near 0 and near ∞ of a q-difference equation of arbitrary order. In this simple case, one can even write $p(x)$ explicitly with the use of Theta-functions.

EXERCISE 1.12. (i) Check directly on the above formula that $p(qx) = p(x)$.

(ii) Write $q = e^{2i\pi\tau}$, $x = e^{2i\pi z}$ and $P(z) := p(x)$. Show that P is meromorphic over $\mathbf{C}$ and Λ_τ-periodic, where Λ_τ is the lattice $\mathbf{Z} + \mathbf{Z}\tau$ (whence the relation to elliptic functions).

(iii) Using $e_a := e_{q,a}$ and the Theta-function θ_q, all introduced in Chapter 2, give an explicit expression for $p(x)$.

(iv) Look at the poles and zeroes of $p(x)$: how are they organized? Where do they come from?

(v) How does it behave whan $q \to 1$? (Guess what you can, then complete after reading Chapter 2 and, later, Section 12.3.)

1.5. Interlude: the q-exponentials

Although these objects are not directly related to Cauchy (and it actually seems hard to trace their precise historical origins), we have already been propelled into the themes of q-analogues and of q-Pochhammer symbols (i.e., $(a;q)_n$ and $(a;q)_\infty$); and q-exponentials are very strongly related to both of them (and also, by the way, to the q-binomial theorem). So this place seems fit for a first approach to them.

1.5.1. We shall write here:

$$f_q(x) := (x;q)_\infty := \prod_{n\geq 0}(1-q^n x) = \sum_{n\geq 0} \frac{(-1)^n q^{n(n-1)/2}}{(q;q)_n} x^n$$

and

$$g_q(x) := \frac{1}{(x;q)_\infty} = \prod_{n\geq 0} \frac{1}{(1-q^n x)} = \sum_{n\geq 0} \frac{1}{(q;q)_n} x^n.$$

In usual form, they respectively satisfy the q-difference equations:

$$f_q(qx) = \frac{1}{1-x} f_q(x) \text{ and } g_q(qx) = (1-x) g_q(x),$$

which we rewrite into the q-derivative form, i.e., using the *q-derivative operator* $D_q := \dfrac{1-\sigma_q}{(1-q)x}$, which sends $f(x)$ to $\dfrac{f(x)-f(qx)}{(1-q)x}$:

$$D_q f_q(x) = \frac{f_q(x)-f_q(qx)}{(1-q)x} = \frac{f_q(x)-f_q(x)/(1-x)}{(1-q)x} = \frac{-1}{1-q}\frac{1}{1-x} f_q(x)$$

and

$$D_q g_q(x) = \frac{g_q(x)-g_q(qx)}{(1-q)x} = \frac{g_q(x)-(1-x)g_q(x)}{(1-q)x} = \frac{1}{1-q} g_q(x).$$

Because of the occurrence of $1-q$ in the denominator, it seems hard to obtain a meaningful degeneracy when $q \to 1$. However, this is easily fixed by *rescaling* the variable x by a factor $\pm(1-q)$; so we introduce[5]:

$$E_q(x) := f_q(-(1-q)x) = (-(1-q)x;q)_\infty = \sum_{n\geq 0} q^{n(n-1)/2} \frac{(1-q)^n}{(q;q)_n} x^n,$$

$$e_q(x) := g_q((1-q)x) = \frac{1}{((1-q)x;q)_\infty} = \sum_{n\geq 0} \frac{(1-q)^n}{(q;q)_n} x^n.$$

Remember the introduction of q-factorials in 1.3. Since:

$$[n]!_q := \prod_{i=1}^{n} [i]_q = \prod_{i=1}^{n} \frac{1-q^i}{1-q} = \frac{(q;q)_n}{(1-q)^n},$$

---

[5]In **[93]**, as well as in much of the q-litterature, the notations e_q, E_q are given for the nonrescaled functions $g_q(x)$ and $f_q(-x)$, as well as the name "q-exponential". We rather use the same convention as in **[130]**: indeed, we prefer to consider q-exponentials as functions which degenerate into the classical exponential function.

we find the nicer forms:

$$E_q(x) = \sum_{n \geq 0} \frac{q^{n(n-1)/2}}{[n]!_q} x^n \text{ and } e_q(x) = \sum_{n \geq 0} \frac{1}{[n]!_q} x^n.$$

Since obviously $\lim\limits_{q \to 1} [n]!_q = n!$, the above form indicates that, termwise, the series $e_q(x)$ and $E_q(x)$ both degenerate into e^x. For that reason, they are considered as q-analogues of the exponential function and called *q-exponentials*. We intend to discuss now if that degeneracy is something more than a formal phenomenon, i.e., if the functions $e_q(x)$ and $E_q(x)$ have e^x as a limit. Note that, since $E_q(x) = 1/e_q(-x)$, it is enough to prove it for one of them.

1.5.2. We now proceed to prove our contention under the assumption that $0 < q < 1$. In Chapter 2, we shall give a more general proof. Since we already know that $e_q \to \exp$ termwise, it is enough to show that the convergence is dominated over any disk $\mathring{D}(0, R)$, i.e., that, for any $R > 0$, there is a sequence (M_n) independent of q such that $\left| \frac{1}{[n]!_q} \right| \leq M_n$ for all n and $\sum M_n R^n < \infty$: we say that the series e_q is dominated by $\sum M_n x^n$.

So let $R > 1$ (this does not restrict generality); and then let $n_0 \in \mathbf{N}$ such that $n_0 > R$. Then there is a unique $q_0 \in]0; 1[$ such that $1 + q_0 + \cdots + q_0^{n_0 - 1} = R$ (since the left-hand side continuously increases from 0 to $n_0 > R$ when q_0 goes from 0 to 1). We may (and will) assume that $1 > q > q_0$. Then, for every $n \geq n_0$:

$$[n]!_q = [n_0]!_q [n_0 + 1]_q \cdots [n]_q \geq [n_0]!_q R^{n - n_0} \geq [n_0]!_{q_0} R^{n - n_0}.$$

So the series $e_q = \sum\limits_{n \geq 0} \frac{x^n}{[n]!_q}$ is dominated by $\sum\limits_{n \leq n_0} \frac{R^n}{[n]!_q} + \sum\limits_{n > n_0} \frac{R^{n_0}}{[n_0]!_{q_0}} (x/R)^n$. This implies that e_q converges to exp uniformly over compact subsets of $\mathbf{C}$.

1.5.3. We return to the q-derivative form of the q-difference equation hoping to observe on it a nice degeneracy. From the usual form of the q-difference equations:

$$e_q(qx) = (1 - (1 - q)x) \, e_q(x) \text{ and } E_q(qx) = \frac{1}{1 + (1 - q)x} E_q(x),$$

we deduce:

$$D_q e_q(x) = e_q(x) \text{ and } D_q E_q(x) = \frac{1}{1 + (1 - q)x} E_q(x),$$

which respectively degenerate into the differential equations $f' = f$ and $g' = g$. Along with the initial conditions $e_q(0) = E_q(0) = 1$, this is again an argument to consider $e_q(x)$ and $E_q(x)$ as q-analogues of the exponential funxction.

1.5.4. Can we compare $e_q(x+y)$ and $e_q(x)e_q(y)$? Hoping for equality, we compare first the homogeneous components and guess that $\frac{(x+y)^n}{[n]!_q} = \sum\limits_{i+j=n} \frac{x^i}{[i]!_q} \frac{y^j}{[j]!_q}$. Using the q-binomial coefficients encountered in 1.3, this would be written:

$$(x + y)^n = \sum_{i+j=n} \begin{bmatrix} n \\ i \end{bmatrix}_q x^i y^j.$$

For $n = 0$ or 1, this is all right. For $n = 2$, this would imply:

$$\frac{x^2 + xy + yx + y^2}{[2]!_q} = \frac{x^2}{[2]!_q [0]!_q} + \frac{xy}{[1]!_q [1]!_q} + \frac{y^2}{[0]!_q [2]!_q} \Longleftrightarrow yx = qxy.$$

Conversely, we shall see in Chapter 2 that such noncommuting variables do satisfy the above equality of the homogeneous components, thus $e_q(x + y) = e_q(x)e_q(y)$. We shall not explore noncommutative algebra (apart from the ring of q-difference operators in Chapter 2) although it is important in that part of q-calculus involved in the theory of "quantum groups". However, since we already encountered Gauß' q-analogue of the expansion of $(1 + x)^n$, we may wonder if the noncommutative q-expansion above can be deduced from it. Instead we shall derive Gauß' formula from the noncommutative expansion.

So let $P_n(x) := (1 + x)_q^n$, so that $P_0 = 1$ and $P_n(x) = (1 + x)P_{n-1}(qx)$, i.e., $P_n = (1 + x)\sigma_q P_{n-1}$, where σ_q is the q-difference operator $P(x) \mapsto P(qx)$. This way, we see that $P_n = (x + x\sigma_q)^n P_0$: we use algebra on operators and x stands for the operator $P \mapsto xP$. Here, x and σ_q do not commute: $\sigma_q x$ (this, of course, denotes the compositum of operators x and σ_q) sends $P(x)$ to $\sigma_q(xP) = qxP(qx)$, so $\sigma_q x = qx\sigma_q$; and similarly $\sigma_q(x\sigma_q) = q(x\sigma_q)\sigma_q$ and the noncommutative q-expansion implies:

$$(x + x\sigma_q)^n = \sum_{i+j=n} \begin{bmatrix} n \\ i \end{bmatrix}_q (\sigma_q x)^i x^j.$$

EXERCISE 1.13. Prove that $(\sigma_q x)^i = q^{i(i-1)/2} x^i {\sigma_q}^i$.

Using the exercise, we deduce that:

$$(x + x\sigma_q)^n = \left(\sum_{i+j=n} q^{i(i-1)/2} \begin{bmatrix} n \\ i \end{bmatrix}_q x^i \right) {\sigma_q}^n.$$

Applying that operator to $P_0 = 1$, we immediately obtain Gauß' formula.

1.6. Heine (Eduard Heine, German, 1821–1881)

In the article **[116]** "Über die Reihe

$$1 + \frac{(q^\alpha - 1)(q^\beta - 1)}{(q - 1)(q^\gamma - 1)} x + \frac{(q^\alpha - 1)(q^{\alpha+1} - 1)(q^\beta - 1)(q^{\beta+1} - 1)}{(q - 1)(q - 1)(q^\gamma - 1)(q^{\gamma+1} - 1)} x^2 + \cdots \text{"}$$

published in 1846, Heine used[6] the letter q for the "basic parameter" and the letter x for the "supplementary parameter". This seems to be the christening of q-calculus under that name. The letter q reappeared in XXth century referring to *quantum* calculus, but *that* seems to be a mere coincidence; see subsection 1.12.2 . The section concentrates on Heine's main contribution to our domain, *q-hypergeometric series* (also called *Basic hypergeometric series*); we shall return to that theme in Section 4.4 of Chapter 4 and then again in [CV]. Other contributions of Heine will be recalled in Sections 1.9 and 1.11 and in Chapter 2. Basic hypergeometric series were later extensively studied by Thomae, but we shall not go that far.

[6] Although Berndt seems to imply the opposite in **[18]**.

1.6.1. Basic hypergeometric series. A geometric series $(a_n)_{n\in\mathbf{N}}$ is one for which a_{n+1}/a_n is constant. If it is a rational function of n, and if the a_n are the coefficients of a power series, it gives rise to a (generalized) hypergeometric series; plain hypergeometric series, going back to Euler and Gauß, are those for which a_{n+1}/a_n is a quotient of quadratic polynomials in n (see Appendix C). They had a historical importance in the birth of Riemann's theory of ordinary differential equations: they are indeed the simplest possible example of equations with a noncommutative monodromy group [**125**, **222**].

In 1846, Heine introduced their q-analogue; with a slight change of notations, it is:

$$ {}_2\Phi_1(a,b;c;q,x) := \sum_{n\geq 0} \frac{(a;q)_n(b;q)_n}{(c;q)_n(q;q)_n} x^n. $$

The parameters a, b, c are nonzero complex numbers. The indices 2, 1 respectively refer to the numbers of "up" parameters, a, b, and "down" parameters, c. The special parameter q is the *base*, whence the terminology "basic". There are more general basic hypergeometric series ${}_{k+1}\Phi_k$ and even ${}_\ell\Phi_k$ with $\ell \neq k+1$, but the latter ones are *irregular*. We shall study ${}_2\Phi_1$ at some depth in 4.4 and ${}_\ell\Phi_k$ in [CV]; but the bible is of course [**93**] (also see [**84**, **235**, **236**]).

The standard convergence test involves:

$$ \frac{a_{n+1}}{a_n} = \frac{(a;q)_{n+1}(b;q)_{n+1}}{(c;q)_{n+1}(q;q)_{n+1}} \frac{(c;q)_n(q;q)_n}{(a;q)_n(b;q)_n} = \frac{(1-aq^n)(1-bq^n)}{(1-q^{n+1})(1-cq^n)}, $$

which tends to 1 as $n \to +\infty$, so the radius of convergence is 1. Note that a_{n+1}/a_n is the quotient of two quadratic polynomials in q^n, explaining the "hypergeometric" denomination.

Moreover, ${}_2\Phi_1$ satisfies a linear q-difference equation of order 2. Indeed, starting from:

$$ \frac{a_{n+1}}{a_n} = \frac{(1-aq^n)(1-bq^n)}{(1-q^{n+1})(1-cq^n)} \Longrightarrow (1-q^{n+1})(1-cq^n)a_{n+1} = (1-aq^n)(1-bq^n)a_n, $$

which we rewrite as $(1-(1+c/q)q^{n+1}+cq^{2n+1})a_{n+1} = (1-(a+b)q^n+abq^{2n})a_n$, then multiply by x^{n+1} and sum for $n \geq 0$; we thus find that ${}_2\Phi_1$ satisfies the relation:

$$ \Phi(x) - (1+c/q)\Phi(qx) + c/q\Phi(q^2x) = x\left(\Phi(x) - (a+b)\Phi(qx) + ab\Phi(q^2x)\right). $$

(The term in x^0 which should have been missing on the left-hand side is actually 0.) We rewrite this as the q-difference equation:

$$ (c/q - abx)\Phi(q^2x) - ((1+c/q)-(a+b)x)\Phi(qx) + (1-x)\Phi(x) = 0. \tag{1.1} $$

From now on, we keep the lighter notation $\Phi(x)$ for ${}_2\Phi_1(a,b;c;q,x)$. We know a priori that Φ is holomorphic over $\mathring{\mathrm{D}}(0;1)$. But the functional equation can be used to provide a meromorphic continuation, as follows. Let $r := |q| < 1$ and define, for $x \in \mathring{\mathrm{D}}(0;r^{-1})$:

$$ \Phi(x) := \frac{-1}{1-x}\left((c/q-abx)\Phi(q^2x) - ((1+c/q)-(a+b)x)\Phi(qx)\right). $$

The right-hand side makes sense because $qx, q^2x \in \mathring{D}(0;1)$ whenever $x \in \mathring{D}(0;r^{-1})$. Note that $\mathring{D}(0,1) \subset \mathring{D}(0;r^{-1})$, so, if $x \in \mathring{D}(0,1)$, this could conflict the earlier definition of Φ; but it does not precisely because Φ satisfies the q-difference equation. So now we have an extended function Φ on $\mathring{D}(0;r^{-1})$, and, by construction, it still satisfies the q-difference equation. The main difference is that it is meromorphic with a simple pole[7] at $x = 1$. Iterating the process, we successively get meromorphic continuations of Φ to the disks $\mathring{D}(0;r^{-n})$ for all $n \geq 0$; all solutions of the same q-difference equation. Since $r < 1$, this eventually leads to a meromorphic continuation of Φ to the whole of $\mathbf{C}$, with simple poles over $\{q^{-n} \mid n \in \mathbf{N}\} = q^{-\mathbf{N}}$.

Dealing with a linear equation of order 2, we could hope for another solution, "independent" from Φ in some sense. This "independence" will be precisely defined and studied in Chapter 5. Inspired by the theory of classical hypergeometric series, we might look for that new solution in the form $x^\lambda\Psi$, with Ψ a power series such that $\Psi(0) = 1$. Substituting $x^\lambda\Psi$ for Φ in (1.1), we find (after division by x^λ):

$$(c/q - abx)q^{2\lambda}\Psi(q^2x) - ((1 + c/q) - (a + b)x)q^\lambda\Psi(qx) + (1 - x)\Psi(x) = 0.$$

Evaluation at $x = 0$ then implies that the *exponent* $s := q^\lambda$ (terminology to be explained in Chapter 5) must be solution of the *characteristic equation*:

$$(c/q)s^2 - (1 + c/q)s + 1 = 0.$$

Solution $s = q^\lambda = 1$ obviously corresponds to our original function $\Phi = {}_2\Phi_1$. The other solution is $s = q/c$ does provide an independent solution, as we shall see in Section 4.4. Of course, there is a degenerate case when $c = q$ and $s = 1$ is a "double exponent". This also will be detailed in Section 4.4.

Note however that the only property of the factor $e(x) := x^\lambda$ that was used is that $e(qx) = se(x)$, where $s := q^\lambda$. It so happens that such functions $e_s(x)$ can be found that are not multivalued (as x^λ generically is) but rather uniform, meromorphic over $\mathbf{C}^*$. (At 0 they usually have an essential singularity.) Such functions will be defined in Chapter 2 and extensively used in the sequel of the book. Indeed, as explained in the introduction, it is one of the big directions of this work to avoid the use of multivalued functions.

On the other hand, resorting systematically to uniform meromorphic functions, as we shall do in this book, has the consequence that there is no more ambiguity of analytic continuation. But that ambiguity, encoded in the monodromy group and representation, was an essential organizing factor in Riemann's theory (and actually the source of Riemann-Hilbert correspondence). So we shall have to find a substitute for it.

EXERCISE 1.14. By taking a or b or c to be 0, or by requiring that $a = qc$, or $c = qa$, etc, find special (possibly degenerated) cases of interest of basic hypergeometric series. This should include examples tackled in the previous sections.

1.6.2. Why it is a q-analogue of hypergeometric series. Apart from the form of the coefficients, similar to that of the classical coefficients, we would like to see basic hypergeometric series as an "unfolding" of classical ones: this meaning that the basic degenerates to the classical when $q \to 1$. We shall check that first on

[7] In special cases, the right factor could vanish at $x = 1$ and the pole be missing; we shall not detail that point here.

the coefficients, then on the functional equation. However, as the theory develops in the book, we shall see that there is much more to the q-analogy.

Inspired by the previous sections, and above all by Heine himself, we set $a := q^\alpha$, $b := q^\beta$ and $c := q^\gamma$. Note that this implies:

$$\frac{(a;q)_n}{(1-q)^n} = \frac{1-a}{1-q} \cdots \frac{1-q^{n-1}a}{1-q} = \frac{1-q^\alpha}{1-q} \cdots \frac{1-q^{\alpha+n-1}}{1-q} = [\alpha]_q \cdots [\alpha+n-1]_q,$$

and similarly for $(b;q)_n$ and $(c;q)_n$. Therefore, when $q \to 1$:

$$\frac{(a;q)_n(b;q)_n}{(c;q)_n(q;q)_n} = \frac{[\alpha]_q \cdots [\alpha+n-1]_q[\beta]_q \cdots [\beta+n-1]_q}{[n]!_q[\gamma]_q \cdots [\gamma+n-1]_q} \to \frac{(\alpha)_n(\beta)_n}{n!(\gamma)_n}.$$

Therefore ${}_2\Phi_1$ degenerates coefficientwise to

$${}_2F_1(\alpha,\beta;\gamma;x) := \sum_{n\geq 0} \frac{(\alpha)_n(\beta)_n}{n!(\gamma)_n} x^n,$$

the classical hypergeometric series. It can be proved, though in a rather indirect way, that there is actually true convergence of the function ${}_2\Phi_1$ to the function ${}_2F_1$ 12.3. Of course, some care must be taken for the poles; see 1.4.5 for a simpler example.

EXERCISE 1.15. Prove directly convergence for $x \in \mathring{\mathrm{D}}(0;1)$.

We now seek a form of the functional equation involving q-derivation. Remember from 1.4.4 that this means the operator $D_q = \dfrac{1-\sigma_q}{(1-q)x}$, so we first rewrite (1.1) using σ_q:

$$(c/q - abx){\sigma_q}^2\Phi(x) - ((1+c/q) - (a+b)x)\sigma_q\Phi(x) + (1-x)\Phi(x) = 0,$$

or, equivalently:

$$L_q\Phi = 0, \text{ where } L_q := (c/q - abx){\sigma_q}^2 - ((1+c/q) - (a+b)x)\sigma_q + (1-x).$$

Then we should replace σ_q by $1-(1-q)xD_q$. Remember that this is in a non-commutative world, since (in the algebra of operators) $\sigma_q x = qx\sigma_q$. However, in studying classical hypergeometric equation, it has often proved useful to use, instead of $D := \dfrac{d}{dx}$ the so-called *Euler's differential operator* $\delta := x\dfrac{d}{dx}$. Accordingly, we introduce its q-analog $\delta_q := xD_q = \dfrac{1-\sigma_q}{(1-q)}$; therefore $\sigma_q = 1-(1-q)\delta_q$, whence ${\sigma_q}^2 = 1-2(1-q)\delta_q + (1-q)^2{\delta_q}^2$ (no more problems of commutation here). We get:

$$\begin{aligned} L_q &= (c/q - abx)\left(1-2(1-q)\delta_q + (1-q)^2{\delta_q}^2\right) \\ &\quad - ((1+c/q)-(a+b)x)(1-(1-q)\delta_q) + (1-x) \\ &= (1-q)^2\left((c/q-abx){\delta_q}^2 + \frac{(1-c/q)-(a+b-2ab)x}{1-q}\delta_q - \frac{(1-a)(1-b)}{(1-q)^2}x\right). \end{aligned}$$

So, for every fixed q:

$$(c/q-abx){\delta_q}^2\,{}_2\Phi_1(a,b;c;q,x) + \frac{(1-c/q)-(a+b-2ab)x}{1-q}\delta_q\,{}_2\Phi_1(a,b;c;q,x) -$$
$$\frac{(1-a)(1-b)}{(1-q)^2}x\,{}_2\Phi_1(a,b;c;q,x) = 0.$$

If we substitute $a := q^\alpha$, etc, we find that, when $q \to 1$:

$$c/q - abx \to 1 - x, \quad \frac{(1-a)(1-b)}{(1-q)^2} \to \alpha\beta,$$

$$\frac{(1-c/q)-(a+b-2ab)x}{1-q} \to (\gamma-1)-(\alpha+\beta)x,$$

so our q-difference equation degenerates into:

$$(1-x)\delta^2 F(x) + ((\gamma-1)-(\alpha+\beta)x)\,\delta F(x) - \alpha\beta x F(x) = 0.$$

This is indeed one of the forms of the hypergeometric equation; see for instance [**222**, p 151, 11.3].

EXERCISE 1.16. Compute $(1-(1-q)xD_q)^2$.

1.7. Jacobi (Carl Gustav Jakob Jacobi, German, 1804–1851)

1.7.1. Theta functions. The theory of elliptic functions, started in the eighteenth century by Euler, Gauß, Legendre ..., was one of the most beautiful realizations of nineteenth century mathematics. Jacobi founded it anew by his theory of Theta functions, which had been anticipated by Gauß (unpublished!) and would be developed again, seemingly from scratch (and in a very different style), by Ramanujan. We shall present a (very small!) part of Jacobi's theory in Chapter 2. For a historical view, see for instance [**112, 255**] but, above all [**69**].

The standard presentation of Theta functions introduces them as holomorphic functions over $\mathbf{C}$ (thus entire functions) which transform nicely under the action of a lattice. One can suppose that the lattice is $\Lambda = \Lambda_\tau := \mathbf{Z} + \mathbf{Z}\tau$, where $\tau \in \mathcal{H}$, the Poincaré half plane, meaning that $\Im\tau > 0$. Elliptic functions (relative to Λ) are meromorphic functions f such that $f(z+1) = f(z+\tau) = f(z)$, i.e., f is Λ-periodic. Such a function, if not constant, necessarily has poles. However, if one merely asks for an "automorphic" behaviour, such as:

$$f(z+1) = f(z), \quad f(z+\tau) = e^{-2i\pi z} f(z),$$

one can easily come up with holomorphic solutions[8], such as:

$$\Theta_\tau(z) := \Theta(z,\tau) := \sum_{n\in\mathbf{Z}} \exp(n^2 i\pi\tau + ni\pi(2z-\tau)) = \sum_{n\in\mathbf{Z}} e^{(n(n-1)/2)(2i\pi\tau)} e^{n.2i\pi z},$$

seen, of course, as a function of z. The theory then says (see Appendix A) that every elliptic function is (up to a nonzero constant factor) a product of translated Theta functions $\Theta_\tau(z-\lambda)$, possibly with negative exponents.

1.7.2. The "multiplicative" model. There is a "multiplicative" form of Theta functions that is particularly well suited for q-calculus[9]. Let $q := e^{2i\pi\tau}$ (so that $0 < |q| < 1$) and $x := e^{2i\pi z} \in \mathbf{C}^*$. We define:

$$\theta_q(x) := \Theta_\tau(z) = \sum_{n\in\mathbf{Z}} q^{n(n-1)/2} x^n.$$

[8]Note that there are many other possible functional equations and corresponding Theta functions; see for instance [**174, 255**].

[9]Another merit of the "multiplicative" form is that it extends smoothly to p-adic analysis and algebraic geometry (theory of elliptic curves), which the "additive" form above does not; this is due to John Tate; see [**210**].

This is well defined for $x \in \mathbf{C}^*$, since a complex $z \in \mathbf{C}$ such that $x := e^{2i\pi z}$ is defined modulo $\mathbf{Z}$ while Θ_τ is $\mathbf{Z}$-periodic. Now, θ_q is holomorphic over $\mathbf{C}^*$ and satisfies:

$$\theta_q(qx) = \frac{1}{x}\theta_q(x).$$

That relation is the basis of the importance of θ_q in this book.

EXERCISE 1.17. (i) What about $\theta_q(x^{-1})$?

(ii) Let $c_1, \dots, c_k, d_1, \dots, d_l \in \mathbf{C}^*$ and let $f(x) := \dfrac{\theta_q(x/c_1)\cdots\theta_q(x/c_k)}{\theta_q(x/d_1)\cdots\theta_q(x/d_l)}$. Under what assumption on the c_i, d_j does f satisfy: $f(qx) = f(x)$?

Call *q-elliptic* or *q-constant* a meromorphic function f on $\mathbf{C}^*$ such that $f(qx) = f(x)$. This is equivalent to say that $\phi(z) := f(e^{2i\pi z})$ is Λ-elliptic, so, by the general theory, all q-elliptic functions are generated (up to nonzero constant factors) by "translated" functions $\theta_q(x/c)$, i.e., they all come in the form found in the exercise above.

Along with the functional equation $\theta_q(qx) = \dfrac{1}{x}\theta_q(x)$, there are two much more difficult properties that make θ_q useful in q-calculus and in q-difference theory. One is *modularity*, we shall state and prove it in Chapter 2 (see also Appendix A). It, by the way, plays a central role in wide parts of mathematics, even nowadays; it also played a very important role in Ramanujan's contributions to the q-world. We shall have use for it specially to study degeneracies when $q \to 1$ (Chapter 2 and Sections 11.2 and 12.3). For the moment, we do not go into detail.

1.7.3. Triple Product Formula. The other "more difficult property", Jacobi's celebrated *Triple Product Formula*, although not very central in mathematics, happened to play an important role in q-calculus and q-difference theory. It was, as uual, anticipated by Gauß and rediscovered by Ramanujan. It states:

$$\theta_q(x) = (q;q)_\infty(-x;q)_\infty(-q/x;q)_\infty = \prod_{n\geq 1}(1-q^n)\prod_{n\geq 0}(1+q^n x)(1+q^{n+1}/x).$$

First let us see how one could ever guess it.

EXERCISE 1.18. Show that $\theta_q(-1) = 0$. (Reindex the sum by $n \leftarrow -n$ and by $n \leftarrow n+1$.)

Thus, from the functional equation, all $\theta_q(-q^n) = 0$ $(n \in \mathbf{Z})$ and one can try to factor out $\theta_q(x) = \prod\limits_{n\geq 0}(1+q^n x)(1+q^{n+1}/x)f(x)$, f holomorphic over $\mathbf{C}^*$. Then direct calculation shows that $f(qx) = f(x)$, so f is holomorphic and q-elliptic, thus constant.

EXERCISE 1.19. (i) Show that $T(x) := (-x;q)_\infty(-q/x;q)_\infty$ satisfies $T(qx) = x^{-1}T(x)$.

(ii) Deduce that, in the power series expansion $T(x) = \sum\limits_{n\in\mathbf{Z}} t_n x^n$, one has $t_{n+1} = q^n t_n$, so $t_n = q^{n(n-1)/2}t_0$, so $T = t_0\theta_q$.

So far we have proved by two methods that $\theta_q(x) = C_q T(x)$ for some nonzero constant C_q. The latter is trickier to determine. We shall give a complete proof of the Triple Product Formula in Chapter 2, but we propose here another path to the determination of C_q. (I found it in [**70**, Exercise 28 p 357]; I don't know where Dieudonné found it.) Indeed it is traditional in this domain to give various elementary proofs of a given q-identity.

We first define Laurent's polynomials $\phi_n(x) := \prod_{i=1}^{n} (1 + q^{2i-1}x)(1 + q^{2i-1}/x)$. One can derive an explicit form from Gauß' q-binomial theorem, but we prefer to proceed directly. Obviously:

$$\phi_n(x) = c_0 + c_1(x + x^{-1}) + \cdots + c_n(x^n + x^{-n}) = \sum_{i \in \mathbf{Z}} c_i x^i,$$

where all $c_i \in \mathbf{C}$, $c_{-i} = c_i$ and $c_i = 0$ for $|i| > n$. Also, clearly, the dominant term has coefficient $c_n = q.q^3 \cdots q^{2n-1} = q^{n^2}$. On the other hand, from the relation $\phi_n(x) = (-qx; q^2)_n(-q/x; q^2)_n$, one draws:

$$\phi_n(q^2x) = \frac{1 + q^{2n+1}x}{1 + qx} \frac{1 + q^{-1}x^{-1}}{1 + q^{2n-1}x^{-1}} \phi_n(x) = \frac{1 + q^{2n+1}x}{qx + q^{2n}} \phi_n(x),$$

whence:

$$(qx + q^{2n})\phi_n(q^2x) = (1 + q^{2n+1}x)\phi_n(x) \Longrightarrow qq^{2i-2}c_{i-1} + q^{2n}q^{2i}c_i = c_i + q^{2n+1}c_{i-1},$$

and then the recursive relation:

$$(1 - q^{2n+2i})c_i = (q^{2i-1} - q^{2n+1})c_{i-1}.$$

EXERCISE 1.20. Check that it implies that $c_i = 0$ for $|i| > n$.

For $i = 1, \ldots, n$, we deduce, iterating the recursive relation:

$$c_i = q^{2i-1} \frac{1 - q^{2n-2i+2}}{1 - q^{2n+2i}} c_{i-1} = \cdots = q^{i^2} \frac{(1 - q^{2n-2i+2}) \cdots (1 - q^{2n})}{(1 - q^{2n+2i}) \cdots (1 - q^{2n+2})} c_0,$$

i.e.:

$$c_i = q^{i^2} \frac{(q^2; q^2)_n}{(q^2; q^2)_{n-i}} \frac{(q^2; q^2)_n}{(q^2; q^2)_{n+i}} c_0.$$

Since we already know the value $c_n = q^{n^2}$, we deduce first that $c_0 = \dfrac{(q^2; q^2)_{2n}}{(q^2; q^2)_n^2}$, then that:

$$c_i = q^{i^2} \frac{(q^2; q^2)_{2n}}{(q^2; q^2)_{n-i}(q^2; q^2)_{n+i}}.$$

This was proved for $i = 0, 1, \ldots, n$, but, since it is invariant by $i \leftarrow -i$, it actually holds for $i = -n, \ldots, n$. Note that one can recognize there Gauß' q-binomial coefficients.

We now slightly change notations: let $\phi_n(x) = \sum c_i(n)x^i$, where $c_i(n) = q^{i^2} d_i(n)$, $d_i(n) := \dfrac{(q^2; q^2)_{2n}}{(q^2; q^2)_{n-i}(q^2; q^2)_{n+i}}$ (and 0 if $|i| > n$). Clearly, for any fixed i, $d_i(n) \to \dfrac{1}{(q^2; q^2)_\infty}$ when $n \to +\infty$, so $\phi_n(x) \to \sum_{n \in \mathbf{Z}} q^{n^2} x^n$ at least termwise. So, formally, i.e., without justification of the interversion of a limit with an infinite sum, we get:

$$\prod_{n \geq 1} (1 + q^{2n-1}x)(1 + q^{2n-1}/x) = \frac{1}{(q^2; q^2)_\infty} \sum_{n \in \mathbf{Z}} q^{n^2} x^n.$$

THEOREM 1.1.

$$\sum_{n \in \mathbf{Z}} q^{n^2} x^n = (q^2; q^2)_\infty (-qx; q^2)_\infty (-q/x; q^2)_\infty.$$

EXERCISE 1.21. Deduce the Triple Product Formula from the above.

Now we sketch a justification of the above interversion $\lim \sum = \sum \lim$. For that, we prove that the convergence is dominated. For $-n \leq i \leq n$:

$$c_i(n) = q^{i^2} \frac{(q^2;q^2)_{2n}}{(q^2;q^2)_{n-i}(q^2;q^2)_{n+i}} = \frac{q^{i^2}}{(q^2;q^2)_\infty} \frac{(q^{2n-2i+2};q^2)_\infty (q^{2n+2i+2};q^2)_\infty}{(q^{4n+2};q^2)_\infty},$$

using the relation $(a;q)_n = \dfrac{(a;q)_\infty}{(q^n a;q)_\infty}$ (proof straightforward). We rewrite the above as:

$$c_i(n) = \frac{q^{i^2} f(q^{2n-2i+2}) f(q^{2n+2i+2})}{f(q^2) f(q^{4n+2})}, \text{ where } f(x) := (x;q^2)_\infty.$$

The function f is holomorphic over $\mathring{\mathrm{D}}(0,|q^2|)$ and continuous over $\overline{\mathrm{D}}(0,|q^2|)$, and it does not vanish there. So there are $m, M > 0$ such that $m \geq |f(x)| \leq M$ for all $x \in \overline{\mathrm{D}}(0,|q^2|)$. This implies that, for $-n \leq i \leq n$:

$$|c_i(n)| \leq |q|^{i^2} \frac{M^2}{m^2},$$

which obviously also holds for $|i| > n$. So the series $\sum_i c_i(n) x^i$ is dominated by $\sum_i q^{i^2} x^i$, which is enough to conclude.

1.8. Rogers (Leonard James Rogers, English, 1862–1933)

At about the same time the theory of q-difference equations was busy being born, along very different lines—somewhere between Jacobi and Ramanujan—Rogers made important discoveries based on "pure ingenuity" (according to Hardy; see [**109**]). We describe the two most famous of them. They bear the name "Rogers-Ramanujan" because, although their discovery (definitions, statements, and complete proofs) was entirely Rogers's work[10], they became famous when Ramanujan rediscovered them (also see [**18**]): first Ramanujan guessed them without proof; then, after reading Rogers, he devised proofs of his own.

1.8.1. Rogers-Ramanujan continued fraction.

It is usually defined as:

$$R(q) := q^{1/5} \cfrac{1}{1 + \cfrac{q}{1 + \cfrac{q^2}{1 + \cdots}}}.$$

The factor $q^{1/5}$ is motivated by "modular properties", which we shall not discuss here (see [**186**] and Chapters 2, 4, as well as [CV]), so we shall consider the fraction without said factor. Evaluating the fraction involves (again!) a supplementary parameter, i.e., we set for instance:

$$f(x) := \cfrac{x}{1 + \cfrac{qx}{1 + \cfrac{q^2 x}{1 + \cdots}}},$$

[10] According to Hardy ([**187**, p 344]), the identities were also rediscovered by Schur; Schur's approach is detailed in [**130**, Chapters 10,11]. They also were independently rediscovered (but without proofs) by Baxter; see Subsection 1.12.2.

and we look for $f(1)$. Here we work formally (we shall later return to a more rigorous study). Clearly:

$$f(x) = \frac{x}{1+f(qx)}.$$

This is a nonlinear q-difference equation and we shall not study such general problems in this book. But we shall see in 4.3.5 how to linearize it: we look for $f = \dfrac{u}{v}$ and solve the functional equation through the path:

$$f(x) = \frac{x}{1+f(qx)} \Longleftrightarrow \frac{u(x)}{v(x)} = \frac{xv(qx)}{u(qx)+v(qx)} \Longleftarrow \begin{cases} u(x) = xv(qx), \\ v(x) = u(qx)+v(qx). \end{cases}$$

We shall see in 4.3.5 that indeed (in spite of the one-way logical arrow at the last step) all solutions can indeed be found that way. Now this system of q-difference equations of rank 1 can be reduced to a single q-difference equation of rank 2:

$$\begin{cases} u(x) = xv(qx), \\ v(x) = u(qx)+v(qx), \end{cases} \Longleftrightarrow \begin{cases} qxv(q^2x)+v(qx)-v(x) = 0, \\ u(x) = xv(qx). \end{cases}$$

The q-difference equation $qxv(q^2x)+v(qx)-v(x) = 0$ admits a unique power series solution:

$$v(x) := \sum_{n\geq 0} \frac{q^{n^2}}{(q;q)_n} x^n.$$

EXERCISE 1.22. Prove it by identification of coefficients and resolution of the recursive relation. (Or else wait till 4.3.5.)

So in the end:

$$f(x) = \frac{xv(qx)}{v(x)} = \frac{\sum_{n\geq 0} \dfrac{q^{n^2+n}}{(q;q)_n} x^{n+1}}{\sum_{n\geq 0} \dfrac{q^{n^2}}{(q;q)_n} x^n} \Longrightarrow \cfrac{1}{1+\cfrac{q}{1+\cfrac{q^2}{1+\cdots}}} = \frac{\sum_{n\geq 0} \dfrac{q^{n^2+n}}{(q;q)_n}}{\sum_{n\geq 0} \dfrac{q^{n^2}}{(q;q)_n}}.$$

1.8.2. Rogers-Ramanujan identities. However, the really impressing part of Rogers' discoveries was the evaluation[11] of the q-series appearing as numerator and denominator of the continued fraction in the form of infinite products:

$$\sum_{n\geq 0} \frac{q^{n^2}}{(q;q)_n} = \frac{1}{(q;q^5)_\infty (q^4;q^5)_\infty}, \quad \sum_{n\geq 0} \frac{q^{n(n+1)}}{(q;q)_n} = \frac{1}{(q^2;q^5)_\infty (q^3;q^5)_\infty},$$

from wich it follows that:

$$\cfrac{1}{1+\cfrac{q}{1+\cfrac{q^2}{1+\cdots}}} = \frac{(q;q^5)_\infty (q^4;q^5)_\infty}{(q^2;q^5)_\infty (q^3;q^5)_\infty}.$$

Now remember that, by Jacobi's Triple Product Formula (1.7.3), we have:

$$\theta_{q^5}(x) = (q^5;q^5)_\infty (-x;q^5)_\infty (-q^5/x;q^5)_\infty,$$

whence:

$$\theta_{q^5}(-q) = (q^5;q^5)_\infty (q;q^5)_\infty (q^4;q^5)_\infty \text{ and } \theta_{q^5}(-q^2) = (q^5;q^5)_\infty (q^2;q^5)_\infty (q^3;q^5)_\infty,$$

[11]Not to be proved in this book, we return to it in [CV].

so that:

$$\cfrac{1}{1+\cfrac{q}{1+\cfrac{q^2}{1+\cdots}}} = \frac{\theta_{q^5}(-q)}{\theta_{q^5}(-q^2)}.$$

In Chapter 2 we shall study the asymptotic behaviour of θ_q when $q \to 1$. The above formula will then provide an asyptotic estimation of Rogers-Ramanujan's continuous fraction when $q \to 1$. (It will also indirectly give a very slight idea of its modular properties.)

1.9. Jackson (Frank Hilton Jackson, English, 1870–1960)

Here we speak of Jackson from the point of view of q-calculus and q-special functions. We shall say a word of his initiation of q-difference equation theory a bit later in 1.12.1.

1.9.1. Jackson's transformation formula. According to [**18**, p 6], "Jackson was contemporaneous with Rogers and even less appreciated for his work". Although the name is most remembered for "Jackson's q-integral" (see below), he did a lot of work on q-series. For instance, he deduced *Jackson's transformation formula* ([**93**]):

$${}_2\Phi_1(a, b; c; q, x) = \frac{(ax; q)_\infty}{(x; q)_\infty} {}_2\Phi_2(a, c/b; c, ax; q, bx)$$

from Heine's transformation formula (see Corollary 4.68 in 4.4.1.4). Hopefully (though we did not define it) you can guess what ${}_2\Phi_2$ is. According to Gasper-Rahman, Jackson more generally started a systematic study of the basic hypergeometric series and, for that, developed the first tools of the domain (q-derivation, q-integration ...).

1.9.2. The search for q-antiderivatives. Remember we define the q-derivation operator D_q by setting $D_q f(x) := \dfrac{f(x) - f(qx)}{(1-q)x}$, so that, for a fixed f, and under rather general conditions, $\lim\limits_{q\to 1} D_q f = f'$. As in the early years of infinitesimal analysis, it is natural to try to invert the process. So we call q-antiderivative of f any F such that $D_q F = f$. Such an F (if it exists) is determined only up to a q-constant, i.e., any $F + C$ is also a q-antiderivative of f if $D_q C = 0$, i.e., if $C(qx) = C(x)$. Such a C may be a "true constant" (a constant function) but it might as well be a q-elliptic function (those were defined in 1.7.2). For instance any Laurent power series $\sum a_n x^n$ such that $a_{-1} = 0$ has as q-antiderivative $\sum a_n x^{n+1}/[n+1]_q$, but $1/x$ has no Laurent power series q-antiderivative.

EXERCISE 1.23. Suppose you want to solve $D_q F(x) = x^n$. Show that, for all $k \geq 0$, one has $F(x) - F(q^k x) = (1 - q^{k(n+1)}) x^{n+1}/[n+1]_q$. What happens when $k \to +\infty$?

According to [**2**, p 364], q-antiderivatives had already been considered by Goursat in the article "Sur un problème relatif à la théorie des équations aux dérivées partielles du second ordre" (1904), where he discussed the equation $f(qx) - f(x) = b(x)$ for a given $b \in \mathbf{C}\{x\}$ (Goursat allowed for $|q| = 1$).

According to [**93**, 1.11 p 23], Thomae obtained a formula for $F(1) - F(0)$ when F is a q-antiderivative of f. Anticipating a notation to be introduced soon, Thomae's formula amounts to:

$$F(1) - F(0) = \int_0^1 f(t)\, d_q t := \sum_{n \geq 0} q^n f(q^n).$$

1.9.3. Jackson's q-integral. From 1910 to 1917, Jackson systematically studied "q-integration". Taking $q \in]0;1[$, here are two of the most standard forms of Jackson's q-integral:

$$\int_0^a f(t)\, d_q t := (1-q) \sum_{n \geq 0} q^n a f(q^n a),$$

$$\int_0^\infty f(t)\, d_q t := (1-q) \sum_{n \in \mathbf{Z}} q^n f(q^n).$$

Those notations admit natural extensions:

$$\int f(x)\, d_q x := \int_0^x f(t)\, d_q t,$$

$$\int_a^b f(t)\, d_q t := \int_0^b f(t)\, d_q t - \int_0^a f(t)\, d_q t,$$

$$\int_{-\infty}^{\infty} f(t)\, d_q t := \int_0^\infty (f(t) + f(-t))\, d_q t.$$

The first one is the indefinite q-integral, with the usual initial condition $F(0) = 0$. Let us justify it formally, as an example of the possible arguments. First, if $D_q F = f$, then $F(x) - F(qx) = (1-q)xf(x)$ and more generally $F(q^n x) - F(q^{n+1} x) = (1-q)q^n x f(q^n x)$. Summing for $n \geq 0$ and assuming that $F(q^n x) \to F(0) = 0$ when $n \to +\infty$, we get:

$$F(x) = \sum_{n \geq 0} (1-q) q^n x f(q^n x) = \int_0^x f(t)\, d_q t = \int f(x)\, d_q x.$$

Conversely, defining $F(x)$ as $\int f(x)\, d_q x$, we get:

$$F(x) - F(qx) = \sum_{n \geq 0} (1-q) q^n x f(q^n x) - \sum_{n \geq 1} (1-q) q^n x f(q^n x) = (1-q) x f(x),$$

whence $D_q F = f$ (and of course $F(0) = 0$).

1.9.4. Classical q-integrals. [**133**, Chapters 18–20] gives elementary properties of q-integrals, some kind of first steps into q-integral calculus; for instance q-analogs of the so-called "fundamental theorem of calculus", of integration by parts, etc.

EXERCISE 1.24. (i) Look at the definition of indefinite q-integrals in [**133**] and prove that:

$$\int f(x) D_q g(x)\, d_q x = f(x) g(x) - \int g(x) D_q f(x)\, d_q x.$$

(ii) Prove that $D_q \int f(x)\, d_q x = f(x)$.
(iii) Discuss equality $\int D_q f(x)\, d_q x = f(x)$.

[133] also discusses degeneracy of $\int f(x)\,d_qx$ to $\int f(x)\,dx$ when $q \to 1$ (for continuous functions at least), an example of which we saw in Section 1.1 on Fermat. An example of q-special function taking the form of a q-integral is:

$$B_q(x, y) := \int_0^1 t^{x-1} \frac{(tq;q)_\infty}{(tq^y;q)_\infty}\, d_qt.$$

Using what we saw in 1.4, we find that the second factor in the q-integral degenerates, under appropriate conditions, to $(1-t)^{y-1}$ when $q \to 1$, so the above is a q-analog of $B(x,y) := \int_0^1 t^{x-1}(1-t)^{y-1}\,dt$, the classical Euler's Beta integral of the theory of special functions **[255]**. The degeneracy is proved in **[93]** for $\Re x, \Re y > 0$.

1.10. Ramanujan (Srinivasa Ramanujan, Indian, 1887–1920)

Ramanujan was a supernova in the firmament of mathematics; see **[7, 18, 109, 185, 186, 253]** for more on his life and work[12]. Much of Ramanujan's contributions rather relate to q-series, but not all to q-difference equations. The *Lost Notebooks* edited by Andrews and Berndt put some order in the near infinite number of extraordinary formulas by Ramanujan that were discovered after his death. Of the four thick volumes, a large part of the second is devoted to q-calculus. We can only show a glimpse of those results. At least two of them were already alluded to: the asymptotics of partitions in 1.2; and Rogers-Ramanujan identities on 1.8. Yet another (Mock Theta Functions) will be mentioned in 1.11.

1.10.1. Asymptotics of partitions. The history of the asymptotic formula is nicely told in **[109]** (but also see the movie *The Man Who Knew Infinity*). Hardy gives an elementary proof of the coarser estimation $\log p(n) \sim \pi\sqrt{2n/3}$. It goes as follows. Let $g(q) := (1-q)\mathcal{P}(q) = \sum\limits_{n\geq 0} (p(n) - p(n-1))q^n$ (with the usual convention that $p(-1) = 0$). There is a "Tauberian theorem"[13] by Hardy that says that, if $f(x) = \sum\limits_{n\geq 0} a_n x^n$ has positive coefficients and $\log f(x) \sim \dfrac{C}{1-x}$ when $x \to 1^-$, then $\log(a_0 + \cdots + a_n) \sim 2\sqrt{Cn}$ when $n \to +\infty$.

We apply it to our g above; the coefficients are clearly positive and we are going to check the other assumption for $C := \pi^2/6$. Since $a_0 + \cdots + a_n = p(n)$, the conclusion is immediate. So let's look at:

$$\begin{aligned}\log g(q) = \log \frac{1-q}{(q;q)_\infty} = \sum_{k\geq 2} -\log(1-q^k) = \sum_{k\geq 2, l\geq 1} \frac{q^{kl}}{l} &= \sum_{l\geq 1} \frac{1}{l(1-q^l)} \\ &= \frac{1}{1-q} \sum_{l\geq 1} \frac{1}{l[l]_q}.\end{aligned}$$

For $|q| < 1$, the computation is perfectly rigorous. Now, when $q \to 1-$, we know that $[l]_q \to 1$, so the series in the last expression converges termwise to $\sum 1/l^2 = \pi^2/6$. We must just show that really $\sum\limits_{l\geq 1} \dfrac{1}{l[l]_q} \to \sum\limits_{l\geq 1} \dfrac{1}{l^2}$.

---

[12] Also see the nice movie *The Man Who Knew Infinity*.

[13] Abelian theorems deduce from the known behaviour of coefficients a_n when $n \to +\infty$ information about the behaviour of $\sum a_n x^n$ when $x \to x_0$, a possible singularity; Abel's radial theorem can be interpreted that way, whence the name. Tauberian theorems are converse to abelian theorems; they usually require stronger assumptions (in the example here, sign of the coefficients). See for instance **[139]**.

EXERCISE 1.25. Justify it rigorously.

Actually, Hardy and Ramanujan proved[14] in 1918 the much more difficult fact that $p(n) \sim \dfrac{1}{4\sqrt{3}n} e^{\pi\sqrt{2n/3}}$. They used the analysis of the singularities of $\mathcal{P}(q) = \dfrac{1}{(q;q)_\infty}$. But that function of q has infinitely many singularities (all roots of unity in $\mathbf{C}$) and they all have infinite order! So they invented a powerful method, the "circle method", that Hardy and Littlewood developed further, with many applications to analytic number theory. Hardy and Ramanujan obtained more than an equivalent: a complete asymptotic expansion, so efficient that using six terms only they were able to compute $p(200) = 3972999029388$ with an error < 0.004. That performance did a lot to convince English mathematicians in Oxford that Ramanujan was not just a freak. Later, Rademacher modified slightly the general term of the asymptotic expansion so that it became a convergent series [**12, 183**]. An interesting discussion by Selberg suggesting a possible reason why Hardy and Ramanujan did not themselves obtain that better formula may be found in [**19**].

The asymptotic formula of Hardy and Ramanujan rests on the "modular behaviour" of the generating series $\mathcal{P}(q)$. Among Euler's formulas around *partitio numerorum* appears the following:

$$f(q) := \mathcal{P}(q)^{-1} = \sum_{n\geq 0}(-1)^n q^{(3n^2-n)/2} \Longrightarrow q^{1/24} f(q) = \sum_{n\geq 0}(-1)^n q^{(6n-1)^2/24},$$

the latter being considered as a "Theta series" (it is Dedekind's Eta function; see Chapter 2). The terminology refers to expressions of the form $\sum q^{Q(n)}$, Q a quadratic polynomial. Some will be considered in Chapter 2. The "modular behaviour" means that, if one writes $q = e^{2i\pi\tau}$, the corresponding functions of τ behave nicely when τ is replaced by $-1/\tau$. Since it is invariant by $\tau \mapsto \tau + 1$, and since those two transformations together generate the group of all homographies $\tau \mapsto \dfrac{a\tau + b}{c\tau + d}$, $a, b, c, d \in \mathbf{Z}$, $ad - bc = 1$, we see that our functions behave nicely under all such transformations. But the orbit of 0 under that group is $\mathbf{Q}$ and this implies that, from the asymptotic behaviour of such a function near $q = 1$, one could deduce its asymptotic behaviour near all roots of 1. This, combined with the circle method, is what Hardy and Ramanujan did. In [**109**, p 118], Hardy summarizes the main steps as follows.

(1) When $q \to 1$, $\mathcal{P}(q)$ "behaves roughly like" $\exp(\pi^2/6(1-q))$. We summarized Hardy's Tauberian argument above, but using the modular behaviour of the Eta function when τ is replaced by $-1/\tau$, the estimation can be made more precise[15].

(2) The other singularities are the $\omega_{l,m} := e^{2i\pi l/m}$ for $0 \leq l \leq m$, $l \wedge m = 1$. When $q \to \omega_{l,m}$, $\mathcal{P}(q)$ "behaves roughly like" $\exp(\pi^2/6m^2(1-|q|))$. This can again be proved from the modular behaviour, but this time using a transformation $\tau \leftarrow (a\tau+b)/(c\tau+d)$, where $a, b, c, d \in \mathbf{Z}$, $ad - bc = 1$ and the homography sends 0 to l/m.

[14] In the historical notes at the end of the chapter, Hardy (p. 130) recalls that a slightly weaker formula was independently obtained by the russian mathematician Uspensky, but unduly ignored because of later publication.

[15] A similar method will be used in Chapter 2 for the asymptotic behaviour of Theta functions, which we intend to apply to the study of degeneracies when $q \to 1$ Section 11.2.

(3) One can then hope for an expansion $p(n) = P_1(n) + \cdots + P_m(n) + R_m(n)$, where each P_m is related to the contributions of terms coming from the $\omega_{l,m}$, and where $R_m(n)$ is a kind of remainder term.

More specifically, we find that, for $k \in \mathbf{N}^*$ and $h, h' \in \mathbf{Z}$ such that $h, h' \equiv -1 \pmod{k}$:

$$f\left(e^{2i\pi(h/k+z)}\right) = \omega_{h,k}\sqrt{z/\mathrm{i}}\exp\left((\pi\mathrm{i}/12)(z + 1/k^2z)\right) f\left(e^{2i\pi(h'/k-1/k^2z)}\right),$$

where $\omega_{h,k}$ is some explicitly computable root of unity. That can be used to deduce the asymptotic formula; for every fixed $\alpha > 0$:

$$p(n) = \frac{1}{\pi\sqrt{2}} \sum_{1 \geq k \geq \alpha\sqrt{n}} A_k(n)C_k(n) + O(n^{-1/4}),$$

where $A_k(n) := \sum_{i=0}^{k-1} \omega_{h,k}e^{-2i\pi hn/k}$ and C_k is the following derivative:

$$C_k(x) := \frac{d}{dx} \frac{\cosh\left(\pi\sqrt{2/3}\sqrt{x - 1/24}/k\right)}{\sqrt{x - 1/24}}.$$

It is a decisive ingredient of the proof that, near all its singularities, which are the roots of unity, $f(q)$ has a "simple" asymptotic expansion related to its "modular behaviour", which itself comes from the properties of "Theta series". We shall see in 1.11 how that fact probably influenced Ramanujan's excitement around "Mock Theta Functions".

1.10.2. Rogers-Ramanujan continued fraction again. Although the analytic theory of continuous fractions [**251**] has numerous applications in the domain of special functions, including q-series, it has not yet interferred a lot with q-difference equations. So we consider the following (already alluded to in 1.8.1, and to be examined again in Chapters 2, 4, and then again in [CV]) as a very interesting, maybe important example, but still lying at the border of our preoccupations.

The theme orignates in the first contact of Ramanujan with western mathematicians, through Hardy [**109**]. Among a wealth of extraordinary formulas, quoted by Hardy from Ramanujan's first letter, the following "defeated [him] completely" (numbering comes from Hardy):

(1.10) If $u = \cfrac{x}{1 + \cfrac{x^5}{1 + \cfrac{x^{10}}{1 + \cfrac{x^{15}}{1 + \cdots}}}}$ and $v = \cfrac{x^{1/5}}{1 + \cfrac{x}{1 + \cfrac{x^2}{1 + \cfrac{x^3}{1 + \cdots}}}}$,

$$\text{then } v^5 = u\frac{1 - 2u + 4u^2 - 3u^3 + u^4}{1 + 3u + 4u^2 + 2u^3 + u^4}.$$

$$(1.11) \quad \cfrac{1}{1 + \cfrac{e^{-2\pi}}{1 + \cfrac{e^{-4\pi}}{1 + \cdots}}} = \left\{\sqrt{\left(\frac{5 + \sqrt{5}}{2}\right)} - \frac{\sqrt{5} + 1}{2}\right\} \times e^{2\pi/5}$$

$$(1.12)\quad \cfrac{1}{1+\cfrac{e^{-2\pi\sqrt{5}}}{1+\cfrac{e^{-4\pi\sqrt{5}}}{1+\cdots}}} = \{\text{rather complicated algebraic expression}\} \times e^{2\pi/\sqrt{5}}$$

All of this reminds us of $R(q)$ encountered in 1.8.1 and to be met again; so we shall not insist any more, except to say that Rogers-Ramanujan continued fraction leads to a lot more than the Rogers-Ramanujan identities.

1.10.3. The "near equality" (Ramanujan, Andrews-Berndt, Zagier). In the Notebooks, part V [**186**, p 36 entry 13], Ramanujan writes that;

$$\cfrac{q}{x+\cfrac{q^4}{x+\cfrac{q^8}{x+\cfrac{q^{12}}{x+\ddots}}}} \quad \text{"nearly equals"} \quad 1-\cfrac{qx}{1+\cfrac{q^2}{1-\cfrac{q^3x}{1+\cfrac{q^4}{1-\ddots}}}}$$

As Don Zagier remarks in [**260**], it is not clear what Ramanujan means by that. However, he quotes numerical experiments by Henri Cohen showing both sides to be very close. Note that for $q = 1$ the left-hand side, say $g(x)$, satisfies formally $g = 1/(x+g)$, whence $g^2 + xg = 1$, while the right-hand side, say $f(x)$, satisfies formally $f = 1 - x/(1+1/f)$, whence $f^2 + xf = 1$, so one could guess that maybe $f = g = \dfrac{-x+\sqrt{x^2+4}}{2}$. More generally, Zagier shows that both sides (without the assumption $q = 1$ are solutions of the same q^2-difference equation: $h(q^2x) = \dfrac{1}{h(x)+qx}$. This is very easy for g, but for f it is rather difficult to guess why (we'll try to in [CV]).

Zagier also obtains asymptotic information on the difference $f_q(x) - g_q(x)$ of both expressions under various assumptions, explaining to a large extent Ramanujan's contention. For instance, with $q \in]0;1[$ arbitrary:

$$f_q(1) - g_q(1) = O\left(\exp \frac{\pi^2/5}{\log q}\right). \quad \text{(There are more general estimations.)}$$

We shall return to this theme in [CV], with methods more in the spirit of this book.

1.10.4. "A remarkable formula with many parameters". This will again be just a glimpse into one of the many marvels of the q-world; maybe this one is not too far from the world of q-difference equations (also see 4.1.3.3 in Chapter 4, where we prove it). In q-litterature [**7, 18, 93, 130**] it is known as *the* ${}_1\psi_1$ *summation formula*:

$$ {}_1\psi_1(a;b;q,x) := \sum_{n=-\infty}^{+\infty} \frac{(a;q)_n}{(b;q)_n} x^n = \frac{(ax;q)_\infty (q/(ax);q)_\infty (q;q)_\infty (b/a;q)_\infty}{(b;q)_\infty (b/(ax);q)_\infty (x;q)_\infty (q/a;q)_\infty}.$$

Note that the series is *bilateral*: n goes from $-\infty$ to $+\infty$; so, while it looks a bit like a simplified q-hypergeometric series, it is really wilder (it has an essential singularity at 0).

In the last chapter of [**109**], Hardy (who suggests it can be proved from what we called here Cauchy's binomial formula, which he attributes to Euler) calls it "a remarkable formula with many parameters", which has caused some ironical comments from the specialists of q-series (sometimes themselves nicknamed "q-wizards"). Indeed, it seems that it came to play a crucial role in q-analysis; see for instance **[7, 93, 130]** and Askey's article in **[19]**. Andrews and Berndt, in volume II of **[186]**, list 22 proofs; and Berndt considers the ${}_1\psi_1$ summation formula as Ramanujan's most famous theorem in the domain of q-series[16].

EXERCISE 1.26. Deduce Cauchy's q-binomial theorem and Jacobi's Triple Product Formula from the ${}_1\psi_1$ summation formula.

1.11. Watson (George Neville Watson, English, 1886–1965)

1.11.1. The final problem. Why Watson gave that title to his lecture on Mock Theta Functions, a theme invented by Ramanujan while on his deathbed, is best explained by himself in **[253]**. He himself gave the theme so much publicity and developed it so far that it seems fit to present it in this section. In a letter to Hardy, three months before his untimely death, Ramanujan remarked that "ϑ-functions in transformed Eulerian form[17] usually have simple asymptotic behaviour at their singularity $q =$ root of unity." He gave examples:

$$\sum_{n\geq 0} \frac{q^{n^2}}{(q;q)_n^2} = \sqrt{t/2\pi}\,\exp(\pi^2/6t - t/24) + o(1),$$

$$\sum_{n\geq 0} \frac{q^{n^2}}{(q;q)_n} = \sqrt{2/(5-\sqrt{5})}\,\exp(\pi^2/15t - t/60) + o(1),$$

where $q = e^{-t}$ and $t \to 0^+$. However some similarly looking series behave less nicely:

$$\sum_{n\geq 0} \frac{q^{n(n+1)/2}}{(q;q)_n^2} = \sqrt{t/(2\pi\sqrt{5})}\,\exp(\pi^2/5t + a_1 t + \cdots + O(a_k t^k)),$$

"where $a_1 = 1/(8\sqrt{5})$ and so on". The latter "false Theta function" behaves "in an unclosed form at the singularities".

The first function above can be derived from a formula of Heine:

$${}_2\Phi_1(a,b;c;q,c/ab) = \frac{(c/a;q)_\infty (c/b;q)_\infty}{(c;q)_\infty (c/ab;q)_\infty}$$

by having $a, b \to \infty$ and $c \to q$. The second is one of the pair appearing in the Rogers-Ramanujan identities. Both do share with standard Theta functions a modular behaviour. The asymptotic behaviour of the third one must have been more complicated to obtain, because that one has no modular behaviour. Ramanujan asked if a function having an "Eulerian form" as above together with, asymptotically, a "closed form at the singularities" would necessarily be a Theta function (in his sense, whatever it was).

---

[16] I personnally would rather have rated higher either the asymptotic expansion of $p(n)$ or formulas around Rogers-Ramanujan identities and continued fraction; but, in both cases, Ramanujan cannot be considered the sole discoverer, not even the first one.

[17] We saw in 1.10.1 what "Theta functions" could mean, although some have interpreted it as meaning (for Ramanujan) any function presenting "modular behaviour"; as for the "Eulerian form", it clearly refers to denominators involving $(q;q)_n$ or similar expressions.

He thought that the answer was no, and proposed a list of conjectureal counterexamples. The list was later extended by Watson, then by other mathematicians. The first one in the list was:

$$f(q) := \sum_{n\geq 0} \frac{q^{n^2}}{(-q;q)_n^2}.$$

Its asymptotic behaviour, which we do not reproduce (it is shown in [**253**, p 329]) can be found by comparison with the true Theta function $\dfrac{(q;q)_\infty}{(-q;q)_\infty^2}$ and, Ramanujan says, "it is unconceivable that a single ϑ-function could be found to cut out the singularities of $f(q)$".

After many decades of hard calculations, a conceptual breakthrough was provided by the thesis of Zwegers [**262**], who related mock Theta functions to a vast generalization of modular forms (which normally pertain to complex analysis) into real analysis, the so-called "harmonic Maass forms". For some time, the subject was thought to be completely understood and closed, although work by Rhoades [**201, 202**] suggested that the "modern" definition of mock Theta functions deduced from that point of view did not coincide with what Ramanujan had in mind.

More recently, work by Changgui Zhang [**267, 268**] concentrated on asymptotic properties of expansions "in Eulerian form"; the new approach clearly indicates that the subject is not closed and, since it involves techniques that are essential in the theory of q-difference equations, we shall return to it in [CV].

1.11.2. Barnes-Mellin-Watson formulas. In the theory of ordinary differential equations, connection formulas for hypergeometric functions played a historical role since they provided explicit representation for the simplest nontrivial examples of monodromy (nontrivial meaning nonabelian; see [**125, 255**], and [**222**, Chapter 11]). Riemann had obtained them in a simple form but relative to unknown fundamental systems of solutions. Then Barnes[18], starting from work of Mellin, found integral representations of the natural solutions leading to explicit connection formulas relative to the natural fundamental systems of solutions. Mellin-Barnes integral representation of the (classical) hypergeometric series (which was defined in 1.6.2) reads as follows:

$${}_2F_1(\alpha,\beta;\gamma;x) = \frac{1}{2\mathrm{i}\pi}\frac{\Gamma(\gamma)}{\Gamma(\alpha)\Gamma(\beta)}\int_C \frac{\Gamma(\alpha+s)\Gamma(\beta+s)\Gamma(-s)}{\Gamma(\gamma+s)}(-x)^s\,ds,$$

for some explicit contour C described in the given references (for instance [**222**, p 157]). The corresponding integral representation for basic hypergeometric series was obtained by Watson:

$${}_2\phi_1(a,b;c;q,x) = \frac{-1}{2\mathrm{i}\pi}\frac{(a;q)_\infty(b;q)_\infty}{(c;q)_\infty(q;q)_\infty}\int_C \frac{(q^{1+s};q)_\infty(cq^s;q)_\infty}{(aq^s;q)_\infty(bq^s;q)_\infty}\frac{\pi(-x)^s}{\sin\pi s}\,ds,$$

for some explicit contour C described in [**93**, §4.2 pp 115–117].

[18]The connection formulas are often called Barnes formulas, but sometimes, and maybe more properly, Mellin-Barnes or Barnes-Mellin.

Barnes derived, from the above Mellin-Barnes integral representation, the connection formula[19]:

$$
\begin{aligned}
{}_2F_1(\alpha,\beta;\gamma;x) &= \frac{\Gamma(\gamma)\Gamma(\beta-\alpha)}{\Gamma(\beta)\Gamma(\gamma-\alpha)}(-x)^{-\alpha}\,{}_2F_1(\alpha,1+\alpha-\gamma;1+\alpha-\beta;x) \\
&+ \frac{\Gamma(\gamma)\Gamma(\alpha-\beta)}{\Gamma(\alpha)\Gamma(\gamma-\beta)}(-x)^{-\beta}\,{}_2F_1(\beta,1+\beta-\gamma;1+\beta-\alpha;x).
\end{aligned}
$$

Watson derived from his integral representation the connection formula[20]:

$$
\begin{aligned}
{}_2\Phi_1(a,b;c;q,x) &= \frac{(b;q)_\infty(c/a;q)_\infty(ax;q)_\infty(q/ax;q)_\infty}{(c;q)_\infty(b/a;q)_\infty(x;q)_\infty(q/x;q)_\infty}\,{}_2\Phi_1(a,aq/c;aq/b;q,cq/abx) \\
&+ \frac{(a;q)_\infty(c/b;q)_\infty(bx;q)_\infty(q/bx;q)_\infty}{(c;q)_\infty(b/a;q)_\infty(x;q)_\infty(q/x;q)_\infty}\,{}_2\Phi_1(b,bq/c;bq/a;q,cq/abx).
\end{aligned}
$$

Why this is indeed a connection formula will be seen in Chapter 12 and its degeneracy when $q \to 1$ to classical Mellin-Barnes connection formula will be explained in Section 12.3.

1.12. The separation of q-calculus and q-difference equation theory

1.12.1. The bifurcation. It seems that q-difference equation theory started with the 1910 article **[126]** in which Jackson started the study of Heine's q-hypergeometric series through the introduction of the operator of "q-derivation" $\Delta : f(x) \mapsto \dfrac{f(qx)-f(x)}{(q-1)x}$ (which, as we saw on a few examples, degenerates into $\dfrac{d}{dx}$ when $q \to 1$. From there, the story of our theory is that of what I called the "Founding Fathers", in alphabetical order: Adams, Birkhoff, Carmichael, (and some others of course).

A completely different thread should however be mentioned here. According to Kuczma [**143**, 6.2.1], while functional equations (not only differential ones) had been studied by Abel, Cauchy, Euler, Riemann ..., it was Poincaré who started a systematic study. For example, in a 1886 note to the Académie des Sciences, he considers what Kuczma calls "Poincaré equation":

$$\psi(sx) = f(\psi(x)).$$

The unknown function is ψ and if one replaces s by q, one has a (probably nonlinear) q-difference equation. Kuczma mentions more general equations where sx is replaced by $\phi(x) := qx + \cdots$ (a power series expansion; we return here to letter q); but a theorem of Koenigs [**143**, 6.3] allows to reduce it to a plain q-difference equation as follows. (This theme will appear again in work of Bézivin; see Section 6.2 of Chapter 6.)

EXERCISE 1.27. Let $\phi(x) := qx + \cdots$ a formal power series, $q \neq 0$.

(i) Show that there is a formal power series $\lambda(x) := x + \cdots$ such that $\phi(x) = \lambda(q\lambda^{-1}(x))$.

(ii) Admitting convergence, show that this allows to reduce the functional equation $\psi(\phi(x)) = f(\psi(x))$ to a q-difference equation $\gamma(qx) = g(\gamma(x))$.

---

[19]This particular formula connects 0 to ∞; there are others connecting 0 to 1 and 1 to ∞. The only one which matters to us is this one.

[20]It connects 0 to ∞, and, in the present state of q-difference theory, this is enough for our purposes.

Yet another thread comes from the importance of projective geometry in nineteenth century. All analytic automorphisms of the projective line have the form of a Mobius transformation (we Frenchmen prefer to call it homography) $R(x) := \frac{ax+b}{cx+d}$, $ad - bc \neq 0$. It can be considered as natural to look at functions transformed a specified way by $R(x)$, i.e., at functional equations $\psi(R(x)) = f(\psi(x))$. Now, if R is nontrivial (i.e., distinct from the identity), it has either one or two fixed points. In the latter case, which is generic, it can be conjugated to a map $x \mapsto qx$, $q \neq 0, 1$; then, as in the exercise, the functional equation boils down to a q-difference equation.

REMARK 1.28. In the former case, which can thus be considered as degenerate (the two fixed points have merged into a double one), R is conjugated to $x \mapsto x+1$ and the functional equation boils down to a difference equation $\gamma(x+1) = g(\gamma(x))$. The fact that this comes through a degeneracy process may explain why the theory of difference equations is so much more difficult than the theory of q-difference equations.

1.12.2. The return of the letter q. To take in account the noncommutative character of quantum mechanics, the physicist Dirac introduced "q-numbers" and "c-numbers". Th c-numbers are classical numbers such as those occurring in classical physics (results of measurements for instance, or their predicted values). The q-numbers, on the other hand, seem to represent what, in Heisenberg's presentation of the theory, was operators. By the way, according to Dirac himself, letter "c" is for "commuting", not for "classical"; and letter "q" is for "queer", not for "quantum". So they are not really related to our q-numbers $[n]_q$, and maybe not even at all to the q-world (at least, not in an obvious way).

Later in the twentieth century, the physicist Baxter, studying the "hard hexagon model" (this belongs to statistical physics) rediscovered the Rogers-Ramanujan identities and a whole lot of similar identities. This is well recounted in [**7**]. So it seems that the q-world came to physics through statistical physics, not quantum physics. However, when mathematicians such as Jimbo and Drinfeld tried to build a conceptual framework for quantum Yang-Baxter equation, which pertains to quantum statistical physics, they discovered q-deformations of classical objects, to be called "quantum groups"; and the q-world then inherited a new meaning for the letter q: quantum group involve q-calculus, renamed "quantum calculus". The fact that the letter q reappeared is therefore purely coincidental. Note that the theory of quantum groups seems to use almost no q-difference equations, and therefore we shall say little about it in this book.

However, q-calculus does appear in modern physics with the letter q sometimes standing for $e^{2i\pi h}$, where h denotes Planck's constant: since it is very small, q is very nearly 1 and the limit $q \to 1$ in quantum physics formulas is expected to lead to classical physics; so here the "continuous limit", our "confluence", may be called "classical limit".

CHAPTER 2

Elementary Special and q-Special Functions

This chapter rests on the classical theory of some fundamental special functions (as recalled for instance in Appendix A), applying it to q-difference theory; involved special functions are mainly the elliptic and modular functions, along with Theta, Gamma and Zeta functions (the last two to a much lesser extent).

First we shall define and study the "multiplicative" (or "basic", in the sense of linked to a "base" q) model of Theta function, the function θ_q. Then we shall apply it to the production of "basic bricks", actually our favorite q-analogs of multivalued functions x^α and $\log x$, which allow us to found the whole theory on *uniform* meromorphic functions. Last we shall begin to give a rather precise meaning to the notion of q-degeneracy. All that is essentially based on the theory of Theta functions in its q-form.

2.1. Classical Theta functions

2.1.1. Elliptic Function Theory on $\mathbf{C}^*/q^{\mathbf{Z}}$. Let $q \in \mathbf{C}^*$ such that $|q| < 1$. We choose once for all $\tau \in \mathcal{H}$ such that $q = e^{2i\pi\tau}$. Many notations will mention q only although the corresponding constructions really involve τ: for instance $q^\alpha := e^{2i\pi\tau\alpha}$, when $\alpha \in \mathbf{C}$ (of course, if $\alpha \in \mathbf{Z}$, the particular choice of τ does not matter). So our notations could cause ambiguities. We shall try to resolve them when they might cause problems.

2.1.1.1. Let σ_q the automorphism of $\mathscr{M}(\mathbf{C}^*)$ defined by: $(\sigma_q f)(x) := f(qx)$. (It is clear that $f \in \mathscr{M}(\mathbf{C}^*) \Longrightarrow \sigma_q f \in \mathscr{M}(\mathbf{C}^*)$ and that $f \mapsto \sigma_q f$ is indeed a field automorphism.) We are here interested in its fixed subfield (see 0.4):

$$\mathscr{M}(\mathbf{C}^*)^{\sigma_q} := \{f \in \mathscr{M}(\mathbf{C}^*) \mid \sigma_q f = f\}.$$

This of course contains $\mathbf{C}$, but it contains much more. First note that the holomorphic map $\underline{e} : \mathbf{C} \to \mathbf{C}^*$, $z \mapsto e^{2i\pi z}$, (also a surjective group morphism) dually defines a field morphism $\mathscr{M}(\mathbf{C}^*) \to \mathscr{M}(\mathbf{C})$, $f \mapsto f \circ \underline{e}$ and that, as every field morphism, it is injective; in other words, the change of variable $x = \underline{e}(z)$ allows us to consider $f(x)$ (defined on $\mathbf{C}^*$) as $g(z) := f(e^{2i\pi z})$ (defined on $\mathbf{C}$). The image of $f \mapsto g := f \circ \underline{e}$ consists in the subfield of 1-periodic meromorphic functions on $\mathbf{C}$, whence an identification:

$$\mathscr{M}(\mathbf{C}^*) \simeq \mathscr{M}(\mathbf{C})^{\mathbf{Z}} := \text{ fixed subfield under the action of } \mathbf{Z} \text{ (according to 0.4).}$$

Through this identification, the transformation of $f(x)$ to $f(qx)$ corresponds to the transformation of $g(z)$ to $g(z+\tau)$, whence a commutative diagram:

$$\begin{array}{ccc} \mathscr{M}(\mathbf{C}^*) & \xrightarrow{\sigma_q} & \mathscr{M}(\mathbf{C}^*). \\ \downarrow & & \downarrow \\ \mathscr{M}(\mathbf{C})^{\mathbf{Z}} & \xrightarrow{g(z)\mapsto g(z+\tau)} & \mathscr{M}(\mathbf{C})^{\mathbf{Z}} \end{array}$$

As a consequence $f = \sigma_q f \Leftrightarrow g(z) = g(z+\tau)$ and since we know in advance that $g(z) = g(z+1)$, this is equivalent to $g \in \mathscr{M}(\mathbf{E}_{\Lambda_\tau})$ (remember that Λ_τ is the lattice $\mathbf{Z} + \mathbf{Z}\tau$ and $\mathscr{M}(\mathbf{E}_{\Lambda_\tau})$ the field of Λ_τ-elliptic functions; see 0.4 and Appendix A).

PROPOSITION 2.1. *The change of variable map $f(x) \mapsto f(e^{2i\pi z})$ induces isomorphisms $\mathscr{M}(\mathbf{C}^*) \simeq \mathscr{M}(\mathbf{C})^{\mathbf{Z}}$ and $\mathscr{M}(\mathbf{C}^*)^{\sigma_q} \simeq \mathscr{M}(\mathbf{C})^{\Lambda_\tau} = \mathscr{M}(\mathbf{E}_{\Lambda_\tau})$.*

2.1.1.2. In the same spirit as for the identification $\mathscr{M}(\mathbf{C})^{\Lambda_\tau} = \mathscr{M}(\mathbf{E}_{\Lambda_\tau})$ (see Subsection A.1.2 of Appendix A), we can give a geometric interpretation for $\mathscr{M}(\mathbf{C}^*)^{\sigma_q}$. Let $f \in \mathscr{M}(\mathbf{C}^*)$ be interpreted as a mapping with target $\mathbf{C} \cup \{\infty\}$. First, in pure set-theoretic (or group-theoretic) terms, we see that the equality $f(qx) = f(x)$ means that f can be factored through the quotient map $\pi : a \mapsto \overline{a} := a \pmod{q^{\mathbf{Z}}}$, from $\mathbf{C}^*$ to the quotient group $\mathbf{C}^*/q^{\mathbf{Z}}$:

$$\begin{array}{ccc} \mathbf{C}^* & \xrightarrow{\quad f \quad} & \mathbf{S} = \mathbf{C} \cup \{\infty\} \\ & {\scriptstyle\pi}\searrow \quad \nearrow & \\ & \mathbf{C}^*/q^{\mathbf{Z}} & \end{array}$$

Next, we note that, as groups, $\mathbf{C}/\Lambda_\tau \simeq \mathbf{C}^*/q^{\mathbf{Z}}$. Indeed, the surjective group morphism $\underline{e} : \mathbf{C} \to \mathbf{C}^*$ induces an isomorphism $\mathbf{C}/\underline{e}^{-1}(q^{\mathbf{Z}}) \simeq \mathbf{C}^*/q^{\mathbf{Z}}$; now:

$$\begin{aligned} \underline{e}(z) \in q^{\mathbf{Z}} &\Longleftrightarrow \exists k \in \mathbf{Z} \, : \, e^{2i\pi z} = e^{2i\pi k\tau} \\ &\Longleftrightarrow \exists k, l \in \mathbf{Z} \, : \, 2i\pi z = 2i\pi k\tau + 2i\pi l \\ &\Longleftrightarrow z \in \Lambda_\tau, \end{aligned}$$

so $\underline{e}^{-1}(q^{\mathbf{Z}}) = \Lambda_\tau$. We actually get a commutative diagram:

$$\begin{array}{ccc} \mathbf{C} & \xrightarrow{\underline{e}} & \mathbf{C}^* \\ {\scriptstyle p}\downarrow & & \downarrow{\scriptstyle \pi} \\ \mathbf{C}/\Lambda_\tau & \xrightarrow{\simeq} & \mathbf{C}^*/q^{\mathbf{Z}}. \end{array}$$

The lower horizontal arrow is a group morphism, but we can use it to carry over the analytic structure on $\mathbf{C}/\Lambda_\tau$ to one on $\mathbf{C}^*/q^{\mathbf{Z}}$, making the latter a compact Riemann surface, which we denote $\mathbf{E}_q$: so we just defined an isomorphism of Riemann surfaces:

$$\mathbf{E}_{\Lambda_\tau} := \mathbf{C}/\Lambda_\tau \simeq \mathbf{E}_q := \mathbf{C}^*/q^{\mathbf{Z}}.$$

There is a direct description of the analytic structure on $\mathbf{E}_q$ in Appendix B. From now on, we call $\mathbf{E}_q$ a complex torus or an elliptic curve. Meromorphic functions on $\mathbf{E}_q$ are called *q-elliptic* functions; when seen as functions on $\mathbf{C}^*$ such that $\sigma_q f = f$, they are also called *q-invariant* or *q-constant functions.*

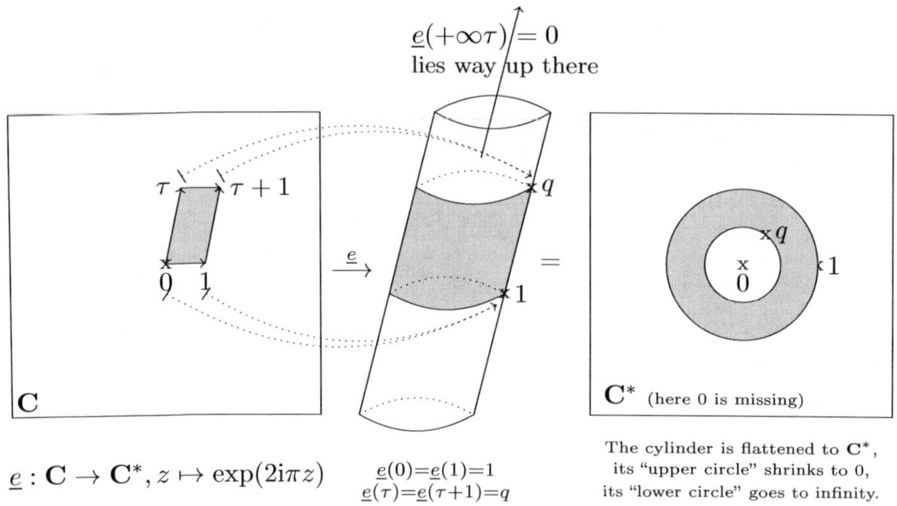

FIGURE 2.1. $\mathbf{C}^*$ as a cylinder and the fundamental annulus (in gray)

2.1.1.3. The map $\underline{e}$ identifies z to $z+1$ and the line $\mathbf{R}\tau$ with the translated line $1+\mathbf{R}\tau$, so that $\mathbf{C}^*$ is essentially a cylinder (see figure 2.1). Every segment $[a;a+1]$ becomes a circle $b\mathbf{U}$, where $b := \underline{e}(a)$ and:

$$\mathbf{U} := \underline{e}([0;1]) = \{x \in \mathbf{C}^* \mid |x| = 1\}$$

is the unit circle in $\mathbf{C}^*$: so the fundamental parallelogram $[0;1] + [0;1]\tau$ becomes the compact annulus $|q| \leq |x| \leq 1$. More generally, any fundamental parallelogram $a + [0;1] + [0;1]\tau$ is transformed into a compact annulus $|q\underline{e}(a)| \leq |x| \leq |\underline{e}(a)|$. To be complete, if $0 < r < R$, set:

$$\overline{\mathcal{C}}(r,R) := \overline{\mathrm{D}}(0;R) \setminus \mathring{\mathrm{D}}(0;r) = \{x \in \mathbf{C}^* \mid r \leq |x| \leq R\},$$

$$\mathring{\mathcal{C}}(r,R) := \mathring{\mathrm{D}}(0;R) \setminus \overline{\mathrm{D}}(0;r) = \{x \in \mathbf{C}^* \mid r < |x| < R\},$$

$$\mathcal{C}(r,R) := \overline{\mathrm{D}}(0;R) \setminus \overline{\mathrm{D}}(0;r) = \{x \in \mathbf{C}^* \mid r < |x| \leq R\}.$$

Then, if $r = R\,|q|$:

- $\pi : \overline{\mathcal{C}}(r,R) \to \mathbf{E}_q$ is surjective;
- $\pi : \mathring{\mathcal{C}}(r,R) \to \mathbf{E}_q$ is injective;
- $\pi : \mathcal{C}(r,R) \to \mathbf{E}_q$ is bijective.

2.1.1.4. The results above can now be translated as follows. Let $f \in \mathscr{M}(\mathbf{C}^*)^{\sigma_q}$, $f \neq 0$. Seen as a q-invariant function on $\mathbf{C}^*$, its zeroes are organized in discrete q-spirals $[a_i;q] := a_i q^{\mathbf{Z}}$, $i = 1, \ldots, k$; and all zeroes on a particular $[a_i;q]$ have the same multiplicity m_i. Similarly, its poles are organized in discrete q-spirals $[b_j;q] := b_j q^{\mathbf{Z}}$, $j = 1, \ldots, l$; and all poles on a particular $[b_j;q]$ have the same multiplicity n_j. Now, seen as a q-elliptic function $f \in \mathscr{M}(\mathbf{E}_q)$, it has zeroes $\overline{a_i} := p(a_i)$ with multiplicities m_i and poles $\overline{b_j} := p(b_j)$ with multiplicities n_j. So its divisor is:

$$\mathrm{div}_{\mathbf{E}_q}(f) = \sum_{i=1}^{k} m_i[\overline{a_i}] - \sum_{j=1}^{l} n_j[\overline{b_j}].$$

The degree of this divisor is $\deg \operatorname{div}_{\mathbf{E}_q}(f) = \sum m_i - \sum n_j = 0$; but its *evaluation*[1] must be computed multiplicatively; it is:

$$\begin{aligned}\operatorname{ev}_{\mathbf{E}_q}(\operatorname{div}_{\mathbf{E}_q}(f)) &= \frac{\prod_{i=1}^k \overline{a_i}^{m_i}}{\prod_{j=1}^l \overline{b_j}^{n_j}} \\ &= \pi\left(\frac{\prod_{i=1}^k a_i^{m_i}}{\prod_{j=1}^l b_j^{n_j}}\right) \in \mathbf{E}_q.\end{aligned}$$

It is supposed to be trivial (as the evaluation of a principal divisor; see Subsection A.1.5 of Appendix A) in $\mathbf{E}_q$, i.e., $\prod a_i^{m_i} \equiv \prod b_j^{n_j} \pmod{q^{\mathbf{Z}}}$ (meaning that the quotient is in $q^{\mathbf{Z}}$). The converse is true; we summarize:

THEOREM 2.1 (q-version of Abel-Jacobi). *Let $f \in \mathscr{M}(\mathbf{E}_q)$, $f \neq 0$, seen as a q-invariant meromorphic function on $\mathbf{C}^*$. Let $\partial D(0,R)$ a circle centered at 0 and containing no zero or pole of f (this holds for all radii R out of a denumerable subset of $\mathbf{R}_+^*$; and then any $|q|^k R$, $k \in \mathbf{Z}$, will have the same property). Let $r := |q|\, R$ and let a_i, resp. b_j, be the zeroes, resp. the poles of f in $\mathring{C}(r,R)$, with multiplicities m_i, resp. n_j. Then $\sum m_i = \sum n_j$ and $\prod a_i^{m_i} \equiv \prod b_j^{n_j} \pmod{q^{\mathbf{Z}}}$.*

Conversely, any data a_i, b_j, m_i, n_j, satisfying those relations come from a q-elliptic function, which is unique up to a factor in $\mathbf{C}^*$.

A direct effective proof of the converse using Theta functions will be given in Section 2.2.2. The total number of zeroes (counted with multiplicities) in an annulus, as described above, is the *order* of the elliptic function; same thing for poles. Recall that the order is 0 for a nonzero constant function, and ≥ 2 for a nonconstant elliptic function (see Appendix A.2).

2.1.2. The real Theta function. This is the function $\vartheta(t) := \sum_{n\in\mathbf{Z}} e^{-\pi n^2 t}$. It is indefinitely differentiable over $\mathbf{R}_+^*$ and satisfies a functional equation which links it to the modular world (cf. below proposition 2.3).

2.1.2.1. *Schwartz class and Fourier transformation.* We say that a function $f : \mathbf{R} \to \mathbf{C}$ has *fast decay (at infinity)* if, for every $N > 0$, one has $|f(x)| = O\left(x^{-N}\right)$ as $|x| \to +\infty$. We say that f *belongs to the Schwartz class* if it is indefinitely differentiable and all its derivatives $f^{(n)}$ have fast decay. The Schwartz class is denoted $\mathcal{S}(\mathbf{R})$. The *Fourier transform* of $f \in \mathcal{S}(\mathbf{R})$ is defined by the formula:

$$\hat{f}(x) := \int_{-\infty}^{+\infty} e^{-2i\pi xy} f(y)\, dy.$$

From the general properties of Lebesgue integration, one proves easily the following facts:

(1) $\hat{f} \in \mathcal{S}(\mathbf{R})$.
(2) If $a > 0$ and $g(x) := f(ax)$, then $g \in \mathcal{S}(\mathbf{R})$ and $\hat{g}(x) = (1/a)\hat{f}(x/a)$.
(3) $f' \in \mathcal{S}(\mathbf{R})$ and $\widehat{f'}(x) = 2i\pi x \hat{f}(x)$.
(4) If $h(x) := xf(x)$, then $h \in \mathcal{S}(\mathbf{R})$ and $\hat{h}(x) = (-1/2i\pi)(\hat{f})'x)$.

[1] As a divisor on a Riemann surface which is a group, it has an "evaluation"; see A.1.3 in Appendix A.

THEOREM 2.2 (Poisson summation formula). *For any* $f \in \mathcal{S}(\mathbf{R})$, *one has* $\sum_{n\in\mathbf{Z}} f(x+n) = \sum_{n\in\mathbf{Z}} \hat{f}(n)e^{2i\pi nx}$.

PROOF. Let $F(x) := \sum_{n\in\mathbf{Z}} f(x+n)$. Since $f \in \mathcal{S}(\mathbf{R})$, the function F is indefinitely differentiable. Clearly it has period 1, whence a Fourier expansion $F(x) = \sum_{n\in\mathbf{Z}} c_n e^{2i\pi nx}$. We compute the coefficients:

$$c_k = \int_0^1 F(x)e^{-2i\pi kx}\,dx = \sum_{n\in\mathbf{Z}} \int_0^1 f(x+n)e^{-2i\pi kx}\,dx =$$

$$\sum_{n\in\mathbf{Z}} \int_0^1 f(x+n)e^{-2i\pi k(x+n)}\,d(x+n) = \sum_{n\in\mathbf{Z}} \int_n^{n+1} f(x)e^{-2i\pi kx}\,dx =$$

$$\int_{-\infty}^{+\infty} f(x)e^{-2i\pi kx}\,dx = \hat{f}(k).$$

□

2.1.2.2. *Application to the real Theta function.*

LEMMA 2.2. (i) *Let* $f(x) := e^{-\pi x^2}$. *Then* $f \in \mathcal{S}(\mathbf{R})$ *and* $\hat{f} = f$.
(ii) *Let* $a > 0$ *and* $f_a(x) := e^{-ax^2}$. *Then* $f_a \in \mathcal{S}(\mathbf{R})$ *and* $\hat{f_a} = \sqrt{\pi/a}\, f_{\pi^2/a}$.
(iii) $\forall x \in \mathbf{R}$, $\sum_{n\in\mathbf{Z}} e^{-a(x+n)^2} = \sqrt{\pi/a} \sum_{n\in\mathbf{Z}} e^{-(\pi^2/a)n^2} e^{2i\pi nx}$.

PROOF. (1) From $f' = -2\pi x f$, using the facts above, one gets $(\hat{f})'(x) = -2\pi x\hat{f}(x)$, whence $\hat{f} = Cf$ for some constant C. One has $f(0) = 1$ (obvious) and $\hat{f}(0) = 1$ (this is the second part of Corollary A.22), so $C = 1$.
(2) comes from $f_a(x) = f(\sqrt{a/\pi}x)$ using the facts above.
(3) results from Poisson summation formula applied to f_a.

□

PROPOSITION 2.3. $\forall t > 0$, $\vartheta(t) = \sqrt{1/t}\,\vartheta(1/t)$.

PROOF. Take $a := \pi t$ and $x := 0$ in the third statement of the lemma. □

We shall give stronger applications of Lemma 2.2.

2.1.3. Classical Theta functions. Our ultimate goal (in 2.2) is the "basic" Theta function $\theta_q(x)$, $|q| < 1$, $x \in \mathbf{C}^*$; however, it is more efficient to start with the classical $\Theta(z, \tau)$, $\tau \in \mathcal{H}$, $z \in \mathbf{C}$. Actually, there are four fundamental such functions, $\vartheta_{i,j}$, $i, j = 0, 1$; see **[174]**. But we shall need only one; the relation to Mumford's notations is:

$$\Theta(z,\tau) = \vartheta\left(z + \frac{1-\tau}{2}, \tau\right) = \vartheta_{0,0}\left(z + \frac{1-\tau}{2}, \tau\right).$$

Theta functions have found use in mechanics (the spinning top), physics (the heat equation), algebraic geometry (abelian varieties), number theory ... (see **[174, 231]**); we shall touch upon the latter application but say nothing about the first three.

2.1.3.1. So we define:

$$\forall z \in \mathbf{C}\,,\ \tau \in \mathcal{H}\,,\ \Theta(z,\tau) := \sum_{n \in \mathbf{Z}} (-1)^n e^{n(n-1)\mathrm{i}\pi\tau} e^{2\mathrm{i}\pi n z} = \sum_{n \in \mathbf{Z}} (-1)^n q^{n(n-1)/2} e^{2\mathrm{i}\pi n z},$$

with our now usual notation $q := e^{2\mathrm{i}\pi\tau}$. Since $\Im\tau > 0 \Rightarrow |q| < 1$, the quadratic exponent $q^{n^2/2}$ dominates and (for fixed τ) the series converges normally over every compact subset of $\mathbf{C}$, so $\Theta_\tau : z \mapsto \Theta(z,\tau)$ is an entire function.

EXERCISE 2.4. Give an argument to justify that it is not the zero function.

PROPOSITION 2.5. *One has:*

$$\Theta_\tau(z+1) = \Theta_\tau(z) \text{ and } \Theta_\tau(z+\tau) = -e^{-2\mathrm{i}\pi z}\Theta_\tau(z).$$

PROOF. The first relation is obvious since Θ_τ depends on z only through $e^{2\mathrm{i}\pi z}$. The second comes easily:

$$\begin{aligned}
\Theta_\tau(z+\tau) &= \sum_{n \in \mathbf{Z}} (-1)^n e^{n(n-1)\mathrm{i}\pi\tau} e^{2\mathrm{i}\pi n(z+\tau)} \\
&= \sum_{n \in \mathbf{Z}} (-1)^n e^{n(n-1)\mathrm{i}\pi\tau + 2\mathrm{i}\pi n\tau} e^{2\mathrm{i}\pi n z} \\
&= \sum_{n \in \mathbf{Z}} (-1)^n e^{n(n+1)\mathrm{i}\pi\tau} e^{2\mathrm{i}\pi n z} \\
&= \sum_{n \in \mathbf{Z}} (-1)^{n-1} e^{n(n-1)\mathrm{i}\pi\tau} e^{2\mathrm{i}\pi(n-1)z} \\
&= -e^{-2\mathrm{i}\pi z} \sum_{n \in \mathbf{Z}} (-1)^n e^{n(n-1)\mathrm{i}\pi\tau} e^{2\mathrm{i}\pi n(z+\tau)},
\end{aligned}$$

as wanted; the penultimate equality comes by shifting indices $n \leftarrow n-1$. □

So Θ_τ functions look quite like our function T in A.1.2 and can therefore be used to produce elliptic functions.

COROLLARY 2.6. *Let $a_j \in \mathbf{C}$, $n_j \in \mathbf{Z}$ for $j = 1, \ldots, k$. Then:*

$$f(z) := \prod_{j=1}^{k} \Theta_\tau(z - a_j)^{n_j} \in \mathscr{M}(\mathbf{E}_{\Lambda_\tau}) \Longleftrightarrow \left(\sum_{j=1}^{k} n_j = 0 \text{ and } \sum_{j=1}^{k} n_j a_j \in \mathbf{Z} \right).$$

PROOF. The function f is obviously meromorphic on $\mathbf{C}$ (this is where we use the above exercise) and satisfies $f(z+1) = f(z)$. Then $\dfrac{f(z+\tau)}{f(z)}$ is equal to:

$$\prod_{j=1}^{k} \frac{\Theta_\tau(z+\tau-a_j)^{n_j}}{\Theta_\tau(z-a_j)^{n_j}} = \prod_{j=1}^{k} (-e^{-2\mathrm{i}\pi(z-a_j)})^{n_j} = (-e^{-2\mathrm{i}\pi z})^{\sum_{j=1}^{k} n_j} e^{2\mathrm{i}\pi \sum_{j=1}^{k} n_j a_j},$$

and the conclusion flows. □

COROLLARY 2.7. *The function Θ_τ has zeroes at all points of Λ_τ with the same multiplicities.*

PROOF. Because of the functional relations, it is enough to verify that $\Theta_\tau(0) = 0$. But in the sum $\Theta_\tau(0) = \sum_{n \in \mathbf{Z}} (-1)^n e^{n(n-1)\mathrm{i}\pi\tau}$, shifting indices by $n \leftarrow 1-n$ gives $\Theta_\tau(0) = -\Theta_\tau(0)$. □

We did not say that those are the only zeroes, although it is true, as will follow from Jacobi's Triple product formula in 2.2.1. Admitting this fact for the moment, we deduce:

COROLLARY 2.8. *Every nonzero* $g \in \mathscr{M}(\mathbf{E}_{\Lambda_\tau})$ *has (up to a nonzero constant factor) the form* $f(z)$ *shown above.*

PROOF. Let g have zeroes and poles a_j with orders n_j in the interior of a fundamental parallelogram such that its boundary contains no zero or pole and let f as shown above. Then g/f is a nonzero Λ_τ elliptic function without zeroes or poles, so it is a nonzero constant. □

2.1.3.2. Now we extend to the complex setting the modular functional equation[2] for the real Theta function proved in the lemma of 2.1.2.2.

THEOREM 2.3 (Modular functional equation for $\Theta(z,\tau)$).

$$\forall z \in \mathbf{C}\,,\ \forall \tau \in \mathcal{H}\,,\ \Theta(z,\tau) = \sqrt{\frac{i}{\tau}}\, e^{-\frac{i\pi}{\tau}\left(z+\frac{1-\tau}{2}\right)^2}\, \Theta\left(\frac{z}{\tau}, \frac{-1}{\tau}\right).$$

PROOF. (Recall from 0.4 that $\mathcal{H}$ denotes the Poincaré half plane $\Im\tau > 0$.) Note first that $\Im\dfrac{-1}{\tau} = \Re\dfrac{\mathrm{i}}{\tau} = \dfrac{\Im\tau}{|\tau|^2} > 0$, so $\dfrac{-1}{\tau} \in \mathcal{H}$ and the Theta factor on the right-hand side is well defined. Also note that, since $\Re\dfrac{\mathrm{i}}{\tau} > 0$, we need only give a meaning to $\sqrt{u}$ for arguments in the right half-plane $\Re u > 0$: we take the main determination on $\mathbf{C} \setminus \mathbf{R}_-$, i.e., the one which restricts to the usual square root on $\mathbf{R}_+^*$ (more explicitly, it is given by the formula $\sqrt{re^{\mathrm{i}\alpha}} = \sqrt{r}e^{\mathrm{i}\alpha/2}$ when $r > 0$ and $-\pi < \alpha < \pi$).

To begin with, we suppose that τ is pure imaginary, so that $a := -\mathrm{i}\pi\tau > 0$. Then, using the lemma from 2.1.2.2, we see that for $z \in \mathbf{C}$:

$$\begin{aligned}\sum_{n\in\mathbf{Z}} e^{\mathrm{i}\pi\tau(z+n)^2} = \sum_{n\in\mathbf{Z}} e^{-a(z+n)^2} &= \sqrt{\frac{\pi}{a}} \sum_{n\in\mathbf{Z}} e^{-(\pi^2/a)n^2} e^{2\mathrm{i}\pi nz} \\ &= \sqrt{\frac{\mathrm{i}}{\tau}} \sum_{n\in\mathbf{Z}} e^{-(\mathrm{i}\pi/\tau)n^2} e^{2\mathrm{i}\pi nz}.\end{aligned}$$

Actually, the lemma gives that for z real, but the extreme sides are analytic functions of z, so, by the principle of analytic continuation, it holds for complex z.

We now replace in the above equality (i.e., the one relating the first expression to the last one) z by $\frac{1}{\tau}\left(z + \frac{1-\tau}{2}\right)$. Then $e^{\mathrm{i}\pi\tau(z+n)^2}$ (from the leftmost expression) is replaced by:

$$e^{\frac{\mathrm{i}\pi}{\tau}\left(z+\frac{1-\tau}{2}\right)^2} e^{2n\mathrm{i}\pi\left(z+\frac{1-\tau}{2}\right)} e^{\mathrm{i}\pi\tau n^2} = e^{\frac{\mathrm{i}\pi}{\tau}\left(z+\frac{1-\tau}{2}\right)^2} (-1)^n e^{n(n-1)\mathrm{i}\pi\tau} e^{2\mathrm{i}\pi nz}.$$

Likewise, $e^{-(\mathrm{i}\pi/\tau)n^2} e^{2\mathrm{i}\pi nz}$ (from the rightmost expression) is replaced by:

$$\begin{aligned}e^{-(\mathrm{i}\pi/\tau)n^2} e^{(2\mathrm{i}\pi n/\tau)\left(z+\frac{1-\tau}{2}\right)} &= e^{-(\mathrm{i}\pi/\tau)n^2} e^{2\mathrm{i}\pi nz/\tau} e^{\mathrm{i}\pi n/\tau} e^{-\mathrm{i}\pi n} \\ &= (-1)^n e^{n(n-1)\mathrm{i}\pi\tau'} e^{2\mathrm{i}\pi nz'},\end{aligned}$$

[2]In modular function theory, there is someting called "modular equation" and which is very different from what is being considered here; see [**109**, lecture 12], [**186**, II ch 9], [**21**]. We call "modular functional equation" the functional equation related to modularity, i.e., the one that involves $-1/\tau$.

where we set for short $\tau' := -1/\tau$ and $z' := z/\tau$. So, after these replacments, our formula becomes:

$$e^{\frac{i\pi}{\tau}\left(z+\frac{1-\tau}{2}\right)^2}\Theta(z,\tau) = \sqrt{\frac{\mathrm{i}}{\tau}}\,\Theta(z',\tau').$$

This is exactly the wanted formula, but proved here under the assumption that τ is pure imaginary. However, for every fixed z, both sides are analytic functions of $\tau \in \mathcal{H}$, so the result comes again from the principle of analytic continuation. □

2.2. "Basic" Theta functions

2.2.1. Generalities. Of course "basic" (as in "basic hypergeometric series", 1.6; also see Chapter 4) refers here to the *base* q, still assumed to satisfy $0 < |q| < 1$. Also we write $q = e^{2i\pi\tau}$ for some chosen $\tau \in \mathcal{H}$.

2.2.1.1. Although we could have done something similar at the beginning of A.1.2, we shall take advantage of the introduction of one of the most important functions in this book to give some motivation on how its defining formula could have been found. A similar discussion but involving the classical model (with z and τ) is to be found at the very beginning of [**174**].

We know from 2.1.1 that any q-invariant holomorphic function on $\mathbf{C}^*$ must be constant. So we may try to relax condition and look for a Laurent power series $f(x) = \sum_{n\in\mathbf{Z}} a_n x^n$ satisfies $f(qx) = cx^k f(x)$ for some $c \in \mathbf{C}^*$ and $k \in \mathbf{Z}$. This is plainly equivalent to a relation among coefficients:

$$f(qx) = cx^k f(x) \Longleftrightarrow \forall n \in \mathbf{Z}\ ,\ q^n a_n = c a_{n-k}.$$

Exploring the consequences, we find that there are three main cases to consider. In all the discussion we exclude a priori the trivial solution $f = 0$.

$\underline{k > 0}$. Then the sequence (a_n) grows, in modulus, grossly as $q^{-n^2/(2k)}$, so, either on the side $n \to +\infty$ or on the side $n \to -\infty$ it diverges.

EXERCISE 2.9. To make the estimation more precise, fix some $r \in \{0,\dots,k-1\}$ and consider the subsequence $b_m := a_{mk+r}$. Find a recursive relation, then an explicit formula and conclude.

$\underline{k = 0}$. The relation boils down to $(c - q^n)a_n = 0$. There are two possibilities: either $c \notin q^{\mathbf{Z}}$, and then all $a_n = 0$, which was excluded; or $c = q^m$, $m \in \mathbf{Z}$, and then all $c_n = 0$ for $n \neq m$, so $f(x) = ax^m$ for some constant a, an uninteresting case that we do not pursue (as it will be of little help in trying to produce q-elliptic functions).

$\underline{k < 0}$. We could, as in the exercise, consider subsequences, but we do something else (which is actually equivalent): letting $l := -k$, we write $f(x) = f_0(x^l) + xf_1(x^l) + \cdots + x^{l-1}f_{l-1}(x^l)$. Then:

$$\begin{aligned} f(qx) = cx^k f(x) &\Longleftrightarrow \forall r = 0,\dots,l-1\ ,\ q^r x^r f_r(q^l x^l) = cx^{-l}x^r f_r(x^l) \\ &\Longleftrightarrow \forall r = 0,\dots,l-1\ ,\ f_r(q^l x^l) = (cq^{-r})x^{-l} f_r(x^l), \end{aligned}$$

i.e., with simplified notations $Q := q^l$, $X := x^l$, equations of the form $F_r(QX) = C_r X^{-1} F_r(X)$. In other words, up to the change of notations, we are considering equations of the form $f(qx) = cx^{-1}f(x)$ (but see the exercise below). So now the relation for coefficients is $q^n a_n = ca_{n+1}$ and we easily solve it: $a_n = c^{-n}q^{n(n-1)/2}a_0$, whence $f(x) = a_0\theta(x/c)$ where $\theta(x) := \sum_{n\in\mathbf{Z}} q^{n(n-1)/2}x^n$. This function θ is the one we shall now study.

EXERCISE 2.10. Having solved the particular case $k = -1$, go back and use it to solve the general case $k < 0$.

2.2.1.2. We shall use very systematically the nontrivial Laurent series:

$$\theta_q(x) := \sum_{n\in\mathbf{Z}} q^{n(n-1)/2}x^n.$$

EXERCISE 2.11. Retrieve much of what follows as a special case $c = 1$ of 2.2.1.1.

Due to the quadratic exponent in $q^{n^2/2}$, the series converges normally on every compact subset of $\mathbf{C}^*$. The following properties are immediately inferred:

(i) $\theta_q \in \mathscr{O}(\mathbf{C}^*)$ (recall from 0.4 that $\mathscr{O}(\mathbf{C}^*)$ is the algebra of holomorphic functions on $\mathbf{C}^*$).

(ii) $\theta_q(qx) = \theta_q(1/x) = \dfrac{1}{x}\theta_q(x)$ (and therefore $\theta_q(q/x) = \theta_q(x)$).

(iii) $\theta_q(-1) = 0$ (shifting indices by $n \leftarrow 1 - n$ gives $\theta_q(-1) = -\theta_q(-1)$).

So more generally from (ii) and (iii), $\theta_q(-q^n) = 0$ for all $n \in \mathbf{Z}$, i.e., θ_q has zeroes at all points of the discrete q-spiral $[-1; q] = -q^{\mathbf{Z}}$. It will follow later (2.2.1.3) from Jacobi's Triple product formula that those zeroes are simple and that they are the only ones.

Compared with the classical Theta function of 2.1.3, we have, for $x = e^{2i\pi z}$:

$$\theta_q(x) = \Theta(z \pm 1/2, \tau), \text{ equivalently: } \theta_q(-x) = \Theta(z, \tau).$$

We shall meet later (in 2.2.4 and then in 2.4) the consequences for $\theta_q(x)$ of the modular functional equation of $\Theta(z, \tau)$.

Before going on, we recall some important q-notations. We write $[a; q] := aq^{\mathbf{Z}}$ the discrete q-spirals; and we use the q-Pochhammer symbols:

$$(a; q)_\infty := \prod_{n\geq 0}(1 - aq^n), \quad (a; q)_n := \frac{(a; q)_\infty}{(aq^n; q)_\infty},$$

so that in particular:

$$\forall n \in \mathbf{N}\ ,\ (a; q)_n = \prod_{0\leq i<n}(1 - aq^i).$$

The analytic properties af these expressions *qua* functions of a will be studied in Chapter 3; but, anticipating that study, we leave you the possibility to prove some of them yourself - as we intend to use them next.

EXERCISE 2.12. (i) $x \mapsto (x; q)_\infty$ is an entire function. (Its power series expansion was actually given in 1.2, also see Chapter 3.)

(ii) That function has simple zeroes over $q^{-\mathbf{N}}$ and no other zeroes.

(iii) It satisfies the functional equation $(qx; q)_\infty = \dfrac{1}{1-x}(x; q)_\infty$.

(iv) We have for all $a \in \mathbf{C}$ a particular case of a "distribution formula": $(a; q)_\infty(-a; q)_\infty = (a^2; q^2)_\infty$.

2.2.1.3. *Jacobi's Triple product formula.*

THEOREM 2.4 (Jacobi's Triple product formula).

$$\theta_q(x) = (q;q)_\infty(-x;q)_\infty(-q/x;q)_\infty.$$

PROOF. A proof was already given in 1.7. The following one comes from **[112]**. There is a different one in **[174]**.

Let $T_q(x) := (-x;q)_\infty(-q/x;q)_\infty$. This function is holomorphic over $\mathbf{C}^*$ (by the rules about infinite products of holomorphic functions; see **[211]**), it has simple zeroes over $[-1;q]$ and no other zeroes. Moreover, from the relation $(x;q)_\infty = (1-x)(qx;q)_\infty$ applied to each of the two factors of T_q, one draws that $T_q(qx) = \dfrac{1+1/x}{1+x}T_q(x) = \dfrac{1}{x}T_q(x)$. This implies that $\theta_q(x) = CT_q(x)$ for some constant $C \in \mathbf{C}^*$. Indeed, there are two possible arguments:

(1) The function θ_q/T_q is q-elliptic and has no pole, so it must be constant (and nonzero since θ_q is a nontrivial Laurent series).
(2) Being holomorphic over $\mathbf{C}^*$, T_q has a Laurent power series expansion $T_q(x) = \sum\limits_{n\in\mathbf{Z}} t_n x^n$, and the functional equation $T_q(qx) = \dfrac{1}{x}T_q(x)$ implies that $q^n t_n = t_{n+1}$, whence $t_n = q^{n(n-1)/2}t_0$ for all $n \in \mathbf{Z}$ (the induction goes both ways, from 0 to $+\infty$ and from 0 to $-\infty$), whence $T_q = t_0\theta_q$ with $t_0 \neq 0$ (that is because $T_q \neq 0$) and one can take $C := t_0^{-1}$.

We must now prove that $C = (q;q)_\infty$. To do that we make explicit the dependence on q and rather write $\theta_q(x) = C(q)T_q(x)$. We shall first use the fact that, for every $x_0 \in \mathbf{C}^*$ such that $x_0 \notin [-1;q]$, we have $C(q) = \dfrac{\theta_q(x_0)}{T_q(x_0)}$. Note that the condition on x_0 also implies that $x_0 \notin [-1;q^N]$ for all $N \in \mathbf{N}^*$, so, for such N,we also have $C(q^N) = \dfrac{\theta_{q^N}(x_0)}{T_{q^N}(x_0)}$; using explicit expressions for the numerator and denominator, we conclude that:

$$\lim_{N\to+\infty} C(q^N) = C(0) = 1.$$

Now we set $D(q) = (q;q)_\infty$, which is holomorphic in $q \in \overset{\circ}{\mathrm{D}}(0,1)$ (same rules on infinite products from **[211]**) and so in particular continuous. We are going to prove that $\dfrac{C(q)}{D(q)} = \dfrac{C(q^4)}{D(q^4)}$. Admitting this fact for a moment, we deduce by iteration and going to the limit:

$$\frac{C(q)}{D(q)} = \frac{C(q^4)}{D(q^4)} = \cdots = \frac{C(q^{4^n})}{D(q^{4^n})} = \cdots = \lim_{n\to+\infty} \frac{C(q^{4^n})}{D(q^{4^n})} = \frac{C(0)}{D(0)} = 1,$$

so indeed $C(q) = D(q)$, which will conclude the proof.

So we go for our identity. First, we choose a square root $\sqrt{-q}$ of $-q$, for instance $\sqrt{-q} := e^{\mathrm{i}\pi(\tau+1/2)}$, but all we shall use is that $\left(\sqrt{-q}\right)^2 = -q$. Then we compute:

$$\theta_q\left(\sqrt{-q}\right) = \sum_{n\in\mathbf{Z}} q^{n(n-1)/2}\left(\sqrt{-q}\right)^n = \sum_{n\in\mathbf{Z}} q^{n(2n-1)}(-q)^n + \sqrt{-q}\sum_{n\in\mathbf{Z}} q^{n(2n+1)}(-q)^n.$$

The summation was split into even index, resp. odd index terms; then both parts were reindexed, the first by $n \leftarrow 2n$, the second $n \leftarrow 2n+1$. The first sum thus obtained is equal to:

$$\sum_{n\in\mathbf{Z}} q^{2n^2}(-1)^n = \sum_{n\in\mathbf{Z}} (q^4)^{n(n-1)/2}(-q^2)^n = \theta_{q^4}(-q^2).$$

The second term (again in the the last expression) vanishes because:

$$\sum_{n\in\mathbf{Z}} q^{n(2n+1)}(-q)^n = \sum_{n\in\mathbf{Z}} (q^4)^{n(n-1)/2}(-q^4)^n = \theta_{q^4}(-q^4).$$

So we are led to equality:

$$\theta_q\left(\sqrt{-q}\right) = \theta_{q^4}(-q^2).$$

Now we turn to T_q; since $-q/\sqrt{-q} = \sqrt{-q}$, we have, using twice the distribution formula (exercise at the end of 2.2.1.2):

$$T_q\left(\sqrt{-q}\right) = \left(-\sqrt{-q};q\right)_\infty\left(\sqrt{-q};q\right)_\infty = (-q;q^2)_\infty = \frac{(q^2;q^4)_\infty}{(q;q^2)_\infty}.$$

Similarly:

$$T_{q^4}(-q^2) = (q^2;q^4)_\infty(q^2;q^4)_\infty = \left((q^2;q^4)_\infty\right)^2.$$

Combining all that:

$$C(q) = \frac{\theta_q\left(\sqrt{-q}\right)}{T_q\left(\sqrt{-q}\right)} = \frac{\theta_{q^4}(-q^2)(q;q^2)_\infty}{(q^2;q^4)_\infty} = \frac{\theta_{q^4}(-q^2)}{T_{q^4}(-q^2)}(q;q^2)_\infty(q^2;q^4)_\infty.$$

Therefore, since $C(q^4) = \dfrac{\theta_{q^4}(-q^2)}{T_{q^4}(-q^2)}$, we have $\dfrac{C(q)}{C(q^4)} = (q;q^2)_\infty(q^2;q^4)_\infty$. Now comes another trick of q-expansions; the last expression is equal to:

$$\prod_{n\geq 0}(1-q^{2n+1})\prod_{n\geq 0}(1-q^{4n+2}) = \prod_{\substack{n\geq 0\\ n\not\equiv 0 \pmod 4}}(1-q^n) = \frac{\prod_{n\geq 0}(1-q^n)}{\prod_{n\geq 0}(1-q^{4n})} = \frac{(q;q)_\infty}{(q^4;q^4)_\infty}.$$

So altogether we have proved that $\dfrac{C(q)}{C(q^4)} = \dfrac{D(q)}{D(q^4)}$, i.e., that $\dfrac{C(q)}{D(q)} = \dfrac{C(q^4)}{D(q^4)}$ as stated. □

The following corollaries are immediate in their θ_q-version; the Θ_τ-version then follows from $\theta_q(x) = \Theta_\tau(z+1/2)$, or $\Theta_\tau(z) = \theta_q(-x)$ (under the standard convention $x = e^{2i\pi z}$).

COROLLARY 2.13. (i) *The only zeroes of* θ_q *over* $\mathbf{C}^*$ *are the points of* $[-1;q]$ *and they are simple.*
(ii) *The only zeroes of* Θ_τ *over* $\mathbf{C}$ *are the points of* Λ_τ *and they are simple.*

COROLLARY 2.14. (i) *When* $x \to -1$, $\theta_q(x) \sim (1+x)(q;q)_\infty^3$.
(ii) *When* $z \to 0$, $\Theta_\tau(z) \sim -2i\pi z(q;q)_\infty^3$.

PROOF.

$$\frac{\theta_q(x)}{1+x} = (q;q)_\infty(-qx;q)_\infty(-q/x;q)_\infty \underset{x\to -1}{\to} (q;q)_\infty^3.$$

□

REMARK 2.15. The asymptotic behaviour of Theta functions for $|x| \to +\infty$ is tackled (for instance) in **[191]**.

The following will be useful more than once.

EXERCISE 2.16. (i) Using Jacobi's Triple Product Formula, show that:

$$\theta_q{}'(-1) = (q;q)_\infty^3.$$

(ii) Deduce that the residue of $1/\theta_q$ at -1 is $1/(q;q)_\infty^3$. What is the residue at $-q^n$, $n \in \mathbf{Z}$?

We can make more precise the role of those residues as follows.

PROPOSITION 2.17. *We have the "infinite partial fraction decomposition" valid over* $U := \mathbf{C}^* \setminus [-1;q]$*:*

$$\frac{1}{\theta_q(t)} = \frac{1}{\theta_q{}'(-1)} \sum_{m\in\mathbf{Z}} \frac{(-1)^m q^{m(m+1)/2}}{1+q^m t}.$$

PROOF. Let $f(t) := \sum_{m\in\mathbf{Z}} \dfrac{(-1)^m q^{m(m+1)/2}}{1+q^m t}$ and $g(t) := f(t)\theta_q(t)$. The series defining $f(t)$ converges normally on every compact subset of U, so $f \in \mathcal{O}(U)$. Also, f has simple poles and θ_q has simple zeroes over $[-1;q]$, so $g \in \mathcal{O}(\mathbf{C}^*)$. Normal convergence justifies the following calculation:

$$\begin{aligned}
g(qt) &= \sum_{m\in\mathbf{Z}} \frac{(-1)^m q^{m(m+1)/2}\theta_q(qt)}{1+q^{m+1}t} = \sum_{m\in\mathbf{Z}} \frac{(-1)^m q^{m(m+1)/2}\theta_q(t)/t}{1+q^{m+1}t} \\
&= \sum_{m\in\mathbf{Z}} \frac{(-1)^{m-1} q^{m(m-1)/2}\theta_q(t)/t}{1+q^m t} = \sum_{m\in\mathbf{Z}} \frac{(-1)^{m-1} q^{m(m+1)/2}\theta_q(t)}{q^m t(1+q^m t)} \\
&= \sum_{m\in\mathbf{Z}} \frac{(-1)^{m-1} q^{m(m+1)/2}\theta_q(t)}{q^m t} - \sum_{m\in\mathbf{Z}} \frac{(-1)^{m-1} q^{m(m+1)/2}\theta_q(t)}{1+q^m t},
\end{aligned}$$

where both series converge normally, the first one vanishes (permutation $m \mapsto 1-m$ of $\mathbf{Z}$ transforms it into its opposite) and the second one is $-g(t)$: therefore $g(qt) = g(t)$. Since $g \in \mathcal{O}(\mathbf{C}^*)$, this implies that g is constant: $g = g(-1)$. Since $f(t)$ is the sum of the $m = 0$ term $1/(1+t)$ and a function holomorphic at -1, we can evaluate $g(-1)$ as the limit of $\theta_q(t)/(1+t)$ when $t \to -1$, i.e., $g = g(-1) = \theta_q{}'(-1)$. The conclusion is then immediate. □

2.2.1.4. *A definite integral.* We are going to prove a formula (apparently due to Changgui Zhang; see **[265]**) which will be of some importance in Chapter 14. That formula holds for real q.

PROPOSITION 2.18. *We assume here that* $0 < q < 1$*. Then:*

$$I := \int_0^{+\infty} \frac{1}{\theta_q(t)} \frac{dt}{t} = \ln(1/q).$$

PROOF. From the obvious inequality $\theta_q(t) > 1 + qt^2$ (for $t > 0$) we draw that the integral converges at infinity. We split the integral as $\int_0^{+\infty} = \int_0^{\sqrt{q}} + \int_{\sqrt{q}}^{+\infty}$ and use the the change of variable $t \leftarrow q/t$ along with the formula $\theta_q(q/t) = \theta_q(t)$ from 2.2.1.2 to get that $\int_0^{\sqrt{q}} = \int_{\sqrt{q}}^{+\infty}$, whence:

$$I = 2\int_0^{\sqrt{q}} \frac{1}{\theta_q(t)} \frac{dt}{t} = 2\int_0^{1/\sqrt{q}} \frac{1}{\theta_q(t)} \frac{dt}{t},$$

the last equality flowing from the change of variable $t \leftarrow qt$ along with the formula $\theta_q(qt) = \theta_q(t)/t$, again from 2.2.1.2. Now using Proposition 2.17, we deduce (invoking normal convergence):

$$\int_0^{1/\sqrt{q}} \frac{1}{\theta_q(t)} \frac{dt}{t} = \frac{1}{\theta_q{}'(-1)} \sum_{m\in\mathbf{Z}} (-1)^m q^{m(m+1)/2} \int_0^{1/\sqrt{q}} \frac{dt}{1+q^m t}$$
$$\Longrightarrow I = \frac{2}{\theta_q{}'(-1)} \sum_{m\in\mathbf{Z}} (-1)^m q^{m(m-1)/2} \ln(1+q^{m-1/2}).$$

We write this as $I = \dfrac{2}{\theta_q{}'(-1)} S$ and expand the logarithms in S; we also group pairs of terms with index $m \geq 1$ and with index $1-m \leq 0$ (note that $m(m-1)/2$ is left invariant by the symmetry $m \mapsto 1-m$):

$$S = \sum_{m\geq 1} q^{m(m-1)/2} \Big((-1)^m \ln(1+q^{m-1/2}) + (-1)^{1-m} \ln(1+q^{1/2-m}) \Big)$$
$$= \sum_{m\geq 1} (-1)^m q^{m(m-1)/2} \ln \frac{1+q^{m-1/2}}{1+q^{1/2-m}} = \sum_{m\geq 1} (-1)^m q^{m(m-1)/2} \ln q^{m-1/2}$$
$$= \sum_{m\geq 1} (-1)^m (m-1/2) q^{m(m-1)/2} \ln q.$$

On the other hand, $\theta_q{}'(x) = \sum_{m\in\mathbf{Z}} q^{m(m-1)/2} m x^{m-1}$ (normal convergence of the derived series justifies termwise derivation of θ_q), so, using the same grouping of pairs:

$$\theta_q{}'(-1) = \sum_{m\in\mathbf{Z}} q^{m(m-1)/2} m(-1)^{m-1} = \sum_{m\geq 1} q^{m(m-1)/2} (m(-1)^{m-1} + (1-m)(-1)^m)$$
$$= \sum_{m\geq 1} (-1)^m q^{m(m-1)/2} (1-2m) = -2S/\ln q.$$

The conclusion $I = \dfrac{2}{\theta_q{}'(-1)} S = -\ln q = \ln 1/q$ immediately follows. □

EXERCISE 2.19. We propose here various formulas due to Changgui Zhang [**269**]. They require some understanding on functions on the universal covering $\widetilde{\mathbf{C}^*}$ of $\mathbf{C}^*$, see 14.3.1 in Chapter 14 for some explanations. We assume q real, $0 < q < 1$, and write it as $q = e^{-2\pi/\kappa}$ (so $\kappa > 0$). We shall denote $\theta(q, x) := \theta_q(x)$ to alleviate typography. Then we set $q^* := e^{-2\pi\kappa}$ and $x^* := x^{-\kappa\mathrm{i}}$ for $x \in \widetilde{\mathbf{C}^*}$. (All this is obviously related to 2.2.4.) Last, we let $e(q,x) := \exp\left(-\frac{\log^2(x/\sqrt{q})}{2\ln q}\right)$. Now prove the following:

$$\theta(q, \sqrt{q}x) = \sqrt{\kappa} e(q, \sqrt{q}x) \theta_{q^*}(q^*, \sqrt{q^*}x^*),$$
$$\theta(q, -x) = \sqrt{\kappa} e(q, -x) \theta_{q^*}(q^*, -x^*),$$
$$\frac{\theta_q(-\lambda x)}{\theta_q(-x)} = \frac{\lambda^{1-m}\theta_q(-\lambda)}{(q;q)_\infty^3} \sum_{\ell\in\mathbf{Z}} \frac{(q^{1-m}x)^2}{1-\lambda q^\ell}.$$

In the last equation, we assume $\lambda \in \mathbf{C}^* \setminus q^{\mathbf{Z}}$, $m \in \mathbf{Z}$, and $q^m < |x| < q^{m-1}$.

2.2.2. Application to q-elliptic functions.

2.2.2.1. Let $a_1, \ldots, a_k, b_1, \ldots, b_l \in \mathbf{C}^*$ and let $f(x) := \dfrac{\theta_q(-x/a_1)\cdots\theta_q(-x/a_k)}{\theta_q(-x/b_1)\cdots\theta_q(-x/b_l)}$. We assume that, for all $i = 1, \ldots, k$ and $j = 1, \ldots l$, $a_i \not\equiv b_j \pmod{q^{\mathbf{Z}}}$, so there is no cancellation of zeroes between the numerator and the denominator. So the zeroes, resp. the poles, of f are the points of the q-spirals $[a_i; q]$, resp. $[b_j; q]$. If a q-spiral occurs m times, that is if m among the a_i are in the same class modulo $q^{\mathbf{Z}}$, then the corresponding zeroes have multiplicity m; and similarly for the poles.

LEMMA 2.20. *The above function f is q-elliptic if, and only if, $k = l$ and $a_1 \cdots a_k = b_1 \cdots b_l$.*

PROOF. From equality $\theta_q(-qx/c) = (-c/x)\theta_q(-qx/c)$, one readily computes $\dfrac{f(qx)}{f(x)} = (-x)^{l-k}\dfrac{a_1\cdots a_k}{b_1\cdots b_l}$. □

Then $\mathrm{div}_{\mathbf{E}_q}(f) = [\overline{a_1}] + \cdots + [\overline{a_k}] - [\overline{b_1}] - \cdots - [\overline{b_l}]$ has degree $k - l = 0$ and evaluation $[\overline{1}]$ as it should by 2.1.1. Note however that the conclusion is not strictly identical to that of the Abel-Jacobi theorem. Indeed, the theorem would guarantee that $a_1 \cdots a_k \equiv b_1 \cdots b_l \pmod{q^{\mathbf{Z}}}$, while here we have equality. But the form of the function is not exactly the same: we did not group parameters a_i or b_j giving rise to the same zero or pole. Instead, we could have something like $(\theta_q(a)\theta_q(qa))/(\theta_q(b)\theta_q(c))$ for instance.

EXAMPLE 2.21. Let $f(x) := \dfrac{\theta_q(x)\theta_q(x/q)}{\theta_q\left(x/\sqrt{q}\right)^2}$, which does satisfy the conditions of the lemma. Its divisor is $2[\overline{-1}] - 2[\overline{-\sqrt{q}}]$, which has zero degree and zero evaluation. (Not the same zero though!) As an element of $\mathscr{M}(\mathbf{E}_q)$, it has a double zero and a double pole. However, f, which is equal to $\dfrac{x}{q}\left(\dfrac{\theta_q(x)}{\theta_q\left(x/\sqrt{q}\right)}\right)^2$, cannot be not the square of a meromorphic function on $\mathbf{C}^*$; for, if it was, x would be too and we know that there is no uniform analytic determination of $\sqrt{x}$ on $\mathbf{C}^*$. In particular, f cannot be put in the form $\dfrac{\theta_q(x/a)^2}{\theta_q(x/b)^2}$, even up to a constant.

2.2.2.2.

THEOREM 2.5. *Every nonzero q-elliptic function has, up to a $\mathbf{C}^*$-factor, the form shown in the lemma.*

PROOF. Let g be such a function. We may (and shall) assume that g is not constant; its divisor is:

$$\mathrm{div}_{\mathbf{E}_q}(g) = [\alpha_1] + \cdots + [\alpha_k] - [\beta_1] - \cdots - [\beta_k]$$

for some $k \geq 1$ and $\alpha_1, \ldots, \alpha_k, \beta_1, \ldots, \beta_k \in \mathbf{E}_q$ such that $\prod \alpha_i = \prod \beta_i$ (an equality in the group $\mathbf{C}^*/q^{\mathbf{Z}}$). We lift these elements to $\mathbf{C}^*$: $\alpha_i = \overline{a_i}$ $\beta_i = \overline{b_i}$, $i = 1, \ldots, k$. Then $\prod a_i = q^m \prod b_i$ for some $m \in \mathbf{Z}$. We replace a_1 by $q^m a_1$, meaning that we can (and will) assume that $m = 0$. Then $f(x) := \dfrac{\theta_q(-x/a_1)\cdots\theta_q(-x/a_k)}{\theta_q(-x/b_1)\cdots\theta_q(-x/b_k)} \in \mathscr{M}(\mathbf{E}_q)$ and $\mathrm{div}_{\mathbf{E}_q}(f) = \mathrm{div}_{\mathbf{E}_q}(g)$, so $g/f \in \mathscr{M}(\mathbf{E}_q)$ has no zeroes and no poles, so it is a nonzero constant. □

Of course, the point where we could not get exactly one factor by zero or pole was when we "poured" q^m among the a_i: this can usually not be done evenly.

COROLLARY 2.22. *The group* $\mathscr{M}(\mathbf{E}_q)^*$ *of nonzero q-elliptic functions is generated by* $\mathbf{C}^*$ *and the functions of the form* $\dfrac{\theta_q(x/a)\theta_q(x/b)}{\theta_q(x/c)\theta_q(x/d)}$, $a, b, c, d \in \mathbf{C}^*$ *such that* $ab = cd$.

PROOF. Let G the subgroup of $\mathscr{M}(\mathbf{E}_q)^*$ generated by such elements. Let $f \in \mathscr{M}(\mathbf{E}_q)^*$ not a constant; we may (and shall) assume that it has the form of the lemma with order $k = l \geq 2$ (indeed we saw in 2.1.1 (also in A.1.3) that a nonconstant elliptic function has at least order 2). Let g the function obtained by replacing $\theta_q(-x/a_{k-1})\theta_q(-x/a_k)$ by $\theta_q(-x/c)\theta_q(-x/b_k)$, where $c := a_{k-1}a_k/b_k$. So $f \equiv g \pmod{G}$. But in g occurs at least a simplification (that of $\theta_q(-x/b_k)$) so it has the same form with strictly smaller order. Iterating, we arrive at a constant C such that $f \equiv C \pmod{G}$. □

EXERCISE 2.23. Prove that, in the corollary, one may fix some arbitrary d_0 and require that all generators are with $d = d_0$. (Something similar will be done with "q-characters" in section 2.3.)

EXERCISE 2.24. We introduce the *Sigma and Zeta functions of Weierstraß*, which will briefly appear in Chapter 12. We follow [**144**, Chapter 18]. Let $\Lambda \subset \mathbf{C}$ a lattice. As usual, $\prod', \sum'$ mean that we operate over $\Lambda \setminus \{0\}$.

(i) Let $\sigma_\Lambda(x) := x\prod'(1 - x/\omega)e^{x/\omega + x^2/2\omega^2}$. Show that this defines an analytic function on $\mathbf{C}$ and that $\sigma_{\lambda\Lambda}(\lambda x) = \sigma_\Lambda(x)$. From now on we write σ for σ_Λ.

(ii) Show that $\zeta := \sigma'/\sigma$ admits the expansion $\zeta(x) = \dfrac{1}{x} + \sum'\left(\dfrac{1}{x-\omega} + \dfrac{1}{\omega} + \dfrac{x}{\omega^2}\right)$ and that its derivative is $-\mathfrak{P}_\Lambda$.

(iii) Deduce that $\zeta(x + \omega) = \zeta(x) + \phi(\omega)$, where $\phi : \Lambda \to \mathbf{C}$ is $\mathbf{Z}$-linear.

(iv) Prove that $\sigma(x + \omega) = \sigma(x)e^{\phi(\omega)(x+\omega/2)} \times \begin{cases} +1 \text{ if } \omega/2 \in \Lambda, \\ -1 \text{ if } \omega/2 \notin \Lambda. \end{cases}$ Therefore σ is "a Theta-function" in the sense that it satisfies a functional relation of the form $\vartheta(x + \omega) = \vartheta(x) \times e^{L(x,\omega)+C(\omega)}$, where $L(x, u)$ is $\mathbf{C}$-linear in x and $\mathbf{R}$-linear in u and where $C : \Lambda \to \mathbf{C}$ is arbitrary.

(v) Let $L = [\omega, \omega']$, $\omega/\omega' \in \mathcal{H}$, and write for short $\eta := \phi(\omega)$ and $\eta' := \phi(\omega')$. prove *Legendre's relation*: $\omega\eta' - \omega'\eta = 2i\pi$.

2.2.3. Almost q-elliptic functions and what to do with them. We shall now give a glimpse at how "vector bundles" will appear in our story. Here we do it in an intuitive informal way, in order to ease the adaptation of the reader unfamiliar with that notion. Moreover, we only deal with "line bundles". The basics of the general theory of vector bundles on a Riemann surface, as far as they will be necessary for us (which is not very far), are presented in Appendix B. Generalities on their use in our context will appear in 7.3 and 9.3, but what we begin here will be completed and generalized in 4.5. (Note however that, in those two places, we will start from the more flexible point of view of *sheaves*.)

2.2.3.1. Let $f \in \mathscr{M}(\mathbf{C}^*)^{\sigma_q}$ a q-elliptic function, i.e., a q-invariant meromorphic function on $\mathbf{C}^*$. In order to see it as a map in the set-theoretic sense, in spite of the poles where it is not defined, we consider it as going to $\mathbf{S} = \mathbf{C} \cup \{\infty\}$ and sending poles to ∞. Since, for every $x \in \mathbf{C}^*$, the points x and qx have the same image by f, we can identify them and go to the quotient of $\mathbf{C}^*$ by the equivalence relation generated by such identifications: this is of course $\mathbf{E}_q = \mathbf{C}^*/q^{\mathbf{Z}}$ and f factors through a map $\overline{f} : \mathbf{E}_q \to \mathbf{S}$. Moreover we were able to endow $\mathbf{E}_q$ with an analytic structure so that $\overline{f}$ is actually a meromorphic function, an element of $\mathscr{M}(\mathbf{E}_q)$.

Now we have met in 2.2.1 functions that are not quite q-elliptic: instead, they obeyed a seemingly innocuous transformation law such as $f(qx) = cx^n f(x)$. We explicitly built nontrivial holomorphic instances under the conditions $c \in q^{\mathbf{Z}}$ and $n \leq 0$, but there are meromorphic instances without those conditions: as for the sign of n, we just take inverses (if $c = 1$, just take θ_q^{-n}); as for the constant, we shall study in 2.3.1 functions such that $f(qx) = cf(x)$ for general $c \in \mathbf{C}^*$. Can we have a geometric model for such behaviour? Yes, and it is provided by line bundles (and, for more general q-difference equations, vector bundles) over $\mathbf{E}_q$.

For the moment, we shall take the following setting: the q-difference equation is $f(qx) = u(x)f(x)$, where $u(x)$ is a given holomorphic nonvanishing function over $\mathbf{C}^*$; for instance, $u(x) = cx^n$ as above. The more general case when u can have singularities (i.e., as we shall then see, zeroes or poles) in $\mathbf{C}^*$ will be tackled in 4.5. So if $f(x) = X$, then $f(qx) = u(x)X$. Since generically X and $u(x)X$ differ, we cannot brutally identify x and qx lest we want to meet a contradiction. But what we can do is to identify the pair (x, X) with the pair $(qx, u(x)X)$. So we proceed as follow.

First, we consider $f : \mathbf{E}_q \to L$, where L is either $\mathbf{S}$ (as above) or $\mathbf{C}$; in the latter case, f is only partially defined: but it comes along with some growth conditions at singularities expressing meromorphy. We shall usually make the latter choice because then L is a vector space, but in litterature with a more differential geometric flavour, the former choice is frequent. (If L was taken to be $\mathbf{S}$, one would speak of a "projective bundle".)

Then we consider f as an object related to $\mathbf{C}^* \times L$, actually a *section* of the first projection map $p : \mathfrak{F} := \mathbf{C}^* \times L \to \mathbf{C}^*$, precisely the map $s : \mathbf{C}^* \to \mathfrak{F}$, $x \mapsto (x, f(x))$ (it is called a section because $p \circ s = \mathrm{Id}_{\mathbf{C}^*}$).

The map $p : \mathfrak{F} \to \mathbf{C}^*$ or, more abusively, the space $\mathfrak{F}$ itself, is called a "trivial line bundle" over $\mathbf{C}^*$. We consider it as a family of lines L, one for each $x \in \mathbf{C}^*$. And the section s sends each x to a point of its own line $\mathfrak{F}_x := p^{-1}(x) = \{x\} \times L$.

2.2.3.2. Now we express the q-difference equation $f(qx) = u(x)f(x)$ in this language. Let $s(x) = (x, X) = (x, f(x))$ and let $x' := qx$. Then $s(x') = (x', X') = (qx, u(x)X)$ is obtained from $s(x)$ by the transformation $T : (x, X) \mapsto (qx, u(x)X)$. So we should quotient $\mathfrak{F}$ by such identifications (i.e., by the equivalence relation generated by the identifications). But before we do so, we note that T is a bijection $\mathfrak{F} \to \mathfrak{F}$ and that the equivalence relation is really induced by a group action, that of the group $T^{\mathbf{Z}} := \{T^n \mid n \in \mathbf{Z}\}$; or, if we prefer, of the group $\mathbf{Z}$.

EXERCISE 2.25. Give an explicit formula for T^n, $n \in \mathbf{Z}$.

We are going to enrich a little the situation by considering at the same time the action of T on $\mathfrak{F}$ and the action of the map $t : x \mapsto qx$ on $\mathbf{C}^*$. They are compatible and give rise to a commutative diagram (left); and the fact that T transforms $s(x)$

to $s(x')$ is also encoded in a commutative diagram (right):

$$\begin{array}{ccc} \mathfrak{F} & \xrightarrow{\ p\ } & \mathbf{C}^* \\ {\scriptstyle T}\big\downarrow & & \big\downarrow{\scriptstyle t} \\ \mathfrak{F} & \xrightarrow{\ p\ } & \mathbf{C}^* \end{array} \qquad\qquad \begin{array}{ccc} \mathfrak{F} & \xleftarrow{\ s\ } & \mathbf{C}^* \\ {\scriptstyle T}\big\downarrow & & \big\downarrow{\scriptstyle t} \\ \mathfrak{F} & \xleftarrow{\ s\ } & \mathbf{C}^*. \end{array}$$

Since this holds not only for T and t but for all pairs T^n, t^n, we see that there are actions of the same group $\mathbf{Z}$ (incarnated by $q^{\mathbf{Z}}$, or $t^{\mathbf{Z}}$ for $\mathbf{C}^*$; and by $T^{\mathbf{Z}}$ for $\mathfrak{F}$) on the bundle $\mathfrak{F}$ and on its "base" $\mathbf{C}^*$, and that these actions are compatible: we speak of an "equivariant line bundle" (and, if s is as above, an "equivariant section").

2.2.3.3. Now, the action of $\mathbf{Z}$ (or $q^{\mathbf{Z}}$, or $t^{\mathbf{Z}}$) on $\mathbf{C}^*$ by the $t^n : x \mapsto q^n x$ has nice topological properties, actually, the same as those stated in Appendix A.3.2 for the action of the modular group that allowed us to make the quotient $\mathbf{E}_q$ into a Riemann surface. The same properties hold for the action of $\mathbf{Z}$ (or $T^{\mathbf{Z}}$) on $\mathfrak{F}$ by the T^n, allowing us to give to the quotient:

$$\mathfrak{F}' := \frac{\mathbf{C}^* \times L}{(x, X) \sim_u (qx, u(x)X)}$$

an analytic structure, in such a way that the induced map $p' : \mathfrak{F}' \to \mathbf{E}_q$ sending the class of (x, X) to the class $\overline{x}$ of X is, in some sense, holomorphic. We thus get a commutative diagram (left); and s also goes to the quotient as an analytic section s' of p', also encoded in a commutative diagram (right):

$$\begin{array}{ccc} \mathfrak{F} & \xrightarrow{\ p\ } & \mathbf{C}^* \\ \big\downarrow & & \big\downarrow{\scriptstyle \pi} \\ \mathfrak{F}' & \xrightarrow{\ p'\ } & \mathbf{E}_q \end{array} \qquad\qquad \begin{array}{ccc} \mathfrak{F} & \xleftarrow{\ s\ } & \mathbf{C}^* \\ \big\downarrow & & \big\downarrow{\scriptstyle \pi} \\ \mathfrak{F}' & \xleftarrow{\ s'\ } & \mathbf{C}^*. \end{array}$$

Note that, although $\mathfrak{F}'$ cannot be put in the form $\mathbf{E}_q \times L'$ for some L', it does share some properties of that kind with $\mathfrak{F} = \mathbf{C}^* \times L \to \mathbf{C}^*$:

(1) For every x' in $\mathbf{E}_q$, its preimage by p' is in bijection with the line L. So in some sense $\mathfrak{F}'$ is a family of lines over the "base" $\mathbf{E}_q$.
(2) More globally (although not completely globally), the base $\mathbf{E}_q$ can be covered by open sets V such that the preimage of V by p' is naturally in bijection with $V \times L$. (This is not completely global because, as just said, one cannot take $V = \mathbf{E}_q$.)

There is something more to be said about those "natural bijections", this will come in a moment. First we make those two contentions a bit more explicit.

For the first one, we first select a fundamental annulus, say $|q| < |x| \leq 1$. Let $x' \in \mathbf{E}_q$. Let x_0 the unique element of the fundamental annulus such that $\pi(x_0) = x'$. The "fibre" ${p'}^{-1}(x')$ is the set of classes modulo $\sim_u$ of pairs (x, X) such that $\pi(x) = x'$. In any such class, there is a unique (x_0, X_0) where x_0 is the element previously selected but $X_0 \in V$ is arbitrary. Therefore the map $X_0 \mapsto (x_0, X_0)$ (mod $T^{\mathbf{Z}}$) is a bijection from L to ${p'}^{-1}(x')$, explaining in which way the fibre is a line.

For the second one, the argument is more or less the same. We choose an arbitrary "small" open set $V \subset \mathbf{E}_q$. Recall from A.1.4 that this means that $\pi^{-1}(V)$

has the form of a disjoint union $\bigsqcup_{n\in\mathbf{Z}} q^nU_0$, where $U_0 \subset \mathbf{C}^*$ is open and each restriction $\pi : q^nU_0 \to V$ is a homeomorphism. Then the preimage ${p'}^{-1}(V)$ is the set of classes modulo $\sim_u$ of pairs (x, X) such that $\pi(x) \in V$. In any such class, there is a unique (x, X) such that $x \in U_0$. Therefore the map $(x, X) \mapsto (x, X) \pmod{T^{\mathbf{Z}}}$ is a bijection from $U_0 \times L$ to ${p'}^{-1}(V)$. Combining with the homeomorphism from V to U_0 inverting π, we get a bijection from $V \times L$ to ${p'}^{-1}(V)$. It has two properties: it is an isomorphism for the analytic structures appearing on $V \times L$ and on ${p'}^{-1}(V)$; and it gives rise to a commutative diagram:

$$\begin{array}{ccc} V \times L & \xrightarrow{\quad\sim\quad} & {p'}^{-1}(V). \\ & {\scriptstyle\text{first projection}}\searrow \qquad \swarrow{\scriptstyle p'} & \\ & V & \end{array}$$

So the restricted mapping ${p'}^{-1}(V) \to V$ is in some sense isomorphic to the trivial line bundle $V \times L \to V$ and $\mathscr{F}' \to \mathbf{E}_q$ is "locally isomorphic to a trivial line bundle".

2.2.3.4. We call "holomorphic sections" of $\mathfrak{F}'$ over an open subset V of $\mathbf{E}_q$ the holomorphic maps $f' : V \to \mathfrak{F}'$ such that $p' \circ f'$ is the identity of V (and therefore f' actually has image in ${p'}^{-1}(V)$). In a general way, we can see (Appendix B) that they are in natural one-to-one correspondence with holomorphic solutions of $f(qx) = u(x)f(x)$ on $\pi^{-1}(V) \subset \mathbf{C}^*$. In particular, in many cases, there are no nontrivial *global* holomorphic sections, i.e., over $\mathbf{E}_q$ itself: this is for instance the case for $u(x) = cx^n$ if $n > 0$ or if $n = 0$ and $c \notin q^{\mathbf{Z}}$.

On the other hand, there is a notion of "meromorphic sections" over V: those should be defined in $V \setminus \Sigma$ where Σ is some discrete subset of V, and have, in some sense, moderate growth at the points of Σ. They are in natural one-to-one correspondence with meromorphic solutions of $f(qx) = u(x)f(x)$ on $\pi^{-1}(V) \subset \mathbf{C}^*$. It can be proved that there are always nontrivial global meromorphic sections (a very special instance of a theorem of Praagman; see section 7.3). For instance, if $u(x) = cx^n$, many functions of the form $\dfrac{\prod \theta_q(a_i x)}{\prod \theta_q(b_j x)}$ provide such global sections.

EXERCISE 2.26. State precise conditions on the a_i and b_j for the above to define such a section.

Now for a meromorphic section of a line bundle, there is a general way to define zeroes, poles, and multiplicities, but in our case there is a direct way. From the equation $f(qx) = u(x)f(x)$ and the fact that u is regular (i.e., defined and nonvanishing) on $\mathbf{C}^*$, it follows readily that, if $a \in \mathbf{C}^*$ is a zero, resp. a pole, then so are all the q^na, $n \in \mathbf{N}$; and with the same multiplicity. Accordingly we declare $\overline{a}$ to be a zero, resp. a pole, of the corresponding meromorphic section f' of $\mathfrak{F}'$; and with the same multiplicity.

So let us call $\alpha_i := \overline{a_i}$, $i = 1, \ldots, k$, the zeroes of the section f' (and m_i their respective multiplicities); and $\beta_j := \overline{b_j}$, $j = 1, \ldots, l$ its poles (and n_j their respective multiplicities). We can define the divisor of the section f':

$$\operatorname{div}_{\mathbf{E}_q}(f') := \sum_{i=1}^{k} m_i[\alpha_i] - \sum_{j=1}^{l} n_j[\beta_j].$$

By slight abuse, we also denote $\mathrm{div}_{\mathbf{E}_q}(f) := \mathrm{div}_{\mathbf{E}_q}(f')$ for f the corresponding function.

Contrary to the case of functions, it is not generally true that the divisor of a meromorphic section has degree $0 \in \mathbf{Z}$ and trivial evaluation in $\mathbf{E}_q$. However, something strong holds.

PROPOSITION 2.27. (i) *Any two nontrivial meromorphic sections of* $\mathfrak{F}'$ *are "linearly equivalent", i.e., their difference is a principal divisor (an element of* $P(\mathbf{E}_q)$*, see* A.1.5*).*

(ii) *All divisors of nontrivial meromorphic sections of* $\mathfrak{F}'$ *have the same degree and evaluation.*

PROOF. (i) Let f', g' two nontrivial meromorphic sections of $\mathfrak{F}'$ and f, g the corresponding meromorphic solutions of $f(qx) = u(x)f(x)$. Then $\phi = f/g$ is q-elliptic and clearly $\mathrm{div}_{\mathbf{E}_q}(f') - \mathrm{div}_{\mathbf{E}_q}(g') = \mathrm{div}_{\mathbf{E}_q}(\phi) \in P(\mathbf{E}_q)$.

(ii) is an immediate consequence.

□

The *degree* of $\mathfrak{F}'$ is, by definition, the common degree of all its nontrivial meromorphic sections.

EXAMPLE 2.28. Let $u(x) = cx^n$. A nontrivial meromorphic solution of $f(qx) = u(x)f(x)$ is, for instance, $f(x) := \dfrac{\theta_q(x/c)}{{\theta_q}^{n+1}(x)}\cdot$ Its divisor is:

$$\mathrm{div}_{\mathbf{E}_q}(f) = [\overline{-c}] - (n+1)[\overline{-1}].$$

The degree of that divisor is $-n$, and it is the degree of $\mathfrak{F}'$. Its evaluation is $[\overline{(-1)^n c}] \in \mathbf{E}_q$, it is the evaluation of all divisors of nontrivial meromorphic sections of $\mathfrak{F}'$.

2.2.4. Application to modular functions.

2.2.4.1. To navigate between the (z,τ) and the (q,x) model, also taking advantage of the modular functional equation, we fix the following conventions and notations: $z \in \mathbf{C}$, $\tau \in \mathcal{H}$, $q := e^{2\mathrm{i}\pi\tau}$, $x := e^{2\mathrm{i}\pi z}$, $\tau' := -1/\tau$, $z' := z/\tau$, $q' := e^{2\mathrm{i}\pi\tau'}$ and $x' := e^{2\mathrm{i}\pi z'}$. So we have:

$$\theta_q(-x) = \Theta(z,\tau), \quad \theta_{q'}(-x') = \Theta(z',\tau'), \quad \Theta(z,\tau) = \sqrt{\frac{\mathrm{i}}{\tau}}\, e^{-\frac{\mathrm{i}\pi}{\tau}\left(z+\frac{1-\tau}{2}\right)^2}\,\Theta(z',\tau').$$

Also recall from Appendix A that we defined the Eisenstein series as $G_k(\tau) := \sum'(m+n\tau)^{-2k}$ (meaning summation over $\mathbf{Z}^2 \setminus \{(0,0)\}$); then we set $g_2 := 60G_2$, $g_3 := 140G_3$ and $\Delta := g_2^3 - 27g_3^2$. We know by A.3.4 that Δ is holomorphic over $\mathcal{H}$, that it satisfies the functional equations $\Delta(\tau+1) = \Delta(\tau)$ and $\Delta(-1/\tau) = \tau^{12}\Delta(\tau)$ and that it has the expansion at infinity: $\Delta = (2\pi)^{12}(q + \cdots)$.

THEOREM 2.6 (Jacobi).

$$\Delta = (2\pi)^{12}\, q(q;q)_\infty^{24}.$$

PROOF. Set $D(\tau) := q(q;q)_\infty^{24}$: this is obviously a nonvanishing holomorphic function on $\mathcal{H}$ such that $D(\tau+1) = D(\tau)$. Also it has expansion at infinity $D(\tau) = q(1 + \cdots)$.

We fix τ (so also τ') and let $z \to 0$, so $z' \to 0$, $x \to 1$ and $x' \to 1$. Based on the corollaries of Jacobi's Triple Product Formula, at the end of 2.2.1.3, we have:

$\Theta(z,\tau) \sim -2\mathrm{i}\pi z(q;q)_\infty^3$; so, similarly $\Theta(z',\tau') \sim -2\mathrm{i}\pi z'(q';q')_\infty^3$, and using the modular functional equation:

$$\frac{-2\mathrm{i}\pi z(q;q)_\infty^3}{-2\mathrm{i}\pi z'(q';q')_\infty^3} \sim \sqrt{\frac{\mathrm{i}}{\tau}}\, e^{-\frac{\mathrm{i}\pi}{\tau}\left(z+\frac{1-\tau}{2}\right)^2}.$$

The left-hand side is equal to $\tau \dfrac{(q;q)_\infty^3}{(q';q')_\infty^3}$. The right-hand side tends to:

$$\sqrt{\frac{\mathrm{i}}{\tau}}\, e^{-\frac{\mathrm{i}\pi}{\tau}\left(\frac{1-\tau}{2}\right)^2} = \sqrt{\frac{\mathrm{i}}{\tau}}\, e^{\mathrm{i}\pi/2}\, e^{-\mathrm{i}\pi\tau/4}\, e^{\mathrm{i}\pi\tau'/4} = \mathrm{i}\sqrt{\frac{\mathrm{i}}{\tau}} q^{-1/8} q'^{1/8}.$$

So we get the nice identity (the usual conventions for square roots apply):

$$\tau\frac{(q;q)_\infty^3}{(q';q')_\infty^3} = \mathrm{i}\sqrt{\frac{\mathrm{i}}{\tau}} q^{-1/8} q'^{1/8} \Longrightarrow q^{1/8}\,(q;q)_\infty^3 = \frac{\mathrm{i}}{\tau}\sqrt{\frac{\mathrm{i}}{\tau}}\, q'^{1/8}\,(q';q')_\infty^3.$$

Raising the last identity to the eighth power yields $D(\tau) = \tau^{-12} D(-1/\tau)$. Combined with the properties of D stated at the beginning of the proof and with those of Δ recalled above, we see that $C := \Delta/D$ is holomorphic over $\mathcal{H}$, with expansion at infinity $C(\tau) = (2\pi)^{12}(1+\cdots)$ and that it satisfies $C(\tau+1) = C(-1/\tau) = C(\tau)$. Altogether, those properties imply that C is constant: $C = (2\pi)^{12}$ (same argument as in the proposition at the end of A.3.5), which concludes the proof. □

EXERCISE 2.29. Let $\tau \in \mathcal{H}$, $q := e^{2\mathrm{i}\pi\tau}$, $\tau' := -1/\tau$ and $q' := e^{2\mathrm{i}\pi\tau'}$. Define:

$$f(q) := 16 q^{1/2} \prod_{n\geq 1}\left(\frac{1+q^n}{1+q^{n-1/2}}\right)^8, \quad g(q) := \prod_{n\geq 1}\left(\frac{1-q^{n-1/2}}{1+q^{n-1/2}}\right)^8,$$

and $h(q) := -f(q)/g(q)$. Prove the following formulas from [**255**, p 481, 21.71]:

$$f(q) + g(q) = 1, \quad f(-q) = h(q), \quad f(q') = g(q).$$

2.2.4.2. Strongly related to the above construction is *Dedekind's Eta function*:

$$\eta(\tau) := e^{\mathrm{i}\pi\tau/12}(q;q)_\infty.$$

Clearly η^{24} is the function $D(\tau)$ introduced in the course of the proof above, so we must have relations of the form:

$$\eta(g.\tau) = \omega_g \sqrt{c\tau+d}\,\eta(\tau),$$

for every $g \in G$ with matrix chosen such that $c > 0$ or $c = 0$ and $d = 1$, and where ω_g denotes some 24th root of unity in $\mathbf{C}^*$.

EXERCISE 2.30. Compute $\eta(\tau+1)$ and $\eta(-1/\tau)$.

The precise definition of the automorphy factors ω_g and $\sqrt{c\tau+d}$ is given in [**12, 183**], along with application of η to the proof of the asymptotic expansions of the partition number by Hardy-Ramanujan and by Rademacher (see 1.2).

EXERCISE 2.31. Prove that $\dfrac{\eta(\tau)^2}{\eta(2\tau)} = \sum\limits_{n\in\mathbf{Z}} (-1)^n q^{n^2}$. (From the product formula, it is equal to $\prod\limits_{n\geq 1} \dfrac{(1-q^n)^2}{1-q^{2n}}$; playing with factorisations and using Jacobi's Triple Product Formula, this is $\theta_{q^2}(-q)$.) For other formulas of this kind, see [**178**, p. 17].

There are many proofs of the modularity of the Eta function [**12**, **231**]. We sketch one proof of the formula for $\eta(-1/\tau)$ based on A.3.7. Remembering that $dq/d\tau = 2\mathrm{i}\pi q$, we obtain the logarithmic derivative:

$$\frac{d \log \eta}{d\tau} = 2\mathrm{i}\pi q \left(\frac{1}{24q} + \sum_{n\geq 1} \frac{-nq^{n-1}}{1-q^n} \right) = \frac{-1}{4\mathrm{i}\pi} G_1(\tau).$$

Using the formula $G_1(-1/\tau) = \tau^2 G_1(\tau) - 2\mathrm{i}\pi\tau$, setting $\phi(\tau) := \dfrac{\eta(-1/\tau)}{\eta(\tau)}$, we find that $\dfrac{d \log \phi}{d\tau} = \dfrac{1}{2\tau}$, so $\phi(\tau)$ has the form $\sqrt{\tau_0/\tau}$. Now, $\phi(\mathrm{i}) = \dfrac{\eta(-1/\mathrm{i})}{\eta(\mathrm{i})} = 1$, so we get:

$$\eta(-1/\tau) = \sqrt{\mathrm{i}/\tau}\, \eta(\tau).$$

2.3. Application of Theta functions to the theory of q-difference equations

The classical theory of complex analytic linear differential equations uses extensively power series (convergent, or even divergent in the case of "irregular" equations), but these are not sufficient. Some "transcendents" are required. They are mainly the "characters" z^α together with logarithms (in the case of "multiple exponents"). Those are enough for "fuchsian" equations but, for irregular equations some expressions such as $e^{1/P(z)}$ (P a polynomial) are also required. At any rate, the use of multivalued functions is unavoidable and actually, ever since Riemann, has been a pillar of the theory (see [**169**, **222**] for all that).

The Founding Fathers of q-difference theory (Carmichaël, Adams, Birkhoff, Jackson ...) replaced z^α, $\log z$, e^{1/z^n}, etc, by similar multivalued functions (some explicit examples will be shown) but it turned out that this caused some ambiguities in matters of classification and Galois theory in the work of Birkhoff; it seems that they were noted first by Deligne, and explained in [**249**]. Anyhow, it is possible to stick to uniform functions. Birkhoff himself proposed using Weierstraß's Sigma function (see the exercise at the end of 2.2.2.2).

Rather, following a suggestion of Ramis [**190**], a program based on the use of Theta functions was launched at the end of the last millenium. It proved succesfull and is one of the pillars of this book. The main features of this approach is that it almost[3] entirely avoids the use of multivalued functions. All phenomena of branching and ramification are replaced by the apparition of poles (essentially coming from the zeroes of Theta functions).

In this section, we present q-analogues of z^α and $\log z$ built with the help of θ_q. Then, in section 2.4, we shall see how, when $q \to 1$, they degenerate to their classical counterparts, i.e., we pursue the theme (already explored in 1.4 and 1.6) of *confluence of a q-difference equation to a differential equation*: this theme will be pursued in other parts of the book, notably 11.2 and 12.3. In Chapter 11 we shall use those q-analogues to solve "fuchsian" q-difference equations; still later on, when studying irregular equations, we shall see how to use Theta functions to replace the classical exponentials $e^{1/P}$.

[3]There will be a limited need of "ramification" (and isogenies of complex tori) for irregular equations with nonintegral slopes.

2.3.1. q-characters.

2.3.1.1. We want to solve equations of the form $f(qx) = cf(x)$ for some fixed $c \in \mathbf{C}^*$. General properties of such equations will be studied in later chapters; here we want to build specific solutions and detail their basic properties.

The Founding Fathers took $f(x) := x^\gamma$, where $q^\gamma = c$. This involves a double ambiguity. First, such a γ is not unique, it is defined only modulo a term in $\mathbf{Z}\tau^{-1}$: indeed we want $e^{2i\pi\tau\gamma} = c$, i.e., $2i\pi\tau\gamma \in \log c$, where $\log c$ here denotes the set $\log_0 c + 2i\pi\mathbf{Z}$ for any particular determination $\log_0$ of the logarithm. Second, when γ has been chosen, $x^\gamma = e^{\gamma \log x}$ is determined only up to a factor in $e^{2i\pi\gamma\mathbf{Z}}$. Such ambiguities may be overcome either going to a covering of $\mathbf{C}^*$ (the Riemann surface of the logarithm) or by cutting lines in $\mathbf{C}^*$ in such a way that what is left is a simply connected domain. At some points (when dealing with confluence in chapters 11 and 12), we shall use such cuts for solutions of differential equations.

DEFINITION 2.32. A *q-character for the exponent $c \in \mathbf{C}^*$* is a nonzero meromorphic function f on $\mathbf{C}^*$ such that $f(qx) = cf(x)$.

2.3.1.2. Our official examples of q-character will be $\theta_q(x)/\theta_q(cx)$ and $\theta_q(x/c)/\theta_q(x)$ (check that they fit!) but we want first to draw general consequences of the definition (and also keep some freedom of choice in some occasions, like confluence or Galois theory); explicit constructions will come in 2.3.1.4. Actually, part of the arguments follow from generalities on the sections of a line bundle (as in 2.2.3.4), but we still detail them in pedestrian style because this is a chapter of initiation.

EXERCISE 2.33. Show that there are no holomorphic q-characters with exponent $c \notin q^{\mathbf{Z}}$. What if $c = q^m$, $m \in \mathbf{Z}$?

Since the set Σ of zeroes and poles of a given q-character f is discrete in $\mathbf{C}^*$, thus denumerable, so is the set of their moduli and therefore, for a generic $r > 0$, the circle $\partial\mathrm{D}(0, r)$ does not meet Σ. We select such a generic radius r. Also the functional equation $f(qx) = cf(x)$ implies that $\Sigma = q\Sigma$. The intersection Σ_0 of Σ with the compact annulus $\overline{\mathcal{C}}(r\,|q|\,, r)$ is finite and lies in its interior $\mathring{\mathcal{C}}(r\,|q|\,, r)$; and $\Sigma = q^{\mathbf{Z}}\Sigma_0$. So let $a_1, \ldots, a_k$ the zeroes, with multiplicities $m_1, \ldots, m_k$; and $b_1, \ldots, b_l$ the poles, with multiplicities $n_1, \ldots, n_l$. Again, from the functional equation, we draw:

- The zeroes of f in $\mathbf{C}^*$ are the points of the finite union of discrete q-spirals $\bigcup_{i=1}^{k} [a_i; q]$, and all those on a given $[a_i; q]$ have multiplicity m_i;
- the poles of f in $\mathbf{C}^*$ are the points of the finite union of discrete q-spirals $\bigcup_{j=1}^{l} [b_j; q]$, and all those on a given $[b_j; q]$ have multiplicity n_j.

It therefore makes sense to say that f has a divisor of zeroes and poles over $\mathbf{E}_q$:

$$\mathrm{div}_{\mathbf{E}_q}(f) := \sum_{i=1}^{k} m_i[\overline{a_i}] - \sum_{j=1}^{l} n_j[\overline{b_j}].$$

EXAMPLE 2.34. Let $f := \theta_q(x)/\theta_q(cx)$ and $g := \theta_q(x/c)/\theta_q(x)$. Then $\mathrm{div}_{\mathbf{E}_q}(f) = [\overline{-1}] - [\overline{-c^{-1}}]$ and $\mathrm{div}_{\mathbf{E}_q}(g) = [\overline{-c}] - [\overline{-1}]$.

We note that both have degree $1 - 1 = 0$ and evaluation:

$$\mathrm{ev}_{\mathbf{E}_q}(\mathrm{div}_{\mathbf{E}_q}(f)) = \mathrm{ev}_{\mathbf{E}_q}(\mathrm{div}_{\mathbf{E}_q}(g)) = \overline{c}.$$

EXERCISE 2.35. Let $h := \theta_q(ax)/\theta_q(bx)$. For which $c \in \mathbf{C}^*$ is it a q-character for the exponent c? Find $\mathrm{div}_{\mathbf{E}_q}(h)$. Conjecture a relation between the exponent and the divisor.

Now let f and g two q-characters for the same exponent $c \in \mathbf{C}^*$ and let $\phi := f/g$. Clearly f is meromorphic over $\mathbf{C}^*$ and:

$$\phi(qx) = \frac{f(qx)}{g(qx)} = \frac{cf(x)}{cg(x)} = \frac{f(x)}{g(x)} = \phi(x),$$

so ϕ is q-elliptic. Now the usual rules on the zeroes and poles of a product or quotient of a meromorphic function immediately imply:

$$\mathrm{div}_{\mathbf{E}_q}(\phi) = \mathrm{div}_{\mathbf{E}_q}(f) - \mathrm{div}_{\mathbf{E}_q}(g).$$

But we know by 2.1.1.4 that $\mathrm{div}_{\mathbf{E}_q}(\phi)$ has zero degree and trivial evaluation; so $\mathrm{div}_{\mathbf{E}_q}(f)$ and $\mathrm{div}_{\mathbf{E}_q}(g)$ have the same degree and the same evaluation. And we know the degree and evaluation of the q-character $\theta_q(x)/\theta_q(cx)$ or $\theta_q(x/c)/\theta_q(x)$, so we conclude that all q-characters of exponent c have degree 0 and evaluation $\overline{c}$.

EXERCISE 2.36. Draw the same conclusion using Cauchy's residue theorem to the boundary of an adequate annulus.

2.3.1.3. Referring again to 2.2.3.4, we see that a q-character of exponent c can be thought of as a section of a line bundle $\mathfrak{F}_c := \dfrac{\mathbf{C}^* \times \mathbf{C}}{(x, X) \sim_c (qx, cX)}$. More precisely we make $\mathbf{Z}$ act on the trivial bundle $\mathbf{C}^* \times \mathbf{C} \to \mathbf{C}^*$ at the same time on the base $\mathbf{C}^*$: there, $(n, x) \mapsto q^n x$; and on the total space $\mathbf{C}^* \times \mathbf{C}$; there, $(n, (x, X)) \mapsto (q^n x, c^n X)$. The actions are compatible (we have an equivariant bundle) and go to the quotient as $\mathfrak{F}_c \to \mathbf{E}_q$, class of $(x, X) \mapsto$ class of x, whence a commutative diagram:

$$\begin{array}{ccc}
\mathbf{C}^* \times \mathbf{C} & \xrightarrow{(x,X)\mapsto x} & \mathbf{C}^*. \\
\Big\downarrow & & \Big\downarrow \pi \\
\mathfrak{F}_c & \longrightarrow & \mathbf{E}_q
\end{array}$$

A meromorphic section of the bundle $\mathbf{C}^* \times \mathbf{C} \to \mathbf{C}^*$ is a map $x \mapsto (x, f(x))$ where $f \in \mathscr{M}(\mathbf{C}^*)$; and it goes to the quotient as a meromorphic section $\mathbf{E}_q \to \mathfrak{F}_c$ if, and only if, f is a q-character with exponent c.

PROPOSITION 2.37. *(i) Let $a_1, \ldots, a_k, b_1, \ldots, b_l \in \mathbf{C}^*$ pairwise nonequivalent* (mod $q^{\mathbf{Z}}$) *and let $m_1, \ldots, m_k, n_1, \ldots, n_l \in \mathbf{N}^*$. Then:*

$$f_0(x) := \frac{\theta_q(-x/a_1)^{m_1} \cdots \theta_q(-x/a_k)^{m_k}}{\theta_q(-x/b_1)^{n_1} \cdots \theta_q(-x/b_l)^{n_l}}$$

is a q-character with exponent c if, and only if,:

$$\sum_{i=1}^{k} m_i = \sum_{j=1}^{l} n_j \text{ and } \frac{a_1^{m_1} \cdots a_k^{m_k}}{b_1^{n_1} \cdots b_l^{n_l}} = c.$$

(ii) The divisor over $\mathbf{E}_q$ *of the above q-character is:*

$$div_{\mathbf{E}_q}(f_0) = \sum_{i=1}^{k} m_i[\overline{a_i}] - \sum_{j=1}^{l} n_j[\overline{b_j}].$$

(iii) Every q-character of exponent c is, up to a $\mathbf{C}^*$ *factor, of the above form.*

PROOF. (i) Use the same computation as in Lemma 2.20.

(ii) is then obvious.

(iii) Let f be a q-character of exponent c. Put its divisor, as described in 2.3.1.2, into the form $\sum m_i[\overline{a_i}] - \sum n_j[\overline{b_j}]$ as above and consider the corresponding explicit q-character f_0. Then f/f_0 is q-elliptic and has no zeroes or poles, so it belongs to $\mathbf{C}^*$.

□

2.3.1.4. We now define our "official" q-characters for this book.

DEFINITION 2.38. For $c \in \mathbf{C}^*$, "the" q-character with exponent c is:

$$e_{q,c}(x) := \frac{\theta_q(x)}{\theta_q(cx)}.$$

The following properties follow immediately:

(1) If $c = q^m$, $m \in \mathbf{Z}$, then $e_{q,c}(x) = q^{m(m-1)/2}x^m$. (This follows from $\theta_q(q^m x) = q^{-m(m-1)/2}x^{-m}\theta_q(x)$, which is easily proved by induction.)

(2) If $c \notin q^{\mathbf{Z}}$, then $\text{div}_{\mathbf{E}_q}(e_{q,c}) = [\overline{-1}] - [\overline{-c^{-1}}]$.

(3) For every $c, d \in \mathbf{C}^*$, the function $\phi_{c,d}(x) := \dfrac{e_{q,c}(x)e_{q,d}(x)}{e_{q,cd}(x)}$ is q-elliptic: $\phi_{c,d}(x) \in \mathscr{M}(\mathbf{E}_q)$.

2.3.1.5. The following important fact is less immediate:

PROPOSITION 2.39. *Together with* $\mathbf{C}^*$*, the functions* $\phi_{c,d}(x)$*,* $c, d \in \mathbf{C}^*$*, generate the group* $\mathscr{M}(\mathbf{E}_q)^*$ *of nontrivial q-elliptic functions.*

PROOF. Note that $\phi_{c,d}(x) = \dfrac{\theta_q(cx)\theta_q(x)}{\theta_q(dx)\theta_q(cdx)}$, so it easily follows from Corollary 2.22. □

The importance of this fact lies in the conjunction of two other facts:

- To solve q-difference equations other than the regular ones (i.e., those for which power series are sufficient; see Chapters 3, 4) we shall need q-characters.
- The "constants" of our theory, either from the point of view of classification or for Galois theory, are the q-elliptic functions.

Hence, as soon as we solve the simplest q-difference equations, we automatically produce a "constant field" $\mathscr{M}(\mathbf{E}_q)$ which is a transcendental extension of $\mathbf{C}$. Etingof in **[82]** did not meet the problem because he considered only regular q-difference equations (and then all exponents are equal to 1). Van der Put and Singer in **[249]** got around the problem by using, instead of the meromorphic functions $e_{q,c}$, formal symbols $e(c)$ constrained to satisfy relations $e(c)e(d) = e(cd)$. This allows classification and Galois theory but much of the power of function theory is thereby lost. (We return to that part of the early history in 13.1.2). So we can wonder if our choice of official q-characters was particularly unhappy. Actually, *no choice is good*:

PROPOSITION 2.40. *Let $(e_c)_{c\in\mathbf{C}^*}$ an arbitrary family of q-characters, i.e., $e_c \in \mathscr{M}(\mathbf{C}^*)^*$ and $e_c(qx) = ce_c(x)$ for all $c \in \mathbf{C}^*$. Then the functions $\psi_{c,d} := \dfrac{e_c e_d}{e_{cd}}$, $c, d \in \mathbf{C}^*$, generate (over $\mathbf{C}$) a transcendental extension of $\mathbf{C}$.*

PROOF. Since $\mathbf{C}$ is algebraically closed, the subfield of $\mathscr{M}(\mathbf{E}_q)$ generated (over $\mathbf{C}$) by the $\psi_{c,d}$ is either a transcendental extension of $\mathbf{C}$ or $\mathbf{C}$ itself. In the latter case, we would have $\mathrm{div}_{\mathbf{E}_q}\psi_{c,d} = 0$, whence $\mathrm{div}_{\mathbf{E}_q}(e_{cd}) = \mathrm{div}_{\mathbf{E}_q}(e_c) + \mathrm{div}_{\mathbf{E}_q}(e_d)$, so the map $c \mapsto \mathrm{div}_{\mathbf{E}_q}(e_c)$ would be a group morphism from $\mathbf{C}^*$ to $\mathrm{Div}(\mathbf{E}_q)$; but such a morphism is necessarily trivial (see lemma below). So we should have $\mathrm{div}_{\mathbf{E}_q}(e_c) = 0$ for all c. But we know that this can only happen for $c \in q^{\mathbf{Z}}$ (by 2.3.1.2), whence a contradiction. □

LEMMA 2.41. *Any morphism from a divisible group (like $\mathbf{C}^*$) to a free abelian group (like $Div(\mathbf{E}_q)$) is trivial.*

PROOF. Let a in the source group have image $\sum n_i x_i$, the x_i a basis of the target group and the $n_i \in \mathbf{Z}$. Let $n \in \mathbf{N}^*$ arbitrary and let b such that $b^n = a$ (since the source group is divisible). The image of b is $\sum p_i x_i$ and $n_i = np_i$ for all i. So the coefficients n_i are multiple of all $n \in \mathbf{N}^*$, which is possible only if they are all 0. □

EXERCISE 2.42. (i) Prove, using Riemann-Hurwitz formula (Appendix B.3.4, also see [**48, 241**]) that the extension generated by the $\psi_{c,d}$ has genus 0 or 1. (*I think it always has genus* 1 *but do not know how to prove it.*)

(ii) Describe the generated extension for the following choices: $e_c(x) := \theta_q(x/c)/\theta_q(x)$; $e_c := e_{\sqrt{q},\sqrt{c}}$. (First check that they are indeed q-characters of exponent c.)

2.3.2. q-logarithms.

2.3.2.1. For q-difference systems having multiple exponents (Chapters 4, 5), we shall have to solve auxiliary q-difference equations of the form $f(qx) = f(x) + 1$. This is impossible with a power series $\sum f_n x^n$, whether in $\mathbf{C}((x))$, in $\mathbf{C}(\{x\})$, or in $\mathscr{O}(\mathbf{C}^*, 0)$, since it would imply $f_0 = f_0 + 1$. We shall not go for a general theory of such equations beyond the simple remark that, if f is a meromorphic solution, then the meromorphic solutions are exactly the $f + \phi$, $\phi \in \mathscr{M}(\mathbf{E}_q)$; also, $f(cx)$ is also a solution for every $c \in \mathbf{C}^*$.

EXERCISE 2.43. Consider equation ${\sigma_q}^2 f - 2c\sigma_q f + c^2 f = 0$. Show that $f_0 := e_c$ is a solution. Look for another solution in the form $f = f_0 g$. Find the q-difference equation satisfied by g and discuss its possible solutions.

DEFINITION 2.44. We shall take $\ell_q(x) := -x{\theta_q}'(x)/\theta_q(x)$ as *the ("official") q-logarithm.*

Indeed, taking logarithmic derivatives on both sides of equation $\theta_q(qx) = \dfrac{1}{x}\,\theta_q(x)$, we find:

$$q{\theta_q}'(qx)/\theta_q(qx) = -\frac{1}{x} + {\theta_q}'(x)/\theta_q(x) \Longrightarrow \ell_q(qx) = \ell_q(x) + 1.$$

Clearly, $\ell_q \in \mathscr{M}(\mathbf{C}^*)$ with simple poles on $[-1;q]$ and no other poles. Its zeroes are somewhat more mysterious. Actually, taking logarithmic derivative of Jacobi's Triple Product Formula, we get a kind of expansion in simple fractions:

$$\ell_q(x) = -\sum_{n\geq 0} \frac{q^n x}{1+q^n x} + \sum_{n\geq 1} \frac{q^n/x}{1+q^n/x}.$$

EXERCISE 2.45. So it has the form $\ell_q = -A + B$ (A, B denoting both sums). Compute separately $A(qx) - A(x)$ and $B(qx) - B(x)$ to verify that $\ell_q(qx) = \ell_q(x) + 1$.

2.3.2.2. We shall more generally have use for the following "higher q-logarithms":

$$\ell_q^{(n)} := \binom{\ell_q}{n} = \frac{\ell_q(\ell_q - 1)\cdots(\ell_q - n + 1)}{n!} \in \mathscr{M}(\mathbf{C}^*).$$

According to the usual conventions on empty products, $\ell_q^{(0)} = 1$; and of course $\ell_q^{(1)} = 1$. Since $\ell_q(qx) = \ell_q(x) + 1$, using Pascal's rule on binomial coefficients, we find that $\ell_q^{(n)}(qx) = \ell_q^{(n)}(x) + \ell_q^{(n-1)}(x)$ (for $n \geq 1$). Indeed:

$$\ell_q^{(n)}(qx) = \binom{\ell_q(qx)}{n} = \binom{\ell_q(x)+1}{n} = \binom{\ell_q(x)}{n} + \binom{\ell_q(x)}{n-1} = \ell_q^{(n)}(x) + \ell_q^{(n-1)}(x).$$

LEMMA 2.46. (i) *The family* $(\ell_q^{(n)})_{n\in\mathbf{N}}$ *is linearly independent over* $\mathbf{C}(\{x\})$.
(ii) *The function* ℓ_q *is transcendental over* $\mathbf{C}(\{x\})$ *(thus over* $\mathbf{C}(x)$ *as well).*

PROOF. (i) To simplify notations, we shall by anticipation resort to a systematic use of the dilatation operator σ_q (of central importance in all the book) which sends a function $f(x)$ to the function $f(qx)$. Its algebraic properties (e.g., it induces a field automorphism of $\mathscr{M}(\mathbf{C}^*)$) will be discussed later but they appear here implicitly as simply justifiable rules of calculation. To begin with, by the formula immediately above:

$$\forall n \in \mathbf{N}\,,\ \sigma_q \ell_q^{(n)} = \ell_q^{(n)} + \ell_q^{(n-1)}.$$

(Of course we take $\ell_q^{(-1)} := 0$.) Also $\sigma_q(f\ell_q^{(n)}) = (\sigma_q f)(\ell_q^{(n)} + \ell_q^{(n-1)})$ for every $f \in \mathbf{C}(\{x\})$. Last, the relation $\sigma_q f - f = 1$ is impossible for $f \in \mathbf{C}(\{x\})$ (indeed the coefficient of x^0 in the expansion of $\sigma_q f - f$ is necessarily 0).

Assume a nontrivial linear relation:

$$f_0\ell_q^{(0)} + \cdots + f_n\ell_q^{(n)} = 0, \quad f_0, \ldots, f_n \in \mathbf{C}(\{x\}), \quad f_n \neq 0.$$

We may (and shall) assume that it is as short as possible, i.e., that n is the smallest possible. It cannot be $n = 0$ because that would mean that $f_0 = 0$. We apply σ_q and obtain a new relation:

$$\sum_{k=0}^{n} (\sigma_q f_k)(\ell_q^{(k)} + \ell_q^{(k-1)}) = 0 \Longrightarrow \sum_{k=0}^{n} (\sigma_q f_k + \sigma_q f_{k+1})\ell_q^{(k)} = 0,$$

with the convention that $f_{n+1} = 0$. We eliminate $\ell_q^{(n)}$ between the two relations by multiplying the first by $(\sigma_q f_n + \sigma_q f_{n+1}) = \sigma_q f_n$, the second by f_n and then substracting. Since the resulting relation is strictly shorter, it must be trivial, and in particular the coefficient of $\ell_q^{(n-1)}$ (for instance) must vanish:

$$f_n(\sigma_q f_{n-1} + \sigma_q f_n) - f_{n-1}\sigma_q f_n = 0 \Longrightarrow \sigma_q(f_{n-1}/f_n) - f_{n-1}/f_n = -1,$$

i.e., $f := -f_{n-1}/f_n \in \mathbf{C}(\{x\})$ satisfies $\sigma_q f - f = 1$, which we know to be impossible.

(ii) It is well known that the Newton polynomials $\binom{X}{k}$, $k = 0, \ldots, n$ have the same span that $1, \ldots, X^n$ and each of the two families can be expressed from the other by linear relations with rational coefficients. So the same is true for the families $\ell_q^{(k)}$, $k = 0, \ldots, n$, and $1, \ell_q, \ldots, \ell_q^{(n)}$. By (i) they have rank $n+1$, so the family $1, \ell_q, \ldots, \ell_q^{(n)}$ is free over $\mathbf{C}(\{x\})$ for all n i.e., ℓ_q is transcental over $\mathbf{C}(\{x\})$ (and therefore over $\mathbf{C}(x)$).

□

EXERCISE 2.47. Following an analogy with the formula $\left(\dfrac{d}{d\alpha} x^\alpha\right)_{|\alpha=0} = \log x$, Anne Duval proposes in **[78]** to look at $\dfrac{d}{dc} e_c(x)_{|c=1}$. Try it (and then read that nice paper).

2.3.3. Algebraic relations (absence of). The results here come from the appendix of **[217]**. We go on using the operator σ_q as we did in the proof of the previous lemma.

2.3.3.1. We shall have use later (Chapter 11) for some properties of algebraic independence among q-characters and q-logarithms, in the line of the lemma in 2.3.2.2. They are related to the structure of the field inside which we solve fuchsian q-difference equations (Chapters 5, 11), namely the subfield of $\mathscr{M}(\mathbf{C}^*, 0)$ generated over $\mathbf{C}(\{x\})$ by ℓ_q and the $e_{q,c}$, $c \in \mathbf{C}^*$.

As a preliminary remark, we note that the constant subfield of that field of resolution is, by nature, contained in $\mathscr{M}(\mathbf{C}^*, 0)^{\sigma_q} = \mathscr{M}(\mathbf{E}_q)$ (the field of q-elliptic functions); but, by the proposition at the beginning of 2.3.1.5, it generates $\mathscr{M}(\mathbf{E}_q)$; so it is actually $\mathscr{M}(\mathbf{E}_q)$.

EXERCISE 2.48. Check that $\mathscr{M}(\mathbf{C}^*, 0)^{\sigma_q} = \mathscr{M}(\mathbf{E}_q)$. (A germ $f \in \mathscr{M}(\mathbf{C}^*, 0)$ such that $\sigma_q f = f$ automatically extends uniquely to a function in $\mathscr{M}(\mathbf{C}^*)$; the argument is detailed in Proposition 3.8, Subsection 3.1.1 of the next chapter.)

2.3.3.2. We already know that the only q-characters in $\mathbf{C}(\{x\})$ are the ax^n, $a \in \mathbf{C}^*$, $n \in \mathbf{Z}$ (and the corresponding exponents are the q^n).

LEMMA 2.49. *The only q-characters in $\mathbf{C}(\{x\})(\ell_q)$ are the ax^n, $a \in \mathbf{C}^*$, $n \in \mathbf{Z}$.*

PROOF. Let $f \in \mathbf{C}(\{x\})(\ell_q)$, $f \neq 0$. Since ℓ_q is transcendental over $\mathbf{C}(\{x\})$, we can write in a unique way $f = u \dfrac{A(\ell_q)}{B(\ell_q)}$, where $u \in \mathbf{C}(\{x\})^*$ and A, B are coprime unitary polynomials over $\mathbf{C}(\{x\})$. Assume that $\sigma_q f = cf$, $c \in \mathbf{C}^*$, i.e.:

$$(\sigma_q u) \frac{A^{\sigma_q}(\ell_q + 1)}{B^{\sigma_q}(\ell_q + 1)} = (cu) \frac{A(\ell_q)}{B(\ell_q)}.$$

(A^{σ_q}, B^{σ_q} denote the result of applying σ_q to the coefficients of A, B.) From the uniqueness mentioned above, we deduce that $\sigma_q u = cu$, $A^{\sigma_q}(\ell_q + 1) = A(\ell_q)$ and $B^{\sigma_q}(\ell_q + 1) = B(\ell_q)$. From $\sigma_q u = cu$ we deduce that $c = q^n$ and $u = ax^n$, $a \in \mathbf{C}^*$, $n \in \mathbf{Z}$.

Then write $A = X^d + a_1 X^{d-1} + \cdots + a_d$. If $d \geq 1$, from $A^{\sigma_q}(X+1) = A(X)$ (recall that ℓ_q is transcendental), we see by identification of coefficients (at degree $d-1$) that $d + \sigma_q a_1 = a_1$, i.e., $g := -a_1/d \in \mathbf{C}(\{x\})$ satisfies $\sigma_q g - g = 1$, which is impossible. Therefore $d = 0$, i.e., $A = 1$; and likewise $B = 1$, and therefore $f = u = ax^n$. □

LEMMA 2.50. *Let $f_1, \dots, f_r \in \mathscr{M}(\mathbf{C}^*)$ be q-characters with exponents $c_1, \dots, c_r \in \mathbf{C}^*$ supposed pairwise noncongruent modulo $q^{\mathbf{Z}}$. Then $f_1, \dots, f_r$ are linearly independent over $\mathbf{C}(\{x\})(\ell_q)$.*

PROOF. Let $A_1 f_1 + \cdots + A_r f_r = 0$, $A_1, \dots, A_r \in \mathbf{C}(\{x\})(\ell_q)$, a nontrivial linear relation. Assume it has been chosen with the smallest possible length r, so that all $A_i \neq 0$ and, of course, $r \geq 2$. Applying σ_q yields $(\sigma_q A_1) c_1 f_1 + \cdots + (\sigma_q A_r) c_r f_r = 0$. The two relations must be proportional, otherwise by elimination we would get a nontrivial shorter one. Therefore, in particular:

$$\frac{(\sigma_q A_1) c_1}{A_1} = \frac{(\sigma_q A_r) c_r}{A_r} \Longrightarrow \sigma_q A = cA,$$

where $A := A_r / A_1 \in \mathbf{C}(\{x\})(\ell_q)^*$ and $c := c_1 / c_r \in \mathbf{C}^*$. By the proposition, this implies that $c \in q^{\mathbf{Z}}$, so $c_1 \equiv c_r \pmod{q^{\mathbf{Z}}}$, contradicting the assumption on exponents[4]. □

2.3.3.3. In essence, the above lemmas reduce algebraic dependencies of q-characters (over $\mathbf{C}(\{x\})(\ell_q)$) to multiplicative dependencies, which we now describe. So assume:

$$\lambda e_{q,c_1} \cdots e_{q,c_r} = e_{q,d_1} \cdots e_{q,d_s}, \quad \lambda, c_1, \dots, c_r, d_1, \dots, d_s \in \mathbf{C}^*.$$

Since $e_{q,1} = 1$, we may assume that all $c_i, d_j \neq 1$. Also, up to simplifications, we may assume that $c_i \neq d_j$ for all i, j. Now we add a simplifying assumption (but see the exercise below): all c_i, d_j are to be taken in $\mathbf{C}^* \setminus q^{\mathbf{Z}}$. Comparing divisors, we get:

$$\sum_{i=1}^{r} \left([\overline{-1}] - [\overline{-c_i^{-1}}] \right) = \sum_{j=1}^{s} \left([\overline{-1}] - [\overline{-d_j^{-1}}] \right).$$

Since (by the "simplifying assumption") all $\overline{-c_i^{-1}} \neq \overline{-1}$ and likewise all $\overline{-d_j^{-1}} \neq \overline{-1}$, we must have $r = s$ and, up to reindexing, $\overline{-c_i^{-1}} = \overline{-d_i^{-1}}$, i.e., $d_i = q^{m_i} c_i$ for some $m_i \in \mathbf{Z}$. From the fact that $e_{q,q^m c} = q^{m(m-1)/2} c^m x^m e_{q,c}$ (which itself follows from $\theta_q(q^m c x) = q^{-m(m-1)/2} c^{-m} x^{-m} \theta_q(x)$, which in turn can be checked by induction), we obtain:

$$\lambda q^{\sum m_i(m_i-1)/2} c^{\sum m_i} x^{\sum m_i} = 1 \Longrightarrow \sum m_i = 0 \text{ and } \lambda = q^{-\sum m_i^2/2}.$$

Conversely, if $\sum m_i = 0$ and $\lambda = q^{-\sum m_i^2/2}$, then $\prod e_{q,d_i} = \lambda \prod e_{q,c_i}$.

EXERCISE 2.51. Relax the "simplifying assumption": how is the conclusion affected?

PROPOSITION 2.52. *θ_q is transcendental over the field generated by $\mathbf{C}(\{x\})$, ℓ_q and the $e_{q,c}$, $c \in \mathbf{C}^*$.*

PROOF. Let f an element of that field and $x_0 \in \mathbf{C}^*$ such that f has no pole over $[x_0; q]$ (such an x_0 is generic by a denumerability argument). Then the sequence $(f(q^n x_0))_{n \geq 0}$ has at most simply exponential growth. On the other hand, $\theta_q(q^n x_0)$ behaves like $q^{-n^2/2}$. So writing an algebraic relation ${\theta_q}^d + f_1 {\theta_q}^{d-1} + \cdots + f_d = 0$ in the equivalent form $\theta_q = -(f_1 {\theta_q}^{-1} + \cdots + f_d {\theta_q}^{-d})$, we would get a contradiction. □

[4]This corollary is number 5 in [**217**, p 1065] but the proof given there is incorrect: on line 3 of p. 1066, operator σ_q should be applied to the coefficients of $P_1, \dots, P_r$.

2.3.4. Universal families of (q, μ)-characters. We propose here a generalisation of characters which is useful in the study of *families* of q-difference equations, as for instance in Subsection 12.2.4, also see **[196]**. Note however that this subsection is not logically necessary in the sequel.

So for $\mu \in \mathbf{N}^*$ and $c \in \mathbf{C}^*$ we define:

$$V_{\mu,c} := \{f \in \mathscr{O}(\mathbf{C}^*) \mid \sigma_q f = cx^{-\mu} f\}$$

and we nickname (q, μ)*-character* any function belonging to some $V_{\mu,c}$.

EXERCISE 2.53. Show that $V_{\mu,c}$ is trivial whenever $\mu < 0$, or $\mu = 0$ and $c \in q^{\mathbf{Z}}$.

2.3.4.1. *Canonical bases of (q, μ)-characters.*

LEMMA 2.54. *The following family is a basis of $V_{\mu,c}$ (the* canonical basis*):*

$$\mathcal{B}_{\mu,c} := \left(\theta_{q^\mu}(x^\mu/c), \ldots, x^i \theta_{q^\mu}(q^i x^\mu/c), \ldots, x^{\mu-1}\theta_{q^\mu}(q^{\mu-1}x/c)\right).$$

PROOF. Write (temporarily) $\mathscr{O}(\mathbf{C}^*)^{(\mu)}$ the fixed subalgebra of the automorphism $f(x) \mapsto f\left(e^{2i\pi/\mu}x\right)$ of $\mathscr{O}(\mathbf{C}^*)$, i.e., $\mathscr{O}(\mathbf{C}^*)^{(\mu)} := \{g(x^\mu) \mid g \in \mathscr{O}(\mathbf{C}^*)\}$. The map $\sigma_q - cx^{-\mu}$ leaves stable each component of the natural splitting $\mathscr{O}(\mathbf{C}^*) = \bigoplus\limits_{i=0}^{\mu-1} x^i \mathscr{O}(\mathbf{C}^*)^{(\mu)}$, so writing accordingly $f = \sum\limits_{i=0}^{\mu-1} x^i g_i(x^\mu)$, and denoting (temporarily) $Q := q^\mu$ and $X := x^\mu$, we have:

$$\begin{aligned}\sigma_q f = cx^{-\mu} f &\Longleftrightarrow \forall i = 0, \ldots, \mu - 1\ ,\ q^i x^i g_i(q^\mu x^\mu) = cx^{-\mu} x^i g_i(x^\mu)\\ &\Longleftrightarrow \forall i = 0, \ldots, \mu - 1\ ,\ g_i(QX) = c_i X^{-1} g_i(X),\end{aligned}$$

where $c_i := cq^{-i}$. The only holomorphic solutions are the $g_i(X) := \alpha_i \theta_Q(X/c_i)$, $\alpha_i \in \mathbf{C}$. □

We now patch all $V_{\mu,c}$ together into a sort of *universal family* of (q, μ)-characters:

$$\forall \mu \in \mathbf{N}^*\ ,\ V_\mu := \{(c, f) \mid c \in \mathbf{C}^*, f \in V_{\mu,c}\}.$$

PROPOSITION 2.55. *The natural projection $(c, f) \mapsto c$, $V_\mu \to \mathbf{C}^*$, makes it a trivial holomorphic vector bundle of rank μ.*

PROOF. Look at Appendix B for complete explanations. We have an isomorphism of fibrations $\mathbf{C}^* \times \mathbf{C}^\mu \simeq V_\mu$ given by:

$$(c, (\alpha_0, \ldots, \alpha_{\mu-1})) \mapsto \left(c, \sum_{i=0}^{\mu-1} \alpha_i x^i \theta_{q^\mu}(q^i x^\mu/c)\right).$$

This means that the following diagram, in which downward arrows are first projections and the horizontal arrow is the one we just defined, is commutative:

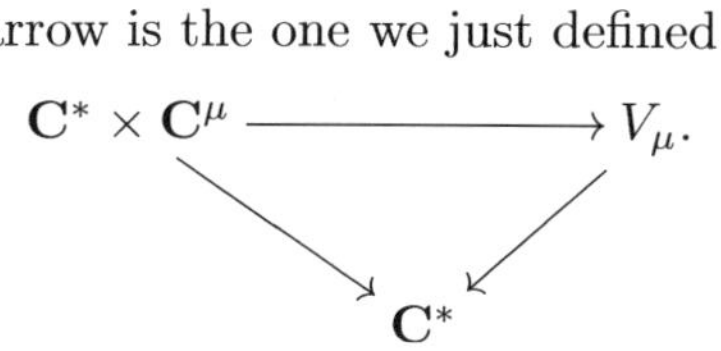

□

So, for a fixed $\mu \geq 1$, we may consider $\mathcal{B}_{\mu,c}$ as a "universal basis" for the "universal space" of (q, μ)-characters. For each particular value of c, it is a basis of the corresponding fiber $V_{\mu,c}$.

2.3.4.2. *An equivariant automorphism of V_μ.* We fix $c \in \mathbf{C}^*$. If $\sigma_q f = cx^{-\mu} f$ and if $g := xf$, then $\sigma_q g = qcx^{-\mu} g$. We thus obtain an isomorphism of fibers $f \mapsto xf$ from $V_{\mu,c}$ to $V_{\mu,qc}$ above the automorphism $c \mapsto qc$ of the base $\mathbf{C}^*$ of the bundle. This is the meaning of "equivariant" in the title of this subsubsection (see Subsection B.5.5 of the quoted appendix). Note that, as a consequence, there is a "quotient bundle" over $\mathbf{E}_q$ when $V_{\mu,c}$ is identified with $V_{\mu,qc}$ through that isomorphism and the result is the fiber over $\overline{c}$; see same Subsection and **[196]**.

LEMMA 2.56. *The action of the above isomorphism relative to bases $\mathcal{B}_{\mu,c}, \mathcal{B}_{\mu,qc}$ is given by the formula:*

$$x\mathcal{B}_{\mu,c} = \mathcal{B}_{\mu,qc} A_{\mu,c}, \text{ where } A_{\mu,c} := \begin{pmatrix} 0 & 0 & \dots & 0 & cq \\ 1 & 0 & \dots & 0 & 0 \\ \vdots & \vdots & \ddots & \vdots & \vdots \\ 0 & 0 & \dots & 0 & 0 \\ 0 & 0 & \dots & 1 & 0 \end{pmatrix} \in GL_\mu(\mathbf{C}).$$

PROOF. Let us write $\mathcal{B}_{\mu,c} = (f_i)_{i=0,\dots,\mu-1}$ and $\mathcal{B}_{\mu,qc} = (g_i)_{i=0,\dots,\mu-1}$ (i.e., $f_i(x) := x^i \theta_Q(q^i x^\mu / c)$ and $g_i(x) := x^i \theta_Q(q^i x^\mu / qc)$, where $Q := q^\mu$). We want equalities of the form $x f_i(x) = \sum_{j=0}^{\mu-1} a_{i,j}(c) g_j(x)$, where the $a_{i,j}$ should depend on c but not on x. From the basic properties of $\theta_Q(x)$, one draws easily:

$$g_i(x) = x^i \theta_Q(q^{i-1} x^\mu / c) = \begin{cases} x f_{i-1}(x) \text{ if } i \geq 1, \\ \dfrac{1}{cq} x f_{\mu-1}(x) \text{ if } i = 0, \end{cases}$$

$$\Rightarrow x f_i(x) = \begin{cases} g_{i+1}(x) \text{ if } i \leq \mu - 2, \\ cq g_0(x) \text{ if } i = \mu - 1. \end{cases}$$

Therefore the isomorphism $f \mapsto xf$ sends $\mathcal{B}_{\mu,c}$ to $x\mathcal{B}_{\mu,c} = \mathcal{B}_{\mu,qc} A_{\mu,c}$, with matrix $A_{\mu,c}$ as described above. □

Denote A_μ the matrix function $c \mapsto A_{\mu,c}$. Anticipating on Subsection B.5 of Chapter 7 or appealing to Subsection B.5.5 of Appendix B, we deduce:

COROLLARY 2.57. *The vector bundle obtained when quotienting V_μ by the above equivariant action is isomorphic to $\mathfrak{F}_{A_\mu}$.*

2.4. Degeneracies when $q \to 1$

The degeneracy of q-object when $q \to 1$ is often called "the continuous limit" in litterature (mostly in physics-oriented work). We saw examples of such degeneracies in 1.4 and 1.6 and will study their theory more deeply[5] in 11.2. But the first step is the study of the behaviour of the "basic bricks", q-characters and q-logarithm:

(1) The q-difference equation $f(qx) = cf(x)$ defining q-characters can be written:

$$D_q f = \frac{1-c}{1-q} \frac{1}{x} f,$$

[5] For a variant combining various degeneracies among q-difference equations, difference equations and differential equations, see the article of Anne Duval and Julien Roques **[79]**.

using the q-derivation operator $D_q : f(x) \mapsto \dfrac{f(x) - f(qx)}{x - qx}$, which formally degenerates to usual derivation when $q \to 1$. So if c depends on q in such a way that $\dfrac{1-c}{1-q} \to \gamma \in \mathbf{C}$ (for instance if $c = q^\gamma$), then the q-difference equation degenerates into $f' = \dfrac{\gamma}{x} f$, and we could hope that $e_{q,c}$ tends to something like x^γ.

(2) On the other hand, the q-difference equation $l_q(qx) = l_q(x) + 1$ can be written:

$$D_q l_q = \frac{-1}{(1-q)x} \Longrightarrow D_q (1-q) l_q = \frac{-1}{x},$$

where we had to rescale ℓ_q by factor $1-q$ to get a reasonable degeneracy to $f' = -1/x$: this makes us hope that $(1-q)l_q$ could tend to a logarithm.

Both those phenomena will be confirmed in 2.4.3. Their proofs rest on consequences of the modular functional equation for Theta functions (and also, in the case of the q-logarithm, on Jacobi's Triple Product Formula), so we shall start by a study of the asymptotic behaviour of θ_q when $q \to 1$. Before going to our true goal (the above mentioned degeneracies), and as a kind of interlude, we apply it to Rogers-Ramanujan continued fraction.

All this section originates in **[217]**. Some differences of signs are due either to the convention on $|q|$ (< 1 here, > 1 there); or on the choice of Theta function (with coefficients involving $(-1)^n$ there, but not here).

2.4.1. Asymptotic behaviour of some Theta expressions.

2.4.1.1. Recall the notations for the modular functional equation of Θ: first $z \in \mathbf{C}$ and $\tau \in \mathcal{H}$; then $z' := z/\tau \in \mathbf{C}$ and $\tau' := -1/\tau \in \mathcal{H}$; last, the equation itself:

$$\Theta(z,\tau) = \sqrt{\frac{\mathrm{i}}{\tau}}\, e^{-\frac{\mathrm{i}\pi}{\tau}\left(z + \frac{1-\tau}{2}\right)^2} \Theta(z', \tau').$$

When applying to θ_q, we add the notations $x := e^{2\mathrm{i}\pi z} \in \mathbf{C}^*$ and $q := e^{2\mathrm{i}\pi\tau}$, $0 < |q| < 1$; together with the relation $\Theta(z,\tau) = \theta_q(-x)$.

Our immediate goal here is the asymptotic behaviour of $\theta_q(q^\alpha x)$ when $q \to 1$ (of course $q^\alpha := e^{2\mathrm{i}\pi\tau\alpha}$; it is understood that x is fixed). For clarity of notations we shall rather consider an expression $\theta_q(-q^\alpha x_0)$, where x_0 is fixed in $\mathbf{C}^*$ and some choice of $z_0 \in \mathbf{C}$ such that $x_0 := e^{2\mathrm{i}\pi z_0}$ has been made; the choice is important and will be explained in 2.4.1.2. At the end of our calculation, we shall replace z_0 by $z_0 + 1/2$, hence x_0 by $-x_0$ (see the statement of the theorem).

We specialize the modular functional equation for $x := q^\alpha x_0 = e^{2\mathrm{i}\pi z}$, where $z = z_0 + \alpha\tau$, so that $z' = \alpha + z_0/\tau = \alpha - z_0\tau'$. To enforce $q \to 1$, we impose $\tau \to 0$ (of course, inside $\mathcal{H}$). As we observed when we studied the degeneracy of Cauchy's q-binomial theorem 1.4, we do not want q-spirals $aq^{\mathbf{Z}}$ to spread all over $\mathbf{C}^*$ (this especially concerns q-spirals of poles), so we force them to condense by setting $\tau = \tau_0\epsilon$ for a fixed $\tau_0 \in \mathcal{H}$ and then letting $\epsilon \to 0^+$ (this means that $\epsilon \to 0$ while staying > 0). In other words, if $q_0 := e^{2\mathrm{i}\pi\tau_0}$, we have $q = q_0^\epsilon$ and the discrete q-spiral $q^{\mathbf{Z}}$ condenses into the continuous logarithmic spiral $q_0^{\mathbf{R}} = e^{2\mathrm{i}\pi\tau_0\mathbf{R}}$. See the figure of condensation in 1.4. Since the points of the discrete spiral seem to merge, by analogy (maybe a little sloppy) with the merging of singularities in so-called "confluent hypergeometric series" (see for instance [**255**, Chapter XVI] or [**123**, Chapter XX]), we nickname "confluence" that phenomenon.

REMARK 2.58. It is conjectured in [**217**] that all the results still hold if one merely requires that $\tau = \tau_0\epsilon + o(\epsilon)$, but it does not seem that this kind of generalisation has found proof or application, so we shall not consider it.

2.4.1.2. We write $\tau_0 = \tau_1 + \mathrm{i}\tau_2$, $\tau_2 > 0$; also $z_0 = u + \mathrm{i}v$. Note that, x_0 being given, z_0 and therefore u is only determined modulo 1, so we may (and shall) assume that:

$$v\tau_1/\tau_2 - 1 < u \leq v\tau_1/\tau_2, \text{ equivalently } -1/2 < u + 1/2 - v\tau_1/\tau_2 \leq 1/2.$$

The equality (at right) is attained if, and only if, $u = v\tau_1/\tau_2$, i.e., $u\tau_2 - v\tau_1 = 0$, i.e., z_0/τ_0 is real, i.e., $x_0 \in q_0^{\mathbf{R}}$. This will be considered as "the special case", while $x_0 \notin q_0^{\mathbf{R}}$ will be called "the generic case". (At the end of 2.4.1.3, there will even appear a "super special case".)

The modular functional equation reads:

$$\theta_q(-q^\alpha x_0) = (\text{automorphy factor}) \times (\text{Theta factor}),$$

where

$$(\text{automorphy factor}) = \sqrt{\frac{\mathrm{i}}{\tau}}\, e^{-\frac{\mathrm{i}\pi}{\tau}\left(z + \frac{1-\tau}{2}\right)^2} \text{ and } (\text{Theta factor}) = \Theta(z', \tau').$$

The automorphy factor is the simplest to estimate; recalling that $z = z_0 + \alpha\tau$, it reads:

$$\begin{aligned}
\sqrt{\frac{\mathrm{i}}{\tau}}\, e^{-\frac{\mathrm{i}\pi}{\tau}\left(z + \frac{1-\tau}{2}\right)^2} &= \sqrt{\frac{\mathrm{i}}{\tau}}\, e^{-\frac{\mathrm{i}\pi}{\tau}(z_0 + 1/2 + \tau(\alpha - 1/2))^2} \\
&= \sqrt{\frac{\mathrm{i}}{\tau}}\, e^{-\frac{\mathrm{i}\pi}{\tau}(z_0+1/2)^2} e^{-2\mathrm{i}\pi(z_0+1/2)(\alpha-1/2)} e^{-\mathrm{i}\pi\tau(\alpha-1/2)^2} \\
&\sim \sqrt{\frac{\mathrm{i}}{\tau}}\, e^{-\frac{\mathrm{i}\pi}{\tau}(z_0+1/2)^2} e^{-2\mathrm{i}\pi(z_0+1/2)(\alpha-1/2)} \text{ when } \tau \to 0.
\end{aligned}$$

2.4.1.3. The Theta factor (using the fact that $z' = \alpha - z_0\tau'$) is:

$$\Theta(z', \tau') = \sum_{n \in \mathbf{Z}} (-1)^n e^{n(n-1)\mathrm{i}\pi\tau'} e^{2\mathrm{i}\pi n z'} = \sum_{n \in \mathbf{Z}} (-1)^n e^{2\mathrm{i}\pi n\alpha} \phi_n(\epsilon),$$

where $\phi_n(\epsilon) := e^{n(n-1)\mathrm{i}\pi\tau'} e^{-2\mathrm{i}\pi n z_0\tau'} = e^{\mathrm{i}\pi\tau'(n^2 - 2n(z_0+1/2))}$. (The dependency on ϵ comes through $\tau' = -1/\tau = -1/(\tau_0\epsilon)$.) So we shall first study $\lim\limits_{\epsilon\to 0^+} \phi_n(c)$ for all n, and then prove the normal convergence of the above series to justify that $\lim\limits_{\epsilon\to 0^+} \sum\limits_{n\in\mathbf{Z}} = \sum\limits_{n\in\mathbf{Z}} \lim\limits_{\epsilon\to 0^+}$. As for the former question, we note that $|\phi_n(\epsilon)| = \exp \Re\, \mathrm{i}\pi\tau'(n^2 - 2n(z_0+$

$1/2)$), so we first compute:

$$\begin{aligned}\Re\, \mathrm{i}\pi\tau'(n^2 - 2n(z_0 + 1/2)) &= -\pi\, \Im\, \frac{-1}{\tau_0\epsilon}(n^2 - 2n(z_0 + 1/2)) \\ &= \frac{\pi}{\epsilon}\, \Im\, \frac{n^2 - 2n(z_0 + 1/2)}{\tau_0} \\ &= \frac{\pi}{\epsilon\,|\tau_0|^2}\, \Im\, (n^2 - 2n(z_0 + 1/2))\overline{\tau_0} \\ &= \frac{\pi}{\epsilon\,|\tau_0|^2}\, \Im\, ((n^2 - n - 2nu) - 2nv\mathrm{i})(\tau_1 - \mathrm{i}\tau_2) \\ &= \frac{-\pi}{\epsilon\,|\tau_0|^2}\, (2n\tau_1 v + \tau_2(n^2 - n - 2nu)) \\ &= \frac{-\pi\tau_2}{\epsilon\,|\tau_0|^2}\, (n^2 - 2nw),\end{aligned}$$

where $w := 1/2 + u - v\tau_1/\tau_2 \in]-1/2, 1/2]$.

The generic case (i.e., $x_0 \notin q_0^{\mathbf{R}}$) occurs when $|w| < 1/2$, so that $n^2 - 2nw > 0$ for all $n \neq 0$. So, in that case, $\lim\limits_{\epsilon \to 0^+} \phi_n(\epsilon) = 0$ for all $n \neq 0$. In the special case, $w = 1/2$ and the limit is still 0 for all $n \neq 0, 1$. So in the generic case only one term remains in the limit (the one with $n = 0$), while in the special case two terms remain (the ones with $n = 0$ and $n = 1$).

As for the normal convergence, it is justified as follows. We let $A := \left|e^{2\mathrm{i}\pi\alpha}\right|$ and $B_\epsilon := \left|e^{-\frac{\pi\tau_2}{\epsilon|\tau_0|^2}}\right|$; clearly $B_\epsilon < B_1$ as soon as $\epsilon < 1$. So:

$$\left|(-1)^n e^{n(n-1)\mathrm{i}\pi\tau'} e^{2\mathrm{i}\pi n z'}\right| \leq A^n B_1^{n^2-n}$$

as soon as $\epsilon < 1$. Since $B_1 < 1$, normal convergence follows immediately, and we confirm:

$$\lim_{\epsilon \to 0^+} \sum_{n \in \mathbf{Z}} (-1)^n e^{2\mathrm{i}\pi n\alpha} \phi_n(\epsilon) = \sum_{n \in \mathbf{Z}} \lim_{\epsilon \to 0^+} (-1)^n e^{2\mathrm{i}\pi n\alpha} \phi_n(\epsilon).$$

Since $\phi_0(\epsilon) = 1$ for all $\epsilon > 0$, when $n = 0$:

$$\lim_{\epsilon \to 0^+} (-1)^n e^{2\mathrm{i}\pi n.\alpha} \phi_n(\epsilon) = \lim_{\epsilon \to 0^+} (-1)^0 e^{2\mathrm{i}\pi 0.\alpha} \phi_0(\epsilon) = \lim_{\epsilon \to 0^+} 1 = 1.$$

Thus we conclude that, in the generic case, the Theta factor tends to 1 and we reach a first conclusion:

$$\forall x_0 \notin q_0^{\mathbf{R}}\,,\ \theta_q(-q^\alpha x_0) \sim \sqrt{\frac{\mathrm{i}}{\tau}}\, e^{-\frac{\mathrm{i}\pi}{\tau}(z_0+1/2)^2} e^{-2\mathrm{i}\pi(z_0+1/2)(\alpha-1/2)} \text{ when } \epsilon \to 0^+.$$

In the special case, the above formula is not really meaningful as such, since the term with $n = 1$ generally does not tend to a limit. All we can write a priori is:

$$\Theta(z', \tau') = 1 - e^{2\mathrm{i}\pi\alpha} \phi_1(\epsilon) + o(1) \text{ when } \epsilon \to 0^+.$$

We are in the case where $x_0 = q_0^s$ and $z_0 = s\tau_0$ for some $s \in \mathbf{R}$. Then:

$$\phi_1(\epsilon) = e^{\mathrm{i}\pi\tau'(1^2 - 2(z_0+1/2))} = e^{-2\mathrm{i}\pi z_0\tau'} = e^{2\mathrm{i}\pi s/\epsilon}.$$

Except in the "super special case" $s = 0$, this is a very quickly oscillating factor. We conclude:

$$\Theta(z', \tau') = 1 - e^{2\mathrm{i}\pi(\alpha + s/\epsilon)} + o(1),$$

and of course in particular $\Theta(z', \tau') \to 1 - e^{2\mathrm{i}\pi\alpha}$ if $s = 0$.

2.4.1.4. We now adapt our results for $x = -x_0$ and $z = z_0 + 1/2$, so $x = e^{2i\pi z}$:

THEOREM 2.7. (i) *If $x \notin -q_0^{\mathbf{R}}$ ("generic case"), then:*

$$\theta_q(q^\alpha x) \sim \sqrt{\frac{i}{\tau}}\, e^{-\frac{i\pi}{\tau}z^2} e^{-2i\pi z(\alpha-1/2)}.$$

(ii) *If $x = -q_0^s$, $s \in \mathbf{R}$ ("special case"), then:*

$$\theta_q(q^\alpha x) \sim \sqrt{\frac{i}{\tau}}\, e^{-\frac{i\pi}{\tau}z^2} e^{-2i\pi z(\alpha-1/2)} \times \left(1 - e^{2i\pi(\alpha+s/\epsilon)} + o(1)\right).$$

□

COROLLARY 2.59. *In the generic case,* $\lim \dfrac{\theta_q(q^\alpha x)}{\theta_q(q^\beta x)} = x^{\beta-\alpha}$.

PROOF. The common factor $\sqrt{\frac{i}{\tau}}\, e^{-\frac{i\pi}{\tau}z^2}$ disappears and we get the limit $e^{2i\pi z(\beta-\alpha)} = x^{\beta-\alpha}$. However, since a choice of z was involved and since we are dealing with a multivalued function, the determination must be specified. Here x lies in $\mathbf{C}^* \setminus (-q_0^{\mathbf{R}})$, a simply connected domain of $\mathbf{C}^*$ containing 1. (If 1 belonged to $-q_0^{\mathbf{R}}$, one would have $|q_0| = 1$.) There is a unique principal determination of the logarithm (i.e., one that maps 1 to 0) on that domain and our choice of z (flowing from the choice of z_0 in 2.4.1.1) is such that $2i\pi z = \log x$. So in our formula above, $x^{\beta-\alpha} = \exp(\beta-\alpha)\log x$. □

REMARK 2.60. If no simplification occurs, i.e., if $\alpha - \beta \notin \mathbf{Z}$, the function $\dfrac{\theta_q(q^\alpha x)}{\theta_q(q^\beta x)}$ has a discrete q-spiral of zeroes $-q^{-\alpha+\mathbf{Z}}$ and a discrete q-spiral of poles $-q^{-\beta+\mathbf{Z}}$. When $q \to 1$ along the "direction" $q_0^{\mathbf{R}_+^*}$, both discrete q-spirals condense into the cut $-q_0^{\mathbf{R}}$.

In the special, but not super-special case $x = -q_0^s$, $s \in \mathbf{R}^*$, the oscillating factor $1 - e^{2i\pi(\alpha+s/\epsilon)}$ makes estimation difficult. The super-special case is nicer:

COROLLARY 2.61. *In the super-special case $x = -1$, we have (assuming $\beta \notin \mathbf{Z}$)* $\lim \dfrac{\theta_q(q^\alpha x)}{\theta_q(q^\beta x)} = \lim \dfrac{\theta_q(-q^\alpha)}{\theta_q(-q^\beta)} = \dfrac{\sin \pi\alpha}{\sin \pi\beta}$.

PROOF. Here, $x_0 = 1$, $z_0 = 0$, $z = 1/2$, $s = 0$. We compute:

$$\frac{\theta_q(-q^\alpha)}{\theta_q(-q^\beta)} = \frac{e^{-i\pi(\alpha-1/2)}}{e^{-i\pi(\beta-1/2)}} \frac{1 - e^{2i\pi\alpha} + o(1)}{1 - e^{2i\pi\beta} + o(1)} \to \frac{e^{-i\pi\alpha} - e^{i\pi\alpha}}{e^{-i\pi\beta} - e^{i\pi\beta}} = \frac{\sin \pi\alpha}{\sin \pi\beta}.$$

□

Note that in the super-special case the limit depends on α and β but not on the direction τ_0. This is as it should be since $-1 \in -q_0^{\mathbf{R}}$ for all q_0, so in that case all possible directions of condensation are alike.

2.4.2. Interlude: application to Rogers-Ramanujan continued fraction. Before going to the true application of 2.4.1 to the subject of this book, we consider here an interesting (indeed, historical) example of the "super special case".

2.4.2.1. We saw in 1.8 (and we shall return to those matters in 4.3.5) various expressions for Rogers-Ramanujan continued fraction:

$$F_q(x) := \cfrac{1}{1+\cfrac{q}{1+\cfrac{q^2}{1+\cdots}}} = \frac{\sum_{n\geq 0} \frac{q^{n^2+n}}{(q;q)_n}}{\sum_{n\geq 0} \frac{q^{n^2}}{(q;q)_n}} = \frac{(q;q^5)_\infty (q^4;q^5)_\infty}{(q^2;q^5)_\infty (q^3;q^5)_\infty} = \frac{\theta_{q^5}(-q)}{\theta_{q^5}(-q^2)}.$$

To evaluate its behaviour when $q \to 1$, we replace q by $q^{1/5}$ in the last expression; in other words, we say that the above is $f(q^5)$, where $f(q) := \dfrac{\theta_q(-q^{1/5})}{\theta_q(-q^{2/5})}\cdot$ This is an instance of the expression at the end of 2.4.1.4, with $\alpha = 1/5$ and $\beta = 1/5$, so we deduce:

$$\lim_{q\to 1} \cfrac{1}{1+\cfrac{q}{1+\cfrac{q^2}{1+\cdots}}} = \lim_{q\to 1} \frac{\theta_q(-q^{1/5})}{\theta_q(-q^{2/5})} = \frac{\sin \pi/5}{\sin 2\pi/5} = \frac{-1+\sqrt{5}}{2}.$$

(The latter can be found by saying that $\sin \pi/5 = \sin 4\pi/5$ and by noting that, if $\omega := e^{2i\pi/5}$, then $u := \omega + \omega^{-1}$ satisfies $u^2 + u - 1 = 0$.) We recognize the golden number, which quite is natural since:

$$\cfrac{1}{1+\cfrac{1}{1+\cfrac{1}{1+\cdots}}} = \frac{-1+\sqrt{5}}{2}.$$

EXERCISE 2.62. Obtain the same results by using the analytic theory of continued fractions (criterion of Worpitzky in **[251]**).

2.4.2.2. Although they are not directly related to Theta functions, we sketch two other possible approaches to the above asymptotic behaviour that fall under the general techniques of q-calculus. We know that $F_q(x) = 1 - qx + \cdots = H_q(qx)/H_q(x) = G_q(x)/H_q(x)$, where $H_q(x) := \sum_{n\geq 0} \dfrac{q^{n^2}}{(q;q)_n} x^n = 1 + \dfrac{q}{1-q} x + \cdots$. It seems reasonable to guess that, when $q \to 1$:

$$F_q(x) \to F_1(x) = \cfrac{1}{1+\cfrac{x}{1+\cfrac{x}{1+\cdots}}}.$$

The right-hand side (if defined) is solution of the algebraic equation $f = \dfrac{1}{1+xf}$, from which we can deduce that $F_1(x) = \dfrac{-1+\sqrt{1+4x}}{2x}$ (the square root is the principal determination on $\mathring{D}(1,1)$). Since $F_1(0) = H_q(0) = 1$, it makes sense to study (for small values of x) $\log F_q(x) = \log H_q(qx) - \log H_q(x)$. This is the kind of relation that can be inverted with the use of Jackson's q-integral:

$$\log H_q(x) = -\sum_{n\geq 0} \log F_q(q^n x) \Longrightarrow (1-q)\log H_q(x) = -\int_0^x \frac{\log F_q(t)}{t}\, d_q t.$$

When $q \to 1$ there are, hopefully, two simultaneous degeneracies: that of F_q to F_1; and that of the q-integral to a plain Riemann's integral. So, heuristically:

$$\begin{aligned}\lim_{q\to 1}(1-q)\log H_q(x) &= -\int_0^x \frac{\log F_1(t)}{t}\,dt \\ &= -\operatorname{Li}_2\left(\frac{1}{2}-\sqrt{x+\frac{1}{4}}\right)+\frac{1}{2}\log^2\left(\frac{1}{2}+\sqrt{x+\frac{1}{4}}\right).\end{aligned}$$

We used the dilogarithm function defined by:

$$\operatorname{Li}_2(x) := \int_0^x \frac{-\log(1-t)}{t}\,dt = \sum_{n\geq 1}\frac{x^n}{n^2}.$$

Using [**261**, p 7], we obtain:

$$\lim_{q\to 1}(1-q)\log H_q(1) = -\operatorname{Li}_2\left(\frac{1-\sqrt{5}}{2}\right)+\frac{1}{2}\log^2\left(\frac{1+\sqrt{5}}{2}\right)=\frac{\pi^2}{15}.$$

Since $H_q(1) = \dfrac{1}{(q^2;q^5)_\infty (q^3;q^5)_\infty}$, we obtain the estimation:

$$\log\left((q^2;q^5)_\infty (q^3;q^5)_\infty\right) \sim \frac{-\pi^2/15}{1-q}.$$

REMARK 2.63. In the product defining $H_q(1)$, resp. $G_q(1)$, appear the factors $1-q^n$ where $n \equiv 2$ or $3 \pmod 5$, resp. $n \equiv 1$ or $4 \pmod 5$, from which we draw:

$$G_q(1)H_q(1) = \frac{(q^5;q^5)_\infty}{(q;q)_\infty}.$$

EXERCISE 2.64. Find an equivalent of $\log G_q(1)$ along the same lines and confront your result with the above remark.

We mentioned two approaches but only sketched one. The other one is the use of the q-Gamma function:

$$\Gamma_q(x) := \frac{(q;q)_\infty}{(q^x;q)_\infty}(1-q)^{1-x}.$$

According to [**93**, p. 20] or [**264**], $\Gamma_q \to \Gamma$ when $q \to 1$.

EXERCISE 2.65. Use that for an alternative approach. (We'll meet again Γ_q in 3.3.2.)

2.4.3. Application to the degeneracy of basic bricks.

2.4.3.1. An immediate consequence of 2.4.1.3 (generic case) is that, under the usual conventions (well, now we can call them "usual" ...) $q = q_0^\epsilon$, $\epsilon \to 0^+$:

$$\forall x \notin -q_0^{\mathbf{R}},\ \lim_{q\to 1} e_{q,q^\gamma}(x) = x^\gamma,$$

where we define x^γ using the principal determination of the logarithm on the simply connected domain $\mathbf{C}^* \setminus (-q_0^{\mathbf{R}})$. This is deduced setting $\alpha := 0$ and $\beta := \gamma$. Yet our hope was that whenever $\dfrac{1-c}{1-q} \to \gamma$ such a limit holds for $e_{q,c}(x)$, of which the above is but a particular case. We now verify the stronger statement.

Observe that the assumption $\dfrac{1-c}{1-q} \to \gamma$ can be written in the form $c = 1 + 2\mathrm{i}\pi\tau_0\gamma\epsilon + o(\epsilon)$, since $1-q = 1-e^{2\mathrm{i}\pi\tau_0\gamma\epsilon} \sim -2\mathrm{i}\pi\tau_0\gamma\epsilon$ when $\epsilon \to 0^+$. So the extension to this more general case will follow easily from:

PROPOSITION 2.66. *Let $x_0 \in \mathbf{C}^* \setminus (-q_0^{\mathbf{R}})$ and let $x_1 = x_1(\epsilon)$ and $x_2 = x_2(\epsilon)$ both tend to x_0 when $\epsilon \to 0^+$, in such a way that $x_2/x_1 = 1 + o(\epsilon)$ (equivalently $|x_1 - x_2| = o(\epsilon)$). Then $\lim \dfrac{(x_2;q)_\infty}{(x_1;q)_\infty} = 1$.*

PROOF. When ϵ approaches 0, both x_1 and x_2 approach x_0, so they stay out of $-q_0^{\mathbf{R}}$ and the statement makes sense. We have:

$$\frac{(x_2;q)_\infty}{(x_1;q)_\infty} = \prod_{n\geq 0} \frac{1-q^n x_2}{1-q^n x_1} = \prod_{n\geq 0}(1+u_n(\epsilon)), \text{ where } u_n(\epsilon) := q^n \frac{x_1 - x_2}{1-q^n x_1}.$$

From [**211**, Theorem 15.4 in Chapter 15], it follows that, if $\sum\limits_{n\geq 0} |u_n(\epsilon)| \to 0$ when $\epsilon \to 0^+$, then the above infinite product tends to 1, concluding the proof. So we proceed to check the condition. From Lemma 2.67, we know that there is a $\delta > 0$ such that, for ϵ small enough (and so x_1 near enough x_0), $|1 - q^n x_1| \geq \delta$ for all $n \geq 0$. Then:

$$\sum_{n\geq 0} |u_n(\epsilon)| \leq \frac{1}{\delta} \sum_{n\geq 0} |q|^n \, |x_1 - x_2| = \frac{1}{\delta} \frac{|x_1 - x_2|}{1-|q|} = \frac{1}{\delta} \frac{|x_1 - x_2|}{1-|q_0|^\epsilon}.$$

Now, from $q_0 = e^{2\mathrm{i}\pi\tau_0}$, $\tau_0 = \tau_1 + \mathrm{i}\tau_2$, $\tau_2 > 0$, we have $|q_0| = \exp \Re(2\mathrm{i}\pi\tau_0) = e^{-2\pi\tau_2}$, whence $1 - |q_0|^\epsilon \sim 2\pi\tau_2\epsilon$. Since $|x_1 - x_2| = o(\epsilon)$, the conclusion follows. □

LEMMA 2.67. *Let $x_0 \in \mathbf{C}^* \setminus q_0^{-\mathbf{R}_+}$. Then there is an open disk $\mathring{D}(x_0, r)$, $r > 0$, and a $\delta > 0$ such that:*

$$\forall x \in \mathring{D}(x_0, r)\ ,\ \forall \lambda \geq 0\ ,\ \left|1 - q_0^\lambda x\right| \geq \delta.$$

PROOF. Since $x_0 \notin q_0^{-\mathbf{R}_+}$, $x_0^{-1} \notin q_0^{\overline{\mathbf{R}}_+} = q_0^{\mathbf{R}_+} \cup \{0\}$; so there is an $r'' > 0$ such that $\mathring{\mathrm{D}}(x_0^{-1}, r'')$ does not meet the compact set $q_0^{\overline{\mathbf{R}}_+}$. For $0 < r' < r''$, the compact $\overline{\mathrm{D}}(x_0^{-1}, r')$ does not meet it either, so their distance δ' is strictly positive. Therefore:

$$\forall x' \in \mathring{\mathrm{D}}(x_0^{-1}, r')\ ,\ \forall \lambda \geq 0\ ,\ \left|x' - q_0^\lambda\right| \geq \delta'.$$

The image of $\mathring{\mathrm{D}}(x_0^{-1}, r')$ by $x \mapsto x^{-1}$ is an open neighborhood of x_0, so it contains an open disk $\mathring{\mathrm{D}}(x_0, r)$, $r > 0$, and $\left|x^{-1} - q_0^\lambda\right| \geq \delta'$ for x in that disk and $\lambda \geq 0$. We may if necessary shrink r and the disk so that $|x|$ has a strictly positive minimum $m > 0$ in that disk. Then, setting $\delta := m\delta' > 0$:

$$\forall x \in \mathring{\mathrm{D}}(x_0, r)\ ,\ \forall \lambda \geq 0\ ,\ \left|1 - q_0^\lambda x\right| \geq |x| \left|x^{-1} - q_0^\lambda\right| \geq \delta.$$

□

THEOREM 2.8. *If $\dfrac{1-c}{1-q} \to \gamma$ then $e_{q,c}(x) \to x^\gamma$. (All this under the usual conventions.)*

PROOF. We apply the proposition with $x_1 = -q^\gamma x$ and $x_2 = -cx$; then again to $x'_1 := q/x_1$ and $x'_2 := q/x_2$. Also, from Jacobi's Triple Product Formula, $\dfrac{\theta_q(x_2)}{\theta_q(x_1)} = \dfrac{(x_2;q)_\infty (x'_2;q)_\infty}{(x_1;q)_\infty (x'_1;q)_\infty}$. □

2.4.3.2. In order to obtain a useful estimation for $(1-q)\ell_q(x)$ when $q \to 1$, we shall start again (as in 2.4.1.1, of which we keep the notations) from the modular functional equation for $\theta_q(-x) = \Theta(z,\tau)$, but now taking the logarithmic derivative with respect to z; but first, taking in account the fact that $dx/dz = 2\mathrm{i}\pi x$, we see that:

$$-2\mathrm{i}\pi x {\theta_q}'(-x) = D\Theta(z,\tau) \Longrightarrow -2\mathrm{i}\pi \ell_q(-x) = D\Theta(z,\tau)/\Theta(z,\tau) =: L(z,\tau),$$

where we denote temporarily by D the derivation with respect to z, i.e., $Df := df/dz$. Thus, from:

$$\Theta(z,\tau) = \sqrt{\frac{\mathrm{i}}{\tau}}\, e^{-\frac{\mathrm{i}\pi}{\tau}\left(z+\frac{1-\tau}{2}\right)^2}\, \Theta(z',\tau'),$$

we draw (since $dz'/dz = 1/\tau$):

$$L(z,\tau) = -\frac{\mathrm{i}\pi}{\tau}(2z+1-\tau) + \frac{1}{\tau} L(z',\tau').$$

From $1-q \sim -2\mathrm{i}\pi\tau$, we now deduce:

$$(1-q)\ell_q(-x) \sim -2\mathrm{i}\pi\tau\ell_q(-x) = \tau L(z,\tau) = -\mathrm{i}\pi(2z+1-\tau) + L(z',\tau').$$

The denominator $\Theta(z',\tau')$ in $L(z',\tau') = D\Theta(z',\tau')/\Theta(z',\tau')$ has a definite nonzero limit in the generic case by 2.4.1. Its numerator tends to 0 because the only term that had a nonzero limit in the series defining $\Theta(z',\tau')$ has vanished under the derivation (and the other terms are essentially bounded the same way). So we conclude that $(1-q)\ell_q(-x) \to -\mathrm{i}\pi(2z+1)$ (remember that $\tau \to 0$).

THEOREM 2.9. *Under the usual conventions, if* $x \notin -q_0^{\mathbf{R}}$*:*

$$\lim_{q\to 1} (q-1)\ell_q(x) = \log x.$$

PROOF. Here $2\mathrm{i}\pi z = \log x$, the principal determination of the logarithm over the simply connected domain $\mathbf{C}^* \setminus (-q_0^{\mathbf{R}})$. The link with the previous formula comes as follows. Write them using x_0, z_0 instead of x, z, then write $x := -x_0$, $x_0 = e^{2\mathrm{i}\pi z_0}$, $x = e^{2\mathrm{i}\pi z}$, $z = z_0 + 1/2$. Then:

$$(1-q)\ell_q(-x_0) \to -\mathrm{i}\pi(2z_0+1) \Longleftrightarrow (1-q)\ell_q(x) \to -2\mathrm{i}\pi z.$$

□

CHAPTER 3

Basic Notions and Tools

In all this book, except in few explicitly stated exceptional occurences, we take $q \in \mathbf{C}^*$, $|q| < 1$: the *base* q is a nonzero complex number with modulus strictly smaller than 1. The case $|q| > 1$ is essentially equivalent through easy changes of formulas, $q \leftrightarrow q^{-1}$: instead of writing (say) $f(qx) = u(x)f(x)$, write $f(q'x) = u'(x)f(x)$, where $q' := q^{-1}$ and $u'(x) := u(q^{-1}x)^{-1}$. Note however that some formulas and statements do change a bit and one must sometimes be careful in handling those changes. Also note that this is not quite the same thing as the change of variables $y := x^{-1}$ which will allow us to replace the local study at $x = \infty$ by a study at $y = 0$. At any rate, there will be some examples and exercises to get a feeling for the change $q \leftarrow q^{-1}$.

On the other hand, the case $|q| = 1$ is *very* different if one takes in account analytic properties. If q has finite order, $q \in \mu_\infty$ (see General Notations 0.4), then the study of the automorphism $f(x) \mapsto f(qx)$, say in $\mathbf{C}(\{x\})$ or $\mathbf{C}((x))$, is a purely algebraic problem: indeed classical Galois theory of cyclic extensions. As such, it usually interests specialists of quantum groups [**135**].

If $|q| = 1$ but $q \notin \mu_\infty$, we are in q-difference theory proper. The purely algebraic part of the theory is the same as if $|q| \neq 1$; but in this book we are dealing with *complex analytic* equations, and there, things go wild (if $|q| = 1$). Much less results have been obtained; see however [**60**], and also [**162, 237**] for motivation.

We choose once for all $\tau \in \mathbf{C}$ such that $q = e^{2i\pi\tau}$; thus assumption $|q| < 1$ translates into $\tau \in \mathcal{H}$ (Poincaré half-plane) i.e., $\Im\, \tau > 0$. Then we set $q^\alpha := e^{2i\pi\tau\alpha}$ for every $\alpha \in \mathbf{C}$. In particular, we let $q_r := q^{1/r} = e^{2i\pi\tau/r}$ for every $r \in \mathbf{N}^*$.

EXERCISE 3.1. Suppose we just need a family of q_r, $r \in \mathbf{N}^*$ such that $q_1 = q$ and $q_{rs}^s = q_r$ for all $r, s \in \mathbf{N}^*$. Does this really require choosing a complex number τ?

3.1. Some q-difference algebra

This is also called "quantum algebra" or "quantum calculus", see for instance [**133**], probably because quantum groupists use it a lot; but this is *not* the way we view it. We give here enough material for our basic needs: for some complements, look at [**51, 61, 81, 130, 133, 147**], and at the references therein.

3.1.1. Various q-difference fields and rings.

3.1.1.1. The traditional goal of the theory is *global*: it deals with functional equations over the Riemann sphere $\mathbf{S}$, i.e., with coefficients in $\mathscr{M}(\mathbf{S}) = \mathbf{C}(x)$, that is, rational coefficients. However the standard approach is to start with a *local*[1]

[1]This means local at 0, as we are only able to localize at 0 and ∞ (see Chapter 11, section 11.1 to undertand why) and the latter boils down to the former by change of variable $x \leftarrow 1/x$.

study, that is, using coefficients in $\mathbf{C}(\{x\})$. And this local study, in turn, usually begins with a *formal* study over $\mathbf{C}((x))$, featuring the more algebraic properties; then it follows with the more analytic parts, involving convergence, domains of definition, approximation ... This is the path that was followed in the theory of ordinary differential equations (see for instance **[222]**) and we shall try to follow it here.

3.1.1.2. In general difference theory, one deals with fields or rings endowed with automorphisms[2]. Our setting here, that of q-difference theory, is entirely concentrated on automorphisms of the form $f(x) \mapsto f(qx)$.

We shall consider a field K, usually the fraction field of a ring R; in some sense R will be the "ring of integers" of K, but we don't want to formalize this. We shall denote σ_q the mapping $f(x) \mapsto f(qx)$. In all cases considered, it will be an automorphism of field or ring (actually, an automorphism of $\mathbf{C}$-algebra). More precisely, here are our main fields and rings of interest (see General Notations 0.4):

$$K = \mathbf{C}(x), R = \mathbf{C}[x]; \quad K = \mathbf{C}(\{x\}), R = \mathbf{C}\{x\}; \quad K = \mathbf{C}((x)), R = \mathbf{C}[[x]];$$
$$K = \mathscr{M}(\mathbf{C}), R = \mathscr{O}(\mathbf{C}); \quad K = \mathscr{M}(\mathbf{C}^*), R = \mathscr{O}(\mathbf{C}^*); \quad K = \mathscr{M}(\mathbf{C}^*, 0), R = \mathscr{O}(\mathbf{C}^*, 0).$$

More generally, if $U \subset \mathbf{C}^*$ is *q-invariant*, i.e., $qU = U$, one can take $K := \mathscr{M}(U)$ and $R := \mathscr{O}(U)$, with the same definition of σ_q. Also possible are $K := \mathscr{M}(U, 0)$ and $R := \mathscr{O}(U, 0)$ (germs at 0). Note that in all cases, the automorphism σ_q of K restricts to the automorphism σ_q of R. Actually, we shall deal mostly with q-difference fields.

REMARK 3.2. In practice, the q-difference fields $\mathbf{C}((x))$, $\mathbf{C}(\{x\})$, $\mathbf{C}(x)$ will usually be the domains of the *coefficients* of our q-difference equations, while $\mathscr{M}(\mathbf{C})$, $\mathscr{M}(\mathbf{C}^*)$, $\mathscr{M}(\mathbf{C}^*, 0)$ ... will rather be the domains where we seek *solutions*. Such q-difference fields as $\mathscr{M}(U)$ appear when we study associated sheaves.

EXERCISE 3.3. For every $K \subset K'$ in that list, compare $K \cap R'$ and R.

DEFINITION 3.4. We call (K, σ_q), resp. (R, σ_q) a q-difference field, resp. a q-difference ring.

Note that in each case that $K \subset K'$ above, the automorphism σ_q of K' restricts to the automorphism σ_q of K (thus justifying a posteriori the fact that we used the same notation).

DEFINITION 3.5. We then say that (K', σ_q) is a *q-difference extension* of (K, σ_q).

3.1.1.3. *q-constants.* Let:

$$K^{\sigma_q} := \{f \in K \mid \sigma_q f = f\}, \quad \text{resp. } R^{\sigma_q} := \{f \in R \mid \sigma_q f = f\},$$

clearly a subfield of K, resp. a subring of R (and in both cases sub-$\mathbf{C}$-algebras). Then $K^{\sigma_q} \cap R = R^{\sigma_q}$.

DEFINITION 3.6. Elements of K^{σ_q} are called *q-constants*, sometimes simply *constants*.

[2] A more general (and more recent) approach **[63, 180, 256]** considers this as the *inversive* difference theory and requires only injective endomorphisms, thus allowing, for instance, for Mahler equations (where the endomorphism is $f(x) \mapsto f(x^p)$) **[175]**.

So we have the *field of constants* K^{σ_q} of K, resp. the *ring of constants* R^{σ_q} of R. Their name is motivated by the fact that, in many aspects of the theory (though not all!), they play the role of constants in the theory of ordinary differential equations. Actually, the basic equation $\frac{df}{dx} = 0$ is replaced [3] here by $\sigma_q f = f$. It is clear that in all cases complex numbers are q-constants:

$$\mathbf{C} \subset R^{\sigma_q} \subset K^{\sigma_q}.$$

The fact that inclusions can be strict is the source of many (interesting!) problems.

For every q-difference extension $K \subset K'$, one has $K'^{\sigma_q} \cap K = K^{\sigma_q}$ (similarly for q-difference rings). This is just because various automorphisms σ_q are restrictions of each other. It follows that, if $K'^{\sigma_q} = \mathbf{C}$, then $K^{\sigma_q} = R'^{\sigma_q} = R^{\sigma_q} = \mathbf{C}$. This is true for most of the examples of our list:

PROPOSITION 3.7. *If K is one of $\mathbf{C}(x)$, $\mathbf{C}(\{x\})$, $\mathbf{C}((x))$ and $\mathscr{M}(\mathbf{C})$, one has $K^{\sigma_q} = R^{\sigma_q} = \mathbf{C}$.*

PROOF. Since $\mathbf{C}(x) \subset \mathscr{M}(\mathbf{C}) \subset \mathbf{C}(\{x\}) \subset \mathbf{C}((x))$ (all q-difference extensions), we need only consider (by the previous remark) the case of $\mathbf{C}((x))$. So let $f := \sum\limits_{n \gg -\infty} a_n x^n$ be such that $\sigma_q f = f$. Of course $\sigma_q f = \sum\limits_{n \gg -\infty} a_n q^n x^n$, so $(q^n - 1)a_n = 0$ for all n. Since $q^n \neq 1$ for $n \neq 0$ (according to our assumption that $|q| < 1$), this means that $f = a_0$, whence $f \in \mathbf{C}$. □

The cases of $\mathscr{M}(\mathbf{C}^*)$ and $\mathscr{M}(\mathbf{C}^*, 0)$ are more exciting[4]

PROPOSITION 3.8. *One has $\mathscr{M}(\mathbf{C}^*)^{\sigma_q} = \mathscr{M}(\mathbf{C}^*, 0)^{\sigma_q} = \mathscr{M}(\mathbf{E}_q)$ (q-elliptic functions).*

PROOF. This will be our first theoretical instance of the use of the functional equation (here $\sigma_q f = f$) as a way to extend the domain of definition of a function f (we did meet some particular cases in Chapter 1.) So we detail the (rather elementary) proof a bit more that absolutely necessary.

Note first that, if $f \in \mathscr{M}(\mathbf{C}^*, 0)$, there is a small punctured open disk $D := \dot{\mathrm{D}}(0, r)$, $r > 0$, over which f is defined and meromorphic. Consider the function f' over $D' := q^{-1}\dot{\mathrm{D}}(0, r)$ defined by $f'(x) := f(qx)$: this is right because $x \in D' \Rightarrow qx \in D$.

On the other hand, $D' = \dot{\mathrm{D}}(0, |q|^{-1} r) \supset D$ because $|q|^{-1} > 1$. Also, the restriction of f' to D is f because this is just what the functional equation $\sigma_q f = f$ says ($f(x) = f(qx) = f'(x)$). Last, f' also satisfies $\sigma_q f' = f'$ on D' (we leave it to you to figure out why).

So what we did was to extend the q-constant f on D to a q-constant f' on D'. Iterating, we see that it can be extended to a q-constant on $q^{-n}D$; moreover these extensions are mutually compatible (i.e., restrictions of each other) so in the end we get a q-constant on $\bigcup\limits_{n \geq 0} q^{-n} D = \mathbf{C}^*$. So we already conclude that $\mathscr{M}(\mathbf{C}^*, 0)^{\sigma_q} = \mathscr{M}(\mathbf{C}^*)^{\sigma_q}$.

---

[3]Note that even in the classical theory of complex differential equations, the condition $\frac{df}{dx} = 0$ does not always imply that f is a complex number; on a nonconnected set, all that can be said is that f is locally constant function. Here (in q-difference theory) things will be still slightly more complicated.

[4]The cases of $\mathscr{M}(U)$ and $\mathscr{M}(U, 0)$, U a q-invariant open subset of $\mathbf{C}^*$, are for technical use only, see the sections on sheaves 7.3, 9.3, and 15.3.

Now $f \in \mathscr{M}(\mathbf{C}^*)^{\sigma_q}$ means that $f(x) = f(qx)$, whence $f(x) = f(q^n x)$ for every $n \in \mathbf{Z}$, so the meromorphic map f seen as a mapping $\mathbf{C}^* \to \mathbf{S}$ is actually a compositum $\mathbf{C}^* \to \mathbf{C}^*/q^{\mathbf{Z}} = \mathbf{E}_q \to \mathbf{S}$, i.e., a q-elliptic function: this is the way we identified $\mathscr{M}(\mathbf{C}^*)^{\sigma_q}$ with $\mathscr{M}(\mathbf{E}_q)$ in Chapter 2. □

EXERCISE 3.9. Prove that $\mathscr{M}(U)^{\sigma_q} = \mathscr{M}(U,0)^{\sigma_q}$ (usual notations).

3.1.1.4. *q-derivations.* To better emphasize some of the analogies of q-difference theory with classical calculus, we introduce here *q-difference operators.* Note however that in this book we shall rather use systematically q-difference automorphism σ_q. For f in any of the q-difference fields listed above, we set first:

$$\delta_q := \frac{1-\sigma_q}{1-q} : f(x) \mapsto \frac{f(x)-f(qx)}{1-q},$$

a *σ_q-twisted derivation*; this is because of:

EXERCISE 3.10. Prove that δ_q satisfies the *twisted Leibniz formula*:

$$\delta_q(fg) = \delta_q(f)\,g + \sigma_q(f)\delta_q(g).$$

The q-derivation δ_q formally degenerates into so-called Euler's derivation $\delta := x\,\dfrac{d}{dx}$ (Euler's derivation) when $q \to 1$. In the easiest sense, that means that, for a fixed f:

$$\lim_{q\to 1} \delta_q f = f.$$

However in section 11.2 on confluence, we shall meet more significant instances of such degeneracies, with a very precise analytical meaning; also see examples in Sections 1.6 and 4.4.

Another σ_q-twisted derivation usually met in books of q-calculus is:

$$D_q := \frac{1-\sigma_q}{(1-q)x} : f(x) \mapsto \frac{f(x)-f(qx)}{(1-q)x},$$

This one formally degenerates when $q \to 1$ into usual derivation $D := d/dx$.

EXERCISE 3.11. Find eigenvectors and eigenvalues of σ_q, δ_q, and δ acting on our various q-difference fields and q-difference rings.

The following exercise is a good preliminary training for the section 3.3 on q-calculus (the notation $\begin{bmatrix} n \\ i \end{bmatrix}_q$ is explained in 3.3.1).

EXERCISE 3.12. Prove the generalized twisted Leibniz formula:

$$D_q^n(fg) = \sum_{i=0}^{n} \begin{bmatrix} n \\ i \end{bmatrix}_q \sigma_q^i D_q^{n-i} f D_q^i g.$$

EXERCISE 3.13. Prove the relations relating powers of σ_q and powers of D_q (thus allowing conversion of q-difference equations into D_q-equations and conversely):

$$\sigma_q^n = \sum_{i=0}^{n} \begin{bmatrix} n \\ i \end{bmatrix}_q (q-1)^i q^{i(i-1)/2} x^i D_q^i,$$

$$D_q^n = \left(\frac{1}{(1-q)x}\right)^n \sum_{i=0}^{n} \begin{bmatrix} n \\ i \end{bmatrix}_{q^{-1}} q^{-i(i-1)/2} \sigma_q^i.$$

3.1.2. Extensions, ramification.

3.1.2.1. In all cases that $K \subset K'$ and the distinguished automorphism σ_q of K' restricts to that of K, we say that the q-difference field (K', σ_q) is a q-difference extension of (K, σ_q) (3.1.1.2), Definition 3.5. We thus have a lattice of q-difference extensions:

$$\begin{array}{ccccccc} \mathbf{C}(x) & \hookrightarrow & \mathscr{M}(\mathbf{C}) & \hookrightarrow & \mathscr{M}(\mathbf{C}, 0) = \mathbf{C}(\{x\}) & \hookrightarrow & \mathbf{C}((x)). \\ & & \downarrow & & \downarrow & & \\ & & \mathscr{M}(\mathbf{C}^*) & \hookrightarrow & \mathscr{M}(\mathbf{C}^*, 0) & & \end{array}$$

3.1.2.2. We now describe another type of q-difference extension, involving other bases $q' \neq q$. We shall write here for short $K := \mathbf{C}(\{x\})$ and $\hat{K} := \mathbf{C}((x))$ (the so-called "x-adic completion" of K). For every $r \in \mathbf{N}^*$, there is a unique (up to isomorphism) algebraic extension of degree r of K, resp. of $\hat{K}$; these respectively are:

$$K_r := K[x^{1/r}] = \mathbf{C}(\{x^{1/r}\}) \quad \text{and} \quad \hat{K}_r := \hat{K}[x^{1/r}] = \mathbf{C}((x^{1/r})).$$

We endow K_r, resp. $\hat{K}_r$ with the automorphism:

$$\sigma_{q_r} : f(x^{1/r}) \mapsto f(q_r x^{1/r}) = f(q^{1/r} x^{1/r}).$$

We thus obtain q-difference extensions (K_r, σ_{q_r}) of (K, σ_q) and $(\hat{K}_r, \sigma_{q_r})$ of $(\hat{K}, \sigma_q)$. We name them q-difference extensions for simplicity although the bigger field is rather a q_r-difference field.

As algebraic extensions, K_r/K and $\hat{K}_r/\hat{K}$ are cyclic extensions, with Galois groups $\mathrm{Gal}(K_r/K)$ and $\mathrm{Gal}(\hat{K}_r/\hat{K})$ both isomorphic to μ_r (the group of rth roots of unity in $\mathbf{C}^*$): for $j \in \mu_r$, the corresponding automorphism is $\tau_j : f(x^{1/r}) \mapsto f(jx^{1/r})$, i.e.:

$$f_0 + f_1 x^{1/r} + \cdots + f_{r-1} x^{(r-1)/r} \mapsto f_0 + f_1 j x^{1/r} + \cdots + f_{r-1} j^{r-1} x^{(r-1)/r},$$

where $f_0, \ldots, f_{r-1} \in K$ or $\hat{K}$. (Remember that $(x^{i/r})_{0 \leq i < r}$ is a basis of the K-linear space K_r, resp. of the $\hat{K}$-linear space $\hat{K}_r$).

An important fact is the following:

LEMMA 3.14. *The σ_{q_r}-action and the Galois action commute to each other.*

PROOF. Indeed the equality $\tau_j \sigma_{q_r} = \sigma_{q_r} \tau_j$ just means that $f(jq_r x^{1/r}) = f(q_r j x^{1/r})$. □

3.1.2.3. By a theorem of Puiseux [**145**] the algebraic closures of K and $\hat{K}$ are:

$$\overline{K} = \bigcup_{r \geq 1} K_r \quad \text{and} \quad \overline{\hat{K}} = \bigcup_{r \geq 1} \hat{K}_r.$$

They can be formally described as generated over the base of K, resp. $\hat{K}$, by a family $(x_r)_{r \geq 1}$ such that $x_1 = x$ and $x_{rs}^s = x_r$. Of course the x_r are just formal abstractions for $x^{1/r}$.

Since for all $r, s \geq 1$ the q-difference field $(K_{rs}, \sigma_{q_{rs}})$ is a q-difference extension of (K_r, σ_{q_r}), σ_q can be extended to $\overline{K}$ to make it a q-difference extension of K; and similarly for $\overline{\hat{K}}$ and $\hat{K}$.

3.1.3. The K-algebra $\mathcal{D}_{K,q}$ of q-difference operators.

3.1.3.1. *q-difference operators.* We fix a q-difference field (K, σ_q) (although much of what follows would still hold for a q-difference ring). Then $\sigma_q, \sigma_q^{-1} \in \mathscr{L}_{\mathbf{C}}(K)$, the $\mathbf{C}$-algebra of $\mathbf{C}$-linear endomorphisms of the $\mathbf{C}$-linear space K. (Thus elements of $\mathscr{L}_{\mathbf{C}}(K)$ are not required to respect the multiplication of K.)

Although $\mathscr{L}_{\mathbf{C}}(K)$ is not in any natural way a K-algebra (can you see why?), we can embed K as a sub-$\mathbf{C}$-algebra by identifying $a \in K$ with the $\mathbf{C}$-linear map $L_a : x \mapsto ax, K \to K$. Indeed $a \mapsto L_a$ is an injective morphism of $\mathbf{C}$-algebras, so an acceptable identification of K with its image in $\mathscr{L}_{\mathbf{C}}(K)$. In this way, we can multiply, i.e., compose, elements of K with powers $\sigma_q^n, n \in \mathbf{Z}$. For instance $a\sigma_q^n$ is the map $x \mapsto a\sigma_q^n(x)$.

CAUTION 3.15. It is customary to write, for short, by juxtaposition fx the application of a morphism f to an element of its source x: so fx means the same as $f(x)$. This convention would lead us to write $\sigma_q a$ for $\sigma_q(a)$. However, if we see σ_q and a as elements of $\mathscr{L}_{\mathbf{C}}(K)$, the notation $\sigma_q a$ also represents their composition $x \mapsto \sigma_q(ax)$, thus an element of $\mathscr{L}_{\mathbf{C}}(K)$. We shall try to avoid ambiguities, but better be careful with this.

EXAMPLE 3.16. The element $\sigma_q a \in \mathscr{L}_{\mathbf{C}}(K)$ is the map $x \mapsto \sigma_q(ax) = \sigma_q(a)\sigma_q(x)$, so it is the same as $\sigma_q(a)\sigma_q$:

$$\sigma_q a = \sigma_q(a)\sigma_q.$$

EXERCISE 3.17. Put $\sigma_q^n a \in \mathscr{L}_{\mathbf{C}}(K)$ in the form $b\sigma_q^n$, $b \in K$.

DEFINITION 3.18. A *q-difference operator* over K is an element of $\mathscr{L}_{\mathbf{C}}(K)$ of the form $\sum\limits_{n \in \mathbf{Z}} a_n \sigma_q^n$, where the $a_n \in K$ are almost all 0 (i.e., all but a finite number of them). The set of all q-difference operators is written:

$$\mathcal{D}_{K,q} := \left\{ \sum_{n \in \mathbf{Z}} a_n \sigma_q^n \mid (a_n) \in K^{(\mathbf{Z})} \right\}.$$

Thus $\mathcal{D}_{K,q}$ is the additive subgroup of $\mathscr{L}_{\mathbf{C}}(K)$ generated by all $a\sigma_q^n$. Since K is a sub K-linear space, $\mathcal{D}_{K,q}$ is one too. Actually, as a vector space:

$$\mathcal{D}_{K,q} = \bigoplus_{n \in \mathbf{Z}} K \sigma_q^n. \tag{3.1}$$

PROPOSITION 3.19. *$\mathcal{D}_{K,q}$ is the sub $\mathbf{C}$-algebra of $\mathscr{L}_{\mathbf{C}}(K)$ generated by K, σ_q and σ_q^{-1}.*

PROOF. Any sub $\mathbf{C}$-algebra of $\mathscr{L}_{\mathbf{C}}(K)$ containing K, σ_q and σ_q^{-1} obviously must contain $\mathcal{D}_{K,q}$, so it is enough to show that the latter is a sub $\mathbf{C}$-algebra. Since it is a sub K-linear space, we just have to check that it is closed under multiplication (that is, composition in $\mathscr{L}_{\mathbf{C}}(K)$). Using the formula $\sigma_q^n a = \sigma_q^n(a)\sigma_q^n$, $a \in K, n \in \mathbf{Z}$, we compute:

$$\left(\sum_{i \in \mathbf{Z}} a_i \sigma_q^i\right)\left(\sum_{j \in \mathbf{Z}} b_j \sigma_q^j\right) = \sum_{i,j \in \mathbf{Z}} a_i \sigma_q^i b_j \sigma_q^j = \sum_{i,j \in \mathbf{Z}} a_i \sigma_q^i(b_j) \sigma_q^i \sigma_q^j = \sum_{k \in \mathbf{Z}} c_k \sigma_q^k,$$

where $c_k := \sum\limits_{i+j=k} a_i \sigma_q^i(b_j)$. □

EXERCISE 3.20. Check that the c_k are almost all 0.

EXERCISE 3.21. Show that the sub **C**-algebra of $\mathscr{L}_{\mathbf{C}}(K)$ generated by K and σ_q is the same as the sub **C**-algebra of $\mathscr{L}_{\mathbf{C}}(K)$ generated by K and δ_q. (Beware this is not $\mathcal{D}_{K,q}$ since we did not take along σ_q^{-1}.)

3.1.3.2. *q-difference operators as noncommutative polynomials.* The multiplication rule found in the proof of the proposition suggests that we define in a purely algebraic, formal way[5] an algebra $K\langle S, S^{-1}\rangle$ of *noncommutative Laurent polynomials*: "Laurent" because the inverse S^{-1} of the indeterminate S is included; "noncommutative" for reasons to appear clearly below.

Elements of $K\langle S, S^{-1}\rangle$ are, by definition, formal linear combinations $\sum\limits_{i\in\mathbf{Z}} a_i S^i$ where $(a_i) \in K^{\mathbf{Z}}$, i.e., all $a_i \in K$ and almost all $a_i = 0$. "Formal" here means that by convention $\sum a_i S^i = \sum b_i S^i$ if, and only if, $a_i = b_i$ for all $i \in \mathbf{Z}$. As is well known we thus obtain a K-linear space:

$$\sum a_i S^i + \sum b_i S^i := \sum (a_i + b_i) S^i,$$
$$\lambda \sum a_i S^i := \sum \lambda a_i S^i.$$

The null element is of course $\sum 0 S^i$. Also we identify the monomial S^k with the $\sum \delta_{k,i} S^i$ (where as usual $\delta_{k,k} = 1$ and $\delta_{k,i} = 0$ whenever $k \neq i$). Then the family $(S^i)_{i\in\mathbf{Z}}$ of all monomials is a basis of the linear space $K\langle S, S^{-1}\rangle$ (its so-called *canonical* basis).

We now define a multiplication on $K\langle S, S^{-1}\rangle$ by setting:

$$\Big(\sum a_i S^i\Big)\Big(\sum b_j S^j\Big) := \sum c_k S^k, \quad \text{where} \quad \forall k \in \mathbf{Z}\,,\ c_k := \sum_{i+j=k} a_i \sigma_q^i(b_j).$$

It is a particular case of the theory of *Ore polynomials*, [**40, 179**]: we thus obtain a ring, indeed a **C**-algebra. The neutral element for multiplication is $S^0 = \sum \delta_{0,i} S^i$. We imbed K into $K\langle S, S^{-1}\rangle$ by sending a to aS^0, i.e., the linear combination $S^0 = \sum a\delta_{0,i} S^i$ (you are invited to check that this is an injective morphism of rings, thus an isomorphism of K to its image).

EXERCISE 3.22. Check directly that the multiplication is associative.

Although $K\langle S, S^{-1}\rangle$ is a ring and a K-linear space, it is not a K-algebra. Indeed, one of the axioms to be satisfied would be that $a(xy) = (ax)y = x(ay)$ for $a \in K$ and $x, y \in K\langle S, S^{-1}\rangle$ (K-bilinearity of the multiplication), but the second equality fails in general. For instance, taking $x = y = S$ and $a \in K$ arbitrary, we should compare aS^2 with $S(aS)$, but the latter is $\sigma_q(a)S^2$.

However, if $a \in K^{\sigma_q}$, then a is *central*:

$$\Big(\sum b_j S^j\Big)a = \sum b_j \sigma_q^j(a) S^j = \sum b_j a S^j = \sum a b_j S^j = a\Big(\sum b_j S^j\Big),$$

and from that it is easy to deduce that $K\langle S, S^{-1}\rangle$ is a K^{σ_q}-algebra.

EXERCISE 3.23. Deduce it. (The only nontrivial part is the bilinearity of the multiplication and it follows from the fact that K^{σ_q} is central.)

The relevance of $K\langle S, S^{-1}\rangle$ to our study of $\mathcal{D}_{K,q}$ lies in the following:

---

[5]The way we abstract here $\mathcal{D}_{K,q}$ as the ring $K\langle S, S^{-1}\rangle$ of "noncommutative Laurent polynomials" is akin to going from *polynomial functions* (concrete objects mapping numbers to numbers) to *polynomials* (abstract, formal objects).

PROPOSITION 3.24. *The map* $\sum a_i S^i \mapsto \sum a_i \sigma_q^i$ *from* $K\langle S, S^{-1}\rangle$ *to* $\mathcal{D}_{K,q}$ *is an isomorphism of* **C***-algebras:* $K\langle S, S^{-1}\rangle \simeq \mathcal{D}_{K,q}$*.*

PROOF. The map is well defined because an element of the source can be uniquely written as $\sum a_i S^i$ and because by definition $\sum a_i \sigma_q^i \in \mathcal{D}_{K,q}$; also by definition of the target it is onto.

The map is obviously a morphism of **C**-linear spaces; and it is a morphism of rings because the rules from multiplication are the same in $K\langle S, S^{-1}\rangle$ and in $\mathcal{D}_{K,q}$ (the former was built for that) and because it sends $1 \in K\langle S, S^{-1}\rangle$ to $1 \in \mathcal{D}_{K,q}$ (you should check it!).

It remains to prove that the map is injective, or, what comes down to the same, that the σ_q^i, $i \in \mathbf{Z}$, are K-linearly independent[6]; but this is a particular case of the Artin-Dedekind lemma [**145**, VIII §4] and [**33**, V §6.2]; however, due to the importance of this fact, we give here a direct proof. Note that the essential point is that the σ_q^i, $i \in \mathbf{Z}$ are pairwise distinct, which in turn comes from the assumption that q is not a root of unity. (Indeed, $\sigma_q^i = \sigma_q^j$ means that $q^i = q^j$.) So assume by contradiction that there is a nontrivial linear relation $\sum a_i \sigma_q^i = 0$. Write it more precisely as $\sum\limits_{i=k}^{l} a_i \sigma_q^i = 0$, where $k \leq l$ and $a_k a_l \neq 0$. Call *length* of such a relation the number $l-k$. Among all such nontrivial linear relations, assume we have chosen one with minimal length. Of course it cannot be 0 (that would mean $a_k \sigma_q^k = 0$, implying $a_k = 0$, a contradiction). So $k < l$, whence $\sigma_q^k \neq \sigma_q^l$: so let $a \in K$ be such $\sigma_q^k(a) \neq \sigma_q^l(a)$. From assumption $\sum\limits_{i=k}^{l} a_i \sigma_q^i = 0$, we draw:

$$\forall x \in K\ ,\ \sum_{i=k}^{l} a_i \sigma_q^i(x) = 0 \Longrightarrow \begin{cases} \forall x \in K\ ,\ \sum_{i=k}^{l} a_i \sigma_q^i(ax) = 0 \text{ and} \\ \forall x \in K\ ,\ \sigma_q^k(a) \sum_{i=k}^{l} a_i \sigma_q^i(x) = 0 \end{cases}$$

$$\Longrightarrow \begin{cases} \forall x \in K\ ,\ \sum_{i=k}^{l} a_i \sigma_q^i(a) \sigma_q^i(x) = 0 \text{ and} \\ \forall x \in K\ ,\ \sum_{i=k}^{l} a_i \sigma_q^k(a) \sigma_q^i(x) = 0 \end{cases}$$

$$\Longrightarrow \forall x \in K\ ,\ \sum_{i=k}^{l} a_i (\sigma_q^i(a) - \sigma_q^k(a)) \sigma_q^i(x) = 0$$

$$\Longrightarrow \forall x \in K\ ,\ \sum_{i=k+1}^{l} a_i (\sigma_q^i(a) - \sigma_q^k(a)) \sigma_q^i(x) = 0,$$

a strictly shorter nontrivial relation (the term of index $i = l$ is nonzero), contradicting the minimality of the length. □

EXERCISE 3.25. Find a similar formal model for the sub-**C**-algebra of $\mathcal{D}_{K,q}$ generated by K and σ_q, i.e., $K\langle \sigma_q \rangle := \left\{ \sum\limits_{n \geq 0} a_n \sigma_q^n \mid (a_n) \in K^{(\mathbf{N})} \right\}$ and also a formal model for the sub-**C**-algebra of $\mathcal{D}_{K,q}$ generated by K and δ_q i.e., $K\langle \delta_q \rangle := \left\{ \sum\limits_{n \geq 0} a_n {\delta_q}^n \mid (a_n) \subset K^{(\mathbf{N})} \right\}$; and then an explicit isomorphism between these two models.

[6]The attentive reader will note that this was implicitly admitted in (3.1), page 78.

We shall now study the structure of the ring $\mathcal{D}_{K,q}$, in particular (as is customary) its center, its unit group and its ideals. By proposition 3.24, it can be identified with $K\langle S, S^{-1}\rangle$, which we shall do in order to emphasize the purely algebraic character of this study. However, when it comes to applications, we shall return to the more down-to-earth model of $\mathcal{D}_{K,q}$ as consisting in q-difference operators, seen as mappings from K to itself.

3.1.3.3. *Center.*

PROPOSITION 3.26. *The center of $K\langle S, S^{-1}\rangle$ is K^{σ_q}; therefore, so is the center of $\mathcal{D}_{K,q}$.*

PROOF. First let $\sum a_i S^i$ belong to the center. Then $S \sum a_i S^i = (\sum a_i S^i)S$, that is $\sum \sigma_q(a_i) S^{i+1} = \sum a_i S^{i+1}$, whence all $a_i \in K^{\sigma_q}$. Also, $a \sum a_i S^i = (\sum a_i S^i)a$ for all $a \in K$, that is $\sum a a_i S^i = \sum a_i \sigma_q^i(a) S^i$. Therefore, for all i such that $a_i \neq 0$, one has $a = \sigma_q^i(a)$ for all $a \in K$, which is only possible if $i = 0$. This means that the only nonzero coefficient is a_0 and our central element is $a_0 \in K^{\sigma_q}$.

The verification that conversely every $a \in K^{\sigma_q}$ is central is easy and left to the reader. □

COROLLARY 3.27. *$K\langle S, S^{-1}\rangle$ is a K^{σ_q}-algebra; therefore, so is $\mathcal{D}_{K,q}$.*

PROOF. See remark about bilinearity of the multiplication in 3.1.3.2. □

3.1.3.4. *Absolute degree, units.* In order to go further, we introduce a useful tool. To motivate its definition, note that, if $P = \sum_{i=k}^{l} a_i S^i$, then SP (which encodes an equivalent q-difference equation) involves powers with exponents going from $k+1$ to $l+1$. Therefore k and l have no really "absolute" meaning, while $l-k$ maybe has.

DEFINITION 3.28. The *(absolute) degree* of $P := \sum_{i=k}^{l} a_i S^i$, where $k \leq l$ and $a_k a_l \neq 0$ is $l - k \in \mathbf{N}$. We write it $\deg P$. The absolute degree of 0 is $\deg 0 := -\infty$. (Usual rules for addition and order are understood to hold on $\mathbf{N} \cup \{-\infty\}$.)

EXAMPLE 3.29. The elements of degree 0 are the aS^n, $a \in K^*$, $n \in \mathbf{Z}$.

PROPOSITION 3.30. *The degree is additive: $\forall P, Q \in K\langle S, S^{-1}\rangle$, $\deg(PQ) = \deg P + \deg Q$.*

PROOF. If $P = 0$ or $Q = 0$, the formula obviously holds. Otherwise, we may write $P := \sum_{i=k}^{l} a_i S^i$ and $Q := \sum_{j=m}^{l} b_n S^j$, where $k \leq l$, $m \neq n$ and $a_k a_l \neq 0$, $b_m b_n \neq 0$, whence:

$$PQ = a_k \sigma_q^k(b_m) S^{k+m} + \cdots + a_l \sigma_q^k(b_n) S^{l+n}$$

(increasing exponents), where $a_k \sigma_q^k(b_m) a_l \sigma_q^k(b_n) \neq 0$. This implies that $\deg(PQ) = (l+n) - (k+m) = \deg P + \deg Q$. □

EXERCISE 3.31. Give an explicit formula for the intermediate coefficients.

Note that there is no formula for $\deg(P+Q)$ (neither equality nor inequality).

COROLLARY 3.32. *$K\langle S, S^{-1}\rangle$ is an integral ring; therefore, so is $\mathcal{D}_{K,q}$. Its invertible elements are the aS^n, $a \in K^*$, $n \in \mathbf{Z}$. The group of unit of $K\langle S, S^{-1}\rangle$ is therefore $K^* S^{\mathbf{Z}}$. The group of units of $\mathcal{D}_{K,q}$ is therefore $K^* \sigma_q^{\mathbf{Z}}$.*

PROOF. The calculation above show that if $P, Q \in K\langle S, S^{-1}\rangle$ and $P, Q \neq 0$, then $PQ \neq 0$.

If P has inverse Q, then $\deg P + \deg Q = \deg PQ = \deg 1 = 0$. The only possibility in $\mathbf{N} \cup \{-\infty\}$ is $\deg P = \deg Q = 0$, whence the form of P (see example above). Conversely, from $(aS^n)(bS^p) = (a\sigma_q^n(b))S^{n+p}$, one draws that aS^n has inverse bS^p if, and only if, $a\sigma_q^n(b) = 1$ and $n + p = 0$, i.e., that $(aS^n)^{-1} = \sigma_q^{-n}(a^{-1})S^{-n}$. □

EXERCISE 3.33. This suggests endowing $K^* \times \mathbf{Z}$ (which is in one-to-one correspondence with $K^* S^{\mathbf{Z}}$) with a group law. Describe explicitly that law. Show that you get a semidirect product.

DEFINITION 3.34. Elements $P, Q \in K\langle S, S^{-1}\rangle$ are said to be *left associated*, resp. *right associated*, if there exists $u \in K^* S^{\mathbf{Z}}$ such that $Q = uP$, resp. $Q = Pu$. We also say that Q is an *associate* of P.

Both relations (left and right association) are equivalence relations and both imply $\deg P = \deg Q$. Moreover, $K\langle S, S^{-1}\rangle P = K\langle S, S^{-1}\rangle Q$ if, and only if, P and Q are left associated and $PK\langle S, S^{-1}\rangle = QK\langle S, S^{-1}\rangle$ if, and only if, P and Q are right associated

EXAMPLE 3.35. Let $a, b \in K^*$. Then $S - a$ and $S - b$ are left associates if, and only if, for some $c \in K^*$ and $k \in \mathbf{Z}$ one has $S - b = cS^k(S - a) = cS^{k+1} - caS^k$. The only possibility is $c = 1$, $k = 0$, that is $a = b$.

On the other hand, $S - a$ and $S - b$ are right associates if, and only if, for some $c \in K^*$ and $k \in \mathbf{Z}$ one has $S - b = (S - a)cS^k = \sigma_q(c)S^{k+1} - caS^k$. Again this implies $a = b$.

EXERCISE 3.36. Let $a, b, a', b' \in K^*$. Show that $aS + b$ and $a'S + b'$ are left associates, resp. right associates, if, and only if, $a'/a = b'/b$, resp. if, and only if, $a'/a = \sigma_q(b'/b)$.

DEFINITION 3.37. We call *standard q-difference operators* or *q-difference operators in standard form* elements $a_0 + \cdots + a_n S^n \in K\langle S, S^{-1}\rangle$, or $a_0 + \cdots + a_n \sigma_q^n \in \mathcal{D}_{K,q}$, where $n \in \mathbf{N}$, $a_0, \ldots, a_n \in K$ and $a_0 a_n \neq 0$. We call *standard unitary q-difference operators* those such that moereover $a_n = 1$.

PROPOSITION 3.38. *Any nonzero q-difference operator is left associate of a unique standard unitary q-difference operator; same thing for right association.*

PROOF. Let $P := \sum_{i=k}^{l} a_i S^i$, where $k \leq l$ and $a_k a_l \neq 0$. Then its only standard unitary left associate is $(a_l S^k)^{-1} P$; its only standard unitary right associate is $P(bS^k)^{-1}$ for some $b \in K^*$. □

EXERCISE 3.39. Check these facts and find b.

Note that these associates are not the same. For instance:

$$aS + b = a(S + b/a) = (S + b/\sigma_q^{-1}(a))\sigma_q^{-1}(a).$$

EXERCISE 3.40. Another useful equivalence relation is *conjugacy*: the conjugates of P are the $u^{-1}Pu$, u invertible, to be used quite a few times in the sequel. Compute $(aS^k)^{-1}(\sum a_i S^i)(aS^k)$.

3.1.3.5. *Euclideanity.*

PROPOSITION 3.41. *$K\langle S, S^{-1}\rangle$ is left and right euclidean; therefore, so is $\mathcal{D}_{K,q}$.*

PROOF. We shall prove that if $A, B \in K\langle S, S^{-1}\rangle$ and if $B \neq 0$, then there exist $Q, R \in K\langle S, S^{-1}\rangle$ such that $A = QB + R$ and $\deg R < \deg B$ (left euclideanity). We leave to the reader to find the corresponding statement and proof for right euclideanity.

So let $A := a_k S^k + \cdots + a_l S^l$ (the case that $A = 0$ would be obvious) and $B := b_m S^m + \cdots + b_n S^n$, with the usual conventions, so that $\deg A = l - k$ and $\deg B = n - m$. The proof is by induction on $\deg A$. If $\deg A < \deg B$, we can take $Q := 0$ and $R := A$.

So assume $\deg A \geq \deg B$, i.e., $l - k \geq n - m$. First we search $c \in K$ and $p \in \mathbf{Z}$ such that $A' := A - cS^p B$ has degree $\deg A' < \deg A$ (see further below). Then we apply induction hypothesis to write $A' = Q'B + R'$, with $\deg R' < \deg B$. Then we have $A = A' + cS^p B = (Q' + cS^p)B + R'$, so we can set $Q := Q' + cS^p$ and $R := R'$: we are done.

Now the search for c, p: we choose to eliminate the highest degree term $a_l S^l$ in A, i.e., we want $(cS^p)(b_n S^n) = a_l S^l$, i.e., $c\sigma_q^p(b_n) = a_l$ and $p + n = l$. Thus we take $p := l - n$ and $c := a_l / sq^p(b_n)$ (recall that $b_n \neq 0$). The operation $A - cS^p B$ therefore eliminates the term in S^l from A, but it may create terms of degrees $j + p$, $m \leq j \leq n$. These degrees all are $\geq m + p = m + l - n \geq k$ since $l - k \geq n - m$. So the terms of A' all have degrees among $k, \ldots, l - 1$ and $\deg A' < \deg A$ as wanted. □

EXERCISE 3.42. (i) Do the corresponding calculation for right division. Note however that there will be a small twist in choosing the coefficient for elimination of the highest degree term, since $(cS^p)(b_n S^n) \neq (b_n S^n)(cS^p)$.

(ii) Do the corresponding calculation eliminating lowest degree term $a_k S^k$ instead of $a_l S^l$.

CAUTION 3.43. There is no uniqueness in this property. Take $A := a + bS$ and $B := S - 1$, then:

$$A = bB + (a + b) = -aB + (a + b)S$$

both are valid euclidean divisions.

EXERCISE 3.44. Which one of those would have been produced by the process described in the proof of the proposition?

COROLLARY 3.45. *$K\langle S, S^{-1}\rangle$ is left and right principal. In particular, every nonzero left ideal (resp. right ideal) of $K\langle S, S^{-1}\rangle$ has the form $K\langle S, S^{-1}\rangle P$ (resp. $PK\langle S, S^{-1}\rangle$) for a unique unitary standard P.*

PROOF. The first statement is standard (!), the second follows from proposition 3.38 □

The following consequence makes sense for "true", "concrete" q-difference operators, those in $\mathcal{D}_{K,q}$.

COROLLARY 3.46. *Let $P \in \mathcal{D}_{K,q}$ and let $f \in K^*$ such that $P(f) = 0$. Set $u := \sigma_q(f)/f$. Then there exists a unique $Q \in \mathcal{D}_{K,q}$ such that $P = Q(\sigma_q - u)$.*

PROOF. We do a euclidean division $P = Q(\sigma_q - u) + R$, where $\deg R < \deg(\sigma_q - u) = 1$, so either $R = 0$ or $R = c\sigma_q^p$ for some $c \in K^*$ and $p \in \mathbf{Z}$. Note that:

$$(\sigma_q - u)(f) = \sigma_q(f) - uf = 0$$

by our definition of u, so $Q(\sigma_q - u)(f) = 0$ so $R(f) = P(f) = 0$. This is not possible in the case $R = c\sigma_q^p$, so $R = 0$.

The uniqueness of Q comes from the fact that $\mathcal{D}_{K,q}$ is an integral ring. □

EXERCISE 3.47. Show how the complete calculation runs. Let $L := a_0 + \cdots + a_n S^n$ and $L' := a'_0 + \cdots + a'_{n-1} S^{n-1}$, $a_0 a_n \neq 0$, $a'_0 a'_n \neq 0$. Then, for $u, b \in K$, translate relation $L = L'(S - u) + b$ first by expressing the a_i in function of the a'_j (and u and b) then by doing the converse.

3.1.3.6. *"Outer automorphisms".* It is customary to call "inner automorphism" of a ring (or group) a map of the form $f \mapsto u^{-1} f u$. We shall have later many uses for those; there was an exercise about them in 3.1.3.4. Here we look at a slightly extended notion.

First suppose K' is a q-difference extension of K and that there is $\theta \in K'^*$ such that $u := \sigma_q\theta/\theta \in K^*$. Note that, from $\sigma_q\theta = u\theta$, one deduces by induction:

$$\forall i \in \mathbf{Z}\ ,\ \sigma_q^i\theta = u^{[i]}\theta, \text{ where } u^{[i]} := \begin{cases} u\sigma_q u \cdots \sigma_q^{i-1} u \text{ if } i \geq 0, \\ (\sigma_q^{-1} u \cdots \sigma_q^i u)^{-1} \text{ if } i < 0. \end{cases}$$

(By the usual convention on empty products, the right-hand side in the first case is 1 if $i = 0$.) From this one draws:

$$\theta^{-1} S^i \theta = \theta^{-1} \sigma_q^i(\theta) S^i = u^{[i]} S^i.$$

Let $P := \sum a_i S^i \in K\langle S, S^{-1}\rangle$. Then:

$$\theta^{-1} P \theta = \theta^{-1} \sum a_i S^i \theta = \sum a_i \theta^{-1} S^i \theta = \sum a_i u^{[i]} S^i.$$

We shall denote, in all generality, for $u \in K^*$ and $P := \sum a_i S^i \in K\langle S, S^{-1}\rangle$:

$$\forall i \in \mathbf{Z}\ ,\ u^{[i]} := \begin{cases} u\sigma_q u \cdots \sigma_q^{i-1} u \text{ if } i \geq 0, \\ (\sigma_q^{-1} u \cdots \sigma_q^i u)^{-1} \text{ if } i < 0, \end{cases}$$

$$P^{[u]} := \sum a_i u^{[i]} S^i.$$

From the above, we see that, if $P, Q \in K\langle S, S^{-1}\rangle$:

$$(P + Q)^{[u]} = P^{[u]} + Q^{[u]} \text{ and } (PQ)^{[u]} = P^{[u]} Q^{[u]}.$$

Also, using $\gamma \in K'^*$ such that $v := \sigma_q\gamma/\gamma \in K^*$:

$$P^{[uv]} = (P^{[u]})^{[v]}.$$

Now, the really interesting point is that the formulas above make sense for every $u, v \in K^*$ independently of a hypothetic q-difference extension K'.

PROPOSITION 3.48. *Formulas $(P+Q)^{[u]} = P^{[u]} + Q^{[u]}$, $(PQ)^{[u]} = P^{[u]} Q^{[u]}$ and $P^{[uv]} = (P^{[u]})^{[v]}$ remain valid in all generality.*

PROOF. Either build a q-difference extension by formally adjoining θ, γ, or do direct computations. □

EXERCISE 3.49. (i) Do those direct computations.
(ii) For which $u \in K^*$ is it true that $P \mapsto P^{[u]}$ is the identity map?

3.1.3.7. *Modules.* We shall prove in Chapter 8 that the left $\mathcal{D}_{K,q}$-modules of finite length are those which are finitely generated and torsion, and also those which are finite-dimensional over K. We shall call them *q-difference modules* and find that they are monogeneous ("cyclic vector lemma"), whence of the form $\mathcal{D}_{K,q}/\mathcal{D}_{K,q}P$ where P is standard unitary. The latter having basis $1, \ldots, S^{n-1}$ (where $n := \deg P$, we see that:

$$\dim_{\mathbf{C}} \mathcal{D}_{K,q}/\mathcal{D}_{K,q}P = \deg P.$$

Clearly this still holds if P is not standard unitary.

EXERCISE 3.50. Compare $\mathcal{D}_{K,q}/\mathcal{D}_{K,q}QR$ to $\mathcal{D}_{K,q}/\mathcal{D}_{K,q}Q$ and $\mathcal{D}_{K,q}/\mathcal{D}_{K,q}R$. (Think either of isomorphism theorems for modules over a ring, or of an exact sequence.)

3.2. Some preliminary generalities on solutions of q-difference equations

This is just a preliminary approach, to be completed in next few chapters. From now on we generally consider q-difference operators as concrete elements $\sum a_i\sigma_q^i \in \mathcal{D}_{K,q} \subset \mathscr{L}_{\mathbf{C}}(K)$ (and not as formal objects $\sum a_i S^i \in K\langle S, S^{-1}\rangle$).

The simplest of all ordinary differential equations is $y' = 0$, with solutions the "constants" (meaning the locally constant functions if the domain of study is not connected). The most similar q-difference equation would be $\sigma_q f = 0$ but it is trivial (the only solution is $f = 0$) and it has no interest at all. So the simplest meaningful q-difference equation is actually $\sigma_q f = f$, which defines the field K^{σ_q} of q-constants (see 3.1.1). In this section, we go a little bit further.

3.2.1. Spaces of solutions.

3.2.1.1. Let $P := \sum a_i\sigma_q^i \in \mathcal{D}_{K,q}$, $P \neq 0$. Then $P : K \to K, f \mapsto P(f) = \sum a_i\sigma_q^i f$ is K^{σ_q}-linear. Therefore its kernel:

$$\mathrm{Sol}(P, K) := \{f \in K \mid P(f) = 0\} = \left\{f \in K \mid \sum a_i\sigma_q^i f = 0\right\},$$

the *space of solutions of P in K*, is a K^{σ_q}-linear space. Let $u \in {\mathcal{D}_{K,q}}^* = K^*\sigma_q^{\mathbf{Z}}$ (Corollary 3.32), so that $u(g) = 0 \Leftrightarrow g = 0$. Then:

$$\mathrm{Sol}(P, K) = \mathrm{Sol}(uP, K).$$

Thanks to 3.1.3.4, we therefore can (and always will) assume that P is standard unitary: $P = \sigma_q^n + a_1\sigma_q^{n-1} + \cdots + a_n$, the $a_i \in K$, $a_n \neq 0$, so that:

$$\begin{aligned}\mathrm{Sol}(P, K) &= \left\{f \in K \mid \sigma_q^n f + a_1\sigma_q^{n-1} f + \cdots + a_n f = 0\right\}\\ &= \left\{f \in K \mid f(q^n x) + a_1(x)f(q^{n-1}x) + \cdots + a_n(x)f(x) = 0\right\},\end{aligned}$$

the space of solutions of a *scalar q-difference equation of order n*.

REMARK 3.51. If $K = \mathbf{C}((x))$ the meaning of x in $f(x)$, etc, is purely formal. For all other cases of our favorite list of q-difference fields, the above relation is meant to be quantified as $\forall x \in U$, etc, where U is some punctured neighborhood[7] of 0 in $\mathbf{C}^*$. By the principle of analytic continuation, it does not matter which particular U is used.

Actually, we shall usually take $P \in \mathcal{D}_{K,q}$ for some fixed "base field" (or "field of coefficients") K, but look for solutions in some q-difference extension K'. This way we obtain a kind of "functor of solutions" $K' \rightsquigarrow \mathrm{Sol}(P, K')$, a K'^{σ_q}-linear space.

[7] When dealing with sheaves, other cases may occur.

3.2.1.2. Casorati, in "Il calcolo delle differenze finite, Annali di Matematica pura ed applicata, Series II, 10, (1880–1882), 10–43, introduced a very useful tool, the analogue in difference algebra of the wronskian of the classical theory of differential equations **[222]**. Many authors call it "Casoratian" or "Casorati matrix" (or determinant). We shall however call it here q-wronskian (matrix or determinant).

Let $f_1, \ldots, f_n \in K'$, some q-difference extension of K. Then their q-wronskian matrix, resp. determinant, is defined as:

$$W_n(f_1, \ldots, f_n) := (\sigma_q^i f_j)_{\substack{0 \leq i \leq n-1 \\ 1 \leq j \leq n}}$$
$$= \begin{pmatrix} f_1 & f_2 & \cdots & f_j & \cdots & f_n \\ \sigma_q f_1 & \sigma_q f_2 & \cdots & \sigma_q f_j & \cdots & \sigma_q f_n \\ \vdots & \vdots & \ddots & \vdots & \ddots & \vdots \\ \sigma_q^i f_1 & \sigma_q^i f_2 & \cdots & \sigma_q^i f_j & \cdots & \sigma_q^i f_n \\ \vdots & \vdots & \ddots & \vdots & \ddots & \vdots \\ \sigma_q^{n-1} f_1 & \sigma_q^{n-1} f_2 & \cdots & \sigma_q^{n-1} f_j & \cdots & \sigma_q^{n-1} f_n \end{pmatrix} \in M_n(K'),$$

resp. $w_n(f_1, \ldots, f_n) := \det W_n(f_1, \ldots, f_n) \in K'$.

We sometimes omit the (redundant) index and write $W(f_1, \ldots, f_n)$ for $W_n(f_1, \ldots, f_n)$, resp. $w(f_1, \ldots, f_n)$ for $w_n(f_1, \ldots, f_n)$. The q-wronskian matrix can be usefully expanded in lines or columns as follows. For $f \in K'$, set:

$$X_f := \begin{pmatrix} f \\ \sigma_q f \\ \vdots \\ \sigma_q^{n-1} f \end{pmatrix} \in K'^n$$

(here viewed as a space of column vectors). Also set:

$$L := (f_1, \ldots, f_n) \in K'^n$$

(this time viewed as a space of line vectors). Then (see 0.4):

$$W_n(f_1, \ldots, f_n) = \begin{bmatrix} X_{f_1} & \cdots & X_{f_n} \end{bmatrix} = \begin{bmatrix} L \\ \sigma_q L \\ \vdots \\ \sigma_q^{n-1} L \end{bmatrix}.$$

EXERCISE 3.52. Write an explicit relation (involving a triangular matrix) between $W_n(f_1, \ldots, f_n)$ and $({\delta_q}^i f_j)_{\substack{0 \leq i \leq n-1 \\ 1 \leq j \leq n}} \in M_n(K')$. In particular, compare their determinants and ranks (the operator δ_q was defined in 3.1.1.4).

3.2.1.3. Let $P \in \mathcal{D}_{K,q}$, $P \neq 0$. We want information on the dimension on the space $\mathrm{Sol}(P, K')$ of solutions over the field $C' := K'^{\sigma_q}$. Relying either on basic examples (there are many in Chapter 4, but see the table in 3.2.1.4) or on our knowledge of linear ordinary differential equations, we expect something like this:

(1) For small q-difference extensions, e.g., $K' = K$, there are maybe not enough solutions and the dimension is low, maybe 0 (think of $\sigma_q f = cf$).
(2) Increasing K' should increase the number of independent solutions, whence the dimension.

(3) At best, for a solution field K' big enough, we expect that an equation of order n has n independent solutions.

We are going to confirm this views: the first two points in this chapter, the third one in Chapter 5. Note however a tricky aspect of the context: while, in the theory of ordinary differential equations, dimensions were computed over $\mathbf{C}$ (the field of complex numbers), here we compute them over a field $C' := K'^{\sigma_q}$ which varies.

We shall for the moment do most of the study over K; of course results can then be applied to an arbitrary q-difference extension K'. As a preliminary remark, note that, if $f_1, \ldots, f_k \in K$ are linearly dependent over $C := K^{\sigma_q}$, then so are the column vectors $X_{f_1}, \ldots, X_{f_k}$. Indeed, if $\sum \lambda_j f_j = 0$ and all $\lambda_j \in C$, then $\sigma_q^i \lambda_j = \lambda_j$ for all $i \leq 0$, whence: $\sum \lambda_j \sigma_q^i f_j = \sigma_q^i (\sum \lambda_j f_j) = 0$; this in turn implies that $\sum \lambda_j X_{f_j} = 0$.

LEMMA 3.53. *Let $f_1, \ldots, f_k \in Sol(P, K)$. Then $f_1, \ldots, f_k$ are C-linearly dependent if, and only if, $X_{f_1}, \ldots, X_{f_k}$ are K-linearly dependent.*

PROOF. Direct implication is immediate: if $f_1, \ldots, f_k$ are C-linearly dependent, then so are $X_{f_1}, \ldots, X_{f_k}$, which of course implies that they are K-linearly dependent.

Suppose conversely that $X_{f_1}, \ldots, X_{f_k}$ are K-linearly dependent. If they are all null, there is nothing to prove, so assume the contrary. Up to a reordering of the f_j, one can assume that for some $r \in \{1, \ldots, k-1\}$ the columns $X_{f_1}, \ldots, X_{f_r}$ are K-free and that, for every $j = r+1, \ldots, k$, setting for short $g := f_j$, one has a relation $X_g = \lambda_1 X_{f_1} + \cdots + \lambda_r X_{f_r}$, where $\lambda_1, \ldots, \lambda_r \in K$; this implies in particular that $g = \lambda_1 f_1 + \cdots + \lambda_r f_r$. We are going to prove that actually $\lambda_1, \ldots, \lambda_r \in C$, from which the conclusion flows.

The relation $X_g = \lambda_1 X_{f_1} + \cdots + \lambda_r X_{f_r}$ translates into n equations:

$$E_0 : g = \lambda_1 f_1 + \cdots + \lambda_r f_r,$$

$$\ldots$$

$$E_i : \sigma_q^i g = \lambda_1 \sigma_q^i f_1 + \cdots + \lambda_r \sigma_q^i f_r,$$

$$\ldots$$

$$E_{n-1} : \sigma_q^{n-1} g = \lambda_1 \sigma_q^{n-1} f_1 + \cdots + \lambda_r \sigma_q^{n-1} f_r.$$

Write $P = \sigma_q^n + a_1 \sigma_q^{n-1} + \cdots + a_n$. Since g and the f_i are solutions of $P(f_i) = 0$, the linear combination $(-a_n)E_0 + \cdots + (-a_1)E_{n-1}$ yields:

$$E_n : g = \lambda_1 \sigma_q^n f_1 + \cdots + \lambda_r \sigma_q^n f_r.$$

On the other hand, call E'_i the equation obtained while applying σ_q to E_{i-1}:

$$E'_1 : \sigma_q g = \sigma_q \lambda_1 \sigma_q f_1 + \cdots + \sigma_q \lambda_r \sigma_q f_r,$$

$$\ldots$$

$$E'_i : \sigma_q^i g = \sigma_q \lambda_1 \sigma_q^i f_1 + \cdots + \sigma_q \lambda_r \sigma_q^i f_r,$$

$$\ldots$$

$$E'_n : \sigma_q^n g = \sigma_q \lambda_1 \sigma_q^n f_1 + \cdots + \sigma_q \lambda_r \sigma_q^n f_r.$$

Now substracting $E_i' - E_i$ for $i = 1, \dots, n$ gives:

$$E_1' - E_1 : 0 = (\sigma_q\lambda_1 - \lambda_1)\sigma_q f_1 + \cdots + (\sigma_q\lambda_r - \lambda_r)\sigma_q f_r,$$

$$\dots$$

$$E_i' - E_i : 0 = (\sigma_q\lambda_1 - \lambda_1)\sigma_q^i f_1 + \cdots + (\sigma_q\lambda_r - \lambda_r)\sigma_q^i f_r,$$

$$\dots$$

$$E_n' - E_n : 0 = (\sigma_q\lambda_1 - \lambda_1)\sigma_q^n f_1 + \cdots + (\sigma_q\lambda_r - \lambda_r)\sigma_q^n f_r.$$

Altogether, this means that:

$$(\sigma_q\lambda_1 - \lambda_1)\sigma_q X_{f_1} + \cdots + (\sigma_q\lambda_r - \lambda_r)\sigma_q X_{f_r} = 0,$$

whence $\sigma_q\lambda_1 - \lambda_1 = \cdots = \sigma_q\lambda_r - \lambda_r = 0$ because the independence assumption on the X_{f_i} implies the same for the $\sigma_q X_{f_i}$; so indeed all the $\lambda_i \in C$. □

EXERCISE 3.54. Why does the independence assumption on the X_{f_i} implies the same for the $\sigma_q X_{f_i}$?

THEOREM 3.1 (The q-wronskian lemma). *Let* $C := K^{\sigma_q}$ *and* $P := \sigma_q^n + a_1\sigma_q^{n-1} + \cdots + a_n \in \mathcal{D}_{K,q}$.

(i) *Let* $f_1, \dots, f_n \in$ *Sol*(P, K). *Then:*

$$\dim_C \mathit{Vect}_C(f_1, \dots, f_n) = \mathit{rank}\, W_n(f_1, \dots, f_n).$$

(ii) *As a consequence:*

$$\dim_C \mathit{Sol}(P, K) \leq n.$$

PROOF. (i) Let $r := \mathrm{rank} W_n(f_1, \dots, f_n)$ the rank of the q-wronskian matrix. Up to reordering, we may assume that $(X_{f_1}, \dots, X_{f_r})$ is a basis of the space $\mathrm{Vect}_K(X_{f_1}, \dots, X_{f_n})$ spanned by the columns of $W_n(f_1, \dots, f_n)$. Then, by the previous lemma, $(f_1, \dots, f_r)$ is a basis of the space $\mathrm{Vect}_C(f_1, \dots, f_n)$ spanned by the f_j, $j = 1, \dots, n$.

(ii) It follows that no family in $\mathrm{Sol}(P, K)$ may have C-rank greater than n.

□

3.2.1.4. The main interest of the q-wronskian lemma is that it the rank criterion does not depend on the q-difference extension considered: the rank of $W_n(f_1, \dots, f_n)$ is the size of the biggest extracted square matrix with nonzero determinant, this calculation does not depend on the field (understood to contain K). So, in what follows, $P \in \mathcal{D}_{K,q}$ is a standard unitary q-difference operator of degree n, K' a q-difference extension of K and C, C' are the respective subfields of constants.

COROLLARY 3.55. $\dim_C \mathit{Sol}(P, K) \leq \dim_{C'} \mathit{Sol}(P, K') \leq n$.

PROOF. The bound $\dim_C \mathrm{Sol}(P, K) \leq n$ is immediate from the theorem and the bound $\dim_{C'} \mathrm{Sol}(P, K') \leq n$ is the same thing but applied to K'.

The majoration $\dim_C \mathrm{Sol}(P, K) \leq \dim_{C'} \mathrm{Sol}(P, K')$ comes from the fact that C-linear independence implies C'-linear independence thanks to the theorem, so a C-basis of $\mathrm{Sol}(P, K)$ is a C'-free family in $\mathrm{Sol}(P, K')$. □

Of course, if $K' \subset K''$ are successive q-difference extensions, then $\dim_{C'} \mathrm{Sol}(P, K') \leq \dim_{C''} \mathrm{Sol}(P, K'')$ all the same.

This leads to the *dilemna* (already illustrated in 2.3.1.5): if we want to have "a full complement of solutions", i.e., $\dim_{C'} \mathrm{Sol}(P, K') = n$, the maximum possible dimension, then we must take K' "big enough" and as a consequence the field of constants C' will be big too. In particular, we might have to compute invariants over something bigger than the complex numbers.

EXAMPLE 3.56. Let $P := \sigma_q - c \in \mathcal{D}_{\mathbf{C}(x),q}$. We tabulate $C := K^{\sigma_q}$ and $r := \dim_C \mathrm{Sol}(P, K)$ for various q-difference extensions K of $\mathbf{C}(x)$. Below is the table when $c \notin q^{\mathbf{Z}}$.

$K =$	$\mathbf{C}(x)$	$\mathbf{C}(\{x\})$	$\mathbf{C}((x))$	$\mathscr{M}(\mathbf{C})$	$\mathscr{M}(\mathbf{C}^*)$	$\mathscr{M}(\mathbf{C}^*, 0)$
$C =$	$\mathbf{C}$	$\mathbf{C}$	$\mathbf{C}$	$\mathbf{C}$	$\mathscr{M}(\mathbf{E}_q)$	$\mathscr{M}(\mathbf{E}_q)$
$r =$	0	0	0	0	1	1

EXERCISE 3.57. Do the same thing for $P := \sigma_q - c \in \mathcal{D}_{\mathbf{C}(x),q}$ when $c \in q^{\mathbf{Z}}$ and then for $P := \sigma_q - (1 + x) \in \mathcal{D}_{\mathbf{C}(x),q}$.

3.2.1.5. Suppose now $f_1, \dots, f_r \in K$ are C-linearly independent ($C := K^{\sigma_q}$ as usual). The set of all q-difference equations they all satisfy, $\{P \in \mathcal{D}_{K,q} \mid P(f_1) = \cdots = P(f_r) = 0\}$, is clearly a left ideal of $\mathcal{D}_{K,q}$. So it is either 0 or a principal ideal $\mathcal{D}_{K,q}P$ for a unique standard unitary P. By 3.2.1.4, such a P would have degree $\geq r$. Moreover, if the degree is r, then $W_r(f_1, \dots, f_r)$ must be invertible. Now consider, for a variable $f \in K$, the q-wronskian determinant:

$$
\begin{aligned}
w_{r+1}(f, f_1, \dots, f_r) &= \det \begin{pmatrix} f & f_1 & \dots & f_r \\ \sigma_q f & \sigma_q f_1 & \dots & \sigma_q f_r \\ \vdots & \vdots & \ddots & \vdots \\ \sigma_q^r f & \sigma_q^r f_1 & \dots & \sigma_q^r f_r \end{pmatrix} \\
&= (-1)^{r-1} w_r(f_1, \dots, f_r)(\sigma_q^r f + a_1 \sigma_q^{r-1} f + \cdots + a_r f),
\end{aligned}
$$

for some $a_1, \dots, a_r \in K$ (we used the fact that $w_r(f_1, \dots, f_r) \neq 0$ to factor it out). Clearly this vanishes if f is one of the f_i (because the determinant then has two identical columns), so setting:

$$
P(f) := \frac{w_{r+1}(f, f_1, \dots, f_r)}{(-1)^{r-1} w_r(f_1, \dots, f_r)},
$$

we define a standard unitary $P = \sigma_q^r + a_1 \sigma_q^{r-1} + \cdots + a_r \in \mathcal{D}_{K,q}$ which yields the minimal q-difference equation satisfied by $f_1, \dots, f_r$.

EXERCISE 3.58. Compute a_r.

3.2.2. Fundamental systems of solutions. As usual, we write $C := K^{\sigma_q}$, $C' := K'^{\sigma_q}$, etc.

3.2.2.1. Let $P := \sigma_q^n + a_1 \sigma_q^{n-1} + \cdots + a_n \in \mathcal{D}_{K,q}$ a standard unitary q-difference operator of degree n.

DEFINITION 3.59. A *fundamental system of solutions* of P over K is a family $(f_1, \dots, f_n)$ in $\mathrm{Sol}(P, K)$ such that $w_n(f_1, \dots, f_n) \neq 0$.

Immediate consequences are (using 3.2.1):

(1) A fundamental system of solutions $(f_1, \dots, f_n)$ is a C-basis of $\mathrm{Sol}(P, K)$.
(2) Let K' a q-difference extension of K; then $(f_1, \dots, f_n)$ is a fundamental system of solutions of P over K', whence a C'-basis of $\mathrm{Sol}(P, K')$.

(3) P is the minimal equation of $(f_1, \dots, f_n)$; in particular:

$$P(f) = (-1)^n \frac{w_{n+1}(f, f_1, \dots, f_n)}{w_n(f_1, \dots, f_n)}.$$

3.2.2.2.

PROPOSITION 3.60. *P admits a fundamental system of solutions over K if, and only if, $\dim_C Sol(P, K) = n$. In that case, every basis of $Sol(P, K)$ is a fundamental system of solutions.*

PROOF. The first statement is immediate using 3.2.1.

Let $(f_1, \dots, f_n)$ a fundamental system of solutions and $(g_1, \dots, g_n)$ another basis, so that $(g_1, \dots, g_n) = (f_1, \dots, f_n)M$ for some $M \in \mathrm{GL}_n(C)$. Since the coefficients of M are q-constants, $(\sigma_q^i g_1, \dots, \sigma_q^i g_n) = (\sigma_q^i f_1, \dots, \sigma_q^i f_n)M$ for $i = 0, \dots, n-1$, so $W_n(g_1, \dots, g_n) = W_n(f_1, \dots, f_n)M$, so $w_n(g_1, \dots, g_n) = w_n(f_1, \dots, f_n) \det M \neq 0$. □

3.2.2.3. To get an anticipated feeling of things to come in Chapter 7, note that, for every $f \in \mathrm{Sol}(P, K)$:

$$\sigma_q X_f = \begin{pmatrix} \sigma_q f \\ \sigma_q^2 f \\ \vdots \\ \sigma_q^n f \end{pmatrix} = \begin{pmatrix} 0 & 1 & 0 & \dots & 0 \\ 0 & 0 & 1 & \dots & 0 \\ \vdots & \vdots & \vdots & \ddots & \vdots \\ 0 & 0 & 0 & \dots & 1 \\ -a_n & -a_{n-1} & -a_{n-2} & \dots & -a_1 \end{pmatrix} \begin{pmatrix} f \\ \sigma_q f \\ \vdots \\ \sigma_q^{n-1} f \end{pmatrix} = A_P X_f$$

where $X_f := \begin{pmatrix} f \\ \sigma_q f \\ \vdots \\ \sigma_q^{n-1} f \end{pmatrix}$ and $A_P := \begin{pmatrix} 0 & 1 & 0 & \dots & 0 \\ 0 & 0 & 1 & \dots & 0 \\ \vdots & \vdots & \vdots & \ddots & \vdots \\ 0 & 0 & 0 & \dots & 1 \\ -a_n & -a_{n-1} & -a_{n-2} & \dots & -a_1 \end{pmatrix}$. Applying this successively to arbitrary elements $f_1, \dots, f_n \in K$ and remembering that $W_n(f_1, \dots, f_n)$ has columns $X_{f_1}, \dots, X_{f_n}$, we obtain:

$$\sigma_q W_n(f_1, \dots, f_n) = A_P W_n(f_1, \dots, f_n).$$

Taking determinants:

$$\sigma_q w_n(f_1, \dots, f_n) = (-1)^n a_n w_n(f_1, \dots, f_n).$$

(We use the fact that $\sigma_q \det M = \det \sigma_q M$ for every M, and also the fact that $\det A_P = (-1)^n a_n$.)

3.2.3. Equations with constant coefficients.

3.2.3.1. Whatever the field K in which we look for solutions, σ_q is a $\mathbf{C}$-linear automorphism and we can, to some extent, apply usual linear algebra. The only limitation comes from the fact that K is not finite dimensional over $\mathbf{C}$, but many classical results hold the same. In particular, let $F := X^n + a_1 X^{n-1} + \dots + a_n \in \mathbf{C}[X]$ a unitary polynomial such that moreover $a_n \neq 0$. Let F admit the decomposition:

$$F = (X - c_1)^{m_1} \cdots (X - c_k)^{m_k}, c_1, \dots, c_k \in \mathbf{C}^* \text{ pairwise distinct}, m_1, \dots, m_k \in \mathbf{N}^*.$$

Then for any **C**-linear endomorphism ϕ of K:

$$\mathrm{Ker} F(\phi) = \bigoplus_{i=1}^{k} \mathrm{Ker}(\phi - c_i \mathrm{Id}_K)^{m_i}.$$

Applying this to $\phi := \sigma_q$, and writing $P := F(\sigma_q) = \sigma_q^n + a_1 \sigma_q^{n-1} + \cdots + a_n \in \mathcal{D}_{K,q}$ and $P_i := (\sigma_q - c_i)^{m_i} \in \mathcal{D}_{K,q}$, $i = 1, \ldots, k$, we get:

$$\mathrm{Sol}(P, K) = \bigoplus_{i=1}^{k} \mathrm{Sol}(P_i, K).$$

Note that this is actually a direct sum of K^{σ_q}-spaces.

3.2.3.2. We are therefore led to look for solutions of $(\sigma_q - c)^m f = 0$, where $c \in \mathbf{C}^*$ and $m \in \mathbf{N}^*$.

EXERCISE 3.61. Show that, if $c \notin q^{\mathbf{Z}}$, there are no power series solutions.

We will look for solutions in the form $f = e_c g$, where $\sigma_q e_c = c e_c$, $e_c \neq 0$. We can take $e_c := e_{q,c}$ (Chapter 2) but the calculation that follows does not really depend on that particular choice of q-character. We begin by using a trick:

$$(\sigma_q - c)^m (e_c g) = 0 \Longleftrightarrow \left(e_c^{-1} (\sigma_q - c)^m e_c\right) g = 0,$$

while the conjugate (remember, we discussed conjugacy at the end of 3.1.3.4) can be computed as follows:

$$e_c^{-1} (\sigma_q - c)^m e_c = \left(e_c^{-1} (\sigma_q - c) e_c\right)^m = c^m (\sigma_q - 1)^m.$$

EXERCISE 3.62. Check this by computing the conjugate $e_c^{-1} (\sigma_q - c) e_c$.

It follows from the preceding argument that, if $e_c \in K$:

$$\mathrm{Sol}((\sigma_q - c)^m, K) = e_c \mathrm{Sol}((\sigma_q - 1)^m, K).$$

Let us denote $E_m := \mathrm{Sol}((\sigma_q - 1)^m, K)$. Combining with the previous result:

$$\mathrm{Sol}(P, K) = \bigoplus_{i=1}^{k} e_{c_i} E_{m_i}.$$

3.2.3.3. We wish to find m independent elements $g_0, \ldots, g_{m-1} \in E_m$. Since the inclusions $E_1 \subset \cdots \subset E_m$ are strict, it is enough to construct a basis adapted to this "flag", i.e., each E_i has basis $(g_0, \ldots, g_{i-1})$, $i = 1, \ldots, m$, which in turn is equivalent to: $g_0 \in E_1 \setminus \{0\}$, $g_1 \in E_2 \setminus E_1$, $\ldots$, $g_{m-1} \in E_m \setminus E_{m-1}$. One way to do this (for motivation, you can look at the theory of reduction of nilpotent endomorphisms in linear algebra) is to solve iteratively $(\sigma_q - 1) g_i = g_{i-1}$.

Of course $E_1 = K^{\sigma_q}$, so we take $g_0 = 1$ (the simplest possible). Then we take for g_1 an element $\ell \in K$ such that $\sigma_q \ell - \ell = 1$. So we have to admit that there is such an element. It can be, for instance, the q-logarithm $\ell_q = -x {\theta_q}'/\theta_q$ (Chapter 2). Then we set:

$$\ell^{(n)} := \binom{\ell}{n} = \frac{1}{n!} \ell(\ell - 1) \cdots (\ell - n + 1).$$

By convention $\ell^{(0)} = 1$ and of course $\ell^{(1)} = \ell$. All these elements belong to K since ℓ is assumed to do so.

EXERCISE 3.63. Check that $\sigma_q \ell^{(n)} = \ell^{(n)} + \ell^{(n-1)}$.

So we see that we can take $(\ell^{(0)}, \ldots, \ell^{(m-1)})$ as a basis of E_m.

3.2.3.4. To summarize, assuming that ℓ and the e_{c_i} are in K, $i = 1, \ldots, k$, we found for $\mathrm{Sol}(P, K)$ the basis made of all $e_{c_i}\ell^{(j)}$, $i = 1, \ldots, k$, $j = 0, \ldots, m_i - 1$. This basis has $m_1 + \cdots + m_k = n$ elements, so it is actually a fundamental system of solutions.

EXERCISE 3.64. Let $\mathcal{B}$ be this basis ordered in such a way that $e_{c_1}E_1$ comes first, then $e_{c_2}E_2$, etc. Show that $\sigma_q\mathcal{B} = \mathcal{B}M$, where the matrix M has k blocks of the form $c_iJ_{m_i}$, where $J_m := I_m + N_m$, N_m the square $m \times m$ matrix with 1 just above the diagonal and 0 elesewhere (thus a Jordan block).

3.2.4. q-holonomic functions. There are various approaches to holonomy, we here look at a particular q-adaptation based only on finite dimensionality conditions (about spaces of solutions). This is intended to be elementary and ad hoc and seems to be sufficient for our needs (a single variable, equations over fields rather than rings). For a more ambitious framework and a thorough study (involving the structure of the algebra of q-difference operators); see the report by Sabbah **[214]** (itself inspired to some extent by Aomoto's approach **[10]**). What is studied here is what Sabbah calls a *rational holonomic system of q-difference equations.*

3.2.4.1. Let K a q-difference field and $K' \supset K$ a q-difference extension. For $f \in K'$, we shall denote (in 3.2.4 only):

$$V_f := \mathrm{Vect}_K\left(\{\sigma_q^k f \mid k \in \mathbf{Z}\}\right).$$

Obviously V_f is a K-vector subspace of K' such that $\sigma_q V_f = V_f$.

EXERCISE 3.65. (i) As a preliminary remark to all that follows, note that, if $V \subset K'$ is a finite dimensional K-vector subspace, so is $\sigma_q V$ (you are invited to check that). As a consequence, $\sigma_q V = V$ if, and only if, $\sigma_q V \subset V$.

(ii) Redo all that follows using the alternative definition $\mathrm{Vect}_K\left(\{\sigma_q^k f \mid k \in \mathbf{N}\}\right)$ for V_f instead of the former one.

PROPOSITION 3.66. (i) *The following three properties are equivalent:*

(a) $\dim_K V_f < \infty$.

(b) *There is a sub K-vector space $V \subset K'$ such that $f \in V$, $\dim_K V < \infty$ and $\sigma_q V \subset V$ (equivalently $\sigma_q V = V$ by the exercise above).*

(c) *There exists a q-difference operator $L \in \mathcal{D}_{K,q}$, $L \neq 0$, such that $Lf = 0$.*

(ii) *If those properties hold, $\dim_K V_f$ is the minimal degree of such a q-difference operator.*

PROOF. (i) Assuming (a), (b) follows since V_f satisfies all three stated properties. Assuming (b), the family of all $\sigma_q^n f$, $n \in \mathbf{Z}$, is included in V, so it is not free, so there is a nontrivial linear relation $\sum a_i \sigma_q^{n_i} f = 0$ (the $a_i \in K$ not all 0, the $n_i \in \mathbf{Z}$ pairwise distinct). Then $L := \sum a_i S^{n_i} \in \mathcal{D}_{K,q}$, $L \neq 0$ and $Lf = 0$; so (c) follows.

Now assume (c). If necessary dividing L at left by some invertible in $\mathcal{D}_{K,q}$, one can assume that it is standard unitary: $L = S^n + a_1 S^{n-1} + \cdots + a_n$, all $a_i \in K$, $a_n \neq 0$. We shall show that V_f is spanned by the $\sigma_q^i f$, $i = 0, \ldots, n-1$. Let V', the subspace generated by those n elements. Equation $Lf = 0$, i.e., $\sigma_q^n f = -(a_n f + \cdots + a_1 \sigma_q^{n-1} f)$ directly implies that $\sigma_q^n f \in V'$. Applying σ_q^{p-n} at left shows that every $\sigma_q^p f$ lies in the span of $\sigma_q^{p-1} f, \ldots, \sigma_q^{p-n+1} f$; so, by induction on $p \geq n$, it shows that all $\sigma_q^p \in V'$ for $p \geq 0$. Now we also have $f = (-1/a_n)(\sigma_q^n f + a_1 \sigma_q^{n-1} f + \cdots + a_{n-1}\sigma_q f)$. Applying σ_q^{-p} shows that each $\sigma_q^{-p} f$ lies in the span of $\sigma_q^{-p+1}f, \ldots, \sigma_q^{-p+n}f$, so, again by induction, all $\sigma_q^{-p}f$, $p \geq 1$, belong to V'. Therefore $V_f \subset V'$, i.e., $V_f = V'$, since the inverse inclusion is obvious. Therefore $\dim_K V_f \leq n$ and (a) follows.

(ii) From the proof that (c) implies (a) we draw that $\dim_K V_f \leq \deg L$ for any nonzero L such that $Lf = 0$. On the other hand, letting $n := \dim_K V_f$, the family $f, \sigma_q f, \ldots, \sigma_q^n f$ is not free, so there is a nonzero L of degree $\leq n$ such that $Lf = 0$. □

COROLLARY 3.67. *Let $n := \dim V_f$. Then there is a unique standard unitary q-difference operator $L \in \mathcal{D}_{K,q}$ of degree n such that $Lf = 0$.*

PROOF. By (ii) of the proposition, there is such a q-difference operator, say L. For any other q-difference operator P such that $Pf = 0$, the euclidean division $P = QL + R$ would imply $Rf = 0$, so by minimality of the degree of L, we would have $R = 0$, so $P = QL$; if P is standard unitary of degree n, this implies $P = L$. □

DEFINITION 3.68. We then say that f is *q-holonomic* (or simply *holonomic*) w.r.t. K, with *degree* $\dim_K V_f$. We write $\mathrm{Hol}_K(K')$ the set of elements of K' which are holonomic w.r.t. K.

EXAMPLE 3.69. If $\sigma_q f = uf$, $u \in K^*$, then f is holonomic of degree 1 (except of course if $f = 0$). This includes in particular all elements of K^* and of K'^{σ_q} (except for 0).

3.2.4.2.

LEMMA 3.70. *Let V, V' K-subspaces of K'.*

(i) *If V, V' are finite dimensional, then so are $V + V'$ and VV'. (The latter is the K-span of $\{vv' \mid v \in V, v' \in V'\}$.) Moreover $\dim_K(V+V') \leq \dim_K V + \dim_K V'$ and $\dim_K(VV') \leq \dim_K V \dim_K V'$.*

(ii) *If V, V' are σ_q-stable, then so are $V + V'$ and VV'.*

PROOF. (i) Let (e_i), resp. (e'_j) a basis of V, resp. V'. Then the e_i and e'_j together generate $V + V'$; and the $e_i e'_j$ together generate VV'.

(ii) is immediate. □

PROPOSITION 3.71. *Holonomic elements form a subring of K' containing K and K'^{σ_q}.*

PROOF. It is enough to prove that, if $f, g \in K'$ are holonomic w.r.t. K, then so are $f + g$ and fg. But $f + g$ belongs to $V_f + V_g$ and fg to $V_f V_g$, and those spaces are finite dimensional and σ_q-stable by the lemma. □

3.2.4.3. We shall now see that the inverse of a holonomic element needs not be holonomic. We take $K := \mathbf{C}(\{x\})$ and $K' := \mathscr{M}(\mathbf{C}^*, 0)$. It was proved in 2.3.2 that the element $l_q := -x\theta_q'/\theta_q$ of K' is transcendental over K. It satisfies $\sigma_q l_q = l_q + 1$, whence $\sigma_q^2 l_q - 2\sigma_q l_q + l_q = 0$, so it is holonomic of degree at most 2. But since $\sigma_q l_q / l_q = 1 + 1/l_q \notin K$, it has degree exactly 2.

Since l_q is transcendental, a relation $\sum a_i \sigma_q^i(1/l_q) = 0$, i.e., $\sum a_i \dfrac{1}{l_q + i} = 0$ (all $a_i \in K$) is equivalent to a relation $\sum a_i \dfrac{1}{X + i} = 0$ in the field $K(X)$ of rational fractions. It is an easy exercise in algebra that this implies that all $a_i = 0$. So $1/l_q$ is not holonomic.

3.3. A bit of q-calculus

Powers of q naturally appear in q-difference theory. The main reason is arguably the fact that they are the eigenvalues of σ_q acting on the fields of interest[8].

However, when we emphasize q-analogies, especially degeneracies (when $q \to 1$) to ordinary differential equations and to classical special functions, one is naturally led to use the q-derivation $\delta_q := \dfrac{1-\sigma_q}{1-q}$ and this has as eigenvalues the so-called "q-numbers" (actually, "q-integers") $\dfrac{1-q^n}{1-q}$.

Other special numbers in q-calculus appear in the work of Euler on partitions. The main one are connected to so-called "q-Pochhammer symbols", to be defined soon. The universe of formulas involving all these objects, as well as some important "q-special functions", is the domain called *q-calculus* (sometimes *q-analysis*). It is a huge domain and, in this book, we barely touch it, centering on what is of direct interest for q-difference equations. (See however the references given at the beginning of this chapter.)

3.3.1. The q-Pochhammer symbol.

This is directly related to *Partitio Numerorum* (1.2) and to basic hypergeometric series (1.6) and 4.4. Recall that, in this book, $0 < |q| < 1$ (except at some exceptional places).

3.3.1.1.

DEFINITION 3.72. Let $a \in \mathbf{C}$ and $n \in \mathbf{N}$. Then:

$$(a;q)_n := \prod_{i=0}^{n-1}(1-aq^i) \text{ and } (a;q)_\infty := \prod_{n\geq 0}(1-aq^n).$$

We shall sometimes also use the condensed notation:

$$(a_1,\ldots,a_r;q) := (a_1;q)\cdots(a_r;q).$$

Thus $(a;q)_0 = 1$ and $(a;q)_{n+1} = (1-aq^n)(a;q)_n$. The infinite product $(a;q)_\infty$ converges because, since $0 < |q| < 1$, the series $\sum\limits_{n\geq 0} aq^n$ is absolutely convergent. This also implies that $(a;q)_\infty$ if, and only if, one of the factors vanishes, i.e., if $a \in q^{-\mathbf{N}}$. Moreover:

$$(a;q)_\infty = \prod_{i=0}^{n-1}(1-aq^i)\prod_{i\geq n}(1-aq^i) = (a;q)_n(aq^n;q)_\infty, \Longrightarrow (a;q)_n = \frac{(a;q)_\infty}{(aq^n;q)_\infty},$$

provided the denominator does not vanish. This allows for a more general definition of the q-Pochhammer symbol:

$$\forall \alpha \in \mathbf{C}\,,\ (a;q)_\alpha := \frac{(a;q)_\infty}{(aq^\alpha;q)_\infty}.$$

(Recall from the beginning of this chapter that $q^\alpha = e^{2i\pi\tau\alpha}$.)

[8] We are here speaking of the *fields of coefficients* $\mathbf{C}(x)$, $\mathbf{C}(\{x\})$, $\mathbf{C}((x))$: in the q-difference extensions $\mathscr{M}(\mathbf{C}^*)$ and $\mathscr{M}(\mathbf{C}^*,0)$ where we look for solutions, all $c \in \mathbf{C}^*$ are eigenvalues.

EXAMPLE 3.73. For $\alpha \in \mathbf{N}$, this definition does coincide with the previous one. For $\alpha = -n$, $n \in \mathbf{N}^*$, setting $b := aq^\alpha = aq^{-n}$, we find:

$$(a;q)_{-n} = (a;q)_\alpha = \frac{(a;q)_\infty}{(aq^\alpha;q)_\infty} = \frac{(bq^n;q)_\infty}{(b;q)_\infty} = \frac{1}{(b;q)_n} = \frac{1}{(aq^{-n};q)_n}.$$

Note that extending the notation to allow for a basis with modulus > 1, one also has:

$$(a;q)_{-n} = \frac{1}{(aq^{-1};q^{-1})_n}.$$

EXERCISE 3.74. (i) Check all the formulas above.
(ii) Prove the "distribution formula" $(a^k;q^k)_n = \prod_{\omega \in \mu_k} (a\omega;q)_n$.

In [**133**], Kac and Cheung introduce the following "suggestive notations" for the "q-twisted powers":

$$(x+a)_q^n := (x+a)\cdots(x+q^{n-1}a) \text{ and } (x-a)_q^n := (x-a)\cdots(x-q^{n-1}a).$$

We shall not use those notations, but the reader is encouraged to state and prove for them formulas of the same kind as those given here for the q-Pochhammer symbols.

EXERCISE 3.75. For $n \in \mathbf{N}$ check that $(a;q)_{-n} = (-1)^n q^{-n(n+1)/2}(a-q)_q^n$.

3.3.1.2. *Function theoretic aspects of the q-Pochhammer symbols.*

PROPOSITION 3.76. *The function $x \mapsto (x;q)_\infty$ is an entire function with simple zeroes on $q^{-\mathbf{N}}$ (and nowhere else). It is the only solution in $\mathbf{C}[[x]]$ of the q-difference equation $\sigma_q f = \dfrac{1}{1-x} f$ with initial condition $f(0) = 1$. Its power series expansion is:*

$$(x;q)_\infty = \sum_{n\geq 0} (-1)^n \frac{q^{n(n-1)/2}}{(q;q)_n} x^n.$$

PROOF. By [**211**, Theorem 15.4], on every compact subset of $\mathbf{C}$, the infinite product $\prod(1+u_n(x))$, where $u_n(x) := -q^n x$, converges uniformly since $\sum u_n(x)$ converges normally. Therefore it defines an entire function (because each factor is holomorphic). The statement on the zeroes also follows from convergence.

This function is obviously a solution in $\mathbf{C}[[x]]$ of $\sigma_q f = \dfrac{1}{1-x} f$ with initial condition $f(0) = 1$ and any other $g \in \mathbf{C}((x))$ with the same constraints will be such that $g/f \in \mathbf{C}(x)^{\sigma_q} = \mathbf{C}$ and $(g/f)(0) = 1$, whence $g/f = 1$, whence the uniqueness.

Writing $f(x) = \sum_{n\geq 0} a_n x^n$ the power series expansion, we have $a_0 = 1$ and, by the identification $(1-x)f(qx) = f(x)$, the recursive relation:

$$\forall n \geq 1 \,,\; q^n a_n - q^{n-1} a_{n-1} = a_n \Longrightarrow a_n = -\frac{q^{n-1}}{1-q^n} a_{n-1}.$$

Easy induction (or telescoping $a_n/a_0 = a_n/a_{n-1} \times \cdots \times a_1/a_0$) then yields $a_n = (-1)^n \dfrac{q^{n(n-1)/2}}{(q;q)_n}$. □

COROLLARY 3.77. *The function* $x \mapsto \dfrac{1}{(x;q)_\infty}$ *is meromorphic over* $\mathbf{C}$, *with simple poles on* $q^{-\mathbf{N}}$ *(and nowhere else). It is the only solution in* $\mathbf{C}((x))$ *of the* q*-difference equation* $\sigma_q f = (1-x)f$ *with initial condition* $f(0) = 1$. *Its power series expansion is:*

$$\frac{1}{(x;q)_\infty} = \sum_{n\geq 0} \frac{1}{(q;q)_n} x^n.$$

PROOF. Everything follows from the proposition, except the power series expansion. Write it $\sum\limits_{n\geq 0} b_n x^n$; then from the functional equation we draw the recursive relation:

$$q^n b_n = b_n - b_{n-1} \Longrightarrow b_n = \frac{1}{1-q^n} b_{n-1},$$

and the conclusion is immediate. □

COROLLARY 3.78. *The following holds for* $|x| < 1$:

$$\frac{1}{(x;q)_\infty} = \exp \sum_{p\geq 1} \frac{x^p}{p(1-q^p)}.$$

PROOF. Let $h(x)$ the series inside the exponential; then $h(0) = 0$ and $h(x) - h(qx) = \sum\limits_{p\geq 1} \dfrac{x^p}{p}$, the well known power series expansion of $-\log(1-x)$. Therefore $H := \exp h$ satisfies the q-difference equation considered in the previous corollary. □

Note that a more direct but less rigorous proof would be as follows:

$$\log \frac{1}{(x;q)_\infty} = \sum_{n\geq 0} \log \frac{1}{1-q^n x} = \sum_{n\geq 0} \sum_{p\geq 1} \frac{(q^n x)^p}{p},$$

and interversion of the summations quickly leads to $\sum_{p\geq 1} \dfrac{x^p}{p(1-q^p)}$.

3.3.1.3. *Degeneracy ("confluence") of the q-Pochhammer symbols.* Two interesting variants of the previous functions are the following:

$$f_q(x) := (-(1-q)x;q)_\infty = \sum_{n\geq 0} \frac{q^{n(n-1)/2}}{[n]_q!} x^n,$$

$$g_q(x) := \frac{1}{((1-q)x;q)_\infty} = \sum_{n\geq 0} \frac{1}{[n]_q!} x^n,$$

where we have (by anticipation on 3.3.2) introduced the notation:

$$[n]_q! := \frac{(q;q)_n}{(1-q)^n} = \prod_{i-1}^{n} \frac{1-q^i}{1-q}.$$

Note that when $q \to 1$, the "q-factorial" $[n]_q!$ tends to $n!$ and both the above series converge coefficientwise to $\sum x^n/n! = \exp x$. In this sense we can say that they are "q-deformations" or "q-analogues" of the exponential function.

Another way to see it is the following. These functions repectively satisfy the q-difference equations:

$$\begin{aligned}\sigma_q f_q &= \frac{1}{1+(1-q)x} f_q \Longrightarrow D_q f_q = \frac{1}{1+(1-q)x} f_q,\\ \sigma_q g_q &= (1-(1-q)x) g_q \Longrightarrow D_q g_q = g_q,\end{aligned}$$

where we have used the q-derivation $D_q := \frac{1-\sigma_q}{(1-q)x}$. Since D_q formally degenerates into $D := \frac{d}{dx}$ when $q \to 1$, the q-difference equations above formally degenerate respectively to $f' = f$ and to $g' = g$, giving yet another meaning to the idea of q-deformation of the exponential. Actually, this and the results of section 11.2 together imply what we are now going to prove directly.

THEOREM 3.2. *When $q \to 1$ while staying in a domain where $|1-q| = O(1-|q|)$ (and of course $0 < |q| < 1$), $f_q \to \exp$ and $g_q \to \exp$.*

PROOF. Due to the relation $f_q(-x)g_q(x) = 1$, it is enough to do the proof for g_q. Recall from Corollary 3.78 that $g_q = \exp h_q$, where:

$$h_q(x) := \sum_{p\geq 1} \frac{(1-q)^p x^p}{p(1-q^p)}.$$

So we just have to prove that $h_q \to x$. This is obvious termwise, since the first term is x and all the terms for $p \geq 2$ have coefficient multiple of $1-q$. To make this more precise, we write $h_q(x) = x + \psi_q(x)$, where:

$$\psi_q(x) := \sum_{n\geq 2} \frac{(1-q)^n x^n}{p(1-q^n)} = \sum_{n\geq 2} (1-q)^{n-1} \frac{x^n}{n[n]_q},$$

where (again by anticipation on 3.3.2) we write $[n]_q := \frac{1-q^n}{1-q}$, so that $[n]_q \to n$ when $q \to 1$. Now, for any $n \geq 2$:

$$|1-q^n| \geq 1 - |q^n| = 1 - |q|^n \geq 1 - |q|.$$

On a domain in which $|1-q| \leq M(1-|q|)$ for some fixed $M > 0$ (that is the meaning of our assumption in the statement of the theorem), one therefore has:

$$\left|\frac{1}{n[n]_q}\right| = \frac{1}{n}\frac{|1-q|}{|1-q^n|} \leq \frac{|1-q|}{1-|q|} \leq M.$$

Therefore $|\psi_q(x)|$ is dominated by the series:

$$\sum_{n\geq 2} |1-q|^{n-1} M |x|^n = M|x| \sum_{n\geq 1} |(1-q)x|^n = M|x| \frac{|(1-q)x|}{1-|(1-q)x|}.$$

The last equality is valid on an arbitrary disk $\overset{\circ}{\mathrm{D}}(0,R)$ as soon as $|1-q| < 1/R$, so we have dominated convergence on any such disk and the conclusion follows. □

EXERCISE 3.79. Let $\phi_0 \in [0, \pi/2[$. Prove that if q stays in the sector:

$$\{1 - re^{\mathrm{i}\phi} \mid r > 0 \text{ and } -\phi_0 \leq \phi \leq \phi_0\}$$

while $q \to 1$, then $|1-q| = O(1-|q|)$. You should also draw this sector to visualize the scope of the proposition.

3.3.1.4. *More on degeneracy.* Another way to look at the behaviour of $(x;q)_\infty$ when $q \to 1$ is the following.

PROPOSITION 3.80. *Let $\alpha \in \mathbf{C}$ and $x \in \overset{\circ}{D}(0,1)$. Then:*

$$\lim_{q \to 1} \frac{(q^\alpha x;q)_\infty}{(x;q)_\infty} = (1-x)^{-\alpha}.$$

PROOF. By Cauchy's binomial theorem (see section 1.4):

$$\frac{(q^\alpha x;q)_\infty}{(x;q)_\infty} = \sum_{n \geq 0} \frac{(q^\alpha;q)_n}{(q;q)_n} x^n.$$

Each individual coefficient $\dfrac{(q^\alpha;q)_n}{(q;q)_n} \prod_{i=0}^{n-1} \dfrac{1-q^{\alpha+i}}{1-q^{1+i}}$ tends to $\prod_{i=0}^{n-1} \dfrac{\alpha+i}{1+i} =: \dfrac{(\alpha)_n}{n!}$ when $q \to 1$. (Recall that $(\alpha)_n := \alpha(\alpha+1)\cdots(\alpha+n-1)$ is the classical Pochhammer symbol). So, using the generalized Newton binomial theorem:

$$(1-x)^{-\alpha} = \sum_{n \geq 0} \binom{-\alpha}{n} (-x)^n = \sum_{n \geq 0} \frac{(\alpha)_n}{n!} x^n,$$

(because $\binom{-\alpha}{n} = (-1)^n \alpha)_n)$, we see that indeed $\dfrac{(q^\alpha x;q)_\infty}{(x;q)_\infty}$ tends to $(1-x)^{-\alpha}$ termwise; and we are again going to prove that the convergence is dominated.

Note first that, if the formula is proved for a given α, it is also proved for $-\alpha$: just replace x by $q^{-\alpha}x$ and then take inverses. Also, if proved for α, it is proved for $\alpha+1$ because $\dfrac{(q^{\alpha+1}x;q)_\infty}{(q^\alpha x;q)_\infty} = (1-q^\alpha x)^{-1}$ obviously tends to $(1-x)^{-1}$. So we may (and shall) assume that $\alpha = a + ib$ with $a > 0$ and $b \geq 0$.

Under those assumptions, we note that, if $|u| < 1$, by the (complex) mean value inequality (applied to the function $v \mapsto v^{1+\alpha/n}$ between 1 and u):

$$\frac{\left|u^{1+\alpha/n} - 1\right|}{|u-1|} \leq \sup_{|v| \leq 1} |1+\alpha/n| \left|v^{\alpha/n}\right| \leq |1+\alpha/n| \sup_{|v| \leq 1} |v|^{a/n} e^{(b/n)\arg v},$$

where the argument is taken so that $|\arg v| \leq \pi$. We apply that bound to $u := q^k$, $b = 1, \ldots, n$, and get:

$$\left|\frac{(q^\alpha;q)_n}{(q;q)_n}\right| \leq \prod_{k=1}^{n} |1+\alpha/k|\, e^{\pi b/k} \leq \prod_{k=1}^{n} (1+|\alpha/k|)\, e^{\pi b H_n} \leq \exp(|\alpha| + \pi b) H_n,$$

where $H_n := 1 + \cdots + 1/n = \log n + \gamma + o(1)$. Therefore $\left|\dfrac{(q^\alpha;q)_n}{(q;q)_n}\right| = O(n^d)$ for some $d > 0$, *uniformly in q*. This means that the series at stake is dominated by $\sum n^d x^n$, i.e., we have dominated convergence in every $\overline{\mathrm{D}}(0,R)$ with $R < 1$ and the conclusion follows. □

The behaviour of $\dfrac{(q^\alpha x;q)_\infty}{(x;q)_\infty}$ out of $\overset{\circ}{\mathrm{D}}(0,1)$ when $q \to 1$ depends heavily on the path followed by q. This is of course to be guessed since there is no unique determination of $(1-x)^{-\alpha}$, one must at least choose a cut from 0 to ∞. Another reason is that the zeroes and poles of $\dfrac{(q^\alpha x;q)_\infty}{(x;q)_\infty}$, respectivily lying on the discrete logarithmic half spirals $q^{-\mathbf{N}-\alpha}$ and $q^{-\mathbf{N}}$, may behave wildly. One way of enforcing a smooth behaviour is to fix some $q_0 = e^{2i\pi\tau_0}$ and then to take $q := q_0^\epsilon = e^{2i\pi\tau_0\epsilon}$, $\epsilon > 0$ tending to 0. Then the discrete half spirals "condense" into the continuous logarithmic half spiral $q_0^{-\mathbf{R}_+}$, which provides a natural cut from 0 to ∞ and a natural determination of $(1-x)^{-\alpha}$. See section 11.2 for details.

3.3.1.5. *Combinatorial properties of the q-Pochhammer symbols.* They come in two guises. The first one is related to *Partitio Numerorum.* As seen in 1.2, the power series expansion:

$$\frac{1}{(q;q)_\infty} =: \sum_{n\geq 0} p(n)q^n$$

provides the partition numbers $p(n)$. More generally, for every $k \geq 0$:

$$\frac{1}{(q;q)_k} =: \sum_{n\geq 0} p_k(n)q^n,$$

where coefficient $p_k(n)$ has three distinct equivalent meanings:

- number of solutions in $\mathbf{N}^k$ of $x_1 + 2x_2 + \cdots + kx_k = n$;
- number of partitions of n in parts $\leq k$;
- number of partitions of n in k parts.

Note that in the third interpretation, parts with value 0 are allowed, although they are not allowed in the general definition of the number of partitions. If we insist on sticking to the usual convention, we just add 1 to each part and get:

- number of partitions of $n+k$ in k (nonzero) parts.

For a tour in the universe of partitions, see **[112]** (for a pleasant elementary initiation) or **[8]** for a true immersion.

3.3.1.6. The second combinatorial interpretation of q-Pochhammer symbol is even more out of the scope of this book, or so it seems; in flavour if not in difficulty. Here *we totally shift our basic assumption* and take for q a "primary" number: $q := p^d$, p a prime number, $d \geq 1$ (so $q > 1$). Thus there exists a (unique up to isomorphism) finite field $\mathbf{F}_q$ having $|\mathbf{F}_q| = q$ elements (a "Galois field" in the Anglo-Saxon world). Then a $\mathbf{F}_q$-linear space of dimension n has q^n elements.

LEMMA 3.81. *The number of free families $(x_1, \ldots, x_r)$ in $\mathbf{F}_q^n$ is:*

$$(-1)^r q^{r(r-1)/2}(q^{n-r+1};q)_r = (-1)^r q^{r(r-1)/2}\frac{(q;q)_n}{(q;q)_{n-r}}.$$

PROOF. There are $q^n - 1$ choices for $x_1 \in \mathbf{F}_q^n \setminus \{0\}$; then $q^n - q$ choices for $x_2 \in \mathbf{F}_q^n \setminus \mathbf{F}_q x_1$; then $q^n - q^2$ choices for $x_3 \in \mathbf{F}_q^n \setminus (\mathbf{F}_q x_1 + \mathbf{F}_q x_2)$ (this is because, by construction, x_1 and x_2 are independent, so $\mathbf{F}_q x_1 + \mathbf{F}_q x_2$ has dimension 2, so it has q^2 elements). Iterating, we see that the total number of choices is:

$$(q^n-1)(q^n-q)\cdots(q^n-q^{r-1}) = 1\times q\times\cdots\times q^{r-1}\times(q^n-1)(q^{n-1}-1)\cdots(q^{n-r+1}-1).$$

Transformation into either of the stated formulas is then easy. □

EXERCISE 3.82. Convert these formulas into formulas involving symbols $(a;q^{-1})_k$.

THEOREM 3.3. (i) *The cardinal of the linear group $GL_n(\mathbf{F}_q)$ is:*

$$|GL_n(\mathbf{F}_q)| = \prod_{i=0}^{n}(q^n - q^i) = (-1)^n q^{n(n-1)/2}(q;q)_n = q^{n^2}(q^{-1};q^{-1})_n.$$

(ii) *The number of subspaces of dimension r in $\mathbf{F}_q^n$ is:*

$$\begin{bmatrix} n \\ r \end{bmatrix}_q := \frac{(q;q)_n}{(q;q)_r (q;q)_{n-r}}.$$

PROOF. (i) An invertible $n \times n$ matrix over $\mathbf{F}_q$ is made of n independent columns in $\mathbf{F}_q^n$, so we apply the lemma with $r = n$.

(ii) To a free family of r elements in $\mathbf{F}_q^n$, associate the subspace of $\mathbf{F}_q^n$ it spans; this is surjective onto the set of subspaces of dimension r. Two families yield the same subspace if, and only if, they are brought into each other by $\mathrm{GL}_r(\mathbf{F}_q)$. So combining the lemma with (i) (applied with n replaced by r) we find:

$$\frac{1}{(-1)^r q^{r(r-1)/2} (q;q)_r} \times (-1)^r q^{r(r-1)/2} \frac{(q;q)_n}{(q;q)_{n-r}},$$

which gives the stated formula. The notation $\begin{bmatrix} n \\ r \end{bmatrix}_q$ is commented below. □

Anticipating 3.3.3, the q-binomial number $\begin{bmatrix} n \\ r \end{bmatrix}_q$ will be defined as $\dfrac{[n]_q!}{[r]_q![n-r]_q!}$ (q-factorials were already presented in 3.3.1.3). But clearly $[n]_q! = (q;q)_n/(1-q)^n$ and $\begin{bmatrix} n \\ r \end{bmatrix}_q$ simplifies to $\dfrac{(q;q)_n}{(q;q)_r (q;q)_{n-r}}$ as wanted.

We shall see in 3.3.3 that the q-binomial number (also called q-binomial coefficient) $\begin{bmatrix} n \\ r \end{bmatrix}_q$ is somehow a q-analogue of the usual binomial number $\dbinom{n}{r}$. Thus, *the number of subspaces of dimension r in $\mathbf{F}_q^n$ is somehow a q-analogue of the number of subsets of cardinal r in* $\{1, \dots, n\}$; a sobering thought ...This, and other strange facts (more sophisticated though) lead to even stranger speculations about "the field with 1 element" (see Wikipedia for references).

EXERCISE 3.83. Calculate $|\mathrm{SL}_n(\mathbf{F}_q)|$. Also see exercises in [**145**, Chapter XIII].

More properties of $(a;q)_\alpha$, $\alpha \in \mathbf{C}$ may be found for instance in [**133**, §14]; and combinatorial interpretations in [**18, 133**].

EXERCISE 3.84. Adapt the formulas of this section to the case $|q| > 1$.

3.3.2. q-numbers, q-factorials.

3.3.2.1. *q-numbers.*

DEFINITION 3.85. Let $\alpha \in \mathbf{C}$. The corresponding *q-number* is:

$$[\alpha]_q := \frac{1 - q^\alpha}{1 - q}.$$

(Remember that $q^\alpha := e^{2i\pi\tau\alpha}$ was defined in the General Notations 0.4 and again at the beginning of this chapter.) Clearly:

$$\lim_{q \to 1} [\alpha]_q = \alpha.$$

We shall actually prefer the name of *q-integer* to those coming from natural integers:

$$\forall n \in \mathbf{N}, \ [n]_q = \frac{1 - q^n}{1 - q} = 1 + q + \cdots + q^{n-1}.$$

EXERCISE 3.86. (i) Let $n \in \mathbf{N}$. Check that $[-n]_q = -q^{-n}[n]_q$.
(ii) let moreover $p \in \mathbf{N}$. Check that $[n+p]_q = [n]_q + q^n[p]_q$.

3.3.2.2. *q-factorials.*

DEFINITION 3.87. Let $n \in \mathbf{N}$. The corresponding *q-factorial* is:

$$[n]_q! := \prod_{i=1}^{n} [i]_q.$$

The usual convention on empty products says that $[0]_q! = 1$. If $n > 0$, then $[n]_q! = [n]_q[n-1]_q!$. And of course, for every $n \in \mathbf{N}$:

$$\lim_{q \to 1} [n]_q! = n!.$$

In general:

$$[n]_q! = \prod_{i=1}^{n} \frac{1-q^i}{1-q} = \frac{(q;q)_n}{(1-q)^n}.$$

For asymptotics when $q \to 1$, we shall sometimes use the above equality in the form:

$$(q;q)_n = [n]_q!(1-q)^n \sim n!\,(1-q)^n.$$

We already used $(1-q)$ as a *scaling factor* when $q \to 1$ while studying q-analogues of the exponential in 3.3.1.3. The following result was used when studying degeneracy of basic hypergeometric series in 1.6

EXERCISE 3.88. Set $a := q^\alpha$. Then prove:

$$(a;q)_n = (1-q)^n[\alpha]_q[\alpha+1]_q \cdots [\alpha+n-1]_q,$$

and deduce that, when $q \to 1$:

$$(a;q)_n \sim (\alpha)_n(1-q)^n.$$

(We encountered classical Pochhammer symbol $(\alpha)_n$ in 3.3.1.4.)

REMARK 3.89. The analogy between the classical and q-factorial is less clear under our convention that $|q| < 1$. Indeed, under the opposite convention, both grow quickly as $n \to +\infty$ and serve as touchstones for so-called "Gevrey" and "q-Gevrey" divergence (see Chapter 14 for the latter).

EXERCISE 3.90. Compare asymptotic behaviours of all our q-things above when $n \to +\infty$ under both conventions for $|q|$.

3.3.2.3. *The q-Gamma function.* We already met it in 2.4.2 and we shall study it more closely in [CV].

DEFINITION 3.91. The *q-Gamma function* is defined as:

$$\Gamma_q(x) := \frac{(q;q)_\infty}{(q^x;q)_\infty}(1-q)^{1-x}.$$

We already know how to define q^x; as for $(1-q)^{1-x} = \exp\left((1-x)\log(1-q)\right)$, this is canonically defined since $|q| < 1$.

EXERCISE 3.92. Check that Γ_q is meromorphic and find its poles.

Using equalities $(q^x;q)_\infty = (1-q^x)(q^{x+1};q)_\infty$ and $(1-q)^{1-x} = (1-q)(1-q)^{1-(x+1)}$, we obtain:

$$\Gamma_q(x+1) = \frac{1-q^x}{1-q}\Gamma_q(x) = [x]_q\Gamma_q(x),$$

which is one of the reasons it is called q-Gamma. Since obviously $\Gamma_q(1) = 1$, we find the value at positive integers:

$$\forall n \in \mathbf{N}^* \;,\; \Gamma_q(n) = [n-1]_q!.$$

We mentioned in Subsection 2.4.2 that $\lim_{q\to 1} \Gamma_q(x) = \Gamma(x)$.

REMARK 3.93. Actually, since Γ_q satisfies no particular q-difference equation, it is not quite clear that it really falls within the scope of this book; however, it is closely related to some important characters of our story. Look at a q-analogue by Heine in [**18**, p. 13].

3.3.3. q-binomial coefficients.

3.3.3.1.

DEFINITION 3.94. The *q-binomial coefficients* (or *q-binomial numbers* are the numbers:

$$\begin{bmatrix} n \\ i \end{bmatrix}_q := \frac{[n]_q!}{[i]_q![n-i]_q!}, i, n \in \mathbf{N} \;,\; 0 \le i \le n.$$

The following properties are immediate:

$$\begin{bmatrix} n \\ 0 \end{bmatrix}_q = \begin{bmatrix} n \\ n \end{bmatrix}_q = 1, \quad \begin{bmatrix} n \\ 1 \end{bmatrix}_q = \begin{bmatrix} n \\ n-1 \end{bmatrix}_q = [n]_q, \quad \begin{bmatrix} n \\ i \end{bmatrix}_q = \begin{bmatrix} n \\ n-i \end{bmatrix}_q,$$

$$\begin{bmatrix} n \\ i \end{bmatrix}_q = \frac{[n]_q[n-1]_q\cdots[n-i+1]_q}{[i]_q!}, \quad \lim_{q\to 1}\begin{bmatrix} n \\ i \end{bmatrix}_q = \binom{n}{i}.$$

PROPOSITION 3.95 (Pascal's rule for q-binomial coefficients). *For $1 \le i \le n$:*

$$\begin{bmatrix} n \\ i \end{bmatrix}_q = \begin{bmatrix} n-1 \\ i-1 \end{bmatrix}_q + q^i \begin{bmatrix} n-1 \\ i \end{bmatrix}_q = q^{n-i}\begin{bmatrix} n-1 \\ i-1 \end{bmatrix}_q + \begin{bmatrix} n-1 \\ i \end{bmatrix}_q.$$

PROOF. We use equality $[n]_q = [i]_q + q^i[n-i]_q$ (see exercise at the end of 3.3.2.1):

$$\begin{aligned}
\begin{bmatrix} n \\ i \end{bmatrix}_q &= \frac{[n]_q!}{[i]_q![n-i]_q!} = \frac{([i]_q + q^i[n-i]_q)[n-1]_q!}{[i]_q![n-i]_q!} = \frac{[i]_q[n-1]_q!}{[i]_q![n-i]_q!} \\
&\quad + q^i\frac{[n-i]_q[n-1]_q!}{[i]_q![n-i]_q!} \\
&= \frac{[n-1]_q!}{[i-1]_q![n-i]_q!} + q^i\frac{[n-1]_q!}{[i]_q![n-i-1]_q!} = \begin{bmatrix} n-1 \\ i-1 \end{bmatrix}_q + q^i\begin{bmatrix} n-1 \\ i \end{bmatrix}_q.
\end{aligned}$$

The second formula then comes by changing i to $n-i$. □

COROLLARY 3.96. $\begin{bmatrix} n \\ i \end{bmatrix}_q$ *is a polynomial in q with coefficients in* $\mathbf{N}$.

PROOF. Obvious for $n = i = 0$, then use q-Pascal's rule for a double induction on i and n. □

EXERCISE 3.97. Check that the leading coefficient is 1 and the degree is $i(n-i)$.

3.3.3.2. *Gauß' binomial theorem.* There are quite a few versions of it and quite a few proofs. In the version here, the left-hand side can be written $(1+x)_q^n$ in the notation of Kac and Cheung, and also $(-x;q)_n$. At any rate, it is a q-deformation of $(1+x)^n$, whence the name of the theorem.

THEOREM 3.4. *For $x \in \mathbf{C}$ and $n \in \mathbf{N}$:*

$$(1+x)\cdots(1+q^{n-1}x) = \sum_{i=0}^{n} q^{i(i-1)/2} \begin{bmatrix} n \\ i \end{bmatrix}_q x^i.$$

PROOF. The left-hand side is clearly a polynomial $P_n(x)$ of degree n:

$$P_n(x) := (1+x)\cdots(1+q^{n-1}x) = a_{0,n} + \cdots + a_{n,n}x^n,$$

where $a_{0,n} = 1$ and $a_{n,n} = q^{n(n-1)/2}$. From the equality $P_{n+1}(x) = (1+x)P_n(qx)$, we deduce that $a_{i,n+1} = q^i a_{i,n} + q^{i-1} a_{i-1,n}$. So setting $a_{i,n} = q^{i(i-1)/2} b_{i,n}$, this translates to $b_{i,n+1} = q^i b_{i,n} + b_{i-1,n}$. Then we use q-Pascal's rule for a double induction on i and n to conclude that the $b_{i,n}$ are the q-binomial coefficients. □

COROLLARY 3.98. *With the same assumptions and moreover $a \in \mathbf{C}$:*

$$(x;q)_n = \sum_{i=0}^{n} (-1)^i q^{i(i-1)/2} \begin{bmatrix} n \\ i \end{bmatrix}_q x^i,$$

$$(x+a)_q^n = \sum_{i=0}^{n} q^{i(i-1)/2} \begin{bmatrix} n \\ i \end{bmatrix}_q a^i x^{n-i}.$$

In the second formula, we of course use the notation of Kac and Cheung.

EXERCISE 3.99. (i) Give another proof of Gauß q-binomial theorem by instantiating Cauchy's q-binomial theorem 1.4.

(ii) Use Cauchy's binomial theorem to prove the formula:

$$\frac{1}{(x-a)_q^{n+1}} = \sum_{i \geq 0} \begin{bmatrix} n+i \\ i \end{bmatrix}_q x^i.$$

EXERCISE 3.100. (i) Prove the following and explain in which way it is related to a classical formula (i.e., not from the q-world):

$$(ab;q)_n = \sum_{i=0}^{n} \begin{bmatrix} n \\ i \end{bmatrix}_q b^i (a;q)_i (b;q)_{n-i}.$$

(ii) Generalize it to a formula for $(a_1 \cdots a_m; q)_n$ involving "q-multinomial coefficients" (to be defined). (See [**93**, exercise 1.3 p 25].)

Note that all the formulas shown here degenerate, when $q \to 1$, to classical formulas involving binomial coefficients.

3.3.3.3. *Miscellani.* There are more uses of q-numbers, q-factorials and q-binomial coefficients than those we just saw. Here are some, without any attempt to exhaustivity (or logical order).

Recall that $D_q := \dfrac{1-\sigma_q}{(1-q)x}$ is a q-deformation of the usual derivation $D := \dfrac{d}{dx}$. Let $0 \leq i \leq n$. Then:

$$D_q(x^n) = \frac{x^n - q^n x^n}{(1-q)x} = [n]_q x^{n-1},$$

so, iterating:

$$D_q^i(x^n) = [n]_q D_q^{i-1}(x^{n-1}) = \cdots = [n]_q \cdots [n-i+1]_q x^{n-i} = \frac{[n]_q!}{[n-i]_q!} x^{n-i}.$$

So, setting $P_n := \dfrac{x^n}{[n]_q!}$, we see that $D_q^i(P_n) = P_{n-i}$.

EXERCISE 3.101. (i) Prove that (using the notation of Kac and Cheung):

$$D_q(x-a)_q^n = [n]_q (x-a)_q^{n-1}.$$

(ii) Check that $(x-a)_q^{n+m} = (x-a)_q^n (x - q^n a)_q^m$ for $m, n \in \mathbf{N}$ and propose a definition for negative "powers" so that this stays true for $m, n \in \mathbf{Z}$.

3.3.3.4.

PROPOSITION 3.102. *Let x, y be two "q-commuting" indeterminates, i.e., $yx = qxy$. (Note that this holds if we replace y by $S \in K\langle S, S^{-1}\rangle$.) Then:*

$$(x+y)^n = \sum_{i=0}^{n} \begin{bmatrix} n \\ i \end{bmatrix}_q x^i y^{n-i}.$$

It is understood that associativity holds and complex numbers are central (i.e., commute with everything).

PROOF. For $n = 0$ or $n = 1$, this is obvious. Assume that it holds for n. We are going to use the fact that $y^j x = q^j x y^j$, which is easy to prove (do it!) by induction:

$$\begin{aligned}
(x+y)^{n+1} &= (x+y)^n x + (x+y)^n y \\
&= \sum_{i=0}^{n} \begin{bmatrix} n \\ i \end{bmatrix}_q x^i y^{n-i} x + \sum_{i=0}^{n} \begin{bmatrix} n \\ i \end{bmatrix}_q x^i y^{n-i+1} \\
&= \sum_{i=0}^{n} q^{n-i} \begin{bmatrix} n \\ i \end{bmatrix}_q x^i y^{n-i} + \sum_{i=0}^{n} \begin{bmatrix} n \\ i \end{bmatrix}_q x^i y^{n-i+1} \\
&= x^{n+1} + y^{n+1} + \sum_{i=1}^{n} (q^{n+1-i} \begin{bmatrix} n \\ i-1 \end{bmatrix}_q + \begin{bmatrix} n \\ i \end{bmatrix}_q) x^i y^{n+1-i} \\
&= \sum_{i=0}^{n+1} \begin{bmatrix} n+1 \\ i \end{bmatrix}_q x^i y^{n+1-i},
\end{aligned}$$

achieving the proof by induction. (The penultimate line was obtained from the previous one by isolating one term in each of the two sums and by shifting index in the first sum). □

3.3.4. One more proof of Jacobi's Triple Product Formula. This one is due to Starcher [**240**].

3.3.4.1. It is sometimes useful to express solutions of q-difference equations as "q-factorial series"[9]. We shall not define them formally but use them in a few

[9]The terminology apparently comes from [**191**], in analogy with the "factorial series" of the theory of difference equations.

examples. Here are, however, some useful properties. We let $f \in \mathbf{C}\{x\}$ admit the expansion into a q-factorial series:

$$f(x) = \sum_{n\geq 0} \frac{a_n x^n}{(x;q)_n} = a_0 + \frac{a_1 x}{1-x} + \frac{a_2 x^2}{(1-x)(1-qx)} + \cdots,$$

Clearly $f(0) = a_0$ and we can write $f = a_0 + \dfrac{x}{1-x}\sigma_q g$, $g \in \mathbf{C}\{x\}$.

LEMMA 3.103. *(i) The function g admits the expansion $g = \sum\limits_{n\geq 0} \dfrac{b_n x^n}{(x;q)_n}$, where $b_n := q^{-n} a_{n+1}$.*

(ii) The function $h := \dfrac{1}{1-x}\sigma_q f$ admits the expansion $h = \sum\limits_{n\geq 0} \dfrac{c_n x^n}{(x;q)_n}$, where $c_n := a_n q^n + a_{n-1} q^{2n-2}$ (with the usual convention $a_{-1} = 0$).

PROOF. These are simple computations and left to you. □

It follows from (i) that, when it exists, a q-factorial series expansion is unique: a_0 can be recovered as $f(0)$, then g is given as Statement (ii) will be used in a moment.
3.3.4.2.

PROPOSITION 3.104. *The Euler partition function admits the q-factorial series expansion:*

$$\frac{1}{(x;q)_\infty} = \sum_{n\geq 0} \frac{q^{n^2-n}}{(q;q)_n} \frac{x^n}{(x;q)_n}.$$

PROOF. The right-hand side converges normally in every closed subdisk $\overline{\mathrm{D}}(0,r) \subset \mathring{\mathrm{D}}(0,1)$ and therefore defines on $\mathring{\mathrm{D}}(0,1)$ a holomorphic function f such that $f(0) = 1$. Its coefficients as a q-factorial series $a_n := \dfrac{q^{n^2-n}}{(q;q)_n}$ satisfy $a_0 = 1$ and, for $n \geq 1$:

$$a_n q^n + a_{n-1} q^{2n-2} = \frac{q^{n^2}}{(q;q)_n} + \frac{q^{(n-1)(n-2)+2n-2}}{(q;q)_{n-1}} = \frac{q^{n^2} + q^{n(n-1)}(1-q^n)}{(q;q)_n} = \frac{q^{n^2-n}}{(q;q)_n} = a_n.$$

By (ii) of the lemma above, $f = \dfrac{1}{1-x}\sigma_q f$, from which, since $f(0) = 1$, we have indeed $f = \dfrac{1}{(x;q)_\infty}$. □

Substituting $x := q$:

COROLLARY 3.105.

$$\frac{1}{(q;q)_\infty} = \sum_{n\geq 0} \frac{q^{n^2}}{(q;q)_n^2}.$$

EXERCISE 3.106. In [**186**, II, p 24, entry 1.6.1], the following formula is given:

$$(aq;q)_\infty \sum_{n\geq 0} \frac{b^n q^{n^2}}{(q;q)_n (aq;q)_n} = \sum_{n\geq 0} \frac{(b/a;q)_n a^n q^{n(n+1)/2}}{(q;q)_n}.$$

Deduce from it the above corollary.

3.3.4.3. Let $T(x) := (x;q)_\infty (q/x;q)_\infty$. This is holomorphic over $\mathbf{C}^*$, so it admits a Laurent power series expansion $T(x) = \sum_{n\in\mathbf{Z}} t_n x^n$. It satisfies a q-difference equation of order 1:

$$T(qx) = (qx;q)_\infty (1/x;q)_\infty = \frac{1}{1-x}(x;q)_\infty (1-1/x)(q/x;q)_\infty =$$
$$= \frac{1-1/x}{1-x}(x;q)_\infty (q/x;q)_\infty = \frac{-1}{x}T(x),$$

from which one deduces that $q^n t_n = -t_{n+1}$ for all $n \in \mathbf{Z}$, which in turn implies that $t_n = (-1)^n q^{n(n-1)/2} t_0$ for all n, whence:

$$(x;q)_\infty (q/x;q)_\infty = t_0 \sum_{n\in\mathbf{Z}} (-1)^n q^{n(n-1)/2} x^n.$$

We found in 3.3.1.2 the power series expansion:

$$(x;q)_\infty = \sum_{n\geq 0} (-1)^n \frac{q^{n(n-1)/2}}{(q;q)_n} x^n,$$

which implies:

$$(q/x;q)_\infty = \sum_{n\geq 0} (-1)^n \frac{q^{n(n+1)/2}}{(q;q)_n} x^{-n}.$$

Multiplying both expansions, selecting the term with x^0 and using 3.3.4.2:

$$t_0 = \sum_{n\geq 0} (-1)^n \frac{q^{n(n-1)/2}}{(q;q)_n} (-1)^n \frac{q^{n(n+1)/2}}{(q;q)_n} = \sum_{n\geq 0} \frac{q^{n^2}}{(q;q)_n^2} = \frac{1}{(q;q)_\infty}.$$

We have proved one form of Jacobi's Triple Product Formula:

THEOREM 3.5.

$$\sum_{n\in\mathbf{Z}} (-1)^n q^{n(n-1)/2} x^n = (q;q)_\infty (x;q)_\infty (q/x;q)_\infty.$$

CHAPTER 4

Equations of Low Order, Elementary Approach

Here, "elementary" means (almost) no modules and no commutative algebra, categories, functional analysis ... However, in the last section we do attempt a smooth initiation to the use of sheaves and vector bundles (here only line bundles) in the study of q-difference equations.

The aim of the chapter is to experiment with basic tricks and techniques, so it is mostly made of examples and, for the most part, not logically necessary for the sequel. However, section 4.4 on basic hypergeometric series is to be taken as fundamental knowledge in the domain.

4.1. Homogeneous equations of order 1

These are equations $f(qx) = u(x)f(x)$, for short $\sigma_q f = uf$, $u \in K^*$, for a q-difference field K taken as "field of coefficients" (solutions will usually be searched in a q-difference extension of K). The case $u = 1$ defining, through equation $\sigma_q f = f$, the subfield of constants, has already been discussed at length in Chapters 2 and 3.

4.1.1. Gauge equivalence and classification.

4.1.1.1. We start from a q-difference field (K, σ_q) where lies the equation: $u \in K^*$. However, we shall usually look for solutions in a q-difference extension K', involving more complicated functions, maybe so-called "new transcendents"[1]. So we consider the change of unknown function $g = pf$ as innocuous if $p \in K^*$.

EXAMPLE 4.1. Let $K := \mathbf{C}(\{x\})$ and $u := ax + b$, $a, b \in \mathbf{C}^*$. To solve $\sigma_q f = uf$, we can look for $g := pf$, where $p := (-ax/b; q)_\infty \in \mathbf{C}(\{x\})$ satisfies $\sigma_q p = \dfrac{1}{1 + ax/b} p$. Then we should solve $\sigma_q g = bg$, but this cannot usually be done in $\mathbf{C}(\{x\})$, we need "new transcendents": in this case, q-characters (see Chapter 2) involving theta functions.

This our first example of *gauge transformation* (they will be systematically studied in subsequent chapters). We write:

$$\sigma_q f = uf \Longleftrightarrow \sigma_q g = (\sigma_q p)(\sigma_q f) = (\sigma_q p)(uf) = \frac{\sigma_q p}{p} ug,$$

so that eventually:

$$\sigma_q f = uf \Longleftrightarrow \sigma_q g = vg, \text{ where } v := \frac{\sigma_q p}{p} u.$$

[1]This old terminology is related to the following question: can we solve a particular functional equation using only already known special functions or do we have to introduce new ones? This is one of the main motivations of differential algebra and differential Galois theory; see **[134]**.

In other words, defining $p[u] := u^{[p]} := \dfrac{\sigma_q p}{p} u$, we get an isomorphism of K^{σ_q}-linear spaces:

$$f \mapsto pf, \quad \mathrm{Sol}(u, K') \to \mathrm{Sol}(v, K').$$

REMARK 4.2. For higher-order equations, we shall use a matricial notation and u, p will be replaced by invertible square matrices U, P. We shall then have to distinguish the left gauge action $U \mapsto P[U] := (\sigma_q P) U P^{-1}$ from the right gauge action $U \mapsto U^{[P]} := (\sigma_q P)^{-1} U P$.

4.1.1.2. We are thus led to consider u and v, or the corresponding q-difference equations $\sigma_q f = uf$ and $\sigma_q g = vg$, as *gauge equivalent (over K)* if there exists such a gauge transformation from the former to the latter:

$$u \underset{K}{\sim} v \Longleftrightarrow \exists p \in K^* \ : v = \frac{\sigma_q p}{p} u.$$

In the present case of homogeneous equation of order 1, the gauge equivalence relation is actually congruence in the abelian group K^* modulo the subgroup $\left\{ \dfrac{\sigma_q p}{p} \mid p \in K^* \right\}$, i.e., $u \underset{K}{\sim} v$ if, and only if, v/u belongs to that subgroup; and so the set of gauge equivalence classes is the quotient group, i.e., the cokernel of the group morphism:

$$\lambda : K^* \to K^*, p \mapsto \frac{\sigma_q p}{p}.$$

Since the kernel of λ is clearly C^*, where $C := K^{\sigma_q}$, we obtain an exact sequence of abelian groups:

$$1 \to C^* \to K^* \xrightarrow{\lambda} K^* \to \mathrm{Pic}(K^*, \sigma_q) \to 1.$$

We use multiplicative notation, so the trivial group is $\{1\}$, shortened into 1; and we write $\mathrm{Pic}(K^*, \sigma_q)$ (or as well $\mathrm{Pic}(K, \sigma_q)$) for the "Picard group" $\mathrm{Coker}\lambda$, the quotient of K^* by the subgroup $\mathrm{Im}\lambda := \left\{ \dfrac{\sigma_q p}{p} \mid p \in K^* \right\}$. This is our classifying group for the classification problem we are interested in. We shall see later that it is in essence the group of classes of q-difference modules of rank 1 for the tensor product; and already in 4.5 below that it is the group of classes of line bundles over $\mathbf{E}_q$, also for the tensor product. Either identification justifies the name "Picard group".

4.1.1.3. We shall now analyze λ in various important cases. Each time, we will determine $\mathrm{Pic}(K^*, \sigma_q) := \mathrm{Coker}\lambda$, but also, for every $u \in K^*$, its inverse image:

$$\lambda^{-1}(u) = \left\{ p \in K^* \mid \frac{\sigma_q p}{p} = u \right\} = \{ p \in K^* \mid \sigma_q p = up \} = \mathrm{Sol}(u, K) \setminus \{0\}.$$

If $p \in \mathrm{Sol}(u, K) \setminus \{0\}$, every element of $\mathrm{Sol}(u, K)$ has the form $p\phi$ where $\phi \in K^{\sigma_q}$, so that actually $\mathrm{Sol}(u, K) = Cp$. Therefore we conclude that

- either $u \not\in \mathrm{Im}\lambda$ and $\mathrm{Sol}(u, K) = \{0\}$;
- or $u \in \mathrm{Im}\lambda$ and $\mathrm{Sol}(u, K)$ has dimension 1 over C.

In this particular setting (homogeneous equations of order 1) resolution and classification come hand in hand.

4.1.2. Classification over K and resolution in K. We distinguish the three main possibilities for K: $\mathbf{C}((x))$, $\mathbf{C}(\{x\})$ and $\mathbf{C}(x)$; in all three cases, $C := K^{\sigma_q} = \mathbf{C}$. Equations that are *regular at* 0 play a special role: these are the q-difference equations $\sigma_q f = uf$ such that u is defined at 0 and $u(0) = 1$. So we fix a notation for the corresponding subgroups of K^* (the subscripts are "f" for "formal", "a" for "(local) analytic" and "g" for "global" (on $\mathbf{S}$, i.e., rational):

$$\begin{aligned} U_f &:= 1 + x\mathbf{C}[[x]] = \{u \in \mathbf{C}[[x]] \mid u(0) = 1\}, \\ U_a &:= 1 + x\mathbf{C}\{x\} = \{u \in \mathbf{C}\{x\} \mid u(0) = 1\}, \\ U_g &:= U_f \cap \mathbf{C}(x) = U_a \cap \mathbf{C}(x). \end{aligned}$$

4.1.2.1. *Formal case.* Here $K := \mathbf{C}((x))$. Each $u \in K^*$ has a unique decomposition $u = cx^\mu \epsilon$, where $c \in \mathbf{C}^*$; $\mu \in \mathbf{Z}$ and $\epsilon = 1 + e_1 x + \cdots \in U_f$. This actually corresponds to a decomposition of $\mathbf{C}((x))^*$ into a direct product of groups:

$$\mathbf{C}((x))^* = \mathbf{C}^* \times x^{\mathbf{Z}} \times U_f.$$

For $u = cx^\mu \epsilon \in \mathbf{C}((x))^*$, one has $\lambda(u) := \dfrac{\sigma_q u}{u} = q^\mu \phi$, where $\phi := \dfrac{\sigma_q \epsilon}{\epsilon} \in U_f$.

LEMMA 4.3. *λ induces an automorphism of U_f.*

PROOF. We already know that it is a morphism. Clearly, if $\epsilon \in U_f$, then $\sigma_q \epsilon \in U_f$, whence $\phi := \dfrac{\sigma_q \epsilon}{\epsilon} \in U_f$. Conversely, let $\phi := 1 + f_1 x + \cdots \in U_f$. Solving $\lambda(\epsilon) = \phi$ for $\epsilon = 1 + e_1 x + \cdots \in U_f$ leads to a system with unknowns the e_k:

$$\sigma_q \epsilon = \phi \epsilon \Longleftrightarrow \forall k \geq 1 \;,\; q^k e_k = 1.f_k + e_1 f_{k-1} + \cdots + e_{k-1} f_1 + e_k.1,$$

which amounts to the recursive relation:

$$\forall k \geq 1 \;,\; e_k := -\frac{1}{1 - q^k}(f_k + e_1 f_{k-1} + \cdots + e_{k-1} f_1).$$

Therefore, ϵ is uniquely determined, i.e., λ is bijective. □

EXERCISE 4.4. Solve equation $\lambda(\epsilon) = \phi$ as a fixpoint equation $\epsilon = \Phi(\epsilon)$, where $\Phi : \epsilon \mapsto \phi^{-1} \sigma_q \epsilon$ is x-adically contracting. Also check that the unique solution is the x-adically converging infinite product $\phi^{-1}(\sigma_q \phi^{-1})(\sigma_q^2 \phi^{-1}) \cdots$.

The endomorphism λ of the group $\mathbf{C}((x))^* = \mathbf{C}^* \times x^{\mathbf{Z}} \times U_f$ sends $\mathbf{C}^*$ to $\{1\}$; it sends $x^{\mathbf{Z}}$ bijectively to $q^{\mathbf{Z}} \subset \mathbf{C}^*$ (this is because $\lambda(x^m) = q^m$); and it sends U_f bijectively to itself. Therefore $\mathrm{Im}\lambda = q^{\mathbf{Z}} \times U_f$ and:

$$\mathrm{Coker}\lambda = \frac{\mathbf{C}^* \times x^{\mathbf{Z}} \times U_f}{q^{\mathbf{Z}} \times \{1\} \times U_f} = \frac{\mathbf{C}^*}{q^{\mathbf{Z}}} \times x^{\mathbf{Z}} \simeq \mathbf{E}_q \times \mathbf{Z}.$$

In this way we identify the image of $u = cx^\mu \epsilon \in \mathbf{C}((x))^*$ in the quotient group $\mathrm{Coker}\lambda = \mathbf{C}((x))^*/\mathrm{Im}\lambda \simeq \mathbf{E}_q \times \mathbf{Z}$ to $(\overline{c}, \mu)$. (Recall that $\overline{c}$ denotes the image in $\mathbf{C}^*/q^{\mathbf{Z}} = \mathbf{E}_q$ of $c \in \mathbf{C}^*$.) The first component $\overline{c}$ is the *exponent* of equation $\sigma_q f = uf$, the second component μ is related to its *slope* (see Chapters 5 and 10).

Now, if $u, v \in \mathbf{C}((x))^*$ encode two q-difference equations, they are gauge equivalent (over $\mathbf{C}((x))$) if, and only if, $u/v \in \mathrm{Im}\lambda$, i.e., if, and only if, they have the same image in $\mathrm{Pic}(\mathbf{C}((x)), \sigma_q)$, i.e., if, and only if, they have same slope and same exponent: *the pair $(\overline{c}, \mu)$ is a complete set of invariants for the gauge classification problem.* Of course, for higher orders, sets of invariants will be more complicated!

Last, we see that $\dim_{\mathbf{C}} \mathrm{Sol}(u, \mathbf{C}((x))) \leq 1$ and that it is 1 if, and only if, $u \in \mathrm{Im}\lambda$, i.e., if it has trivial image in $\mathrm{Pic}(\mathbf{C}((x)), \sigma_q)$, i.e., if, and only if, its slope is $0 \in \mathbf{Z}$ and its exponent is $\overline{1} \in \mathbf{E}_q$. (Note that $\mathbf{E}_q = \mathbf{C}^*/q^{\mathbf{Z}}$ is written multiplicatively and $\overline{1}$ is its neutral element; of course the underlying group of a complex torus $\mathbf{C}/\Lambda_\tau$ is more often written additively, and its neutral element is then written 0.) More concretely, the q-difference equation $\sigma_q f = uf$ has a nontrivial solution in $\mathbf{C}((x))$ if, and only if, $u = q^m \phi$ and such a solution is then $x^\mu \epsilon$ (with the above notations).

4.1.2.2. *Local analytic case.* Here $K := \mathbf{C}(\{x\})$. Similar to the previous case, each $u \in K^*$ has a unique decomposition $u = cx^\mu \epsilon$, where $c \in \mathbf{C}^*$, $\mu \in \mathbf{Z}$ and $\epsilon \in U_a$. This corresponds to a decomposition of $\mathbf{C}(\{x\})^*$ into a product of groups:

$$\mathbf{C}(\{x\})^* = \mathbf{C}^* \times x^{\mathbf{Z}} \times U_a.$$

Also, for $u = cx^\mu \epsilon \in \mathbf{C}(\{x\})^*$, one has $\lambda(u) := \dfrac{\sigma_q u}{u} = q^\mu \phi$, where $\phi := \dfrac{\sigma_q \epsilon}{\epsilon} \in U_a$.

LEMMA 4.5. *λ induces an automorphism of U_a.*

PROOF. Clearly $\epsilon \in U_a \Rightarrow \dfrac{\sigma_q \epsilon}{\epsilon} \in U_a$. Taking into account the proof of Lemma 4.3 and the notations there, we are reduced to prove that, if $\sum f_n x^n$ converges , then so does $\sum e_n x^n$. This is a particular case of a general result (Cauchy's so-called "method of majorants"), so we leave it as an exercise to the reader (also see the exercise following this proof).

We rather show here that the infinite product solution converges. Let $r > 0$ such that $r < 1$ and $r <$ the radius of convergence of $1/\phi$. Write $1/\phi = 1 + x\psi$, $\psi \in \mathbf{C}\{x\}$. Then $M := \sup\limits_{x \in \overline{\mathrm{D}}(0,r)} |\psi(x)| < \infty$. This implies that $\left|\sigma_q^n(x\psi)\right| \leq M r^n$ on $\mathring{\mathrm{D}}(0, r)$, so the series $\sum \sigma_q^n(x\psi)$ converges normally there, so $\prod(1 + \sigma_q^n(x\psi))$ converges uniformly there and defines a holomorphic function [**211**]. But this product is $1/(\phi \sigma_q \phi \cdots)$ and its image by λ is ϕ. □

EXERCISE 4.6. Show that $D := \inf\limits_{k \geq 0} \left|1 - q^k\right| > 0$ and that, if $|f_k| < D' r^{-k}$ for all k, then $|f_k| \leq ((1 + DD')r)^{-k}$ for all k.

From there on, the argument follows the same road as in 4.1.2.1. The image of $\mathbf{C}(\{x\})^* = \mathbf{C}^* \times x^{\mathbf{Z}} \times U_a$ by λ is $q^{\mathbf{Z}} \times U_a$, so:

$$\mathrm{Pic}(\mathbf{C}(\{x\}), \sigma_q) = \mathrm{Coker}\lambda = \mathbf{C}^*/q^{\mathbf{Z}} \times x^{\mathbf{Z}} \simeq \mathbf{E}_q \times \mathbf{Z}.$$

The q-difference equation $\sigma_q f = uf$, $u = cx^\mu \epsilon \in \mathbf{C}(\{x\})^*$, is completely classified over $\mathbf{C}(\{x\})$ (that is for analytic gauge equivalence) by its exponent $\overline{c} \in \mathbf{E}_q$ and its slope $\mu \in \mathbf{Z}$. And $\mathrm{Sol}(u, \mathbf{C}(\{x\}))$ is trivial, except if $c = q^m$ and $\mu = 0$, in which case it is generated by $x^m \phi$, where $\phi \in U_a$ is explicit.

EXERCISE 4.7. Find examples of this in Chapter 1.

EXERCISE 4.8. Discuss resolution in $\mathscr{M}(\mathbf{C})$.

4.1.2.3. *Global (or rational) case.* Here $K := \mathbf{C}(x)$. The usual decomposition becomes $u = cx^\mu \epsilon$, $c \in \mathbf{C}^*$, $\mu \in \mathbf{Z}$ and $\epsilon \in U_g$, whence:

$$\mathbf{C}(x)^* = \mathbf{C}^* \times x^{\mathbf{Z}} \times U_g.$$

The action of λ on factors $\mathbf{C}^*$ and $x^{\mathbf{Z}}$ is the same, but the case of U_g is maybe more subtle. Elements of U_g can be uniquely written:

$$\epsilon = \prod (1 - x/a_i)^{m_i},$$

where the $a_i \in \mathbf{C}^*$ and the $m_i \in \mathbf{Z}$. ("Uniquely" means of course up to putting in or out trivial factors with $m_i = 0$.) The divisor $\mathrm{div}_{\mathbf{C}^*}(\epsilon)$, sometimes also denoted (ϵ), of zeroes and poles of ϵ (see Chapter 2) is:

$$\mathrm{div}_{\mathbf{C}^*}(\epsilon) = \sum m_i [a_i] \in \mathrm{Div}(\mathbf{C}^*).$$

Recall that $\mathrm{Div}(\mathbf{C}^*)$ is the free abelian group with basis the symbols $[a]$, $a \in \mathbf{C}^*$. The map $\epsilon \mapsto (\epsilon)$ is a group isomorphism from U_g to $\mathrm{Div}(\mathbf{C}^*)$, so:

$$\mathbf{C}(x)^* \simeq \mathbf{C}^* \times \mathbf{Z} \times \mathrm{Div}(\mathbf{C}^*).$$

The action of λ on the first factor is trivial ($\mathbf{C}^*$ is sent to $\{1\}$) and on $\mathbf{Z}$ it is given by $m \mapsto q^m$. It sends each generator $1 - x/a$ of U_g to $(1 - qx/a)/(1 - x/a) \in U_g$, so it induces an endomorphism of U_g, whence an endomorphism Δ of $\mathrm{Div}(\mathbf{C}^*)$, and we should find:

$$\mathrm{Coker}\lambda = \frac{\mathbf{C}^* \times \mathbf{Z}}{q^{\mathbf{Z}} \times \{0\}} \times \frac{U_g}{\lambda(U_g)} \simeq \mathbf{E}_q \times \mathbf{Z} \times \mathrm{Coker}\Delta.$$

To understand Δ, we note that σ_q sends $1 - x/a$ to $1 - qx/a$ and the corresponding action on divisors (the "pullback" of σ_q, here abreviating the restriction $\sigma_{q|U_g}$ of σ_q to U_g) is the map:

$$\sigma_q^* : \mathrm{Div}(\mathbf{C}^*) \to \mathrm{Div}(\mathbf{C}^*), \sum m_i [a_i] \mapsto \sum m_i [q^{-1} a_i];$$

Then the map corresponding to λ (here abreviating the restriction $\lambda_{|U_g}$ of λ to U_g) is $\Delta := \lambda^* = \sigma_q^* - \mathrm{Id}$:

$$\Delta : \sum m_i [a_i] \mapsto \sum m_i [q^{-1} a_i] - \sum m_i [a_i] = \sum m_i \left([q^{-1} a_i] - [a_i] \right).$$

This can be expressed by a commutative diagram:

$$\begin{array}{ccc} U_g & \xrightarrow{\ \lambda\ } & U_g \\ {\scriptstyle \mathrm{div}} \downarrow & & \downarrow {\scriptstyle \mathrm{div}} \\ \mathrm{Div}(\mathbf{C}^*) & \xrightarrow[\ \Delta\ ]{} & \mathrm{Div}(\mathbf{C}^*). \end{array}$$

Since the vertical arrows are isomorphisms, this provides an identification of $\mathrm{Coker}\lambda$ with $\mathrm{Coker}\Delta$ and of $\mathrm{Ker}\lambda$ with $\mathrm{Ker}\Delta$.

LEMMA 4.9. *Δ is an injective endomorphism with cokernel:*

$$\mathit{Coker}\Delta = \mathit{Div}(\mathbf{C}^*/q^{\mathbf{Z}}) = \mathit{Div}(\mathbf{E}_q) = \mathit{Div}(\mathcal{C}_q).$$

PROOF. The fact that Δ is a morphism of groups is obvious.

Note that, for every $D \in \mathrm{Div}(\mathbf{C}^*)$, the support Σ of $\sigma_q^* D$ is $q^{-1}\Sigma$. So, if $D \in \mathrm{Ker}\Delta$, i.e., $\sigma_q^* D = D$, then $\Sigma = q\Sigma$. This is only possible if Σ is empty or infinite. The second case is excluded for a divisor on $\mathbf{E}_q$ and the first case means that $D = 0$. So $\mathrm{Ker}\Delta$ is trivial and Δ is injective.

In particular, the image $\{[q^{-1}a] - [a] \mid a \in \mathbf{C}^*\} = \{[a] - [qa] \mid a \in \mathbf{C}^*\}$ of the basis $\{[a] \mid a \in \mathbf{C}^*\}$ of $\mathrm{Div}(\mathbf{C}^*)$ by Δ is a basis of $\mathrm{Im}\Delta$. Pick a set X of representatives in $\mathbf{C}^*$ of the congruence modulo $\mathrm{Im}\Delta$. This amounts to picking a fundamental domain for the projection $\pi : \mathbf{C}^* \to \mathbf{C}^*/q^{\mathbf{Z}} = \mathbf{E}_q$, i.e., a subset $X \subset \mathbf{C}^*$ which is mapped bijectively by π to $\mathbf{E}_q$; for instance, the fundamental annulus $\mathcal{C}_q : |q| < |x| \leq 1$ would do. Then $\{[a] \mid a \in X\}$ is the basis of the subspace $\text{Div}(X)$ of $\text{Div}(\mathbf{C}^*)$ and we have the direct sum decomposition:

$$\text{Div}(\mathbf{C}^*) = \text{Div}(X) \oplus \text{Im}\Delta.$$

Therefore $\text{Coker}\Delta = \text{Div}(\mathbf{C}^*)/\text{Im}\Delta \simeq \text{Div}(X)$. The latter is moreover isomorphic to $\text{Div}(\mathbf{E}_q)$ through the map $[a] \mapsto [\overline{a}]$, whence the conclusion. □

Practically, the image of $\sum m_i[a_i] \in \text{Div}(\mathbf{C}^*)$ in $\text{Coker}\Delta$ as identified to $\text{Div}(\mathbf{E}_q)$ is $\sum m_i[\overline{a_i}]$; but the latter is no longer a free combination, simplifications may occur.

EXERCISE 4.10. Prove rigorously that $\text{Div}(\mathbf{C}^*) = \text{Div}(X) \oplus \text{Im}\Delta$.

We conclude by the same arguments as before that:

$$\text{Pic}(\mathbf{C}(x), \sigma_q) \simeq \mathbf{E}_q \times \mathbf{Z} \times \text{Div}(\mathbf{E}_q).$$

For instance, equation $\sigma_q f = uf$, $u \in \mathbf{C}(x)^*$, has a nontrivial solution in $\mathbf{C}(x)$ if, and only if, $c \in q^{\mathbf{Z}}$, $\mu = 0$ and $\sum m_i[\overline{a_i}] = 0$ (with the previous notation $\epsilon = \prod(1 - x/a_i)^{m_i}$).

EXERCISE 4.11. Prove that the third condition is equivalent to: ϵ is a product of elements of the form $(1 - xq/a)/(1 - x/a)$ (with maybe negative exponents); and give an expression for a solution.

4.1.3. Meromorphic solutions.

4.1.3.1. We now look for solutions in $K := \mathscr{M}(\mathbf{C}^*, 0)$, resp. in $K := \mathscr{M}(\mathbf{C}^*)$ for q-difference equations $\sigma_q f = uf$, where $u \in K_0 := \mathbf{C}(\{x\})$, resp. $K_0 := \mathbf{C}(x)$. So we distinguish the q-difference field K_0 *of coefficients* from the q-difference field K *of solutions.* Note that $K_0 := \mathbf{C}((x))$ is not considered here, since it does not naturally embed into a field of meromorphic functions[2].

In both cases mentioned above, K is a q-difference extension of K_0 and $K_0^{\sigma_q} = \mathbf{C}$ while $K^{\sigma_q} = \mathscr{M}(\mathbf{E}_q)$. We shall find that in both cases nontrivial solutions always exist so that $\text{Sol}(u, K)$ is a $\mathscr{M}(\mathbf{E}_q)$-linear space of dimension 1. We write $u = cx^\mu\epsilon$ as usual. Solving independently for the three factors, we find a solution $f = f_1 f_2 f_3$, where (with notations from Chapter 2):

- $\sigma_q f_1 = cf_1$: we can take $f_1 := e_{q,c} \in \mathscr{M}(\mathbf{C}^*)$;
- $\sigma_q f_2 = x^\mu f_2$: we can take $f_2 := \theta_q^{-\mu} \in \mathscr{M}(\mathbf{C}^*)$;
- $\sigma_q f_3 = \epsilon f_3$: we can take $f_3 := 1/(\epsilon\sigma_q\epsilon \cdots) \in U_a$ by 4.1.2.2.

So at any rate we are certain that $f \in \mathscr{M}(\mathbf{C}^*, 0)$ (because it contains $\mathscr{M}(\mathbf{C}^*)$ and U_a). However, in the "global" case $K_0 = \mathbf{C}(x)$, writing $\epsilon = \prod(1 - x/a_i)^{m_i}$, we find that $f_3 := \prod(x/a_i; q)_\infty^{-m_i}$ is a valid choice, and it belongs to $\mathscr{M}(\mathbf{C}^*)$, so there actually is a solution $f \in \mathscr{M}(\mathbf{C}^*)$. To summarize:

$$f(x) := e_{q,c}(x)\theta_q(x)^{-\mu} \prod (x/a_i; q)_\infty^{-m_i}.$$

Then all solutions have the form fg, $g \in \mathscr{M}(\mathbf{E}_q) \subset \mathscr{M}(\mathbf{C}^*)$ and thus belong to $\mathscr{M}(\mathbf{C}^*)$.

---

[2]In the algebraic approach of **[249]**, it is possible to find a so-called "Picard-Vessiot extension" (not a field, though) but that method does not seem fit for a Riemann-Hilbert approach such as ours.

EXERCISE 4.12. Prove directly that in the global case any solution in $\mathscr{M}(\mathbf{C}^*, 0)$ actually lays in $\mathscr{M}(\mathbf{C}^*)$ by using the functional equation $f = u^{-1}\sigma_q f$ to iteratively extend the domain of definition.

4.1.3.2. *A first look at Birkhoff's connection matrix.* Here, $u \in K_0 := \mathbf{C}(x)$, which is included in $\mathbf{C}(\{x\})$ but also in $\mathbf{C}(\{1/x\})$, so we can "solve at infinity" the equation $\sigma_q f = uf$. To do that, we set $y := 1/x$ and $f(x) = g(1/x) = g(y)$, so that our q-difference equation is equivalent to:

$$g(qy) = f(1/qy) = f(x/q) = u(x/q)^{-1} f(x) = u(1/qy)^{-1} g(y),$$

i.e., to $g(qy) = v(y)g(y)$ where $v(y) := u(1/qy)^{-1}$. So, if $u(x) = cx^\mu \prod (1 - x/a_i)^{m_i}$:

$$\begin{aligned} v(y) &= \left(c(1/qy)^\mu \prod (1 - 1/qa_i y)^{m_i}\right)^{-1} \\ &= c^{-1} q^\mu y^\mu \left(\prod (-1/qa_i y)^{m_i} \prod (1 - qa_i y)^{m_i}\right)^{-1} \\ &= dy^\nu \prod (1 - y/b_i)^{n_i}, \end{aligned}$$

where $d := c^{-1} q^\mu \prod (-qa_i)^{m_i}$, $\nu := \mu + \sum m_i$, $n_i := -m_i$ and $b_i := 1/qa_i$.

By the method of 4.1.3.1, using $e_{q,c}(x) := \theta_q(x)/\theta_q(cx)$, we find the solution:

$$g(y) := e_{q,d}(y)\theta_q(y)^{-\nu} \prod (y/b_i; q)_\infty^{-n_i}.$$

Recall from Chapter 2 the relations $\theta_q(qx) = \theta_q(x^{-1}) = x^{-1}\theta_q(x)$ and $\theta_q(x) = a(-x; q)_\infty(-q/x; q)_\infty$ for some explicit $a \in \mathbf{C}^*$; then:

$$g(y) = \frac{\theta_q(x^{-1})}{\theta_q(dx^{-1})} \theta_q(x^{-1})^{-\nu} \prod (qa_i/x; q)_\infty^{m_i} = \frac{1}{d} x^{-\nu} \frac{\theta_q(x)^{-\nu+1}}{\theta_q(x/d)} \prod (qa_i/x; q)_\infty^{m_i}.$$

Now write $f^{(0)}(x) := e_{q,c}(x)\theta_q(x)^{-\mu} \prod (x/a_i; q)_\infty^{-m_i}$ the solution f found in 4.1.3.1; and $f^{(\infty)}(x)$ the solution g just found. We consider them respectively as solutions *local at* 0 and *local at* ∞, although neither needs be defined at 0 or at ∞ (they generally have essential singularities there). So "local at 0", resp. "at ∞" rather refers to the process of construction. Since $f^{(0)}, f^{(\infty)} \in \mathrm{Sol}(u, K)$, the quotient $p := f^{(\infty)}/f^{(0)}$ is q-elliptic: $p \in \mathscr{M}(\mathbf{E}_q)^*$. This p is the rank 1 version of *Birkhoff's connection matrix*; see Chapter 12. Let's compute it explicitly to be sure:

$$\begin{aligned} p(x) &= \frac{1}{d} x^{-\nu} \frac{\theta_q(cx)\theta_q(x)^{-\nu+\mu}}{\theta_q(x/d)} \prod (x/a_i; q)_\infty^{m_i} \prod (qa_i/x; q)_\infty^{m_i} \\ &= a' x^{-\nu} \frac{\theta_q(cx)\theta_q(x)^{-\sum m_i}}{\theta_q(x/d)} \prod \theta_q(-x/a_i)^{m_i} \end{aligned}$$

for some $a' \in \mathbf{C}^*$. Therefore:

$$\frac{p(qx)}{p(x)} = q^{-\nu} \frac{1/cx}{1/(x/d)} x^{\sum m_i} \prod (-a_i/x)^{m_i} = \frac{q^{-\nu}}{cd} \prod (-a_i)^{m_i} = 1.$$

So indeed $p \in \mathscr{M}(\mathbf{C}^*)^{\sigma_q} = \mathscr{M}(\mathbf{E}_q)$.

EXERCISE 4.13. Find the divisor of the elliptic function p and try to understand how it can be inferred from $u(x)$. (The link between the singularities of the connection matrix and those of the equation is an important theme.)

4.1.3.3. *An "application": Ramanujan's summation for* ${}_1\psi_1$. We already met it in 1.10. Called by Hardy [**109**, p 222] "a remarkable formula with many parameters", which was source of ironical comments by specialists of q-calculus, who see it as central, it goes as follows (precise conditions of validity will be detailed in the theorem further below):

$$\sum_{n\in\mathbf{Z}} \frac{(a;q)_n}{(b;q)_n} x^n = \frac{(q;q)_\infty (b/a;q)_\infty (ax;q)_\infty (q/(ax);q)_\infty}{(b;q)_\infty (q/a;q)_\infty (x;q)_\infty (b/(ax);q)_\infty}.$$

The left-hand side is one of the simplest of the family of *bilateral series* (see [**93**, Chapter 5] and [**186**, II, ch 3]):

$${}_r\psi_s \begin{pmatrix} a_1,\ldots,a_r \\ b_1,\ldots,b_s \end{pmatrix}; q, x\Big) = \sum_{n\in\mathbf{Z}} \frac{(a_1,\ldots,a_r;q)}{(b_1,\ldots,b_s;q)} (-1)^{s-r} q^{(s-r)n(n-1)/2} x^n.$$

The word "bilateral" refers of course to the summation over $n \in \mathbf{Z}$. (Also recall from Chapter 3 the shortened notation $(a_1,\ldots,a_r;q) := (a_1;q)\cdots(a_r;q)$.)

EXERCISE 4.14. Discuss the domain of convergence of ${}_r\psi_s$.

We shall prove the formula by a method due to Andrews and Askey (there are many more, look for them in [**93, 186**]!). We introduce the following function:

$$f(x) := \sum_{n\in\mathbf{Z}} b^n \frac{(a;q)_n}{(x;q)_n}.$$

We already saw (Chapter 2) that for negative values of the index, the q-Pochhammer symbols have the following values: if $n = -m$ and $m \geq 1$, then $(a;q)_n = \dfrac{(-1)^m q^{m(m+1)/2}}{(a-q)\cdots(a-q^m)}$. This entails:

$$f(x) = \sum_{n\in\mathbf{N}} b^n \frac{(a;q)_n}{(x;q)_n} + \sum_{m\in\mathbf{N}^*} b^{-m} \frac{(x-q)\cdots(x-q^m)}{(a-q)\cdots(a-q^m)}.$$

So, for all the terms of $f(x)$ to have a meaning, we assume that $x \notin q^{-\mathbf{N}}$, $b \neq 0$ and $a \notin q^{\mathbf{N}^*}$; and for both series to converge, we assume $|b| < 1$ (for the first one) and $|x| < |ab|$ (for the second one). (This is done by finding the limit of the quotient of two successive terms in each series.) As a series of functions of x, we may more precisely say that it (the second term) converges normally in any closed disk $\overline{\mathrm{D}}(0,r)$ where $0 < r < r_0$, where $r_0 := \min(1, |ab|)$; so it defines a holomorphic function in $\mathring{\mathrm{D}}(0,r_0)$. We now find a q-difference equation for f. First, from the relation $(x;q)_{n+1} - (1-x)(qx;q)_n$, we draw:

$$f(qx) = \sum_{n\in\mathbf{Z}} b^n \frac{(a;q)_n}{(qx;q)_n} = (1-x) \sum_{n\in\mathbf{Z}} b^n \frac{(a;q)_n}{(x;q)_{n+1}}.$$

Second, assuming for a moment that $x \neq 0, 1$ and shifting the summation index:

$$\begin{aligned} f(x) &= \sum_{n\in\mathbf{Z}} b^{n+1}\frac{(a;q)_{n+1}}{(x;q)_{n+1}} = b\sum_{n\in\mathbf{N}} b^n \frac{(a;q)_n}{(x;q)_n}\frac{1-aq^n}{1-xq^n} \\ &= b\sum_{n\in\mathbf{N}} b^n\frac{(a;q)_n}{(x;q)_n}\left(\frac{1-a/x}{1-xq^n}+\frac{a/x-aq^n}{1-xq^n}\right) \\ &= b(1-a/x)\sum_{n\in\mathbf{Z}} b^n\frac{(a;q)_n}{(x;q)_{n+1}} + ba/x\sum_{n\in\mathbf{N}} b^n\frac{(a;q)_n}{(x;q)_n} \\ &= b(1-a/x)\frac{f(qx)}{1-x}+\frac{ba}{x}f(x), \end{aligned}$$

whence the functional equation:

$$f(x) = \frac{b(1-a/x)}{(1-x)(1-ba/x)}f(qx) = \frac{1-x/a}{(1-x)(1-x/ab)}f(qx).$$

In the latter form, the relation obviously remains valid for $x = 0$. Note that this q-difference equation allows us to extend f to a meromorphic function over $\mathbf{C}$.

EXERCISE 4.15. Use the q-difference equation to find all its zeroes and poles.

Third, iterating the relation, we find that:

$$f(x) = C\frac{(x/a;q)_\infty}{(x;q)_\infty(x/ab;q)_\infty},$$

where $C := f(0)$ is a constant to be determined. To do so, we evaluate at $x = q$. By Cauchy's q-binomial formula (1.4 or Chapter 3):

$$f(q) = \sum_{n\geq 0}\frac{(a;q)_n}{(q;q)_n}b^n + \sum_{m\geq 1} 0 = \frac{(ab;q)_\infty}{(b;q)_\infty}, \text{ whence } C = \frac{(q;q)_\infty(q/ab;q)_\infty(ab;q)_\infty}{(q/a;q)_\infty(b;q)_\infty}$$

and at last:

$$f(x) = \frac{(q;q)_\infty(q/ab;q)_\infty(ab;q)_\infty(x/a;q)_\infty}{(q/a;q)_\infty(b;q)_\infty(x;q)_\infty(x/ab;q)_\infty}.$$

We can now deduce:

THEOREM 4.1 (Ramanujan's ${}_1\psi_1$ summation formula). *Let $a, b \in \mathbf{C}$, $a \notin q^{\mathbf{N}^*}$, $b \notin q^{-\mathbf{N}}$. Then, for $x \in \mathbf{C}$ such that $|b/a| < |x| < 1$:*

$${}_1\psi_1\left(\begin{matrix} a \\ b\end{matrix}; q, x\right) = \frac{(q, b/a, ax, q/(ax); q)_\infty}{(b, q/a, x, b/(ax); q)_\infty}.$$

PROOF. Just interchange b and x in the above discussion. □

4.2. Nonhomogeneous equations of order 1

These are equations of the form:

(4.1) $\quad \sigma_q f = uf + v, u, v \in$ some q-difference field ("coefficient field").

We shall barely touch upon these equations, because, in their most general form, their "true" theory already exhibits the need for more powerful techniques (q-asymptotics, resummation, etc: Chapter 14 and [CV]).

We consider here equations $\sigma_q f = uf + v$, $u, v \in K^*$. We shall distinguish the cases that $u = 1$ or $u \neq 1$. We mostly tackle examples.

4.2.1. Equations $\sigma_q f = f + v$, $v \neq 0$.

4.2.1.1. Suppose $v = \sum_{n\geq 0} v_n x^n$. Then a power series solution $f = \sum_{n\geq 0} f_n x^n$ must satisfy $\forall n$, $(q^n - 1)f_n = v_n$. If $v_0 = 0$, there is a unique such solution $\sum \dfrac{v_n}{q^n - 1} x^n$. Moreover, it is convergent if, and only if, v is (and they have the same radius of convergence). So, for $K = \mathbf{C}((x))$ or $K = \mathbf{C}(\{x\})$, we get the following exact sequence (the so-called "(small) additive q-de Rham complex", see 8.3.2 and 10.3.5):

$$0 \to \mathbf{C} \to K \xrightarrow{\sigma_q - \mathrm{Id}} K \xrightarrow{v \mapsto v_0} \mathbf{C} \to 0.$$

On the left side, this expresses that $\mathrm{Ker}(\sigma_q - \mathrm{Id}) = K^{\sigma_q} = \mathbf{C}$ as we already know. On the right-hand side (following from the previous calculation) we see that $\mathrm{Coker}(\sigma_q - \mathrm{Id}) = \mathbf{C}$, but, more precisely, that $K = \mathrm{Im}(\sigma_q - \mathrm{Id}) \oplus \mathbf{C}$, where the decomposition of $v \in K$ is $(v - v_0) + v_0$.

EXERCISE 4.16. If $v \in \mathscr{O}(\mathbf{C}^*, 0)$, it has a power series expansion $v = \sum v_n x^n$ with indexes (and exponents) not bounded below. Then going from $\sum v_n x^n$ to $\sum \dfrac{v_n}{q^n - 1} x^n$ changes the radius of convergence on one side by a factor $|q|$. What does the "small additive De Rham complex" become in that case?

4.2.1.2. Equation $\sigma_q f = f + v$ can be seen as a fixpoint equation $f = \Phi(f) := -v + \sigma_q f$ and we can try to solve it by iterating Φ. First we get:

$$f = -(v + \sigma_q v + \cdots + \sigma_q^{k-1} v) + \sigma_q^k f.$$

Then assuming $f(0) = 0$ (which requires $v(0) = 0$, i.e., $v_0 = 0$), we obtain:

$$f = -\sum_{k\geq 0} \sigma_q^k v.$$

However this only gives a meaningful result if $v(0) = 0$, i.e., if $v = \sum\limits_{n>0} v_n x^n$. Then we get:

$$f = -\sum_{k\geq 0}\sum_{n>0} q^{kn} v_n x^n = \sum_{n>0} \frac{v_n}{q^n - 1} x^n.$$

If $v = \sum\limits_{n<0} v_n x^n$ we rather write as follows the fixpoint equation and iteration:

$$f = sq^{-1} v + \sigma_q^{-1} f = \cdots = sq^{-1} v + \cdots + sq^{-(k-1)} v + \sigma_q^{-k} f.$$

Now we can make $k \to \infty$ because $\lim\limits_{k\to\infty} \sigma_q^{-k} f(x) = f(\infty)$, which is meaningful (and vanishes) if f has the form $\sum\limits_{n<0} f_n x^n$.

EXERCISE 4.17. Finish the computation of f by that method.

4.2.1.3. There is an interesting way to understand the "obstruction" v_0 to the resolution of $\sigma_q f = f + v$. Let Γ a simple analytic contour in $\mathbf{C}^*$, for instance a circle centered at 0. Then $q\Gamma$ is also a simple contour and, for any function f analytic in the neighborhood of $q\Gamma$:

$$\int_\Gamma \sigma_q f(x) \, \frac{dx}{x} = \int_{q\Gamma} f(x) \, \frac{dx}{x}.$$

(You should check it using a parameterization $\gamma : [a, b] \to \mathbf{C}^*$ and the definition $\int_\Gamma g(x)\,dx := \int_a^b g(\gamma(t))\gamma'(t)\,dt$.) Now for any two contours Γ_1, Γ_2, use the abreviation $\int_{\Gamma_2 - \Gamma_1} := \int_{\Gamma_2} - \int_{\Gamma_1}$. We deduce from the above (assuming hypotheses on f left for you to state):

$$\int_{q\Gamma - \Gamma} f(x)\,\frac{dx}{x} = \int_\Gamma (\sigma_q f - f)(x)\,\frac{dx}{x} = \int_\Gamma v(x)\,\frac{dx}{x}.$$

By Cauchy theorem, the former is $2i\pi\times$ the sum of the residues of f/x in the domain delimited by the contours Γ and $q\Gamma$ (an annulus if Γ is a circle) while the latter is $2i\pi\times$ the sum of the residues of f/x in the domain delimited by the contour Γ. So, if we take for Γ a circle small enough that v is holomorphic in the punctured disk, writing a_i the poles of f in the corresponding annulus and r_i the residues at these poles, we get:

$$\sum r_i/a_i = v_0.$$

In particular if $v_0 \neq 0$ there is no hope for a solution without poles.

EXERCISE 4.18. Apply this to $v = 1$ to get information on all possible q-logarithms and see what it gives in particular for $-x\theta_q{}'/\theta_q$.

EXERCISE 4.19. If g is another solution: $\sigma_q g - g = v$, writing b_j its poles and s_j the corresponding residues, we see that $\sum r_i/a_i = \sum s_j/b_j$. Check that this actually is the residue formula for the q-elliptic function $g - f$.

4.2.1.4. Here is a multiplicative application of the above. Suppose that $\sigma_q F = VF$. Then $f := xf'/f$ satisfies $\sigma_q f = f + v$, where $v := xV'/V$. (Taking $V := 1/x$ gives you the example of q-logarithms.) As a consequence, if F has zeroes and poles a_i with orders r_i in the fundamental annulus (orders of poles are counted negatively, i.e., they are actually valuations), then:

$$\int_\Gamma \frac{V'}{V}(x)\,dx = 2i\pi \sum r_i.$$

Taking for example $V := c \in \mathbf{C}^*$, we see that a q-character has as many zeroes as poles (counting multiplicities).

EXERCISE 4.20. Do the calculation when $F = (x; q)_\infty$.

4.2.1.5. *When v is rational.* Using partial fraction decomposition $v = \sum v_i$ of $v \in \mathbf{C}(x)$ and solving independently $\sigma_q f_i - f_i = v_i$ (so that $f := \sum f_i$ solves $\sigma_q f - f = v$), we can assume that either v is a polynomial or $v = (x-a)^{-r}$, $r \in \mathbf{N}^*$.

- If $v = v_0 + \cdots + v_n x^n$, then we can take $f := v_0 l_q + \dfrac{v_1}{q-1} x + \cdots + \dfrac{v_n}{q^n - 1} x^n$.
- If $v = x^{-r}$, $r \in \mathbf{N}^*$, we can take $f := \dfrac{1}{q^{-r} - 1} x^{-r}$.
- If $v = (x-a)^{-r}$, $r \in \mathbf{N}^*$, $a \in \mathbf{C}^*$, we can use the second iteration process and take $f := \sum\limits_{n \geq 1} \dfrac{1}{(q^{-n}x - a)^{-r}}$.

EXERCISE 4.21. Solve the last case above by expanding v in $\mathbf{C}\{x\}$.

4.2.2. Equations $\sigma_q f = uf + v$, $u \neq 1$, $v \neq 0$. If $u := \sum u_n x^n$ and $v := \sum v_n x^n$ are power series and we look for a power series solution $f := \sum f_n x^n$, we are led to the relations $q^n f_n = v_n + \sum_{k+l=n} u_k f_l$. This a recursive system if $u_k = 0$ for $k \leq 0$, i.e., if $v_0(u) > 0$ (recall from General Notations 0.4 that v_0 denotes the x-adic valuation, in other terms the order at 0). There will be more information on this kind of process in Chapter 5, although section 4.3 will provide some indirect information via the associated homogeneized q-difference equation of order 2. Here we explore other possible approaches on some simple minded examples.

4.2.2.1. *Iteration.* We shall only look at the "natural" iteration, that which involves $f(q^n x)$ for $n \to +\infty$. So we rather write our q-difference equation as $f = a\sigma_q f + b$, $a := 1/u$, $b := -v/u$. Iteration of $f = b + a\sigma_q f$ gives (easy induction is left to you):

$$\forall m \geq 1\,,\; f = b + a\sigma_q b + \cdots + a\sigma_q a \cdots \sigma_q^{m-1} a \sigma_q^m b + a\sigma_q a \cdots \sigma_q^m a \sigma_q^{m+1} f.$$

It is sensible to write $a := cx^k \epsilon$ and $b := dx^l \eta$ (conventions similar to those from 4.1.2.1). Then:

$$a\sigma_q a \cdots \sigma_q^{m-1} a = q^{km(m-1)/2} c^m x^{km} \epsilon_m, \text{ where } \epsilon_m := \epsilon \sigma_q \epsilon \cdots \sigma_q^{m-1} \epsilon.$$

If we hope for formal convergence ("in the x-adic sense"), we want to only have $k > 0$ in present terms, i.e., $v_0(a) > 0$. The choice of b seems less determining. So we now examine an instance of each of the following signs for $v_0(a)$.

EXAMPLE 4.22 ($a = x, b = 1$ i.e., $\sigma_q f = x^{-1} f - x^{-1}$). Then iteration and going to the limit $n \to +\infty$ gives:

$$f = 1 + x\sigma_q f + \cdots + q^{m(m-1)/2} x^m + q^{m(m+1)/2} x^{m+1} \sigma_q^{m+1} f = \cdots = \sum_{m \geq 0} q^{m(m-1)/2} x^m.$$

This gives an entire series with quickly convergent coefficients and an infinite radius of convergence.

EXAMPLE 4.23 ($a = x^{-1}, b = -x^{-1}$ i.e., $\sigma_q f = xf + 1$). Here we find:

$$f = -\sum_{m \geq 1} q^{-m(m-1)/2} x^{-m},$$

a (very) divergent series in x^{-1}. We can also consider iterating the equivalent equation $f = \sigma_q^{-1}(xf + 1) = 1 + q^{-1} x \sigma_q^{-1} f$; this gives an element of $\mathbf{C}[[x]]$, but divergent (see exercise below).

EXAMPLE 4.24 ($a \in \mathbf{C}^*, b = x^l$). Iteration gives $f := (\sum\limits_{m \geq 0} a^m q^{ml}) x^l$, which makes sense if, and only if, $l > 0$. However a direct look gives in all cases $f := \dfrac{1}{1 - aq^l}$, except if $a = q^{-l}$ which is in essence the case of the q-logarithm.

EXERCISE 4.25. (i) Solve those three q-difference equations with recursive relations on coefficients.

(ii) Solve the second one by rather iterating $f = \sigma_q^{-1}(xf + 1)$.

(iii) Complete the exceptional case in the third example.

4.2.2.2. *Straight Lagrange method.* Now we try to solve $\sigma_q f = uf + v$ in $\mathscr{M}(\mathbf{C}^*, 0)$. In analogy with Lagrange method for ordinary differential equations, we first solve the associated homogeneous equation $\sigma_q p = up$; the unknown function is called p because we now use a solution $p_0 \neq 0$ as a gauge transformation $f = p_0 g$, which leads us to a q-difference equation for g; our equation transforms to:

$$\sigma_q(p_0 g) = up_0 g + v \Longleftrightarrow up_0 \sigma_q g = up_0 g + v \Longleftrightarrow \sigma_q g = g + v/up_0,$$

which we already studied in 4.2.1.

EXAMPLE 4.26. To solve $\sigma_q f = xf + 1$, we take $p_0 := 1/\theta_q$ and are led to q-difference equation $\sigma_q g - g = \theta_q / x = \sum\limits_{n \in \mathbf{Z}} t_n x^n$. (You may replace the t_n by their values from 2.2.1.) So we isolate the term t_0 which requires a q logarithm and take:

$$f := \frac{1}{\theta_q} \left(t_0 l_q + \sum_{n \neq 0} \frac{t_n}{q^n - 1} x^n \right).$$

Note that the series has quickly decaying coefficients for $n \to \pm\infty$, so it defines a function in $\mathscr{O}(\mathbf{C}^*)$.

EXERCISE 4.27. The example failed while trying to stick to plain power series. Prove that there is no solution without essential singularities at 0 and at ∞.

4.2.2.3. *Twisted Lagrange method.* This is intended to avoid q-logarithms. These are caused by obstructions to $\sigma_q g - g = w$ in the form of a nonzero constant term in w. More generally, an equation $\sigma_q g - cg = w$, $c \in \mathbf{C}^*$, would find, if $c = q^n$, a similar obstruction in the form of a nonzero x^n term in w. This encourages us to try the gauge transform $f = p_0 g$ for p_0 a nontrivial solution of the "twisted" associated homogeneous equation $\sigma_q p = cup$ (read: the product of three factors!), $c \in \mathbf{C}^* \setminus q^{\mathbf{Z}}$. This process will be generalized under the name of "algebraic summation" in Chapter 15.

EXAMPLE 4.28. To solve $\sigma_q f = xf + 1$, we first solve $\sigma_q p = (x/c)p$, taking for instance $p_0(x) := \dfrac{1}{\theta_q(x/c)}$. We are led to:

$$\sigma_q g - cg = \frac{c}{x} \theta_q(x/c) = \sum_{n \in \mathbf{Z}} t_n (x/c)^n,$$

using the same notation as in the example of 4.2.2.2. We eventually reach the solution:

$$f(x) := \frac{1}{\theta_q(x/c)} \sum_{n \in \mathbf{Z}} \frac{t_n c^{-n}}{q^n - c} x^n.$$

EXERCISE 4.29. Show that f is the only solution in $\mathscr{M}(\mathbf{C}^*, 0)$ of $\sigma_q f = xf + 1$ having at most a simple pole at $-c$.

4.3. Homogeneous equations of order 2

4.3.1. Generalities. These are equations of the following form (or reducible to it):

$$\sigma_q^2 f + a \sigma_q f + bf = 0, \tag{4.2}$$

where $a, b \in K$, some q-difference field of coefficients that we shall usually take to be $\mathbf{C}(\{x\})$ or $\mathbf{C}((x))$, sometimes $\mathbf{C}(x)$; and for which we shall usually look for solutions in some q-difference extension of K, usually $\mathscr{M}(\mathbf{C}^*,0)$ or $\mathscr{M}(\mathbf{C}^*)$. Existence of solutions in the base field is considered exceptional, as we already saw for equations of order 1. However, solutions involving *divergent power series* appear spontaneously.

As with complex linear ordinary differential equations, homogeneous equations of order ≥ 2 generically admit such *divergent solutions*; yet, since Euler, it has been possible to use them as analytic objects. In truth, this was not done rigorously until much later (beginning with Poincaré). A similar theory exists for q-difference equations; we shall superficially touch upon it in this section, then more seriously later in Chapters 14 and 15. However, in sharp contrast with classical theory, q-difference equations always have (also) nontrivial *convergent solutions.* This seems to be due to Adams and Birkhoff and will be shown, in the particular case studied here, in a direct way; then in full generality in Chapter 5.

We begin this section by exploring the relation to the previous one.

4.3.1.1. *From* (4.1) *to* (4.2). First we start from a nonhomogeneous equations of order 1 (4.1) $\sigma_q f = uf + v$ (we assume $v \neq 0$). This can be written as:

$$(\sigma_q - u)f = v.$$

Setting $u' := \sigma_q v/v$, so that $(\sigma_q - u')v = 0$, we conclude that:

$$\sigma_q f = uf + v \Longrightarrow (\sigma_q - u')(\sigma_q - u)f = 0,$$

i.e., it implies (4.2) with $a := -(u' + \sigma_q(u))$ and $b := uu'$.

Conversely, if f is a solution of the above homogeneous equation, we see that $f' := (\sigma_q - u)f$ satisfies $\sigma_q f' = u'f'$; so $f' = vg$, where g is q-invariant (a σ_q-constant). Then $(\sigma_q - u)(f/g) = ((\sigma_q - u)f)/g = f'/g = v$, i.e., f/g is a solution of (4.1). We conclude that, *up to a q-constant factor*, every solution of (4.2) is a solution of (4.1).

4.3.1.2. *From* (4.2) *to* (4.1). Suppose conversely that we start from a homogeneous equation of order 2 (4.2) and assume that we found a particular nontrivial solution $f_0 \neq 0$. Then setting $u := \sigma_q f_0/f_0$, we have $(\sigma_q - u)f_0 = 0$. Euclidean division of the order 2 q-difference operator by this one writes $\sigma_q^2 + a\sigma_q + b = (\sigma_q - u')(\sigma_q - u) + c$ (Chapter 3) and since both $\sigma_q^2 + a\sigma_q + b$ and $(\sigma_q - u)$ kill f_0, we have $cf_0 = 0$, whence $c = 0$, whence:

$$\sigma_q^2 + a\sigma_q + b = (\sigma_q - u')(\sigma_q - u), \text{i.e. } u' + \sigma_q(u) = -a \text{ and } uu' = b.$$

Either of these two relations allow us to compute u' from u.

EXERCISE 4.30. Why do they give the same value for u'?

Once we have found this factorisation (from the knowledge of the particular nontrivial solution f_0), we can solve (4.2) in two step: first look for v such that $\sigma_q v = u'v$ (thus, an homogeneous equation of order 1); second look for f such that $(\sigma_q - u)f = v$, i.e., $\sigma_q f = uf + v$ (thus, a nonhomogeneous equation of order 1).

4.3.1.3.

EXAMPLE 4.31. Quickly convergent series $f_0 := 1 + x + qx^2 + q^3x^3 + \cdots = \sum_{n\geq 0} q^{n(n-1)/2}x^n$ is solution of $f = 1 + x\sigma_q f$, so $(x\sigma_q - 1)f = -1$, so $(\sigma_q - 1)(x\sigma_q - 1)f = 0$, i.e.:

$$qx\sigma_q^2 f - (1+x)\sigma_q f + f = 0.$$

Looking for power series solutions $f = \sum a_n x^n$, we arrive at the recursive relations $q^{2n-1}a_{n-1} - q^n a_n - q^{n-1}a_{n-1} + a_n = 0$, i.e.:

$$(1-q^n)a_n = (1-q^n)q^{n-1}a_{n-1} \Longrightarrow \forall n \neq 0\ ,\ a_n = q^{n-1}a_{n-1}.$$

The recursion splits f into two parts, respectively with positive and negative exponents:

$$f = a_0 f_0 + a_{-1} f_1, \text{ where } f_1 := q^{-1}(qx^{-1} + q^3x^{-2} + \cdots) = q^{-1} \sum_{n\geq 1} q^{n(n+1)/2} x^{-n}.$$

The series f_1 is also quickly convergent but in x^{-1}. So one part defines a function in $\mathscr{O}(\mathbf{C})$, the other in $\mathscr{O}(\mathbf{S} \setminus \{0\}$: if $a_0, a_{-1} \neq 0$, both parts are present and the solution is in $\mathscr{O}(\mathbf{C}^*)$.

EXERCISE 4.32. Relate $f_1(x)$ to $f_0(q/x)$ and $f_0 + qf_1$ to θ_q. Also look at what $(x\sigma_q - 1)$ does to these three functions.

4.3.1.4.

EXAMPLE 4.33. Strongly divergent series $f_0 := 1 + q^{-1}x + q^{-3}x^2 + \cdots = \sum\limits_{n\geq 0} q^{-n(n+1)/2}x^n$ is solution of $\sigma_q f = xf + 1$, i.e., of $(\sigma_q - x)f = 1$, whence of $(\sigma_q - 1)(\sigma_q - x)f = 0$, i.e., $\sigma_q^2 f - (1+qx)\sigma_q f + xf = 0$. We return to this in 5.3 under the name of Tshakaloff.

EXERCISE 4.34. Check that f_0 is, up to a constant factor, the only power series solution of this equation. Also check all nondetailed calculations in this number.

As we shall see later, it is useful to look for solutions in the form $\theta_q^{-\mu}e_{q,c}g$, g a power series[3]; we can as well take $g = 1 + \cdots$ because any power of x can be taken care of in the q-character $e_{q,c}$ (using the fact that $e_{q,qc} = xe_{q,c}$ times a constant, see Chapter 2) and constant factors in solutions are inessential. From $f = \theta_q^{-\mu}e_{q,c}g$, we draw

$$\sigma_q f = cx^\mu \theta_q^{-\mu} e_{q,c}\sigma_q g \text{ and then } \sigma_q^2 f = c^2 q^\mu x^{2\mu}\theta_q^{-\mu}e_{q,c}\sigma_q^2 g,$$

so, substituting in our equation and dividing by common factors $\theta_q^{-\mu}e_{q,c}$, we are led to:

$$c^2q^\mu x^{2\mu}\sigma_q^2 g - (1+qx)cx^\mu\sigma_q g + xg = 0.$$

Evaluating at 0, we see that the three terms must be such that two of them must have the same valuation, the third must have a higher one. Indeed, if one term only has the smallest valuation $\nu := \inf(2\mu, \mu, 1)$, then after division by x^ν and evaluation at 0 we get an impossible equation $0 \neq 0$. For instance, if $\mu = -1$, the three terms have valuations $-2, -1, 1$ and multiplying by x^2 then evaluating yields $c^2q^\mu = 0$, so this case must be excluded.

One can visualize the condition on valuations by saying that $(2\mu, \mu, 1)$ must look like one of the following configurations: $\circ\ \circ\ \circ$ or $^\circ\ \circ\ \circ$ or $\circ\ ^\circ\ \circ$ or $\circ\ \circ\ ^\circ$, but not one of the following: $^\circ\ ^\circ\ \circ$ or $^\circ\ \circ\ ^\circ$ or $\circ\ ^\circ\ ^\circ$. (More on this "visualisation tool" in 4.3.2.)

If $\mu > 1$, $(2\mu, \mu, 1)$ rather looks like $^\circ\ ^\circ\ \circ$ while if $\mu < 0$ it looks like $\circ\ ^\circ\ ^\circ$. At any

[3]This is to be compared with solutions of ordinary differential equations obtained as a product of an exponential, a "character" z^α and a power series; so exponentials are replaced here by Theta powers, but the former are more varied and can be wilder that the latter.

rate, we conclude that $\mu = 0$ or $\mu = 1$. We leave it to the reader to complete the case $\mu = 0$ by checking that it entails $c = 1$ and corresponds to our series above.

If $\mu = 1$, then we go for a solution $\theta_q^{-1} e_{q,c} g$ such that:

$$c^2 qx\sigma_q^2 g - (1 + qx)c\sigma_q g + g = 0.$$

Evaluating at 0 yields $-c + 1 = 0$, whence $c = 1$, i.e., a solution $\theta_q^{-1} g$ such that:

$$qx\sigma_q^2 g - (1 + qx)\sigma_q g + g = 0.$$

The recursive relation on coefficients of $g = \sum_{n \geq 0} b_n x^n$ is:

$$\forall n \geq 1\,,\; q^{2n-1} b_n - q^n b_n - q^n b_{n-1} + b_n = 0,$$

i.e., $(1 - q^n) b_n = q^n (1 - q^{n-1}) b_{n-1}$, with initialisation $b_0 := 1$. It is readily checked (by you!) that this implies that $\forall n \geq 1\,,\; b_n = 0$, i.e., $g = 1$ and we are left with solution θ_q^{-1}. Of course we could have guessed it because of the right factor $(\sigma_q - x)$ in $(\sigma_q - 1)(\sigma_q - x)f = 0$!

4.3.1.5. Note that in the two examples we know in advance one particular nontrivial solution f_0 of the q-difference equation, so, setting $u := (\sigma_q f_0)/f_0$, we can factor the q-difference operator as $(\sigma_q - u')(\sigma_q - u)$ where $u' := -a - \sigma_q(u) = b/u$, thus reducing the work to the resolution of two equations of order 1. In 4.3.1.6, we shall describe another approach allowing the same reduction (but leading to an important generalisation). Here we consider convergence matters. But this is not the way we proceeded.

In the first example, we obtained the q-difference operator from a convergent factorisation (i.e., involving convergent series): $qx\sigma_q^2 - (1 + x)\sigma_q + 1 = (\sigma_q - 1)(x\sigma_q - 1)$. In the second example, the q-difference operator was factored as $\sigma_q^2 - (1 + qx)\sigma_q + x = (\sigma_q - 1)(\sigma_q - x)$. If we had used the factorisation $(\sigma_q - u')(\sigma_q - u)$ indicated above, the result would have been different since neither $\sigma_q - x^{-1}$ nor $\sigma_q - x$ is the corresponding $\sigma_q - u$. But at least, in the first example, the factorisation would have been convergent (because $u := (\sigma_q f_0)/f_0$ was, so also u'); while in the second example, it would not have been convergent (because $u := (\sigma_q f_0)/f_0$ cannot be convergent if f_0 is not). So we see that in both cases the q-difference operator admits *at least* one convergent factorisation.

EXERCISE 4.35. Try to guess how the convergent and divergent factorisations of $\sigma_q^2 - (1 + qx)\sigma_q + x$ are respectively related to the two parts of the configuration $\circ\,\circ\,{}^{\circ}$.

4.3.1.6. Let f, g two solutions of (4.2) and let $w := f\sigma_q g - g\sigma_q f$ their q-*wronskian* (this was defined in 3.2.1.2). Then $\sigma_q^2 f + a\sigma_q f + bf = \sigma_q^2 g + a\sigma_q g + bg = 0$, which entails:

$$\sigma_q w = \sigma_q f \sigma_q^2 g - \sigma_q g \sigma_q^2 f = \sigma_q f(-a\sigma_q g - bg) - \sigma_q g(-a\sigma_q f - bf) = b(f\sigma_q g - g\sigma_q f) = bw.$$

So it is anyway useful to look for a nontrivial solution w of $\sigma_q w = bw$. Then if by chance one finds a particular solution f of (4.2), solving the nonhomogeneous equation of order 1: $f\sigma_q g - g\sigma_q f = w$ (the unknown is g), one gets a second independent solution.

EXERCISE 4.36. Show that g is independent of f in the sense that g/f is not a q-constant.

4.3.2. Resolution by power series and some q-special functions (i.e., "basic bricks"). All that we are going to do now for order 2 q-difference equations will be subsequently generalized to arbitrary orders; so this here is more for initiation and motivation, and it will be illustrated on the case of "basic hypergeometric series" in section 4.4. We first look for formal solutions. Convergence matters will be discussed in 4.3.3.

4.3.2.1. We slightly change the form of the equation we are adressing and instead look at:

$$a\sigma_q^2 f + b\sigma_q f + cf = 0, a, b, c \in K, ac \neq 0. \tag{4.3}$$

Here again $K = \mathbf{C}(\{x\})$ or $\mathbf{C}((x))$. We write $a = \alpha x^m u$, $b = \beta x^n v$ and $c = \gamma x^p w$, where $\alpha, \beta, \gamma \in \mathbf{C}$, $\alpha\gamma \neq 0$, $m, n, p \in \mathbf{Z}$ and $u, v, w \in U_a$ or U_f (depending on K). If $b = 0$, we take $\beta := 0$, $n = \infty$ and $v = 1$ (for instance). Up to multiplying our equation by $x^{-\min(m,n,p)}$, we may and will assume that $\min(m, n, p) = 0$.

Our arguments rest upon a visualisation process of which ones, among m, n, p, are positive. This will be generalized by the "Newton polygon" in Chapter 10. The possible configurations are the following:

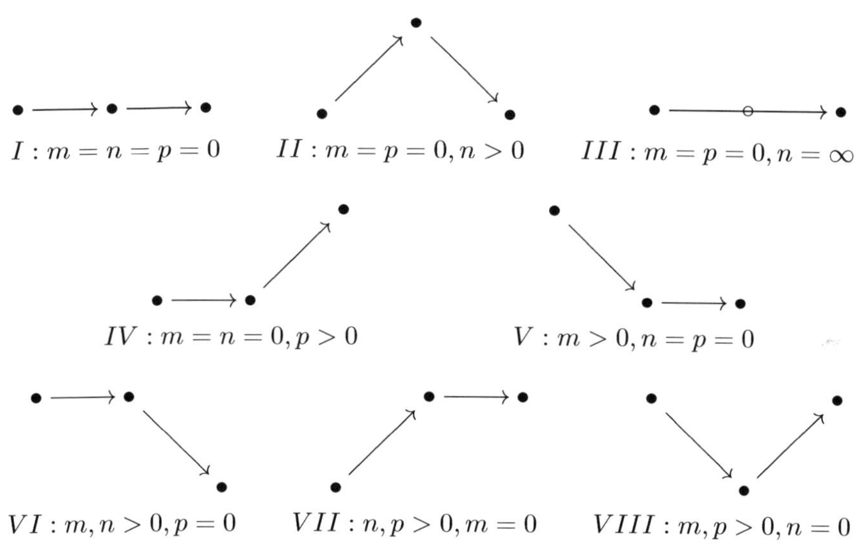

Here in 4.3.2.1, we look for conditions under which the q-difference equation (4.3) admits a *regular power series solution*, that is a solution $f \in K$ such that $v_0(f) = 0$, i.e., f has no pole at 0 and $f(0) \neq 0$. (Later we shall enlarge our domain of search.) Note that we can as well assume that $f(0) = 1$. Evaluating equation $a\sigma_q^2 f + b\sigma_q f + cf = 0$ at $x = 0$, we get the relation:

- In case I: $\alpha + \beta + \gamma = 0$.
- In cases II and III: $\alpha + \gamma = 0$.
- In case IV: $\alpha + \beta = 0$.
- In case V: $\beta + \gamma = 0$.
- In case VI: $\gamma = 0$.
- In case VII: $\alpha = 0$.
- In case VIII: $\beta = 0$.

Since we know that $\alpha\gamma \neq 0$, cases VI and VII are not allowed. Case VIII is not allowed either because $\beta = 0$ occurs only when $b = 0$, which is case III. So we see that only cases I to V are allowed, and to each case corresponds a particular relation. We can describe the five allowed configurations by noting that, in each case, there are at least two points in the bottom line: the configuration lays on a segment. We can also describe the associated equation by noting that, in each case, it reads:

$$a(0) + b(0) + c(0) = 0.$$

4.3.2.2. To understand better the condition we found, in particular the equation $a(0)+b(0)+c(0) = 0$, we enlarge slightly our domain of search and look for a power series solution f such that $v_0(f) = k$, i.e., $f = x^k g$, where g is a regular power series; we can as well assume that $g(0) = 1$, so $f = x^k + \cdots$, $\sigma_q f = q^k x^k + \cdots$ and $\sigma_q^2 f = q^{2k} x^k + \cdots$. Then evaluating the equation $a\sigma_q^2 f + b\sigma_q f + cf = 0$ at $x = 0$, we get the relation:

- In case I: $\alpha q^{2k} + \beta q^k + \gamma = 0$.
- In cases II and III: $\alpha q^{2k} + \gamma = 0$.
- In case IV: $\alpha q^{2k} + \beta q^k = 0$.
- In case V: $\beta q^k + \gamma = 0$.

In other words, in each case:

$$a(0)q^{2k} + b(0)q^k + c(0) = 0.$$

We are thus led to attach to our lower horizontal segment a *characteristic equation*:

$$a(0)s^2 + b(0)s + c(0) = 0,$$

and to say that there is a power series solution f such that $v_0(f) = k$ only if q^k is a root of the characteristic equation. This is a necessary condition, we have not proved that it is sufficient. This will be discussed later. We call these roots *exponents*, so our necessary condition here is that q^k is an exponent.

For a q-difference operator $L := a\sigma_q^2 + b\sigma_q + c$ such that $\min(v_0(a), v_0(b), v_0(c)) = 0$, we shall write:

$$\overline{L}(s) := a(0)s^2 + b(0)s + c(0).$$

4.3.2.3. Now we look for solutions of the form $f = e_{q,s}g$, $s \in \mathbf{C}^*$, g a regular power series. This case is actually a generalisation of the previous one since, up to a constant factor, x^k is e_{q,q^k}. Since $\sigma_q(e_{q,s}g) = se_{q,s}\sigma_q g$ and $\sigma_q^2(e_{q,s}g) = s^2 e_{q,s}\sigma_q^2 g$, we get (after division by $e_{q,s}$) the equation $as^2\sigma_q^2 g + bs\sigma_q g + cg = 0$, whence, evaluating at $x = 0$:

$$a(0)s^2 + b(0)s + c(0) = 0.$$

Therefore, for equation (4.3) to have a solution of the form $e_{q,s}g$, g a regular power series, it is necessary that s be a root of the *characteristic equation*, i.e., an *exponent*. Of course, since by definition the exponent $s \neq 0$, the characteristic equation must have at least two terms, and this precisely means that we are in one of the cases I to V.

EXERCISE 4.37. Let $L' := e_{q,s_0}^{-1} L e_{q,s_0}$. Compare $\overline{L}(s)$ and $\overline{L'}(s)$ and determine the logical relationship between the findings of 4.3.2.1, 4.3.2.2, and 4.3.2.3.

4.3.2.4. We shall now conversely try to determine if the above conditions (on the characteristic equation) are sufficient for the existence of a solution of the form indicated. We assume that we are in one of the cases I to V and write $a := \sum_{n\geq 0} a_n x^n$, and similarly for b and c. We do not anymore use the letters $m, n, p, \alpha, \beta, \gamma$ in the way we did.

We first look for a regular power series solution $f = \sum_{n\geq 0} f_n x^n$, $f_0 \neq 0$. So we must assume that $a_0 + b_0 + c_0 = \overline{L}(1) = 0$. We solve:

$$\sum_{n\geq 0} a_n x^n \sum_{n\geq 0} f_n q^{2n} x^n + \sum_{n\geq 0} b_n x^n \sum_{n\geq 0} f_n q^n x^n + \sum_{n\geq 0} c_n x^n \sum_{n\geq 0} f_n x^n = 0,$$

which is equivalent to:

$$\forall n \geq 0\ , \ \sum_{l+m=n} a_l q^{2m} f_m + \sum_{l+m=n} b_l q^m f_m + \sum_{l+m=n} c_l f_m = 0.$$

For $n = 0$, this boils down to $(a_0 + b_0 + c_0) f_0 = 0$, which (since we assumed that $a_0 + b_0 + c_0 = 0$) leaves us free to take any $f_0 \neq 0$. For $n \geq 1$, this becomes $\sum_{l+m=n} (a_l q^{2m} + b_l q^m + c_l) f_m = 0$, which in turn is equivalent to:

$$(a_0 q^{2n} + b_0 q^n + c_0) f_n = \text{ some explicit linear combination of } f_0, \ldots, f_{n-1}.$$

This is a valid recursive definition of the sequence (f_n) if no $a_0 q^{2n} + b_0 q^n + c_0$ vanishes for $n \geq 1$, i.e., if no q^n, $n \geq 1$, is an exponent. So, in the present state of our knowledge, for (4.3) to have a regular power series solution:

- a *necessary* condition is that $\overline{L}(1) = 0$;
- a *sufficient* condition is that $\overline{L}(1) = 0$ and $\overline{L}(q^n) \neq 0$ for all $n \geq 1$.

Now if we look for solutions of the form $f = e_{q,s} g$, g a regular power series, it is useful to introduce:

$$L' := e_{q,s}^{-1} L e_{q,s} = s^2 a \sigma_q^2 + s b \sigma_q + c,$$

so that $L(f) = 0 \Leftrightarrow L'(g) = 0$. The corresponding necessary condition on L' for such a solution g becomes:

$$\overline{L'}(1) = 0 \Longleftrightarrow a_0 s^2 + b_0 s + c_0 = 0 \Longleftrightarrow L(s) = 0.$$

Similarly, you can (and should) check that $\overline{L'}(q^n) = \overline{L}(q^n s)$ and we conclude that, for (4.3) to have a solution $f = e_{q,s} g$, g a regular power:

- a *necessary* condition is that $\overline{L}(s) = 0$;
- a *sufficient* condition is that $\overline{L}(s) = 0$ and $\overline{L}(q^n s) \neq 0$ for all $n \geq 1$.

We then say that s is a *nonresonant exponent*, i.e., a *nonresonant root* of the characteristic equation.

REMARK 4.38. In cases IV and V, the characteristic equation is either $\alpha s^2 + \beta s = 0$ or $\beta s + \gamma = 0$ (with the previous notations), so it has exactly one nonzero root. So resonancy could only occur in cases I, II, III. (Those will be later recognized as the *fuchsian* case, see Chapter 11.)

EXERCISE 4.39. Show that resonancy cannot occur in cases II and III.

4.3.3. Convergence.

4.3.3.1. So far, we said nothing about the convergence of the power series appearing in the solutions. Suppose for instance 1 is a nonresonant exponent. The recursion relation on the coefficients of f then has the form:

$$f_n := \frac{1}{a_0 q^{2n} + b_0 q^n + c_0} \times \text{ a linear combination of } f_0, \ldots, f_{n-1}.$$

This suggests that $|f_n|$ will grow more slowly if $c_0 \neq 0$ (if $c_0 = 0$, the leading factor is a fraction with denominator tending to 0). It will be shown in great generality in Chapter 5 that, if $c_0 \neq 0$, the resulting series has indeed a positive radius of convergence. This leaves as possible configurations I, II, III and V: we want to have the horizontal segment at right. This, in Chapter 10, will be expressed in terms of *Newton polygon* and *slopes*.

4.3.3.2. We shall now look at a few examples to illustrate the various possibilities. We start from q-difference operators given in factorised form. You are invited to look at the way the "slopes" of the resulting configurations can be read off the factors. This will be detailed afterward. You should also check the calculations, which will be less detailed than before ...

In the following examples, $a \in \mathbf{C}^*$ and $u \in U_a$, so that there is a unique $g \in U_a$ such that $u = \sigma_q g/g$.

EXAMPLE 4.40. $L := (x\sigma_q - a)(\sigma_q - u) = x\sigma_q^2 - (a + x\sigma_q(u))\sigma_q + au$: this is configuration V $^{\circ}$ o o. The characteristic equation $-as + a = 0$ has root 1, unique whence nonresonant(recall that resonancy is only possible in the case of configuration I o o o), so there is a unique regular power series solution with constant term 1. This is of course g (it anihilates the right factor $\sigma_q - u$).

EXAMPLE 4.41. $L := (\sigma_q - u)(x\sigma_q - a) = qx\sigma_q^2 - (a + ux)\sigma_q + au$: this is again configuration V $^{\circ}$ o o and again the characteristic equation is $-as + a = 0$. However, no regular power series f can anihilate the right factor. Instead we look at it so that $x\sigma_q f - af$ anihilates the left factor; so we are led to solve $x\sigma_q f - af = g$. We choose to do so by considering the fixpoint equation $f = \dfrac{-1}{a} g + \dfrac{x}{a}\sigma_q f$. Iteration and going to the limit yield:

$$f = \sum_{k=0}^{n-1} \left(\frac{x}{a}\sigma_q\right)^k \left(\frac{-1}{a} g\right) + \left(\frac{x}{a}\sigma_q\right)^n (f) = \frac{-1}{a} \sum_{n \geq 0} a^{-n} (x\sigma_q)^n (g).$$

That series converges normally on some disk. Indeed, choose $R, M > 0$ such that $|g| \leq M$ on $\overline{\mathrm{D}}(0, R)$ (it is possible since $g \in U_a$). Then, since $(x\sigma_q)^n = q^{n(n-1)/2} x^n \sigma_q^n$, we have $|(x\sigma_q)^n(g)| \leq q^{n(n-1)/2} R^n M$ on $\overline{\mathrm{D}}(0, R)$.

EXERCISE 4.42. Solve $x\sigma_q f - af = g$ by power series expansion; the recursive relation $q^{n-1} f_{n-1} - af_n = g_n$ can be made telescopic by dividing by $q^{n(n-1)/2}$.

Here the power series solution is related to the left factor; it converges, but we shall see that this is not true as a general rule.

EXAMPLE 4.43. $L := (a\sigma_q - x)(\sigma_q - u) = a\sigma_q^2 - (x + a\sigma_q(u))\sigma_q + ux$: this is configuration IV o o $^{\circ}$. The characteristic equation $as^2 - as = 0$ has only 1 as root in $\mathbf{C}^*$, so there is a unique regular power series solution with constant term 1. This is of course g.

EXERCISE 4.44. Compute $L' := (\theta_q e_{q,a}) L (\theta_q e_{q,a})^{-1}$ and then look for a solution in the form $(\theta_q e_{q,a})^{-1} \times$ a regular power series.

EXAMPLE 4.45. $L := (\sigma_q - u)(a\sigma_q - x) = asq^2 - (qx + au)\sigma_q + xu$: this is again configuration IV o o $^{\circ}$, with characteristic equation $as^2 - as = 0$. We look at the regular power series solution with constant term 1 by solving $a\sigma_q(f) - xf = g$. The recursive relation is $aq^n f_n - f_{n-1} = g_n$, with initialisation $af_0 = g_0 = 1$, whence $f_0 = 1/a$. The recursion is simplified (made telescopic) by a trick which is really an instance of so-called "q-Borel-Ramis" transformation[4]. We set:

$$\tilde{f}_n := q^{n(n+1)/2} f_n.$$

Then our recursion becomes:

$$a\tilde{f}_n - \tilde{f}_{n-1} = q^{n(n-1)/2} g_n \Longrightarrow a^n \tilde{f}_n - a^{n-1}\tilde{f}_{n-1} = q^{n(n-1)/2} a^{n-1} g_n.$$

Taking in account initial conditions, this entails:

$$a^n \tilde{f}_n = \sum_{k=0}^{n} q^{k(k-1)/2} a^{k-1} g_k \Longrightarrow \tilde{f}_n = a^{-n} \sum_{k=0}^{n} q^{k(k-1)/2} a^{k-1} g_k.$$

Now, a *necessary* condition for $\sum f_n x^n$ to have a positive radius of convergence is that $a^n \tilde{f}_n = a^n q^{n(n+1)/2} f_n \to 0$ when $n \to +\infty$; that is:

$$\sum_{n\geq 0} q^{n(n-1)/2} a^{n-1} g_n = 0.$$

We now discuss that condition, considering the convergent regular power series g as fixed, but not the parameter a. Since g has a positive radius of convergence, the value of the series $\sum\limits_{n\geq 0} q^{n(n-1)/2} g_n \xi^n$ is well defined for every ξ: it defines an entire function of ξ (the use of the letter ξ for the variable in the "Borel plane" is traditional in this context). This entire function does not vanish identically since its constant term is $g_0 = 1$. So its set of zeroes in the plane is discrete, thus denumerable. So for almost all values of a (meaning here: all but for a denumerable set), this sum is not 0 so the power series solution f found above does not converge.

So convergence of the solution attached to the left segment seems to be rather exceptional.

Note that the above argument provided us with a numerical criterion for the convergence of the formal power series solution f, that is: $\sum\limits_{n\geq 0} q^{n(n-1)/2} g_n a^n = 0$.

EXERCISE 4.46. Write $\mathcal{B}_q f(\xi) := \sum\limits_{n\geq 0} \tilde{f}_n \xi^n$. Translate the simplified (telescopic) recursion for $\tilde{f}_n$ into a condition on $\mathcal{B}_q f$.

EXERCISE 4.47. Study convergence and divergence of solutions of $\sigma_q^2 f + q\sigma_q f + xf = 0$ and of $(1-qx^2)\sigma_q^2 f - (1+x-qx^2+q^2x^3)\sigma_q f + x(1-q^3x^2)f = 0$ (see [**1**]).

We shall now look at two degenerate situations: first with a resonancy; then with a double exponent.

EXAMPLE 4.48. $L := (\sigma_q - (1-x))(\sigma_q - q) = \sigma_q^2 - (1+q-x)\sigma_q + q(1-x)$: this is configuration I ∘ ∘ ∘. The characteristic equation is $s^2 - (1+q)s + q = 0$, so the exponents are 1 and q. Because of the exponent 1, we look for a regular power series solution f. This cannot satisfy $(\sigma_q - q)(f) = 0$ (can you see why?) so it must be such that $\sigma_q f - qf = g$, a regular power series solution of $(\sigma_q - (1-x))(g) = 0$. We take $g := \dfrac{1}{(x;q)_\infty}$ and go on to solve:

$$\sigma_q f - qf = \sum_{n\geq 0} \frac{1}{(q;q)_n} x^n,$$

---

[4]This notion will be more formally introduced in Sections 6.1.3 and 14.1.3.

but this is impossible for a power series f: indeed, the q-difference operator $\sigma_q - q$ would kill the term of degree 1 and $\sigma_q f - qf$ would have no such term. This impossibility does not contradict our criteria, because 1 is here a resonant exponant and there is no guarantee in that case! So we first exploit the exponent q, which is nonresonant. It yields a solution $f := x$. To look for a second independent solution g, we shall use the q-wronskian $w := f\sigma_q g - g\sigma_q f = x(\sigma_q g - qg)$. By 4.3.1.6, it should satisfy $\sigma_q w = q(1-x)w$. So we take $w := \dfrac{x}{(x;q)_\infty}$ and proceed to solve:

$$x(\sigma_q g - q) = \frac{x}{(x;q)_\infty} \iff \sigma_q g - qg = \frac{x}{(x;q)_\infty},$$

which indeed is not possible with a power series, again because of the term in degree 1.

EXERCISE 4.49. Noting that $\sigma_q(g/f) - (g/f) = \dfrac{1}{qx(x;q)_\infty}$, obtain g in the form $g_1 + g_2$ where g_1 is a power series and g_2 a constant multiple of fl_q. Compute g_1 and g_2.

EXAMPLE 4.50. $L := (1-\alpha x)\sigma_q^2 - (2-\beta x)\sigma_q + (1-\gamma x)$, $\alpha, \beta, \gamma \in \mathbf{C}$: this is again configuration I o o o. The characteristic equation is $s^2 - 2s + 1 = 0$, so 1 is a double exponent. It is not resonant so there is a regular power series solution $f = \sum f_n x^n$. We initialize $f_0 := 1$ and write the recursive relation on coefficients:

$$(q^{2n} f_n - \alpha q^{2n-2} f_{n-1}) - (2q^n f_n - \beta q^{n-1} f_{n-1}) + (f_n - \gamma f_{n-1}) = 0,$$

whence:

$$f_n = \frac{\alpha q^{2n-2} - \beta q^{n-1} + \gamma}{(1-q^n)^2} f_{n-1}.$$

We suppose now that $\alpha\gamma \neq 0$ (but you should by yourself tackle the case that $\alpha\gamma = 0$). Then there are $\lambda, \mu \in \mathbf{C}^*$ such that $\lambda + \mu = \beta/\alpha$ and $\lambda\mu = \gamma/\alpha$, so that the numerator (in the recursive relation) writes $\gamma(1-\lambda q^{n-1})(1-\mu q^{n-1})$; and we get:

$$f_n = \gamma^n \frac{(\lambda;q)_n (\mu;q)_n}{(q;q)_n^2}.$$

Actually, f is essentially a "basic hypergeometric series"; see Section 4.4. To find a second independent solution g, we could again use the q-wronskian, but we rather do it as suggested in 4.3.1.2. Let $u := (\sigma_q f)/f = 1 + \cdots$. Then:

$$L = (1-\alpha x)(\sigma_q - u')(\sigma_q - u), \text{ where } u' := \frac{1-\gamma x}{1-\alpha x}\frac{1}{u} = 1 + \cdots$$

So now we solve $\sigma_q f' = u'f'$, getting $f' = 1 + \cdots$; and then $\sigma_q g - ug = f'$, equivalently $\sigma_q(g/f) - (g/f) = f'/(\sigma_q f) = 1 + \cdots$, which necessarily involves a q-logarithm.

EXERCISE 4.51. Find g using the q-wronskian and confirm the results.

4.3.4. Algorithm for fuchsian equations of order 2.

4.3.4.1. We now give a slightly more systematic presentation for the resolution of scalar q-difference equations of order 2 when they are "fuchsian", i.e., in the case of configuration I, II or III. Such an equation can be written $L(f) = 0$, where:

$$L := \sigma_q^2 - \lambda\sigma_q + \mu, \lambda, \mu \in K.$$

(Recall that $K = \mathbf{C}((x))$ or $\mathbf{C}(\{x\})$); and the fuchsian condition says that $v_0(\lambda) \geq 0$ and $v_0(\mu) = 0$, i.e., λ has no pole at 0 and μ is a regular power series. The *characteristic equation* then has the form:

$$\overline{L}(s) = s^2 - \lambda(0)s + \mu(0) = (s-c)(s-d) = 0,$$

where $c, d \in \mathbf{C}^*$ are the *exponents*. All that follows can be either deduced, or at least intuited from the above. Since at any rate these facts are particular cases of Chapter 5, we shall state them without proofs. The reader may consider it a useful exercise to find out which facts are not proved rigorously—and then to provide the lacking proofs.

The main criterion in the discussion is whether $c \not\equiv d \pmod{q^{\mathbf{Z}}}$ ("strong nonresonancy"), $c \equiv d \pmod{q^{\mathbf{Z}}}$ but $c \neq d$ ("resonancy") or $c = d$ (double exponent).

4.3.4.2. *Strongly nonresonant case:* $c \not\equiv d \pmod{q^{\mathbf{Z}}}$. Then there are unique solutions in the form $f := e_{q,c}\phi$ and $g := e_{q,d}\psi$ where ϕ, ψ are regular power series such that $\phi(0) = \psi(0) = 1$. If $K = \mathbf{C}(\{x\})$, then ϕ and ψ are convergent. In that case, $f, g \in \mathscr{M}(\mathbf{C}^*, 0)$. Moreover f and g are a basis of the $\mathscr{M}(\mathbf{C}^*, 0)^{\sigma_q}$-linear space $\mathrm{Sol}(L, \mathscr{M}(\mathbf{C}^*, 0))$.

EXERCISE 4.52. Prove that f, g are linearly independent over the constant field $\mathscr{M}(\mathbf{C}^*, 0)^{\sigma_q} = \mathscr{M}(\mathbf{E}_q)$. (Setting $h := g/f$, $(\sigma_q h)/h = (d/c)\times$ a regular power series, so $h \notin \mathscr{M}(\mathbf{C}^*, 0)^{\sigma_q}$.)

4.3.4.3. *Double exponent case:* $c = d$. Otherwise said, $\overline{L}(s) = (s-c)^2$, i.e., $\lambda(0) = 2c$ and $\mu(0) = c^2$. We shall first reduce this to the case that $c = d = 1$ by a *gauge transformation* $f = e_{q,c}f'$. We see that $L(f) = 0 \Longrightarrow (e_{q,c}^{-1} L e_{q,c})(f') = 0$ and readily check that $e_{q,c}^{-1} L e_{q,c} = c^2\sigma_q^2 - c\lambda\sigma_q + \mu$; so we set:

$$L' := c^{-2} e_{q,c}^{-1} L e_{q,c} = \sigma_q^2 - c^{-1}\lambda\sigma_q + c^{-2}\mu \Longrightarrow \overline{L'}(s) = s^2 - 2s + 1 = (s-1)^2.$$

So L' has the double exponent 1 and solutions of L are solutions of L' multiplied by $e_{q,c}$. So we may as well from start assume that $c = d = 1$ and let the reader eventually translate our findings back to the general case.

EXERCISE 4.53. Let L an arbitrary fuchsian q-difference operator of order 2. Prove that for any $c \in \mathbf{C}^*$ $e_{q,c}^{-1} L e_{q,c}$ is fuchsian and compare their characteristic equations.

Since (by our new assumption) 1 is a double exponent, it is not resonant; so, by 4.3.2.4, there is a unique regular power series solution such that $f(0) = 1$. It can, in principle, be found using the recursion relation for the coefficients. We now have two paths to complete resolution, i.e., to finding a second independent solution g.

- Setting $u := (\sigma_q f)/f$, we obtain a fatorisation:

  $$L = (\sigma_q - u')(\sigma_q - u), \text{ where } u' := \mu/u = \lambda - \sigma_q(u).$$

  Note that $u(0) = 1$, implying that $u'(0) = 1$. As a consequence, there is a unique f' such that $(\sigma_q f')/f' = u'$ and $f'(0) = 1$. (Indeed, $f' = 1/(u\,\sigma_q u\,\sigma_q^2 u \cdots)$.) Now we look for g by solving:

  $$\sigma_q g - ug = f' \Longleftrightarrow \sigma_q h - h = f'/\sigma_q(f), \text{ where } h := g/f.$$

Now, since the right-hand side $f'/\sigma_q(f) = 1 + \cdots$, this cannot be solved with a power series. One has to take $h := \ell_q + h'$ where h' is some power series such that $h'(0) = 0$ (and h' is uniquely determined). So, slightly changing notations we end with a solution $g = f l_q + g'$, g' a power series such that $g'(0) = 0$.

- Calling $w := f\sigma_q g - g\sigma_q f$ the q-wronskian of the looked for solution, we must have $\sigma_q w = \mu w$, so we take for instance $w := 1/(\mu\sigma_q\mu\sigma_q^2\mu\cdots) = 1 + \cdots$. Then we solve:

$$f\sigma_q g - g\sigma_q f = w \Longleftrightarrow \sigma_q h - h = w/(f\sigma_q(f)), \text{ where } h := g/f.$$

Again the right-hand side is of the form $1 + \cdots$ and the end of the story is the same. Actually, we obtain exactly the same solution by both methods: indeed, from $\sigma_q(ff')/(ff') = uu' = \mu$ and from the initial conditions $ff'(0) = w(0) = 1$, we draw that $w = ff'$, so $w/(f\sigma_q(f)) = f'/\sigma_q(f)$.

EXERCISE 4.54. (i) Compute $L(fl_q)$ and deduce that g' is found by solving nonhomogeneous q-difference equation: $L(g') = \lambda\sigma_q f + \mu f$. Translate this into a recursion relation for coefficients of g'.

An example of this kind will be treated in great detail in section 4.4.

4.3.4.4. *Resonant case:* $d = q^{-k}c$, $k \geq 1$. Again, up to a gauge transformation $f = e_{q,c}f'$, we may as well assume from start that $c = 1$ and $d = q^{-k}$. So 1 is a nonresonant exponent and there is a unique regular power series solution f with $f(0) = 1$. To find the other independent solution g, we use their q-wronskian w.

Since $cd = \mu(0)$, we also see that $\mu = q^{-k} \times \mu'$, where μ' is a regular power series with constant term $\mu'(0) = 1$. So we take $w := x^{-k}w'$, where $\sigma_q w' = \mu' w'$; and we can take $w' = 1 + \cdots$. Then we put $h := g/f$ and we solve:

$$f\sigma_q g - g\sigma_q f = w \Longleftrightarrow \sigma_q h - h = w/(f\sigma_q(f)) = x^{-k}h', \text{ where } h' = 1 + h'_1 x + \cdots$$

Now there are two possibilities: either $h'_k = 0$, in which case there is a power series solution $h = x^{-k}(1 + \cdots)$, and it can be made unique by requiring that it has no term of degree 0; or $h'_k \neq 0$ and then it is necessary to incorporate a q-logarithm in h, whence in g.

EXERCISE 4.55. In both cases, prove that f and g are K^{σ_q}-independent.

4.3.5. Vectorialisation (from scalar equations to systems).

4.3.5.1. A standard trick in the classical theory of linear ordinary differential equations is to transform a scalar order 2 equation $y'' + uy' + vy = 0$ into a vectorial equation $Y' = AY$, where $Y := \begin{pmatrix} y \\ y' \end{pmatrix}$ and $A := \begin{pmatrix} 0 & 1 \\ -v & -u \end{pmatrix}$. Moreover, the "cyclic vector lemma" conversely guarantees that any vectorial equation $Y' = AY$, where $Y := \begin{pmatrix} f \\ g \end{pmatrix}$ is a vector of functions and $A := \begin{pmatrix} a & b \\ c & d \end{pmatrix}$ is a matrix of functions, can somehow be reduced to a scalar equation of order 2. (Actually, this remains true for any order n, not only for $n = 2$.) This is due to Birkhoff.

The same trick is available for q-difference equations (indeed for general difference equations). Starting from a *scalar q-difference equation* of order 2:

$$\sigma_q^2 f + u\sigma_q f + vf = 0, u, v \in K, v \neq 0,$$

we go to[5]:

$$\sigma_q X = AX, \text{ where } X := \begin{pmatrix} f \\ \sigma_q f \end{pmatrix} \text{ and } A := \begin{pmatrix} 0 & 1 \\ -v & -u \end{pmatrix}.$$

Note that $\det A = v \neq 0$, expressing that the order is really 2.

4.3.5.2. The cyclic vector lemma in this setting (and in this elementary case) runs as follows. We start from a rank 2 *vectorial q-difference equation* $\sigma_q X = AX$, where $X := \begin{pmatrix} f \\ g \end{pmatrix}$ and $A := \begin{pmatrix} a & b \\ c & d \end{pmatrix}$, $a, b, c, d \in K$. Of course, we do not know where to look for the solutions f, g. This vectorial equation is equivalent to a q-difference system $\begin{cases} \sigma_q f = af + bg, \\ \sigma_q g = cf + dg. \end{cases}$ Assuming for instance that $b \neq 0$ (a more systematic study will come in Chapter 7), this is equivalent to the system:

$$g = \frac{\sigma_q f - af}{b} \text{ and } \sigma_q \left(\frac{\sigma_q f - af}{b} \right) = cf + d \frac{\sigma_q f - af}{b},$$

which, in turn, transforms to:

$$g = \frac{\sigma_q f - af}{b} \text{ and } \sigma_q^2 f - \frac{b\sigma_q a + d\sigma_q b}{b} \sigma_q f + \frac{(da - bc)\sigma_q b}{b} f = 0.$$

So we are indeed reduced to solve $\sigma_q^2 f + u\sigma_q f + vf = 0$, where $u := -\frac{b\sigma_q a + d\sigma_q b}{b}$ and $v := \frac{(da - bc)\sigma_q b}{b}$. If $\det A = ad - bc \neq 0$, then $v \neq 0$ and this q-difference equation really has order 2.

EXERCISE 4.56. (i) Treat similarly the case $c \neq 0$, writing $f = \cdots$.
(ii) Try the case $b = c = 0$. (This is trickier.)

4.3.5.3. *Rogers-Ramanujan's continued fraction.* Here is an example of a q-difference system arising "naturally". The continued fraction $\dfrac{1}{1 + \dfrac{1}{1 + \cdots}}$ has value the *golden number* $\phi := \dfrac{-1 + \sqrt{5}}{2} \simeq 0,618$. As we saw in 1.10, Ramanujan introduces the following q-deformation[6] of ϕ:

$$\phi_q := \cfrac{1}{1 + \cfrac{q}{1 + \cfrac{q^2}{1 + \cfrac{q^3}{1 + \cdots}}}}.$$

To study it, we reproduce Euler's trick from 1.2 and introduce:

$$f(x) := \cfrac{x}{1 + \cfrac{qx}{1 + \cfrac{q^2 x}{1 + \cfrac{q^3 x}{1 + \cdots}}}},$$

[5]There are other possible ways of vectorialisation; see further below 4.3.5.5 and, more significantly, 12.3.2.3.

[6]Actually, Ramanujan considers $q^{1/5}\phi_q$; see section 1.10 and [CV].

so that $\phi_q = f(1)$. General properties of analytic continued fractions **[251]** guarantee that f is analytic and vanishes at 0, and more precisely that $f = x + \cdots \in \mathbf{C}\{x\}$. We shall study more closely Rogers-Ramanujan's continued fraction in [CV] (we already met it in 1.10). Here, we just note that $f(x)$ is characterized by the conditions:

$$f(0) = 0 \text{ and } f = \frac{x}{1 + \sigma_q f}.$$

Heuristically, we look for it in the form $f = \dfrac{u}{v}$, with:

$$\frac{u}{v} = \frac{x}{1 + \dfrac{\sigma_q u}{\sigma_q v}} = \frac{x\sigma_q v}{\sigma_q u + \sigma_q v},$$

which we decide to transform into the q-difference system:

$$\begin{cases} u &= x\sigma_q v, \\ v &= \sigma_q u + \sigma_q v, \end{cases} \Longleftrightarrow \sigma_q \begin{pmatrix} u \\ v \end{pmatrix} = \begin{pmatrix} -1/x & 1 \\ 1/x & 0 \end{pmatrix} \begin{pmatrix} u \\ v \end{pmatrix}.$$

Of course this is not equivalent: the latter implies the former; we shall straighten this up soon. Anyway, that system is equivalent to:

$$u = x\sigma_q v,$$
$$v = qx\sigma_q^2 v + \sigma_q v.$$

Therefore we should look for $f := \dfrac{x\sigma_q v}{v}$ with v submitted to condition $qx\sigma_q^2 v + \sigma_q v - v = 0$.

Now that we know what we *should* look for, we justify it more rigorously. From $f = x + \cdots \in \mathbf{C}\{x\}$, we deduce that $g := f/x \in U_a$, so there exists a unique $v \in U_a$ such that $g = (\sigma_q v)/v$, i.e., $f = \dfrac{x\sigma_q v}{v}$. Moreover, an easy computation (left for your pleasure) gives:

$$f = \frac{x}{1 + \sigma_q f} \Longleftrightarrow \frac{x\sigma_q v}{v} = \frac{x}{1 + \dfrac{qx\sigma_q^2 v}{\sigma_q v}} = \frac{x\sigma_q v}{\sigma_q v + qx\sigma_q^2 v} \Longleftrightarrow v = \sigma_q v + qx\sigma_q^2 v$$

So this confirms our reduction to q-difference equation $qx\sigma_q^2 v + \sigma_q v - v = 0$.

The latter corresponds to configuration V $^{\circ}$ o o. Its characteristic equation is $s - 1 = 0$, so it admits a unique regular power series solution v such that $v(0) = 1$. Setting $v = \sum\limits_{n \geq 0} v_n x^n$, our q-difference equation leads to the recursive equation:

$$\forall n \geq 1 \,,\; q^{2n-1} v_{n-1} + (q^n - 1) v_n = 0,$$

which, along with initialisation $v_0 := 1$, solves into:

$$\forall n \geq 0 \,,\; v_n = \frac{q^{n^2}}{(q;q)_n},$$

whence $v = \sum_{n\geq 0} \frac{q^{n^2}}{(q;q)_n} x^n$, which entails $\sigma_q v = \sum_{n\geq 0} \frac{q^{n^2+n}}{(q;q)_n} x^n$ and at last $f(x) = x \frac{\sum_{n\geq 0} \frac{q^{n^2+n}}{(q;q)_n} x^n}{\sum_{n\geq 0} \frac{q^{n^2}}{(q;q)_n} x^n}$. In the end, we conclude:

$$\phi_q = \frac{\sum_{n\geq 0} \frac{q^{n^2+n}}{(q;q)_n}}{\sum_{n\geq 0} \frac{q^{n^2}}{(q;q)_n}}.$$

EXERCISE 4.57. (i) Find a solution corresponding to the left segment $\circ$ o of the configuration. To that end look for it in the form $v'(x) = \theta_q(-x)v''(x)$. (The motivation for this will become clear in Chapter 5.)

(ii) Look for v' using the q-wronskian $w := v\sigma_q v' - v'\sigma_q v$.

4.3.5.4. We now take as example a simplified version of [**44**]. The intent is to show how one tackles a q-difference system directly, not as above through a q-difference equation. The system is:

$$\begin{cases} \sigma_q f = af + bg, \\ \sigma_q g = cf + dg, \end{cases} \qquad a, b, c, d \in \mathbf{C}\{x\}, ad - bc \neq 0,$$

equivalently:

$$\sigma_q X = AX \text{ where } X := \begin{pmatrix} u \\ v \end{pmatrix} \text{ and } A := \begin{pmatrix} a & b \\ c & d \end{pmatrix}.$$

We write $a := \sum_{n\geq 0} a_n x^n$, all $a_n \in \mathbf{C}$; and similarly for b, c, d. Also $A := \sum_{n\geq 0} A_n x^n$, all $A_n \in \mathrm{Mat}_2(\mathbf{C})$.

We look for conditions for the system to have regular power series solutions, which means $X \in \mathbf{C}\{x\}^2$ such that $X(0) \neq 0$. So we write $f := \sum_{n\geq 0} f_n x^n$ and $g := \sum_{n\geq 0} g_n x^n$, all $f_n, g_n \in \mathbf{C}$; and $X := \sum_{n\geq 0} X_n x^n$, all $X_n \in \mathbf{C}^2$. Initialisation comes from evaluating at $x = 0$: $\begin{cases} f_0 = a_0 f_0 + b_0 g_0, \\ g_0 = c_0 f_0 + d_0 g_0, \end{cases}$, otherwise said $X_0 = A_0 X_0$. The desired property $X_0 \neq 0$ then imposes on us the *necessary* condition that 1 be an eigenvalue of A_0.

The recursion relation then is that, for all $n \geq 1$:

$$\begin{cases} q^n f_n = (a_0 f_n + \cdots + a_n f_0) + (b_0 g_n + \cdots + b_n g_0) = \sum_{i+j=n} a_i f_j + \sum_{i+j=n} b_i g_j, \\ q^n g_n = (c_0 f_n + \cdots + c_n f_0) + (d_0 g_n + \cdots + d_n g_0) = \sum_{i+j=n} c_i f_j + \sum_{i+j=n} d_i g_j. \end{cases}$$

In a more compact form: $q^n X_n = A_0 X_n + \cdots + A_n X_0$. The truly recursive form goes as follows:

$$\forall n \geq 1\,,\ (q^n I_2 - A_0) X_n = A_1 X_{n-1} + \cdots + A_n X_0.$$

We certainly can solve the recursion if all $q^n I_2 - A_0$, $n \geq 1$ are invertible; i.e., if no q^n, $n \geq 1$ is an eigenvalue of A_0. Along with the condition that 1 be an eigenvalue of A_0, this gives a *sufficient* condition for the existence of a regular power series solution.

Actually, we only obtained condition for a *formal* solution (i.e., $f, g \in \mathbf{C}[[x]]$) but we shall prove in all generality in Chapter 5 that they do converge.

EXERCISE 4.58. (i) In **[44]** the system really has the form $\sigma_q X = x^\mu AX$. Solve it using some Theta power (along with power series).

(ii) More generally, look for solutions involving Theta powers and q-characters (the same for f and g). You should get conditions on the eigenvalues of A_0.

4.3.5.5. If we want to consider a scalar q-difference equation or a q-difference system as a q-deformation of a differential scalar equation or system, it is convenient to use q-derivation $\delta_q = \dfrac{1-\sigma_q}{1-q}$ or $D_q = \dfrac{1-\sigma_q}{(1-q)x}$. An important example will be detailed in section 4.4, so here we just touch upon the process.

EXAMPLE 4.59. Let $f := \sum\limits_{n \geq 0} [n]_q! x^{n+1}$, a q-analogue of Euler series. Recalling that $\delta_q x^n = [n]_q x^n$, we see that:

$$\delta_q f = \sum_{n \geq 0} [n+1]_q! x^{n+1} \Longrightarrow x\delta_q f = \sum_{n \geq 0} [n+1]_q! x^{n+2} = f - x \Longrightarrow f = x + x\delta_q f.$$

We can rewrite this as:

$$f = x + x\frac{f - \sigma_q f}{1-q} \Longrightarrow \frac{x}{1-q}\sigma_q f + \left(1 - \frac{x}{1-q} f\right) = x \Longrightarrow L(f) = 0,$$

where $L := (\sigma_q - q)\left(\dfrac{x}{1-q}\sigma_q + 1 - \dfrac{x}{1-q}\right)$.

EXERCISE 4.60. (i) Treat this scalar equation and the associated q-difference system by the methods shown in this section. Among other thing you will get a vectorial equation $\sigma_q X = AX$.

(ii) Deduce an equation $\delta_q X = A'X$ (with the same X).

Another approach is to stay entirely within the δ_q world. Since $\delta_q x = x$, we have:

$$(x\delta_q - 1)f = -x \Longrightarrow L'(f) = 0.$$

where $L' := (\delta_q - 1)(x\delta_q - 1)$. Using the noncommutation rule $\delta_q x = qx\delta_q + x$, this transforms into:

$$L' = qx{\delta_q}^2 + x\delta_q - x\delta_q - \delta_q + 1 = qx{\delta_q}^2 - \delta_q + 1 \Longrightarrow qx{\delta_q}^2 f - \delta_q f + f = 0.$$

Setting $X' := \begin{pmatrix} f \\ \delta_q f \end{pmatrix}$, this can be vectorialized as:

$$\delta_q X' = A'X', \text{ where } A' := \begin{pmatrix} 0 & 1 \\ -1/qx & 1/qx \end{pmatrix}.$$

EXERCISE 4.61. Compare A and A' and try to find a general rule to go from one representation to the other.

When $q \to 1$, the equation $\delta_q X' = A'X'$ degenerates into $\delta Y = BY$, where $\delta := x\dfrac{d}{dx}$ (Euler's differential operator) and $Y := \begin{pmatrix} f \\ \delta f \end{pmatrix}$. The matrix is $B := \begin{pmatrix} 0 & 1 \\ -1/x & 1/x \end{pmatrix}$. So we are led to solve: $\delta^2 f = (-1/x)f + (1/x)\delta f$, i.e., $x\delta^2 f - \delta f + f = 0$.

EXERCISE 4.62. Check that the termwise limit $\sum_{n\geq 0} n!x^{n+1}$ of $\sum_{n\geq 0} [n]_q!x^{n+1}$ when $q \to 1$ is solution of the above differential equation.

4.4. Heine's basic hypergeometric series

General basic hypergeometric series ${}_r\Phi_s$ (with many parameters), in particular *confluent* ones, will be discussed in [CV]. Here we consider only ${}_2\Phi_1$, abreviated as Φ. Note however that some instances of the irregular case will later be discussed in 14.2.5.

4.4.1. The series and the equation.

4.4.1.1. *The series.* Let $a, b, c \in \mathbf{C}^*$. We define the *basic hypergeometric series* with parameters[7] a, b, c as:

$$\Phi(a, b; c; q, x) := {}_2\Phi_1(a, b; c; q, x) := \sum_{n\geq 0} \frac{(a;q)_n(b;q)_n}{(c;q)_n(q;q)_n} x^n. \tag{4.4}$$

The word *basic* refers to the *base* q. Other notations exist: see for instance **[18]** and the bible **[93]**.

If $c \in q^{-\mathbf{N}}$, a denominator vanishes, so we exclude that possibility. If a or $b \in q^{-\mathbf{N}}$, a numerator vanishes and we get a polynomial; this case is less interesting and we exclude it too. So we assume $a, b, c \in \mathbf{C}^* \setminus q^{-\mathbf{N}}$. Also, if a or b is in $q^{\mathbf{Z}}$ or in $cq^{\mathbf{Z}}$, many simplifications occur: we retrieve series already encountered around Cauchy's q-binomial theorem.

EXERCISE 4.63. Recognize those series and write explicit formulas for their sums.

4.4.1.2. *The functional equation.* Let $\phi_n := \dfrac{(a;q)_n(b;q)_n}{(c;q)_n(q;q)_n}$ the general coefficient of Φ. For $n \geq 1$, we have $\phi_n = \dfrac{(1-aq^{n-1})(1-bq^{n-1})}{(1-cq^{n-1})(1-q^n)}\phi_{n-1}$, whence:

$$(c/q)q^{2n}\phi_n - (1+c/q)q^n\phi_n + \phi_n = abq^{2n-2}\phi_{n-1} - (a+b)q^{n-1}\phi_{n-1} + \phi_{n-1}.$$

Multiplying both sides of the equality by x^n and summing for $n \geq 1$, we get:

$$(c/q)(\sigma_q^2\Phi - 1) - (1+c/q)(\sigma_q\Phi - 1) + (\sigma_q\Phi - 1) = x(ab\sigma_q^2\Phi - (a+b)\sigma_q\Phi + \Phi),$$

where we once more abreviated Φ for $\Phi(a, b; c; q, x)$. After some cancellations, this boils down to:

$$(c/q - abx)\sigma_q^2\Phi - ((1+c/q) - (a+b)x)\sigma_q\Phi + (1-x)\Phi = 0. \tag{4.5}$$

This is the *basic hypergeometric equation* (again with base q and parameters a, b, c). We shall also write this q-difference equation as:

$$\sigma_q^2\Phi - \lambda\sigma_q\Phi + \mu\Phi = 0, \text{ where } \lambda := \frac{(1+c/q) - (a+b)x}{c/q - abx} \text{ and } \mu := \frac{1-x}{c/q - abx}. \tag{4.6}$$

So we have an equation with rational coefficients, i.e., the coefficient field is $\mathbf{C}(x)$.

---

[7]In the original article, Heine had $a = q^\alpha$, $b = q^\beta$, $c = q^\gamma$. We shall take up that convention in 4.4.4.

4.4.1.3. *Meromorphic continuation of the series.* Since all ϕ_n are defined and nonzero, we find the radius of convergence by looking at the limit when $n \to +\infty$ of $\frac{\phi_n}{\phi_{n-1}} = \frac{(1 - aq^{n-1})(1 - bq^{n-1})}{(1 - cq^{n-1})(1 - q^n)}$; that limit is obviously 1, so Φ defines an analytic function in $\overset{\circ}{\mathrm{D}}(0, 1)$. This is of course called a basic hypergeometric function, and, as in the classical theory of complex ordinary differential equations, we next go for an analytic continuation; however, contrary to the classical theory, we shall see that there is a unique meromorphic continuation to the whole complex plane: no phenomenon of multivaluedness here **[222]**. For that, we rewrite our q-difference equation in the form:

$$\Phi(x) = \frac{1}{1-x}\left[((1 + c/q) - (a + b)x)\Phi(qx) - (c/q - abx)\Phi(q^2x)\right]. \tag{4.7}$$

First let $|x| < R := \frac{1}{|q|}$ (so $R > 1$), implying that $qx, q^2x \in \overset{\circ}{\mathrm{D}}(0, 1)$. Using relation (4.7) as a definition, we see that it allows us to give a meaning to $\Phi(x)$, except of course if $x = 1$. So we have extended Φ to a function over $\overset{\circ}{\mathrm{D}}(0, R)$ that still satisfies our q-difference equation. Indeed:

- the fact that it extends Φ, i.e., that its restriction to $\overset{\circ}{\mathrm{D}}(0, 1)$ is our "old" Φ just expresses the fact that the latter satisfies the q-difference equation;
- the fact that it satisfies the q-difference equation is obvious on $\overset{\circ}{\mathrm{D}}(0, 1)$ and, over $\overset{\circ}{\mathrm{D}}(0, R) \setminus \overset{\circ}{\mathrm{D}}(0, 1)$, comes from the way we defined it there.

Now we take $x \in \overset{\circ}{\mathrm{D}}(0, R^2)$, so that $qx, q^2x \in \overset{\circ}{\mathrm{D}}(0, R)$, and again extend Φ the same way. This time, Φ has pole 1 and also those x such that qx is a pole of Φ on $\overset{\circ}{\mathrm{D}}(0, R)$, that is $qx = 1$. So Φ on $\overset{\circ}{\mathrm{D}}(0, R^2)$ is meromorphic with simple poles at 1 and q^{-1}.

Iterating the process, we get a meromorphic extension over $\overset{\circ}{\mathrm{D}}(0, R^p)$, still satisfying our q-difference equation, and with simple poles at the q^{-i}, $i = 0, \dots, p - 1$. Last, noting that the union of all $\overset{\circ}{\mathrm{D}}(0, R^p)$, $p \in \mathbf{N}$, is $\mathbf{C}$, we conclude:

PROPOSITION 4.64. *The basic hypergeometric function Φ on $\overset{\circ}{\mathrm{D}}(0, 1)$ admits a unique meromorphic extension to $\mathbf{C}$, with simple poles over $q^{\mathbf{N}}$. This extension is a solution of the q-difference equation* (4.5).

PROOF. Uniqueness comes from the principle of analytic continuation. Existence is the content of the discussion above. □

EXERCISE 4.65. Write a formal proof, successively defining functions Φ_p over all $\overset{\circ}{\mathrm{D}}(0, R^p)$, $p \in \mathbf{N}$ and proving inductively that each one extends the previous one and satisfies the q-difference equation.

EXERCISE 4.66. Redo it all when $|q| > 1$. (You may look at **[217]**.)

4.4.1.4. *Heine's transformation formulas.* These are not strongly related to q-difference theory (see however exercise 4.77 at the end of Subsection 4.4.3), but they are an important feature of the q-hypergeometric series.

THEOREM 4.2 (Heine's first transformation formula).

$$\Phi(a,b;c;q,x) = \frac{(b;q)_\infty (ax;q)_\infty}{(c;q)_\infty (x;q)_\infty} \Phi(c/b, x; ax; q, b).$$

PROOF. We shall use two facts: the obvious equality $(a;q)_\infty = (a;q)_n (aq^n;q)_\infty$; and Cauchy's q-binomial theorem (from 1.4) $\dfrac{(ax;q)_\infty}{(x;q)_\infty} = \sum_{n\geq 0} \dfrac{(a;q)_n}{(q;q)_n} x^n$. The reader is invited to spot where they intervene in the following calculation:

$$\begin{aligned}
\Phi(a,b;c;q,x) &= \sum_{n\geq 0} \frac{(a;q)_n (b;q)_n}{(c;q)_n (q;q)_n} x^n = \frac{(b;q)_\infty}{(c;q)_\infty} \sum_{n\geq 0} \frac{(a;q)_n}{(q;q)_n} \frac{(cq^n;q)_n}{(bq^n;q)_n} x^n \\
&= \frac{(b;q)_\infty}{(c;q)_\infty} \sum_{n\geq 0} \frac{(a;q)_n}{(q;q)_n} x^n \sum_{m\geq 0} \frac{(c/b;q)_m}{(q;q)_m} (bq^n)^m \\
&= \frac{(b;q)_\infty}{(c;q)_\infty} \sum_{m\geq 0} \frac{(c/b;q)_m}{(q;q)_m} b^m \sum_{n\geq 0} \frac{(a;q)_n}{(q;q)_n} (q^m x)^n \\
&= \frac{(b;q)_\infty}{(c;q)_\infty} \sum_{m\geq 0} \frac{(c/b;q)_m}{(q;q)_m} b^m \frac{(aq^m x;q)_\infty}{(q^m x;q)_\infty} \\
&= \frac{(b;q)_\infty}{(c;q)_\infty} \frac{(ax;q)_\infty}{(x;q)_\infty} \sum_{m\geq 0} \frac{(c/b;q)_m}{(q;q)_m} b^m \frac{(x;q)_m}{(ax;q)_m} \\
&= \frac{(b;q)_\infty (ax;q)_\infty}{(c;q)_\infty (x;q)_\infty} \sum_{m\geq 0} \frac{(c/b;q)_m (x;q)_m}{(q;q)_m (ax;q)_m} b^m \\
&= \frac{(b;q)_\infty (ax;q)_\infty}{(c;q)_\infty (x;q)_\infty} \Phi(c/b, x; ax; q, b).
\end{aligned}$$

The reader should also justify the interversion of summations $\sum_n$ and $\sum_m$. □

EXERCISE 4.67. Deduce the more symmetric form:

$$\frac{(by;q)_\infty}{(y;q)_\infty} \Phi(a,y;by;q,x) = \frac{(ax;q)_\infty}{(x;q)_\infty} \Phi(b,x;ax;q,y)\cdot$$

Iterating directly the formula leads us back to the original expression (it is an "involutive" transformation). However, combining it with the symmetric role of the first and second argument (a and b in $\phi(a,b;c;q,x)$) does lead us somewhere else.

COROLLARY 4.68 (Heine's second transformation formula).

$$\Phi(a,b;c;q,x) = \frac{(abx/c;q)_\infty}{(x;q)_\infty} \Phi(c/a, c/b; c; q, abx/c).$$

PROOF. Apply the first transformation; then exchange the first and second argument; then apply again the first transformation; then exchange again the first and second argument; then apply for the last time the first transformation. Many simplifications occur among the q-Pochhammer factors. □

EXERCISE 4.69. Write down explicitly the conditions of validity of the proof and formula ($|x| < 1$, etc).

ALTERNATIVE PROOF OF HEINE'S SECOND TRANSFORMATION FORMULA. Although I do not know how the first transformation formula can be proved with the help of q-difference equations, there is a simple such derivation for the second. We set $g(x) := \Phi(c/a, c/b; c; q, abx/c)$ and $f(x) := \dfrac{(abx/c;q)_\infty}{(x;q)_\infty}$, thus the right-hand side of the wanted equality. Adapting to g the q-difference equation for $\phi(a,b;c;q,x)$, we get:

$$g(q^2x) - \frac{(1+c/q) - (c/a + c/b)(abx/c)}{c/q - (c/a)(c/b)(abx/c)} g(qx) + \frac{1 - abx/c}{c/q - (c/a)(c/b)(abx/c)} g(x) = 0,$$

equivalently:

$$g(q^2x) - \frac{(1+c/q) - (a+b)x}{(c/q)(1-qx)} g(qx) + \frac{1 - abx/c}{(c/q)(1-qx)} g(x) = 0.$$

Now, from $g(x) = \dfrac{(x;q)_\infty}{(abx/c;q)_\infty} f(x)$, one successively draws:

$$g(qx) = \frac{1 - abx/c}{1 - x} \frac{(x;q)_\infty}{(abx/c;q)_\infty} f(qx),$$

then:

$$g(q^2x) = \frac{(1 - abqx/c)(1 - abx/c)}{(1-qx)(1-x)} \frac{(x;q)_\infty}{(abx/c;q)_\infty} f(q^2x).$$

We plug this into the q-difference equation for g and simplify the common $\dfrac{(x;q)_\infty}{(abx/c;q)_\infty}$ factor, getting:

$$f(q^2x) - \lambda(x) f(qx) + \mu(x) f(x) = 0,$$

the q-difference equation for $\phi(a,b;c;q,x)$. But f is plainly holomorphic at 0 and satisfies $f(0) = 1$. Since we know that $\Phi(a,b;c;q,x)$ is the unique such solution of the q-difference equation (remember why?), it is indeed equal to f. □

EXERCISE 4.70. (i) In both transformation formulas, set $a := q^\alpha$, etc, and see what happens when $q \to 1$.

(ii) Prove the following consequence of Heine's second transformation formula, due to Jackson [**18**, Theorem 2.4]:

$$\Phi(a,b;c;q,x) = \frac{(b;q)_\infty (ax;q)_\infty}{(c;q)_\infty (x;q)_\infty} \Phi(c/b, x; ax; q, b).$$

4.4.1.5. *Other approaches.* There are still two more ways (at least!) to write q-difference equation (4.6):

(1) Matricial (vectorialization): let $X_\Phi := \begin{pmatrix} \Phi \\ \sigma_q \Phi \end{pmatrix}$. Then:

$$\sigma_q X_\Phi = A X_\Phi, \text{ where } A := \begin{pmatrix} 0 & 1 \\ -\mu & \lambda \end{pmatrix}.$$

(2) With q-derivation; substituting $1 + (1-q)\delta_q$ for σ_q and simplifying, we get:

$$\delta_q{}^2 \Phi - \frac{\lambda - 2}{q - 1} \delta_q \Phi + \frac{\mu - \lambda + 1}{(q-1)^2} \Phi = 0.$$

We shall encounter these approaches again 12.3.

EXERCISE 4.71. (i) Combine these approaches to an equivalent equation $\delta_q Y = BY$.

(ii) Take $a = q^\alpha$, $b = q^\beta$, $c = q^\gamma$; what become the equations in the three approaches when $q \to 1$?

4.4.2. Local resolution at 0 of the q-hypergeometric equation.

4.4.2.1. Equation (4.5) corresponds to configuration I of Subsection 4.3.2, i.e., $\bullet \longrightarrow \bullet \longrightarrow \bullet$. It has characteristic equation:

$$s^2 - (q/c + 1)s + q/c = 0.$$

The exponents are 1 and q/c; since $q/c \notin q^{\mathbf{N}^*}$, the exponent 1 is nonresonant, so there exists a unique solution in $1 + x\mathbf{C}\{x\}$:

$$f_c(x) := \Phi(a, b; c; q, x).$$

4.4.2.2. *Second solution, generic case.* If $c \notin q^{\mathbf{Z}}$ (generic case), exponent q/c is nonresonant, whence a second independent solution:

$$g_c(x) := e_{q/c}(x)\Phi(aq/c, bq/c; q^2/c; q, x).$$

EXERCISE 4.72. Check that by looking for a solution in the form $e_{q/c}(x) \times f$, where f is a power series satisfying an appropriate q-hypergeometric equation.

Note that solutions $f_c, g_c \in \mathscr{M}(\mathbf{C}^*)^{\sigma_q} = \mathscr{M}(\mathbf{E}_q)$ are independent because they correspond to different exponents. This means that they are linearly independent over $\mathscr{M}(\mathbf{E}_q)$ and therefore form a basis of $\mathrm{Sol}((4.5), \mathscr{M}(\mathbf{C}^*))$ by 3.2.

4.4.2.3. *Second solution, degenerate case.* If $c = q$ (degenerate case)[8], 1 is a double exponent. Thus, by general results from section 4.3, $\Phi := f_q = \Phi(a, b; q; q, x)$ is the unique power series solution (up to constant factor); and there is an independent solution in the form:

$$g_q := \Phi l_q + \Psi, \tag{4.8}$$

with Ψ a power series without constant term; with notation $L := \sigma_q^2 - \lambda\sigma_q + \mu$:

$$L(\Phi l_q) = L(\Phi) l_q + 2\sigma_q^2\Phi - \lambda\sigma_q\Phi = \lambda\sigma_q\Phi - 2\mu\Phi,$$

whence:

$$L(\Phi l_q + \Psi) = 0 \Longleftrightarrow L(\Psi) = -L'(\Phi),$$

where $L' := \lambda\sigma_q - 2\mu$.

Direct computation of Ψ. We solve $L(\Psi) = -L'(\Phi)$ (the unknown is Ψ), first for $\Phi := \Phi(a, b; c; q, x)$, i.e., without specializing c to q (we shall do it at the end of the calculation). We write $\Phi := \sum\limits_{n \geq 0} \phi_n(c)x^n$ and $\Psi := \sum\limits_{n \geq 0} \psi_n x^n$ where $\phi_n(c) := \dfrac{(a; q)_n (b; q)_n}{(c; q)_n (q; q)_n}$, so that:

$$\forall n \geq 1\,,\ (1 - cq^{n-1})(1 - q^n)\phi_n(c) = (1 - aq^{n-1})(1 - bq^{n-1})\phi_{n-1}(c). \tag{4.9}$$

Chasing denominators in our equation, we have:

$$\begin{aligned}(c/q - abx)\sigma_q^2\Psi - ((c/q + 1) - (a + b)x)\sigma_q\Psi + (1 - x)\Psi \\ = -((c/q + 1) - (a + b)x)\sigma_q\Phi + (1 - x)\Phi.\end{aligned}$$

By identification of coefficients:

$$\begin{aligned}\forall n \geq 1\,,\ (cq^{2n-1} - (c + q)q^{n-1} + 1)\psi_n - (abq^{2n-2} - (a + b)q^{n-1} + 1)\psi_{n-1} = \\ (2 - (q + c)q^{n-1})\phi_n(c) - (2 - (a + b)q^{n-1})\phi_{n-1}(c).\end{aligned}$$

[8] We shall not consider the other degenerate cases, i.e., $c \in q^{-\mathbf{N}}$.

This can also be written:

$$(1-cq^{n-1})(1-q^n)\psi_n - (1-aq^{n-1})(1-bq^{n-1})\psi_{n-1} =$$
$$((1-cq^{n-1})+(1-q^n))\phi_n(c) - ((1-aq^{n-1})+(1-bq^{n-1}))\phi_{n-1}(c).$$

Now we use equality (4.9) and divide both sides by the quantity shown there, yielding a telescopic relation:

$$\frac{\psi_n}{\phi_n(c)} - \frac{\psi_{n-1}}{\phi_{n-1}(c)} = \frac{1}{1-cq^{n-1}} + \frac{1}{1-q^n} - \frac{1}{1-aq^{n-1}} - \frac{1}{1-bq^{n-1}}.$$

Since $\phi_0(c)=1$:

$$\psi_n = \psi_0 + \phi_n(c)\sum_{k=1}^{n}\left(\frac{1}{1-cq^{k-1}} + \frac{1}{1-q^k} - \frac{1}{1-aq^{k-1}} - \frac{1}{1-bq^{k-1}}\right).$$

Now we specialize $c := q$, which allows us to require $\psi_0 := 0$ (the reader is invited to check why!) and we conclude:

$$\Psi(x) = \sum_{n\geq 1} A_n\phi_n x^n, \text{ where } A_n := \sum_{k=1}^{n}\left(\frac{2}{1-q^k} - \frac{1}{1-aq^{k-1}} - \frac{1}{1-bq^{k-1}}\right).$$

EXERCISE 4.73. Taking $a := q^\alpha$ and $b := q^\beta$, find an asymptotic equivalent of A_n when $q \to 1$ (compare with [**98**, p. 41]).

Method of Goursat-Zhang. This is the adaptation by Changgui Zhang (personal communication) of a method apparently due to Goursat [**98**, III, no 17, pp 40–41].

From now on λ is written λ_c, etc; so the degenerate case above is λ_q, etc.

We use the notation[9]: $DF_c := \left[\dfrac{\partial F_c}{\partial c}\right]_{|c=q}$, so it is clear that, when $F_c = G(q/c)$, $DF_c = (-1/q)\left[\dfrac{\partial G(t)}{\partial t}\right]_{|t=1}$. For instance $De_{q/c} = (-1/q)l_q$.

Recall g_c is the second solution in the generic case:

$$Dg_c = (De_{q/c})\left[\Phi(aq/c, bq/c; q^2/c; q, x)\right]_{|c=q} + \left[e_{q/c}(x)\right]_{|c=q} D\Phi(aq/c, bq/c; q^2/c; q, x)$$
$$= (-1/q)l_q\Phi + D\Phi(aq/c, bq/c; q^2/c; q, x).$$

Let ϕ_c a solution of the q-difference equation depending analytically on c (e.g., f_c or g_c). From

$$\phi_c(q^2x) - \lambda_c\phi_c(qx) + \mu_c\phi_c(x) = 0,$$

applying derivation D, we find:

$$(D\phi_c)(q^2x) - \lambda_q(D\phi_c)(qx) + \mu_q(D\phi_c)(x) = (D\lambda_c)\phi_q(qx) - (D\mu_c)\phi_q(x).$$

We apply that to $\phi_c = f_c$ and to $\phi_c = g_c$, noting that $f_q = g_q = \Phi$:

$$\sigma_q^2 Df_c - \lambda_q\sigma_q Df_c + \mu_q Df_c = \sigma_q^2 Dg_c - \lambda_q\sigma_q Dg_c + \mu_q Dg_c = (D\lambda_c)\sigma_q\Phi - (D\mu_c)\Phi.$$

So now we take $h := -q(Dg_c - Df_c)$ and have a solution:

$$\sigma_q^2 h - \lambda\sigma_q h + \mu h = 0.$$

Besides:

$$h = l_q\Phi - qD\Phi(aq/c, bq/c; q^2/c; q, x) + qD\Phi(a, b; c; q, x) = l_q\Phi + \Psi,$$

[9]This notation will be used in Subsubsection 4.4.2.3 only and must not be confused with the notation D for the usual derivation d/dx used elsewhere in the book.

where $\Psi := qD\Phi(a,b;c;q,x) - qD\Phi(aq/c, bq/c; q^2/c; q, x)$. So h has the form mentioned in (4.8).

End of the computation of Ψ. We keep the notation

$$\Phi(a,b;c;q,x) = \sum_{n\geq 0} \phi_n(c)x^n$$

and set moreover $\Phi(aq/c, bq/c; q^2/c; q, x) = \sum_{n\geq 0} \phi_n'(c)x^n$. Note that $\phi_n(q) = \phi_n'(q)$ $= \dfrac{(a;q)_n(b;q)_n}{(q;q)_n^2} =: \phi_n$. Thus, taking into account that:

$$D\phi_n(c) = \phi_n(q)\left[\frac{\partial \log \phi_n(c)}{\partial c}\right]_{|q=c},$$

$$D\phi_n'(c) = \phi_n(q)\left[\frac{\partial \log \phi_n'(c)}{\partial c}\right]_{|q=c},$$

we write:

$$D\Phi(a,b;c;q,x) = \sum_{n\geq 0}\phi_n(q)\left[\frac{\partial \log \phi_n(c)}{\partial c}\right]_{|q=c} x^n,$$

$$D\Phi'(a,b;c;q,x) = \sum_{n\geq 0}\phi_n(q)\left[\frac{\partial \log \phi_n'(c)}{\partial c}\right]_{|q=c}.$$

On the other hand:

$$\left[\frac{\partial \log \phi_n(c)}{\partial c}\right]_{|q=c} = -\sum_{i=0}^{n-1}\left[\frac{-q^i}{1-q^ic}\right]_{|q=c} = \sum_{i=1}^{n}\frac{q^{i-1}}{1-q^i},$$

$$\left[\frac{\partial \log \phi_n'(c)}{\partial c}\right]_{|q=c} = (-1/q)\left[\frac{\partial \log \phi_n''(t)}{\partial t}\right]_{|t=1}$$

$$= -(1/q)\sum_{i=0}^{n-1}\left(\left[\frac{-q^ia}{1-q^iat}\right]_{|t=1} s\right.$$

$$\left. + \left[\frac{-q^ib}{1-q^ibt}\right]_{|t=1} - \left[\frac{-q^{i+1}}{1-q^{i+1}t}\right]_{|t=1}\right).$$

We have used $\phi_n''(t) := \dfrac{(at;q)_n(bt;q)_n}{(qt;q)_n(q;q)_n}$, so that $\phi_n'(c) = \phi_n''(q/c)$. We conclude from the above:

$$q\left[\frac{\partial \log \phi_n'(c)}{\partial c}\right]_{|q=c} = \sum_{i=1}^{n}\left(\frac{q^{i-1}a}{1-q^{i-1}a} + \frac{q^{i-1}b}{1-q^{i-1}b} - \frac{q^i}{1-q^i}\right).$$

Now, replacing $\dfrac{z}{1-z}$ by $-1+\dfrac{1}{1-z}$ and eliminating all occurrences of -1:

$$q\left[\frac{\partial \log \phi_n(c)}{\partial c}\right]_{|q=c} - q\left[\frac{\partial \log \phi_n'(c)}{\partial c}\right]_{|q=c} = \sum_{i=1}^{n}\left(2\frac{q^i}{1-q^i} - -\frac{q^{i-1}a}{1-q^{i-1}a} - \frac{q^{i-1}b}{1-q^{i-1}b}\right)$$

$$= \sum_{i=1}^{n}\left(2\frac{1}{1-q^i} - -\frac{1}{1-q^{i-1}a} - \frac{1}{1-q^{i-1}b}\right).$$

So, if we set:

$$A_n := \sum_{i=1}^{n} \left(2\frac{1}{1-q^i} - -\frac{1}{1-q^{i-1}a} - \frac{1}{1-q^{i-1}b} \right),$$

we find that $\Psi = \sum_{n \geq 1} A_n \phi_n x^n$.

4.4.3. Global resolution of the q-hypergeometric equation, connection formulas.

4.4.3.1. What has been done up to now can be seen as "local study near 0": although most solutions are meromorphic over $\mathbf{C}^*$ with an essential singularity at 0 (coming from the q-characters $e_{q,c}$ or, in the degenerate cases, from the q-logarithm l_q), they are built from basic q-special functions along with elements of $\mathbf{C}(\{x\})$.

For the corresponding "local study at infinity"[10] we think of $\mathbf{C}$ as embedded in the Riemann sphere $\mathbf{S} = \mathbf{C} \cup \{\infty\}$ (see **[222]**) and use the local coordinate $y := 1/x$; thus $y = 0$ corresponds to $x = \infty$. So our coefficient field (for the study of basic hypergeometric series) will be $K := \mathbf{C}(y) = \mathbf{C}(x)$, while our solution field will be $\mathcal{M}(\mathbf{C}^*)$ (meromorphy in y is equivalent to meromorphy in y). As for the functions, their power series part will be very different since $\mathbf{C}(\{y\}) = \mathbf{C}(\{1/x\}) \neq \mathbf{C}(\{x\})$.

EXERCISE 4.74. Show that the q-character and q-logarithmic parts are not "very different". (Find and use formulas for $e_{q,c}(1/x)$ and $l_q(1/x)$.)

So we start from equation (4.6), written as $f(q^2x) - \lambda(x)f(qx) + \mu(x)f(x) = 0$, where $\lambda(x) := \dfrac{(1+c/q)-(a+b)x}{c/q - abx}$ and $\mu(x) := \dfrac{1-x}{c/q-abx}$. We temporarily set $g(y) := f(x) = f(1/y)$, so that $f(qx) = g(1/qx) = g(y/q)$ and $f(q^2x) = g(1/q^2x) = g(y/q^2)$. Therefore, f is a solution of (4.6) if, and only if,:

$$g(y/q^2) - \lambda(1/y)g(y/q) + \mu(1/y)g(y) = 0.$$

It is understood that this is true for all y, so we replace y by q^2y and then divide to get the standardized form of the equation:

$$\text{(4.6)} \quad o \Longleftrightarrow g(y) - \lambda(1/q^2y)g(qy) + \mu(1/q^2y)g(q^2y) = 0 \Longleftrightarrow g(q^2y) - \lambda^\infty(y)g(qy) + \mu^\infty(y)g(y) = 0,$$

where:

$$\lambda^\infty(y) = \frac{\lambda(1/q^2y)g(qy)}{\mu(1/q^2y)} = \frac{(a+b)-(qc+q^2)y}{1-q^2y} \text{ and } \mu^\infty(y) = \frac{1}{\mu(1/q^2y)} = \frac{ab-qcy}{1-q^2y}.$$

This corresponds to configuration I o o o (or exceptionally to configuration II or III, but it does not really matter here), with characteristic equation $t^2-(a+b)t+ab = 0$: the exponents are a and b.

[10]Contrary to the case of differential equations, 0 and ∞ are a priori the only places at which a local study makes sense; this is explained by the discussion in the introduction, in 5.1 and then again in 7.2.2.

4.4.3.2. We now assume that $a \not\equiv b \pmod{q^{\mathbf{Z}}}$, so that the exponents are neither resonant nor double. Then there are unique solutions $e_{q,a}(y)\phi(y)$ and $e_{q,b}(y)\psi(y)$, where $\phi, \psi \in \mathbf{C}\{y\}$ and $\phi(0) = \psi(0) = 1$. We do the following calculations for the first case, the other case is immediately deduced. After substituting $g(y) := e_{q,a}(y)\phi(y)$ and dividing by $a^2 e_{q,a}(y)$, the equation becomes:

$$\phi(q^2y) - \frac{(1+b/a) - (q(c+q)/a)y}{1-q^2y}\phi(qy) + \frac{b/a - (qc/a^2)y}{1-q^2y}\phi(y) = 0.$$

We try to get it closer to a q-hypergeometric equation:

$$\phi(q^2y) - \frac{(1+a/b) - (q(c+q)/b)y}{a/b - (q^2a/b)y}\phi(qy) + \frac{1 - (qc/ab)y}{a/b - (q^2a/b)y}\phi(y) = 0.$$

To get right there, we change variable: $y' := (qc/ab)y$ and rename the unknown function accordingly: $\phi'(y') := \phi(y)$, whence $\phi'(qy') := \phi(qy)$ and $\phi'(q^2y') := \phi(q^2y)$. Also, $\phi' \in \mathbf{C}(\{y'\})$ and $\phi'(0) = 1$. Then:

$$\phi'(q^2y') - \frac{(1+a/b) - (a(c+q)/c)y'}{a/b - (qa^2/c)y'}\phi'(qy') + \frac{1-y'}{a/b - (qa^2/c)y'}\phi'(y') = 0.$$

Setting $a' := a$, $b' := qa/c$ and $c' := qa/b$, this becomes:

$$\phi'(q^2y') - \frac{(1+a/b) - (a(c+q)/c)y'}{a/b - (qa^2/c)y'}\phi'(qy') + \frac{1-y'}{a/b - (qa^2/c)y'}\phi'(y') = 0.$$

Setting $a' := a$, $b' := qa/c$ and $c' := qa/b$, this becomes:

$$\phi'(q^2y') - \frac{(1+c'/q) - (a'+b')y'}{c'/q - a'b'y'}\phi'(qy') + \frac{1-y'}{c'/q - a'b'y'}\phi'(y') = 0.$$

We conclude that $\phi'(y') = \Phi(a', b'; c'; q, y')$, i.e., $\phi(y) = \Phi(a, qa/c; qa/b; q, (qc/ab)y)$.

Combining with the similar results for exponent b, we conclude that equation (4.6) has a basis of solutions at infinity:

$$e_{q,a}(y)\Phi(a, qa/c; qa/b; q, (qc/ab)y) \text{ and } e_{q,b}(y)\Phi(b, qb/c; qb/a; q, (qc/ab)y).$$

EXERCISE 4.75. Check that these two solutions are independent (i.e., that their quotient is not q-invariant).

4.4.3.3. We now assume for simplicity that the situation is generic in every possible way, that is:

$$a, b, c, a/b, a/c, b/c \not\in q^{\mathbf{Z}} \Longleftrightarrow \overline{a}, \overline{b}, \overline{c}, \overline{q} \in \mathbf{E}_q \text{ are pairwise distinct.}$$

Then we have two independent solutions $f^{(0)}, g^{(0)}$ "local at 0":

$$\begin{cases} f^{(0)}(x) := \Phi(a, b; c; q, x), \\ g^{(0)}(x) := e_{q,q/c}(x)\Phi(qa/c, qb/c; q^2/c; q, x), \end{cases}$$

and two independent solutions $f^{(\infty)}, g^{(\infty)}$ "local at ∞":

$$\begin{cases} f^{(\infty)}(x) := e_{q,a}(x^{-1})\Phi(a, qa/c; qa/b; q, (qc/ab)x^{-1}), \\ g^{(\infty)}(x) := e_{q,b}(x^{-1})\Phi(b, qb/c; qb/a; q, (qc/ab)x^{-1}). \end{cases}$$

We note that the q-characters $e_{q,q/c}, e_{q,a}, e_{q,b}$ used here can be replaced at will by arbitrary q-characters $e'_{q/c}, e'_a, e'_b$ in $\mathscr{M}(\mathbf{C}^*)$, as their particular choice was used nowhere in the resolution of (4.5). This may be made necessary if we want to control singularities, see 4.4.4.

All these solutions belong to $\mathrm{Sol}((4.5), \mathscr{M}(\mathbf{C}^*))$ and "independent" here means "linearly independent over $\mathscr{M}(\mathbf{C}^*)^{\sigma_q} = \mathscr{M}(\mathbf{E}_q)$". From the general theory of section 3.2, we know that $\mathcal{B}^{(0)} := (f^{(0)}, g^{(0)})$ and $\mathcal{B}^{(\infty)} := (f^{(\infty)}, g^{(\infty)})$ are bases of the $\mathscr{M}(\mathbf{E}_q)$-linear space $\mathrm{Sol}((4.5), \mathscr{M}(\mathbf{C}^*))$. Therefore, there exist unique coefficients $a, b, c, d \in \mathscr{M}(\mathbf{E}_q)$ such that:

$$\begin{cases} f^{(0)} = af^{(\infty)} + cg^{(\infty)}, \\ g^{(0)} = bf^{(\infty)} + dg^{(\infty)}, \end{cases} \iff \mathcal{B}^{(0)} = \mathcal{B}^{(\infty)}P, \text{ where } P := \begin{pmatrix} a & b \\ c & d \end{pmatrix} \in \mathrm{GL}_2(\mathscr{M}(\mathbf{E}_q)).$$

The matrix P is invertible because it connects two bases. It is made of elements of $\mathscr{M}(\mathbf{E}_q)$ because these are bases of solutions of the same q-difference equation. It is an instance of *Birkhoff's connection matrix*, which plays a fundamental role in the global theory of q-difference equations, somewhat analogous to the role of the monodromy in the theory of linear differential equations.

In Chapter 12 we shall introduce Birkhoff's connection matrix in a slightly different way. Note that, a and b being q-elliptic:

$$f^{(0)} = af^{(\infty)} + cg^{(\infty)} \Longrightarrow \sigma_q f^{(0)} = a\sigma_q f^{(\infty)} + c\sigma_q g^{(\infty)} \Longrightarrow X^{(0)} = aX^{(\infty)} + cY^{(\infty)},$$

where we introduced $X^{(0)} := X_{f^{(0)}} = \begin{pmatrix} f^{(0)} \\ \sigma_q f^{(0)} \end{pmatrix}$, $X^{(\infty)} := X_{f^{(\infty)}} = \begin{pmatrix} f^{(\infty)} \\ \sigma_q f^{(\infty)} \end{pmatrix}$ and $Y^{(\infty)} := X_{g^{(\infty)}} = \begin{pmatrix} g^{(\infty)} \\ \sigma_q g^{(\infty)} \end{pmatrix}$. Similarly, writing $Y^{(0)} := X_{g^{(0)}} = \begin{pmatrix} g^{(0)} \\ \sigma_q g^{(0)} \end{pmatrix}$, we have $Y^{(0)} = bX^{(\infty)} + dY^{(\infty)}$. So, introducing the q-wronskian matrices:

$$\mathcal{X}^{(0)} := \left[X^{(0)}, Y^{(0)}\right] \text{ and } \mathcal{X}^{(\infty)} := \left[X^{(\infty)}, Y^{(\infty)}\right],$$

we have $\mathcal{X}^{(0)} = \mathcal{X}^{(\infty)}P$, which allows us to compute the connection matrix as:

$$P = \left(\mathcal{X}^{(\infty)}\right)^{-1} \mathcal{X}^{(0)}.$$

Matrices $\mathcal{X}^{(0)}, \mathcal{X}^{(\infty)}$ are actually *fundamental matricial solutions* of the same equation, which here has the form $\mathcal{X} = M\mathcal{X}$ for some $M \in \mathrm{GL}_2(\mathbf{C}(x))$ (actually the companion matrix for q-difference operator $\sigma_q^2 - \lambda\sigma_q + \mu$), so that:

$$\begin{aligned} \sigma_q P &= \sigma_q \left(\left(\mathcal{X}^{(\infty)}\right)^{-1} \mathcal{X}^{(0)}\right) \\ &= \left(\sigma_q \mathcal{X}^{(\infty)}\right)^{-1} \sigma_q \mathcal{X}^{(0)} \\ &= \left(M\mathcal{X}^{(\infty)}\right) M\mathcal{X}^{(0)} = \sigma_q \left(\mathcal{X}^{(\infty)}\right)^{-1} \mathcal{X}^{(0)} \\ &= P, \end{aligned}$$

a confirmation of the q-invariance of P.

For instance, solving the system $\begin{cases} af^{(\infty)} + cg^{(\infty)} = f^{(0)}, \\ a\sigma_q f^{(\infty)} + c\sigma_q g^{(\infty)} = \sigma_q f^{(0)}, \end{cases}$ yields:

$$a = \frac{f^{(0)}\sigma_q g^{(\infty)} - g^{(\infty)}\sigma_q f^{(0)}}{f^{(\infty)}\sigma_q g^{(\infty)} - g^{(\infty)}\sigma_q f^{(\infty)}} \text{ and } c = \frac{f^{(\infty)}\sigma_q f^{0)} - f^{(0)}\sigma_q f^{(\infty)}}{f^{(\infty)}\sigma_q g^{(\infty)} - g^{(\infty)}\sigma_q f^{(\infty)}}.$$

There are actually very beautiful explicit expressions of the coefficients of P, called the *Barnes-Mellin-Watson* formulas (see 1.11.2), they will be proved in [CV].

EXERCISE 4.76. (i) Write the matrix M.

(ii) Check directly that a and c are q-elliptic and find their poles. You can use the fact that the numerators and the denominator all are q-wronskians.

EXERCISE 4.77. Try to use Heine's transformation formulas from Subsubsection 4.4.1.4 to derive connection formulas.

4.4.4. Degeneracy when $q \to 1$. This is just a first look at this important domain. The general theory will be studied in Sections 11.2 and 12.3.

4.4.4.1. In accordance with Heine's original notations in **[116]**, we now set $a := q^\alpha$, $b := q^\beta$ and $c := q^\gamma$, where α, β, γ are *fixed*: this precision matters since we are going to let q vary (and tend to 1). For simplicity, we restrict to the most generic case: $a, b, c, a/b, a/c, b/c \not\in q^{\mathbf{Z}}$; said otherwise, we assume that $\alpha, \beta, \gamma, \alpha - \beta, \alpha - \gamma, \beta - \gamma \not\in \mathbf{Z}$.

We intend to look at the behaviour of $\Phi(a, b; c; q, x)$ when $q \to 1$.The coefficient of x^n is:

$$\phi_n := \phi_n(q) := \frac{(q^\alpha; q)_n (q^\beta; q)_n}{(q^\gamma; q)_n (q; q)_n},$$

which can be rewritten, using q-numbers $[\delta]_q := \dfrac{1 - q^\delta}{1 - q}$, as:

$$\phi_n(q) = \prod_{i=0}^{n-1} \frac{[\alpha + i]_q [\beta + i]_q}{[\gamma + i]_q [1 + i]_q},$$

from which we readily obtain the limit coefficientwise:

$$\lim_{q \to 1} \phi_n(q) = \prod_{i=0}^{n-1} \frac{(\alpha + i)(\beta + i)}{(\gamma + i)(1 + i)} = \frac{(\alpha)_n (\beta)_n}{(\gamma)_n (1)_n}.$$

Recall the classical Pochhammer symbol $(\delta)_n := \delta(\delta+1) \cdots (\delta+n-1)$, in particular $(1)_n = n!$. So we see that coefficientwise, the q-hypergeometric series $\phi(a, b; c; q, x)$ tends to the classical hypergeometric series:

$$F(\alpha, \beta; \gamma; x) := \sum_{n \geq 0} \frac{(\alpha)_n (\beta)_n}{(\gamma)_n (1)_n} x^n = \sum_{n \geq 0} \frac{(\alpha)_n (\beta)_n}{(\gamma)_n n!} x^n.$$

We are now going to make this statement more precise to a certain extent.

4.4.4.2.

LEMMA 4.78. *For every positive $R < 1$, the convergence of the series*

$$\phi(q^\alpha, q^\beta; q^\gamma; q, x)$$

to the series $F(\alpha, \beta; \gamma; x)$ when $q \to 1$ is dominated in the domain $x \in \mathring{D}(0, R)$ if one restricts q to a domain where $|1 - q| = O(1 - |q|)$ (and for q near enough to 1, see the proof).

Recall that such restricted domains for q were already considered in 3.3.1.

PROOF. Fix such a $R \in]0; 1[$ and also a $M > 0$, restriciting the domain of variation of q so that $|1 - q| \leq M(1 - |q|)$. We have:

$$\frac{\phi_{n+1}}{\phi_n} = \frac{[n + \alpha]_q [n + \beta]_q}{[n + \gamma]_q [n + 1]_q}.$$

We are going to ensure that, under the said restrictions for q, as soon as q is near enough to 1, there is an integer $n_0 \in \mathbf{N}$ (not depending on q) such that:

$$\forall n \geq n_0\,, \left|\frac{\phi_{n+1}}{\phi_n}\right| \leq R^{-1},$$

whence $|\phi_n| = O(R^{-n})$, which indeed entails domination by $\sum(x/R)^n$, whence dominated convergence in $\mathring{\mathrm{D}}(x, R)$.

From the general equality $[n+\delta]_q = [n]_q + q^n[\delta]_q$ (see 3.3.2), we deduce the bounds:

$$|[n]_q| - |[\delta]_q| \leq |[n+\delta]_q| \leq |[n]_q| + |[\delta]_q|\,.$$

The function $q \mapsto [\delta]_q$ is continuous on $\overline{\mathrm{D}}(0,1) \cap (\Re q \geq 1/2)$ (of course we extend it by the value δ at $q = 1$), whence bounded above; let $A > 0$ a common upper bound for $[\alpha]_q, [\beta]_q, [\gamma]_q$ and $[1]_q = 1$. Then, writing $u := |[n]_q|$:

$$\left|\frac{\phi_{n+1}}{\phi_n}\right| \leq \left(\frac{u+A}{u-A}\right)^2,$$

at least if $u > A$. Since $\left(\dfrac{u+A}{u-A}\right)^2 \to 1 < R^{-1}$ when $u \to +\infty$, there exists a $K > 0$ such that, as we would like:

$$|[n]_q| \geq K \Longrightarrow \left|\frac{\phi_{n+1}}{\phi_n}\right| \leq R^{-1}.$$

Now we arbitrarily choose $r_0 \in \,]0,1[$ and $n_0 \in \mathbf{N}$ such that:

$$1 + r_0 + \cdots + r_0^{n_0-1} \geq KM.$$

Now let $|q| \geq r_0$ (this obviously holds for q near enough to 1); then, since:

$$|[n]_q| = \left|\frac{1-q^n}{1-q}\right| \geq \frac{1-|q^n|}{|1-q|} = \frac{1-|q|^n}{|1-q|} = \frac{1-|q|^n}{1-|q|}\,\frac{1-|q|}{|1-q|},$$

we have:

$$\begin{aligned}|[n]_q| \geq \left(1+|q|+\cdots+|q|^{n-1}\right)\frac{1-|q|}{|1-q|} &\geq \left(1+r_0+\cdots+r_0^{n_0-1}\right)\frac{1-|q|}{|1-q|} \\ &\geq (KM)M^{-1},\end{aligned}$$

i.e., $|[n]_q| \geq K$ as wanted. □

4.4.4.3. From the above lemma, we conclude that the q-hypergeometric *series* $\phi(q^\alpha, q^\beta; q^\gamma; q, x)$ degenerates, when $q \to 1$, into the classical hypergeometric series $F(\alpha, \beta; \gamma; x)$, at least inside their common disk of convergence $\mathring{\mathrm{D}}(0,1)$. In order to better understand the degeneracy of the q-hypergeometric *functions*, the following topics should be adressed:

(1) We saw that the power series $\phi(a, b; c; q, x)$ admits a uniform meromorphic continuation to the whole of $\mathbf{C}$. So what of the degeneracy of $\phi(q^\alpha, q^\beta; q^\gamma; q, x)$ over $\mathbf{C} \setminus \mathring{\mathrm{D}}(0,1)$? Since, by 4.4.1.3, said function admits a half q-spiral $q^{-\mathbf{N}}$ of simple poles, we cannot hope for any regular behaviour if q is allowed to vary freely. The most natural solution seems to force $q^{-\mathbf{N}}$ to *condense* into a continuous half spiral. This can be done as follows: we fix $q_0 := e^{2i\pi\tau_0}$, $\Im\tau_0 > 0$, and then set $q := q^\epsilon = e^{2i\pi\tau_0\epsilon}$, where

$\epsilon > 0$. Then if $\epsilon \to 0^+$, we see that $q \to 1$ along a "loxodromic"[11] direction. More visually, since $-\mathbf{N}\epsilon$ condenses into $\mathbf{R}_-$, the discrete half spiral $q^{-\mathbf{N}} = q_0^{-\mathbf{N}\epsilon}$ condenses into $q_0^{\mathbf{R}_-}$, a continuous half logarithmic spiral; see Figure 1.1, page 12 in Chapter 1.

(2) Solutions of the form $e_{q,q/c}(x)\Phi(qa/c, qb/c; q^2/c; q, x)$ must be adressed too. This involves the asymptotic behaviour of the q-characters e_{q,q^δ} when $q \to 1$ and can be deduced easily from Chapter 2. If we want a simple geometric figure, we may have to use characters with q-spiral of poles condensating to $q_0^{\mathbf{R}_-}$, i.e., built with Theta functions having zeroes at $[1; q] = q^{\mathbf{Z}}$ instead of $[-1; q] = -q^{\mathbf{Z}}$.

(3) If we understand properly the degeneracy of solutions at 0 and also at ∞, we may arrive at the same for connection formulas. At the limit, we should recover connection formulas for classical hypergeometric functions between 0 and ∞ [**222**]. But the latter have multivalued continuation to the whole plane and the coefficients of their connection formulas are related to their monodromy matrices. So we expect Birkhoff's connection matrix (the coefficients of which come from the connection formulas for q-hypergeometric functions) to represent some kind of q-analogue of monodromy (although all our q-hypergeometric functions are uniform over $\mathbf{C}^*$).

(4) Last, we saw in Sections 1.5 and 3.3 that the q-difference equations for q-hypergeometric functions should also be written in terms of δ_q to find how they degenerate into the ordinary differential equations for the classical hypergeometric functions.

More details and the solution to most remaining mysteries will be given in 12.3.2.

EXERCISE 4.79. Attack the program sketched in the last point: in equation (4.6), replace σ_q by $1 + (1-q)\delta_q$; then expand and simplify by some power of $1-q$ to get an equation ${\delta_q}^2\phi - \lambda'(q)\delta_q\phi + \mu'(q)\phi = 0$. Then, taking $a := q^\alpha$, $b := q^\beta$ and $c := q^\gamma$, make $q \to 1$ to, hopefully, obtain a second order linear ordinary differential equation satisfied by $F(\alpha, \beta; \gamma; x)$. You may, if you prefer, replace σ_q by $1 + (1-q)xD_q$. (You may also do both ...)

4.5. A first look at associated sheaves and bundles

> By sheafifying one suddenly was able to obtain results one had not dared to dream of in 1950. "Il faut faisceautiser les problèmes."
> —H. Grauert and R. Remmert[12] [**100**, Chapter 2 §5 3]

Ever since Riemann, the dialectics of local *vs* global behaviour has been a central paradigm in the study of complex analytic functions, in particular of solutions of differential equations. Analytic continuation provided the main ingredient. For q-difference equations, our choice to stick to uniform solutions makes that ingredient useless. However, using sheaves (as we did for differential equations in [**222**] for example) is a convenient substitute.

[11] The word "loxodromy" refers to an old problem of navigation: to follow a route that cuts all meridians at a fixed angle. Up to the standard projection of the (Riemann) sphere to the complex plane, such routes are logarithmic spirals.

[12] I would guess that the part in French is a quotation of Henri Cartan.

It then appears that all sheaves that we meet are *locally free*, hence they are sheaves of sections of vector bundles (for rank 1 q-difference equations: of line bundles), so we thus retrieve the geometric approach that was initiated in 2.2.3, of which the present section is a (partial) generalisation by other means. Truly general study will come in 7.3 and 9.3.

4.5.1. Local solutions.

4.5.1.1. *General setting.* The basic remark is the following. If f is to be considered as a "local" solution of a q-difference equation $f(qx) = u(x)f(x)$, it should be defined on an open set $U \subset \mathbf{C}$ such that $qU = U$. If $0 \in U$ this implies that $U = \mathbf{C}$ (see the exercise) and this is not really local, so we shall assume that $U \subset \mathbf{C}^*$. In 4.5.1.3 we shall be more precise on how to choose U to make the study even more local.

EXERCISE 4.80. Check that $qU = U$ together with $0 \in U$ entails $U = \mathbf{C}$.

Now let a particular q-difference equation $\sigma_q f = uf$, where $u \in \mathbf{C}(x)^*$. Iterating (or using induction), we see that it implies, for all $n \in \mathbf{Z}$, the relation $\sigma_q^n f = u_n f$, where $u_n \in \mathbf{C}(x)^*$ is given by $u_0 := 1$ and in general:

$$u_n(x) := \begin{cases} u(x) \cdots u(q^{n-1}x) \text{ if } n > 0, \\ \left(u(q^{-1}x) \cdots u(q^{-p}x)\right)^{-1} \text{ if } n = -p, p > 0. \end{cases} \qquad \text{(So, naturally, } u_1 = u.\text{)}$$

4.5.1.2. *Singularities of a rational q-difference equation of rank* 1. Since we just intend here an initiation, we shall restrict to a rather simple (if meaningful) class of examples, where $u \in \mathbf{C}(x)^*$ is taken in the form:

$$u(x) := cx^m \frac{(x - a_1) \cdots (x - a_k)}{(x - b_1) \cdots (x - b_l)}, \quad m \in \mathbf{Z},\ c, a_1, \ldots, a_k, b_1, \ldots, b_l \in \mathbf{C}^*.$$

(So the case discussed in 2.2.3 corresponds to $k = l = 0$). For simplicity, we assume that $a_1, \ldots, a_k, b_1, \ldots, b_l$ are pairwise non congruent modulo $q^{\mathbf{Z}}$; equivalently, the $(k+l)$ q-spirals $[a_i; q]$, $[b_j; q]$, $i = 1, \ldots, k$, $j = 1, \ldots, l$, are pairwise disjoint. In some sense, this is a kind of "fuchsianity" condition at the corresponding singularities, but we cannot delve into that. For reasons which will appear soon, we consider as *singular set* of our equation, either:

$$\Sigma := \{a_1, \ldots, a_k, b_1, \ldots, b_l\} \subset \mathbf{C}^* \text{ or } \overline{\Sigma} := \bigcup_{i=1,\ldots,k} [a_i; q] \cup \bigcup_{j=1,\ldots,l} [b_j; q].$$

(Those are the so-called *intermediate singularities* as we do not care here whether 0 and ∞ are to be seen as singular.) Those singularity properties are expressed in a condensed form in the divisor:

$$\mathrm{div}_{\mathbf{C}^*}(u) = [a_1] + \cdots + [a_k] - [b_1] - \cdots - [b_l].$$

4.5.1.3. *Coverings by small open sets.* As in Chapter 2 (and Appendix A), we consider as "small" a (nonempty) open subset $V \subset \mathbf{E}_q$ such that:

$$\pi^{-1}(V) = \bigsqcup_{n \in \mathbf{Z}} q^n U_0 \quad \text{(i.e., a } \textit{disjoint} \text{ union)}$$

for some open subset $U_0 \subset \mathbf{C}^*$ which is mapped homeomorphically onto V by π (so those are the trivializing open subsets for the etale covering π).

We leave it to you to check that $\mathbf{E}_q$ can be covered by such small V and that we may moreover for each of them choose U_0 such that $U_0 \cap \overline{\Sigma}$ is either empty, or equal to some $\{a_i\}$, or to some $\{b_j\}$. In the second and third case, the choice of U_0 is moreover uniquely determined by the condition. In what follows, we shall stick to those notations and speak of the first, second and third case. If we restrict u to such an open set U_0, we find that $\mathrm{div}_{U_0}(u)$ is 0 in the first case, $[a_i]$ in the second case, $-[b_j]$ in the third case.

4.5.1.4. *Meromorphic solutions.* Let U_0 and V as above. If f is a solution meromorphic on U, its restriction f_0 to U_0 is meromorphic on U_0. Conversely, any $f_0 \in \mathscr{M}(U_0)$ uniquely determines such a solution f. Indeed, since the open sets $U_n := q^n U_0$ are pairwise disjoint, we may define independently the restrictions $f_n \in \mathscr{M}(U_n)$ and we just have to check that f_n and f_{n+1} are correctly related:

$$\forall n \in \mathbf{Z} \;,\; \forall x \in U_n \;,\; f_{n+1}(qx) = u(x) f_n(x).$$

The function f_0 being given, this obviously defines uniquely (by ascending and descending induction) all the $f_n \in \mathscr{M}(U_n)$, and, glueing them, $f \in \mathscr{M}(U)$ such that $\sigma_q f = uf$. At any rate, we conclude that the restriction map $f \mapsto f_0 := f_{|U_0}$ is a $\mathbf{C}$-linear isomorphism:

$$\mathrm{Sol}(u, \mathscr{M}(U)) := \{\text{Solutions of } \sigma_q f = uf \text{ in } \mathscr{M}(U)\} \longrightarrow \mathscr{M}(U_0).$$

We just proved bijectivity and $\mathbf{C}$-linearity is obvious.

Actually, a little more holds that mere $\mathbf{C}$-linearity. Indeed, if f is a solution and if $g \in \mathscr{M}(U)$ is σ_q-invariant, then gf is a solution. This makes $\mathrm{Sol}(u, \mathscr{M}(U))$ a $\mathscr{M}(U)^{\sigma_q}$-module. Recall that $\mathscr{M}(U)^{\sigma_q}$ can be identified with $M(U/q^{\mathbf{Z}})$, i.e., to $\mathscr{M}(V)$. On the other hand, $\mathscr{M}(U_0)$ too can be identified with $\mathscr{M}(V)$ since $\pi : U_0 \to V$ is an homeomorphism and an analytic chart. Therefore, we conclude:

PROPOSITION 4.81. *The restriction map $f \mapsto f_0 := f_{|U_0}$ is an isomorphism of $\mathscr{M}(V)$-modules from Sol$(u, \mathscr{M}(U))$ to $\mathscr{M}(U_0) = \mathscr{M}(V)$.*

□

4.5.2. Conditions of holomorphy.

We are going to define various subspaces of $\mathrm{Sol}(u, \mathscr{M}(U))$ by putting conditions of holomorphy on solutions.

4.5.2.1. *Holomorphic solutions.* Here, we want to characterize those $f_0 \in \mathscr{M}(U_0)$ such that the corresponding $f \in \mathrm{Sol}(u, \mathscr{M}(U))$ is holomorphic all over U, i.e., such that $f \in \mathrm{Sol}(u, \mathscr{O}(U))$.

We begin with a preliminary remark; if $f_n \in \mathscr{O}(U_n)$ and $f_{n+1} \in \mathscr{O}(U_{n+1})$ are related by $\sigma_q f_{n+1} = u f_n$ (this makes sense on U_n since $U_{n+1} = qU_n$), then $f_n \in \mathscr{O}(U_n) \Leftrightarrow f_{n+1} \in \mathscr{O}(U_{n+1})$, except if u has a pole or a zero on U_n. So, if $u \in \mathscr{O}(U_n)^*$, i.e., if it has neither zeroes nor poles on U_n, let us say that U_n and U_{n+1} are "linked" (this is but a provisional terminology).

In the first case (as defined in 4.5.1.3), this holds true for all $n \in \mathbf{Z}$. So here, all U_n, $n \in \mathbf{Z}$, are linked; so that, in the notations of 4.5.1.4, either all f_n are holomorphic or none of them is.

More generally, in any of the three cases, all the U_n with $n \geq 1$ are linked and all the U_n with $n \leq 0$ are linked. The only possible problem would arises between U_0 and U_1 in the second or third case.

PROPOSITION 4.82. *Let $f \in$ Sol$(u, \mathscr{M}(U))$ and $f_0 := f_{|U_0}$. Then:*

(i) *In the first and second case, $f \in$ Sol$(u, \mathscr{O}(U)) \Leftrightarrow f_0 \in \mathscr{O}(U_0)$.*

(ii) *In the third case, $f \in$ Sol$(u, \mathscr{O}(U)) \Leftrightarrow f_0 \in (x - b_j)\mathscr{O}(U_0)$. (Recall that $\{b_j\} := U_0 \cap \overline{\Sigma}$.)*

PROOF. The first case has already been taken care of. For the other cases note that, from the preliminary remark, f is holomorphic (on U) if, and only if, f_0 and f_1 are (respectively on U_0 and U_1); and that f_1 is holomorphic on U_1 if, and only if, uf_0 is holomorphic on U_0. So we already have:

$$f \in \mathrm{Sol}(u, \mathscr{O}(U)) \Longleftrightarrow f_0 \in \mathscr{O}(U_0) \cap u^{-1}\mathscr{O}(U_0).$$

Now, on U_0, we have $u = (x - a_i) \times$ an element of $\mathscr{O}(U_0)^*$ (in the second case) or $u = \frac{1}{x-b_j} \times$ an element of $\mathscr{O}(U_0)^*$ (in the third case), so that:

$$\mathscr{O}(U_0) \cap u^{-1}\mathscr{O}(U_0) = \begin{cases} \mathscr{O}(U_0) \text{ in the second case,} \\ (x - b_j)\mathscr{O}(U_0) \text{ in the third case.} \end{cases}$$

□

REMARK 4.83. The apparent dissymmetry between the second and third case, and between the a_i and the b_j, can be explained as follows. It is not really the open set U_0 that is special, it is rather the "unlinked pair" (U_0, U_1). If we had decided to use $f_1 \in \mathscr{M}(U_1)$ as a touchstone instead of $f_0 \in \mathscr{M}(U_0)$, the second case would have involved a_i and the third no b_j.

EXERCISE 4.84. Check that contentions by using conditions on f_1 instead of conditions on f_0.

4.5.2.2. *Solutions holomorphic near* 0 *or near* ∞. In future chapters, we shall often concentrate on local study. In that context, global phenomena on $\mathbf{C}^*$ are not meaningful. So we shall be interested by solutions that are holomorphic on $U := \pi^{-1}(V)$ *in some (punctured) neighborhood of* 0, which we shall write: *in* $(\mathbf{C}^*, 0)$; and similarly near ∞.

So, for a precise definition, we shall say that $f \in \mathrm{Sol}(u, \mathscr{M}(U))$ is *holomorphic on* U *in* $(\mathbf{C}^*, 0)$ if $f \in \mathscr{O}(U \cap \overset{\bullet}{\mathrm{D}}(0, R))$ for some $R > 0$. We then write[13] $f \in \mathscr{O}(U, 0)$. Likewise, we shall say that it is *holomorphic on* U *in* $(\mathbf{C}^*, \infty)$, and then write $f \in \mathscr{O}(U, \infty)$, if $f \in \mathscr{O}(U \setminus \overline{\mathrm{D}}(0, R))$ for some $R > 0$. (Punctured disks $\overset{\bullet}{\mathrm{D}}(0, R)$ and closed disks $\overline{\mathrm{D}}(0, R)$ are defined in 0.4).

From the previous remarks we see that holomorphy on U in $(\mathbf{C}^*, 0)$ is equivalent to the holomorphy of f_1 on U_1; that holomorphy on U in $(\mathbf{C}^*, \infty)$ is equivalent to the holomorphy of f_0 on U_0; and that both are equivalent in the first case. So the reader will easily complete the proof of the following. As before, depending on the case, $U_0 \cap \overline{\Sigma}$ is either empty, or $\{a_i\}$, or $\{b_j\}$.

PROPOSITION 4.85. *Let* $f \in \mathit{Sol}(u, \mathscr{M}(U))$ *and* $f_n := f_{|U_n}$ *for* $n = 0, 1$.

(i) *In the first case:*

$$f \in \mathscr{O}(U, 0) \Longleftrightarrow f \in \mathscr{O}(U, \infty) \Longleftrightarrow f \in \mathscr{O}(U) \Longleftrightarrow f_0 \in \mathscr{O}(U_0) \Longleftrightarrow f_1 \in \mathscr{O}(U_1).$$

(ii) *In the second case:*

$$f \in \mathscr{O}(U, 0) \Longleftrightarrow f_1 \in \mathscr{O}(U_1) \Longleftrightarrow f_0 \in \frac{1}{x - a_i}\mathscr{O}(U_0),$$

$$f \in \mathscr{O}(U, \infty) \Longleftrightarrow f_0 \in \mathscr{O}(U_0) \Longleftrightarrow f_1 \in (x - a_i q)\mathscr{O}(U_1).$$

---

[13]Beware that we do not really define $\mathscr{O}(U, 0)$ here, nor $\mathscr{O}(U, \infty)$: they are really sets of *germs*, see the remark at the beginning of 4.5.3.1.

(iii) In the third case:

$$f \in \mathscr{O}(U,0) \Longleftrightarrow f_1 \in \mathscr{O}(U_1) \Longleftrightarrow f_0 \in (x-b_j)\mathscr{O}(U_0),$$

$$f \in \mathscr{O}(U,\infty) \Longleftrightarrow f_0 \in \mathscr{O}(U_0) \Longleftrightarrow f_1 \in \frac{1}{x-b_jq}\mathscr{O}(U_1).$$

□

4.5.3. Sheaves and line bundles. We now shall "glue" together the local information above in the form of sheaves, then of line bundles.

4.5.3.1. *Sheaves of solutions.* For every open subset V of $\mathbf{E}_q$, the open subset $U := \pi^{-1}(V)$ is q-invariant: $qU = U$. Conversely, every q-invariant open subset U of $\mathbf{C}^*$ is equal to $\pi^{-1}(V)$, where $V := \pi(U)$ is an open subset of $\mathbf{E}_q$. So it is natural to parameterize our sets of local solutions by such open subsets V of $\mathbf{E}_q$, and we set:

$$\mathscr{F}_u(V) := \{f \in \mathscr{O}(\pi^{-1}(V)) \mid \sigma_q f = uf\},$$
$$\mathscr{F}_u^{(0)}(V) := \{f \in \mathscr{O}(\pi^{-1}(V),0) \mid \sigma_q f = uf \text{ near } 0\},$$
$$\mathscr{F}_u^{(\infty)}(V) := \{f \in \mathscr{O}(\pi^{-1}(V),\infty) \mid \sigma_q f = uf \text{ near } \infty\}.$$

REMARK 4.86. The restrictions "near 0", "near ∞" are of course necessary here since the corresponding solutions f are a priori not defined everywhere. It is time to be more precise. The elements of $\mathscr{O}(U,0)$ are *germs*, i.e., functions defined and holomorphic in *some* punctured neighborhood of 0, and two germs are considered equal if the representative functions coincide in *some* (maybe smaller) punctured neighborhood of 0. Similar definitions are made at ∞. In case of rational q-difference equations (as here), we have unique meromorphic continuation to the whole of U so this definition is equivalent to the approach in 4.5.2 (see the exercise).

EXERCISE 4.87. Check for instance that one can identify $\mathscr{F}_u^{(0)}(V)$ as defined above with $\{f \in \mathrm{Sol}(u,\mathscr{M}(U)) \mid f \text{ is holomorphic on } U \text{ in } (\mathbf{C}^*,0)\}$.

What matters for us is that:

(1) $\mathscr{F}_u, \mathscr{F}_u^{(0)}, \mathscr{F}_u^{(\infty)}$ are sheaves on $\mathbf{E}_q$: being a section on any of them is a local property;
(2) they are sheaves of modules over the sheaf $\mathscr{O}_{\mathbf{E}_q}$ of holomorphic functions on $\mathbf{E}_q$: clearly all $\mathscr{F}_u(V), \mathscr{F}_u^{(0)}(V), \mathscr{F}_u^{(\infty)}(V)$ are $\mathbf{C}$-linear spaces, but moreover, for each $f \in \mathscr{O}_{\mathbf{E}_q}(V)$ and each $g \in \mathscr{F}_u(V)$ (for instance), $fg \in \mathscr{F}_u(V)$;
(3) they are *locally free sheaves of rank* 1, i.e., they are locally isomorphic to the sheaf $\mathscr{O}_{\mathbf{E}_q}$: this directly follows from 4.5.2.

4.5.3.2. *The associated line bundles.* It is a general fact that, X being a Riemann surface, every sheaf of modules over $\mathscr{O}_X$ which is locally free of rank r (i.e., locally isomorphic to $\mathscr{O}_X^r$) can be identified with the sheaf of sections of a vector bundle $\mathfrak{F}$, a line bundle if the rank is $r = 1$. So we should be able to define line bundles $\mathfrak{F}_u, \mathfrak{F}_u^{(0)}, \mathfrak{F}_u^{(\infty)}$ corresponding to the sheaves above. Following 2.2.3, we would set for instance:

$$\mathfrak{F}_u := \frac{\mathbf{C}^* \times \mathbf{C}}{(x,X) \sim (qx,u(x)X)}.$$

However this presents difficulty since here (contrary to what was the case in 2.2.3), $u(x)$ may be undefined at some values of x; or the relation $(qx, u(x)X)$ may lack symmetry if $u(x) = 0$.

We shall overcome that difficulty for $\mathfrak{F}_u^{(0)}, \mathfrak{F}_u^{(\infty)}$ in the following way. We do it for $\mathfrak{F}_u^{(0)}$ (see exercise as regards ∞). We consider $(\mathbf{C}^*, 0)$ as a "germ of space". This can be defined and studied formally; see [**85**, p. 16] or [**16**, Chapter 0, §III]. But we can take it pragmatically by saying that all constructions and statements involving $(\mathbf{C}^*, 0)$ are understood to hold in some punctured neighborhood of 0, and are to be identified (or considered equivalent) if they coincide in a (maybe smaller) punctured neighborhood of 0. Then, given that $u \in \mathbf{C}(x)^*$ has neither zeroes nor poles in a small enough punctured neighborhood of 0, we can set:

$$\mathfrak{F}_u^{(0)} := \frac{(\mathbf{C}^*, 0) \times \mathbf{C}}{(x, X) \sim (qx, u(x)X)}.$$

One then checks easily that $\mathfrak{F}_u^{(0)}$ is a line bundle over $\mathbf{E}_q$ and that its sheaf of sections is naturally identified with $\mathscr{F}_u^{(0)}$.

EXERCISE 4.88. (i) Do the same at ∞.
(ii) How would you adapt the construction to take care of $\mathfrak{F}_u$?

4.5.3.3. *An alternative construction.* It rests on the fact that we have here explicit way of producing solutions (this remains true for all "fuchsian equations", more generally for all those with "integral slopes", see Chapters 11 and 15). We rewrite u in two ways:

$$u = \begin{cases} c'x^m u_0, \text{ where } c' := (-1)^{k-l} c \dfrac{a_1 \cdots a_k}{b_1 \cdots b_l} \text{ and } u_0 := \frac{(1-x/a_1)\cdots(1-x/a_k)}{(1-x/b_1)\cdots(1-x/b_l)}, \\ cx^{m'} u_\infty, \text{ where } m' := m + k - l \text{ and } u_\infty = \frac{(1-a_1/x)\cdots(1-a_k/x)}{(1-b_1/x)\cdots(1-b_l/x)}. \end{cases}$$

This leads (by the methods of this chapter) to a solution $f^{(0)}(x) := \dfrac{\theta_q(x/c')}{\theta_q(x)^{m+1}} g^{(0)}(x)$ and to a solution $f^{(\infty)}(x) := \dfrac{\theta_q(x/c)}{\theta_q(x)^{m'+1}} g^{(\infty)}(x)$, where $\sigma_q f^{(0)} = u_0 f^{(0)}$ and $\sigma_q f^{(\infty)}$ $= u_\infty f^{(\infty)}$, so we can take:

$$g^{(0)}(x) := \frac{(x/b_1;q)_\infty \cdots (x/b_l;q)_\infty}{(x/a_1;q)_\infty \cdots (x/a_k;q)_\infty} \text{ and } g^{(\infty)}(x) := \frac{(qa_1/x;q)_\infty \cdots (qa_k/x;q)_\infty}{(qb_1/x;q)_\infty \cdots (qb_l/x;q)_\infty}.$$

EXERCISE 4.89. Recall why $f^{(0)}/f^{(\infty)}$ is q-elliptic. Check it by expressing with Theta functions. (You may invoke Jacobi's Triple Product Formula.)

Now let f be either $f^{(0)}$ or $f^{(\infty)}$, at any rate, an explicit function of which we know the zeroes and poles. Then all meromorphic solutions of our equation on $\mathbf{C}^*$ have the form $f\phi$, where $\phi \in \mathscr{M}(\mathbf{C}^*)^{\sigma_q} = \mathscr{M}(\mathbf{E}_q)$. More generally, for $V \subset \mathbf{E}_q$, $U := \pi^{-1}(V)$:

$$\mathscr{F}_u(V) = \{f\phi \mid \phi \in \mathscr{M}(U)^{\sigma_q}, \phi f \in \mathscr{O}(U)\} \simeq \{\phi \mid \phi \in \mathscr{M}(V), \phi f \in \mathscr{O}(U)\}.$$

Note that identifying $\mathscr{M}(V)$ with $\mathscr{M}(U)^{\sigma_q}$ means that , if $a \in \mathbf{C}^*$ and $p := \pi(a) \in \mathbf{E}_q$, we set $\phi(a) := \phi(p)$. So the condition $\phi f \in \mathscr{O}(U)$ reads:

$$\forall a \in U \ , \ v_{\pi(a)}(\phi) + v_a(f) \geq 0, \text{ i.e. } \forall p \in V \ , \ \forall a \in \pi^{-1}(p) \ , \ v_p(\phi) + v_a(f) \geq 0.$$

This is equivalent to a condition on divisors:

$$\phi f \in \mathscr{O}(U) \iff \mathrm{div}_V(\phi) \geq -\sum_{p\in V} n_p[p], \text{ where } n_p := \min_{\pi(a)=p} v_a(f).$$

So call $D := \mathrm{div}_{\mathbf{C}^*}(f)$ the divisor of f and $D_{|U}$ its restriction to U (i.e., if $D = \sum_{a\in\mathbf{C}^*} v_a(f)[a]$, then $D_U := \mathrm{div}_U(f) = \sum_{a\in U} v_a(f)[a]$). We introduce a general operation, say $\tilde{\pi}$, on divisors:

$$\tilde{\pi} : \sum_{a\in U} n_a[a] \mapsto \sum_{p\in V} \tilde{n}_p[p], \text{ where } \tilde{n}_p := \min_{\pi(a)=p} n_a.$$

Then we got an isomorphism of $\mathscr{M}(V)$ modules:

$$\mathscr{F}_u(V) \simeq \{\phi \in \mathscr{M}(V) \mid \mathrm{div}_V(\phi) \geq \tilde{D}_V\}, \text{ where } D := \tilde{\pi}\mathrm{div}_{\mathbf{C}^*}(f).$$

EXERCISE 4.90. (i) Compute $\tilde{\pi}\left(\mathrm{div}_{\mathbf{C}^*}(f^{(0)})\right)$ and $\tilde{\pi}\left(\mathrm{div}_{\mathbf{C}^*}(f^{(\infty)})\right)$.

(ii) Adapt the whole argument to describe $\mathscr{F}_u^{(0)}$ and $\mathscr{F}_u^{(\infty)}$.

CHAPTER 5

Resolution of (General) Scalar Equations and Factorisation of q-Difference Operators

General references for this chapter are [**20, 65, 87, 191, 197, 221**].

5.1. General considerations

Classically, in the theory of functional equations, one is primarily interested in the *global* study over the Riemann sphere $\mathbf{S}$: this means equations with *rational* coefficients. Indeed most special functions are solutions of linear functional equations with rational (or polynomial, this amounts to the same) coefficients. Of course, the most well known functional equations are ordinary differential equations, but these are not the only ones that matter: just think of the Zeta, Gamma and Theta functions!

It has been a successful strategy from the beginning to start with a *local* study, i.e., taking coefficients in $\mathbf{C}(\{x\})$ or $\mathbf{C}\{x\}$ and looking for solutions near 0 (say); and often one starts with a *formal* study, i.e., taking coefficients in $\mathbf{C}((x))$ or $\mathbf{C}[[x]]$, thus emphasizing the more algebraic aspects. Then of course, in the end, one must adress the convergence of solutions.

When studying (complex analytic) ordinary differential equations, the local study at 0 keeps making sense when going to other points of the Riemann sphere, respectively replacing $\mathbf{C}(\{x\})$ and $\mathbf{C}((x))$ by $\mathbf{C}(\{x-a\})$, $\mathbf{C}((x-a))$ (at $a \in \mathbf{C}^*$) or $\mathbf{C}(\{1/x\})$, $\mathbf{C}((1/x))$ (at ∞). However, for q-difference equations, it is not easy to give a meaning to a local study at $a \in \mathbf{C}^*$, since the q-dilatation operator central to the functional equation sends a to qa, etc. So "local" for us will mean at 0 or at ∞.

In this section of generalities, the q-difference field K will be one of the coefficient fields mentioned above and we will look for solutions in some q-difference extension K' of K, to be precised later. Then, for $L \in \mathcal{D}_{K,q}$, the space $\mathrm{Sol}(L, K')$ will be a linear space over K'^{σ_q}, although we shall sometimes write $\mathrm{Ker}L$ to maintain some undefiniteness.

5.1.1. Homogeneous or not. In most of the chapter, except from some examples (and in contrast to the next Chapter 6), we shall restrict ourselves to homogeneous equations; but some words about nonhomogeneous equations are necessary.

5.1.1.1. Let K a q-difference field (one of the coefficient fields mentioned above). Classically, one distinguishes *homogeneous* q-difference equations:

$$Lf = 0, \text{ where } L \in \mathcal{D}_{K,q}, \tag{5.1}$$

from *nonhomogeneous* q-difference equations:

$$Lf = b, \text{ where } L \in \mathcal{D}_{K,q} \text{ and } b \in K^*. \tag{5.2}$$

Clearly one can assume that L is standard unitary (why?):

$$L = \sigma_q^n + a_1\sigma_q^{n-1} + \cdots + a_n, \text{ where } a_1, \ldots, a_n \in K \text{ and } a_n \neq 0.$$

We make (5.2) homogeneous by introducing:

$$L' := \left(\sigma_q - \frac{\sigma_q(b)}{b}\right) L, \text{ so that } Lf = b \Longrightarrow L'f = 0.$$

Conversely suppose that $L'f = 0$ and set $g := Lf$. If $g = 0$, then f is solution of the homogeneous equation (5.1). However, we often expect that the inclusion $\mathrm{Ker}L \subset \mathrm{Ker}L'$ would usually be strict, so we hope to be able find to such f that $g \neq 0$. In that case:

$$\left(\sigma_q - \frac{\sigma_q(b)}{b}\right) g = \left(\sigma_q - \frac{\sigma_q(b)}{b}\right) Lf = L'f = 0 \Longrightarrow \sigma_q(g/b) = g/b.$$

So we write $g = b\epsilon$, where $\epsilon \neq 0$ is σ_q-invariant, that is q-elliptic. This implies that $\epsilon L = L\epsilon$ (section 3.1) so:

$$L(f/\epsilon) = (Lf)/\epsilon = g/\epsilon = b.$$

Said otherwise, f may not be itself a solution of (5.2) but some f/ϵ is, ϵ being q-elliptic.

5.1.1.2. There is another view on the logical relation of nonhomogeneous to homogeneous equations. Let f_0 a particular solution of (5.2). Then, for every other solution f:

$$Lf = b = Lf_0 \Longrightarrow L(f - f_0) = 0.$$

So the set $L^{-1}(b)$ of solutions of (5.2) is either empty, or equal to $f_0 + \mathrm{Ker}L$ for some f_0, that is, an affine space with underlying linear space $\mathrm{Ker}L$. The obstruction to existence of such a f_0 is measured by the cokernel $K'/\mathrm{Im}L$ and a deeper understanding will lead us, in Chapter 6, to introduce the "small additive q-de Rham complex":

$$0 \to \mathrm{Ker}L \to K' \xrightarrow{L} K' \to \mathrm{Coker}L \to 0.$$

5.1.2. About ramification.

5.1.2.1. Consider the equation $\sigma_q^2 f = qf$; equivalently: $\sigma_{q^2} f = qf$. It can be solved, for instance, by the multivalued function $f := x^{1/2}$. For complex linear ordinary differential equations, the use of multivalued functions cannot be avoided **[222]**. However, for linear q-difference equations, *everything can be done with uniform functions.* This is due in principle to Praagman, see the theorem 7.4 in Section 7.3; it can be made explicit for fuchsian equations, see Chapter 11; or for irregular equations with integral slopes, see Chapters 14 and 15.

There are perhaps some good reasons to avoid multivalued functions. But our reasons to do so are more positive: taking $\mathscr{M}(\mathbf{C}^*)$ or $\mathscr{M}(\mathbf{C}^*, 0)$ as solution field, we have $\mathscr{M}(\mathbf{E}_q)$ as constant field and the whole theory of elliptic functions is at our disposal. The idea to do so was suggested by Birkhoff at the end of **[23]** and revived by Ramis **[190]**.

5.1.2.2. Considered as a q-difference equation, $\sigma_q^2 f = q$ has a basis of solutions (in $\mathscr{M}(\mathbf{C}^*)$) made, for instance, of $\dfrac{\theta_q(x)}{\theta_q(\sqrt{q}x)}$ and $\dfrac{\theta_q(x)}{\theta_q(-\sqrt{q}x)}\cdot$ Considered as a q^2-difference equation $\sigma_{q^2} f = qf$, its solutions are q^2-characters and a basis consists for instance in $\dfrac{\theta_{q^2}(x)}{\theta_{q^2}(qx)}\cdot$

Note that these are bases over different constant fields: $\mathscr{M}(\mathbf{E}_q)$ in the first case, $\mathscr{M}(\mathbf{E}_{q^2})$ in the second case; and of course, the latter is an extension of degree 2 of the former (they are isogeneous by Chapter 2).

At any rate, the reader can check that there is no nontrivial holomorphic solution. While avoiding the multivaluedness of the solution $x^{1/2}$, we lost its holomorphic character and gained instead discrete spirals of poles.

EXERCISE 5.1. (i) Verify that there is no nontrivial holomorphic solution in $\mathscr{M}(\mathbf{C}^*)$ through the following argument: such a solution would be of the form $\dfrac{\theta_{q^2}(x)}{\theta_{q^2}(qx)}\phi$ for a q^2-elliptic function ϕ and the divisor of such a ϕ cannot compensate the poles.

(ii) Do it again with an argument based on Cauchy residue theorem as in 4.2.1.

5.1.2.3. Now consider $\sigma_q^2 f = xf$; equivalently: $\sigma_{q^2} f = xf$. In its second form, it admits the obvious solution $\dfrac{1}{\theta_{q^2}(x)}\cdot$ However, this means that we are going to use linear spaces over the constant field $\mathscr{M}(\mathbf{E}_{q^2})$; for classification problems, including q-difference Galois theory[1], we may wish to stay in the q-world and use only spaces over $\mathscr{M}(\mathbf{E}_q)$.

One obvious solution is to reintroduce multivalued functions, like $\sqrt{\dfrac{1}{\theta_q}}$ or $\dfrac{1}{\theta_q(\sqrt{x})}\cdot$ This will complicate a little the matter, we shall see how when studying irregular equations, in particular their Stokes phenomenonin Chapters 14, 15.

5.1.3. "Truly constant" coefficients.

5.1.3.1. Here we suppose that the coefficients in (5.1) are complex numbers: $a_1, \ldots, a_n \in \mathbf{C}$. ("Truly constant" is, of course, meant in contrast to "q-constant".) Therefore the operator L belongs to the sub-$\mathbf{C}$-algebra $\mathbf{C}[\sigma_q]$ of $\mathcal{D}_{K,q}$, which is commutative. Actually, factoring the polynomial:

$$X^n + a_1 X^{n-1} + \cdots + a_n = \prod_{i=1}^{k} (X - c_i)^{r_i}, k \in \mathbf{N}^*, c_1, \ldots, c_k \in \mathbf{C}^*, r_1, \ldots, r_k \in \mathbf{N}^*,$$

(it is understood that the c_i are pairwise distinct), we see that:

$$L = L_1 \cdots L_k, \text{ where } L_i = (\sigma_q - c_i)^{r_i}, i = 1, \ldots, k.$$

Then we can apply the rules of usual linear algebra over $\mathbf{C}$ and deduce:

$$\mathrm{Ker} L = \mathrm{Ker} L_1 \oplus \cdots \oplus \mathrm{Ker} L_k.$$

[1] We shall see in Chapters 13 and 15 that for us Galois theory is, among other things, a classification theory (by representations of an algebraic group).

5.1.3.2. The above argument allows us to reduce the problem to the case that $L = (\sigma_q - c)^n$, $n \in \mathbf{N}^*$, $c \in \mathbf{C}^*$. We use once more the trick of gauge transformation $f = e_{q,c}f'$ (already used in Chapters 3 and 4), i.e., we introduce:

$$L' := e_{q,c}^{-1} L e_{q,c}, \text{ so that } Lf = 0 \Longleftrightarrow L'f' = 0.$$

Now L' is computed as follows. As operators, $(\sigma_q - c)e_{q,c} = c e_{q,c}(\sigma_q - 1)$, whence:

$$\begin{aligned} e_{q,c}^{-1}(\sigma_q - c)e_{q,c} = c(\sigma_q - 1) \Longrightarrow L' &= e_{q,c}^{-1}(\sigma_q - c)^n e_{q,c} \\ &= \left(e_{q,c}^{-1}(\sigma_q - c)e_{q,c}\right)^n = c^n(\sigma_q - 1)^n. \end{aligned}$$

Therefore, if $f = e_{q,c}f'$:

$$(\sigma_q - c)^n f = 0 \Longleftrightarrow (\sigma_q - 1)^n f' = 0.$$

5.1.3.3. The above argument allows us to reduce the problem to the case that $L = (\sigma_q - 1)^n$, $n \in \mathbf{N}^*$. For $n = 1$, this defines q-constants, a basic solution is 1. For $n = 2$, we know that l_q is a solution:

$$(\sigma_q - 1)^2 l_q = (\sigma_q - 1)(\sigma_q l_q - l_q) = (\sigma_q - 1)1 = 0.$$

In general, remember from 2.3.2 that we defined:

$$l_q^{(n)} := \binom{\ell_q}{n} = \frac{1}{n!} l_q(l_q - 1)\cdots(l_q - n + 1),$$

and proved that $(\sigma_q - 1)l_q^{(n)} = l_q^{(n-1)}$. This includes the cases $l_q^{(0)} = 1$ and $l_q^{(1)} = l_q$. We deduce that $1, l_q, \ldots, l_q^{(n-1)}$ is a basis of solutions of equation $(\sigma_q - 1)^n f = 0$.

5.1.3.4. We summarize the results as follows.

THEOREM 5.1. *Let* $L := (\sigma_q - c_1)^{r_1} \cdots (\sigma_q - c_k)^{r_k}$, $k \geq 1$, $r_1, \ldots, r_k \geq 1$, $c_1, \ldots, c_k \in \mathbf{C}^*$. *Then a basis of KerL is the family of all* $e_{q,c_i} l_q^{(j)}$, $i = 1, \ldots, k$, $j = 0, \ldots, r_i - 1$.

5.1.4. Convergent or divergent solutions and factorisation.

5.1.4.1. Irregular (complex, linear) ordinary differential equations have solutions involving basic functions: "characters" z^α, logarithms, exponentials of some rational functions; and power series. For so-called "fuchsian" equations **[222]**, there is no exponential factor and the power series are convergent; but in general, the power series may diverge. More precisely, some differential equations have no nontrivial solution involving convergent power series. Also the exponential factors may have a rather complicated structure **[150]**.

For fuchsian q-difference equations (to be defined in Chapter 11), the basic functions are q-characters and q-logarithms and the associated power series always converge: so this is rather like the corresponding case for differential equations. However, for irregular q-difference equations (Chapters 14 and 15) there are some big differences. The main one is that there is always a nontrivial solution whose power series factor is convergent. This is Adams's lemma, see 5.3.2. Also, exponential factors are here replaced by Theta factors whose structure is rather simple.

5.1.4.2. An important consequence of Adams's lemma is that analytic q-difference operators (those over $\mathbf{C}(\{x\})$) admit an analytic factorisation; this is not true for differential operators, which generally admit only a formal factorisation **[150, 159, 204]** (for the case of difference operators, see for instance **[75]**).

The reason is that, if f is a nontrivial "convergent solution" of $L(f) = 0$, i.e., one whose power series factor converges, then $L = L'(\sigma_q - \sigma_q(f)/f)$ and L' is analytic. To be true, maybe some solutions will be "ramified", i.e., multivalued as we saw in 5.1.2.3; but grouping them in some way will yield nonramified factors. This will be detailed in 5.4.2.

5.2. Formal solutions and the Newton polygon

The method of resolution shown in this section is, in essence, due to Adams [**1**, **2**]. The path followed is: formal resolution, then discussion of convergence.

5.2.1. Formal power series solutions.

5.2.1.1. We slightly change the indexing of coefficients of q-difference operators in order to ease calculations. So let:

$$L := a_n\sigma_q^n + \cdots + a_0 \in \mathcal{D}_{K,q}, a_0a_n \neq O.$$

To begin with, we take $K := \mathbf{C}((x))$. Clearly, left-multiplying L by an invertible element of $\mathcal{D}_{K,q}$ does not change the space of solutions of q-difference equation $L(f) = 0$. But here, instead of taking L unitary (i.e., such that $a_n = 1$), we shall assume that the coefficients a_i are mutually coprime in $\mathbf{C}[[x]]$. This amounts to require that $\min\{v_0(a_i), i = 0, \ldots, n\} = 0$, equivalently:

$$a_0, \ldots, a_n \in \mathbf{C}[[x]] \text{ and } a_i(0) \neq 0 \text{ for some } i.$$

We look for a power series solution $f := \sum f_k x^k \in \mathbf{C}((x))$; more precisely, we look for conditions of existence of a nontrivial power series solution. First, we reduce the problem to the case of a *regular* power series solution (0.4 and 4.3.2). Writing $f := x^l g$ and using the fact that $x^{-l}\sigma_q^i x^l = q^{il}\sigma_q^i$, we see that:

$$L(f) = 0 \Longleftrightarrow L'(g) = 0 \text{ where } L' := x^{-l}Lx^l = a_nq^{nl}\sigma_q^n + \cdots + a_1q^l\sigma_q + a_0.$$

We are therefore going to look for conditions of existence of a regular power series solution; then, if necessary, we shall translate them back into conditions of existence of a nontrivial power series solution.

5.2.1.2. So let $f := \sum\limits_{k\geq 0} f_k x^k \in \mathbf{C}[[x]]$, $f_0 \neq 0$. To study in detail the q-difference equation, we set $a_i = a_{i,0} + a_{i,1}x + \cdots$, $i = 0, \ldots, n$. The condition assumed above says that $a_i(0) = a_{i,0} \neq 0$ for at least one i. We compute the coefficients $(Lf)_k$ of $L(f)$ as follows (we index the coefficients of f by l):

$$L(f) = \sum_{i=0}^{n}\sum_{j\geq 0}\sum_{l\geq 0} a_{i,j}x^j\sigma_q^i(f_lx^l) = \sum_{i=0}^{n}\sum_{j\geq 0}\sum_{l\geq 0} a_{i,j}q^{il}f_lx^{j+l} = \sum_{l\geq 0}\left(\sum_{\substack{i=0,\ldots,n\\ j+l=k}} a_{i,j}q^{il}f_l\right)x^k.$$

We deduce that $(Lf)_0 = \sum\limits_{i=0}^{n} a_{i,0}$ and, for $k \geq 1$:

$$\begin{aligned}(Lf)_k &= \sum_{\substack{i=0,\ldots,n\\ j=0,\ldots,k}} a_{i,j}q^{i(k-j)}f_{k-j}\\ &= \left(\sum_{i=0,\ldots,n} a_{i,0}q^{ik}\right)f_k + \text{some explicit linear combination of } f_0, \ldots, f_{k-1}.\end{aligned}$$

We can make that formula more explicit and more useful by introducing some auxiliary notations. We write $L = L_0 + xL_1 + \cdots$, where $L_j := \sum\limits_{i=0}^{n} a_{i,j}\sigma_q^i \in \mathbf{C}[\sigma_q]$

for all $j \geq 0$. Moreover we associate to each such constant coefficient q-difference operator L_j a kind of "characteristic polynomial":

$$P_j(s) := \sum_{i=0}^{n} a_{i,j} s^i \in \mathbf{C}[s].$$

Then, the above formula is made precise as follows:

$$\forall k \geq 0\,,\; (Lf)_k = \sum_{j+l=k} P_j(q^l) f_l = P_0(q^k) f_k + \cdots + P_k(q^0) f_0.$$

EXERCISE 5.2. Make explicit the relation to the characteristic equation $\overline{L}(s) = 0$ introduced in 4.3.2.2.

5.2.1.3. The above calculation allows us to formulate conditions for the existence of a regular power series solution.

PROPOSITION 5.3. (i) *A* necessary *condition is that* $P_0(1) = 0$.
(ii) *A* sufficient *condition is that* $P_0(1) = 0$ *and* $P_0(q^k) \neq 0$ *for all* $k \geq 1$.

PROOF. (i) From 5.2.1.2, evaluation at 0 yields the constant term $(Lf)(0) = P_0(1) f_0$. If $L(f) = 0$ with $f_0 \neq 0$, then $P_0(1) = 0$.
(ii) Equation $L(f) = 0$ is equivalent to $(Lf)_k = 0$ for all $k \geq 0$. Relation $(Lf)_0 = 0$ boils down to $P_0(1) f_0 = 0$, which allows any f_0 by the first assumption of (ii). Then relation $(Lf)_k = 0$ reduces, for $k \geq 1$, to equality:

$$P_0(q^k) f_k + \cdots + P_k(q^0) f_0 = 0 \Longleftrightarrow f_k = -\frac{P_1(q^{k-1}) f_{k-1} + \cdots + P_k(q^0) f_0}{P_0(q^k)},$$

which is a valid recursive definition by the second assumption of (ii).

□

We are led to the following definitions. Recall from 5.2.1.1 that L has coefficients in $\mathbf{C}[[x]]$ and coprime.

DEFINITION 5.4. (i) We denote $\overline{L} := P_0$. Equation $\overline{L}(s) = 0$ is the *characteristic equation* attached to the q-difference equation $L(f) = 0$ (or to the q-difference operator L).
(ii) Nonzero roots of the polynomial $\overline{L}$ are called *exponents* of the q-difference equation or of the q-difference operator.
(iii) A *nonresonant exponent* is a root $s \in \mathbf{C}^*$ of the characteristic equation $\overline{L}(s) = 0$ such that no $q^k s$, $k \geq 1$, is a root: $\forall k \geq 1\,,\; \overline{L}(q^k s) \neq 0$.

This definition will be generalized in 5.2.2.6. The proposition can there be reformulated:

COROLLARY 5.5. (i) *A* necessary *condition for q-difference equation $L(f) = 0$ to have a regular power series solution is that* 1 *be an exponent.*
(ii) *A* sufficient *condition is that* 1 *be a nonresonant exponent.*

Taking in account the discussion in 5.2.1.1, we deduce:

COROLLARY 5.6. *A* necessary and sufficient *condition for q-difference equation $L(f) = 0$ to have a nontrivial power series solution is that some q^k, $k \in \mathbf{Z}$, be an exponent.*

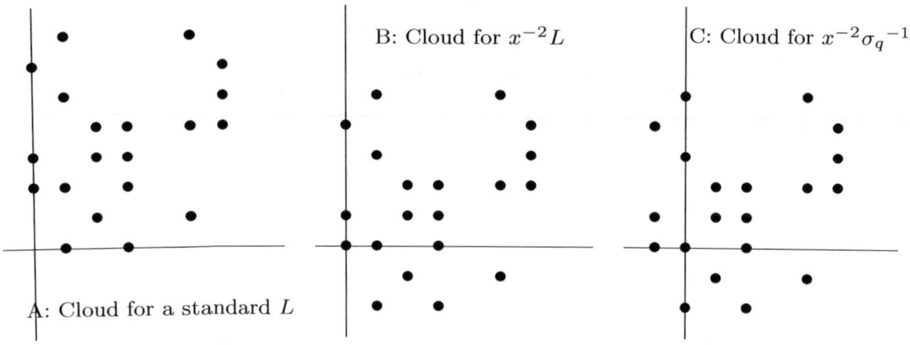

FIGURE 5.1. Clouds: prelude to the Newton polygon

PROOF. Necessity is a direct translation of the previous corollary and of the discussion (here k is of course $v_0(f)$, i.e., x^{-k} f is a regular power series). Now, if q^k is an exponent, there is an exponent q^l with l minimal and this one is the nonresonant exponent to which the previous corollary and the discussion apply. □

These results which be generalized below in two ways:

(1) Arbitrary exponents $c \in \mathbf{C}^*$ will be related to solutions $e_{q,c}f$, f regular (the case $c = q^k$ corresponds with power series solution such that $v_0(f) = k$).
(2) To get solutions $\theta_q^\mu e_{q,c}f$, we shall introduce exponents "attached to slope μ" (the above corresponds to slope 0).

EXERCISE 5.7. (i) To appreciate the strategy just outlined, first give conditions for the q-difference equation $L(f) = 0$ to have a solution $e_{q,c}f$, f a regular power series. Hint: consider $e_{q,c}^{-1}Le_{q,c}$.
(ii) Now give conditions for a solution $\theta_q^\mu e_{q,c}f$.

5.2.1.4. *Graphical interpretation: configurations and clouds.* We try to visualize the possibility of the conditions in the proposition. With the above exercise and the discussion on "configurations" in Chapter 4, we hope that the following will help you find natural the constructions to come next (in 5.2.2).

We want the characteristic equation $\overline{L}(s) = 0$ to have the root 1 (this is exactly the condition in the first corollary of the proposition); or some root q^k (this is the condition in the second corollary); or actually any nonzero exponent c (this is the condition that should appear if you solve question (i) in the above exercise). We use the notations of 5.2.1.3. For the equation $\sum\limits_{i=0}^{n} a_{i,0}s^i = 0$ To have a root in $\mathbf{C}^*$, it is necessary and sufficient that it has at least two nonvanishing terms, i.e., that there exist at least two indexes $i_0 < i_1$ in $\{0, \ldots, n\}$ such that $a_{i_0,0}a_{i_1,0} \neq 0$. We visualize this condition by drawing all points $(i, j) \in \mathbf{N}^2$ such that $0 \leq i \leq n$ and $a_{i,j} \neq 0$. This should look like figure 5.1 A, which was drawn for a standard q-difference operator $L := a_0 + \cdots + a_6\sigma_q^6$ and two of its "translates" (see herebelow) $x^{-2}L$ and $x^{-2}\sigma_q^{-1}L$, where we took:

$$a_0 := x^2 + x^3 + x^6, a_1 := 1 + x^2 + x^5 + x^7, a_2 := x + x^3 + x^4,$$
$$a_3 := 1 + x^2 + x^3 + x^4, a_4 := 0, a_5 := x + x^4 + x^7, a_6 := x^4 + x^5 + x^6.$$

What matters is the part of the *cloud of points* sitting on the x-axis, the lower horizontal segment.

If L was not required to be standard from start, it would be so up to some $\sigma_q^{i_0}$ factor on the right or on the left (this does not give the same q-difference operator but it gives the same cloud): see figure 5.1 B. If coefficients of L were not required to be in $\mathbf{C}[[x]]$ and coprime, this could be arranged by a x^{j_0} factor on the left on or the right: see 5.1 C. What is invariant under these possible changes is the following property of the cloud: it has a nontrivial lower horizontal segment (on our figures: two points).

EXERCISE 5.8. For an arbitrary q-difference operator L of absolute degree ≥ 1, show that there is a unique pair (i_0, j_0) such that $L' := x^{j_0}\sigma_q^{i_0} L$ is standard and with coefficients in $\mathbf{C}[[x]]$ and coprime. Show that then $L'' := L x^{j_0}\sigma_q^{i_0}$ has the same properties. Compare the clouds and the characteristic equations of L' and L''.

5.2.2. The Newton polygon.

5.2.2.1. Here, $K := \mathbf{C}((x))$. Let:

$$L := \sum_{i=i_0}^{i_1} a_i \sigma_q^i \in \mathcal{D}_{K,q}, i_0 \leq i_1, a_{i_0} a_{i_1} \neq 0,$$

so that $L \neq 0$. We usually assume that $i_1 > i_0$, i.e., $\deg L > 0$, i.e., L is not invertible. We often let $L := \sum a_i \sigma_q^i$ written as a finite sum but without precision on the range of the index i. Each $a_i \in \mathbf{C}((x))$ is written as $a_i = \sum a_{i,j} x^j$, where of course the indices j are bounded below. The following definitions rest on elementary properties of convex hulls in $\mathbf{R}^2$ that we do not bother to prove. (But some of them are nice exercises, although not within the main themes of this book!)

We imagine drawn in the plane the "cloud of points" $\{(i, v_0(a_i)) \mid i \in \mathbf{Z}\}$ attached to coeficients of L (if $a_i = 0$, $v_0(a_i) = \infty$, there is no corresponding point). Replacing each point with the upward vertical half line starting from it, we obtain the set $\{(i,j) \in \mathbf{Z}^2 \mid j \geq v_0(a_i)\}$. We are interested in its convex hull (figure 5.2).

The boundary of $NP(L)$ is made of two upward vertical half-lines with lower extremities (i_0, j_0) and (i_1, j_1), where $j_0 := v_0(a_{i_0})$ and $j_1 := v_0(a_{i_1})$; and of a finite number of segments, which we make into vectors by orienting them from left to right; we write them (r_i, d_i), $i = 1, \ldots, k$, each $(r_i, d_i) \in \mathbf{N}^* \times \mathbf{Z}$, so that their *slopes* $\mu_i := \dfrac{d_i}{r_i} \in \mathbf{Q}$ are ordered: $\mu_1 < \cdots < \mu_k$. Clearly, the width of the corresponding figure is:

$$r_1 + \cdots + r_k = i_1 - i_0 = \deg L.$$

DEFINITION 5.9 (Adams). The *Newton polygon* of the q-difference operator L, written $NP(L)$, is the convex hull of the set $\{(i,j) \in \mathbf{Z}^2 \mid j \geq v_0(a_i)\}$. Its *slopes* are the slopes $\mu_i := \dfrac{d_i}{r_i} \in \mathbf{Q}$, $i = 1, \ldots, k$ of the vectors $(r_i, d_i) \in \mathbf{N}^* \times \mathbf{Z}$ constituting the nonvertical part of its boundary, oriented and indexed from left to right (so that $\mu_1 < \cdots < \mu_k$). The set of slopes is denoted $S(L) := \{\mu_1, \ldots, \mu_k\} \subset \mathbf{Q}$.

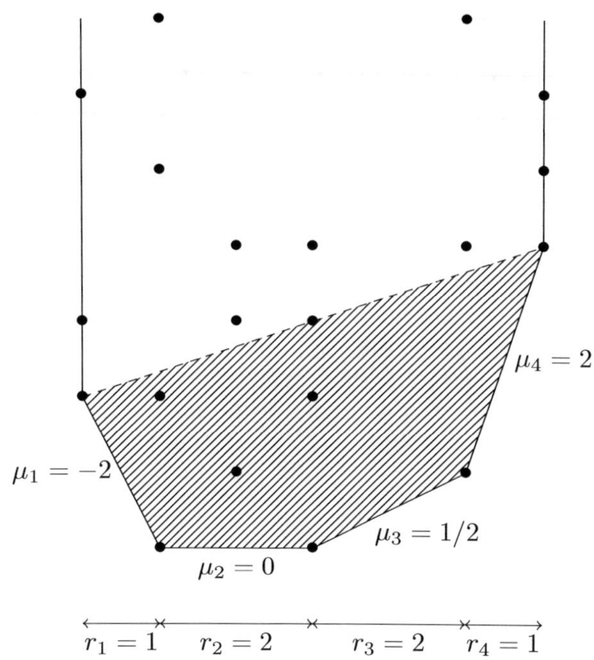

FIGURE 5.2. The Newton polygon for the clouds in figure 5.1

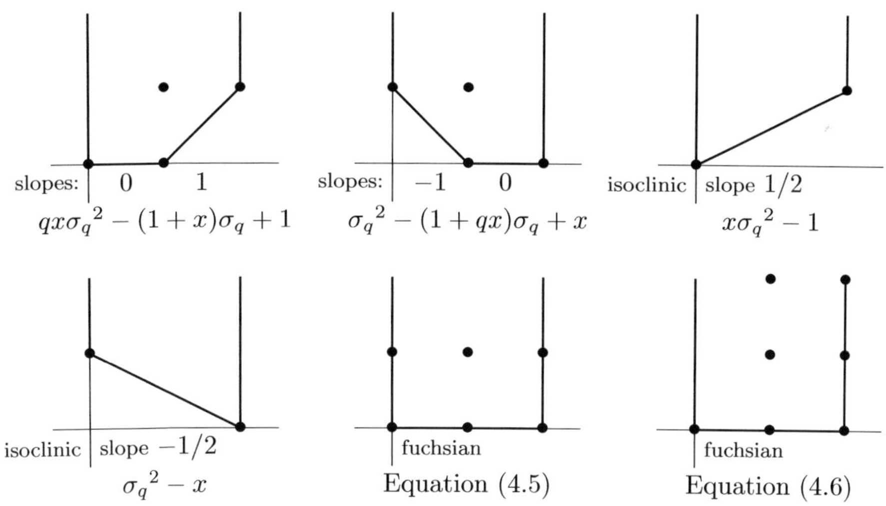

FIGURE 5.3. Some particular Newton polygons (all in standard position)

CAUTION 5.10. We have indexed the slopes from left to right, so that $\mu_1 < \cdots < \mu_k$. For systems and modules (in the forthcoming chapters), the convention will be different.

EXERCISE 5.11. (i) Compute $j_1 - j_0$ from the data r_i, μ_i. (We shall soon see that thay are the truly relevant data.)

(ii) In Chapter 15, we shall have use for the quantity $\sum_{1\leq i<j\leq k} (r_i\mu_j - \mu_i r_j)$ (up to the change of convention on slopes). Show that it is a positive integer and interpret it geometrically. (Look for instance at Figure 5.2.)

5.2.2.2. *Some examples.* We have collected in Figures 5.3 and 5.4 some examples taken from the previous chapter and this one. recall that we are particularly interested in the horizontal lower segment, that is slope 0, because it is the one with a characteristic equation and, if some exponents are in $q^{\mathbf{Z}}$, nontrivial power series solutions.

As regards nonzero slopes, at least if they are integral, there may be a way of "straightening them up" (or down) by a gauge transformation of a very simple kind: replacing equation $L(f) = 0$ by $L'(g) = 0$ where $f = ug$ amounts to take $L' := u^{-1}Lu$. (We shall see that a similar trick can change exponents, but this is not our problem here.) So actually $L' = \sum a'_i\sigma_q^i$, where (with the usual calculus of q-difference operators):

$$a'_i\sigma_q^i = u^{-1}a_i\sigma_q^i u \Longrightarrow a'_i = a_i\frac{\sigma_q^i(u)}{u}.$$

In particular, if $\sigma_q u = x^\mu u$ (for instance if $u = \theta_q^{-\mu}$), then $v_0(a'_i) = v_0(a_i) + \mu i$. This means each point (i, j) of the cloud of L gives rise to the point $(i, j + \mu i)$ of the cloud of L', whence a similar transformation from $NP(L)$ to $NP(L')$; we come back to this in greater generality in 5.2.2.6.

But first, here is a small case study in degree 2 (we leave it to you to draw the corresponding clouds and Newton polygons). Let:

$$L := \alpha_0(x^{j_0} + \cdots) + \alpha_1(x^{j_1} + \cdots)\sigma_q + \alpha_2(x^{j_2} + \cdots)\sigma_q^2,$$

where $\alpha_0, \alpha_1, \alpha_2 \in \mathbf{C}^*$.

First we suppose that $(j_0, j_1, j_2) = (1, -1, 0)$. Then $S(L) = \{-2, 1\}$.

(1) Let $L' := u^{-1}Lu$, where $\sigma_q u = x^2u$. Point (i, j) of the cloud of L gives rise to point $(i, j + 2i)$ of the cloud of L'. The figure shows that slope are increased by 2 and $S(L') = \{0, 3\}$, so the left slope -2 was straightened to 0.

(2) Let $L'' := v^{-1}Lv$, where $\sigma_q v = x^{-1}v$. Point (i, j) of the cloud of L gives rise to point $(i, j - i)$ of the cloud of L''. The figure shows that slope are decreased by 1 and $S(L') = \{-3, 0\}$, so the right slope 1 was straightened to 0.

Now suppose that $(j_0, j_1, j_2) = (-1, 0, 0)$. Then $S(L) = \{1/2\}$. Transforming L to $u^{-1}Lu$, where $\sigma_q u = x^\mu u$ we transform the ordinates (j_0, j_1, j_2) of the lower points of the cloud to $(j_0 + \mu.0, j_1 + \mu.1, j_2 + \mu.2) = (-1, \mu, 2\mu)$. Since the ordinate of the middle point is strictly greater that the mean of the ordinates of the left and right points the lower segment of the Newton polygon lonks these two extreme points and the unique slope is $\mu + 1/2$. We cannot straighten it to 0. (Ramification will solve the puzzle!)

5.2.2.3. We would like information that does not depend on nonessential features; in particular it should be the same for L and for uL, $u \in \mathcal{D}_{K,q}^* = K^*\sigma_q^{\mathbf{Z}}$. We see easily that, if $L' = \alpha\sigma_q^i L$, $\alpha \in \mathbf{C}((x))^*$, $i \in \mathbf{Z}$, then the cloud of L' is that of L translated by $(i, v_0(\alpha))$; therefore:

$$NP(\alpha\sigma_q^i L) = NP(L) + (i, v_0(\alpha)).$$

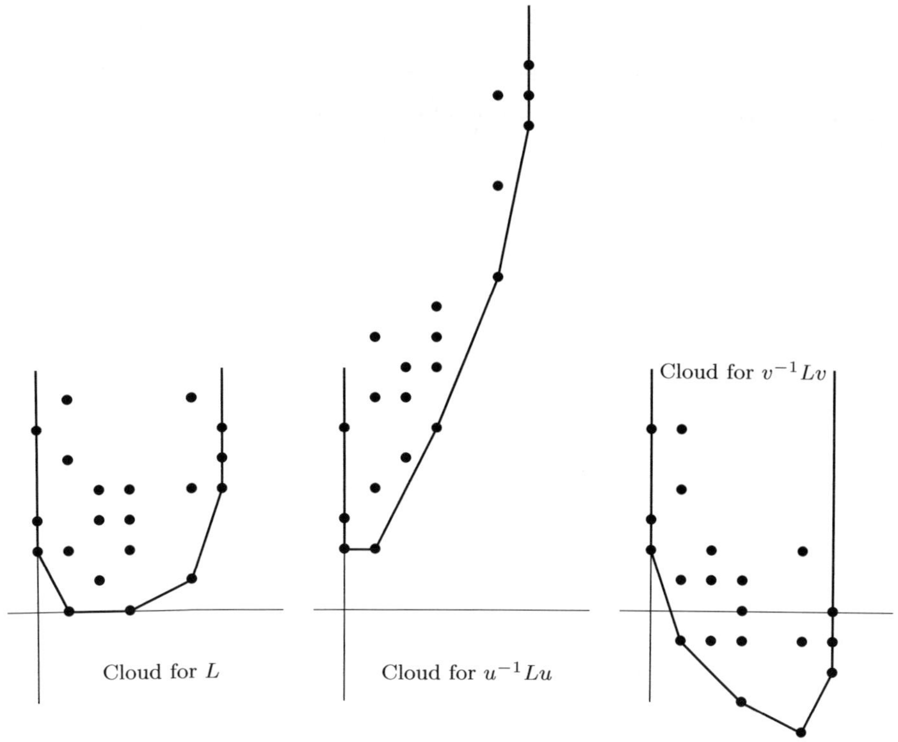

FIGURE 5.4. Clouds, Newton polygons and conjugacy

So we want an information that does not distinguish a Newton polygon from its translates.

DEFINITION 5.12. The *Newton function* of L is the map $r_L : \mathbf{Q} \to \mathbf{N}$ with support $S(L)$ and such that, with the above notations, $r_L(\mu_i) = r_i$, $i = 1, \ldots, k$. We call $r_L(\mu)$ the *multiplicity* of the slope μ.

It is clear that, if $u \in \mathcal{D}^*_{K,q}$, then $r_{uL} = r_L$.

REMARK 5.13. Instead of the function $r_L : \mathbf{Q} \to \mathbf{N}$, we could equivalently use the *multiset* (or *bag*):

$$SS(L) := \{\{\mu_1, \ldots, \mu_1, \ldots, \mu_k, \ldots, \mu_k\}\},$$

in which each μ_i occurs r_i times.

EXERCISE 5.14. (i) Prove that any finitely supported function $r : \mathbf{Q} \to \mathbf{N}$ such that $\forall \mu \in \mathbf{Q}$, $\mu r(\mu) \in \mathbf{Z}$ is the Newton function of some q-difference operator.

(ii) Prove that two Newton polygons give rise to the same Newton function if, and only if, they are translates of each other.

(iii) Compare the sum of two Newton polygons (seen as subsets of the group $\mathbf{R}^2$) to the sum of the corresponding Newton functions.

DEFINITION 5.15. The q-difference operator L is called *pure isoclinic of slope* μ if $S(L) = \{\mu\}$. It is called *fuchsian* if it is pure isoclinic of slope 0.

EXERCISE 5.16. (i) Recognize all pure isoclinic, resp. all fuchsian operators in the examples given.

(ii) Show that a standard unitary q-difference operator $\sigma_q^n + a_1\sigma_q^{n-1} + \cdots + a_n$ is fuchsian if, and only if, $a_1, \ldots, a_n \in \mathbf{C}[[x]]$ and $a_n(0) \neq 0$.

5.2.2.4. *Elementary gauge transformations (or "outer automorphisms").* Recall from Chapter 3, no 3.1.3.6 the notation $L^{[u]}$ and its properties: let $u \in \mathbf{C}((x))^*$ and let $\phi \neq 0$ in some q-difference extension be such that $\sigma_q\phi = u\phi$. Then $\phi^{-1}L\phi$ can be expressed in terms of L and u alone (no explicit occurrence of ϕ) and the resulting expression is written $L^{[u]}$. To the properties listed in *loc. cit.*, we add the following:

LEMMA 5.17. *If $L = \sum a_i\sigma_q^i$ and $L^{[u]} = \sum b_i\sigma_q^i$, then:*

$$\forall i\ ,\ v_0(b_i) = v_0(a_i) + iv_0(u).$$

PROOF. It is proved in 3.1.3.6 that $b_i = a_iu^{[i]}$ where $u^{[i]} = u\sigma_q u \cdots \sigma_q^{i-1}u$ if $i \geq 0$ and $u^{[i]} = (\sigma_q^{-1}u \cdots \sigma_q^{i}u)^{-1}$ if $i < 0$, so that in all cases $v_0(u^{[i]}) = iv_0(u)$. (Because of course σ_q preserves the valuation v_0.) □

Elementary gauge transformations mentioned here are those of the form $f = \phi g$, which ensure that $L(f) = 0 \Leftrightarrow L^{[u]}(g) = 0$; thus it is interesting to compare properties of L to those of $L^{[u]}$ in terms of resolution. The following proposition is an immediate consequence of the lemma.

PROPOSITION 5.18. *The cloud of points attached to $L^{[u]}$ is the image of the cloud of points attached to L by the transformation $(x, y) \mapsto (x, y + v_0(u)x)$.*

COROLLARY 5.19. (i) *The Newton polygon of $L^{[u]}$ is the image of the Newton polygon of L by the transformation $(x, y) \mapsto (x, y + v_0(u)x)$.*

(ii) *Let $\mu_0 := v_0(u)$. Then $r_{L^{[u]}}(\mu) = r_L(\mu - \mu_0)$.*

(iii) *One has $S(L^{[u]}) = S(L) + \mu_0$.*

PROOF. (i) is immediate from the proposition.

(ii) follows from the fact that the linear map $(x, y) \mapsto (x, y+v_0(u)x)$ sends the vector (r_i, d_i) in the boundary of $NP(L)$ to the vector $(r_i, d_i+\mu_0 r_i)$ in the boundary of $NP(L^{[u]})$; so the slope $\mu_i := d_i/r_i$ of L, which has multiplicity $r_i = r_L(\mu_i)$, gives rise to the slope $(d_i + \mu_0 r_i)/r_i = \mu_i + \mu_0$ of $L^{[u]}$, which also has multiplicity r_i, so that $r_{L^{[u]}}(\mu_i + \mu_0) = r_L(\mu_i)$.

(iii) follows immediately.

□

EXERCISE 5.20. (i) Show that $L^{[u]}$ is pure isoclinic if, and only if, L is.

(ii) For which u does $L^{[u]}$ admit the slope 0, i.e., $0 \in S(L^{[u]})$?

5.2.2.5. *Ramification.* Let $r \geq 2$ and write $x' := x^{1/r}$, generating the extension $K' := K_r := \mathbf{C}((x^{1/r}))$ of $K := \mathbf{C}((x))$. Remember we consider K' as a q-difference extension by setting $\sigma_q(x') := q'x'$, where $q' := q_r = q^{1/r}$. For a q-difference operator L' with coefficients in K', the constructions leading to the Newton polygon will be defined with respect to x': the coefficients of L' are expanded in power series of L', the cloud of points involves points (i, j) such that the monomial $x'^j\sigma_q^i$ is present, the valuations are taken with respect to x'. Therefore:

PROPOSITION 5.21. *(i) If L' comes from $L \in \mathcal{D}_{K,q}$ by substituting x'^r for x, the cloud of points attached to L' is the image of the cloud of points attached to L by the transformation $(x, y) \mapsto (x, ry)$.*

(ii) The Newton polygons obey the same law: $NP(L')$ is the image of $NP(L)$ by the same transformation.

(iii) One has $r_{L'}(\mu) = r_L(\mu/r)$ and $S(L') = rS(L)$.

PROOF. (i) flows from the conventions of construction described above, (ii) and (iii) are immediate consequences.

□

So if $S(L) = \{\mu_1, \ldots, \mu_k\}$, then $S(L') = \{r\mu_1, \ldots, r\mu_k\}$: so taking for r a common denominator of the slopes allows one to reduce many problems to the case where all slopes are integral. Then 5.2.2.4 is used to have a null slope.

5.2.2.6. *Characteristic equations, exponents.* The characteristic equation $\overline{L}(s) = 0$ was defined in 5.2.1.3 in the case of a q-difference operator $L = \sum_{i=0}^{n} a_i \sigma_q^i$, $a_0 a_n \neq 0$, with coefficients a_i in $\mathbf{C}[[x]]$ and coprime (Definition 5.4).

DEFINITION 5.22. Let $L = \sum a_i \sigma_q^i \in \mathcal{D}_{K,q}$ arbitrary nontrivial. Let $j_0 := \min v_0(a_i)$ and $i_0 := \min\{i \mid a_i \neq 0\}$, so that the q-difference operator $L' := x^{-j_0} L \sigma_q^{-i_0})$ satisfies the assumptions of Definition 5.4. We define the *characteristic equation attached to slope* 0 *of* L as $\overline{L} := \overline{L'}$ and the *exponents attached to slope* 0 *of* L as the nonzero roots of $\overline{L}$. An exponent s is *nonresonant* if no $q^k s$, $k \geq 1$, is an exponent.

The characteristic equation is considered as defined modulo a factor in $\mathbf{C}^* s^{\mathbf{Z}}$.

In dealing with exponents, we shall frequently use the notation from 0.4:

$$c = q^{\epsilon(c)} \overline{c}, \epsilon(c) \in \mathbf{Z}, \overline{c} \in \mathcal{C}_q.$$

Note that, if 0 is not a slope of L, $0 \notin S(L)$, then $\overline{L}$ consists of just one monomial and can be identified to 1 (modulo a factor in $\mathbf{C}^* s^{\mathbf{Z}}$).

DEFINITION 5.23. Let $L = \sum a_i \sigma_q^i \in \mathcal{D}_{K,q}$ arbitrary nontrivial. Let $\mu \in \mathbf{Z}$ and let $L' := \theta_q^{\mu} L \theta_q^{-\mu} = L^{[x^{-\mu}]}$, so that $S(L') = S(L) - \mu$. We define the *characteristic equation attached to slope* μ *of* L as the characteristic equation attached to slope 0 of L', and the exponents, etc, in accordance.

If μ is not a slope, see the remark following the previous definition. *We have no definition for the case of* $\mu \in \mathbf{Q} \setminus \mathbf{Z}$.

EXERCISE 5.24. (i) Find the exponents for all integral slopes of the examples in the figure.

(ii) Suggest a definition for the cases of nonintegral slopes.

We can summarize the calculations in this section in the following:

PROPOSITION 5.25. *Let $L \in \mathcal{D}_{K,q}$. For the equation $L(f) = 0$ to have a solution $f = \theta_q^{-\mu} e_{q,c}\, g$, g a regular power series, it is necessary that c be an exponent attached to the slope μ of L; and sufficient that it be a nonresonant such exponent.*

EXERCISE 5.26. Draw the Newton polygon of irregular basic hypergeometric equations (those where one parameter is 0).

5.2.3. Factorisation and the Newton polygon.

5.2.3.1. Let again $K := \mathbf{C}((x))$ and $L \in \mathcal{D}_{K,q}$. Everytime we find a nontrivial solution f of $L(f) = 0$, we get a factorisation (see Corollary 3.46 in Section 3.1 or the lemma at the beginning of Section 5.4):

$$L = L_1(\sigma_q - \sigma_q(f)/f).$$

If moreover $f = \theta_q^{-\mu} e_{q,c} g$, $g \in \mathbf{C}((x))^*$, then $\sigma_q(f)/f = cx^\mu \sigma_q(g)/g \in \mathbf{C}((x))$, so the factorisation is inside $\mathcal{D}_{K,q}$. We saw how to find the solution f using the Newton polygon of L, let us see what it tells us about the Newton polygon of L_1. Then we shall attempt to iterate the process.

EXERCISE 5.27. (i) Try to prove before reading what follows that L_1 has just once less than L the slope μ, everything being equal besides; more precisely, $r_{L_1} = r_L$, except that $r_{L_1}(\mu) = r_L(\mu) - 1$.

(ii) Compute the Newton polygon of $\sigma_q - cx^\mu h$, $h \in \mathbf{C}[[x]]^*$, and try to deduce a generalisation of (i) (again before reading the sequel).

So we formulate a general version of what is needed.

THEOREM 5.2 (Additivity of the Newton function). *Let $L_1, L_2 \in \mathcal{D}_{K,q} \setminus \{0\}$. Then:*

$$\forall \mu \in \mathbf{Q}\ ,\ r_{L_1 L_2}(\mu) = r_{L_1}(\mu) + r_{L_2}(\mu).$$

In order to prove the theorem, we shall have to introduce some useful auxiliary notions. Note that nowhere in 5.2.3.2 is it used or assumed that μ be an integer.

5.2.3.2. *Twisted valuations and initial forms.* We fix some $\mu \in \mathbf{R}$ (although only values $\mu \in \mathbf{Q}$ will be useful) and define[2] a *twisted valuation* as follows:

$$\forall L = \sum a_i \sigma_q^i \in \mathcal{D}_{K,q}\ ,\ v_\mu(L) := \min(v_0(a_i) - \mu i).$$

In particular $v_\mu(a\sigma_q^l) = v_0(a) - \mu l$ for $a \in \mathbf{C}(\{x\})^*$ and $l \in \mathbf{Z}$. We say that L is *homogeneous* (with respect to v_μ) if all terms $a_i\sigma_q^i$ present in L have the same (twisted) valuation $v_\mu(a_i\sigma_q^i) = v_0(a_i) - \mu i$; this common is of course the (twisted) valuation $v_\mu(L)$.

Then we define the corresponding *twisted initial form* of $L = \sum a_i \sigma_q^i$:

$$\mathrm{In}_\mu(L) := \sum_{v_\mu(a_i\sigma_q^i) = v_\mu(L)} a_i \sigma_q^i.$$

We leave to the reader the (easy) proof of the following properties:

(1) $\forall L_1, L_2 \in \mathcal{D}_{K,q}\ ,\ v_\mu(L_1 + L_2) \geq \min(v_\mu(L_1) + v_\mu(L_2))$.

(2) $\mathrm{In}_\mu(L)$ is homogeneous with respect to v_μ.

(3) $v_\mu(L - \mathrm{In}_\mu(L)) > v_\mu(L)$.

Note that the last two properties together completely determine $\mathrm{In}_\mu(L)$.

LEMMA 5.28. (i) *If L_1 and L_2 are homogeneous, so is L_1L_2 and its twisted valuation is:*

$$v_\mu(L_1L_2) = v_\mu(L_1) + v_\mu(L_2).$$

(ii) *The initial form is multiplicative:*

$$\forall L_1, L_2 \in \mathcal{D}_{K,q}\ ,\ In_\mu(L_1L_2) = In_\mu(L_1)In_\mu(L_2).$$

[2] Do not confuse the notation v_μ introduced here with the notation $v_a(f)$ for the order of a function f at $a \in \mathbf{C}$ in the General Notations 0.4.

PROOF. (i) There are d_1, d_2 such that all terms present in L_1 have the form $a_i\sigma_q^i$ with $v_0(a_i) - \mu i = d_1$ and terms present in L_2 have the form $b_j\sigma_q^j$ with $v_0(b_j) - \mu j = d_2$. Then all terms in L_1L_2 are sums of terms of the form $a_i\sigma_q^i b_j\sigma_q^j = a_i\sigma_q^i(b_j)\sigma_q^{i+j}$, so they have the form $c_k\sigma_q^k$, where $c_k := \sum\limits_{i+j=k} a_i\sigma_q^i(b_j)$. Now, if $c_k \neq 0$, since:

$$v_0(a_i\sigma_q^i(b_j)) = v_0(a_i) + v_0(b_j) = d_1 + \mu i + d_2 + \mu j = d_1 + d_2 + \mu k,$$

one has $v_0(c_k) = d_1 + d_2 + \mu k$ and all terms of L_1L_2 have twisted valuation $d_1 + d_2$. So L_1L_2 is homogeneous.

(ii) Set $I_1 := \mathrm{In}_\mu(L_1)$, $I_2 := \mathrm{In}_\mu(L_2)$, $L_1' := L_1 - I_1$ and $L_2' := L_2 - I_2$; also let $d_1 := v_\mu(L_1)$ and $d_2 := v_\mu(L_2)$. Then:

$$L_1L_2 = I_1I_2 + L', \text{ where } L' := I_1L_2' + L_1'I_2 + L_1'L_2'.$$

By (i), I_1I_2 is homogeneous of valuation d_1+d_2. By the properties before the lemma, $v_\mu(L') > d_1 + d_2$. By the remark following those properties, $I_1I_2 = \mathrm{In}_\mu(L_1L_2)$.

□

LEMMA 5.29. *The multiplicity of slope μ of L is the absolute degree of the twisted initial form:*

$$\forall L \in \mathcal{D}_{K,q}\ ,\ \forall \mu \in \mathbf{Q}\ ,\ r_L(\mu) = \deg \mathit{In}_\mu(L).$$

PROOF. Let $L = \sum a_i\sigma_q^i$ and $d := v_\mu(L)$. Write $I := \mathrm{In}_\mu(L)$ the initial form and $L' := L - I$; we write it $I = \sum\limits_{i=i_0}^{i_1} a_i\sigma_q^i$, $i_0 \leq i_1$, $a_{i_0}a_{i_1} \neq 0$.

In the $\mathbf{R}^2$ plane with coordinates (i, j), we draw the cloud of points of L and the straight line $\Delta_d : j - \mu i = d$. By definition, for all $a_i\sigma_q^i$ present in I, the point $(i, j) = (i, v_0(a_i))$ is on Δ_d and it comes from the lowest degree term $a_{i,j} \neq 0$ of the power series $a_i = a_{i,j}x^j + \cdots$; and all other terms of a_i as well as all terms coming from L' give rise to points strictly above Δ_d.

Therefore Δ_d *touches* the convex set $NP(L)$, i.e., either their intersection is either a single point on the boundary, or it is one of the segments of said boundary. In the second case, the slope of that segment is μ and the abcissa of the corresponding left-to-right vector is $i_1 - i_0$, that is the absolute degree of I: this is the relation wanted for.

We leave it to the reader to check that the first case corresponds to $r_L(\mu)$ and $\deg \mathrm{In}_\mu(L)$ both 0. □

PROOF OF THEOREM 5.2. It follows immediately from the two lemmas above.

□

5.2.3.3. *Application to the characteristic equations.* We now assume that $\mu \in \mathbf{Z}$. The coefficients of $L' := L^{[x^{-\mu}]}$, are related to those of L by equality $a_i' = q^{-i(i-1)/2}x^{-\mu i}a_i$.

EXERCISE 5.30. Call $\overline{L}^{(\mu)}$ the operator L' defined in 5.2.2.6.

(i) How does it depend on the particular Theta function used?

(ii) What is $\overline{L_1L_2}^{(\mu)}$? Particular case: $L = u(\sigma_q - v)$.

(iii) How is $\overline{L}^{(\mu)}$ related to $\mathrm{In}_\mu(L)$?

5.3. Adams's lemma and analytic solutions

5.3.1. Discussion of examples.

5.3.1.1. We now take as coefficient field $K := \mathbf{C}(\{x\})$ (this will be further restricted to $\mathscr{M}(\mathbf{C})$ in 5.3.2.2) and discuss the existence and calculation of *convergent* solutions. Before stating and proving in 5.3.2 the key result, Adams's lemma, we discuss some more examples.

First, recall from Chapter 4 that for equations of order 2 having only the slope 0 (this was called the cases of configurations I, II and III), we found generically the existence of two independent solutions of the form: a q-character times a regular convergent power series. The degenerate cases came when there were resonant exponents, and then we had to use moreover a q-logarithm; but still all power series appearing in the solutions were convergent. The whole theory of q-hypergeometric equations belongs to those cases.

The case of a pure isoclinic equation of integral slope can be reduced to the previous ones by an elementary gauge transform, so we expect the solutions to involve Theta powers, q-characters, maybe a q-logarithm; and only convergent power series. For a nonintegral slope, one has to use some ramification but again only convergent power series may appear.

So herebelow we discuss two examples in order two having two different slopes. Note that the slopes are by force integral, because the boundary vectors are necessarily $(1, d_1)$ and $(1, d_2)$ with $d_1, d_2 \in \mathbf{Z}$, $d_1 < d_2$.

5.3.1.2. We introduce so-called Tshakaloff series[3], one of the few q-analogues of Euler's series $\sum_{n \geq 0} n! x^n$, itself a standard touchstone for all theories on divergent series:

$$Ч(x) := \sum_{n \geq 0} q^{-n(n+1)/2} x^n.$$

(The cyrillic letter Ч is pronounced "tch".) Note that, when $n \to \infty$, $\log n! \sim n \log n$, while $\log \left| q^{-n(n+1)/2} \right| \sim a n^2$ for some $a > 0$, so Tshakaloff series diverges much more quickly that Euler's series. It is a formal solution of q-difference equation:

$$(5.3) \quad \sigma_q f = 1 + x f \Longrightarrow L(f) = 0, \text{ where } L := (\sigma_q - 1)(\sigma_q - x) = \sigma_q^2 - (1 + qx)\sigma_q + x.$$

The slopes of L are -1 and 0 and the characteristic equation for slope 0 is $s^2 - s = 0$, whence a unique exponent 1 and a unique regular formal power series solution with constant term 1: of course, this is Ч.

To take in account slope -1, we compute $L^{[x^{-1}]} = xL'$, where $L' := qx\sigma_q^2 - (1 + qx)\sigma_q + 1$ has slopes $0, 1$ and characteristic equation (for slope 0) $1 - s = 0$, so again a unique regular formal power series solution with constant term 1: and actually, this is the constant 1. The corresponding solution for (5.3.1.2) (second equation) is $1/\theta_q$, which is actually solution of the homogeneous associated equation $\sigma_q f = 1 + x f$ as expected.

EXERCISE 5.31. Write the recursive relation for the coefficients in both cases above and confirm that solving these recursions we respectively find Ч and 1.

[3]There are various q-analogues of Euler series beyond Tshakaloff. Moreover, depending on whether $|q| < 1$ or $|q| > 1$, summation and degeneracy questions lead to some subtle phenomena; see **[66]**.

We now look at q-difference equation $L(f) = 0$, $L := x\sigma_q^2 - \sigma_q + x$. Its slopes are -1 and 1. So we shall successively consider $L' := L^{[x^{-1}]}$ and $L'' := L^{[x]}$.

The q-difference operator $L' = x(1 - \sigma_q + qx^2\sigma_q^2)$ has slopes 0 and 2. The characteristic equation for slope 0 is $1 - s = 0$. The recursion relation for the coefficients of the unique regular formal power series solution with constant term 1 is $(1 - q^n)f_n + q^{2n-1}f_{n-2} = 0$ (so we already predict quick convergence). With initialisation $f_0 := 1$, we find that all $f_{2m+1} = 0$ and $f_{2m} = (-1)^m \dfrac{q^{m(2m+1)}}{(q^2;q^2)_m}$. The corresponding convergent solution is therefore:

$$\frac{1}{\theta_q(x)} \sum_{m \geq 0} (-1)^m \frac{q^{m(2m+1)}}{(q^2;q^2)_m} x^{2m}.$$

The q-difference operator $L'' = q^{-1}x^{-1}(qx^2 - q\sigma_q + \sigma_q^2)$ has slopes -2 and 0 and characteristic equation (for slope 0) $s^2 - qs = 0$, so we look for a power series solution $\sum f_n x^n$ starting at $f_1 = 1$. The recursive relation is $(q^{2n} - q^{n+1})f_n + qf_{n-2} = 0$, so we predict strong divergence.

EXERCISE 5.32. (i) Show that the power series is $f = \sum\limits_{m \geq 0} \dfrac{q^{-m(m+2)}}{(q^2;q^2)_m} x^{2m+1}$ and give the corresponding solution for the initial equation $L(f) = 0$.

(ii) Which part of the Newton polygon seems to always yield convergent solutions?

5.3.2. Adams's lemma.

5.3.2.1. *Statement and proof.* We now state and prove Adams's lemma for q-difference operators with analytic coefficients: here $K := \mathbf{C}(\{x\})$. We assume integral slopes; application to q-difference operators and q-difference equations with arbitrary slopes will be formulated when needed.

THEOREM 5.3 (Adams's lemma). *Let $L \in \mathcal{D}_{K,q}$ have slopes $S(L) = \{\mu_1, \ldots, \mu_k\}$, $\mu_1 < \cdots < \mu_k$. Assume the smallest (or leftmost) slope is integral $\mu_1 \in \mathbf{Z}$. Let $\theta_q^{-\mu_1} e_{q,c} f$ a solution attached to that slope by the method of* 5.2.1, *i.e., c is a non-resonant exponent for slope μ_1 and $f \in \mathbf{C}((x))$ is a regular power series. Then f converges: $f \in \mathbf{C}\{x\}$.*

PROOF. Up to an elementary gauge transformation we may assume from start that $\mu_1 = 0$ and $c = 1$. We can also assume that L is standard unitary: $L = a_0 + \cdots + a_n\sigma_q^n$, $a_0, \ldots, a_n \in \mathbf{C}(\{x\})$, $a_0 a_n \neq 0$; also we can assume that $a_0, \ldots, a_n$ are coprime, so that actually $a_0, \ldots, a_n \in \mathbf{C}\{x\}$ and not all $a_i(0)$ are 0. We write as usual $a_i = \sum\limits_{n \geq 0} a_{i,j}x^j$. Recall from 5.2.1.3 the notation $P_j(s) := \sum\limits_{i=0}^{n} a_{i,j}s^i \in \mathbf{C}[s]$ and the recursive relation for the coefficients of f:

$$\forall l \geq 1\,,\; f_l = -\frac{P_1(q^{l-1})f_{l-1} + \cdots + P_l(q^0)f_0}{P_0(q^l)}.$$

Now we interpret the assumption that $\mu_1 = 0$ and $\mu_2, \ldots, \mu_k > 0$. It exactly means that $a_{0,0} \neq 0$ and that there is a $i_0 \in \{1, \ldots, n\}$ such that $a_{i_0,0} \neq 0$. Indeed, if we moreover choose the biggest such i_0, the boundary vector yielding slope 0 is the one going from $(0,0)$ to $(i_0, 0)$. As a consequence, $P_0(s) = a_{0,0} + \cdots + a_{i_0,0}s^{i_0}$, the important facts being that:

- $P_0(1) = 0$ since $c = 1$ is an exponent attached to slope 0 (this is the fact that allows the initialisation $f_0 := 1$, see 5.2.1.3);
- $P_0(q^l) \neq 0$ for $l \geq 1$ since $c = 1$ is a nonresonant exponent (this is the fact that makes the above recursive formula meaningful);
- $P_0(0) = a_{0,0} \neq 0$ (this expresses the fact that we are considering the leftmost slope and it is also the crucial fact in proving the convergence of f).

Since $P_0(q^l) \neq 0$ for $l \geq 1$ and $\lim\limits_{l \to +\infty} P_0(q^l) = P_0(0) \neq 0$, there is a $C_0 > 0$ such that:

$$\forall l \geq 1\,, \quad \left| \frac{1}{P_0(q^l)} \right| < C_0.$$

By assumption all a_i converge, so, being in finite number, they have a common nontrivial disk of convergence, so there exists $C_1, R > 0$ such that:

$$\forall i = 0; \ldots, n\,,\ \forall j \geq 0\,,\ |a_{i,j}| < C_1 R^{-j}.$$

By construction, all polynomials P_j have degree $\leq n$; for any $s \in \mathbf{C}$ such that $|s| < 1$, one has:

$$|P_j(s)| \leq \sum_{i=1}^{n} |a_{i,j}|\, |s|^i \leq n C_1 R^{-j}.$$

This holds in particular for $s := q^l$, $l \geq 1$. From those inequalities and the recursive relation, one draws, for $l \geq 1$:

$$|f_l| \leq C_2 \sum_{j=1}^{l} R^{-j}\, |f_{l-j}|\,, \text{ where } C_2 := n C_0 C_1.$$

We can now achieve our goal with elementary explicit arguments. Let $g_j := R^j\, |f_j|$. Then $g_0 = 1$ and $g_l \leq C_2(g_0 + \cdots + g_{l-1})$ for $l \geq 1$. Introducing the sequence (h_j) defined by $h_0 = 1$ and $h_l := C_2(h_0 + \cdots + h_{l-1})$ for $l \geq 1$, we see by induction that $0 \leq g_l \leq h_l$ for all l; and, for $l \geq 2$, that:

$$h_l - h_{l-1} = C_2(h_0 + \cdots + h_{l-1}) - C_2(h_0 + \cdots + h_{l-2}) = C_2 h_{l-1} \Longrightarrow C_l = (1 + C_2) h_{l-1}.$$

Also $h_1 = C_2 h_0$, so $h_l = C_2(1 + C_2)^{l-1}$, so $g_l \leq C_2(1 + C_2)^{l-1}$, so:

$$\forall l \geq 1\,,\ |f_l| \leq C_2(1 + C_2)^{l-1} R^{-l},$$

and we conclude that f converges: $f \in \mathbf{C}\{x\}$. □

For resonant exponents attached to the leftmost slope, some of the solutions obtained by the method of 5.2.1 have the form $\theta_q^{-\mu_1} e_{q,c} f$ times a q-logarithmic factor. We shall see later that their power series factor f also converges.

5.3.2.2. *Meromorphic continuation.* In 5.3.2.1, we took coefficients $a_i \in K := \mathbf{C}(\{x\}) = \mathscr{M}(\mathbf{C}, 0)$. Now we shall assume that $K := \mathscr{M}(\mathbf{C})$. Of course, this will be applied in particular to the most important case of *rational* coefficients, see 12.1.

So let $L := a_0 + \cdots + a_n \sigma_q^n$, all the $a_i \in \mathscr{M}(\mathbf{C})$, $a_0 a_n \neq 0$. As we did for power series coefficients, we can chase denominators and factor out common divisors, so we assume that actually $a_0, \ldots, a_n \in \mathscr{O}(\mathbf{C})$. (In the particular case of rational coefficients, we would take the a_i in $\mathbf{C}[x]$.) We intend to study the analytic continuation of a convergent solution $\theta_q^{\mu} e_{q,c} f$, where $f \in \mathbf{C}\{x\}$ is regular. To begin with, we shall assume that $\mu = 0$ and $c = 1$. We introduce what we consider to be the "singular locus" of the q-difference equation $L(f) = 0$:

$$\Sigma := a_0^{-1}(0) \cap \mathbf{C}^* = \{x \in \mathbf{C}^* \mid a_0(x) = 0\}.$$

Snce $a_0 \neq 0$, the singular locus Σ is discrete in $\mathbf{C}^*$; and since 0 is not an essential singularity, Σ does not accumulate at 0. Thus Σ is discrete in $\mathbf{C}$. Let $r_0 := \min\limits_{x \in \Sigma} |x|$.

The convergent power series solution f is holomorphic over some open disk $\mathring{\mathrm{D}}(0, r)$ and satisfies there the functional equation:

$$f(x) = \frac{-1}{a_0(x)} \left(a_1(x) f(qx) + \cdots + a_n(x) f(q^n x)\right). \tag{5.4}$$

LEMMA 5.33. (i) *Let* $r' := \dfrac{r}{|q|}$, *so that* $r' > r$. *The function* f *admits a unique analytic continuation* g *to* $\mathring{\mathrm{D}}(0, r')$ *satisfying the functional equation* (5.4).

(ii) *If* $r' < r_0$, *the function* g *is holomorphic.*

PROOF. (i) Uniqueness will follow from the principle of analytic continuation. As for existence, we first define provisonally a function g on $\mathring{\mathrm{D}}(0, r')$ by setting:

$$g(x) := \frac{-1}{a_0(x)} \left(a_1(x) f(qx) + \cdots + a_n(x) f(q^n x)\right). \tag{5.5}$$

This makes sense since for $x \in \mathring{\mathrm{D}}(0, r')$ we have $qx, \ldots, q^n x \in \mathring{\mathrm{D}}(0, r)$ and f is well defined there. Moreover, if $x \in \mathring{\mathrm{D}}(0, r)$, the functional equation (5.4) says that $g(x) = f(x)$, so g is indeed an extension of f. It follows that we may replace f by g in the right-hand side of (5.5), i.e., g indeed satisfies (5.4).

(ii) By construction, g is holomorphic on $\mathring{\mathrm{D}}(0, r)$ (since it is equal to f there). If $x \in \mathring{\mathrm{D}}(0, r') \setminus \mathring{\mathrm{D}}(0, r)$, then g is given by the functional equation as a product; the second factor is holomorphic because it involves arguments within $\mathring{\mathrm{D}}(0, r)$; the first factor is holomorphic because the denominator a_0 is nonzero inside $\mathring{\mathrm{D}}(0, r_0)$ and *a fortiori* inside $\mathring{\mathrm{D}}(0, r')$.

□

REMARK 5.34. The fact that g satisfies (5.4) is actually an immediate consequence of the principle of analytic continuation. However, the scheme of proof above is interesting in that it allows one to obtain extension of solutions of some functional equations even when these are not analytic and the principle does not apply.

THEOREM 5.4. (i) *The solution* f *admits a unique meromorphic continuation to* $\mathbf{C}$.

(ii) *It is holomorphic over* $\mathring{\mathrm{D}}(0, r_0)$ *and all its poles lie over* $q^{-\mathbf{N}}\Sigma$.

PROOF. (i) Starting from $\mathring{\mathrm{D}}(0, r)$ defined above, extend f iteratively to $\mathring{\mathrm{D}}(0, r/|q|^k)$ for all k. Since the union of these disks is $\mathbf{C}$, this proves the fisrst statement.

(ii) Up to shrinking r if necessary, one can assume that $r = |q|^N r_0$ for some $N \geq 1$. Applying N times the lemma (in particular its second statement) one gets a continuation f holomorphic over $\overset{\circ}{\mathrm{D}}(0, r_0)$.

Now let f_0 be this solution holomorphic over $D_0 := \overset{\circ}{\mathrm{D}}(0, r_0)$ and let f_k, $k \geq 1$, the solution on $D_k := \overset{\circ}{\mathrm{D}}(0, |q|^{-k} r_0)$ obtained by k successive applications of the lemma. Let S_k the set of poles of f_k and let $\Sigma_k := \Sigma \cap D_k$ the set of zeroes of a_0 over $D_k \setminus \{0\}$. from equality:

$$f_{k+1}(x) := \frac{-1}{a_0(x)} \left(a_1(x) f_k(qx) + \cdots + a_n(x) f_k(q^n x) \right),$$

we deduce that x is a pole of f_{k+1} only if it belongs to Σ_{k+1} or if one of $qx, \ldots, q^n x$ belong to S_k:

$$S_{k+1} \subset \Sigma_{k+1} \cup q^{-1} S_k \cup \cdots \cup q^{-n} S_k \subset \Sigma \cup q^{-\mathbf{N}^*} S_k.$$

Now it is easy to check by induction that all S_k are included in $q^{-\mathbf{N}}\Sigma$.

□

EXERCISE 5.35. (i) Give a more precise upper bound for S_k computed from all the Σ_j, $j \leq k$. You may want to use instead the sets $S'_k := S_k \setminus S_{k-1}$ and $\Sigma'_k := \Sigma_k \setminus \Sigma_{k-1}$.

(ii) Compute all those singular sets for the equation $\sigma_q f = \dfrac{1 - ax}{1 - bx} f$.

5.4. Factorisation

In this section, $K := \mathbf{C}((x))$ to begin with (in 5.4.1 and 5.4.2) then $K := \mathbf{C}(\{x\})$ (in 5.4.3).

Solutions of a q-difference equation and right factors of the corresponding q-difference operator L are related as follows. Let f in some q-difference extension of K, $f \neq 0$ and suppose that $u := \sigma_q(f)/f \in K$. Then the q-difference operator $\sigma_q - u \in \mathcal{D}_{K,q}$ has unique slope $\mu := -v_0(u)$; also, it satisfies $(\sigma_q - u)(f) = 0$. Now let $L' \in \mathcal{D}_{K,q}$ an arbitrary q-difference operator and let $L := L'(\sigma_q - u) \in \mathcal{D}_{K,q}$. Then $L(f) = 0$ and $\mu \in S(L)$ (this is because of Theorem 5.2). This has a converse:

LEMMA 5.36. *Let $L := a_0 + \cdots + a_n \sigma_q^n \in \mathcal{D}_{K,q}$, $a_0 a_n \neq 0$. Let $f \neq 0$ a solution of $L(f) = 0$ such that $u := \sigma_q(f)/f \in K$. Then there is a unique $L' \in \mathcal{D}_{K,q}$ such that $L = L'(\sigma_q - u)$. One has $L' = a'_0 + \cdots + a'_{n-1}\sigma_q^{n-1}$, $a'_0 a'_{n-1} \neq 0$. As a consequence, $\mu := -v_0(u) \in S(L)$.*

PROOF. The argument is similar to that of Corollary 3.46 in Section 3.1. Euclidean division $L = L'(\sigma_q - u) + b$ has a remainder b either null or with absolute degree 0, and such that $b(f) = 0$, so, since $f \neq 0$, by necessity $b = 0$. All the other statements are immediate consequences (again using theorem 5.2 for the last one). □

EXERCISE 5.37. Compare $\overline{L}^{(\mu)}$ and $\overline{L'}^{(\mu)}$ (notation from the exercise in 5.2.3.3).

The discussion strongly suggests that, in order to extract first degree factor from a q-difference operator L, we assume first that $S(L) \subset \mathbf{Z}$.

5.4.1. Formal factorisation: integral slopes.

5.4.1.1. Here we take $K := \mathbf{C}((x))$ and otherwise keep the above notations.

PROPOSITION 5.38. *Let $L \in \mathcal{D}_{K,q}$, $L \neq 0$, be such that $S(L) \subset \mathbf{Z}$. Then there is a factorisation $L = \alpha(\sigma_q - u_1)\ldots(\sigma_q - u_n)$, $\alpha, u_1, \ldots, u_n \in K^*$ and the bag of slopes (see remark 5.13, page 165) is:*

$$SS(L) = \{\{-v_0(u_1), \ldots, -v_0(u_n)\}\}.$$

PROOF. A solution f as used in the previous lemma can be built in the form $\theta_q^\mu e_{q,c} g$, g a regular power series, using any slope $\mu \in S(L)$ and any nonresonant exponent c attached to μ. Then $u = cx^{-\mu}h$, h a regular power series and we can apply the lemma. Iterating yields the conclusion. □

The above is obviously impossible if $S(L) \not\subset \mathbf{Z}$. In that case, one can obtain a factorisation through ramification to a degree r such that $S(L) \subset (1/r)\mathbf{Z}$. However, something can be done in the general case, in accordance with the model of the following corollary. We shall write $S(L) = \{\mu_1, \ldots, \mu_k\}$, with multiplicities $r_i := r_L(\mu_i)$, $i = 1, \ldots, k$. Here and in 5.4.2, *we do not require that the indexing respects the order of the slopes.*

COROLLARY 5.39. *The q-difference operator L admits a factorisation $L = L_1 \cdots L_k$, $L_1, \ldots, L_k \in \mathcal{D}_{K,q}$, such that $S(L_i) = \{\mu_i\}$ for $i = 1, \ldots, k$. For each such factorisation, $\deg L_i = r_i$ for $i = 1, \ldots, k$.*

PROOF. The first statement follows from the proposition by grouping factors. The second statement follows from theorem 5.2. □

5.4.1.2. Now we prepare for dealing, in 5.4.2, with the case of arbitrary (possibly nonintegral) slopes.

LEMMA 5.40. *Let $L \in \mathcal{D}_{K,q}$ admit two factorisations $L = L'L'' = M'M''$, where $S(L'') = S(M'') = \{\mu\}$ and $S(L'), S(M') \not\ni \mu$. Then there exists $\alpha \in \mathcal{D}_{K,q}^* = K^*\sigma_q^{\mathbf{Z}}$ such that $M' = L'\alpha$ and $M'' = \alpha L''$.*

PROOF. Recall that we still work under the assumption that $S(L) \subset \mathbf{Z}$. Therefore, there is, in some q-difference extension of K, a solution f of $L(f) = 0$ attached to slope μ, i.e., such that $u := \sigma_q(f)/f$ has valuation $v_0(u) = -\mu$. We make two euclidean divisions:

$$L'' = L'''(\sigma_q - u) + b \text{ and } M'' = M'''(\sigma_q - u) + c.$$

From $L(f) = (\sigma_q - u)(f) = 0$, we draw that $(L'b)(f) = (M'c)(f) = 0$. If for instance $b \neq 0$, then $\deg b = 0$, whence $S(L') = S(L'b)$; but this does not contain μ so it cannot have solution f by Lemma 5.36. So in fact $b = 0$ and similarly $c = 0$, whence $L'' = L'''(\sigma_q - u)$ and $M'' = M'''(\sigma_q - u)$, whence $L'L''' = M'M'''$. We then iterate the process until the right factor (with unique slope μ) is entirely peeled off: the corresponding $L^{(i)}, M^{(i)}$ are then invertible and the conclusion is easy. □

EXERCISE 5.41. Write down rigorously that "easy" conclusion.

COROLLARY 5.42. *Let $L = L_1 \cdots L_k = L'_1 \cdots L'_k$, $L_1, \ldots, L_k, L'_1, \ldots, L'_k \in \mathcal{D}_{K,q}$ two factorisations of the kind considered in* 5.4.1.1. *Then there exist $\alpha_1, \ldots, \alpha_{k-1} \in \mathcal{D}_{K,q}^* = K^*\sigma_q^{\mathbf{Z}}$, such that, setting $\alpha_0, \alpha_k := 1$:*

$$\forall i = 1, \ldots, k\ ,\ L'_i = \alpha_{i-1} L_i \alpha_i^{-1}.$$

PROOF. Easy by iteration of the lemma. □

EXERCISE 5.43. Write down rigorously that "easy" proof.

5.4.2. Formal factorisation: arbitrary slopes. Our goal is to extend the above results to the case of arbitrary slopes. However, we shall first establish some auxiliary results for q-difference operators with integral slopes. (And, as previously announced, $K := \mathbf{C}((x))$.)

5.4.2.1. Let $r \geq 2$ and $\rho \in \mu_r^*$ a primitive rth root of unity. Define $\tau_\rho : K \to K$, $a(x) \mapsto a(\rho x)$, thus a field automorphism commuting with σ_q.

LEMMA 5.44. *Let $f \in \mathbf{C}[[x]]$, $f(0) = 1$. Suppose $f\tau_\rho f \cdots \tau_\rho^{r-1} f = 1$. Then there exists $g \in \mathbf{C}[[x]]$, $g(0) = 1$, such that $f = \tau_\rho(g)/g$. Moreover, if f converges, g can be taken convergent.*

PROOF. This is a form of Hilbert's theorem 90 **[145]** but we give a direct argument. We use the maps $\exp : x\mathbf{C}[[x]] \to 1 + x\mathbf{C}[[x]]$ and $\log : 1 + x\mathbf{C}[[x]] \to x\mathbf{C}[[x]]$, which are bijective and mutually reciprocal. In the convergent case, we use the analogous maps $x\mathbf{C}\{x\} \leftrightarrow 1 + x\mathbf{C}\{x\}$. Those maps commute with τ_ρ. So we may write $f = \exp\phi$, $\phi(0) = 0$ (and ϕ converges if f does); and the assumption on f translates into $\phi + \tau_\rho\phi + \cdots + \tau_\rho^{r-1}\phi = 0$; and the wanted conclusion can be written $\phi = \tau_\rho\gamma - \gamma$ (and γ converges if ϕ does), thus yielding $g := \exp\gamma$.

So let $\phi = \sum \alpha_n x^n$. We look for $\gamma = \sum \beta_n x^n$ such that $\alpha_n = (\rho^n - 1)\beta_n$ for all n. From the well known formulas:

$$\sum_{s=0}^{r-1} \tau_\rho^s(x^n) = \begin{cases} 0 \text{ if } r \text{ does not divide } n, \\ rx^n \text{ if } r \text{ divides } n, \end{cases}$$

we see (from the assumption on ϕ) that $\alpha_n = 0$ for all $n \in r\mathbf{Z}$. Then we can take $\beta_n := \dfrac{\alpha_n}{\rho^n - 1}$ when $n \notin r\mathbf{Z}$, $\beta_n := 0$ when $n \in r\mathbf{Z}$. (Convergence of γ is obvious if ϕ converges.) □

We make τ_ρ operate on $\mathcal{D}_{K,q}$ by acting on coefficients:

$$\tau_\rho\big(\sum a_i\sigma_q^i\big) := \sum \tau_\rho(a_i)\sigma_q^i.$$

This is an automorphism of the $\mathbf{C}$-algebra $\mathcal{D}_{K,q}$.

EXERCISE 5.45. Prove it. This uses the fact that τ_ρ commutes with σ_q.

5.4.2.2.

LEMMA 5.46. *Let $L \in \mathcal{D}_{K,q}$, $L \neq 0$, such that $S(L) \subset \mathbf{Z}$. Suppose $\tau_\rho(L) = \alpha L$ for some $\alpha \in \mathcal{D}_{K,q}^*$. Then there is some $m \in \{0, \ldots, r-1\}$ and some regular $g \in \mathbf{C}[[x]]$, $g(0) = 1$, such that $L = x^m g M$, where $\tau_\rho(M) = M$. If L has convergent coefficients, g can be taken in $\mathbf{C}\{x\}$.*

PROOF. Write $L = \sum\limits_{i=i_0}^{i_1} a_i\sigma_q^i$, $i_0 \leq i_1$, $a_{i_0}a_{i_1} \neq 0$. Also write $\alpha = u\sigma_q^l$, $u \in K^*$, $l \in \mathbf{Z}$. From $\tau_\rho(L) = \alpha L$, i.e.:

$$\sum_{i=i_0}^{i_1} \tau_\rho(a_i)\sigma_q^i = \sum_{i=i_0}^{i_1} (ua_i)\sigma_q^{i+l},$$

we deduce first that $l = 0$, whence $\tau_\rho(a_i) = ua_i$ for all i, which in turn implies $v_0(u) = 0$ (since $v_0 \circ \tau_\rho = v_0$ and since some a_i are nonzero). So $u = cf$, $c \in \mathbf{C}^*$ and $f \in \mathbf{C}[[x]]$ such that $f(0) = 1$; and $\tau_\rho(a_i) = cfa_i$ for all i.

Iterating this relation (seen as a functional relation for a_i) we obtain $\tau_\rho^k(a_i) = c^k f\tau_\rho(f)\cdots\tau_\rho^{k-1}(f)a_i$ for all i,k, whence, taking $k := r$ (and since τ_ρ^r is the identity) $a_i = c^r f\tau_\rho(f)\cdots\tau_\rho^{r-1}(f)a_i$ for all i, whence $c^r f\tau_\rho(f)\cdots\tau_\rho^{r-1}(f) = 1$ (since some a_i are nonzero).

Evaluating at 0 yields $c^r = 1$, whence $c = \rho^m$ for some $m \in \{0,\ldots,r-1\}$; and $f\tau_\rho(f)\cdots\tau_\rho^{r-1}(f) = 1$; and $f = \tau_\rho(g)/g$ for some $g \in \mathbf{C}\{x\}$, $g(0) = 1$, by Lemma 5.44. Then $cf = \dfrac{\tau_\rho(x^m g)}{x^m g}$ and, writing $a_i = x^m g b_i$, we see that $b_k = \tau_\rho(b_i)$ for all i, so we can take $M := \sum\limits_{i=i_0}^{i_1} b_i\sigma_q^i$. Statements on convergence also come from lemma 5.44. □

COROLLARY 5.47. *Let $L = L_1\cdots L_k$ (same conventions as before) such that $\tau_\rho(L) = L$. Then there is a decomposition $L = M_1\cdots M_k$ such that $\tau_\rho(M_i) = M_i$ and $L_i = \alpha_{i-1}M_i\alpha_i^{-1}$, $i = 1,\ldots,k$, $\alpha_0 = \alpha_k = 1$. One can require that $\alpha_i = x^{m_i}g_i$, $g_i(0) = 1$ for $i = 1,\ldots,k$. If all the L_i have convergent coefficients, one can also require that the M_i have convergent coefficients and that the $g_i \in \mathbf{C}\{x\}$.*

5.4.2.3. Now comes the reward! recall that $K = \mathbf{C}((x))$.

THEOREM 5.5. *Let $K := \mathbf{C}((x))$ and let $L \in \mathcal{D}_{K,q}$, $L\ /0$. Let $S(L) = \{\mu_1,\ldots,\mu_k\}$ with multiplicity $r_1,\ldots,r_k$. We do not assume that $S(L) \subset \mathbf{Z}$ and we do not assume either that the slopes μ_i are ordered. Then L admits a factorisation $L = L_1\cdots L_k$ with each $L_i \in \mathcal{D}_{K,q}$, $S(L_i) = \{\mu_i\}$, for $i = 1,\ldots,k$. This factorisation is unique up to the possible transformations described in* 5.4.2.2

As before, it follows that $\deg L_i = r_i$ for $i = 1,\ldots,k$. The important point in the theorem is that, although the slopes may be nonintegral and therefore the degree 1 factors need ramification, the factors L_i live in $\mathcal{D}_{K,q}$, they do not need ramification.

PROOF. Existence: We operate a ramification of degree r any common denominator of the slopes. Call L' the q-difference operator L seen as an operator over the q-difference extension $K' := K_r$. Since it has integral slopes (by 5.2.2.5), L' admits a factorisation $L' = L'_1\cdots L'_k$ with $S(L'_i) = \{\mu_i\}$ for $i = 1,\ldots,k$ (formal factorisation, first case, see 5.4.1). Then apply 5.4.2.2 to the automorphism τ_ρ of K', yielding a factorisation by τ_ρ-invariant q-difference operators, that is q-difference operators over K.

Uniqueness: In the uniqueness lemma (Corollary 5.42), the α_i should a priori be defined over K'. However, equality $L'_1 = L_1\alpha_1^{-1}$ (since $\alpha_0 = 1$) implies that $\alpha_1 \in K$; then equalities $L'_i = \alpha_{i-1}L_i\alpha_i^{-1}$ allow us to inductively deduce that all $\alpha_i \in K$. □

5.4.3. Analytic factorisation. Although this is one of the most important results of the theory, it will come for free, since all the hard work has already been done. Here of course $K := \mathbf{C}(\{x\})$.

THEOREM 5.6. *Let $L \in \mathcal{D}_{K,q}$. We write $S(L) = \{\mu_1,\ldots,\mu_k\}$ with $\mu_1 < \cdots < \mu_k$.*

(i) *If $S(L) \subset \mathbf{Z}$, there is a factorisation as in* 5.4.1.1, *with all $u_i \in \mathbf{C}(\{x\})$ and the $v_0(u_i)$ in increasing order.*

(ii) *Without the assumption* $S(L) \subset \mathbf{Z}$, *there is a factorisation as in* 5.4.2.3 *(so with factors of increasing slopes) with all* L_i *defined over* $\mathbf{C}(\{x\})$ *(i.e., having convergent coefficients).*

PROOF. (i) comes by repeated iteration of Adams's lemma (theorem 5.3). (ii) comes from 5.4.2.3 taking in account that $\mathbf{C}(\{x^{1/r}\}) \cap \mathbf{C}((x)) = \mathbf{C}(\{x\})$. □

5.5. Application to the explicit resolution of scalar equations

5.5.1. Outline.

5.5.1.1. In this section, we intend to produce a "full complement" (ancient terminology) of solutions of the scalar equation $Lf = 0$, where $L := \sigma_q^n + a_1\sigma_q^{n-1} + \cdots + a_n$, all $a_i \in K$; that is, in nowadays terminology, a fundamental system of solutions, i.e., a family $(f_1, \ldots, f_n)$ in some q-difference extension such that $Lf_i = 0$, $i = 1, \ldots, n$, and $w_n(f_1, \ldots, f_n) \neq 0$ (recall the q-wronskian determinant was defined in Chapter 3, where the formal properties of a fundamental system of solutions have been explored).

We shall do that for $K := \mathbf{C}(\{x\})$ and then solutions will be found in $\mathscr{M}(\mathbf{C}^*, 0)$. When specialized to equations with coefficients in $\mathscr{M}(\mathbf{C})$ or $\mathbf{C}(x)$, the process will yield solutions in $\mathscr{M}(\mathbf{C}^*)$. The process can also be extended to $\mathbf{C}((x))$: then a more algebraic approach is required; such an approach is proposed in **[249]** in relation with its main motivation, Galois theory in Picard-Vessiot style.

5.5.1.2. We shall be able to produce enough explicit solutions only when all the slopes of the Newton polygon of L are integral: $S(L) \subset \mathbf{Z}$. In the most general case, let $r \in \mathbf{N}^*$ a common denominator of the slopes: $S(L) \subset \frac{1}{r}\mathbf{Z}$. Then our method provides *ramified solutions* $f(x^{1/r})$, $f \in \mathscr{M}(\mathbf{C}^*, 0)$. The reason is that we use complete factorisation of L, as obtained in section 5.4, to reduce the resolution of $Lf = 0$ to a succession of resolutions of nonhomogeneous equations of order 1: $\sigma_q f = uf + g$. So, if the $\sigma_q - u_i$, $i = 1, \ldots, n$, are the degree 1 factors of L, we need a q-difference extension (field or ring) of K such that all the $\sigma_q - u_i$ act surjectively and not injectively in that extension (more precise assumptions are stated in 5.5.2).

In 7.3 we shall prove that every $L \in \mathcal{D}_{K,q}$ of degree at least 1 actually has a full complement of solutions inside $\mathscr{M}(\mathbf{C}^*, 0)$ (so, without ramification). However, the proof (due to Praagman) is rather indirect and by no way explicit. (It uses properties of holomorphic vector bundles usually proved with the help of functional analysis.)

As for explicit resolution of q-difference *systems*, we shall tackle it in Chapters 11 and (for the case of integral slopes) 15.

5.5.2. Iterative resolution.

5.5.2.1. We assume that L can be completely factored over K:

$$L = (S - u_1) \cdots (S - u_n), \text{ all } u_i \in K^*.$$

We let K' a q-difference extension of K in which no $\sigma_q - u_i$, $i = 1, \ldots, n$, acts injectively; and in which all $\sigma_q - u_i$, $i = 2, \ldots, n$, acts surjectively (see 5.5.3 for such an extension).

We shall successively solve equations $L_m f = 0$ for $m = 1, \ldots, n$, where:

$$L_m := (S - u_1) \cdots (S - u_m);$$

i.e., we shall find for each $m = 1, \ldots, n$ a fundamental system of solutions $(f_1^{(m)}, \ldots, f_m^{(m)})$: all $L_m f_i^{(m)} = 0$, $i = 1, \ldots, m$, and $D_m := w_m(f_1^{(m)}, \ldots, f_m^{(m)}) \neq 0$.

Initialisation is obvious: since $\sigma_q - u_1$ is not injective in K', there exists $f_1^{(1)} \in K'$, $f_1^{(1)} \neq 0$, such that $L_1 f_1^{(1)} = (\sigma_q - u_1) f_1^{(1)} = 0$; and then $D_1 = f_1^{(1)}$.

5.5.2.2. The iteration step relies on a lemma.

LEMMA 5.48. *Let $P \in \mathcal{D}_{K,q}$, $P \neq 0$, and let $f_1, \ldots, f_m$ solutions of $Pf = 0$ in a q-difference extension K' of K. We set $f_{m+1} := 0$. Let $u \in K^*$ and let $g_i \in K'$, $i = 1, \ldots, m+1$, such that $(\sigma_q - u) g_i = f_i$.*

(i) *The g_i are solutions of $Qg = 0$, where $Q := P(\sigma_q - u)$.*

(ii) *$w_{m+1}(g_1, \ldots, g_{m+1}) = (-1)^m \, g_{m+1} \, w_m(f_1, \ldots, f_m)$.*

PROOF. (i) is clear: $Qg_i = P(\sigma_q - u) g_i = P f_i = 0$.

(ii) Let the line vectors $\Phi := \begin{pmatrix} f_1 & \ldots & f_{m+1} \end{pmatrix}$ and $\Gamma := \begin{pmatrix} g_1 & \ldots & g_{m+1} \end{pmatrix}$, so that:

$$\sigma_q \Gamma = u\Gamma + \Phi;$$

indeed, that equality summarizes the relations $(\sigma_q - u) g_i = f_i$, equivalently: $\sigma_q g_i = u g_i + f_i$. Also, the wronskian matrix:

$$W_{m+1}(g_1, \ldots, g_{m+1}) = \begin{pmatrix} \Gamma \\ \sigma_q \Gamma \\ \vdots \\ \sigma_q^m \Gamma \end{pmatrix}.$$

In order to start standard "operations on lines and columns" (actually, only lines here), we write $L_0, \ldots, L_m$ those lines (do not confuse with q-difference operators); so $L_i = \sigma_q^i \Gamma$. We successively perform operations $L_i \leftarrow L_i - (\sigma_q^{i-1} u) L_{i-1}$ for $i = m, \ldots, 1$.

Those operations do not change the determinant and they transform the matrix into

$$\begin{pmatrix} \Gamma \\ \Phi \\ \sigma_q \Phi \\ \vdots \\ \sigma_q^{m-1} \Phi \end{pmatrix} \text{ the last column of which is } \begin{pmatrix} g_{m+1} \\ 0 \\ 0 \\ \vdots \\ 0 \end{pmatrix} \text{ since } f_{m+1} = 0.$$

Moreover the block complementary to g_{m+1} in that matrix (i.e., all lines except the first, all columns except the last) is the q-wronskian matrix $W_m(f_1, \ldots, f_m)$. The stated formula then follows from usual expansion of the determinant along its last column.

□

5.5.2.3. We now apply the lemma to the following situation: for $m < n$, assuming a fundamental system of solutions $(f_1^{(m)}, \ldots, f_m^{(m)})$ with wronskian D_m has been obtained in K' for the equation $L_m f = 0$, we let (notations of the lemma) $P := L_m$, the $f_i := f_i^{(m)}$ for $i = 1, \ldots, m$ and $u := u_{m+1}$ (so that $Q = L_{m+1}$). Then we take the family $(g_1, \ldots, g_{m+1})$ as in the lemma (possible under the assumptions made on K' since $\sigma_q - u = \sigma_q - u_{m+1}$ is surjective not injective) and rewrite that family as $(f_1^{(m+1)}, \ldots, f_{m+1}^{(m+1)})$. By the lemma, its wronskian is $D_{m+1} = (-1)^m f_{m+1}^{(m+1)} D_m \neq 0$, so $(f_1^{(m+1)}, \ldots, f_{m+1}^{(m+1)})$is indeed a fundamental system of solutions of equation $L_{m+1} f = 0$.

THEOREM 5.7. *The above algorithm produces a fundamental system of solutions* $(f_1^{(n)}, \ldots, f_n^{(n)})$ *of equation* $Lf = 0$, *with wronskian:*

$$w_n(f_1^{(n)}, \ldots, f_n^{(n)}) = (-1)^{n(n-1)/2} f_1^{(1)} \cdots f_n^{(n)}.$$

PROOF. The statement on the fundamental system of solutions is immediate by inuction. The formula for the wronskian is a particular case of formula $D_m = (-1)^{m(m-1)/2} f_1^{(1)} \cdots f_m^{(m)}$, also immediate by inuction. □

EXERCISE 5.49. (i) Show that $\dfrac{\sigma_q w_n(f_1^{(n)}, \ldots, f_n^{(n)})}{w_n(f_1^{(n)}, \ldots, f_n^{(n)})} = u_1 \cdots u_n$.

(ii) From which previously known fact can that formula be deduced?

5.5.3. Constructing an appropriate extension. From now on, the base q-difference field is $K := \mathbf{C}(\{x\})$, but we keep the notation K for short. We proceed to build an appropriate q-difference extension $K' \subset \mathscr{M}(\mathbf{C}^*, 0)$.

5.5.3.1. *A first extension.* Let $L := (S - u_1) \cdots (S - u_n)$, where, for $i = 1, \ldots, n$, $u_i = c_i c^{-\mu_i} \epsilon_i$, $c_i \in \mathbf{C}^*$, $\mu_i \in \mathbf{Z}$, $\epsilon_i \in 1 + x\mathbf{C}\{x\}$. The slopes of L are the $\mu_i = -v(u_i)$ and, by 5.4.3, we may assume that $\mu_1 > \cdots > \mu_n$.

Since we want to get nontrivial solutions for all $\sigma_q - u_i$, we need to put into K' many q-characters and Theta powers (Chapters 2 and 4). The q-characters are best obtained as quotients of twisted Theta functions $\theta_{q,c}(x) := \theta_q(x/c)$; the standard choice is $e_{q,c} := \theta_{q,c}/\theta_q$. So we introduce a first extension *ring* (notation is provisional):

$$R_0 := K\left[\text{all } \theta_{q,c}, {\theta_{q,c}}^{-1}, c \in \mathbf{C}^*\right].$$

We know from Chapter 4 that all equations $\sigma_q f = uf$, $u \in \mathbf{C}(\{x\})^*$, have a nontrivial solution in R_0.

EXERCISE 5.50. Show that $R_0^{\sigma_q} = \mathscr{M}(\mathbf{E}_q)$.

5.5.3.2. *A motivating example.* In order to motivate the next step, let us look at the example $L := (\sigma_q - u)(\sigma_q - x)$, where $u := (\sigma_q f_0)/f_0$, $f_0 \in \mathbf{C}(\{x\})^*$. Our fundamental system of solutions (f_1, f_2) as obtained along the lines of 5.5.2 will have (if we take $f_1^{(1)} := f_0$ at the initialisation step) $\sigma_q f_1 - x f_1 = f_0$, $\sigma_q f_2 = x f_2$. We take $f_2 := 1/\theta_q$ (the simplest choice). As for f_1, we know that, except for exceptional choices of f_0 (see "q-Borel-Ramis" in 4.3.3) it cannot be a convergent power series. (A generically divergent power series solution can however be obtained by starting from 0 and iterating $f \mapsto \sigma_q^{-1}(f_0 + xf)$.)

EXERCISE 5.51. What condition on f_0 would ensure convergence?

To solve $\sigma_q f_1 - x f_1 = f_0$, we can try the Lagrange method. Since the associated homogeneous equation has the nontrivial solution $f_2 = 1/\theta_q$, we look for f_1 as $f_2 g = g/\theta_q$, so we arrive to the equation $\sigma_q g - g = x^{-1} f_0 \theta_q$. The right-hand side belongs to $\mathscr{O}(\mathbf{C}^*, 0)$, so it admits a Laurent power series expansion $\sum \gamma_n x^n$ which converges on some punctured open disk $\overset{\bullet}{\mathrm{D}}(0, r) := \overset{\circ}{\mathrm{D}}(0, r) \setminus \{0\}$, $r > 0$. The term γ_0 is usually present, it is an obstruction to having a power series solution, but we can use a q-logarithm ℓ_q (recall from Chapter 2 that $\ell_q := -x\theta_q'/\theta_q$ satisfies $\sigma_q \ell_q = \ell_q + 1$). Eventually we get:

$$g = \gamma_0 \ell_q + \sum_{n \neq 0} \frac{\gamma_n}{q^n - 1} x^n \Longrightarrow f = \frac{1}{\theta_q} \left(\gamma_0 \ell_q + \sum_{n \neq 0} \frac{\gamma_n}{q^n - 1} x^n \right).$$

So we get a solution in the form h/θ_q^2, $h \in \mathscr{O}(\mathbf{C}^*, 0)$.

However, we can do a little better and avoid the double poles by using *twisted* Lagrange method. We look for f_1 in the form $f_1 = g/\theta_{q,c}$, whence equation $c^{-1}\sigma_q g - g = x^{-1} f_0 \theta_{q,c}$. The right-hand side is in $\mathscr{O}(\mathbf{C}^*, 0)$, write it $\sum \gamma_n x^n$. If c was taken out of $q^{\mathbf{Z}}$, we can then solve:

$$g = \sum \frac{\gamma_n}{c^{-1} q^n - 1} x^n \Longrightarrow f = \frac{1}{\theta_{q,c}} \sum \frac{\gamma_n}{c^{-1} q^n - 1} x^n.$$

This time the sum is in $\mathscr{O}(\mathbf{C}^*, 0)$ so the only possible poles are the zeroes of $\theta_{q,c}$, they are simple.

Thinking about how the process may be iterated (along the lines of 5.5.2), we see that, generally speaking, we can probably solve equations $\sigma_q f - uf = g$ with f a quotient of an element of $\mathscr{O}(\mathbf{C}^*, 0)$ by a product of twisted Theta factors.

REMARK 5.52. It is likely that not every element of $\mathscr{O}(\mathbf{C}^*, 0)$ can appear as the numerator of such a solution, some constraints must hold; i.e., not every quotient of an element of $\mathscr{O}(\mathbf{C}^*, 0)$ by a product of twisted Theta factors is holonomic with respect to $\mathbf{C}(\{x\})$.

5.5.3.3. *Back to the theory.* So we are led to introduce the following q-difference ring:

$$R := \mathscr{O}(\mathbf{C}^*, 0) \left[\text{all } {\theta_{q,c}}^{-1}, c \in \mathbf{C}^* \right] \subset \mathscr{M}(\mathbf{C}^*, 0).$$

EXERCISE 5.53. Check that $\sigma_q R = R$.

An element of T has the form $\dfrac{\gamma}{T}$, $\gamma \in \mathscr{O}(\mathbf{C}^*, 0)$, T a product of twisted Theta functions $\theta_{q,c}$. Therefore, in an appropriate punctured open disk $\dot{\mathrm{D}}(0, r)$, $r > 0$, over which γ is holomorphic, the poles of γ/T are organized in half q-spirals $cq^{\mathbf{N}}$; more precisely, the set of poles is included in a finite union of such half q-spirals. This does not hold for arbitrary elements of $\mathscr{M}(\mathbf{C}^*, 0)$. For instance, $1/\sin \notin R$, although $\sin \in R$. So R is not a field and the inclusion $R \subset \mathscr{M}(\mathbf{C}^*, 0)$ is strict.

THEOREM 5.8. *For every $u \in K^*$ and $g \in R$, the equation $\sigma_q f = uf + g$ admits a solution in R.*

PROOF. Let $u := cx^m \epsilon$, $c \in \mathbf{C}^*$, $m \in \mathbf{Z}$, $\epsilon \in 1 + x\mathbf{C}\{x\}$, so that, by Chapters 3 and 4, $\epsilon = (\sigma_q \eta)/\eta$ for some $\eta \in \mathbf{C}\{x\}^*$. Setting $u' := cx^m$, $f' := f/\eta$ and $g' := g/(\sigma_q \eta)$, equation $\sigma_q f = uf + g$ is equivalent to equation $\sigma_q f' = u' f' + g'$; also $f' \in R \Leftrightarrow f \in R$ and $g' \in R \Leftrightarrow g \in R$, so we may as well assume from start that $u = cx^m$.

Write $g = \dfrac{\gamma}{T}$, $\gamma \in \mathscr{O}(\mathbf{C}^*, 0)$, T a product of functions $\theta_{q,c}$, so that $\sigma_q T = dx^n T$ for some $n \in -\mathbf{N}$ and $d \in \mathbf{C}^*$. Setting $f' := fT$, $c' := cd$, $m' := m + n$ and $g' := dx^n g T = dx^n \gamma \in \mathscr{O}(\mathbf{C}^*, 0)$, equation $\sigma_q f = uf + g$ is equivalent to equation $\sigma_q f' - c' x^{m'} f' = g'$, so we may as well assume from start that $u = cx^m$ and $g \in \mathscr{O}(\mathbf{C}^*, 0)$. Now, there are three cases to consider:

<u>$m > 0.$</u> We use a twisted Lagrange method: we look for $f = \phi/T$, $\phi \in \mathscr{O}(\mathbf{C}^*, 0)$ and T a product of twisted Theta functions such that $\sigma_q T = dx^n T$ for some $n \in -\mathbf{N}$ and $d \in \mathbf{C}^*$. So we must solve $\sigma_q \phi - cdx^{m+n}\phi = dx^n gT$. Since $T \in \mathscr{O}(\mathbf{C}^*, 0)$, the right-hand side $dx^n gT$ is also in $\mathscr{O}(\mathbf{C}^*, 0)$. Clearly (this means: you check it!), we can choose T such that $m + n = 0$ and $cd \notin q^{\mathbf{Z}}$. Then we fall under the next case, first subcase.

<u>$m = 0.$</u> There are two subcases!

- If $c \notin q^{\mathbf{Z}}$, we solve $\sigma_q f - cf = g$, where g has the Laurent power series expansion $g = \sum g_n x^n$, by taking $f := \sum f_n x^n$, where $f_n := \dfrac{g_n}{q^n - c}$; then $f \in \mathscr{O}(\mathbf{C}^*, 0)$.
- If $c \in q^{\mathbf{Z}}$, say $c = q^k$, the above fails for term f_k. Using the fact that $(\sigma_q - q^k)(q^{-k}x^k \ell_q) = x^k$, we just replace the missing term $f_k x^k$ by $g_k q^{-k} x^k \ell_q \in R$.

<u>$m < 0.$</u> We write $g(x) = g_1(x) + g_2(x^{-1})$, $g_1 \in \mathbf{C}\{x\}$, $g_2 \in \mathscr{M}(\mathbf{C})$ (that decomposition is not unique since the term of degree 0 can go either way). We look for f in the same form: $f(x) = f_1(x) + f_2(x^{-1})$, $f_1 \in \mathbf{C}\{x\}$, $f_2 \in \mathscr{M}(\mathbf{C})$, where $\sigma_q f_1 - cx^m f_1 = g_1$ and (check it!) $f_2(q^{-1}x^{-1}) - cx^m f_2(x^{-1}) = g_2(x^{-1})$, i.e., $f_2(x) - cq^{-m}x^{-m} f_2(qx) = g_2(qx)$, i.e., $f_2(qx) - c^{-1}q^m x^m f_2(x) = -c^{-1}q^m x^m g_2(qx)$. So this reduces our problem to the following proposition. □

Let $C(R) := \mathscr{O}(\mathring{\mathrm{D}}(0, R)) \cap \mathscr{C}(\overline{\mathrm{D}}(0, R), \mathbf{C})$ (it will appear again in 6.2.1).

PROPOSITION 5.54. *Let $n \in \mathbf{N}^*$ and $d \in \mathbf{C}^*$. Equation $\phi - dx^n \sigma_q \phi = \gamma$ admits*
(i) *a solution $\phi \in C(R)$ if $\gamma \in C(R)$;*
(ii) *a solution $\phi \in \mathbf{C}\{x\}$ if $\gamma \in \mathbf{C}\{x\}$;*
(iii) *a solution $\phi \in \mathscr{O}(\mathbf{C})$ if $\gamma \in \mathscr{O}(\mathbf{C})$.*

PROOF. (i) We solve our equation by considering ϕ as a fixed point of $\phi \mapsto \gamma + dx^n \sigma_q \phi$. It suffices to check that the series $\sum\limits_{k \geq 0} (dx^n \sigma_q)^k \gamma$ converges normally on $\overline{\mathrm{D}}(0, R)$; for that, just observe that $(dx^n \sigma_q)^k = q^{nk(k-1)/2} d^k x^{nk} \sigma_q^k$, so that, on $\overline{\mathrm{D}}(0, R)$:

$$\left|((dx^n \sigma_q)^k)(x)\right| \leq \left|q^{n/2}\right|^{k^2} \left(|q|^{-n/2} R^n |d|\right)^k \|\gamma\|_R .$$

The conclusion is then obvious.
(ii) and (iii) follow immediately.

□

Similar results will be proved (and greatly improved) in Chapter 6.

EXERCISE 5.55. In the proof of the theorem, track down where the Theta denominators appear. What can be deduced about the solutions of $Lf = 0$?

CHAPTER 6

Further Analytic Properties of Solutions: Index Theorems, Growth

In the last chapter of his book on the history of Functional Analysis, Dieudonné **[68]** enumerates the applications to functional equations, mainly differential and partial differential equations. While he mentions the use of fixed point theorems, of universal interest, he says nothing of index theory. There are only a few lines on Fredholm operators. Actually, index theory progressively outgrew fixed point theorems, at least in the area of complex differential equations; next to our subject here are the works of Deligne, Katz, Malgrange, Ramis, Sibuya ....

The power of index theory is even greater when it is used in association to sheaf theory and homological algebra. This will be done in Chapter 9. In this chapter, we shall use some *moderate*[1] functional analysis (index theorems, mostly in Banach spaces; all prerequisites are recalled in Appendix D); and some very elementary homological algebra (around exact sequences). All significant results given here come from the papers **[20]** by Jean-Paul Bézivin (Section 6.2) and **[191]** by Jean-Pierre Ramis (Section 6.3).

The orders of growth of coefficients of divergent series, or of entire functions, give rise to notions of q-Gevrey classes where (with respect to the Gevrey classification in the theory of ordinary differential equations) factorials $n!$ are replaced by more rapidly growing expressions like $|q|^{-n^2/2}$: the corresponding exponents are the so-called orders. More refined classification takes in account types (coming e.g. in the form of a factor A^n). We do not go into these more precise results. See **[87]**.

6.1. Introductory examples

6.1.1. Preliminary discussion.

6.1.1.1. We are interested in properties of solutions of nonhomogeneous equations $Lf = g$, where L is a q-difference operator, g a given analytic function (maybe a germ, maybe meromorphic) and f is the unknown function. Of course, part of the preliminary study is formal, so maybe at some places our objects are formal rather than analytic. Also, by chasing denominators, one can usually assume that L has coefficients in $\mathbf{C}[[x]]$ or $\mathbf{C}[x]$ or even $\mathscr{O}(\mathbf{C})$; and that the right-hand side g lies in the same ring. So, contrary with what is done in most of this book, we shall here work with q-difference *rings*.

6.1.1.2. Call R either one of the q-difference rings quoted above. We are interested in the $\mathbf{C}$-linear map $R \to R$, $f \mapsto Lf$ (note that here $R^{\sigma_q} = \mathbf{C}$, so $\mathbf{C}$-linearity is the best that we can hope for). Indeed,

[1]There is some more fancy functional analysis involving "FN" and "DFN" spaces, this can be seen as a black box, just taking in faith the corresponding results in Appendix D.

- Ker$L$ is the linear space of solutions of the associated homogeneous equation $Lf = 0$, thus of the affine space of solutions of the nonhomogeneous equation $Lf = g$ (except of course if the said space is empty);
- and Im L is the linear space of those g such that $Lf = g$ has solutions.

The existence of solutions is guaranteed if, and only if, L is onto, i.e., Im $L = R$, i.e., Coker $L = \{0\}$ (remember that the cokernel of L is defined as Coker $L := R/(\text{Im } L)$). The uniqueness is guaranteed if, and only if, Ker $L = \{0\}$.

REMARK 6.1. We shall use the same notation for q-difference operator L acting on various spaces of functions R, so it would be more rigorous to write $\text{Ker}(L, R)$ and $\text{Coker}(L, R)$. We shall do so every time it is needed for clarity.

6.1.1.3. Therefore $\dim_{\mathbf{C}} \text{Coker } L$ measures the defect of guarantee of existence; and likewise $\dim_{\mathbf{C}} \text{Ker}L$ measures the defect of uniqueness. We shall find out that, in all the cases of interest, the kernel and cokernel have finite dimension: L acting on R is then said to *have an index* and called a *Fredholm operator* (Appendix D). Then the number:

$$\chi(L, R) := \dim_{\mathbf{C}} \text{Ker}(L, R) - \dim_{\mathbf{C}} \text{Coker}(L, R)$$

is a well defined rational integer, the *index of L (operating on R)*. This number is important because it has very convenient properties (Appendix D).

6.1.1.4. *Examples.*

(1) The q-difference operator $L := \sigma_q$ acting on $R := \mathbf{C}[[x]]$ or $R := \mathbf{C}\{x\}$ or $R := \mathscr{O}(\mathbf{C})$ is bijective so $\dim_{\mathbf{C}} \text{Ker}L = \dim_{\mathbf{C}} \text{Coker } L = 0$ so $\chi(L, R) = 0$.
(2) The q-difference operator $L := x$ (i.e., the map $f \mapsto xf$) acting on $R := \mathbf{C}[[x]]$ or $R := \mathbf{C}\{x\}$ or $R := \mathscr{O}(\mathbf{C})$ is injective and its image has codimension 1 because the decomposition $f = (f - f(0)) + f(0)$ yields in all cases the splitting $R = \text{Im}(L, R) \oplus \mathbf{C}$. Therefore $\dim_{\mathbf{C}} \text{Ker}L = 0$ and $\dim_{\mathbf{C}} \text{Coker } L = 1$ so $\chi(L, R) = -1$.
(3) Let more generally $L_m := x^m$, also be injective on $R := \mathbf{C}[[x]]$ or $R := \mathbf{C}\{x\}$ or $R := \mathscr{O}(\mathbf{C})$. From the splitting $R = \text{Im}(L_m, R) \oplus \mathbf{C} \oplus \cdots \oplus \mathbf{C}x^{m-1}$ (which follows for instance from Taylor expansion at order m) we draw that $\dim_{\mathbf{C}} \text{Coker } L = m$ whence $\chi(L, R) = -m$. This illustrate the general additivity formula $\chi(L_1 L_2) = \chi(L_1) + \chi(L_2)$ (Appendix D).
(4) Here $R := \mathbf{C}[[x]]$ or $R := \mathbf{C}\{x\}$. Any nonzerio $u \in R$ is uniquely written $u = x^m \epsilon$, where $m := v(u) \in \mathbf{N}$ and $\epsilon \in R^*$. As in the first example, $\chi(\epsilon, R) = 0$ and, using the additivity formula, we get $\chi(L, R) = -m$.

6.1.1.5. *Another example.* This one is treated separately because it features nicely the *stability properties* of the index. It is somehow similar to the classical "Fredholm alternative" **[234]** which was one of the startings point of index theory. We take $c \in \mathbf{C}^*$ and $L := \sigma_q - c$. Solving $Lf = g$ where $f := \sum f_n x^n$ and $g := \sum g_n x^n$ reduces to solving the equations $(q^n - c) f_n = g_n$ for all $n \geq 0$. Clearly there are two cases to consider:

- If $c \notin q^{\mathbf{N}}$, then there is a unique solution $f := \sum x^n/(q^n - c) \in R$ (convergence is checked easily if $R = \mathbf{C}\{x\}$), L has bijectively and has index 0.
- If $c \notin q^{\mathbf{N}}$, say $c = q^m$, then $\text{Ker}(L, R) = \mathbf{C}x^m$. On the other hand, only the $(q^n - c) f_n = g_n$ for $n := m$ may have no solution, if $g_m = 0$, so

there is a splitting $R = \mathrm{Im}(L, R) \oplus \mathbf{C}x^m$ and we end with $\dim_{\mathbf{C}} \mathrm{Ker} L = \dim_{\mathbf{C}} \mathrm{Coker}\, L = 1$ so $\chi(L, R) = 0$.

The fact that the index is 0 in all cases, although kernel and cokernel vary, is an example of the "stability" mentioned above.

6.1.2. Operators of degree 1.

6.1.2.1. In all Subsection 6.1.2 we shall take $R := \mathbf{C}[[x]]$ or $R := \mathbf{C}\{x\}$ and denote $v(f) := v_0(f)$ the x-adic valuation.

LEMMA 6.2. *Let $w \in R \setminus \{0\}$. Then $\chi(wL, R) = \chi(L, R) - v(w)$.*

PROOF. As in all similar cases, it is implicitly understood that the complete statement would be: "if L has an index, then so does wL and one has ...". We write $w = \epsilon x^m$ with $\epsilon \in R^*$ and $m := v(w)$. From one of the examples above we see that the equality to be proved flows from the additivity of the index. Yet we want a direct proof. (Yes, this is not necessary; it just aims at acclimatization!) There are two independent steps.

The formula $\chi(\epsilon L, R) = \chi(L, R)$ comes from the obvious fact that ϵL has the same kernel as L; and from the less obvious but easy fact that $\epsilon R = R$ and $\mathrm{Im}(\epsilon L) = \epsilon \mathrm{Im} L$, so left multiplication by ϵ induces an isomorphism $\mathrm{Coker}\, L \simeq \mathrm{Coker}\, \epsilon L$. The formula $\chi(x^m L, R) = \chi(L, R) - m$ comes by iteration from the case $m = 1$. This one rests first on the obvious fact that xL has the same kernel as L. Then, noting that $\mathrm{Im}(xL) = x\mathrm{Im} L$, the formula boils down to the following: if the subspace E of R has codimension d, then xE has codimension $d + 1$. This comes as follows:

$$\dim_{\mathbf{C}} R/xE = \dim_{\mathbf{C}} R/xR + \dim_{\mathbf{C}} xR/xE = 1 + \dim_{\mathbf{C}} R/E.$$

□

EXERCISE 6.3. (i) Explain why $\dim_{\mathbf{C}} R/xR = 1$ and $\dim_{\mathbf{C}} xR/xE = \dim_{\mathbf{C}} R/E$.

(ii) Explain why we did not go through formula $\dim_{\mathbf{C}} E/xE = 1$. (Take $E := \mathrm{Im}(\sigma_q - q)$.)

So we are left with the study of the special case $L = u\sigma_q + v$, $u, v \in R \setminus \{0\}$. And since in R (a valuation ring in both cases considered!) either $u|v$ or $v|u$, we may even assume that u or v is invertible.

If u is invertible, up to one more division (can you see which?) we may write $L = \epsilon\sigma_q - cx^m$, $\epsilon \in R^*$, $\epsilon(0) = 1$, $c \in \mathbf{C}^*$ and $m \in \mathbf{N}$. Now recall from Chapter 4 that $\epsilon = \eta/\sigma_q\eta$ for some $\eta \in R^*$, so $L = \eta(\sigma_q - cx^m)\eta^{-1}$ and again (same type of argument as before; please fill in the details!) we can reduce to the case $L = \sigma_q - cx^m$. Similarly, if v is invertible we may assume that $L = x^m\sigma_q - c$ (fill in the details!).

So in the end we are left with three cases: either $L = \sigma_q - c$; or $L = x^m\sigma_q - c$, $m \geq 1$; or $L = \sigma_q - cx^m$, $m \geq 1$. The first case is the example treated in 6.1.1.5, we saw that $\chi(L, \mathbf{C}[[x]]) = \chi(L, \mathbf{C}\{x\}) = 0$. The operator $\sigma_q - c$ does not "see the difference formal/analytic". We look at $L = x^m\sigma_q - c$ and $L = \sigma_q - cx^m$ further below.

6.1.2.2. $L = x^m\sigma_q - c$. Equation $x^m\sigma_q f - cf = g$ reduces to the sequence of recursive relations:

$$\forall n \geq 0 \,,\; q^{n-m} f_{n-m} - cf_n = g_n.$$

(We adopt here and everywhere the natural convention that $f_n = 0$ whenever $n < 0$.)

The kernel $\mathrm{Ker}(L, R)$ is trivial since it is characterized by the relations $f_n = c^{-1}q^{n-m}f_{n-m}$ which imply $f_n = 0$ directly for $n = 0, \dots, m-1$ and then by induction for $n \geq m$.

We shall show that $L : R \to R$ is onto, so the cokernel is trivial. Indeed, the above sequence of recursive relations can be solved by setting $f_n := -c^{-1}g_n$ for $n = 0, \dots, m-1$ and then inductively $f_n := c^{-1}(q^{n-m}f_{n-m} - g_n)$ for $n \geq m$. This certainly proves surjectivity in the formal case, so $\mathrm{Coker}(L, \mathbf{C}[[x]]) = \{0\}$ and then $\chi(L, \mathbf{C}[[x]]) = 0$. In the case of $\mathbf{C}\{x\}$, we shall prove convergence, so $\mathrm{Coker}(L, \mathbf{C}\{x\}) = \{0\}$ and then $\chi(L, \mathbf{C}\{x\}) = 0$: the operator $x^m\sigma_q - c$, $m \geq 1$, does not either "see the difference formal/analytic".

First method. Fix $j \in \{0, \dots, m-1\}$ and set $\phi_i := f_{mi+j}$, $\gamma_i := g_{mi+j}$ for all $i \geq 0$. Then $\phi_i = aQ^i\phi_{i-1} + b\gamma_i$ for some $a, b \in \mathbf{C}^*$ (to be computed by *you*) and $Q := q^m$. We make this telescopic by setting:

$$\tilde{\phi}_i := a^{-i}Q^{-i(i+1)/2}\phi_i \text{ and } \tilde{\gamma}_i := a^{-i}Q^{-i(i+1)/2}\gamma_i,$$

thus yielding:

$$\tilde{\phi}_i = \tilde{\phi}_{i-1} + b\tilde{\gamma}_i.$$

Second method. Consider the equation as a fixpoint problem and iterate the relation $f = c^{-1}(-g + x^m\sigma_q f)$ within some closed disc $\overline{\mathrm{D}}(0, r)$.

EXERCISE 6.4. In both methods, use previous calculations of the same kind to complete the estimation to obtain convergence of f (assuming convergence of g).

6.1.2.3. $L = \sigma_q - cx^m$. Equation $\sigma_q f - cx^m f = g$ amounts to the sequence of recursive relations:

$$\forall n \geq 0\ ,\ q^n f_n - cf_{n-m} = g_n.$$

The kernel $\mathrm{Ker}(L, R)$ is trivial since it is characterized by the relations $f_n = cq^{-n}f_{n-m}$ which imply $f_n = 0$ directly for $n = 0, \dots, m-1$ and then by induction for $n \geq m$.

Over $R := \mathbf{C}[[x]]$, the operator is onto, since the f_n can be recursively defined by relations $f_n := q^{-n}(cf_{n-m} + g_n)$. So $\mathrm{Coker}(L, \mathbf{C}[[x]]) = \{0\}$, whence $\chi(L, \mathbf{C}[[x]]) = 0$.

EXERCISE 6.5. Alternative proof of surjectivity: to solve the fixpoint equation $f = \sigma_q^{-1}(g + cx^m f)$, initialize $f^{(0)} := 0$, iterate $f^{(n+1)} := \sigma_q^{-1}(g + cx^m f^{(n)})$; prove x-adic convergence in $\mathbf{C}[[x]]$.

However the presence of a factor q^{-n} in $f_n = cq^{-n}f_{n-m} + \cdots$ gives little hope of (classical) convergence in $\mathbf{C}\{x\}$ if $g \in \mathbf{C}\{x\}$ is arbitrary. We shall prove that L is *not* onto. Actually, $\mathrm{Im}(L, \mathbf{C}\{x\})$ has codimension m in $\mathbf{C}\{x\}$, so $\dim_{\mathbf{C}} \mathrm{Coker}(L, \mathbf{C}\{x\}) = m$ and $\chi(L, \mathbf{C}\{x\}) = -m$. The operator $\sigma_q - cx^m$ does "see the difference formal/analytic". The proof will come in two steps of different flavourr.

PROPOSITION 6.6. *The operator $\sigma_q - cx$ acting on $\mathbf{C}\{x\}$ has index -1; equivalently, it has image of codimension* 1.

COROLLARY 6.7. *The operator $\sigma_q - cx^m$ acting on $\mathbf{C}\{x\}$ has index $-m$; equivalently, it has image of codimension m.*

6.1.2.4. *Proof of the corollary.* So let $L : f \mapsto \sigma_q f - x^m f$, $\mathbf{C}\{x\} \to \mathbf{C}\{x\}$. Write $E_i := x^i \mathbf{C}\{x^m\}$ for $i = 0, \ldots, m-1$, so that $\mathbf{C}\{x\} = E_0 \oplus \cdots \oplus E_{m-1}$. Moreover, each component E_i is stable under L, i.e., it induces L_i on E_i, so that:

$$\mathrm{Ker}(L, \mathbf{C}\{x\}) = \mathrm{Ker}(L_0, E_0) \oplus \cdots \oplus \mathrm{Ker}(L_{m-1}, E_{m-1})$$

and

$$\mathrm{Coker}(L, \mathbf{C}\{x\}) = \mathrm{Coker}(L_0, E_0) \oplus \cdots \oplus \mathrm{Coker}(L_{m-1}, E_{m-1}),$$

whence $\chi(L, \mathbf{C}\{x\}) = \chi(L_0, E_0) + \cdots + \chi(L_{m-1}, E_{m-1})$.

Let $f(x) = x^i F(x^m)$. We have $Lf = q^i x^i F(q^m x^m) - c x^{i+m} F(x^m)$. We can express this relation as a commutative diagram:

$$\begin{array}{ccc} \mathbf{C}\{X\} & \xrightarrow{L'_i} & \mathbf{C}\{X\} \\ \downarrow & & \downarrow \\ E_i & \xrightarrow[L_i]{} & E_i. \end{array}$$

Both vertical arrows stand for $F(X) \mapsto x^i F(x^m)$, so they are isomorphisms; and L'_i is the Q-difference operator $q^i \sigma_Q - cX$, where $Q := q^m$. The factor q^i is innocuous so applying the lemma we get that $\chi(L_i, E_i) = -1$, and adding up that $\chi(L, \mathbf{C}\{x\}) = -m$.

6.1.2.5. *Proof of the proposition.* Studying equation $\sigma_q f - cxf = g$, we are actually dealing with the recursice relation:

$$\forall n \geq 0 \,,\; q^n f_n - c f_{n-1} = g_n, \tag{6.1}$$

with the convention: $f_n = 0$ for $n < 0$, whence the initialisation: $f_0 := g_0$. The assumption that $g \in \mathbf{C}\{x\}$ means that there exist $C, A > 0$ such that:

$$\forall n \geq 0 \,,\; |g_n| < CA^n.$$

We make (6.1) telescopic by multiplying it by $c^{-n} q^{n(n-1)/2}$; so setting $\overline{f}_n := c^{-n} q^{n(n+1)/2} f_n$ and $\overline{g}_n := c^{-n} q^{n(n-1)/2} g_n$ (again a "q-Borel-Ramis" type transformation, see 6.1.3; but beware the exponent on q is not the same for f and g), we get:

$$\forall n \geq 0 \,,\; \overline{f}_n - \overline{f}_{n-1} = \overline{g}_n. \tag{6.2}$$

Therefore:

$$\overline{f}_n = \sum_{i=0}^n \overline{g}_i \Longrightarrow f_n = c^n q^{-n(n-1)/2} \sum_{i=0}^n \overline{g}_i.$$

By itself, this computation ensures bijectivity in the formal case and injectivity in the convergent case. To conclude about surjectivity (or the defect of it) in the convergent case, we must check if the f_n given above define a $f \in \mathbf{C}\{x\}$. First note that:

$$|\overline{g}_n| \leq C \, |q|^{n(n-1)/2} \, (A / |c|)^n,$$

that is a very fast decay of the coefficients of $\overline{g}(x) := \sum\limits_{n \geq 0} \overline{g}_n x^n$, thus making the latter the expansion of an entire function $\overline{g} \in \mathscr{O}(\mathbf{C})$. Similarly, if $f \in \mathbf{C}\{x\}$ then the $\overline{f}_n$ define a $\overline{f} := \sum\limits_{n \geq 0} \overline{f}_n x^n \in \mathscr{O}(\mathbf{C})$; but this is conditional.

On the other hand, (6.2) implies that:

$$(1 - x)\overline{f} = \overline{g}.$$

Therefore, if $f \in \mathbf{C}\{x\}$, we can substitute $x := 1$ and conclude that $\overline{g}(1) = 0$. This is a *necessary* condition for the convergence of f. We will now see that it is sufficient. At any rate, we already know that the image of $L := \sigma_q - cx$ operating on $\mathbf{C}\{x\}$ is contained in the kernel of the linear map:

$$\lambda : g \mapsto \sum_{n \geq 0} c^{-n} q^{n(n-1)/2} g_n.$$

This is a nontrivial linear form on $\mathbf{C}\{x\}$, so its kernel has codimension 1. So we have found part of our answer:

$$\operatorname{codim}_{\mathbf{C}} \operatorname{Im}(L, \mathbf{C}\{x\}) \geq \operatorname{codim}_{\mathbf{C}} \operatorname{Ker} \lambda = 1 \Longrightarrow \chi(L, \mathbf{C}\{x\}) \leq -1.$$

So we are now up to proving that condition $\overline{g}(1) = 0$ conversely implies that f defined by relations $\overline{f}_n = \sum_{i=0}^n \overline{g}_i$ has a positive radius of convergence; it is is the content of the following lemma. It will implies that $\operatorname{Im}(L, \mathbf{C}\{x\}) \operatorname{Ker} \lambda$, so $\operatorname{codim}_{\mathbf{C}} \operatorname{Im}(L, \mathbf{C}\{x\}) = 1$, so $\chi(L, \mathbf{C}\{x\}) = -1$, thus achieving the proof of the proposition.

6.1.2.6.

LEMMA 6.8. *Let* $g := \sum g_n x^n \in \mathbf{C}\{x\}$ *and let* $\overline{g} := \sum c^{-n} q^{n(n-1)/2} g_n x^n$, *so that* $\overline{g} \in \mathscr{O}(C)$. *Assume* $\overline{g}(1) = 0$, *so that* $\overline{f} := \dfrac{1}{1-x}\overline{g} \in \mathscr{O}(C)$ *also. Let* $\overline{f} = \sum \overline{f}_n x^n$ *its power series expansion. Define* $f_n := c^n q^{-n(n+1)/2} \overline{f}_n$ *and* $f(x) := \sum f_n x^n$. *Then* $f \in \mathbf{C}\{x\}$.

PROOF. By assumption, $\overline{g}(1) = 0 \Rightarrow \sum_{n \geq 0} \overline{g}_n = 0$, where $\overline{g}_n := c^{-n} q^{n(n-1)/2} g_n$. Therefore:

$$\overline{f}_n = \sum_{i=0}^{n} \overline{g}_i = -\sum_{i>n} \overline{g}_i = -\sum_{i>n} c^{-i} q^{i(i-1)/2} g_i.$$

From the assumption $g \in \mathbf{C}\{x\}$, one has $|g_n| \leq C A^n$ for some $C, A > 0$ and all $n \geq 0$, whence:

$$|\overline{f}_n| \leq \sum_{i>n} \left| c^{-i} q^{i(i-1)/2} g_i \right| \leq \sum_{i>n} \left| c^{-i} q^{i(i-1)/2} C A^i \right| = C \sum_{i>n} |q|^{i^2/2} \left(\frac{A}{|c|\,|q|^{1/2}} \right)^i.$$

That upper bound can be rewritten:

$$C\,|q|^{n^2/2} \left(\frac{A}{|c|\,|q|^{1/2}} \right)^n \sum_{i \geq 1} |q|^{ni + i^2/2} \left(\frac{A}{|c|\,|q|^{1/2}} \right)^i \leq C'\,|q|^{(n^2+n)/2} \left(\frac{A}{|cq|} \right)^n,$$

where $C' := C \sum_{i \geq 1} |q|^{i^2/2} \left(\dfrac{A}{|c|\,|q|^{1/2}} \right)^i < \infty$. Thus, setting $A' := A/|q|$, we get:

$$|f_n| = \left| c^n q^{-n(n+1)/2} \overline{f}_n \right| \leq C' A'^n,$$

whence $f \in \mathbf{C}\{x\}$. □

Note that f may have a greater radius of convergence that g.

EXERCISE 6.9. If $g \in \mathscr{O}(\mathbf{C})$ prove that $f \in \mathscr{O}(\mathbf{C})$ and deduce the value of $\chi(L, \mathscr{O}(\mathbf{C}))$.

6.1.2.7. We summarize the information obtained so far:

L	$\sigma_q - c$	$x^m\sigma_q - c$	$\sigma_q - cx^m$
$\chi(L, \mathbf{C}[[x]])$	0	0	0
$\chi(L, \mathbf{C}\{x\})$	0	0	$-m$
look of the Newton polygon $NP(L)$	$(0,0) \longrightarrow (1,0)$	$(0,0) \nearrow (1,m)$	$(0,m) \searrow (1,0)$
slope	0	m	$-m$

We note the important fact that $\chi(L, \mathbf{C}[[x]]) \neq \chi(L, \mathbf{C}\{x\})$ whenever the slope is strictly negative and only in that case.

EXERCISE 6.10. Equation $Lf = g$ implies $L'f = 0$, where $L' := (\sigma_q - u)L$, $u := \sigma_q(g)/g$. Draw the Newton polygon of L' in each of the above three cases. Relate the results obtained here with those in Chapter 5.

6.1.3. Further remarks about the examples.

6.1.3.1. Anticipating notations to be introduced later (see 6.2), we now introduce *q-Borel-Ramis transforms*, here at level 1:

$$\mathcal{B}_{q;1} : \sum_{n\geq 0} f_n x^n \mapsto \sum_{n\geq 0} q^{n(n+1)/2} f_n x^n.$$

a priori, it operates bijectively on $\mathbf{C}[[x]]$ and makes divergent series less divergent. It makes some divergent series convergent; for instance it realizes a bijection from the space

$$\mathbf{C}[[x]]_{q,1} := \left\{ \sum_{n\geq 0} f_n x^n \mid \exists C, A > 0 \mid \forall n \geq 0\,,\ |f_n| \leq q^{-n(n+1)/2} C A^n \right\}$$

of "q-Gevrey series of level 1" to $\mathbf{C}\{x\}$. It also makes convergent series more convergent; for instance it realizes a bijection from $\mathbf{C}\{x\}$ to the space

$$\mathbf{C}\{x\}_{q,-1} := \left\{ \sum_{n\geq 0} f_n x^n \mid \exists C, A > 0 \mid \forall n \geq 0\,,\ |f_n| \leq q^{n(n+1)/2} C A^n \right\}$$

of "q-Gevrey series of level -1", which is obviously included in the space of power series expansions of entire functions: $\mathbf{C}[[x]]_{q,-1} \subset \mathscr{O}(\mathbf{C})$.

CAUTION 6.11. Although square brackets $[[-]]$ are used, $\mathbf{C}[[x]]_{q,-1}$ is made up of convergent series.

Slightly modifying some previous calculations, we start from the q-difference equation $\sigma_q f - cxf = g$, or equivalently from the recursive relations $q^n f_n - c f_{n-1} = g_n$ and multiply the latter by $q^{n(n-1)/2}$; setting $\overline{f}_n := q^{n(n+1)/2} f_n$ and $\overline{g}_n :=$

$q^{n(n+1)/2}g_n$, we get $\overline{f}_n - c\overline{f}_{n-1} = q^{-n}\overline{g}_n$. Now setting $\overline{f} := \mathcal{B}_{q;1}\, f = \sum_{n\geq 0} \overline{f}_n x^n$ and $\overline{g} := \mathcal{B}_{q;1} g = \sum_{n\geq 0} \overline{g}_n x^n$, we see that $(1-cx)\overline{f}(x) = \overline{g}(x/q)$, i.e., $\sigma_q \circ (1-cx)\overline{f} = \overline{g}$. We summarize this formula by commutative diagrams:

$$\begin{array}{ccc} \mathbf{C}[[x]] & \xrightarrow{\sigma_q - cx} & \mathbf{C}[[x]] \\ \downarrow{\scriptstyle \mathcal{B}_{q;1}} & & \downarrow{\scriptstyle \mathcal{B}_{q;1}} \\ \mathbf{C}[[x]] & \xrightarrow{\sigma_q\circ(1-cx)} & \mathbf{C}[[x]] \end{array} \qquad \begin{array}{ccc} \mathbf{C}[[x]]_{q,1} & \xrightarrow{\sigma_q - cx} & \mathbf{C}[[x]]_{q,1} \\ \downarrow{\scriptstyle \mathcal{B}_{q;1}} & & \downarrow{\scriptstyle \mathcal{B}_{q;1}} \\ \mathbf{C}\{x\} & \xrightarrow{\sigma_q\circ(1-cx)} & \mathbf{C}\{x\} \end{array} \qquad \begin{array}{ccc} \mathbf{C}\{x\} & \xrightarrow{\sigma_q - cx} & \mathbf{C}\{x\} \\ \downarrow{\scriptstyle \mathcal{B}_{q;1}} & & \downarrow{\scriptstyle \mathcal{B}_{q;1}} \\ \mathbf{C}\{x\}_{q,-1} & \xrightarrow{\sigma_q\circ(1-cx)} & \mathbf{C}\{x\}_{q,-1}. \end{array}$$

Since all vertical arrows are linear bijections, information on the lower horizontal arrows (e.g. index computation) will automatically provide information on the upper horizontal arrows.

EXERCISE 6.12. Before reading the rest of this chapter:

(i) compute the index of the lower horizontal arrow in each of the three diagrams above;

(ii) compute separately the conjugates by $\mathcal{B}_{q;1}$ of σ_q and of x (i.e., the operator $f \mapsto xf$). This means you should complete the following commutative diagrams:

$$\begin{array}{ccc} \mathbf{C}[[x]] & \xrightarrow{\sigma_q} & \mathbf{C}[[x]] \\ \downarrow{\scriptstyle \mathcal{B}_{q;1}} & & \downarrow{\scriptstyle \mathcal{B}_{q;1}} \\ \mathbf{C}[[x]] & \xrightarrow[?]{} & \mathbf{C}[[x]] \end{array} \qquad \begin{array}{ccc} \mathbf{C}[[x]] & \xrightarrow{x} & \mathbf{C}[[x]] \\ \downarrow{\scriptstyle \mathcal{B}_{q;1}} & & \downarrow{\scriptstyle \mathcal{B}_{q;1}} \\ \mathbf{C}[[x]] & \xrightarrow[?]{} & \mathbf{C}[[x]] \end{array}$$

6.1.3.2. The index computations in 6.1.2 look a little abstract (what could mean a possibly negative *difference* of dimensions?) but it actually can provide us with very concrete information on *the rate of growth of coefficients of formal solutions* of the nonhomogeneous q-difference equation $Lf = g$, when $g \in \mathbf{C}\{x\}$. We shall observe that for $m = 1$, i.e., $L := \sigma_q - x$. We keep notations from 6.1.2.4.

First, for the right-hand side g taken as the constant function 1, write it for clarity e, we have $\overline{e}(1/cq) = 1 \neq 0$, so $\overline{e} \notin \mathrm{Im}(L, \mathbf{C}\{x\})$. Therefore $\mathbf{C}e$ is a supplementary space for $\mathrm{Im}(L, \mathbf{C}\{x\})$ in $\mathbf{C}\{x\}$:

$$\mathbf{C}\{x\} = \mathrm{Im}(L, \mathbf{C}\{x\}) \oplus \mathbf{C}e.$$

An arbitrary right-hand side function $g \in \mathbf{C}\{x\}$ decomposes as $g = g' + \alpha e$, where $\alpha := \overline{e}(1)$ and $g' := g - \alpha e \in \mathrm{Im}(L, \mathbf{C}\{x\})$. So the unique $f \in \mathbf{C}[[x]]$ such that $Lf = g$ decomposes as $f' + \alpha E$, where $f' \in \mathbf{C}\{x\}$ is such that $Lf' = g'$ and $E \in \mathbf{C}[[x]]$ is such that $LE = e$, that is, $E = \sum c^{-n} q^{-n(n+1)/2} x^n$.

EXERCISE 6.13. Check that and relate E to a function previously encountered. (Hint: it is denoted by a cyrillic letter ...)

Therefore we see that $f_n = c^{-n} q^{-n(n+1)/2} + f'_n$, where $f'_n = O(A^n)$ for some $A > 0$. The resulting alternative says that the rate of growth of the coefficients f_n of the formal solution f is

- either tame for exceptional cases (f lies in a hyperplane of $\mathbf{C}\{x\}$),
- or wild, of the same order as $|q|^{-n(n+1)/2}$ (later to be nicknamed as "q-Gevrey of order 1").

This alternative will be generalized and refined in this chapter. The relevant relation here, "of the same order" means the following:

$$(u_n) \approx (v_n) \underset{def}{\Longleftrightarrow} \exists A, B > 0 \; : \; u_n = O(v_n A^n) \text{ and } v_n = O(u_n B^n).$$

6.1.3.3. Suppose we want to tackle directly the equation $\sigma_q f - cx^m f = g$, without resorting to the change of variable $X = x^m$ used to prove the corollary in 6.1.2.4. We want a telescopic version of the recursive equation $q^n f_n - c f_{n-m} = g_n$; we do that multiplying each such equation by a $t_n \in \mathbf{C}^*$, for an adequately chosen family (t_n). More precisely, we shall assume that:

$$\forall n \; , \; t_n = q^n t_{n-m}.$$

Compare this to the case $m = 1$. Note that we do not restrict to positive values of the index. Setting $\overline{f}_n := t_n f_n$, $\overline{f} := \sum \overline{f}_n x^n$ and similarly $\overline{g}_n := t_n g_n$, $\overline{g} := \sum \overline{g}_n x^n$.

EXERCISE 6.14. Show that relation $t_n = q^n t_{n-m}$ implies very fast decay of the $|t_n|$ as $n \to +\infty$, so that $\overline{g}$ is automatically an entire function and $\overline{f}$ also is if $f \in \mathbf{C}\{x\}$.

Then our recursive relation becomes:

$$t_n(q^n f_n - c f_{n-m}) = t_n g_n \Longleftrightarrow q^n \overline{f}_n - c q^n t_{n-m} f_{n-m} = \overline{g}_n \Longleftrightarrow \overline{f}_n - c \overline{f}_{n-m} = q^{-n} \overline{g}_n.$$

Multiplying by x^n and summing over $n \geq 0$, we see that it is in turn equivalent to:

$$(1 - cx^m)\overline{f}(x) = \overline{g}(x/q).$$

So we get a *necessary* condition for f to be convergent: that $\overline{g}$ vanish at all the $1/qc_i$, $i = 1, \ldots, m$, where the c_i are the mth roots of c. Concluding the argument shows that the image of $\sigma_q - cx^m$ is the kernel of the linear map $g \mapsto (g(1/qc_1), \ldots, g(1/qc_m))$ from $\mathbf{C}\{x\}$ to $\mathbf{C}^m$. This map being onto, the image has codimension m.

EXERCISE 6.15. (i) Fill in details (surjectiveness, convergence ...).
(ii) Express the above calculation through a commutative diagram and state its formal properties.

6.1.3.4. We make more precise the decay property stated in the penultimate exercise above. Let $n = am + b$ with $a \in \mathbf{N}$ and $0 \leq b \leq m - 1$. Then, setting $Q := q^m$ and writing provisionnaly $T_a^{(b)} := t_{am+b}$, we have:

$$t_n = q^n t_{n-m} \Longrightarrow T_a^{(b)} = q^b Q^a T_{a-1}^{(b)}.$$

This entails that, for a fixed b, the family of $T_a^{(b)}$ has the same order of decay that the family of the $Q^{a^2/2}$. Replacing a by $(n-b)/m \sim n/m$, we get:

$$t_n \approx q^{sn^2/2}, \text{ where } s := 1/m.$$

So we see as in 6.1.3.2 that either the coefficients f_n of the solution f of $Lf = g$ are tame (exceptional case, valid on a subspace of codimension m), or they have the order of growth of $q^{-sn^2/2}$: they form a "q-Gevrey sequence of order s" (see sections 6.2 and 6.3). Note how the q-Gevrey level $s = 1/m$ is related to the slope m.

EXERCISE 6.16. Let $T := \sum t_n x^n$. Show that the recursive relation characterizing the t_n is equivalent to $\sigma_q(x^m T) = T$ and explain how one can produce such families (t_n) using Theta functions.

6.2. Index theorems and the growth of coefficients of divergent formal solutions (Bézivin)

This section is mostly devoted to the results of J.-P. Bézivin [**20**].

6.2.1. Index theory for q-difference equations.

6.2.1.1. As explained at the beginning of this chapter, index theory turned out an efficient tool in the theory of complex differential equations. Bézivin started its adaptation to q-difference equations theory. He actually considered more general functional equations (following e.g. [**143**]) of the form $\sum a_n \sigma^n f = g$, where the "difference operator" σ has the form $f \mapsto f \circ \psi$, where $\psi := qx + \cdots$, $q \neq 0$, is a power series (formal and convergent cases may be considered). Actually, conjugating ψ to qx through a so-called "Schröder series" (*loc. cit.*) allows, under rather general conditions, to reduce the study to the case where $\sigma := \sigma_q$, i.e., to the case of q-difference equations (we already spoke of that linearisation process in 1.12.1). This is not litterally true since the left-hand side in the equation above is allowed to involve infinitely many terms.

For technical reasons related to the use of a "q-Borel-Ramis transformation", even starting with a plain q-difference equation involving a finite number of terms, one may end with a generalized one as above. Since such operators and equations will appear in this section only, We shall give them a temporary name: GBQDO for "generalized Bézivin q-difference operator". They are even more general since, dealing with irregular equations and possibly nonintegral slopes, we shall use non-integral powers of σ_q; for any $k \in \mathbf{R}$ we define, in the obvious way:

$$\sigma_q^k : f(x) \mapsto f(q^k x), \text{ where, as usual: } q^k := e^{2\mathrm{i}\pi\tau k}.$$

The precise definition of GBQDOs is given at the beginning of 6.2.2.

6.2.1.2. This section and the following one will rely on some standard functional analysis in Banach spaces, and also on some slightly more sophisticated facts about direct and inverse limits of such spaces. They are summarized in Appendix D. Following Bézivin, we shall denote, for any $R > 0$:

$$C(R) := \mathscr{O}(\mathring{\mathrm{D}}(0,R)) \cap \mathscr{C}(\overline{\mathrm{D}}(0,R), \mathbf{C}),$$

the algebra of complex continuous functions on $\overline{\mathrm{D}}(0,R)$ which are holomorphic in $\mathring{\mathrm{D}}(0,R)$. There are more customary notations for this space in analysis, but anyway those spaces will only appear in this section. We endow $C(R)$ with the uniform norm:

$$\forall f \in \mathbf{C}(R)\,,\ \|f\|_R := \sup_{x \in \overline{\mathrm{D}}(0,R)} |f(x)| = \max_{x \in \overline{\mathrm{D}}(0,R)} |f(x)| = \max_{x \in \partial \mathrm{D}(0,R)} |f(x)|\,.$$

(Last equality follows from the maximum principle for analytic functions [**211**].) Recall from complex analysis that $C(R)$ with that norm is a Banach space [**212**]. Clearly:

$$\mathbf{C}\{x\} = \bigcup_{R>0} C(R) \text{ and } \mathscr{O}(\mathbf{C}) = \bigcap_{R>0} C(R).$$

One can also write this as:

$$\mathbf{C}\{x\} = \lim_{\longrightarrow R \searrow 0} C(R) \text{ and } \mathscr{O}(\mathbf{C}) = \lim_{\longleftarrow R \nearrow \infty} C(R),$$

that is a direct, resp. an inverse limit; and endow $\mathbf{C}\{x\}$, $\mathscr{O}(\mathbf{C})$ with the corresponding topologies. This will be further explained in the following and in the appendix.

6.2.1.3. In 6.2.2 we shall derive index theorems for GBQDOs acting on $\mathbf{C}[[x]]$, $\mathbf{C}\{x\}$ and $\mathscr{O}(\mathbf{C})$. For $\mathbf{C}[[x]]$, this will be done algebraically. For $\mathbf{C}\{x\}$ and $\mathscr{O}(\mathbf{C})$, this will rest on auxiliary theorems for $C(R)$ and going to the limit.

In 6.2.3, we shall transport those theorems to "q-Gevrey spaces" using q-Borel-Ramis transformation. Then we shall apply these results to obtain information on the growth of coefficients of divergent power series solutions of usual q-difference equations $Lf = g$.

6.2.2. Index theorems in $\mathbf{C}[[x]]$, $C(R)$, $\mathbf{C}\{x\}$ and $\mathscr{O}(\mathbf{C})$.

6.2.2.1. We consider, following **[20]**, a "generalized q-difference operator", or GBQDO, defined by an infinite series:

$$L := \sum_{i\geq 0} a_i \sigma_q^{k_i}, \text{ where } \forall i \geq 0 \ , \ a_i \in \mathbf{C}[[x]] \text{ and } k_i \in \mathbf{R}.$$

To have L operate on $\mathbf{C}[[x]]$, we shall moreover assume:

$$\lim_{i\to\infty} v(a_i) = +\infty. \tag{6.3}$$

In other words, for all $N \geq 0$, the set $\{i \in \mathbf{N} \mid v(a_i) \leq N\}$ is finite. We shall moreover denote:

$$v(L) := \min_{i\geq 0} v(a_i).$$

If $L \neq 0$, $v(L) \in \mathbf{N}$; and $v(0) = +\infty$.

Proposition 6.17. *Such a GBQDO defines a* $\mathbf{C}$*-linear endomorphism of* $\mathbf{C}[[x]]$.

Proof. We must prove that $Lf \in \mathbf{C}[[x]]$ is well defined whenever $f \in \mathbf{C}[[x]]$ (then $\mathbf{C}$-linearity is obvious). Formally, this means that the series $\sum_{i\geq 0} a_i \sigma_q^{k_i} f$ converges in the x-adic topology. In more down-to-earth terms, we are to check that each term in x^m in Lf receives finitely many contributions from the infinitely many terms $a_i \sigma_q^{k_i} f$. But only those $a_i \sigma_q^{k_i} f$ with $v(a_i) \leq m$ can contribute; and any particular such term is just the usual product of two power series, so it involves finitely terms contributing to x^m. □

We shall have use for more precise information. So let $f := \sum f_k x^k$ and, for all i, let $a_i := \sum a_{i,j} x^j$. Then:

$$\sum_{i\geq 0} a_i \sigma_q^{k_i} f = \sum_{i,j,k\geq 0} a_{i,j} x^j q^{k_i k} f_k x^k = \sum_{n\geq 0} \left(\sum_{\substack{i,j,k\geq 0 \\ j+k=n}} a_{i,j} q^{k_i k} f_k \right) x^n.$$

Note that each coefficient $\sum a_{i,j} q^{k_i k} f_k$ is actually a finite sum. Indeed:

$$\begin{aligned}\sum_{\substack{i,j,k\geq 0 \\ j+k=n}} a_{i,j} q^{k_i k} f_k &= \left(\sum_{i\geq 0} a_{i,0} q^{n k_i} \right) f_n + \cdots + \left(\sum_{i\geq 0} a_{i,n} q^{0 k_i} \right) f_0 \\ &= P_0(q^n) f_n + \cdots + P_n(q^0) f_0,\end{aligned}$$

where we define the "polynomials":

$$P_j(s) := \sum_{i \geq 0} a_{i,j} s^{k_i}.$$

Thanks to condition (6.3), such an L_j has but a finite number of terms; however, it has nonintegral exponents! We summarize our result:

LEMMA 6.18. *If f has expansion $\sum f_k x^k$, then $g := L(f)$ has expansion $\sum g_n x^n$, where:*

$$\forall n \geq 0\,,\ g_n = P_0(q^n) f_n + \cdots + P_n(q^0) f_0 = \sum_{i+j=n} P_j(q^i) f_i. \tag{6.4}$$

COROLLARY 6.19. *If no q^n, $n \geq 0$, is a root of P_0, then L is bijective from $\mathbf{C}[[x]]$ to itself.*

PROOF. The equation $Lf = g$ is equivalent to the sequence of recursive relations $P_0(q^n) f_n + \cdots + P_n(q^0) f_0 = g_n$ which is uniquely solved by taking:

$$f_n := \frac{g_n - \left(P_1(q^{n-1}) f_{n-1} + \cdots + P_n(q^0) f_0\right)}{P_0(q^n)}.$$

□

Note the similarity of the above with the calculations in 5.2.1. The only difference is that the "polynomials" P_j here have nonintegral coefficients.

6.2.2.2. We now assume more about the GBQDO L:

$$\text{The sequence } (k_i) \text{ is strictly increasing and } \lim_{i \to \infty} k_i = +\infty. \tag{6.5}$$

PROPOSITION 6.20. *The operator L acting on $\mathbf{C}[[x]]$ has index:*

$$\chi(L, \mathbf{C}[[x]]) = -v(L).$$

PROOF. We know that the operator x (which maps any f to xf) acting on $\mathbf{C}[[x]]$ has index $\chi(L, \mathbf{C}[[x]]) = -1$ (it has trivial kernel and its image $\mathrm{Im}(x, \mathbf{C}[[x]]) = x\mathbf{C}[[x]]$ has codimension 1 because $\mathbf{C}[[x]] = \mathbf{C} \oplus x\mathbf{C}[[x]]$). So we write $L = x^m L'$ where $m := v(L)$ and $v(L') = 0$, and, by the additivity of the index (Appendix D):

$$\chi(L, \mathbf{C}[[x]]) + v(L) = \chi(L', \mathbf{C}[[x]]) - m + m = \chi(L', \mathbf{C}[[x]]),$$

so it is enough to prove that $\chi(L', \mathbf{C}[[x]]) = 0$ knowing that $v(L') = 0$, and we may as well assume from start that $v(L) = 0$.

We introduce $L^{(N)} := x^{-N} L x^N$, also a GBQDO:

$$L^{(N)} = \sum_{i \geq 0} a_i x^{-N} \sigma_q^{k_i} x^N = \sum_{i \geq 0} a_i q^{N k_i} \sigma_q^{k_i}.$$

From the commutative diagram:

$$\begin{array}{ccc}
\mathbf{C}[[x]] & \xrightarrow{\ L^{(N)}\ } & \mathbf{C}[[x]] \\
{\scriptstyle x^N} \big\downarrow & & \big\downarrow {\scriptstyle x^N} \\
\mathbf{C}[[x]] & \xrightarrow[\ L\ ]{} & \mathbf{C}[[x]]
\end{array}$$

we conclude, by additivity of the index, that $\chi(L, \mathbf{C}[[x]]) = \chi(L^{(N)}, \mathbf{C}[[x]])$ (meaning implicitly that one has index if, and only if, the other has). Now, with obvious adaptation of the notations of 6.2.2.1:

$$P_0^{(N)}(s) = \sum_{i \geq 0} q^{Nk_i} a_{i,0} s^{k_i} = P_0(q^N s).$$

Now condition (6.5), together with the fact that there is at least one $a_{i,0} \neq 0$ (since $v(L) = 0$), implies that P_0 does not vanish for small enoug s:

$$\exists \epsilon > 0 \ : \ \forall s \ , \ |s| < \epsilon \Rightarrow P_0(s) \neq 0.$$

We take N such that $|q^N| < \epsilon$ (possible since $|q| < 1$). From the equality above, we get:

$$\forall s \ , \ |s| < 1 \Rightarrow P_0^{(N)}(s) \neq 0.$$

In particular, no q^n is a root of $P_0^{(N)}$. From Corollary 6.19 we then conclude that $L^{(N)}$ acts bijectively on $\mathbf{C}[[x]]$, whence $\chi(L^{(N)}, \mathbf{C}[[x]]) = 0$. □

6.2.2.3. In order to compute the index of L in $C(R)$, $\mathbf{C}\{x\}$ and $\mathscr{O}(\mathbf{C})$, we formulate new conditions. They will not all be required all the time, so we give them names and shall state explicitly for each proposition which conditions are required. Notations for L, the a_i and the k_i are the same as before.

- H1(R): $R > 0$, all the $a_i \in C(R)$ and $\sum \|a_i\|_R < \infty$.
- H1: H1(R) holds for some $R > 0$.
- H2: H1(R) holds for every $R > 0$.

These conditions will respectively serve for the study in $C(R)$, in $\mathbf{C}\{x\}$ and in $\mathscr{O}(\mathbf{C})$.

6.2.2.4.

Proposition 6.21. *Under condition H1(R) and assuming moreover that $k_0 = 0$, the GBQDO L acts continuously on $C(R)$.*

Proof. Let $f \in C(R)$ and let $M := \|f\|_R$. Since $k_0 = 0$ implies that all $k_i \geq 0$, one has $q^{k_i}\overline{\mathrm{D}}(0, R) \subset \overline{\mathrm{D}}(0, R)$ for all i, implying that each $\sigma_q^{k_i} f \in C(R)$ and that $\|\sigma_q^{k_i} f\|_R \leq M$. Therefore Lf is well defined (absolutely convergent series) and $\|Lf\|_R \leq M \sum \|a_i\|_R$. Linearity being obvious, L is a continuous linear operator of subordinate norm $\leq \sum \|a_i\|_R$. □

Let $a \in C(R)$ vanishing nowhere on the boundary $\partial\mathrm{D}(0, R)$. Then a has a finite number of zeroes $x_1, \ldots, x_k$ in $\mathring{\mathrm{D}}(0, R)$: indeed, $\overline{\mathrm{D}}(0, R)$ is compact and the zeroes of a nontrivial analytic function cannot accumulate. Let $m_1, \ldots, m_k$ the multiplicities of those zeroes and $m := m_1 + \cdots + m_k$.

Lemma 6.22. *The operator $a : f \mapsto af$ on $C(R)$ has index $\chi(a, C(R)) = -m$.*

Proof. Obviously, the kernel is trivial. The image is made of functions vanishing at each x_i with multiplicity at least m_i: one inclusion is obvious and conversely, each such function divided by a clearly yields an element of $C(R)$. The image is therefore the kernel of the map $C(R) \to \mathbf{C}^m$ sending $f \in C(R)$ to $\left(f^{(j)}(x_i)\right)_{\substack{i=1,\ldots,k \\ j=0,\ldots,m_i-1}}$. This linear map is surjective (it already is when restricted to polynomials) so the $\mathrm{Im}(a, C(R))$ has codimension m. □

EXERCISE 6.23. Prove that polynomials of degree $\leq m-1$ form a supplementary space to $\operatorname{Im}(a, C(R))$ in $C(R)$.

LEMMA 6.24. *For every $k > 0$, the operator $L := \sigma_q^k$ from $C(R)$ to itself is compact.*

PROOF. Let $R' := R\,|q|^{-k} > R$. Then L is the compositum of the isomorphism $\sigma_q^k : C(R) \to C(R')$ followed by the restriction map $C(R') \to C(R)$. The latter map is compact (Appendix D). □

PROPOSITION 6.25. *Let $L := \sum a_i \sigma_q^{k_i}$ a GBQDO, with the same notations as before, satisfying H1(R) and such that moreover $k_0 = 0$. Suppose that a_0 vanishes nowhere on $\partial D(0, R)$ and let m the number of its zeroes in $\overset{\circ}{D}(0, R)$ (counted with multiplicity). Then L acting on $C(R)$ has index $\chi(L, C(R)) = -m$.*

PROOF. Write $L = L_0 + L_1$, $L_0 := a_0$, and write $L_1 = \sigma_q^{k_1} \circ L_1'$. Then L_1' is a GBQDO satisfying H1(R), so it is continuous from $C(R)$ to itself. Therefore, since $\sigma_q^{k_1}$ is compact by the above lemma, L_1 is compact. By Appendix D, L has the same index as L_0 and the latter is $-m$ by the penultimate lemma. □

6.2.2.5. Let now the GBQDO L satisfy H1. We shall use Appendix D.2.2.

PROPOSITION 6.26. *The GBQDO L defines a continuous operator from $\mathbf{C}\{x\}$ to itself, with index $\chi(L, \mathbf{C}\{x\}) = -v(a_0)$.*

PROOF. First note that, for every $k > 0$, the operator σ_q^k from $C(R)$ to itself is continuous, so the $(\sigma_q^k)_{|C(R)}$ for $R > 0$ form an inductive family of continuous maps and their direct limit $(\sigma_q^k)_{|\mathbf{C}\{x\}}$ is therefore continuous; moreover it is bijective, so it has null index. Writing $L := \sigma_q^{k_0} \circ L'$, we are reduced to prove that $L' = a_0' + \cdots$ is continuous on $\mathbf{C}\{x\}$ and that its index is $-v(a_0') = -v(a_0)$ (this is because $a_0' = \sigma_q^{-k_0} a_0 \sigma_q^{k_0} = \sigma_q^{-k_0}(a_0)$). In other words, we may as well assume from the beginning that $k_0 = 0$.

Let $R' > 0$ such that H1(R') holds. and $R'' > 0$ such that a_0 does not vanish on $\overset{\circ}{D}(0, R'')$, except maybe at 0 ($L$ is of course supposed nontrivial; and a_0 is supposed nontrivial, otherwise we could reindex the terms: so the zeroes of a_0 cannot accumulate). Let $R_0 < \min(R', R'')$, $R > 0$. Then 6.2.2.4 applies for all $R < R_0$and the restriction L_R of L to $C(R)$ has index $-v(a_0)$. The lemma in D.2.2 then allows us to conclude. □

6.2.2.6. Now suppose that L satisfies H2. We use D.2.2 again.

PROPOSITION 6.27. (i) *The GBQDO L induces a continuous map from $\mathcal{O}(\mathbf{C})$ to itself.*

(ii) *Assume that a_0 has a finite number N of zeroes in $\mathbf{C}$ (counted with multiplicities). Then L has index $\chi(L, \mathcal{O}(\mathbf{C})) = -N$.*

PROOF. As in 6.2.2.5 we may assume that $k_0 = 0$. The proof of (i) is straightforward.

Let $R_0 > 0$ such that all zeroes of a_0 lie in $Do(0, R_0)$. Then for all $R > R_0$ the restriction L_R of L to $C(R)$ is continuous and has index $-N$ by 6.2.2.4, so D.2.2 allows us to conclude. □

COROLLARY 6.28. *If a_0 is a polynomial, $\chi(L, \mathcal{O}(\mathbf{C})) = -\deg a_0$.*

6.2.2.7. Here we suppose that L satisfies H1. We intend to investigate "how L distinguishes formal from convergent solutions". More precisely, L acts on $\mathbf{C}[[x]]$ and on $\mathbf{C}\{x\}$, so it acts on their quotient $\mathbf{C}[[x]]/\mathbf{C}\{x\}$: we want information on that action.

LEMMA 6.29. *There is an N such that Im $L \supset x^N\mathbf{C}[[x]]$.*

PROOF. Write $L = x^m L'$ with $m := v(L)$, so that $v(L') = 0$. If we prove that Im $L' \supset x^{N'}\mathbf{C}[[x]]$ for some N', then $N := m + N'$ will do. So we may as well assume from start that $v(L) = 0$.

In the course of the proof the proposition in 6.2.2.2, we saw that there is an N such that $x^{-N}Lx^N$ is bijective, so $x^{-N}Lx^N\mathbf{C}[[x]] = \mathbf{C}[[x]]$, so $Lx^N\mathbf{C}[[x]] = x^N\mathbf{C}[[x]]$, and since the left-hand side $Lx^N\mathbf{C}[[x]]$ contains Im L we are done. □

PROPOSITION 6.30. *The GBQDO L acts surjectively on $\mathbf{C}[[x]]/\mathbf{C}\{x\}$ and its kernel has dimension $v(a_0) - v(L)$:*

$$\mathit{Coker}(L, \mathbf{C}[[x]]/\mathbf{C}\{x\}) = \{0\} \text{ and } \dim \mathit{Ker}(L, \mathbf{C}[[x]]/\mathbf{C}\{x\}) = v(a_0) - v(L).$$

PROOF. The first statement says that every $g \in \mathbf{C}[[x]]$ can be written as $Lf+g'$, where $f \in \mathbf{C}[[x]]$ and $g' \in \mathbf{C}\{x\}$. Using the lemma, it is enough to take for g' the polynomial of degree $< N$ obtained by truncation T_N (see 0.4) at order N (so obviously $g' \in \mathbf{C}\{x\}$); and then $g - g' \in x^N\mathbf{C}[[x]]$ so $g - g' \in L\mathbf{C}[[x]]$ by the lemma.

Now we use the following commutative diagram:

$$\begin{array}{ccccccccc} 0 & \longrightarrow & \mathbf{C}\{x\} & \longrightarrow & \mathbf{C}[[x]] & \longrightarrow & \mathbf{C}[[x]]/\mathbf{C}\{x\} & \longrightarrow & 0 \\ & & \downarrow{\scriptstyle L} & & \downarrow{\scriptstyle L} & & \downarrow{\scriptstyle L} & & \\ 0 & \longrightarrow & \mathbf{C}\{x\} & \longrightarrow & \mathbf{C}[[x]] & \longrightarrow & \mathbf{C}[[x]]/\mathbf{C}\{x\} & \longrightarrow & 0. \end{array}$$

From Appendix D.1.3, Prop., (i), we conclude that:

$$\chi(L, \mathbf{C}[[x]]/\mathbf{C}\{x\}) = \chi(L, \mathbf{C}[[x]]) - \chi(L, \mathbf{C}\{x\}) = v(a_0) - v(L).$$

Together with the triviality of the cokernel (proved above), this allows us to conclude. □

COROLLARY 6.31. *Let L as before satisfying H1. Then there is a logical equivalence:*

$$\{\forall f \in \mathbf{C}[[x]]\ ,\ Lf \in \mathbf{C}\{x\} \Rightarrow f \in \mathbf{C}\{x\}\} \Longleftrightarrow \{\forall i \geq 0\ ,\ v(a_i) \geq v(a_0)\}$$

PROOF. The left-hand side statement says that L acts injectively on $\mathbf{C}[[x]]/\mathbf{C}\{x\}$, i.e., that $\mathrm{Ker}(L, \mathbf{C}[[x]]/\mathbf{C}\{x\}) = 0$. The right-hand side statement says that $v(a_0) = \min_{i\geq 0} v(a_i) = v(L)$, i.e., that $v(a_0) - v(L) = 0$. □

DEFINITION 6.32. The *irregularity of L* is the number:

$$\mathrm{Irr}(L) := \chi(L, \mathbf{C}[[x]]/\mathbf{C}\{x\}).$$

The motivation for this name is that "regular q-difference operators" (actually singular regular ones, to be studied in Chapter 11) "see no difference between formal and analytic". We shall relate $\mathrm{Irr}(L) = v(a_0) - v(L)$ to the Newton polygon further below in 6.4.2.3 and explain there in what sense it measures the defect of regularity.

6.2.2.8. Here we suppose that L satisfies H2, and also that a_0 has a finite number of zeroes in $\mathbf{C}$ (counted with multiplicities). Since, when a_0 is a polynomial, this number is $\deg a_0$ we (somewhat improperly) write it $\deg a_0$ even if a_0 is not. So, by 6.2.2.6, we have: $\chi(L, \mathcal{O}(\mathbf{C})) = -\deg a_0$.

PROPOSITION 6.33. *The GBQDO L acts surjectively on* $\mathbf{C}[[x]]/\mathcal{O}(\mathbf{C})$ *and its kernel has dimension* $\deg a_0 - v(L)$*:*

$$\textit{Coker}(L, \mathbf{C}[[x]]/\mathcal{O}(\mathbf{C})) = \{0\} \textit{ and } \dim \textit{Ker}(L, \mathbf{C}[[x]]/\mathcal{O}(\mathbf{C})) = \deg a_0 - v(L).$$

PROOF. The proof is almost identical to that of the proposition in 6.2.2.7. □

COROLLARY 6.34. *Let L as before satisfying H2. Then there is a logical equivalence:*

$$\{\forall f \in \mathbf{C}[[x]] \; , \; Lf \in \mathcal{O}(\mathbf{C}) \Rightarrow f\mathcal{O}(\mathbf{C}) \in\} \Longleftrightarrow \left\{ \begin{array}{l} \forall i \geq 0 \; , \; v(a_i) \geq v(a_0) \\ a_0 \textit{ has no zeroes in } \mathbf{C}^* \end{array} \right\}$$

PROOF. The left-hand side statement says that L acts injectively on

$$\mathbf{C}[[x]]/\mathcal{O}(\mathbf{C}).$$

This is equivalent to $\deg a_0 - v(L) = 0$. Since $\deg a_0 - v(L) = (\deg a_0 - v(a_0)) + (v(a_0) - v(L))$, a sum of nonnegative numbers, its vanishing is equivalent to $\deg a_0 = v(a_0)$ and $v(a_0) = v(L)$; the latter equality has already been commented. The former says thay a_0 has no zeroes in $\mathbf{C}^*$. □

EXERCISE 6.35. Check the very last statement. (Recall that $\deg a_0$ has an extended definition here.)

6.2.3. Index theorems in q-Gevrey spaces. We saw in 6.2.1 that, even for equations with analytic coefficients, divergent formal power series solution naturally arise with a peculiar type of growth of the coefficients. Before giving a precise definition, recall that we introduced in Subsubsection 6.1.3.2 a notation for sequences having "the same order of growth *up to a geometric factor*":

$$(a_n) \approx (b_n) \underset{def}{\Longleftrightarrow} \exists A, B > 0 \; : \; b_n = O(a_n A^n) \text{ and } a_n = O(b_n B^n).$$

6.2.3.1. *q-Gevrey spaces.* Let $s > 0$. Following Bézivin[2] we introduce the power series:

$$\eta_{q,s} := \sum_{n \geq 0} q^{sn(n+1)/2} \text{ and } \eta^*_{q,s} := \sum_{n \geq 0} q^{-sn(n+1)/2}.$$

Obviously, the former defines an entire function while the latter is wildly divergent. Recall the *Hadamard product* of two power series is defined as:

$$\left(\sum a_n x^n\right) \square \left(\sum b_n x^n\right) := \sum a_n b_n x^n.$$

It is associative and has neutral element the series $\sum x^n$. Moreover, the series $\eta_{q,s}$ and $\eta^*_{q,s}$ above are inverse to each other for this product.

DEFINITION 6.36 (q-Gevrey spaces). We say that $f \in \mathbf{C}[[x]]$ has *q-Gevrey order* $\leq s$ if $f \square \eta_{q,s} \in \mathbf{C}\{x\}$. The space of series having q-Gevrey order $\leq s$ is written $\mathbf{C}[[x]]_{q,s}$.

We say that $f \in \mathbf{C}[[x]]$ has *q-Gevrey-Beurling order* $\leq s$ if $f \square \eta_{q,s} \in \mathcal{O}(\mathbf{C})$. The space of series having q-Gevrey-Beurling order $\leq s$ is written $\mathbf{C}[[x]]_{q,(s)}$.

[2]There is a small sign mistake in *loc.cit.* when defining $\eta^*_{q,s}$.

In either case, we speak of *exact order* s if the order is $\leq s$ but is not $\leq s'$ for any $s' < s$.

In other words:

$$\mathbf{C}[[x]]_{q,s} = \{\sum a_n x^n \in \mathbf{C}[[x]] \mid \exists A > 0 \ : \ a_n = O(q^{-sn(n+1)/2} A^n)\},$$
$$\mathbf{C}[[x]]_{q,(s)} = \{\sum a_n x^n \in \mathbf{C}[[x]] \mid \forall A > 0 \ , \ a_n = O(q^{-sn(n+1)/2} A^n)\}.$$

The obvious extension of those definitions to the case $s = 0$ immediately yields $\eta_{q,0} = \eta^*_{q,0} = \sum x^n$ and:

$$\mathbf{C}[[x]]_{q,0} = \mathbf{C}\{x\} \text{ and } \mathbf{C}[[x]]_{q,(0)} = \mathscr{O}(\mathbf{C}).$$

(Note that this will be further extended to the case $s < 0$ in Section 6.3, with much stronger consequences.)

Here are some easy consequences of the above definitions; let $0 \leq s' < s$:

$$\mathbf{C}[[x]]_{q,s'} \subset \mathbf{C}[[x]]_{q,(s)} \subset \mathbf{C}[[x]]_{q,s}.$$

Obviously, $\mathbf{C}\{x\} \subset \bigcap_{s>0} \mathbf{C}[[x]]_{q,s}$ and $\bigcup_{s>0} \mathbf{C}[[x]]_{q,s} \subset \mathbf{C}[[x]]$.

EXERCISE 6.37. Prove that both inclusions are strict by producing adequate counterexamples. (So the "q-Gevrey filtration" is not exhaustive[3], be it downward or upward.)

6.2.3.2. *The q-Borel-Ramis transformation.* Let $s > 0$. We denote $\mathcal{B}_{q;s}$ the map $f \mapsto \eta_{q,s} \square f$ from $\mathbf{C}[[x]]$ to itself and call it[4] *q-Borel-Ramis transformation of order s*. It admits the map $f \mapsto \eta^*_{q,s} \square f$ as inverse, so it is bijective. It is clearly $\mathbf{C}$-linear. The definitions of the q-Gevrey and q-Gevrey-Beurling spaces above amount to say that $\mathcal{B}_{q;s}$ induces isomorphisms of vector spaces:

$$\mathcal{B}_{q;s} : \mathbf{C}[[x]]_{q,s} \simeq \mathbf{C}\{x\} \text{ and } \mathcal{B}_{q;s} : \mathbf{C}[[x]]_{q,(s)} \simeq \mathscr{O}(\mathbf{C}).$$

This allows us to endow $\mathbf{C}[[x]]_{q,s}$ and $\mathbf{C}[[x]]_{q,(s)}$ with the topologies respectively transported from $\mathbf{C}\{x\}$ and $\mathscr{O}(\mathbf{C})$ (thus a DFN and a FN topology, see the appendix), so that the above maps are actually isomorphisms of topological vector spaces.

For any GBQDO operator L, we shall denote $\tilde{L}$ its conjugates by $\mathcal{B}_{q;s}$:

$$\tilde{L} := \mathcal{B}_{q;s} \circ L \circ \mathcal{B}_{q;s}^{-1}.$$

So beware the notation does not explicitly mention the order s: hopefully, this will not be confusing. The plural on "conjugates" is motivated by the multiple possible commutative diagrams (depending on the assumptions made about L):

$$\begin{array}{ccc} \mathbf{C}[[x]] & \xrightarrow{L} & \mathbf{C}[[x]] \\ {\scriptstyle \mathcal{B}_{q;s}} \downarrow & & \downarrow {\scriptstyle \mathcal{B}_{q;s}} \\ \mathbf{C}[[x]] & \xrightarrow{\tilde{L}} & \mathbf{C}[[x]] \end{array} \qquad \begin{array}{ccc} \mathbf{C}[[x]]_{q,s} & \xrightarrow{L} & \mathbf{C}[[x]]_{q,s} \\ {\scriptstyle \mathcal{B}_{q;s}} \downarrow & & \downarrow {\scriptstyle \mathcal{B}_{q;s}} \\ \mathbf{C}\{x\} & \xrightarrow{\tilde{L}} & \mathbf{C}\{x\} \end{array} \qquad \begin{array}{ccc} \mathbf{C}[[x]]_{q,(s)} & \xrightarrow{L} & \mathbf{C}[[x]]_{q,(s)} \\ {\scriptstyle \mathcal{B}_{q;s}} \downarrow & & \downarrow {\scriptstyle \mathcal{B}_{q;s}} \\ \mathscr{O}(\mathbf{C}) & \xrightarrow{\tilde{L}} & \mathscr{O}(\mathbf{C}). \end{array}$$

[3]To state this more rigorously: the family of the $\mathbf{C}[[x]]_{q,s}/\mathbf{C}\{x\}$ provides an ascending and a descending filtration of $\mathbf{C}[[x]]/\mathbf{C}\{x\}$, neither of which is exhaustive.

[4]The relation to classical Borel transformation is detailed in Chapter 14, where the name q-Borel-Ramis is also explained.

For any of these diagrams, the vertical maps being isomorphisms of topological vector spaces, L is continuous if, and only if, $\tilde{L}$ is; and L has an index if, and only if, $\tilde{L}$ has, and then the index is the same.

EXERCISE 6.38. Why is $\mathcal{B}_{q;s} : \mathbf{C}[[x]] \to \mathbf{C}[[x]]$ an isomorphism of topological vector spaces?

6.2.3.3. We shall apply the conjugacy by q-Borel-Ramis transformation to a GBQDO $L := \sum a_i \sigma_q^{k_i}$; recall that $(k_i)_{i \geq 0}$ is a strictly increasing sequence with limit $+\infty$ and that the sequence $(v(a_i))_{i \geq 0}$ also has limit $+\infty$. As a guide to the computations that follow, we shall use notations such that $Lf = g$, $\tilde{f} := \mathcal{B}_{q;s} f$ and $\tilde{g} := \mathcal{B}_{q;s} g$, so that the conjugacy simply means that $\tilde{L}\tilde{f} = \tilde{g}$ and we have a commutative diagram:

$$\begin{array}{ccc} f & \overset{L}{\longmapsto} & g. \\ {\scriptstyle \mathcal{B}_{q;s}} \big\downarrow & & \big\downarrow {\scriptstyle \mathcal{B}_{q;s}} \\ \tilde{f} & \overset{\tilde{L}}{\longmapsto} & \tilde{g} \end{array}$$

LEMMA 6.39. *Let $L_{k,m} := x^m \sigma_q^k$ for some $m \in \mathbf{N}$ and $k \in \mathbf{R}$. Then:*

$$\tilde{L}_{k,m} = q^{sm(m+1)/2} x^m \sigma_q^{k+sm}.$$

PROOF. Let $f := x^n$, so that $\tilde{f} = q^{sn(n+1)/2} x^n$ and $g = x^m \sigma_q^k f = q^{kn} x^{m+n}$. Then:

$$\begin{aligned} \tilde{g} &= q^{kn+s(m+n)(m+n)/2} x^{m+n} = q^{sm(m+1)/2} x^m q^{kn+smn+sn(n+1)/2} x^n \\ &= q^{sm(m+1)/2} x^m \sigma_q^{k+sm} (q^{sn(n+1)/2} x^n) = q^{sm(m+1)/2} x^m \sigma_q^{k+sm} \tilde{f}. \end{aligned}$$

So $\tilde{L}_{k,m}$ and $q^{sm(m+1)/2} x^m \sigma_q^{k+sm}$ coincide on all elements of the form x^n. So by linearity, they coincide on all polynomials. A continuity argument then entails equality for all $f \in \mathbf{C}[[x]]$. □

EXERCISE 6.40. Explain rigorously the "continuity argument" necessary to go from polynomials to the whole of $\mathbf{C}[[x]]$. (Recall that the topology on $\mathbf{C}[[x]]$ is defined in a purely algebraic way.)

Formally, an arbitrary GBQDO L is an "infinite linear combination" of operators $L_{k,m}$ so we can argue as follows. Let $a_i = \sum a_{i,m} x^m$, so that $L = \sum a_i \sigma_q^{k_i} = \sum a_{i,m} L_{k_i,m}$ and:

$$\tilde{L} = \sum_{i,m \geq 0} a_{i,m} \tilde{L}_{k_i,m} = \sum_{i,m \geq 0} a_{i,m} q^{sm(m+1)/2} x^m \sigma_q^{k_i+sm}.$$

We must be careful manipulating this doubly indexed infinite sum. First note that, for any fixed $M > 0$, the inequality $k_i + sm \leq M$ implies $k_i \leq M$ and $m \leq M/s$, so only a finite number of pairs (i,m) can ensure it (recall that $\lim k_i = +\infty$). As a consequence the set $\{k_i + sm \mid a_{i,m} \neq 0\}$ of exponents of σ_q which really appear can be organized into a strictly increasing sequence $l_0 < l_1 < \cdots$ having limit $\lim l_i = +\infty$.

EXERCISE 6.41. The set E of really appearing exponents is a priori a denumerable subset of $\mathbf{R}$. Generally study under which condition such a set can be organized into a strictly increasing sequence as above. (Note for instance that the set $\mathbf{Q}$ could not.)

So with the above notation, we can write:

$$\tilde{L} = \sum_{j \geq 0} \tilde{a}_j \sigma_q^{l_j}, \text{ where: } \tilde{a}_j := \sum_{\substack{i,m \geq 0 \\ k_i + sm = l_j}} a_{i,m} q^{sm(m+1)/2} x^m.$$

For any fixed $K > 0$, the inequality $v(\tilde{a}_j) \leq K$ implies that there is some pair (i, m) such that $a_{i,m} \neq 0$, $m \leq k$ and $k_i + sm = l_j$. We leave it to the reader to check that this is only possible for a finite number of m, which implies that $v(\tilde{a}_j)$ tends to $+\infty$ as $j \to +\infty$. So $\tilde{L}$ is a well defined GBQDO.

Last but not least, note that, since $k_i + sm \leq l_j$ implies $m \leq (l_j - k_i)/s \leq (l_j - k_0)/s$, each $\tilde{a}_j$ is actually a polynomial of degree $\leq (l_j - k_0)/s$.

6.2.3.4. We shall need the following elementary fact from complex analysis: let $R > 0$ and $\phi(x) := \sum \alpha_n x^n \in C(R)$. Then $|\alpha_n R^n| \leq \|\phi\|_R$ for all $n \geq 0$. Indeed, by Cauchy residue formula:

$$\alpha_n = \frac{1}{2i\pi} \int_\Gamma \frac{\phi(x)}{x^{n+1}}\, dx,$$

where Γ is the circle $\partial \mathrm{D}(0, R)$, run through once counterclockwise. If $\phi \leq M$ on $\partial \mathrm{D}(0, R)$, for instance if $M := \|\phi\|_R$, this immediately implies that $|\alpha_n| \leq M R^{-n}$.

PROPOSITION 6.42. *If L satisfies H1, then $\tilde{L}$ satisfies H2.*

PROOF. Let $R > 0$ such that L satisfies H1(R), i.e., $\sum M_i < \infty$, where we write $M_i := \|a_i\|_R$ for all $i \geq 0$. From the "elementary fact" recalled above, and with the usual notations, $|a_{i,m} R^m| \leq M_i$ for all $m \geq 0$.

Now let $R' > 0$ arbitrary and let $|x| \leq x$. We look for an upper bound $|\tilde{a}_j(x)| \leq \tilde{M}_j$, independent of x and in such a way that $\sum \tilde{M}_j < \infty$. Using the above inequality along with thee definition of $\tilde{a}_j$ given in 6.2.3.3:

$$|\tilde{a}_j(x)| \leq \sum_{k_i + sm = l_j} |a_{i,m}| \, |q|^{sm(m+1)/2} R'^m \leq \sum_{k_i + sm = l_j} M_i R^{-m} \, |q|^{sm(m+1)/2} R'^m.$$

So we set $\tilde{M}_j := \sum_{k_i + sm = l_j} M_i R^{-m} \, |q|^{sm(m+1)/2} R'^m$, whence $\|\tilde{a}_j|_{R'} \leq \tilde{M}_j$, and:

$$\begin{aligned} \sum_{j \geq 0} \|\tilde{a}_j|_{R'} \leq \sum_{j \geq 0} \tilde{M}_j &= \sum_{i,j \geq 0} M_i R^{-m} \, |q|^{sm(m+1)/2} R'^m \\ &= \left(\sum_{i \geq 0} M_i \right) \left(\sum_{j \geq 0} |q|^{sm(m+1)/2} (R'/R)^m \right). \end{aligned}$$

The sum in the second factor of the last expression obviously converges since the exponent of $|q| < 1$ grows like m^2. □

6.2.3.5. Thanks to that proposition, we may apply the results of 6.2.2 to compute the index of $\tilde{L}$, whence that of L. By 6.2.2.5, $\tilde{L}$ acting on $\mathbf{C}\{x\}$ has index $-v(\tilde{a}_0)$, so we have:

$$\chi(L, \mathbf{C}[[x]]_{q,s}) = \chi(\tilde{L}, \mathbf{C}\{x\}) = -v(\tilde{a}_0).$$

We can also use 6.2.2.6 with $N := \deg \tilde{a}_0$ since $\tilde{a}_0$ is a polynomial:

$$\chi(L, \mathbf{C}[[x]]_{q,(s)}) = \chi(\tilde{L}, \mathscr{O}(\mathbf{C})) = -\deg \tilde{a}_0.$$

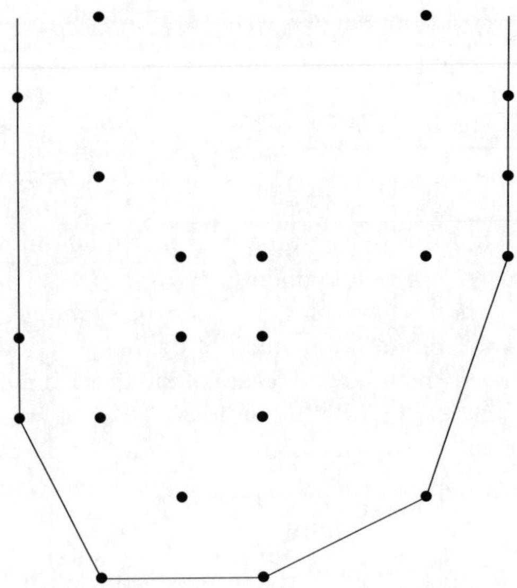

FIGURE 6.1. Clouds for GBQDOs

It remains to express $v(\tilde{a}_0)$ and $\deg \tilde{a}_0$ directly in terms of L. Recall that:

$$\tilde{a}_0 = \sum_{\substack{i,m \geq 0 \\ k_i + sm = l_0}} a_{i,m} q^{sm(m+1)/2} x^m,$$

and also that l_0 is the smallest possible value of $k_i + sm$ for all pairs (i, m) such that $a_{i,m} \neq 0$. This leads us to an adaptation of the Newton polygon to the present setting (that of GBQDOs).

The cloud of points associated to L is now the set of points (k_i, m) for $a_{i,m} \neq 0$. Therefore, conditions of the form $k_i + sm = l$ will be intepreted in the (k, m)-plane as meaning that the point (k_i, m) belongs to some line $k + sm = l$. In the "standard case" of true q-difference operators, we take $k_i := i$ and $a_0 \neq 0$, thus recovering the usual cloud of points used to define the Newton polygon in Chapter 5. The illustration below is drawn under this convention[5].

For any $s > 0$, the lines $k + sm = l$ $(l \in \mathbf{R})$ are parallel to each other, with the same slope $-1/s < 0$, so they look decreasing. If $l < l'$, the line $k + sm = l$ is left and down (or south-west) from the line $k + sm = l'$. So the line $k + sm = l_0$ is the one in the family that is leftmost while touching the cloud. In the example above:

- For $s < 1/2$, we get $l_0 = 3$ and the line $k + sm = l_0$ meets the cloud at point $(0, 3)$ only;
- For $s = 1/2$, we get $l_0 = 3/2$ and the line $k + sm = l_0$ meets the cloud at points $(0, 3)$ and $(1, 1)$.

So more generally we see that there are some exceptional values of s (here above: $s = 1/2$) such that the line $k + sm = l_0$ meets the cloud at more than one point. The corresponding points lie on the boundary of the convex hull of the

[5]The presentation of this construction in [**20**, **191**] is more formal, relying on inequalities and computations. We give here a more intuitive geometric picture; this can easily made rigorous by appealing to plane geometry, in particular properties of convex subsets of the plane.

cloud, more precisely on the left down (or south-west) part of the Newton polygon, where slopes are negative; and these slomes are the corresponding $\mu := -1/s$.

Let $s < 0$, with value exceptional or not. Recall that:

$$\tilde{a}_0 = \sum_{\substack{i,m \geq 0 \\ k_i + sm = l_0}} a_{i,m} q^{sm(m+1)/2} x^m.$$

The index of L in $\mathbf{C}[[x]]_{q,s}$, resp. in $\mathbf{C}[[x]]_{q,(s)}$ is related to $v(\tilde{a}_0)$, resp. to $\deg \tilde{a}_0$, that is, to the lowest, resp. to the highest value of m really present in the sum. These are the ordinates of the lowest, resp. the highest point of intersection of the line $k + sm = l_0$ with the cloud. For an exceptional value of s, these points are the extremities of a segment on the Newton polygon; for anonexceptional value, there is only one point, which is one of the vertices. So we introduce the following notation; for $s > 0$ and $\mu := -1/s$:

- $I^+(s)$ is the ordinate of the highest point of intersection of the line $k+sm = l_0$ with the cloud;
- $I^-(s)$ is the ordinate of the lowest point of intersection of the line $k+sm = l_0$ with the cloud.

We summarize the discussion.

PROPOSITION 6.43. *For any $s > 0$:*

$$\chi(L, \mathbf{C}[[x]]_{q,s}) = -I^-(s) \text{ and } \chi(L, \mathbf{C}[[x]]_{q,(s)}) = -I^+(s).$$

We note some more easy facts about those numbers.

(1) We have $I^+(s) \geq I^-(s)$; equality occurs if, and only if, $\mu := -1/s$ is one slope of the Newton polygon.
(2) Both $I^+(s)$ and $I^-(s)$ are decreasing functions of s, with jumps at the exceptional values.
(3) For s small enough, the common value of $I^+(s)$ and $I^-(s)$ is $v(a_0)$.
(4) For s big enough, the common value of $I^+(s)$ and $I^-(s)$ is $v(L)$.

EXERCISE 6.44. Study $I^+(s)$ and $I^-(s)$ in the case of the cloud shown in the figure above (Figure 6.1 on page 202).

6.2.3.6. Although the results of Bézivin are valid for his extended clas of operators (the ones we nicknamed GBQDOs), we shall state the most important ones for usual q-difference operators, which are the main theme of this book[6].

So let $L := a_0 + \cdots + a_n \sigma_q^n$, $a_0, \ldots, a_n \in \mathbf{C}\{x\}$, $a_0 a_n \neq 0$. Keeping the previous notations, we have here $k_i = i$ for all i and the cloud and Newton polygon are the usual ones from Chapter 5.

THEOREM 6.1. *Let $\{\mu_1, \ldots, \mu_k\} := S(L) \cap \mathbf{R}^*_-$ the negative slopes of L ordered as usual: $\mu_1 < \cdots < \mu_k < 0$. Let $s_i := -1/\mu_i$, so that $0 < s_1 < \cdots < s_k$. For any $s \in [0, +\infty]$ define $I^+(s)$ and $I^-(s)$ as the ordinates of the highest and the lowest point of intersection of $NP(L)$ with the contact line of slope $\mu := -1/s$. (For limit cases $s = 0$ and $s = +\infty$, see explanations in the proof.) Then:*

$$\chi(L, \mathbf{C}[[x]]_{q,s}) = -I^-(s) \text{ and } \chi(L, \mathbf{C}[[x]]_{q,(s)}) = -I^+(s).$$

[6]Note however that even if L is a "usual" q-difference operator, $\tilde{L}$ will in general not be, so the introduction of GBQDOs is by no means arbitrary.

PROOF. The "normal" cases $s \in]0, +\infty[$ have been detailed in 6.2.3.5. They can be extended to the limit values of s as follows.

First let $s = 0$. We then take the convention that $\mu = -\infty$ and that the contact line is the leftmost vertical line touching $NP(L)$. The points of intersection are the $(0, m)$ such that $a_{0,m} \neq 0$, so (draw yourself a picture to see why):

$$I^+(0) = \deg a_0 \text{ and } I^-(0) = v(a_0).$$

Also note that we took the convention (or the limiting application of the definition) that $\mathbf{C}[[x]]_{q,0} = \mathbf{C}\{x\}$ and $\mathbf{C}[[x]]_{q,(0)} = \mathcal{O}(\mathbf{C})$, so the values given in the theorem are indeed those found in 6.2.2.5 and 6.2.2.8.

As for the value $s = +\infty$, it is natural to consider it as representing all $s > s_k$ and this case has also been discussed in 6.2.3.5. The corresponding part of the Newton polygon is the lowest point or horizontal segment; this provides for the limit cases $\mathbf{C}[[x]]_{q,\infty} = \mathbf{C}[[x]]_{q,(\infty)} = \mathbf{C}[[x]]$ by 6.2.2.2. □

EXAMPLES 6.45. (i) Let $L := \sigma_q - cx$, $c \in \mathbf{C}^*$.

s	0	$]0,1[$	1	$]1,+\infty[$	$+\infty$
$I^-(s)$	1	1	1	0	0
$I^+(s)$	1	1	0	0	0

(ii) Let $L := x\sigma_q - 1$. There are no exceptional values (the unique slope is $+1$) and $I^+(s) = I^-(s) = 0$ for all $s \in [0, +\infty]$.

EXERCISE 6.46. Check all these values for the examples above and for Figure 6.1 on page 202.

6.2.3.7.

COROLLARY 6.47. *For all* $s \in [0, +\infty]$*:*

$$\chi(L, \mathbf{C}[[x]]_{q,s}/\mathbf{C}[[x]]_{q,(s)}) = I^-(s) - I^+(s).$$

So this index vanishes at all nonexceptional values of q.

PROOF. The proof is almost identical to that of the proposition in 6.2.2.7. □

The above argument can be extended to more general circumstances. Recall that, if $0 < s < t$, we have inclusion $\mathbf{C}[[x]]_{q,s} \subset \mathbf{C}[[x]]_{q,(t)}$. Then:

$$\chi(L, \mathbf{C}[[x]]_{q,(t)}/\mathbf{C}[[x]]_{q,s}) = I^-(s) - I^+(t).$$

If moreover $]s, t[$ contains no exceptional values, $I^-(s) = I^+(t)$ and the above index is 0. We may then go to the limits appealing to the lemma in Appendix D.2.3, with $E_1 = F_1 = \mathbf{C}[[x]]_{q,s}$ and $E_2 = F_2 = \mathbf{C}[[x]]_{q,(t)}$ (clearly all injections $\mathbf{C}[[x]]_{q,s} \subset \mathbf{C}[[x]]_{q,(t)}$ have dense images, since $\mathbf{C}[x]$ is dense in both source and target by Weierstraß's theorem). We conclude:

LEMMA 6.48. *If* $]s, t[\cap \{s_1, \ldots, s_k\} = \emptyset$, *then* $Ker(L, \mathbf{C}[[x]]_{q,(t)}/\mathbf{C}[[x]]_{q,s}) = \{0\}$.

PROOF. The quoted lemma says that the cokernel is trivial, and we just saw that the index is 0. □

6.2.3.8. *An application to nonhomogeneous equations.* Let $g \in \mathbf{C}\{x\}$ and let L a q-difference operator with coefficients in $\mathbf{C}\{x\}$. What can be said about the formal solutions of the equation $Lf = g$?

THEOREM 6.2. *Let $f \in \mathbf{C}[[x]]$ such that $Lf \in \mathbf{C}\{x\}$. Then, either $f \in \mathbf{C}\{x\}$, or there exixts $s > 0$ such that $f \in \mathbf{C}[[x]]_{q,s} \setminus \mathbf{C}[[x]]_{q,(s)}$, i.e., f has exact q-Gevrey order s.*

PROOF. Recall from the end of 6.2.3.5 that, for s big enough, $I^+(s) = I^-(s) = v(a_0)$ and that, for s small enough, $I^+(s) = I^-(s) = v(L)$. Therefore, for s big enough, $\chi(L, \mathbf{C}[[x]]_{q,s}) = \chi(L, \mathbf{C}[[x]])$; and, for s small enough, $\chi(L, \mathbf{C}[[x]]_{q,s}) = \chi(L, \mathbf{C}\{x\})$. This entails that, for s big enough, by the same argument as in 6.2.2.7, $\mathrm{Ker}(L, \mathbf{C}[[x]]/\mathbf{C}[[x]]_{q,s}) = 0$; for such an s, $f \in \mathbf{C}[[x]]_{q,s}$. If $f \in \mathbf{C}[[x]]_{q,(s)}$, we apply 6.2.3.7 and conclude that there is some $s' < s$ such that $f \in \mathbf{C}[[x]]_{q,s'} \setminus \mathbf{C}[[x]]_{q,(s')}$. If s' small enough, that means that $f \in \mathbf{C}\{x\}$. □

EXAMPLE 6.49. Taking $L := \sigma_q - cx$ and using the table at the end of 6.2.3.6, we recover the conclusion of 6.1.3.2.

6.3. Application to entire functions (Ramis)

If $s < 0$, the definitions 6.36 applied to a power series imply a very fast decay of the coefficients, whence give rise to entire functions (see 6.3.1). On the other hand, there is a duality between $\mathbf{C}[[x]]_{q,s}$ and $\mathbf{C}[[x]]_{q,-s}$ (6.3.2) arising from the general theory of topological vector spaces. In **[191]**, J.-P. Ramis used these facts together with Bézivin's theory to obtain q-analogues of more ancient results on ordinary differential equations. The q-analogy mainly stems on the fact that, for q-difference equations, the part of the factorial $n!$ is played (at least when $|q| > 1$, see the remark in 3.3.2.2) by such quantities as $q^{n(n+1)/2}$ and the like.

We shall here present some of Ramis's results. The whole section comes from **[191]**, but, since in this book, $0 < |q| < 1$, there are some changes of notations.

6.3.1. Growth of entire functions and asymptotic behaviour of their Taylor coefficients.

6.3.1.1. Let $s \in \mathbf{R}$. Then $f \in \mathbf{C}[[x]]$ is said to be q-Gevrey, resp. q-Gevrey-Beurling, of order s if:

$$\exists A, C > 0 \ : \ \forall n \geq 0 \, , \ |a_n| \leq C A^n \, |q|^{-sn(n+1)/2} \, ,$$

resp. if:

$$\forall A > 0 \, , \ \exists C > 0 \ : \ \forall n \geq 0 \, , \ |a_n| \leq C A^n \, |q|^{sn(n+1)/2} \, .$$

The space of q-Gevrey, resp. q-Gevrey-Beurling, series of order s is written $\mathbf{C}[[x]]_{q,s}$, resp. $\mathbf{C}[[x]]_{q,(s)}$. Elements of $\mathbf{C}[[x]]_{q,s} \setminus \bigcup\limits_{s'<s} \mathbf{C}[[x]]_{q,s'}$ are said to have exact q-Gevrey order s. When $s > 0$, these definitions coincide with definitions 6.36; and when $s = 0$, with their extension given in 6.2.3.1. The formal properties stated there keep their validity. In this section, we are interested in the case $s < 0$. Clearly, one then has $\mathbf{C}[[x]]_{q,s} \subset \mathscr{O}(\mathbf{C})$.

6.3.1.2. *Orders of growth.*

DEFINITION 6.50. Let $k > 0$. A function $f \in \mathcal{O}(\mathbf{C})$ is said to have *q-exponential growth of order k and of finite type* if:

$$\exists K > 0\,,\ \alpha \in \mathbf{R}\ :\ \forall x \in \mathbf{C}\,,\ |f(x)| \leq K\, |q|^{-\frac{k}{2}\left(\frac{\log|x|}{\log|q|}\right)^2} |x|^\alpha\,.$$

Of course, it is implicitly understood here that "order k" means "order at most k" rather than "exact order k".

LEMMA 6.51. *Let $f := \sum a_n x^n \in \mathcal{O}(\mathbf{C})$ and let $a > 0$.*

(i) *If $\exists C > 0\,,\ \alpha \in \mathbf{R}\ :\ \forall n \geq 0\,,\ |a_n| \leq Ce^{(n-\alpha)^2/2a}$, then $\exists C' > 0\,,\ \alpha' > \alpha$: $\forall x \in \mathbf{C}\,,\ |f(x)| \leq C' e^{\frac{a}{2}(\log|x|)^2} |x|^{\alpha'}$.*

(ii) *If $\forall x \in \mathbf{C}\,,\ |f(x)| \leq Ce^{\frac{a}{2}(\log|x|)^2} |x|^\alpha$, then $\forall n \geq 0\,,\ |a_n| \leq Ce^{-(n-\alpha)^2/2a}$.*

PROOF. (ii) We proved in 6.2.3.4 that if $|f| \leq M$ on $\overline{\mathrm{D}}(0, R)$, then all $|a_n R^n| \leq M$. Thus in our case:

$$|a_n| \leq Ce^{\frac{a}{2}(\log R)^2} R^{\alpha-n}.$$

The function $\phi(R) := e^{\frac{a}{2}(\log R)^2} R^{\alpha-n} = \exp\left(\frac{a}{2}(\log R)^2 + (\alpha - n)\log R\right)$ has a derivative ϕ' of the same sign as $a\log R+\alpha-n$; its minimum, reached at $R := e^{(n-\alpha)/a}$, has value $e^{-(n-\alpha)^2/2a}$. This entails the conclusion.

(i) Let $R := |x|$ and $b := \alpha + a\log R$. Then:

$$|a_n x^n| \leq Ce^{-(n-b)^2/2a} R^a e^{\frac{a}{2}(\log R)^2}.$$

So we want an upper bound for $\sum e^{-(n-b)^2/2a}$. Combining a comparison series-integral with decomposition of the domain as $\mathbf{R}_+ = [0, b] \cup [b, +\infty[$ implies that this sum is $O(b)$ and the conclusion flows.

□

EXERCISE 6.52. Check the swift computations in (i) and (ii).

THEOREM 6.3. *Let $f := \sum a_n x^n \in \mathcal{O}(\mathbf{C})$ and let $s < 0$; write $k := -1/s > 0$. Then the power series f has q-Gevrey order s if, and only if, the entire function f has q-exponential growth of order k.*

PROOF. We apply the lemma, noticing that, for $a > 0$:

- $|a_n| \leq Ce^{-(n-\alpha)^2/2a}$ means that the series has q-Gevrey order $s := 1/a\log|q| < 0$.
- $|f(x)| \leq Ce^{\frac{a}{2}(\log|x|)^2} |x|^\alpha$ means that the function has q-exponential growth of order $k := -a\log|q|$.

□

6.3.1.3. *An alternative characterization of q-exponential growth.* Various alternative characterizations of q-exponential growth are possible using special functions from the q-world. Ramis, in [**191**, Proposition 5.6] proposes one related to the q-exponential $\dfrac{1}{((1-q)x; q)_\infty}$, but requires q to be real. We shall present here another one from [**197**, 5.1.2 p 63] and more closely related to matters of asymptotics and summation; see Chapter 14.

Let $T_q(x) := \theta_{|q|}(|x|)$, which is continuous on $\mathbf{C}^*$ and vanishes nowhere (since the zeroes of $\theta_{|q|}$ are in $-|q|^{\mathbf{Z}}$). It also satisfies:

$$T_q(qx) = \theta_{|q|}(|q|\,|x|) = \frac{1}{|x|}\theta_{|q|}(|x|) = \frac{1}{|x|}T_q(x).$$

This is called $e(x;q)$ in **[197]** but this could cause confusion with other q-functions called e in this book and in the litterature.

LEMMA 6.53. *There are constants $A, B > 0$, $a, b \in \mathbf{R}$ such that:*

$$\forall x \in \mathbf{C}^* \;,\; A\,|x|^a\,|q|^{-\frac{1}{2}\left(\frac{\log|x|}{\log|q|}\right)^2} \leq T(x) \leq B\,|x|^b\,|q|^{-\frac{1}{2}\left(\frac{\log|x|}{\log|q|}\right)^2}.$$

PROOF. Let $C \subset \mathbf{C}^*$ the compact annulus $|q| \leq |x| \leq 1$ and $m, M > 0$ such that $m \leq T \leq T$ on C. Write $x = q^n\overline{x}$ with $n \in \mathbf{Z}$ and $\overline{x} \in C$, so that $|q|^{n+1} \leq |x| \leq |q|^n$. Writing $\lambda := \dfrac{\log|x|}{\log|q|}$, one therefore has $n \leq \lambda \leq n+1$.

From the functional relation $T_q(qx) = \dfrac{1}{|x|}T_q(x)$ we get by the usual induction that:

$$T_q(x) = T_q(q^n\overline{x}) = |q|^{-n(n-1)/2}\,|x|^{-n}\,T(\overline{x}),$$

whence:

$$m\,|q|^{-n(n-1)/2} \leq T(x) \leq M\,|q|^{-n}\,|q|^{-n(n-1)/2}.$$

Using the fact that $n \in [\lambda - 1, \lambda]$, the conclusion comes immediately. □

EXERCISE 6.54. Express A, B, a, b explicitly in function of $|q|$, m and M.

COROLLARY 6.55. *The entire function f has q-exponential growth of order k and of finite type if, and only if, there exists $C, c > 0$ such that:*

$$\forall x \in \mathbf{C}^* \;,\; |f(x)| \leq CT(cx)^k.$$

EXERCISE 6.56. Explain how the "external" factor $|x|^\alpha$ of the definition was replaced here by the "internal" factor c.

REMARK 6.57. One has clearly $\theta_q(x) \leq T_q(x)$. A reciprocal domination relation is valid away from the zeroes of θ_q; see **[265]** and Chapter 14.

6.3.2. Dualities.

6.3.2.1. As explained in Appendix D.2.3, there is a duality between the DFN space $E := \mathbf{C}\{x\}$ and the FN space[7] $F := \mathscr{O}(\mathbf{C})$. It comes from the mapping $E \times F \to \mathbf{C}$, written as $(f,g) \mapsto \langle f, g\rangle$ and defined as follows:

$$\langle \sum a_n x^n, \sum b_n x^n \rangle := \sum a_n b_n.$$

The isomorphisms $\mathbf{C}[[x]]_{q,s} \simeq \mathbf{C}\{x\}$ and $\mathbf{C}[[x]]_{q,(-s)} \simeq \mathscr{O}(\mathbf{C})$ induced by the q-Borel-Ramis transformations $f \mapsto \eta_{q,s}\Box f$ and $g \mapsto \eta^*_{q,s}\Box g$ allow us to transport the above into a pairing $\mathbf{C}[[x]]_{q,s} \times \mathbf{C}[[x]]_{q,(-s)} \to \mathbf{C}$; then latter is given by exactly the same formulas, since:

$$\sum a_n b_n = \sum \left(q^{sn(n+1)/2}a_n\right)\left(q^{-sn(n+1)/2}b_n\right).$$

[7]Instead of $\mathscr{O}(\mathbf{C})$, Ramis uses the FN space $x^{-1}\mathscr{O}(\mathbf{C}^* \cup \{\infty\})$, with elements $\sum a_n x^{-n-1}$. The present formulation is equivalent and suffices for our needs.

All that keeps sense and validity when extended to the cases $s = 0$ and $s = \infty$. The former gives back the pairing we started from; the latter, from the conventions in 6.2.3.6, gives a pairing of the FN space $\mathbf{C}[[x]]$ with the DFN space $\mathbf{C}[x]$.

By the usual mechanism $f \mapsto (g \mapsto \langle f, g\rangle)$ (and symmetrically), each such pairing $E \times F \to \mathbf{C}$ provides us with isomorphisms of topological vector spaces $E \simeq F'$ and $F \simeq E'$ (continuous duals). By transport of structures, we thus obtain, for every $s > 0$, isomorphisms:

$$\mathbf{C}[[x]]_{q,s} \simeq \left(\mathbf{C}[[x]]_{q,(-s)}\right)', \qquad \mathbf{C}[[x]]_{q,-s} \simeq \left(\mathbf{C}[[x]]_{q,(s)}\right)',$$
$$\mathbf{C}[[x]]_{q,(s)} \simeq \left(\mathbf{C}[[x]]_{q,-s}\right)', \qquad \mathbf{C}[[x]]_{q,(-s)} \simeq \left(\mathbf{C}[[x]]_{q,s}\right)'.$$

6.3.2.2. Again appealing to Appendix D.2.3, those dualities will be used to compute the index of a q-difference operator L acting on one of the spaces E above from the index of its adjoint L' acting on E':

$$\chi(L, E) = -\chi(L', E').$$

The left-hand side space E will be one of the spaces of entire functions defined by Ramis, the right-hand side E' will be identified (thanks to the above dualities) to one of the q-Gevrey spaces introduced by Bézivin. It is just a matter of relating L' to a q-difference operator.

EXERCISE 6.58. Before reading the sequel, find the adjoints of σ_q and of x acting for instance on $\mathbf{C}\{x\}$.

REMARK 6.59. There is a small mistake in Lemma 4.3 of **[191]**: the operator x on the space $x^{-1}\mathcal{O}(\mathbf{C}^* \cup \{\infty\})$ does not exist. The difficulty is rooted in the fact that Bézivin, and then Ramis, work with rings of series, negative exponents are not allowed. For the same reason, only q-difference operators with polynomial coefficients allow for complete result (see the use of *upper Newton polygon* by Ramis). The flaw can be easily fixed in various ways and all the results in *loc.cit* remain true. One way is to use the interpretation of the dual as a quotient, as is done a bit further in the paper of Ramis (see the remark at the end of 6.3.2.3). We shall choose a more down to earth description.

EXAMPLE 6.60. In this example we do not assume polynomial coefficients, but the exercise below will show you that this is not at the heart of the matter. We take $E := \mathbf{C}\{x\}$ and identify E' with $\mathcal{O}(\mathbf{C})$ in accordance with the formulas above. Let $L := \sigma_q - \dfrac{1}{1-x}\cdot$ We use notations $f = \sum f_n x^n \in E$, $g = \sum g_n x^n \in E'$, $\overline{f} := Lf$, $\overline{g} := L'g$, etc, so that we want $\langle f, \overline{g}\rangle = \langle \overline{f}, g\rangle$, i.e., $\sum f_n \overline{g}_n = \sum \overline{f}_n g_n$. Now, from the definition of L:

$$\overline{f}_n = q^n f_n - (f_0 + \cdots + f_n) \Longrightarrow \sum \overline{f}_n g_n = \sum q^n f_n g_n - \sum_{m \le n} f_m g_n$$
$$\Longrightarrow \sum f_n \overline{g}_n = \sum f_n \Big(q^n g_n - \sum_{m \ge n} g_m\Big)$$
$$\Longrightarrow \overline{g}_n = q^n g_n - \sum_{m \ge n} g_m,$$
$$\Longrightarrow \overline{g}(x) = g(qx) - \sum_{m \ge n} g_m x^n,$$

i.e., $\overline{g}(x) = g(qx) - \sum g_m \dfrac{1-x^{m+1}}{1-x}$, whence $L'g = \sigma_q g + \dfrac{x}{1-x} g - \dfrac{1}{1-x} g(1)$. So in the end $L' = L^* + L''$, where $L^* := \sigma_q + \dfrac{x}{1-x}$ is a q-difference operator (but it does not leave E' stable!) and the compensation term L'' acts as $g \mapsto -\dfrac{1}{1-x} g(1)$ (actually a compact map).

EXERCISE 6.61. Check that in the example $\chi(L,E) = \chi(L',E') = 0$. For the latter, use the splitting $E' = \mathbf{C} \oplus (1-x)E'$.

6.3.2.3. So we let $L := a_0 + \cdots + a_n \sigma_q^n$, $a_0, \ldots, a_n \in \mathbf{C}[x]$, $a_0 a_n \neq 0$. We write as usual $a_i = \sum a_{i,j} x^j$; also we set $N_0 := \max(\deg a_0, \ldots, \deg a_n) < \infty$. For any $s > 0$, we let V denote either the DFN space $\mathbf{C}[[x]]_{q,-s}$ or the FN space $\mathbf{C}[[x]]_{q,(-s)}$ and identify the continuous dual V' accordingly with the FN space $\mathbf{C}[[x]]_{q,(s)}$ or the DFN space $\mathbf{C}[[x]]_{q,s}$. Since the operator L is a finite linear combination $\sum a_{i,j} x^j \sigma_q^i$ (acting on V), its adjoint is readily decomposed as $L' = \sum a_{i,j} (x^j \sigma_q^i)'$ (acting on V').

LEMMA 6.62. *In either case (regarding the choice of V):*

(i) *the adjoint of σ_q^i (acting on V), is σ_q^i (acting on V');*

(ii) *the adjoint of x^j (acting on V as $f \mapsto x^i f$) is the compositum $\pi \circ x^{-j}$, where, N denoting any integer $\geq N_0$, the map $x^{-j} : V' \to x^{-N} V' = V' \oplus (\mathbf{C}x^{-1} \oplus \cdots \oplus \mathbf{C}x^{-N})$ is $g \mapsto x^{-j} g$ and where $\pi : V' \oplus (\mathbf{C}x^{-1} \oplus \cdots \oplus \mathbf{C}x^{-N}) \to V'$ is the projection to the first component (i.e., the truncation of negative powers).*

PROOF. In either case $L = \sigma_q^i$ or $L = x^j$, we use the same notations as in the example above: $f = \sum f_n x^n \in E$, $g = \sum g_n x^n \in E'$, $\overline{f} := Lf$, $\overline{g} := L'g$, etc, so that we want $\langle f, \overline{g} \rangle = \langle \overline{f}, g \rangle$, i.e., $\sum f_n \overline{g}_n = \sum \overline{f}_n g_n$. This should uniquely define the $\overline{g}_n$, whence $\overline{g}$, whence L'.

(i) Since in general $(L_1 L_2)' = L_2' L_1'$, we have $(\sigma_q^i)' = ({\sigma_q}')^i$, so we can as well take $L := \sigma_q$. This yields $\overline{f}_n = q^n f_n$ and the adjointness relation is written $\sum f_n \overline{g}_n = \sum q^n f_n g_n$, whence immediately $\overline{g}_n = q^n g_n$, whence $\overline{g} = \sigma_q g$, i.e., $L' = \sigma_q$ (acting on V').

(ii) Here $\overline{f}_n = f_{n-j}$ (with the usual convention that this is 0 if $n - j < 0$), so we want:

$$\sum f_n \overline{g}_n = \sum f_{n-j} g_n = \sum f_n g_{n+j} \Longrightarrow \overline{g}_n = g_{n+j} \Longrightarrow \overline{g} = \sum_{n \geq j} g_n x^{n-j}.$$

The latter is exactly the effect of truncating negative powers from $x^{-j} g$ so indeed $L' = \pi \circ x^{-j}$ (acting on V').

□

REMARK 6.63. Following Ramis **[191]**, (ii) could also be interpreted as describing an action on the quotient space $\dfrac{V' \oplus x^{-1}\mathbf{C}[x^{-1}]}{x^{-1}\mathbf{C}[x^{-1}]}$.

6.3.2.4. We introduce for commodity a mock adjoint operator:

$$L^* := \sum a_{i,j} \sigma_q^i x^{-j} = \sum a_{i,j} q^{-ij} x^{-j} \sigma_q^i : V' \to V'' := V' \oplus (\mathbf{C}x^{-1} \oplus \cdots \oplus \mathbf{C}x^{-N}).$$

As in the lemma above, $\pi : V'' \to V'$ is the first projection (truncation of negative powers).

PROPOSITION 6.64. $L' = \pi \circ L^*$.

PROOF. Linearity of $L \mapsto L'$ and the rule $(L_1 \circ L_2)' = L_2' \circ L_1'$, along with the above lemma, imply:

$$L' = \sum a_{i,j}(x^j \sigma_q^i)' = \sum a_{i,j}(\sigma_q^i)'(x^j)' = \sum a_{i,j}\sigma_q^i \circ \pi \circ x^{-j}.$$

Then we note that $\sigma_q^i \circ \pi = \pi \circ \sigma_q^i$ (obvious) and we are done. □

COROLLARY 6.65. $\chi(L, V) = -N - \chi(x^N L^*, V')$

PROOF. Note that the right-hand side makes sense since all exponents of x are ≥ 0 in $x^N L^* := \sum\sum a_{i,j} q^{-ij} x^{N-j} \sigma_q^i$, so it does send V' into itself. By Appendix D.2.3, $\chi(L, V) = -\chi(L', V')$ and, by the proposition (and additivity of the index), $\chi(L', V') = \chi(L^*, V', V'') + \chi(\pi, V'', V')$.

Since π is onto with kernel $\mathbf{C}x^{-1} \oplus \cdots \oplus \mathbf{C}x^{-N}$, we get $\chi(\pi, V'', V') = N$. On the other hand, $L^* : V' \to V''$ is a compositum:

$$V' \overset{x^N L^*}{\longrightarrow} V' \overset{x^{-N}}{\longrightarrow} V'',$$

the right arrow being an isomorphism. The formula follows. □

EXERCISE 6.66. Show directly that $-N - \chi(x^N L^*, V')$ does not depend on a particular choice of $N \geq N_0$.

EXAMPLE 6.67. $L := \sigma_q - x$, then $L^* = \sigma_q - x^{-1}$ and $\chi(L, V) = -1 - \chi(x\sigma_q - 1, V')$. Whether V' be $\mathbf{C}[[x]]_{q,s}$ or $\mathbf{C}[[x]]_{q,(s)}$, we have $\chi(x\sigma_q - 1, V') = 0$, so $\chi(L, V) = -1$ whether V be $\mathbf{C}[[x]]_{q,(-s)}$ or $\mathbf{C}[[x]]_{q,-s}$. We verify that directly. In both cases $V \subset \mathbf{C}\{x\}$, so $\mathrm{Ker}(\sigma_q - x, V) = \{0\}$; and equality $\sigma_q f - xf = 1$ is impossible, so the cokernel has at least dimension 1.

EXERCISE 6.68. Use arguments as in 6.1.3.2 to show that the dimension is exacly 1.

EXAMPLE 6.69. $L = x\sigma_q - 1$, then $L^* = q^{-1}x^{-1}\sigma_q - 1$ and $\chi(L, V) = -1 - \chi(\sigma_q - qx, V')$. We shall only consider the case $s = -1$ (the most interesting by the example at the end of 6.2.3.6). We expect:

$$\chi(x\sigma_q - 1, \mathbf{C}[[x]]_{q,(-1)}) = -1 - \chi(\sigma_q - x, \mathbf{C}[[x]]_{q,1}) = -1,$$
$$\chi(x\sigma_q - 1, \mathbf{C}[[x]]_{q,-1}) = -1 - \chi(\sigma_q - x, \mathbf{C}[[x]]_{q,(1)}) = 0.$$

Since in any case the kernel is trivial, we should check that $x\sigma_q - 1$ is onto in the second case but not in the first. We slightly adapt q-Borel-Ramis transformation used by Bézivin and Ramis, setting $T(\sum f_n x^n) := \sum q^{-n(n-1)/2} f_n x^n$; we then find that $g = x\sigma_q f - f$ is equivalent to $Tg = (x-1)Tf$, whence coomutative diagrams:

$$\begin{array}{ccc} \mathbf{C}[[x]]_{q,-1} & \xrightarrow{x\sigma_q - 1} & \mathbf{C}[[x]]_{q,-1} \\ {\scriptstyle T}\downarrow & & \downarrow{\scriptstyle T} \\ \mathbf{C}\{x\} & \xrightarrow{x-1} & \mathbf{C}\{x\} \end{array} \qquad \begin{array}{ccc} \mathbf{C}[[x]]_{q,(-1)} & \xrightarrow{x\sigma_q - 1} & \mathbf{C}[[x]]_{q,(-1)} \\ {\scriptstyle T}\downarrow & & \downarrow{\scriptstyle T} \\ \mathscr{O}(\mathbf{C}) & \xrightarrow{x-1} & \mathscr{O}(\mathbf{C}). \end{array}$$

In both diagrams, vertical arrows are isomorphisms so upper horizontal arrows are conjugated to lower ones. On the left diagram, $x - 1$ being invertible in $\mathbf{C}\{x\}$, we have an isomorphism. On the right diagram, the image is the hyperplane of those g such that $g(1) = 0$, whence the cokernel has dimension 1.

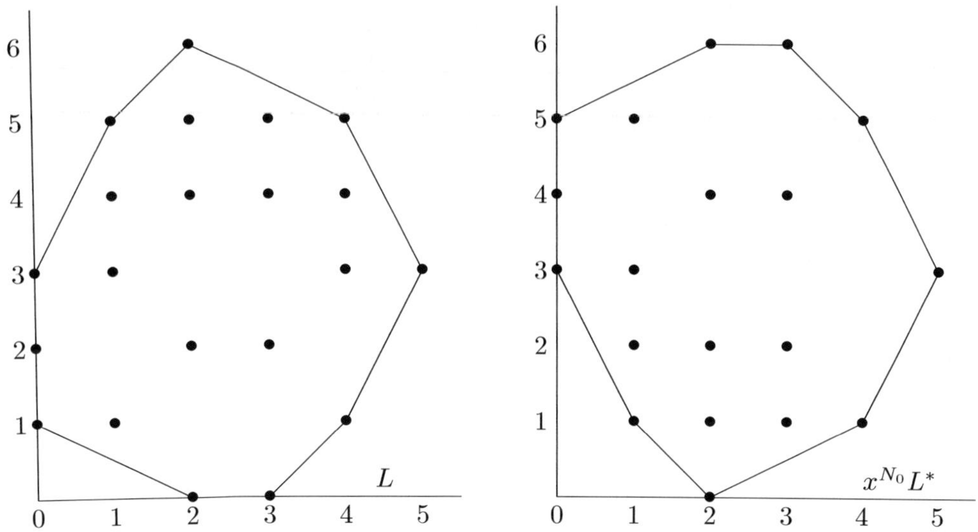

FIGURE 6.2. First appearance of the upper Newton polygon

EXAMPLE 6.70. We show on an example what $x^{N_0}L^*$ looks like. On the left and right drawing (see Figure 6.2 on page 211) are respectively shown the cloud of points of L and of $x^{N_0}L^*$.

Since $x^{N_0}L^* := \sum\sum a_{i,j}q^{-ij}x^{N_0-j}\sigma_q^i$, the cloud for $x^{N_0}L^*$ is the image of the cloud for L by the axial symmetry $(i,j) \mapsto (i, N_0 - j)$ (w.r.t. the horizontal line at ordinate $N_0/2$).

In the example, the relevant slopes for computing the index of $x^{N_0}L^*$ are those of the two down left segments (slopes: $-2, -1$; special values of s: $1/2$, 1). They come by symmetry from the two upper left slopes 2, 1 for L, thus explaining Ramis's introduction of the *upper Newton polygon.*

6.3.3. Index of q-difference operators operating on spaces of entire functions.

6.3.3.1. In order to apply the formula $\chi(L,V) = -N - \chi(x^N L^*, V')$ obtained in 6.3.2.4, we shall make a few preliminary remarks. We keep the notation $L = \sum a_{i,j}x^j\sigma_q^i$ and moreover write the mock adjoint operator L^* in the form $L^* = \sum a_{i,j}^* x^j \sigma_q^i$, where $a_{i,j}^* := q^{ij}a_{i,-j}$. The corresponding clouds of points $C := \{(i,j) \mid a_{i,j} \neq 0\}$ and $C^* := \{(i,j) \mid a_{i,j}^* \neq 0\}$ are therefore symmetric to each other with respect to the horizontal axis $j = 0$. Then the cloud for $x^N L^*$ is $(0,N) + C^*$, resulting of a vertical translation of C^*. So it is the symmetric of C with respect to the horizontal axis $j = N/2$. (You are encouraged to check this by a drawing of your own.)

The relevant features of $x^N L^*$ used while computing its index in $\mathbf{C}[[x]]_{q,s}$ or $\mathbf{C}[[x]]_{q,(s)}$ for some $s \in [0, +\infty]$ are:

- The segments with slope $\mu := -1/s < 0$ on the boundary of the convex hull of the cloud (this is for $0 < s < +\infty$); plus the lower horizontal segment or single lowest point (this is for $\mu = 0$, $s = +\infty$); plus the left vertical segment or single leftmost point (this is for $\mu = -\infty$, $s = 0$).

- More precisely, the ordinates $I^+(s)$, $I^-(s)$ of the highest and lowest points of said segments (or isolated points).

Also note that $I^\pm(s)$ computed for $x^N L^*$ is equal to $N + I^\pm(s)$ computed for L^*.

6.3.3.2. Following Ramis, we therefore introduce the *upper Newton polygon* $N_u(L)$, defined as the boundary of the convex hull of

$$C + \{0\} \times \mathbf{R}_- = \{(i,j) \mid j \leq \deg a_i\}.$$

Taking in account the preceding symmetry arguments, we are interested in its leftmost vertical segment (or single point); its positive slopes; and its upper horizontal segment (or single point). We order as follows the positive slopes of $N_u(L)$:

$$+\infty > \nu_1 > \cdots > \nu_k > 0.$$

The corresponding slopes for L^* are:

$$-\infty < \mu_1 := -\nu_1 < \cdots < \mu_k := -\nu_k < 0.$$

The corresponding special values of s are:

$$0 < s_1 := -1/\mu_1 = 1/\nu_1 < \cdots < s_k := -1/\mu_k = 1/\nu_k < +\infty.$$

For every $\nu > 0$, $\mu := -\nu$, $s := -1/\mu = 1/\nu$, we set (following Ramis):

$$j_1(\nu) := I^+(s) \text{ and } j_2(\nu) := I^-(s),$$

where $I^\pm(s)$ is taken with respect to C^*. We can now state the main result of this section:

THEOREM 6.4. *Let $0 \leq s \leq +\infty$ and $\nu := 1/s$. Then:*

$$\chi(L, \mathbf{C}[[x]]_{q,-s}) = -j_1(\nu) \text{ and } \chi(L, \mathbf{C}[[x]]_{q,(-s)}) = -j_2(\nu).$$

The special case $s = +\infty$, $\nu = 0$ (denoted 0_u by Ramis) corresponds to the upper horizontal segment or single point. In either case, $j_1(0) = j_2(0) =: j(0)$ and $\mathbf{C}[[x]]_{q,-s}) = \mathbf{C}[[x]]_{q,(-s)} = \mathbf{C}[x]$.
6.3.3.3.

EXAMPLE 6.71. Let $L := 1 - x\sigma_q$; then $S_u(L) = \{1\}$ (set of slopes of $N_u(L)$). Following the method above, we tabulate:

s	0	1	$+\infty$
ν	$+\infty$	1	0
$j_1(\nu)$	0	0	1
$j_2(\nu)$	0	1	1

So, for instance, $\chi(L, \mathbf{C}[[x]]_{q,-1}) = 0$ while $\chi(L, \mathbf{C}[[x]]_{q,(-1)}) = -1$. We are going to confirm these values by direct arguments (not through duality).

Note that for $s \geq 0$, the equation $f = x\sigma_q f$ has no solution in $\mathbf{C}[[x]]_{q,-s} \subset \mathbf{C}\{x\}$ so in all cases $\mathrm{Ker}(L, \mathbf{C}[[x]]_{q,-s}) = \mathrm{Ker}(L, \mathbf{C}[[x]]_{q,(-s)}) = \{0\}$. For the image, we refer to the incomplete diagrams:

$$\begin{array}{ccc} \mathbf{C}[[x]]_{q,-s} & \xrightarrow{L} & \mathbf{C}[[x]]_{q,-s} \\ {\scriptstyle \mathcal{B}_{q,-s}}\downarrow & & \downarrow{\scriptstyle \mathcal{B}_{q,-s}} \\ \mathbf{C}\{x\} & \xrightarrow[\tilde{L}]{} & \mathbf{C}\{x\} \end{array} \qquad \begin{array}{ccc} \mathbf{C}[[x]]_{q,(-s)} & \xrightarrow{L} & \mathbf{C}[[x]]_{q,(-s)} \\ {\scriptstyle \mathcal{B}_{q,-s}}\downarrow & & \downarrow{\scriptstyle \mathcal{B}_{q,-s}} \\ \mathcal{O}(\mathbf{C}) & \xrightarrow[\tilde{L}]{} & \mathcal{O}(\mathbf{C}). \end{array}$$

We lack an explicit form of $\tilde{L}$. The simplest way to find it (in concrete cases and by hand) is to write $Lf = g$ as a recursive relation $f_n - q^{n-1} f_{n-1} = g_n$; then (with the usual conventions) $\tilde{f}_n := q^{-sn(n+1)/2} f_n$ and $\tilde{g}_n := q^{-sn(n+1)/2} g_n$ satisfy $\tilde{f}_n - q^{(1-s)n-1} \tilde{f}_{n-1} = \tilde{g}_n$.

For $s = 1$, this boils down to $\tilde{f}_n - q^{-1} \tilde{f}_{n-1} = \tilde{g}_n$, i.e., to $(1 - x/q)\tilde{f} = \tilde{g}$ so $\tilde{L} = 1 - x/q$. This is an automorphism in $\mathbf{C}\{x\}$ but in $\mathcal{O}(\mathbf{C})$ the image has codimension 1 (hyperplane defined by $g(q) = 0$). We do recover the values of χ above.

In general, $\tilde{L} = 1 - \frac{x}{q}\sigma_q^{1-s}$. The cases $s = 0$ and $s = 2$ are easily reduced to those of $x\sigma_q - c$ and $\sigma_q - cx$, which have already been treated.

EXERCISE 6.72. Show that the case $s = +\infty$ can be reduced to the study of multiplication by $1 - x/q$ in $\mathbf{C}[x]$.

6.3.3.4. The following consequence is mentioned in view of a classical application.

COROLLARY 6.73. *The q-difference operator L acting on $\mathbf{C}[[x]]_{q,-s}/\mathbf{C}[x]$, resp. on $\mathbf{C}[[x]]_{q,(-s)}/\mathbf{C}[x]$, is onto and its kernel has dimension $j(0) - j_1(\nu)$, resp. $j(0) - j_2(\nu)$.*

PROOF. Same argument as in 6.2.3.7. □

THEOREM 6.5. *Let $L := a_0 + \cdots + a_n\sigma_q^n$, $a_0, \ldots, a_n \in \mathbf{C}[x]$, $a_0 a_n \neq 0$ and let $g \in \mathbf{C}[x]$. Let $f \in \mathcal{O}(\mathbf{C})$ such that $Lf = g$. Then either f is a polynomial, or it has q-exponential growth of order $s_i := 1/\nu_i$ for one of the special slopes ν_i.*

PROOF. Argument similar to the argument in 6.2.3.8. □

In **[191]**, Ramis also discusses common solutions to algebraic q-difference equations and ordinary differential equations (also the nonvalidity of the same results for nonlinear q-difference equations).

6.4. A more elementary approach to index computations

Here, we shall obtain index values for q-difference operators operation over q-difference *fields*, in accordance with the general framework of this book rather than the more ring centered framework of the present chapter. The interested reader is invited to adapt the elementary methods here to the case of q-difference rings. Some indications are sketched in 6.4.3.

We shall let the letter K denote any of $\mathbf{C}((x)) = \mathbf{C}[[x]][1/x]$, $\mathbf{C}(\{x\}) = \mathbf{C}\{x\}[1/x]$ or $\mathbf{C}((x))_{q,s} := \mathbf{C}[[x]]_{q,s}[1/x]$: in each case, the fraction field of the corresponding ring (this is well known for $\mathbf{C}((x))$ and $\mathbf{C}(\{x\})$ but deserved to be verified for $\mathbf{C}((x))_{q,s}$). Also in each case the constant field is $K^{\sigma_q} = \mathbf{C}$.

EXERCISE 6.74. Prove that the fraction field of $\mathbf{C}[[x]]_{q,s}$ is indeed $\mathbf{C}[[x]]_{q,s}[1/x]$. You could start by proving that the group of units is $\mathbf{C}[[x]]_{q,s}^* = \mathbf{C}[[x]]_{q,s} \cap \mathbf{C}[[x]]^*$. Extend those facts to $\mathbf{C}[[x]]_{q,(s)}$

We shall restrict to q-difference operators with coefficients in $\mathbf{C}(\{x\})$. Since many arguments are similar in the case of fields, we shall at some places refer to calculations previously done in this chapter: necessary adaptations are easy and left to the reader. Our main tool is the factorisation from Chapter 5, whence the plan followed.

6.4.1. Operators of degree 1.

6.4.1.1. *Preliminary reductions.* Since any $u \in K^*$ operates bijectively on K, its index is $\chi(u, K) = 0$. Thus, if $L' = uLv$, $u, v \in K^*$, then $\chi(L, K) = \chi(L', K)$. Now take $L := a\sigma_q + b$, $a, b \in K^*$, so that $L' = a'\sigma_q + b'$, $a' := u\sigma_q(v)a$, $b' := uvb$. The unique slope of L' is $v(a'/b') = v((a/b)(v/\sigma_q v)) = v(a/b)$, the unique slope of L. So we can assume as was done before that $L = \sigma_q - cx^m$, $c \in \mathbf{C}^*$, $m \in \mathbf{Z}$, the slope being $\mu := -m$.

Note however that c is only determined modulo $q^{\mathbf{Z}}$, so the true invariants are the slope $\mu := -m$ and the exponent $\overline{c} \in \mathbf{E}_q$.

EXERCISE 6.75. (i) Check that requiring $u(a\sigma_q + b)v = \sigma_q - cx^m$ amounts to solving $\begin{cases} ua\sigma_q(v) = 1, \\ ubv = -cx^m, \end{cases}$ and solve it. (The unknowns are u, v, c, m.)

(ii) Check that $\sigma_q - cx^m = u(\sigma_q - dx^n)v$ is possible if, and only if, $m = n$ and $\overline{c} = \overline{d}$.

6.4.1.2. *Kernels.*

LEMMA 6.76. *If* $(\mu, \overline{c}) = (0, \overline{1})$, *then* $\dim \mathit{Ker}(L, K) = 1$. *Otherwise,* $\dim \mathit{Ker}(L, K) = 0$.

PROOF. If $m \neq 0$, equation $Lf = 0$ is equivalent to the recursive relations $q^n f_n = c f_{n-m}$. For power series with indices bounded below (that is, such that $f_n = 0$ for $n << 0$), this implies that all $f_n = 0$, i.e., $f = 0$.

If $m = 0$, the equations gives $(q^n - c) f_n = 0$. If $c = q^{n_0}$ for some $n_0 \in \mathbf{Z}$, we get $\mathrm{Ker}(L, K) = \mathbf{C}x^{n_0}$; otherwise, $\mathrm{Ker}(L, K) = \{0\}$. □

6.4.1.3. *Cokernels and indices: the formal and convergent cases.* The proof of the following lemma is exactly the same as in the case of rings and the proposition is an immediate consequence.

LEMMA 6.77. (i) *Let* $K := \mathbf{C}((x))$; *then* L *acting on* K *is onto if, and only if,* $(m, \overline{c}) \neq (0, \overline{1})$; *otherwise, its image has codimension* 1.

(ii) *Let* $K := \mathbf{C}(\{x\})$. *If* $m < 0$ *or* $m = 0$ *and* $\overline{c} \neq \overline{1}$, *then* l *is onto. If* $(m, \overline{c}) = (0, \overline{1})$, *its image has codimension* 1. *If* $m > 0$, *its image has codimension* m.

PROPOSITION 6.78. *For any* q*-difference operator* L *of degree* 1*:*

$$\chi(L, \mathbf{C}((x))) = 0,$$

$$\chi(L, \mathbf{C}(\{x\})) = -\max(0, m) = \min(0, \mu) = \begin{cases} \mu \text{ if } \mu \leq 0, \\ 0 \text{ if } \mu \geq 0. \end{cases}$$

6.4.1.4. *Cokernels and indices: the q-Gevrey case.* Here we consider $L := \sigma_q - cx^m$, $c \in \mathbf{C}^*$, $m \in \mathbf{Z}$, acting on $\mathbf{C}((x))_{q,s}$. Clearly $\mathcal{B}_{q;s}$ realizes an isomorphism $\mathbf{C}((x))_{q,s} \simeq \mathbf{C}(\{x\})$, so we want to complete the commutative diagram i.e., to find $\tilde{L}$):

$$\begin{array}{ccc} \mathbf{C}((x))_{q,s} & \xrightarrow{L} & \mathbf{C}((x))_{q,s} \\ {\scriptstyle \mathcal{B}_{q,s}}\big\downarrow & & \big\downarrow{\scriptstyle \mathcal{B}_{q,s}} \\ \mathbf{C}(\{x\}) & \xrightarrow[\tilde{L}]{} & \mathbf{C}(\{x\}). \end{array}$$

So we let $f := \sum f_n x^n$ (but here n can be < 0), $g := \sum g_n x^n$, $Lf = g$ $\tilde{f} := \mathcal{B}_{q,s} f$, $\tilde{g} := \mathcal{B}_{q,s} g$, etc, and want to express $\tilde{g}$ as $\tilde{L}\tilde{f}$. From $g_n = q^n f_n - c f_{n-m}$, we get:

$$\begin{aligned} \tilde{g}_n &= q^{sn(n+1)/2}(q^n f_n - c f_{n-m}) \\ &= q^n \tilde{f}_n - c q^{sn(n+1)/2 - s(n-m)(n-m+1)/2} \tilde{f}_{n-m} \\ &= q^n \tilde{f}_n - c q^{sm(m+1)/2} q^{sm(n-m)} \tilde{f}_{n-m}, \end{aligned}$$

that is, $\tilde{g} = \sigma_q \tilde{f} - d x^m \sigma_q^{sm} \tilde{f}$, where $d := c q^{sm(m+1)/2}$, i.e.:

$$\tilde{L} = \sigma_q - d x^m \sigma_q^{sm}.$$

Since σ_q and σ_q^{sm} act bijectively on $\mathbf{C}(\{x\})$, their index is 0; the same for operator dx^m. So we get three cases:

$1 > sm$: We look for the index of $\sigma_q^{1-sm} - dx^m = \sigma_{q'} - dx^m$, where $q' := q^{1-sm}$ falls under our general assumptions for q. By 6.4.1.3, the index is $\max(0, -m)$.

$1 = sm$: We look for the index of $1 - dx^m$. This is 0 because $m = 1/s \neq 0$ so $1 - dx^m$ is invertible.

$1 < sm$: We look for the index of $\sigma_q^{sm-1} - \frac{1}{d} x^{-m}$. This is $-\max(0, -m)$ which is 0 since here $m > 1/s > 0$.

PROPOSITION 6.79. *For $s > 0$ and $L = \sigma_q - cx^m$:*

$$\chi(L, \mathbf{C}((x))_{q,s}) = \begin{cases} -m \text{ if } 0 \leq m < 1/s, \\ 0 \text{ otherwise.} \end{cases}$$

6.4.2. Operators of arbitrary degree.

6.4.2.1. From Chapter 5 we know that a q-difference operator with integral slopes can be factored into degee 1 q-difference operators: let $L \in \mathcal{D}_{\mathbf{C}(\{x\}),q}$ with slopes (maybe counted multiply) $\nu_1 \geq \ldots \geq \nu_n$; then $L = L_1 \cdot L_n$, each L_i of degree 1 with slope ν_i. Up to the elementary transformations allowed in 6.4.1.1, let L be equivalent to $\sigma_q - c_i x^{m_i}$: then $\nu_i = -m_i$. On the other hand, for any q-difference operator of degree 1 and slope μ, we saw in 6.4.1 how to compute its index as a function $I_K(\mu)$ defined as follows:

$$\text{If } K = \mathbf{C}((x)) \text{ then } I_K(\mu) := 0,$$
$$\text{If } K = \mathbf{C}(\{x\}) \text{ then } I_K(\mu) := \min(0, \mu),$$
$$\text{If } K = \mathbf{C}((x))_{q,s} \text{ then } I_K(\mu) := \begin{cases} \mu \text{ if } -1/s < \mu \leq 0, \\ 0 \text{ otherwise.} \end{cases}$$

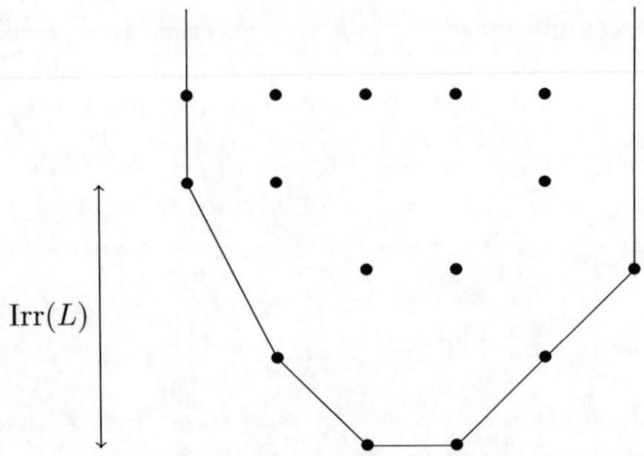

FIGURE 6.3. Irregularity of a q-difference operator

6.4.2.2. Using the additivity of the index, we know that $\chi(L, K) = \sum \chi(L_i, K)$. Returning to the usual notation $\mu_1 > \cdots > \mu_k$ for the slopes of L (and $r_1, \ldots, r_k \in \mathbf{N}^*$ their multiplicities), so that $\chi(L, K) = \sum r_i I_K(\mu_i)$, we summarize:

THEOREM 6.6. (i) $\chi(L, \mathbf{C}((x))) = 0$.
(ii) $\chi(L, \mathbf{C}(\{x\})) = \sum\limits_{\mu_i<0} r_i\mu_i$.
(iii) $\chi(L, \mathbf{C}((x))_{q,s}) = \sum\limits_{-1/s<\mu_i<0} r_i\mu_i$.

PROOF. If L has integral slopes, this follows from the discussion in 6.4.2.1 and 6.4.2.2. In the general case, we invoke the usual ramification argument (note that the $r_i\mu_i$ are rational integers). □

EXERCISE 6.80. Write down properly the "usual ramification argument".

6.4.2.3. *The* irregularity *of* L.

COROLLARY 6.81. $\chi(L, \mathbf{C}((x))\mathbf{C}(\{x\})) = -\sum\limits_{\mu_i<0} r_i\mu_i$.

Similar to [158], the number $\mathrm{Irr}(L) := -\sum\limits_{\mu_i<0} r_i\mu_i$ could be called the *irregularity* of the q-difference operator L: in some sense, it expresses how indices "see" the difference between formal and convergent. It has a geometrical interpretation, see Figure 6.3 page 216.

However, we shall see in 15.2.3.3 and in 15.2.4.4 that the true measure of irregularity is something that would best be denoted $\mathrm{Irr}((L^\vee \otimes L)^{>0})$, which also has a geometrical interpretation and which says more precisely how far the formal class is from determining the analytic class.

6.4.3. Adaptation to the case of rings. We sketch here the way the elementary methods of 6.4.1, 6.4.2 (but applied there to the case of q-difference fields) can be adapted to suit the case of q-difference rings as in 6.2 and 6.3. The reader is invited to fill in the details. Note that some terminology ("integral", "primitive") is purely local, it will not be used elsewhere in the book.

6.4.3.1. *How to adapt factorisation.* Let K the fraction field of R (among our usual q-difference fields and q-difference rings). We say that $L \in \mathcal{D}_{K,q}$ is "integral" if it belongs to $\mathcal{D}_{R,q}$. If $L = \sum a_i \sigma_q^i$, recall that $v(L) = \min v(a_i)$ and since in all our cases R is the valuation ring of v within the valued field K, we see that L is integral if, and only if, $v(L) \geq 0$. Also, any $L \in \mathcal{D}_{K,q}$ can be uniquely written as $L = x^k L'$, where $k := v(L)$ and $v(L') = 0$. We say that such a L' is "primitive": it means that $L' = \sum a' i \sigma_q^i$ where the $a'_i \in R$ are coprime. (Compare to Gauß' theory of polynomials over a unique factorisation domain.)

Facts to be proved in order.

(1) If L_1 and L_2 are primitive, so is $L_1 L_2$.
(2) $v(L_1 L_2) = v(L_1) + v(L_2)$ (this uses the previous fact).
(3) Let $L' := x^{-m} L x^m$, $m \in \mathbf{Z}$; then $v(L') = v(L)$. In particular, if L is integral (resp. primitive), so is L'.
(4) Let $L = L_1 L_2$ be integral. Then $L = L'_1 L'_2$, where $L'_i = x^{a_i} L_i x^{b_i}$ for some $a_i, b_i \in \mathbf{Z}$, $i = 1, 2$.
(5) Suppose that $L = u_0 L_1 u_1 L_2 \cdots u_{k-1} L_k u_k$, all $u_i \in K^*$, all $L_i \in \mathcal{D}_{K,q} \setminus \{0\}$, and L is integral. Then $L = u'_0 L'_1 u'_1 L'_2 \cdots u'_{k-1} L'_k u'_k$, all $u'_i \in K^*$, all $L'_i \in \mathcal{D}_{R,q} \setminus \{0\}$, $L'_i = x^{a_i} L_i x^{b_i}$ for some $a_i, b_i \in \mathbf{Z}$, $i = 1, \ldots, k$ (in particular, they have the same degree).

EXERCISE 6.82. For $L = \sum a_i \sigma_q^i$ primitive, write $\overline{L} := \sum a_i(0) \sigma_q^i$ (in essence, the associated characteristic equation for slope 0: see Chapter 5). Then $\overline{L_1 L_2} = \overline{L_1}\, \overline{L_2}$.

6.4.3.2. *Formal and convergent cases.* Thanks to 6.4.3.1, to relate $\chi(L, R)$ to the factorisation of L, we just have to compute: $\chi(u, R)$ for $u \in R \setminus \{0\}$; $\chi(\sigma_q - c, R)$ for $c \in \mathbf{C}^*$; $\chi(\sigma_q - c x^m, R)$ for $c \in \mathbf{C}^*$ and $m \in \mathbf{N}^*$; $\chi(x^m \sigma_q - c, R)$ for $c \in \mathbf{C}^*$ and $m \in \mathbf{N}^*$.

EXERCISE 6.83. Strictly speaking, the previous reduction would only be valid for q-difference operators with integral slopes. Show how to deal with arbitrary slopes.

Here, we tackle the cases $R = \mathbf{C}[[x]]$ and $R = \mathbf{C}\{x\}$.

- $L = u$ has index $-v(u)$ (trivial kernel and the image uR has codimension $v(u)$).
- $L = \sigma_q - c$ is injective if, and only if, $c \notin q^{\mathbf{N}}$ and then it is bijective. If $c = q^k$, $k \in \mathbf{N}$, then the kernel is $\mathbf{C} x^k$ and the image has supplementary space $\mathbf{C} x^k$. Therefore $\chi(L, R) = 0$ both in $\mathbf{C}[[x]]$ and $\mathbf{C}\{x\}$.
- $L = x^m \sigma_q - c$ is bijective, thus has index 0, in both $\mathbf{C}[[x]]$ and $\mathbf{C}\{x\}$ by the same arguments as in 6.3.
- $L = \sigma_q - c x^m$ is injective both in $\mathbf{C}[[x]]$ and $\mathbf{C}\{x\}$.
- $L = \sigma_q - c x^m$ is surjective in $\mathbf{C}[[x]]$: so it has index 0.
- The image of $L = \sigma_q - c x^m$ has codimension m in $\mathbf{C}\{x\}$, so the index is $-m$; indeed the image is characterized by the vanishing of some q-Borel-Ramis transform at m points.

REMARK 6.84. One can have $LR \subset R$ even if L is not integral, as shown by the example $L = x^{-1}(\sigma_q - 1)$. For such a "quasi-integral" L, the index $\chi(L, R)$ makes sense and can be computed from the above since we know $\chi(x, R) = -1$.

EXERCISE 6.85. Let $L = \sum b_i {\delta_q}^i$ (for the notation of q-derivation, see section 3.3). The $LR \subset R$ if, and only if, all $b_i \in R$.

6.4.3.3. *q-Gevrey case.* We use the q-Borel-Ramis conjugate $\tilde{L} := \mathcal{B}_{q;s} L \mathcal{B}_{q;s}^{-1}$, whence the usual commutative diagram:

$$\begin{array}{ccc} \mathbf{C}[[x]]_{q,s} & \xrightarrow{L} & \mathbf{C}[[x]]_{q,s} \\ \downarrow{\scriptstyle \mathcal{B}_{q,s}} & & \downarrow{\scriptstyle \mathcal{B}_{q,s}} \\ \mathbf{C}\{x\} & \xrightarrow[\tilde{L}]{} & \mathbf{C}\{x\}. \end{array}$$

- If $L = \sigma_q - cx^m$, then $\tilde{L} = \sigma_q - cq^{sm(m+1)/2} x^m \sigma_q^{sm}$.
- If $L = x^m \sigma_q - c$, then $\tilde{L} = q^{-sm(m+1)/2} x^m \sigma_q^{1-sm} - c$.

Then we can apply 6.4.3.2.

CHAPTER 7

Equations and Systems

In the two years at Chicago, [the author] came to the conclusion that it was better to study a single matrix differential equation

$$Y'(x) = A(x)Y(x)$$

(that is, n equations of the first order) than a single linear differential equation of the nth order. Perhaps it was the conspicuous analogy between the first and nth order cases in this symbolic form which attracted me to the matrix notation.

—George D. Birkhoff (1938)[1]

Two important advances of early twentieh century mathematics were: matrix, and matrix algebra, to deal with linear operations and equations; and, more deeply, the *linearisation* of many problems.

EXAMPLE 7.1. An algebraic number $x \in \mathbf{C}$ satisfies a polynomial equation $P(x) := x^n + a_1 x^{n-1} + \cdots + a_n = 0$ with all coefficients $a_i \in \mathbf{Q}$. This is equivalent to saying that the spanned space $V := \mathrm{Vect}_{\mathbf{Q}}(1, x, \ldots, x^n, \ldots)$ has finite dimension; and then that dimension is the degree n of the shortest possible equation satisfied by x. Then $\mathcal{B} := (1, x, \ldots, x^{n-1})$ is a basis of V and the map $y \mapsto xy$ an endomorphism of V, the matrix of which (relative to basis $\mathcal{B}$) is companion to the polynomial P; so its eigenvalues are x and its conjugates. To deal with an algebraic integer x, we just require the a_i to be rational integers and take V to be a $\mathbf{Z}$-module.

In **[23]**, Birkhoff initiated the use of matrices for linear ordinary differential equations, but also for difference and q-difference equations. Let $L := \sigma_q^n + a_1 \sigma_q^{n-1} + \cdots + a_n$ a q-difference operator. The scalar equation $Lf = 0$ can be written $\sigma_q X = AX$, where:

$$X := \begin{pmatrix} f \\ \sigma_q f \\ \vdots \\ \sigma_q^{n-2} f \\ \sigma_q^{n-1} f \end{pmatrix} \text{ and } A := \begin{pmatrix} 0 & 1 & \ldots & 0 & 0 \\ 0 & 0 & \ldots & 0 & 0 \\ \vdots & \vdots & \ddots & \vdots & \vdots \\ 0 & 0 & \ldots & 0 & 1 \\ -a_n & -a_{n-1} & \ldots & -a_2 & -a_1 \end{pmatrix}.$$

For a general $X := {}^t(f_1, \ldots, f_n)$ and $A := (a_{i,j})_{1 \leq i,j \leq n}$, the relation $\sigma_q X = AX$ is but a shorthand for the system $\begin{cases} \sigma_q f_1 = a_{1,1} f_1 + \cdots + a_{1,n} f_n, \\ \cdots\cdots\cdots\cdots \\ \sigma_q f_n = a_{n,1} f_1 + \cdots + a_{n,n} f_n, \end{cases}$ and we shall make no difference between vectorial equations of rank n (using matrices) and systems of scalar equations with n unknowns. So the trick of Birkhoff translates a general

[1] In §7 of *Fifty years of American Mathematics, Collected Mathematical Papers*, vol. 3.

scalar equation of order n (with one unknown) into a vectorial equation of rank n. However, Birkhoff also proved that general vectorial equations $\sigma_q X = AX$ are not really more general than scalar equations (see 7.1.4), but this fact is by no way tautological and we shall prove it twice, here and in the next chapter. (Actually, Birkhoff proved the same thing for linear ordinary differential equations and for difference equations.)

In this chapter, we study the relations between scalar equations and vectorial equations (or systems) and start to show the power of the latter. Throughout, (K, σ_q) denotes (unless otherwise specified) an arbitrary q-difference field.

7.1. From equations to systems and back

7.1.1. An introductory example. This is intended to help you understand the more general facts that will follow.

7.1.1.1. We start with a scalar equation of order 2:

(7.1)
$\sigma_q^2 f + a\sigma_q f + bf = 0, a, b \in K, b \neq 0.$ (Do you remember why the condition on b?)

It is equivalent to the vectorial equation of rank 2:

$$\sigma_q X_f = AX_f, \text{ where } X_f := \begin{pmatrix} f \\ \sigma_q f \end{pmatrix} \text{ and } A := \begin{pmatrix} 0 & 1 \\ -b & -a \end{pmatrix}. \tag{7.2}$$

It is indeed obvious that (7.1) implies (7.2), but the converse must be understood in a strong sense: any solution of $\sigma_q X = AX$ has the form X_f *and moreover* f is a solution of (7.1). Indeed:

$$\left(\sigma_q X = AX \text{ with } X = \begin{pmatrix} f \\ g \end{pmatrix}\right) \iff \begin{cases} \sigma_q f = g, \\ \sigma_q g = -bf - ag, \end{cases} \iff \begin{cases} \sigma_q f = g, \\ \sigma_q^2 f = -bf - ag. \end{cases}$$

So for A a companion matrix as in (7.2), the vectorial equation amounts to a scalar equation.

7.1.1.2. Start conversely with an *arbitrary* rank 2 vectorial equation:

$$\sigma_q X = AX, \text{ where } X := \begin{pmatrix} f \\ g \end{pmatrix} \text{ and } A := \begin{pmatrix} a & b \\ c & d \end{pmatrix} \in \mathrm{GL}_2(K). \tag{7.3}$$

The reason to assume that A is invertible will appear later. However, in the example (7.2), note that $\det A = b \neq 0$.

EXERCISE 7.2. Show that if $\det A = 0$ then (7.3) can be reduced to a scalar equation of order 1.

The vectorial equation (7.4) is actually equivalent to the system:

$$\begin{cases} \sigma_q f = af + bg, \\ \sigma_q g = cf + dg. \end{cases} \tag{7.4}$$

We proceed to show how to reduce it to a scalar equation of order 2. We solve it by substitution, with a few cases to be considered.

Assume $b \neq 0$. Then we look for $g := \dfrac{\sigma_q f - af}{b}$ such that:

$$\sigma_q \frac{\sigma_q f - af}{b} = cf + d\frac{\sigma_q f - af}{b} \iff \sigma_q^2 f + \alpha\sigma_q f + \beta f = 0.$$

EXERCISE 7.3. (i) Compute α and β explicitly.

(ii) Do the calculation again by noting that the second equation is $(\sigma_q - d)g = cf$ so we want to solve $Lf = 0$, where $L := (\sigma_q - d)\left(\dfrac{\sigma_q - a}{b}\right) - c$.

Assume $c \neq 0$.

EXERCISE 7.4. Do the (symmetrical) calculations.

Assume $b = c = 0$. Then $ad \neq 0$ and we have decoupled equations $\sigma_q f = af$, $\sigma_q g = dg$, which we know how to solve. However, without waiting for the general process in 7.1.4, you can train yourself with the following.

EXERCISE 7.5. (i) Let $h := f + g$. Show abstractly that $h, \sigma_q h$ are linearly independent over K but $h, \sigma_q h, \sigma_q^2 h$ are not.

(ii) Find explicit $\alpha, \beta \in K$ such that $\sigma_q^2 h + \alpha\sigma_q h + \beta h = 0$. (You may have to assume something about a and d.)

(iii) Show how to recover f, g from h and $\sigma_q h$.

7.1.1.3. We return to the case $b \neq 0$ and try to extract concepts[2] from the grubby calculations above. How are related the original system $\sigma_q X = AX$ and the particular system $\sigma_q Y = BY$ encoding the scalar equation $\sigma_q^2 f + \alpha\sigma_q f + \beta f = 0$? On the one hand, the latter must come from:

$$Y := \begin{pmatrix} f \\ \sigma_q f \end{pmatrix} \text{ and } B := \begin{pmatrix} 0 & 1 \\ -\beta & -\alpha \end{pmatrix}.$$

The relation reducing one system to the other was $g := \dfrac{\sigma_q f - af}{b}$, which translates to:

$$X := \begin{pmatrix} f \\ g \end{pmatrix} = FY, \text{ where } F := \begin{pmatrix} 1 & 0 \\ -a/b & 1/b \end{pmatrix}.$$

Relation $X = FY$ can be inverted to:

$$Y = GX \text{ where } G := F^{-1} = \begin{pmatrix} 1 & 0 \\ a & b \end{pmatrix}.$$

So we can recover X from Y or Y from X indifferently. Also we understand that $\sigma_q X = AX$ is equivalent to $\sigma_q Y = BY$. This can lead us to the following calculation:

$$\sigma_q X = AX \iff \sigma_q(FY) = AFY \iff \sigma_q Y = \left((\sigma_q F)^{-1} AF\right) Y.$$

EXERCISE 7.6. Check that $(\sigma_q F)^{-1} AF = (\sigma_q G) AG^{-1} = B$.

The whole process of setting $X = FY$ to go from $\sigma_q X = AX$ to $\sigma_q Y = BY$, where $B := (\sigma_q F)^{-1} AF$ is called *gauge transformation* and is crucial. We are led to the following notations:

$$G[A] := (\sigma_q G) AG^{-1} \text{ and } A^F := (\sigma_q F)^{-1} AF,$$

so that our gauge transformation writes $B = G[A] = A^F$.

[2] Hilbert's motto was "replace calculations by ideas".

EXERCISE 7.7. (i) Deduce from those equalities the converse equalities: $A = F[B] = B^G$.

(ii) More generally, and before reading the sequel, try to state and prove rules about gauge transformations. For instance, what can be said about $F_2[F_1[B]]$, etc.

7.1.2. Vocabulary of systems. The theory proper will start in 7.2. For now, we shall consider systems with matrices $A = (a_{i,j}) \in \mathrm{GL}_n(K)$, so:

$$\left(\sigma_q X = AX, \text{ where } X := \begin{pmatrix} f_1 \\ \vdots \\ f_n \end{pmatrix}\right) \iff \begin{cases} \sigma_q f_1 = a_{1,1} f_1 + \cdots + a_{1,n} f_n, \\ \dots\dots\dots\dots\dots\dots\dots\dots \\ \sigma_q f_n = a_{n,1} f_1 + \cdots + a_{n,n} f_n. \end{cases}$$

EXERCISE 7.8. Suppose (in this exercise only) that $\det A = 0$, so there exists $X_1 \in K^n \setminus \{0\}$ such that $AX_1 = 0$. Complete X_1 into a basis $X_1, \dots, X_n$ of K^n (column vectors). Look for X, solution of $\sigma_q X = AX$, in the form $g_1 X_1 + \cdots + g_n X_n$: show that this leads to a system of rank $n-1$ in $g_2, \dots, g_n$, along vith an explicit expression of g_1 in function of $g_2, \dots, g_n$. So we do not really have a system of rank n. (The same reduction will be performed again later using gauge transforms.)

7.1.2.1. *Some rules of computation.* For any rectangular matrix

$$M := (m_{i,j})_{\substack{1 \le i \le n \\ 1 \le \le p}}$$

in $\mathrm{Mat}_{n,p}(K)$, define $\sigma_q M := (\sigma_q m_{i,j})_{\substack{1 \le i \le n \\ 1 \le \le p}} \in \mathrm{Mat}_{n,p}(K)$. Then $\sigma_q(M_1 + M_2) = \sigma_q M_1 + \sigma_q M_2$; and $\sigma_q(fM) = (\sigma_q f)(\sigma_q M)$ for $f \in K$. We say that the action of σ_q on $\mathrm{Mat}_{n,p}(K)$ is σ_q*-linear* (or sometimes with less precision *skew linear*). We shall meet that notion again in Chapter 8. In particular, if $f \in K^{\sigma_q}$, we have $\sigma_q(fM) = f(\sigma_q M)$, so the action is K^{σ_q}-linear.

If now N is such that M and N can be multiplied, we find that $\sigma_q(MN) = (\sigma_q M)(\sigma_q N)$. So σ_q is an automorphism of the K^{σ_q}-algebra $\mathrm{Mat}_n(K)$ (not of the K-algebra though!).

DEFINITION 7.9. The space of solutions of $A \in \mathrm{Mat}_n(K)$ in K^n (column vectors) is:

$$\mathrm{Sol}(A, K) := \mathrm{Ker}(\sigma_q - A, K^n) = \{X \in K^n \mid \sigma_q X = AX\}.$$

It is a K^{σ_q}-linear space. We shall see in 7.2 that, as such, it has dimension $\le n$.

We shall usually find that there are "not enough" solutions over K (i.e., in K^n): the space $\mathrm{Sol}(A, K)$ will have K^{σ_q}-dimension $< n$. So we shall enlarge the q-difference field in hope to find enough. That means we shall consider a q-difference extension K' of K, so that $\sigma_q - A$ will act on K'^n and $\mathrm{Sol}(A, K)$ will naturally imbed into $\mathrm{Sol}(A, K')$; however, the latter will be a linear space over K'^{σ_q}, which may be strictly larger than K^{σ_q}. This, by the way, is one of the main difficulties in the study of q-difference equations: even studying complex analytic q-difference equations, one falls upon spaces of solutions over fields bigger than $\mathbf{C}$.

However, when studying q-difference equations over some q-difference field K, even if we admit solutions over a q-difference extension K', we shall only allow for gauge transformations over K, i.e., with matrix in $\mathrm{GL}_n(K)$. To understand why, think of differential equations with rational coefficients. Maybe their solutions involve bigger fields than $\mathbf{C}(x)$, i.e., they define "new transcendants". However, solving them, one usually does not allow transformations of the unknowns involving nonrational coefficients, otherwise the transformations themselves would introduce new transcendants and one would not know exactly which are mandatory.

7.1.2.2. *Gauge transformations.* Let $F \in \mathrm{GL}_n(K)$. We set:

$$F[A] := (\sigma_q F)AF^{-1} \text{ and } A^F := (\sigma_q F)^{-1}AF.$$

So actually $A^F = F^{-1}[A]$. Some rules are easy to prove (this means you should prove them):

$$F_2[F_1[A]] = (F_2F_1)[A] \text{ and } I_n[A] = A,$$

in other words, $(F, A) \mapsto F[A]$ defines a left action of $\mathrm{GL}_n(K)$. Similarly:

$$(A^{F_1})^{F_2} = A^{F_1F_2} \text{ and } A^{I_n} = A,$$

in other words, $(F, A) \mapsto A^F$ defines a right action of $\mathrm{GL}_n(K)$. These formal properties justify the following definition.

DEFINITION 7.10. Let K' a q-difference extension of K. Matrices $A, B \in \mathrm{GL}_n(K')$ are said to be *gauge equivalent* or simply *equivalent (over K)* if there is a *gauge transformation* $F \in \mathrm{GL}_n(K)$ such that $B = F[A]$. We write $A \sim_K B$ or simply $A \sim B$.

For instance, gauge equivalence over $\mathbf{C}(x)$, resp. over $\mathbf{C}(\{x\})$, resp. over $\mathbf{C}((x))$, is called *rational*, resp. *analytic*, resp. *formal* equivalence.

PROPOSITION 7.11. *Let $F[A] = B$, where $F, A, B \in GL_n(K)$. Then, for any q-difference extension K' of K, the map $X \mapsto FX$ from K'^n to itself induces an isomorphism $Sol(A, K') \simeq Sol(B, K')$.*

PROOF. The map $X \mapsto FX$ is clearly a K'^{σ_q}-automorphism of K'^n. The conclusion then follows from the equivalence $\sigma_q X = AX \Leftrightarrow \sigma_q(FX) = B(FX)$, which comes from the following calculation:

$$\sigma_q X = AX \Longleftrightarrow \sigma_q F \sigma_q X = \sigma_q FAX = ((\sigma_q F)AF^{-1})(FX) = B(FX).$$

□

Of course the inverse isomorphism $\mathrm{Sol}(B, K') \simeq \mathrm{Sol}(A, K')$ is $Y \mapsto F^{-1}Y$.

EXERCISE 7.12. Suppose that $F \in \mathrm{Mat}_n(K)$ is such that $(\sigma_q F)A = BF$. How can you relate $\mathrm{Sol}(A, K')$ and $\mathrm{Sol}(B, K')$?

EXERCISE 7.13. Suppose that $\det A = 0$. Find $F \in \mathrm{GL}_n(K)$ such that the first column of $B := F[A]$ is 0. Then show that $\sigma_q Y = BY \Leftrightarrow \sigma_q Z = CZ$ where $Z = GY$, for some explicit C and G of size $n - 1$.

7.1.3. Vectorialisation. We generalize the process exhibited in example 7.1.1. Let $L := \sigma_q^n + a_1\sigma_q^{n-1} + \cdots + a_n \in \mathcal{D}_{K,q}$, $a_n \neq 0$, a q-difference operator in standard unitary form. Let the companion matrix:

$$A_L := \begin{pmatrix} 0 & 1 & \dots & 0 & 0 \\ 0 & 0 & \dots & 0 & 0 \\ \vdots & \vdots & \ddots & \vdots & \vdots \\ 0 & 0 & \dots & 0 & 1 \\ -a_n & -a_{n-1} & \dots & -a_2 & -a_1 \end{pmatrix}.$$

Then $\det A_L = (-1)^n a_n$, so $A_L \in \mathrm{GL}_n(K)$. For $f \in K'$, an arbitrary q-difference extension of K, let:

$$X_f := \begin{pmatrix} f \\ \sigma_q f \\ \vdots \\ \sigma_q^{n-2} f \\ \sigma_q^{n-1} f \end{pmatrix}.$$

Then, clearly, $Lf = 0 \Leftrightarrow \sigma_q X_f = A_L X_f$.

EXERCISE 7.14. Detail the computations leading to that equivalence.

PROPOSITION 7.15. *The map $f \mapsto X_f$ from K' to K'^n induces an isomorphism of K'^{σ_q}-linear spaces $Sol(L, K') \simeq Sol(A_L, K')$.*

PROOF. Since the map is clearly K'^{σ_q}-linear and injective, we are reduced to prove that any element of $\mathrm{Sol}(A_L, K')$ has the form X_f for some $f \in K'$; and then the equivalence in the above exercise will give the wanted conclusion.

But if $X := {}^t(f_1, \ldots, f_n)$ satisfies $\sigma_q X = AX$, then the $n-1$ first among the n equations of the corresponding system say that $\sigma_q f_1 = f_2, \ldots, \sigma_q f_{n-1} = f_n$, i.e., $X = X_{f_1}$. □

The inverse isomorphism is clearly induced by the map ${}^t(f_1, \ldots, f_n) \mapsto f_1$ from K'^n to K'.

EXERCISE 7.16. Show that the proposition can be deduced from consideration of the kernels in the following commutative diagram:

$$\begin{array}{ccc} K' & \xrightarrow{\;L\;} & K' \\ \downarrow & & \downarrow \\ K'^n & \xrightarrow[\sigma_q - A_L]{} & K'^n, \end{array}$$

where the left vertical arrow is $f \mapsto X_f$ and the right vertical arrow is $g \mapsto {}^t(g, 0, \ldots, 0)$. (Cokernels will be interesting too; see 7.1.5.)

EXERCISE 7.17. Show how to vectorialize a q-difference equation written with δ_q. Compare the two systems thus obtained. (See 4.4.)

7.1.4. Cyclic vector lemma. In essence, it says that going from scalar equations to systems (or vectorial equations) does not really enlarge the class of objects at stake. More precisely:

THEOREM 7.1 (Birkhoff). *Let $A \in GL_n(K)$. Then there is a standard unitary operator $L \in \mathcal{D}_{K,q}$ such that $A \sim_K A_L$, i.e., A is equivalent (or gauge equivalent) to A_L over K.*

PROOF. We give here a proof in the style of Birkhoff; it is valid only in the analytic case $K := \mathbf{C}(\{x\})$. A general (and more algebraic) proof due to Lucia Di Vizio and inspired by Katz-Deligne will be given in the next chapter: that one will be valid for any q-difference field of characteristics 0.

So we start from $A \in \mathrm{GL}_n(K)$. First we fix some $x_0 \in \mathbf{C}^*$ such that:

$$\forall i = 0, \ldots, n-1\,,\ A(q^i x_0) \in \mathrm{GL}_n(\mathbf{C}).$$

In other words, the set $\{x_0, qx_0, \ldots, q^{n-1}x_0\}$ is included in the common domain of definition of A and A^{-1}. This x_0 will serve to fix a kind of "initial condition".

We are simultaneously looking for $L = \sigma_q^n + a_1\sigma_q^{n-1} + \cdots + a_n \in \mathcal{D}_{K,q}$ and for $F \in \mathrm{GL}_n(K)$ such that $F[A] = A_L$, i.e., $(\sigma_q F)A = A_L F$. We write F as a stack of lines: $F = \begin{bmatrix} L_1 \\ \vdots \\ L_n \end{bmatrix}$ (so beware the L_i are not q-difference operators, they are line vectors in K^n). Then, denoting $\Psi(X) := (\sigma_q X)A$ (again, X is a line vector), we easily get:

$$(\sigma_q F)A = A_L F \iff \begin{cases} \Psi(L_1) = L_2, \\ \cdots\cdots\cdots\cdots\cdots\cdots \\ \Psi(L_{n-1}) = L_n, \\ \Psi(L_n) = -(a_n L_1 + \cdots + a_1 L_n). \end{cases}$$

We moreover require $F \in \mathrm{GL}_n(K)$, i.e., $\det(L_1, \ldots, L_n) \neq 0$. Altogether, these conditions boil down to the fact that L_1 is a line vector in K^n such that *setting* $L_2 := \Psi(L_1)$, $\ldots$, $L_n := \Psi(L_{n-1})$, we get a basis $(L_1, \ldots, L_n)$ of K^n: indeed then the a_i will be uniquely determined from the last equation in the system above. The vector L_1 may be considered as a *cyclic vector* for Ψ (in the terminology of reduction theory in linear algebra; except that here Ψ is not linear).

Setting $F := \begin{bmatrix} L_1 \\ \vdots \\ L_n \end{bmatrix}$ the gauge matrix we intend to define, we shall now see that we can choose L_1 in such a way that $F(x_0) = I_n$. This will guarantee thet $\det(L_1, \ldots, L_n) \neq 0$ (since, as an element of $\mathbf{C}(\{x\})$, it has value 1 at $x = x_0$); and so that F is an invertible matrix.

Note that iterating Ψ we find, for an arbitrary line vector X:

$$\Psi(X) = (\sigma_q X)A, \Psi^2(X) = (\sigma_q^2 X)(\sigma_q A)A, \ldots, \Psi^i(X) = (\sigma_q^i X)A_i,$$

where $A_i := (\sigma_q^{i-1}A)\cdots(\sigma_q A)A$. (Actually $A_0 = I_n$ and $A_{i+1} = (\sigma_q A_i)A$.) So in particular, $L_i = \Psi^{i-1}(L_1) = (\sigma_q^{i-1}L_1)A_{i-1}$.

Write $e_1, \ldots, e_n$ the lines of I_n; they are also the canonical basis of the space of line vectors. Then we can rewrite condition $F(x_0) = I_n$ as:

$$\begin{cases} L_1(x_0) = e_1, \\ \cdots\cdots\cdots \\ L_n(x_0) = e_n \end{cases} \iff \begin{cases} L_1(x_0) = e_1, \\ \cdots\cdots\cdots\cdots\cdots\cdots \\ (\sigma_q^{n-1}L_1)(x_0)A_{n-1}(x_0) = e_n \end{cases}$$

$$\iff \begin{cases} L_1(x_0) = e_1, \\ \cdots\cdots\cdots\cdots\cdots \\ L_1(q^{n-1}x_0) = e_n A_{n-1}^{-1}(x_0) \end{cases}$$

All the right-hand sides $e_i A_{i-1}^{-1}(x_0)$ are well defined by our initial choice of x_0. So we are reduced to find a line vector L_1 of analytic functions taking prescribed line values in $\mathbf{C}^n$ at n distinct points. Using Lagrange interpolation, that can even be done with a vector of polynomials, each having degree $\leq n - 1$. $\square$

EXERCISE 7.18. State and prove a similar theorem valid for $A \in \mathrm{GL}_n(\mathbf{C}((x)))$, assuming $A(0) \in \mathrm{GL}_n(\mathbf{C})$.

To go further, recall (or anticipate) that, if A is rational, $\Sigma := \mathrm{Sing}(A)$ (see the General Notations 0.4 or 7.2.3.3) is the singular locus of A (poles of A and of A^{-1}).

COROLLARY 7.19. *If $A \in GL_n(\mathbf{C}(x))$, the gauge transformation matrix F can be chosen with coefficients in $\mathbf{C} \oplus \cdots \oplus \mathbf{C}x^{n-1}$ and such that $F(x_0) = I_n$, where x_0 is any point in the punctured disk $\overset{\bullet}{D}(0,r)$, where $r := \min\{|x| \mid x \in \Sigma\}$.*

EXERCISE 7.20. State and prove a similar corollary valid for $A \in \mathrm{GL}_n(\mathbf{C}(\{x\}))$.

EXAMPLE 7.21. Let $A := \begin{pmatrix} a & b \\ c & d \end{pmatrix}$, so $\Delta := \det A = ad - bc$ and $A^{-1} = \frac{1}{\Delta}\begin{pmatrix} d & -b \\ -c & a \end{pmatrix}$. We look for $L_1 := (f,g)$ such that $(f(x_0), g(x_0)) = (1,0)$ and $(f(qx_0), g(qx_0)) = (-c/\Delta, a/\Delta)(x_0)$. For instance, if $b = c = 0$, let $r := \min\{|x| \mid x \neq 0 \text{ and } ad(x) \text{ defined and } \neq 0\}$; we take $0 < |x_0| < r$. Then we can set $f := \dfrac{x - qx_0}{x_0 - qx_0}$ and $g := \dfrac{1}{d(x_0)} \dfrac{x - x_0}{qx_0 - x_0}$. Now let $(f', g') := (\sigma_q f, \sigma_q g)A$, i.e.:

$$f' = a\frac{qx - qx_0}{x_0 - qx_0} + c\frac{1}{d(x_0)}\frac{qx - x_0}{qx_0 - x_0}, \quad g' = b\frac{qx - qx_0}{x_0 - qx_0} + d\frac{1}{d(x_0)}\frac{qx - x_0}{qx_0 - x_0}.$$

Then $F := \begin{pmatrix} f & g \\ f' & g' \end{pmatrix}$.

EXERCISE 7.22. There are good algorithms to find the cyclic vector L_1. However, taking $L_1 := e_1$ (or any line vector chosen at random) will work with probability 1 (bad luck cannot be excluded though). Explain why.

7.1.5. Indices related to scalar, resp. vectorial equations.

7.1.5.1. *Solutions and cosolutions.* Let L a q-difference operator and A the matrix of a q-difference system. Recall the definitions of their spaces of solutions in K as $\mathrm{Sol}(L,K) := \mathrm{Ker}(L,K)$ and $\mathrm{Sol}(A,K) := \mathrm{Ker}(\sigma_q - A, K^n)$. We shall also be interested in so-called *cosolutions*:

$$\mathrm{Cosol}(L,K) := \mathrm{Coker}(L,K) \text{ and } \mathrm{Cosol}(A,K) := \mathrm{Coker}(\sigma_q - A, K^n).$$

Those spaces fit in exact sequences:

$$0 \longrightarrow \mathrm{Sol}(L,K) \longrightarrow K \xrightarrow{L} K \longrightarrow \mathrm{Cosol}(L,K) \longrightarrow 0, \tag{7.5}$$

$$0 \longrightarrow \mathrm{Sol}(A,K) \longrightarrow K^n \xrightarrow{\sigma_q - A} K^n \longrightarrow \mathrm{Cosol}(A,K) \longrightarrow 0. \tag{7.6}$$

In that terminology, the index of L acting over K, resp. of $\sigma_q - A$ acting over K^n, is:

$$\chi(L,K) = \dim_K \mathrm{Sol}(L,K) - \dim_K \mathrm{Cosol}(L,K),$$
$$\chi(\sigma_q - A, K^n) = \dim_K \mathrm{Sol}(A,K) - \dim_K \mathrm{Cosol}(A,K).$$

We are going to relate indexes for q-difference equations with indexes for q-difference systems. We begin by a useful auxiliary result.

PROPOSITION 7.23. *Let $F \in GL_n(K)$ and $B := F[A]$. Then the following square is commutative:*

$$\begin{array}{ccc} K^n & \xrightarrow{\sigma_q - A} & K^n \\ {\scriptstyle F}\downarrow & & \downarrow{\scriptstyle \sigma_q F} \\ K^n & \xrightarrow[\sigma_q - B]{} & K^n \end{array}$$

PROOF. Let $X \in K^n$. Using the fact that $(\sigma_q F)A = BF$, we find:

$$(\sigma_q - B)FX = \sigma_q(FX) - BFX = (\sigma_q F)(\sigma_q X) - (\sigma_q F)AX = (\sigma_q F)(\sigma_q X - AX).$$

□

COROLLARY 7.24. *F induces an isomorphism $Sol(A, K) \simeq Sol(B, K)$ and $\sigma_q F$ an isomorphism $Cosol(A, K) \simeq Cosol(B, K)$. In particular:*

$$\chi(\sigma_q - A, K^n) = \chi(\sigma_q - B, K^n).$$

7.1.5.2. *Comparing solutions and cosolutions.* The companion matrix A_L to a q-difference operator $L := \sigma_q^n + a_1\sigma_q^{n-1} + \cdots + a_n$ was defined in 7.1.3 as:

$$A_L := \begin{pmatrix} 0 & 1 & \ldots & 0 & 0 \\ 0 & 0 & \ldots & 0 & 0 \\ \vdots & \vdots & \ddots & \vdots & \vdots \\ 0 & 0 & \ldots & 0 & 1 \\ -a_n & -a_{n-1} & \ldots & -a_2 & -a_1 \end{pmatrix}.$$

Since we shall look at the kernel and cokernel of the K^{σ_q}-linear map $\sigma_q - A_L$, we introduce a notation for it:

$$\Delta_L := \sigma_q - A_L : \begin{pmatrix} f_1 \\ \vdots \\ f_n \end{pmatrix} \mapsto \begin{pmatrix} \sigma_q f_1 - f_2 \\ \vdots \\ \sigma_q f_{n-1} - f_n \\ \sigma_q f_n + a_1 f_n + \cdots + a_n f_1 \end{pmatrix}.$$

In order to formalize the relation of L to A_L, and in particular of the way their solutions and cosolutions relate, we also use the following maps from K to K^n:

$$X : f \mapsto \begin{pmatrix} f \\ \sigma_q f \\ \vdots \\ \sigma_q^{n-1} f \end{pmatrix}, \quad U : g \mapsto \begin{pmatrix} 0 \\ \vdots \\ 0 \\ g \end{pmatrix}.$$

PROPOSITION 7.25. *The following square is commutative:*

$$\begin{array}{ccc} K & \xrightarrow{L} & K \\ {\scriptstyle X}\downarrow & & \downarrow{\scriptstyle U} \\ K^n & \xrightarrow[\Delta_L]{} & K^n. \end{array}$$

PROOF. Let $f \in K$. Then:

$$\Delta_L X_f = \begin{pmatrix} \sigma_q f - \sigma_q f \\ \vdots \\ \sigma_q \sigma_q^{n-2} f - \sigma_q^{n-1} f \\ \sigma_q \sigma_q^{n-1} + a_1 \sigma_q^{n-1} f + \cdots + a_n f \end{pmatrix} = \begin{pmatrix} 0 \\ \vdots \\ 0 \\ Lf \end{pmatrix} = ULf.$$

□

COROLLARY 7.26. *We have two morphisms Sol*$(L,K) \to$ *Sol*(A,K) *and Cosol*$(L,K) \to$ *Cosol*(A,K) *respectively induced by* X *and by* U*.*

However, the vertical arrows are obviously not isomorphisms, so we cannot directly infer that the above morphisms (in the corollary) are isomorphisms, which they actually are. We shall give two proofs of it, one through the snake lemma, the other through a small excursion in the theory of complexes (cochain complexes to be precise).

7.1.5.3. *Through the snake lemma.* We introduce two maps from K^n to K^{n-1}:

$$Y : \begin{pmatrix} f_1 \\ \vdots \\ f_n \end{pmatrix} \mapsto \begin{pmatrix} \sigma_q f_1 - f_2 \\ \vdots \\ \sigma_q f_{n-1} - f_n \end{pmatrix}, \quad V : \begin{pmatrix} g_1 \\ \vdots \\ g_n \end{pmatrix} \mapsto \begin{pmatrix} g_1 \\ \vdots \\ g_{n-1} \end{pmatrix}.$$

Those maps fit into a natural extension of the commutative square in 7.1.5.2:

$$\begin{array}{ccc} 0 & \longrightarrow & 0 \\ \downarrow & & \downarrow \\ K & \xrightarrow{L} & K \\ {\scriptstyle X}\downarrow & & \downarrow{\scriptstyle U} \\ K^n & \xrightarrow[\Delta_L]{} & K^n \\ {\scriptstyle Y}\downarrow & & \downarrow{\scriptstyle V} \\ K^{n-1} & \xrightarrow[\mathrm{Id}]{} & K^{n-1} \\ \downarrow & & \downarrow \\ 0 & \longrightarrow & 0. \end{array}$$

It will be more convenient, in order to apply the snake lemma, to rotate the diagram by $\pi/2$ (also some repositioning):

$$\begin{array}{ccccccccc} 0 & \longrightarrow & K & \xrightarrow{X} & K^n & \xrightarrow{Y} & K^{n-1} & \longrightarrow & 0 \\ & & {\scriptstyle L}\downarrow & & {\scriptstyle \Delta_L}\downarrow & & {\scriptstyle \mathrm{Id}}\downarrow & & \\ 0 & \longrightarrow & K & \xrightarrow[U]{} & K^n & \xrightarrow[V]{} & K^{n-1} & \longrightarrow & 0. \end{array}$$

PROPOSITION 7.27. *The diagram above is a morphism of exact sequences.*

PROOF. (i) First we prove that the upper sequence is exact. Injectivity of $X : K \to K^n$ is obvious. Surjectivity of $Y : K^n \to K^{n-1}$ is easy because an antecedent of an element of the target can be found for instance by imposing $f_n := 0$ and then solving inductively (but backward) for $f_{n-1}, \dots, f_1$. Last an element of Ker$Y$ is charecterized by the relations $f_{i+1} = \sigma_q f_i$, i.e., it is an X_{f_1}, whence exactness at the middle.

(ii) Second we must prove that the lower exact sequence is exact. But this is obvious.

(iii) Last we must prove that the diagram is commutative, that is: $U \circ L = \Delta_L \circ X$; and $Y = V \circ \Delta_L$. The first equality was proved in 7.1.5.2. The second is immediate by definition of the three maps involved.

□

COROLLARY 7.28. *We have isomorphisms $Sol(L, K) \simeq Sol(A_L, K)$ and $Cosol(L, K) \simeq Cosol(A_L, K)$.*

PROOF. From the snake lemma (Appendix D.1.2) we have an exact sequence:

$$0 \to \text{Ker } L \to \text{Ker } \Delta_L \to \text{Ker Id} \to \text{Coker } L \to \text{Coker } \Delta_L \to \text{Coker Id} \to 0.$$

Since Ker Id $= 0$ and Coker Id $= 0$, the consequence follows. Note that by the precise statement of the snake lemma, the isomorphisms are respectively induced by X and by U. □

7.1.5.4. *Through homotopy of q-de Rham complexes.* We introduce a sequence of q-difference operators:

$$L_0 := 1, L_1 := \sigma_q + a_1, \ldots, L_{i+1} := L_i\sigma_q + a_{i+1}, \ldots, L_n := L.$$

Convention: $a_0 := 1$, $g_0 := 0$, all summations with indices $i, j, k, l \geq 0$.

Let $X', U' : K^n \to K$ be defined by:

$$X' : \begin{pmatrix} f_1 \\ \vdots \\ f_n \end{pmatrix} \mapsto f_1, \quad U' : \begin{pmatrix} g_1 \\ \vdots \\ g_n \end{pmatrix} \mapsto L_0 g_n + \cdots + L_n g_0.$$

PROPOSITION 7.29. (i) $X'X = U'U = Id_K$
(ii) $L \circ X' = U' \circ \Delta_L$ *i.e., the following square is commutative:*

$$\begin{array}{ccc} K^n & \xrightarrow{\Delta_L} & K^n \\ {\scriptstyle X'}\downarrow & & \downarrow{\scriptstyle U'} \\ K & \xrightarrow[L]{} & K. \end{array}$$

PROOF. (i) Equality $X'X = \text{Id}_K$ is obvious. Equality $U'U = \text{Id}_K$ is easy because the terms $L_1 g_{n-1} + \cdots + L_n g_0$ vanish.

(ii) We let $\begin{pmatrix} g_1 \\ \vdots \\ g_n \end{pmatrix} := \Delta_L \begin{pmatrix} f_1 \\ \vdots \\ f_n \end{pmatrix}$ and compute $U'\Delta_L \begin{pmatrix} f_1 \\ \vdots \\ f_n \end{pmatrix} = U' \begin{pmatrix} g_1 \\ \vdots \\ g_n \end{pmatrix}$ as:

$$\begin{aligned}
&= L_0 g_n + \cdots + L_n g_0 \\
&= (\sigma_q f_n + a_1 f_n + \cdots + a_n f_1) + L_1(\sigma_q f_{n-1} - f_n) + \cdots + L_{n-1}(\sigma_q f_1 - f_2) \\
&= (a_1 f_n + \cdots + a_n f_1) + (\sigma_q - L_1) f_n + (L_1 \sigma_q - L_2) f_{n-1} \\
&\quad + \cdots + (L_{n-2}\sigma_q - L_{n-1}) f_2 + L_{n-1}\sigma_q f_1 \\
&= (a_1 f_n + \cdots + a_n f_1) + (-a_1) f_n + (-a_2) f_{n-1} + \cdots + (-a_{n-2}) f_2 + L_{n-1}\sigma_q f_1 \\
&= (a_n + L_{n-1}\sigma_q) f_1 = L_n f_1 = LX' \begin{pmatrix} f_1 \\ \vdots \\ f_n \end{pmatrix}.
\end{aligned}$$

□

We now replace our previous diagrams by diagrams involving the *complexes* (Appendix E):

$$0 \longrightarrow \underset{0}{K} \xrightarrow{L} \underset{1}{K} \longrightarrow 0. \tag{7.7}$$

$$0 \longrightarrow \underset{0}{K^n} \xrightarrow{\sigma_q - A} \underset{1}{K^n} \longrightarrow 0. \tag{7.8}$$

Clearly, the homology of the first complex is $H^0 = \mathrm{Sol}(L, K)$, $H^1 = \mathrm{Cosol}(L, K)$; and that of the second complex is $H^0 = \mathrm{Sol}(A, K)$, $H^1 = \mathrm{Cosol}(A, K)$. We want to prove that, when $A = A_L$, they have same homology. The commutative diagrams previously obtained extend to morphisms of complexes (X, U):

$$\begin{array}{ccccccc}
0 & \longrightarrow & K & \xrightarrow{L} & K & \longrightarrow & 0 \\
 & & \downarrow{\scriptstyle X} & & \downarrow{\scriptstyle U} & & \\
0 & \longrightarrow & K^n & \xrightarrow[\Delta_L]{} & K^n & \longrightarrow & 0
\end{array}$$

$$\begin{array}{ccccccc}
0 & \longrightarrow & K^n & \xrightarrow{\Delta_L} & K^n & \longrightarrow & 0 \\
 & & \downarrow{\scriptstyle X'} & & \downarrow{\scriptstyle U'} & & \\
0 & \longrightarrow & K & \xrightarrow[L]{} & K & \longrightarrow & 0.
\end{array}$$

We can compose these morphisms of complexes in two ways. The first:

$$\begin{array}{ccccccc}
0 & \longrightarrow & K & \xrightarrow{L} & K & \longrightarrow & 0 \\
 & & \downarrow{\scriptstyle X'X} & & \downarrow{\scriptstyle U'U} & & \\
0 & \longrightarrow & K & \xrightarrow[L]{} & K & \longrightarrow & 0.
\end{array}$$

Since $X'X = U'U = \mathrm{Id}_{K^n}$, this says that the corresponding composite maps $\mathrm{Sol}(L, K) \to \mathrm{Sol}(A, K) \to \mathrm{Sol}(L, K)$ and $\mathrm{Cosol}(L, K) \to \mathrm{Cosol}(A, K) \to \mathrm{Cosol}(L, K)$ are identities.

Composing the other way gives:

$$\begin{array}{ccccccc} 0 & \longrightarrow & K^n & \xrightarrow{\Delta_L} & K^n & \longrightarrow & 0 \\ & & {\scriptstyle XX'}\big\downarrow & & {\scriptstyle UU'}\big\downarrow & & \\ 0 & \longrightarrow & K^n & \xrightarrow[\Delta_L]{} & K^n & \longrightarrow & 0. \end{array}$$

This will induce composite maps $\mathrm{Sol}(A, K) \to \mathrm{Sol}(L, K) \to \mathrm{Sol}(A, K)$ and $\mathrm{Cosol}(A, K) \to \mathrm{Cosol}(L, K) \to \mathrm{Cosol}(A, K)$. Of course it would be quite unreasonable to hope that XX' and UU' be isomorphisms, since they factor through K. Besides, we can compute it explicitly:

$$XX' : \begin{pmatrix} f_1 \\ \vdots \\ f_n \end{pmatrix} \mapsto \begin{pmatrix} f_1 \\ \vdots \\ \sigma_q^{n-1} f_1 \end{pmatrix}, \quad UU' : \begin{pmatrix} g_1 \\ \vdots \\ g_n \end{pmatrix} \mapsto \begin{pmatrix} 0 \\ \vdots \\ 0 \\ L_0 g_n + \cdots + L_n g_0 \end{pmatrix}.$$

However we shall see that the morphism (XX', UU') is *homotopic to the identity*. To that aim, we introduce a new map H in the following position:

$$\begin{array}{ccccccc} 0 & \longrightarrow & K^n & \xrightarrow{\Delta_L} & K^n & \longrightarrow & 0 \\ & & {\scriptstyle XX'}\big\downarrow & \overset{H}{\swarrow} & \big\downarrow{\scriptstyle UU'} & & \\ 0 & \longrightarrow & K^n & \xrightarrow[\Delta_L]{} & K^n & \longrightarrow & 0. \end{array}$$

It is defined as follows:

$$H \begin{pmatrix} g_1 \\ \vdots \\ g_n \end{pmatrix} := \begin{pmatrix} 0 \\ g_1 \\ g_2 + \sigma_q g_1 \\ \vdots \\ g_{n-1} + \sigma_q g_{n-2} + \cdots + \sigma_q^{n-2} g_1 \end{pmatrix}.$$

LEMMA 7.30. $H \circ \Delta_L = XX' - Id$

PROOF. We apply the left-hand side $H \circ \Delta_L$ with the notations:

$$\begin{pmatrix} g_1 \\ \vdots \\ g_n \end{pmatrix} := \Delta_L \begin{pmatrix} f_1 \\ \vdots \\ f_n \end{pmatrix}, \quad \begin{pmatrix} f'_1 \\ \vdots \\ f'_n \end{pmatrix} := H \begin{pmatrix} g_1 \\ \vdots \\ g_n \end{pmatrix}.$$

We have:

$$g_1 = \sigma_q f_1 - f_2, \ldots, g_{n-1} = \sigma_q f_{n-1} - f_n, g_n = \sigma_q f_n + a_1 f_n + \cdots + a_n f_1,$$

then:

$$f'_1 = 0, f'_2 = g_1 = \sigma_q f_1 - f_2, f'_3 = g_2 + \sigma_q g_1 = (\sigma_q f_2 - f_3) + \sigma_q(\sigma_q f_1 - f_2) = \sigma_q^2 f_1 - f_3,$$

up to:

$$\begin{aligned} f'_n &= g_{n-1} + \sigma_q g_{n-2} + \cdots + \sigma_q^{n-2} g_1 \\ &= (\sigma_q f_{n-1} - f_n) + \sigma_q(\sigma_q f_{n-2} - f_{n-1}) + \cdots + \sigma_q^{n-2}(\sigma_q f_1 - f_2) \\ &= \sigma_q^{n-1} f_1 - f_n, \end{aligned}$$

whence:

$$\begin{pmatrix} f_1' \\ f_2' \\ \vdots \\ f_n' \end{pmatrix} = \begin{pmatrix} f_1 - f_1 \\ \sigma_q f_1 - f_2 \\ \vdots \\ \sigma_q^{n-1} f_1 - f_n \end{pmatrix} = \begin{pmatrix} f_1 \\ \sigma_q f_1 \\ \vdots \\ \sigma_q^{n-1} f_1 \end{pmatrix} - \begin{pmatrix} f_1 \\ f_2 \\ \vdots \\ f_n \end{pmatrix} = XX' \begin{pmatrix} f_1 \\ f_2 \\ \vdots \\ f_n \end{pmatrix} - \begin{pmatrix} f_1 \\ f_2 \\ \vdots \\ f_n \end{pmatrix}.$$

□

LEMMA 7.31. $\Delta_L \circ H = UU' - Id$

PROOF. We apply the left-hand side $\Delta_L \circ H$ with the notations:

$$\begin{pmatrix} g_1' \\ \vdots \\ g_n' \end{pmatrix} := \Delta_L \begin{pmatrix} f_1 \\ \vdots \\ f_n \end{pmatrix}, \quad \begin{pmatrix} f_1 \\ \vdots \\ f_n \end{pmatrix} := H \begin{pmatrix} g_1 \\ \vdots \\ g_n \end{pmatrix}.$$

With the conventions $g_0 = 0$ and only indexes ≥ 0 allowed:

$$f_k = \sum_{i+j=k} \sigma_q^i g_j.$$

For $k = 1, \ldots, n-1$:

$$g_k' = \sigma_q f_k - f_{k+1} = \sigma_q \left(g_{k-1} + \sigma_q g_{k-2} + \cdots + \sigma_q^{k-1} g_0\right) - (g_k + \sigma_q g^{k-1} + \cdots + \sigma_q^k g_0) = -g_k,$$

after eliminations. Now comes the hard part:

$$g_n' = \sigma_q f_n + a_1 f_n + \cdots + a_n f_1 = \sigma_q \left(g_{n-1} + \sigma_q g_{n-2} + \cdots + \sigma_q^{n-1} g_0\right) + \sum_{k=1}^{n} a_k f_{n+1-k},$$

so with the convention $a_0 = 1$:

$$\begin{aligned} g_n + g_n' &= a_0 \sum_{i+j=n} \sigma_q^i g_j + \sum_{k=1}^{n} a_k \sum_{i+j=n-k} \sigma_q^i g_j \\ &= \sum_{i+j+k=n} a_k \sigma_q^i g_j \\ &= \sum_{j=0}^{n} L_{n-j}' g_j, \end{aligned}$$

where we introduced the q-difference operators:

$$L_l' := \sum_{i+k=l} a_k \sigma_q^i = \sigma_q^l + a_1 \sigma_q^{l-1} + \cdots + a_l,$$

that is, $L_l' = L_l$. So in the end $g_n' = \sum\limits_{i+j=n} L_i g_j - g_n$ and:

$$(\Delta_L \circ H) \begin{pmatrix} g_1 \\ \vdots \\ g_n \end{pmatrix} = \begin{pmatrix} -g_1 \\ \vdots \\ -g_{n-1} \\ \sum\limits_{i+j=n} L_i g_j - g_n \end{pmatrix} = (UU' - \mathrm{Id}) \begin{pmatrix} g_1 \\ \vdots \\ g_n \end{pmatrix}.$$

□

PROPOSITION 7.32. *The endomorphism (XX', UU') of the complex $0 \to K^n \to K^n \to 0$ is homotopic to the identity morphism (Id, Id).*

PROOF. It is exactly the content of the lemmas. □

COROLLARY 7.33. *The maps $Sol(L,K) \to Sol(A_L,K)$ and $Cosol(L,K) \to Cosol(A_L,K)$ are isomorphisms.*

PROOF. We already saw that the composition $\mathrm{Sol}(L,K) \to \mathrm{Sol}(A_L,K) \to \mathrm{Sol}(L,K)$ and $\mathrm{Cosol}(L,K) \to \mathrm{Cosol}(A_L,K) \to \mathrm{Cosol}(L,K)$ are identities. From the homotopy in the proposition, we deduce that the compositions $\mathrm{Sol}(A_L,K) \to \mathrm{Sol}(L,K) \to \mathrm{Sol}(A_L,K)$ and $\mathrm{Cosol}(A_L,K) \to \mathrm{Cosol}(L,K) \to \mathrm{Cosol}(A_L,K)$ also are identities. □

7.1.6. Index of $\sigma_q - A$, the direct way. The previous results allow us (in principle at least) to compute the index of $\sigma_q - A$ by first finding an equivalent scalar equation (i.e., applying effectively the cyclic vector lemma). It is traditional in such circumstances to ask whether one can work directly from the system. We show here that, to some extent, this is indeed possible. For that, we shall have to anticipate some results on normal forms (Section 15.2 of Chapter 15). In all this subsection, the letter K denotes either $\mathbf{C}(\{x\})$ or $\mathbf{C}((x))$. We write Δ_A for $\sigma_q - A$.

7.1.6.1.

LEMMA 7.34. *Let A be in block-triangular form: $A = \begin{pmatrix} A_1 & \dots & \star \\ \vdots & \ddots & \vdots \\ 0 & \dots & A_k \end{pmatrix}$, $A_i \in GL_{n_i}(K)$, $n_1 + \dots + n_k = n$, all stars $\star$ denoting arbitrary rectangular matrices. Then $\chi(\Delta_A, K^n) = \sum_{i=1}^{k} \chi(\Delta_{A_i}, K^{n_i})$.*

PROOF. It is of course enough to consider the case of two blocks:

$$A = \begin{pmatrix} A' & \star \\ 0 & A'' \end{pmatrix},$$

and then to iterate. Since (writing lines instead of columns to ease printing):

$\Delta_A(x',0) = (\sigma_q x',0) - A(x',0) = (\sigma_q x',0) - (A'x',0) = (\sigma_q x' - A'x',0) = (\Delta_{A'}x',0)$,

and similarly $\Delta_A(x',x'')$ has the form $(\star, \Delta_{A''}x'')$ (computation left to you), the canonical exact sequence $0 \to K^{n'} \xrightarrow{i} K^n \xrightarrow{p} K^{n'} \to 0$ defined by $i(x') := (x',0)$ and $p(x',x'') := x''$ gives rise to a commutative diagram:

$$\begin{array}{ccccccccc} 0 & \longrightarrow & K^{n'} & \xrightarrow{i} & K^n & \xrightarrow{p} & K^{n''} & \longrightarrow & 0 \\ & & \downarrow{\scriptstyle \Delta_{A'}} & & \downarrow{\scriptstyle \Delta_A} & & \downarrow{\scriptstyle \Delta_{A''}} & & \\ 0 & \longrightarrow & K^{n'} & \xrightarrow{i} & K^n & \xrightarrow{p} & K^{n''} & \longrightarrow & 0. \end{array}$$

The conclusion then follows from the additivity of the index with respect to exact sequences (Appendix D). □

EXERCISE 7.35. Check that $p \circ \Delta_A = \Delta_{A''} \circ p$.

REMARK 7.36. It is however not true that the snake exact sequence (Appendix D) splits into two exact sequences, one for the kernels, one for the cokernels; said otherwise, $\mathrm{Ker}\Delta_A \to \mathrm{Ker}\Delta_{A''}$ needs not be surjective and $\mathrm{Coker}\Delta_{A'} \to \mathrm{Coker}\Delta_A$ needs not be injective. As a consequence, the additivity of the indices cannot be strengthened into the additivity of their components dim Ker and dim Coker.

EXAMPLE 7.37. Let $A := \begin{pmatrix} 1 & 1 \\ 0 & 1 \end{pmatrix}$. It is not true that the canonical map $\mathrm{Ker}(\sigma_q - I_2, K^2) \to \mathrm{Ker}(\sigma_q - 1, K)$ is surjective: Indeed, $1 \in \mathrm{Ker}(\sigma_q - 1, K)$ cannot be lifted to $(f, 1) \in \mathrm{Ker}(\sigma_q - I_2, K^2)$ because that would require to solve $\sigma_q f - f = 1$, which we know to be impossible in $\mathbf{C}(\{x\})$ or $\mathbf{C}((x))$.

EXERCISE 7.38. Discuss the case of $A := \begin{pmatrix} x^m & 1 \\ 0 & 1 \end{pmatrix}$, $m \neq 0$. (You will find that the map is onto, except when $K = \mathbf{C}(\{x\})$ and $m > 0$.)

7.1.6.2.

LEMMA 7.39. *Let $A = x^m A_0$, $A_0 \in GL_n(\mathbf{C})$. Then $\chi(\Delta_A, K^n) = nI_K(-m)$. (Recall that notation I_K was introduced in* 6.4.2.1.*)*

PROOF. Let $A_0 = PTP^{-1}$, where $P \in \mathrm{GL}_n(\mathbf{C})$ and $T \in \mathrm{GL}_n(\mathbf{C})$ is upper triangular; then $A = P(x^m T)P^{-1}$. In order to use the lemma below, we write that equality in the form $A = (\sigma_q P)(x^m T)P^{-1}$. Then the said lemma implies that $\chi(\Delta_A, K^n) = \chi(\Delta_{x^m T}, K^n)$; then we apply the previous lemma along with 6.4.2 to conclude. □

Anticipating slightly on 7.2.2, we consider a "gauge equivalence"

$$A = (\sigma_q P)BP^{-1},$$

where $A, B, P \in \mathrm{GL}_n(K)$. The conjugacy (or similarity) relation is just the special case $P \in \mathrm{GL}_n(\mathbf{C})$.

LEMMA 7.40. *Let $A, B, P \in GL_n(K)$ such that $A = (\sigma_q P)BP^{-1}$. Then $\chi(\Delta_A, K^n) = \chi(\Delta_B, K^n)$.*

PROOF. First, we check that $\Delta_A \circ p = (\sigma_q p) \circ \Delta_B$, where $p : K^n \to K^n$ has matrix P and $(\sigma_q p)$ has matrix $\sigma_q P$; this amounts to:

$$\sigma_q(PX) - APX = (\sigma_q P)(\sigma_q X) - (\sigma_q P)BX = (\sigma_q P)(\sigma_q X - BX).$$

Then we use the following commutative diagram:

$$\begin{array}{ccc} K^n & \xrightarrow{\;p\;} & K^n \\ {\scriptstyle \Delta_B}\big\downarrow & & \big\downarrow{\scriptstyle \Delta_A} \\ K^n & \xrightarrow[\sigma_q p]{} & K^n. \end{array}$$

Since the horizontal arrows are isomorphisms, $\chi(\Delta_A, K^n) = \chi(\Delta_B, K^n)$. □

EXERCISE 7.41. Justify the last statement.

7.1.6.3. In Chapter 5 we introduced the Newton polygon of a q-difference operator; and in 7.1.4 we saw that every $A \in \mathrm{GL}_n(K)$ is equivalent to A_L for some q-difference operator L. In 10.1, we shall see that, A being given, all the q-difference operators L such that A is equivalent to A_L have the same Newton polygon. We thus may speak of the slopes $\mu_1, \ldots, \mu_k$ of A and of their multiplicities $r_1, \ldots, r_k$. Again anticipating on results to come (in 10.1), we shall see that, *if the slopes are integral* and indexed in such a way that $\mu_1 > \cdots > \mu_k$, then there exist a gauge transformation matrix $F \in \mathrm{GL}_n(K)$ such that:

$$A = (\sigma_q F) \begin{pmatrix} x^{\mu_1} A_1 & \ldots & \star \\ \vdots & \ddots & \vdots \\ 0 & \ldots & x^{\mu_k} A_k \end{pmatrix} F^{-1},$$

where all $A_i \in \mathrm{GL}_{r_i}(\mathbf{C})$ and all symbols $\star$ denote rectangular matrices over K. Using 7.1.6.2, we conclude:

LEMMA 7.42. *If A has integral slopes $\mu_1, \ldots, \mu_k$ with multiplicities $r_1, \ldots, r_k$, then* $\chi(\Delta_A, K^n) = \sum\limits_{i=1}^{k} r_i I_K(-\mu_i)$.

□

7.1.6.4. The above result can be extended to the case of nonintegral slopes, but, it seems, not through such direct methods[3]. The natural attack would be as follows.

Let r a common denominator of the slopes and let $K' := K[x']$, $x' := x^{1/r}$, $q' := q^{1/r}$. Over K', the q'-difference system with matrix $A'(x') := A(x)$ has slopes $r\mu_1, \ldots, r\mu_k$ and the above lemma can be applied to it. It will follow from the sheaf-theoretic approach that the corresponding operator $\Delta'_{A'}$ has index $\chi(\Delta'_{A'}, K'^n) = r\chi(\Delta_A, K^n)$.

EXERCISE 7.43. Admitting that fact, give a general formula for $\chi(\Delta_A, K^n)$.

Here is a *failed* attempt at proving the formula. We write $K' = \bigoplus\limits_{i=0}^{r-1} K x'^i$. On each component $(Kx'^i)^n$ of K'^n, $\Delta'_{A'}$ acts according to the following commutative diagram:

$$\begin{array}{ccc} (Kx'^i)^n & \longrightarrow & K^n \\ {\scriptstyle \Delta'_{A'}} \big\downarrow & & \big\downarrow {\scriptstyle \Delta_{A_i}} \\ (Kx'^i)^n & \longrightarrow & K^n, \end{array}$$

where the horizontal maps are easy to guess isomorphisms, and where $A_i := q^{-i/r} A$. So it all comes down to proving that all Δ_{A_i} have the same index.

EXERCISE 7.44. (i) Guess the isomorphisms.
(ii) *Try* to prove directly that all Δ_{A_i} have the same index.

[3]At least, I presently do not know any direct proof without resorting to the properties of the sheaf of solutions; see 7.3 and above all 9.3.

7.1.6.5. We now sketch a possible shortcut of independent interest in the case $A = x^\mu A_0$, $A_0 \in \mathrm{GL}_n(\mathbf{C})$, $\mu \in \mathbf{Z} \setminus \{0\}$: this is "direct" in the sense that it avoids trigonalisation and above all additivity of the index with respect to exact sequences. If $X = \sum X_k x^k$, then:

$$X \in \mathrm{Ker}\Delta_A \Longleftrightarrow \sigma_q X = x^\mu A_0 X \Longleftrightarrow \forall k\ ,\ q^k X_x = A_0 X_{k-\mu}.$$

Since $\mu \neq 0$, this would imply that either all $X_k \neq 0$, which is impossible over K, or that all $X_k = 0$; so we conclude that $\mathrm{Ker}(\Delta_A, K^n) = \{0\}$ in the formal as well as in the convergent case.

Now we go for surjectivity, first in the formal case $K = \mathbf{C}((x))$; we study separately according to whether $\mu > 0$ or < 0.

(1) If $\mu > 0$, we solve $\Delta_A X = Y$ in the form $X = \sigma_q^{-1}(Y + x^\mu A_0 X) = Z + \Phi(X)$, where $Z := \sigma_q^{-1} Y$ and $\Phi := \sigma_q^{-1} x^\mu A_0$. Clearly Φ is x-adically (i.e., for the adic metrics determined by x) contracting and there is a unique formal solution $\sum\limits_{k \geq 0} \Phi^k(Z)$.

(2) If $\mu < 0$, set $m := \mu$ and solve $\Delta_A X = Y$ in the form $X = Z + \Psi(X)$, where $Z := -x^m A_0^{-1} Y$ and $\Psi := x^m A_0^{-1} \sigma_q$. The argument goes on as above.

So the index is $\chi(\Delta_A, K^n) = 0$ whichever the sign of μ.

EXERCISE 7.45. Compute Φ^k and Ψ^k for all $k \geq 0$.

In the analytic case $K = \mathbf{C}((x))$, there is a big difference. Computation of Φ^k, resp. of Ψ^k, makes appear powers $q^{-\mu^2/2}$, resp. $q^{\mu^2/2}$. So $\Delta_A : K^n \to K^n$ is surjective if $\mu < 0$ but probably not if $\mu > 0$. This is confirmed and the dimension of the cokernel can be computed by the following steps:

(1) Writing $K = \bigoplus\limits_{i=0}^{\mu-1} x^i \mathbf{C}(\{x^\mu\})$, one reduces the study to the case $\mu = 1$.

(2) In that case, conjugate $x^{-1}\sigma_q - A_0$ by a transformation of q-Borel-Ramis type as in 6.1.3.1.

EXERCISE 7.46. Complete this program along the following lines:

(i) Find a commutative diagram:

$$\begin{array}{ccc} K^n & \xrightarrow{\mathcal{B}_{q;1}} & L^n \\ {\scriptstyle x^{-1}\sigma_q - A_0}\big\downarrow & & \big\downarrow{\scriptstyle x^{-1} - A_0} \\ K^n & \xrightarrow[\mathcal{B}_{q;1}]{} & L^n \end{array}$$

(ii) Determine exactly the cokernel of $x^{-1} - A_0$ acting on L^n.

You should say what ring is denoted by L. Also, there is an explicit linear form on L^n whose kernel is the image of $x^{-1} - A_0$; and so an explicit linear form on K^n whose kernel is the image of $x^{-1}\sigma_q - A_0$, whence an explicit criterion of analytic solvability of the equation $\Delta_A X = Y$.

7.2. Generalities on systems

We go on with the q-difference field (K, σ_q) and write moreover $C := K^{\sigma_q}$ its subfield of constants. Generally speaking, unless said otherwise, we identify K^n

with $\mathrm{Mat}_{n,1}(K)$ (column vectors). The first approach to systems shown here will be thoroughly completed in the next chapter, with more appropriate tools.

7.2.1. Solutions. Let $A \in \mathrm{GL}_n(K)$ (for the case that $\det A = 0$; see 7.1.2). The space of solutions:

$$\mathrm{Sol}(A, K) := \{X \in K^n \mid \sigma_q X = AX\}$$

is a C-linear space. We shall let the field of solutions vary; if K' is a q-difference extension of K, and $C' := K'^{\sigma_q}$ its subfield of constants, $\mathrm{Sol}(A, K')$ is a C'-linear space.

7.2.1.1. To get a feeling for the (moderate) subtleties of what follows, consider a field extension $K \subset K'$ (no q-difference operators or q-difference equations here) and a family of column vectors $X_1, \ldots, X_m$ in K^n. Then the K-rank r of the family, i.e., $\dim_K \mathrm{Vect}(X_1, \ldots, X_m)$, is equal to the maximum size of a regular (invertible) square matrix extracted from the $n \times m$ rectangular matrix $[X_1, \ldots, X_m]$. In particular, if the X_i are K-free, i.e., if $r = m$, then they are K'-free, because the criterion just mentioned works the same. Conversely, if $X_1, \ldots, X_m$ are K'-free, they are K-free. (This, by the way, is even more obvious since any linear relation over K is a linear relation over K'.)

However, if $X_1, \ldots, X_m$ belong to K'^n (coefficients in the bigger field), the situation is not the same: L'-freeness implies K'-freeness for the same obvious reason as above; but the converse is false, the above criterion does not apply. The classical counter-example is $K = \mathbf{R}$, $K' = \mathbf{C}$ and the family $(1, \mathrm{i})$: it is $\mathbf{R}$-free but not $\mathbf{C}$-free.

We now concentrate on solutions. Let $X_1, \ldots, X_m \in Sol(A, K)$. Then:

$$\dim_K \mathrm{Vect}_K(X_1, \ldots, X_m) = \mathrm{rank}\,[X_1, \ldots, X_m]$$

is the maximum size r of a $r \times r$ matrix with determinant $\neq 0$ extracted from the $n \times m$ matrix. (Extracted means that we choose indices $1 \leq i_1 < \cdots < i_r \leq n$ and $1 \leq j_1 < \cdots < j_r \leq m$ and keep the coefficients with such indices only.)

7.2.1.2. *The q-wronskian lemma.*

LEMMA 7.47. *Assume $X_1, \ldots, X_m \in Sol(A, K)$ are K-linearly independent (as elements of K^n). Let $Y \in Sol(A, K)$ such that $Y = \lambda_1 X_1 + \cdots + \lambda_m X_m$, all $\lambda_i \in K$. Then all $\lambda_i \in C$.*

PROOF. On the one hand, we apply σ_q, on the other we left multiply by A:

$$\sigma_q Y = (\sigma_q \lambda_1) \sigma_q X_1 + \cdots + (\sigma_q \lambda_m) \sigma_q X_m,$$
$$AY = \lambda_1 A X_1 + \cdots + \lambda_m A X_m.$$

Since Y and all the X_i are in $\mathrm{Sol}(A, K)$, equaling:

$$(\sigma_q \lambda_1 - \lambda_1) A X_1 + \cdots + (\sigma_q \lambda_m - \lambda_m) A X_m = 0.$$

Since A is invertible and the X_i are K-free, we deduce:

$$\sigma_q \lambda_1 - \lambda_1 = \cdots = \sigma_q \lambda_m - \lambda_m = 0,$$

whence the conclusion. □

THEOREM 7.2 (q-wronskian lemma).

$$\dim_C \mathit{Vect}_C(X_1, \ldots, X_m) = \mathit{rank}[X_1, \ldots, X_m].$$

PROOF. Let $r := \operatorname{rank}[X_1, \dots, X_m] = \dim_K \operatorname{Vect}_K(X_1, \dots, X_m)$ this rank so that, up to reordering, we can assume that $X_1, \dots, X_r$ are K-linearly independent and $\operatorname{Vect}_K(X_1, \dots, X_m) = \operatorname{Vect}_K(X_1, \dots, X_r)$. Note that K-linear independence trivially implies C-linear independence so $\dim_C \operatorname{Vect}_C(X_1, \dots, X_m) \geq r$. We want to show equality. According to the assumption, each X_j, $j > r$, belongs to $\operatorname{Vect}_K(X_1, \dots, X_r)$, i.e., $X_j = \lambda_1 X_1 + \cdots + \lambda_r X_r$, all $\lambda_i \in K$. But, after the lemma, all $\lambda_i \in C$, whence $X_j \in \operatorname{Vect}_C(X_1, \dots, X_r)$ and at last we may conclude that $\dim_C \operatorname{Vect}_C(X_1, \dots, X_m) = r$. □

EXERCISE 7.48. Deduce from this the q-wronskian lemma for scalar equations (Theorem 3.1 in Chapter 3).

COROLLARY 7.49. (i) $\dim_C$ *Sol*$(A, K) \leq n$.
(ii) *If K' is a q-difference extension of K' with constant subfield C', then* $\dim_C$ *Sol*$(A, K) \leq \dim_{C'}$ *Sol*(A, K').

PROOF. (i) For any $X_1, \dots, X_m$ in K^n, we have $\operatorname{rank}[X_1, \dots, X_m] \leq n$.
(ii) Let $X_1, \dots, X_r \in K^n$ be K-free and a C-basis of $\operatorname{Sol}(A, K)$. Then they belong to $\operatorname{Sol}(A, K')$ and remain free by the rank criterion.

□

However, we must recall the dilemna: if we increase K' in order to obtain more solutions, then the constant subfield $C' := K'^{\sigma_q}$ increases too.

7.2.1.3. *Fundamental systems of solutions.* Suppose now that the situation is "optimal": $\dim_C \operatorname{Sol}(A, K) = n$. Then there is a *fundamental system (or basis) of solutions* $X_1, \dots, X_n \in K^n$, a C-basis of $\operatorname{Sol}(A, K)$. We introduce the invertible square matrix:

$$\mathcal{X} := [X_1, \dots, X_n] \in \mathrm{GL}_n(K).$$

It satisfies:

$$\sigma_q \mathcal{X} = [\sigma_q X_1, \dots, \sigma_q X_n] = [AX_1, \dots, AX_n] = A[X_1, \dots, X_n] = A\mathcal{X},$$

so it is a "fundamental matricial solution", after the following definition:

DEFINITION 7.50. (i) A *matricial solution* (over K) is a rectangular matrix $M \in \operatorname{Mat}_{n,m}(K)$ such that $\sigma_q \mathcal{X} = A\mathcal{X}$.
(ii) A *fundamental matricial solution* is an invertible (thus square) matricial solution.

Clearly $M \in \operatorname{Mat}_{n,m}(K)$ is a matricial solution if, and only if, its columns are in $\operatorname{Sol}(A, K)$; and a matricial solution is fundamental if, and only if, it is square and its columns are a C-basis of $\operatorname{Sol}(A, K)$. So a fundamental matricial solution exists if, and only if, $\dim_C \operatorname{Sol}(A, K) = n$.

PROPOSITION 7.51. *Suppose* $\dim_C$ *Sol*$(A, K) = n$ *and $\mathcal{X}$ is a fundamental matricial solution. Then:*

$$\mathit{Sol}(A, K) = \{\mathcal{X} X_0 \mid X_0 \in C^n\}.$$

More precisely, $X_0 \mapsto \mathcal{X} X_0$ is an isomorphism $C^n \to$ Sol(A, K).

PROOF. Any $X \in K^n$ can be written $X = \mathcal{X} X_0$ for a unique $X_0 \in K^n$. Then:

$$X \in \operatorname{Sol}(A, K) \Longleftrightarrow \sigma_q(\mathcal{X} X_0) = A\mathcal{X} X_0 \Longleftrightarrow (\sigma_q \mathcal{X})(\sigma_q X_0) = A\mathcal{X} X_0$$
$$\Longleftrightarrow A\mathcal{X} \sigma_q X_0 = A\mathcal{X} X_0 \Longleftrightarrow \sigma_q X_0 = X_0 \Longleftrightarrow X_0 \in C^n.$$

□

COROLLARY 7.52. (i) *If $\mathcal{X}$ is a particular fundamental matricial solution, all fundamental matricial solutions are of the form $\mathcal{X}X_0$, $X_0 \in GL_n(C)$.*

(ii) *All bases of Sol(A, K) have the form $(\mathcal{X}X_1, \ldots, \mathcal{X}X_n)$ where $(X_1, \ldots, X_n)$ is a basis of C^n.*

7.2.2. Gauge transformations.

7.2.2.1. *Gauge equivalence.*

DEFINITION 7.53. Let K' a q-difference extension of K and let $A, B \in \mathrm{GL}_n(K')$. We say that A is *gauge equivalent* or *equivalent* to B over K, and we write $A \sim_K B$, if there exists $F \in \mathrm{GL}_n(K)$ such that $F[A] = B$. Equivalence over $\mathbf{C}(x)$, resp. $\mathbf{C}(\{x\})$, resp. $\mathbf{C}((x))$, is called *rational* (or *global*), resp. *analytic*, resp. *formal*; analytic and formal equivalence are *local*.

PROPOSITION 7.54. *Let K' a q-difference extension of K and $F \in GL_n(K)$ such that $F[A] = B$. Then the map $X \mapsto FX$ is a K'^{σ_q}-isomorphism Sol$(A, K') \simeq$ Sol(B, K').*

PROOF. This is just a restatement of things already seen. □

EXAMPLE 7.55. Let $A \in \mathrm{GL}_n(K)$. From the cyclic vector lemma in 7.1.4, we find $F \in \mathrm{GL}_n(K)$ and $L \in \mathcal{D}_{K,q}$ standard unitary of degree n such that $A = F[A_L]$. Combining with 7.1.3, we have an isomorphism $f \mapsto FX_f$ from $\mathrm{Sol}(L, K')$ to $\mathrm{Sol}(A, K')$.

In particular, $(f_1, \ldots, f_n)$ is a K'^{σ_q}-basis of $\mathrm{Sol}(L, K')$ if, and only if, $(FX_{f_1}, \ldots, FX_{f_1})$ is a K'^{σ_q}-basis of $\mathrm{Sol}(A, K')$. Note that:

$$[FX_{f_1}, \ldots, FX_{f_1}] = F\,[X_{f_1}, \ldots, X_{f_1}] = F\mathcal{W},$$

where $\mathcal{W} := [X_{f_1}, \ldots, X_{f_1}]$ is the wronskian matrix of $(f_1, \ldots, f_n)$. The matrix $F\mathcal{W}$ is a fundamental matricial solution of A in K'.

7.2.2.2. *A discussion on classification.* These are just some preliminary remarks, aiming at some clarification on our long way toward Riemann-Hilbert correspondence. The study of q-difference systems over the Riemann sphere $\mathbf{S}$ (or complex projective line $\mathbf{P}^1(\mathbf{C})$), i.e., over $\mathscr{M}(\mathbf{S}) = \mathbf{C}(x)$, involves rational classification, that is:

- effective characterization of rational equivalence;
- description of equivalence classes;
- definition and computation of invariants;
- calculation of normal forms;
- if (very) lucky, description of the quotient space, maybe with a structure making it a moduli space.

See 12.1, 12.2, and 15.3 for significant steps along the above lines.

Global classification is preceded by *local* classification. In q-difference theory, "local" usually means "in a neighborhood of 0" (or maybe of ∞, which amounts to the same thing through the change of coordinate $x \leftarrow 1/x$)[4]. And then local study itself usually begins with *formal* study, followed by *analytic* study. So we deal with a tower of fields: $\mathbf{C}(x) \subset \mathbf{C}(\{x\}) \subset \mathbf{C}((x))$. For instance, gross formal classification

[4]Local study near a point in $\mathbf{C}^*$ poses conceptual difficulties since those are not invariant by the map $x \mapsto qx$. It is progressively tackled in more or less recent works **[176, 207, 209]** but we shall have little to say about it.

will provide us with formal invariants and formal classes; then we shall divide each formal class into analytic classes, requiring more (analytic) invariants to distinguish them[5].

REMARK 7.56. An immediate calculation shows that $F[uA] = uF[A]$ for every $u \in K$, $F, A \in \mathrm{GL}_n(K)$. So, if $u \in K^*$, then $A \sim B \Leftrightarrow uA \sim uB$. Therefore in order to characterize the equivalence class of A, we may replace A by some convenient uA. For instance, if $A \in \mathrm{GL}_n(\mathbf{C}(x))$, we may assume that $A \in \mathrm{Mat}_n(\mathbf{C}[x])$ and that its n^2 coefficients are mutually coprime. (This is done in **[23]** and **[176]**; also see 12.2.)

7.2.3. Systems regular at 0.

7.2.3.1. In the theory of (linear, complex, analytic) ordinary differential equations "ordinary points" are those at which Cauchy theorem (in its complex analytic guise) can be applied, thus yielding a fundamental system of *power series solutions* **[222]**. For differential equations with formal power series coefficients, Cauchy theorem is replaced by the method of indeterminate coefficients. Practically, the point 0 is ordinary if the equation is (or boils down to) a system $X' = AX$ where $A(0) \in \mathrm{Mat}_n(\mathbf{C})$, meaning that 0 is not a pole. We now explain the corresponding notion for q-difference systems.

DEFINITION 7.57. The system $\sigma_q X = AX$, $A \in \mathrm{GL}_n(\mathbf{C}((x)))$, is said to have an *ordinary point at* 0, or to be *regular at* 0, if A is defined (i.e., has no pole) at 0 and $A(0) = I_n$.

This makes sense in particular for analytic or rational systems since $\mathbf{C}(x), \mathbf{C}(\{x\}) \subset \mathbf{C}((x))$. It means that A can be expanded as a power series with matricial coefficients:

$$A = A_0 + A_1 x + \cdots, \text{ where } A_0 = I_n.$$

7.2.3.2. *Local resolution.* Exactly as in the case of ordinary differential equations, such systems can be solved by the "method of indeterminate coefficients" going back to Newton; and then, for analytic systems, convergence of solutions comes by the "method of majorants" going back to Cauchy.

PROPOSITION 7.58. (i) *Let $A \in GL_n(\mathbf{C}((x)))$ be regular at 0. Then there is a unique fundamental matricial solution $\mathcal{X} \in GL_n(\mathbf{C}((x)))$ of $\sigma_q \mathcal{X} = A\mathcal{X}$ subject to the initial condition $\mathcal{X}(0) = I_n$.*

(ii) *If $A \in GL_n(\mathbf{C}(\{x\}))$, i.e., if A converges, then $\mathcal{X} \in GL_n(\mathbf{C}(\{x\}))$, i.e., then $\mathcal{X}$ converges.*

PROOF. (i) We look for $\mathcal{X}$ in the form $\mathcal{X} = \mathcal{X}_0 + \mathcal{X}_1 x + \cdots$. The initial condition reads $\mathcal{X}_0 = I_n$. Likewise expanding A as $A = A_0 + A_1 x + \cdots$, where, by assumption, $A_0 = I_n$, the functional equation, written coefficientwise, says that:

$$\forall k \geq 0 \,,\; q^k \mathcal{X}_k = A_0 \mathcal{X}_k + \cdots + A_k \mathcal{X}_0.$$

For $k = 0$ it is implied by the initial condition. For $k \geq 1$, it transforms to:

$$\mathcal{X}_k = \frac{-1}{1 - q^k} \left(A_1 \mathcal{X}_{k-1} + \cdots + A_k \mathcal{X}_0\right).$$

This is a *bona fide* recurrence definition and it guarantees the existence and uniqueness of $\mathcal{X}$.

[5]Subtleties about *isoformal analytic classes* will be dealt with in Chapter 15.

(ii) Now assume that A converges. We expand A as before. Using an arbitrary matricial norm[6], we know that there exist $C, r > 0$ such that $|||A_k||| < Cr^{-k}$ for all $k \geq 0$. Now $|||\mathcal{X}_0||| = 1$ and:

$$\forall k \geq 1\,,\ |||\mathcal{X}_k||| \leq \frac{1}{|1-q^k|}\left(|||A_1|||\ |||\mathcal{X}_{k-1}||| + \cdots + |||A_k|||\ |||\mathcal{X}_0|||\right).$$

So we let $d := \min\{|1-q^k| \mid k \geq 1\} > 0$ (can you see why $d > 0$?) and deduce:

$$\forall k \geq 1\,,\ |||\mathcal{X}_k||| \leq \frac{C}{d}\left(r^{-1}\ |||\mathcal{X}_{k-1}||| + \cdots + r^{-k}\ |||\mathcal{X}_0|||\right),$$

i.e., setting $y_k := r^k\ |||\mathcal{X}_k|||$ and $D := C/d$:

$$y_0 = 1 \text{ and } \forall k \geq 1\,,\ y_k \leq D(y_0 + \cdots + y_{k-1}).$$

By easy induction, $y_k \leq (1+D)^k$, whence $|||\mathcal{X}_k||| \leq \left(\frac{1+D}{r}\right)^k$ for all $k \geq 0$, from which the convergence of $\mathcal{X}$ follows.

□

COROLLARY 7.59. *In the analytic case:*

$$\mathcal{X} = A^{-1}(\sigma_q A)^{-1} \cdots = \prod_{k \geq 0} (\sigma_q^k A)^{-1} \text{ (ordered infinite product).}$$

PROOF. Let $B_k := A^{-1}(\sigma_q A)^{-1} \cdots (\sigma_q^{k-1} A)^{-1}$ (so by the usual convention on empty products $B_0 = I_n$). Iterating the relation:

$$\mathcal{X} = A^{-1}\sigma_q\mathcal{X} = A^{-1}\sigma_q(A^{-1}\sigma_q\mathcal{X}) = A^{-1}(\sigma_q A)^{-1}\sigma_q^2\mathcal{X} = \cdots,$$

we find that $\mathcal{X} = B_k \sigma_q^k \mathcal{X}$ for all $k \geq 0$. When $k \to +\infty$, $\sigma_q^k \mathcal{X}(x) = \mathcal{X}(q^k x)$ tends to $\mathcal{X}(0) = I_n$ and our relation tends to the one to be proved. □

EXAMPLE 7.60. The system $\begin{cases} \sigma_q f = f + xg, \\ \sigma_q g = g, \end{cases}$ translates in matricial terms to $\sigma_q X = AX$ where $X := \begin{pmatrix} f \\ g \end{pmatrix}$ and $A := \begin{pmatrix} 1 & x \\ 0 & 1 \end{pmatrix} = I_2 + xN$, where $N^2 = 0$ and $\sigma_q N = qN$. Then $\sigma_q^k A^{-1} = (\sigma_q A^{-1})^k = I_2 - q^k xN$ and one finds easily that $B_k = I_2 - \frac{1-q^k}{1-q} xN$, so in the end:

$$\mathcal{X} = \begin{pmatrix} 1 & -\frac{x}{1-q} \\ 0 & 1 \end{pmatrix}.$$

This means that the fundamental system of solutions of our system with the given initial conditions is:

$$\begin{pmatrix} f_1 \\ g_1 \end{pmatrix} = \begin{pmatrix} 1 \\ 0 \end{pmatrix}, \quad \begin{pmatrix} f_2 \\ g_2 \end{pmatrix} = \begin{pmatrix} -x/(1-q) \\ 1 \end{pmatrix}.$$

[6] All norms on $\mathrm{Mat}_n(\mathbf{C})$ are equivalent and can be used to characterize absolute convergence, but we call *matricial norm* one such $|||I_n||| = 1$ and $|||MN||| \leq |||M|||\ |||N|||$.

7.2.3.3. *Meromorphic continuation.* Consider again the system $\sigma_q X = AX$, supposed to be regular at 0: $A(0) = I_n$ and moreover supposed to be rational: $A \in \mathrm{GL}_n(\mathbf{C}(x))$. The local fundamental matricial solution $\mathcal{X}$ such that $\mathcal{X}(0) = I_n$ is well defined (it converges) and regular (with values in $\mathrm{GL}_n(\mathbf{C})$) in some open disk $\mathring{\mathrm{D}}(0,r)$, $r > 0$. Then, using the functional equation in the same way as we did in 4.4.1.3 and in 5.3.2.2, we can extend $\mathcal{X}$ to $\mathring{\mathrm{D}}(0,r')$, $r' := |q|^{-1} r > r$: we first define $\mathcal{X}'$ on $\mathring{\mathrm{D}}(0,r')$ by setting $\mathcal{X}'(x) := A(x)^{-1}\mathcal{X}(qx)$ (it may have poles, we'll see that in a moment). Then from the functional equation $\sigma_q \mathcal{X} = A\mathcal{X}$ we see that $\mathcal{X}' = \mathcal{X}$ on $\mathring{\mathrm{D}}(0,r)$, so it is indeed an extension of $\mathcal{X}$ to $\mathring{\mathrm{D}}(0,r')$.

As for poles, we first note that, if x is a pole of A^{-1}, since $\mathcal{X}$ is regular at qx (i.e., $\mathcal{X}(qx) \in \mathrm{GL}_n(\mathbf{C})$), then x is a pole of $\mathcal{X}'$. But we are also interested in places where $\mathcal{X}$ fails to be regular, i.e., where $\det \mathcal{X}$ vanishes. One of the reasons is that solutions X of $\sigma_q X = AX$ are parameterized by q-elliptic vectors: they have the form $X = \mathcal{X}C$, where $C := \mathcal{X}^{-1}X$ has coefficients in $\mathscr{M}(\mathbf{E}_q)$ (the argument was reproduced in 7.2.1.3). So poles of $\mathcal{X}^{-1}$ also are spots where something happens. So we call *singularities* of $\mathcal{X}'$ its poles and the zeroes of its determinant (equivalently, its poles and those of its inverse) and we are led to introduce the singular locus of A:

$$\mathrm{Sing}(A) := \{\text{poles of } A\} \cup \{\text{poles of } A^{-1}\} = \{\text{poles of } A\} \cup \{\text{zeroes of } \det A\}.$$

It follows that, at our first step of extension, $\mathrm{Sing}(\mathcal{X}') = \mathrm{Sing}(A) \cap \mathring{\mathrm{D}}(0,r)$.

EXERCISE 7.61. Check that:

$$\{\text{poles of } A\} \cup \{\text{poles of } A^{-1}\} = \{\text{poles of } A\} \cup \{\text{zeroes of } \det A\}.$$

Iterating, we obtain a meromorphic extension of $\mathcal{X}$ to the whole of $\mathbf{C}$. We go on denoting it $\mathcal{X}$. At each new step of extension $\mathcal{X}'(x) := A(x)^{-1}\mathcal{X}(qx)$, it may have acquired singularities from A or from the previous extension $\mathcal{X}$; and also some singularities may have disappeared because, if A^{-1} has a pole at x and $\det \mathcal{X}$ has a zero at qx, or the converse, one cannot be sure of the status of $\mathcal{X}'(x) := A(x)^{-1}\mathcal{X}(qx)$. So in the end we conclude:

PROPOSITION 7.62. *If A is regular at 0 and rational, then the unique fundamental matricial solution $\mathcal{X}$ of the system $\sigma_q X = AX$ such that $\mathcal{X}(0) = I_n$ is meromorphic over $\mathbf{C}$ and:*

$$\mathit{Sing}(\mathcal{X}) \subset q^{-\mathbf{N}}\mathit{Sing}(A).$$

7.2.4. Systems with "truly constant" coefficients. This is a first contact, we shall return to them in 11.1.

7.2.4.1. Here, "truly constant" means "in $\mathbf{C}$" in contrast to "q-constant". So we are looking at systems $\sigma_q X = AX$, $A \in \mathrm{GL}_n(\mathbf{C})$. We shall solve such a system by building an explicit fundamental matricial solution $\mathcal{X} \in \mathrm{GL}_n(\mathscr{M}(\mathbf{C}^*))$. The examples given in the previous chapters show that we cannot hope for much better. More precisely, we can find an explicit smaller subfield K of $\mathscr{M}(\mathbf{C}^*)$ in which the coefficients will lie; but we cannot avoid the essential singlarity at 0. Also, we cannot avoid a subfield of constant K^{σ_q} that be transcendental over $\mathbf{C}$.

Recall from Chapter 2, section 2.3, that we defined:

- for each $c \in \mathbf{C}^*$ a *q-character*, i.e., a nonzero function $e_{q,c} \in \mathscr{M}(\mathbf{C}^*)$ such that $\sigma_q e_{q,c} = c e_{q,c}$;
- a *q-logarithm*, i.e., a function $\ell_q \in \mathscr{M}(\mathbf{C}^*)$ such that $\sigma_q \ell_q = \ell_q + 1$.

The choice of those particular functions is, to some extent, immaterial. So, in this section, we shall just assume that we have a family of nonzero functions $e_c \in \mathscr{M}(\mathbf{C}^*)$ such that $\sigma_q e_c = c e_c$; and a function $\ell \in \mathscr{M}(\mathbf{C}^*)$ such that $\sigma_q \ell = \ell + 1$.

We shall use the *multiplicative Dunford decomposition* of A **[222]** (see F.4 in Appendix F): this means that $A = SU = US$, where $S \in \mathrm{GL}_n(\mathbf{C})$ is semi-simple and $U \in \mathrm{GL}_n(\mathbf{C})$ is unipotent. The matrices S and U are uniquely characterized by these conditions and actually are polynomial in A: $S, U \in \mathbf{C}[A]$. We shall denote $N := U - I_n$, which is therefore nilpotent: $N^n = 0$, and also polynomial in A: $N \in \mathbf{C}[A]$.

7.2.4.2. *Solving for the semi-simple part.* Let $S = P\mathrm{Diag}(c_1, \ldots, c_n)P^{-1}$. Then we set $e_S := P\mathrm{Diag}(e_{c_1}, \ldots, e_{c_n})P^{-1}$. Although the diagonalisation of S is not unique, the resulting e_S is unique, so our definition is consistent, thanks to the corresponding discussion in the introduction (see again F.4 in Appendix F).

PROPOSITION 7.63. (i) *The matrix e_S is polynomial in S: $e_S \in \mathscr{M}(\mathbf{C}^*)[S]$.*
(ii) *$\sigma_q e_S = S e_S = e_S S$.*
(iii) *$e_{PSP^{-1}} = P e_S P^{-1}$.*

PROOF. It all follows from the lemma and easy calculations. □

7.2.4.3. *Solving for the unipotent part.* Recall (see for instance **[222]**) that, U being unipotent, we can define its powers with respect to any complex exponent λ:

$$U^\lambda = (I_n + N)^\lambda = \sum_{k=0}^{n-1} \binom{\lambda}{k} N^k = I_n + \lambda N + \cdots \frac{\lambda(\lambda-1)\cdots(\lambda-n+2)}{(n-1)!} N^{n-1}.$$

This Newton binomial series doesn't go further since $N^n = 0$. This operation satisfies the usual rules: $U^{\lambda+\mu} = U^\lambda U^\mu$, $(U^\lambda)^\mu = U^{\lambda\mu}$ and, if U_1 and U_2 commute, $(U_1 U_2)^\lambda = U_1^\lambda U_2^\lambda$. These notations and properties extend immediately to exponents that are complex valued functions (or actually elements of any commutative $\mathbf{C}$-algebra), thus yielding matrix functions (i.e., at will functions with values that are complex matrices or matrices with coefficients that are complex valued functions). So we can set:

$$e_U := U^\ell = \sum_{k=0}^{n-1} \ell^{(k)} N^k, \text{ where } \ell^{(k)} := \binom{\ell}{k}.$$

EXERCISE 7.64. Compute $\sigma_q \ell^{(k)} - \ell^{(k)}$.

Clearly $e_U \in \mathscr{M}(\mathbf{C}^*)[N] = \mathscr{M}(\mathbf{C}^*)[U]$. The following formal computation:

$$\sigma_q e_U = \sigma_q \left(U^\ell\right) = U^{\sigma_q \ell} = U^{\ell+1} = U e_U = e_U U$$

can be directly justified using the expanded formula for e_U and the exercise above. We summarize:

PROPOSITION 7.65. (i) *The matrix e_U is polynomial in*
$U : e_U \in \mathscr{M}(\mathbf{C}^*)[U]$.
(ii) *$\sigma_q e_U = U e_U = e_U U$.*
(iii) *$e_{PUP^{-1}} = P e_U P^{-1}$.*

PROOF. It all follows from the above and some more easy calculations. □

7.2.4.4. *Solving it all.* Now we set $e_A := e_S e_U = e_U e_S$. The commutation comes from the fact that e_S is polynomial in S, whence in A; and similarly for e_U. The same facts allow to complete the arguments for the following.

THEOREM 7.3. (i) *The matrix e_A is polynomial in A: $e_A \in \mathscr{M}(\mathbf{C}^*)[A]$.*
(ii) $\sigma_q e_A = A e_A = e_A A$.
(iii) $e_{PAP^{-1}} = P e_A P^{-1}$.

EXERCISE 7.66. Try to find the form of the coefficients of e_A (involving q-characters and q-logarithms) and compare with the resolution of scalar equations with constant coefficients in Chapter 5.

7.2.4.5. *Bifurcations.* We just mention for the sake of completeness two other possible ways of defining e_A.

Through Theta functions of matrices. You can define, for arbitrary $A \in \mathrm{GL}_n(\mathbf{C})$:

$$\Theta(A) := \sum_{n \in \mathbf{Z}} q^{n(n-1)/2} A^n,$$

a convergent sum in the Banach algebra $\mathrm{Mat}_n(\mathbf{C})$. Then the function $T_A : x \mapsto \Theta(xA)$ from $\mathbf{C}^*$ to $\mathrm{Mat}_n(\mathbf{C})$ is holomorphic and satisfies a nice functional equation w.r.t. σ_q. Combining such functions yields variants of e_A.

Through holomorphic functional calculus. (References for functional calculus, sometimes called operational calculus or symbolic calculus, are [**39**, Chapter I], [**199**, Chapter VII], [**212**, Chapter 10] and [**227**, Chapter 6].) Functions defined by convergent power series are easily extended to Banach algebras as above. A more general process rests on Cauchy integrals. Thus, if a function f is holomorphic on a domain Ω containing the spectrum of A and if Γ is a simple positive contour in Ω enclosing that contour, one may set:

$$f(A) := \frac{1}{2i\pi} \int_\Gamma (xI_n - A)^{-1} f(x)\, dx.$$

Fixing x_0, this can be applied to $f : c \mapsto e_c(x_0)$ to define $e_A(x_0)$.

EXERCISE 7.67. Explore both of those alternative approaches.

7.3. Sheaves of solutions

The reader is advised to read (again?) Subsection 4.5.3 for a first introduction to sheaves of solutions in an elementary setting; and, if necessary, to look at Appendix B.1.3.4 for some basic information about sheaves of holomorphic functions on $\mathbf{E}_q$.

7.3.1. Generalities: local *vs* global.

7.3.1.1. An aspect of Riemann's point of view on complex functions is to combine local information (e.g., behaviour near singularities) with global information (e.g., for differential equations, connection formulas between solutions near distant singularities). This was later encoded by Hilbert in the so-called "monodromy representation", thus giving rise to the Riemann-Hilbert correspondence [**169, 222, 250**]. One most flexible framework to do that is the formalism of sheaves (of solutions).

Direct transposition to q-difference equations of the methods that were successful for differential equations presents two difficulties. First of all, the classical theory rests essentially on analytic continuation of multivalued solutions, while in this book we made a consistent choice to use only uniform solutions. This choice has been stated several times and will be illustrated and discussed again; see in particular section 12.1.1.

Second, it is by no way obvious what a *local study* would look like at a point $a \in \mathbf{C}^*$ which is moved elsewhere by the q-dilatation operator $a \mapsto qa \neq a$ (in contrast, the operator d/dx does not seem[7] to move points). So, to begin with, we shall attempt a local study only at the fixed points of the q-dilatation operator: 0 and ∞; note however that in 7.3.4 we will hint at what can be done at "intermediate singularities", those in $\mathbf{C}^*$.

7.3.1.2. *Dialectics of local vs global.* When zooming at 0; the local/global dialectics manifests itself firstly through the q-expansion of the domain of definition that was decribed in 7.2.3.3 (and of which various examples had already appeared, for instance in Chapter 1). In the case of a rational equation or system, this can be iterated all the way from 0 to ∞ and it results (at least in the fuchsian case) in the *Birkhoff connection matrix* (Chapter 12).

On the other hand, each time the expansion process produces a singularity a, this will also be iterated to all $q^n a$ ($n > 0$ or $n < 0$ as may be the case) and we are led to take in account the whole q-spiral $[a; q]$, or, equivalently, the class $\overline{a} \in \mathbf{E}_q$. Also, starting from a neighborhood $U_0 \ni a$ in $\mathbf{C}^*$ (the reason not to allow $0 \in U_0$ is explained in the exercise below) , we may have to consider all the $U_n := q^n U_0$ as well as the neihborhood $V := \pi(U_0) \ni \overline{a}$ in $\mathbf{E}_q$ and its preimage $U := \pi^{-1}(V) = \bigcup_n U_n$.

EXERCISE 7.68. If $0 \in U_0$ then $U = \mathbf{C}$. (See 14.1.1.)

Then a miracle happens. The fact that $\mathbf{C}^* \to \mathbf{E}_q$ is an etale covering (4.5.1.3) or Appendix B.1.3.4 entails that the neighborhood U_0 can be chosen such that all U_n are pairwise disjoint; and then a solution can be (almost) freely chosen on U_0 (then extended to U), i.e., the sheaf of solutions is *locally free.* But to such a sheaf is attached a *holomorphic vector bundle* on $\mathbf{E}_q$, a very nice geometric structure which—to a large extent—encodes our equation or system. This mere fact is one more sign that the theory of q-difference equations is simpler than the theory of differential equations (also see 0.1.2 in the Introduction).

7.3.1.3. Since it is going to be a central argument, we briefly recall (in a down-to-earth way) why every $a \in \mathbf{C}^*$ admits an open neighborhood $U_0 \subset \mathbf{C}^*$ such that the $U_n := q^n U_0$, $n \in \mathbf{Z}$ are pairwise disjoint (a *trivializing* neighborhood: this implies in particular that each U_n maps homeomorphically to the open neighborhood $V := \pi(U_0)$ of $\overline{a}$ in $\mathbf{E}_q$).

Let $c_q := \min\limits_{n \in \mathbf{Z}} \dfrac{|1 - q^n|}{1 + |q|^n} > 0$ (see the exercise below). Now let $a \in \mathbf{C}^*$ and $r > 0$ such that $r < c_q |a|$. Then the open disks $\overset{\circ}{\mathrm{D}}(q^n a, |q|^n r) = q^n \overset{\circ}{\mathrm{D}}(a, r)$ are pairwise disjoint (exercise also) and we may take $U_0 := \overset{\circ}{\mathrm{D}}(a, r)$.

EXERCISE 7.69. (i) Prove that $c_q := \min\limits_{n \in \mathbf{Z}} \dfrac{|1 - q^n|}{1 + |q|^n}$ exists and that $c_q > 0$.

(ii) Check that the $q^n \overset{\circ}{\mathrm{D}}(a, r)$ are pairwise disjoint.

[7]However, for a geometric view of what happens "inside a point"; see [**55, 151**].

7.3.2. Sheaves of solutions. Here we generalize the constructions displayed in 4.5.3.1, so it may be a good idea (for the reader not accustomed to sheaves) to first read (again?) that subsection.

To begin with, we concentrate on the local analytic setting at 0, i.e., we do everything over $\mathbf{C}(\{x\})$. The local analytic case at ∞ (i.e., over $\mathbf{C}(\{1/x\})$ will be dealt with only as far as definitions are involved, the corresponding facts are left to the reader to state and prove. The formal local case (over $\mathbf{C}((x))$) does not seem[8] to give rise to interesting geometric constructions. The global case (over $\mathbf{C}(x)$) will be tackled in 7.3.4.

7.3.2.1. *Definition of the relevant sheaves.* We consider simultaneously the case of a scalar q-difference operator and that of a system; so we let:

$$L := \sigma_q^n + a_1\sigma_q^{n-1} + \cdots + a_n,\ a_1, \ldots, a_n \in \mathbf{C}(\{x\}),\ a_n \neq 0,$$

and $A \in \mathrm{GL}_n(\mathbf{C}(\{x\}))$ (by no way related to each other). Since either equation $Lf = 0$ or $\sigma_q X = AX$ only has a meaning in some unspecified neighborhood of 0, we consider germs of solutions. Since we cannot hope for a smooth behaviour at 0 itself, we consider punctured neighborhoods. And, of course, they must be q-invariant, whence of the form $\pi^{-1}(V)$. So, for every open subset $V \subset \mathbf{E}_q$, we define:

$$\mathscr{F}_L^{(0)}(V) := \{f \in \mathscr{O}(\pi^{-1}(V), 0) \mid Lf = 0 \text{ near } 0\},$$
$$\mathscr{F}_A^{(0)}(V) := \{X \in \mathscr{O}(\pi^{-1}(V), 0)^n \mid \sigma_q X = AX \text{ near } 0\}.$$

In case L or A has coefficients in $\mathbf{C}(\{1/x\})$, we would use instead:

$$\mathscr{F}_L^{(\infty)}(V) := \{f \in \mathscr{O}(\pi^{-1}(V), \infty) \mid Lf = 0 \text{ near } \infty\},$$
$$\mathscr{F}_A^{(\infty)}(V) := \{X \in \mathscr{O}(\pi^{-1}(V), \infty)^n \mid \sigma_q X = AX \text{ near } \infty\}.$$

See Remark 4.86 (where the case $n = 1$ was explained) to get a precise understanding of the restrictions "near 0", "near ∞".

7.3.2.2. *Local freeness.* From the very same arguments as in 4.5.3.1, we draw that all the above are sheaves of modules over the sheaf $\mathscr{O}_{\mathbf{E}_q}$ of holomorphic functions on $\mathbf{E}_q$, for short $\mathscr{O}_{\mathbf{E}_q}$-modules; and that moreover:

PROPOSITION 7.70. *The $\mathscr{O}_{\mathbf{E}_q}$-modules $\mathscr{F}_L^{(0)}, \mathscr{F}_A^{(0)}$ (resp $\mathscr{F}_L^{(\infty)}, \mathscr{F}_A^{(\infty)}$ as the case may be) are locally free of rank n.*

PROOF. Although the proofs are straightforward transpositions of those given in 4.5.3, we shall sketch them. For an arbitrary $\overline{a} \in \mathbf{E}_q$, let $a \in \mathbf{C}^*$ a representative, U_0 a trivializing open neighborhood of a and $V = \pi(U_0)$ the corresponding neighborhood of $\overline{a}$. We can choose U_0 so that all the coefficients of L (resp. of A) are holomorphic over U_0 and a_n vanishes nowhere over U_0 (resp. A is invertible over U_0): indeed it is enough to take U_0 inside a small enough disk $\mathring{\mathrm{D}}(0, r)$ and this can be done by replacing U_0 by some $q^p U_0$, p big enough.

An element f of $\mathscr{F}_L^{(0)}(V)$ is given by the sequence $(f_p)_{p \geq 0}$ of its restrictions $f_{|U_p}$, where each $f_p \in \mathscr{O}(U_p)$, and equation $Lf = 0$ is satisfied. The latter condition

[8] An apparent contradiction is the article **[17]**, but it deals with fuchsian equations only and, for those, the formal and analytic study are essentially equivalent (Chapter 11).

translates to a recurrence relation of order n among the f_p, in which $f_0, \ldots, f_{n-1}$ are free to choose. This gives an isomorphism:

$$f \mapsto (f_0, \ldots, f_{n-1}), \quad \mathcal{F}_L^{(0)}(V) \to \mathcal{O}_{\mathbf{C}^*}(U_0) \times \cdots \times \mathcal{O}_{\mathbf{C}^*}(U_{n-1}) \simeq \mathcal{O}_{\mathbf{E}_q}(V)^n,$$

the isomorphism being as $\mathcal{O}_{\mathbf{E}_q}(V)$-modules. The family of those isomorphisms (when V varies) is clearly compatible with the restriction maps of the sheaves, so it yields local isomorphisms of $\mathcal{F}_L^{(0)}$ with $\mathcal{O}_{\mathbf{E}_q}^n$ as required.

For systems, choosing a solution X amounts to choosing $X_0 \in \mathcal{O}_{\mathbf{C}^*}(U_0)^n$ freely and the local isomorphy of $\mathcal{F}_A^{(0)}$ with $\mathcal{O}_{\mathbf{E}_q}^n$ flows even more directly. □

EXERCISE 7.71. Write down explicitly the recurrence relation for the sequence (f_p) and check that the resulting map $f \mapsto (f_0, \ldots, f_{n-1})$ is indeed bijective. (The $\mathcal{O}_{\mathbf{E}_q}(V)$-linearity is of course obvious).

7.3.2.3. *Equations and systems.* Suppose that $A = A_L$, the companion system to the q-difference operator. We proved in 7.1.3 that their spaces of solutions in any q-difference extension K' were isomorphic. Exactly the same calculation and argument remain valid if K' is replaced by $\mathcal{O}(\pi^{-1}(V), 0)$ and K'^{σ_q} by $\mathcal{O}(V)$, so we get isomorphisms of $\mathcal{O}(V)$-modules $\mathcal{F}_L^{(0)}(V) \simeq \mathcal{F}_A^{(0)}(V)$. When moreover V varies, those isomorphisms are compatible with restrictions maps. In short:

PROPOSITION 7.72. *If $A = A_L$, then $\mathcal{F}_L^{(0)} \simeq \mathcal{F}_A^{(0)}$ as $\mathcal{O}_{\mathbf{E}_q}$-modules.*

7.3.2.4. *Gauge equivalence.* We saw in 7.1.2.2 that a gauge equivalence $B = F[A]$ gives rise to an isomorphism $X \mapsto FX$ of spaces of solutions in K'. Just as above, this remains valid if K' is replaced by $\mathcal{O}(\pi^{-1}(V), 0)$, etc, and we have compatibility with the restriction maps:

PROPOSITION 7.73. *If $B = F[A]$, $F \in GL_n(\mathbf{C}(\{x\}))$, then the maps $X \mapsto FX$ define an isomorphism $\mathcal{F}_A^{(0)} \xrightarrow{\simeq} \mathcal{F}_B^{(0)}$ of $\mathcal{O}_{\mathbf{E}_q}$-modules.*

Combining those two propositions and the cyclic vector lemma from 7.1.4, we see that the study of the sheaves of solutions can, to a large extent, concentrate on the case of systems. This will be the case in all the sequel of Section 7.3.

7.3.3. The associated vector bundle.

We refer to Appendix B.5 for general facts on holomorphic vector bundles on a Riemann surface.

7.3.3.1. We noted in 4.5.3 the "general fact" that every locally free $\mathcal{O}_X$-module (X a Riemann surface) can be identified with the sheaf of sections of a holomorphic vector bundle over X. We shall now explicitly construct the vector bundle[9] $\mathfrak{F}_A^{(0)}$ over $\mathbf{E}_q$ associated to the sheaf $\mathcal{F}_A^{(0)}$.

EXERCISE 7.74. Assume that A is regular over $\mathbf{C}^*$ (i.e., $A(x)$ is defined and invertible for all $x \in \mathbf{C}^*$). Check that $\dfrac{\mathbf{C}^* \times \mathbf{C}^n}{(x, X) \sim (qx, A(x)X)}$ is a vector bundle over $\mathbf{E}_q = \dfrac{\mathbf{C}^*}{x \sim qx}$ and describe its sheaf of holomorphic sections.

[9]We presently use (slightly) different letters $\mathfrak{F}$ and $\mathcal{F}$ to write the bundle and the sheaf, although a very general custom is to represent them in the same way.

As also noted in *loc. cit.*, we cannot do that in general since $A \in \mathrm{GL}_n(\mathbf{C}(\{x\}))$ cannot be supposed to meet the stated conditions. So here again we consider the germ of space $(\mathbf{C}^*, 0)$ [**85**, p 16], [**16**, Chapter 0 §III], and use the same conventions as those explained in 4.5.3 about functions over unprecised "small enough" punctured neighborhoods of 0. Now we define:

$$\mathfrak{F}_A^{(0)} := \frac{(\mathbf{C}^*, 0) \times \mathbf{C}^n}{(x, X) \sim (qx, A(x)X)}.$$

We then leave it to the reader to prove:

PROPOSITION 7.75. *This defines a holomorphic vector bundle $\mathfrak{F}_A^{(0)} \to \mathbf{E}_q$ whose sheaf of holomorphic sections can be canonically identified with $\mathscr{F}_A^{(0)}$.*

EXERCISE 7.76. Avoid resorting to germs of spaces along the following lines. For every small enough punctured open disk D (centered at 0), there is a well defined sheaf $\mathfrak{F}_A^{(D)}$. The family of all such $\mathfrak{F}_A^{(D)}$ form a direct system in which all morphisms are isomorphisms. Then $\mathfrak{F}_A^{(0)}$ can be taken as the direct limit. (See [**216**, 2.1, p 408].)

7.3.3.2. *Holomorphic triviality.* Here and in the next subsection, we invoke Theorem B.8 of Appendix B.5.6. After statement (i) of that theorem, for every $\overline{a} \in \mathbf{E}_q$, the restriction of the vector bundle $\mathfrak{F}_A^{(0)}$ to $V := \mathbf{E}_q \setminus \{\overline{a}\}$ is trivial; equivalently, the restriction to V of the sheaf $\mathscr{F}_A^{(0)}$ is free. This means that there is a family of holomorphic sections $s_1, \ldots, s_n \in \mathscr{F}_A^{(0)}(V)$ such that every $\mathscr{F}_A^{(0)}(V')$, $V' \subset V$, is freely generated by the restrictions of the s_i to V'. Interpreting those sections as holomorphic (vector valued) functions X_i on $(U, 0)$, where $U := \pi^{-1}(V) = \mathbf{C}^* \setminus [a; q]$ ($a \in \mathbf{C}^*$ any representative of $\overline{a}$), such that $\sigma_q X_i = AX_i$, then setting $\mathcal{X}^{(0)} := [X_1, \ldots, X_n]$, we draw:

PROPOSITION 7.77. *The system $\sigma_q X = AX$ admits a fundamental matricial solution $\mathcal{X}^{(0)} \in GL_n(\mathscr{O}(\mathbf{C}^* \setminus [a; q], 0))$.*

However, we do not consider the above to be a very useful result as it says nothing of the singular behaviour of $\mathcal{X}$ on $[a; q]$: this can be expected to be very wild and, truly, the interest of *meromorphic* solutions (poles) is much greater for us. So the next result is much more spectacular.

7.3.3.3. *Meromorphic triviality and Praagman's theorem.* After statement (ii) of theorem B.8, the vector bundle is $\mathfrak{F}_A^{(0)}$ *meromorphically trivial.* This means that there is a family of *global meromorphic sections* $s_1, \ldots, s_n$. Interpreting those sections as meromorphic (vector valued) functions X_i on $(\mathbf{C}^*, 0)$ such that $\sigma_q X_i = AX_i$, then setting $\mathcal{X}^{(0)} := [X_1, \ldots, X_n]$, we draw:

THEOREM 7.4 (Praagman). *The system $\sigma_q X = AX$ admits a global meromorphic fundamental matricial solution $\mathcal{X}^{(0)} \in GL_n(\mathscr{M}(\mathbf{C}^*, 0))$.*

This striking result, due to Praagman [**182, 249**], corresponds to nothing in the case of differential equations, since it is well known that even very simple equations such as $xf' = \alpha f$ admit no nontrivial uniform solutions unless $\alpha \in \mathbf{Z}$. Here "global" of course means that it is valid over a whole (unprecised) punctured neighborhood of 0, no "cuts" are necessary as in the case of x^α (see [**222**]). A truly global result (over the whole of $\mathbf{C}^*$) would require that $A = (\sigma_q \mathcal{X}^{(0)})(\mathcal{X}^{(0)})^{-1}$ also be globally defined; and conversely this is sufficient thanks to the expansion of the disks of definition:

COROLLARY 7.78. *If* $A \in GL_n(\mathcal{M}(\mathbf{C}))$ *then* $\mathcal{X}^{(0)} \in GL_n(\mathcal{M}(\mathbf{C}^*))$.

REMARK 7.79. There are many such fundamental solutions and (a particular one being given) all have the form $\mathcal{X}^{(0)}P$, where $P \in \mathrm{GL}_n(\mathcal{M}(\mathbf{E}_q))$: this is because $\mathcal{M}(\mathbf{C}^*, 0)^{\sigma_q} = \mathcal{M}(\mathbf{E}_q)$ so P, which is a priori q-invariant, is automatically defined all over $\mathbf{C}^*$. Therefore the conclusion of the corollary would apply either to all fundamental solutions or to none.

7.3.3.4. *Matricial divisors.* We take $A \in \mathrm{GL}_n(\mathbf{C}(\{x\}))$ and choose a particular fundamental matricial solution $\mathcal{X}^{(0)} \in \mathrm{GL}_n(\mathcal{M}(\mathbf{C}^*, 0))$ as above. Let $V \subset \mathbf{E}_q$ an open subset. Then the usual argument shows that there is an isomorphism of $\mathcal{O}_{\mathbf{E}_q}(V)$-modules (taking again $U := \pi^{-1}(V)$):

$$C \mapsto \mathcal{X}^{(0)}C, \quad \{C \in \mathcal{M}(\mathbf{E}_q)^n \mid \mathcal{X}^{(0)}C \text{ is holomorphic over } (U, 0)\} \to \mathcal{F}_A^{(0)}(V)$$

This is reminiscent of the notion of matricial divisors developed by Weil in **[254]**. That line of thought does not seem to have been pursued, although some related ideas will be displayed in 7.3.4.

EXERCISE 7.80. Describe in this way $\mathfrak{F}_u$ and $\mathfrak{F}_u^{(0)}$ (defined in 4.5.3) for $u = cx^m$, $c \in \mathbf{C}^*$, $m \in \mathbf{Z}$. You may begin with the cases $u = c$ and $u = x$.

7.3.4. Intermediate singularities. Proofs are left as exercises or to be looked for in **[209]**, where all this comes from; beware that there $|q| > 1$.

7.3.4.1. When $A \in \mathrm{GL}_n(\mathbf{C}(x))$ ("global" or "rational setting"), the local study may be performed at 0 and at ∞ and gives rise to two sheaves $\mathcal{F}_A^{(0)}$ and $\mathcal{F}_A^{(\infty)}$. Using fundamental matricial solutions $\mathcal{X}^{(0)}$ and $\mathcal{X}^{(\infty)}$, both in $\mathrm{GL}_n(\mathcal{M}(\mathbf{C}^*))$ as seen above, we have $\mathcal{X}^{(0)} = \mathcal{X}^{(\infty)}P$ for some $P \in \mathrm{GL}_n(\mathbf{E}_q)$ (the so-called "connection matrix" of Birkhoff; see Chapter 12). Then a solution $\mathcal{X}^{(0)}C$ "local at 0" may be viewed as a solution $\mathcal{X}^{(\infty)}D$, $D := PC$ "local at ∞".

EXERCISE 7.81. Show that $C \mapsto PC$ can be interpreted as a "meromorphic isomorphism" of bundles from $\mathfrak{F}_A^{(0)}$ to $\mathfrak{F}_A^{(\infty)}$.

In some sense, the connection matrix P (or more intrinsically the above meromorphic isomorphism) carries information on how the local structure at 0 of q-difference equation A is transformed into the local structure at ∞. Although this is rather vague, such local structures will be examined more closely in the coming chapters, starting from Chapter 11.

Here we are rather interested in what happens during the transformation. It is a natural guess (more than 150 years after Riemann!) that everything significant should be related to *intermediate singularities*, those in $\mathbf{C}^*$. Such singularities are *priori* the points $a \in \mathbf{C}^*$ where A fails to be regular, i.e., where $A(a)$ fails to be defined and invertible (the singular locus $\mathrm{Sing}(A)$ was precisely defined in 7.2.3.3); and a posteriori all points of the corresponding q-spirals $[a; q]$, because of the long range effects of q-dilatations $x \mapsto q^{\pm 1}x$. We shall give some (very partial) information on what is known and understood (which is mighty little).

7.3.4.2. So let $A \in \mathrm{GL}_n(\mathbf{C}(x))$. Along with the sheaves $\mathscr{F}_A^{(0)}$ and $\mathscr{F}_A^{(\infty)}$ on $\mathbf{E}_q$ introduced in 7.3.2, we introduce two new ones; let V and U be as usual:

$$\mathscr{F}_A(V) := \{\text{solutions of } \sigma_q X = AX \mid X \text{ holomorphic over } U\},$$
$$\mathscr{F}'_A(V) := \{\text{solutions of } \sigma_q X = AX \mid X \text{ almost holomorphic over } U\},$$

where "almost holomorphic over U" means: having at most finitely many poles on any q-spiral $[a;q] \subset U$. Clearly $\mathscr{F}_A \subset \mathscr{F}'_A$. There are good reasons for the seemingly unnatural definition of $\mathscr{F}'_A$. The first one is the following.

PROPOSITION 7.82. (i) $\mathscr{F}'_A = \mathscr{F}_A^{(0)} \cap \mathscr{F}_A^{(\infty)}$.
(ii) *If* $B = F[A]$, $F \in GL_n(\mathbf{C}(x))$, *then* $F\mathscr{F}'_A = \mathscr{F}'_B$.

The proposition corresponding to (ii) in 7.3.2.4 used the fact that, if $F \in \mathrm{GL}_n(\mathbf{C}(\{x\}))$, then left multiplication by F strictly preserves holomorphy in a punctured neighborhood of 0 (and of course that it sends solutions of A to solutions of B, a general property of gauge transformations). Here we use the fact that, if $F \in \mathrm{GL}_n(\mathbf{C}(x))$, then left multiplication by F strictly preserves the finiteness condition defining $\mathscr{F}'_A$. However, it does not preserve the holomorphy condition defining $\mathscr{F}_A$:

EXAMPLE 7.83. Let $n = 1$ $A = 1$, $F = \dfrac{1}{x-1}$, so that $B := F[A] = \dfrac{x-1}{qx-1}$. Then $\mathscr{F}_A = \mathscr{O}_{\mathbf{E}_q}$. The comparison with $\mathscr{F}_B$ goes as follows: if $\overline{1} \notin V$, then $\mathscr{F}_A(V) = \mathscr{F}_B(V)$. However, if $\overline{1} \in V$, then:

$$\mathscr{F}_B(V) = \frac{1}{x-1}\mathscr{F}_A(V) \cap \mathscr{O}_{\mathbf{C}^*}(U) \neq \frac{1}{x-1}\mathscr{F}_A(V),$$

since for instance $\dfrac{1}{x-1} \in \dfrac{1}{x-1}\mathscr{F}_A(V) \setminus \mathscr{O}_{\mathbf{C}^*}(U)$.

However, although less intrinsic, $\mathscr{F}_A$ is easier to use than $\mathscr{F}'_A$. The following facts will make everything smoother. Say that A is "nonresonant"[10] if $\mathrm{Sing}(A) \cap q^{\mathbf{N}^*}\mathrm{Sing}(A) = \emptyset$, i.e., any two distinct singularities are non congruent modulo $q^{\mathbf{Z}}$.

PROPOSITION 7.84. (i) *If A is nonresonant, then* $\mathscr{F}_A = \mathscr{F}'_A$.
(ii) *Every matrix in* $GL_n(\mathbf{C}(x))$ *is rationally equivalent to a nonresonant one.*

7.3.4.3. To understand the specific contribution of the intermediate singularities amounts to separate it from the local descriptions at 0 and ∞. One way to do that is to describe the quotient sheaves $\mathscr{F}_A^{(0)}/\mathscr{F}'_A$ and $\mathscr{F}_A^{(\infty)}/\mathscr{F}'_A$. To do that, we can assume that A is nonresonant and $\mathscr{F}_A = \mathscr{F}'_A$. In what follows, we take the convention that either $a \in \mathrm{Sing}(A)$ or $[a;q]$ does not meet $\mathrm{Sing}(A)$ (so if $\overline{a}$ is singular, a is its unique representative in $\mathrm{Sing}(A)$).

PROPOSITION 7.85. *The sheaves* $\mathscr{F}_A^{(0)}/\mathscr{F}_A$ *and* $\mathscr{F}_A^{(\infty)}/\mathscr{F}_A$ *are torsion sheaves, whence finite sums of skyscraper sheaves (see Appendix* H*); more precisely, they have trivial stalks at nonsingular points and their stalks at a singular* $\overline{a}$ *are finite length* $\mathscr{O}_{\mathbf{E}_q,\overline{a}}$*-modules (thus finite-dimensional complex vector spaces) completely described by the local structure of* A *at* a *(i.e., its elementary divisors).*

[10]This terminology has a purely local scope. In this book, we shall rather use this expression in relation with exponents, as in Chapters 4, 5, and 11.

Fix $\overline{a} \in \mathbf{E}_q$ and a representative $a \in \mathbf{C}^*$ with the above convention. Then the local ring $R := \mathscr{O}_{\mathbf{E}_q,\overline{a}} = \mathscr{O}_{\mathbf{C}^*,a}$ is the discrete valuation ring $\mathbf{C}\{u\}$, where $u := x - a$ (uniformizing parameter). A first lemma rests on the conditions that ensure holomorphy of the solutions:

LEMMA 7.86. *The stalks, as R-modules, are isomorphic to:*

$$\left(\mathscr{F}_A^{(0)}/\mathscr{F}_A\right)_{\overline{a}} \simeq R^n/(R^n \cap A^{-1}R^n) \text{ and } \left(\mathscr{F}_A^{(\infty)}/\mathscr{F}_A\right)_{\overline{a}} \simeq R^n/(R^n \cap AR^n)$$

The theorem of elementary divisors ([**33**, VII] or [**128**]) then gives us a decomposition $A = CDE$, with $C, E \in \mathrm{GL}_n(R)$ and $D = \mathrm{Diag}(u^{d_1}, \ldots, u^{d_n})$, where $d_1 \leq \cdots \leq d_n$ are in $\mathbf{Z}$. Now the following is pure linear algebra:

LEMMA 7.87. *As R-modules:*

$$R^n/(R^n \cap AR^n) \simeq \prod_{d_i>0} R/u^{d_i}R \text{ and } R^n/(R^n \cap A^{-1}R^n) \simeq \prod_{d_i<0} R/u^{-d_i}R.$$

Clearly the stalks at a nonsingular point are trivial and, at a singular point, as stated in the proposition.

CHAPTER 8

Systems and Modules

This chapter and the following present most useful algebraic tools for the study of q-difference equations, so we shall try to motivate carefully the basic constructions. In essence, we gain freedom - coordinate freedom. But first, why (and how) should one go coordinate free? In linear algebra and in geometry, one finds that change of coordinates may be useful. When doing so, what remains true, what keeps meaning? The geometric properties themselves.

Moreover, applying linear methods in some situation, the use of coordinates may be not so easy. For instance, $A \mapsto PAP^{-1}$ is a linear map from $\text{Mat}_n(K)$ to itself and its spectral properties are interesting (7.2.4) or **[222]**. Yet, writing its matrix in the canonical basis $(E_{i,j})_{1 \leq i,j \leq n}$ of $\text{Mat}_n(K)$ is complicated and maybe not necessary. So it is often more convenient to think of a linear map $f : V \to V'$ than to write down its explicit description in coordinates or matrix form.

A homogeneous linear system of equations $AX = 0$ (where $A \in \text{Mat}_{m,n}(K)$ and X is to be found in K^n identified, as often, to $\text{Mat}_{n,1}(K)$) can be thought of geometrically, its solutions form a linear subspace of K^n, the kernel of the linear maps $X \mapsto AX$. To some extent, something similar can be done with differential equations and systems. The system $X' = AX$ can be modelled or encoded by a pair (K^n, Δ_A), where $\Delta_A : K^n \to K^n$ is the map $X \mapsto X' - AX$ (note however that it is not K-linear, only C-linear, where C is the constant subfield of elements such that $f' = 0$) **[222]**. The solutions of the differential system are the "horizontal vectors". To describe the space $\text{Ker}\Delta_A$ in coordinate free style, one replaces (K^n, Δ_A) by a pair (V, Δ) where V is a K-vector space of finite dimension and Δ a "twisted-linear" map, meaning that it satisfies Leibniz formula: $\Delta(fX) = f'X + f\Delta(X)$ (you should check that Δ_A satisfies that formula too). The formal frame underlying system $X' = AX$ is that of *differential module* (V, Δ).

Changes of coordinates (i.e., change of unknowns), be it for linear systems of equations or for differential linear systems, lead to the notion of *isomorphism*. For differential systems, this comes in the form $X = FY$, $F \in \text{GL}_n(K)$, leading to $Y' = BY$ where $B := F^{-1}(AF - F')$ ("gauge transformation"). We already saw the corresponding notion for q-difference systems in 7.1.1.3 of Chapter 7. Before describing the more abstract notion for differential modules, we note that the above relation is equivalent to $F' = AF - FB$; in that form it remains meaningful even if F is not invertible, or not even a square matrix. This leads to the notion of *morphism* and this widened scope is just as fruitful as going from isomorphisms of vector spaces to more general linear maps.

This chapter is devoted to q-difference modules, which are the q-analogues of differential modules, their morphisms and their structure. In the next chapter,, we look at more sophisticated properties of q-difference modules that will be useful in classification and in Galois theory: tensor properties in 9.1 and homological properties in 9.2. The reader not conversant with those subjects may either admit on faith the main results, or come back to that chapter only when needed.

REMARK 8.1. In the theory of differential equations, there is another coordinate involved. For instance, when studying differential equation $y' = u(x)y$, one may wish to see what happens "at infinity". The practical way is to use a new variable $z := 1/x$ and to look at $z = 0$. More intrinsically, one may consider the Riemann sphere **S** as a surface, for which we know there is a coordinate free theory: (complex) differential geometry. This includes a possibility of formulating differential systems independently of the variable x: the corresponding notion is that of connection on a *vector bundle*[1]. This has been explored for q-difference equations by Tarasov and Varchenko in [**243**] but we shall not go into that in this book.

CAUTION 8.2. We are going to deal with categories and morphisms. After some hesitations, and in accordance with the pedestrian style of this book, I chose to denote $\mathrm{Hom}_{\mathscr{C}}(X,Y)$ (instead of the more standard $\mathrm{Mor}_{\mathscr{C}}(X,Y)$) the set of all morphisms from an object X to an object Y in a category $\mathscr{C}$.

8.1. Generalities on q-difference modules

Let us emphasize right now that the new objects we are going to introduce - q-difference modules, that is pairs (V,Φ) of a vector space with an additional structure (a σ_q-linear automorphism) - "are" (i.e., may be identified with) the same thing as modules over the ring $\mathcal{D}_{K,q}$ of q-difference operators, constrained with an additional property (finite length). In essence, we shall construct the category q-diffMod of q-difference modules over the q-difference field (K,σ_q) so that it can be identified with the full subcategory of $\mathcal{D}_{K,q}$Mod (this is the category of $\mathcal{D}_{K,q}$-modules) having as objects the $\mathcal{D}_{K,q}$-modules of finite length. As a consequence, it will automatically be a K^{σ_q}-linear abelian category. So we shall go back and forth between the two models. However, there is an important difference: the tensor structure that we will put on the category of q-difference modules in Chapter 9 is *not* induced by the tensor structure of $\mathcal{D}_{K,q}$Mod.

Our idea, in generalizing the operator Δ_A mentioned above, is to consider solutions of the q-difference system $\sigma_q X = AX$, $A \in \mathrm{GL}_n(K)$, as fixed points[2] of the map: $X \mapsto A^{-1}\sigma_q X$. While the solutions and the fixed points are usually to be searched in q-difference extensions of K, the structural properties of the q-difference system $\sigma_q X = AX$ are already completely determined by the action of that map on K^n. So our relevant object will be:

$$\Phi_A : K^n \to K^n, X \mapsto A^{-1}\sigma_q X.$$

The pair (K^n, Φ_A) should be seen as a q-analogue of the pair (K^n, Δ_A).

Recall that for a differential module, we abstracted Leibniz formula:

$$\Delta_A(fX) = (fX)' - AfX = f'X + fX' - AfX = f'X + f\Delta_A(X),$$

itself coming from the derivation property $(fg)' = f'g + fg'$, into an axiom $\Delta(fX) = f'X + f\Delta(X)$. The automorphism property $\sigma_q(fg) = (\sigma_q f)(\sigma_q g)$ which replaces

[1] Vector bundles appear in this book but with a quite different use; and the word "connection" is used too, but with a totally different meaning.

[2] In Section 7.1.5 of Chapter 7, we considered said solutions as zeroes of the map $X \mapsto \sigma_q X - AX$. That linear approach is well adapted to some questions (in particular homological questions) but do not properly account for the very *multiplicative* character of q-difference theory.

the derivation property implies the "q-Leibniz formula":

$$\Phi_A(fX) = A^{-1}\sigma_q(fX) = A^{-1}(\sigma_q f)(\sigma_q X) = (\sigma_q f)(A^{-1}\sigma_q X) = (\sigma_q f)\Phi_A(X).$$

This will motivate our axiom for q-difference modules.

The change of unknown $X = FY$ or $Y = FX$ was already studied in Section 7.1.2 of Chapter 7 and led to the notion of gauge transformation, defined either by $A^F := (\sigma_q F)^{-1}AF$ or by $F[A] := (\sigma_q F)AF^{-1}$. If we consider for instance equality $B = (\sigma_q F)AF^{-1}$ as an isomorphism from A to B (we give it an orientation!), we can rewrite it as $(\sigma_q F)A = BF$, which still has a meaning if F is not invertible, and see it as a morphism from A to B.

If we want to express morphisms in terms of our new operator Φ_A, we are led to the following computation (it will motivate the definition of morphisms of q-difference modules in 8.2.2):

$$\begin{aligned}(\sigma_q F)A = BF \Longleftrightarrow B^{-1}\sigma_q F = FA^{-1} &\Longleftrightarrow \forall X \in K^n\ ,\ B^{-1}\sigma_q F\sigma_q X = FA^{-1}\sigma_q X \\ &\Longleftrightarrow \forall X \in K^n\ ,\ B^{-1}\sigma_q(FX) = F(A^{-1}\sigma_q X) \\ &\Longleftrightarrow \forall X \in K^n\ ,\ \Phi_B(FX) = F\Phi_A(X) \\ &\Longleftrightarrow \Phi_B \circ F = F \circ \Phi_A.\end{aligned}$$

EXERCISE 8.3. If $\Phi_B \circ F = F \circ \Phi_A$, show how F relates fixed points of Φ_A, Φ_B.

8.1.1. Basic definitions.

DEFINITION 8.4. A *q-difference module* over a q-difference field K is a pair (V, Φ) consisting of a finite-dimensional K-linear space V and a *σ_q-linear* group automorphism Φ of K. Its *rank*, denoted rank M, is the dimension $\dim_K V$ of the underlying space. Here, "σ_q-linear" means that Φ satisfies the following q-analogue of Leibniz rule:

$$\forall f \in K\ ,\ \forall X \in V\ ,\ \Phi(fX) = \sigma_q(f)\Phi(X).$$

Note that if $\sigma_q f = f$ the above equality becomes $\Phi(fX) = f\Phi(X)$, so σ_q-linearity implies K^{σ_q}-linearity. We write $\mathscr{L}_{\sigma_q}(V, W)$ the set of σ_q-linear maps $V \to W$ between two K-vector spaces.

EXAMPLE 8.5. If $A \in \mathrm{GL}_n(K)$, the pair $M_A := (K^n, \Phi_A)$ is a q-difference module. The *trivial* q-difference module has trivial underlying space: $V := \{0\}$. The *unit* q-difference module (Tannakian terminology) is $\underline{\mathbf{1}} := (K, \sigma_q)$.

In order to show that any q-difference module is, in essence, of the form (K^n, Φ_A), we shall now introduce a notion that will be generalized later (see the definition of morphisms in 8.2.2).

DEFINITION 8.6. An *isomorphism of q-difference modules* $(V, \Phi) \to (W, \Psi)$ is an isomorphism of K-linear spaces $F : V \to W$ intertwining Φ and Ψ, i.e., such that the diagram

$$\begin{array}{ccc} V & \xrightarrow{\ \Phi\ } & V \\ {\scriptstyle F}\big\downarrow & {\scriptstyle \Psi\circ F = F\circ\Phi} & \big\downarrow{\scriptstyle F} \\ W & \xrightarrow[\ \Psi\ ]{} & W. \end{array}$$

commutes. (Thus F is K-linear, not σ_q-linear.)

EXAMPLE 8.7. Let F an isomorphism $(K^n, \Phi_A) \to (K^p, \Phi_B)$. As an isomorphism $K^n \to K^p$ it implies that $n = p$ and it can be interpreted as a matrix in

$\mathrm{GL}_n(K)$. The intertwining condition then reads:

$$\forall X \in K^n \ , \ FA^{-1}\sigma_q X = B^{-1}\sigma_q(FX) = B^{-1}(\sigma_q F)(\sigma_q X),$$

equivalently, $FA^{-1} = B^{-1}(\sigma_q F)$, i.e., $B = F[A]$. In other words, for modules of the form (K^n, Φ_A), isomorphy is the same thing as gauge equivalence.

PROPOSITION 8.8. *Every q-difference module is isomorphic to one of the form (K^n, Φ_A).*

PROOF. Let (V, Φ) a q-difference module and let $\mathcal{B} = (e_i)$ a basis of V. Then $\Phi(\mathcal{B}) = (\Phi(e_i))$ is a basis of V; indeed:

(1) If $\sum \alpha_i \Phi(e_i) = 0$, then by applying Φ^{-1}, which is σ_q^{-1}-linear, we get $\sum \sigma_q^{-1}(\alpha_i) e_i = 0$, whence all $\sigma_q^{-1}(\alpha_i) = 0$, whence all $\alpha_i = 0$ and $\Phi(\mathcal{B})$ is a free family.
(2) Let $x \in V$ and write $\Phi^{-1}(x)$ as $\sum \alpha_i e_i$. Then $x = \Phi(\Phi^{-1}(x)) = \sum \sigma_q(\alpha_i)\Phi(e_i)$ and $\Phi(\mathcal{B})$ is a generating family.

Therefore, there exists a matrix $A \in \mathrm{GL}_n(K)$ such that $\mathcal{B} = \Phi(\mathcal{B})A$. Then the isomorphism $F : X \mapsto \mathcal{B}X$ from K^n to V fits in a commutative diagram:

$$\begin{array}{ccc} K^n & \xrightarrow{\ ?\ } & K^n \\ {\scriptstyle F}\big\downarrow & & \big\downarrow{\scriptstyle F} \\ V & \xrightarrow{\ \Phi\ } & V. \end{array}$$

The left vertical arrow should send X to an Y such that $\Phi(F(X)) = F(Y)$, i.e., $\Phi(\mathcal{B}X) = \mathcal{B}Y$, i.e., $\Phi(\mathcal{B})\sigma_q(X) = \mathcal{B}Y$ (here we used σ_q-linearity of Φ), i.e., $\mathcal{B}A^{-1}\sigma_q(X) = \mathcal{B}Y$, i.e., $Y = A^{-1}\sigma_q(X) = \Phi_A(X)$. Therefore the missing arrow is Φ_A and the diagram shows that $(V, \Phi) \simeq (K^n, \Phi_A)$. □

REMARK 8.9. If $\mathcal{B} = (e_1, \ldots, e_n)$ (a family in a K-linear space V) and $X = {}^t(x_1, \ldots, x_n)$ (a column vector in $\mathrm{Mat}_{n,1}(K)$, we denote $\mathcal{B}X := \sum\limits_{i=1}^{n} x_i e_i \in V$. If $X \in \mathrm{Mat}_{n,p}(K)$ has columns $X_1, \ldots, X_p$, then $\mathcal{B}X \in V^p$ has components $\mathcal{B}X_1, \ldots, \mathcal{B}X_p$. If one admits that $ex = xe$ when $x \in K$, $e \in V$, that calculus works smoothly (associativity, bilinearity ...).

8.1.2. From q-difference modules to $\mathcal{D}_{K,q}$-modules of finite length and back. Recall that, in linear algebra, the reduction theory of endomorphisms can be approached through the classification of pairs (V, ϕ), V a vector space, ϕ an endomorphism of V. Such a pair can be identified with a $K[X]$-module with underlying group V and with external multiplication defined as $(P, v) \in K[X] \times V \mapsto P.v := P(\phi)(v) \in V$. Finite dimensionality of V is equivalent to that $K[X]$-module being a finitely generated torsion module; or, equivalently, to having finite length. A similar approach is available for differential modules **[222]** and we shall mimick it for q-difference modules; however, there will be a twist ...

From the moment we shall have properly detailed the correspondence between q-difference modules and $\mathcal{D}_{K,q}$-modules of finite length, we shall identify them when convenient, or else (if we want to compare both points of view) say that "(V, Φ) corresponds to M".

8.1.2.1. *From q-difference modules to $\mathcal{D}_{K,q}$-modules.* So let (V,Φ) a q-difference module. We begin by a heuristic computation. For a polynomial $P := \sum a_i X^i$ and an element $v \in V$, define:

$$P.v := \sum a_i \Phi^i(v).$$

Note that, since Φ is bijective, we may allow negative exponents i, so we may as well take for P a Laurent polynomial: $P \in K[X,X^{-1}]$. For the above definition to define a module, we should verify some axioms. Actually, all of them come smoothly (the computations are obvious, do them!) except from the "external associativity" which should write: $P.(Q.v) = (PQ).v$. So we compute the left-hand side of that wished-for equality; taking $Q := \sum b_j X^j$:

$$P.(Q.v) = \sum a_i \Phi^i\left(\sum b_j \Phi^j(v)\right) = \sum a_i \sigma_q^i(b_j) \Phi^{i+j}(v) = \sum c_k \Phi^k(v),$$

where we set $c_k := \sum\limits_{i+j=k} a_i \sigma_q^i(b_j)$. This means that we should define a new product of Laurent polynomial by formula $PQ := R$, where $R := \sum c_k X^k$. For instance, we may compute $(aX)(bX)$ either by that definition or by redoing the above calculation in that particular case. We do the latter:

$$a\Phi(b\Phi(v)) = a\sigma_q(b)\Phi^2(v) \Longrightarrow (aX)(bX) = a\sigma_q(b)X^2.$$

So this multiplication is *not commutative.* The basic case deserves special notice:

EXERCISE 8.10. Show that $\forall a \in K\ ,\ Xa = \sigma_q(a)X$.

We recognize the noncommutative polynomials of Chapter 3, i.e., we are actually defining a structure of $\mathcal{D}_{K,q}$-module. In what follows, we shall take up the notations from *loc. cit.*, in particular the letter S for the formal noncommutative indeterminate generating $K\langle S,S^{-1}\rangle$.

PROPOSITION 8.11. *For $P := \sum a_i S^i \in K\langle S,S^{-1}\rangle = \mathcal{D}_{K,q}$ and $v \in V$, set $P.v := \sum a_i \Phi^i(v)$. This gives the group V a structure of left $\mathcal{D}_{K,q}$-module.*

PROOF. Most of the proof is trivial. We detail it mainly to make clear that there are axioms to be satisfied. With obvious notations, we must check:

$$P.(v+w) = P.v + P.w, \quad (P+Q).v = P.v + Q.v, \quad 1.v = v, \quad P.(Q.v) = (PQ).v.$$

Note that they amount to the following: each map $P(\Phi)$ is an endomorphism of the group V: (this translates formula $P.(v+w) = P.v + P.w$ and it is obvious since $P(\Phi)$ actually is an endomorphism of the K^{σ_q}-vector space V); and the map $P \mapsto P(\Phi)$ is a morphism of rings from $\mathcal{D}_{K,q}$ to the ring of endomorphisms of the group V (this translates the three other formulas).

Equalities $(P+Q).v = P.v+Q.v$ and $1.v = v$ respectively say that $(P+Q)(\Phi) = P(\Phi) + Q(\Phi)$ and that $1(\Phi) := \Phi^0 = \mathrm{Id}$: like the first formula, they are particular cases of classical facts about polynomial of endomorphisms of the K^{σ_q}-space V.

Equality $P.(Q.v) = (PQ).v$ says that $P(\Phi) \circ Q(\Phi)) = (PQ)(\Phi)$, i.e., with obvious notations, that $(\sum a_i \Phi^i) \circ (\sum b_j \Phi^j) = \sum c_k \Phi^k$, where $\sum c_k S^k := (\sum a_i S^i) \circ (\sum b_j S^j)$ (multiplication in $\mathcal{D}_{K,q}$). Using the distributivities of composition (of endomorphisms) and of multiplication (of non commutative polynomials) with respect to additions, this boils down to formula $(a\Phi^i) \circ (b\Phi^j) = c\Phi^k$, where $cS^k := (aS^i)(bS^j) = a\sigma_q^i(b)S^{i+j}$. So we are left to check that $(a\Phi^i) \circ (b\Phi^j) = a\sigma_q^i(b)\Phi^{i+j}$, i.e., that $a\Phi^i(b\Phi^j)(v) = a\sigma_q^i(b)\Phi^{i+j}(v)$ for all $v \in V$. This follows easily from the σ_q-linearity of Φ. □

8.1.2.2. *q-difference modules give rise to finite length $\mathcal{D}_{K,q}$-modules.* Thus any q-difference module (V, Φ) can be seen as a left $\mathcal{D}_{K,q}$-module; let us write it M. Of course, the underlying group of M is V, but it has additional structure. Conversely, let M a $\mathcal{D}_{K,q}$-module. Since K is a subring of $\mathcal{D}_{K,q}$, we can see M as a K-module ("restriction of scalars"), i.e., as a K-vector space: let us write V that space. Define:

$$\Phi : V \to V, v \mapsto S.v,$$

that is, the left multiplication by $S \in V$. From properties of modules, this is a group automorphism (it is bijective because S is invertible in $\mathcal{D}_{K,q}$). Moreover:

$$\forall a \in K \ , \ \forall v \in V \ , \ \Phi(av) = Sav = \sigma_q(a)Sv = \sigma_q(a)\Phi(v)$$

so Φ is σ_q-linear. This almost makes (V, Φ) a q-difference module: we should require that V be finite-dimensional, which is not automatic, and actually not true for an arbitrary $\mathcal{D}_{K,q}$-module. For instance, taking $M := \mathcal{D}_{K,q}$ (considered as a left module over itself) would not work since it has infinite dimension over K (the S^i, $i \in \mathbf{Z}$, are an infinite basis).

Actually, there is an obvious necessary condition on M: for any strictly increasing sequence of submodules $0 = M_0 \subset \cdots \subset M_l = M$ (thus $M_i \neq M_{i+1}$ for $i = 0, \ldots, l-1$), one must have $l \leq \dim_K V$: indeed, each submodule M_i is *a fortiori* a subspace, so we must have $\dim_K M_{i+1} \geq \dim_K M_i + 1$, whence $l \leq \dim_K M_l = \dim_K V$. Therefore, the *length* of the module M, that is the greatest possible value of the length l of such a subsequence, is bounded above by $\dim_K V$: the module M *has finite length.* It can be shown **[145]** that all maximal possible subsequences have the same length l, which is of course the length of M. Also, clearly, such a subsequence is maximal if, and only if, there is no intermediate module between M_i and M_{i+1} for $i = 0, \ldots, l-1$. (See 8.2.1.5 on that.)

PROPOSITION 8.12. *In the above construction of (V, Φ) from M, the K-vector space V has finite dimension if, and only if, the $\mathcal{D}_{K,q}$-module M has finite length.*

PROOF. This could actually be deduced from 8.1.5 and from the study of exact sequences in 8.2, however we prefer to give a direct proof, because it initiates us smoothly to $\mathcal{D}_{K,q}$-modules reasoning.

First, if we start from (V, Φ) and assume that V has finite dimension n, we just proved that M has finite length $l \leq \dim_K V$.

Conversely suppose that M has finite length l and let $0 = M_0 \subset \cdots \subset M_l = M$ a maximal sequence as above. Since the M_i are K-subspaces, $\dim_K M = \dim_K V$ is equal to the sum of the $\dim_K M_{i+1} - \dim_K M_i = \dim_K M_{i+1}/M_i$, $i = 0, \ldots, l-1$ (usual properties of dimension for vector spaces). So it is enough to show that each M_{i+1}/M_i has finite dimension. But we know that it is a *simple* $\mathcal{D}_{K,q}$-module, i.e., it is nontrivial and has no other submodule other than 0 and itself (this is because there is no intermediate module between M_i and M_{i+1}).

So we must show that a simple $\mathcal{D}_{K,q}$-module N is finite-dimensional as a K-vector space. Let $v \in N$, $v \neq 0$. Since $\mathcal{D}_{K,q}v$ is a nontrivial submodule, it must be equal to N (this is because N is simple). So there is a surjective morphism of $\mathcal{D}_{K,q}$-modules:

$$\mathcal{D}_{K,q} \to N, P \mapsto Pv.$$

Let I its kernel, so that $N \simeq \mathcal{D}_{K,q}/I$. The left ideal I is not 0 because $\mathcal{D}_{K,q}$ is not a simple $\mathcal{D}_{K,q}$-module (see the exercise below). So by euclideanity (3.1.3), I is principal: $I = \mathcal{D}_{K,q}P$ for some $P \neq 0$; and we may assume P to be standard unitary (this was defined in 3.1.3), so, thanks to euclidean division $\mathcal{D}_{K,q}/I = \mathcal{D}_{K,q}/\mathcal{D}_{K,q}P$ has finite dimension $\deg P$ as a K-vector space (3.1.3 again). □

The q-difference module $\mathcal{D}_{K,q}/\mathcal{D}_{K,q}P$ will be denoted M_P.

EXERCISE 8.13. Show that $\mathcal{D}_{K,q}$ is not a simple $\mathcal{D}_{K,q}$-module: choose a noninvertible element P in $\mathcal{D}_{K,q}$ (you must check that this exists) and consider $\mathcal{D}_{K,q}P$, a submodule of $\mathcal{D}_{K,q}$.

8.1.2.3. *From $\mathcal{D}_{K,q}$-modules back to q-difference modules.* Let M a $\mathcal{D}_{K,q}$-module of finite length, and write V its underlying K-vector space. By 8.1.2.2, we know that $\dim_K V < \infty$. Define:

$$\Phi : V \to V, v \mapsto S.v,$$

thus using the external multiplication $S.v$ due to the $\mathcal{D}_{K,q}$-module structure. Obviously Φ is an automorphism of the group V (it is bijective because S is invertible). Moreover, if $a \in K$ and $v \in V$:

$$\Phi(av) = S.(av) = (Sa).v = (\sigma_q(a)S).v = \sigma_q(a)(S.v) = \sigma_q(a)\Phi(v).$$

Therefore, Φ is σ_q-linear and (V, Φ) is a q-difference module. The following is almost tautological and we shall not care to prove it:

PROPOSITION 8.14. *The process in* 8.1.2.1 *associating to a q-difference module* (V, Φ) *a* $\mathcal{D}_{K,q}$*-module* M *of finite length (by* 8.1.2.2*) and the process in* 8.1.2.3 *associating to a* $\mathcal{D}_{K,q}$*-module of finite length* M *a q-difference module* (V, Φ) *are inverse of each other.*

From now on, we shall identify a q-difference module (V, Φ) with the corresponding finite length $\mathcal{D}_{K,q}$-module M; we shall alternate notations without further notice.

EXERCISE 8.15. We have defined isomorphy of q-difference modules in 8.1.2.1 and there is a standard notion of isomorphy for $\mathcal{D}_{K,q}$-modules. Show that there are equivalent in the case of $\mathcal{D}_{K,q}$-modules of finite length. (This will be generalized to all morphisms in Section 8.2.)

It will be shown in 8.1.5 that every q-difference module (V, Φ) admits a "cyclic vector", that is an element $v \in V$ such that $(v, \Phi(v), \ldots, \Phi^{n-1}(v))$ is a basis of V (here $n := \dim_K V$). Equivalently, seen as a $\mathcal{D}_{K,q}$-module (of finite length by necessity), it is generated by an element. Therefore, by the same argument as in the proof above, there is a standard unitary $P \in \mathcal{D}_{K,q}$ such that $M \simeq \mathcal{D}_{K,q}/\mathcal{D}_{K,q}P$. However, P is not unique, and it is rather complicated to determine whether two standard unitary elements P, Q of $\mathcal{D}_{K,q}$ are such that $\mathcal{D}_{K,q}/\mathcal{D}_{K,q}P \simeq \mathcal{D}_{K,q}/\mathcal{D}_{K,q}Q$. Of course they must have the same degree (since it is the K-dimension of the underlying vector space). Examples of this difficulty are shown in 8.1.4.

So we have two computational models for q-difference modules: (K^n, Φ_A), using matrices; or $\mathcal{D}_{K,q}/\mathcal{D}_{K,q}P$, using noncommutative Laurent polynomials. Both are useful and powerful. However, in this book, we shall rely more on the matricial model, for its wonderful flexibility. Examples in Subsection 8.1.4 are meant to illustrate the interplay of these two models, but we first examine some general facts.

8.1.3. Correspondence between the models (K^n, Φ_A) and $\mathcal{D}_{K,q}/\mathcal{D}_{K,q}L$. In this subsection, we use the letter S (rather than σ or σ_q) to denote the generator of $\mathcal{D}_{K,q}$, in order to emphasize the formal, algebraic character of the arguments. We have so far introduced two special notations for q-difference modules:

$$M_A := (K^n, \Phi_A), A \in \mathrm{GL}_n(K); \text{ and } M_L := \mathcal{D}_{K,q}/\mathcal{D}_{K,q}L, L \in \mathcal{D}_{K,q}, L \neq 0.$$

8.1.3.1. Let $L := S^n + a_1 S^{n-1} + \cdots + a_n \in \mathcal{D}_{K,q}$. Using euclidean division, we saw that, by 3.1.3, as a K-vector space V:

$$V := \mathcal{D}_{K,q}/\mathcal{D}_{K,q}L = K \oplus \cdots \oplus K s^{n-1}, \text{ where } s := S \pmod{\mathcal{D}_{K,q}L},$$

i.e., it has basis $(v_1, \ldots, v_n) := (1, s, \ldots, s^{n-1})$. Call $\Phi : V \to V$ the left multiplication by S (thus involving its structure of left $\mathcal{D}_{K,q}$-module). Clearly:

$$\Phi(1) = s, \ldots, \Phi(s^{n-2}) = s^{n-1} \text{ and } \Phi(s^{n-1}) = s^n = -(a_n + a_{n-1}s + \cdots + a_1 s^{n-1}),$$

so the matrix of the basis $(\Phi(v_1), \ldots, \Phi(v_n))$ in the basis $(v_1, \ldots, v_n)$ is:

$$A' := \begin{pmatrix} 0 & 0 & 0 & \ldots & 0 & -a_n \\ 1 & 0 & 0 & \ldots & 0 & -a_{n-1} \\ 0 & 1 & 0 & \ldots & 0 & -a_{n-2} \\ \vdots & \vdots & \vdots & \ddots & \vdots & \vdots \\ 0 & 0 & 0 & \ldots & 0 & -a_2 \\ 0 & 0 & 0 & \ldots & 1 & -a_1 \end{pmatrix} \tag{8.1}$$

This is the transpose matrix ${}^t\!A_L$ (recall that matrix A_L was defined in Subsection 7.1.3). Then, the σ_q-linearity of Φ (which simply translates the fact that $\Phi(av) = (Sa)v = \sigma_q(a)Sv = \sigma_q(a)\Phi(v)$) implies:

$$\begin{aligned} \Phi(x_1 v_1 + \cdots + x_n v_n) &= \sigma_q(x_1)\Phi(v_1) + \cdots + \sigma_q(x_n)\Phi(v_n) \\ &= \sigma_q(x_1)v_2 + \cdots + \sigma_q(x_{n-1})v_n + \sigma_q(x_n)(-a_n v_1 - \cdots - a_1 v_n) \\ &= y_1 v_1 + \cdots + y_n v_n, \end{aligned}$$

where $y_1 := -a_n \sigma_q(x_1)$ and $y_i := \sigma_q(x_{i-1}) - a_{n-i+1}\sigma_q(x_n)$ for $i \geq 2$, i.e., $Y = {}^t\!A_L \sigma_q X$, where $X := \begin{pmatrix} x_1 \\ \vdots \\ x_n \end{pmatrix}$ and $Y := \begin{pmatrix} y_1 \\ \vdots \\ y_n \end{pmatrix}$. This means that the q-difference module (V, Φ) is isomorphic to the q-difference module $(K^n, \Phi_{A^\vee})$, where $A^\vee$ denotes the *contragredient matrix* $A^\vee := {}^t\!A^{-1}$. More precisely, the diagram:

$$\begin{array}{ccc} K^n & \xrightarrow{F} & V \\ {\scriptstyle \Phi_{A^\vee}} \downarrow & & \downarrow {\scriptstyle \Phi} \\ K^n & \xrightarrow[F]{} & V \end{array}$$

is commutative, where F is the isomorphism $X \mapsto \sum x_i v_i$ from K^n to V. This is a first hint that the vectorialisation (K^n, Φ_{A_L}) of the scalar equation $Lf = 0$ and the $\mathcal{D}_{K,q}$-module $\mathcal{D}_{K,q}/\mathcal{D}_{K,q}L$ are in some sense *dual* to each other. Thus if (K^n, Φ_{A_L}) can be put in the form $\mathcal{D}_{K,q}/\mathcal{D}_{K,q}L'$, the q-difference operators L and L' should also be dual to each other. We shall meet that phenomenon again in 8.1.3.3 and obtain a conceptual explanation in Appendix G (duality in $\otimes$-categories).

EXERCISE 8.16. If $L := \sigma_q^n - u$ and $L' := \sigma_q^n - u^{-1}$, show that $A_L^\vee = A_{L'}$.

8.1.3.2. Conversely, starting from $M := (K^n, \Phi_B)$, where $B := A_L^\vee$, let us see how to recover the form $\mathcal{D}_{K,q}/\mathcal{D}_{K,q}L$. We saw that $\Phi_B(X)$ is the map $X \mapsto A'\sigma_q X$, where A' is the matrix shown in (8.1). The image of the canonical basis $(e_1, \ldots, e_n)$ is therefore given by:

$$\Phi_B(e_1) = e_2, \ldots, \Phi_B(e_{n-1}) = e_n, \Phi_B(e_n) = -(a_n e_1 + \cdots + a_1 e_n).$$

The first $(n-1)$ relation show that $v := e_1$ is a cyclic vector: $\Phi_B^k(x) = e_{k+1}$ for $k = 0, \ldots, n-1$. Then:

$$\Phi_B^n(v) = -(a_n e_1 + \cdots + a_1 e_n) = -(a_n v + \cdots + a_1 \Phi_B^{n-1}(v)),$$

whence:

$$(S^n + a_1 S^{n-1} + \cdots + a_n).v = (\Phi_B^n + a_1 \Phi_B^{n-1} + \cdots + a_n)(v) = 0,$$

and this is the shortest possible relation since $v, \ldots, \Phi_B^{n-1}(v)$ are free. More precisely, if $P.v = 0$, from the euclidean division $P = QL + R$ we get $R.v = 0$, which is a linear relation for this free family, whence trivial, whence $P = QL$. It follows that the $\mathcal{D}_{K,q}$-linear map $\mathcal{D}_{K,q} \to M$, $P \mapsto P.v$ is onto with kernel $\mathcal{D}_{K,q}L$ and that $M \simeq \mathcal{D}_{K,q}/\mathcal{D}_{K,q}L$. This argument shows more generally, under the assumption that the cyclic vector lemma holds, that every q-difference module is isomorphic to some $\mathcal{D}_{K,q}/\mathcal{D}_{K,q}L$ (using a cyclic vector v just as we did). Alternatively, this can be shown using Birkhoff's version of the cyclic vector lemma given in 7.1.4.

EXERCISE 8.17. (i) Show that $B = F[A] \Rightarrow B^\vee = F^\vee[A^\vee]$.
(ii) Deduce that every q-difference module is isomorphic to some $\mathcal{D}_{K,q}/\mathcal{D}_{K,q}L$.

8.1.3.3. Now we start from a q-difference operator L, go to its vectorialization (K^n, Φ_{A_L}) and then put it in the form $\mathcal{D}_{K,q}/\mathcal{D}_{K,q}L'$; more precisely we find an explicit L' such that $(K^n, \Phi_{A_L}) \simeq \mathcal{D}_{K,q}/\mathcal{D}_{K,q}L'$.

EXERCISE 8.18. Check that:

$$A_L^{-1} = \begin{pmatrix} -a_{n-1}/a_n & -a_{n-2}/a_n & \ldots & -a_1/a_n & -1/a_n \\ 1 & 0 & \ldots & 0 & 0 \\ \vdots & \vdots & \ddots & 0 & 0 \\ 0 & 0 & \ldots & 0 & 0 \\ 0 & 0 & \ldots & 1 & 0 \end{pmatrix}.$$

Letting again $(e_1, \ldots, e_n)$ the canonical basis of the column vectors, we get the relations $\Phi_{A_L}(e_i) = b_i e_1 + e_{i+1}$, $i = 1, \ldots, n-1$ and $\Phi_{A_L}(e_n) = b_n e_1$. Let $v := e_1$, so that $e_{i+1} = \Phi_{A_L}(e_i) - b_i v$ for $i = 1, \ldots, n-1$.

EXERCISE 8.19. Check by induction that, for $i = 1, \ldots, n$:

$$e_i = \Phi_{A_L}^{i-1}(v) - \sum_{j=2}^{i} (\sigma_q^{i-j} b_{j-1}) \Phi_{A_L}^{i-j}(v).$$

In particular, $e_n = \Phi_{A_L}^{n-1}(v) - \sum_{j=2}^{n} (\sigma_q^{n-j} b_{j-1}) \Phi_{A_L}^{n-j}(v)$. Then using $\Phi_{A_L}(e_n) = b_n e_1$, we get:

$$\Phi_{A_L}^n(v) - (\sigma_q^{n-1} b_1) \Phi_{A_L}^{n-1}(v) - \cdots - (\sigma_q b_{n-1}) \Phi_{A_L}(v) - b_n v = 0,$$

i.e., $L'v = 0$, where $L' := S^n - (\sigma_q^{n-1} b_1) S^{n-1} - \cdots - (\sigma_q b_{n-1}) S - b_n$. We write this as:

$$L' = S^n + a_1' S^{n-1} + \cdots + a_n', \text{ where } a_i' := -\sigma_q^{n-i} b_i = \sigma_q^{n-i} \frac{a_{n-i}}{a_n}.$$

Reasoning as before, we see that v is a cyclic vector with minimal q-difference equation L' and that $(K^n, \Phi_{A_L}) \simeq \mathcal{D}_{K,q}/\mathcal{D}_{K,q}L'$. We consider L' as some sort of dual of L.

EXERCISE 8.20. How would the "bidual" $(L')'$ be related to L?

8.1.3.4. Let K' a q-difference extension of the q-difference field K and let $f \in K'$ be holonomic (3.2.4) with respect to K. Let $L := S^n + a_1 S^{n-1} + \cdots + a_n$ a minimal equation for f. Recall that $n = \dim_K V_f$, where V_f is the K-subspace of K' spanned by all the $\sigma_q^k f$, $k \in \mathbf{Z}$.

Since $\sigma_q V_f = V_f$, that space can be considered as a q-difference module, with $\Phi : g \mapsto \sigma_q g$. Then f is a cyclic vector, the minimal equation is L and our q-difference module is isomorphic to $\mathcal{D}_{K,q}/\mathcal{D}_{K,q}L$. On the other hand, the matrix of Φ in the basis $(f, \ldots, \sigma_q^{n-1} f)$ is the matrix A' shown in (8.1), so $\Phi = \Phi_{A_L^\vee}$.

8.1.4. Examples.

8.1.4.1. *q-difference modules of rank* 1. (Recall that rank M was defined in 8.1.1.) In the form (V, Φ), a q-difference module of rank 1 can be identified as (K, Φ_a) for some $a \in K^*$. An isomorphism $(K, \Phi_a) \to (K, \Phi_b)$ is a $f \in K^*$ such that $(\sigma_q f)a = bf$, so:

$$(K, \Phi_a) \simeq (K, \Phi_b) \Longleftrightarrow b/a \in \{(\sigma_q f)/f \mid f \in K^*\}.$$

So, with the notations of 4.1.1, the set of isomorphism classes of q-difference modules of rank 1 is $\mathrm{Pic}(K, \sigma_q)$. (The Picard group $\mathrm{Pic}(K, \sigma_q)$ is defined in 4.1.1.)

Now let M a $\mathcal{D}_{K,q}$-module which is of rank 1 as a q-difference module, i.e., as a K-vector space. Any generator v of that space is *a fortiori* a generator of the $\mathcal{D}_{K,q}$-module, so that the $\mathcal{D}_{K,q}$-linear map $P \mapsto P.v$ from $\mathcal{D}_{K,q}$ to M is onto. We have seen (e.g. in Subsubsection 8.1.3.2) that the kernel is a nontrivial left ideal $\mathcal{D}_{K,q}L$ and $M \simeq \mathcal{D}_{K,q}/\mathcal{D}_{K,q}L$. Since $\dim_K \mathcal{D}_{K,q}/\mathcal{D}_{K,q}L = \deg L$, we may therefore assume that $M = \mathcal{D}_{K,q}/\mathcal{D}_{K,q}(S - a)$ for some $a \in K^*$: this is the $\mathcal{D}_{K,q}$-module form of a q-difference module of rank 1.

Let us now determine when $\mathcal{D}_{K,q}/\mathcal{D}_{K,q}(S-a) \simeq \mathcal{D}_{K,q}/\mathcal{D}_{K,q}(S-b)$, where $a, b \in K^*$. Generally speaking, any morphism $\mathcal{D}_{K,q}/\mathcal{D}_{K,q}L \to \mathcal{D}_{K,q}/\mathcal{D}_{K,q}L'$ is completely determined by the image of the generator $u := 1 \pmod{\mathcal{D}_{K,q}L}$ of $\mathcal{D}_{K,q}/\mathcal{D}_{K,q}L$. Let $v = P \pmod{\mathcal{D}_{K,q}L'}$ that image. The morphism is then given by the formula $F \pmod{\mathcal{D}_{K,q}L} \mapsto FP \pmod{\mathcal{D}_{K,q}L'}$. For that formula to be consistent it is necessary and sufficient that $L \pmod{\mathcal{D}_{K,q}L}$ be mapped to 0, i.e., that $LP \in \mathcal{D}_{K,q}L'$, i.e., that $LP = QL'$ for some $Q \in \mathcal{D}_{K,q}$. Another way to express it is that, in the commutative diagram:

$$\begin{array}{ccc} \mathcal{D}_{K,q} & \longrightarrow & \mathcal{D}_{K,q} \\ \downarrow & & \downarrow \\ \mathcal{D}_{K,q}/\mathcal{D}_{K,q}L & \longrightarrow & \mathcal{D}_{K,q}/\mathcal{D}_{K,q}L' \end{array}$$

(the vertical maps are the canonical projections), we want the upper horizontal map $F \mapsto FP$ to go to the quotient, which is equivalent to $(\mathcal{D}_{K,q}LP) \subset \mathcal{D}_{K,q}L'$, which in turn is equivalent to $LP \in \mathcal{D}_{K,q}L'$. For general L, L', this is rather difficult to check. So we do it for our special case above.

Note that the P defining the morphism is only determined modulo $\mathcal{D}_{K,q}(S-b)$, so we can take it in K (just by replacing it by its remainder in the division by $S-b$); and even in K^*, lest the morphism be trivial while we want an isomorphism. We want that $(S-a)P = Q(S-b)$. The left-hand side is $\sigma_q(P)S - aP = \sigma_q(P)(S-b) + \sigma_q(P)b - aP$, so it belongs to $\mathcal{D}_{K,q}(S-b)$ if, and only if, $\sigma_q(P)b - aP = 0$. So we conclude that:

$$\mathcal{D}_{K,q}/\mathcal{D}_{K,q}(S-a) \simeq \mathcal{D}_{K,q}/\mathcal{D}_{K,q}(S-b) \Longleftrightarrow b/a \in \{P/\sigma_q(P) \mid P \in K^*\}.$$

So we conclude again that the set of isomorphism classes of q-difference modules of rank 1 is $\mathrm{Pic}(K, \sigma_q)$.

EXERCISE 8.21. Explain the similitude of the conclusions by proving that $\mathcal{D}_{K,q}/\mathcal{D}_{K,q}(S - a^{-1}) \simeq (K, \Phi_a)$.

EXAMPLE 8.22. Let $V := K$ and $\Phi := \Phi_1 = \sigma_q$. As a $\mathcal{D}_{K,q}$-module, the left multiplication in K is defined by $P.f = P(\sigma_q)(f)$. Any nonzero element f of K is a cyclic vector. Since $S.f = \sigma_q(f)$, its minimal equation is $S - \sigma_q(f)/f$:

$$(K, \sigma_q) \simeq \mathcal{D}_{K,q}/\mathcal{D}_{K,q}(S - \sigma_q(f)/f).$$

Its simplest form is of course $\mathcal{D}_{K,q}/\mathcal{D}_{K,q}(S-1)$. This is compatible with the above study, taking $a := 1$. It also comes through the vectorialisation of the scalar equation $\sigma_q f = f$.

We saw (at the beginning of Subsection 8.1.1) and we shall meet again (in Section 9.1) this special q-difference module under the name of *unit object* and the notation $\underline{\mathbf{1}}$ (not to be confused with the *trivial object* $(0, -)$; see Section 8.2).

8.1.4.2. *q-difference modules of rank* 2. We start from (K^2, Φ_A), where $A := \begin{pmatrix} a & b \\ c & d \end{pmatrix} \in \mathrm{GL}_2(K)$, whence $\Delta := \det A = ad - bc \neq 0$. We shall write for short $\Phi := \Phi_A$. For every $u \in K^2$, the elements $u, \Phi u, \Phi^2 u$ are linearly dependent; we look for u such that u, Φu are independent and therefore $(u, \Phi u)$ will be a basis; that condition amounts to $u \neq 0$ and $\Phi u \notin Ku$.

EXERCISE 8.23. Prove that the family $(\Phi^n u)_{n \geq 0}$ spans K^2 if, and only if, $u \neq 0$ and $\Phi u \notin Ku$.

For such a *cyclic vector* u, we have $\Phi^2 u = -a_2 u - a_1 \Phi u$ for some $a_1, a_2 \in K$, i.e., (for the $\mathcal{D}_{K,q}$-module structure) $L.u = 0$, where $L := S^2 + a_1 S + a_2$. By the usual argument, the mapping $P \mapsto Pu$ from $\mathcal{D}_{K,q}$ to M (the name of our q-difference module as a $\mathcal{D}_{K,q}$-module) is onto and has kernel $\mathcal{D}_{K,q}L$, so $M \simeq \mathcal{D}_{K,q}/\mathcal{D}_{K,q}L$.

The trick: we shall look for u in $\mathbf{C}^2$ (or in C^2, $C := K^{\sigma_q}$ for a general q-difference field K), so that $\sigma_q u = u$ and $\Phi u = A^{-1}u$. This trick does not always work. For instance, if $A = I_2$ this choice will give $\Phi u = u$ and certainly not lead to a cyclic vector. This is true more generally if A is a scalar matrix, and, as we shall see in a moment, only in that case.

Note that the family $(u, A^{-1}u)$ is free if, and only if, the family (u, Au) is (apply A to the former) and that, since $\Phi^2 u = A^{-1}\sigma_q A^{-1} u$:

$$\begin{aligned}\Phi^2 u + a_1 \Phi u + a_2 u = 0 &\Longleftrightarrow A^{-1}\sigma_q A^{-1} u + a_1 A^{-1} u + a_2 u = 0 \\ &\Longleftrightarrow u + a_1 (\sigma_q A) u + a_2 (\sigma_q A) A u = 0.\end{aligned}$$

Letting $u := \begin{pmatrix} \alpha \\ \beta \end{pmatrix}$, we see that $(u, \Phi u)$ is free if, and only if, $\beta(a\alpha+b\beta)-\alpha(c\alpha+d\beta) \neq 0$, i.e., $c\alpha^2 + (a-d)\alpha\beta + b\beta^2 \neq 0$. If $b = c = a - d = 0$, that is if A is a scalar matrix, this is impossible and our trick is doomed. If A is not scalar, this defines the complement of a degenerate conic in $\mathbf{C}^2$ (a pair of lines, may be a double line) so it is dense in $\mathbf{C}^2$ and a vector u chosen *at random* will almost surely be a cyclic vector. Moreover, if A is not diagonal, we can take u in the canonical basis.

Before pursuing the analysis, we mention a preliminary computation:

$$(\sigma_q A)A = \begin{pmatrix} \sigma_q a & \sigma_q b \\ \sigma_q c & \sigma_q d \end{pmatrix} \begin{pmatrix} a & b \\ c & d \end{pmatrix} = \begin{pmatrix} a\sigma_q a + c\sigma_q b & b\sigma_q a + d\sigma_q b \\ a\sigma_q c + c\sigma_q d & b\sigma_q c + d\sigma_q d \end{pmatrix}$$

Case $c \neq 0$. By the quadratic condition above, we can take $u := \begin{pmatrix} 1 \\ 0 \end{pmatrix}$. Then $\Phi^2 u + a_1 \Phi u + a_2 u = 0$ is equivalent to the first column of $a_1(\sigma_q A) + a_2(\sigma_q A)A$ being $-u$, i.e.:

$$\begin{cases} a_1\sigma_q a + a_2(a\sigma_q a + c\sigma_q b) = -1, \\ a_1\sigma_q c + a_2(a\sigma_q c + c\sigma_q d) = 0 \end{cases}$$

So we find $a_1 = -\dfrac{a\sigma_q c + c\sigma_q d}{c\sigma_q \Delta}$ and $a_2 = \dfrac{\sigma_q c}{c\sigma_q \Delta}$. Therefore, $M \simeq \mathcal{D}_{K,q}/\mathcal{D}_{K,q}P$, where $P := S^2 - \dfrac{a\sigma_q c + c\sigma_q d}{c\sigma_q \Delta} S + \dfrac{\sigma_q c}{c\sigma_q \Delta}$.

Case $b \neq 0$. By the quadratic condition above, we can take $u := \begin{pmatrix} 0 \\ 1 \end{pmatrix}$. Then $\Phi^2 u + a_1 \Phi u + a_2 u = 0$ is equivalent to the second column of $a_1(\sigma_q A) + a_2(\sigma_q A)A$ being $-u$, i.e.:

$$\begin{cases} a_1\sigma_q b + a_2(b\sigma_q a + d\sigma_q b) = 0, \\ a_1\sigma_q d + a_2(b\sigma_q c + d\sigma_q d) = -1 \end{cases}$$

So we find $a_1 = -\dfrac{b\sigma_q a + d\sigma_q b}{b\sigma_q \Delta}$ and $a_2 = \dfrac{\sigma_q b}{b\sigma_q \Delta}$. Therefore, $M \simeq \mathcal{D}_{K,q}/\mathcal{D}_{K,q}Q$, where $Q := S^2 - \dfrac{b\sigma_q a + d\sigma_q b}{b\sigma_q \Delta} S + \dfrac{\sigma_q b}{b\sigma_q \Delta}$.

Note that, if $bc \neq 0$, then $M \simeq \mathcal{D}_{K,q}/\mathcal{D}_{K,q}P \simeq \mathcal{D}_{K,q}/\mathcal{D}_{K,q}Q$ although P and Q are rather different and in particular $\mathcal{D}_{K,q}P \neq \mathcal{D}_{K,q}Q$.

EXERCISE 8.24. (i) Find a cyclic vector in $\mathbf{C}^2$ when $b = c = 0$ (but of course $a \neq d$).

(ii) Apply the previous study to A_L, $L := S^2 + pS + q$, and recover (for the special case of rank 2) results from 8.1.3.

8.1.5. Cyclic vectors.

8.1.5.1.

LEMMA 8.25. *Let (V, Φ) a q-difference module of rank n and let $v \in V$. Then, the following are equivalent:*

(a) *the family $(\Phi^k(v))_{k \geq 0}$ spans V;*
(b) *the family $(\Phi^k(v))_{k \in \mathbf{Z}}$ spans V;*
(c) *the finite family $(\Phi^k(v))_{k=0,\ldots,n-1}$ is a basis of V.*

PROOF. Implications (c) $\Rightarrow$ (a) $\Rightarrow$ (b) are obvious. Suppose (b) holds. Since $\dim V < \infty$, there is a nontrivial linear relation $\sum_{i=l}^{m} a_i \Phi^i(v) = 0$, with $0 \leq l \leq m$ and $a_l a_m \neq 0$. Applying Φ^{-l} to that equality yields nontrivial linear relation $\sum_{i=0}^{m-l} (\sigma_q^{-l} a_i) \Phi^i(v) = 0$, so we see that we may assume from the beginning that $l = 0$. So we take a nontrivial linear relation $\sum_{i=0}^{m} a_i \Phi^i(v) = 0$, with $0 \leq m$ and $a_0 a_m \neq 0$ and assume moreover that it has been chosen with the smallest possible value of m. In particular, the $\Phi^i(v)$ with $0 \leq i \leq m-1$ are independent, so $m \leq n$. We are going to prove that they generate V, so they are a basis and $m = n$.

Let V_k the subspace generated by the $\Phi^i(v)$, $i = 0, \ldots, k-1$. From equality $\Phi^m(v) = \dfrac{-1}{a_m}(a_0 v + \cdots + a_{m-1}\Phi^{m-1}(v))$, we deduce that $\Phi^m(v) \in V_{m-1}$, so $V_m = V_{m-1}$. Applying Φ^k, $k \geq 0$, to the same equality, we see that $\Phi^{m+k}(v) \in V_{m+k-1}$, so, by induction, all $\Phi^l(v) \in V_{m-1}$ for $l \geq m$ (and obviously for all $l \geq 0$).

On the other side of 0, we use equality $v = \dfrac{-1}{a_0}(a_1 \Phi(v) + \cdots + a_m \Phi^m(v))$ and deduce by repeated application of Φ^{-1} that $\Phi^l(v) \in V_{m-1}$ for $l \leq 0$. This ends the proof. □

DEFINITION 8.26. A vector v satisfying the conditions of the lemma is called a *cyclic vector.*

LEMMA 8.27. *Let v a cyclic vector of q-difference module $M := (V, \Phi)$. The map $P \mapsto P.v$ from $\mathcal{D}_{K,q}$ to M is onto and its kernel is $\mathcal{D}_{K,q} L$ for a unique q-difference operator:*

$$L = S^n + a_1 S^{n-1} + \cdots + a_n = \sigma_q^n + a_1 \sigma_q^{n-1} + \cdots + a_n.$$

As a consequence, $M \simeq \mathcal{D}_{K,q}/\mathcal{D}_{K,q} L$.

PROOF. The fact that the map is onto is an immediate consequence of the definition. It is clearly a morphism of left $\mathcal{D}_{K,q}$-modules, so its kernel is a left submodule of $\mathcal{D}_{K,q}$, i.e., a left ideal $\mathcal{D}_{K,q} L$ (since $\mathcal{D}_{K,q}$ is principal; see Chapter 3. Since V has finite dimension over K while $\mathcal{D}_{K,q}$ has infinite dimension (it has the denumerable basis $(S^k)_{k \in \mathbf{Z}}$), the map cannot be injective, so $L \neq 0$ and the conclusion follows (a generator can always be chosen standard unitary). □

THEOREM 8.1 (Cyclic vector lemma). *Every q-difference module (V, Φ) admits a cyclic vector.*

The proof requires some "exterior algebra" which will be recalled in 8.1.5.2; then we will give the proof in 8.1.5.3.

COROLLARY 8.28. *As a $\mathcal{D}_{K,q}$-module, every q-difference module of rank n is isomorphic to some $\mathcal{D}_{K,q}/\mathcal{D}_{K,q} L$, L standard unitary of degree n.*

8.1.5.2. *Some exterior algebra.* See **[37, 145]** for more general information. Here we deal only with finite-dimensional spaces over a fixed commutative field K. This allows us to take as definition what is really a provable property in the more general setting.

So let V a K-vector space of dimension n. Recall that the space $\mathcal{A}^r(V)$ of alternating r-linear forms[3] on V has dimension $\binom{n}{r}$. Its dual is called the *rth*

[3]Recall that those are maps $\Phi : V^r \to K$ which are separately linear in each of their arguments; and which vanish whenever two of their arguments are equal. This implies that they are

exterior power of V and denoted:

$$\wedge^r(V) := \mathcal{A}^r(V)^*.$$

Let $v_1, \ldots, v_r \in V$. They define a particular linear form on $\mathcal{A}^r(V)$:

$$v_1 \wedge \cdots \wedge v_r : \phi \mapsto \phi(v_1, \ldots, v_r), \quad \mathcal{A}^r(V) \to K,$$

so we have a map:

$$(v_1, \ldots, v_r) \mapsto v_1 \wedge \cdots \wedge v_r, \quad V^r \to \wedge^r(V).$$

That map has the following properties:

(1) The above mapping $V^r \to \wedge^r(V)$ is r-linear and alternating.
(2) For every $v_1, \ldots, v_r \in V$, the family $(v_1, \ldots, v_r)$ is free if, and only if, $v_1 \wedge \cdots \wedge v_r \neq 0$.
(3) Let $(e_1, \ldots, e_n)$ a basis of V. Then the $e_{i_1} \wedge \cdots \wedge e_{i_r}$ with $1 \leq i_1 < \cdots < e_{i_r} \leq n$ form a basis of $\wedge^r(V)$. (You can check that there is the right number $\binom{n}{r}$ of elements in that basis.)
(4) If $r + s \leq n$ there is a unique bilinear mapping from $\wedge^r(V) \times \wedge^s(V)$ to $\wedge^{r+s}(V)$ sending $(v_1 \wedge \cdots \wedge v_r, v'_1 \wedge \cdots \wedge v'_s)$ to $v_1 \wedge \cdots \wedge v_r \wedge v'_1 \wedge \cdots \wedge v'_s$. We write it[4] $(U, V) \mapsto U \wedge V$.
(5) The above mapping is nondegenerate, i.e., if $U \wedge U' = 0$ for all U', then $U = 0$; and symmetrically.

8.1.5.3. *The proof of the Cyclic vector lemma.*

Proof. The mainmotivation for using the formalism of exterior algebra is that, given a q-difference module (V, Φ) of rank n, we are looking for v such that $(v, \Phi(v), \ldots, \Phi^{n-1}(v))$ is free, that is, such that:

$$v \wedge \Phi(v) \wedge \cdots \wedge \Phi^{n-1}(v) \neq 0.$$

For $v \in V$, let V_v the subspace spanned by $v, \Phi(v), \ldots$; from the proof of Lemma 8.25, we see that, denoting $m := \dim_K V_v$ (so that $m \leq n$), the family $(v, \Phi(v), \ldots, \Phi^{m-1}(v))$ is free, actually, a basis of V_v. We deduce that m is maximal with the property that $v \wedge \Phi(v) \wedge \cdots \wedge \Phi^{m-1}(v) \neq 0$.

Now let m the maximum possible dimension of such a V_v and let v realizing it: $\dim_K V_v = m$. We want to prove that $m = n$, so, by contradiction, we assume that $m < n$. We will show that the element $U := v \wedge \Phi(v) \wedge \cdots \wedge \Phi^{m-1}(v) \in \wedge^m(V)$ is such that $U \wedge U' = 0$ for all $U' \in V$, so, by the nondegeneracy property in 8.1.5.2, this will be the contradiction.

We introduce an arbitrary $v' \in V$, and $\lambda \in \mathbf{C}$ and $s \in \mathbf{Z}$; and set $w := v + \lambda x^s v'$. By the definition of m, we have $w \wedge \Phi(w) \wedge \cdots \wedge \Phi^m(w) = 0$. Using the multilinearity of the exterior product, we see that $w \wedge \Phi(w) \wedge \cdots \wedge \Phi^m(w)$ is a polynomial expression in λ. Its constant term is $v \wedge \Phi(v) \wedge \cdots \wedge \Phi^m(v)$, which is 0, but this is no information. Its term of degree 1 is:

$$\sum_{i=0}^{m} v \wedge \cdots \wedge \Phi^i(\lambda x^s v') \wedge \cdots \wedge \Phi^m(v) = \lambda x^s \sum_{i=0}^{m} q^{is} v \wedge \cdots \wedge \Phi^i(v') \wedge \cdots \wedge \Phi^m(v),$$

---

antisymmetrical, i.e., $\phi(v_{\tau(1)}, \ldots, v_{\tau(r)}) = \operatorname{sgn}(\tau)\phi(v_1, \ldots, v_r)$ for all permutations τ of $\{1, \ldots, r\}$ (sgn denotes the signature). The converse (antisymmetrical implies alternating) is true if K has characteristics $\neq 2$.

[4]The notation is consistent because there is some kind of "associativity".

so we deduce that:

$$\sum_{i=0}^{m} q^{is} v \wedge \cdots \wedge \Phi^i(v') \wedge \cdots \wedge \Phi^m(v) = 0.$$

The expression $\sum_{i=0}^{m} X^i v \wedge \cdots \wedge \Phi^i(v') \wedge \cdots \wedge \Phi^m(v)$ can be seen as a polynomial in the indeterminate X with coefficients in the finite-dimensional space $\wedge^{m+1}(V)$. The above formula says that it has all the q^s, $s \in \mathbf{Z}$, as roots. Since q is not a root of unity (actually the only assumption necessary for this proof), those are infinitely many roots, so said polynomial is zero. In particular, its coefficient in X^m vanishes: $v \wedge \cdots \wedge \Phi^{m-1}(v) \wedge \Phi^m(v') = 0$. This is true for every v'; but, Φ (and therefore Φ^m) being bijective, we see that $U \wedge U' = v \wedge \cdots \wedge \Phi^{m-1}(v) \wedge U' = 0$ for all $U' \in V$, whence $U = 0$, the contradiction looked for. □

8.2. Morphisms of q-difference modules

We discuss usual structures of the linear part of commutative algebra, as applied to q-difference theory.

8.2.1. Submodules, quotients.

8.2.1.1. *Submodules.* Let (V, Φ) a q-difference module and M the corresponding $\mathcal{D}_{K,q}$-module. So V is the K-vector space underlying M (that is, obtained by "restriction of scalars" from $\mathcal{D}_{K,q}$ to K). Additional structures on V are, on the q-difference module side, the σ_q-linear automorphism Φ; and, on the other side, left multiplication by elements of $\mathcal{D}_{K,q}$.

A sub $\mathcal{D}_{K,q}$-module $M' \subset M$ is a subgroup which is moreover stable by left multiplication; stability under left multiplication by K makes it a sub K-vector space $V' \subset V$; then, stability under left multiplication by S and S^{-1} mean that $\Phi(V') = V'$. Conversely, a subspace $V' \subset V$ such that $\Phi(V') = V'$ is a subgroup $M' \subset M$ stable under left multiplication by K, S and S^{-1}, whence by the whole of $\mathcal{D}_{K,q}$: it is a sub $\mathcal{D}_{K,q}$-module.

EXERCISE 8.29. Show that $\Phi(V') \subset V'$ implies $\Phi(V') = V'$. (From an argument about the image by Φ of a basis used in 8.1.1, deduce that $\Phi(V')$ is a K-subspace of V having the same dimension as V'.)

DEFINITION 8.30. A *sub-q-difference module* of the q-difference module (V, Φ) is a pair (V', Φ') where V' is a Φ-stable subspace of V and $\Phi' := \Phi_{|V'}$. (Here Φ-stable means indifferently that $\Phi(V') \subset V'$ or that $\Phi(V') = V'$ since, by the above exercise, those relations are equivalent.) We shall also call (V', Φ') a *submodule* of (V, Φ).

EXAMPLE 8.31. Let $M = (V, \Phi)$ and $v \in V$. The smallest sub-q-difference module of V containing v is (as a $\mathcal{D}_{K,q}$-module) $\mathcal{D}_{K,q}v$ and, as a K-linear space, the span of the family $(\Phi^n(v))_{n \in \mathbf{Z}}$. If $v \neq 0$, the annihilator ideal $\{P \in \mathcal{D}_{K,q} \mid P.v = 0\}$ admits a unique standard unitary generator, the *minimal polynomial* P of v, and $M \simeq M_P$.

In order to have compatibility with the notations for $\mathcal{D}_{K,q}$-modules, we shall then write $(V', \Phi') \subset (V, \Phi)$ and speak of inclusion (of strict inclusion if $V' \neq V$).

A sub-q-difference module automatically inherits a structure of q-difference module and corresponds to a $\mathcal{D}_{K,q}$-submodule. (And inherited structure of q-difference module corresponds to the inherited structure of $\mathcal{D}_{K,q}$-module, but we shall not always mention explicitly such trivialities - although they are not tautologies!)

EXAMPLE 8.32. Let $V := K^n$ and $\Phi := \Phi_A$, where A has the following block structure: $A = \begin{pmatrix} A' & \star \\ 0 & A'' \end{pmatrix}$, $A' \in \mathrm{GL}_{n'}(K)$, $A'' \in \mathrm{GL}_{n''}(K)$ and $\star$ denotes an arbitrary $n'' \times n'$ matrix (of course $n' + n'' = n$). Then $A^{-1} = \begin{pmatrix} A'^{-1} & \star \\ 0 & A''^{-1} \end{pmatrix}$, so that the subspace $V' := K^{n'} \times \{0\} \subset K^n$ is Φ_A-stable and defines a submodule. The induced action of Φ_A on $K^{n'} \times \{0\}$ is $\Phi_{A'} \times \mathrm{Id}_{\{0\}}$, so our submodule is $(K^{n'} \times \{0\}, \Phi_{A'} \times \mathrm{Id}_{\{0\}})$. Clearly, it is canonically isomorphic to the q-difference module $(K^{n'}, \Phi_{A'})$ through the map $v' \mapsto \begin{pmatrix} v' \\ 0 \end{pmatrix}$ from $K^{n'}$ to $K^{n'} \times \{0\}$, so we shall identify them and consider $(K^{n'}, \Phi_{A'})$ as a submodule of (K^n, Φ_A).

EXAMPLE 8.33. It follows from 8.1.5 that any q-difference module of rank n corresponds to some $\mathcal{D}_{K,q}$-module $\mathcal{D}_{K,q}/\mathcal{D}_{K,q}P$, where $P \in \mathcal{D}_{K,q}$ is a (nonunique) standard unitary q-difference operator of degree n. The submodules of $\mathcal{D}_{K,q}/\mathcal{D}_{K,q}P$ are the $\mathcal{D}_{K,q}R/\mathcal{D}_{K,q}P$ with R standard unitary such that $\mathcal{D}_{K,q}P \subset \mathcal{D}_{K,q}R$, i.e., $P \in \mathcal{D}_{K,q}R$, i.e., $P = QR$ with Q standard unitary. So submodules of $\mathcal{D}_{K,q}/\mathcal{D}_{K,q}P$ are in one-to-one correspondence with decompositions $P = QR$ with $Q, R \in \mathcal{D}_{K,q}$ standard unitary. (Recall that, if $\mathcal{D}_{K,q}R_1 = \mathcal{D}_{K,q}R_2$, where R_1 and R_2 are standard unitary, then $R_1 = R_2$.)

There is a converse to example 8.32, it comes by as follows. Let (V', Φ') a submodule of (V, Φ). Let $\mathcal{B}'$ a basis of V', giving rise to an isomorphism of vector spaces $K^{n'} \simeq V'$ and to an isomorphism of q-difference modules $(K^{n'}, \Phi_{A'}) \simeq (V', \Phi')$. Complete $\mathcal{B}'$ into a basis $\mathcal{B}$ of V, giving rise to an isomorphism of vector spaces $K^n \simeq V$ and to an isomorphism of q-difference modules $(K^n, \Phi_A) \simeq (V, \Phi)$. The isomorphism $K^n \simeq V$ restricts to an isomorphism $K^{n'} \times \{0\} \simeq V'$, which can be identified with $K^{n'} \simeq V'$ above. Similarly, up to this identification, the isomorphism $(K^n, \Phi_A) \simeq (V, \Phi)$ restricts to the isomorphism $(K^{n'}, \Phi_{A'}) \simeq (V', \Phi')$. Actually A has the same block structure as in the example: $A = \begin{pmatrix} A' & \star \\ 0 & A'' \end{pmatrix}$ for some $\star$ and A'' and so, up to compatible isomorphisms, every occurrence of submodules (V', Φ') of (V, Φ) has the form given in the example. Anticipating on the definition of morphisms of q-difference modules (to come soon in 8.2.2), we have a commutative diagram:

$$\begin{array}{ccc} (K^{n'}, \Phi_{A'}) & \xrightarrow{\simeq} & (V', \Phi') \\ {\scriptstyle j}\downarrow & & \downarrow{\scriptstyle i} \\ (K^n, \Phi_A) & \xrightarrow{\simeq} & (V, \Phi). \end{array}$$

The left vertical map is $j : v' \mapsto \begin{pmatrix} v' \\ 0 \end{pmatrix}$ and yields an isomorphism with the corresponding submodule. The right vertical map is simply the inclusion map i of V' into V. Both vertical maps are "injective morphisms". The word "morphism" expresses their compatibility with the q-difference module structure. For instance, the fact that Φ restricted to V' is Φ' can be translated into: $\Phi \circ i = i \circ \Phi'$. Likewise, the fact that $\Phi_{A'}$ is induced by Φ_A can be translated into: $\Phi_A \circ j = j \circ \Phi_{A'}$.

8.2.1.2. *Quotients.* We now describe the quotient of a q-difference module by a submodule. We leave to the reader the proof of the following.

PROPOSITION 8.34. *Let (V, Φ) a q-difference module and M the corresponding $\mathcal{D}_{K,q}$-module. Let (V', Φ') a submodule of (V, Φ) and $M' \subset M$ the corresponding $\mathcal{D}_{K,q}$-submodule. Let $V'' := V/V'$ (quotient vector space) and $M'' := M/M'$ (quotient $\mathcal{D}_{K,q}$-module).*

(i) *The map $\Phi'' : V'' \to V''$ sending $v \pmod{V'}$ to $\Phi(v) \pmod{V'}$ is well defined and it is a σ_q-linear automorphism of V''.*

(ii) *The resulting q-difference module (V'', Φ'') corresponds to the $\mathcal{D}_{K,q}$-module M''.*

PROOF. Instead of giving a demonstration, we just give a more precise meaning to statement (i). Calling i the inclusion map from V' to V, we saw at the end of 8.2.1.1 that the compatibility property of Φ and Φ' (which makes V' a submodule) can be translated as $\Phi \circ i = i \circ \Phi'$, i.e., by a commutative diagram:

$$\begin{array}{ccc} V' & \xrightarrow{i} & V \\ {\scriptstyle \Phi'}\downarrow & & \downarrow{\scriptstyle \Phi} \\ V' & \xrightarrow{i} & V. \end{array}$$

As we shall see in 8.2.2, this says that i is a morphism of q-difference modules from (V', Φ') to (V, Φ) (an injective morphism too). So we shall write $i : (V', \Phi') \to (V, \Phi)$.

Now let $p : V \to V''$ the canonical projection $v \mapsto v \pmod{V'}$. The condition in (i) says that $\Phi'' \circ p = p \circ \Phi$, whence a commutative diagram:

$$\begin{array}{ccc} V & \xrightarrow{p} & V'' \\ {\scriptstyle \Phi}\downarrow & & \downarrow{\scriptstyle \Phi''} \\ V & \xrightarrow{p} & V''. \end{array}$$

In other words, p is a (surjective) morphism of q-difference modules from (V, Φ) to (V'', Φ''), and we shall write $p : (V, \Phi) \to (V'', \Phi'')$.

Last, we note that i and p combine into an exact sequence of vector spaces:

$$0 \to V' \xrightarrow{i} V \xrightarrow{p} V'' \to 0$$

which can itself be inserted in a bigger commutative diagram:

$$\begin{array}{ccccccccc} 0 & \longrightarrow & V' & \xrightarrow{i} & V & \xrightarrow{p} & V'' & \longrightarrow & 0 \\ & & {\scriptstyle \Phi'}\downarrow & & {\scriptstyle \Phi}\downarrow & & \downarrow{\scriptstyle \Phi''} & & \\ 0 & \longrightarrow & V' & \xrightarrow{i} & V & \xrightarrow{p} & V'' & \longrightarrow & 0. \end{array}$$

We consider it as an exact sequence of q-difference modules:

$$0 \to (V', \Phi') \xrightarrow{i} (V, \Phi) \xrightarrow{p} (V'', \Phi'') \to 0.$$

□

We shall return to exact sequences in 8.3.

DEFINITION 8.35. The q-difference module (V'', Φ'') is called the *quotient (q-difference module)* of (V, Φ) by (V', Φ').

EXAMPLE 8.36. Let $A := \begin{pmatrix} A' & \star \\ 0 & A'' \end{pmatrix}$, $A' \in \mathrm{GL}_{n'}(K)$, $A'' \in \mathrm{GL}_{n''}(K)$, as in example 8.32. Recall that the map $\begin{pmatrix} v' \\ v" \end{pmatrix} \mapsto v''$ from K^n (here identified with $K^{n'} \times K^{n''}$) to $K^{n''}$ allows us to identify $K^n/K^{n'}$ with $K^{n''}$; thus, there is an exact sequence $0 \to K^{n''} \to K^n \to K^{n''} \to 0$. We can insert it into a commutative diagram:

$$\begin{array}{ccccccccc} 0 & \longrightarrow & K^{n'} & \longrightarrow & K^n & \longrightarrow & K^{n''} & \longrightarrow & 0 \\ & & \Big\downarrow{\scriptstyle \Phi_{A'}} & & \Big\downarrow{\scriptstyle \Phi_A} & & \Big\downarrow{\scriptstyle \Phi_{A''}} & & \\ 0 & \longrightarrow & K^{n'} & \longrightarrow & K^n & \longrightarrow & K^{n''} & \longrightarrow & 0. \end{array}$$

Then the natural identification of $K^n/K^{n'}$ with $K^{n''}$ results in an identification of the quotient q-difference module $(K^n, \Phi_A)/(K^{n'}, \Phi_{A'})$ with $(K^{n''}, \Phi_{A''})$.

EXAMPLE 8.37. Let $M := \mathcal{D}_{K,q}/\mathcal{D}_{K,q}P$ as in example 8.33 and let $P = QR$, where Q and R are standard unitary; therefore $M' := \mathcal{D}_{K,q}R/\mathcal{D}_{K,q}P$ is a submodule of M. From the standard isomorphism theorems for modules, we know that the quotient module $M/M' = \dfrac{\mathcal{D}_{K,q}/\mathcal{D}_{K,q}P}{\mathcal{D}_{K,q}R/\mathcal{D}_{K,q}P}$ is isomorphic to $M'' := \mathcal{D}_{K,q}/\mathcal{D}_{K,q}R$.

We can give a better look to the resulting exact sequence by noting that the map $X \mapsto XR$ (right multiplication by R in $\mathcal{D}_{K,q}$) yields an isomorphism $\mathcal{D}_{K,q} \to \mathcal{D}_{K,q}R$, which in turn restricts to an isomorphism $\mathcal{D}_{K,q}Q \to \mathcal{D}_{K,q}QR = \mathcal{D}_{K,q}P$, and then induces an isomorphism:

$$\mathcal{D}_{K,q}/\mathcal{D}_{K,q}Q \xrightarrow{\simeq} \mathcal{D}_{K,q}R/\mathcal{D}_{K,q}P \subset \mathcal{D}_{K,q}/\mathcal{D}_{K,q}P.$$

Eventually we get an exact sequence:

$$0 \to \mathcal{D}_{K,q}/\mathcal{D}_{K,q}Q \xrightarrow{\times R} \mathcal{D}_{K,q}/\mathcal{D}_{K,q}P \to \mathcal{D}_{K,q}/\mathcal{D}_{K,q}R \to 0.$$

Our contention that this is a kind of model for our expected exact sequence $0 \to M' \to M \to M'' \to 0$ can be encoded in a commutative diagram:

$$\begin{array}{ccccccccc} 0 & \longrightarrow & \mathcal{D}_{K,q}/\mathcal{D}_{K,q}Q & \longrightarrow & \mathcal{D}_{K,q}/\mathcal{D}_{K,q}P & \longrightarrow & \mathcal{D}_{K,q}/\mathcal{D}_{K,q}R & \longrightarrow & 0 \\ & & \Big\downarrow & & \Big\downarrow & & \Big\downarrow & & \\ 0 & \longrightarrow & M' & \longrightarrow & M & \longrightarrow & M'' & \longrightarrow & 0. \end{array}$$

We let to the reader the pleasant task of describing precisely the arrows and check commutativities.

EXERCISE 8.38. What can be deduced if $P = P_1 \cdots P_k$?

8.2.1.3. *Intersections, sums, direct sums, products.* The notions of intersection, sum, direct sum and product are well known for modules over a ring, here we briefly translate them from the frame of $\mathcal{D}_{K,q}$-modules to that of q-difference modules. We leave to the reader the (mechanical) verification of that translation.

Let M a $\mathcal{D}_{K,q}$-module and M_1, M_2 submodules. Let (V, Φ), and $(V_1, \Phi_1), (V_2, \Phi_2)$ their q-difference module counterparts, so that $V_i \subset V$ and $\Phi_i = \Phi_{|V_i}$ for $i = 1, 2$.

Then $V_1 \cap V_2$ is stable under any Φ, Φ_1, Φ_2 and thus defines a submodule of any one of $(V, \Phi), (V_1, \Phi_1), (V_2, \Phi_2)$. That submodule corresponds to $M_1 \cap M_2$.

With the same notations, we see that $V_1 + V_2$ is stable under Φ, thus defines a submodule of (V, Φ). That submodule corresponds to $M_1 + M_2$. The sum of those $\mathcal{D}_{K,q}$-submodules is direct if, and only if, $M_1 \cap M_2 = 0$, which is equivalent to $V_1 \cap V_2 = 0$. In particular, $M = M_1 + M_2$ is equivalent to $V = V_1 + V_2$ and $M = M_1 \oplus M_2$ is equivalent to $V = V_1 \oplus V_2$. Note however that a vector space decomposition $V = V_1 + V_2$, resp. $V = V_1 \oplus V_2$, does not necessarily corresponds to a $\mathcal{D}_{K,q}$-module decomposition $M = M_1 + M_2$, resp. $M = M_1 \oplus M_2$: we must moreover require Φ-stability of V_1, V_2.

EXERCISE 8.39. Let p a projector of V (i.e., an idempotent endomorphism) and $V_1 := \text{Im } p$, $V_2 := \text{Ker } p$, whence a vector space decomposition $V = V_1 \oplus V_2$. Show that it corresponds to a q-difference module decomposition if, and only if, $p \circ \Phi = \Phi \circ p$.

Now let (V_1, Φ_1), (V_2, Φ_2) arbitrary q-difference modules (i.e., not assumed to be submodules of some q-difference module) and let M_1, M_2 the corresponding $\mathcal{D}_{K,q}$-modules. Then the map $\Phi_1 \times \Phi_2 : (v_1, v_2) \mapsto (\Phi_1(v_1), \Phi_2(v_2))$ from $V_1 \times V_2$ to itself is σ_q-linear, so it defines a product q-difference module $(V_1, \Phi_1) \times (V_2, \Phi_2) := (V_1 \times V_2, \Phi_1 \times \Phi_2)$. This corresponds to the product $\mathcal{D}_{K,q}$-module $M_1 \times M_2$.

Coming back to the situation when $M = M_1 \oplus M_2$, we know that there is an isomorphism $(x_1, x_2) \mapsto x_1 + x_2$ from $M_1 \times M_2$ to M. On the q-difference module side, we have $V = V_1 \oplus V_2$, whence an isomorphism of vector spaces $(v_1, v_2) \mapsto v_1 + v_2$ from $V_1 \times V_2$ to V: this is an isomorphism of q-difference modules. It is therefore customary in some situations, and for purely psychological reasons[5], to write $M_1 \oplus M_2$ instead of $M_1 \times M_2$ and, more generally, $\bigoplus_{i=1}^{k} M_i$ instead of $\prod_{i=1}^{k} M_i$.

More generally, again anticipating 8.2.2, assuming only that (V_1, Φ_1) an (V_2, Φ_2) are submodules of (V, Φ), the map $(v_1, v_2) \mapsto v_1 + v_2$ is a morphism from $(V_1, \Phi_1) \times (V_2, \Phi_2)$ to (V, Φ) with image the sum $(V_1, \Phi_1) + (V_2, \Phi_2)$.

EXERCISE 8.40. Check that statement, i.e., writing s that morphism, check that $s \circ (\Phi_1 \times \Phi_2) = \Phi \circ s$.

8.2.1.4. *Flags.* We shall call *flag*[6] *of length* k in a q-difference module $M = (V, \Phi)$ a finite sequence of submodules $0 = M_0 \subset M_1 \subset \cdots \subset M_{k-1} \subset M_k = M$ such that each inclusion $M_{i-1} \subset M_i$, $i = 1, \ldots, k$, is strict. We call M together with a flag a *filtered q-difference module.* Its *associated graded q-difference module* is the direct sum $\bigoplus_{i=1}^{k} M_i / M_{i-1}$.

According to 8.2.1.1, we have two possible models:

---

[5]An infinite direct sum of $\mathcal{D}_{K,q}$-modules can be defined and it is different from the corresponding product if they are nontrivial (actually, the former is a strict submodule of the latter); but this does not occur in the frame of q-difference modules because of the condition of finite length or dimension.

[6]This is not exactly the correct terminology but it is concise and it will be convenient for us. We shall come to a more precise formulation (distinguishing flags, or towers, from filtrations), in Chapter 10.

- $M=(K^n,\Phi_A)$ with $A=\begin{pmatrix} A_1 & \star & \star & \dots & \star & \star \\ 0 & A_2 & \star & \dots & \star & \star \\ 0 & 0 & A_3 & \dots & \star & \star \\ \vdots & \vdots & \vdots & \ddots & \vdots & \vdots \\ 0 & 0 & 0 & \dots & A_{k-1} & \star \\ 0 & 0 & 0 & \dots & 0 & A_k \end{pmatrix}$, $A_i \in \mathrm{GL}_{n_i}(K)$, $n_1+\cdots+n_k = n$ and each $M_i = (K^{m_i}, \Phi_{A_{\leq i}})$, where $m_i := n_1+\cdots+n_i$ and $A_{\leq i}$ is the upper left block of A of size m_i: $A_{\leq i} = \begin{pmatrix} A_1 & \star & \dots & \star \\ 0 & A_2 & \dots & \star \\ \vdots & \vdots & \ddots & \vdots \\ 0 & 0 & \dots & A_i \end{pmatrix}$.

  The associated graded q-difference module is (K^n, Φ_{A_0}), where:

$$A_0 := \begin{pmatrix} A_1 & 0 & 0 & \dots & 0 & 0 \\ 0 & A_2 & 0 & \dots & 0 & 0 \\ 0 & 0 & A_3 & \dots & 0 & 0 \\ \vdots & \vdots & \vdots & \ddots & \vdots & \vdots \\ 0 & 0 & 0 & \dots & A_{k-1} & 0 \\ 0 & 0 & 0 & \dots & 0 & A_k \end{pmatrix}.$$

- $M = \mathcal{D}_{K,q}/\mathcal{D}_{K,q}P$ where $P = R_1 \cdots R_k$ and each $M_i = \mathcal{D}_{K,q}R_{i+1} \cdots R_n/\mathcal{D}_{K,q}P \simeq \mathcal{D}_{K,q}/\mathcal{D}_{K,q}R_1 \cdots R_i$ (usual conventions: $R_{i+1} \cdots R_n = \mathcal{D}_{K,q}$ if $i = n$ and $R_1 \cdots R_i = \mathcal{D}_{K,q}$ if $i = 0$).

  The associated graded q-difference module is $\bigoplus_{i=1}^{k} \mathcal{D}_{K,q}/\mathcal{D}_{K,q}R_i$.

8.2.1.5. *Composition series.* A flag $0 = M_0 \subset M_1 \subset \cdots \subset M_{k-1} \subset M_k = M$ can be *refined* if there is an i such that M_i/M_{i-1} is not simple, so that one can find an intermediate module N making:

$$0 = M_0 \subset M_1 \subset \cdots \subset M_{i-1} \subset N \subset M_i \subset \cdots \subset M_{k-1} \subset M_k = M$$

a flag of length $k+1$ (of course, the submodules beyond N should be reindexed). If the flag cannot be refined, i.e., if all quotients $P_i := M_i/M_{i-1}$ are simple, then the flag is called a *composition series*.

The theorem of Jordan-Holder **[37, 145]** says that all the composition series have the same length, which is actually the length of the module (we already used that fact in 8.1.2). However, it says more: if $0 = M_0 \subset M_1 \subset \cdots \subset M_{k-1} \subset M_k = M$ and $0 = M'_0 \subset M'_1 \subset \cdots \subset M'_{k-1} \subset M'_k = M$ are two composition series and if $P_i := M_i/M_{i-1}$, $P'_i := M'_i/M'_{i-1}$ for $i = 1, \ldots, k$, then there is a permutation τ of $\{1, \ldots, k\}$ such that $P'_i \simeq P_{\tau(i)}$ for $i = 1, \ldots, k$. As a consequence, the associated graded modules $\bigoplus P_i$ and $\bigoplus P'_i$ are isomorphic. (In the theory of modules over noncommutative rings, they are called "semi-simple modules".)

8.2.2. Morphisms of q-difference modules.

8.2.2.1. *Definition of morphisms.* In 8.1.1 we defined an isomorphism of q-difference modules $(V,\Phi) \xrightarrow{\sim} (W,\Psi)$ as a K-linear isomorphism of vector spaces $F : V \xrightarrow{\sim} W$ such that $\Psi \circ F = F \circ \Phi$ (so F "intertwines" Φ and Ψ), in other words making the following diagram commute:

$$\begin{array}{ccc} V & \xrightarrow{\Phi} & V \\ {\scriptstyle F}\downarrow & & \downarrow{\scriptstyle F} \\ W & \xrightarrow[\Psi]{} & W. \end{array}$$

Relaxing the requirement that F be bijective, we get the following:

DEFINITION 8.41. A *morphism* $F : (V, \Phi) \to (W, \Psi)$ is a K-linear map of vector spaces $F : V \to W$ intertwining Φ and Ψ, i.e., such that $\Psi \circ F = F \circ \Phi$.

EXAMPLE 8.42. Let $(V, \Phi) := (K^n, \Phi_A)$ and $(W, \Psi) := (K^p, \Phi_B)$. A K-linear map $K^n \to K^p$ is, up to usual identifications, a matrix $F \in \mathrm{Mat}_{p,n}(K)$. The intertwining condition says that

$$\forall X \in K^n,\ B^{-1}\sigma_q(FX) = FA^{-1}\sigma_q X \iff \forall X \in K^n,\ B^{-1}(\sigma_q F)(\sigma_q X) = FA^{-1}\sigma_q X$$
$$\iff B^{-1}(\sigma_q F) = FA^{-1} \iff (\sigma_q F)A = BF.$$

So we shall say that, up to a slight abuse, a morphism from $A \in \mathrm{GL}_n(K)$ to $B \in \mathrm{GL}_p(K)$ is a matrix $F \in \mathrm{Mat}_{p,n}(K)$ such that $(\sigma_q F)A = BF$.

EXERCISE 8.43. Recall the q-difference module $\underline{1}$ defined in 8.1.4.1 as (K, σ_q).

(i) Show that morphisms from $\underline{1}$ to (V, Φ) can be identified with elements $v \in V$ such that $\Phi(v) = v$. (Linear maps $K \to V$ have the form $a \mapsto av$ for some v and the intertwining condition translates into $\Phi(av) = (\sigma_q a)v$.)

(ii) Show that morphisms from $\underline{1}$ to (K^n, Φ_A) can be identified to elements $X \in K^n$ such that $\sigma_q X = AX$. (Matrices in $\mathrm{Mat}_{p,1}(K)$ are such column vectors X and condition found above is just $(\sigma_q X)1 = AX$.)

EXAMPLE 8.44. Let M, N the $\mathcal{D}_{K,q}$-modules respectively corresponding to (V, Φ) and (W, Ψ). Let F a K-linear map from V to W. The intertwining condition, written in terms of $\mathcal{D}_{K,q}$-modules, says that $F(Sx) = SF(x)$ for all $x \in M$. From that it follows by induction (plus a separate argument for S^{-1}) that $F(S^k x) = S^k F(x)$ for all $k \in \mathbf{Z}$; and then by K-linearity that $F(Px) = PF(x)$ for all $P \in \mathcal{D}_{K,q}$, i.e., F is a morphism of $\mathcal{D}_{K,q}$-modules. Conversely every morphism of $\mathcal{D}_{K,q}$-modules $F : M \to N$ is K-linear and satisfies $F(Sx) = SF(x)$, i.e., $\Psi \circ F = F \circ \Phi$, so it is a morphism of q-difference modules.

For instance, a morphism $\mathcal{D}_{K,q}/\mathcal{D}_{K,q}P \to \mathcal{D}_{K,q}/\mathcal{D}_{K,q}P'$ is obtained from a morphism (of left $\mathcal{D}_{K,q}$-modules) $\mathcal{D}_{K,q} \to \mathcal{D}_{K,q}$ by going to the quotient. The latter morphism has the form $Q \mapsto QP_0$, where P_0 is the image of 1. The condition that it goes to the quotient is that it sends $\mathcal{D}_{K,q}P$ in $\mathcal{D}_{K,q}P'$, i.e., that $PP_0 \in \mathcal{D}_{K,q}P'$. Of course P_0 is determined only modulo $\mathcal{D}_{K,q}P'$.

EXERCISE 8.45. Recall that in 8.1.4.1, the q-difference module $\underline{1}$ was also described as $\mathcal{D}_{K,q}/\mathcal{D}_{K,q}(S-1)$. Show (using the example above) that morphisms from $\mathcal{D}_{K,q}/\mathcal{D}_{K,q}P$ to $\underline{1}$ are in one-to-one correspondence with elements $f \in K$ such that $P(f) = 0$. (With the notations of the example, $P' = S-1$ and, thanks to euclidean division, P_0 can be uniquely chosen in K: this is f; now we want $P.f \in \mathcal{D}_{K,q}(S-1)$, but this is equivalent to $P(f) = 0$.)

The usual vocabulary of morphisms in linear algebra applies here: a monomorphism, resp. an epimorphism, is an injective, resp. a surjective morphism (and an isomorphism[7] is both); an endomorphism, resp. an automorphism, is a morphism, resp. an isomorphism from a q-difference module to itself.

8.2.2.2. *Basic properties of morphisms.* What we are going to show could be deduced mechanically from general properties of the morphisms of modules over a ring (here $\mathcal{D}_{K,q}$); however, it is useful to be able to compute and reason with the model (V, Φ).

Let $M := (V, \Phi)$ a q-difference module. Then the identity map of V is an automorphism of M. Indeed, the intertwining condition here writes $\Phi \circ \mathrm{Id}_V = \mathrm{Id}_V \circ \Phi$.

Let $M := (V, \Phi)$, $M' := (V', \Phi')$ and $M'' := (V'', \Phi'')$ three q-difference modules and let $F : M \to M'$ and $F' : M' \to M''$ two morphisms. Then $F' \circ F$ is a morphism from M to M''. Indeed, it is obviously a K-linear map from V to V'' and:

$$\Phi'' \circ (F' \circ F) = (\Phi'' \circ F') \circ F = (F' \circ \Phi') \circ F = F' \circ (\Phi' \circ F) = F' \circ (F \circ \Phi) = (F' \circ F) \circ \Phi,$$

and the equality $\Phi'' \circ (F' \circ F) = (F' \circ F) \circ \Phi$ is just the required intertwining condition.

Let us write, just for now, $\mathrm{Hom}((V, \Phi), (W, \Psi))$ the set of all morphisms of q-difference modules from (V, Φ) to (W, Ψ); this notation will be refined soon (in 8.2.2.3). First, we show that it is a group, actually a vector space.

Clearly $\mathrm{Hom}((V, \Phi), (W, \Psi))$ is a subset of the space $\mathscr{L}_K(V, W)$ of K-linear maps. The intertwining relation shows that it is actually the kernel of the morphism $F \mapsto \Psi \circ F - F \circ \Phi$. But why can we say that this map is a morphism? Note that $\Psi \circ F$ and $F \circ \Phi$ are not K-linear, they are σ_q-linear. (See 8.1.1 for the definition of "σ_q-linear".)

EXERCISE 8.46. (i) Show that composing a K-linear with a σ_q-linear map produces a σ_q-linear map.

(ii) Show that the set $\mathscr{L}_{\sigma_q}(V, W)$ of σ_q-linear maps is, in a natural way, a K-vector space.

So the map $F \mapsto \Psi \circ F - F \circ \Phi$ goes from $\mathscr{L}_K(V, W)$ to $\mathscr{L}_{\sigma_q}(V, W)$. It is clearly a morphism of groups:

$$\Psi \circ (F_1 + F_2) - (F_1 + F_2) \circ \Phi = (\Psi \circ F_1 - F_1 \circ \Phi) + (\Psi \circ F_2 - F_2 \circ \Phi).$$

However, since $F \mapsto \Psi \circ F$ is σ_q-linear while $F \mapsto F \circ \Phi$ is K-linear, so the best we can get is that they are both K^{σ_q}-linear and therefore so is the map $F \mapsto \Psi \circ F - F \circ \Phi$. So we see in the end that its kernel $\mathrm{Hom}_{\mathscr{C}}((V, \Phi), (W, \Psi))$ is a K^{σ_q}-vector space. There is a last property:

EXERCISE 8.47. Show that the composition map $(F, F') \mapsto F' \circ F$:

$$\mathrm{Hom}((V, \Phi), (V', \Phi')) \times \mathrm{Hom}_{\mathscr{C}}((V', \Phi'), (V'', \Phi'')) \to \mathrm{Hom}((V, \Phi), (V'', \Phi''))$$

is K^{σ_q}-bilinear.

---

[7] Strictly speaking, "isomorphism" means *invertible morphism*; but, as almost always in linear algebra, this is here equivalent to *bijective morphism*.

8.2.2.3. *Categories.* From the fact that identities are left and right neutral for the composition of maps and that said composition is associative, we deduce that q-difference modules over a fixed q-difference field (K, σ_q), together with their morphisms, form a category q-diffMod. For simplicity of notations, we shall temporarily denote it $\mathscr{C}$. So, according to the general formalism of categories, we write $\mathrm{Hom}_{\mathscr{C}}((V, \Phi), (W, \Psi))$ the set of all morphisms from (V, Φ) to (W, Ψ). We summarize all its properties displayed in 8.2.2.2 by saying that $\mathscr{C}$ (i.e., q-diffMod) is a K^{σ_q}-linear category.

Sticking to the temporary notation $\mathscr{C}$, let $\mathscr{C}_0$ the full subcategory of objects (K^n, Φ_A). Thus, by definition:

$$\mathrm{Hom}_{\mathscr{C}_0}((K^n, \Phi_A), (K^p, \Phi_B)) = \mathrm{Hom}_{\mathscr{C}}((K^n, \Phi_A), (K^p, \Phi_B)).$$

From the fact that any q-difference module is isomorphic to some (K^n, Φ_A), we deduce that $\mathscr{C}$ and $\mathscr{C}_0$ are equivalent categories: indeed, the inclusion functor from $\mathscr{C}_0$ to $\mathscr{C}$ is obviously fully faithful and, because of the property just recalled, essentially surjective.

In the other direction, we have no canonically defined functor[8]. A very non-canonical one shall be defined as follows. To each K-vector space V, attach an arbitrary basis, say $\mathcal{B}_V$. Once this super-choice has been made, there is a well defined isomorphism $\phi_V : V \xrightarrow{\sim} K^{n_V}$, where $n_V := \dim_K V$. Now here comes our functor: to the object (V, Φ) it associates (K^{n_V}, Φ_A), where Φ_A is defined as usual from Φ and $\mathcal{B}_V$ (in other words, $\mathcal{B}_V = \Phi(\mathcal{B}_V)A$); and to the morphism $F : (V, \Phi) \to (W, \Psi)$ it associates $\phi_W \circ F \circ \phi_V^{-1}$.

EXERCISE 8.48. Check that $\phi_W \circ F \circ \phi_V^{-1}$ is indeed a morphism from (K^{n_V}, Φ_A) to the (K^{n_W}, Φ_B) associated to (W, Ψ); and then check that what was defined above is indeed a functor from $\mathscr{C}$ to $\mathscr{C}_0$.

Now let $\mathcal{D}_{K,q}$Mod the category of all left $\mathcal{D}_{K,q}$-modules and $\mathcal{D}_{K,q}$flMod its full subcategory consisting of $\mathcal{D}_{K,q}$-modules of finite length. Again for typographical simplicity, we respectively denote $\mathscr{C}'$ and $\mathscr{C}'_0$ the former one and the latter one, so that, for any M, N in $\mathscr{C}'_0$:

$$\mathrm{Hom}_{\mathscr{C}'_0}(M, N) = \mathrm{Hom}_{\mathscr{C}'}(M, N) = \{\mathcal{D}_{K,q}\text{-linear maps } M \to N\}.$$

(Be aware however that, whereas $\mathscr{C}$ and $\mathscr{C}_0$ are equivalent, $\mathscr{C}'$ and $\mathscr{C}'_0$ are not.) From the general theory of modules over a ring, this set is known to be an additive group for the obvious addition defined by the formula $F_1 + F_2 : x \mapsto F_1(x) + F_2(x)$; and actually a module over the center of $\mathcal{D}_{K,q}$ for the external multiplication defined by the formula $aF : x \mapsto aF(x)$.

EXERCISE 8.49. Check that, when verifying that aF is $\mathcal{D}_{K,q}$-linear, one uses the fact that a belongs to the center of $\mathcal{D}_{K,q}$.

We already saw in 3.1.3 that the center of $\mathcal{D}_{K,q}$ is K^{σ_q}, so $\mathrm{Hom}_{\mathscr{C}'_0}(M, N)$ is actually a K^{σ_q}-vector space. Again one finds that $\mathscr{C}'_0$ is a K^{σ_q}-linear category. This is not by chance, since actually $\mathscr{C}'_0$ and $\mathscr{C}$ are very much "the same" category, in the following sense. Recall that in 8.1.2 we associated to every q-difference module a $\mathcal{D}_{K,q}$-module of finite length; and that we proved in 8.2.2.1 that, under that identification, morphisms of q-difference modules correspond to morphisms of

[8]The whole situation is entirely similar to that of the category of all finite-dimensional vector space *versus* its full subcategory of objects K^n; there is just some additional structure here.

$\mathcal{D}_{K,q}$-modules. This means that we defined a functor from $\mathscr{C}$ to $\mathscr{C}'_0$. From what we saw in 8.1.2, the process can be reversed, so that our functor admits an inverse: it is an isomorphism of categories. In particular, it is an equivalence.

REMARK 8.50. One seldom meets isomorphisms of categories and it is actually not a very useful property. The truly relevant notion to express that two categories are "structurally similar" is that of equivalence.

The very important fact to note is that, if M, N are two q-difference modules, and so can be seen as $\mathcal{D}_{K,q}$-modules of finite length, the K^{σ_q}-vector spaces $\mathrm{Hom}_{\mathscr{C}}(M,N)$, $\mathrm{Hom}_{\mathscr{C}'}(M,N)$, and $\mathrm{Hom}_{\mathscr{C}'_0}(M,N)$ are the same; and if $M = M_A$, $N = M_B$, they are also the same as $\mathrm{Hom}_{\mathscr{C}_0}(M,N)$.

Practically, in all that follows, we shall most of the time identify:

$$\mathscr{C} = \mathscr{C}_0 = \mathscr{C}'_0.$$

8.2.3. Some important functors.

8.2.3.1. *A first look at some functors.* To motivate some of the subtleties which might appear somewhat artificial, we begin by some remarks on functorial arguments in the context basic linear algebra.

So let V a finite-dimensional vector space over a (commutative) field K and let $V^\vee := \mathscr{L}_K(V,K)$ its dual. We know that V and $V^\vee$ are isomorphic, but we also know that there is no *canonical* isomorphism between them: the standard way to produce an isomorphism between them is to *choose* a basis (e_i) of V, and the resulting isomorphism (which maps the basis (e_i) to its dual basis) heavily depends on that arbitrary choice. Of course, this is different if V comes along with a canonical basis, like K^n or the space of polynomials of degree $\leq n$; or if V comes equipped with a nondegenerate symmetric bilinear form.

On the opposite side, there is a canonical isomorphism from V to its *bidual* $V^{\vee\vee} := (V^\vee)^\vee$: it sends $x \in V$ to the linear form $\phi \mapsto \phi(x)$ on $V^\vee$. But it is also known that this isomorphism is *natural.* This related to *morphisms* of vector spaces, i.e., to linear maps. Indeed, for every linear map $f : V \to W$ we can define a linear map $f^{\vee\vee} : V^{\vee\vee} \to W^{\vee\vee}$ and the this process is *functorial*: $(\mathrm{Id}_V)^{\vee\vee} = \mathrm{Id}_{V^{\vee\vee}}$; and, whenever this composition makes sense, $(g \circ f)^{\vee\vee} = g^{\vee\vee} \circ f^{\vee\vee}$. We say that $V \rightsquigarrow V^{\vee\vee}$, $f \rightsquigarrow f^{\vee\vee}$ is a *covariant functor.*

EXERCISE 8.51. Show that there is no way to define $f^\vee$ in such a way that $V \rightsquigarrow V^\vee$, $f \rightsquigarrow f^\vee$, is a covariant functor.

Recall that actually, $f^\vee$ is well defined as a map from $W^\vee$ to $V^\vee$ (it is the map $\phi \mapsto \phi \circ f$) and that it satisfies $(\mathrm{Id}_V)^\vee = \mathrm{Id}_{V^\vee}$; and, whenever this composition makes sense, $(g \circ f)^\vee = f^\vee \circ g^\vee$ (the order is reversed!). We say that $V \rightsquigarrow V^\vee$, $f \rightsquigarrow f^\vee$ is a *contravariant functor.* Then $f^{\vee\vee}$ used above is just $(f^\vee)^\vee$.

Then the *naturality* of the isomorphism $V \xrightarrow{\sim} V^{\vee\vee}$ is expressed by the commutativity of the following diagram:

$$\begin{array}{ccc} V & \longrightarrow & V^{\vee\vee} \\ {\scriptstyle f}\big\downarrow & & \big\downarrow{\scriptstyle f^{\vee\vee}} \\ W & \longrightarrow & W^{\vee\vee}. \end{array}$$

In the language of categories, one says that *the functor "bidual" is isomorphic to the identity functor.* Nothing similar can be designed using duals instead of biduals.

EXERCISE 8.52. (i) We fix a space V_0. Define a covariant functor $V \rightsquigarrow \mathscr{L}_K(V_0, V)$. (You are supposed to say what it does to $f : V \to W$.)

(ii) Show that the functor $V \rightsquigarrow \mathscr{L}_K(K, V)$ is isomorphic to the identity functor.

(iii) Define a contravariant functor $V \rightsquigarrow \mathscr{L}_K(V, V_0)$.

Now we adapt those constructions to the category $\mathscr{C}$, in a generalized form. There will be however an important difference: in the above constructions, everything occurred within the category of finite-dimensional vector spaces; this is because, if V and W are finite-dimensional vector spaces, then so is $\mathscr{L}_K(V, W)$. In contrast, if M, N are objects of $\mathscr{C}$ (i.e., q-difference modules), $\mathrm{Hom}_{\mathscr{C}}(M, N)$ is not: it is a K^{σ_q}-vector space. So we are going to define functors from $\mathscr{C}$ to the category of K^{σ_q}-vector spaces. Of course, the process cannot be iterated. We are not really defining here the dual of a q-difference module: that will be done in 9.1.

First we fix a q-difference module M_0. To every q-difference module M we associate the K^{σ_q}-vector space $\mathrm{Hom}_{\mathscr{C}}(M_0, M)$. Temporarily, to have light notations, we set $\tilde{M} := \mathrm{Hom}_{\mathscr{C}}(M_0, M)$. If $F : M \to N$ is a morphism of q-difference modules, then we deduce a morphism $\tilde{F} : \tilde{M} \to \tilde{N}$: it sends $G \in \tilde{M} = \mathrm{Hom}_{\mathscr{C}}(M_0, M)$, i.e., $G : M_0 \to M$, to $F \circ G : M_0 \to N$, i.e., $F \circ G \in \mathrm{Hom}_{\mathscr{C}}(M_0, N) = \tilde{N}$. The reader is encouraged to prove (it is easy) the formulas: $\widetilde{\mathrm{Id}_M} = \mathrm{Id}_{\tilde{M}}$ and $\widetilde{F' \circ F} = \tilde{F'} \circ \tilde{F}$. We conclude that $M \rightsquigarrow \tilde{M}$, $F \rightsquigarrow \tilde{F}$ is a covariant functor from $\mathscr{C}$ to the category of K^{σ_q}-vector spaces. We shall write it more completely as $\mathrm{Hom}_{\mathscr{C}}(M_0, -)$.

EXERCISE 8.53. Define a contravariant functor $\mathrm{Hom}_{\mathscr{C}}(-, M_0)$ from $\mathscr{C}$ to the category of K^{σ_q}-vector spaces.

REMARK 8.54. These are actually very general results. In any category $\mathscr{C}$, each object M_0 provides a covariant functor $\mathrm{Hom}_{\mathscr{C}}(M_0, -)$ and a contravariant functor $\mathrm{Hom}_{\mathscr{C}}(-, M_0)$ to the category of sets; and if $\mathscr{C}$ is C-linear, to the category of C-vector spaces (or C-modules if C is a ring, not a field). Also if $\mathscr{C}$ is the category of modules over a ring $\mathcal{D}$, then it is C-linear, where C denotes the center of $\mathcal{D}$.

8.2.3.2. *Solutions and cosolutions.* What follows is related to the examples and exercises in 8.2.2.1. First let $M := (V, \Phi)$. Then we found a bijection:

$$\mathrm{Hom}_{\mathscr{C}}(\underline{1}, M) \to V^{\Phi} := \{v \in V \mid \Phi(v) = v\}, F \mapsto F(1).$$

Moreover, the map $F \mapsto F(1)$ is obviously K^{σ_q}-linear, so the above is actually an isomorphism of K^{σ_q}-vector spaces: $\mathrm{Hom}_{\mathscr{C}}(\underline{1}, M) \simeq V^{\Phi}$. Note that $\mathrm{Hom}_{\mathscr{C}}(\underline{1}, -)$ is a covariant functor by 8.2.3.1, but actually there also is a functor $(V, \Phi) \rightsquigarrow V^{\Phi}$: a morphism of q-difference modules $F : (V, \Phi) \to (W, \Psi)$ sends V^{Φ} into W^{Ψ}, thus defines a morphism of vector spaces $V^{\Phi} \to W^{\Psi}$; indeed, if $v \in V^{\Phi}$, i.e., if $\Phi(v) = v$, then:

$$\Psi(F(v)) = (\Psi \circ F)(v) = (F \circ \Phi)(v) = F(\Phi(v)) = F(v),$$

so $F(v) \in V^{\Phi}$. Other properties making it a functor are immediately verified. We then check easily that there is a commutative diagram:

$$\begin{array}{ccc}
\mathrm{Hom}_{\mathscr{C}}(\underline{1}, (V, \Phi)) & \longrightarrow & V^{\Phi} \\
\big\downarrow & & \big\downarrow \\
\mathrm{Hom}_{\mathscr{C}}(\underline{1}, (W, \Psi)) & \longrightarrow & W^{\Psi}.
\end{array}$$

We say that the functors $\mathrm{Hom}_{\mathscr{C}}(\underline{1}, -)$ and $(V, \Phi) \rightsquigarrow V^{\Phi}$ are isomorphic (as functors). (It is customary, as in the second case here, to denote a functor by writing explicitly only its action on objects, omitting morphisms.)

In the particular case that $M = (K^n, \Phi_A)$, an elements $v \in V$ can be seen as a column vector $X \in K^n$ and $\Phi(v) = v \Leftrightarrow \sigma_q X = AX$, so that V^{Φ} is actually identified with $\mathrm{Sol}(A, K)$, whence an isomorphism of K^{σ_q}-vector spaces: $\mathrm{Hom}_{\mathscr{C}}(\underline{1}, M) \simeq \mathrm{Sol}(A, K)$. We shall write:

$$\Gamma(M) := \mathrm{Hom}_{\mathscr{C}}(\underline{1}, M), \text{ seen as a } K^{\sigma_q}\text{-vector space.}$$

The same argument as above (*mutatis mutandis*) shows that the functor Γ restricted to $\mathscr{C}_0$ is isomorphic to the functor $(K^n, \Phi_A) \rightsquigarrow \mathrm{Sol}(A, K)$.

REMARK 8.55. The notation Γ is traditionally used to denote the group of global sections of a sheaf; our present use will be shown to be consistent with the usual convention in Subsection 9.3.1. However, the idea of considering $\mathrm{Hom}_{\mathscr{C}}(\underline{1}, M)$ as a kind of group of global sections has a wider scope; see for instance [**52**, Example 5.5].

DEFINITION 8.56. The *functor of solutions* is the covariant functor:

$$\Gamma := \mathrm{Hom}_{\mathscr{C}}(\underline{1}, -) : M \rightsquigarrow \Gamma(M) := \mathrm{Hom}_{\mathscr{C}}(\underline{1}, M)$$

from $\mathscr{C}$ to the category of finite-dimensional K^{σ_q}-vector spaces. In particular:

$$\Gamma(V, \Phi) = V^{\Phi}.$$

We shall find that the functor of solutions in K, by itself, does not provide enough information about the q-difference module M. It will be useful to consider solutions in a q-difference extension K' (whence a functor that could be denoted $\Gamma_{K'}(M) := \mathrm{Sol}(M, K')$ and obtained in a similar way as above; see 8.3.3.2); and even, in order to get a *sheaf of solutions*, functors of solutions in various q-difference *rings*; see 9.3.

If $L \in \mathcal{D}_{K,q}$ is standard unitary, it follows from 8.1.3 that $\mathrm{Sol}(L, K)$ can be identified to $\Gamma(M_L)$, where $M_L := (K^n, \Phi_{A_L})$. But from Exercise 8.45 in 8.2.2.1, we know that $\mathrm{Sol}(L, K)$ can be also be identified to $\mathrm{Hom}_{\mathscr{C}}(\mathcal{D}_{K,q}/\mathcal{D}_{K,q}L, \underline{1})$. This motivates the following:

DEFINITION 8.57. The *functor of cosolutions* is the contravariant functor:

$$\Gamma^{\vee} := \mathrm{Hom}_{\mathscr{C}}(-, \underline{1}) : M \rightsquigarrow \Gamma^{\vee}(M) := \mathrm{Hom}_{\mathscr{C}}(M, \underline{1})$$

from $\mathscr{C}$ to the category of finite-dimensional K^{σ_q}-vector spaces.

We shall define in 9.1.1 the dual q-difference module $M^{\vee}$ and obtain an identification $\Gamma^{\vee}(M) = \Gamma(M^{\vee})$.

So in the end we have two functorial ways to express solutions, in some sense dual with each other; setting $M_A := (K^n, \Phi_A)$ and $M_L := (K^n, \Phi_{A_L})$:

$$\mathrm{Sol}(A, K) = \Gamma(M_A), \quad \mathrm{Sol}(L, K) = \Gamma(M_L);$$

and setting $M'_L := \mathcal{D}_{K,q}/\mathcal{D}_{K,q}L$:

$$\mathrm{Sol}(L, K) = \Gamma^{\vee}(M'_L).$$

8.3. Abelian properties of q-difference modules

The word "abelian" will be explained and motivated in the course of the text.

8.3.1. Kernels, images, exact sequences.

CAUTION 8.58. In general, the theory of abelian categories [**94, 96, 104**] is more abstract, for instance exact sequences are not defined by brutal *equalities* $\text{Im } f = \text{Ker} g$. Here we adopt the more down-to-earth approach of [**32, 145**] which works perfectly well for categories of modules over a ring (although [**145**] only considers commutative rings).

8.3.1.1. *Prelude.* We shall now briefly recall what it concretely means that "the category $\mathscr{C}'$ of $\mathcal{D}_{K,q}$-modules is abelian". (Modules, submodules, etc, are implicitly understood to be left $\mathcal{D}_{K,q}$-modules, submodules, etc). We already saw in 8.2.2 that it is C-linear, where C is the center of $\mathcal{D}_{K,q}$: all sets of morphisms $\text{Hom}_{\mathscr{C}'}(M,N)$ are C-modules (actually C-vector spaces, since $C = K^{\sigma_q}$ is here a field); and the composition is C-bilinear.

For every morphism $f : M \to N$, the kernel $\text{Ker} f$ is a submodule of M and and the image $\text{Im } f$ is a submodule of N; and there is a canonical isomorphism $M/\text{Ker} f \xrightarrow{\sim} \text{Im } f$. One also has use for the *cokernel* $\text{Coker} f := N/\text{Im } f$. To make all this more flexible, one uses *exact sequences*. We say that $M' \xrightarrow{f} M \xrightarrow{g} M''$ is *exact* if $\text{Im } f = \text{Ker} g$. This implies that $g \circ f = 0$ but it is a stronger property. If we have a longer sequence of morphisms, say:

$$\cdots \to M_{i-1} \to M_i \to M_{i+1} \to \cdots$$

we say that it is *exact at* M_i if the extracted small sequence[9] $M_{i-1} \to M_i \to M_{i+1}$ is exact. The same sequence is *exact* if it is exact at each "intermediate" module. (At extremities, nothing is required.)

There are important special cases. We write 0 the trivial module.

(1) The sequence $0 \to M' \xrightarrow{f} M$ is exact if, and only if, f is injective (so f is a monomorphism).
(2) The sequence $M \xrightarrow{g} M'' \to 0$ is exact if, and only if, g is surjective (so g is an epimorphism).
(3) The sequence $0 \to M' \xrightarrow{f} M \xrightarrow{g} M'' \to 0$ is exact if, and only if, f is injective, g is surjective and $\text{Im } f = \text{Ker} g$. Such a sequence is christened *short exact sequence.*

Recall that in $\mathscr{C}'$, a bijective morphism has an inverse which is a morphism[10], so it is an isomorphism. (The converse is true!)

EXERCISE 8.59. Show that $0 \to M \xrightarrow{f} N \to 0$ is exact if, and only if, f is an isomorphism.

EXAMPLES 8.60. (1) Every submodule $N \subset M$ gives rise to a short exact sequence $0 \to N \to M \to M/N \to 0$.
(2) Every morphism $f : M_1 \to M_2$ gives rise to a short exact sequence: $0 \to \text{Ker} f \to M_1 \to \text{Im } f \to 0$.
(3) The morphism $f : M_1 \to M_2$ also gives rise to the short exact sequence $0 \to \text{Im } f \to M_2 \to \text{Coker} f \to 0$.

[9] "Short" exact sequences will denote something slightly different; see further below.

[10] This is not tautological: in the category of topological spaces, the morphisms are the continuous maps and the inverse of a bijective continuous map need not be continuous.

The first two examples are actually prototypes for all short exact sequences, i.e., every short exact sequence is isomorphic to one of each type. This means the following. Let $0 \to M' \to M \to M'' \to 0$ a short exact sequence; then:

(1) There is a submodule $N \subset M$ and a commutative diagram in which the vertical arrows are isomorphisms and the lower horizontal arrows are the obvious ones:

$$\begin{CD} 0 @>>> M' @>>> M @>>> M'' @>>> 0 \\ @. @VVV @VV{\mathrm{Id}_M}V @VVV @. \\ 0 @>>> N @>>> M @>>> M/N @>>> 0. \end{CD}$$

(2) There is a morphism $f : M \to N$ a commutative diagram in which the vertical arrows are isomorphisms and the lower horizontal arrows are the obvious ones:

$$\begin{CD} 0 @>>> M' @>>> M @>>> M'' @>>> 0 \\ @. @VVV @VV{\mathrm{Id}_M}V @VVV @. \\ 0 @>>> \mathrm{Ker} f @>>> M @>>> \mathrm{Im}\ f @>>> 0. \end{CD}$$

EXERCISE 8.61. (i) Construct the first diagram by taking for N the kernel of $M' \to M$.

(ii) Construct the second diagram by taking for f the map $M \to M''$.

More generally, the short exact sequences $0 \to M_1' \to M_1 \to M_1'' \to 0$ and $0 \to M_2' \to M_2 \to M_2'' \to 0$ are isomorphic if they can be inserted in a commutative diagram in which the vertical arrows are isomorphisms:

$$\begin{CD} 0 @>>> M_1' @>>> M_1 @>>> M_1'' @>>> 0 \\ @. @VVV @VVV @VVV @. \\ 0 @>>> M_2' @>>> M_2 @>>> M_2'' @>>> 0. \end{CD}$$

This definition extends to isomorphisms of arbitrary exact sequences having the same number of terms.

8.3.1.2. *Back to q-difference modules.* From the explicit description of submodules and quotients of q-difference modules in 8.2.1, and from the two prototypes in 8.3.1.1, we find two models for all short exact sequences in $\mathscr{C}$:

(1) Let $A = \begin{pmatrix} A' & \star \\ 0 & A'' \end{pmatrix}$ and let $M_A := (K^n, \Phi_A)$, etc. Then we have an exact sequence:

$$0 \to M_{A'} \to M_A \to M_{A''} \to 0 \tag{8.2}$$

and every short exact sequence is isomorphic to an exact sequence of that form.

(2) Let $Q, R \in \mathcal{D}_{K,q}$ standard unitary and let $P := QR$. Then we gave an exact sequence:

$$0 \to \mathcal{D}_{K,q}/\mathcal{D}_{K,q}Q \to \mathcal{D}_{K,q}/\mathcal{D}_{K,q}P \to \mathcal{D}_{K,q}/\mathcal{D}_{K,q}R \to O \tag{8.3}$$

and every short exact sequence is isomorphic to an exact sequence of that form. (Morphisms have been described in Example 8.42.)

8.3.1.3. *Split exact sequences.* Among short exact sequences, a special class is related to direct sums; if M_1 and M_2 are q-difference modules, let $i : M_1 \to M_1 \oplus M_2$ be the natural injection $x_1 \mapsto x_1 + 0 = x_1$ and let $p : M_1 \oplus M_2 \to M_2$ be the natural projection $x_1 + x_2 \mapsto x_2$. Then we have a short exact sequence:

$$0 \to M_1 \xrightarrow{i} M_1 \oplus M_2 \xrightarrow{p} M_2 \to 0$$

A short exact sequence is said to be *split* if it is isomorphic to one of the above form. One then proves in module theory [**37, 46, 145**] that, given a short exact sequence $0 \to M' \xrightarrow{f} M \xrightarrow{g} M'' \to 0$, the following four conditions are equivalent:

(1) The sequence is split.
(2) The module M admits a direct sum decomposition $M = M_1 \oplus M_2$ such that f is an isomorphism $M' \tilde{\to} M_1$ and the restriction of g is an isomorphism $M_2 \tilde{\to} M''$.
(3) The surjective morphism g admits a *section*, i.e., a morphism $s : M'' \to M$ such that $g \circ s = \mathrm{Id}_{M''}$.
(4) The injective morphism f admits a *retraction*, i.e., a morphism $r : M \to M'$ such that $r \circ f = \mathrm{Id}_{M'}$.

EXERCISE 8.62. With the above notations, show that:

(i) The projections of M on M_1 and M_2 can be given by $f \circ r$ and $s \circ g$;
(ii) $0 \to M'' \xrightarrow{s} M \xrightarrow{r} M' \to 0$ is a split short exact sequence.

All the above is general formalism for modules over a ring. However there is no easy criterion to decide if a given short exact sequence of the form (8.3) as in 8.3.1.2 is split.

EXAMPLE 8.63. Let $P = QR$. Then a section of the epimorphism $g : \mathcal{D}_{K,q}/\mathcal{D}_{K,q}P \to \mathcal{D}_{K,q}/\mathcal{D}_{K,q}R$ must have the form $s : X \pmod{\mathcal{D}_{K,q}R} \mapsto XL \pmod{\mathcal{D}_{K,q}P}$, where $\mathcal{D}_{K,q}RL \subset \mathcal{D}_{K,q}P$ (for s to be well defined) and where $\forall X \in \mathcal{D}_{K,q}\ ,\ , XL - X \in \mathcal{D}_{K,q}R$ (for s to be a section of g). The first condition says that $RL = AP$ for some A; the second that $L - 1 = BR$ for some B. So we should take $L := 1 + BR$, where $R(1+BR) = AP = AQR$ (for some A), i.e., where $1 + RB = AQ$. The latter condition, written $AQ - RB = 1$, looks like a bizarre form of Bézout condition in Q, R: it says that $1 \in \mathcal{D}_{K,q}Q + R\mathcal{D}_{K,q}$, but it is not easy to test (the Euclid algorithm uses either left or right division, here they would be mixed).

EXERCISE 8.64. The example was based on the third criterion. What can be said using the fourth?

However, for a short exact sequence of the form (8.2), there is a rather explicit test. We shall write $A = \begin{pmatrix} A' & U \\ 0 & A'' \end{pmatrix}$, where $A' \in \mathrm{GL}_{n'}(K)$, $A'' \in \mathrm{GL}_{n''}(K)$ and $U \in \mathrm{Mat}_{n',n''}(K)$; also $M := (K^n, \Phi_A)$, $M' := (K^{n'}, \Phi_{A'})$ and $M'' := (K^{n''}, \Phi_{A''})$. Last, under the identification of K^n with $K^{n'} \times K^{n''}$ (recall we do not distinguish such a cartesian product from tha corresponding direct sum), we write $f : M' \to M$, $x' \mapsto (x', 0)$ and $g : M \to M''$, $(x', x'') \mapsto x''$.

PROPOSITION 8.65. *With those notations, the short exact sequence* $0 \to M' \xrightarrow{f} M \xrightarrow{g} M'' \to 0$ *is split if, and only if, there exists*[11] $F \in Mat_{n',n''}(K)$ *such that* $(\sigma_q F)A'' - A'F = U$.

PROOF. Writing elements of K^n, $K^{n'}$, $K^{n''}$ as column vectors, the map $f : K^{n'} \to K^n$ and $g : K^n \to K^{n'}$ respectively have the form $X' \mapsto \begin{pmatrix} X' \\ 0_{n''} \end{pmatrix}$ (here $0_{n''}$ denotes a column vector of size n'') and $\begin{pmatrix} X' \\ X'' \end{pmatrix} \mapsto X''$. So they are respectively given by matrices $\begin{pmatrix} I_{n'} \\ 0_{n'',n'} \end{pmatrix}$ (here $0_{n'',n'}$ denotes a matrix of format $n'' \times n'$) and $\begin{pmatrix} 0_{n'',n'} & I_{n''} \end{pmatrix}$.

Using for instance the third criterion, we look for the section $s : K^{n''} \to K^{n'} \times K^{n''}$ in the form of a matrix $H := \begin{pmatrix} F \\ G \end{pmatrix}$, $F \in \mathrm{Mat}_{n',n''}(K)$, $G \in \mathrm{Mat}_{n'',n''}(K)$. The condition $g \circ s = \mathrm{Id}_{M''}$ translates into:

$$\begin{pmatrix} 0_{n'',n'} & I_{n''} \end{pmatrix} \begin{pmatrix} F \\ G \end{pmatrix} = I_{n''},$$

i.e., $G = I_{n''}$. The condition that s be a morphism of q-difference modules translates into $(\sigma_q H)A'' = AH$, i.e.:

$$\begin{pmatrix} \sigma_q F \\ I_{n''} \end{pmatrix} A'' = \begin{pmatrix} A' & U \\ 0 & A'' \end{pmatrix} \begin{pmatrix} F \\ I_{n''} \end{pmatrix},$$

i.e., $(\sigma_q F)A'' = A'F + U$. The matrix F is the one we look for. □

EXERCISE 8.66. Derive a proof using the fourth criterion (with a retraction).

EXAMPLE 8.67. Letting $A := \begin{pmatrix} 1 & 1 \\ 0 & 1 \end{pmatrix}$, we get a short exact sequence $0 \to \underline{\mathbf{1}} \to M \to \underline{\mathbf{1}} \to 0$. It is split if, and only if, there is $f \in K$ such that $\sigma_q f - f = 1$. If $K = \mathbf{C}((x))$ or $\mathbf{C}(\{x\})$ or $\mathbf{C}(x)$, this is impossible.

8.3.2. Extensions. Extensions of q-difference modules are a useful tool in the classification of irregular systems and in Galois theory. The fundamentals about extensions can be found in [**46**, Chapter 14 §1] and in [**32**, §5]. The application to q-difference modules is rather mechanical, and it is pretty much detailed in [**197**, Appendix A, Section 4]. So we shall give no proofs here. (But of course the reader is warmly encouraged to write them down!)

8.3.2.1. *Generalities.* So, from the general theory of extensions of modules:

- An *extension* of a module M'' by a module M' is a short exact sequence $0 \to M' \to M \to M'' \to 0$.
- A morphism from extension $0 \to M' \to M_1 \to M'' \to 0$ to extension $0 \to M' \to M_2 \to M'' \to 0$ is a morphism $h : M_1 \to M_2$ such that the

[11] The notation F (and, in the proof, G) is not meant to express a particular relation to morphisms f, g.

following diagram commutes:

$$\begin{CD} 0 @>>> M' @>>> M_1 @>>> M'' @>>> 0 \\ @. @V{\mathrm{Id}_{M'}}VV @V{h}VV @VV{\mathrm{Id}_{M''}}V \\ 0 @>>> M' @>>> M_2 @>>> M'' @>>> 0. \end{CD}$$

In this context, one rather uses the diagram:

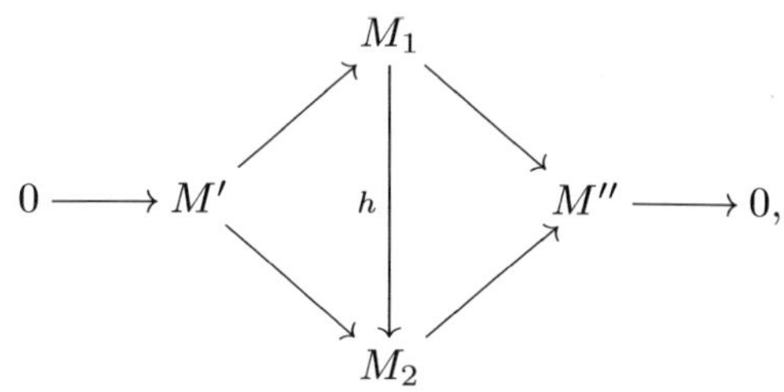

or even its simplified form:

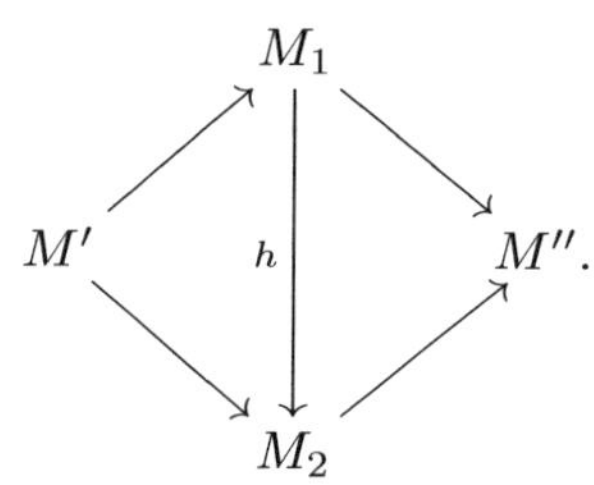

One sometimes (often?) says improperly, for short, that M is an extension of M'' by M'.

EXAMPLES 8.68. (i) The trivial extension is $0 \to M' \to M' \oplus M'' \to M'' \to 0$.

(ii) A split (or trivial) extension is one that is split as an exact sequence. It is equivalent to say that it is isomorphic (see definition below) to the trivial extension.

(iii) The example at the end of 8.3.1.3 shows a nonsplit extension of $\underline{1}$ by itself.

In the following, it is tacitly understood that M' and M'' are fixed and that "extensions" are extensions of M'' by M'. A morphism of extensions as above is automatically an isomorphism $M_1 \tilde{\to} M_2$ and then h^{-1} is also a morphism of extensions. Also, composing morphisms of extension yields morphisms of extensions. And of course, identities are morphisms of extensions. So clearly isomorphy of extensions is an equivalence relation.

DEFINITION 8.69. One writes $\mathrm{Ext}(M'', M')$ the set[12] of classes of extensions.

By **[46]** and specially **[32]**, $\mathrm{Ext}(M'', M')$ actually has a structure of module over the center of $\mathcal{D}_{K,q}$, that is of a K^{σ_q}-vector space. Its zero element is the class formed by trivial extensions. Operations will be described soon.

[12]It is by no way tautological that such classes do form a set, but, in the case of q-difference modules, it will follow from the concrete description.

8.3.2.2. *The* (V, Φ) *model.* Now we take $M' := (V', \Phi')$ and $M'' := (V'', \Phi'')$. Let $u \in \mathscr{L}_{\sigma_q}(V'', V')$, i.e., $u : V'' \to V'$ is a σ_q-linear map. We are going to build from u an extension M_u of M'' by M'. Let $V := V' \times V''$ and let $\Phi_u : V \to V$ be given by the formula:

$$\forall x \in V', y \in V'' \ , \ \Phi_u(x, y) := (\Phi'(x) + u(y), \Phi''(y)).$$

It is easily checked (meaning: you should do it) that:

- Φ_u is a σ_q-linear automorphism of V, so $M_u := (V, \Phi_u)$ is a q-difference module;
- $x \mapsto (x, 0)$ and $(x, y) \mapsto y$ are morphisms $M' \to M$ and $M \to M''$;
- $0 \to M' \to M \to M'' \to 0$ is a short exact sequence.

So we defined an extension M_u of M'' by M'. The crucial fact (to be included in the next theorem) is that *every extension of* M'' *by* M' *is isomorphic to* M_u *for some* $u \in \mathscr{L}_{\sigma_q}(V'', V')$. So the map $u \mapsto$ isomorphism class of extension M_u is surjective from $\mathscr{L}_{\sigma_q}(V'', V')$ to $\mathrm{Ext}(M'', M')$.

Of course we must discuss when such extensions are isomorphic; so we take $u, v \in \mathscr{L}_{\sigma_q}(V'', V')$ and look for possible vertical arrows in the diagram

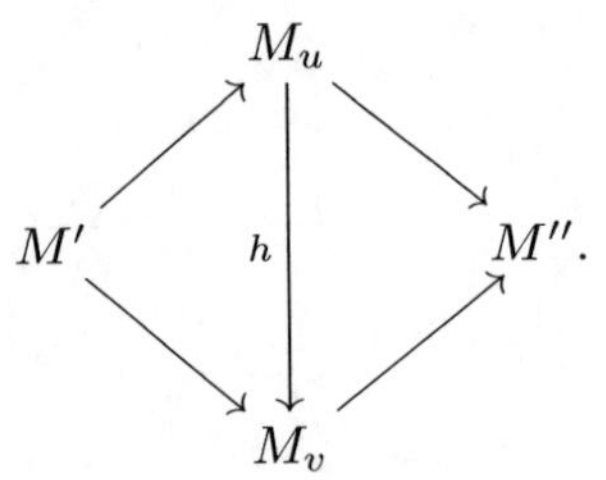

So we look for a linear map $h : V' \times V'' \to V' \times V''$ such that:

- h is a morphism: $\Phi_v \circ h = h \circ \Phi_u$;
- the left triangle commutes: $h(x, 0) = (x, 0)$;
- the right triangle commutes: the second component of $h(x, y)$ is y.

From the second and third condition, we see that $h(x, y) = (x + f(y), y)$, where $f \in \mathscr{L}_K(V'', V')$. Then the first condition writes:

$$\forall x \in V', y \in V'' \ , \ (\Phi'(x+f(y))+v(y), \Phi''(y))) = (\Phi'(x)+u(y)+f(\Phi''(y)), \Phi''(y))).$$

The above equation boils down to: $\Phi'(f(y)) + v(y) = u(y) + f(\Phi''(y))$, i.e.:

$$v - u = f \circ \Phi'' - \Phi' \circ f.$$

Note that this is consistent since both sides are σ_q-linear maps $V'' \to V'$. The above analysis is summarized in the following (see [**197**, Theorem A.4.1 p 126]):

THEOREM 8.2. (i) *The map* $u \mapsto$ *isomorphism class of extension* M_u *is surjective from* $\mathscr{L}_{\sigma_q}(V'', V')$ *to* $\mathrm{Ext}(M'', M')$.

(ii) $u, v \in \mathscr{L}_{\sigma_q}(V'', V')$ *have the same image in* $\mathrm{Ext}(M'', M')$ *if, and only if,* $v - u$ *belongs to the image of the* K^{σ_q}*-linear map* $t_{M'',M'} : f \mapsto f \circ \Phi'' - \Phi' \circ f$ *from* $\mathscr{L}_K(V'', V')$ *to* $\mathscr{L}_{\sigma_q}(V'', V')$.

The surjection $\mathscr{L}_{\sigma_q}(V'', V') \to \mathrm{Ext}(M'', M')$ can therefore be used to endow $\mathrm{Ext}(M'', M')$ with a unique structure of K^{σ_q}-vector space identifying it with the cokernel of the map $t_{M'',M'}$. Noting that the latter has kernel $\mathrm{Hom}_{\mathscr{C}}(M'',M')$, we obtain an exact sequence of K^{σ_q}-vector spaces:

$$0 \to \mathrm{Hom}_{\mathscr{C}}(M'',M') \to \mathscr{L}_K(V'',V') \overset{t_{M'',M'}}{\longrightarrow} \mathscr{L}_{\sigma_q}(V'',V') \to \mathrm{Ext}(M'',M') \to 0.$$

In the language of cochain complexes (Appendix E), we can say that the (new avatar of the) q-de Rham complex[13]:

$$\underset{0}{\mathscr{L}_K(V'',V')} \xrightarrow{t_{M'',M'}} \underset{1}{\mathscr{L}_{\sigma_q}(V'',V')}$$

(concentrated in degrees 0 and 1) has cohomology spaces $\mathrm{Hom}_{\mathscr{C}}(M'',M')$ and $\mathrm{Ext}(M'',M')$.

EXERCISE 8.70. Check that the neutral element of $\mathrm{Ext}(M'',M')$ is the trivial class.

REMARK 8.71. In [**32**,**46**] $\mathrm{Ext}(M'',M')$ is given a structure of module over the center of $\mathcal{D}_{K,q}$, i.e., K^{σ_q}; this is done in a very general setting, independent of the constructions above (which are of course specific to q-difference modules since they involve the model (V,Φ)). It is proved in [**197**, Theorem A.4.1, p 126] that *the two structures are the same* (for conciseness, we shall admit that result). So we shall freely use the general theorems involving that K^{σ_q}-structure.

8.3.2.3. *The* (K^n,Φ_A) *model.* The above constructions are readily transposed to the case that $V' := K^{n'}$, $\Phi' := \Phi_{A'}$, $V'' := K^{n''}$, $\Phi'' := \Phi_{A''}$. Then $V := V' \times V''$ is identified with K^n, where $n := n' + n''$. The map σ_q-linear map $u : V'' \to V'$ has the form $X \mapsto T\sigma_q X$ for some matrix $T \in \mathrm{Mat}_{n',n''}(K)$ and $\Phi_u = \Phi_A$ where $A := \begin{pmatrix} A' & U \\ 0 & A'' \end{pmatrix}$ for some $U \in \mathrm{Mat}_{n',n''}(K)$.

EXERCISE 8.72. Check (or admit) that $U = A'TA''$.

Let $U,V \in \mathrm{Mat}_{n',n''}(K)$ and let $f \in \mathscr{L}_K(V'',V')$ correspond to matrix $F \in \mathrm{Mat}_{n',n''}(K)$. Then :

$$v - u = f \circ \Phi'' - \Phi' \circ f \Longleftrightarrow V - U = (\sigma_q F)A'' - A'F.$$

So we get an exact sequence:

$$0 \to \mathrm{Hom}_{\mathscr{C}}(M_{A''},M_{A'}) \to \mathrm{Mat}_{n',n''}(K) \overset{T_{A',A''}}{\longrightarrow} \mathrm{Mat}_{n',n''}(K) \to \mathrm{Ext}(M_{A''},M_{A'}) \to 0,$$

where $T_{A',A''} : \mathrm{Mat}_{n',n''}(K) \to \mathrm{Mat}_{n',n''}(K)$ is the map $F \mapsto (\sigma_q F)A'' - A'F$.

EXAMPLE 8.73. Let $V' = V'' = K$ and $A' = A'' = 1$. Then we obtain $\mathrm{Ext}(\underline{\mathbf{1}},\underline{\mathbf{1}})$ as the cokernel of:

$$K \xrightarrow{f \mapsto \sigma_q f - f} K.$$

So the extensions corresponding to matrices $\begin{pmatrix} 1 & u \\ 0 & 1 \end{pmatrix}$ and $\begin{pmatrix} 1 & v \\ 0 & 1 \end{pmatrix}$ are equivalent if, and only if, $\sigma_q f - f = v - u$ can be solved in K.

EXERCISE 8.74. Here, $K := \mathbf{C}(\{x\})$ or $\mathbf{C}((x))$.

---

[13] Other approaches to the q-de Rham complexes are possible; see, e.g., that of Aomoto in [**11**] and subsequent works.

(i) Discuss the equivalence of matrices $\begin{pmatrix} 1 & 0 \\ 0 & x^m \end{pmatrix}$ and $\begin{pmatrix} 1 & 1 \\ 0 & x^m \end{pmatrix}$ in both cases. (The sign of m may matter.)

(ii) Interpret in terms of extensions. (This example will be generalized when we study the formal and analytic classification of irregular systems in Chapter 15.)

8.3.2.4. *Weakening the equivalence.* We now discuss the seemingly more natural question: when are two modules M_1, M_2 obtained as extensions of M'' by M' iso morphic as modules? The equivalence defined above is about the isomorphism of two short exact sequences:

$$0 \to M' \xrightarrow{i_1} M_1 \xrightarrow{p_1} M'' \to 0 \text{ and } 0 \to M' \xrightarrow{i_2} M_2 \xrightarrow{p_2} M'' \to 0$$

as extensions, i.e., requiring identity arrows $M' \to M'$ and $M'' \to M''$. It allowed us to define $\mathrm{Ext}(M'', M')$ as the quotient of the set[14] $\mathcal{E}(M'', M')$ of all such exact sequences by the equivalence relation, say $\sim$, described in 8.3.2.1. So let us write $\mathrm{Ext}(M'', M') := \mathcal{E}(M'', M')/\sim$. Also, in order to shorten notations, we shall denote (i_1, p_1) and (i_2, p_2) the above elements of $\mathcal{E}(M'', M')$ and say that $M_1 := \mathrm{Im} i_1 = \mathrm{Ker} p_1$ and $M_2 := \mathrm{Im} i_2 = \mathrm{Ker} p_2$ are there respective middle terms.

We also noticed that the equivalence of such extensions implies that their middle terms are isomorphic: $M_1 \simeq M_2$. So we have a weaker equivalence relation on $\mathcal{E}(M'', M')$ and, correspondingly, a surjective mapping:

$$\mathrm{Ext}(M'', M') = \frac{\mathcal{E}(M'', M')}{\sim} \longrightarrow \widetilde{\mathrm{Ext}}(M'', M') := \frac{\mathcal{E}(M'', M')}{\simeq}.$$

We shall see that it is actually a quotient map under the group action of $\mathrm{Aut}(M') \times \mathrm{Aut}(M'')$. To define that action, which actually comes from an action on $\mathcal{E}(M'', M')$, let $f' \in \mathrm{Aut}(M')$ and $f'' \in \mathrm{Aut}(M'')$ and let $(i, p) \in \mathcal{E}(M'', M')$ with middle term M. One checks immediately that $(i \circ f'^{-1}, f'' \circ p) \in \mathcal{E}(M'', M')$ with same middle term M; and that this indeed defines a group action of $\mathrm{Aut}(M') \times \mathrm{Aut}(M'')$ on $\mathcal{E}(M'', M')$.

PROPOSITION 8.75. (i) *That action goes to the quotient as an action on the quotient set* $\mathit{Ext}(M'', M') = \mathcal{E}(M'', M')/\sim$.

(ii) *The map* $\mathit{Ext}(M'', M') \to \widetilde{\mathit{Ext}}(M'', M')$ *is the quotient by that action.*

PROOF. (i) We must show that if $(i_1, p_1) \sim (i_2, p_2)$ in $\mathcal{E}(M'', M')$ (equivalence of extensions), then $(i_1 \circ f'^{-1}, f'' \circ p_1) \sim (i_2 \circ f'^{-1}, f'' \circ p_2)$. But the same $f : M_1 \to M_2$ such that $f \circ i_1 = i_2$ and $p_2 \circ f = p_1$ also satisfies $f \circ i_1 \circ f'^{-1} = i_2 \circ f'^{-1}$ and $f'' \circ p_2 \circ f = f'' \circ p_1$.

(ii) We must show that two extensions (i_1, p_1) and (i_2, p_2) in $\mathcal{E}(M'', M')$ have isomorphic middle terms if, and only if, they are equivalent to extensions in the same orbit. One way is easy (guess which). Conversely, assume that $f : M_1 \to M_2$ is an isomorphism of their middle terms. It induces automorphisms f' of M' and f'' of M'' and we can see that f is an equivalence of the extensions $(i_1 \circ f'^{-1}, f'' \circ p_1)$ and (i_2, p_2).

□

[14] This is not really a set, it is too "big" for that. But this could easily be fixed by requiring that all involved modules M have as underlying set the set $M' \times M''$. So we do no take care of that subtlety.

We apply the proposition to the matrix model and take $M' := (K^{n'}, \Phi_{A'})$ and $M'' := (K^{n''}, \Phi_{A''})$, so that $\mathrm{Ext}(M'', M')$ can be identified with the cokernel of the map $T_{A',A''} : F \mapsto (\sigma_q F)A'' - A'F$ from $\mathrm{Mat}_{n',n''}(K)$ to itself. Here $\mathrm{Aut}(M')$ and $\mathrm{Aut}(M'')$ can respectively be identified to the subgroups $\mathrm{Aut}(A')$ and $\mathrm{Aut}(A'')$ of $\mathrm{GL}_{n'}(K)$ and $\mathrm{GL}_{n''}(K)$ defined by conditions $F'[A'] = A'$ and $F''[A''] = A''$. We have a commutative diagram:

$$\begin{array}{ccc} \mathrm{Mat}_{n',n''}(K) & \xrightarrow{T_{A',A''}} & \mathrm{Mat}_{n',n''}(K) \\ \downarrow & & \downarrow \\ \mathrm{Mat}_{n',n''}(K) & \xrightarrow{T_{A',A''}} & \mathrm{Mat}_{n',n''}(K). \end{array}$$

The vertical arrows are $F \mapsto F'FF''$ and $F \mapsto (\sigma_q F')FF''$. This way, we get an automorphism of $\mathrm{Coker}\, T_{A',A''}$ and, in the end, an action of $\mathrm{Aut}(A') \times \mathrm{Aut}(A'')$ on that space. The quotient may be identified with $\widetilde{\mathrm{Ext}}(M'', M')$.

EXAMPLE 8.76. Taking $n' = n'' = 1$, $A' := ax$, $A'' := b$, $a, b \in \mathbf{C}^*$, the cokermel is isomorphic to $\mathbf{C}$, and $\mathrm{Aut}(A') = \mathrm{Aut}(A'') = \mathbf{C}^*$. The action of $\mathbf{C}^* \times \mathbf{C}^*$ on $\mathbf{C}$ gives a nonseparated quotient $\{0, *\}$. Here, 0, the class of $0 \in \mathbf{C}$, corresponds to the class of all split extensions, that of all the $A_u := \begin{pmatrix} ax & u \\ 0 & b \end{pmatrix}$ such that $u = b\sigma_q f - axf$ can be solved for some $f \in K$; and $*$ is the class of all the other A_u, which are mutually gauge equivalent. The point 0 is closed in the quotient while the point $*$ is dense.

EXERCISE 8.77. Make more precise the example in case $K = \mathbf{C}(\{x\})$, resp. $K = \mathbf{C}((x))$. Caution: in the latter case the conclusion must be *corrected*. In the former case, a criterion using the q-Borel-Ramis transform $\mathcal{B}_{q;1}$ from Chapter 6 may be invoked.

8.3.3. Base change (or extension of scalars) for q-difference modules. Our hesitation on the title of this subsection comes from the possible confusion with both choices: "base change" refers to a change of the base field K (for us usually $\mathbf{C}(\{x\})$ or $\mathbf{C}((x))$), although in q-difference theory, q is classically referred as the base; and "extension of scalars" refers to the extension of the field of scalars of a vector space, although our base fields K are rather fields of functions and our field of scalars is rather $\mathbf{C}$.

One main use of the formalism will be for ramification (see 3.1.2 in Chapter 3 and 5.2.2.5 in Chapter 5) as a tool for defining and studying the Newton polygon of a q-difference module in Section 10.1, i.e., replacing x by $x^{1/r}$ and q by $q^{1/r}$. The other main use is "formalization", i.e., going from a q-difference module over $\mathbf{C}(\{x\})$ to one over $\mathbf{C}((x))$. We should also mention the use of "Picard-Vessiot extensions" in the alternative Picard-Vessiot approach to Galois theory. So it seems convenient to first describe the more general mechanism of replacing a q-difference field K by a q-difference extension K'.

8.3.3.1. *Base change for $\mathcal{D}_{K,q}$-modules.* There is a general formalism of base change for K-algebras, unhappily it cannot be directly applied here since $\mathcal{D}_{K,q}$ is not a K-algebra. There is a more powerful formalism using so-called bimodules that works, it is used for instance in [**4**]. We shall proceed in a more pedestrian way that is sufficient for our needs.

So let $(K', \sigma_{q'})$ a q-difference extension of (K, σ_q). For the moment, we consider $\mathcal{D}_{K,q}$ and $\mathcal{D}_{K',q'}$ as algebras of noncommutative polynomials, with basis $(S^n)_{n\in\mathbf{Z}}$ (see 3.1.3.2 in Chapter 3). The K'-space $K' \otimes_K \mathcal{D}_{K,q}$ also has basis the family $(S^n)_{n\in\mathbf{Z}}$, which allows us to identify as K'-spaces:

$$\mathcal{D}_{K',q'} = K' \otimes_K \mathcal{D}_{K,q}.$$

The difference of interpretation of S as σ_q or as $\sigma_{q'}$ comes when we consider modules. Recall that a $\mathcal{D}_{K,q}$-module M is the same thing as a pair (V, Φ), where V is a K-vector space and Φ a σ_q-linear automorphism of V. The action of $\mathcal{D}_{K,q}$ on an element x of M, that is of V, is defined as $(\sum a_n S^n).x := \sum a_n \Phi^n(x)$.

LEMMA 8.78. *Let $V' := K' \otimes_K V$. There is a unique $\sigma_{q'}$-linear automorphism Φ' of V' such that Φ' restricts to Φ on V.*

PROOF. Let (e_i) a K-basis of V, thus a K'-basis of V'. The only possible definition of Φ' is $\sum \lambda_i e_i \mapsto \sum \sigma_{q'}(\lambda_i)\Phi(e_i)$. It is obviously additive. It is σ_q-linear because:

$$\Phi'\left(\lambda \sum \lambda_i e_i\right) = \sum \sigma_{q'}(\lambda\lambda_i)\Phi(e_i) = \sigma_{q'}(\lambda) \sum \sigma_{q'}(\lambda_i)\Phi(e_i) = \sigma_{q'}(\lambda)\Phi'\left(\sum \lambda_i e_i\right).$$

It is bijective with inverse $\sum \lambda_i e_i \mapsto \sum \sigma_{q'}^{-1}(\lambda_i)\Phi^{-1}(e_i)$. □

So (V', Φ') is a $\mathcal{D}_{K',q'}$-module, we denote it $K' \otimes_K M$ (the operators σ_q and $\sigma_{q'}$ are omitted in the notation). To obtain a functor $M \rightsquigarrow K' \otimes_K M$, we must define the effect of morphism. So let $f : M \to N$ a morphism of $\mathcal{D}_{K,q}$-modules. Identifying M as (V, Φ) and N as (W, Ψ), we see f as a K-linear map $V \to W$ such that $\Psi \circ f = f \circ \Phi$. By Appendix F, it extends to a K'-linear map f' from $V' := K' \otimes_K V$ to $W' := K' \otimes_K W$. We define Φ', Ψ' as in the above lemma.

LEMMA 8.79. $\Psi' \circ f' = f' \circ \Phi'$.

PROOF. Let (e_i) a K-basis of V and let the $\lambda_i \in K'$. Then:

$$\begin{aligned}(\Psi' \circ f')\left(\sum \lambda_i e_i\right) = \Psi'\left(f'\left(\sum \lambda_i e_i\right)\right) = \Psi'\left(\sum \lambda_i f(e_i)\right) = \sum \sigma_{q'}(\lambda_i)\Psi(f(e_i)) = \\ \sum \sigma_{q'}(\lambda_i) f(\Phi(e_i)) = f'\left(\sum \sigma_{q'}(\lambda_i)\Phi(e_i)\right) = f'\left(\Phi'\left(\sum \lambda_i e_i\right)\right) \\ = (f' \circ \Phi')\left(\sum \lambda_i e_i\right).\end{aligned}$$

□

So we have a functor from $\mathcal{D}_{K,q}$-modules to $\mathcal{D}_{K',q'}$-modules.

PROPOSITION 8.80. *That functor is additive, exact and faithful. If $K' \otimes_K M$ is trivial, so is M.*

PROOF. Properties such as $M = M_1 \oplus M_2$ and the exactness of $M_1 \to M_2 \to M_3$ are checked on the underlying K-vector spaces as $V = V_1 \oplus V_2$ and the exactness of $V_1 \to V_2 \to V_3$, so it all boils down to the corresponding statements in Appendix F; same thing for the vanishing of $K' \otimes_K M$ or M. A similar argument applies to the vanishing of a morphism, which is equivalent to the vanishing of the underlying linear map. □

EXAMPLE 8.81. From the exact sequence:

$$0 \to \mathcal{D}_{K,q} \xrightarrow{-\times P} \mathcal{D}_{K,q} \to \mathcal{D}_{K,q}/\mathcal{D}_{K,q}P \to 0,$$

we deduce (identifying $P \in \mathcal{D}_{K,q}$ to $P \in \mathcal{D}_{K',q'}$) the exact sequence:

$$0 \to \mathcal{D}_{K',q'} \xrightarrow{-\times P} \mathcal{D}_{K',q'} \to \mathcal{D}_{K',q'}/\mathcal{D}_{K',q'}P \to 0$$

In particular, $K' \otimes (\mathcal{D}_{K,q}/\mathcal{D}_{K,q}P) = \mathcal{D}_{K',q'}/\mathcal{D}_{K',q'}P$.

8.3.3.2. *Base change for q-difference modules.* Again let $(K', \sigma_{q'})$ a q-difference extension of (K, σ_q) and let $M = (V, \Phi)$ a q-difference module over K. The only difference with 8.3.3.1 is that now we assume $\dim_K V < \infty$, so defining $M' := K' \otimes_K M = (V', \Phi')$ as above, we have $\dim_{K'} V' = \dim_K V < \infty$ and M' is a q-difference module over K' (maybe more properly a q'-difference module). All constructions from 8.3.3.1, and in particular, if $M = \mathcal{D}_{K,q}/\mathcal{D}_{K,q}P$ then $M' = \mathcal{D}_{K',q'}/\mathcal{D}_{K',q'}P$. Now we deal with the other standard form of a q-difference module.

EXAMPLE 8.82. let $M = (K^n, \Phi_A)$. Then $M' = (K'^n, \Phi'_A)$, where $\Phi'_A(X) := A^{-1}\sigma_{q'}X$. Indeed, let $(e_1, \ldots, e_n)$ the canonical basis of K^n over K and also of K'^n over K' and let $\lambda_1, \ldots, \lambda_n$ the components of X, so that $X = \sum \lambda_i e_i$. Then $\Phi'_A(X) = \sum \sigma_{q'}(\lambda_i)\Phi(e_i)$, and the $\Phi(e_i)$ are the columns of A^{-1}, whence the conclusion.

Note that if $F \in \mathrm{Mat}_{p,n}(K)$ is a morphism from (K^n, Φ_A) to (K^p, Φ_B), i.e., if $(\sigma_q F)A = BF$, then $F \in \mathrm{Mat}_{p,n}(K')$ is a morphism from (K'^n, Φ'_A) to (K'^p, Φ'_B) as it should.

Taking $n = 1$, $A := (1)$, we see that $K' \otimes_K \underline{1}$ is the unit object over K', also denoted $\underline{1}$ (with the hope that this will cause no confusion). This can also be seen from the identification $K' \otimes (\mathcal{D}_{K,q}/\mathcal{D}_{K,q}P) = \mathcal{D}_{K',q'}/\mathcal{D}_{K',q'}P$, where $P := S - 1$.

The functor $M \rightsquigarrow K' \otimes_K M$ is additive, exact and faithful: this is a particular case of the similar facts stated in 8.3.3.1. Likewise, $K' \otimes_K M$ is trivial only if M is.

Recall in particular from 8.2.3.2 that a solution, resp. a cosolution of M is a morphism $\underline{1} \to M$, resp. a morphism $M \to \underline{1}$. By base extension, we therefore get maps $\Gamma(M) \to \Gamma(K' \otimes_K M)$ and $\Gamma^\vee(M) \to \Gamma^\vee(K' \otimes_K M)$. Of course $\Gamma(K' \otimes_K M)$, also denoted $\Gamma_{K'}(M)$, must be understood in the K'-category, i.e., as $\mathrm{Hom}(\underline{1}', K' \otimes_K M)$ where $\underline{1}' := K' \otimes_K \underline{1}$ (temporary notation!).

Those maps are injective, but, in general, not surjective. Indeed:

- If $M = (K^n, \Phi_A)$ then its solutions are the $X \in K^n$ such that $\sigma_q X = AX$ and the map $\Gamma(M) \to \Gamma(K' \otimes_K M)$ sends such a X to itself, an element of K'^n such that $\sigma_{q'} X = AX$. There may of course be more solutions in K'^n.
- If $M = \mathcal{D}_{K,q}/\mathcal{D}_{K,q}P$ then its cosolutions are the $f \in K$ such that $P(\sigma_q)(f) = 0$ and the map $\Gamma^\vee(M) \to \Gamma^\vee(K' \otimes_K M)$ sends such a f to itself, an element of K' such that $P(\sigma_{q'})(f) = 0$. There may of course be more solutions in K'.

For examples where $\Gamma(K' \otimes_K M)$, resp. $\Gamma^\vee(K' \otimes_K M)$, is not trivial while $\Gamma(M)$, resp. $\Gamma^\vee(M)$ is; see below in 8.3.3.3. We return to those matters in 9.2.

8.3.3.3. *First example of base change: formalization.* First, let $K := \mathbf{C}(\{x\})$ and $K' := \mathbf{C}((x))$. Starting from an "analytic" q-difference module, i.e., a q-difference module over $\mathbf{C}(\{x\})$, we define its *formalization* as $\hat{M} := \mathbf{C}((x)) \otimes_{\mathbf{C}(\{x\})} M$. In particular, writing for short K and $\hat{K}$ for $\mathbf{C}(\{x\})$ and $\mathbf{C}((x))$ and $\mathcal{D}$ and $\hat{\mathcal{D}}$ the corresponding algebras of q-difference operators:

$$\widehat{(K^n, \Phi_A)} = (\hat{K}^n, \Phi_A) \text{ if } A \in \mathrm{Mat}_n(K) \text{ and } \widehat{\mathcal{D}/\mathcal{D}P} = \hat{\mathcal{D}}/\hat{\mathcal{D}}P \text{ if } P \in \mathcal{D}.$$

EXAMPLE 8.83. Let us write for short $\mathcal{D}$ and $\hat{\mathcal{D}}$ the corresponding rings of q-difference operators. Let $P := (\sigma_q - 1)(\sigma_q - x) = \sigma_q^2 - (1 + qx)\sigma_q + x$ and $M := \mathcal{D}/\mathcal{D}P$. Then $\Gamma^\vee(M)$ consists of solutions of $Pf = 0$ in $\mathbf{C}(\{x\})$, that is, $\Gamma^\vee(M)$ is trivial; while the formal solution $\sum\limits_{n\geq 0} q^{-n(n+1)/2} x^n$ generates $\Gamma^\vee(\hat{M})$ (we recognize the Tshakaloff q-difference operator from 4.3.1 and 5.3.1).

EXERCISE 8.84. Let $A := \begin{pmatrix} x & 1 \\ 0 & 1 \end{pmatrix}$ and $M := (\mathbf{C}(\{x\})^n, \Phi_A)$, so that $\hat{M} = (\mathbf{C}((x))^n, \Phi_A)$. Check that $\Gamma(M)$ is trivial while $\Gamma(\hat{M})$ is not.

As usual, it is easier to deal with solutions in the matricial model and with cosolutions in the q-difference operator model.

8.3.3.4. *Second example of base change: ramification.* Let $K := \mathbf{C}(\{x\})$ or $K := \mathbf{C}((x))$ at will, and the q-difference extension $K' := K[x']$ endowed with $\sigma_{q'}$, where $x' := x^{1/r}$ and $q' := q^{1/r}$ (here $r \in \mathbf{N}$, $r \geq 2$). So, if $M := (K^n, \Phi_A)$, then $M' = (K'^n, \Phi_A)$ with the same matrix $A \in \mathrm{GL}_n(K) \subset \mathrm{GL}_n(K')$.

EXAMPLE 8.85. Let $A := \begin{pmatrix} 0 & q \\ 1 & 0 \end{pmatrix}$ and $M := (K^2, \Phi_A)$, so that the elements of $\Gamma(M)$ are the $\begin{pmatrix} f \\ g \end{pmatrix}$ such that $f, g \in K$ satisfy $\sigma_q f = qg$, $\sigma_q g = f$. Since $\sigma_q^2 f = qf$ is impossible in K (remember why?), $\Gamma(M)$ is trivial. But $\Gamma(K' \otimes_K M)$ is generated by $\begin{pmatrix} x' \\ x'/q' \end{pmatrix}$. Actually, A is conjugated with $\mathrm{Diag}(q', -q')$ in $\mathrm{GL}_n(\mathbf{C})$, so they are *a fortiori* gauge-equivalent over K, which implies that M splits as $M = M_1 \oplus M_2$, where $M_1 \simeq (K, \Phi_{q'})$ and $M_2 \simeq (K, \Phi_{-q'})$. Clearly, $\Gamma(M_1)$ and $\Gamma(M_2)$ both are trivial. However, setting $M_i' := K' \otimes_K M_i$, we see that $M_1' \simeq (K', \Phi_{q'})$, which (as a q-difference module over K') is trivial, so $\Gamma(K' \otimes_K M_1)$ has dimension 1.

EXERCISE 8.86. What about $\Gamma(K' \otimes_K M_2)$?

EXAMPLE 8.87. Let $A := \begin{pmatrix} 0 & 1 \\ x & 0 \end{pmatrix}$ and $M := (K^2, \Phi_A)$, so that $\Phi_A \begin{pmatrix} f \\ g \end{pmatrix} = \begin{pmatrix} x^{-1}\sigma_q g \\ \sigma_q f \end{pmatrix}$. The q-difference module M is simple: indeed, any element as above generating an invariant line would have (f, g) and $(x^{-1}\sigma_q g, \sigma_q f)$ colinear, whence $x^{-1} g \sigma_q g = f \sigma_q f$, which is only possible in K for $f = g = 0$ (taking valuations). On the other hand, over K':

$$A = \begin{pmatrix} 0 & 1 \\ x'^2 & 0 \end{pmatrix} \sim_{K'} \mathrm{Diag}(cx', -cx'),$$

where $c^2 = 1/q'$. This can be seen as a splitting of $M' := K' \otimes_K M$ as $M' = M_1' \oplus M_2'$, where M_1' is generated by $e_1' := \begin{pmatrix} 1 \\ cx' \end{pmatrix}$ and M_2' by $e_2' := \begin{pmatrix} 1 \\ -cx' \end{pmatrix}$. Indeed, direct calculation shows that $\Phi'(e_1') = (cx')^{-1} e_1'$ and $\Phi'(e_2') = (cx')^{-1} e_2'$. So here M is simple but $K' \otimes_K M$ splits.

EXERCISE 8.88. Find the gauge transformation diagonalizing A.

8.3.4. Restriction of scalars for q-difference modules.

8.3.4.1. Let $(K',\sigma_{q'})$ a q-difference extension of (K,σ_q) and let $M'=(V',\Phi')$ a q'-difference module over $(K',\sigma_{q'})$. Let $V:=V'_{(K)}$, so that V' is the same as V but seen as a K-space. Let Φ be Φ' seen as a group automorphism of V. Then:

$$\forall a\in K, v\in V\ ,\ \Phi(av)=\Phi'(av)=\sigma_{q'}(a)\Phi'(v)=\sigma_q(a)\Phi(v).$$

In the last equality, we used the fact that, by definition of a q-difference extension, $\sigma_{q'|K}=\sigma_q$. Therefore Φ is σ_q-linear and $M:=(V,\Phi)$ is a q-difference module over (K,σ_q), obtained by *restriction of scalars* and denoted (if necessary) by $M'_{(K)}$. The adjunction formula from Appendix F still makes sense and holds true in this context:

$$\mathrm{Hom}_{\mathscr{C}}(M,N'_{(K)})=\mathrm{Hom}_{\mathscr{C}'}(K'\otimes_K M,N).$$

8.3.4.2. *An important example (I).* This is due to van der Put end Reversat [**248**] and will be used in Subsection 10.1.5 and then applied in 11.1.4 in relation with formal classification.

We shall need a particularly useful special case of restriction of scalars. We let $K=\mathbf{C}(\{x\})$ or $\mathbf{C}((x))$, and, for some $r\geq 2$, $x':=x^{1/r}$, $q':=q^{1/r}$ and $K':=K[x']$. We consider the q'-difference module $M':=(V',\Phi')$, where $V':=K'e'$, a line with basis e', and Φ' is defined by equality $\Phi'(e'):=ax'^de'$ for some fixed $a\in\mathbf{C}^*$, $d\in\mathbf{Z}$. Clearly then:

$$M'\simeq\mathcal{D}_{K,q'}/\mathcal{D}_{K,q'}L',\ \text{where}\ L':=\sigma_{q'}-ax'^d,$$

so L' has (as a q'-difference operator) is pure isoclinic of slope d (section 5.2. As we shall see in Chapter 10, one can then say that the q'-difference module M' is pure isoclinic of slope d.

We now look for the structure of $M:=M'_{(K)}$. The underlying space $V:=V'_{(K)}$ has K-basis $e_0,\dots,e_{r-1}$, where $e_i:=x'^ie'$, $i=0,\dots,r-1$. Its structural σ_q-linear automorphism is defined by:

$$\Phi(e_i)=\Phi'(x'^ie')=\sigma_{q'}(x'^i)\Phi'(e')=q'^ix'^iax'^de'=aq'^ix'^{(\alpha(i)r+\beta(i))}e'=aq'^ix^{\alpha(i)}e_{\beta(i)},$$

where we have effected the euclidean division $i+d=\alpha(i)r+\beta(i)$, $\alpha(i)\in\mathbf{Z}$, $\beta(i)\in\{0,\dots,r-1\}$. Clearly β is a permutation of $\{0,\dots,r-1\}$ (expressing in terms of representatives the translation by d in $\mathbf{Z}/r\mathbf{Z}$).

Since Φ^j sends Ke_0 to $Ke_{\beta^j(0)}$, we see that e_0 is a cyclic vector if, and only if, β is an r-cycle; this, of course is equivalent to $d\wedge r=1$ (d, r are coprime). *We now assume that this condition is satisfied.* We compute:

$$\Phi^r(e_0)=\Phi'^r(e')=a^rq'^{dr(r-1)/2}x'^{rd}e'=a^rq^{d(r-1)/2}x^de_0.$$

Therefore $M\simeq\mathcal{D}_{K,q}/\mathcal{D}_{K,q}L$, where $L:=\sigma_q^r-a^rq^{d(r-1)/2}x^d$ is pure isoclinic of slope d/r; and so is $M:=M'_{(K)}$, according to Section 5.2 and Chapter 10. Actually, $M\simeq(K^r,\Phi_A)$, where (by the exercise at the end of 8.1.3.1):

$$A:=\begin{pmatrix}0&1&0&\dots&0&0\\0&0&1&\dots&0&0\\0&0&0&\dots&0&0\\\vdots&\vdots&\vdots&\ddots&\vdots&\vdots\\0&0&0&\dots&0&1\\u^{-1}&0&0&\dots&0&0\end{pmatrix}.$$

EXERCISE 8.89. (i) Check the formula given for $\Phi'^r(e')$.

(ii) More generally, compute $\Phi^j(e_0)$, $j = 0, \ldots, r-1$.

8.3.5. Galois descent.

8.3.5.1. Let K'/K a field extension, V a linear space over K, and let $V' := K' \otimes_K V$. Every K-subspace $W \subset V$ gives rise to a K'-subspace $W' := K' \otimes_K W \subset V'$. Using a basis of W and completing it into a basis of V one can easily see that $W = W' \cap V$. Conversely, how does one recognize if an arbitrary K'-subspace $W' \subset V'$ comes in this way from a K-subspace $W \subset V$?

This can be effectively answered if K'/K is a finite Galois extension and the process belongs to the theory of *Galois descent*; see [**33**, Chapter V §10.4] or [**239**, 11.1]. (At `https://kconrad.math.uconn.edu/blurbs/galoistheory/galoisdescent.pdf` one can find a nice introduction by Keith Conrad.) Indeed, the Galois group $G := Gal(K'/K)$ acts upon V' according to the formula:

$$\Psi_\tau\left(\sum \lambda_i v_i\right) := \sum \tau(\lambda_i) v_i.$$

EXERCISE 8.90. Prove that the fixed subset under this action is ${V'}^G = V$.

Note that, while Ψ_τ is not K'-linear[15] but merely τ-linear, for every K'-subspace W' of V', the image $\Psi_\tau(W')$ is a K'-subspace. Clearly, if W' is of the form $K' \otimes_K W$, it is stable under the action of G, i.e., $\Psi_\tau(W') = W'$ for every $\tau \in G$. It is then a theorem that the converse holds: every K'-subspace W' of V' which is G-stable is of the form $K' \otimes_K W$. Of course, then $W = {W'}^G = W' \cap {V'}^G = W' \cap V$.

8.3.5.2. To apply this result in 10.2, we shall need two auxiliary results. The context is the following: $K = \mathbf{C}(\{x\})$ or $K = \mathbf{C}((x))$, at will; $r \in \mathbf{N}$, $r \geq 2$ being given, let $x' := x^{1/r}$, $q' := q^{1/r}$ and $K' := K[x']$ endowed with $\sigma_{q'}$. Then K'/K is a cyclic Galois extension, its Galois group G being generated by the K-automorphism τ_r such that $x' \mapsto jx'$, where $j := e^{2i\pi/r}$. It is an important fact that τ_r and $\sigma_{q'}$ commute: indeed, $\tau_r \circ \sigma_{q'}$ and $\sigma_{q'} \circ \tau_r$ both induce σ_q on K and both send x' to $q'jx'$. In the following calcultaiton, we write $\tau := \tau_r$.

Let $M = (V, \Phi)$ a q-difference module over K and let $M' := K' \otimes_K M = (V', \Phi')$. Then there is an action Ψ_τ on V' and it commutes with Φ'; indeed, if the $\lambda_i \in K'$ and the $v_i \in V$:

$$\Psi_\tau \circ \Phi'\left(\sum \lambda_i v_i\right) = \Psi_\tau\left(\sum \sigma_{q'}(\lambda_i)\Phi(v_i)\right) = \sum \tau\sigma_{q'}(\lambda_i)\Phi(v_i) =$$

$$\sum \sigma_{q'}\tau(\lambda_i)\Phi(v_i) = \Phi'\left(\sum \tau(\lambda_i) v_i\right) = \Phi' \circ \Psi_\tau\left(\sum \lambda_i v_i\right).$$

Now let N' a sub module of M. Then, by definition, its underlying linear space W' is Φ'-stable. Since $\Psi_\tau \circ \Phi' = \Phi' \circ \Psi_\tau$, the image $\Psi_\tau(W')$ is also Φ'-stable. Therefore it defines a submodule $\Psi_\tau(N')$.

LEMMA 8.91. *Write for short $\mathcal{D}'$ the ring of q-difference operators over K'. If $N' \simeq \mathcal{D}'/\mathcal{D}'P$, then $\Psi_\tau(N') \simeq \mathcal{D}'/\mathcal{D}'P^\tau$, where P^τ is obtained by applying τ to the coefficients of P.*

PROOF. Let v a cyclic vector of N' such that $P(\Phi')(v) = 0$. Direct calculation (left to you) shows that for any $Q \in \mathcal{D}'$, one has:

$$\Psi_\tau\left(Q(\Phi')(v)\right) = Q^\tau(\Phi')\left(\Psi_\tau(v)\right).$$

[15] Contrary to what is stated in [**220**]. Luckily the rest of the argument there is correct.

So, on the one hand, since W' is the set of all $Q(\Phi')(v)$, we see that $\Psi_\tau(W')$ is the set of all $Q^\tau(\Phi')(\Psi_\tau(v))$, i.e., $\Psi_\tau(v)$ is a cyclic vector for $\Psi_\tau(W')$ (indeed, when Q spans $\mathcal{D}'$, so does Q^τ); on the other hand, $P^\tau(\Phi')(\Psi_\tau(v)) = 0$ and this is a minimal equation. □

Note that $\tau(f)$ has the same valuation as f for any $f \in K$, so the operators P and P^τ have the same Newton polygon.

CHAPTER 9

Further Algebraic Properties of q-Difference Modules

In this chapter, we complete the algebraic foundations of q-difference module theory regarding *tensorial*, *homological* and *sheaf theoretical* properties.

9.1. Tensor properties

Although this section will have immediate use for classification, its conclusions will come to full power when we speak of "Tannakian categories" in Chapter 13 and in Section 15.4.

Our primary goal is, of course, applications to q-difference modules. However, some make sense (and are useful) in the broader context of arbitrary $\mathcal{D}_{K,q}$-modules. So it seems fit to slightly generalize[1] the constructions of 8.1, in particular the equivalence:

$$\left\{ \begin{matrix} q\text{-difference modules} = \text{pairs } (V, \Phi) \\ \dim_K V < \infty \end{matrix} \right\} \longleftrightarrow \left\{ \begin{matrix} \mathcal{D}_{K,q} - \text{modules } M \\ \text{length } M < \infty \end{matrix} \right\}$$

into a more general one:

$$\left\{ \begin{matrix} \text{pairs } (V, \Phi) \\ \dim_K V \text{ arbitrary} \end{matrix} \right\} \longleftrightarrow \left\{ \begin{matrix} \mathcal{D}_{K,q} - \text{modules } M \\ \text{length } M \text{ arbitrary} \end{matrix} \right\}.$$

Just use arbitrary K-vector spaces V and the same conditions on Φ. In the same way, if $M = (V, \Phi)$ and $N = (W, \Psi)$ are seen as $\mathcal{D}_{K,q}$-modules, the same definition of morphisms as in 8.2.2 is equivalent to that of morphisms of $\mathcal{D}_{K,q}$-modules.

EXERCISE 9.1. Check that this generalisation works[2] smoothly.

9.1.1. Internal Hom.

9.1.1.1. Let $M_1 := (V_1, \Phi_1)$ and $M_2 := (V_2, \Phi_2)$ two q-difference modules over the q-difference field K. On the K-vector space $V_{1,2} := \mathscr{L}_K(V_1, V_2)$, we define the following map (the notation is by no way definitive):

$$\Phi_{1,2} : f \mapsto \Phi_2 \circ f \circ \Phi_1^{-1}.$$

[1]This is the case in Subsections 9.1.1 and 9.1.4, but not in Subsections 9.1.2 and 9.1.3where only q-difference modules occur.

[2]This does *not* apply to the cyclic vector lemma, which holds only for q-difference modules.

To see that $\Phi_2 \circ f \circ \Phi_1^{-1} \in \mathscr{L}_K(V_1, V_2)$, first note that Φ_1^{-1} is ${\sigma_q}^{-1}$-linear (verification left to you); then, if $a, b \in K$ and $x, y \in V_1$:

$$\begin{aligned}
\left(\Phi_2 \circ f \circ \Phi_1^{-1}\right)(ax + by) &= \left(\Phi_2 \circ f\right)\left({\sigma_q}^{-1}(a)\Phi_1^{-1}(x) + {\sigma_q}^{-1}(b)\Phi_1^{-1}(y)\right) \\
&= \Phi_2\left({\sigma_q}^{-1}(a)(f \circ \Phi_1^{-1}(x)) + {\sigma_q}^{-1}(b)(f \circ \Phi_1^{-1}(y))\right) \\
&= \sigma_q({\sigma_q}^{-1}(a))\left(\Phi_2 \circ f \circ \Phi_1^{-1}\right)(x) \\
&\quad + \sigma_q({\sigma_q}^{-1}(b))\left(\Phi_2 \circ f \circ \Phi_1^{-1}\right)(y) \\
&= a\left(\Phi_2 \circ f \circ \Phi_1^{-1}\right)(x) + b\left(\Phi_2 \circ f \circ \Phi_1^{-1}\right)(y).
\end{aligned}$$

The map $\Phi_{1,2}$ is obviously bijective (its inverse is $f \mapsto \Phi_2^{-1} \circ f \circ \Phi_1$). It is also σ_q-linear because Φ_2 is: since $(af + bg) \circ \Phi_1^{-1} = a(f \circ \Phi_1^{-1}) + b(g \circ \Phi_1^{-1})$ without any assumption on Φ_1, we have, using the σ_q-linearity:

$$\begin{aligned}
\Phi_2 \circ (af + bg) \circ \Phi_1^{-1} &= \Phi_2\left(a(f \circ \Phi_1^{-1}) + b(g \circ \Phi_1^{-1})\right) \\
&= \sigma_q(a)\Phi_2 \circ f \circ \Phi_1^{-1} + \sigma_q(b)\Phi_2 \circ g \circ \Phi_1^{-1}.
\end{aligned}$$

DEFINITION 9.2. The q-difference module $(V_{1,2}, \Phi_{1,2})$, where $V_{1,2} := \mathscr{L}_K(V_1, V_2)$ and $\Phi_{1,2}$ is as above, is called the *internal Hom* of $M_1 := (V_1, \Phi_1)$ and $M_2 := (V_2, \Phi_2)$ and denoted $\underline{\mathrm{Hom}}(M_1, M_2)$.

Examples will be given in 9.1.1.3. Beware that while the "usual Hom" $\mathrm{Hom}(M_1, M_2)$ is a K^{σ_q}-vector space, the internal Hom is a q-difference module. Their relationship is encoded as follows.

PROPOSITION 9.3. *Morphisms are solutions of the internal Hom. More explicitly, up to a natural identification:*

$$Hom(M_1, M_2) = \Gamma(\underline{Hom}(M_1, M_2)).$$

PROOF. We saw in 8.2.3.2 that $\Gamma(V, \Phi)$ can be identified with V^{Φ} (fixed points of Φ in V). For $(V_{1,2}, \Phi_{1,2}) := \underline{\mathrm{Hom}}(M_1, M_2)$, a fixed point is an $f \in \mathscr{L}_K(V_1, V_2)$ such that $\Phi_2 \circ f \circ \Phi_1^{-1} = f$, i.e., such that $\Phi_2 \circ f = f \circ \Phi_1$, so it is the same thing as a morphism of q-difference modules $M_1 \to M_2$. □

9.1.1.2. *Functoriality of Internal Hom.* Now let $M_3 := (V_3, \Phi_3)$ yet another q-difference module and let $F : M_2 \to M_3$ a morphism, thus a linear map $V_2 \to V_3$ such that $\Phi_3 \circ F = F \circ \Phi_2$. Let us denote $(V_{1,2}, \Phi_{1,2}) := \underline{\mathrm{Hom}}(M_1, M_2)$ and $(V_{1,3}, \Phi_{1,3}) := \underline{\mathrm{Hom}}(M_1, M_3)$. (Recall that $V_{1,2} := \mathscr{L}_K(V_1, V_2)$ and $V_{1,3} := \mathscr{L}_K(V_1, V_3)$.) As a K-linear map $V_2 \to V_3$, F induces a linear map $\mathscr{L}_K(V_1, V_2) \to \mathscr{L}_K(V_1, V_3)$, $f \mapsto F \circ f$ (covariant functoriality). That map should be denoted $\mathscr{L}_K(V_1, F)$, we abreviate it into $\tilde{F}$. We compute:

$$\Phi_{1,3}(F \circ f) = \Phi_3 \circ (F \circ f) \circ \Phi_1^{-1} = (\Phi_3 \circ F \circ \Phi_2^{-1}) \circ (\Phi_2 \circ f \circ \Phi_1^{-1}) = F \circ \Phi_{1,2}(f),$$

since, F being a morphism of q-difference modules, $\Phi_3 \circ F \circ \Phi_2^{-1} = F$. This means that:

$$\Phi_{1,3} \circ \tilde{F} = \tilde{F} \circ \Phi_{1,2}.$$

We therefore have a commutative diagram:

$$\begin{array}{ccc} V_{1,2} & \xrightarrow{\tilde{F}} & V_{1,3} \\ {\scriptstyle \Phi_{1,2}}\downarrow & & \downarrow{\scriptstyle \Phi_{1,3}} \\ V_{1,2} & \xrightarrow{\tilde{F}} & V_{1,3}. \end{array}$$

Therefore $\tilde{F}$ is a morphism of q-difference modules from $\underline{\mathrm{Hom}}(M_1, M_2)$ to $\underline{\mathrm{Hom}}(M_1, M_3)$. It is formally denoted $\underline{\mathrm{Hom}}(M_1, F)$.

PROPOSITION 9.4. (i) *We thus define for any q-difference module* M_0 *a covariant functor* $M \rightsquigarrow \underline{Hom}(M_0, M)$, $F \rightsquigarrow \underline{Hom}(M_0, F)$, *from q-difference modules to q-difference modules.*

(ii) *Similarly, we have a contravariant functor* $M \rightsquigarrow \underline{Hom}(M, M_0)$, $F \rightsquigarrow \underline{Hom}(F, M_0)$, *from q-difference modules to q-difference modules.*

PROOF. (i) The two properties to be verified (about identities and about composition) are exactly those satisfied by the analogous covariant functor on vector space $V \rightsquigarrow \mathscr{L}_K(V_1, V)$.

(ii) Proof entirely similar; of course, $F : M := (V, \Phi) \to N := (W, \Psi)$ being given, $\underline{\mathrm{Hom}}(F, M_0)$ is the map $f \mapsto f \circ F$ from $\mathscr{L}_K(W, V_0)$ to $\mathscr{L}_K(V, V_0)$ (where $M_0 = (V_0, \Phi_0)$).

□

9.1.1.3. *Examples.* Recall the unit object $\underline{\mathbf{1}} := (K, \sigma_q)$. Let $M := (V, \Phi)$. Then $\mathrm{Hom}(\underline{\mathbf{1}}, M)$ is the vector space $\mathscr{L}_K(K, V)$ endowed with the σ_q-linear automorphism $\Psi : f \mapsto \Phi \circ f \circ {\sigma_q}^{-1}$. Recall the canonical isomorphism of vector spaces $V \tilde{\to} \mathscr{L}_K(K, V)$, $v \in\mapsto f_v$, where $f_v : K \to V$ is the linear map $a \mapsto av$. Then $\Psi(f_v) = \Phi \circ f_v \circ {\sigma_q}^{-1}$ is defined by:

$$a \mapsto \Phi \circ f_v \circ {\sigma_q}^{-1}(a) = \Phi(f_v({\sigma_q}^{-1}(a))) = \Phi({\sigma_q}^{-1}(a)v) = a\Phi(v),$$

so $\Psi(f_v) = f_{\Phi(v)}$. Therefore, we have a commutative diagram:

$$\begin{array}{ccc} V & \longrightarrow & \mathscr{L}_K(K, V) \\ {\scriptstyle \Phi}\downarrow & & \downarrow{\scriptstyle \Psi} \\ V & \longrightarrow & \mathscr{L}_K(K, V) \end{array}$$

in which both horizontal arrows are the isomorphism $V \tilde{\to} \mathscr{L}_K(K, V)$, $v \mapsto f_v$. This means that it is an isomorphism of q-difference modules $M \to \underline{\mathrm{Hom}}(\underline{\mathbf{1}}, M)$. In the same way one usually identifies V with $\mathscr{L}_K(K, V)$, we shall identify M with $\underline{\mathrm{Hom}}(\underline{\mathbf{1}}, M)$. Not only is that isomorphism canonical (i.e., free of arbitrary choices), it is also functorial:

EXERCISE 9.5. Let $F : M \to N$ a morphism of q-difference modules, thus inducing a morphism of q-difference modules $\tilde{F} : \underline{\mathrm{Hom}}(\underline{\mathbf{1}}, M) \to \underline{\mathrm{Hom}}(\underline{\mathbf{1}}, N)$ (by the proposition in 9.1.1.2). Verify the commutativity of the following diagram, in which the horizontal arrows are the canonical identifications just defined:

$$\begin{array}{ccc} M & \longrightarrow & \underline{\mathrm{Hom}}(\mathbf{1}, M) \\ {\scriptstyle F}\downarrow & & \downarrow{\scriptstyle \tilde{F}} \\ N & \longrightarrow & \underline{\mathrm{Hom}}(\mathbf{1}, N). \end{array}$$

Now consider $M^\vee := \underline{\mathrm{Hom}}(M, \mathbf{1})$, the *dual q-difference module* of M. Its underlying vector space is the usual dual $V^\vee := \mathscr{L}_K(V, K)$. It is endowed with the σ_q-linear automorphism $\Phi^\vee : \lambda \mapsto \sigma_q \circ \lambda \circ \Phi^{-1}$. By 9.1.1.2, we have a contravariant functor $M \rightsquigarrow M^\vee$. Another way to see it is to write the action of $V^\vee$ on V as a pairing: $\langle \lambda, v \rangle := \lambda(v)$. Then the above formula for $\Phi^\vee$ can be written:

$$\langle \Phi^\vee(\lambda), \Phi(v) \rangle = \sigma_q \langle \lambda, v \rangle.$$

If $F : M := (V, \Phi) \to N := (W, \Psi)$ is a morphism, $F^\vee$ seen as a linear map $W^\vee \to V^\vee$ is $f \mapsto f \circ F$, so it is the usual transpose map ${}^tF$ of linear algebra.

EXERCISE 9.6. Show that the canonical isomorphism $V \tilde{\to} V^{\vee\vee}$ actually defines an isomorphism of q-difference modules $M \tilde{\to} M^{\vee\vee}$.

Suppose in particular that $M = (K^n, \Phi_A)$, $A \in \mathrm{GL}_n(K)$. We identify $(K^n)^\vee$ with K^n the usual way: the element $(a_i)_{1 \leq i \leq n} \in K^n$ defines the linear form $(x_i)_{1 \leq i \leq n} \mapsto \sum a_i x_i$, an element of $(K^n)^\vee$. (This actually corresponds to the isomorphism from K^n to $(K^n)^\vee$ sending the canonical basis to its dual basis.) Under that identifications, writing elements of K^n as column vectors, the pairing $(K^n)^\vee \times K^n \to K$ is given by $\langle X, Y \rangle =^t XY$ (usual scalar product). Therefore the definition of $\Phi_A^\vee$ should be such that:

$$\langle \Phi_A^\vee X, \Phi_A Y \rangle = \sigma_q({}^tXY) \Longleftrightarrow {}^t(\Phi_A^\vee X)(A^{-1}\sigma_q Y) = {}^t(\sigma_q X)(\sigma_q Y).$$

That being true for all Y:

$${}^t(\Phi_A^\vee X)A^{-1} = {}^t(\sigma_q X) \Longleftrightarrow {}^tA^{-1}(\Phi_A^\vee X) = \sigma_q X \Longleftrightarrow \Phi_A^\vee X = {}^tA\sigma_q X.$$

This means that $\Phi_A^\vee = \Phi_{A^\vee}$, where $A^\vee := {}^tA^{-1}$ is the *contragredient* matrix of A. We already saw in 8.1.3 and again in 8.2.3 that it plays somehow a dual role to that of A.

More generally, let $M = (K^n, \Phi_A)$, $A \in \mathrm{GL}_n(K)$ and $N = (K^p, \Phi_B)$, $B \in \mathrm{GL}_p(K)$. We identify $\mathscr{L}_K(K^n, K^p)$ with $\mathrm{Mat}_{p,n}(K)$, rectangular matrices acting on column vectors by left multiplication. So $\underline{\mathrm{Hom}}(M, N) = (\mathrm{Mat}_{p,n}(K), \Phi_{A,B})$ for some σ_q-linear automorphism $\Phi_{A,B}$ that we must characterize. For an arbitrary matrix $F \in \mathrm{Mat}_{p,n}(K)$, we should have:

$$\forall X \in K^n,\ (\Phi_{A,B}(F))X = \Phi_B(F\Phi_A^{-1}(X)) = B^{-1}\sigma_q\left(F{\sigma_q}^{-1}(AX)\right) = \left(B^{-1}(\sigma_q F)A\right)X,$$

so $\Phi_{A,B}(F) = B^{-1}(\sigma_q F)A$.

EXERCISE 9.7. Deduce from that the previous description of $(K^n, \Phi_A)^\vee$.

9.1.2. Tensor product. For the general theory of tensor products of modules; see [**37, 145**]. Note however that the tensor product of two q-difference modules is *not* their tensor products as $\mathcal{D}_{K,q}$-modules, it is a very specific construction on its own right. In our setting, it is however based on the tensor product of vector spaces, of which we recall the basic properties in Appendix F.

EXERCISE 9.8. After reading what follows, try to describe the effect of a base change on tensorial constructions (including internal Hom and dual).

9.1.2.1. *Tensor product of q-difference modules.* Let $M := (V, \Phi)$ and $N := (W, \Psi)$ two q-difference modules. The general formalism of tensor product does not by itself allow us to define $\Phi \otimes \Psi$ as a σ_q-linear automorphism of $V \otimes_K W$, because it only applies to linear maps. We shall however sketch a way to get around that difficulty (see Remark 9.11). But we first explore a more direct approach.

Let (e_i) a basis of V and (f_j) a basis of W, so that $(e_i \otimes f_j)$ is a basis of $V \otimes_K W$. The following map $\Gamma : V \otimes_K W \to V \otimes_K W$ is therefore well defined:

$$\Gamma\left(\sum_{i,j} \lambda_{i,j} e_i \otimes f_j\right) := \sum_{i,j} \sigma_q(\lambda_{i,j}) \Phi(e_i) \otimes \Psi(f_j).$$

It is σ_q-linear as shows the following calculation:

$$\begin{aligned}
\Gamma\left(a \sum_{i,j} \lambda_{i,j} e_i \otimes f_j\right) &= \Gamma\left(\sum_{i,j} a\lambda_{i,j} e_i \otimes f_j\right) = \sum_{i,j} \sigma_q(a\lambda_{i,j}) \Phi(e_i) \otimes \Psi(f_j) \\
&= \sum_{i,j} \sigma_q(a)\sigma_q(\lambda_{i,j}) \Phi(e_i) \otimes \Psi(f_j) \\
&= \sigma_q(a) \sum_{i,j} \sigma_q(\lambda_{i,j}) \Phi(e_i) \otimes \Psi(f_j) \\
&= \sigma_q(a) \Gamma\left(\sum_{i,j} \lambda_{i,j} e_i \otimes f_j\right).
\end{aligned}$$

It is bijective with inverse the map:

$$\sum_{i,j} \lambda_{i,j} e_i \otimes f_j \mapsto \sum_{i,j} {\sigma_q}^{-1}(\lambda_{i,j}) \Phi^{-1}(e_i) \otimes \Psi^{-1}(f_j).$$

LEMMA 9.9. *There is a unique σ_q-linear automorphism of $V \otimes_K W$ such that $x \otimes y \mapsto \Phi(x) \otimes \Psi(y)$ for all $x \in V$, $y \in W$.*

PROOF. Such a map is necessarily defined by the formula given above for Γ. We already saw that Γ is indeed a σ_q-linear automorphism of $V \otimes_K W$. Now, letting $x = \sum \lambda_i e_i$ and $y = \sum \mu_j f_j$, we have:

$$\begin{aligned}
\Gamma(x \otimes y) &= \Gamma\left(\sum_{i,j} \lambda_i \mu_j e_i \otimes f_j\right) = \sum_{i,j} \sigma_q(\lambda_i \mu_j) \Phi(e_i) \otimes \Psi(f_j) \\
&= \sum_{i,j} \sigma_q(\lambda_i \mu_j) \Phi(e_i) \otimes \Psi(f_j) = \sum_{i,j} \sigma_q(\lambda_i) \sigma_q(\mu_j) \Phi(e_i) \otimes \Psi(f_j) \\
&= \left(\sum_i \sigma_q(\lambda_i) \Phi(e_i)\right) \otimes \left(\sum_j \sigma_q(\mu_j) \Phi(e_i) \Psi(f_j)\right) = \Gamma(x) \otimes \Gamma(y).
\end{aligned}$$

□

DEFINITION 9.10. We write $\Phi \otimes \Psi$ the σ_q-linear automorphism of $V \otimes_K W$ defined by the lemma. The q-difference module $(V \otimes_K W, \Phi \otimes \Psi)$ is denoted $M \otimes N$ and called the *tensor product* of M and N.

REMARK 9.11. We sketch an alternative construction. Details can be found in [**197**, Appendix A2.4]. Let ${\sigma_q}^* V$ the space with the same underlying group as V but with external multiplication by K defined by $a \star v := \sigma_q(a)v$. A σ_q-linear automorphism of V is the same thing as a linear isomorphism $V \xrightarrow{\sim} {\sigma_q}^* V$. So given Φ and Ψ as above we have a linear isomorphism $V \otimes_K W \xrightarrow{\sim} {\sigma_q}^* V \otimes_K {\sigma_q}^* W$. On the other hand, there is a canonical isomorphism ${\sigma_q}^* V \otimes_K {\sigma_q}^* W \xrightarrow{\sim} {\sigma_q}^* V \otimes_K W$, which, by composition, gives the linear isomorphism $V \otimes_K W \xrightarrow{\sim} {\sigma_q}^* V \otimes_K W$, i.e., the σ_q-linear automorphism of $V \otimes_K W$.

It follows from the corresponding properties of $V \otimes_K W$ that $M \otimes N \simeq N \otimes M$ and that $M_1 \otimes (M_2 \otimes M_3) \simeq (M_1 \otimes M_2) \otimes M_3$. We also check that $\underline{1} \otimes M \simeq M \otimes \underline{1} \simeq \underline{1}$. This is obvious at the level of the underlying spaces since:

$$K \otimes_K V \simeq V \otimes_K K \simeq V,$$

but we should also check that, for instance, the isomorphism $K \otimes_K V \xrightarrow{\sim} V$, $a \otimes x \mapsto ax$, is compatible with the corresponding strctural σ_q-linear automorphism. On $K \otimes_K V$, by definition, $a \otimes x \mapsto \sigma_q(a) \otimes \Phi(x)$. Since $\sigma_q(a) \otimes \Phi(x) = \Phi(ax)$, this works all right.

PROPOSITION 9.12. *Under the conventions of Appendix* F.3*:*

$$(K^n, \Phi_A) \otimes (K^p, \Phi_B) = (K^{np}, \Phi_{A \otimes B}).$$

PROOF. We let the canonical bases (e_i) and (f_j) as before, and (e'_i), (f'_j) their respective images by Φ_A and Φ_B, so that the basis $(e_i \otimes f_j)$ of $K^n \otimes K^p = K^{np}$ has image $(e'_i \otimes f'_j)$. Under the implicit reordering, $(e_i \otimes f_j)$ is the canonical basis of K^{np}. We have $(e_i) = (e'_i)A$ and $(f_j) = (f'_j)B$ (this was how we "discovered" Φ_A and Φ_B in 8.1) so $(e'_i \otimes f'_j) = (e_i \otimes f_j)(A \otimes B)$, which precisely means that $\Phi_A \otimes \Phi_B = \Phi_{A \otimes B}$. □

Specializing to rank 1 q-difference modules (K, Φ_a) and remembering that $(K, \Phi_a) \simeq (K, \Phi_b)$ is equivalent to $b/a = (\sigma_q f)/f$ for some $f \in K^*$ (8.1.4), we see that the isomorphism classes of rank 1 q-difference modules form a group for the tensor product and that this is the Picard group $\mathrm{Pic}(K, \sigma_q)$ defined in 4.1.1.

EXAMPLE 9.13. We take M and N of rank 1. By the above, $(K, \Phi_a) \otimes (K, \Phi_b) = (K, \Phi_{ab})$. In a description as $\mathcal{D}_{K,q}$-modules, $(K, \Phi_a) \otimes (K, \Phi_b) = (K, \Phi_{ab})$. In a description as $\mathcal{D}_{K,q}$-modules, $(K, \Phi_a) \simeq \mathcal{D}_{K,q}/\mathcal{D}_{K,q}(S - a')$, $(K, \Phi_b) \simeq \mathcal{D}_{K,q}/\mathcal{D}_{K,q}(S - b')$, where $a' := a^{-1}$, $b' := b^{-1}$, so:

$$\mathcal{D}_{K,q}/\mathcal{D}_{K,q}(S - a') \otimes \mathcal{D}_{K,q}/\mathcal{D}_{K,q}(S - b') \simeq \mathcal{D}_{K,q}/\mathcal{D}_{K,q}(S - a'b').$$

This bears little relation to the tensor product as $\mathcal{D}_{K,q}$-modules[3]

Let $M := \mathcal{D}_{K,q}/\mathcal{D}_{K,q}P$ and $N := \mathcal{D}_{K,q}/\mathcal{D}_{K,q}Q$. We assume P and Q are standard unitary with respective degrees n and p, so, writing u and v the respective images of 1 in M and N, those modules have bases $(u, \ldots, S^{n-1}u)$ and $(v, \ldots, S^{p-1}v)$. Therefore $M \otimes N$ admits $(S^i u \otimes S^j v)_{\substack{i=0,\ldots,n-1 \\ j=0,\ldots,p-1}}$ as a basis.

[3]Actually, over a noncommutative ring, one can only speak of the tensor product of a left module by a right module. But we may compare with the tensor product $(R/I) \otimes_R (R/J) = R/(I + J)$, where I, J are ideals of a commutative ring R. Obviously what we have here is quite different.

9.1.2.2. *Functoriality of tensor product.* Let $M := (V, \Phi)$, $N := (W, \Psi)$, $M' := (V', \Phi')$, $N' := (W', \Psi')$, and let $F : M \to M'$ and $G : N \to N'$ be morphisms. In particular, they are linear maps $V \to V'$ and $G : W \to W'$, so $F \otimes G$ is a linear map $V \otimes_K W \to V' \otimes_K W'$. Now $(\Phi' \otimes \Psi') \circ (F \otimes G)$ and $(F \otimes G) \circ (\Phi \otimes \Psi)$ both are σ_q-linear maps from $V \otimes_K W$ to $V' \otimes_K W'$ (since they are compositions of a linear and a σ_q-linear map, although not in the same order). Also, for all $x \in V, y \in W$:

$$((\Phi' \otimes \Psi') \circ (F \otimes G))\,(x \otimes y) = (\Phi' \otimes \Psi')\,(F(x) \otimes G(y)) = \Phi'(F(x)) \otimes \Psi'(G(y)),$$

while $((F \otimes G \circ (\Phi \otimes \Psi)))\,(x \otimes y) = (F \otimes G)\,(\Phi(x) \otimes \Psi(y)) = F(\Phi(x)) \otimes G(\Psi(y))$,

so that, since $\Phi'(F(x)) = F(\Phi(x))$ and $\Psi'(G(y) = G(\Psi(y))$, those maps coincide on the generating system $(x \otimes y)_{x \in V, y \in W}$ of $V \otimes_K W$, so they are equal: $F \otimes G$ is a morphism of q-difference modules from $M \otimes N$ to $M' \otimes N'$.

PROPOSITION 9.14. *The tensor product of q-difference modules is a bifunctor covariant in each argument $M, N \rightsquigarrow M \otimes N$, $F, G \rightsquigarrow F \otimes G$. it is K^{σ_q}-bilinear.*

PROOF. We just saw that $F, G \rightsquigarrow F \otimes G$ sends pairs of morphisms to morphisms. Good behaviour with respect to identities and compositions follows from the corresponding statement for the tensor products of vector spaces and linear maps, and K^{σ_q}-bilinearity is immediate. □

PROPOSITION 9.15. *With similar notations as above, let $M' := K' \otimes_K M$ and $N' := K' \otimes_K N$. Then:*

$$M' \otimes_{K'} N' = K' \otimes_K (M \otimes_K N).$$

PROOF. According to [**37**, II, §5, no 1, Proposition 3], if $V' := K' \otimes_K V$ and $W' := K' \otimes_K W$, then:

$$V' \otimes_{K'} W' = K' \otimes_K (V \otimes_K W).$$

The part about the structural skew linear automorphisms is then immediate. □

9.1.2.3. *Application to solutions and cosolutions.* Remember the covariant solution functor defined as $\Gamma(M) := \mathrm{Hom}_{\mathscr{C}}(\underline{1}, M)$. Given such a morphism $F : \underline{1} \to M$ and likewise an element of $\Gamma(N)$, i.e., a morphism $G : \underline{1} \to N$, we have a morphism $F \otimes G : \underline{1} \otimes \underline{1} \to M \otimes N$. Under the identification of $\underline{1} \otimes \underline{1}$ with $\underline{1}$, this gives us an element of $\Gamma(M \otimes N)$. So we have defined the mapping:

$$\Gamma(M) \times \Gamma(N) \to \Gamma(M \otimes N).$$

It will follow from the concrete description below that this mapping is K^{σ_q}-bilinear, so it yields:

$$\Gamma(M) \otimes_{K^{\sigma_q}} \Gamma(N) \to \Gamma(M \otimes N).$$

We now proceed to an explicit description. For $M := (V, \Phi)$, an element of $\Gamma(M)$ is a linear map $K \to V$, completely determined by the image $f \in V^{\Phi}$ of $1 \in K$. The corresponding bijection from $\Gamma(M)$ to V^{Φ} is a K^{σ_q}-linear isomorphism. Let similarly $N := (W, \Psi)$ and $g \in W^{\Psi}$. Then:

$$(\Psi \otimes \Phi)(f \otimes g) = \Psi(f) \otimes \Phi(g) = f \otimes g,$$

so $f \otimes g \in (V \otimes_K W)^{\Phi \otimes \Psi}$. Clearly the map $(f, g) \mapsto f \otimes g$ from $V^{\Phi} \times W^{\Psi}$ to $(V \otimes_K W)^{\Phi \otimes \Psi}$ is K^{σ_q}-bilinear, and so is therefore the map $\Gamma(M) \times \Gamma(N) \to \Gamma(M \otimes N)$, as stated above.

EXAMPLE 9.16. Let $M := (K^n, \Phi_A)$ and $N := (K^p, \Phi_B)$. An element of $\Gamma(M) = (K^n)^{\Phi_A}$ (equality here and below is understood under the usual identifications) is a column vector $X \in K^n$ such that $\sigma_q X = AX$. An element of $\Gamma(N) = (K^p)^{\Phi_B}$ is a column vector $Y \in K^p$ such that $\sigma_q Y = BY$. Then $Z := X \otimes Y \in K^n \otimes K^p = K^{np}$ is a column vector such that $\sigma_q Z = CZ$, where $C := A \otimes B \in \mathrm{GL}_{np}(K)$.

If $n = p = 1$, writing $a := A$, $b := B$, $c := C$, $f := X$, $g := Y$ and $h := Z$, we see that $c = ab$ and $h = fg$: a solution f of $\sigma_q f = af$ and a solution g of $\sigma_q g = bg$ give rise to a solution fg of $\sigma_q h = abh$. All this happens in K.

Note that taking $b := a^{-1}$ so $c = 1$, we have $\Gamma(M \otimes N) = K^{\sigma_q}$ while $\Gamma(M)$ and $\Gamma(N)$ may be both trivial (if $\sigma_q f = af$ has no nontrivial solution: for instance, if $K = \mathbf{C}(\{x\})$ or $\mathbf{C}((x))$ and $a = x$), so our map $\Gamma(M) \otimes_{K^{\sigma_q}} \Gamma(N) \to \Gamma(M \otimes N)$ needs not be surjective.

EXERCISE 9.17. (i) Show that $\Gamma(M) \otimes_{K^{\sigma_q}} \Gamma(N) \to \Gamma(M \otimes N)$ is injective. (This relies on the properties of the tensor product of vector spaces.) Also discuss the functoriality of that map.

(ii) State and prove something similar for cosolutions.

9.1.2.4. *Exterior power, determinant, wronskian.* Exterior algebra was introduced in Subsection 8.1.5. Here we shall only define the nth exterior power $\bigwedge^n M$ of a q-difference module M of rank n. This construction will only be used in Section 10.4, so we describe it succinctly, leaving it to the reader to fill the missing details (which should be mechanical). If $M = (V, \Phi)$, the underlying space of $\bigwedge^n M$ will be $W := \bigwedge^n V$ (thus a vector space of rank 1). If $\mathcal{B} := (e_1, \ldots, e_n)$ is a basis of V, so that $e_1 \wedge \cdots \wedge e_n$ is a basis of W, there is a unique σ_q-linear automorphism Ψ of W such that $\Psi(e_1 \wedge \cdots \wedge e_n) = \Phi(e_1) \wedge \cdots \wedge \Phi(e_n)$. As a formal argument or the following explicit computations allow to show, this Ψ does not depend on the choice of the basis $\mathcal{B}$. We then set $\bigwedge^n M := (W, \Psi)$.

We now proceed to give two concrete descriptions of $\bigwedge^n M$. We start with the matricial model $M := M_A = (K^n, \Phi_A)$, $A \in \mathrm{GL}_n(K)$. Standard multilinear algebra shows that, if $X_1, \ldots, X_n$ are column vectors in K^n, then $(BX_1) \wedge \cdots \wedge (BX_n) = (\det B)(X_1 \wedge \cdots \wedge X_n)$ for every square matrix B. In particular, if $B = A^{-1}$:

$$\begin{aligned}\Psi_A(X_1 \wedge \cdots \wedge X_n) &= \Phi_A(X_1) \wedge \cdots \wedge \Phi_A(X_n) = (B\sigma_q X_1) \wedge \cdots \wedge (B\sigma_q X_n)\\ &= (\det B)(\sigma_q X_1 \wedge \cdots \wedge \sigma_q X_n) = (\det A)^{-1}\sigma_q(X_1 \wedge \cdots \wedge X_n).\end{aligned}$$

So $\bigwedge^n (K^n, \Phi_A) = (K, \Phi_{\det A})$.

Now we use the polynomial model $M := M_P = \mathcal{D}_{K,q}/\mathcal{D}_{K,q}P$, where $P = \sum_{i=0}^{n} a_i S^i$, $a_0 a_n \neq 0$, which we write as (V, Φ), where V has as basis the classes $e_0, \ldots, e_{n-1}$ of $1, S, \ldots, S^{n-1}$; we have $\Phi(e_0) = e_1, \ldots, \Phi(e_{n-2}) = e_{n-1}$ and $\Phi(e_{n-1}) = \dfrac{-1}{a_n}(a_0 e_0 + \cdots + a_{n-1}e_{n-1})$. Therefore $M \simeq (K^n, \Phi_A)$, where the matrix A of basis $e_0, \ldots, e_{n-1}$ relatively to basis $\Phi(e_0), \ldots, \Phi(e_{n-1})$ (we let you write it down) has determinant $\det A = (-1)^n a_n/a_0$. From the matricial description, we draw $\bigwedge^n M \simeq (K, \Phi_{\det A})$. The kernel of the unique morphism of left $\mathcal{D}_{K,q}$-modules $\mathcal{D}_{K,q} \to M$ sending 1 to $e_0 \wedge e_1 \wedge \cdots \wedge e_{n-1}$ is therefore the left ideal generated by $S - (-1)^n \dfrac{a_0}{a_n}$, so we may for instance write $\bigwedge^n M$ as $\mathcal{D}_{K,q}/\mathcal{D}_{K,q}Q$, where $Q := a_0 + (-1)^{n-1} a_n S$.

EXERCISE 9.18. Make explicit the relation with the wronskian mentioned in the title of this subsubsection.

9.1.3. Adjunction formulas. Such formulas hold in many categories. We shall explain each one first in the case of vector spaces. Then we extend them to q-difference modules, sometimes with and sometimes without a complete verification (and then you are supposed to fill the details). For the general linear algebra involved; see Appendix F and, of course, **[37]**.

9.1.3.1. Let V, W two finite-dimensional K-vector spaces. The map $V^\vee \times W \to \mathscr{L}_K(V, W)$ sending (λ, w) to the linear map $v \mapsto \lambda(v)w$ is bilinear and thus induces a linear map $V^\vee \otimes_K W \to \mathscr{L}_K(V, W)$, which is actually an isomorphism[4]. Moreover, this isomorphism is functorial (one also says: natural). This means the following. Note that the left-hand side as well as the right-hand side are K^{σ_q}-bilinear bifunctors, contravariant in V and covariant in W. So let $f : V' \to V$ and $g : W \to W'$ two linear maps, whence a diagram:

$$\begin{array}{ccc} V^\vee \otimes_K W & \longrightarrow & \mathscr{L}_K(V, W). \\ {\scriptstyle {}^t f \otimes g} \Big\downarrow & & \Big\downarrow {\scriptstyle \mathscr{L}_K(f,g)} \\ V'^\vee \otimes_K W' & \longrightarrow & \mathscr{L}_K(V', W') \end{array}$$

Said functoriality means that the diagram is commutative. So from now on we identify $V^\vee \otimes_K W$ with $\mathscr{L}_K(V, W)$ and write $V^\vee \otimes_K W = \mathscr{L}_K(V, W)$.

EXERCISE 9.19. In the equality $V^\vee \otimes_K V = \mathscr{L}_K(V, V)$, which element corresponds on the left-hand side to element Id_V on the right-hand side?

PROPOSITION 9.20. *For any two q-difference modules, $M^\vee \otimes N = \underline{Hom}(M, N)$.*

PROOF. Let $M := (V, \Phi)$ and $N := (W, \Psi)$. We just saw the isomorphism $V^\vee \otimes_K W \xrightarrow{\sim} \mathscr{L}_K(V, W)$ of the underlying vector spaces. We must check that it is compatible with the structural σ_q-linear automorphisms Γ on $V^\vee \otimes_K W$ and Δ on $\mathscr{L}_K(V, W)$. So let $\lambda \in V^\vee$ and $w \in W$. The image under Γ of $\lambda \otimes w$ is $(\sigma_q \circ \lambda \circ \Phi^{-1}) \otimes \Psi(w)$, which corresponds in $\mathscr{L}_K(V, W)$ to the map $v \mapsto \sigma_q(\lambda(\Phi^{-1}(v)))\Psi(w)$. We just followed the south-then-east path in the diagram:

$$\begin{array}{ccc} V^\vee \otimes_K W & \longrightarrow & \mathscr{L}_K(V, W) \\ {\scriptstyle \Gamma} \Big\downarrow & & \Big\downarrow {\scriptstyle \Delta} \\ V^\vee \otimes_K W & \longrightarrow & \mathscr{L}_K(V, W). \end{array}$$

As for the east-then-south map, it first sends $\lambda \otimes w$ to the linear map $v \mapsto \lambda(v)w$, then composes the latter on the left with Ψ and on the right with Φ^{-1}, yielding the linear map $v \mapsto \Psi(\lambda(\Phi^{-1}(v))w)$. So we must verify:

$$\Psi(\lambda(\Phi^{-1}(v))w) = \sigma_q(\lambda(\Phi^{-1}(v)))\Psi(w),$$

which is immediate. This proves that we have an isomorphism of q-difference modules $M^\vee \xrightarrow{\sim} N = \underline{\mathrm{Hom}}(M, N)$. Then we must prove its functoriality, i.e., that if we

[4]If V and W are not finite-dimensional, only injectivity is guaranteed.

have morphisms of q-difference modules $F : M' \to M$ and $G : N \to N'$, then the diagram:

$$\begin{array}{ccc} M^\vee \otimes N & \longrightarrow & \underline{\mathrm{Hom}}(M,N) \\ \downarrow & & \downarrow \\ M'^\vee \otimes N' & \longrightarrow & \underline{\mathrm{Hom}}(M',N') \end{array}$$

is commutative. But it is actually *identical* (spaces and maps) with the corresponding diagram for vector spaces seen before the proposition. □

COROLLARY 9.21. *We have a functorial isomorphism, whence an identification:*

$$\Gamma(M^\vee \otimes N) = Hom_{\mathscr{E}}(M,N).$$

PROOF. Indeed, $\Gamma(M^\vee \otimes N) = \Gamma(\underline{\mathrm{Hom}}(M,N)) = \mathrm{Hom}_{\mathscr{E}}(M,N)$. □

EXAMPLES 9.22. If we take $M := \underline{\mathbf{1}}$, taking in account the fact that $\underline{\mathbf{1}}^\vee = \underline{\mathbf{1}}$, we obtain the natural identifications $\underline{\mathbf{1}} \otimes N = \underline{\mathrm{Hom}}(\underline{\mathbf{1}},N) = N$.
If we take $N := \underline{\mathbf{1}}$, we obtain the natural identification $M^\vee = \underline{\mathrm{Hom}}(M,\underline{\mathbf{1}})$.
If we take $M = N$, we obtain $M^\vee \otimes M = \underline{\mathrm{Hom}}(M,M)$.

COROLLARY 9.23. *We have a functorial isomorphism, whence an identification (recall $\Gamma^\vee$ was defined in* 8.2.3.2*):*

$$\Gamma^\vee(M) = \Gamma(M^\vee).$$

PROOF. Indeed, $\Gamma^\vee(M) = \mathrm{Hom}(M,\underline{\mathbf{1}}) = \Gamma(\underline{\mathrm{Hom}}(M,\underline{\mathbf{1}}))$. □

9.1.3.2. For any three K-vector spaces V_1, V_2, V_3, there is an isomorphism from $\mathscr{L}_K(V_1, \mathscr{L}_K(V_2,V_3))$ to $\mathrm{Bil}_K(V_1,V_2;V_3)$ which sends the linear map $f : V_1 \to \mathscr{L}_K(V_2,V_3)$ to the bilinear map $\phi V_1 \times V_2 \to V_3$ defined by $\phi(x_1,x_2) := (f(x_1))(x_2)$. The inverse map sends the bilinear map ϕ to the linear map $f : x_1 \mapsto (x_2 \mapsto \phi(x_1,x_2))$. That isomorphism is functorial (contravariant in V_1, V_2 and covariant in V_3), and so we write:

$$\mathscr{L}_K(V_1, \mathscr{L}_K(V_2,V_3)) = \mathrm{Bil}_K(V_1,V_2;V_3) = \mathscr{L}_K(V_1 \otimes_K V_2, V_3).$$

The second equality is justified by the very definition of tensor product in F.2.

PROPOSITION 9.24. *For any three q-difference modules M_1, M_2, M_3, we have:*

$$\underline{Hom}(M_1, \underline{Hom}(M_2,M_3)) = \underline{Hom}(M_1 \otimes M_2, M_3).$$

PROOF. Write $M_i := (V_i, \Phi_i)$, $i = 1,2,3$. The equality at the level of the underlying vector spaces was explained before. Functoriality will be automatic. All we have to prove is compatibility with the structural σ_q-linear automorphisms Γ and Δ, i.e., commutativity of the following diagram:

$$\begin{array}{ccc} \mathscr{L}_K(V_1, \mathscr{L}_K(V_2,V_3)) & \longrightarrow & \mathscr{L}_K(V_1 \otimes_K V_2, V_3) \\ \Gamma\downarrow & & \downarrow\Delta \\ \mathscr{L}_K(V_1, \mathscr{L}_K(V_2,V_3)) & \longrightarrow & \mathscr{L}_K(V_1 \otimes_K V_2, V_3). \end{array}$$

We write $\Gamma : f \mapsto f'$ and $\Delta : g \mapsto g'$ the vertical arrows and $f \mapsto g$ the upper horizontal arrow, i.e., $g(x_1 \otimes x_2) = (f(x_1))(x_2)$. As for Γ and Δ:

$$f'(x_1) : x_2 \mapsto \Phi_3\left((f(\Phi_1^{-1})(x_1))(\Phi_2^{-1}(x_2))\right) \text{ and } g'(x_1 \otimes x_2) = \Phi_3\left((\Phi_1 \otimes \Phi_2)^{-1}(x_1 \otimes x_2)\right).$$

So we should prove that $g'(x_1 \otimes x_2) = (f'(x_1))(x_2)$, i.e., that:

$$\Phi_3\left((\Phi_1 \otimes \Phi_2)^{-1}(x_1 \otimes x_2)\right) = \Phi_3\left((f(\Phi_1^{-1})(x_1))(\Phi_2^{-1}(x_2))\right).$$

This is left to the reader.

(You should use the fact that $(\Phi_1 \otimes \Phi_2)^{-1}(x_1 \otimes x_2) = \Phi_1^{-1}(x_1) \otimes \Phi_2^{-1}(x_2)$; after proving it, of course!) □

COROLLARY 9.25. *We have the "adjunction formula":*

$$Hom(M_1, \underline{Hom}(M_2, M_3)) = Hom(M_1 \otimes M_2, M_3).$$

PROOF. Just apply the solution functor Γ. □

As usual, the meaning of the equalities in the proposition and in the corollary is that there is a canonical isomorphism and that it is functorial in all three arguments.

REMARK 9.26. An *adjunction formula* in category theory has the form:

$$\mathrm{Hom}(F(X), Y) = \mathrm{Hom}(X, G(Y)),$$

where F and G are two functors. Here, M_2 is considered as fixed, $F(X) := X \otimes M_2$ and $G(Y) := \underline{\mathrm{Hom}}(M_2, Y)$.

EXERCISE 9.27. What becomes of the proposition and the corollary if one replaces one of the arguments by $\underline{1}$? If one applies Γ?

9.1.3.3. *Some more formulas.* Here we just give the construction of isomorphisms in the case of vector spaces, then the corresponding statements for q-difference modules (of course, so-called "equalities" have the same meaning as above); verification of the compatibility with structural σ_q-linear automorphisms is left to you. The maps are isomorphisms because the vector spaces are finite-dimensional.

PROPOSITION 9.28. *Let $M, N, M_1, M_2, M_3, N_1, N_2$ q-difference modules. Then:*
(i) $M^\vee \otimes N^\vee = (M \otimes N)^\vee$.
(ii) $\underline{Hom}(M_1, M_2) \otimes M_3 = \underline{Hom}(M_1, M_2 \otimes M_3)$.
(iii) $\underline{Hom}(M_1, M_2) \otimes \underline{Hom}(N_1, N_2) = \underline{Hom}(M_1 \otimes N_1, M_2 \otimes N_2)$.

PROOF. (i) The linear map $V^\vee \otimes_K W^\vee \to (V \otimes_K W)^\vee$ is defined as: $\lambda \otimes \mu \mapsto (x \otimes y \mapsto \lambda(x)\mu(y))$.

(ii) The linear map $\mathscr{L}_K(V_1, V_2) \otimes_K V_3 \to \mathscr{L}_K(V_1, V_2 \otimes_K V_3)$ is defined as $f \otimes x_3 \mapsto (x_1 \mapsto f(x_1) \otimes x_3)$.

(iii) The linear map $\mathscr{L}_K(V_1, V_2) \otimes_K \mathscr{L}_K(W_1, W_2) \to \mathscr{L}_K(V_1 \otimes W_1, V_2 \otimes W_2)$ is defined as $f \otimes g \mapsto (x_1 \otimes y_1 \mapsto f(x_1) \otimes g(y_1))$.

□

9.1.4. Extension of some results to arbitrary $\mathcal{D}_{K,q}$-modules. We use here the (V, Φ) model for arbitrary $\mathcal{D}_{K,q}$-modules, as explained at the beginning of 9.1.

9.1.4.1. The definitions of the internal Hom and of the tensor product of q-difference modules in 9.1.1 and in 9.1.2 generalize immediately to the case of arbitrary $\mathcal{D}_{K,q}$-modules, as well as the functoriality properties (including the definition of the tensor product of morphisms of q-difference modules). We must insist however on the fact that the tensor product of $\mathcal{D}_{K,q}$-modules, as defined this way, is not a particular case of the definition of the tensor product of modules over a ring. Indeed, if it was, one would find that $\mathcal{D}_{K,q} \otimes \underline{1} = \underline{1}$ while actually $M \otimes \underline{1} = \underline{1}$ for every M. (Also see Example 9.13 and its footnote.)

9.1.4.2. The adjunction formulas given in 9.1.3 rest on corresponding formulas for vector spaces over a commutative field. Hereafter, we write equalities for canonical functorial isomorphisms, all taken from [**30**, §4]:

$$\begin{aligned}
\mathscr{L}_K(V_1, \mathscr{L}_K(V_2, V_3)) &= \mathscr{L}_K(V_1 \otimes_K V_2, V_3),\\
\mathscr{L}_K(V_1, V_2) &= V_1^\vee \otimes_K V_2,\\
\mathscr{L}_K(V_1, V_2) \otimes_K V_3 &= \mathscr{L}_K(V_1, V_2 \otimes_K V_3),\\
\mathscr{L}_K(V_1, V_2) \otimes_K \mathscr{L}_K(W_1, W_2) &= \mathscr{L}_K(V_1 \otimes W_1, V_2 \otimes W_2),\\
V^\vee \otimes_K W^\vee &= (V \otimes_K W)^\vee.
\end{aligned}$$

In our application to q-difference modules, we used the fact that K is a commutative field (so K-modules are automatically projective) and all our spaces were finite-dimensional. However, according to *loc. cit.*, the first equality is valid without any condition on dimension; the second requires only that V_1 or V_2 be finite-dimensional and the third that V_1 or V_3 be finite-dimensional.

EXERCISE 9.29. Check in *loc. cit.* what of the fourth and fifth and what can be deduced for q-difference modules and $\mathcal{D}_{K,q}$-modules.

9.1.4.3. With the same arguments as in 9.1.3, we can then deduce the following adjunction formulas for $\mathcal{D}_{K,q}$-modules:

$$\begin{aligned}
\mathrm{Hom}(M_1, \underline{\mathrm{Hom}}(M_2, M_3)) &= \mathrm{Hom}(M_1 \otimes M_2, M_3),\\
M^\vee \otimes N &= \underline{\mathrm{Hom}}(M, N),\\
\underline{\mathrm{Hom}}(M_1, M_2) \otimes M_3 &= \underline{\mathrm{Hom}}(M_1, M_2 \otimes M_3).
\end{aligned}$$

The first one is valid without any condition. The second (resp. the third) one requires that M_1 or M_2 (resp. M_1 or M_3) be a q-difference module.

9.2. Homological properties

The crucial fact is that the category of q-difference modules is a K^{σ_q}-linear abelian subcategory of the category of $\mathcal{D}_{K,q}$-modules; in particular, kernels and images of morphisms of q-difference modules being q-difference modules, exact sequences of q-difference modules are just exact sequences of $\mathcal{D}_{K,q}$-modules. This allows us to use elementary homological algebra in the style of [**32**] or [**145**] (although the latter allows only modules over commutative rings).

For any ring R with center C, the category of left R-modules is C-linear abelian. We know from 3.1.3 that the center of $\mathcal{D}_{K,q}$ is K^{σ_q}, so, for all this section, we shall set $C := K^{\sigma_q}$.

9.2.1. Generalities on functors and exactness.

9.2.1.1. A C-linear functor (this includes all the functors on q-difference modules we have encountered) transforms the trivial q-difference module 0 into either the trivial q-difference module 0 (if it is a functor to q-difference modules) or the trivial vector space 0 (if it is a functor to vector spaces). So it transforms a short exact sequence $0 \to M' \to M \to M'' \to 0$ into a short sequence, either $0 \to F(M') \to F(M) \to F(M'') \to 0$ (if F is covariant) or $0 \to F(M'') \to F(M) \to F(M') \to 0$ (if F is contravariant).

Moreover, since a C-linear functor transforms the trivial morphism 0 into the trivial morphism 0, the composite map $F(M') \to F(M'')$ or $F(M'') \to F(M')$ is 0, i.e., the sequence we obtained is a *complex* (see Chapter 6 and Appendix E.1). Discussion of the possible exactness of such sequences at M' or M or M'' has been found to be of utmost importance in many domains of mathematics, and is actually one of the foundations of homological algebra. This is the subject matter of this section.

We are going to examine the "exactness" of some functors on q-difference modules. An elementary but crucial fact is the following: the sequence $(V', \Phi') \overset{F}{\to} (V, \Phi) \overset{G}{\to} (V'', \Phi'')$ is exact if, and only if, the underlying sequence of vector spaces $V' \overset{F}{\to} V \overset{G}{\to} V''$ is exact (8.3.1).

Also note that sequences of vector spaces are always split but sequences of q-difference modules are not necessarily (see the part on extensions in 8.3.1). So, while additive functors (those which send direct sums to direct sums) on linear spaces are exact, this is no longer true for q-difference modules.

9.2.1.2. The covariant C-linear functor F is said to be *exact*, resp. *left exact*, resp. *right exact* if, starting from the short exact sequence $0 \to M' \to M \to M'' \to 0$, one gets:

- a short exact sequence: $0 \to F(M') \to F(M) \to F(M'') \to 0$,
- resp. an exact sequence: $0 \to F(M') \to F(M) \to F(M'')$,
- resp. an exact sequence: $F(M') \to F(M) \to F(M'') \to 0$.

The contravariant C-linear functor F is said to be *exact*, resp. *left exact*, resp. *right exact* if, starting from the short exact sequence $0 \to M' \to M \to M'' \to 0$, one gets:

- a short exact sequence: $0 \to F(M'') \to F(M) \to F(M') \to 0$,
- resp. an exact sequence $0 \to F(M'') \to F(M) \to F(M')$,
- resp. an exact sequence $F(M'') \to F(M) \to F(M') \to 0$.

Thus in both cases (co or contravariant), left exactness implies that there is an injective map at the left of the *target sequence*, while right exactness means that there is a surjective map at the right of the *target sequence*. One can prove that:

- if F is exact covariant, then:

$$M_1 \to \cdots \to M_n \text{ exact } \Longrightarrow F(M_1) \to \cdots \to F(M_n) \text{ exact;}$$

- if F is left exact covariant, then:

$$0 \to M_1 \to \cdots \to M_n \text{ exact } \Longrightarrow 0 \to F(M_1) \to \cdots \to F(M_n) \text{ exact;}$$

- if F is right exact covariant, then:

$$M_1 \to \cdots \to M_n \to 0 \text{ exact } \Longrightarrow F(M_1) \to \cdots \to F(M_n) \to 0 \text{ exact;}$$

- if F is exact contravariant, then:

$$M_1 \to \cdots \to M_n \text{ exact } \Longrightarrow F(M_n) \to \cdots \to F(M_1) \text{ exact;}$$

- if F is left exact contravariant, then:

$$M_1 \to \cdots \to M_n \to 0 \text{ exact } \Longrightarrow 0 \to F(M_n) \to \cdots \to F(M_1) \text{ exact;}$$

- if F is right exact contravariant, then:

$$0 \to M_1 \to \cdots \to M_n \text{ exact } \Longrightarrow F(M_n) \to \cdots \to F(M_1) \to 0 \text{ exact.}$$

9.2.1.3. *The case of vector spaces.* From general linear algebra, we know that all the following functors are exact and additive; we just denote them by their effect on objects, omitting the effect on morphisms:

- for V fixed, the covariant functor $W \rightsquigarrow \mathscr{L}_K(V, W)$;
- for W fixed, the contravariant functor $V \rightsquigarrow \mathscr{L}_K(V, W)$;
- the contravariant functor $V \rightsquigarrow V^\vee$;
- for V fixed, the covariant functor $W \rightsquigarrow V \otimes_K W$;
- for W fixed, the covariant functor $V \rightsquigarrow V \otimes_K W$.

Note that all this is true because we consider vector spaces over a base field K. If we considered modules over a ring, we would have the first two functors additive and left exact; and the last two functors additive and right exact. Modules ensuring, when fixed, exactness in the other argument bear special names (projective, injective, flat and flat). The case of the dual for modules over a ring is more complex (see Bourbaki or Lang).

9.2.2. Exactness of functors on q-difference modules. As noted at the beginning of this section, we shall rely on the category of $\mathcal{D}_{K,q}$-modules. A noteworthy exception is that of the tensor constructions, which are not at all the same for $\mathcal{D}_{K,q}$-modules and for q-difference modules.

9.2.2.1. It immediately follows from 9.2.1 that the following functors from q-difference modules to q-difference modules are exact and additive:

- for M fixed, the covariant functor $N \rightsquigarrow \underline{\mathrm{Hom}}(M, N)$;
- for N fixed, the contravariant functor $M \rightsquigarrow \underline{\mathrm{Hom}}(M, N)$;
- the contravariant functor $M \rightsquigarrow M^\vee$;
- for M fixed, the covariant functor $N \rightsquigarrow M \otimes N$;
- for N fixed, the covariant functor $M \rightsquigarrow M \otimes N$.

9.2.2.2. The following is a general fact (for functors $M \rightsquigarrow \mathrm{Hom}(M_0, M)$ in a category of modules) but a direct proof maybe gives more intuition of how to measure the defect of right exactness, in other words, of what the Ext functors look like (see 9.2.3).

PROPOSITION 9.30. *The functor $\Gamma : M \rightsquigarrow Hom(\underline{1}, M)$ from q-difference modules to C-vector spaces is left exact and additive. (Recall that Γ was defined in 8.2.3.)*

PROOF. Recall from 8.2.3 that Γ is actually the functor $(V, \Phi) \rightsquigarrow V^\Phi$, sending a morphism $F : (V, \Phi) \to (W, \Psi)$ to its (well defined and C-linear) restriction $V^\Phi \to W^\Psi$. So we are actually discussing the exactness and of the short sequence: $0 \to V'^{\Phi'} \to V^\Phi \to V''^{\Phi''} \to 0$ (additivity of the functor Γ is obvious and left to the reader).

We said nothing and therefore prove nothing about exactness at right (see 9.2.2.3). At left, we must prove injectivity, which is obvious since $V'^{\Phi'} \to V^{\Phi}$ is the restriction of $V' \to V$, which is assumed injective.

At the middle, the inclusion Im $\subset$ Ker is automatic, because it means that the compositum is trivial, which directly follows from the fact that the compositum $M' \to M''$ is (and the fact that Γ transforms trivial morphisms into trivial morphisms).

So we are left with the following: let $x \in V^{\Phi}$ with trivial image in V''; then show that it comes from $V'^{\Phi'}$. But it comes from $x' \in V'$ (since it has trivial image in V'' and the sequence $V' \to V \to V''$ is exact). Now, the image of $\Phi'(x')$ in V is $\Phi(x) = x$, i.e., the image of x'; since $V' \to V$ is injective, $\Phi'(x') = x'$, so $x' \in V'^{\Phi'}$, so x is the image of an element of $V'^{\Phi'}$, which was our goal. □

COROLLARY 9.31. *The functors $M \rightsquigarrow Hom(M,N)$ and $N \rightsquigarrow Hom(M,N)$ are left exact and additive.*

PROOF. They are obtained by composing the left exact additive functor Γ by the exact additive functor $M \rightsquigarrow \underline{\mathrm{Hom}}(M,N)$ resp. $N \rightsquigarrow \underline{\mathrm{Hom}}(M,N)$. □

9.2.2.3. Let $K := \mathbf{C}((x))$ or $\mathbf{C}(\{x\})$ indifferently and let $A := \begin{pmatrix} 1 & 1 \\ 0 & 1 \end{pmatrix}$, so that $\Phi_A \begin{pmatrix} x \\ y \end{pmatrix} = \begin{pmatrix} \sigma_q x - \sigma_q y \\ \sigma_q y \end{pmatrix}$. Let $M := (K^2, \Phi_A)$. We have an extension: $0 \to \underline{\mathbf{1}} \to M \to \underline{\mathbf{1}} \to 0$ (see 8.3.2.3). Applying Γ we get, at the right, a mapping:

$$\left\{ \begin{pmatrix} x \\ y \end{pmatrix} \mid \begin{cases} \sigma_q x - \sigma_q y = x, \\ \sigma_q y = y \end{cases} \right\} \to \{y \mid \sigma_q y = y\}, \quad \begin{pmatrix} x \\ y \end{pmatrix} \mapsto y.$$

Then $y := 1$ is in the target but not in the image, because equation $\sigma_q x - 1 = x$ has no solution in either $\mathbf{C}(\{x\})$ or $\mathbf{C}((x))$. So the functor is not right exact.

EXERCISE 9.32. Describe completely the image sequence and verify its left exactness.

The defect of exactness may be related to the fact that the sequence is not split. Indeed, additive functors transform split extensions or sequences into split extensions or sequences. As we shall see soon, for q-difference modules, the defect of right exactness is exactly measured by the existence of nontrivial classes of extensions, i.e., of nonsplit extensions.

9.2.3. Right derived functors. Our main goal here is to complete on the right some of the incomplete exact sequences of 9.2.1.2 in the case of Hom functors.

9.2.3.1. In "bifunctors" like Hom and Ext^1, we shall use letter N at left (contravariant side) and M at right (covariant side).

First we resort to [**32**, §5,§8]. In particular, [**32**, §8.6] together with the principality of $\mathcal{D}_{K,q}$ (see 3.1.3) entail the following properties for left $\mathcal{D}_{K,q}$-modules:

- Every exact sequence $0 \to M' \to M \to M'' \to 0$ gives rise to an exact sequence of C-vector spaces (N is an arbitrary left $\mathcal{D}_{K,q}$-module):

$$0 \to \mathrm{Hom}(N,M') \to \mathrm{Hom}(N,M) \to \mathrm{Hom}(N,M'') \to$$
$$\to \mathrm{Ext}^1(N,M') \to \mathrm{Ext}^1(N,M) \to \mathrm{Ext}^1(N,M'') \to 0.$$

In other words, using the (slightly) more sophisticated language of [**46, 94, 104**], for fixed N, the covariant functor $\mathrm{Hom}(N,-) : M \rightsquigarrow \mathrm{Hom}(N,M)$ admits as right derived functors:

$$R^1\mathrm{Hom}(N,-) = \mathrm{Ext}^1(N,-) \text{ and } R^i\mathrm{Hom}(N,-) = 0 \text{ for } i \geq 2.$$

- Every exact sequence $0 \to N' \to N \to N'' \to 0$ gives rise to an exact sequence of C-vector spaces (M is an arbitrary left $\mathcal{D}_{K,q}$-module):

$$0 \to \mathrm{Hom}(N'',M) \to \mathrm{Hom}(N,M) \to \mathrm{Hom}(N',M) \to$$
$$\to \mathrm{Ext}^1(N'',M) \to \mathrm{Ext}^1(N,M) \to \mathrm{Ext}^1(N',M) \to 0.$$

In other words, for fixed M, the contravariant functor $\mathrm{Hom}(-,M) : N \rightsquigarrow \mathrm{Hom}(N,M)$ admits as right derived functors:

$$R^1\mathrm{Hom}(-,M) = \mathrm{Ext}^1(-,M) \text{ and } R^i\mathrm{Hom}(-,M) = 0 \text{ for } i \geq 2.$$

We also know from [**32**] that the C-spaces $\mathrm{Ext}^1(N,M)$ can be realized as the C-spaces of extension classes $\mathrm{Ext}(N,M)$, as was illustrated concretely in 8.3.2. We return to this in 9.2.4.

9.2.3.2. We now focus on q-difference modules, i.e., we let our $\mathcal{D}_{K,q}$-modules have finite length. We introduce a new functor Γ^1 from q-difference modules to C-vector spaces by setting:

$$\Gamma^1(M) := \mathrm{Ext}(\underline{1}, M).$$

Letting $N := \underline{1}$ in 9.2.3.1, we see that an exact sequence $0 \to M' \to M \to M'' \to 0$ gives rise to an exact sequence of C-vector spaces:

$$0 \to \Gamma(M') \to \Gamma(M) \to \Gamma(M'') \to \Gamma^1(M') \to \Gamma^1(M) \to \Gamma^1(M'') \to 0.$$

In other words, the right derived functors of the left exact functor Γ are $R^1\Gamma = \Gamma^1$ and $R^i\Gamma = 0$ for $i \geq 2$. Actually the general exact sequence with M, M', M'', N as in 9.2.3.1 may be recovered from this particular one. Indeed, recall from 9.1.3 the "equalities" (see 9.1.3.3 about our use of that word):

$$\mathrm{Hom}(N,M) = \Gamma(\underline{\mathrm{Hom}}(N,M)) = \Gamma(N^\vee \otimes M).$$

Then the general exact sequence follows if we replace M by $N^\vee \otimes M$ etc, in the special one, thanks to the following lemma.

LEMMA 9.33.

$$\mathit{Ext}^1(N,M) = \Gamma^1(N^\vee \otimes M).$$

PROOF. We take for N a q-difference module but let M denote an arbitrary $\mathcal{D}_{K,q}$-module (so we shall have to use some of the generalised formulas from 9.1.4). The covariant functor $M \rightsquigarrow \mathrm{Hom}(N,M)$ is obtained by composing the covariant functor $M \rightsquigarrow N^\vee \otimes M = \underline{\mathrm{Hom}}(N,M)$ (which is exact by 9.2.2.1 with the left exact covariant functor Γ. Moreover, the former functor sends injective modules (those M such that $\mathrm{Hom}(-,M)$ is exact) to injective modules thanks to the adjunction formula $\mathrm{Hom}(P, \underline{\mathrm{Hom}}(N,M)) \simeq \mathrm{Hom}(P \otimes N, M)$. One can then compute the derived functors of $M \rightsquigarrow \mathrm{Hom}(N,M)$ using [**94**, Chapter III, §7, Theorem 1]. □

EXERCISE 9.34. After 8.2.3.2, we have $\Gamma^\vee(N) = \Gamma(\underline{\mathrm{Hom}}(N,\underline{1})) = \Gamma(N^\vee)$. What can be drawn from that in the same spirit as above?

9.2.4. The q-de Rham complex (aka: complex of solutions). We study ways to explicitly compute the right derived functors defined in 9.2.3.

9.2.4.1. We shall now make precise the above constructions within the (V, Φ) model. According to 8.3.2, given q-difference modules $M = (V, \Phi)$ and $N = (W, \Psi)$, there is an exact sequence of C-vector spaces:

$$0 \to \mathrm{Hom}(N, M) \to \mathscr{L}_K(W, V) \xrightarrow{t_{N,M}} \mathscr{L}_{\sigma_q}(W, V) \to \mathrm{Ext}(N, M) \to 0.$$

The meaning of $\mathscr{L}_K(W, V)$ and we recall that elements of $\mathscr{L}_{\sigma_q}(W, V)$ are those goup morphisms $f : W \to V$ that are σ_q-linear, i.e., such that $f(\lambda x) = \sigma_q(\lambda) f(x)$ for $\lambda \in K$, $x \in V$. Another way to see it is to consider the K-linear space $\sigma_q^* V$ defined with underlying additive group V and external multiplication $(\lambda, x) \mapsto \sigma_q(\lambda) x$. Then clearly:

$$\mathscr{L}_{\sigma_q}(W, V) = \mathscr{L}_K(W, \sigma_q^* V).$$

From this, we see easily that $M \rightsquigarrow \mathscr{L}_{\sigma_q}(W, V)$ is a covariant exact functor from q-difference modules to C-vector spaces.

The central map $t_{N,M}$ was defined (up to notations) as $f \mapsto f \circ \Psi - \Phi \circ f$. Its C-linearity is obvious, as the fact that its target is indeed $\mathscr{L}_{\sigma_q}(W, V)$. What was not quite obvious (and which we admitted, with reference to [**197**]) is that the rightmost map $\mathscr{L}_{\sigma_q}(W, V) \to \mathrm{Ext}(N, M)$ is C-linear when its target is endowed with the C-linear structure on $\mathrm{Ext}(N, M) = \mathrm{Ext}^1(N, M)$ as described in [**32**]. Now we interpret our exact sequence above in the language of Appendix E.

COROLLARY 9.35. *The q-de Rham complex concentrated in degrees* 0 *and* 1 *(also called* complex *of solutions in* [**197**]*):*

$$\underset{0}{\mathscr{L}_K(W, V)} \xrightarrow{t_{N,M}} \underset{1}{\mathscr{L}_{\sigma_q}(W, V)}$$

has cohomology spaces $H^0 = Hom(N, M)$ and $H^1 = Ext(N, M)$.

Now fix N and provisionally denote $K(M)$ this complex. From the remarks above, we have a covariant exact functor $M \rightsquigarrow K(M)$ from q-difference modules to the category of complexes of vector spaces.

EXERCISE 9.36. (i) Using the long exact sequence for complexes (Example E.2.3 in Appendix E), recover one of the exact sequences displayed in 9.2.3.1.
(ii) Redo all the arguments for the contravariant argument N.

9.2.4.2. *Avatars of the complex of solutions.* Composing on the right $t_{N,M} : \mathscr{L}_K(W, V) \longrightarrow \mathscr{L}_{\sigma_q}(W, V)$ with the C-linear isomorphism $g \mapsto \Phi^{-1} \circ g$, we get a new complex and a diagram of complexes:

$$\begin{array}{ccc} \mathscr{L}_K(W, V) & \xrightarrow{t_{N,M}} & \mathscr{L}_{\sigma_q}(W, V) \\ \downarrow & & \downarrow \\ \mathscr{L}_K(W, V) & \xrightarrow[\Gamma - \mathrm{Id}]{} & \mathscr{L}_K(W, V), \end{array}$$

where $\Gamma : f \mapsto \Phi^{-1} \circ f \circ \Psi$ is actually the structural σ_q-linear automorphism of $\underline{\mathrm{Hom}}(N, M)$ and the left (resp. the right) vertical map is identity (resp. right composition $g \mapsto \Phi^{-1} \circ g$). This is an isomorphism of complexes, so we can compute $H^0 = \mathrm{Hom}(N, M)$ as $\mathrm{Ker}(\Gamma - \mathrm{Id})$ (which anyway is obvious!) and $H^1 = \mathrm{Ext}(N, M)$ as $\mathrm{Coker}(\Gamma - \mathrm{Id})$.

Next, a matricial description (which will turn out extremely useful in Chapter 15) of the q-de Rham complex and of $\mathrm{Ext}(N, M)$ was also given in 8.3.2.3. We take

$M := M_A := (K^n, \Phi_A)$ and $N := M_B := (K^p, \Phi_B)$. Then the complex takes the form $\mathrm{Mat}_{n,p}(K) \to \mathrm{Mat}_{n,p}(K)$, $F \mapsto (\sigma_q F)B - AF$. Again its cohomology identify with $\mathrm{Hom}(N, M), \mathrm{Ext}(N, M)$. If we compose the map on the right with $G \mapsto GB^{-1}$, we obtain again $\mathrm{Mat}_{n,p}(K) \to \mathrm{Mat}_{n,p}(K)$ but with map $F \mapsto \sigma_q F - AFB^{-1}$, a complex isomorphic to the former one (whence with isomorphic cohomologies). A specially interesting case is when $p = 1$, $B = (1)$, which yields tha complex:

$$\underset{0}{K^n} \xrightarrow{\sigma_q - A} \underset{1}{K^n},$$

which has cohomology $H^0 = \Gamma(M_A)$, $H^1 = \Gamma^1(M_A)$.

9.2.4.3. *A homotopy.* Let a q-difference operator $L := {\sigma_q}^n + a_1 {\sigma_q}^{n-1} + \cdots + a_n \in \mathcal{D}_{K,q}$ and let A_L its companion matrix (as in 7.1.3). In 7.1.5 we defined two diagrams of complexes (recall that $\Delta_L := \sigma_q - A_L$):

$$\begin{array}{ccc} K & \xrightarrow{L} & K \\ {\scriptstyle X}\downarrow & & \downarrow{\scriptstyle U} \\ K^n & \xrightarrow[\Delta_L]{} & K^n \end{array} \qquad \begin{array}{ccc} K & \xrightarrow{L} & K \\ {\scriptstyle X'}\uparrow & & \downarrow{\scriptstyle U'} \\ K^n & \xrightarrow[\Delta_L]{} & K^n. \end{array}$$

Moreover $X'X = U'U = \mathrm{Id}_K$ while both XX' and UU' are homotpic to Id_{K^n}. So the two complexes are homotopic, and, according to Example E.3 in Appendix A, have same cohomology. Therefore the associated q-difference module (K^n, Φ_{A_L}) has cohomology $H^0 = ker L$, $H^1 = \mathrm{Coker} L$.

9.2.4.4. *The $\mathcal{D}_{K,q}/\mathcal{D}_{K,q}P$-model.* The following facts are proved in [**197**, A7], we leave them to you as exercises.

EXERCISE 9.37. (i) Let $N := \mathcal{D}_{K,q}/\mathcal{D}_{K,q}P$ and $M := (V, \Phi)$ two difference modules. Then there is an exact sequence of C-vector spaces:

$$0 \to \mathrm{Hom}(N, M) \to V \overset{P(\Phi)}{\longrightarrow} V \to \mathrm{Ext}^1(N, M) \to 0.$$

(Hint: use the presentation $\mathcal{D}_{K,q} \overset{\times P}{\longrightarrow} \mathcal{D}_{K,q} \to N$.)

(ii) For any dual $P^\vee$ of P the complex $K \overset{P^\vee(\sigma_q)}{\longrightarrow} K$ has cohomology

$$\Gamma(N), \Gamma^1(N).$$

(iii) Look at the special case $P := \sigma_q - u$.

9.2.5. Indexes of q-difference modules.

9.2.5.1. *Definition and additivity.* According to Corollary 7.49 to the q-wronskian lemma in 7.2.1, for a q-difference module M of rank n, the C-vector space $\Gamma(M) = \mathrm{Sol}(M, K)$ has dimension at most n, and, at any rate, $< \infty$. Therefore all spaces $\mathrm{Hom}(N, M)$ are finite-dimensional of dimension $\leq$ (rank M)(rank N). We introduce $h^0(M) := \dim_C \Gamma(M)$ and, when defined (i.e., finite):

$$h^1(M) := \dim_C \Gamma^1(M) \in \mathbf{N} \text{ and } \chi(M) := h^0(M) - h^1(M) \in \mathbf{Z}.$$

Those notations extend (when possible) as follows:

$$h^i(N, M) := h^i(N^\vee \otimes M) = h^i(\underline{\mathrm{Hom}}(N, M)) \text{ and}$$

$$\chi(N, M) := h^0(N, M) - h^1(N, M) = \chi(N^\vee \otimes M) = \chi(\underline{\mathrm{Hom}}(N, M)).$$

In the same vein of (simplifying?) overloading, with the notations of 9.2.4.3:

$$h^i(A) := h^i(M_A), h^i(L) := h^i(A_L) \text{ and, in all cases, } \chi := h^0 - h^1.$$

By Chapter 6, in particular section 6.4, those are indeed finite whenever $K = \mathbf{C}(\{x\})$ or $K = \mathbf{C}((x))$. We shall return to that case in Chapter 10.

Admitting the finiteness of all involved h^1, the general properties of indexes (Chapter 6 and Appendix D) then allow us to state:

PROPOSITION 9.38. *If $0 \to M' \to M \to M'' \to 0$ is an exact sequence of q-difference modules:*

$$\chi(M) = \chi(M') + \chi(M'').$$

9.2.5.2. *Effect of base change.* Let K' a q-difference extension of K. The natural definition of solutions in K' of the q-difference module M over K is:

$$\Gamma_{K'}(M) := \mathrm{Hom}(K' \otimes \underline{1}, K' \otimes M) \text{ so that indeed } \Gamma_{K'}(M_A) = \mathrm{Sol}(A, K').$$

We now from 3.2.1 that, K and M being fixed, $\Gamma_{K'}(M)$ grows with K' (but with dimension over $C' := K'^{\sigma_q}$ bounded above by rank M) (see a precision in (i) of the exercise herebelow). On the other hand, this is not true of Γ^1 ((ii) of the exercise), so that the behaviour of χ by base extension is rather irregular.

EXERCISE 9.39. (i) Define a natural injection $C' \otimes_C \Gamma_K(M) \to \Gamma_{K'}(M)$.

(ii) Let $K := \mathbf{C}(\{x\})$, $K' := \mathbf{C}((x))$. Show that $\mathrm{Ext}(K' \otimes \underline{1}, K' \otimes M_x) = 0$ while $\mathrm{Ext}(\underline{1}, M_x) \neq 0$. (Use the matricial version of the q-de Rham complex.)

9.3. Sheaves of solutions of q-difference modules over $\mathbf{C}(\{x\})$

This section is intended to complete section 7.3 in a more intrinsic setting, while taking advantage of the properties of q-difference modules from Chapters 8 and 9. So, for all of 9.3 (apart from a specified exception in Subsubsection 9.3.1.2), the base field is $K := \mathbf{C}(\{x\})$. Once the basic formalism is established, most statements follow from mechanical verifications which will be left to the reader.

In order to go beyond pure formalism, we sometimes anticipate results from Chapters 10 and 11 so as to get interesting examples. More meaningful applications appear in **[209, 216, 219]** and Chapter 15.

CAUTION 9.40. If $\mathscr{F}$ is a sheaf over the space X, the set of sections over an open subset U is denoted $\Gamma(U, \mathscr{F})$ or $\mathscr{F}(U)$. If $\mathscr{F}$ is moreover a sheaf of groups, the zeroth cohomology group $H^0(U, \mathscr{F})$ is naturally identified with $\mathscr{F}(U) = \Gamma(U, \mathscr{F})$. So we shall generally not distinguish them.

9.3.1. Construction of the sheaf of solutions. Here, we extend 7.3.2; the corresponding extension of 7.3.3 is automatic and we shall not discuss it.

9.3.1.1. Every q-difference module M is isomorphic to some $M_A = (K^n, \Phi_A)$, where A is unique up to a gauge equivalence. And if $F[A] = B$ is such an equivalence, inducing $M_A \simeq M_B$, then we saw that $X \mapsto FX$ defines an isomorphism $\mathscr{F}_A^{(0)} \simeq \mathscr{F}_B^{(0)}$ of sheaves of $\mathcal{O}_{\mathbf{E}_q}$-modules. So, starting from the q-difference module M, we could define $\mathscr{F}_M^{(0)}$ as any of the sheaves $\mathscr{F}_A^{(0)}$ such that $M \simeq M_A$, this being well defined up to isomorphism. However it would be more elegant to define directly the sheaf $\mathscr{F}_M^{(0)}$ on $\mathbf{E}_q$ by:

$$\mathscr{F}_M^{(0)}(V) := \{\text{solutions of } M \text{ in } \mathcal{O}_{\mathbf{E}_q}\left(\pi^{-1}(V), 0\right)\}.$$

For that we need to know what "solutions of M in $\mathcal{O}_{\mathbf{E}_q}\left(\pi^{-1}(V), 0\right)$" are.

9.3.1.2. Just for 9.3.1.2 we return to the general case of a q-difference field (K, σ_q) and we call *q-difference-algebra* a commutative K-algebra $\mathcal{A}$ endowed with a σ_q-linear automorphism extending σ_q. If $\mathcal{A}$ is a field K', we recover the notion of q-difference extension. Remember that in that case we had defined the K'^{σ_q}-space of solutions:

$$\mathrm{Sol}(M, K') = \Gamma_{K'}(M) = \mathrm{Hom}(K' \otimes \underline{1}, K' \otimes M).$$

This works exactly the same for an arbitrary q-difference-algebra $\mathcal{A}$, as will the following generalisations:

- A q-difference module over $\mathcal{A}$ is a finite rank $\mathcal{A}$-module endowed with a σ_q-linear automorphism.
- Extension of scalars $M \rightsquigarrow \mathcal{A} \otimes_K M$ is an exact functor from q-difference modules over K to q-difference modules over $\mathcal{A}$ (exactness is automatic for tensor products over a field).
- There is a well defined $\mathcal{A}^{\sigma_q}$-module:

$$\mathrm{Sol}(M, \mathcal{A}) := \Gamma_{\mathcal{A}}(M) := \mathrm{Hom}(\mathcal{A} \otimes \underline{1}, \mathcal{A} \otimes M).$$

Of course, we then have $\mathrm{Sol}(M_A, \mathcal{A}) = \{X \in \mathcal{A}^n \mid \sigma_q X = AX\}$. This defines a left exact covariant functor from q-difference modules over K to $\mathcal{A}^{\sigma_q}$-modules.

EXERCISE 9.41. (i) Check the validity of those foundations; in particular that $\mathrm{Sol}(M_A, \mathcal{A})$ has the desired meaning of *space of solutions of M in $\mathcal{A}$*.

(ii) Discuss functoriality with respect to A (if $(\sigma_q F)A = BF$ then $X \mapsto FX$ is a morphism $\mathrm{Sol}(M_A, \mathcal{A}) \to \mathrm{Sol}(M_B, \mathcal{A})$) and with respect to $\mathcal{A}$ (define morphisms of q-difference-algebras and say what happens to solutions if $\mathcal{A} \to \mathcal{A}'$ is one).

(iii) Discuss the q-wronskian lemma when $\mathcal{A}$ is not integral.

9.3.1.3. We return (until the end of 9.3) to the case of interest $K := \mathbf{C}(\{x\})$. Let $V \subset \mathbf{E}_q$ an open subset and $U := \pi^{-1}(V) \subset \mathbf{C}^*$. Then $\mathcal{A}^{(0)}(V) := \mathscr{O}_{\mathbf{C}^*}(U, 0)$ has an obvious structure of q-difference-algebra over $\mathbf{C}(\{x\})$ and its q-constant subalgebra is $\mathcal{A}^{(0)}(V)^{\sigma_q} := \mathscr{O}_{\mathbf{E}_q}(V)$. Therefore, for M a q-difference module over $\mathbf{C}(\{x\})$:

$$\Gamma_{\mathcal{A}^{(0)}(V)}(M) = \mathrm{Sol}(M, \mathcal{A}^{(0)}(V)) = \mathrm{Hom}(\mathcal{A}^{(0)}(V) \otimes \underline{1}, \mathcal{A}^{(0)}(V) \otimes M)$$

is a $\mathscr{O}_{\mathbf{E}_q}(V)$-module. The restriction maps $\mathcal{A}^{(0)}(V) \to \mathcal{A}^{(0)}(V')$ (when $V' \subset V$) make the mapping:

$$\mathscr{F}_M^{(0)} : V \mapsto \Gamma_{\mathcal{A}^{(0)}(V)}(M) =: H^0(V, \mathscr{F}_M^{(0)})$$

a sheaf of modules over $\mathscr{O}_{\mathbf{E}_q}$. Clearly, if $M = M_A$ then $\mathscr{F}_M^{(0)}$ can be naturally identified with $\mathscr{F}_A^{(0)}$ and we have obtained the desired intrisic construction. In particular, $\mathscr{F}_M^{(0)}$ is locally free of rank rank M over $\mathscr{O}_{\mathbf{E}_q}$.

That construction is easily adapted to obtain a sheaf $\mathscr{F}_M^{(\infty)}$ from a q-difference module M over $\mathbf{C}(\{1/x\})$.

EXERCISE 9.42. Let $M := M_A$, where $A := (x^\mu)$, $\mu \in \mathbf{Z}$ (so rank $M = 1$). Compare the $\mathbf{C}$-space $H^0(\mathbf{E}_q, \mathscr{F}_M^{(0)})$ of global sections to $\Gamma(M)$.

9.3.2. Abelian and homological properties of the functor $M \rightsquigarrow \mathscr{F}_M^{(0)}$.

9.3.2.1. *Basic properties.* Let $F : M \to N$ a morphism of q-difference modules. This extends first to $\mathcal{A} \otimes M \to \mathcal{A} \otimes N$, then to $\mathrm{Sol}(M, \mathcal{A}) \to \mathrm{Sol}(N, \mathcal{A})$; last, letting $\mathcal{A} := \mathcal{A}^{(0)}(V)$ for varying $V \subset \mathbf{E}_q$, we get a morphism of sheaves $\mathscr{F}_M^{(0)} \to \mathscr{F}_N^{(0)}$. One checks readily that this defines a covariant functor $M \rightsquigarrow \mathscr{F}_M^{(0)}$ from q-difference modules over $\mathbf{C}(\{x\})$ to sheaves of $\mathbf{C}$-vector spaces over $\mathbf{E}_q$. Moreover, this functor has image in the subcategory of locally free $\mathscr{O}_{\mathbf{E}_q}$-modules of finite rank and can be seen as a functor $M \rightsquigarrow \mathfrak{F}_M^{(0)}$ to holomorphic vector bundles over $\mathbf{E}_q$.

Temporarily denote $\tilde{F} : \mathscr{F}_M^{(0)} \to \mathscr{F}_N^{(0)}$ the morphism of sheaves arising from a morphism of q-difference modules$F : M \to N$. Direct computation in the case of M_A, M_B or a more general argument left to you shows that, if $H = \alpha F + \beta G$, $\alpha, \beta \in \mathbf{C}$, then $\tilde{H} = \alpha\tilde{F} + \beta\tilde{G}$. In other words the map $F \mapsto \tilde{F}$ from $\mathrm{Hom}(M, N)$ to $\mathrm{Hom}(\mathscr{F}_M^{(0)}, \mathscr{F}_N^{(0)})$ is $\mathbf{C}$-linear. Again sticking to the case of M_A, M_B, we see that $\tilde{F} = 0$ implies that all maps $X \mapsto FX$ on all $\mathscr{F}_A^{(0)}(V)$ are trivial, whence $F = 0$.

PROPOSITION 9.43. *The functor $M \rightsquigarrow \mathscr{F}_M^{(0)}$ is additive, $\mathbf{C}$-linear and faithful.*

PROOF. Its $\mathbf{C}$-linearity was just proven and faithfulness follows because, having a trivial kernel, the map $\mathrm{Hom}(M, N) \to \mathrm{Hom}(\mathscr{F}_M^{(0)}, \mathscr{F}_N^{(0)})$ is injective.

Additivity means that, if $P = M \oplus N$, then $\mathscr{F}_P^{(0)} = \mathscr{F}_M^{(0)} \oplus \mathscr{F}_N^{(0)}$ (up to natural identifications). This follows from $\mathrm{Sol}(P, \mathcal{A}) = \mathrm{Sol}(M, \mathcal{A}) \oplus \mathrm{Sol}(N, \mathcal{A})$ which is easy either in the M_A, M_B model (using 8.2.1) or through their definition as $\mathrm{Hom}(\mathcal{A} \otimes \underline{1}, \mathcal{A} \otimes M)$, etc. □

However the functor is not *fully* faithful, i.e., in general the natural linear map from $\mathrm{Hom}(M, N)$ to $\mathrm{Hom}(\mathscr{F}_M^{(0)}, \mathscr{F}_N^{(0)})$ is not onto:

EXAMPLE 9.44. Let $A := (x)$ and $B := (1)$ (thus in rank 1). Recall from Chapter 2 that $\sigma_q\theta_q = x^{-1}\theta_q$, so $\sigma_q f = xf \Rightarrow \sigma_q(\theta_q f) = \theta_q f$. This means that $f \mapsto \theta_q f$ defines a map $\mathrm{Sol}(A, \mathcal{A}^{(0)}(V)) \to \mathrm{Sol}(B, \mathcal{A}^{(0)}(V))$ for every open subset $V \subset \mathbf{E}_q$, all those maps glueing into a nontrivial morphism of sheaves $\mathscr{F}_{M_A}^{(0)} = \mathscr{F}_A^{(0)} \to \mathscr{F}_{M_B}^{(0)} = \mathscr{F}_B^{(0)}$. On the other hand, we know that $\mathrm{Hom}(M_A, M_B) = 0$, and $\mathrm{Hom}(M_A, M_B) \to \mathrm{Hom}(\mathscr{F}_{M_A}^{(0)}, \mathscr{F}_{M_B}^{(0)})$ is certainly not onto.

The problem is that θ_q is almost a morphism $A \to B$ since $(\sigma_q\theta_q)A = B\theta_q$, but for the fact that it behave wildly at 0 instead of being meromorphic there. So the possibility of lifting a morphism $\mathscr{F}_M^{(0)} \to \mathscr{F}_N^{(0)}$ to a morphism $M \to N$ is related to questions of asymptotic behaviour at 0. We shall return to that in Chapter 11. (Also see 9.3.4 for some special cases.)

9.3.2.2. *Exactness.*

THEOREM 9.1. *The functor $M \rightsquigarrow \mathscr{F}_M^{(0)}$ is exact.*

PROOF. Let $0 \to M' \to M \to M'' \to 0$ an exact sequence of q-difference modules. By 8.3.1 it is isomorphic to an exact sequence of the form $0 \to M_{A'} \to M_A \to M_{A''} \to 0$, where A is a block upper triangular matrix $\begin{pmatrix} A' & N \\ 0 & A'' \end{pmatrix}$. The sequence of sheaves $0 \to \mathscr{F}_{M'}^{(0)} \to \mathscr{F}_M^{(0)} \to \mathscr{F}_{M''}^{(0)} \to 0$ is in turn isomorphic to $0 \to \mathscr{F}_{A'}^{(0)} \to \mathscr{F}_A^{(0)} \to \mathscr{F}_{A''}^{(0)} \to 0$ and we shall prove the exactness of the latter.

We must first prove that $0 \to \mathcal{F}^{(0)}_{A'}(V) \to \mathcal{F}^{(0)}_{A}(V) \to \mathcal{F}^{(0)}_{A''}(V)$ is exact for every open subset $V \subset \mathbf{E}_q$. This boils down to proving the exactness of sequences $0 \to \mathrm{Sol}(A', \mathcal{A}) \to \mathrm{Sol}(A, \mathcal{A}) \to \mathrm{Sol}(A'', \mathcal{A})$, where the two main maps have the form $X' \mapsto (X', 0)$ and $(X', X'') \mapsto X''$ (understanding implicitly relations $\sigma_q X' = A'X'$, $\sigma_q(X', 0) = A(X', 0)$ for the first map; and $\sigma_q(X', X'') = A(X', X'')$, $\sigma_q X'' = A''X''$ for the second map). We leave it to the reader to check such exactness properties, this is very easy. So $M \rightsquigarrow \mathcal{F}^{(0)}_M$ is left exact.

Right exactness is (slightly) more tricky since the exactness of the sequence of sheaves $\mathcal{F}^{(0)}_A \to \mathcal{F}^{(0)}_{A''} \to 0$ does *not* mean that all maps $\mathcal{F}^{(0)}_A(V) \to \mathcal{F}^{(0)}_{A''}(V)$ are onto. It means that the maps of *stalks* $(\mathcal{F}^{(0)}_A)_x \to (\mathcal{F}^{(0)}_{A''})_x$ are onto for every $x \in \mathbf{E}_q$. We proceed to prove it.

An element of $(\mathcal{F}^{(0)}_{A''})_x$ is a germ of solution, represented in some neighborhood V of x by a $X'' \in \mathcal{F}^{(0)}_{A''}(V)$; and we want to lift it to a germ of solution $X := (X', X'') \in \mathcal{F}^{(0)}_A(V')$ in some possibly smaller neighborhood $V' \subset V$ of x. Since:

$$\sigma_q X = AX \iff \{\sigma_q X' = A'X' + NX'' \text{ and } \sigma_q X'' = A''X''\}$$

and we know in advance that $\sigma_q X'' = A''X''$, we are left to solve (for X'' given) the nonhomogeneous equation $\sigma_q X' = A'X' + C$, where $C := NX''$. Practically, X'' is holomorphic over $(\pi^{-1}(V), 0)$ and, shrinking V if necessary, we can assume that it is trivializing:

$$U := \pi^{-1}(V) = \bigsqcup_{k \in \mathbf{Z}} U_k, \text{ where each } U_k := q^k U_0 \text{ projects homeomorphically to } V.$$

Also, letting $r > 0$ such that A is regular (i.e., holomorphic vith invertible values) over $\dot{\mathrm{D}}(0, r)$ (and therefore A', A'' are also regular and N is holomorphic there) we can choose U_0 such that $q^{\mathbf{N}} U_0 \subset \dot{\mathrm{D}}(0, r)$. Last changing r if necessary, we can assume that X'', whence C, is holomorphic over $q^{\mathbf{N}} U_0$.

With all those choices made, we find the solution X' by taking it *arbitrary* over U_0 then defining it iteratively over all the $q^k U_0$, $k \leq 1$, with the help of the functional equation $\sigma_q X' = A'X' + C$. This solves our problem. □

EXERCISE 9.45. (i) Prove carefully left exactness.
(ii) Let $A := \begin{pmatrix} 1 & 1 \\ 0 & 1 \end{pmatrix}$. Show that $\mathcal{F}^{(0)}_A(\mathbf{E}_q) \to \mathcal{F}^{(0)}_{A''}(\mathbf{E}_q)$ is not onto.

COROLLARY 9.46. (i) *Suppose that M is given a filtration $0 = M_0 \subset M_1 \subset \cdots \subset M_k = M$ with successive quotients $P_i := M_i / M_{i-1}$, $i = 1, \ldots, k$. Then there is a corresponding filtration by (locally free) subsheaves $0 = \mathcal{F}_0 \subset \mathcal{F}_1 \subset \cdots \subset \mathcal{F}_k = \mathcal{F}^{(0)}_M$ with successive quotients $\mathcal{F}_i / \mathcal{F}_{i-1} = \mathcal{F}^{(0)}_{P_i}$, $i = 1, \ldots, k$.*

(ii) *The same is true for the associated bundles: there is a filtration by subbundles $0 = \mathfrak{F}_0 \subset \mathfrak{F}_1 \subset \cdots \subset \mathfrak{F}_k = \mathfrak{F}^{(0)}_M$ with successive quotients $\mathfrak{F}_i / \mathfrak{F}_{i-1} = \mathfrak{F}^{(0)}_{P_i}$, $i = 1, \ldots, k$.*

PROOF. (i) is an immediate consequence of the theorem. Then (ii) follows from the special circumstance that the quotients are locally free, so that the short exact sequences $0 \to \mathcal{F}_{i-1} \to \mathcal{F}_i \to \mathcal{F}^{(0)}_{P_i} \to 0$ are locally split.

□

EXAMPLE 9.47. It will follow from Chapters 10 and 11 that any $A \in \mathrm{GL}_n(\mathbf{C}(\{x\}))$ can be taken (up to analytic gauge equivalence) in block upper triangular form with "pure" diagonal blocks $B_i \in \mathrm{GL}_{r_i}(\mathbf{C}(\{x\}))$, $i = 1, \ldots, k$; and that, in the case that the Newton polygon of A has integral slopes $\mu_1 > \cdots > \mu_k$, then those blocks can be taken in the form $B_i = x^{\mu_i} A_i$, $A_i \in \mathrm{GL}_{r_i}(\mathbf{C})$. Then $\mathscr{F}_A^{(0)}$ has a filtration with successive quotients $\mathscr{F}_{x^{\mu_i} A_i}^{(0)}$. (The case of nonintegral slopes is rather more complicated; see 11.1.4 and [CV], and also [**209**].)
Sheaves $\mathscr{F}_{x^m C}^{(0)}$, $m \in \mathbf{Z}$, $C \in \mathrm{GL}_r(\mathbf{C})$ are (using terminology to come soon in 9.3.3 and to be completed in Chapter 11) tensor products $\mathscr{F}_{x^m}^{(0)} \otimes \mathscr{F}_C^{(0)}$, where $\mathscr{F}_{x^m}^{(0)}$ is locally free of rank 1 and $\mathscr{F}_C^{(0)}$ is "flat". (Also see 9.3.4 for more results.)

9.3.2.3. *Cohomology.* We shall be concise, giving here the basic useful information, and hoping that it will be further illustrated, e.g., in 9.3.4 and again in Chapter 15. The basic theory is recalled in Appendix H.

For every coherent sheaf $\mathscr{F}$ over an analytic manifold X, which includes in particular locally free sheaves of finite rank over $\mathbf{E}_q$, there are *cohomology spaces* $H^i(X, \mathscr{F})$, $i \in \mathbf{N}$, which are finite-dimensional complex vector spaces. In particular $H^0(X, \mathscr{F}) = \Gamma(X, \mathscr{F})$, the space of global sections, and $H^i(X, \mathscr{F}) = 0$ for $i > \dim X$. Every short exact sequence of sheaves $0 \to \mathscr{F}' \to \mathscr{F} \to \mathscr{F}'' \to 0$ gives rise to a *cohomology long exact sequence*:

$$0 \to H^0(X, \mathscr{F}') \to H^0(X, \mathscr{F}) \to H^0(X, \mathscr{F}'') \to$$
$$\to H^1(X, \mathscr{F}') \to H^1(X, \mathscr{F}) \to H^1(X, \mathscr{F}'') \to \cdots$$

The above facts imply that the index:

$$\chi(X, \mathscr{F}) := \sum_{i \geq 0} (-1)^i \dim_{\mathbf{C}} H^i(X, \mathscr{F}) \in \mathbf{Z}$$

(actually the *Euler characteristics*, also abreviated as $\chi(\mathscr{F}) := \chi(X, \mathscr{F})$) is well defined and additive i.e., for an exact sequence as above:

$$\chi(\mathscr{F}) = \chi(\mathscr{F}') + \chi(\mathscr{F}'').$$

For our purposes, we deduce from the previous results that, for every q-difference module M over $\mathbf{C}(\{x\})$, the spaces $H^i(\mathscr{F}_M^{(0)}, \mathbf{E}_q)$ are finite-dimensional; that they are trivial for $i \geq 2$; that an exact sequence of q-difference modules $0 \to M' \to M \to M'' \to 0$ gives rise to a cohomology exact sequence:

$$0 \to H^0(X, \mathscr{F}_{M'}^{(0)}) \to H^0(X, \mathscr{F}_M^{(0)}) \to H^0(X, \mathscr{F}_{M''}^{(0)}) \to$$
$$\to H^1(X, \mathscr{F}_{M'}^{(0)}) \to H^1(X, \mathscr{F}_M^{(0)}) \to H^1(X, \mathscr{F}_{M''}^{(0)}) \to 0.$$

Also $\chi(\mathscr{F}_M^{(0)}) = \dim_{\mathbf{C}} H^0(X, \mathscr{F}_M^{(0)}) - \dim_{\mathbf{C}} H^1(X, \mathscr{F}_M^{(0)})$ and additivity formula:

$$\chi(\mathscr{F}_M^{(0)}) = \chi(\mathscr{F}_{M'}^{(0)}) + \chi(\mathscr{F}_{M''}^{(0)}).$$

In particular, in case of a filtration as in the Corollary 9.46:

$$\chi(\mathscr{F}_M^{(0)}) = \sum_{i=1}^{k} \chi(\mathscr{F}_{P_i}^{(0)}).$$

9.3.3. Tensor properties of the functor $M \rightsquigarrow \mathscr{F}_M^{(0)}$.

9.3.3.1. *Tensor product.* Let $A \in \mathrm{GL}_n(\mathbf{C}(\{x\}))$, $B \in \mathrm{GL}_p(\mathbf{C}(\{x\}))$ and $\mathcal{A}$ a q-difference-algebra over $\mathbf{C}(\{x\})$. Then there is a map $\mathrm{Sol}(A,\mathcal{A}) \times \mathrm{Sol}(B,\mathcal{A}) \to \mathrm{Sol}(A \otimes B,\mathcal{A})$ sending (X,Y) to $X \otimes Y$; indeed:

$$\sigma_q(X \otimes Y) = (\sigma_q X) \otimes (\sigma_q Y) = (AX) \otimes (BY) = (A \otimes B)(X \otimes Y).$$

That map is $\mathcal{A}^{\sigma_q}$-bilinear and therefore induces a $\mathcal{A}^{\sigma_q}$-linear map $\mathrm{Sol}(A,\mathcal{A}) \otimes \mathrm{Sol}(B,\mathcal{A}) \to \mathrm{Sol}(A \otimes B,\mathcal{A})$. Setting $\mathcal{A} := \mathcal{A}^{(0)}(V)$ for a varying $V \subset \mathbf{E}_q$, and $M := M_A$, $N := M_B$ (so that $M \otimes N$ is naturally identified with $M_{A\otimes B}$), we get a morphism of sheaves of $\mathscr{O}_{\mathbf{E}_q}$-modules:

$$\mathscr{F}_M^{(0)} \otimes \mathscr{F}_N^{(0)} \to \mathscr{F}_{M\otimes N}^{(0)}.$$

This extends naturally to any pair q-difference modules M, N and it is functorial in the obvious sense.

THEOREM 9.2. *The natural morphism* $\mathscr{F}_M^{(0)} \otimes \mathscr{F}_N^{(0)} \to \mathscr{F}_{M\otimes N}^{(0)}$ *is an isomorphism, so we write:*

$$\mathscr{F}_{M\otimes N}^{(0)} = \mathscr{F}_M^{(0)} \otimes \mathscr{F}_N^{(0)}.$$

PROOF. We may assume that $M := M_A$, $N := M_B$, and it is then enough to check that $\mathscr{F}_A^{(0)}(V) \otimes \mathscr{F}_B^{(0)}(V) \to \mathscr{F}_{A\otimes B}^{(0)}(V)$ is bijective for all V in an open covering of $\mathbf{E}_q$, so we may (and shall) take V trivializing and choose a connected component U_0 of $\pi^{-1}(V)$ such that A and B (whence $A \otimes B$) are regular over a punctured disk $\overset{\bullet}{\mathrm{D}}(0,r)$ containing $q^{\mathbf{N}}U_0$. We know that restricting a solution X to U_0 yields isomorphisms $\mathscr{F}_A^{(0)}(V) \to \mathscr{O}_{\mathbf{C}^*}(U_0)^n \to \mathscr{O}_{\mathbf{E}_q}(V)^n$, same for the two other sheaves, so our morphism above translates to:

$$\mathscr{O}_{\mathbf{E}_q}(V)^n \otimes \mathscr{O}_{\mathbf{E}_q}(V)^n \to \mathscr{O}_{\mathbf{E}_q}(V)^{np},$$

which is of course an isomorphism. □

9.3.3.2. *Dual and internal Hom.* Since the solutions of $\underline{1}$ in $\mathcal{A}$ are the elements of $\mathcal{A}^{\sigma_q}$, it is rather clear that we have $\mathscr{F}_{\underline{1}}^{(0)} = \mathscr{O}_{\mathbf{E}_q}$. For what follows, we assume the reader to be at ease with tensor operations on sheaves (otherwise look at classical treatises).

PROPOSITION 9.48. *We have a natural identification:*

$$\mathscr{F}_{M^\vee}^{(0)} = (\mathscr{F}_M^{(0)})^\vee.$$

PROOF. We use the M_A model. Recall that $M_A^\vee = M_{A^\vee}$, where $A^\vee = {}^tA^{-1}$, the contragredient of A. Let $X \in \mathrm{Sol}(A,\mathcal{A})$ and $Y \in \mathrm{Sol}(A^\vee,\mathcal{A})$. Then:

$$\{\sigma_q X = AX \text{ and } \sigma_q Y = A^\vee Y\} \Longrightarrow \sigma_q\left({}^tYX\right) = {}^t(\sigma_q Y)\sigma_q X = {}^tYA^{-1}AX = {}^tYX,$$

whence ${}^tYX \in \mathrm{Sol}(\underline{1},\mathcal{A}) = \mathcal{A}^{\sigma_q}$. Letting as usual $\mathcal{A} := \mathcal{A}^{(0)}(V)$, etc, we get a pairing: $\mathscr{F}_{M^\vee}^{(0)} \times \mathscr{F}_M^{(0)} \to \mathscr{O}_{\mathbf{E}_q}$. Restricting as above to trivializing open sets (and using the corresponding identifications with $\mathscr{O}_{\mathbf{E}_q}(V)^n$, etc, we see that it is a perfect pairing. □

COROLLARY 9.49. *We have a natural identification:*

$$\mathscr{F}_{\underline{Hom}(M,N)}^{(0)} = \underline{Hom}(\mathscr{F}_M^{(0)}, \mathscr{F}_N^{(0)}).$$

PROOF. We use the identifications $\underline{\mathrm{Hom}}(M,N) = M^\vee \otimes N$ (for q-difference modules) and $\underline{\mathrm{Hom}}(\mathscr{F},\mathcal{G}) = \mathscr{F}^\vee \otimes \mathcal{G}$ (for sheaves of modules). □

EXERCISE 9.50. Prove the corollary directly from the definitions of internal Hom, without reference to dual and tensor product.

9.3.3.3. *An example.* This is related to the Stokes sheaf, for further details see Chapter 15 and **[216, 219]**. In 15.3.2 we shall consider a situation which we now summarize. Let $M_0 := P_1 \oplus \cdots \oplus P_k$, each $P_i := M_{B_i}$, $B_i \in \mathrm{GL}_{r_i}(\mathbf{C}(\{x\}))$, $i = 1, \ldots, k$, so that $M_0 = M_{A_0}$ where $A_0, = \mathrm{Diag}(B_1, \ldots, B_k)$ (a block diagonal matrix). Then the internal End $\underline{\mathrm{End}}(M_0) = \bigoplus_{i,j=1,\ldots,k} \underline{\mathrm{Hom}}(P_j, P_i)$ admits a q-difference-submodule $\underline{\mathrm{End}}^{>0}(M_0) := \bigoplus_{1\leq i<j\leq k} \underline{\mathrm{Hom}}(P_j, P_i)$. Let $\lambda^{i,j} := \mathscr{F}^{(0)}_{\underline{\mathrm{Hom}}(P_j,P_i)}$ the sheaf associated to $\underline{\mathrm{Hom}}(P_j, P_i)$:

$$\lambda^{i,j}(V) = \{f_{i,j} \in \mathrm{Mat}_{r_i,r_j}(\mathcal{A}^{(0)}(V)) \mid (\sigma_q f_{i,j})B_j = B_i f_{i,j}\}.$$

The sheaf associated to $\underline{\mathrm{End}}(M_0)$ is $\bigoplus_{i,j=1,\ldots,k} \lambda^{i,j}$. Its sections over V are matrices F holomorphic over $(\pi^{-1}(V), 0)$ such that $(\sigma_q F)A_0 = A_0 F$, they have a block decomposition $(f_{i,j})_{i,j=1,\ldots,k}$ where the $f_{i,j}$ are as above. The sheaf associated to $\underline{\mathrm{End}}^{>0}(M_0)$ is $\bigoplus_{1\leq i<j\leq k} \lambda^{i,j}$. Its sections over V are strict upper triangular matrices F with a block decomposition $(f_{i,j})_{1\leq i<j\leq k}$ where the $f_{i,j}$ are as above.

9.3.4. Some complements depending on Chapters 10 and 11. Here we shall use by anticipation some facts related to slopes: first in 9.3.4.1 to make precise the relation of $\Gamma(M)$ with $H^0(\mathbf{E}_q, \mathscr{F}_M^{(0)})$ in the so-called "fuchsian case"; second in 9.3.4.2 to complete and relativize some pessimistic views from 9.3.2.1 around essential surjectivity of the functor $M \rightsquigarrow \mathscr{F}_M^{(0)}$; last in 9.3.4.2 to illustrate the advantage of combining sheaf theoretic methods with robust algebraic foundations.

We take from Chapter 10 that to every q-difference module M over $\mathbf{C}(\{x\})$ is attached a set of slopes $S(M) = \{\mu_1, \ldots, \mu_k\} \subset \mathbf{Q}$. The usual convention is that $\mu_1 > \cdots > \mu_k$.

9.3.4.1. *Global sections in the case of fuchsian q-difference modules.* Let $M = M_A$, $A \in \mathrm{GL}_n(\mathbf{C}(\{x\}))$, a q-difference module in matricial form. The elements of $\Gamma(M)$ are the $X \in \mathbf{C}(\{x\})^n$ such that $\sigma_q X = AX$. The elements of $H^0(\mathbf{E}_q, \mathscr{F}_M^{(0)}) = H^0(\mathbf{E}_q, \mathscr{F}_A^{(0)})$ are the $X \in \mathscr{O}(\mathbf{C}^*, 0)^n$ such that $\sigma_q X = AX$. Therefore we have an inclusion of $\mathbf{C}$-linear spaces $\Gamma(M) \subset H^0(\mathbf{E}_q, \mathscr{F}_M^{(0)})$. According to the Exercise **??**, the inclusion may be strict. However:

PROPOSITION 9.51. *If $S(M) = \{0\}$, then $\Gamma(M) \subset H^0(\mathbf{E}_q, \mathscr{F}_M^{(0)})$.*

PROOF. It will be proved in 11.1.1 that, if $S(M) = \{0\}$, then M is isomorphic to M_A for some $A \in \mathrm{GL}_n(\mathbf{C})$, $n = \mathrm{rank}\ N$. Let $X \in \mathscr{O}(\mathbf{C}^*, 0)^n$ such that $\sigma_q X = AX$. Let $\dot{\mathrm{D}}(0, r)$ a punctured disk on which X is holomorphic. Then equality $X(q^m x) = A^m X(x)$ holds on $\dot{\mathrm{D}}(0, r)$.

Now let $r' < r$ and $C < \infty$ the maximum of $\|X(x)\|$ on the compact annulus $r'\,|q| \leq |x| \leq r'$ and let $D := |||A|||$ the matricial norm of A. If $D \leq 1$, it follows immediately that $\|X(x)\| \leq C$ on $\overset{\bullet}{\mathrm{D}}(0, r')$. Otherwise:

$$\forall x \in \overset{\bullet}{\mathrm{D}}(0, r')\ ,\ \log \|X(x)\| \leq \log C + D \frac{\log(|x|/r')}{\log|q|}.$$

Therefore, in any case, X has at most polynomial growth near 0, so it is meromorphic: $X \in \Gamma(M)$. □

EXERCISE 9.52. (i) Prove the bound on $\|X(x)\|$.

(ii) After reading the next chapters, try to find other cases where equality $\Gamma(M) = H^0(\mathbf{E}_q, \mathscr{F}_M^{(0)})$ holds.

9.3.4.2. *Essential surjectivity in special cases.* We saw in 9.3.2.1 that the functor $M \rightsquigarrow \mathscr{F}_M^{(0)}$ is not fully faithful: the map $\mathrm{Hom}(M, N) \to \mathrm{Hom}(\mathscr{F}_M^{(0)}, \mathscr{F}_N^{(0)})$ needs not be onto. We now show a case where it is.

First "recall" from 10.1.4.3 that, if $S(M) = \{\mu\}$ and $S(N) = \{\nu\}$, then $S\left(\underline{\mathrm{Hom}}(M, N)\right) = \{\nu - \mu\}$.

PROPOSITION 9.53. *If $S(M) = S(N) = \{\mu\}$, then the map $Hom(M, N) \to Hom(\mathscr{F}_M^{(0)}, \mathscr{F}_N^{(0)})$ is onto.*

PROOF. Using the fact "recalled" above and the formulas $\mathrm{Hom}(M, N) = \Gamma(\underline{\mathrm{Hom}}(M, N))$ and $\mathrm{Hom}(\mathscr{F}_M^{(0)}, \mathscr{F}_N^{(0)}) = H^0\left(\mathbf{E}_q, \underline{\mathrm{Hom}}(\mathscr{F}_M^{(0)}, \mathscr{F}_N^{(0)})\right)$, this boils down to the proposition in 9.3.4.1. □

For more general statements involving the slope filtration; see **[216]**.

9.3.4.3. *Computation of the Euler characteristics of $\mathscr{F}_A^{(0)}$.* We shall do that under the assumption that the slopes are integral: $S(M_A) \subset \mathbf{Z}$. Then it is proved in 11.1.1.6 that there is an analytic gauge equivalence:

$$A \underset{\mathbf{C}(\{x\})}{\sim} \begin{pmatrix} x^{\mu_1} A_1 & \star & \dots & \star \\ 0 & x^{\mu_2} A_2 & \dots & \star \\ \vdots & \vdots & \ddots & \vdots \\ 0 & 0 & \dots & x^{\mu_k} A_k \end{pmatrix}, \text{ where } A_i \in \mathrm{GL}_{r_i}(\mathbf{C}), i = 1, \dots, k.$$

PROPOSITION 9.54. *If $S(M_A) \subset \mathbf{Z}$, then $\chi\left(\mathscr{F}_A^{(0)}\right) = v_0(\det A)$.*

PROOF. First note that the statement is invariant under analytic equivalence. For the left-hand side, this is clear from previous considerations; for the right-hand side, it flows from:

$$\begin{aligned} v_0(\det F[A]) &= v_0((\det \sigma_q F)(\det A)(\det F^{-1})) = v_0(\det A) + v_0\left(\frac{\det \sigma_q F}{\det F}\right) \\ &= v_0(\det A) \end{aligned}$$

since $v_0(\det \sigma_q F) = v_0(\sigma_q \det F) = v_0(\det F)$.

Now, setting $B_i := x^{\mu_i} A_i$, we know from 9.3.2.3 that

$$\chi\left(\mathscr{F}_A^{(0)}\right) = \sum_{i=1,\dots,k} \chi\left(\mathscr{F}_{B_i}^{(0)}\right),$$

and, of course, $v_0(\det A) = \sum\limits_{i=1,\dots,k} v_0(\det B_i)$. So we are led to prove the proposition for a matrix $B = x^\mu C$, $\mu \in \mathbf{Z}$, $C \in \mathrm{GL}_r(\mathbf{C})$. Moreover, we can suppose that C

is upper triangular and then the same additivity argument reduces the problem to the case that the rank is 1 and $B = cx^\mu$, $c \in \mathbf{C}^*$.

Now we have an explicit meromorphic global section of the sheaf $\mathscr{F}^{(0)}_{cx^\mu}$, for instance the function $\theta_q(x)^{\mu+1}/\theta_q(cx)$. Using the theorem of Riemann-Roch from Appendix B.6, that case follows easily. □

EXERCISE 9.55. Write in detail the last part of the argument.

The case of arbitrary slopes is a bit more complicated; see [**209**] or [CV].

CHAPTER 10

Newton Polygons and Slope Filtrations

10.1. Newton polygon

In this section (on the Newton polygon of a q-difference module) and the following (on the slope filtration), the base q-difference field is[1] $K = \mathbf{C}(\{x\})$ or $K = \mathbf{C}((x))$. The results in this section for $\mathbf{C}(\{x\})$ are actually easy consequences of those for $\mathbf{C}((x))$ (and they are essentially the same), but this will not be the case in the next section 10.2. All the results come from **[220]** but the presentation will be a little different.

10.1.1. Outline of this section. Our introduction here of the Newton polygon of a q-difference module (and later of its slope filtration) will be somewhat more detailed than that of **[220]**, because it allows us to unveil some structural properties. The risk is that the path and the results appear less clearly, so we begin with an outline.

10.1.1.1. *The main results.* The main result is the following: a q-difference module M being given, all q-difference operators $P \in \mathcal{D}_{K,q}$ such that $M \simeq M_P := \mathcal{D}_{K,q}/\mathcal{D}_{K,q}P$ (remember notations M_A, M_P from 8.1.3) have the same Newton polygon[2]. This way we define the Newton polygon $NP(M) := NP(P)$ of M, the Newton function $r_M := r_P$ and their support $S(M) := S(P)$. (Note that M and P appear at the same place in the notation, but the chances of confusing a q-difference module with a q-difference operator are really small!)

The main "abelian" property of the Newton function is additivity for exact sequences: if $0 \to M' \to M \to M'' \to 0$ is an exact sequence, then $r_M = r_{M'} + r_{M''}$. It implies that $S(M) = S(M') \cup S(M'')$, whence a wealth of nice properties (10.1.4).

10.1.1.2. *The path.* The strategy consists in "peeling off", or "unscrewing"[3] the q-difference module M into its subquotients. More precisely, we endow it with a "composition series", i.e., a sequence of submodules $0 = M_0 \subset \cdots \subset M_n = M$ such that each $P_i := M_i/M_{i-1}$ is a simple q-difference module. Hopefully, each P_i has rank 1 (we then say that M is "split") and for them the theorem comes smoothly; we add those elementary Newton polygons and the theorem of Jordan-Hölder ensures us that the result depends on M only (not on the particular sequence of submodules used). This is where some formalism around composition series and "Grothendieck

[1] At the beginning of the chapter, we deal with the category $\mathscr{C} := \mathcal{D}flMod$ of finite length modules over a nonnecessarily commutative ring $\mathcal{D}$; then we specialize $\mathcal{D}$ to $\mathcal{D}_{K,q}$, so (according to Chapter 8) $\mathscr{C}$ specializes to the category of q-difference modules over K.

[2] Of course this is only true up to translation, so what is really at stake is the Newton function; see 5.2.2 in Chapter 5.

[3] Under the name "dévissage", Grothendieck systematized this way of reducing the complexity of problems; here is of course an ultrasimple version of the process.

groups", while delaying a bit the conclusion, may shed some light on structural properties of q-difference modules.

Maybe some of the subquotients P_i from the composition series do not have rank 1; then the rules of base change from 8.3.3 along with the properties of ramification detailed in 5.2.2.5, Chapter 5, guarantee that some extension $K' \otimes_K M$ will do better, and that all possible K' will yield the same result.

Those structural theorems will then allow us to prove the nice abelian properties of the Newton polygon. They will be completed by "tensor properties" in 10.1.4.3.

10.1.2. Split modules.

10.1.2.1. *Modules of finite length and their Grothendieck group.* We first recall some facts about (left) modules over an arbitrary (i.e., nonnecessarily commutative, but, as always in this book, endowed with a unit element) ring, say $\mathcal{D}$; the main references here are [**145**, Chapter IV, §3] and [**40**, §2,§3,§11]. A $\mathcal{D}$-module M is simple if $M \neq \{0\}$ and its only submodules are $\{0\}$ and itself. A $\mathcal{D}$-module M has finite length if there is a finite sequence of submodules $\{0\} = M_0 \subset \cdots \subset M_l = M$ such that $P_i := M_i/M_{i-1}$ is simple for $i = 1, \ldots, l$. We call such a sequence a *composition series* ("simple filtration" in Lang, "Jordan-Hölder sequence" in Bourbaki) and l is its length. We already mentioned in 8.1.2 (also see 3.1.3) that all composition series have the same length l, it is the length of M.

The theorem of Jordan-Hölder says more precisely that for any two composition series of M, the multisets of isomorphism classes of the simple subquotients are the same; more formally, if $\{0\} = M'_0 \subset \cdots \subset M'_l = M$ is another composition series, there is a permutation τ of $\{1, \ldots, l\}$ such that $M'_i/M'_{i-1} \simeq P_{\tau(i)}$ for $i = 1, \ldots, l$. This leads us to introduce the following notations:

- Let $\mathcal{D}Mod$ the category of (left) $\mathcal{D}$-modules and $\mathcal{D}flMod$ its full subcategory with objects the $\mathcal{D}$-modules of finite length. Write $[M]$ the isomorphism class of a $\mathcal{D}$-module M. Note that, if $0 \to M' \to M \to M'' \to 0$ is an exact sequence, M is in $\mathcal{D}flMod$ if, and only if, M' and M'' are. We shall say that *$\mathcal{D}flMod$ has the Euler-Poincaré property* (Lang's terminology in *loc. cit.*) or that it is a *thick subcategory of $\mathcal{D}Mod$* ("épaisse" in Grothendieck's terminology in [**104**]).
- Let $K(\mathcal{D}flMod)$ the free abelian group generated by the isomorphism classes $[P]$ of all simple $\mathcal{D}$-modules P. The elements of $K(\mathcal{D}flMod)$ are the formal linear combinations $\sum n_i[P_i]$, all P_i simple, all $n_i \in \mathbf{Z}$, almost all $n_i = 0$. This is a kind of "Grothendieck group" (see [**40, 145**]).
- If $\{0\} = M_0 \subset \cdots \subset M_l = M$ is a composition series of M, the element $\sum[M_i/M_{i-1}]$ of $K(\mathcal{D}flMod)$ is well defined by Jordan-Hölder. We write it $\phi(M)$.

REMARK 10.1. We consider ϕ as a "map" $\mathrm{Ob}(\mathcal{D}flMod) \to K(\mathcal{D}flMod)$, the word "map" being quoted because the source $\mathrm{Ob}(\mathcal{D}flMod)$ is not really a set, only a class. We may get around this superficial difficulty by proving that isomorphism classes of $\mathcal{D}$-modules of finite length form a set and call it the "skeleton" $\mathrm{sk}(\mathcal{D}flMod)$ of the category $\mathcal{D}flMod$. This does not seem to be necessary here.

In the terminology of Lang, ϕ is an *Euler-Poincaré map*; it is additive for exact sequences:

$$0 \to M' \to M \to M'' \to 0 \text{ is exact } \Longrightarrow \phi(M) = \phi(M') + \phi(M'').$$

Moreover, it is universal for this property: if Γ is an abelian group and ϕ' is a "map" $\mathrm{Ob}(\mathcal{D}flMod) \to \Gamma$, and if moreover ϕ' is additive for exact sequences, then there is a unique group morphism $f : K(\mathcal{D}flMod) \to \Gamma$ such that $\phi' = f \circ \phi$.

EXAMPLE 10.2. The length map is such a map ϕ', with $\Gamma = \mathbf{Z}$ as target; the corresponding f is the "degree map" (defined on any free abelian group) deg : $K(\mathcal{D}flMod) \to \mathbf{Z}$, $\sum n_i[P_i] \mapsto \sum n_i$.

10.1.2.2. *Split q-difference modules.* We now specialize $\mathcal{D}$ to $\mathcal{D}_{K,q}$, where (K, σ_q) is (to begin with) any q-difference field. So our modules of finite length are exactly q-difference modules and $\mathcal{D}flMod = \mathscr{C}$ (notation from 8.2.2). The simple objects in $\mathscr{C}$ are the pairs (V, Φ) where $V \neq \{0\}$ and the only Φ-stable subspaces of V are $\{0\}$ and itself; equivalently: every $x \in V \setminus \{0\}$ is a cyclic vector. Among them are all q-difference modules of rank $\dim_K V = 1$, i.e., those isomorphic to some (K, Φ_a), $a \in K^*$; equivalently, those isomorphic to some $M_P := \mathcal{D}_{K,q}/\mathcal{D}_{K,q}P$, $\deg P = 1$ (and we can take $P = \sigma_q - a$ for some $a \in K^*$). By 8.1.4.1 we know that isomorphism classes of q-difference modules of rank 1 are in bijection with the Picard group $\mathrm{Pic}(K, \sigma_q)$.

DEFINITION 10.3. We say[4] that a q-difference module is *split* if all the simple subquotients P_i from a composition series have rank 1. (This of course does not depend on the particular composition series used.)

Clearly, if the sequence $0 \to M' \to M \to M'' \to 0$ is exact, then M is split if, and only if, M' and M'' are split (the multiset of simple subquotients of M, i.e., the support of $\phi(M)$, is the union of those of M' and M''). Therefore the full subcategory $\mathscr{C}_1$ of $\mathscr{C}$ having as objects the split q-difference modules is thick (the term was introduced in 10.1.2.1).

LEMMA 10.4. (i) *Let $P \in \mathcal{D}_{K,q}$, $P \neq 0$. Then the module $\mathcal{D}_{K,q}/\mathcal{D}_{K,q}P$ is split if, and only if, P is split, i.e., P is a product of q-difference operators of degree 1.*

(ii) *Let $A \in GL_n(K)$. Then the q-difference module (K^n, Φ_A) if, and only if, A is equivalent to an upper triangular matrix in $GL_n(K)$.*

PROOF. (i) We know that the rank r of $\mathcal{D}_{K,q}/\mathcal{D}_{K,q}P$ is the degree of P. If it is 0 or 1, the conclusion is immediate. We proceed by induction on r and assume $r \geq 2$.

If P is split, then $P = QR$ where Q and R are split and both have degree $< r$. By induction, $\mathcal{D}_{K,q}/\mathcal{D}_{K,q}Q$ and $\mathcal{D}_{K,q}/\mathcal{D}_{K,q}R$ are split. From the thickness of $\mathscr{C}_1$ and from the exact sequence in Example 8.37 of 8.2.1.2, we deduce that $\mathcal{D}_{K,q}/\mathcal{D}_{K,q}P$ is split.

Conversely, if $\mathcal{D}_{K,q}/\mathcal{D}_{K,q}P$ is split, it cannot be simple (for it would then have rank 1), so it admits a nontrivial proper submodule $\mathcal{D}_{K,q}R/\mathcal{D}_{K,q}P$, where $P = QR$ and Q and R both have degree $< r$. By induction, Q and R are split and therefore so is P.

(ii) Similar argument, using example 8.36 of 8.2.1.2 and induction on n.

□

[4]We hope there will be no confusion with the notion of *split exact sequence* from 8.3.1. Here the notion is rather related to the completely factored q-difference operators of 5.5.2, so the use of the word is reminiscent of *splitting fields* in field theory.

Since isomorphic q-difference modules have the same rank, we can define the subgroup $K_1(\mathscr{C})$ of $K(\mathscr{C})$ generated by the classes of q-difference modules of rank 1. Our split modules are those such that $\phi(M) \in K_1(\mathscr{C})$. We identified above those classes with the elements of the Picard group $\mathrm{Pic}(K, \sigma_q)$.

10.1.2.3. *From $K_1(\mathscr{C})$ to Newton polygons for split modules.* From now on, $K = \mathbf{C}(\{x\})$ or $K = \mathbf{C}((x))$. We then know that the q-difference operator P is split if, and only if, $S(P) \subset \mathbf{Z}$, i.e., all the slopes of P are integers (one implication comes from Chapter 5, the other from the obvious fact that a q-difference operator of degree 1 has integral slope, combined with the additivity of Newton polygon). Also, as a group, $\mathrm{Pic}(K, \sigma_q) \simeq \mathbf{E}_q \times \mathbf{Z}$. Keeping in mind that, if $a = a_0 x^n(1 + \cdots)$, then $\sigma_q - a$ has slope $-n$ and exponent $\overline{a_0}$, we select such an isomorphism as follows: we map the isomorphism class of $M_{\sigma_q - a}$ to $(\overline{a_0}, -n)$.

We shall therefore identify $K_1(\mathscr{C})$ as the free abelian group with basis $\mathbf{E}_q \times \mathbf{Z}$. Under this identification, when $M \in \mathrm{Ob}(\mathscr{C}_1)$, we shall use the notation $\psi(M)$ for the corresponding $\mathbf{Z}$-linear combination of elements of $\mathbf{E}_q \times \mathbf{Z}$.

EXERCISE 10.5. What is the image under ψ of (K^n, Φ_A) if A is upper triangular? (Recall that $(K, \Phi_a) \simeq M_{a^{-1}}$.)

Note that for any set X, the free abelian group generated by X can be seen either as the set of all formal $\mathbf{Z}$-linear combinations or as the set $\mathbf{Z}^{(X)}$ of maps $X \to \mathbf{Z}$ with finite support. The linear combination $\sum n_i[x_i]$ (the $x_i \in X$ pairwise distinct, the $n_i \in \mathbf{Z}$, almost all $n_i = 0$) corresponds to the map $X \to \mathbf{Z}$ such that each $x_i \mapsto n_i$ and all other $x \in X$ map to 0.

In this way, if M is split, the element $\psi(M) = \sum n_i[(\alpha_i, m_i)]$ (the $n_i \in \mathbf{Z}$, the (α_i, m_i) in $\mathbf{E}_q \times \mathbf{Z}$ pairwise distinct) can be seen as a function $\psi_M : \mathbf{E}_q \times \mathbf{Z} \to \mathbf{Z}$ such that each $(\alpha_i, m_i) \mapsto n_i$ and all other $(\alpha, m) \mapsto 0$.

PROPOSITION 10.6. *Let $M = \mathcal{D}/\mathcal{D}P$ (so P is split and $S(P) \subset \mathbf{Z}$, i.e., r_P has support in $\mathbf{Z}$). Then:*

$$\forall \mu \in \mathbf{Z}\,,\ r_P(\mu) = \sum_{\alpha \in \mathbf{E}_q} \psi_M(\alpha, \mu).$$

PROOF. Let $\sigma_q - a_j$, $j = 1, \ldots, \deg P$, the degree 1 factors in some decomposition of P and let $(\alpha_j, m_j) := \psi(\sigma_q - a_j)$, so that the unique slope of $\sigma_q - a_j$ is m_j. Then $r_P(\mu)$ is the number of factors such that $m_j = \mu$. (Here, we use additivity of the Newton function for products of q-difference operators, theorem 5.2 in 5.2.3.1, Chapter 5.) Each such factor appears $\psi_M(\alpha_j, m_j)$ times and the conclusion follows. □

If we put (temporarily) $\psi_P := \psi_M$, where $M = \mathcal{D}/\mathcal{D}P$, we can therefore see ψ_P as a kind of enriched Newton function (it keeps track of the multiplicities of the exponents, not only of the slopes).

DEFINITION 10.7. Let M a split module. We define the *Newton function* r_M of M as:

$$\forall \mu \in \mathbf{Z}\,,\ r_M(\mu) = \sum_{\alpha \in \mathbf{E}_q} \psi_M(\alpha, \mu).$$

The *set of slopes* $S(M)$ of M is the support of r_M (so $S(M) \subset \mathbf{Z}$).

It follows that all q-difference operators P such that $M = \mathcal{D}/\mathcal{D}P$ (a given split q-difference module) have the same Newton function $r_P = r_M$.

COROLLARY 10.8. *The Newton function is additive with respect to exact sequences in $\mathscr{C}_1$.*

PROOF. One can either use the fact that ψ is additive with respect to exact sequences and easily deduce the same property for $M \mapsto \left(\mu \mapsto \sum\limits_{\alpha \in \mathbf{E}_q} \psi_M(\alpha,\mu)\right)$; or appeal again to additivity of the Newton function for products of q-difference operators. □

REMARK 10.9. Generally speaking, a map $f : X \to X'$ induces a group morphism $\mathbf{Z}^{(X)} \to \mathbf{Z}^{(X')}$ which sends $\sum n_i[x_i]$ to $\sum n_i[f(x_i)]$ and the corresponding finitely supported function $\phi : X \to \mathbf{Z}$ to $\phi' : X' \to \mathbf{Z}$, $x' \mapsto \sum\limits_{f(x)=x'} \phi(x)$. The above transformation from ψ_M to r_M is thus obtained from the second projection $\mathbf{E}_q \times \mathbf{Z} \to \mathbf{Z}$ (forget the exponent, keep the slope).

10.1.3. Ramification and splitting. Here again $K = \mathbf{C}(\{x\})$ or $K = \mathbf{C}((x))$. We fix q and will consider the behaviour of our previous constructions under ramification at varying levels. Along with q are given the operator σ_q, the algebra $\mathcal{D} := \mathcal{D}_{K,q}$ (abreviation to simplify notations), the category $\mathscr{C}$, its full subcategory $\mathscr{C}_1$ of split q-difference modules, and the Grothendieck groups $K(\mathscr{C})$ and $K_1(\mathscr{C})$ together with maps $\phi : \mathrm{Ob}\mathscr{C} \to K(\mathscr{C})$ and $\psi : \mathrm{Ob}\mathscr{C}_1 \to K_1(\mathscr{C})$ (here identified with the free abelian group with basis $\mathrm{Pic}(K,\sigma_q) = \mathbf{E}_q \times \mathbf{Z}$).

10.1.3.1. *Ramification of split modules.* First we consider a given level of ramification $r \geq 2$. We set $x' := x^{1/r}$ and $q' := q^{1/r}$. We denote K', $\mathcal{D}'$, $\mathscr{C}'$ the corresponding q-difference extension, algebra of q'-difference operators and category of q'-difference modules. From 8.3.3 we have a base change functor $M \rightsquigarrow K' \otimes_K M$ from $\mathscr{C}$ to $\mathscr{C}'$. It is additive, exact, and it sends $\mathcal{D}/\mathcal{D}P$, $P \in \mathcal{D}$, to $\mathcal{D}'/\mathcal{D}'P$; and (K^n, Φ_A), $A \in \mathrm{GL}_n(K)$, to (K'^n, Φ'_A).

We write $\phi' : \mathrm{Ob}(\mathscr{C}') \to K(\mathscr{C}')$ the universal map defined in 10.1.2. The composed map $M \mapsto \phi'(K' \otimes_K M)$ from $\mathrm{Ob}(\mathscr{C})$ to $K(\mathscr{C}')$ is additive on exact sequences (because ϕ' is so and $M \rightsquigarrow K' \otimes_K M$ is exact). Therefore it factors uniquely through a group morphism $K(\mathscr{C}) \to K(\mathscr{C}')$.

Note that $\dim_{K'}(K' \otimes_K M) = \dim_K M$, so our functor sends q-difference modules of rank 1 to q'-difference modules of rank 1 and (by exactness) $\mathscr{C}_1$ to $\mathscr{C}'_1$ (split modules in $\mathscr{C}'$). Hence our group morphism $K(\mathscr{C}) \to K(\mathscr{C}')$ restricts to $K_1(\mathscr{C}) \to K_1(\mathscr{C}')$. We shall now describe this morphism in terms of their bases $\mathrm{Pic}(K,\sigma_q) = \mathbf{E}_q \times \mathbf{Z}$ and $\mathrm{Pic}(K',\sigma_{q'}) = \mathbf{E}_{q'} \times \mathbf{Z}$. Let $M := \mathcal{D}/\mathcal{D}(\sigma_q - a)$, $a = a_0 x^m(1 + \cdots)$, so that $\psi(M) = [(\overline{a_0}, -m)]$. Recall that $\overline{a_0}$ denotes the class of $a_0 \in \mathbf{C}^*$ in $\mathbf{E}_q := \mathbf{C}^*/q^{\mathbf{Z}}$. Then $M' := K' \otimes_K M = \mathcal{D}'/\mathcal{D}(\sigma_q - a')$, $a' = a_0 x'^{mr}(1 + \cdots)$, so that its image $\psi(M')$ is $[(\overline{a_0}', -mr)]$, where this time $\overline{a_0}'$ denotes the class of a_0 in $\mathbf{E}_{q'} := \mathbf{C}^*/q'^{\mathbf{Z}}$. Therefore we define:

$$R_r : \mathbf{E}_q \times \mathbf{Z} \to \mathbf{E}_{q'} \times \mathbf{Z},\ (\alpha,\mu) \mapsto (\alpha', r\mu),$$

where $\alpha' \in \mathbf{E}_{q'}$ is the image of $\alpha \in \mathbf{E}_q$ by the natural isogeny $\mathbf{C}^*/q^{\mathbf{Z}} \to \mathbf{C}^*/q'^{\mathbf{Z}}$ induced by the inclusion $q^{\mathbf{Z}} \subset q'^{\mathbf{Z}}$. We conclude:

LEMMA 10.10. *If M is split in $\mathscr{C}$, then $M' := K' \otimes_K M$ is split in $\mathscr{C}'$ and:*

$$\psi_M([\alpha,\mu]) = \psi_{M'}([R_r(\alpha), r\mu]).$$

PROOF. It follows from the above calculation in the base case using exactness and additivity. □

PROPOSITION 10.11. *let M split and $M' := K' \otimes_K M$. Then $r_M(\mu) = r_{M'}(r\mu)$.*

PROOF. It follows from the proposition by summing on all exponents. □

COROLLARY 10.12. *With the same notations, $S(M') = rS(M)$.*

We now agree to extend the source of the Newton function for a split module to all $\mu \in \mathbf{Q}$, by setting $r_M(\mu) := 0$ whenever $\mu \in \mathbf{Q} \setminus \mathbf{Z}$. All the above statements then hold without change.

10.1.3.2. *General definition of the Newton polygon.* Let $M \in \mathrm{Ob}(\mathscr{C})$. If $M = \mathcal{D}/\mathcal{D}P$ is not split, there is a $r \geq 2$ such that $S(P) \subset r^{-1}\mathbf{Z}$ and we take on the notations of 10.1.3.1; if M is split, we take $K' := K$, etc. In any case, $M' := K' \otimes_K M$ is split in $\mathscr{C}'$.

Now let $s \geq 2$ and let K''/K' a further q-difference extension associated with $x'' := x'^{1/s} = x^{1/rs}$ and $q'' := q'^{1/s} = q^{1/rs}$, and all other notations flowing from that. Then $M'' := K'' \otimes_{K'} M' = K'' \otimes_K M$ (by the last rule in 8.3.3.1) is split in $\mathscr{C}''$. From 10.1.3.1 we have:

$$\forall \mu \in \mathbf{Q}\,,\; r_{M''}(rs\mu) = r_{M'}(r\mu).$$

Therefore the following definition makes sense.

DEFINITION 10.13. Let M a q-difference module in $\mathscr{C}$ and let r such that (with the notations of 10.1.3.1) $M' := K' \otimes_K M$ is split in $\mathscr{C}'$. We define the *Newton function r_M of M* by

$$\forall \mu \in \mathbf{Q}\,,\; r_M(\mu) := r_{M'}(r\mu).$$

Clearly isomorphic q-difference modules have the same Newton function and polygon. But more is true. Let M an analytic q-difference module i.e., over $\mathbf{C}(\{x\})$, and let $\hat{M} := \mathbf{C}((x)) \otimes_{\mathbf{C}(\{x\})} M$ its formalization (this was defined in 8.3.3.3). Write for short $\mathcal{D} := \mathcal{D}_{\mathbf{C}(\{x\}),q}$ and $\hat{\mathcal{D}} := \mathcal{D}_{\mathbf{C}((x)),q}$. We saw in the same place that, if $M \simeq \mathcal{D}/\mathcal{D}P$, then $\hat{M} \simeq \hat{\mathcal{D}}/\hat{\mathcal{D}}P$. Since the x-adic valuations of the coefficients of P are the same whether computed in $\mathbf{C}(\{x\})$ or in $\mathbf{C}((x))$, the q-difference operator P has the same Newton function whether computed in $\mathcal{D}$ or in $\hat{\mathcal{D}}$. Therefore, $r_M = r_{\hat{M}}$; more generally, if $\hat{M} \simeq \hat{N}$ then $r_M = r_N$:

PROPOSITION 10.14. *The Newton polygon and function are formal invariants.*

10.1.3.3. *Additivity.*

THEOREM 10.1. *The Newton function is additive with respect to exact sequences.*

PROOF. Let $0 \to M_1 \to M_2 \to M_3 \to 0$ an exact sequence in $\mathscr{C}$. Let r a common level of splitting by ramification for all three modules (for instance the least common multiple of such levels for each one). Write $M_i' := K' \otimes_K M_i$, $i = 1, 2, 3$. Then, by exactness of base change and additivity of the Newton function for split modules:

$$r_{M_2}(\mu) = r_{M_2'}(r\mu) = r_{M_1'}(r\mu) + r_{M_3'}(r\mu) = r_{M_1}(\mu) + r_{M_3}(\mu).$$

□

COROLLARY 10.15. *Let* $M \simeq \mathcal{D}_{K,q}/\mathcal{D}_{K,q}P$, *where* $P = P_1 \cdots P_k$. *Then* $r_M = r_{P_1} + \cdots + r_{P_k}$.

PROOF. From 8.2.1.4, we have a tower $0 = M_0 \subset M_1 \subset \cdots \subset M_k = M$ such that $M_i/M_{i-1} \simeq \mathcal{D}_{K,q}/\mathcal{D}_{K,q}P_i$ for $i = 1, \ldots, k$. From the exact sequence $0 \to M_{i-1} \to M_i \to M_i/M_{i-1} \to 0$, we deduce that:

$$r_{M_i} = r_{M_{i-1}} + r_{M_i/M_{i-1}} = r_{M_{i-1}} + r_{P_i},$$

whence the conclusion by induction. □

10.1.3.4. *The Newton polygon of* (K^n, Φ_A). We give a few easy yet useful rules of computation. Let us write M_A for (K^n, Φ_A) and r_A for r_{M_A}.

From 8.2.1.4, we draw (as in the corollary above):

$$A = \begin{pmatrix} A_1 & \star & \star & \ldots & \star & \star \\ 0 & A_2 & \star & \ldots & \star & \star \\ 0 & 0 & A_3 & \ldots & \star & \star \\ \vdots & \vdots & \vdots & \ddots & \vdots & \vdots \\ 0 & 0 & 0 & \ldots & A_{k-1} & \star \\ 0 & 0 & 0 & \ldots & 0 & A_k \end{pmatrix} \Longrightarrow r_A = \sum_{i=1}^{k} r_{A_i}.$$

If $B = F[A]$, then $M_B \simeq M_A$, whence $r_B = r_A$.

In particular, if $A \in \mathrm{GL}_n(\mathbf{C})$, then $S(M_A) = \{0\}$ (i.e., M_A is *fuchsian*; see 10.1.4.2). Indeed, trigonalisation $A = CTC^{-1}$, $C, T \in \mathrm{GL}_n(\mathbf{C})$, amounts to a gauge transformation $A = C[T]$; we then apply the two rules above and reduce the problem to $S(M_a) = \{0\}$ when $a \in \mathbf{C}^*$: but we know by 8.1.4 that $(K, \Phi_a) \simeq \mathcal{D}_{K,q}/\mathcal{D}_{K,q}(S - a^{-1})$, so its (unique) slope is indeed 0.

REMARK 10.16. We shall see in 11.1 that, more generally, if A is regular at 0, i.e., $A(0)$ is defined and invertible, which we write for short $A(0) \in \mathrm{GL}_n(\mathbf{C})$, then M_A is fuchsian. The opposite implication is easier.

EXERCISE 10.17. Prove that, if $S(M) = 0$, then there is a regular A such that $M \simeq M_A$. (First take $P = a_0 + \cdots + a_n S^n$ such that $a_0, a_n \in \mathbf{C}[[x]]^*$, all $a_i \in \mathbf{C}[[x]]$, and $M \simeq M_P$; Then take from 8.1.5 a particular A such that $M_A \simeq M_P$.)

If $A = fB$, $f \in K^*$, then $\Phi_B = f\Phi_A$, which entails $\Phi_B^k = f^{(k)}\Phi_A^k$, where $f^{(k)} := f\sigma_q f \cdots {\sigma_q}^{k-1} f$. It immediately follows that Φ_A and Φ_B have the same cyclic vectors. Also, if v is such a cyclic vector and $P := \sum a_i S^i$ is a minimal polynomial such that $P(\Phi_A)(v) = 0$, then setting $Q := \sum b_i S^i$ with $b_i := a_i/f^{(i)}$, we see that Q is a minimal polynomial such that $Q(\Phi_B)(v) = 0$. So $M_A \simeq \mathcal{D}_{K,q}/\mathcal{D}_{K,q}P$ and $M_B \simeq \mathcal{D}_{K,q}/\mathcal{D}_{K,q}Q$. Now let $\alpha := v(f)$, so that $v(f^{(k)}) = k\alpha$. We conclude that the cloud for P is the image of the cloud for Q by the transformation $(i,j) \mapsto (i, j + i\alpha)$. It follows that $S(A) = S(B + \alpha)$, multiplicities being respected, i.e., more precisely: $r_B(\mu) = r_A(\mu + \alpha)$. We conclude:

$$r_{fA}(\mu) = r_A(\mu + v(f)).$$

10.1.4. Algebraic properties. A wealth of nice consequences! Beware we shall no longer use the "prime" in M' to denote base change.

10.1.4.1. *Set of slopes.* We define $S(M)$ to be the support of r_M and call its elements the *slopes of* M. It is empty if, and only if, $M = \{0\}$. The *multiplicity* of slope μ is $r_M(\mu)$.

COROLLARY 10.18. *If the sequence* $0 \to M' \to M \to M'' \to 0$ *is exact,* $S(M) = S(M') \cup S(M'')$. *In particular,* $S(M') \subset S(M)$ *and* $S(M/M') \subset S(M)$.

COROLLARY 10.19. *Let* $f : M \to N$ *a morphism of q-difference modules. Then* $S(Im\ f) \subset S(M) \cap S(N)$. *In particular, if* $S(M) \cap S(N) = \emptyset$, *then* f *is trivial.*

COROLLARY 10.20. (i) *Let* M_1, M_2 *two submodules of the q-difference module* M. *Then* $S(M_1 + M_2) = S(M_1) \cup S(M_2)$.
(ii) *If* $S(M_1) \cap S(M_2) =$, *the sum is direct.*

PROOF. (i) Use first the exact sequence $0 \to M_1 \to M_1 \oplus M_2 \to M_2 \to 0$ to conclude that $S(M_1 \oplus M_2) = S(M_1) \cup S(M_2)$; then the surjection $M_1 \oplus M_2 \to M_1 + M_2$ to conclude that $S(M_1 + M_2) \subset S(M_1) \cup S(M_2)$; last the inclusions $S(M_1) \subset S(M_1 + M_2)$, $S(M_2) \subset S(M_1 + M_2)$, for the reciprocal inclusion.
(ii) Indeed then $M_1 \cap M_2 = \{0\}$.

□

10.1.4.2. *Pure isoclinic modules.*

DEFINITION 10.21. The q-difference module M is *pure isoclinic of slope* μ if $S(M) = \{\mu\}$.

The particular case $\mu = 0$ ("fuchsian" q-difference modules) will be studied in Chapter 11. The more general "pure modules" (not necessary isoclinic) will appear in the next section 10.2.

COROLLARY 10.22. (i) *A submodule, a quotient module of a pure isoclinic module is either trivial or pure isoclinic of the same slope.*
(ii) *An extension of pure isoclinic modules of same slope* μ *is pure isoclinic of slope* μ.

EXAMPLE 10.23. Let $A := \begin{pmatrix} 0 & a \\ b & 0 \end{pmatrix}$, $a, b \in K^*$ (K is either $\mathbf{C}(\{x\})$ or $\mathbf{C}((x))$) and let $M := (K^2, \Phi_A)$ and $e := \begin{pmatrix} 1 \\ 0 \end{pmatrix}$. Then $\Phi_A(e) = \begin{pmatrix} 0 \\ a^{-1} \end{pmatrix}$ and $\Phi_A^2(e) = c^{-1}e$, where $c := b\sigma_q a$. Therefore e is cyclic and $M \simeq \mathcal{D}_{K,q}/\mathcal{D}_{K,q}P$, where $P := cS^2 - 1$. It follows that M is pure isoclinic with slope $v_0(c)/2$.

If $v_0(c)$ is odd, M is simple: indeed a proper nontrivial submodule should have rank 1 and slope $v_0(c)/2$, which is impossible since rank 1 q-difference modules have integral slopes.

Assume $v_0(c)$ is even. A proper nontrivial submodule N would be generated by some nonzero vector $v := \begin{pmatrix} f \\ g \end{pmatrix}$ which is colinear with $\Phi_A(v) = \begin{pmatrix} b^{-1}\sigma_q g \\ a^{-1} f\sigma_q \end{pmatrix}$; that is equivalent to $fb^{-1}\sigma_q g = ga^{-1}f$, which in turns is equivalent to $h\sigma_q h = c$, where $h := ga/f$ (neither g nor f can be 0). Conversely, if h can be found, $f := a$, $g := h$ will do. By the following exercise, if $v_0(c)$ is even, then the equation $h\sigma_q h = c$ has a solution in K (the converse is obvious) so M is not simple in that case.

Note that $h' := -h$ is also a solution, yielding a nonproportional vector v', whence another rank 1 submodule N', whence $M = N \oplus N'$: the q-difference module M is not indecomposable, which is stronger that not simple.

EXERCISE 10.24. In $\mathbf{C}(\{x\})$ or $\mathbf{C}((x))$, the equation $h\sigma_q h = c$ has a solution if, and only if, $v_0(c)$ is even. (Write $c = c_1 c_2$, $c_1 = c_0 x^{2r}$, $c_2 = 1 + \cdots$ and solve separately $h_1 \sigma_q h_1 = c$, which is easy, and $h_2 \sigma_q h_2 = c_2$, with $h_2 := \prod {\sigma_q}^{2i} c_2 / \prod {\sigma_q}^{2i+1} c_2$; then take $h := h_1 h_2$.)

COROLLARY 10.25. (i) *Let M a q-difference module. The sum $M(\mu)$ of all its pure isoclinic submodules of slope μ is trivial if $\mu \notin S(M)$. If $\mu \in S(M)$ it is the biggest pure isoclinic submodule of slope μ of M.*

(ii) *The rank of $M(\mu)$ is at most $r_M(\mu)$.*

(iii) *The submodules $M(\mu)$ of M are in direct sum (i.e., their sum as submodules of M is a direct sum).*

Our main task, in section 10.2, will be to find out when $M(\mu)$ has the maximum possible rank $r_M(\mu)$.

10.1.4.3. *Tensor product.* According to Example 8.73, we have:

$$(K, \Phi_a) \otimes (K, \Phi_b) = (K, \Phi_{ab}),$$
$$\mathcal{D}_{K,q}/\mathcal{D}_{K,q}(S - a') \otimes \mathcal{D}_{K,q}/\mathcal{D}_{K,q}(S - b') \simeq \mathcal{D}_{K,q}/\mathcal{D}_{K,q}(S - a'b').$$

We are going to deduce the general formula:

$$r_{M_1 \otimes M_2}(\mu) = \sum_{\mu_1 + \mu_2 = \mu} r_{M_1}(\mu_1) r_{M_2}(\mu_2).$$

If M_1 and M_2 have rank 1, it immediately follows from either one of the two formulas above. To make more transparent the "unscrewing" argument leading to the general case, we shall introduce a nice auxiliary tool. It actually looks a lot like "Hilbert-Samuel series" used in commutative algebra and algebraic geometry [**36**]. So we write:

$$\mathcal{R}_M(T) := \sum_{\mu \in \mathbf{Q}} r_M(\mu) T^{\mu}.$$

This is a "Puiseux-Laurent polynomial": a polynomial because the sum is finite (only $\mu \in S(M)$ can appear); Laurent because negative exponents are allowed; Puiseux because nonintegral exponents are allowed[5]. In this language, Theorem 10.1 writes:

The sequence $0 \to M' \to M \to M'' \to 0$ is exact $\Longrightarrow \mathcal{R}_M(T) = \mathcal{R}_{M'}(T) + \mathcal{R}_{M''}(T)$.

and the above formula to be proved writes:

$$\mathcal{R}_{M_1 \otimes M_2}(T) = \mathcal{R}_{M_1}(T) \mathcal{R}_{M_2}(T).$$

Note that ramification $M' := K[x^{1/r}] \otimes_K M$ entails $\mathcal{R}_{M'}(T) = \mathcal{R}_M(T^r)$. Therefore it is enough to prove our contention for split modules. Then, if for instance the sequence $0 \to M_1' \to M_1 \to M_1'' \to 0$ is exact, then so it the sequence $0 \to M_1' \otimes M_2 \to M_1 \otimes M_2 \to M_1'' \otimes M_2 \to 0$ and by induction on such sequences we reduce the problem to the case when the left factor has rank 1. (Here we used the exactness of the tensor product; see 9.2.2.) Then a similar argument goes for the right factor and we reach the conclusion.

EXERCISE 10.26. (i) Confirm the limit case of the induction (rank 1 case) by expressing $\mathcal{R}_M(T)$ when $M := (K, \Phi_a)$ and when $M := \mathcal{D}_{K,q}/\mathcal{D}_{K,q}(\sigma_q - a')$.

(ii) Using 8.3.3.2, compute $r_{\underline{\mathrm{Hom}}(M_1, M_2)}$.

---

[5] Actually, if we denote $T^{\mathbf{Q}}$ the additive group $\mathbf{Q}$ written multiplicatively, we are here dealing with the group algebra of $T^{\mathbf{Q}}$ over $\mathbf{C}$; see [**145**].

REMARK 10.27. Among the slope filtrations studied in **[5]**, there are mostly two possible behaviours when tensoring: either slopes add up (like here, more generally like in **[213]**); or the slopes of $A \otimes B$ are the $\max(\mu, \nu)$, μ slope of A, ν slope of B (maybe min instead of max according to conventions): this is the case of ordinary differential equations; see **[5]**. We pursue the comparison of the present theory with **[5]** in Section 10.4.

10.1.5. An important example (II). This is a followup of 8.3.4.2; it is also due to van der Put and Reversat. The motivation is an application to formal classification in 11.1.4. For an alternative approach by Virginie Bugeaud, using cyclic vectors; see **[43]**. Here, K is $\mathbf{C}(\{x\})$ or $\mathbf{C}((x))$. Letting $r \in \mathbf{N}^*$, $d \in \mathbf{Z}$ and $a \in \mathbf{C}^*$ and set $L := {\sigma_q}^r - ax^d$. Then the q-difference module $M := \mathcal{D}/\mathcal{D}L$ has rank r and is pure isoclinic with slope d/r. Actually, $M \simeq M_A = (K^r, \Phi_A)$, where (according to 8.1.3.1 and 8.3.4.2 and denoting $v := (ax^d)^{-1}$):

$$A := A_L^\vee = \begin{pmatrix} 0 & 1 & 0 & \dots & 0 & 0 \\ 0 & 0 & 1 & \dots & 0 & 0 \\ 0 & 0 & 0 & \dots & 0 & 0 \\ \vdots & \vdots & \vdots & \ddots & \vdots & \vdots \\ 0 & 0 & 0 & \dots & 0 & 1 \\ v & 0 & 0 & \dots & 0 & 0 \end{pmatrix}.$$

LEMMA 10.28. *M is irreducible (or simple) if, and only if, r and d are coprime: $r \wedge d = 1$.*

PROOF. The "if" part (backwards implication) is directly related to the content of this chapter: if $M' \subset M$ is a nontrivial submodule, then $S(M') = \{d/r\}$; since d/r cannot be put in the form d'/r' with $d' \in \mathbf{Z}$, $r' \in \mathbf{N}^*$ and $r' < r$, the submodule M' cannot have rank $< r$, so it is M.

The "only if" part (direct implication) relies on the explicit computation of a nontrivial factorisation of L. Let $r = ks$, $d = ke$ with $k \geq 2$. Set $\tau := {\sigma_q}^s$ and $y := \alpha x^e$, where $\alpha^k = a$: thus $L = \tau^k - y^k$. Now use the exercise below. □

EXERCISE 10.29. Find $\lambda \in \mathbf{C}^*$ such that $\tau(y) = \lambda y$ and deduce the values of $\alpha_1, \dots, \alpha_{k-1}, \mu \in \mathbf{C}^*$ such that:

$$\tau^k - y^k = (\tau^{k-1} + \alpha_1 y \tau^{k-2} + \cdots + \alpha_{k-1} y^{k-1})(\tau - \mu y).$$

(ii) Find the corresponding nontrivial submodule M' of M.

Let now N the similar q-difference module obtained by replacing a by $b \in \mathbf{C}^*$.

LEMMA 10.30. *Assume $r \wedge d = 1$, so M and N are simple. Then $M \simeq N$ if, and only if, $a \equiv b \pmod{q^{\mathbf{Z}}}$.*

PROOF. Let B the matrix obtained by replacing v by $w := (bx^d)^{-1}$ in A. Comparing determinants, we see that $F[A] = B$ implies $a/b = w/v = \sigma_q(\det F)/\det F$, which is only possible if $a/b \in q^{\mathbf{Z}}$.

Conversely, using the exercise below, we see that L is conjugate to ${\sigma_q}^r - q^{\beta d - \alpha r} a x^d$ for any $\alpha, \beta \in \mathbf{Z}$; if $r \wedge d = 1$ the conclusion follows. □

EXERCISE 10.31. Check the following equalities:

$$x^\alpha L (x^\alpha)^{-1} = q^{\alpha r} {\sigma_q}^r - ax^d \text{ and } {\sigma_q}^\beta L ({\sigma_q}^\beta)^{-1} = {\sigma_q}^r - q^{\beta d} a x^d.$$

10.2. The canonical slope filtration

The slope filtration of analytic and formal q-difference modules is a *very important* structural and structuring property, and it will be heavily used in classification and Galois theory. It essentially rests on factorisation from Chapter 5 and so, ultimately, on Adams's lemma, as well as its extension by Birkhoff and Guenther. The novelty of this property of q-difference theory with respect to the theory of ordinary differential equations was mentioned in 5.1.4, in comparison with [**75**, **150**, **159**, **204**].

We first state and prove the existence and uniqueness of the filtration with a not very rigorous terminology, not distinguishing between flags (towers of submodules) and filtrations proper. In 10.2.2 we give the proper terminology. (Note however that, for instance, [**145**] calls "filtrations" our flags.) We then proceed to study properties related to morphisms and exact sequences. In 10.2.3 we study tensor properties. In Section 10.3, we shall see in what sense the associated graduation is the exact algebraic counterpart of formalization (i.e., forgetting convergence issues).

Part of what follows can be subsumed in the general theory of slope filtrations in [**5**]; see Section 10.4; however, we favor a direct independent treatment.

EXERCISE 10.32. For your edification, compare various notions of filtrations respectively taken from [**145**], [**5**], and [**232**, II §2].

10.2.1. Definition, existence, uniqueness. As in all the chapter, $K = \mathbf{C}(\{x\})$ or $K = \mathbf{C}((x))$. We write for short $\mathcal{D} := \mathcal{D}_{K,q}$.

10.2.1.1.

THEOREM 10.2 (The canonical slope filtration). *Let M a q-difference module over K and $S(M) = \{\mu_1, \ldots, \mu_k\}$, with multiplicities $r_i := r_M(\mu_i)$. If $K = \mathbf{C}(\{x\})$, we assume moreover that $\mu_1 > \cdots > \mu_k$ (otherwise the order is arbitrary). Then there is a unique flag (or tower of submodules) $\{0\} = M_0 \subset M_1 \subset \cdots \subset M_k = M$ such that each M_i/M_{i-1}, $i = 1, \ldots, k$, is pure isoclinic of slope μ_i. Moreover M_i/M_{i-1} has rank r_i.*

PROOF. The statement on the rank of M_i/M_{i-1} directly flows from the fact that the Newton function of M is the sum of the Newton functions of the M_i/M_{i-1}.

We prove uniqueness by induction on k, the limiting case $k = 1$ (of a pure isoclinic module) being easy. If such a filtration exists, then M_1 is the greatest pure isoclinic submodule of slope μ_1 (see Corollary 10.25), so it is uniquely determined. Then from the conditions on the successive quotients M_i/M_{i-1}, $i = 2, \ldots, k$, we draw that $S(M/M_1) = \{\mu_2, \ldots, \mu_k\}$ and $\{0\} = M_1/M_1 \subset M_2/M_1 \subset \cdots \subset M_k/M_1 = M/M_1$ is a flag with similar properties with respect to M/M_1, whence, by induction, uniquely determined.

As for the existence, we write $M \simeq \mathcal{D}/\mathcal{D}P$ and use a factorisation $P = P_1 \cdots P_k$ such that $S(P_i) = \{\mu_i\}$ (Theorem 5.5 in 5.4.2 and Theorem 5.6 in 5.4.3). From 8.2.1.4 we then get a tower $\{0\} = M_0 \subset M_1 \subset \cdots \subset M_k = M$ such that each $M_i/M_{i-1} \simeq \mathcal{D}/\mathcal{D}P_i$. □

COROLLARY 10.33 (Splitting of the canonical slope filtration in the formal case). *If $K = \mathbf{C}(\{x\})$, the filtration splits*[6]*, i.e., there are subsmodules $N_1, \ldots, N_k$*

[6]So we return to the other meaning of "split", that related to exact sequences associated to direct sums.

of M such that $M_i = N_1 \oplus \cdots \oplus N_i$, $i = 1, \ldots, k$, each N_i being pure isoclinic with slope μ_i and rank r_i.

PROOF. Using the theorem with μ_1 any particular slope of M (since in the formal case the order is arbitrary), we see that the greatest submodule of slope μ has rank $r_M(\mu)$. The indexing being given, call N_i that submodule for $\mu = \mu_i$. From 10.1.4.2 we know that the sum of the N_i is direct. The conclusion follows. □

10.2.1.2. *An alternative argument.* The above proof rests, at a critical point (Theorems 5.5 and 5.6) on a complicated galoisian argument to reduce the case of arbitrary slopes to the case of integral slopes. We feel that the argument is more transparent in its "linear form", using Galois descent for q-difference modules from 8.3.5; so we present here a second proof[7].

So let M an arbitrary q-difference module. Clearly (seeing the proof of the theorem above), it is enough to check that its biggest submodule of slope μ_1 has rank $r_1 := r_M(\mu_1)$. If M is split, this is clear using the factorisation of P such that $M \simeq \mathcal{D}/\mathcal{D}P$ (no galoisian argument needed here, just usual resolution and factorisation as in 5.4.1). In the general case, we use ramification of degree r such that $S(M) \subset r^{-1}\mathbf{Z}$. Thus we get a split module $M' := K' \otimes_K M$, which has a pure isoclinic submodule M'_1 of slope $r\mu_1$ and rank r_1. We just want to prove that $M'_1 := K' \otimes_K M_1$ for some $M_1 \subset M$ because then M_1 will have unique slope μ_1 and rank r_1.

Now by Lemma 8.91 (and using notations there), $\Psi_\tau(M'_1)$ is a submodule of same rank and slope as M'_1, so $\Psi_\tau(M'_1) = M'_1$ (uniqueness of the submodule of greatest slope), so Galois descent applies and $M_1 := M'_1 \cap M$ does the job.

10.2.1.3. Combining the theorem with 8.2.1.4, we see that, if $K = \mathbf{C}(\{x\})$, the q-difference module M can be put in the form (K^n, Φ_A), where A is block-triangular with diagonal blocks $A_i \in \mathrm{GL}_{r_i}(K)$ such that each (K^{r_i}, Φ_{A_i}) is pure of slope μ_i. From the corollary we see that, if $K = \mathbf{C}((x))$, then A can be taken in block-diagonal form. So we shall frequently meet such matrices:

$$A_U = \begin{pmatrix} A_1 & U_{1,2} & U_{1,3} & \ldots & U_{1,k-1} & U_{1,k} \\ 0 & A_2 & U_{2,3} & \ldots & U_{2,k-1} & U_{2,k} \\ 0 & 0 & A_3 & \ldots & U_{3,k-1} & U_{3,k} \\ \vdots & \vdots & \vdots & \ddots & \vdots & \vdots \\ 0 & 0 & 0 & \ldots & A_{k-1} & U_{k-1,k} \\ 0 & 0 & 0 & \ldots & 0 & A_k \end{pmatrix},$$

and $A_0 = \mathrm{Diag}(A_1, \ldots, A_k)$ (the block diagonal matrix extracted from A_U). Although in the latter case we can take the order of the slopes arbitrary, we shall usually require that $\mu_1 > \cdots > \mu_k$. Actually, we shall generally derive A_0 as the "graded counterpart" of A_U (see Section 10.3).

The above forms will be refined in Chapters 11 and 15, but we already state without proof what can be said when $S(M) \subset \mathbf{Z}$:

- We can take $A_i = x^{\mu_i} B_i$, $1 \leq i \leq k$, with $B_i \in \mathrm{GL}_{r_i}(\mathbf{C})$;
- we can take the rectangular block $U_{i,j}$, $1 \leq i < j \leq k$, with coefficients in $\sum\limits_{\mu_i \geq d > \mu_j} \mathbf{C}^d$ (this is the so-called Birkhoff-Guenther Normal Form).

[7] It is actually the original proof in **[220]**, but for a minor error there, which we have corrected here (see the footnote in 8.3.5).

At any rate, we can require that $A \in \mathrm{GL}_n(\mathbf{C}[x, x^{-1}])$.

10.2.1.4. *The case of pure q-difference modules.*

DEFINITION 10.34. A *pure* q-difference module is a direct sum of pure isoclinic q-difference modules.

Suppose $M = N_1 \oplus \cdots \oplus N_k$, where each N_i is pure isoclinic of slope $\mu_i \in \mathbf{Q}$ and rank $r_i \geq 1$ (all the μ_i pairwise distinct). Its Newton function is given by $S(M) = \{\mu_1, \ldots, \mu_k\}$ and $r_M(\mu_i) = r_i$, $i = 1, \ldots, k$. Clearly $M^{(\mu_i)} = N_i$. Set $M_i := N_1 \oplus \cdots \oplus N_i$. Then we have a flag $\{0\} = M_0 \subset M_1 \subset \cdots \subset M_k = M$ such that each $M_i/M_{i-1} = N_i$ is pure isoclinic of slope μ_i. If the slopes are indexed so that $\mu_1 > \cdots > \mu_k$, we recover the canonical slope filtration and it splits (as in Corollary 10.33).

It follows from 10.2.1.3 that every formal q-difference module (that is q-difference module over $\mathbf{C}((x))$) is pure. This is not true over $\mathbf{C}(\{x\})$. The relation will be somewhat clarified in section 10.3. We show a typical example.

EXAMPLE 10.35. Let $K := \mathbf{C}(\{x\})$ and $M := (K^2, \Phi_A)$, where $A := \begin{pmatrix} x & 1 \\ 0 & 1 \end{pmatrix}$. This fits into an exact sequence $0 \to M' \to M \to M'' \to 0$, where $M' := (K, \Phi_x)$ and $M'' := (K, \Phi_1)$. Therefore $S(M) = \{0, 1\}$ and $r_M(0) = r_M(1) = 1$.

If M was pure, it would have a unique nontrivial submodule of slope 0 and rank 1 and this would have to be isomorphic to M'' (indeed the projection $M \to M''$ would necessarily send to 0 the component of slope 1). So there would be a nontrivial morphism $M'' \to M$, i.e., a $F := \begin{pmatrix} f \\ g \end{pmatrix} \in \mathrm{Mat}_{2,1}(K)$ such that $(\sigma_q F).1 = AF$, i.e. $\begin{cases} \sigma_q f = xf + g \\ \sigma_q g = g \end{cases}$.

If $g = 0$, knowing that there are no nontrivial power series solution of $\sigma_q f = xf$, we draw that $f = 0$, hence $F = 0$, which is not allowed (we look for an injective morphism).

If $g \neq 0$, then $h := f/g$ satisfies $\sigma_q h = xh + 1$. By Chapter 4, the only power series solution is $\sum\limits_{n \geq 0} q^{-n(n+1)/2} x^n \in \mathbf{C}((x)) \setminus \mathbf{C}(\{x\})$. So there is no solution in $\mathbf{C}(\{x\})$ and no possible nontrivial $F \in \mathrm{Mat}_{2,1}(\mathbf{C}(\{x\}))$, so that M is not pure.

10.2.2. Abelian properties. For a q-difference module M over K, the "usual notations" will mean: the Newton polygon of M, presented as $S(M) = \{\mu_1, \ldots, \mu_k\}$ with order $\mu_1 > \cdots > \mu_k$ and multiplicities $r_1, \ldots, r_k$; and its canonical slope filtration, presented as a flag $\{0\} = M_0 \subset M_1 \subset \cdots \subset M_k = M$ such that each M_i/M_{i-1} is isoclinic pure with slope μ_i and rank r_i.

10.2.2.1. *Descending filtrations.* We explicit the relation of filtrations as defined in **[5, 54, 213]** with flags. A *descending filtration* of the $\mathcal{D}$-module M (for us, always by submodules and always indexed by $\mathbf{Q}$) is a family of submodules $(M^{\geq \mu})_{\mu \in \mathbf{Q}}$ such that $\mu < \mu' \Rightarrow M^{\geq \mu} \supset M^{\geq \mu'}$. Denoting $M^{\geq +\infty} := \bigcap\limits_{\mu \in \mathbf{Q}} M^{\geq \mu}$ and $M^{\geq -\infty} := \bigcup\limits_{\mu \in \mathbf{Q}} M^{\geq \mu}$ (both are submodules), we shall moreover assume that $M^{\geq +\infty} = \{0\}$

and $M^{\geq -\infty} = M$. Let $M^{>\mu} := \bigcup_{\mu' > \mu} M^{\geq \mu'}$, whence a submodule of $M^{\geq \mu}$, and[8] $M^{(\mu)} := M^{\geq \mu}/M^{>\mu}$. We say that there is a *break* or a *jump* at μ if $M^{(\mu)} \neq \{0\}$.

If M has finite length, there is a finite number of breaks, say $\mu_1 > \cdots > \mu_k$. We can therefore associate to our filtration a flag $\{0\} = M_0 \subset M_1 \subset \cdots \subset M_k = M$ such that each $M_i = M^{\geq \mu_i}$. Introducing $\mu_0 := +\infty$ and $\mu_{k+1} := -\infty$, we see that:

$$\forall \mu \in \mathbf{Q}\ ,\ M^{\geq \mu} = M_i \Longleftrightarrow \mu_{i-1} > \mu \geq \mu_i.$$

Conversely, from every flag $\{0\} = M_0 \subset M_1 \subset \cdots \subset M_k = M$ the above formula defines a descending filtration. One has $M^{>\mu_i} = M_{i-1}$ and $M^{(\mu_i)} = M_i/M_{i-1}$, all the other $M^{(\mu)}$ being trivial.

In this language, the *(canonical) slope filtration* on M is the unique descending filtration such that each $M^{(\mu)}$ is either trivial or pure of slope μ; the breaks occur at $\mu \in S(M)$.

EXERCISE 10.36. Prove that $S(M^{\geq \mu}) = S(M) \cap [\mu, +\infty[$ and $S(M^{>\mu}) = S(M) \cap]\mu, +\infty[$.

EXERCISE 10.37. State and prove analogous properties for $M_{\leq \mu} := M/M^{>\mu}$ and $M_{<\mu} := M/M^{\geq \mu}$.

10.2.2.2. We now formulate the main abelian properties of the canonical slope filtration, using sometimes the language of descending filtrations and sometimes the language of flags. They are essentially corollaries of the uniqueness property in the theorem in 10.2.1.1. Also we refer at will to the "usual notations" without caring to recall them. All formulas given hereafter for $M^{\geq \mu}$, etc, imply immediately the corresponding formulas for $M^{>\mu}$, etc.

COROLLARY 10.38. *Let $M' \subset M$ a submodule and $M'' := M/M'$. Then the slope filtrations on M' and M'' are given by $M'^{\geq \mu} = M' \cap M^{\geq \mu}$ and $M''^{\geq \mu} =$ the image of $M^{\geq \mu}$ in M'', i.e., $(M^{\geq \mu} + M')/M' = M^{\geq \mu}/M'^{\geq \mu}$.*

PROOF. Set $N'^{\geq \mu} := M' \cap M^{\geq \mu}$ and $N''^{\geq \mu} := (M^{\geq \mu} + M')/M'$ (the last equality comes from the general isomorphism theorem $(A+B)/B \simeq A/(A \cap B)$). They are descending filtrations respectively on M' and M'' with pure breaks with corresponding slopes (you should of course check that!) and the uniqueness property gives the conclusion. □

COROLLARY 10.39. *Let $f : M \to N$ a morphism of q-difference modules. Then $f(M^{\geq \mu}) = f(M) \cap N^{\geq \mu}$.*

PROOF. Both are equal to $f(M)^{\geq \mu}$ by the previous lemma (taking $M' := \operatorname{Ker} f$ and $M'' := \operatorname{Im} f$). □

Note that one would say that f is compatible with the two filtrations if $f(M^{\geq \mu}) \subset N^{\geq \mu}$. The above equality is stronger and it is called *strict compatibility* or *strictness*. The following result is a general property of strictness [**54**, 1.1] but we give a direct proof.

COROLLARY 10.40. *Let $M' \xrightarrow{f} M \xrightarrow{g} M''$ an exact sequence. Then the induced sequence $M'^{(\mu)} \to M^{(\mu)} \to M''^{(\mu)}$ is exact.*

[8] Be careful not to confuse $M^{(\mu)}$ here with $M(\mu)$ from 10.1.4.2.

PROOF. The image of the left map is:

$$\frac{f\left(M'^{\geq\mu}\right)}{f\left(M'^{>\mu}\right)} = \frac{(f(M') \cap M^{\geq\mu})}{(f(M') \cap M^{>\mu})} = \frac{(\operatorname{Ker} g \cap M^{\geq\mu})}{(\operatorname{Ker} g \cap M^{>\mu})} = \operatorname{Ker}\left(M^{(\mu)} \to M''^{(\mu)}\right).$$

□

As often, the most important application is:

COROLLARY 10.41. *Let $0 \to M' \to M \to M'' \to 0$ an exact sequence. Then the induced sequence $0 \to M'^{(\mu)} \to M^{(\mu)} \to M''^{(\mu)} \to 0$ is exact.*

10.2.2.3. The smallest slope μ_k (also the leftmost slope by the form of the Newton polygon) sometimes plays a special part, so we shall denote it $\mu(M) := \min S(M)$ (compare with [**5**]). We extend the notation by agreeing that $\mu(\{0\}) = +\infty$, which is consistent with the facts that $S(\{0\}) = \emptyset$ and that $\min \emptyset = +\infty$ (the only convention logically compatible with most usual formulas).

Recall that, with the "usual notations", we have: $S(M_i) = \{\mu_1, \dots, \mu_i\}$ and $S(M/M_i) = \{\mu_{i+1}, \dots, \mu_k\}$.

PROPOSITION 10.42. *Let $N \subset M$ a submodule.*

(i) $N \subset M_i \Longleftrightarrow \mu(M) \geq \mu_i \Longleftrightarrow S(N) \subset \{\mu_1, \dots, \mu_i\}$.
(i') $N \subset M^{\geq\mu} \Longleftrightarrow \mu(M) \geq \mu$.
(ii) $\mu(N) = \mu_i \Longleftrightarrow (N \subset M_i$ *and* $N \not\subset M_{i-1})$.
(ii') $\mu(N) = \mu \Longleftrightarrow \left(N \subset M^{\geq\mu}\right.$ *and* $\left.N \not\subset M^{>\mu}\right)$

PROOF. (i') and (ii') respectively are translations of (i) and (ii), and (ii) easily follows from (i).

In (i), the second equivalence and the first direct implication both come from $S(N) \subset S(M)$.

Conversely, Suppose that $S(N) \subset \{\mu_1, \dots, \mu_i\}$. Then $S(N) \cap S(M/M_i) = \emptyset$, so that the compositum $N \to M \to M/M_i$ is 0 by 10.1.4.1, i.e., $N \subset M_i$. □

Let $v \in M$, $v \neq 0$. We define $\mu(v) := \mu(\mathcal{D}v)$. Note that if $v \in N$, a submodule of M, the calculation of $\mu(v)$ in N or M gives the same result. We have $S(\mathcal{D}v) \subset S(M)$ so $\mu(v) \in S(M)$ and $\mu(v) \geq \mu(M)$. If $P(\Phi)$ is the minimal polynomial of v (defined in 8.2.1), then $\mathcal{D}v \simeq \mathcal{D}/\mathcal{D}P$ so $\mu(v)$ is the leftmost slope of the Newton polygon of P.

COROLLARY 10.43. (i) $\mu(v) \geq \mu_i \Longleftrightarrow v \in M_i$.
(i') $\mu(v) \geq \mu \Longleftrightarrow v \in M^{\geq\mu}$.
(ii) $\mu(v) = \mu_i \Longleftrightarrow v \in M_i \setminus M_{i-1}$.
(iii) $\mu(v) = \mu \Longleftrightarrow v \in M^{\geq\mu} \setminus M^{>\mu}$.
(iv) *If v is a cyclic vector, $\mu(v) = \mu_k$.*
(v) *Over the ramified extension $K' \otimes_K M$ of degree r, $\mu(1 \otimes v) = r\mu(v)$.*
(vi) *Over the formalization $\hat{K} \otimes_K M$, $\mu(1 \otimes v) = \mu(v)$.*
(vii) *If $f : M \to N$ is a morphism, $\mu(f(v)) \geq \mu(v)$.*

PROOF. Only the last one is not an immediate consequence of the proposition. It comes from the fact that $\mathcal{D}f(v)$ is a quotient of $\mathcal{D}v$, so $S(\mathcal{D}f(v)) \subset S(\mathcal{D}v)$. □

10.2.2.4. The properties of $\mu(v)$ allow us to introduce a new useful tool to characterize the slope filtration (we shall use it in 10.2.3).

DEFINITION 10.44. Let $M = (V, \Phi)$ a q-difference module, and, taking again the "usual notations", let V_i the underlying space to M_i. A basis $\mathcal{B}$ of V is *adapted (to the canonical slope filtration)* if it can be partitioned into $\mathcal{B}_1, \ldots, \mathcal{B}_k$ such that each $\mathcal{B}_1 \cup \cdots \cup \mathcal{B}_i$, $i = 1, \ldots, k$, is a basis of V_i. (So it is adapted in the sense of standard linear algebra to the flag $\{0\} = V_0 \subset V_1 \subset \cdots \subset V_k = V$.)

The elements of $\mathcal{B}$ can then be bi-indexed as $v_{i,j}$, $i = 1, \ldots, k$, $j = 1, \ldots, r_i$ in such a way that each $v_{i,j} \in M_i$, i.e., $\mu(v_{i,j}) \geq \mu_i$. The converse is true:

LEMMA 10.45. *Let $(v_{i,j})_{\substack{i=1,\ldots,k \\ j=1,\ldots,r_i}}$ a basis of V such that all $\mu(v_{i,j}) \geq \mu_i$. Then it is an adapted basis.*

PROOF. The space spanned by all $v_{i,j}$ with $i \geq i_0$ has dimension $r_1 + \cdots + r_{i_0}$ and is included in M_{i_0} (or more properly its underlying space V_{i_0}, but let's not insist on it) by 10.2.2.3, so it is M_{i_0}. □

EXERCISE 10.46. Deduce from the adapted basis $\mathcal{B}$ a basis for each of the $M^{(\mu_i)}$.

10.2.2.5. If M_0 is pure, $M_0 = P_1 \oplus \cdots \oplus P_k$, each P_i is $M_0(\mu_i)$ so the decomposition is unique and canonical.

If the filtration on M splits (here in the sense that all the short exact sequences are split, not in the sense that the factors have rank 1), then $M_i = M_i' \oplus M_{i-1}$, $i = 1, \ldots, k$, then M_i' is the P_i above so the splitting is unique and canonical.

10.2.3. Tensor properties. We now look at the behaviour of tensor products with respect to the canonical filtrations.

10.2.3.1.

THEOREM 10.3. *Let M_1, M_2 two q-difference modules and let $M := M_1 \otimes M_2$. The slope filtration on M is described by the formulas:*

$$M^{\geq \mu} = \sum_{\mu_1 + \mu_2 = \mu} M_1^{\geq \mu_1} \otimes M_2^{\geq \mu_2},$$

$$M^{> \mu} = \sum_{\mu_1 + \mu_2 = \mu} \left(M_1^{> \mu_1} \otimes M_2^{\geq \mu_2} + M_1^{\geq \mu_1} \otimes M_2^{> \mu_2} \right).$$

PROOF. Note that, if $N_1 \subset M_1$ and $N_2 \subset M_2$ are submodules, then $N_1 \otimes N_2$ canonically embeds into $M_1 \otimes M_2$ (this follows from exactness of the tensor product; see 9.2.1). So it makes sense to define the submodule $N^{\geq \mu} := \sum\limits_{\mu_1 + \mu_2 = \mu} M_1^{\geq \mu_1} \otimes M_2^{\geq \mu_2}$ of M. Clearly, $(N^{\geq \mu})$ is a descending filtration on M. Also, from the rules on the Newton polygon $S(\sum A_i) = \bigcup S(A_i)$ (in 10.1.4.1) and $S(A \otimes B) = S(A) + S(B)$ (in 10.1.4.3), $N^{\geq \mu} \subset M^{\geq \mu}$.

Now let $\mu_1 + \mu_2 = \mu$ and $\mu_1' > \mu_1$, so that $\mu_1' + \mu_2 > \mu$: we deduce that $M_1^{\geq \mu_1'} \otimes M_2^{\geq \mu_2} \subset N^{> \mu}$; taking all $\mu_1' > \mu_1$ and also using symmetrically the $\mu_2' > \mu_2$, we conclude:

$$\sum_{\mu_1 + \mu_2 = \mu} \left(M_1^{> \mu_1} \otimes M_2^{\geq \mu_2} + M_1^{\geq \mu_1} \otimes M_2^{> \mu_2} \right) \subset N^{> \mu}.$$

From that we derive surjective morphisms of q-difference modules:

$$\bigoplus_{\mu_1+\mu_2=\mu} \frac{M_1^{\geq\mu_1} \otimes M_2^{\geq\mu_2}}{M_1^{>\mu_1} \otimes M_2^{\geq\mu_2} + M_1^{\geq\mu_1} \otimes M_2^{>\mu_2}}$$

$$\longrightarrow \frac{\sum_{\mu_1+\mu_2=\mu} M_1^{\geq\mu_1} \otimes M_2^{\geq\mu_2}}{\sum_{\mu_1+\mu_2=\mu} \left(M_1^{>\mu_1} \otimes M_2^{\geq\mu_2} + M_1^{\geq\mu_1} \otimes M_2^{>\mu_2}\right)} \longrightarrow \frac{N^{\geq\mu}}{N^{>\mu}}.$$

The first one comes from the general fact from linear algebra that there is a surjective map $\bigoplus A_i \to \sum A_i$ inducing $\bigoplus A'_i \to \sum A'_i$ if all the $A'_i \subset A_i$. Now, another general fact, if $A' \subset A$ and $B' \subset B$, is that the linear map $A \otimes B \to (A/A') \otimes (B/B')$ is surjective with kernel $A \otimes B' + A' \otimes B$, whence an isomorphism $(A/A') \otimes (B/B') \simeq A \otimes B/(A \otimes B' + A' \otimes B)$. It follows that each of the above $\dfrac{M_1^{\geq\mu_1} \otimes M_2^{\geq\mu_2}}{M_1^{>\mu_1} \otimes M_2^{\geq\mu_2} + M_1^{\geq\mu_1} \otimes M_2^{>\mu_2}} \simeq M_1^{(\mu_1)} \otimes M_2^{(\mu_2)}$ is pure of slope μ, and, by the surjective morphisms above, so is $\dfrac{N^{\geq\mu}}{N^{>\mu}}$. The filtration $(N^{\geq\mu})$ having pure breaks $N^{(\mu)}$ of slope μ is therefore the canonical slope filtration on M, whence the first contention.

Now, comparison of ranks entails that all the above surjective morphisms are actually bijective, whence the second contention. □

COROLLARY 10.47. $M^{(\mu)} = \bigoplus\limits_{\mu_1+\mu_2=\mu} M_1^{(\mu_1)} \otimes M_2^{(\mu_2)}$.

10.2.3.2. We sketch another possible approach to the above theorem. First note that, if $v \in M$ and $w \in N$, then $\mathcal{D}(v \otimes w) \subset \mathcal{D}v \otimes \mathcal{D}w$.

EXERCISE 10.48. Show that the inclusion may be strict. (Take v, w with respective minimal polynomials $\Phi^2 - a$, $\Phi^2 - b$.)

Since generally speaking:

$$\begin{aligned}\mu(M' \otimes N') &= \min S(M' \otimes N') = \min\left(S(M') + S(N')\right)\\ &= \min S(M') + \min S(N') = \mu(M') + \mu(N'),\end{aligned}$$

we have $\mu(v \otimes w) \geq \mu(v) + \mu(w)$.

Now let M, N two q-difference modules with $S(M) = \{\mu_1, \ldots, \mu_k\}$ and $S(N) = \{\nu_1, \ldots, \nu_l\}$ (in decreasing order) and with respective adapted bases $(v_{i,j})_{\substack{i=1,\ldots,k\\ j=1,\ldots,r_i}}$ and $(w_{m,n})_{\substack{m=1,\ldots,l\\ n=1,\ldots,s_n}}$. From:

$$\mu(v_{i,j}) \geq \mu_i \text{ and } \mu(w_{m,n}) \geq \nu_m \Longrightarrow \mu(v_{i,j} \otimes w_{m,n}) \geq \mu_i + \nu_m,$$

it is easy to deduce that the $v_{i,j} \otimes w_{m,n}$ form an adapted basis for $M \otimes N$.

EXERCISE 10.49. (i) Prove it rigorously and deduce from it the theorem.
(ii) Also deduce that, generally speaking, $\mu(v \otimes w) = \mu(v) + \mu(w)$.

10.3. Graduation and formalization

We now describe the (very nice) properties of the functor $\mathrm{gr} : M \rightsquigarrow \bigoplus M^{(\mu)}$. A special feature of q-difference theory is that the difference between filtration and graduation (or, said otherwise, the extent to which the filtration fails to split) reflects rather faithfully the difference between the formal and the analytic category.

10.3.1. Graduation. In this subsection, $K = \mathbf{C}(\{x\})$ or $K = \mathbf{C}((x))$. We first introduce some functorial notations. Then we translate into that language results from Section 10.1 and Section 10.2.

10.3.1.1. Let M a q-difference module over K. We write:

$$F^{\geq\mu}M := M^{\geq\mu}, \quad F^{>\mu}M := M^{>\mu} \text{ and } \mathrm{gr}^{(\mu)}M := \frac{F^{\geq\mu}M}{F^{>\mu}M} = \frac{M^{\geq\mu}}{M^{>\mu}} = M^{(\mu)}.$$

Since any morphism $f : M \to N$ sends $M^{\geq\mu}$ to $N^{\geq\mu}$ and $M^{>\mu}$ to $N^{>\mu}$ and therefore induces $M^{(\mu)} \to N^{(\mu)}$, we see that, for every $\mu \in \mathbf{Q}$, $F^{\geq\mu}$, $F^{>\mu}$ and $\mathrm{gr}^{(\mu)}$ are functors. The formal properties of functors with respect to identities and composition are immediately checked, as well as the fact that all these functors are $\mathbf{C}$-linear and additive.

Let $\mathscr{C}$ the category of q-difference modules over K and let $\mathscr{C}_{p,\mu}$ the full subcategory of pure isoclinic submodules of slope μ; then $\mathscr{C}_{p,\mu}$ is thick (see 10.1.2.1 and 10.1.4.2). On the other hand, the category $\mathscr{C}_p$ of pure q-difference modules (i.e., direct sums of pure isoclinic q-difference modules, but the slope being allowed to vary) has no reason to be thick. It is indeed if $K = \mathbf{C}((x))$ but it is not if $K = \mathbf{C}(\{x\})$, as Example 10.35 plainly shows: it displays a nonpure extension of two pure q-difference modules.

The functors $F^{\geq\mu}$, $F^{>\mu}$ and $\mathrm{gr}^{(\mu)}$ are endofunctors of $\mathscr{C}$ (i.e., functors from $\mathscr{C}$ to itself) but we may as well see $\mathrm{gr}^{(\mu)}$ as a functor from $\mathscr{C}$ to $\mathscr{C}_{p,\mu}$. By Corollary 10.41, it is exact.

10.3.1.2.

DEFINITION 10.50. The *associated graded module* of the q-difference module M is $\mathrm{gr}M := \bigoplus_{\mu\in\mathbf{Q}} M^{(\mu)}$.

Accordingly, we have a *graduation functor* $\mathrm{gr} := \bigoplus_{\mu\in\mathbf{Q}} \mathrm{gr}^{(\mu)}$ from $\mathscr{C}$ to $\mathscr{C}_p$, its full subcategory of pure modules (these are defined in 10.2.1.4).

THEOREM 10.4. *The functor gr is* $\mathbf{C}$*-linear, additive, exact, faithful and* $\otimes$*-compatible. It is a retraction of the inclusion* $\mathscr{C}_p \to \mathscr{C}$*. (The last two statements are explained in the proof.)*

PROOF. As a direct sum of the $\mathbf{C}$-linear, additive, exact functors $\mathrm{gr}^{(\mu)}$, it is exact. It is a retraction of the inclusion $\mathscr{C}_p \to \mathscr{C}$ because, starting from a pure module M, the filtration is split and one has an identity (or canonical isomorphism) $\mathrm{gr}M = M$: in the usual notations, $M_i = N_1 \oplus \cdots \oplus N_i$, where $N_i := M^{(\mu_i)}$ so M_i/M_{i-1} canonically identifies to N_i.

Faithfulness means that, if $f : M \to N$ is a morphism, then $\mathrm{gr}f = 0 \Rightarrow f = 0$ (so by $\mathbf{C}$-linearity the map $f \mapsto \mathrm{gr}f$ is injective). Indeed, the premise means that $f(M^{\geq\mu}) \subset N^{>\mu}$ for all μ. Since $f(M^{\geq\mu}) = f(M) \cap N^{\geq\mu}$ by 10.2.2.2, this implies that $f(M) \cap N^{\geq\mu} = f(M) \cap N^{>\mu}$ for all μ. From that, using $\bigcup N^{\geq\mu} = N$ and $\bigcap N^{\geq\mu} = \{0\}$, it follows easily that $f(M) = \{0\}$.

Tensor compatibility, abbreviated $\otimes$-compatibility, means that there is a natural (and functorial) isomorphism $\mathrm{gr}(M \otimes N) = (\mathrm{gr}M) \otimes (\mathrm{gr}N)$. That natural isomorphism was described and proved in 10.2.3. The fact that it is functorial (understand: in each argument M, N) is easy and tedious to prove, the reader may look for a complete argument in [**213**, IV.2.1] (the proof there rests on assumptions on the filtration that are verified here). □

10.3.2. Formalization. We now write $\mathscr{C}$ and $\mathscr{C}_p$ for the categories of q-difference modules over $\mathbf{C}(\{x\})$ and $\hat{\mathscr{C}}$ and $\hat{\mathscr{C}}_p$ the corresponding categories over $\mathbf{C}((x))$. We shall likewise abbreviate $\mathcal{D} := \mathcal{D}_{\mathbf{C}(\{x\}),q}$ and $\hat{\mathcal{D}} := \mathcal{D}_{\mathbf{C}((x)),q}$.

10.3.2.1. We know by 10.2.1.1 and 10.2.1.4 that all q-difference modules in $\hat{\mathscr{C}}$ are pure and the canonical filtration splits, so $\hat{\mathscr{C}} = \hat{\mathscr{C}}_p$.

COROLLARY 10.51. *Over $\mathbf{C}((x))$, the functor gr is canonically isomorphic to the identity functor.*

The explanation is essentially that, since $\hat{\mathscr{C}} = \hat{\mathscr{C}}_p$, the only retraction of $\hat{\mathscr{C}}_p \to \hat{\mathscr{C}}$ (as in 10.3.1.2) is the identity functor.

10.3.2.2. Now let M a q-difference module over $\mathbf{C}(\{x\})$ and $\hat{M} := \mathbf{C}((x)) \otimes_{\mathbf{C}(\{x\})} M$ its formalization. We adapt some previous notations by setting

$$M_A := (\mathbf{C}(\{x\})^n, \Phi_A),$$

resp. $\hat{M}_A := (\mathbf{C}((x))^n, \Phi_A)$ if $A \in \mathrm{GL}_n(\mathbf{C}(\{x\}))$, resp. $A \in \mathrm{GL}_n(\mathbf{C}((x)))$. Likewise, we set $M_P := \mathcal{D}/\mathcal{D}P$, resp. $\hat{M}_P := \hat{\mathcal{D}}/\hat{\mathcal{D}}P$ if $P \in \mathcal{D}$, resp. if $P \in \hat{\mathcal{D}}$. We then know that:

$$\widehat{M_A} = \hat{M}_A \text{ and } \widehat{M_P} = \hat{M}_P.$$

PROPOSITION 10.52. (i) *$r_M = r_{\hat{M}}$ and $S(M) = S(\hat{M})$.*
(ii) *$\widehat{M^{\geq \mu}} = \hat{M}^{\geq \mu}$ and $\widehat{M^{>\mu}} = \hat{M}^{>\mu}$.*
(iii) *There are canonical isomorphisms $\widehat{M^{(\mu)}} = \hat{M}^{(\mu)}$ and $\widehat{grM} = gr\hat{M}$.*

PROOF. (i) Immediately follows from $\widehat{M_P} = \hat{M}_P$.
(ii) We explain the first statement, the second works the same way. Applying the formalization functor to $M^{\geq \mu} \to M$ defines a canonical injective morphism $\widehat{M^{\geq \mu}} \to \hat{M}$: using Newton functions, we see that its image is $\hat{M}^{\geq \mu}$.
(iii) Is an immediate consequence of (ii).

□

10.3.2.3.

THEOREM 10.5. *Let M and N analytic q-difference modules (i.e., objects of $\mathscr{C}$). Then:*

$$\hat{M} \simeq \hat{N} \Longleftrightarrow grM \simeq grN.$$

PROOF. Since $\widehat{\mathrm{gr}M} = \mathrm{gr}\hat{M} = \hat{M}$, we are reduced to prove that, if M and N are pure analytic q-difference modules, $\hat{M} \simeq \hat{N} \Rightarrow M \simeq N$ (the converse implication being obvious). Thanks to the rules on Newton polygons, it is enough to do that for pure isoclinic q-difference modules of the same slope μ. Taking M, N in the form M_A, M_B, this follows from the stronger statement in the lemma below. □

LEMMA 10.53. *If $A, B \in GL_n(\mathbf{C}(\{x\}))$ are pure isoclinic of slope μ and if $F \in GL_n(\mathbf{C}((x)))$ is such that $F[A] = B$, then $F \in GL_n(\mathbf{C}(\{x\}))$.*

That will be a consequence of our study of fuchsian systems in the next chapter; see in particular 11.1.2. The statement is stronger than what was required because, from the existence of such a $F \in \mathrm{GL}_n(\mathbf{C}((x)))$, we just needed to deduce the existence of a $G \in \mathrm{GL}_n(\mathbf{C}(\{x\}))$ such that $G[A] = B$; and we actually got $G = F$.

EXERCISE 10.54. Using the internal Hom show that looking for F such that $F[A] = B$ amounts to solving a fuchsian equation, so that the principle used really is: any formal solution of a fuchsian equation converges.

10.3.3. Matricial interpretation and application to classification. With the formal and analytic classification in view, we discuss the matricial interpretation of our results. To begin with, we do not state precisely whether we speak of analytic or formal q-difference modules, and the same about analytic or formal isomorphisms: so K is either $\mathbf{C}(\{x\})$ or $\mathbf{C}((x))$. From now on, our favorite model is the $M_A = (K^n, \Phi_A)$, where we suppose moreover that A has the block upper triangular form A_U described in 10.2.1.3; and A_0 denotes its block diagonal part.

10.3.3.1. So let $M := (K^n, \Phi_A)$ and $N := (K^n, \Phi_B)$, both in block upper triangular form. We discuss isomorphisms from M to N. This requires of course that $\mathrm{gr}M \simeq \mathrm{gr}N$, i.e., that the block diagonal matrices A_0, B_0 be isomorphic. Necessary conditions are that $n = p$ and that the Newton functions be the same, so that in particular the diagonal blocks have the same sizes $r_1, \ldots, r_k$.

The strict compatibility of morphisms with the canonical filtrations implies that any isomorphism $f : M \to N$ is represented by a matrix $F \in \mathrm{GL}_n(K)$ in block upper triangular form and $\mathrm{gr}f$ by its block diagonal part F_0:

$$F = \begin{pmatrix} F_{1,1} & F_{1,2} & \ldots & F_{1,k-1} & F_{1,k} \\ 0 & F_{2,2} & \ldots & F_{2,k-1} & F_{2,k} \\ \vdots & \vdots & \ddots & \vdots & \vdots \\ 0 & 0 & \ldots & F_{k-1,k-1} & F_{k-1,k} \\ 0 & 0 & \ldots & 0 & F_{k,k} \end{pmatrix}, \quad F_0 = \begin{pmatrix} F_{1,1} & 0 & \ldots & 0 & 0 \\ 0 & F_{2,2} & \ldots & 0 & 0 \\ \vdots & \vdots & \ddots & \vdots & \vdots \\ 0 & 0 & \ldots & F_{k-1,k-1} & 0 \\ 0 & 0 & \ldots & 0 & F_{k,k} \end{pmatrix},$$

So the relation $F[A] = B$ implies the relation $F_0[A_0] = B_0$. Let us detail them. For that, we set $A := A_U$ (as in 10.2.1.3) and B similar with B_i in place of A_i and $V_{i,j}$ in place of $U_{i,j}$.

- The relation $F_0[A_0] = B_0$ translates into the system:

$$\forall i = 1, \ldots, k \;,\; (\sigma_q F_i) A_i = B_i F_{i,i}.$$

- The relation $F_0[A_0] = B_0$ contains the previous one, and moreover:

$$\forall i, j \;:\; 1 \le i < j \le k \;,\; (\sigma_q F_{i,j}) A_j + \sum_{i<l\le j} (\sigma_q F_{i,l}) U_{l,j} = B_i F_{i,j} + \sum_{i\le l<j} V_{i,l} F_{l,j}.$$

We consider the above as systems with unknowns the $F_{i,j}$, $i \le j$. Then:

- The faithfulness of the functor gr, i.e., the injectivity of $f \mapsto \mathrm{gr}f$, says that, whatever the field K (formal or analytic), solutions $F_{i,i}$ to the first (diagonal) system being given, solutions $F_{i,j}$ to the second (strictly upper diagonal) system are unique if they exist.
- The fact that in the formal category $\hat{\mathcal{C}}$ the map $f \mapsto \mathrm{gr}f$ is bijective says that, formal solutions $F_{i,i}$ to the first system being given, formal solutions $F_{i,j}$ to the second system exist.

Note that, in some sense, the second system is triangular. Indeed, if $1 \le i < j \le k$, we must solve for $F_{i,j}$ the equation:

$$(\sigma_q F_{i,j}) A_j - B_i F_{i,j} = \sum_{i\le l<j} V_{i,l} F_{l,j} - \sum_{i<l\le j} (\sigma_q F_{i,l}) U_{l,j}.$$

But on the right-hand side there are $F_{i',j'}$ with $0 < j' - i' < j - i$. So we can start from the lowest values of $j - i$ (the very lowest ones have no $F_{i',j'}$ at all on the right-hand side) and go "up level".

Actually, we will find in Chapter 15 that it is more natural to define the "level" of the block $F_{i,j}$ as $\mu_i - \mu_j > 0$. Again we find that the corresponding right-hand side has only strictly lower levels $\mu_i - \mu_l$ with $i \leq l < j$ and $\mu_l - \mu_j$ with $i < l \leq j$.

EXERCISE 10.55. It is more complicated to describe the triangular form for arbitrary morphisms between q-difference modules which do not necessarily have the same Newton polygon: the description is somewhat awkward. Try it.

10.3.3.2. We now suppose that M and N are analytic q-difference modules such that $\hat{M} \simeq \hat{N}$ and we seek criteria for analytic isomorphy $M \simeq N$. Since the Newton polygon is a formal invariant, we may put M and N in the form $M_A = (\mathbf{C}(\{x\})^n, \Phi_A)$, $M_B = (\mathbf{C}(\{x\})^n, \Phi_B)$ with A and B in the form described in 10.3.3.1. We therefore have $\hat{F} \in \mathrm{GL}_n(\mathbf{C}((x)))$ in the form above such that $\hat{F}[A] = B$, whence $\hat{F}_0[A_0] = B_0$ and $\hat{F}_0 \in \mathrm{GL}_n(\mathbf{C}(\{x\}))$. So we rather write it F_0.

Note that $A' := F_0[A]$ defines an analytic q-difference module $M_{A'} \simeq M_A$ (analytic isomorphism), so it is another possible matricial representation for M (again in upper block diagonal form). Moreover, $A'_0 = F_0[A_0] = B_0$. If we set $\hat{F}' := \hat{F}F_0^{-1}$, we see that $\hat{F}'[A'] = B$ and that $\hat{F}'_0 = I_n$. So, replacing if necessary A by A', we may assume from the beginning that $A_0 = B_0$ and that $\hat{F}_0 = I_n$.

So we have $A, B \in \mathrm{GL}_n(\mathbf{C}(\{x\}))$ in the upper block diagonal form prescribed above and such that $A_0 = B_0$ and $\hat{F} \in \mathrm{GL}_n(\mathbf{C}((x)))$ such that $\hat{F}[A] = B$ and $\hat{F}_0 = I_n$. And we want to know if there exists $F \in \mathrm{GL}_n(\mathbf{C}(\{x\}))$ such that $F[A] = B$. Its block diagonal part F_0 is a priori arbitrary, but with the same trick as above (replacing A by $F_0[A]$ while taking in account that $F_0[A_0] = B_0$) we find that we may suppose that $F_0 = I_n$. But then $\hat{F}$ and F have the same diagonal part, so, as deduced in 10.3.3.1 from the faitfulness of the functor gr, we must actually have $\hat{F} = F$. So the question has been transformed into the following:

Problem: Let $A, B \in \mathrm{GL}_n(\mathbf{C}(\{x\}))$ in the prescribed upper block diagonal form and such that $A_0 = B_0$ (so really we can write them A_U, A_V with the notation from 10.2.1.3). Let $\hat{F} \in \mathrm{GL}_n(\mathbf{C}((x)))$ such that $\hat{F}[A] = B$ and $\hat{F}_0 = I_n$. That matrix exists and is unique. Give criteria on A, B to ensure $\hat{F} \in \mathrm{GL}_n(\mathbf{C}(\{x\}))$.

In this way the problem of *isoformal analytic classification*, that is recognizing analytic equivalence classes within a formal equivalence class, has been replaced (at least in formulation) by a more algebraic problem: what is fixed instead of the formal class is the associated graded q-difference module.

10.3.3.3. We now go backwards and give that algebraic problem a less matricial and more intrinsic form. We do that within the category $\mathscr{C}$ of analytic q-difference modules. Instead of fixing a formal class (that of $\hat{M}$ or $\hat{N}$ as in 10.3.3.2) we fix a pure q-difference module M_0. We are going to define a classification problem giving rise to a set of "isograded classes" $\mathscr{F}(M_0)$ and then verify that its matricial interpretation is exactly the problem stated at the end of 10.3.3.2.

So let M_0 a pure analytic q-difference module,i.e., an object of $\mathscr{C}_p$. We define $\mathscr{F}(M_0)$ as follows. First we introduce the "class"[9] $\mathscr{E}(M_0)$ (this is a temporary

[9]This is not a set, it is "too big" for that. The quotient $\mathscr{F}(M_0)$ by the equivalence relation $\sim$ will be a *bona fide* set. We shall admit it, however the concrete desciption in terms of matrices obviously yields a set.

notation, valid until the end of the chapter only) of all pairs (M, u), where M is an analytic q-difference module and $u : \mathrm{gr}M \to M_0$ is an isomorphism. We say that $(M, u) \sim (N, v)$ if there exists a morphism $f : M \to N$ such that $u = v \circ \mathrm{gr}f$, i.e., the diagram

$$\begin{array}{ccc} \mathrm{gr}M & \xrightarrow{\mathrm{gr}f} & \mathrm{gr}N \\ & {\scriptstyle u}\searrow \quad \swarrow{\scriptstyle v} & \\ & M_0 & \end{array}$$

commutes.

EXERCISE 10.56. Show that f is automatically an isomorphism. (This is a general property of filtrations and graduations; see [**34**, III.2.4] or [**54**, 1.1].)

We clearly have an equivalence relation, and we define:

$$\mathscr{F}(M_0) := \frac{\mathscr{E}(M_0)}{\sim}.$$

In view of 10.3.4 (also see 10.3.5), it will appear that the equivalence relation used here, which is more rigid than plain analytic isomorphy, is a generalisation of the equivalence of extensions as defined in 8.3.2 (compare in particular 10.3.5.4 to 8.3.2.4).

10.3.4. The "linear" or "additive" case $k = 2$. The reason to call it "linear" or "additive" will appear clearly in Chapter 15. Here, we take $k := 2$ and $M_0 := P_1 \oplus P_2$, where P_i is pure isoclinic of slope μ_i, $i = 1, 2$ and we assume that $\mu_1 > \mu_2$. We shall go in great detail so that the usually implied tedious checking have been completely written down at least once!

10.3.4.1. *Filtered modules with two slopes.* A filtered module is encoded as a pair (M, M_1) where $M_1 \subset M$; we write $i : M_1 \to M$ and $s : M \to M/M_1$ the natural injection and surjection. Since $\mathrm{gr}M = M_1 \oplus (M/M_1)$, an element of $\mathscr{E}(M_0)$ is defined by $u_1 : M_1 \simeq P_1$ and $u_2 : M/M_1 \simeq P_2$; we describe it as (M, M_1, u_1, u_2).

A morphism $f : M \to N$ of filtered modules can be encoded as $(f, f_1, f_2) : (M, M_1, i, p) \to (N, N_1, , j, t)$, where $f_1 : M_1 \to N_1$ and $f_2 : M/M_1 \to N/N_1$ are such that $\begin{cases} (F_1)\ f \circ i = j \circ f_1, \\ (F_2)\ t \circ f = f_2 \circ s. \end{cases}$ (Strictness is not taken in account here.) We say that (f, f_1, f_2) realizes an equivalence of (M, M_1, u_1, u_2) with (N, N_1, v_1, v_2) if $(G_i) v_i \circ f_i = u_i$, $i = 1, 2$. So if we write (temporarily, just in 10.3.4) $F(M_0)$ the class of all such (M, M_1, u_1, u_2) and $\sim$ the above equivalence, we find that $\mathscr{F}(M_0) = F(M_0)/\sim$. Here is the corresponding (pretty) diagram:

$$\begin{array}{ccccccccc} 0 & \longrightarrow & N_1 & \longrightarrow & N & \longrightarrow & N/N_1 & \longrightarrow & 0 \\ & & \nearrow\ \downarrow & & \nearrow\ \downarrow & & \nearrow\ \downarrow & & \\ 0 & \longrightarrow M_1 & \longrightarrow & M & \longrightarrow & M/M_1 & \longrightarrow & 0 & \\ & & \searrow\ \downarrow & & \searrow\ \downarrow & & \searrow\ \downarrow & & \\ 0 & \longrightarrow & P_1 & \longrightarrow & M_0 & \longrightarrow & P_2 & \longrightarrow & 0 \end{array}$$

10.3.4.2. *Relation to extensions.* We consider the class $E(P_2, P_1)$ of all extensions $(M, \alpha_1, \alpha_2) : 0 \to P_1 \overset{\alpha_1}{\to} M \overset{\alpha_2}{\to} P_2 \to 0$; and morphisms of such extensions, i.e., maps $f : M \to N$ such that (with obvious notations) $\begin{cases} (E_1) : \beta_1 = f \circ \alpha_1, \\ (E_2) : \alpha_2 = \beta_2 \circ f \end{cases}$. So, as seen in 8.3.2, $\mathrm{Ext}(P_2, P_1) = E(P_2, P_1)/\sim$, where $\sim$ is the equivalence relation induced by such morphisms.

THEOREM 10.6. *There is a natural bijection $\mathscr{F}(M_0) \simeq Ext(P_2, P_1)$. The equivalence class of split filtered modules corresponds to the equivalence class of trivial extensions.*

PROOF. We first define $F(M_0) \to E(P_2, P_1)$, then prove its properties. To the element (M, M_1, u_1, u_2) of $F(M_0)$, it associates the element (M, α_1, α_2) of $E(P_2, P_1)$, where $\alpha_1 := i \circ u_1^{-1}$ and $\alpha_2 := u \circ s$, i.e., such that the following diagram commutes:

$$\begin{array}{ccccccccc} 0 & \longrightarrow & M_1 & \xrightarrow[i]{} & M & \xrightarrow[s]{} & M/M_1 & \longrightarrow & 0 \\ & & \Big\downarrow{\scriptstyle u_1} & & \Big\downarrow{\scriptstyle \mathrm{Id}} & & \Big\downarrow{\scriptstyle u_2} & & \\ 0 & \longrightarrow & P_1 & \xrightarrow[\alpha_1]{} & M_0 & \xrightarrow[\alpha_2]{} & P_2 & \longrightarrow & 0. \end{array}$$

Now the lower horizontal line is exact, since it is isomorphic to the upper exact sequence, so it is indeed an extension.

Step 1: existence, injectivity. Let as above $(M, M_1, u_1, u_2) \mapsto (M, \alpha_1, \alpha_2)$ and likewise $(N, N_1, v_1, v_2) \mapsto (N, \beta_1, \beta_2)$. We shall prove that:

$$(M, M_1, u_1, u_2) \sim (N, N_1, v_1, v_2) \Longleftrightarrow (M, \alpha_1, \alpha_2) \sim (N, \beta_1, \beta_2)$$

(the former equivalence in $F(M_0)$, the latter in $E(P_2, P_1)$). This will imply that our map defined above induces (while going to the quotients) an injection $\mathscr{F}(M_0) \to \mathrm{Ext}(P_2, P_1)$.

So we prove our contention. This boils down to taking a morphism $f : M \to N$ and checking that $(G_1, G_2) \Leftrightarrow (E_1, E_2)$; this will use (F_1, F_2):

$$\begin{aligned} (E_1) : \beta_1 = f \circ \alpha_1 \Leftrightarrow j \circ v_1^{-1} = f \circ i \circ u_1^{-1} &\overset{(F_1)}{\Longleftrightarrow} j \circ v_1^{-1} = j \circ f \circ u_1^{-1} \\ &\overset{j \text{ inj.}}{\Longleftrightarrow} v_1^{-1} = f \circ u_1^{-1} \Longleftrightarrow (G_1), \\ (E_2) : \alpha_2 = \beta_2 \circ f \Leftrightarrow u_2 \circ s = v_2 \circ t \circ f &\overset{(F_2)}{\Longleftrightarrow} u_2 \circ s = v_2 \circ f \circ s \\ &\overset{s \text{ onto}}{\Longleftrightarrow} u_2 = v_2 \circ f \Longleftrightarrow (G_2). \end{aligned}$$

Step 2: surjectivity. Now let (M, α_1, α_2) in $E(P_2, P_1)$ and set $M_1 := \alpha_1(P_1)$. Then α_1 is an isromorphism $P_1 \simeq M_1 \subset M$, so let $u_1 : M_1 \simeq P_1$ the inverse isomorphism. The sequence $0 \to P_1 \to M \to P_2 \to 0$ being exact, α_2 induces an isomorphism $u_2 : M/\alpha_1(P_1) = M/M_1 \simeq P_2$. So we have an element (M, M_1, u_1, u_2) of $F(M_0)$. One checks easily (usual notations) that $i \circ u_1^{-1} = \alpha_1$ and $u_2 \circ s = \alpha_2$, i.e., the image of (M, M_1, u_1, u_2) in $E(P_2, P_1)$ is (M, α_1, α_2). Therefore our map above is surjective, and so is the quotient map $\mathscr{F}(M_0) \to \mathrm{Ext}(P_2, P_1)$.

Step 3: split filtrations correspond to split extensions. To say that the filtration of (M, M_1, i, s) splits is to say that there is a submodule $M_2' \subset M$ such that $M = M_1 \oplus M_2'$. (Note that, by the basic rules of Newton polygons, M_2' is automatically pure isoclinic with slope μ_2 and rank r_2.) From the standard properties of direct sums of modules in linear algebra, this is equivalent to the existence of a retraction: a morphism $r : M \to M_1$ such that $r \circ i = \mathrm{Id}_{M_1}$.

To say that the extension (M, α_1, α_2) splits is likewise equivalent to the existence of $\alpha_1' : M \to P_1$ such that $\alpha_1' \circ \alpha_1 = \mathrm{Id}_{P_1}$.

So if we have r, we set $\alpha_1' := u_1 \circ r$; and, conversely, if we have α_1', we set $r := u_1^{-1} \circ \alpha_1'$. Then:

$$\alpha_1' \circ \alpha_1 = \mathrm{Id}_{P_1} \Longrightarrow r \circ i = u_1^{-1} \circ \alpha_1' \circ \alpha_1 \circ u_1 = \mathrm{Id}_{M_1},$$

and conversely:

$$r \circ i = \mathrm{Id}_{M_1} \Longrightarrow \alpha_1' \circ \alpha_1 = u_1 \circ r \circ i \circ u_1^{-1} = \mathrm{Id}_{P_1}.$$

So, if (M, M_1, i, s) and (M, α_1, α_2) correspond to each other, the former splits if, and only if, the latter does. □

10.3.4.3. *More functorially.* We complete the above description of the "linear" case by a more functorial point of view. This is motivated by the general principle that classification problems, usually solved in the form of bijections, e.g. $F(M_0)/\sim\ \simeq E(P_2, P_1)/\sim$, are often best understood in terms of isomorphism classes in some category. So we introduce a category $\mathfrak{F}$ (this letter is a gothic F) with objects the triples (M, u, M_0), M in $\mathscr{C}$, M_0 in $\mathscr{C}_p$, $u : \mathrm{gr}M \simeq M_0$; and with morphisms from (M, u, M_0) to (N, v, N_0) the pairs (f, f_0), where $f : M \to N$ is a morphism in $\mathscr{C}$ and $f_0 : M_0 \to N_0$ is a morphism in $\mathscr{C}_p$ such that $v \circ \mathrm{gr} f = f_0 \circ u$, i.e., the following diagram commutes:

$$\begin{array}{ccc} \mathrm{gr}M & \xrightarrow{\ \mathrm{gr}f\ } & \mathrm{gr}N \\ {\scriptstyle u}\downarrow & & \downarrow{\scriptstyle v} \\ M_0 & \xrightarrow{\ f_0\ } & N_0. \end{array}$$

This is an example of a so-called "comma-category"; see **[154]**. So you notice that M_0 here is not fixed.

EXERCISE 10.57. Fix M_0 and define the category $\mathfrak{F}_{M_0}$ really of interest for us.

On the other hand, we fix P_1, P_2 with the usual assumptions (but see the next exercise) and define the category $\mathfrak{E}_{P_1,P_2}$ (a gothic E) with objects the extensions (M, α_1, α_2) of P_2 by P_1 and morphisms $(M, \alpha_1, \alpha_2) \to (N, \beta_1, \beta_2)$ the morphisms $f : M \to N$ such that $f \circ \alpha_1 = \beta_1$ and $\alpha_2 = \beta_2 \circ f$ (we already saw the corresponding commutative diagram).

EXERCISE 10.58. (i) Use the constructions in the theorem to define a functor from $\mathfrak{F}_{M_0}$ to $\mathfrak{E}_{P_1,P_2}$ and prove that it is an equivalence of categories. (Therefore it induces bijections between isomorphism classes in both categories.)

(ii) Define a category $\mathfrak{E}$ with P_1, P_2 not fixed and explain the relation with $\mathfrak{F}$.

It is easy to define a more general category of filtered modules instead of $\mathfrak{F}$ (i.e., without requiring $k = 2$), but it is not clear what we should use in place of $\mathfrak{E}_{P_1,P_2}$ or $\mathfrak{E}$ above.

10.3.5. Classification and matrices, general case $k \geq 2$.

10.3.5.1. Note that if $w : M_0 \to N_0$ is an isomorphism, the map $(M, u) \mapsto (M, w \circ u)$ is a one-to-one correspondence between $\mathscr{E}(M_0)$ and $\mathscr{E}(N_0)$, and it is strictly compatible with the equivalence relations, which means that $(M, u) \sim (N, v)$ in $\mathscr{E}(M_0)$ if, and only if, $(M, w \circ u) \sim (N, w \circ v)$ in $\mathscr{E}(N_0)$. Therefore we have a bijection $\mathscr{F}(M_0) \to \mathscr{F}(N_0)$, class of $(M, u) \mapsto$ class of $(M, w \circ u)$. Since every pure analytic q-difference module is isomorphic to some $(\mathbf{C}(\{x\})^n, \Phi_{A_0})$ with A_0 in the already described block diagonal form, we shall take M_0 of that form and write $\mathscr{F}(M_0) =: \mathscr{F}(A_0)$.

Now let $(M, u) \in \mathscr{E}(M_0)$ and let $f : M \to N$ an arbitrary isomorphism. Setting $v := u \circ (\mathrm{gr} f)^{-1}$, we see that $(M, u) \sim (N, v)$. Since every analytic q-difference module is isomorphic to some $(\mathbf{C}(\{x\})^n, \Phi_B)$, with B in the same block upper triangular form as before, we see that we can restrict $\mathscr{E}(M_0)$ to the subclass of such (N, v). Moreover, using the arguments and calculations of 10.3.3.2, we see that we can require B to have block diagonal part $B_0 = A_0$, that is, B has the form A_U. Also, again using 10.3.3.2, we can require that the isomorphism $v : \mathrm{gr}N \to M_0$, i.e., $\mathrm{gr}(\mathbf{C}(\{x\})^n, \Phi_B) \to (\mathbf{C}(\{x\})^n, \Phi_{A_0})$ be encoded by the identity matrix $I_n : B_0 \to A_0$. So we are led to the following definitions:

$$\mathscr{E}(A_0) := \{A_U \mid (U_{i,j}) \in \prod_{1 \leq i < j \leq k} \mathrm{Mat}_{r_i, r_j}(\mathbf{C}(\{x\}))\}$$

and, for $A_U, A_V \in \mathscr{E}(A_0)$:

$$A_U \sim A_V \underset{def}{\Longleftrightarrow} \exists F = \begin{pmatrix} I_{r_1} & F_{1,2} & \dots & F_{1,k} \\ 0 & I_{r_2} & \dots & F_{2,k} \\ \vdots & \vdots & \ddots & \vdots \\ 0 & 0 & \dots & F_{k,k} \end{pmatrix} \in \mathrm{GL}_n(\mathbf{C}(\{x\})) \; : \; F[A_U] = A_V.$$

The conclusion is now that sending the class of $A_U \in \mathscr{E}(A_0)$ to the class of the pair (M, u), where $M := (\mathbf{C}(\{x\})^n, \Phi_A)$ and u is the identity of $(\mathbf{C}(\{x\})^n, \Phi_{A_0})$, we define a bijective map $\dfrac{\mathscr{E}(A_0)}{\sim} \to \mathscr{F}(A_0)$. So from now on we shall identify $\mathscr{F}(A_0)$ with the quotient set $\dfrac{\mathscr{E}(A_0)}{\sim}$.

EXAMPLE 10.59 (Linear case: $k = 2$). Let $A_0 := \begin{pmatrix} A_1 & 0 \\ 0 & A_2 \end{pmatrix} \in \mathrm{GL}_n(\mathbf{C}(\{x\}))$, $A_i \in \mathrm{GL}_{r_i}(\mathbf{C}(\{x\}))$, $i = 1, 2$. (Also with assume isoclinic purity conditions with slopes $\mu_1 > \mu_2$.) Whenever $U \in \mathrm{Mat}_{r_1, r_2}(\mathbf{C}(\{x\}))$, let $A_U := \begin{pmatrix} A_1 & U \\ 0 & A_2 \end{pmatrix} \in \mathrm{GL}_n(\mathbf{C}(\{x\}))$. Note that, whenever $F \in \mathrm{Mat}_{r_1, r_2}(\mathbf{C}((x)))$:

$$\begin{aligned} \begin{pmatrix} I_{r_1} & F \\ 0 & I_{r_2} \end{pmatrix} [A_U] = A_V &\Longleftrightarrow \begin{pmatrix} I_{r_1} & \sigma_q F \\ 0 & I_{r_2} \end{pmatrix} A_U = A_V \begin{pmatrix} I_{r_1} & F \\ 0 & I_{r_2} \end{pmatrix} \\ &\Longleftrightarrow U + (\sigma_q F) A_2 = A_1 F + V. \end{aligned}$$

There exists a priori a unique such formal F, write it $\hat{F}_{U,V}$. Then:

$$\begin{aligned} A_U \sim A_V &\Longleftrightarrow \exists F \in \mathrm{Mat}_{r_1, r_2}(\mathbf{C}(\{x\})) \; : \; U + (\sigma_q F) A_2 = A_1 F + V \\ &\Longleftrightarrow \hat{F}_{U,V} \in \mathrm{Mat}_{r_1, r_2}(\mathbf{C}((x))). \end{aligned}$$

The latter condition is of analytic nature (some particular power series is required to have positive radius of convergence), we shall extensively study it in Chapters

14 and 15 with the help of so-called "q-Borel-Ramis" transformation. The former condition is algebraic (and even of "homological" nature). Indeed consider:

$$t_{A_1,A_2} : \mathrm{Mat}_{r_1,r_2}(\mathbf{C}(\{x\})) \to \mathrm{Mat}_{r_1,r_2}(\mathbf{C}(\{x\})), F \mapsto (\sigma_q F)A_2 - A_1 F.$$

(This is actually a matricial version of the q-de Rham complex in the form encountered in Chapter 8.3.2.) Then $A_U \sim A_V$ if, and only if, $V - U \in \mathrm{Im}\, t_{A_1,A_2}$, which entails that $\mathscr{F}(A_0)$ can be identified with Coker t_{A_1,A_2}.

EXERCISE 10.60. What is the "meaning" of Ker t_{A_1,A_2}?

10.3.5.2. We draw on the principle stated at the end of 10.3.3.2, viz that isoformal analytic classification boils down to isograded classification. We stick to the matricial model $M = (\mathbf{C}(\{x\})^n, \Phi_A)$ and consistently take A in the form A_U described in 10.2.1.3; so, A_0 denoting its block diagonal part, $M_0 := \mathrm{gr}M$ is taken as $M_0 = (\mathbf{C}(\{x\})^n, \Phi_{A_0})$. Note that this description contains implicitly the structural morphism $u : \mathrm{gr}M \simeq M_0$.

We know a priori that there is a unique formal isomorphism $\hat{M}_0 \simeq \hat{M}$ respecting the structural data. This means that there is a unique $\hat{F} \in \mathrm{GL}_n(\mathbf{C}((x)))$ in block upper triangular form shown in 10.3.3.1, but with all $F_{i,i} = I_{r_i}$ (so this is actually the form shown in 10.3.5.1) and such that $\hat{F}[A_0] = A_U$. We write it $\hat{F}_U$. We say for short that it is unipotent block upper triangular. If we consider A_U and A_V, there is likewise a unique unipotent block upper triangular $\hat{F} \in \mathrm{GL}_n(\mathbf{C}((x)))$ such that $\hat{F}[A_U] = A_V$. We write it $\hat{F}_{U,V}$. Clearly $\hat{F}_V \circ \hat{F}_U^{-1}$ is unipotent block upper triangular and it sends A_U to A_V, so we conclude:

$$\hat{F}_{U,V} = \hat{F}_V \circ \hat{F}_U^{-1}.$$

We summarize (part of) the general discussion in the following:

PROPOSITION 10.61. *The block diagonal matrix A_0 being given in the usual form, let $\mathscr{E}(A_0)$ the set of all analytic matrices*

$$A_U,\ U := (U_{i,j}) \in \prod_{1 \leq i < j \leq k} Mat_{r_i,r_j}(\mathbf{C}(\{x\})).$$

Let the equivalence relation $\sim$ on $\mathscr{E}(A_0)$ be defined by:

$$A_U \sim A_V \underset{def}{\iff} \exists F \in GL_n(\mathbf{C}(\{x\})) : \begin{cases} F[A_U] = A_V, \\ F \text{ unipotent block upper triangular.} \end{cases}$$

(i) *Then $\mathscr{F}(M_0)$ is canonically identified with $\mathscr{E}(A_0)/\sim$;*
(ii) *and $A_U \sim A_V \Leftrightarrow \hat{F}_{U,V} \in GL_n(\mathbf{C}(\{x\}))$.*

□

EXAMPLE 10.62. Let $k = 2$, and write $\mu > \nu$ the slopes, here assumed integral, and r, s their multiplicities. Then, as we shall see in Chapter 11, the pure blocks can be taken in the form $x^\mu B$, $x^\nu C$, where $B \in \mathrm{GL}_r(\mathbf{C})$ and $C \in \mathrm{GL}_s(\mathbf{C})$, so we can take:

$$A_0 := \begin{pmatrix} x^\mu B & 0 \\ 0 & x^\nu C \end{pmatrix} \text{ and } A_U := \begin{pmatrix} x^\mu B & U \\ 0 & x^\nu C \end{pmatrix}, U \in \mathrm{Mat}_{r,s}(\mathbf{C}(\{x\})).$$

Let $U, V \in \mathrm{Mat}_{r,s}(\mathbf{C}(\{x\}))$. There is a unique

$$\hat{F}_{U,V} = \begin{pmatrix} I_r & F \\ 0 & I_s \end{pmatrix},\ F \in \mathrm{Mat}_{r,s}(\mathbf{C}((x)))$$

sending A_U to A_V, i.e., such that $(\sigma_q F)(x^\nu C) - (x^\mu B)F = V - U$, and $A_U \sim A_V$ if, and only if, $F \in \mathrm{Mat}_{r,s}(\mathbf{C}(\{x\}))$. We write the above as a fixed point equation:

$$F = F_0 + x^\delta L(F), \text{ where } F_0 := \sigma_q{}^{-1}\left(x^{-\nu}(V-U)C^{-1}\right), \delta := \mu - \nu \in \mathbf{N}^*,$$

and where we introduce the operator L on $\mathrm{Mat}_{r,s}(\mathbf{C}((x)))$ such that $x^\delta L(F) = \sigma_q{}^{-1}\left(x^\delta BFC^{-1}\right)$, i.e., $L(F) = q^{-\delta}\sigma_q{}^{-1}B(\sigma_q{}^{-1}F)C^{-1}$. So we set $\Lambda := x^\delta L$ and solve by iteration:

$$F = F_0 + \Lambda(F) \Longrightarrow F = \sum_{m\geq 0} \Lambda^m(F_0).$$

Clearly $\Lambda^m(F_0) = x^{m\delta}F_m$, where F_m has the same x-content as F_0 (see the exercise that follow), so that the series converges x-adically and defines an element of $\mathrm{Mat}_{r,s}(\mathbf{C}((x)))$, the unique solution of the fixed point equation.

EXERCISE 10.63. Compute precisely $\Lambda^m(F_0)$.

Note that if the slopes are not integral, the conclusion that the x-content of $\Lambda^m(F_0)$ is $m\delta$ still holds in some sense, as can be deduced from a q-analogue of Jurkat's criterion (see the footnote at the beginning of Chapter 11).

10.3.5.3. The failure to obtain an analytic solution F above comes from the fact that iterating Λ involves iterating $\sigma_q{}^{-1}$ and that $(\sigma_q{}^{-1})^m(a_k x^k) = q^{-mk}a_k x^k$, so we get growing coefficients and, generically, divergent series.

EXAMPLE 10.64. We take the same setting with $r = s = 1$, $\mu = 1$, $\nu = 0$. So here there are $a, b \in \mathbf{C}^*$ such that $A_u := \begin{pmatrix} ax & u \\ 0 & b \end{pmatrix}$ for $u \in \mathbf{C}(\{x\})$. Since in the linear case $A_u \sim A_v \Leftrightarrow A_{u-v} \sim A_0$, we just look for $\hat{F}_u = \begin{pmatrix} 1 & f \\ 0 & 1 \end{pmatrix}$ with $f \in \mathbf{C}((x))$ such that $\hat{F}_u[A_0] = A_u$, i.e.:

$$b\sigma_q f - axf = u \Longleftrightarrow \forall m\ ,\ bq^m f_m - af_{m-1} = u_m,$$

where of course $u = \sum u_m x^m$ and $f = \sum f_m x^m$. Instead of solving that equation by iteration (as in the previous example), we work with coefficients. As already seen (for instance in 6.1.3), we can make the equation more telescopic by introducing its q-Borel transform; so let:

$$\tilde{u}_m := q^{m(m+1)/2}u_m, \tilde{u}(x) := \sum \tilde{u}_m x^m \text{ and } \tilde{f}_m := q^{m(m+1)/2}f_m, \tilde{f}(x) := \sum \tilde{f}_m x^m,$$

so that multiplying the above equation by $q^{m(m-1)/2}$ yields:

$$b\sigma_q f - axf = u \Longleftrightarrow \forall m\ ,\ b\tilde{f}_m - a\tilde{f}_{m-1} = q^{-m}\tilde{u}_m \Longleftrightarrow (b - ax)\tilde{f}(x) = \tilde{u}(x/q).$$

Note that, u being convergent, $\tilde{u}$ is holomorphic on $\mathbf{C}^*$ and meromorphic at 0 (its coefficients $q^{m(m+1)/2}u_m$ have very fast decay). The same holds for $\tilde{f}$ if f is convergent. In that case, one is allowed to substitute $x := b/a$ and we see that a necessary condition for f to be convergent is that $\tilde{u}(qb/a) = 0$. Therefore:

$$A_u \simeq A_0 \Longrightarrow \sum q^{m(m-1)/2}u_m(b/a)^m = 0.$$

We have a well defined linear map $\mathcal{B} : u \mapsto \sum q^{m(m-1)/2}u_m(b/a)^m$ on $\mathbf{C}(\{x\})$ such that the class of A_u is trivial only if $u \in \mathrm{Ker}\mathcal{B}$.

EXERCISE 10.65. (i) Prove that conversely if

$$\tilde{u}(qb/a) = \sum q^{m(m-1)/2} u_m (b/a)^m = 0$$

then the function $\dfrac{\tilde{u}(x/q)}{b - ax}$ (which is clearly holomorphic on $\mathbf{C}^*$ and meromorphic at 0) is $\tilde{f}$ for a unique $f \in \mathbf{C}(\{x\})$ and that, as a consequence, $A_u \simeq A_0$.

(ii) Deduce an isomorphism: class of $A_u \mapsto \mathcal{B}(u)$ from $\mathscr{F}(A_0)$ to $\mathbf{C}$.

This fundamental process will be extended and detailed in Chapters 14 and 15.

10.3.5.4. *Weakening of isograded equivalence.* We somehow generalize here the weakening of equivalence of extensions in 8.3.2. We take up the notations A_0, A_U from the beginning of 10.2.1.3, recalling that each A_i is pure of slope μ_i and that $\mu_1 > \cdots > \mu_k$. We have been studying isograded equivalence, i.e., that induced by gauge transformations $F[A_U] = A_V$ where F is block upper triangular unipotent; see 10.3.5.1. We want to weaken that condition to $G[A_U] = A_V$ with arbitrary $G \in \mathrm{GL}_n(\mathbf{C}(\{x\}))$. We shall then write $A_U \simeq A_V$. From the fact that $\mathscr{F}(M_0) = \mathscr{E}(A_0)/\sim$, and from the obvious fact that $\sim \Rightarrow \simeq$, defining $\widetilde{\mathscr{F}}(M_0) = \mathscr{E}(A_0)/\simeq$, we get a surjective mapping $\mathscr{F}(M_0) \to \widetilde{\mathscr{F}}(M_0)$.

From the functoriality of the filtration and of the graduation, we deduce that any $G \in \mathrm{GL}_n(\mathbf{C}(\{x\}))$ such that $G[A_U] = A_V$ is block upper triangular, with diagonal blocks $G_i \in \mathrm{GL}_{r_i}(\mathbf{C}(\{x\}))$ such that $G_i[A_i] = A_i$; in other words, $G_i \in \mathrm{Aut}(A_i)$ and $G_0 := \mathrm{Diag}(F_1, \ldots, F_k)$ can be seen as an element of $\mathrm{Aut}(A_0) = \prod_{i=1}^{k} \mathrm{Aut}(A_i)$ (natural identification). Now $F := FG_0^{-1}$ is block upper triangular unipotent and, setting $A_{U'} := G_0[A_U] \in \mathscr{E}(A_0)$ (its out diagonal blocks are $U'_{i,j} := (\sigma_q G_i) U_{i,j} G_j^{-1}$), it satisfies $F[A_{U'}] = A_V$; so $G_0[A_U] \sim A_V$.

So we are led to define an action of the group $\mathrm{Aut}(A_0) = \prod_{i=1}^{k} \mathrm{Aut}(A_i)$ on $\mathscr{E}(A_0)$ by gauge equivalence (it is immediate that it sends $\mathscr{E}(A_0)$ to itself). We just saw that $A_U \simeq A_V$ implies that some $A_{U'}$ in the orbit of A_U satisfies $A_{U'} \sim A_V$. The converse is obvious. More generally:

PROPOSITION 10.66. (i) *The group action of Aut(A_0) on $\mathscr{E}(A_0)$ goes to the quotient as a group action on $\mathscr{F}(M_0)$.*

(ii) *The mapping $\mathscr{F}(M_0) \to \widetilde{\mathscr{F}}(M_0)$ is a quotient by that action.*

PROOF. (i) If $A_U \sim A_V$, say through F, then $G_0[A_U] \sim G_0[A_V]$ through $G_0 F G_0^{-1}$ (which is indeed block upper triangular unipotent).

(ii) then follows from the above discussion.

□

A special case is when all slopes are integral, so instead of A_i we write $x^{\mu_i} A_i$, $A_i \in \mathrm{GL}_{r_i}(\mathbf{C})$. If we moreover require (as is possible) that $\mathrm{Sp}\, A_i \subset \mathcal{C}_q$, the fundamental annulus, we see that $\mathrm{Aut}(x^{\mu_i} A_i) = Z(A_i) \subset \mathrm{GL}_{r_i}(\mathbf{C})$, the commutant of A_i. So the group $\mathrm{Aut}(A_0) = \prod Z(A_i) \subset \prod \mathrm{GL}_{r_i}(\mathbf{C})$ is a linear group. We shall see in Chapter 15 that $\mathscr{F}(M_0)$ admits a natural structure of affine space and then $\widetilde{\mathscr{F}}(M_0)$ is the quotient of that affine space by the action of a linear group, often a rather badly behaved space.

10.4. Relation to other theories of slope filtrations

Here, we briefly discuss how the slope filtration of q-difference modules described in this chapter fits into Yves André's general theory of slope filtrations (see **[5]**, where our filtrations are kindly nicknamed "Adams-Sauloy") and also how it relates to the Harder-Narasimhan filtration of vector bundles on a curve. This section will not be needed in the rest of the book. We heavily use notations from *loc. cit.* without reproducing their explanation.

CAUTION 10.67. In **[5]**, references to q-difference modules mention the assumption $|q| > 1$ and speak about *ascending* filtrations. Indeed, our conventions in various papers by Sauloy, Ramis-Sauloy-Zhang and Ramis-Sauloy have evolved with time. Here, as in all the book, our assumption is $0 < |q| < 1$ and the filtrations are descending, so the conventions of **[5]** hold *verbatim*.

In [**5**, I.2] various generalizations of abelian categories are considered. Our categories of q-difference modules are abelian, so that all constructions behave in the "standard" way: epi-monics are isomorphisms (so indeed the first condition of the definition of slope functions in [**5**, I.3.1.1] is empty); in the terms of [**5**, I.2.1], all our subobjects and quotient objects are strict; and, in the terms of [**5**, I.2.6.2], the mapping $(V, \Phi) \mapsto \dim_K V$ is a rank function.

10.4.1. Comparison with [5].

10.4.1.1. *Complements on the Newton polygon.* We begin with some complements to the general notions and notations of this chapter. To a q-difference operator $P = a_0 + \cdots + a_n S^n \in \mathcal{D}_{K,q} = K\langle S, S^{-1}\rangle$, $a_0 a_n \neq 0$, we have associated first the set $\{(i,j) \mid j \geq v(a_i)\}$, then the boundary of the convex hull of that set, which consists in two vertical half lines (going up) and a finite number of vectors (r_i, d_i), $i = 1, \ldots, k$, $r_i \in \mathbf{N}^*$, $d_i \in \mathbf{Z}$. If we number them from right to left, the slopes $\mu_i := d_i/r_i$ satisfy $\mu_1 > \cdots > \mu_k$. We have $r_1 + \cdots + r_k = n = \deg P$ (absolute degree of P) and $\delta(P) := d_1 + \cdots + d_k = v(a_n/a_0)$; see for instance Figure 5.2 in Chapter 5, but beware that we have here adapted the numbering to fit the context of q-difference modules rather than that of q-difference operators.

By direct calculation, one checks that $\delta(PQ) = \delta(P) + \delta(Q)$. The general theory implies that the Newton polygon of P only depends on the isomorphism class of the q-difference module $M := \mathcal{D}_{K,q}/\mathcal{D}_{K,q}P$. As a consequence, this is also true of $\delta(P)$ and we can set $\delta(M) := \delta(P)$. From Subsubsection 9.1.2.4, we then deduce that $\delta\left(\bigwedge^n M\right) = \delta(M)$. We can also define three more invariants of M:

$$\overline{\mu}(M) := \mu_1, \quad \underline{\mu}(M) := \mu_k, \quad \mu(M) := \frac{\delta(P)}{\deg(P)} = \frac{r_1\mu_1 + \cdots + r_k\mu_k}{r_1 + \cdots + r_k}.$$

We respectively call them the *upper slope*, the *lower slope* and the *slope* of M. Note from the last equality that $\underline{\mu}(M) \leq \mu(M) \leq \overline{\mu}(M)$ (strict inequalities except if M is pure isoclinic).

10.4.1.2. *Starting from the slope function in* [**5**, I.3]. If we apply to our function $\mu(M)$ the definition 3.1.1 in [**5**, I.3], we see that what is called there degree of M is what we just denoted $\delta(M)$. Then, from equality $\delta(PQ) = \delta(P) + \delta(Q)$ for q-difference operators, we deduce that δ is additive on short exact sequences. (We use the description of such sequences in Subsection 8.2.1.) So $\mu(M)$ is indeed a slope function in the sense of *loc. cit.*.

Then, looking at the definition of semi-stability in [**5**, I.3.2], we find that semi-stable q-difference modules are exactly the pure isoclinic ones; and that stable q-difference modules are exactly the simple ones. It follows first that the universal destabilizing subobject of M is $M_1 = M^{\geq \mu_1}$, then that the filtration deduced from the slope function in the proof of theorem 4.2.3 of [**5**, I.4] is our slope filtration.

Note that the slope function, as briefly described in [**5**, III.13.3.1], indeed coincides with our $\mu(M)$ thanks to the equality $\delta\left(\bigwedge^n M\right) = \delta(M)$.

10.4.1.3. *Starting from the slope filtration in* [**5**, I.4]. Following notations from [**5**, I.4.1] (and ours above), the descending filtration functor is defined by $F^{\geq \lambda} M = M_i$ for $\mu_i \leq \lambda < \mu_{i-1}$ (extreme cases are taken care of through the conventions $\mu_0 := +\infty$, $\mu_{k+1} := -\infty$). The breaks are therefore the μ_i, $i = 1, \ldots, k$ and the multiplicity of the break μ_i is r_i.

Now following [**5**, I.4.2], what is called there the degree of M is:

$$\mu_1 r_1 + \cdots + \mu_k r_k = d_1 + \cdots + d_k = \delta(M).$$

Since we use the natural rank function, the slope function deduced from the filtration is $\mu := \delta/n$, i.e., our $\mu(M)$ defined above.

The highest break defined in [**5**, I.4.3] is clearly our μ_1, i.e., the upper slope $\overline{\mu}(M)$ of M. Proposition 4.3.2 from *loc. cit.* then states first that M is semi-stable if, and only if, $\overline{\mu}(M'') \geq \overline{\mu}(M)$ for every nonzero quotient M''. Since we saw that semi-stable means pure isoclinic, one implication is clear (and actually equality is implied). On the other hand, if M is not pure then $\overline{\mu}(M) = \mu_1 > \overline{\mu}(M/M_1) = \mu_2$, so the converse holds.

The same proposition also states that (in our language) the function $\overline{\mu}$ completely determines the slope function and filtration, which is clear by iteration from $\mu_2 = \overline{\mu}(M/M_1)$; indeed, the "universal destabilizing subobject" M_1 can be characterized as the single largest subobject such that $\overline{\mu}(M) > \overline{\mu}(M/M_1)$.

Last, in [**5**, I.4.4], there is a definition of the Newton polygon attached to a slope filtration as the convex hull of the points defined in our language as $(0,0)$, (r_1, d_1), $(r_1 + r_2, d_1 + d_2)$, $\ldots$, $(r_1 + \cdots + r_k, d_1 + \cdots + d_k)$. Those points are the images of the vertices of our Newton polygon by the central symmetry $(x, y) \mapsto (n - x, \delta - y)$ (see Figure 10.1) so we obtain the same slopes.

REMARK 10.68. It is stated in [**5**, I.5.3] that the slope filtration of q-difference modules is not split. This is true for q-difference modules over $\mathbf{C}(\{x\})$, but for those over $\mathbf{C}((x))$ the filtration is actually split. We return to that at the end of the next Subsubsection.

10.4.2. The Harder-Narasimhan filtration. As is well known [**5, 233**], holomorphic vector bundles over a compact Riemann surface (or projective curve) admit a canonical slope filtration, the *Harder-Narasimhan filtration.* Hereafter, we shall denote $\lambda(\mathfrak{F})$ the slope of a vector bundle $\mathfrak{F}$. Since every q-difference module M over $\mathbf{C}(\{x\})$ gives rise to a holomorphic vector bundle $\mathfrak{F}_M$ over the compact torus (or elliptic curve) $\mathbf{E}_q := \mathbf{C}^*/q^{\mathbf{Z}}$ (Sections 7.3 and 9.3), it is only natural to compare the filtrations on M and on $\mathfrak{F}_M$. Let us recall the construction of $\mathfrak{F}_M$ in the matricial model $M = (K^n, \Phi_A)$, $A \in \mathrm{GL}_n(K)$ (in all the present Subsection, K stands for $\mathbf{C}(\{x\})$). We first describe a *sheaf* $\mathscr{F}_A$ on $\mathbf{E}_q$; denoting $\pi : \mathbf{C}^* \to \mathbf{E}_q$ the canonical projection, and letting V an open subset of $\mathbf{E}_q$:

$$\mathscr{F}_A(V) := \{X \in \mathscr{O}(\pi^{-1}(V), 0) \mid \sigma_q X = AX\},$$

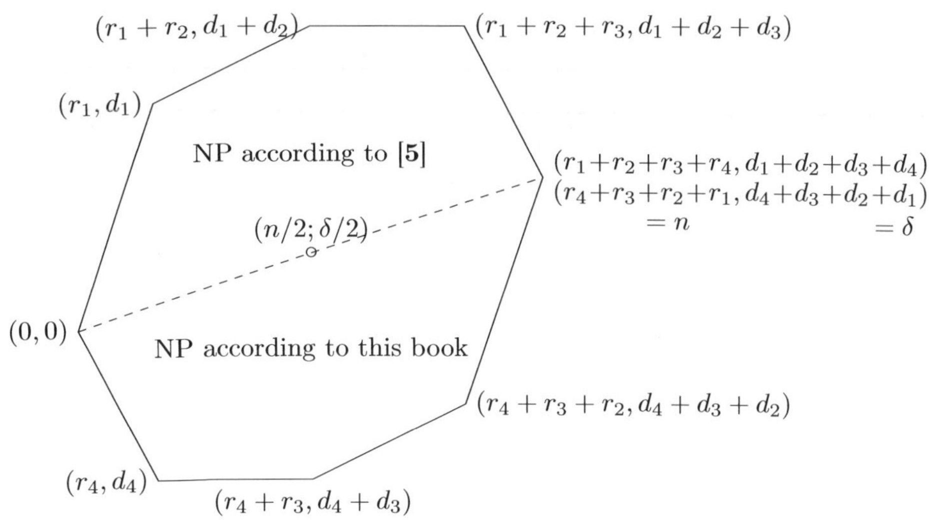

FIGURE 10.1. Newton polygons according to this book and to [5]

the set of germs of solutions of the associated q-difference equation on the punctured neighborhood $\pi^{-1}(V)$ of 0 in $\mathbf{C}^*$. This sheaf is locally free over $\mathbf{E}_q$ an therefore defines a vector bundle $\mathfrak{F}_A = \mathfrak{F}_M$. Moreover (equalities meaning canonical isomorphies) $\bigwedge^n(K^n, \Phi_A) = (K, \Phi_{\det A})$, so that $\bigwedge^n \mathfrak{F}_A = \mathfrak{F}_{\det A}$.

LEMMA 10.69. *We have* $\lambda(\mathfrak{F}_M) = -\mu(M)$.

PROOF. In the terminology of [5], both filtrations are determinantal, meaning that, writing n the rank of M and of $\mathfrak{F}_M$ (they are the same), we have $\mu(M) = \dfrac{\delta(M)}{n}$, where $\delta(M) = \delta\left(\bigwedge^n M\right)$, and $\lambda(\mathfrak{F}_M) = \dfrac{\deg(\mathfrak{F}_M)}{n}$, where $\deg(\mathfrak{F}_M) = \deg\left(\bigwedge^n \mathfrak{F}_M\right)$. Moreover, if $L := \bigwedge^n M$, then $\bigwedge^n \mathfrak{F}_M = \mathfrak{F}_L$ (this will be clear from the explicit description). So we are reduced to compare the slopes (i.e., the degrees) in the case of a q-difference module rank 1.

So we should look for the degree of the line bundle $\mathfrak{F}_u$ for an arbitrary $u \in K^*$ (the degree of the q-difference module (K, Φ_u) is obviously $v(u)$). The degree of a line bundle is the degree of the divisor of any of its nonzero meromorphic sections. In our case, those are the meromorphic solutions of the q-difference equation $\sigma_q f = uf$. Writing $u = cx^m v$, where $c \in \mathbf{C}^*$, $m \in \mathbf{Z}$ and $v \in \mathbf{C}\{x\}$, $v(0) = 1$, we have the typical solution:

$$f(x) := \frac{\theta_q(x)}{\theta_q(cx)} \theta_q^{-m}(x) w(x), \text{ where } w := \prod_{i \geq 0} \frac{1}{\sigma_q{}^i v}.$$

(See Section 4.1; we use the Theta function $\theta_q \in \mathscr{O}(\mathbf{C}^*)$ from Section 2.2.) Since w does not vanish near 0, its divisor as a section of $\mathfrak{F}_u$ is trivial; therefore, using (notations from Appendix B:

$$\operatorname{div}_{\mathbf{E}_q}(f) = (1 - m)[\overline{-1}] - [\overline{-c}] \Longrightarrow \deg \operatorname{div}_{\mathbf{E}_q}(f) = -m,$$

so that $\deg \mathfrak{F}_u = -v(u)$. The conclusion follows. □

It is noted in [**5**, 13.3] that the slope filtration on q-difference modules is not split. For instance, $M := (K^2, \Phi_A)$, $A := \begin{pmatrix} x & 1 \\ 0 & 1 \end{pmatrix}$ has a two-steps filtration with $P_1 = (K, \Phi_x)$ and $P_2 = (K, \Phi_1)$, but it is not isomorphic to $P_1 \oplus P_2$ (the failure to split has been detailed in Sections 8.3 and 10.3). In essence, the impossibility boils down to the fact that the nonhomogeneous "homological" q-difference equation $\sigma_q f = xf + 1$ has no solution $f \in \mathbf{C}(\{x\})$. However, it has a solution in $\mathbf{C}((x))$, the Tshakaloff series (see Section 5.3) $\text{Ч}(x) := \sum_{n \geq 0} q^{-n(n+1)/2} x^n$. More generally, over $\mathbf{C}((x))$, the Adams-Sauloy filtration splits functorially (Section 10.3), so it is actually a *split filtration* in the terminology of [**5**, 5.3]. On the other hand, the Harder-Narasimhan filtration splits, *but not functorially* (see [**5**, 5.3.2]).

CHAPTER 11

Fuchsian q-Difference Equations and Systems: Local Study

The general references for this chapter and the next one are [**217, 218**], to be completed by more recent works to be quoted at specific places. Since here we are dealing with the local study, it is fit to give a specific notation for the category (previously denoted $\mathscr{C}$ in a more general setting) of q-difference modules over $\mathbf{C}(\{x\})$: we write it $\mathscr{E}^{(0)}$, although, from 11.1.3 on, we will mainly be interested in its full subcategory $\mathscr{E}_f^{(0)}$ (and the equivalent categories $\mathscr{P}^{(0)}$ and $\hat{\mathscr{E}}_f^{(0)}$).

By the way, since we saw in Chapter 8 that $\mathscr{E}^{(0)}$ is equivalent to its full subcategory of q-difference modules $M_A := (K^n, \Phi_A)$ (thus here with $A \in \mathrm{GL}_n(\mathbf{C}(\{x\}))$), we shall frequently take q-difference modules in that form without saying so, and even make no difference between M_A and A, thus writing things like $\mathrm{Hom}(A, B) = \mathrm{Hom}(M_A, M_B)$, etc.

For ordinary differential equations [**123, 169, 222, 252**], also see [**3**, Chapter 8], "regular singularities" are defined as those singularities where all solutions have "moderate growth". For scalar equations, a criterion of Fuchs[1] allows one to recognize them at sight: it is a simple algebraic condition on coefficients; it is then completed by Fuchs-Frobenius method of resolution. For differential *systems*, there is no such simple necessary and sufficient condition: there is only a sufficient condition (being "of the first kind") and some algorithms for qualifying the singularity. Then, in the frame of the global study over the Riemann sphere $\mathbf{S}$, "fuchsian equations" are those having only ordinary points and regular singularities.

For q-difference equations, terminology has not (yet) been firmly established[2]. Adams, Carmichael and the Founding Fathers used a Newton polygon for scalar equations, and their results amount to an extension of Fuchs's criterion. Birkhoff, in the famous paper [**23**, p 299] (which is the main inspiration for the results here and in the next chapter), used a characterization by the form of the solutions (see Section 12.1.1.4 in the next chapter) and considered it as equivalent to something analogous to "of the first kind".

We shall introduce some similar terminology and criteria. Two remarks are relevant here. First, on the local side, we shall properly define *regular singularity* at 0 (and, for rational systems, at ∞) and define *fuchsian* (at 0 or ∞) as something equivalent, and *first kind* as something more manageable: the former conditions mean that $S(M) = \{0\}$ (i.e., pure isoclinic of slope 0), while the latter (which is strictly stronger) means that $A(0) \in \mathrm{GL}_n(\mathbf{C})$.

Then, on the global side, we shall define (in the next chapter) a *fuchsian rational system* as one which is fuchsian at 0 and ∞. This is because, at the moment (end

[1] This is Lazarus Fuchs, not to be confused with his less well-known son Richard Fuchs.

[2] Besides, here we slightly change it with respect to [**217, 218**].

2023), we have no really satisfactory local theory at "intermediate singularities", those on $\mathbf{C}^*$: indeed, they are not fixed by the q-dilatation $x \mapsto qx$, and it is not always clear how to localize at a $q^{\mathbf{Z}}$ or $q^{\mathbf{N}}$-orbit. (However we shall tackle intermediate singularities to some extent in section 12.2 of Chapter 12.)

The first section of this chapter deals with the local study at 0: various equivalent criteria, form and properties of solutions, some applications to pure modules[3] and slope filtration. Most important is the reduction to systems with constant coefficients, whence a complementary study of $e_{q,A}$.

The second section deals with degeneracy when $q \to 1$ to the universe of ordinary differential equations. This is known as *confluence* for reasons already hinted at in 1.4, 2.4, 4.4.4 and more precisely developed here. It will be completed in the next chapter by a study of its global consequences (in particular in terms of monodromy) and then in the following chapter, Section 13.3.4, by indications on the degeneracy of the Galois group.

11.1. Local study

In all this section, the field K is either $\mathbf{C}(\{x\})$ or $\mathbf{C}((x))$; and, accordingly, we denote $\mathscr{O}$ its subring $\mathbf{C}\{x\}$ or $\mathbf{C}[[x]]$, a discrete valuation ring (this fact will matter). Whenever we refer to $\mathcal{D}_{K,q}$, we abreviate it $\mathcal{D}$.

11.1.1. Reduction of fuchsian systems.

11.1.1.1. *Systems fuchsian at* 0.

DEFINITION 11.1. (i) The q-difference module M is said to be *fuchsian (at* 0*)* if it is pure isoclinic of slope 0, i.e., if $S(M) = \{0\}$.

(ii) The system matrix $A \in \mathrm{GL}_n(K)$ is said to be *fuchsian (at* 0*)* if the q-difference module (K^n, Φ_A) is.

(iii) The q-difference operator $P \in \mathcal{D}$ is said to be *fuchsian (at* 0*)* if the q-difference module $\mathcal{D}/\mathcal{D}P$ is.

In case $K = \mathbf{C}(\{x\})$, it will follow from the proposition below that it changes nothing to consider A or P as defined over $\mathbf{C}((x))$ (this is simply because $\mathrm{GL}_n(\mathbf{C}(\{x\})) \subset \mathrm{GL}_n(\mathbf{C}((x)))$).

EXAMPLES 11.2. (i) The unit object $\underline{\mathbf{1}}$ as well as all regular systems (7.2.3) are fuchsian.

(ii) Systems with constant coefficients $A \in \mathrm{GL}_n(\mathbf{C})$ are fuchsian (7.2.4).

(iii) The Tshakaloff q-difference operator $L := (\sigma_q - 1)(\sigma_q - x) = \sigma_q^2 - (1 + qx)\sigma_q + x$ (5.3) and the matrix $\begin{pmatrix} x & \star \\ 0 & 1 \end{pmatrix}$ are not fuchsian.

EXERCISE 11.3. (i) Give explicit criteria (probably involving a discussion) to test whether $\sigma_q^2 + u\sigma_q + v \in \mathcal{D}$ and $\begin{pmatrix} a & b \\ c & d \end{pmatrix} \in \mathrm{GL}_2(K)$ are fuchsian.

(ii) Generalized q-hypergeometric systems are those correponding to generalized q-hypergeometric series:

$$ {}_r\Phi_s(a_1, \ldots, a_r; b_1, \ldots, b_s; q, x) := \sum_{n \geq 0} \frac{(a_1;q)_n \cdots (a_r;q)_n}{(q;q)_n (b_1;q)_n \cdots (b_s;q)_n} x^n. $$

[3]Criteria of the kind seen in Corollary 11.23 will be completed in [CV] by some q-analogues of what is known as "Jurkat's criterion" for ordinary differential equations.

Find the corresponding q-difference equation and q-difference system then discuss fuchsianity (at 0).

PROPOSITION 11.4. (i) *If M is fuchsian and $N \simeq M$, then N is fuchsian.*

(i') *If $A \in GL_n(K)$ is fuchsian and if $F \in GL_n(K)$, then $F[A]$ is fuchsian.*

(ii) *The q-difference module M over $\mathbf{C}(\{x\})$ is fuchsian if, and only if, $\hat{M}$ is.*

(ii') *The q-difference module M is fuchsian if, and only if, $M' := K' \otimes_K M$ obtained by ramification is.*

(iii) *Let $0 \to M' \to M \to M'' \to 0$ an exact sequence. Then M is fuchsian if, and only if, M' and M'' are.*

(iv) *If two among M_1, M_2 and $M_1 \otimes M_2$ are fuchsian, so is the third.*

(iv') *If M is fuchsian, so is $M^\vee$.*

(v) *If M, N are pure isoclinic of the same slope μ, then $\underline{Hom}(M,N)$ is fuchsian.*

(vi) *Let $P := \sigma_q^n + a_1\sigma_q^{n-1} + \cdots + a_n \in \mathcal{D}$ standard unitary. Then, P is fuchsian if, and only if, $a_1, \ldots, a_{n-1} \in \mathcal{O}$ and $a_n \in \mathcal{O}^*$. (Said otherwise: $a_1(0), \ldots, a_n(0)$ are well defined and $a_n(0) \in \mathbf{C}^*$.)*

PROOF. (i) is obvious and (i') follows immediately.

(ii) and (ii') follow from the rules $S(\hat{M}) = S(M)$ and $S(M') = rS(M)$ (r the degree of ramification).

(iii) follows from the rule $S(M) = S(M') \cup S(M'')$.

(iv) follows from the rule $S(M_1 \otimes M_2) = S(M_1) + S(M_2)$ and (iv') from the rule $S(M^\vee) = -S(M)$.

(v) follows from $\underline{\mathrm{Hom}}(M,N) \simeq M^\vee \otimes N$ and the same rules used in (iv), (iv').

(vi) is a direct translation of the condition $S(P) = \{0\}$.

□

REMARK 11.5. In spite of (ii), the interplay between analytic and formal setting is important and not tautological.

11.1.1.2. *First kind systems and resonance.*

DEFINITION 11.6. The q-difference system $\sigma_q X = AX$ (or the matrix A itself) is *of the first kind* if $A \in \mathrm{GL}_n(\mathcal{O})$, i.e., if $A(0) \in \mathrm{GL}_n(\mathbf{C})$.

The latter condition of course means that A has no pole at 0, so that $A(0)$ is well defined, and that then $A(0) \in \mathrm{GL}_n(\mathbf{C})$. The fact that such matrices are those in $\mathrm{GL}_n(\mathcal{O})$ is standard.

LEMMA 11.7. *$A \in GL_n(K)$ is of the first kind if, and only if, $\Phi_A(\mathcal{O}^n) = \mathcal{O}^n$.*

PROOF. It is also standard that, letting $B \in \mathrm{GL}_n(K)$ act on K^n, one has $B \in \mathrm{GL}_n(\mathcal{O})$ if, and only if, $B\mathcal{O}^n = \mathcal{O}^n$. On the other hand, $\sigma_q \mathcal{O}^n = \mathcal{O}^n$ (obviously) and we apply the former fact to $B := A^{-1}$. □

CAUTION 11.8. Contrary to fuchsianity, being of the first kind is *not* an intrinsic condition, if A is of the first kind, $F[A]$ needs not be: taking $A := \begin{pmatrix} 1 & 1 \\ 0 & 1 \end{pmatrix}$ (which is of the first kind) and $F := \begin{pmatrix} 1 & 0 \\ 0 & x \end{pmatrix}$ yields $F[A] = \begin{pmatrix} 1 & x^{-1} \\ 0 & q \end{pmatrix}$, which is not of the first kind. The condition "of the first kind" is not about a q-difference module (V, Φ) but about its representation relative to some basis of V. So the closest intrinsic condition would be that Φ preserves some lattice: see 11.1.1.6.

DEFINITION 11.9. Let $A \in \mathrm{GL}_n(K)$ of the first kind. Then:

(i) A is *nonresonant* if any two distinct eigenvalues of $A(0)$ are noncongruent modulo $q^{\mathbf{Z}}$, i.e., if $\forall c, d \in \mathrm{Sp}\ A(0)\ ,\ c \neq d \Rightarrow c \not\equiv d \pmod{q^{\mathbf{Z}}}$. Equivalently, $\mathrm{Sp}\ A(0) \cap q^{\mathbf{N}^*}\mathrm{Sp}\ A(0) = \emptyset$.

(ii) A is *strictly nonresonant* if it is nonresonant and $A(0)$ is separable (i.e., it has n pairwise distinct eigenvalues); equivalently $A(0)$ has n pairwise noncongruent eigenvalues.

(iii) A is *normalized* if all its eigenvalues lie in the fundamental annulus $\mathcal{C}_q$: $\forall c \in \mathrm{Sp}\ A(0)\ ,\ |q| < |c| \leq 1$.

Clearly normalized first kind matrices are nonresonant and those such that $A(0)$ is separable are strictly nonresonant. Strict nonresonance (which holds generically true) was the condition implicitly assumed by Birkhoff in **[23]** under the name "general case". But nonresonance allows more or less to the same theory, except for the presence of q-logarithms in the solution; and, in some sense (as will be shown), it essentially covers all fuchsian theory.

EXAMPLE 11.10. Let $A := \begin{pmatrix} 1 & 1 \\ 0 & 1 \end{pmatrix}$ again and let this time $F := \begin{pmatrix} x & 0 \\ 0 & 1 \end{pmatrix}$ so that $B := F[A] = \begin{pmatrix} q & qx \\ 0 & 1 \end{pmatrix}$. Then B is resonant, A is not, and the solution involves a q-logarithm, which is not obvious by inspection of the separable matrix $B(0)$.

11.1.1.3. *Reduction of first kind to nonresonant.*

PROPOSITION 11.11. *Every system of the first kind can be transformed into a nonresonant system by a succession of gauge transformations either with matrices in $GL_n(\mathbf{C})$ (so they really are conjugacies); or with* shearing matrices*:*

$$S_\mu := \begin{pmatrix} x^{-1}I_\mu & 0_{\mu,n-\mu} \\ 0_{n-\mu,\mu} & I_{n-\mu} \end{pmatrix}.$$

PROOF. It is actually an algorithm. We may partition $\mathrm{Sp}\ A(0)$ into congruence classes modulo $q^{\mathbf{Z}}$. Such a class would look like $\{c, \ldots, cq^m\}$, involving powers q^i with exponents i between 0 and $m \geq 0$ (all values from 0 to m need not be present). Call m the "height" of such a class: our algorithm lowers at each step one of the nonzero heights without changing the other ones, so in the end all heights are 0 and the resulting matrix is nonresonant.

So assume there is a class $\{c, \ldots, cq^m\}$ with $m \geq 1$. We isolate the eigenvalue cq^m of $A(0)$ by using a conjugacy with matrix $C \in \mathrm{GL}_n(\mathbf{C})$ such that $A(0)$ goes to the block diagonal form:

$$CA(0)C^{-1} = \begin{pmatrix} a_0 & 0 \\ 0 & d_0 \end{pmatrix},\ \begin{cases} a_0 \in \mathrm{GL}_\mu(\mathbf{C}), \mathrm{Sp}\ a_0 = \{cq^m\}, \\ d_0 \in \mathrm{GL}_{n-\mu}(\mathbf{C}), \mathrm{Sp}\ d_0 \not\ni cq^m. \end{cases}$$

Let $A' := CAC^{-1} = \begin{pmatrix} a & b \\ c & d \end{pmatrix}$, so a, b, c, d have coefficients in $\mathcal{O}$, and $a(0) = a_0$, $b(0) = 0$, $c(0) = 0$, $d(0) = d_0$. Then $A'' := S_\mu[A'] = \begin{pmatrix} q^{-1}a & q^{-1}x^{-1}b \\ xc & d \end{pmatrix}$ has coefficients in $\mathscr{O}$ since $b(0) = 0$; and:

$$A''(0) = \begin{pmatrix} q^{-1}a_0 & \star \\ 0 & d_0 \end{pmatrix} \Longrightarrow \text{Sp } A''(0) = \text{Sp } a_0 \cup \text{Sp } d_0$$
$$= \{cq^{m-1}\} \cup (\text{Sp } A(0) \setminus \{cq^m\}),$$

so the class of c has height $m-1$ while the other classes are unchanged. □

COROLLARY 11.12. *If A is of the first kind, there exists $F \in GL_n\left(\mathbf{C}[x, x^{-1}]\right)$ such that $F[A]$ is nonresonant.*

PROOF. It is an immediate consequence of the proposition, since $\mathrm{GL}_n(\mathbf{C})$ and all shearing matrices belong to the subgroup $\mathrm{GL}_n\left(\mathbf{C}[x, x^{-1}]\right)$ of $\mathrm{GL}_n(K)$. □

EXERCISE 11.13. The group $\mathrm{GL}_n(\mathbf{C}[x, x^{-1}])$ is generated by $\mathrm{GL}_n(\mathbf{C})$ and shearing matrices.

REMARK 11.14. A similar algorithm could be used to normalize $F[A]$ (i.e., send its eigenvalues into the fundamental annulus), which would automatically enforce nonresonance; we prefer to separate the two processes, which have different use (see the end of 11.1.1.4).

11.1.1.4. *Reduction of nonresonant to constant.*

DEFINITION 11.15. A (formal or analytic) *Birkhoff gauge transformation* $F \in \mathrm{GL}_n(K)$ is one such that $F(0) = I_n$, i.e., one admitting an expansion $F = I_n + xF_1 + \cdots$, all $F_i \in \mathrm{Mat}_n(\mathbf{C})$.

Note that if $F \in \mathrm{Mat}_n(K)$ admits such an expansion, then automatically $F \in \mathrm{GL}_n(K)$.

PROPOSITION 11.16. (i) *Let $A \in GL_n(\mathbf{C}((x)))$ of the first kind and nonresonant and write $A_0 := A(0) \in GL_n(\mathbf{C})$. Then there is a unique Birkhoff gauge transformation $F = I_n + xF_1 + \cdots$ such that $F[A_0] = A$.*
(ii) *If A converges, then F converges.*

PROOF. (i) We shall use the following well known fact (F.4) in Appendix F. Let $M \in \mathrm{Mat}_m(\mathbf{C})$ and $N \in \mathrm{Mat}_n(\mathbf{C})$. Then the endomorphism $\Phi_{M,N} : X \mapsto MX - XN$ of $\mathrm{Mat}_{m,n}(\mathbf{C})$ has spectrum:

$$\text{Sp } \Phi_{M,N} = \text{Sp } M - \text{Sp } N = \{\{\lambda - \mu \mid \lambda \in \text{Sp } M, \mu \in \text{Sp } N\}\},$$

all spectra there being seen as multisets.

We write $A = A_0 + xA_1 + \cdots$ and look for $F = F_0 + xF_1 + \cdots$ with initial condition $F_0 := I_n$ and functional equation $F[A_0] = A$ being equivalent to:

$$(F_0 + qxF_1 + \cdots)A_0 = (A_0 + xA_1 + \cdots)(F_0 + xF_1 + \cdots)$$
$$\Longleftrightarrow \forall k \geq 1\,,\; q^k F_k A_0 = A_0 F_k + \cdots + A_k F_0.$$

Indeed, the condition for $k = 0$ would be $F_0 A_0 = A_0 F_0$ which is automatically satisfied since $F_0 = I_n$. The general condition yields a recursive relation (using the notation from the lemma just quoted):

$$q^k F_k A_0 = A_0 F_k + \cdots + F_k A_0 \Longleftrightarrow \Phi_{A_0, q^k A_0}(F_k) = -\left(A_1 F_{k-1} + \cdots + A_k F_0\right).$$

Since A is nonresonant, $\text{Sp } A_0 \cap \text{Sp } q^k A_0 = \emptyset$ for $k \geq 1$, so, by the lemma, $\Phi_{A_0, q^k A_0}$ is bijective, so the coefficients F_k are indeed uniquely determined by the initial condition $F_0 := I_n$ (which, by the way, guarantees that F will be invertible) and the recursive definition $F_k := -\Phi^{-1}_{A_0,q^kA_0}(A_1F_{k-1} + \cdots + A_kF_0)$.

(ii) So let $A \in \mathrm{GL}_n(\mathbf{C}(\{x\}))$. We choose any multiplicative norm $|||-|||$ on $\mathrm{Mat}_n(\mathbf{C})$, i.e., one such that $|||I_n||| = 1$ and $|||AB||| \leq |||A|||\,|||B|||$. Convergence of A means that there exist $C, R > 0$ such that $|||A_k||| \leq CR^{-k}$ for all $k \geq 0$. On the other hand, in the normed space $\mathrm{End}\,(\mathrm{Mat}_n(\mathbf{C}))$, when $k \to +\infty$:

$$q^kA_0 \to 0 \Longrightarrow \Phi_{A_0,q^kA_0} \to \Phi_{A_0,0} \Longrightarrow \Phi^{-1}_{A_0,q^kA_0} \to \Phi^{-1}_{A_0,0}.$$

Now $\Phi^{-1}_{A_0,0}$ is $X \mapsto A_0^{-1}X$ and it has norm $|||A_0^{-1}|||$, so the norm of $\Phi^{-1}_{A_0,q^kA_0}$ is uniformly bounded: there is a $D > 0$ such that $\left|\left|\left|\Phi^{-1}_{A_0,q^kA_0}\right|\right|\right| \leq D$ for all $k \geq 1$ (note that Φ_{A_0,A_0} is not invertible). In the following, we set $f_k := |||F_k|||$ and $g_k := R^kf_k$; in particular, $g_0 = f_0 = 1$. By the recursive definition (and the triangular inequality), for all $k \geq 1$:

$$\begin{aligned} f_k &= \left|\left|\left|\Phi^{-1}_{A_0,q^kA_0}(A_1F_{k-1} + \cdots + A_kF_0)\right|\right|\right| \\ &\leq D\,(|||A_1|||\,f_{k-1} + \cdots + |||A_k|||\,f_0) \\ &\leq D\left(CR^{-1}f_{k-1} + \cdots + CR^{-k}f_0\right) \\ \Longrightarrow g_k &\leq DC(g_0 + \cdots + g_{k-1}). \end{aligned}$$

So if we define $h_0 := 1$ and $h_k := DC(h_0 + \cdots + h_{k-1})$ for $k \geq 1$, an immediate induction entails that $g_k \leq h_k$ for all k. But an equally easy induction yields the equality $h_k = DC(1+DC)^{k-1}$ for $k \geq 1$, whence:

$$|||F_k||| = f_k = R^{-k}g_k \leq R^{-k}h_k = \frac{DC}{1+DC}\left(\frac{1+DC}{R}\right)^k$$

for all $k \geq 1$, whence the convergence of F.

□

COROLLARY 11.17. *Every first kind matrix A is gauge equivalent (through $GL_n(K)$) to a constant matrix $A^{(0)}$. We shall frequently write $A = M^{(0)}[A^{(0)}]$.*

PROOF. Combine the gauge transformation in $\mathrm{GL}_n(\mathbf{C}[x, x^{-1}])$ from 11.1.1.3 with the Birkhoff gauge transformation just found. □

Note however that the matrix $A^{(0)} \in \mathrm{GL}_n(\mathbf{C})$ is not necessarily $A(0)$ if A is resonant, as there have been previous transformations in the course of the algorithm; and also that $A^{(0)}$ is not unique, even if we require it to be normalized. What can be predicted is that $A^{(0)}$ and $A(0)$ have the same spectrum modulo $q^{\mathbf{Z}}$; see 11.1.1.7.

EXERCISE 11.18. Let $A_0, A_1 \in \mathrm{GL}_n(\mathbf{C})$ having the same spectrum modulo $q^{\mathbf{Z}}$. Show that there exists a first kind matrix A such that $A(0) = A_0$ and $A^{(0)} = A_1$.

11.1.1.5. *More about $e_{q,A}$.* The following subtleties are useful for classification matters. Let $A \in \mathrm{GL}_n(\mathbf{C})$. Write $A = A_sA_u$ (a multiplicative Dunford decomposition) and $A_s = P\mathrm{Diag}(c_1, \ldots, c_n)P^{-1}$. Recall from F.4 in Appendix F that, for any complex valued function f defined at least on $\mathrm{Sp}\ A = \mathrm{Sp}\ A_s = \{c_1, \ldots, c_n\}$, the matrix $f(A_s) := P\mathrm{Diag}(f(c_1), \ldots, f(c_n))P^{-1}$ is well defined (it does not depend on the particular diagonalisation of A_s). And, for any B semi-simple, $f(PBP^{-1}) = Pf(B)P^{-1}$.

We write each $c \in \mathbf{C}^*$ uniquely as $c = Q(c)\overline{c}$, where $Q(c) = q^m$, $m \in \mathbf{Z}$ and $\overline{c} \in \mathcal{C}_q$, the fundamental annulus. Then let $X(c) := x^m$ so $\sigma_q X(c) = Q(c)X(c)$. So we have $A_s = Q\overline{A_s}$, where:

$$Q := Q(A_s) = P\mathrm{Diag}(q^{m_1}, \ldots, q^{m_n})P^{-1} \text{ and } \overline{A_s} := P\mathrm{Diag}(\overline{c_1}, \ldots, \overline{c_n})P^{-1}.$$

More precisely, $A_s = X\left[\overline{A_s}\right]$ and $A = X\left[\overline{A}\right]$, where:

$$X := X(A_s) = P\mathrm{Diag}(x^{m_1}, \ldots, x^{m_n})P^{-1} \in \mathrm{GL}_n(\mathbf{C}[x, x^{-1}])$$

and where we have written $\overline{A} = \overline{A_s}A_u = A_u\overline{A_s}$ (thus also a multiplicative Dunford decomposition).

Corollary 11.19. *Every first kind matrix A is gauge equivalent (through $GL_n(K)$) to a normalized constant matrix $\overline{A^{(0)}}$.*

Morphisms and reduction. We recall from 0.4 the decomposition $c = q^{\epsilon(c)}\overline{c}$, $\epsilon(c) \in \mathbf{Z}$, $\overline{c} \in \mathcal{C}_q$ (the fundamental annulus). We complete it with some useful notations and formulas. If $m := \epsilon(c)$, let $\chi(c) := q^m$. We have with the standard definition of q-characters:

$$\xi(c) := \frac{e_{qc}}{e_c} = q^{m(m-1)/2}\overline{c}^m x^m.$$

Note that, if we had chosen a definition of q-characters such that $e_{qc} = xe_c$ for all c, the constant factor $q^{m(m-1)/2}\overline{c}^m$ would not be required.

Exercise 11.20. Show that relation $e_{qc} = xe_c$ cannot be obtained if we want $c \mapsto e_c$ continuous.

At any rate, applying the above maps to the eigenvalues of $A \in \mathrm{GL}_n(\mathbf{C})$, we draw a decomposition $A_s = \chi(A_s)\overline{A_s} \Longrightarrow \sigma_q X = KX$, where $K := \chi(A_s)$, $X := \xi(A_s)$. All those are polynomial in A, whence mutually commuting. From the above we draw that $A = K\overline{A} = X[\overline{A}]$, where $\overline{A} = \overline{A_s}A_u$ (easy computations left to you). To summarize:

$$A_s = \chi(A_s)\overline{A_s} = \xi(A_s)\left[\overline{A_s}\right].$$

Now, if $S : A \to B$ is a morphism, and if $X : \overline{A} \to A$ is as above and $Y : \overline{B} \to B$ is similar, then $R := Y^{-1}SX$ is a morphism from $\overline{A}$ to $\overline{B}$. Since they are normalized, R has coefficients in $\mathbf{C}$ and $S = YRX^{-1}$ has coefficients in $\mathbf{C}[x, x^{-1}]$.

Morphisms, reduction and e_A. Now, for $A \in \mathrm{GL}_n(\mathbf{C})$, write $A = A_sA_u$ and then $A_s = \chi(A_s)\overline{A_s}$, based on the decompositions $c_i = q^{\epsilon(c_i)}\overline{c_i}$. Letting $\overline{A} = \overline{A_s}A_u$ and $e_c = \xi(c)e_{\overline{c}}$, we may write:

$$e_{A_s} = \xi(A_s)e_{\overline{A_s}} \text{ and } e_A = \xi(A_s)e_{\overline{A}}.$$

Note that all implicit factors are polynomial in A, whence commute.

Letting $X : \overline{A} \to A$ and $Y : \overline{B} \to B$ the gauge transformations, we have $e_A = Xe_{\overline{A}}$, $e_B = Ye_{\overline{B}}$ and if R, S are as above, we deduce:

$$Se_{\overline{A}} = e_{\overline{B}}R \Longrightarrow Se_A = e_BR.$$

11.1.1.6. *Characterisations of fuchsian systems.* Almost all classical criteria (i.e., the q-analogues of those for ODEs) appear here, only the one by the growth of solutions is delayed until 11.1.2.4.

THEOREM 11.1. *Let $A \in GL_n(K)$. The following properties are equivalent:*
(i) *A is fuchsian at 0.*
(ii) *A is gauge equivalent through $GL_n(K)$ to a matrix of the first kind.*
(iii) *A is gauge equivalent through $GL_n(K)$ to some $A^{(0)} \in GL_n(\mathbf{C})$.*

PROOF. (iii) $\Rightarrow$ (i) follows from 11.1.1.1, example, (iii) and proposition, (i'). (i) $\Rightarrow$ (ii) uses a cyclic vector: let $(K^n, \Phi_A) \simeq \mathcal{D}/\mathcal{D}P$ with P as in (v) of the proposition in 11.1.1.1 (since the module is fuchsian); then $\mathcal{D}/\mathcal{D}P \simeq (K^n, \Phi_{A_P})$, where the companion matrix A_P (the notation was introduced in 7.1.3) is of the first kind since its coefficients are either 1 or those of P, which are in $\mathcal{O}$, and its determinant is $\pm a_n \in \mathcal{O}^*$; also $(K^n, \Phi_A) \simeq (K^n, \Phi_{A_P})$ implies $A \underset{K}{\sim} A_P$. (ii) $\Rightarrow$ (iii) follows from Corollary 11.17. □

COROLLARY 11.21. *If A is pure isoclinic with slope $\mu \in \mathbf{Z}$, then $A \underset{K}{\sim} x^\mu A_0$ for some $A_0 \in GL_n(\mathbf{C})$.*

PROOF. Since $(K^n, \Phi_A) \simeq (K, \Phi_{x^\mu}) \otimes (K^n, \Phi_{x^{-\mu}A})$, we have $S(A) = \mu + S(x^{-\mu}A)$, whence $S(x^{-\mu}A) = S(A) - \mu = \{0\}$, whence $x^{-\mu}A \underset{K}{\sim} A_0$ for some $A_0 \in \mathrm{GL}_n(\mathbf{C})$ (by the theorem) and this clearly entails the conclusion (as generally $F[fA_0] = fF[A_0]$ when $f \in K$). □

COROLLARY 11.22 (Normal analytic and forms in case of integral slopes). *Let $A \in GL_n(\mathbf{C}(\{x\}))$, resp. $A \in GL_n(\mathbf{C}((x)))$, with integral slopes $\mu_1 > \cdots > \mu_k$. Then there are $A_i \in GL_{r_i}(\mathbf{C})$, $i = 1, \ldots, k$, such that:*

$$A \underset{\mathbf{C}(\{x\})}{\sim} \begin{pmatrix} x^{\mu_1}A_1 & \star & \ldots & \star \\ 0 & x^{\mu_2}A_2 & \ldots & \star \\ \vdots & \vdots & \ddots & \vdots \\ 0 & 0 & \ldots & x^{\mu_k}A_k \end{pmatrix}, \text{ resp. } A \underset{\mathbf{C}((x))}{\sim} \begin{pmatrix} x^{\mu_1}A_1 & 0 & \ldots & 0 \\ 0 & x^{\mu_2}A_2 & \ldots & 0 \\ \vdots & \vdots & \ddots & \vdots \\ 0 & 0 & \ldots & x^{\mu_k}A_k \end{pmatrix}.$$

PROOF. Combine the corollary above with the results on slope filtration and graduation from Chapter 10. □

In the case of arbitrary slopes, formal normal forms (due to van der Put and Reversat [**248**]) are more complicated; see 11.1.4. The following criterion comes from [**249**]. (Also see footnote on Jurkat's criterion at the beginning of this chapter.)

COROLLARY 11.23. *The q-difference module (V, Φ) is fuchsian if, and only if, there is a lattice $\Lambda \subset V$ (see the proof) such that $\Phi(\Lambda) = \Lambda$.*

PROOF. Recall from [**35**, VII §4.1] or the exercises in [**145**, Chapter 17] that, K being the fraction field of the discrete valuation ring $\mathcal{O}$, a lattice of V is a free $\mathcal{O}$-submodule of maximal rank; equivalently, it is the submodule $\mathcal{O}e_1 \oplus \cdots \oplus \mathcal{O}e_n$ generated by a basis $(e_1, \ldots, e_n)$ of V. The lattices of K^n are the $F\mathcal{O}^n$, $F \in \mathrm{GL}_n(K)$.

So we identify (V, Φ) as (K^n, Φ_A) and notice the following basic fact: if $B = F[A]$, then $\Phi_A(\Lambda) = \Lambda \Leftrightarrow \Phi_B(F\Lambda) = F\Lambda$. The conclusion now flows from the equivalence of (i) and (ii) in the theorem and from Lemma 11.7. □

EXERCISE 11.24. Prove the stated "basic fact".

11.1.1.7. *Exponents.* As we saw, if A is of the first kind, $A \underset{K}{\sim} A^{(0)}$ for some nonunique $A^{(0)} \in \mathrm{GL}_n(\mathbf{C})$. Nondeterministic gauge transformations by shearing matrices may multiply the eigenvalues by integral powers of q. So we cannot hope for better invariant exponents than in $\mathbf{E}_q$. Actually, that much can indeed be achieved.

PROPOSITION 11.25. *Let A, B of the first kind and such that $A \underset{K}{\sim} B$. Then the multisets Sp $A(0) \pmod{q^{\mathbf{Z}}}$ and Sp $B(0) \pmod{q^{\mathbf{Z}}}$ are identical.*

PROOF. Let $F[A] = B$, $F \in \mathrm{GL}_n(K)$. Since K is the fraction field of the discrete valuation ring $\mathscr{O}$, the theorem of elementary divisors ([**33**, VII] or [**128**]) gives us a decomposition $F = CDE$, with $C, E \in \mathrm{GL}_n(\mathscr{O})$ and

$$D = \mathrm{Diag}(x^{m_1} I_{r_1}, \ldots, x^{m_k} I_{r_k}),$$

where $m_1 > \cdots > m_k$ are in $\mathbf{Z}$ and $r_1, \ldots, r_k \in \mathbf{N}^*$. Then $A' := E[A]$ is of the first kind and $A(0), A'(0)$ are conjugated by $E(0) \in \mathrm{GL}_n(\mathbf{C})$, so, as multisets, $\mathrm{Sp}\ A'(0) = \mathrm{Sp}\ A(0)$. Similarly, $B' := C^{-1}[B]$ is of the first kind and $\mathrm{Sp}\ B'(0) = \mathrm{Sp}\ B(0)$. Last, $D[A'] = B'$. So, up to replacing A by A' and B by B', we may as well start from A and B such that $D[A] = B$.

Writing A and B in blocks $(A_{i,j})$ and $(B_{i,j})$ of sizes $r_i \times r_j$, we have the relations $B_{i,j} = q^{m_i} x^{m_i - m_j} A_{i,j}$, which imply (taking in account the order on the m_i and the fact that A, B have coefficients in $\mathscr{O}$):

(1) $B_{i,j}(0) = 0$ for $i < j$, i.e., $B(0)$ is block lower triangular;
(2) $A_{i,j}(0) = 0$ for $i > j$, i.e., $A(0)$ is block upper triangular;
(3) $B_{i,i}(0) = q^{m_i} A_{i,i}(0)$.

Combining those facts, we see that, as multisets:

$$\mathrm{Sp}\ B(0) = \bigcup_{i=1}^{k} \mathrm{Sp}\ B_{i,i}(0) = \bigcup_{i=1}^{k} q^{m_i} \mathrm{Sp}\ A_{i,i}(0), \text{ while } \mathrm{Sp}\ A(0) = \bigcup_{i=1}^{k} \mathrm{Sp}\ A_{i,i}(0),$$

so indeed the multisets $\mathrm{Sp}\ A(0) \pmod{q^{\mathbf{Z}}}$ and $\mathrm{Sp}\ B(0) \pmod{q^{\mathbf{Z}}}$ are identical. □

COROLLARY 11.26. *If A is fuchsian and $A \underset{K}{\sim} B$, B of the first kind, then the multiset Sp $B(0) \pmod{q^{\mathbf{Z}}}$ in $\mathbf{E}_q$ depends on A only.*

DEFINITION 11.27. The multiset $\mathrm{Sp}\ B(0) \pmod{q^{\mathbf{Z}}}$ is the multiset of *exponents* of A.

EXERCISE 11.28. Relate exponents to the discussion on the Grothendieck group in 10.1.2.

11.1.2. Solutions of fuchsian systems. We study analytic solutions (form, singularities, growth ...) of analytic fuchsian systems, so here $K = \mathbf{C}(\{x\})$ all along.

11.1.2.1. *Resolution.*

THEOREM 11.2. *Let $A \in GL_n(\mathbf{C}(\{x\}))$ fuchsian at 0. There exist $M^{(0)} \in GL_n(\mathbf{C}(\{x\}))$ and $A^{(0)} \in GL_n(\mathbf{C})$ such that $\mathcal{X}^{(0)} := M^{(0)} e_{A^{(0)}}$ is a fundamental matricial solution of the system $\sigma_q X = AX$.*

PROOF. From Theorem 11.1, we know that $A = M^{(0)}[A^{(0)}]$ for some $M^{(0)}, A^{(0)}$ as in the statement. The conclusion then follows from the general relation between solutions of A and $F[A]$ (7.2.2). □

COROLLARY 11.29. *There is a fundamental matricial solution:*

$$\mathcal{X}^{(0)} \in GL_n\left(\mathscr{M}(\mathbf{C}^*, 0)\right)$$

and all solutions have the form $X = \mathcal{X}^{(0)}C$, $C \in \mathscr{M}(\mathbf{E}_q)^n$.

REMARK 11.30. (i) We consider the method of resolution used as a q-analogue of the Fuchs-Frobenius method for differential equations **[123, 222]**. The exponents of the classical theory (only defined modulo $\mathbf{Z}$) are here replaced by the eigenvalues of $A^{(0)}$, modulo $q^{\mathbf{Z}}$.

(ii) The "log-car" form described in **[217]** is related to the above as follows. Write $A^{(0)} = PDUP^{-1}$, D diagonal, U unipotent, $DU = UD$; then, in the notations used there, $M := M^{(0)}P$, $N := e_D e_U$.

EXAMPLE 11.31. Suppose, as Birkhoff did in **[23]**, that A is of the first kind and strictly nonresonant. Then we can take $A^{(0)} := \mathrm{Diag}(\rho_1, \ldots, \rho_n)$, $\rho_1, \ldots, \rho_n \in \mathbf{C}^*$. Writing $M^{(0)}$ as $(m_{i,j})_{i,j=1,\ldots,n}$, this gives $\mathcal{X}^{(0)} = (m_{i,j}e_{\rho_j})_{i,j=1,\ldots,n}$. This is Birkhoff's "general" form of the solutions of a fuchsian system, apart from the fact that we replaced his multivalued q-characters $x^{\log_q \rho_j}$ by e_{ρ_j}.

In order to define "normalized" solutions (in relation with the normalisation of the constant matrix in 11.1.1.5), we complete the remarks there, using the same notations, but under the convention that q-characters were chosen such that $e_{qc} = xe_c$. Write $A^{(0)} = A_s^{(0)}A_u^{(0)}$ the Dunford decomposition and $A_s^{(0)} = Q\overline{A_s^{(0)}} = X\boxed{A_s^{(0)}}$, so that $A^{(0)} = X\boxed{A^{(0)}}$. From the relations $e_{qc} = xe_c$, it readily follows that $e_{A^{(0)}} = Xe_{\overline{A^{(0)}}}$. Simplifying notations, we get:

PROPOSITION 11.32. (i) *The fuchsian system* $\sigma_q X = AX$ *admits a* normalized *fundamental matricial solution* $\mathcal{X}^{(0)} := M^{(0)}e_{A^{(0)}}$, *i.e., one such* $Sp\, A^{(0)} \subset \mathcal{C}_q$.

(ii) *Let* $\mathcal{X}^{(0)} := M^{(0)}e_{A^{(0)}}$, $\mathcal{Y}^{(0)} := N^{(0)}e_{B^{(0)}}$ *be two such normalized fundamental matricial solutions. Then there is* $C \in GL_n(\mathbf{C})$ *such that* $N^{(0)} = M^{(0)}C$, $B^{(0)} = C^{-1}A^{(0)}C$, $e_{B^{(0)}} = C^{-1}e_{A^{(0)}}C$ *and* $\mathcal{Y}^{(0)} = \mathcal{X}^{(0)}C$.

PROOF. (i) comes from the discussion above.

(ii) We have $(\sigma_q\mathcal{Y}^{(0)})(\mathcal{Y}^{(0)})^{-1} = A = (\sigma_q\mathcal{X}^{(0)})(\mathcal{X}^{(0)})^{-1}$, which, by replacing each fundamental matricial solution by its defining expression and setting $F := (M^{(0)})^{-1}N^{(0)}$, yields $(\sigma_q F)A^{(0)} = B^{(0)}F$. Expanding $F = \sum F_k x^k$, we get $q^k F_k A^{(0)} = B^{(0)}F_k$ for all k. Because of normalisation of the eigenvalues, this implies that $F_k = 0$ for $k \neq 0$. So we set $C := F_0$ and the rest follows mechanically. □

The argument is actually a particular case of Proposition 11.46.

EXERCISE 11.33. Find all solutions $N^{(0)}e_{B^{(0)}}$ (not necessarily normalized) from a given normalized $M^{(0)}e_{A^{(0)}}$.

11.1.2.2. *Singularities of solutions.* Recall that a *singularity* of a function matrix $M(x)$ is a value of x where either M or M^{-1} has a pole, and that M is *regular* at a if $M(a)$ is defined and invertible. The singular locus of $e_{q,A}$ thus avoids its poles (generically: the q-spirals $[-1/c;q]$ for c in the spectrum of A) and the zeroes of its determinant (the q-spiral $[-1;q]$). ("Generically" because there might be some compensations if A has eigenvalues in $q^{\mathbf{Z}}$.)

In a small enough punctured neighborhood V^* of 0, the meromorphic component $M^{(0)}$ of the fundamental matricial solution $\mathcal{X}^{(0)} = M^{(0)} e_{A^{(0)}}$ is *regular*, i.e., defined and with invertible values. Its singular locus there is therefore $\mathrm{Sing}_{V^*} M^{(0)} = \emptyset$, so $\mathrm{Sing}_{V^*} \mathcal{X}^{(0)} = \mathrm{Sing}_{V^*} e_{A^{(0)}}$. The latter is a finite union of half q-spirals[4] $[c;q]_0 = cq^{\mathbf{N}}$ (indeed recall that $\mathrm{Sing}\, e_{A^{(0)}} = \mathrm{Sing}_{\mathbf{C}^*} e_{A^{(0)}}$ is a finite union of q-spirals $[c;q] = cq^{\mathbf{Z}}$).

An arbitrary solution $\mathcal{X}^{(0)}C$, C elliptic, will moreover inherit the singularities of C. Since the singular locus of an elliptic function in $\mathbf{C}^*$ is a finite union of q-spirals, we conclude from 11.1.2.1:

COROLLARY 11.34. *Let $\sigma_q X = AX$ a fuchsian system. For every small enough punctured neighborhood V^* of 0, the singular locus $\mathrm{Sing}_{V^*}\mathcal{X}$ of every fundamental matricial solution $\mathcal{X}$ is contained in a finite union of half q-spirals. The set V^* can be chosen from the fuchsian system only (it is the same for all fundamental matricial solutions).*

COROLLARY 11.35. *In such a punctured neighborhood, all solutions are holomorphic out of a finite union of half q-spirals.*

In 12.1, we shall take $A \in \mathrm{GL}_n(\mathbf{C}(x))$ and refine the study of the singular locus.

11.1.2.3. *Growth criteria.* In order to define conditions of moderate growth on the solutions of q-difference systems, in particular in order to define a notion of "regular singularity" (in analogy with differential equations), we shall therefore have to avoid finite unions of half q-spirals. Also, due to the multiplicative character of q-difference theory (alluded to in the introduction of the book), we shall have to state precisely, along with conditions of moderate growth, conditions of moderate decay.

EXAMPLE 11.36. Let $f \in \mathscr{M}(\mathbf{C}^*)$, $f \neq 0$, such that $f(qx) = cf(x)$, $c \in \mathbf{C}^*$, $c \neq 1$. Let $x_0 \in \mathbf{C}^*$ such that $f(x_0) \neq 0$ (implicitly we mean that f is defined at x_0). Then $f(q^m x) = c^m f(x)$, which implies that, when $m \to +\infty$, $\log|f(c^m x_0)| \sim \alpha m$ for some $\alpha \neq 0$. Since $\log|q^m x_0| \sim \beta m$ for some $\beta \neq 0$, we see that $\log|f(x)| \sim \gamma \log|x|$ for some $\gamma \neq 0$ when $x \to 0$ along the half q-spiral $[x_0;q]_0 := x_0 q^{\mathbf{N}}$.

DEFINITION 11.37. We say that a complex vector valued function $X(x)$ with coefficients in $\mathscr{M}(\mathbf{C}^*, 0)$ has *moderate growth along half q-spirals near* 0, or simply *moderate growth* (near 0), if, for some punctured neighborhood V^* of 0 and almost all half q-spirals $S \subset V^*$ (meaning: all but a finite number of them), $X(x)$ is defined on S and $\dfrac{\log\|X(x)\|}{|\log|x||}$ is bounded above on S.

This means that for x on S, x near 0, $\|X(x)\| < c\,|x|^N$ where c, N depend on S only. Of course any norm will do.

[4]This is rigorously true if V^* is a punctured disk, or even a punctured star shaped set; otherwise, this is only true as germs.

REMARK 11.38. A slightly different condition (more subtle) is stated in was obtained by Julien Roques (private communication; not yet published). The proof there of the criterion of regular singularity (for fuchsian systems) is totally different from the proof here.

EXERCISE 11.39. Let $A \in \mathrm{GL}_n(\mathbf{C})$. Prove that e_A and e_A^{-1} have moderate growth and decay along half q-spirals near 0.

PROPOSITION 11.40. (i) *Elliptic functions, vectors, matrices have moderate growth.*

(ii) *Meromorphic functions, vectors, matrices have moderate growth.*

(iii) *Linear combinations and (tensor) products preserve moderate growth.*

(iv) *Ramification strictly preserves moderate growth.*

PROOF. All statements are easy and left to the reader. The third one says in particular that there is a $\mathbf{C}$-algebra of germs with moderate growth. The fourth one means that, if $x' := x^{1/r}$ and $X'(x') = X(x)$, then X has moderate growth if, and only if, X' has (but, of course, the latter with respect to half q'-spirals, $q' := q^{1/r}$). □

11.1.2.4. *Regular singularity.*

DEFINITION 11.41. Let $A \in \mathrm{GL}_n(\mathbf{C}(\{x\}))$. Then A *is regular singular at* 0 or *has a regular singularity at* 0 if the system $\sigma_q X = AX$ admits a fundamental matricial solution $\mathcal{X}$ such that $\mathcal{X}$ and $\mathcal{X}^{-1}$ have moderate growth at 0.

By Proposition 11.40, this remains true for every fundamental matricial solution since they all have the form $\mathcal{X}C$, $C \in \mathrm{GL}_n(\mathscr{M}(\mathbf{E}_q))$. For the same reason, all solutions have moderate growth, which in return implies that any fundamental matricial solution $\mathcal{X}$ has moderate growth (since the columns of $\mathcal{X}$ are solutions). Thus, since we know a priori that there are fundamental matricial solutions by the theorem of Praagman in 7.3, the condition on $\mathcal{X}$ is equivalent to: all solutions have moderate growth (here we omit the condition on $\mathcal{X}^{-1}$).

As for the moderate growth of $\mathcal{X}^{-1}$, it is obviously equivalent to the same for its transpose ${}^t\mathcal{X}^{-1} = \mathcal{X}^\vee$, which we can consider as the dual of $\mathcal{X}$.

PROPOSITION 11.42. (i) *For A to be regular singular at 0, it is necessary and sufficient that all solutions of A and its dual $A^\vee$ have moderate growth at* 0.

(ii) *The property is intrinsic: A is singular regular if, and only if, $F[A]$ is (where $F \in GL_n(\mathbf{C}(\{x\}))$).*

(iii) *It is strictly preserved by ramification.*

PROOF. (i) Note that $\sigma_q\mathcal{X} = A\mathcal{X} \Leftrightarrow \sigma_q\mathcal{X}^\vee = A^\vee\mathcal{X}^\vee$ (immediate calculation; inversion and transposition each change the order in the product, but dualisation does not!). So it follows from the discussion just before the proposition.

(ii) A fundamental matricial solution of $F[A]$ is $F\mathcal{X}$ and we apply Proposition 11.40.

(iii) Again by Proposition 11.40.

□

The following properties will follow from the theorem (the above ones also, but we shall need them to prove it!).

EXERCISE 11.43. Prove directly (i.e., without appealing to the theorem):

(i) If two among A, B, $A \otimes B$ are regular singular, so is the third.
(ii) If A and B are regular singular, so is $\underline{\mathrm{Hom}}(A, B) := \underline{\mathrm{Hom}}(M_A, M_B)$.

THEOREM 11.3. *Let $A \in GL_n(\mathbf{C}(\{x\}))$. Then A is fuchsian (at 0) if, and only if, it is singular regular (at 0).*

PROOF. If A is fuchsian, $A \underset{\mathbf{C}(\{x\})}{\sim} A^{(0)} \in \mathrm{GL}_n(\mathbf{C})$ (Theorem 11.1), so, by (ii) of the proposition above, we may as well assume that $A \in \mathrm{GL}_n(\mathbf{C})$. A solution X then satisfies:

$$\begin{aligned}\forall m \geq 0 \,,\; X(q^m x_0) = A^m X(x_0) &\Longrightarrow \forall m \geq 0 \,,\; \log \|X(q^m x_0)\| \leq m \log |||A||| \\ &\qquad\qquad + \log \|X(x_0)\| \\ &\Longrightarrow \log \|X(x)\| \leq \alpha \left|\log |x|\right| + \beta \text{ on } [x_0; q]_0,\end{aligned}$$

where $\alpha := \log |||A||| \dfrac{\log |x|}{\log |q|}$ (which is > 0 when x is near 0) and $\beta := \log \|X(x_0)\| - \dfrac{\log |||A||| \log |x_0|}{\log |q|}$, so all solutions have moderate growth. Now, $A^\vee \in \mathrm{GL}_n(\mathbf{C})$, so the same conclusion applies to it and A is singular regular.

Conversely, if A is singular regular, $S(A), S(A^\vee) \subset \mathbf{Q}_+$ by the lemma below. Since $S(A^\vee) = -S(A)$, this implies that $S(A) = \{0\}$, so A is fuchsian. □

LEMMA 11.44. *If all solutions of $\sigma_q X = AX$ have moderate growth, then $S(A) \subset \mathbf{Q}_+$.*

PROOF. We can assume (by ramification, see the proposition in 11.1.2.3) that $S(A) \subset \mathbf{Z}$. By the corollaries of the theorem in 11.1.1.6, up to a gauge transform (thus using again 11.1.2.3), we can write:

$$A = \begin{pmatrix} x^{\mu_1} A_1 & \star & \ldots & \star \\ 0 & x^{\mu_2} A_2 & \ldots & \star \\ \vdots & \vdots & \ddots & \vdots \\ 0 & 0 & \ldots & x^{\mu_k} A_k \end{pmatrix},$$

all $A_i \in \mathrm{GL}_{r_i}(\mathbf{C})$. Choose a fundamental matricial solution $\mathcal{X}$ (using once more Praagman's theorem in 7.3) and write it in $r_i \times r_j$ blocks $\mathcal{X}_{i,j}$. Then, for $j = 1, \ldots, k$:

$$\sigma_q \mathcal{X}_{k,j} = x^{\mu_k} A_k \mathcal{X}_{k,j} \Longrightarrow \mathcal{X}_{k,j}(q^m x) = q^{\mu_k m(m-1)/2} \mathcal{X}_{k,j}(x).$$

If $\mu_k < 0$, this is compatible with moderate growth only if $\mathcal{X}_{k,j} = 0$. Since this can't be true for all j (because $\mathcal{X}$ is invertible), $\mu_k \geq 0$, whence $S(M) \subset \mathbf{N}$. □

EXERCISE 11.45. Define and study a "moderate decay" condition in such a way as to obtain a criterion of fuchsianity based on moderate growth and decay of a fundamental matricial solution.

11.1.2.5. *Morphisms and classification.* Let A, B be fuchsian, or, more generally pure isoclinic of the same slope μ. Then $\underline{\mathrm{Hom}}(A, B) = \underline{\mathrm{Hom}}(M_A, M_B)$ is fuchsian and any morphism $F : A \to B$ can be seen as a solution in $\mathbf{C}(\{x\})$ of this fuchsian q-difference module. That fact could be used (along with the previous results here) to prove what follows, but we shall rather tackle it directly.

PROPOSITION 11.46. *Let A, B be fuchsian and $F : A \to B$ a morphism.*

(i) *If $A \in GL_n(\mathbf{C})$ and $B \in GL_m(\mathbf{C})$ are normalized (this was defined in 11.1.1.2), then $F \in Mat_{m,n}(\mathbf{C})$.*

(ii) *If $A \in GL_n(\mathbf{C})$ and $B \in GL_m(\mathbf{C})$ but not supposed to be normalized, then $F \in Mat_{m,n}(\mathbf{C}[x, x^{-1}])$.*

PROOF. Expanding $F = \sum F_k x^k$, the relation $(\sigma_q F)A = BF$ boils down to $q^k F_k A_k = B_k F_k$ for all k.

(i) If A, B are normalized, $\mathrm{Sp}(q^k A_k) \cap \mathrm{Sp}(B_k) = \emptyset$ for $k \neq 0$, so, by the usual lemma, $F = F_0$.

(ii) In general, $\mathrm{Sp}(q^k A_k) \cap \mathrm{Sp}(B_k) = \emptyset$ for all but a finite number of k and F has a finite number of terms $F_k x^k$.

□

COROLLARY 11.47. *Let A, B be pure isoclinic of the same slope μ and let $F \in Mat_{m,n}(\mathscr{M}(\mathbf{C}^*, 0))$ such that $(\sigma_q F)A = BF$. Then $F \in Mat_{m,n}(\mathbf{C}(\{x\}))$ (so it is actually a morphism).*

PROOF. We only sketch it. Ramification allows us to take $\mu \in \mathbf{Z}$. Then the relation stated for A, B is equivalent to the same for $x^{-\mu}A, x^{-\mu}B$, so we may assume A, B fuchsian. We write them as $A = M^{(0)}[A^{(0)}]$, $B = N^{(0)}[B^{(0)}]$ and see that $G := {N^{(0)}}^{-1} F M^{(0)}$ satisfies $(\sigma_q G)A^{(0)} = B^{(0)}G$, so the proposition (ii) applies to G. □

EXERCISE 11.48. (i) Do it all using $\underline{\mathrm{Hom}}(A, B)$.

(ii) In **[217]**, a stronger definition of normalisation was used (involving a fixed arbitrary linear order on $\mathbf{E}_q$). Show how it allows for more "rigidity" (uniqueness properties of morphisms).

We apply those facts to the general situation of a morphism $F : A \to B$ between fuchsian systems. Writing first $A = M^{(0)}[A^{(0)}]$ and $B = N^{(0)}[B^{(0)}]$ as above; and then $A^{(0)} = X\left[\overline{A^{(0)}}\right]$ and $B^{(0)} = Y\left[\overline{B^{(0)}}\right]$ as in 11.1.2.1, we get a commutative diagram:

$$\begin{array}{ccccc} \overline{A^{(0)}} & \xrightarrow{X} & A^{(0)} & \xrightarrow{M^{(0)}} & A. \\ {\scriptstyle R}\downarrow & & & & \downarrow{\scriptstyle F} \\ \overline{B^{(0)}} & \xrightarrow{Y} & B^{(0)} & \xrightarrow{N^{(0)}} & B \end{array}$$

The leftmost vertical arrow $R := \left(N^{(0)}Y\right)^{-1} F \left(M^{(0)}X\right)$ was not given, it has been defined to complete the diagram. So it is a morphism from $\overline{A^{(0)}}$ to $\overline{B^{(0)}}$, so, by the proposition, $R \in \mathrm{Mat}_{n,m}(\mathbf{C})$ and actually $R\overline{A^{(0)}} = \overline{B^{(0)}}R$. Then we know we can write the arbitrary morphism F as:

$$F = \left(N^{(0)}Y\right) R \left(M^{(0)}X\right)^{-1},$$

with rather strong constraints on the various factors. In terms of fundamental matricial solutions, this translates as follows. We can a priori write $FM^{(0)}e_{A^{(0)}} = N^{(0)}e_{B^{(0)}}C$, with C elliptic. We also know from 11.1.2.1 that $e_{A^{(0)}} = Xe_{\overline{A^{(0)}}}$ and $e_{B^{(0)}} = Ye_{\overline{B^{(0)}}}$, so:

$$FM^{(0)}Xe_{\overline{A^{(0)}}} = N^{(0)}Ye_{\overline{B^{(0)}}}C \Longleftrightarrow Re_{\overline{A^{(0)}}} = e_{\overline{B^{(0)}}}C.$$

However, since $R\overline{A^{(0)}} = \overline{B^{(0)}}R$ (as we just saw), by 11.1.2.5, we can deduce that $Re_{\overline{A^{(0)}}} = e_{\overline{B^{(0)}}}R$, so $C = R$. So we deduce the much stronger information:

COROLLARY 11.49. *Let $F : A \to B$ a morphism between fuchsian systems and let $\mathcal{X}^{(0)} := M^{(0)}e_{A^{(0)}}$, $\mathcal{Y}^{(0)} := N^{(0)}e_{B^{(0)}}$ fundamental matricial solutions obtained by the Fuchs-Frobenius method. Then $F\mathcal{X}^{(0)} = \mathcal{Y}^{(0)}R$ for some $R \in Mat_{n,m}(\mathbf{C})$.*

This important fact allows to use complex matrices instead of elliptic ones for classification. It was proved in **[217]** by rather different methods.

EXERCISE 11.50. Note that a fundamental matricial solution $\mathcal{X}$ of the q-difference $\sigma_q X = AX$ can be considered as a kind of morphism $I_n \overset{\mathcal{X}}{\to} A$ ("a kind of" because it has coefficients in a q-difference extension). So complete the previous diagram by a vertical arrow $F^{(0)}$ in the middle and on the left by:

$$\begin{array}{ccc}
I_n & \xrightarrow{\;e_{\overline{A^{(0)}}}\;} & \overline{A^{(0)}} \\
{\scriptstyle R}\downarrow & & \downarrow{\scriptstyle R} \\
I_n & \xrightarrow[\;e_{\overline{B^{(0)}}}\;]{} & \overline{B^{(0)}}
\end{array}$$

to deduce that $F^{(0)}e_{\overline{A^{(0)}}} = e_{\overline{B^{(0)}}}R$.

11.1.3. Local structure of fuchsian q-difference modules.

11.1.3.1. We noticed in 11.1.2 that the Grothendieck group of the category of q-difference modules (formal or analytic) involves information about slopes and about exponents. The information about slopes was translated in terms of Newton polygon and then given a structural counterpart in the slope filtration and graduation.

We now restrict to fuchsian modules and do something similar for exponents. One striking difference is that the formal and analytic settings are then, in many ways, equivalent. We shall give no proofs of the following statements, as all the (relatively) hard work has already been done and the deductions are easy.

EXERCISE 11.51. Deduce all that follows from the results of 11.1.1 and 11.1.2.

11.1.3.2. We let $\mathcal{E}_f^{(0)}$ the category of analytic q-difference modules fuchsian at 0 and $\mathcal{P}^{(0)}$ its full subcategory of objects (K^n, Φ_A) with $A \in \mathrm{GL}_n(\mathbf{C})$. Also let $\hat{\mathcal{E}}_f^{(0)}$ the category of formal q-difference modules fuchsian at 0. There is an inclusion functor from $\mathcal{P}^{(0)}$ to $\mathcal{E}_f^{(0)}$ and a formalization functor from $\mathcal{E}_f^{(0)}$ to $\hat{\mathcal{E}}_f^{(0)}$. The following is of course quite specific to the fuchsian setting.

THEOREM 11.4. *Both these functors are equivalence of categories.*

11.1.3.3. From now on, $K = \mathbf{C}(\{x\})$ but, as noted, much applies to $\mathbf{C}((x))$ as well. We write $E(M) \subset \mathbf{E}_q$ the multiset of exponents of a fuchsian q-difference module. When applying Jordan-Hölder theorem, we use the:

LEMMA 11.52. (i) *Simple fuchsian modules have the form (K, Φ_c) and the isomorphism class of such a module is uniquely determined by $\overline{c} \in \mathbf{E}_q$.*

(ii) *As a group[5] for the tensor product, the Picard group of $\mathcal{E}_f^{(0)}$ is $Pic\left(\mathcal{E}_f^{(0)}\right) \simeq \mathbf{E}_q$.*

[5]The Picard group of a tensor category is the group of its isomorphism classes of invertible objects (for the tensor product).

THEOREM 11.5. (i) *Let* $0 \to M' \to M \to M'' \to 0$ *an exact sequence. Then (as multisets, i.e., taking in account multiplicities):*

$$E(M) = E(M') \cup E(M'').$$

(ii) $E(M_1 \otimes M_2) = E(M_1)E(M_2)$ *and* $E(M^\vee) = E(M)^{-1}$.

COROLLARY 11.53. (i) *If* $E(M) \cap E(N) = \emptyset$, *then the only morphism* $M \to N$ *is* 0.

(ii) *If* $M_1, M_2 \subset M$ *and* $E(M_1) \cap E(M_2) = \emptyset$, *then their sum is direct.*

So summing all the submodules of M having only exponent $\overline{c}$ yields a biggest submodule $M_{\overline{c}}$ with the same property, and the sum of all the various $M_{\overline{c}}$ is direct. Actually each such $M_{\overline{c}}$ has maximal possible rank, as follows from the above study:

PROPOSITION 11.54. *M is the direct sum of its submodules* $M_{\overline{c}}$, $\overline{c} \in E(M)$.

11.1.3.4. Call "pseudo-pure" a fuchsian q-difference module with only one exponent, say $\overline{c}$; and "unipotent" a pseudo-pure q-difference module with exponent 1 (those notations are of temporary nature).

PROPOSITION 11.55. (i) *A pseudo-pure module has the form* $(K, \Phi_c) \otimes M'$, *the second factor* M' *being unipotent.*

(ii) *A unipotent q-difference module has the form* (K^n, Φ_U), *where* $U \in GL_n(\mathbf{C})$ *is unipotent and determined up to conjugacy.*

Thus every fuchsian module can be decomposed canonically as:

$$\bigoplus_{\substack{c \pmod{q^{\mathbf{Z}}} \\ U \text{ modulo conjugacy}}} (K, \Phi_c) \otimes (K^n, \Phi_U).$$

EXERCISE 11.56. Translate in terms of scalar fuchsian equations.

11.1.4. Application to the formal classification of q-difference modules. The main result of this subsection (the formal classification of pure q-difference modules with arbitrary slopes) is due to van der Put and Reversat [**248**, 1.4, Corollary 1.6]. Its (rather complicated) proof will be detailed in [CV], along with a more general study of what happens when slopes are nonintegral. Here we describe only the easy parts of the proof.

Recall that a q-difference module M is *indecomposable* if:

$$M = M_1 \oplus M_2 \Longrightarrow (M_1 = 0 \text{ or } M_2 = 0)\,;$$

and that M is *irreducible* or *simple* if:

$$0 \subset M' \subset M \Longrightarrow (M' = 0 \text{ or } M' = M)\,.$$

Obviously the latter condition is stronger than the former. (The fact that it is *strictly* stronger gave rise to the study of extensions in Section 8.3.2.) Also, we shall resort to the notations (in which $m \in \mathbf{N}^*$ and $v \in \mathbf{C}((x))$):

$$U_m := \begin{pmatrix} 1 & 1 & 0 & \dots & 0 & 0 \\ 0 & 1 & 1 & \dots & 0 & 0 \\ 0 & 0 & 1 & \ddots & 0 & 0 \\ \vdots & \vdots & \vdots & \ddots & \ddots & \vdots \\ 0 & 0 & 0 & \dots & 1 & 1 \\ 0 & 0 & 0 & \dots & 0 & 1 \end{pmatrix}, \quad \text{and } C_v := \begin{pmatrix} 0 & 1 & 0 & \dots & 0 & 0 \\ 0 & 0 & 1 & \dots & 0 & 0 \\ 0 & 0 & 0 & \ddots & 0 & 0 \\ \vdots & \vdots & \vdots & \ddots & \ddots & \vdots \\ 0 & 0 & 0 & \dots & 0 & 1 \\ v & 0 & 0 & \dots & 0 & 0 \end{pmatrix}.$$

11.1.4.1. *Usefulness of indecomposable modules in the formal classification.* In Chapter 10, we used the theorem of Jordan-Hölder to ensure the existence and (partial) unicity of a kind of decomposition of a q-difference module into simple ones. Here we shall use the Krull-Remak-Schmidt theorem on indecomposable modules (see [**40**, VIII §2.5, thm. 2] or [**145**, VI §9, thm. 9.5]). In our context, it asserts that any q-difference module is, in a unique way (up to isomorphisms) a direct sum of indecomposable q-difference modules; here, a q-difference module is said to be *indecomposable* if it does not split into the direct sum of two nontrivial q-difference modules. In the form stated in [**40**], the theorem applies because our q-difference modules have finite length as $\mathcal{D}_{K,q}$-modules; and in the form stated in [**145**], it applies because those modules are artinian (every strictly descending sequence of submodules is finite) and noetherian (same for strictly ascending sequences). Note that both conditions (finite length or artinian-noetherian) are actually equivalent.

As a consequence, the formal classification of general q-difference modules is completely reduced to that of the indecomposable ones. Now, it follows from the formal splitting of the canonical slope filtration (proved in 10.2.1) that a formal q-difference module with more than one slope splits, thus that an indecomposable q-difference module is necessarily pure isoclinic.

One of the main results of [**248**] is that every indecomposable q-difference module over $\mathbf{C}((x))$ has the form $M \otimes (K^m, \Phi_{U_m})$ for some irreducible q-difference module M (unique up to isomorphism) and some $m \in \mathbf{N}^*$, where $U_m \in \mathrm{GL}_m(\mathbf{C})$ denotes the unipotent Jordan block U_m (see beginning of section). So we now turn to irreducible q-difference modules.

11.1.4.2. *The structure of irreducible q-difference modules over* $\mathbf{C}((x))$. We refer to the "important example" of Subsubsection 8.3.4.2 (and its follow up in Subsection 10.1.5 but using notations[6] from [**43**]. Let $r \in \mathbf{N}^*$, $d \in \mathbf{Z}$ and $c \in \mathbf{C}^*$; then:

$$E(r,d,c) := \mathcal{D}_{K,q}/\mathcal{D}_{K,q}L, \text{ where } L := \sigma_q^r - u, \quad u := q^{d(r-1)/2}cx^d.$$

The factor $q^{d(r-1)/2}$ is clearly inessential, but it appeared naturally in 8.3.4.2 and it does simplify some formulas in [**43, 248**]. It follows from 8.3.4.2 (or 8.1.3.1) that:

$$E(r,d,c) \simeq (K^r, \Phi_A) \text{ where } A := A_L^\vee = C_v,\ v := u^{-1}.$$

(See beginning of section.) By Subsection 10.1.5, the q-difference module $E(r,d,c)$ is irreducible if, and only if, $r \wedge d = 1$, which we now assume. Then other main result of [**248**] is that every irreducible q-difference module is isomorphic to some $E(r,d,c)$, where $r \in \mathbf{N}^*$, $d \in \mathbf{Z}$ are such that $r \wedge d = 1$ and uniquely determined; and $c \in \mathbf{C}^*$ is unique modulo $q^{\mathbf{Z}}$. The uniqueness of r, d is easy to prove, the statement about c follows directly from Subsection 10.1.5. We summarize the discussion by the following theorem. (Complete proof maybe found in [**248**]; a different approach by Virginie Bugeaud appears in [**43**]; also see [CV].)

THEOREM 11.6 (van der Put,Reversat). *Every q-difference module over* $\mathbf{C}((x))$ *is isomorphic to a direct sum:*

$$\bigoplus E(r_i,d_i,c_i) \otimes (K^{m_i}, \Phi_{U_{m_i}}),$$

all $r_i, m_i \in \mathbf{N}^*$, $d_i \in \mathbf{Z}$ *being uniquely determined and* $c_i \in \mathbf{C}^*$ *up to* $q^{\mathbf{Z}}$. *Thus a* normal form *is defined by prescribing that all* c_i *lie in the fundamental annulus* $\mathcal{C}_q$.

[6]We slightly differ from *loc. cit.* as regards signs, because of the convention $|q| > 1$ there.

EXERCISE 11.57. Using either this section or [**40**, §3.2, thm 1], show that, if M is an irreducible q-difference module, then $\mathrm{End}(M) = \mathbf{C}$.

11.1.5. Parameterized analytic families of q-difference equations. We shall have use (see for instance Subsection 12.2.5 and, above all, in [**196**]) for q-difference equations with coefficients depending analytically on a parameter. The following classical lemma from complex analysis in several variables (see for instance [**119**, Lemma 2.2.11]) will be useful: *If $(a_m(t))_{m\geq 0}$ is a sequence of analytic functions on a polydisk Δ and $F > 0$ is such that $\sup_\Delta |a_m(t)| \leq F^m$ for all m, then $a(t,x) := \sum a_m(t)x^m$ is analytic on $\Delta \times \mathring{D}(0, 1/F)$.* Before stating and proving sufficient conditions for the existence of solutions also depending analytically on that parameter, we return for a moment to resolution by power series.

11.1.5.1. *Plain direct resolution with power series.* We intend to solve the q-difference equation with prescribed initial condition $X(0) := X_0 \in \mathbf{C}^n$:

$$\sigma_q X = AX, \text{ where } A(x) := A_0 + A_1 x + \cdots \in \mathrm{GL}_n(\mathbf{C}\{x\}) \text{ (so that } A_0 \in \mathrm{GL}_n(\mathbf{C})).$$

We hope for a power series solution $X := X_0 + X_1 x + \cdots \in \mathbf{C}\{x\}^n$. This requires of course a compatibility condition $A_0 X_0 = X_0$ and then reduces to the recursive relations $q^m X_m = A_0 X_m + \cdots + A_m X_0$, i.e., $(q^m I_n - A_0)X_m = A_1 X_{m-1} + \cdots + A_m X_0$ for al $m \geq 1$. It is then natural, as in Subsection 11.1.1, to require a nonresonancy condition ensuring that all $q^m I_n - A_0$ are invertible; then the above relation provides a formal iterative resolution and one may hope for convergence.

PROPOSITION 11.58. *Let $A(x) := A_0 + A_1 x + \cdots \in GL_n(\mathbf{C}\{x\})$ be such that $Sp(A_0)$ does not meet $q^{\mathbf{N}^*}$ and let $X_0 \in \mathbf{C}^n$ such that $A_0 X_0 = X_0$. Then the q-difference equation with initial condition $\sigma_q X = AX$, $X(0) := X_0$, admits a unique formal solution $X := X_0 + X_1 x + \cdots \in \mathbf{C}[[x]]^n$ and that solution has a strictly positive radius of convergence, i.e., $X \in \mathbf{C}\{x\}^n$.*

PROOF. Taking in account the previous computation and the nonresonancy assumption, we immediately deduce the existence and uniqueness of the formal solution iteratively defined by formula $X_m := (q^m I_n - A_0)^{-1}(A_1 X_{m-1} + \cdots + A_m X_0)$. Then the convergence is a very particular case of what will be proved with parameter in the next subsubsection, so we leave it as an exercise. (Some care should be necessary when adapting condition (3) of Theorem 11.7.) □

11.1.5.2. *Systems with parameters.* We now assume a parameter $t \in \Delta$, when Δ is (for instance) a polydisk in some $\mathbf{C}^d$ (the reader may, for simplicity, take the classical one parameter case of a disk in $\mathbf{C}$ centered at 0). We consider an invertible functional matrix $A(t,x) := \sum_{m\geq 0} A_m(t)x^m$ supposed analytic in $\Delta \times \mathring{\mathrm{D}}(0,R)$ for some $R > 0$. We want to solve the functional equation $X(t,qx) = A(t,x)X(t,x)$ with help of a power series (in x) solution $X(t,x) := X_0(t) + X_1(t)x + \cdots$ hopefully analytic in some (maybe smaller) domain inside $\Delta \times \mathring{\mathrm{D}}(0,R)$. As always, matrix norms are multiplicative, i.e., such that $|||I_n||| = 1$ and $|||PQ||| \leq |||P|||\,|||Q|||$.

THEOREM 11.7. *We make the following assumptions:*

(1) *There exist positive constants $M_m > 0$ such that:*

$$\forall m \geq 0\,,\ \forall t \in \Delta\,,\ |||A_m(t)||| \leq M_m \text{ and } \sum_{m\geq 0} M_m R^m < \infty.$$

(2) *There is a "uniformly bounded initial condition"* $X_0(t)$ *analytic in* Δ *and with values in* $\mathbf{C}^n$*:*

$$\forall t \in \Delta\ ,\ A_0(t)X_0(t) = X_0(t) \text{ and } \|X_0(t)\| \leq \alpha_0 \text{ (where } \alpha_0 > 0 \text{ is fixed).}$$

(3) *The* q*-difference equation is "uniformly nonresonant", i.e., there is a constant* $C > 0$ *such that:*

$$\forall m \geq 1\ ,\ \forall t \in \Delta\ ,\ \left|\left|\left|(q^m I_n - A_0(t))^{-1}\right|\right|\right| \leq C.$$

Then the q*-difference equation* $X(t, qx) = A(t, x)X(t, x)$ *admits a unique formal solution* $X(t, x) := X_0(t) + X_1(t)x + \cdots \in \mathscr{O}(\Delta)[[x]]$ *and that solution is analytic in* $\Delta \times \overset{\circ}{D}(0, R')$ *for some* $R' \in]0, R]$.

PROOF. The constant term $X_0(t)$ being given, the same formal computation as before yields the recursive relation:

$$\forall m \geq 1\ ,\ \forall t \in \Delta\ ,\ X_m(t) := (q^m I_n - A_0(t))^{-1}(A_1(t)X_{m-1}(t) + \cdots + A_m(t)X_0(t)).$$

Taking in account the given bounds:

$$\|X_m(t)\| \leq C\left(M_1 \|X_{m-1}(t)\| + \cdots + M_m \|X_0(t)\|\right).$$

We define recursively a real sequence with first term α_0 and:

$$\forall m \geq 1\ ,\ \alpha_m := C(M_1\alpha_{m-1} + \cdots + M_m\alpha_0).$$

Immediate induction gives $\|X_m(t)\| \leq \alpha_m$ for all $m \geq 1$ and $t \in \Delta$, so the formal series $X(t)(x)$ is dominated (in the sense of Cauchy's method of majorants) by:

$$\alpha(x) := \sum \alpha_m x^m = \frac{\alpha_0}{1 - CxM(x)}, \text{ where we set } M(x) := \sum \alpha_m x^m.$$

Thus we can take $R' := \min(R, 1/CM(R))$ (recall that $M(R) < \infty$). □

It is clear that the condition of uniform boundedness of initial conditions is automatically satisfied over any compact subset of Δ. This is also the case of the uniform nonresonancy condition. Since we shall detail a similar fact in the case of parameterized reduction (Subsection 12.2.5 of the next chapter), we leave it to you to check it in the simpler framework here. At any rate, since Δ can be covered by relatively compact open subsets, we deduce:

COROLLARY 11.59. *The conclusions above hold even if we do not require uniform boundedness of initial conditions and uniform nonresonancy.*

11.2. Confluence of fuchsian systems (local study)

We summarize here[7] (and, as regards the global consequences, in Section 12.3 of the next chapter) results from **[217]**; with however, at some points, a slight departure from the methods of *loc. cit.*. This will be completed in 13.3.4 by a brief summary of the corresponding facts from **[218]**.

11.2.1. Some generalities. We intend to display some systematic patterns underlying the well known degeneracy of q-difference systems to ordinary differential equations when $q \to 1$, of which we gave many examples in Chapters 1, 2, 4, and again in various places.

[7] There are similar works by Anne Duval and by Julien Roques involving difference equations, or even mixtures of differential, difference and q-difference equations, see **[76, 77, 79, 206]**; also see reference to related work by Thomas Dreyfus in 12.3.

11.2.1.1. *Position of the problem.* The general pattern we shall follow is this. Let $\sigma_q X = A_q X$ a q-difference system, where $A_q(x)$ depends on q in such a way that there is a limit (we shall later be more precise about the nature of the convergence):

$$\tilde{B}(x) := \lim_{q \to 1} \frac{A_q(x) - I_n}{q - 1}.$$

Then we hope that *some* fundamental matricial solution $\mathcal{X}_q(x)$ degenerates, when $q \to 1$, to a fundamental matricial solution $\tilde{\mathcal{X}}(x)$ of the differential equation $\delta X = \tilde{B}X$ (recall the Euler differential operator $\delta := x\frac{d}{dx}$). Of course, since $\mathcal{X}_q$ can be multiplied by arbitrary $C_q \in \mathrm{GL}_n(\mathscr{M}(\mathbf{E}_q))$ (with arbitrary dependence on q), this cannot be true for an arbitrary family $(\mathcal{X}_q)$, some kind of "rigidification" will be needed.

Now we restrict to rational system matrices A_q, supposed fuchsian at 0 and ∞; moreover $\tilde{B}$ is supposed rational, with intermediate poles $\tilde{x}_1, \ldots, \tilde{x}_r \in \mathbf{C}^*$ (there might also be polarity at 0 and ∞). Under adequate assumptions, there would be a canonical choice of solutions $\mathcal{X}_q^{(0)}$ and $\mathcal{X}_q^{(\infty)}$ at 0 and at ∞ and a well defined "connection matrix" $P_q := \left(\mathcal{X}_q^{(\infty)}\right)^{-1} \mathcal{X}_q^{(0)}$ (this will be formally introduced in Chapter 12). The latter being q-elliptic comes with q-spirals of singularities. Then we may hope, and will indeed find under favorable circumstances:

- Those q-spirals *condense* into continuous logarithmic spirals going from 0 to ∞, in $\mathbf{S}$; there is one through each of the poles $\tilde{x}_i$, plus may be one coming from the q-characters and q-logarithms used in the local solutions $\mathcal{X}_q^{(0)}, \mathcal{X}_q^{(\infty)}$. Along those continuous spirals, P_q has no limit.
- The complementary subset $\mathbf{S} \setminus \bigcup\{$ said spirals $\}$ has $(r+1)$ connected components $\tilde{\Omega}_i$, $i = 0, \ldots, r$; and, on each such $\tilde{\Omega}_i$, the connection matrix P_q degenerates into a *constant matrix* $\tilde{P}_i \in \mathrm{GL}_n(\mathbf{C})$.
- the "quotient matrices" $\tilde{P}_i^{-1}\tilde{P}_{i-1}$ are the monodromy matrices of $\tilde{X}$ at the $\tilde{x}_i$ (relative to adequate bases).

A word about the use of the word "confluence". It has been traditional to say that (for instance) the irregular differential equation $y' = \frac{1}{x^2}y$ may be (for instance) obtained from the fuchsian equation $y' = \left(\frac{\epsilon^{-1}}{x} - \frac{\epsilon^{-1}}{x+\epsilon}\right) y$ by letting $\epsilon \to 0$, so the two regular singularities $0, -\epsilon$ merge (in French: *elles confluent*, like rivers) into the irregular singularity 0 (about confluence of singularities, see **[123]**). Thus a "confluent hypergeometric equation" **[156, 169]** may be obtained this way from plain (fuchsian) hypergeometric equations. We nicknamed "confluence" the condensation of discrete spirals of singularities into continuous ones with that image as a justification. Note that this is one more sign that differential equations are more degenerate that q-difference equations.

11.2.1.2. *A first look at general hypotheses.* In accordance with the examples in Chapters 1, 2, 4, and more, we shall have $q \to 1$ along a "continuous spiral direction", so the discrete q-spirals may have a chance to condense. More precisely, we fix such a direction in the form $q_0 := e^{2i\pi\tau_0}$, $\Im\tau_0 > 0$, thus defining a continuous logarithmic spiral $q_0^{\mathbf{R}} := e^{2i\pi\tau_0\mathbf{R}}$. Then we shall set[8] $q := q_0^\epsilon = e^{2i\pi\tau}$, $\tau := \tau_0\epsilon$, where $\epsilon > 0$ and we want to have $\epsilon \to 0$ (abreviation: $\epsilon \to 0^+$). it follows that

[8] Actually, most results hold if $\tau = \tau_0\epsilon + o(\epsilon)$ **[217]** but we see no need to prove this.

all discrete spirals, resp. half-spirals, $q^{\mathbf{Z}}$, resp. $q^{\pm\mathbf{N}}$, lie in $q_0^{\mathbf{R}}$, resp. in $q_0^{\pm\mathbf{R}_+}$. For example $q^{\mathbf{N}} = q_0^{\mathbf{N}\epsilon} \subset q_0^{\mathbf{R}_+}$.

We shall look at q-difference systems in a q-family: $\sigma_q X = A_q X$, where each $A_q \in \mathrm{GL}_n(\mathbf{C}(x))$. We shall also frequently use notations[9] such as:

$$A_\epsilon := A_q,\ B_\epsilon := B_q,\ X_\epsilon := X_q,\ \mathcal{X}_\epsilon^{(0)} := \mathcal{X}_q^{(0)},\ \mathcal{X}_\epsilon^{(\infty)} := \mathcal{X}_q^{(\infty)}, \ldots$$

with, hopefully, little risk of confusion). For $\epsilon > 0$ near enough 0 (abreviation: $\epsilon \approx 0^+$), we moreover assume that A_q is of the first kind at 0 and at ∞; recall from 11.1.1.2 that this means that $A_q(0), A_q(\infty) \in \mathrm{GL}_n(\mathbf{C})$. Actually, we shall also need nonresonancy, but it will flow from the other assumptions.

For the differential equation in the "continuous limit" (this is traditional terminology for the $q \to 1$ degeneracy), we require $\tilde{B}(x) \in \mathrm{Mat}_n(\mathbf{C}(x))$ to be of the first kind and non resonant at 0 and at ∞. It means that $\tilde{B}(0), \tilde{B}(\infty) \in \mathrm{Mat}_n(\mathbf{C})$ and that:

$$\forall \lambda, \mu \in \mathrm{Sp}\ \tilde{B}(0)\ ,\ \lambda - \mu \notin \mathbf{N}^* \text{ and the same for } \tilde{B}(\infty).$$

We briefly recall how the monodromy of the system $\delta X = \tilde{B}X$ at each of the poles $\tilde{x}_i \in \mathbf{C}^*$ can be described **[169, 222]**: if $\tilde{\mathcal{X}}$ is a fundamental matricial solution in a neighborhood of some ordinary point a (one at which $\tilde{\mathcal{X}}(a) \in \mathrm{Mat}_n(\mathbf{C})$) and let γ_i a loop based at a and with winding number $+1$ around $\tilde{x}_i$ and 0 around all other $\tilde{x}_j$. Then analytic continuation along γ_i transforms $\tilde{\mathcal{X}}$ into $\tilde{\mathcal{X}}\tilde{M}_i$ for some $\tilde{M}_i \in \mathrm{GL}_n(\mathbf{C})$, the corresponding monodromy matrix. (Using a different fundamental matricial solution or a different loop yields a conjugate matrix.) We intend to derive the $\tilde{M}_i$, $i = 1, \ldots, r$, for a particular choice of $\tilde{\mathcal{X}}$ and of the loops γ_i.

Now we come to the assumptions on the way of degeneracy. We require that on each compact subset of $\mathbf{S} \setminus \{\tilde{x}_1, \ldots, \tilde{x}_r\}$, the matrix $\dfrac{A_\epsilon - I_n}{q-1}$ tends uniformly to $\tilde{B}$ as $\epsilon \to 0^+$. We shall write $A_\epsilon = I_n + (q-1)B_\epsilon$, so $B_\epsilon \to \tilde{B}$ in the way explained. This implies in particular that $B_\epsilon(0) \to \tilde{B}(0)$ and that $B_\epsilon(\infty) \to \tilde{B}(\infty)$ and this in turn will imply that, for ϵ near enough 0, the q-difference equation $\sigma_q X = A_q X$ will be nonresonant at 0 and ∞. (We already know it is of the first kind, so it does make sense.)

All this together will not be quite enough to guarantee a smooth degeneracy, so one more technical assumption (SDJS for "Smooth degeneracy of the Jordan Structure") will be introduced in due time (Subsubsection 11.2.1.3) regarding the eigenstructures of $A_\epsilon(0)$ and $A_\epsilon(\infty)$.

11.2.1.3. *Strategy.* In the end, we shall get $P_\epsilon := \left(\mathcal{X}_\epsilon^{(\infty)}\right)^{-1} \mathcal{X}_\epsilon^{(0)}$ to degenerate when $q \to 1$, so we should begin with the local degeneracy, that of $\mathcal{X}_\epsilon^{(0)} = M_\epsilon^{(0)} e_{A_\epsilon^{(0)}}$ (same at infinity, where our conclusions are easily transposed). As for the $e_{A_\epsilon^{(0)}}$ component (dealt with in 11.2.3), knowing that $\dfrac{A_\epsilon^{(0)} - I_n}{q-1} = \dfrac{A_\epsilon(0) - I_n}{q-1} = B_\epsilon(0)$ (equality due to the nonresonancy assumption) tends to $\tilde{B}(0)$, we hope to find that $e_{A_\epsilon^{(0)}}$ degenerates to $x^{\tilde{B}(0)}$ (maybe up to some factor), thus generalizing 2.4 (recall that $e_{A_\epsilon^{(0)}}$ involves q-characters and, generically, q-logarithms).

---

[9]Note that we mention *en passant* local conditions at infinity, since they will be needed later (in 12.3) for the global picture.

As for the $M^{(0)}$ component (dealt with in 11.2.4.2), we note that it is characterized by two conditions:

- $M_\epsilon^{(0)}\left[A_\epsilon^{(0)}\right] = A_\epsilon$, i.e., $\sigma_q M_\epsilon^{(0)} = A_\epsilon M_\epsilon^{(0)}\left(A_\epsilon^{(0)}\right)^{-1}$, which can be seen as a q-difference equation $\sigma_q X = \mathcal{A}X$, where X is a vector standing for the unknown matrix $M_\epsilon^{(0)}$ and $\mathcal{A}$ the matrix encoding the linear map $M \mapsto A_\epsilon M\left(A_\epsilon^{(0)}\right)^{-1}$. The vector space $\mathbf{C}(\{x\})^N$, $N := n^2$ will stand for $\mathrm{Mat}_n(\mathbf{C}(x))$ and $\mathcal{A} \in \mathrm{Mat}_N(\mathbf{C}(x))$. That q-difference equation is actually of the first kind and nonresonant.
- For every $\epsilon > 0$, there is an initial condition $M_\epsilon^{(0)}(0) = I_n$, so our q-difference equation also comes with an initial condition $X(0) = X_0$ *independent of* ϵ.
- Of course, $\mathcal{A} = \mathcal{A}_\epsilon$ depends on ϵ. Then, under the assumption (or intermediate conclusion) $\dfrac{\mathcal{A}_\epsilon - I_N}{q-1} \to \tilde{\mathcal{B}}$, we hope that the solution X_ϵ degenerates into some $\tilde{X}$ such that $\sigma_q \tilde{X} = \tilde{\mathcal{B}}\tilde{X}$ and $\tilde{X}(0) = X_0$.

That desired conclusion will indeed be proved in two steps: a local proof (in a neighborhood of 0) by inspection of power series solution and taditional arguments of dominated convergence; then escaping the neighborhood by an argument similar to Euler's discretization of differential equations with variable steps (here a multiplicative avatar).

EXERCISE 11.60. (i) Suggest a natural choice for L.
(ii) Formalize the equation for $M_\epsilon^{(0)}$ in terms of internal Hom.

Practically, in this section, we tackle *local* confluence only (that of $\mathcal{X}^{(0)}$) and begin with two extreme cases: degeneracy of a family of *regular* systems (i.e., those such that all $A_\epsilon(0) = I_n$), but imposing an initial condition $X_\epsilon(0) = X_0$; and degeneracy of a family of *constant coefficient* systems (i.e., those such that all $A_\epsilon \in \mathrm{GL}_n(\mathbf{C})$). Then all this will be put together to study first the confluence of the $M^{(0)}$ component, then that of $\mathcal{X}^{(0)}$ itself.

11.2.2. Degeneracy of regular systems with fixed initial conditions.

11.2.2.1. *Preparation.*

The data. We have $\tilde{B} \in \mathrm{Mat}_n(\mathbf{C}(x))$ such that $\tilde{B}(0) \in \mathrm{Mat}_n(\mathbf{C})$. We assume that $\mathrm{Sp}\ \tilde{B}(0) \cap \mathbf{Z} = \{0\}$. Also we fix $X_0 \in \mathbf{C}^n \setminus \{0\}$ such that $\tilde{B}(0)X_0 = 0$. Next we call $\tilde{x}_1, \ldots, \tilde{x}_r \in \mathbf{C}$ the poles of $\tilde{B}$. Because of our assumption $\tilde{B}(0) \in \mathrm{Mat}_n(\mathbf{C})$, all $\tilde{x}_i \in \mathbf{C}^*$.

We shall assume a family $B_\epsilon \in \mathrm{Mat}_n(\mathbf{C}(x))$ such that $B_\epsilon \to \tilde{B}$ uniformly on every compact of $\mathbf{C} \setminus \{\tilde{x}_1, \ldots, \tilde{x}_r\}$ when[10] $\epsilon \to 0^+$. As a consequence, if $0 < \rho < \min(|\tilde{x}_1|, \ldots, |\tilde{x}_r|)$, then $B_\epsilon \to \tilde{B}$ uniformly on $\mathring{D}(0, \rho)$.

LEMMA 11.61. *Let* $\{\tilde{\mu_1}, \ldots, \tilde{\mu_n}\} :=$ *Sp* $\tilde{B}(0) \subset \mathbf{C}$ *(with repetitions allowed, thus a multiset here). Then, for* $\epsilon \approx 0^+$, *the spectrum Sp* $B_\epsilon(0)$ *may be written as* $\{\mu_1(\epsilon), \ldots, \mu_n(\epsilon)\}$ *in such a way that* $\mu_i(\epsilon) \to \tilde{\mu}_i$ *for* $i = 1, \ldots, n$.

[10] Recall the conventions: $\epsilon \to 0^+$ means that $\epsilon \to 0$ with $\epsilon > 0$; and $\epsilon \approx 0^+$ means that ϵ is near enough 0 with $\epsilon > 0$; also, all limits are understood to hold when $\epsilon \to 0^+$ unless otherwise explicitly stated. Last, we shall often resort without coining it to the *sequential criterion*, that is the fact that: $\lim_{\epsilon \to 0^+} f(\epsilon) = l$ is equivalent to: $\lim_{m \to +\infty} f(\epsilon_m) = l$ for every sequence of $\epsilon_m > 0$ such that $\lim_{m \to +\infty} \epsilon_m = 0$.

This is of course a form of "continuity of roots", iself sometimes proved using the theorem of Rouché. Possible references for Rouché's theorem and its applications are [**3**, 4.5.2], [**211**, 10.4.2], [**244**, 3.4.2]; also see the *Éléments d'analyse*, IX,17, by Dieudonné, which contains a complicated application to polynomials; or, easier to follow, the approach in [**170**, 2.7.5] through linear algebra (polar decomposition of the companion matrix). Last, it can be deduced from topological considerations on etale coverings, see Section A.5 of Appendix A.

PROOF. Changing the indexation w.r.t. the statement of the lemma, we write the characteristic polynomial of $\tilde{B}(0)$ as $\tilde{P}(X) = (X-\tilde{\mu}_1)^{r_1}\cdots(X-\tilde{\mu}_k)^{r_k}$, the $\tilde{\mu}_i$ being pairwise distinct, so that $r := \frac{1}{2}\min_{i\neq j}(|\tilde{\mu}_i-\tilde{\mu}_i|) > 0$ and the disks $D_i := \mathring{\mathrm{D}}(\tilde{\mu}_i, r)$ are pairwise disjoint. For $m \gg 0$, the characteristic polynomial P_m of $B_{\epsilon_m(0)}$ is such that $\left|P_m - \tilde{P}\right| < \left|\tilde{P}\right|$ on each circle $\partial D_i = \partial\mathrm{D}(\tilde{\mu}_i, r)$. Rouché's theorem then ensures that P_m admits r_i zeroes in D_i. Our contention follows. □

Recall that $q = q_0^\epsilon = 1 + 2i\pi\tau_0\epsilon + o(\epsilon)$ and $A_q = A_\epsilon = I_n + (q-1)B_\epsilon$, which (at 0) has spectrum $\mathrm{Sp}\ A_\epsilon(0) = \{\lambda_1(\epsilon), \ldots, \lambda_n(\epsilon)\}$, where for $i = 1, \ldots, n$:

$$\lambda_i(\epsilon) = 1 + (q-1)\mu_i(\epsilon) = 1 + 2i\pi\tau_0\epsilon\mu_i(\epsilon) + o(\epsilon) = 1 + 2i\pi\tau_0\epsilon\tilde{\mu}_i + o(\epsilon).$$

This entails that $\mathrm{Sp}\ A_\epsilon(0) \cap q^{\mathbf{Z}} \subset \{1\}$ for $\epsilon \approx 0^+$ (which will be crucial to justify nonresonancy in 11.2.4.2).

We are now going to solve simultaneously a q-difference system (here given in two equivalent forms) and a differential system, both with initial conditions:

$$\begin{cases}\sigma_q X = A_\epsilon X,\\ X(0) = X_0,\end{cases}, \text{ equivalently } \begin{cases}\dfrac{\sigma_q - I_n}{q-1}X = B_\epsilon X,\\ X(0) = X_0,\end{cases}, \text{ and } \begin{cases}\tilde{X} = \tilde{B}\tilde{X},\\ \tilde{X}(0) = X_0.\end{cases}$$

For that, while shortening notation: $A := A_\epsilon$, etc, we use power series expansions $A = A_0 + A_1x + \cdots$, $B = B_0 + B_1x + \cdots$, $\tilde{B} = \tilde{B}_0 + \tilde{B}_1x + \cdots$, $X = X_0 + X_1x + \cdots$, $\tilde{X} = \tilde{X}_0 + \tilde{X}_1x + \cdots$. In its first form, the q-difference system gibes rise to the inductive formula $X_m := (q^mI_n - A_0)^{-1}(A_1X_m + \cdots + A_mX_0)$ for $m \geq 1$. However, using the second form displays better the similarity of both resolution procedures; of course, $\tilde{X}_0 = X_0$, and, for $m \geq 1$:

$$X_m := ([m]_qI_n - B_0)^{-1}(B_1X_m + \cdots + B_mX_0),$$
$$\tilde{X}_m := (mI_n - \tilde{B}_0)^{-1}(\tilde{B}_1\tilde{X}_m + \cdots + \tilde{B}_m\tilde{X}_0).$$

(The first formulas are well founded for $\epsilon \approx 0^+$.) Since $[m]_q \to m$ for all m, easy induction shows that $X_m = X_m(\epsilon) \to \tilde{X}_m$ for all m, i.e., $X = X_\epsilon \to \tilde{X}$ coefficient-wise. In the first step (11.2.2.2 and 11.2.2.3) we shall show that convergence truly holds within some open disk $\mathring{\mathrm{D}}(0, r)$; and in the second step (11.2.2.4) and 11.2.2.6 we shall extend the domain of that convergence.

11.2.2.2. *First step: near 0.* As in many previous occasions (for instance in 11.1.1.4), we fix some matricial (multiplicative) norm; but, for simplicity, we just write it $\|-\|$, with the same symbol as the vectorial norm it comes from.

LEMMA 11.62. (i) *There exist $R, C > 0$ and $\epsilon_1 > 0$ such that:*

$$0 < \epsilon < \epsilon_1 \Longrightarrow \forall m \geq 0\,,\ \|B_m\| < CR^{-m}.$$

(ii) *There exist $D > 0$ and $\epsilon_2 > 0$ such that:*

$$0 < \epsilon < \epsilon_2 \Longrightarrow \forall m \geq 1\,,\ \left\|([m]_q I_n - B_0)^{-1}\right\| < D.$$

PROOF. (i) Let generally a sequence of entire power series $f_k := \sum a_{k,m}x^m$ converge uniformly to $g := \sum b_m x^m$ on some $\mathring{D}(0,r)$. Then, from Cauchy's estimates [**211**, 10.26] it follows that f_k converges normally to g on every $\mathring{D}(0,r')$, $r' < r$. We apply that fact separately to all of the n^2 components of B_m (implicitly using the sequential criterion).

(ii) This will be follow from a sequence of auxiliary results given in 11.2.2.3. □

Assuming the lemma, we now state and prove the main result of the first step.

PROPOSITION 11.63. *There is a $\mathring{D}(0,r)$, $r > 0$, on which X_ϵ converges normally to $\tilde{X}$ for $\epsilon \approx 0^+$.*

PROOF. Let $\epsilon_0 := \min(\epsilon_1, \epsilon_2)$ (notations coming from the lemma). From the recursive formula in 11.2.2.1, if $0 < \epsilon < \epsilon_0$::

$$\forall m \geq 1\,,\ \|X_m\| \leq CD(R^{-1}\|X_{m-1}\| + \cdots + R^{-}\|X_0\|),$$

so that, setting $u_m := R^m\|X_m\|$ and $U_m := u_0 + \cdots + u_m$:

$$u_m = U_m - U_{m-1} \leq CD(u_0 + \cdots + u_{m-1}) = CDU_{m-1} \Longrightarrow U_m \leq (1+CD)^m u_0$$

$$\Longrightarrow u_m \leq CD(1+CD)^{m-1}u_0 \Longrightarrow \|X_m\| \leq R^{-m}CD(1+CD)^{m-1}\|X_0\|.$$

So we can set $r := R/(1+CD)$. □

11.2.2.3. *Continuation of the first step.* We now go for the proof of (ii) in the previous lemma.

LEMMA 11.64. *There exist $C > 0$ and $\epsilon_0 > 0$ such that:*

$$0 < \epsilon < \epsilon_0 \Longrightarrow \forall m \geq 0\,,\ |[m]_q| \geq Cm.$$

PROOF. First, $|1-q| = \left|1 - e^{2i\pi\tau_0\epsilon}\right| \sim |2i\pi\tau_0\epsilon| = 2\pi|\tau_0|\epsilon$ when $\epsilon \to 0^+$. So letting $a > 2\pi|\tau_0|$, we have $|1-q| < a\epsilon$ when $\epsilon \approx 0^+$. Second, writing $\tau_0 = \tau_1 + i\tau_2$ and setting $r_0 := |q_0| = e^{-2\pi\tau_2}$, so that $|q^m| = r_0^{m\epsilon}$, we have:

$$|1-q^m| \geq 1 - |q^m| = 1 - r_0^{m\epsilon} \Longrightarrow |[m]_q| \geq \frac{1-r_0^{m\epsilon}}{a\epsilon} \geq \frac{b}{a}m,$$

where b is the (strictly positive) minimum of the function $\dfrac{1-r_0^x}{x}$ on $\mathbf{R}_+$ (the function if continuous if one sets $0 \mapsto -\log r_0 > 0$, it is strictly positive on $\mathbf{R}_+$ and has limit $+\infty$ at $+\infty$). □

LEMMA 11.65. *Let $C \in Mat_n(\mathbf{C})$, $C \neq 0$. Then there are constants strictly positive $A = A(C)$, $B = B(C)$, depending continuously on C, such that:*

$$\forall \lambda \in \mathbf{C} \setminus Sp\,C\,,\ \left\|(\lambda I_n - C)^{-1}\right\| \leq \max\left(A, \frac{B}{d(\lambda, Sp\mathbf{C})^n}\right).$$

PROOF. If $|\lambda| \geq 2\|C\|$, we expand:

$$(\lambda I_n - C)^{-1} = \lambda^{-1}\left(I_n + \frac{1}{\lambda}C + \cdots\right) \Longrightarrow \left\|(\lambda I_n - C)^{-1}\right\| \leq \frac{1}{2\|C\|}(1 + 1/2 + \cdots),$$

so we may take $A := \dfrac{1}{\|C\|}\cdot$ If $|\lambda| < 2\|C\|$, using Cramer's formula $M^{-1} = \dfrac{1}{\det M}{}^t comatrix(M)$ and the fact that $\|comatrix(M)\| \leq K\|M\|^{n-1}$ for some universal $K > 0$ (that bound coming from homogeneity), we get (writing χ_C the characteristic polynomial of C):

$$\left\|(\lambda I_n - C)^{-1}\right\| \leq \frac{K}{|\chi_C(\lambda)|}\|\lambda I_n - C\|^{n-1} \leq \frac{K}{d(\lambda, \mathrm{Sp}\mathbf{C})^n}(3\|C\|)^{n-1},$$

so we may take $B := K(3\|C\|)^{n-1}$. □

EXERCISE 11.66. Check the implicitly used inequalities $|\chi_C(\lambda)| \geq d(\lambda, \mathrm{Sp}\mathbf{C})^n$ and $\|\lambda I_n - C\| \leq 3\|C\|$.

LEMMA 11.67. *Let $\tilde{C} \in Mat_n(\mathbf{C})$ such that $Sp\, \tilde{C} \cap \mathbf{N}^* = \emptyset$ and let a family (C_ϵ) in $Mat_n(\mathbf{C})$ have limit $\tilde{C}$ when $\epsilon \to 0^+$. Then, for $\epsilon \approx 0^+$, the $\left\|([m]_q I_n - C_\epsilon)^{-1}\right\|$, $m \geq 1$, are uniformly bounded.*

PROOF. Let $r > 0$ such that the closed disk $\overline{\mathrm{D}}(\lambda, r)$, $\lambda \in \mathrm{Sp}\,\tilde{C}$, are pairwise disjoint and moreover do not meet $\mathbf{N}^*$. Let F the union of those disks, so that $\mathrm{Sp}\,\tilde{C} \subset \overset{\circ}{F}$ (the interior of F) and $\delta := d(F, \mathbf{N}^*) > 0$ (distance of a compact to a closed set). Using again that application of Rouché's theorem to continuity of eigenvalues (as in 11.2.2.1), we deduce that there exists a $\epsilon_1 > 0$ such that:

$$0 < \epsilon < \epsilon_1 \Longrightarrow \mathrm{Sp}\ C_\epsilon \subset \overset{\circ}{F}.$$

Clearly F is bounded, so $F \subset \overset{\circ}{\mathrm{D}}(0, R)$ for some $R > 0$. Let $\delta' := d(F, \mathbf{C} \setminus \overset{\circ}{\mathrm{D}}(0, R))$, so $\delta' > 0$ (again a compact and a closed set). Then, for $0 < \epsilon < \epsilon_1$:

$$|\lambda| > R \Longrightarrow d(\lambda, \mathrm{Sp}\ C_\epsilon) \geq d(\lambda, F) \geq \delta'.$$

Let C, ϵ_0 as in a previous lemma such that $|[m]_q| \geq Cm$ for all $m \geq 0$. Let $m_0 := \lceil R/C \rceil$ (the smallest integer $\geq R/C$), so that $Cm_0 \geq R$. Then:

$$0 < \epsilon < \epsilon_0 \Longrightarrow \forall m \geq m_0\ ,\ d([m]_q, \mathrm{Sp}\ C_\epsilon) \geq \delta'.$$

Now let $m_1 := \max(m_0, \lceil R \rceil) \in \mathbf{N}^*$, so that $m_1 \geq R$ and $m_1 \geq m_0$. Since $[m]_q \to m$ for all m, we have:

$$\lim_{\epsilon \to 0^+} \max_{m=1,\ldots,m_1} |[m]_q - m| = 0.$$

Therefore, for some $\epsilon_2 > 0$, using the δ from the beginning of the proof:

$$0 < \epsilon < \epsilon_0 \Longrightarrow \forall m = 1, \ldots, m_1\ ,\ d([m]_q, \mathrm{Sp}\ C_\epsilon) \geq \delta/2.$$

In conclusion, if $0 < \epsilon < \min(\epsilon_0, \epsilon_1, \epsilon_2)$ and $\delta'' := \min(\delta', \delta/2) > 0$, using the previous lemma:

$$\left\|([m]_q I_n - C_\epsilon)^{-1}\right\| \leq \max\left(A(C_\epsilon), \frac{B(C_\epsilon)}{\delta''^n}\right).$$

Since $C_\epsilon \to \tilde{C}$ and $A(C_\epsilon), B(C_\epsilon)$ are continuous in C_ϵ, this ensures a finite upper bound. □

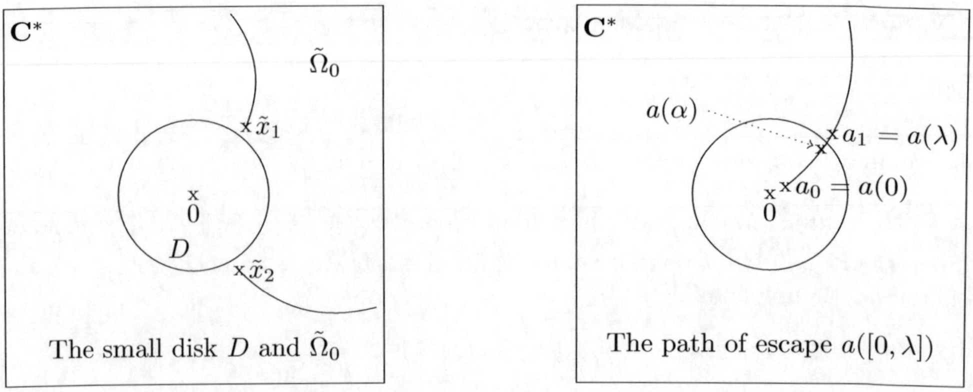

FIGURE 11.1. Escaping from the small disk $D := \mathring{D}(0, r)$ to $\tilde{\Omega}_0$

We now leave it to the reader to check that combining the last two lemmas is enough to prove (ii) in the Lemma of 11.2.2.2.

EXERCISE 11.68. Both lemmas were proved assuming that $C \neq 0$; can you see why? Modify the proof for $C = 0$.

11.2.2.4. *Second step: escaping the neighborhood.* Recall the poles $\tilde{x}_1, \ldots, \tilde{x}_r \in \mathbf{C}$ of $\tilde{B}$; we introduce (see left side of figure 11.1):

$$\tilde{\Omega}_0 := \mathbf{C} \setminus \bigcup_{i=1,\ldots,r} \tilde{x}_i q_0^{\mathbf{R}_-} \subset \mathbf{C} \setminus \{\tilde{x}_1, \ldots, \tilde{x}_r\},$$

thus a simply connected domain such that $B_\epsilon \to \tilde{B}$ uniformly over every compact subset of $\tilde{\Omega}_0$. Canonical solutions X_ϵ of $\sigma_q X = A_\epsilon X$ with initial condition $X(0) = X_0$ have been shown (step 1) to converge uniformly over a disk $\mathring{D}(0, r)$ included in $\tilde{\Omega}_0$ to the canonical solution of $\delta \tilde{X} = \tilde{B} \tilde{X}$ with initial condition $\tilde{X}(0) = X_0$.

In the following, we shall assume that the q_0 spirals $\tilde{x}_i q_0^{\mathbf{R}_-}$ do not overlap:

$$i \neq j \Longrightarrow \tilde{x}_i / \tilde{x}_j \notin q_0^{\mathbf{R}}.$$

REMARK 11.69. Maybe the tiring calculation that follows may be replaced by the theory of product integrals as, for example, in [**71**, 1.3].

Now let $a_1 \in \tilde{\Omega}_0$. We want to show that $X_\epsilon(a_1) \to \tilde{X}(a_1)$. If $a_1 \in \mathring{D}(0, r)$, there is nothing to prove. So we assume that $a_1 \notin \mathring{D}(0, r)$. Note that both $X_\epsilon(a_1)$ and $\tilde{X}(a_1)$ are well defined: the former by meromorphic continuation of solutions (while avoiding poles), as for example in 5.3.2.2 (for q-difference equations) or 7.2.3.3 (for q-difference systems), the latter by standard properties of linear differential equations over a simply connected domain.

We join a_1 to a point $a_0 \in \mathring{D}(0, r)$ via a segment of continuous q_0-spiral (see right side of Figure 11.1). Concretely, let $\lambda > 0$ such that $\left| q_0^\lambda \right| < r / |a_1|$, so that $a_0 := q_0^\lambda a_1 \in \mathring{D}(0, r)$. Also an auxiliary role will be played by $\alpha \in \,]0, \lambda[$ such that $q_0^{-\alpha} a_0 \in \mathring{D}(0, r)$. The (spiralling) path of escape is:

$$a : [0, \lambda] \to \mathbf{C}^*, t \mapsto a_0 q_0^{-t}, \text{ whence } a(0) = a_0, a(\lambda) = a_1 \text{ and } a\left([0, \alpha]\right) \subset \mathring{D}(0, r).$$

Since $a([0,\lambda]) \subset \tilde{\Omega}_0 \subset \mathbf{C} \setminus \{\tilde{x}_1, \ldots, \tilde{x}_r\}$ is compact, we draw that $B_\epsilon \circ a$ converges uniformly to $\tilde{B} \circ a$ on $[0,\lambda]$. This will allow us to transform (next) a problem of complex analysis in $\tilde{\Omega}_0$ into a problem of real analysis over $[0,\lambda]$. After doing that, we shall recall some facts on resolvents and Euler's method with variable steps, then apply them to our problem of real analysis.

EXERCISE 11.70. Let $t > 0$. Check that $|a(t)| < R \Leftrightarrow t > \dfrac{1}{2\pi\tau_2} \log(|a_0|/R)$.

Translation from complex analysis to real analysis. We define vector valued functions over $[0,\lambda]$:

$$Y_\epsilon(t) := X_\epsilon(a(t)) = X_\epsilon(a_0 q_0^{-t}) \text{ and } \tilde{Y}(t) := \tilde{X}(a(t)) = \tilde{X}(a_0 q_0^{-t}).$$

Basic facts are:

(1) $Y_\epsilon \to \tilde{Y}$ uniformly on $[0,\alpha]$ (since $X_\epsilon \to \tilde{X}$ on $a([0,\alpha]) \subset \mathring{\mathrm{D}}(0,r)$).
(2) $Y_\epsilon(t+\epsilon) = (I_n + \epsilon C_\epsilon(t)) Y_\epsilon(t)$, where $C_\epsilon(t) := D_\epsilon(t+\epsilon)$ and:

$$D_\epsilon(t) := -\frac{q-1}{\epsilon} B_\epsilon(a(t)) \left(I_n + (q-1) B_\epsilon(a(t))\right)^{-1}.$$

(3) $\dfrac{d\tilde{Y}}{dt} = \tilde{C}\tilde{Y}$, where $\tilde{C} := -2i\pi\tau_0 \left(\tilde{B} \circ a\right)$ (apply the chain rule to the differential equation satisfied by $\tilde{X}$, noting that $da/dt = -2i\pi\tau_0 a$).
(4) $C_\epsilon \to \tilde{C}$ uniformly on $[0,\lambda]$.

EXERCISE 11.71. (i) Check the second fact above, starting from functional equation $X_\epsilon(q^{-1}a(t)) = A_\epsilon(q^{-1}a(t))^{-1} X_\epsilon(a(t))$ and from identity $(I_n + D)^{-1} = I_n - D(I_n + D)^{-1}$.
(ii) Deduce from it the fourth fact.

11.2.2.5. *Reminders on resolvents and on Euler's method.* Our main reference here is [**31**, IV §2], also see [**22**, **57**]. What we call here resolvent[11] is called "matrizant" in [**22**].

Let $I \subset \mathbf{R}$ an open interval, $E := \mathrm{Mat}_n(\mathbf{C})$ seen as a real vector space and $A : I \to E$ a matrix valued continuous function (in [**31**], E is more general). We denote: $t \mapsto U(t,t_0)$ the unique fundamental matricial solution of $X' = AX$ with initial condition $t_0 \mapsto I_n$. It follows that the unique vectorial solution of $X' = AX$ with initial condition $t_0 \mapsto X_0 \in \mathbf{C}^n$ is $t \mapsto U(t,t_0)X_0$. The *resolvent* $U(t,t_0)$ has the following properties:

(1) $U(s,s) = I_n$, $U(s,t) = U(t,s)^{-1}$, $U(s,t)U(t,u) = U(s,u)$.
(2) $U : I \times I \to \mathrm{GL}_n(\mathbf{C})$ is continuous.
(3) If $B \in \mathrm{Mat}_n(\mathbf{C})$ commutes with all $A(t)$, $t \in I$, then it commutes with all $U(s,t)$, $s,t \in I$.
(4) Let $K \subset I$ be compact and let $k := \max_K \|A\|$. Then:

$$\forall t, t_0 \in I \,,\ \|U(t,t_0) - I_n\| \leq e^{k|t-t_0|} - 1.$$

In our setting, we shall denote $\tilde{U}$ the resolvent for the system $Y' = \tilde{C}Y$. Thus:

$$\tilde{X}(a_1) = \tilde{Y}(\lambda) = \tilde{U}(\lambda,0)\tilde{Y}(0) = \tilde{U}(\lambda,0)\tilde{X}(a_0).$$

Euler's method can be summarized in the following theorem.

---

[11]Not to be confused with the resolvent of functional calculus!

THEOREM 11.8. *let $t_0 < t_1$ in I. For each $p \geq 1$, consider a subdivision of length p:*

$$s_0^{(p)} = t_0 < s_1^{(p)} < \cdots < s_p^{(p)} = t_1, \text{ with steps } \Delta s_i^{(p)} := s_i^{(p)} - s_{i-1}^{(p)}, i = 1, \ldots, p$$

and denote $\epsilon^{(p)} := \max_{i=1,\ldots,p} \Delta s_i^{(p)}$. Assume that $\epsilon^{(p)} \to 0$ when $p \to +\infty$, we have:

$$U(t_1, t_0) = \lim_{p \to +\infty} \left(I_n + A(s_p^{(p)})\Delta s_p^{(p)}\right) \cdots \left(I_n + A(s_1^{(p)})\Delta s_1^{(p)}\right).$$

To apply it to equation $Y' = AY$ with initial condition $Y(t_0) = Y_0$, we choose variable steps $\epsilon_1, \ldots, \epsilon_p > 0$ such that $\epsilon_1 + \cdots + \epsilon_p = t_1 - t_0$ and define inductively $s_0 = t_0$, then $s_i := s_{i-1} + \epsilon_i$, $i = 1, \ldots, p$. Soving inductively the interpolation equations (with initialization $Z(t_0) := Y_0$):

$$\frac{Z(s_{i+1}) - Z(s_i)}{s_{i+1} - s_i} = A(s_i)Z(s_i),$$

we get an approximation $Z(t_1)$ of $Y(t_1)$. Euler's method ensures that having $\max(\epsilon_1, \ldots, \epsilon_p) \to 0$ entails $Z(t_1) \to Y(t_1)$.

11.2.2.6. *Continuation of the second step and conclusion.* We keep notations from the "Translation from complex analysis to real analysis" in 11.2.2.4. Fix $\epsilon \approx 0^+$ and set $N := \lfloor \lambda/\epsilon \rfloor$, $\nu := \lambda/\epsilon - N \in [0, 1[$. Deduce a subdivision of $[0, \lambda]$ with first step $\nu\epsilon$ followed by N steps ϵ (this is consistent since $\nu\epsilon + N\epsilon = \lambda$). then, from 11.2.2.5:

$$\tilde{U}(\lambda, 0) = \lim_{\epsilon \to 0^+} \left(I_n + \epsilon\tilde{C}(\lambda - \epsilon)\right) \cdots \left(I_n + \epsilon\tilde{C}(\lambda - N\epsilon)\right) \left(I_n + \nu\epsilon\tilde{C}(0)\right).$$

Note that the rightmost factor tends to I_n so we may skip it at will. However, this is not the formula we would like: instead of values of $\tilde{C}$, we would prefer values of C_ϵ. This will be taken care of thanks to the following lemma.

LEMMA 11.72. (i) *let $A_1, \ldots, A_m, E_1, \ldots, E_m \in Mat_n(\mathbf{C})$ and set $A_i' := A_i + E_i$, $i = 1, \ldots, m$. Then:*

$$\|(I_n + A_1') \cdots (I_n + A_m') - (I_n + A_1) \cdots (I_n + A_m)\| \leq \left(\sum_{i=1}^{m} \|E_i\|\right) \exp\left(\sum_{i=1}^{m} (\|A_i\| + \|E_i\|)\right).$$

(ii) *Let $(A_{p,i})$, $(A'_{p,i})$, $p \in \mathbf{N}^*$, $i = 1, \ldots, p$, two families in $Mat_n(\mathbf{C})$ and set:*

$$U_p := (I_n + A_{p,1}) \cdots (I_n + A_{p,p}) \text{ and } U_p' := (I_n + A'_{p,1}) \cdots (I_n + A'_{p,p}).$$

Assume that $\sum_{i=1,\ldots,p} \left\|A'_{p,i} - A_{p,i}\right\| \to 0$ when $p \to +\infty$. Then $U_p' - U_p \to 0$ when $p \to +\infty$.

PROOF. (i) is easy using properties of the exponential of matrices and (ii) is then an immediate consequence.

□

COROLLARY 11.73.

$$\tilde{U}(\lambda, 0) = \lim_{\epsilon \to 0^+} (I_n + \epsilon C_\epsilon(\lambda - \epsilon)) \cdots (I_n + \epsilon C_\epsilon(\lambda - N\epsilon)) (I_n + \nu\epsilon C_\epsilon(0)).$$

We now reach the conclusion of our study.

THEOREM 11.9. *$X_\epsilon \to \tilde{X}$ on $\tilde{\Omega}_0$.*

PROOF. The relation $X_\epsilon(a_1) \to \tilde{X}(a_1)$ for an arbitrary $a_1 \in \tilde{\Omega}_0$ either follows from the study near 0 if $|a_1^3| < r$, or translates to $Y_\epsilon(\lambda) \to \tilde{Y}(\lambda) = \tilde{U}(\lambda, 0)\tilde{Y}(0)$, which comes from the first case (applied at a_0) and the above limit formula for the resolvent. (You should check carefully the argument, there is an extra factor to be taken in account.) □

The proof is not precise enough to say if the convergence is uniform over compact subsets of $\tilde{\Omega}_0$. I would guess that this holds true, but it does not seem to be needed in applications.

EXERCISE 11.74. Was the nonoverlapping of q_0-spirals assumed in 11.2.2.4 really necessary?

11.2.3. Degeneracy of a family of constant coefficient systems. Here we suppose that all $A_\epsilon = I_n + (q-1)B_\epsilon \in \mathrm{GL}_n(\mathbf{C})$ and that $B_\epsilon \to \tilde{B} \in \mathrm{Mat}_n(\mathbf{C})$ (plain convergence in $\mathrm{Mat}_n(\mathbf{C})$). As often, we distinguish two complementary cases: A_ϵ unipotent (in 11.2.3.1, it is the easiest one!) and A_ϵ semi-simple (in 11.2.3.2); then we mix them in 11.2.3.3.

We shall of course reduce the problem to *degeneracy of basic bricks* as explained in 2.4.3. This holds under a restriction of the domain of definition: either $q_0^{\mathbf{R}}$ or $-q_0^{\mathbf{R}}$ has to be cut off[12]; and the resulting functions are, as the case may be, $\log \pm x$, resp. $(\pm x)^\gamma$. In any case, log denotes the principal determination of the logarithm (the one such that $1 \mapsto 0$) on the simply connected set $\mathbf{C}^* \setminus (-q_0^{\mathbf{R}})$; and of course $x^\gamma := \exp(\gamma \log x)$, etc.

11.2.3.1. *Unipotent systems.* We assume all A_ϵ unipotent, whence all $B_\epsilon = \dfrac{A_\epsilon - I_n}{q-1}$ nilpotent: $B_\epsilon^n = 0$, whence $\tilde{B} = \lim B_\epsilon$ nilpotent: $\tilde{B}^n = 0$. For unipotent q-difference systems, there is a canonical solution from 7.2.4.1:

$$\mathcal{X}_\epsilon = A_\epsilon^{\ell_q} = (I_n + (q-1)B_\epsilon)^{\ell_q} = \sum_{k=0}^{n} \binom{\ell_q}{k} (q-1)^k B_\epsilon^k.$$

We then write:

$$\binom{\ell_q}{k}(q-1)^k = (q-1)^k \frac{\ell_q(\ell_q - 1)\cdots(\ell_q - k + 1)}{k!} = \frac{1}{k!}(q-1)^k {\ell_q}^k + \text{remaining terms},$$

where all remaining terms are constant multiples of $(q-1)^k {\ell_q}^p$ with $p < k$. Now we know from 2.4.3.2 that $(q-1)\ell_q(x) \to \log x$ if $x \notin -q_0^{\mathbf{R}}$. Therefore all remaining terms tend to 0 and $\binom{\ell_q}{k}(q-1)^k \to \dfrac{1}{k!}(\log x)^k$, whence:

$$\forall x \notin q_0^{\mathbf{R}}\ ,\ \mathcal{X}_\epsilon(x) \to \sum_{k=0}^{n} \frac{1}{k!}(\log x)^k \tilde{B}^k = \exp\left((\log x)\tilde{B}\right).$$

(Recall that $\tilde{B}$ is nilpotent, so the exponential series is truncated.) It is customary to denote $\exp\left((\log x)\tilde{B}\right)$ as $x^{\tilde{B}}$ (principal determination on the simply connected set $\mathbf{C}^* \setminus (-q_0^{\mathbf{R}}) \ni 1$, the one such that $1^{\tilde{B}} = I_n$). It is the canonical fundamental matricial solution of $\delta\tilde{X} = \tilde{B}\tilde{X}$ obtained by the Fuchs-Frobenius method on that domain [**222**]. We summarize:

$$\forall x \notin -q_0^{\mathbf{R}}\ ,\ e_{q,A_\epsilon}(x) \to x^{\tilde{B}}.$$

[12] In [**217**] the cut was $q_0^{\mathbf{R}}$ and the continuous limits were $\log(-x)$, resp. $(-x)^\gamma$.

11.2.3.2. *Semi-simple systems.* We assume all A_ϵ semi-simple:

$$A_\epsilon = P_\epsilon \mathrm{Diag}(c_1(\epsilon), \ldots, c_n(\epsilon))P_\epsilon^{-1}, P_\epsilon \in \mathrm{GL}_n(\mathbf{C}), c_i(\epsilon) = 1 + (q-1)\gamma_i(\epsilon) \in \mathbf{C}^*.$$

Then $B_\epsilon = \dfrac{A - I_n}{q-1} = P_\epsilon \mathrm{Diag}(\gamma_1(\epsilon), \ldots, \gamma_n(\epsilon))P_\epsilon^{-1}$ is also semi-simple. If we write $\mathrm{Sp}\,\tilde{B} = \{\tilde{\gamma}_1, \ldots, \tilde{\gamma}_n\}$, using the arguments of 11.2.2.1 we see that (up to permutation of the indices) the P_ϵ can be chosen so that $\gamma_i(\epsilon) \to \tilde{\gamma}_i$, $i = 1, \ldots, n$. However, there appears in this case a double difficulty: first, $\tilde{B} = \lim B_\epsilon$ needs not be semi-simple (the space of semi-simple matrices is dense in $\mathrm{Mat}_n(\mathbf{C})$, not closed); and even if it was, the diagonalisation matrices P_ϵ may behave very wildly, as shows the following example from **[217]** (a slight sign error there is corrected here).

EXERCISE 11.75. Let $B_\epsilon := \epsilon \begin{pmatrix} \cos^2 \epsilon^{-1} & -\sin \epsilon^{-1} \cos \epsilon^{-1} \\ \sin \epsilon^{-1} \cos \epsilon^{-1} & \sin^2 \epsilon^{-1} \end{pmatrix}$. Show that $\tilde{B} = 0$ but that no diagonalisation of $\tilde{B}$ can be unfolded into a family of diagonalisations of the B_ϵ.

There probably are various ways of escaping that problem and the one we choose is certainly not the most economical. We assume that the matrices P_ϵ (which are of course not unique!) may be chosen so as to have a limit $\tilde{P} \in \mathrm{GL}_n(\mathbf{C})$. Then $\tilde{B}$ is itself semi-simple:

$$\tilde{B} = \lim B_\epsilon = \lim P_\epsilon \mathrm{Diag}(\gamma_1(\epsilon), \ldots, \gamma_n(\epsilon))P_\epsilon^{-1} = \tilde{P}\mathrm{Diag}(\tilde{\gamma}_1, \ldots, \tilde{\gamma}_n)\tilde{P}^{-1}.$$

Now the canonical fundamental matricial solution we chose to associate to equation $\sigma_q X_\epsilon = A_\epsilon X_\epsilon$ in 7.2.4.2 is:

$$\mathcal{X}_\epsilon = P_\epsilon \mathrm{Diag}(e_{q,c_1(\epsilon)}, \ldots, e_{q,c_n(\epsilon)})P_\epsilon^{-1}.$$

Since $\dfrac{c_i(\epsilon) - 1}{q-1} = \gamma_i(\epsilon) \to \tilde{\gamma}_i$, $i = 1, \ldots, n$, it follows from 2.4.3.1 that $e_{q,c_i(\epsilon)} \to x^{\tilde{\gamma}_i}$ for $x \notin -q_0^{\mathbf{R}}$, so that, under the same conditions:

$$\mathcal{X}_\epsilon = P_\epsilon \mathrm{Diag}(e_{q,c_1(\epsilon)}, \ldots, e_{q,c_n(\epsilon)})P_\epsilon^{-1} \to \tilde{P}\mathrm{Diag}(x^{\tilde{\gamma}_1}, \ldots, x^{\tilde{\gamma}_n})\tilde{P}^{-1} = x^{\tilde{B}}.$$

We summarize; again under the stated assumption, on matrices P_ϵ:

$$\forall x \notin -q_0^{\mathbf{R}}\ ,\ e_{q,A_\epsilon}(x) \to x^{\tilde{B}}.$$

11.2.3.3. *General case.* We now have a family of $A_\epsilon = I_n + (q-1)B_\epsilon \in \mathrm{GL}_n(\mathbf{C})$ with $B_\epsilon \to \tilde{B} \in \mathrm{Mat}_n(\mathbf{C})$. We are going to need subscripts s, n, u for the additive and mutiplicative Dunford decompositions (and the letter n for "nilpotent" will even risk to collide with the rank n in I_n, but so be it ...), so we rather write $A := A(\epsilon) := A_\epsilon$ and $B := B(\epsilon) := B_\epsilon$. We shall say that the family has *smooth degeneracy of Jordan structure*, abreviated SDJS, if there is a family of diagonalizing matrices P_ϵ for the semi-simple parts $A_s(\epsilon)$ which has a limit $\tilde{P} \in \mathrm{GL}_n(\mathbf{C})$.

Write $B = B_s + B_n$ and $\tilde{B} = \tilde{B}_s + \tilde{B}_n$ (additive Dunford decompositions) whence the additive Dunford decomposition $A = (I_n + (q-1)B_s) + (I_n + (q-1)B_n)$. Therefore the multiplicative Dunford decomposition $A = A_s A_u$ is given by the formulas:

$$A_s = I_n + (q-1)B_s, \quad A_u = I_n + (q-1)B'_n \text{ where } B'_n := A_s^{-1} B_n.$$

From SDJS it follows that $B_s \to \tilde{B}_s$, whence $B_n \to \tilde{B}_n$ and (since $A \to I_n$) that $B'_n \to \tilde{B}_n$. From 11.2.3.1 and 11.2.3.2 we then draw:

$$\forall x \notin -q_0^{\mathbf{R}}\ ,\ e_{q,A_s}(x) \to x^{\tilde{B}_s} \text{ and } e_{q,A_u}(x) \to x^{\tilde{B}_n}.$$

THEOREM 11.10. *Assume the family (A_ϵ) has SDJS. Then:*

$$\forall x \notin -q_0^{\mathbf{R}}\ ,\ e_{q,A}(x) \to x^{\tilde{B}}.$$

PROOF. By construction $e_{q,A}(x) = e_{q,A_s}(x)e_{q,A_u}(x)$, so, by the above, it tends to $x^{\tilde{B}_s}x^{\tilde{B}_n} = x^{\tilde{B}_s+\tilde{B}_n} = x^{\tilde{B}}$. □

EXERCISE 11.76. Recall why $x^{\tilde{B}_s}x^{\tilde{B}_n} = x^{\tilde{B}_s+\tilde{B}_n}$.

11.2.4. Getting it together.

11.2.4.1. *Summarizing our setting.* We return to the general case of a family $A_\epsilon = I_n + (q-1)B_\epsilon \in \mathrm{GL}_n(\mathbf{C}(x))$ degenerating to $\tilde{B} \in \mathrm{Mat}_n(\mathbf{C}(x))$ (this means in our context that $B_\epsilon \to \tilde{B}$). The general conventions hold: $q = q_0^\epsilon$, all limits are for $\epsilon \to 0^+$; as often as required, $\epsilon \approx 0^+$ (see 11.2.2.1).

The necessary assumptions that appeared throughout the study of special cases are the following:

(1) $\tilde{B}(0) \in \mathrm{Mat}_n(\mathbf{C})$ is nonresonant, i.e., no two of its eigenvalues have difference in $\mathbf{N}^*$.
(2) Let $\tilde{x}, \ldots, \tilde{x}_r \in \mathbf{C}^*$ the poles of $\tilde{B}$. Then $B_\epsilon \to \tilde{B}$ on every compact subset of $\mathbf{C} \setminus \{\tilde{x}, \ldots, \tilde{x}_r\}$.
(3) The family $A_\epsilon(0)$ in $\mathrm{GL}_n(\mathbf{C})$ has SDJS (see 11.2.3.3).

Since we saw that for $\epsilon \approx 0^+$ the q-difference system $\sigma_q X = A_\epsilon X$ is nonresonant, there is a canonical fundamental matricial solution $\mathcal{X}_\epsilon = M_\epsilon^{(0)} e_{q,A_\epsilon(0)}$, where $M_\epsilon^{(0)} = I_n + \cdots \in \mathrm{GL}_n(\mathbf{C}(\{x\}))$ is uniquely determined by the condition that $M_\epsilon^{(0)}[A_\epsilon(0)] = A_\epsilon$; in other words, $M_\epsilon^{(0)}$ is solution of the q-difference equation $\sigma_q M = A_\epsilon M A_\epsilon(0)^{-1}$.

By classical theory **[222]**, the differential system $\delta\tilde{X} = \tilde{B}\tilde{X}$ likewise admits a canonical fundamental matricial solution $\tilde{\mathcal{X}} = \tilde{M}^{(0)}x^{\tilde{B}(0)}$, where $\tilde{M}^{(0)} = I_n + \cdots \in \mathrm{GL}_n(\mathbf{C}(\{x\}))$ is uniquely determined by the condition that $\tilde{M}^{(0)}[\tilde{B}(0)] = \tilde{B}$. However, here, the gauge transformation is taken in the sense of differential systems and it means that $\tilde{M}^{(0)}$ is solution of the differential equation $\delta\tilde{M} = \tilde{B}\tilde{M} - \tilde{M}\tilde{B}(0)$. As for the second factor $x^{\tilde{B}(0)}$, it is defined as $\exp(\log x)\tilde{B}(0)$, the logarithm log being (say) the principal determination on some simply connected subset of $\mathbf{C}^*$ not containing 1.

We shall (in account of 11.2.3) choose the simply connected set $\mathbf{C}^* \setminus (-q_0^{\mathbf{R}})$ and prove that indeed $\mathcal{X}_\epsilon \to \tilde{\mathcal{X}}$ on a large subset of it. As we have already seen in 11.2.3 that $e_{q,A_\epsilon(0)} \to x^{\tilde{B}(0)}$ on $\mathbf{C}^* \setminus (-q_0^{\mathbf{R}})$, we just have to check that $M_\epsilon^{(0)} \to \tilde{M}^{(0)}$.

11.2.4.2. *Confluence of the $M^{(0)}$ component.* Equation $\sigma_q M = A_\epsilon M A_\epsilon(0)^{-1}$ may be seen as a q-difference system if we consider M as a vector $\mathcal{M}$ in a space of dimension n^2 and the map $\mathcal{A}_\epsilon : M \mapsto A_\epsilon M A_\epsilon(0)^{-1}$ as a linear automorphism, whence a matrix of rank $N := n^2$: this just amounts to replacing from the start the q-difference module of interest by the Internal Hom q-difference module (see 9.1.1).

At any rate, the matrix $\mathcal{A}_\epsilon(0)$ encodes the conjugacy automorphism $M \mapsto A_\epsilon(0)MA_\epsilon(0)^{-1}$ of $\mathbf{C}^N = \mathrm{Mat}_n(\mathbf{C})$, so its eigenvalues are the λ/μ, $\lambda, \mu \in \mathrm{Sp}\ A_\epsilon(0)$, and, from the opening remark in 11.2.2.1, we deduce that it is nonresonant for

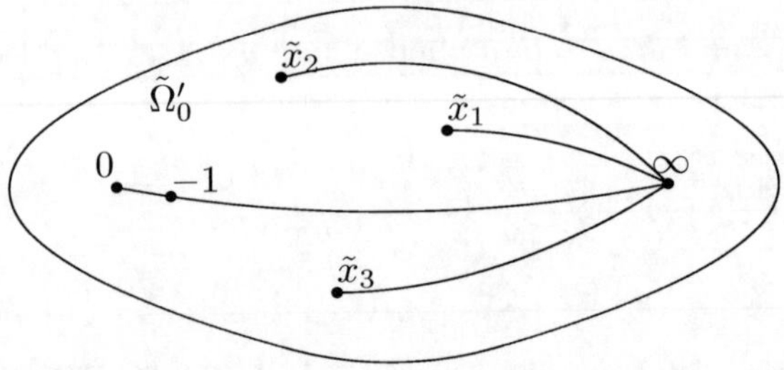

FIGURE 11.2. Domain of the degeneracy $\mathcal{X}_\epsilon^{(0)} \to \tilde{\mathcal{X}}^{(0)}$

$\epsilon \approx 0^+$. On the other hand, equation $\sigma_q M = A_\epsilon M A_\epsilon(0)^{-1}$, written $\sigma_q \mathcal{M} = \mathcal{A}_\epsilon \mathcal{M}$, comes with an initial condition $M(0) = I_n$, whence of the form $\mathcal{M}(0) = \mathcal{M}_0$. So we are in the setting of 11.2.2. We must examine where the degeneracy leads to.

Writing $A_\epsilon = I_n + (q-1)B_\epsilon$, we immediately see that:

$$A_\epsilon M A_\epsilon(0)^{-1} = M + (q-1)(B_\epsilon M - M B_\epsilon(0)) + \text{ multiple of } (q-1)^2,$$

which can be written $N := n^2$) as $\mathcal{A}_\epsilon = I_N + (q-1)\mathcal{B}_\epsilon$, where $\mathcal{B}_\epsilon$ encodes the transformation $M \mapsto B_\epsilon M - M B_\epsilon(0) +$ multiple of $(q-1)$. This tends to the matrix $\tilde{\mathcal{M}}$ encoding $\tilde{M} \mapsto \tilde{B}\tilde{M} - \tilde{M}\tilde{B}(0)$, which is nonresonant. So we are exactly in the setting of 11.2.2. Obviously, the poles of $\tilde{\mathcal{M}}$ are those of $\tilde{B}$ and we may conclude that:

$$\forall x \in \tilde{\Omega}_0 \,,\; M_\epsilon^{(0)}(x) \to \tilde{M}^{(0)}(x).$$

EXERCISE 11.77. Why is $\tilde{\mathcal{M}}$ nonresonant?

11.2.4.3. *Conclusion.* We define (see figure 11.2):

$$\tilde{\Omega}'_0 := \tilde{\Omega}_0 \setminus (-q_0^{\mathbf{R}}) = \mathbf{C}^* \setminus \bigcup_{i=1,\ldots,r} \tilde{x}_i q_0^{\mathbf{R}-} \setminus (-q_0^{\mathbf{R}}).$$

THEOREM 11.11. *Under the conventions and assumptions summarized in* 11.2.4.1*:*

$$\forall x \in \tilde{\Omega}'_0 \,,\; M_\epsilon^{(0)}(x) \to \tilde{M}^{(0)}(x).$$

PROOF. This just combines previous results. □

A beautiful example will be treated at length in Subsection 12.3.2.

CHAPTER 12

Fuchsian q-Difference Equations and Systems: Global Study

This chapter is devoted to the *global* study of rational systems over $\mathbf{S}$. It mainly revolves around rational classification by Birkhoff connection matrix as initiated in [**23**], and some modern variants. For reasons to appear soon, it can be seen as a first initiation to the monodromy and Galois theory of q-difference equations.

The first section is a partly historical exposition of Birkhoff's approach. To stick more closely to *loc. cit.*, we take (there and only there) the convention that $|q| > 1$. The main goal of Birkhoff was classification, and it is here formulated (and proved) in rather old style. Classification in a more modern style but based on Birkhoff connection matrix will be tackled in Chapter 13.

The second section offers a geometric language to deal with some special families of fuchsian q-difference equations. It is completed by a first application to q-Painlevé equations in 12.2.4, also see [**176, 196**] and [CV].

The third section extends to the global setting the phenomenon of confluence which was studied locally in 11.2. It is one (among quite a few) way to understand the idea (rooted in [**23**]) that Birkhoff's matrix is a substitute for monodromy. Since monodromy is closely related to Galois theory [**222**], Chapter 13 will provide other hints in that direction. The general principles are then applied to q-hypergeometric series.

12.1. Global study: the Riemann-Hilbert-Birkhoff correspondence for fuchsian systems

Here, "global" means "over the Riemann sphere $\mathbf{S}$", so we are dealing with q-difference systems $\sigma_q X = AX$ where A is rational: $A \in \mathrm{GL}_n(\mathbf{C}(x))$. This is in straight line of the classical theory of complex ordinary differential equations as initiated by Riemann. Some of the main problems revolve around classification (invariants, normal forms, moduli spaces ...) and their heart is the Riemann-Hilbert correspondence. For an idea of the classical model that we will be trying to follow, see for instance [**121, 125, 222, 250**]. Note however that there also is a "Riemann-Hilbert-Birkhoff" correspondence for irregular q-difference equations; see Chapters 14 and 15; and also a galoisian approach to classification; see 13.3 and 15.4.

For fuchsian equations and systems, Birkhoff initiated in [**23**] a unified approach for ordinary differential equations, difference equations and q-difference equations under the name "generalized Riemann problem". It applies to fuchsian q-difference systems (we shall not tackle usual difference equations, those involving $f(x+1)$). For differential as well as q-difference equations, the standard strategy is to start by a local study (distinguishing ordinary points from singularities) before patching together local information.

As already noted (and more than once), in this book, we mostly do the local study at 0 and then deduce it for ∞ by a change of variables $y := 1/x$. Other points on $\mathbf{S}$, i.e., those on $\mathbf{C}^*$, are moved by the dilatation operator $x \mapsto qx$, so it is complicated to say what "local" means for them. Those "intermediate singularities" will be dealt with in some special cases only. So we consider that the local study of fuchsian q-difference systems has been done in section 11.1.

We shall begin in 12.1.1 by an informal presentation of the article of Birkhoff (the part about q-difference equations). We cover this at some length because it was a crucial step in q-difference theory (indeed, in the history of functional equations), and actually some of the ideas there have been exploited up to recent times (e.g. the content of 12.2.1). Of course, that presentation comes with comments on how those ideas were adapted in modern treatments.

12.1.1. Birkhoff 1913: the "Generalized Riemann Problem".

12.1.1.1. a priori, we should consider rational classification of rational q-difference systems $\sigma_q X = AX$, $A \in \mathrm{GL}_n(\mathbf{C}(x))$. However, as Birkhoff remarks, $A \sim B \Leftrightarrow fA \sim fB$ for any $f \in \mathbf{C}(x)^*$, so we can as well chase denominators and assume that A has polynomial coefficients. As we shall see in 12.2, if we have started from a system fuchsian at 0 and ∞, we may take A in the form:

$$A = A_0 + A_1 x + \cdots + A_\mu x^\mu, \mu \in \mathbf{N}^*, \text{ all } A_i \in \mathrm{Mat}_n(\mathbf{C}), A_0, A_\mu \in \mathrm{GL}_n(\mathbf{C}).$$

This system is of the first kind at 0. At ∞, through the change of variable $y := 1/x$ and the change of unknown $Y(y) := X(x)$, it is transformed into the system $Y(qy) = B(y)Y(y)$, where:

$$B(y) := A(1/qy)^{-1} = q^\mu y^\mu (A_\mu + \cdots + A_0 q^\mu y^\mu)^{-1} = y^\mu (B_0 + B_1 y + \cdots),$$

an expansion starting at $B_0 := q^\mu A_\mu^{-1}$. Therefore the latter system (in variable y) is pure isoclinic of slope μ with exponents the inverses of the eigenvalues of A_μ (taken modulo $q^{\mathbf{Z}}$). However, Birkhoff keeps the variable x for all his calculations.

12.1.1.2. Birkhoff assumes that A_0 and A_μ are diagonalisable; and maybe, implicitly, that they are strictly nonresonant. Let $r_1, \ldots, r_n$, resp. $s_1, \ldots, s_n$ the eigenvalues[1] of A_0, resp. of A_μ. Writing $q = e^{2i\pi\tau}$ (note that Birkhoff takes $|q| > 1$, so $\Im\, \tau < 0$), we introduce ρ_j such that $q^{\rho_j} = r_j$ and σ_i such that $q^{\sigma_i} = s_i^{-1}$, so that $\sigma_q x^{\rho_j} = r_j x^{\rho_j}$ and $\sigma_q x^{-\sigma_i} = s_i x^{-\sigma_i}$; and t such that $x = q^t$, so that $\sigma_q q^{\mu t(t-1)/2} = x^\mu q^{\mu t(t-1)/2}$. Then there are local fundamental matricial solutions "at 0" and "at ∞" in the form:

$$\mathcal{X}^{(0)} = M^{(0)} D^{(0)} \text{ and } \mathcal{X}^{(\infty)} = M^{(\infty)} D^{(\infty)},$$

where $M^{(0)} = (a_{i,j}(x)) \in \mathrm{GL}_n(\mathbf{C}\{x\})$ and $M^{(\infty)} = (b_{i,j}(x)) \in \mathrm{GL}_n(\mathbf{C}\{1/x\})$ and where:

$$D^{(0)} := \begin{pmatrix} x^{\rho_1} & \ldots & 0 \\ \vdots & \ddots & \vdots \\ 0 & \ldots & x^{\rho_n} \end{pmatrix}, \text{ and } D^{(\infty)} := q^{\mu t(t-1)/2} \begin{pmatrix} x^{-\sigma_1} & \ldots & 0 \\ \vdots & \ddots & \vdots \\ 0 & \ldots & x^{-\sigma_n} \end{pmatrix},$$

[1] So, in our terminology, they are the exponents (at least their classes in $\mathbf{E}_q$ are) although maybe the ρ_j, σ_i of Birkhoff, to be introduced soon, better deserve that name.

Of course, in this book, we use the uniform functions $\theta_q^{-\mu}$, e_{q,r_j} and e_{q,σ_i} rather than the multivalued $q^{\mu t(t-1)/2}$, x^{ρ_j} and x^{σ_i}.

EXERCISE 12.1. Admitting the properties of those multivalued functions, use the methods of Section 11.1 to obtain $M^{(0)}$ and $M^{(\infty)}$.

REMARK 12.2. We should already give a look (at least superficially) at so-called *intermediate singularities*. Here, they are the points where $A(x)$ becomes singular. So we compute:

$$\det\ A(x) = (\det\ A_\mu)(x - x_1)\cdots(x - x_N) = (\det\ A_0)(1 - x/x_1)\cdots(1 - x/x_N),$$

where $N := n\mu$ and $x_1, \ldots, x_N \in \mathbf{C}^*$. This implies a q-analogue of Fuchs relations (see [**169**, Part I,4.1.2]):

$$x_1\cdots x_N = (-1)^N \frac{\det\ A_0}{\det\ A_\mu} = (-1)^N \frac{r_1\cdots r_n}{s_1\cdots s_n}.$$

The exponents $r_1, \ldots, r_n$, resp. $s_1, \ldots, s_n$ (along with μ) can be considered as "local monodromy data" at 0, resp. ∞; and $x_1, \ldots, x_N$ play a similar role in $\mathbf{C}^*$. The fact that no supplementary data are given at each x_k stems from the (hidden) fact that we do not (yet) try to analyse what happens locally at them.

12.1.1.3. We "connect" the fundamental matricial solutions $\mathcal{X}^{(0)} = M^{(0)}D^{(0)}$ and $\mathcal{X}^{(\infty)} = M^{(\infty)}D^{(\infty)}$ through *Birkhoff's connection matrix*:

$$P := \left(\mathcal{X}^{(\infty)}\right)^{-1}\mathcal{X}^{(0)} = \left(D^{(\infty)}\right)^{-1} MD^{(0)}, \text{ where } M := \left(M^{(\infty)}\right)^{-1} M^{(0)}.$$

(The second form will be useful later, but as well see already the inner structure of P.) We think of the relation $\mathcal{X}^{(0)} = \mathcal{X}^{(\infty)}P$ as mimicking the analytic continuation of $\mathcal{X}^{(0)}$ from "near 0" to "near ∞", where we express it in terms of $\mathcal{X}^{(\infty)}$. Of course that analytic continuation is virtual since anyhow all our functions are defined over $\mathbf{C}^*$. Anyway, from the fact that $\mathcal{X}^{(0)}$ and $\mathcal{X}^{(\infty)}$ are solutions of the same q-difference equation $\sigma_q X = AX$, we deduce that P is q-invariant:

$$P(qx) = \left(\mathcal{X}^{(\infty)}(qx)\right)^{-1}\mathcal{X}^{(0)}(qx) = \left(A(x)\mathcal{X}^{(\infty)}(x)\right)^{-1} A(x)\mathcal{X}^{(0)}(x) =$$

$$\left(\mathcal{X}^{(\infty)}(x)\right)^{-1}(A(x))^{-1}A(x)\mathcal{X}^{(0)}(x) = \left(\mathcal{X}^{(\infty)}(x)\right)^{-1}\mathcal{X}^{(0)}(x) = P(x).$$

(This can also be proved by applying σ_q to the relation $\mathcal{X}^{(0)} = \mathcal{X}^{(\infty)}P$.) However, P is not q-elliptic in the sense we give to that expression in Chapter 2, since it is not made of uniform functions. The multivaluedness is traditionally expressed in terms of going from x to $e^{2\mathrm{i}\pi}x$; otherwise said, if $x = e^u$, going from u to $u + 2\mathrm{i}\pi$. Since $M^{(0)}$ and $M^{(\infty)}$ are uniform, what matters is the behaviour of $D^{(0)}$ and $D^{(\infty)}$:

$$\begin{cases} D^{(0)}\left(e^{2\mathrm{i}\pi}x\right) = \mathrm{Diag}\left(e^{2\mathrm{i}\pi\rho_1}, \ldots, e^{2\mathrm{i}\pi\rho_n}\right) D^{(0)}(x), \\ D^{(\infty)}\left(e^{2\mathrm{i}\pi}x\right) = \phi(x)\,\mathrm{Diag}\left(e^{-2\mathrm{i}\pi\sigma_1}, \ldots, e^{-2\mathrm{i}\pi\sigma_n}\right) D^{(\infty)}(x), \end{cases}$$

The scalar factor in the second relation comes from $x = q^t = e^{2\mathrm{i}\pi\tau t}$, so that, setting provisionally $\tau' := 1/\tau = 2\mathrm{i}\pi/\log q$, we expect that $e^{2\mathrm{i}\pi}x = q^{t+\tau'}$ and:

$$\phi(x) = \frac{q^{\mu(t+\tau')(t+\tau'-1)/2}}{q^{\mu t(t-1)/2}} = e^{2\mathrm{i}\pi\mu(t+\frac{\tau'-1}{2})} = (-1)^\mu e^{2\mathrm{i}\pi\mu(t+\tau'/2)}$$

So the multivaluedness of the connection matrix takes the form:

$$P\left(e^{2i\pi}x\right)=(-1)^{\mu}e^{-2i\pi\mu(t+\tau'/2)}\mathrm{Diag}\left(e^{2i\pi\sigma_1},\ldots,e^{2i\pi\sigma_n}\right)P(x)\mathrm{Diag}\left(e^{2i\pi\rho_1},\ldots,e^{2i\pi\rho_n}\right).$$

In our treatment in 12.2 (and again in 13.3) concentrating on the "central part" M, we shall see that it is uniform and that the above "automorphy factors" rather come from the transformation $x \mapsto qx$, thus directly relating to usual q-difference equations.

Before pursuing Birkhoff's analysis, let us discuss why P has been, ever since **[23]**, considered as a substitute for monodromy. In classical theory (that of complex differential equations), analytic continuation along a path going from (the neighborhood of) a singularity to another would be followed by continuation along another path going back to the origin. Thus we obtain continuation along loops and the corresponding transformations of the solutions make up the monodromy group **[125, 222, 250]**.

Here, relation $\mathcal{X}^{(0)} = \mathcal{X}^{(\infty)}P$ may be understood as meaning that the solution $\mathcal{X}^{(0)}$ which is "local at 0" can be continued at ∞ and expressed there in function of the solution $\mathcal{X}^{(\infty)}$ which is "local at ∞". So P plays the role of a connection matrix of the classical theory. But here, P depends on x, so we can see it either as a continuous family of paths or even maybe as an indetermined path, that can be determined by choosing an evaluation $P(x_0)$ (in the same way as occurs wave packet reduction in quantum mechanics ...); and maybe the evaluated matrices $P(x_1)^{-1}P(x_0)$ (when defined and nonsingular) could be thought of as monodromy matrices. This was first confirmed by Etingof in **[82]**. We shall develop that point of view in several ways.

12.1.1.4. From the representation $x = q^t$, Birkhoff introduces $\overline{P}(t) := P(q^t)$, with coefficients $\overline{p}_{i,j}(t) = p_{i,j}(q^t)$, $i, j = 1, \ldots, n$. From 12.1.1.2 we draw:

$$\overline{p}_{i,j}(t+1) = \overline{p}_{i,j}(t) \text{ and } \overline{p}_{i,j}(t+\tau') = (-1)^{\mu}e^{-2i\pi\mu(t+\tau'/2)}e^{2i\pi(\rho_j+\sigma_i)}\overline{p}_{i,j}(t).$$

On the other hand, he appeals to the Weierstraß' sigma function attached to the lattice $[1;\tau']$ (see the exercise at the end of 2.2.2), which satisfies:

$$\sigma(t+1) = e^{-\eta(t+1/2)}\sigma(t) \text{ and } \sigma(t+\tau') = e^{-\eta'(t+\tau'/2)}\sigma(t).$$

The next deduction relies on properties of σ that we did not state (let alone prove ...) but we shall do something similar with θ_q in 12.2.1, so the reader may fill in the details. From the above properties of $\overline{p}_{i,j}$, it follows that:

$$\overline{p}_{i,j}(t) = c_{i,j}e^{-(\eta\mu/2)t^2+(\eta(\rho_j+\sigma_i)-\eta'\mu/2)t}\prod_{k=1}^{\mu}\sigma\left(t-\alpha_{i,j}^{(k)}\right).$$

Here, the $c_{i,j}$ and the $\alpha_{i,j}^{(k)}$ all belong to $\mathbf{C}$, the latter being moreover subject to the condition:

$$\forall i,j = 1,\ldots,n\,,\ \sum_{k=1}^{\mu}\alpha_{i,j}^{(k)} = \sigma_i + \rho_j - \mu\tau'/2.$$

12.1.1.5. *Counting.* Birkhoff then counts[2] the "characteristic constants" (we would call them the *invariants*); this amounts to a heuristic calculation of the dimension of a hoped-for space of equivalence classes.

[2] As the "great headman, Ghân-buri-Ghân", who counted "stars in sky, leaves on trees, men in the dark" in *The Lord of the Rings*, mathematicians have, for a long time, based classification theories on counting degrees of freedom.

(1) There are $2n$ complex numbers ρ_j, σ_i, then n^2 complex numbers $c_{i,j}$ and $n^2\mu$ complex numbers $\alpha_{i,j}^{(k)}$.

(2) However, the $\alpha_{i,j}^{(k)}$ satisfy n^2 relations $\sum_k \alpha_{i,j}^{(k)} = \ldots$, thus reducing the dimension by n^2.

(3) Moreover, there is a "gauge freedom" coming from the possibility to replace $\mathcal{X}^{(0)}$ by $\mathcal{X}^{(0)}\mathrm{Diag}(f_1, \ldots, f_n)$ and $\mathcal{X}^{(\infty)}$ by $\mathcal{X}^{(\infty)}\mathrm{Diag}(h_1, \ldots, h_n)$ (all $f_j, h_i \in \mathbf{C}^*$), whence each $c_{i,j}$ by $c_{i,j}f_j/h_i$; the factors f_j/h_i make up for $2n-1$ independent numbers, thus reducing the dimension by $2n-1$.

REMARK 12.3. The general rule of thumb is that, if you start with d degrees of freedom (so a space of dimension d) and you enforce e *independent* constraints (so you go to a subspace defined by e independent equations), you get dimension $d-e$. Now if you choose to identify data equivalent under the action of a group with c *independent* parameters (and which acts freely) , you get dimension $d-e-c$. In linear algebra, this is quite standard (see the exercise below). In algebraic geometry, there are ways to deal with such matters, we shall do that in 12.2.1.

EXERCISE 12.4. Let a linear space V of dimension d. We express the equations by a linear map $g : V \to W$, so the corresponding subspace is $\mathrm{Ker} g$. We likewise express the group action by a linear map $f : U \to V$ (so $u \in U$ represents the action parameters and we want identifications $v \sim v + f(u)$) and assume that this acts inside the subspace, i.e., $\mathrm{Im} f \subset \mathrm{Ker} g$. In the end, we should consider the space $\mathrm{Ker} g/\mathrm{Im} f$. We write $c := \dim U$ and $e := \dim W$.

(i) How would you express the independence of equations, resp. of the action parameters? Check that in that case $\dim \mathrm{Ker} g/\mathrm{Im} f = d - e - c$.

(ii) What if there are dependencies? (A quantitative answer is expected.)

Resuming our counting process, we end up with the following estimation for the total number of degrees of freedom:

$$2n + n^2 + n^2\mu - n^2 - (2n-1) = \mu n^2 + 1.$$

This stands for the *right-hand side* of the Riemann-Hilbert correspondence:

$$\left\{ \begin{array}{c} \text{Rational equivalence classes} \\ \text{of rational } q\text{-difference systems} \end{array} \right\} \longleftrightarrow \left\{ \begin{array}{c} \text{Space of tuples of characteristic} \\ \text{constants satisfying proper relations} \\ \text{modulo relevant equivalence relation} \end{array} \right\}.$$

On the *left-hand side*, we see that our q-difference system is characterized by the matrix $A = A_0 + A_1 x + \cdots + A_\mu x^\mu$, which involves $n^2(\mu+1)$ complex coefficients. However there is a "gauge freedom" which amounts to replace A by CAC^{-1} for an arbitrary $C \in \mathrm{GL}_n(\mathbf{C})$, and this C is generically given up to a scalar factor (because generically the simultaneous commutant of all A_i is $\mathbf{C}^* I_n$), so the conjugacy decreases the dimension by $n^2 - 1$ and the resulting count is $n^2(\mu+1) - (n^2-1) = \mu n^2 + 1$ again.

If we admit (which Birkhoff does not seem to explicitly justify, but which will follow from our version in 12.2) that the stated gauge freedoms are the only possible ones, we deduce that we have an injective map (at least generically injective) between "spaces" of the same "dimension". The quotation marks are because we are actually manipulating raw sets as long as we did not put some structure on them. Anyhow, we may hope that our map is (at least generically) surjective. This is what Birkhoff proves next (see 12.1.1.5), and it is by far the deepest part of his contribution.

REMARK 12.5. Nowadays, we try to distinguish the *local* contributions to the *monodromy data* implicit in that procedure - just as Riemann did by starting from the local behaviour near singularities before connecting them. For us, the ρ_j (or the r_j) stand for the local data at 0 and the σ_i (or the s_i), along with the slope μ, for the local data at ∞. Intermediate singularities $x_1, \ldots, x_N$ (see the remark at the end of 12.1.1.2) are the other loci where something happens. If we fix the ρ_j and the σ_i, we decrease the dimension by $2n$ (frozen degrees of freedom); we cannot play with μ which is a discrete invariant and has no clear geometrical meaning. Fixing $x_1, \ldots, x_N$ amounts to N equations $\det A(x_1) = \cdots = \det A(x_N) = 0$, but only $N-1$ of them are independent because of Fuchs relation. (We'll come back to that in 12.2.) So fixing the local data leaves us with the dimension:

$$\mu n^2 + 1 - 2n - (N-1) = \mu n^2 + 1 - 2n - (\mu n - 1) = (n-1)(\mu n - 2).$$

For $n = 2$, $\mu = 1$, this number vanishes, meaning that the local data (or local isomorphism classes) essentially determine the global isomorphy class: this important property is known as *rigidity*. The q-hypergeometric equations fall under that case. The next interesting case is $n = 2$, $\mu = 2$, for which we have dimension 2, thus a surface of monodromy data. We shall discuss that surface in [CV](also see [**196**]).

EXERCISE 12.6. Explain directly rigidity when $n = 1$.

12.1.1.6. As for surjectivity, which Birkhoff formulates as "the inverse problem" of recovering the system matrix A from the connection matrix P (given the local data at 0 and ∞), it goes as follows. We are given the ρ_j, the σ_i and μ, along with an invertible $n \times n$ matrix P such that $P(qx) = P(x)$ as well as the specified behaviour of $P\left(e^{2i\pi}x\right)$ described in 12.1.1.3. Note first that $D^{(0)}$ and $D^{(\infty)}$ can be recovered from said local data. Then we extract the "central part" of P:

$$M := D^{(\infty)} P \left(D^{(0)}\right)^{-1},$$

which is uniform because of the prescribed behaviour of $P\left(e^{2i\pi}x\right)$. Suppose now that we are able to obtain $M^{(0)} \in \mathrm{GL}_n(\mathscr{M}(\mathbf{C}))$ and $M^{(\infty)} \in \mathrm{GL}_n(\mathscr{M}(\mathbf{C}_\infty))$ (recall that $\mathbf{C}_\infty = \mathbf{S} \setminus \{0\} = \mathbf{C} \cup \{\infty\}$) such that $M = \left(M^{(\infty)}\right)^{-1} M^{(0)}$. Then we set $\mathcal{X}^{(0)} := M^{(0)} D^{(0)}$ and $\mathcal{X}^{(\infty)} := M^{(\infty)} D^{(\infty)}$. By construction, $P = \left(\mathcal{X}^{(\infty)}\right)^{-1} \mathcal{X}^{(0)}$. Moreover:

$$\left(\sigma_q \mathcal{X}^{(0)}\right) \left(\mathcal{X}^{(0)}\right)^{-1} = \left(\sigma_q \mathcal{X}^{(\infty)} P\right) \left(\mathcal{X}^{(\infty)} P\right)^{-1} = \left(\sigma_q \mathcal{X}^{(\infty)}\right) \left(\mathcal{X}^{(\infty)}\right)^{-1},$$

call $A(x)$ that matrix, so that $\mathcal{X}^{(0)}$ and $\mathcal{X}^{(\infty)}$ are fundamental matricial solutions of the q-difference system $\sigma_q X = AX$. On the one hand:

$$A = \left(\sigma_q \mathcal{X}^{(0)}\right) \left(\mathcal{X}^{(0)}\right)^{-1} = \left(\sigma_q M^{(0)}\right) \left(\sigma_q D^{(0)}\right) \left(D^{(0)}\right)^{-1} \left(M^{(0)}\right)^{-1} =$$

$$\left(\sigma_q M^{(0)}\right) \mathrm{Diag}(r_1, \ldots, r_n) \left(M^{(0)}\right)^{-1} \in \mathrm{GL}_n(\mathscr{M}(\mathbf{C})).$$

On the other hand, through a similar calculation, $A \in \mathrm{GL}_n(\mathscr{M}(\mathbf{C}_\infty))$. This means that A is invertible and meromorphic over $\mathbf{C}$ and over $\mathbf{C}_\infty$, so over the whole of $\mathbf{S}$, so $A \in \mathrm{GL}_n(\mathbf{C}(x))$ as wanted.

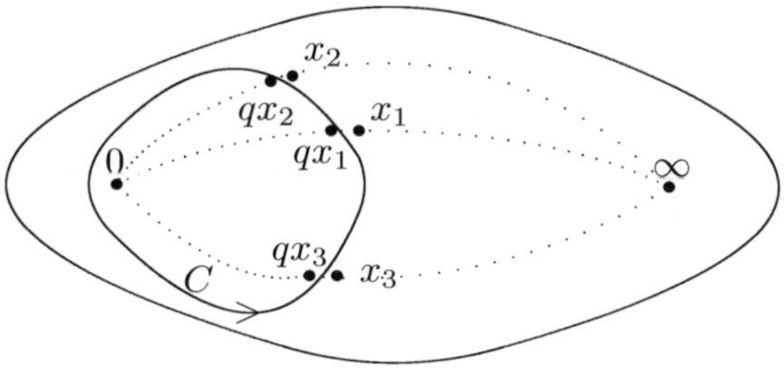

FIGURE 12.1. The analytic contour for Birkhoff's Preliminary Theorem

So we have to justify the crucial step, the existence of an "analytic matrix factorisation":

$$M = \left(M^{(\infty)}\right)^{-1} M^{(0)}.$$

This relies on the *preliminary theorem* of Birkhoff [**23**, pp 266–267], stated here in the basic case of a single simple contour (see Figure 12.1).

THEOREM 12.1 (Birkhoff's Preliminary Theorem). *Let C a simple closed analytic curve on $\mathbf{S}$ separating 0 from ∞ and let $D_0 \ni 0$, $D_\infty \ni \infty$ the connected components of $\mathbf{S} \setminus C$. Let $M(x)$ an analytic invertible matrix in a neighborhood of C (i.e., $x \mapsto M(x)$ is analytic with values in $GL_n(\mathbf{C})$). Then there exists open neighborhoods V_0 of $\overline{D_0}$ and V_∞ of $\overline{D_\infty}$ and analytic matrices M_0 on V_0 and M_∞ on V_∞ such that:*

(1) *$M_0 = M_\infty M$ in a neighborhood of C contained in $V_0 \cap V_\infty$.*
(2) *M_0 is regular (i.e., holomorphic with holomorphic inverse) over V_0.*
(3) *M_∞ is regular over $V_\infty \setminus \{\infty\}$ and holomorphic at ∞.*

Note however that the last condition cannot be sharpened, one cannot in general require that $M_\infty(\infty) \in \mathrm{GL}_n(\mathbf{C})$. Also, it is easy to state variants, where M_0 and M_∞, resp. M_0 and M_0^{-1} exchange their positions, etc.

Since we will be appealing to that theorem in section 12.2 and then again in section 13.3, we do not detail here how Birkhoff applies it. Such factorisation problems play a quite important role in the study of functional equations, sometimes under the name of "Riemann-Hilbert problem". Birkhoff's proof is hard functional analysis, more modern proofs involve a mixture of analysis and of homological algebra: see [**249**] for a setting similar to ours; and [**88**] for an application to Riemann-Hilbert correspondence for differential equations, due to Röhrl [**205**]. In [**222**] another factorisation theorem, due to Cartan, was used to the same aim.

12.2. A geometric approach to Riemann-Hilbert-Birkhoff correspondence

12.2.1. A "space of monodromy data". The following variant of Birkhoff's classification theorem comes from [**176**]. The notational conventions will slightly differ from those introduced in 12.1.1. In particular, we go on with the general convention of this book: $0 < |q| < 1$.

We shall consider the rational classification of matrices $A_0 + \cdots + A_\mu x^\mu$ (see the precise assumptions further below); but, instead of Birkhoff's connection matrix P, we shall rely on its "central component" M. The main underlying reason for that is that M represents the contributions of the intermediate singularities, freed of the local contributions at 0 and ∞. The practical reason is that it works very well, for classification as well as for Galois theory (see **[218]** and 13.3).

As argued before, once fixed the local monodromies, in the guise of the exponents[3] $\rho_1, \ldots, \rho_n, \sigma_1, \ldots, \sigma_n$ and the positions $x_1, \ldots, x_N$ of the intermediate singularities, what we are doing is, in accordance with Riemann's views, classifying equations through their monodromies. In the legacy of Hilbert's views, this was encoded as a monodromy representation, so our classifying space is a space of representations of the fundamental group. It is called a "character variety" **[124, 155]** (also see a work to appear[4] by Emmanuel Paul and Jean-Pierre Ramis) because characters are elementary representations of a group.

However, at present, we are not able to propose an adequate q-analog for the *global* fundamental group[5]. So we shall rather call what we are building a *space of monodromy data.*

The word "space", as opposed to "set", is by no way innocent. We insist on being able to endow our classifying set with some geometric structure (actually, an algebraic variety) which properly reflects the underlying geometry of the space of q-difference systems. As an illustration, we apply our constructions to q-hypergeometric equations and to q-Painlevé equations.

12.2.1.1. *The general setting.* Let $n \geq 2$ and $\mu \geq 1$. We fix the diagonal matrices of exponents at 0 and ∞:

$$R := \mathrm{Diag}(\rho_1, \ldots, \rho_n), \quad S := \mathrm{Diag}(\sigma_1, \ldots, \sigma_n),$$

all $\rho_i, \sigma_j \in \mathbf{C}^*$, moreover subject to the *strict nonresonancy conditions*:

$$(SNR) \quad \forall i, j, i', j' \in \{1, \ldots, n\} \text{ such that } i \neq i', j \neq j', \text{ we have:} \quad \rho_i \not\equiv \rho_{i'}, \ \sigma_j \not\equiv \sigma_{j'}.$$

Recall that all congruences $\equiv$ in $\mathbf{C}^*$ are modulo $q^{\mathbf{Z}}$. We shall consider $n \times n$ matrices of the form $A_0 + \cdots + A_\mu x^\mu$; so, generically, their determinant is a poynomial of degree $N := n\mu$ and we shall denote $x_1, \ldots, x_N$ its roots. So we fix N nonzero complex numbers x_k, $k = 1, \ldots, N$, also subject to a supplementary nonresonancy condition:

$$(SNR) \quad \forall k, k' \in \{1, \ldots, N\} \text{ such that } k \neq k', \text{ we have: } x_k \not\equiv x_{k'}.$$

The N-uple, we abbreviate $\underline{x} := (x_1, \ldots, x_N)$. For reasons to appear soon, we moreover require that they satisfy *Fuchs relation*; beware that this is an equality, not a congruence:

$$(FR) \quad x_1 \cdots x_N = (-1)^N \frac{\rho_1 \cdots \rho_n}{\sigma_1 \cdots \sigma_n}.$$

---

[3]We slightly depart from the notations in **[23]** and in 12.1.1, as we have no use for the logarithms of the eigenvalues.

[4]It should be part of the *Handbook of Geometry and Topology of Singularities VI: Foliations*, edited by Felipe Cano and Jose Seade, to be published by Springer. Meanwhile, see https://arxiv.org/abs/2301.08597.

[5]The situation is different for the *local* study of *irregular* q-difference equations: we then have a good q-analog of the "wild fundamental group" of Ramis; see **[195, 224]** and 15.4.

Now we define the relevant space of q-difference systems, via their matrices:
(12.1)

$$E_{R,S,\underline{x}} := \left\{ A_0 + \cdots + A_\mu x^\mu \mid A_0, \ldots, A_n \in \mathrm{Mat}_n(\mathbf{C}) \text{ s.t. } \begin{cases} A_0 \sim R, A_\mu \sim S, \\ \det A = 0 \text{ on } \underline{x}. \end{cases} \right\}.$$

The last condition means of course that $\forall k \in \{1, \ldots, N\}$, $\det A(x_k) = 0$. The first two conditions are conjugacies in $\mathrm{Mat}_n(\mathbf{C})$. They imply that $\det A_0 = \rho_1 \cdots \rho_n$ and $\det A_\mu = \sigma_1 \cdots \sigma_n$, so, together with the prescribed roots, we have:

$$\det A(x) = \sigma_1 \cdots \sigma_n (x - x_1) \cdots (x - x_N) = \rho_1 \cdots \rho_n (1 - x/x_1) \cdots (1 - x/x_N).$$

Evaluating at $x = 0$ yields $(-1)^N \sigma_1 \cdots \sigma_n = \rho_1 \cdots \rho_n$. Thus, without assumption (FR), the space $E_{R,S,\underline{x}}$ would be empty.

We denote by $\sim$ the equivalence relation induced on $E_{R,S,\underline{x}}$ by rational gauge equivalence:

$$\forall A, B \in E_{R,S,\underline{x}} \; , \; A \sim B \underset{def}{\Longleftrightarrow} (\exists F \in \mathrm{GL}_n(\mathbf{C}(x)) \; : \; B = F[A]) \, . \tag{12.2}$$

Our goal is to describe (and understand!) the set of equivalence classes:

$$\mathscr{E}_{R,S,\underline{x}} := \frac{E_{R,S,\underline{x}}}{\sim}. \tag{12.3}$$

The notations will often be shortened to $E := E_{R,S,\underline{x}}$ and $\mathscr{E} := \mathscr{E}_{R,S,\underline{x}}$.

EXERCISE 12.7. (i) What would become all that if $\mu = 0$?
(ii) Same question for $n = 1$.
(iii) Give an example with $n = 2$, $\mu = 1$, if possible in relation with Chapter 1.

12.2.1.2. *Local reductions.* We fix some $A \in E_{R,S,\underline{x}}$.

LEMMA 12.8. (i) *Let $C \in GL_n(\mathbf{C})$ such that $A_0 = CRC^{-1}$. There exists a unique $M_0 \in GL_n(\mathbf{C}\{x\})$ such that $M_0(0) = C$ and $M_0[R] = A$.*

(ii) *Let $D \in GL_n(\mathbf{C})$ such that $A_\mu = DSD^{-1}$. There exists a unique $M_\infty \in GL_n(\mathbf{C}\{1/x\})$ such that $M_\infty(\infty) = D$ and $M_\infty[Sx^\mu] = A$.*

PROOF. (i) Note that $C[R] = CRC^{-1} = A_0$, so looking at M_0 in the form $F_0 C$, the requirements are equivalent to: $F_0 \in \mathrm{GL}_n(\mathbf{C}\{x\})$ such that $F_0(0) = I_n$ and $F_0[A_0] = A$. This precisely amounts to the reduction of a nonresonant first kind system to a constant one, which was obtained in 11.1.

(ii) The condition $M_\infty[Sx^\mu] = A$ is equivalent to $M_\infty[S] = A_\mu + \cdots + A_0 x^{-\mu}$. So the change of variable $y = x^{-1}$ reduces the second statement to the first one. □

EXERCISE 12.9. Do precisely the reduction of (ii) to (i). (Beware that, if $y = x^{-1}$, then $x \mapsto qx$ does not translate into $y \mapsto qy$.)

LEMMA 12.10. *With the same notations as in the previous lemma, let moreover $C', D' \in GL_n(\mathbf{C})$ such that $A_0 = C'RC'^{-1}$ and $A_\mu = D'SD'^{-1}$. Let M'_0 and M'_∞ be deduced accordingly. Then there exist $\Gamma, \Delta \in \mathbf{D}_n(\mathbf{C})$ such that $C' = C\Gamma$, $D' = D\Delta$; and they satisfy $M'_0 = M_0\Gamma$, $M'_\infty = M_\infty\Delta$.*

PROOF. Let $\Gamma := C^{-1}C'$ and $\Delta := D^{-1}D'$. Then from $C'RC'^{-1} = CRC^{-1}$ and $D'SD'^{-1} = DSD^{-1}$ we deduce that Γ commutes with R and Δ commutes with S. Since R and S are diagonal with pairwise unequal coefficients, Γ and Δ are diagonal. The situation is now encoded in the diagram:

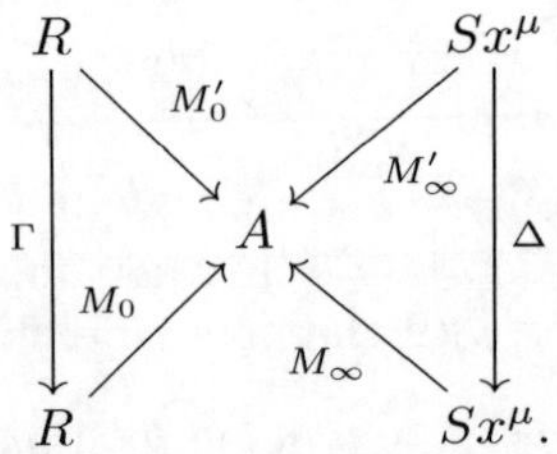

Computationally: $\Gamma[R] = R$, so $(M_0\Gamma)[R] = A$; and $(M_0\Gamma)(0) = C\Gamma = C'$. So $M_0\Gamma$ satisfies the properties which uniquely characterize M_0', so $M_0\Gamma = M_0'$. The same argument at infinity entails $M_\infty\Delta = M_\infty'$. □

EXERCISE 12.11. How does the above relate to the general fact that morphisms $A \to B$ are the solutions of $\underline{\mathrm{Hom}}(A, B) = \underline{\mathrm{Hom}}(M_A, M_B)$?

PROPOSITION 12.12 (Properties of M_0, M_∞). *Let the notations be as in the first lemma.*

(i) $M_0 \in GL_n(\mathbf{C}\{x\})$ *admits a unique extension* $M_0 \in GL_n(\mathscr{M}(\mathbf{C}))$ *such that:*

- M_0 *has simple poles over* $\underline{x}q^{-\mathbf{N}} = \bigcup x_k q^{-\mathbf{N}}$*, and no other poles;*
- M_0^{-1} *is holomorphic all over* $\mathbf{C}$.

(ii) $M_\infty \in GL_n(\mathbf{C}\{1/x\})$ *admits a unique extension* $M_0 \in GL_n(\mathscr{M}(\mathbf{C}_\infty))$ *such that:*

- M_∞ *is holomorphic all over* $\mathbf{C}_\infty$*;*
- M_∞^{-1} *has simple poles over* $\underline{x}q^{\mathbf{N}^*} = \bigcup x_k q^{-\mathbf{N}}$*, and no other poles.*

PROOF. (i) a priori M_0 is regular near 0, so there is a disk $\overset{\circ}{\mathrm{D}}(0, r)$, $r > 0$, over which M_0 and M_0^{-1} take values in $\mathrm{GL}_n(\mathbf{C})$. We are going to iteratively extend the domains of M_0 and M_0^{-1} with the help of the functional equation. A similar process was already displayed, see for instance 4.4.1, 5.3.2, or 7.2.3. Since M_0^{-1} is regular on $\overset{\circ}{\mathrm{D}}(0, r)$, so is $\sigma_q M_0^{-1}$ on $\overset{\circ}{\mathrm{D}}(0, r')$, where $r' := |q|^{-1} r > r$. Then relation $M_0[R] = A$ translates to $M_0^{-1} = R^{-1}\left(\sigma_q M_0^{-1}\right) A$, so M_0^{-1} extends holomorphically to $\overset{\circ}{\mathrm{D}}(0, r')$. The extension satisfies the same functional equation (for instance because of the principle of analytic continuation), so, iterating, we get a holomorphic extension of M_0^{-1} to $\overset{\circ}{\mathrm{D}}\left(|q|^{-m} r\right)$ for all $m \geq 1$. Since $|q|^{-m} \to +\infty$, this gives the holomorphic extension to the whole of $\mathbf{C}$.

From the relation $M_0 = A^{-1}\left(\sigma_q M_0\right) R$, we see that, if M_0 has simple poles at $\Sigma \subset \overset{\circ}{\mathrm{D}}(0, r)$ for some $r > 0$ (and no other poles on that disk), then, on $\overset{\circ}{\mathrm{D}}(0, r')$, $r' := |q|^{-1} r$, it has simple poles at some $\Sigma' \supset q^{-1}\Sigma$, where $\Sigma' \setminus q^{-1}\Sigma$ is made of the zeroes of $\det A$ on $\overset{\circ}{\mathrm{D}}(0, r') \setminus \overset{\circ}{\mathrm{D}}(0, r)$. Again, iterating gives the conclusion.

(ii) The arguments are similar; here, σ_q expands the "open disks centered at ∞", which have the form $\mathbf{C}_\infty \setminus \overline{\mathrm{D}}(0, r)$.

□

EXERCISE 12.13. (i) Find a functional relation satisfied by $\det M_0$, resp. by $\det M_\infty$. Use it to describe the polar loci of $\det M_0^{-1}$ and $\det M_\infty$. Then use the holomorphy statements in (i) and

(ii) of the proposition to deduce anew the polarity statements.

(iii) Deduce a functional equation and polar properties for $\det M_0^{-1}M_\infty$ and then a product expansion in Theta factors.

12.2.1.3. *The connection matrix (central component).* With the notations of Lemmas 12.8 and 12.10 of 12.2.1.2, completed by the information on the extended M_0, M_∞ (and also, as a consequence, on M_0', M_∞'), we can define:

$$M := M_0^{-1}M_\infty \in \mathrm{GL}_n(\mathscr{M}(\mathbf{C}^*,0)) \text{ and } M' := M_0'^{-1}M_\infty' \in \mathrm{GL}_n(\mathscr{M}(\mathbf{C}^*,0))$$

and therefore complete the previous commutative diagram into[6]:

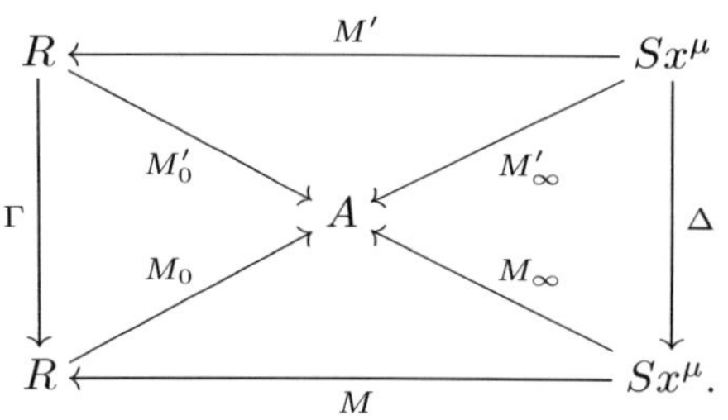

Indeed, from $M_0[R] = M_0'[R] = M_\infty[Sx^\mu] = M_\infty'[Sx^\mu] = A$, it follows readily by direct calculation that $M[Sx^\mu] = M'[Sx^\mu] = R$.

Proposition 12.14. (i) *The matrix* $M \in GL_n(\mathscr{M}(\mathbf{C}^*))$ *(and therefore also the matrix* M'*) has the following properties:*

- $\sigma_q M = RM(Sx^\mu)^{-1}$;
- M *is holomorphic over* $\mathbf{C}^*$;
- M^{-1} *has simple poles at* $[\underline{x};q] = \underline{x}q^{\mathbf{Z}} = \bigcup x_k q^{\mathbf{Z}}$ *and no other poles. (Equivalently:* $\det M$ *has simple zeroes at* $[\underline{x};q]$ *and no other zeroes.)*

(ii) *There exist* $\Gamma, \Delta \in \mathbf{D}_n(\mathbf{C})$ *such that* $M' = \Gamma^{-1}M\Delta$.

Proof. The relations $\sigma_q M = RM(Sx^\mu)^{-1}$ and $M' = \Gamma^{-1}M\Delta$ can either be deduced by computation from the properties of $M_0, M_\infty, M_0', M_\infty'$ given in 12.2.1.2, or read off the above commutative diagram (the lower horizontal arrow says that $M[Sx^\mu] = R$, i.e., that $(\sigma_q M)Sx^\mu = RM$; the outer rectangle says that $\Gamma M' = M\Delta$). The holomorphy and polarity properties immediately flow from 12.2.1.2. □

So it is natural to introduce the following space:

$$F_{R,S,\underline{x}} := \left\{ M \in \mathrm{Mat}_n(\mathscr{O}(\mathbf{C}^*)) \mid \begin{cases} \sigma_q M = RM(Sx^\mu)^{-1}, \\ \det M \text{ has simple zeroes} \\ \text{at } [\underline{x};q] \text{ and nowhere else} \end{cases} \right\}. \tag{12.4}$$

Because of (ii) in the proposition, it is then natural to introduce the following equivalence relation on $F_{R,S,\underline{x}}$:

$$\forall M, N \in F_{R,S,\underline{x}}\,,\ M \sim N \underset{def}{\iff} \left(\exists \Gamma, \Delta \in \mathbf{D}_n(\mathbf{C})\ :\ N = \Gamma^{-1}M\Delta\right). \tag{12.5}$$

Then we are interested in the quotient space:

$$\mathscr{F}_{R,S,\underline{x}} := \frac{F_{R,S,\underline{x}}}{\sim}. \tag{12.6}$$

We shall often shorten those notations into $F := F_{R,S,\underline{x}}$ and $\mathscr{F} := \mathscr{F}_{R,S,\underline{x}}$.

---

[6]To be perfectly rigorous, although $M[Sx^\mu] = M'[Sx^\mu] = R$, the matrices M, M' are not really morphisms from Sx^μ to R since their coefficients lie in $\mathscr{M}(\mathbf{C}^*,0)$. However, the use of such diagram makes sense and eases the discussion.

EXERCISE 12.15. Show that setting $M^{(\Gamma,\Delta)} := \Gamma^{-1}M\Delta$ we define a right action of the group $\mathbf{D}_n(\mathbf{C}) \times \mathbf{D}_n(\mathbf{C})$ on $F_{R,S,\underline{x}}$. (Later on we shall rather use the left action given by $(\Gamma, \Delta) \cdot M := \Gamma M \Delta^{-1}$.)

COROLLARY 12.16. *Associating to $A \in E_{R,S,\underline{x}}$ the class of any of the possible $M \in F_{R,S,\underline{x}}$ as above, we obtain a well defined map $E_{R,S,\underline{x}} \to \mathscr{F}_{R,S,\underline{x}}$.*

We shall next show 1) that the above map goes to the quotient into $\mathscr{E}_{R,S,\underline{x}} \to \mathscr{F}_{R,S,\underline{x}}$ and 2) that the resulting map ("Riemann-Hilbert-Birkhoff correspondence") is a bijection.

12.2.1.4. *The Riemann-Hilbert-Birkhoff correspondence.*

THEOREM 12.2 (Riemann-Hilbert-Birkhoff[7] correspondence). *The map $E_{R,S,\underline{x}} \to \mathscr{F}_{R,S,\underline{x}}$ goes to the quotient and defines a bijection $\mathscr{E}_{R,S,\underline{x}} \to \mathscr{F}_{R,S,\underline{x}}$.*

PROOF. There are three things to prove: 1) that the map goes to the quotient, 2) that the resulting map is injective and 3) that it is surjective. We shall use abreviations E, $\mathscr{E}$, F, $\mathscr{F}$.

The quotient map. We must check that if $A, B \in E$ respectively give rise to $M, N \in F$, then $A \sim B \Rightarrow M \sim N$. So let $A := A_0 + \cdots + A_\mu x^\mu$ and $B := B_0 + \cdots + B_\mu x^\mu$ both in E and $F \in \mathrm{GL}_n(\mathbf{C}(x))$ such that $F[A] = B$. Also let $M_0 : R \to A$ and $M_\infty : Sx^\mu \to A$ as in 12.2.1.2 and, similarly, $N_0 : R \to B$ and $N_\infty : Sx^\mu \to B$. We set $\Gamma := N_0^{-1}FM_0$ and $\Delta := N_\infty^{-1}FM_\infty$, whence a commutative diagram:

$$\begin{array}{ccccc} R & \xrightarrow{M_0} & A & \xleftarrow{M_\infty} & Sx^\mu \\ {\scriptstyle\Gamma}\downarrow & & {\scriptstyle F}\downarrow & & \downarrow{\scriptstyle\Delta} \\ R & \xrightarrow[N_0]{} & B & \xleftarrow[N_\infty]{} & Sx^\mu. \end{array}$$

By construction $\Gamma = (\gamma_{i,j}) \in \mathrm{GL}_n(\mathbf{C}(\{x\}))$ is such that $\Gamma[R] = R$, i.e., $\sigma_q \gamma_{i,j} = \dfrac{\rho_i}{\rho_j}\gamma_{i,j}$ for all $i, j = 1, \ldots, n$. But for $i \neq j$, this implies $\gamma_{i,j} = 0$ (since then $\rho_i/\rho_j \notin q^{\mathbf{Z}}$); and each $\gamma_{i,i} \in \mathbf{C}$ (since it is q-elliptic in $\mathbf{C}(\{x\})$). Also Γ is invertible, so $\Gamma \in \mathbf{D}_n(\mathbf{C})$. We prove in the same way that $\Delta \in \mathbf{D}_n(\mathbf{C})$. The outer rectangle in the diagram (reversing the right horizontal arrows) shows that $N\Delta = \Gamma M$, so $M \sim N$ as hoped for.

Injectivity. We want to show that if $A, B \in E$ give rise to $M, N \in F$ such that $M \sim N$ in F, then $A \sim B$ in E. We use the same diagram as above, but start from the outer rectangle and define $F := N_0 \Gamma M_0^{-1} = N_\infty \Gamma M_\infty^{-1}$, thus producing the middle vertical line. By construction, $F \in \mathrm{GL}_n(\mathscr{M}(\mathbf{C})) \cap \mathrm{GL}_n(\mathscr{M}(\mathbf{C}_\infty)) = \mathrm{GL}_n(\mathscr{M}(\mathbf{S})) = \mathrm{GL}_n(\mathbf{C}(x))$ and $F[A] = B$, so indeed $A \sim B$ as wanted.

Surjectivity. This, of course, is the hard part! We start from $M \in F$ and try to build an antecedent $A \in E$, at least up to rational gauge equivalence. We shall use the following corollary of Birkhoff's Preliminary Theorem (which was stated in 12.1.1):

[7] In the theory of differential equations, that name is usually associated to the extension of the Riemann-Hilbert correspondence to the irregular case, with Stokes operators being used along with monodromy. We refer here to the extension by Birkhoff of the Riemann-Hilbert correspondence to the theory of q-difference equations, **[23]**.

COROLLARY 12.17. *Let $M \in GL_n(\mathscr{M}(\mathbf{C}^*,0)) \cap Mat_n(\mathscr{O}(\mathbf{C}^*,0))$ with singular locus $\Sigma = \det^{-1}(0)$. We assume that $\det M$ has only simple zeroes. Let C as in the preliminary theorem (so C does not meet Σ) and write $\Sigma_0 := \Sigma \cap D_0$, $\Sigma_\infty := \Sigma \cap D_\infty$. Then one has a factorisation $M = M_0^{-1} M_\infty$ over $\mathbf{C}^*$, with:*

(1) *M_0 is regular over $\mathbf{C} \setminus \Sigma_\infty$, M_0^{-1} is holomorphic over $\mathbf{C}$ and $\det M_0^{-1}$ has simple zeroes over Σ_∞.*
(2) *M_∞ is regular over $\mathbf{C}^* \setminus \Sigma_0$, holomorphic over $\mathbf{S} \setminus \{0\}$ and $\det M_\infty$ has simple zeroes over Σ_0.*

PROOF. We apply the theorem with $M \in \mathrm{GL}_n(\mathscr{M}(\mathbf{C}^*,0))$. Then the relations $M_0 = M_\infty M$ and $M_\infty = M_0 M^{-1}$ automatically allow for an invertible meromorphic extension of M_0 and M_∞ over $\mathbf{C}^*$, and we simply write $M_0, M_\infty \in \mathrm{GL}_n(\mathscr{M}(\mathbf{C}^*,0))$. Moreover, the regularity conditions in the conclusion of the theorem then say that M_0 has the same singularities as M over $V_\infty \setminus \{\infty\}$ and that M_∞ as the same singularities as M over V_0. □

To apply the corollary, we draw on the Riemann sphere a simple closed analytic contour C winding once around 0 and separating each x_k from qx_k (see Figure 12.1). This is possible thanks to the (SNR) assumptions. At any rate, we thus obtain a decomposition $M = M_0^{-1} M_\infty$ such that:

- M_0 is regular (i.e., holomorphic with an holomorphic inverse) on $\mathbf{C} \setminus q^{-\mathbf{N}}\underline{x}$ and has simple poles over $q^{-\mathbf{N}}\underline{x}$;
- M_0^{-1} is holomorphic over $\mathbf{C}$ and $\det M_0$ has simple zeroes over $q^{-\mathbf{N}}\underline{x}$ (of course, the latter is redundant with respect to the previous condition);
- M_∞ is regular on $\mathbf{C}^* \setminus q^{\mathbf{N}^*}\underline{x}$, it is holomorphic over $q^{\mathbf{N}^*}\underline{x}$ and $\det M_\infty$ has simple zeroes there; also, it is holomorphic (but may be not regular) at ∞.
- M_∞^{-1} is holomorphic over $\mathbf{C}^*$ and meromorphic at ∞ (again there is some redundancy here).

From the arrow $Sx^\mu \overset{M}{\to} R$, i.e., from the condition $M[Sx^\mu] = R$ (which is part of the assumptions), we deduce thet $M_0[R] = M_\infty[Sx^\mu]$. Call it A, so it is inserted in a diagram $R \overset{M_0}{\to} A \overset{M_\infty}{\leftarrow} Sx^\mu$. We want to show that $A \in E$ and that its class in $\mathscr{E}$ is the preimage of the class of M in $\mathscr{F} = F/\sim$. Clearly:

$$A \in \mathrm{GL}_n(\mathscr{M}(\mathbf{C})) \cap \mathrm{GL}_n(\mathscr{M}(\mathbf{C}_\infty)) = \mathrm{GL}_n(\mathscr{M}(\mathbf{S})) = \mathrm{GL}_n(\mathbf{C}(x)).$$

Actually, more can be said. From the listed properties of M_0 and M_∞, one gets that A is holomorphic over $\mathbf{C}$ and meromorphic at ∞, whence polynomial:

$$A = A_0 + \cdots + x^d A_d, \quad A_0, \ldots, A_d \in \mathrm{Mat}_n(\mathbf{C}),\ A_d \neq 0.$$

Here $A_0 = A(0) = CRC^{-1} \in \mathrm{GL}_n(\mathbf{C})$, where $C = M_0(0) \in \mathrm{GL}_n(\mathbf{C})$.

From the given relations, we also see that $a(x) := \det A(x)$ has simple zeroes at $\underline{x}$ and nowhere else. Thus $a(x) = s(x - x_1) \cdots (x - x_N)$ for some $s \in \mathbf{C}^*$. Then $a(0) = s(-1)^N x_1 \cdots x_N$, but also $a(0) = \det A_0 = \det CRC^{-1} = \det R$, so by Fuchs relation $s = \det S$. Now we use the relation $A = M_\infty[Sx^\mu]$; taking the determinant and setting $f := \det M_\infty \in \mathbf{C}(\{1/x\})^*$, we draw:

$$\frac{\sigma_q f}{f} = \frac{a}{sx^N} = \prod_{i=1}^N (1 - x_i/x) \Longrightarrow f = \phi \prod_{i=1}^N \frac{1}{(x_i/x;q)_\infty},$$

where ϕ is elliptic. But at the same time, $\phi \in \mathbf{C}(\{1/x\})^*$, so that actually $\phi \in \mathbf{C}^*$. This implies that $f(\infty) = \phi$, so that M_∞ is regular at ∞ (which Birkhoff factorisation did not automatically imply). Setting $D := M_\infty(\infty) \in \mathrm{GL}_n(\mathbf{C})$, we see that $A = M_\infty[Sx^\mu]$ is asymptotic to $(DSD^{-1})x^\mu$ at ∞. Since it is also asymptotic to $x^d A_d$, we get that $d = \mu$ and $A_d = DSD^{-1} \in \mathrm{GL}_n(\mathbf{C})$. It is then immediate that $A \in E$ and the fact that its class in $\mathscr{E} = E/\sim$ is the antecedent of the class of M in $\mathscr{F} = F/\sim$ follows from the various equalities we found out during the computation (i.e., the construction of M goes through the M_0 and M_∞ used in the proof).

This concludes the proof of existence and bijectivity of the Riemann-Hilbert-Birkhoff correspondence. □

12.2.1.5. *The return of Ghân-buri-Ghân.* Although we intend to study more closely $\mathscr{F}_{R,S,\underline{x}}$ in Subsection 12.2.2 (in particular, most facts stated here without proper justification, including exercises, will be then proved), it may be instructive to already estimate its putative dimension.

First, let $V := \{M \in \mathrm{Mat}_n(\mathscr{O}(\mathbf{C}^*, 0)) \mid \sigma_q M = RM(Sx^\mu)^{-1}\}$. This is a vector space and the constraints on each component on M being independent, it has a product decomposition:

$$V \simeq \prod_{1 \leq i,j \leq n} V_{i,j}, \text{ where each } V_{i,j} := \{f \in \mathscr{O}(\mathbf{C}^*, 0) \mid \sigma_q f = \frac{\rho_i}{\sigma_j} x^{-\mu} f\}.$$

Each $V_{i,j}$ has dimension μ (see exercise), so V has dimension $n^2\mu$.

EXERCISE 12.18. Check that, if $\rho \in \mathbf{C}^*$ and $\mu \geq 1$, then the linear space $V_{\mu,\rho} := \{f \in \mathscr{O}(\mathbf{C}^*, 0) \mid \sigma_q f = \rho x^{-\mu} f\}$ has dimension μ. (Expand f in a Laurent series and count degrees of freedom.)

Inside V, let $\overline{F} := \{M \in V \mid \det M(x_1) = \cdots = \det M(x_N) = 0\}$. This subspace (caution: not a *linear* subspace!) is defined by $N = n\mu$ equations, so we could expect its dimension to be $n^2\mu - n\mu$. However those N equations are not independent, any one of them is a consequence of the $(N-1)$ others (see exercise), so the dimension count for $\overline{F}$ should rather give $n^2\mu - (n\mu - 1)$.

EXERCISE 12.19. (i) Let $W := \{f \in \mathscr{O}(\mathbf{C}^*, 0) \mid \sigma_q f = \dfrac{\rho_1 \cdots \rho_n}{\sigma_1 \cdots \sigma_n} x^{-N} f\}$. So, using the notations from the previous exercise, $W = V_{N,(\rho_1 \cdots \rho_n)/(\sigma_1 \cdots \sigma_n)}$. Check (as above) that $\dim W = N$. Describe its nonzero elements as products of Theta functions.

(ii) Let $x_1, \ldots, x_N \in \mathbf{C}^*$ pairwise noncongruent modulo $q^{\mathbf{Z}}$. Assume Fuchs relation $x_1 \cdots x_N \equiv (-1)^N \dfrac{\rho_1 \cdots \rho_n}{\sigma_1 \cdots \sigma_n}$. Show that:

$$f(x_1) = \cdots = f(x_{N-1}) = 0 \Longrightarrow f(x_N) = 0.$$

(iii) Assume that the x_k do not satisfy Fuchs relation. Show that:

$$f(x_1) = \cdots = f(x_N) = 0 \Longrightarrow f = 0.$$

(iv) Call Λ_k the linear form $f \mapsto f(x_k)$ on W. Translate (ii) and (iii) in terms of linear (in)dependency relations among the Λ_k, $k = 1, \ldots, N$.

(v) Check that $\det M \in W$ and the conclusions above thus apply to it.

Now F is the open subset of $\overline{F}$ defined by $\det \neq 0$ (i.e., those matrices M such that $\det M$ is not identically zero). Since the equation $\det M = 0$ is polynomial in the coordinates of M (considered as an element of the space V, there are two possibilities: either it vanishes identically on the "algebraic subset" $\overline{F}$ (note that this would indeed happen if the x_k did not satisfy Fuchs relation!), or it defines a nowhere dense closed subset of $\overline{F}$. The first case would imply that F is empty, which cannot be (because E is obviously not!!!). So the second case holds and we expect that F also has dimension $n^2\mu - (n\mu - 1)$.

Last we must take in account the equivalence relation induced by $\mathbf{D}_n(\mathbf{C}) \times \mathbf{D}_n(\mathbf{C})$ on F, the quotient of which is $\mathscr{F}$. We can replace each $m_{i,j}$ in M by $\dfrac{\lambda_i}{\mu_j} m_{i,j}$. The n^2 complex numbers $\dfrac{\lambda_i}{\mu_j}$ are not independent, they really have $(2n-1)$ degrees of freedom (indeed, the diagonal matrices Γ, Δ represent $2n$ degrees of freedom, but they can be replaced by $\alpha\Gamma, \alpha\Delta$ for any $\alpha \in \mathbf{C}^*$ without changing the action) so they decrease the dimension by as much and we should therefore expect that:

$$\dim \mathscr{F}_{R,S,\underline{x}} = n^2\mu - (n\mu - 1) - (2n - 1) = (n - 1)(n\mu - 2).$$

EXERCISE 12.20. Try the corresponding dimension count for $\mathscr{E}_{R,S,\underline{x}}$.

12.2.1.6. *An extended version of the theorem.* Theorem 12.2 is close to the original one of Birkhoff as described in 12.1.1, yet we did not to assume anything about the form of the solutions (in particular: multivalued or not). It will be enough for a nice application to q-Painlevé equation (see references at the very end of 12.2.4.3).

Although we shall have no need for it, we mention a stronger version with more natural hypotheses. In essence, we relax the strong nonresonancy assumption. It is convenient to take data in a slightly different form:

- Let $R, S \in \mathrm{GL}_n(\mathbf{C})$ such that $\mathrm{Sp}R, \mathrm{Sp}S \subset \mathbf{C}$ (the fundamental annulus).
- Let $x_1, \ldots, x_N$ as before (i.e., $x_i/x_j \notin q^{\mathbf{Z}}$ for $i \neq j$, and $x_1 \cdots x_N = (-1)^N(\rho_1 \cdots \rho_n)(\sigma_1 \cdots \sigma_n)$).
- The sets $E_{R,S,\underline{x}}$ and $\mathscr{E}_{R,S,\underline{x}}$ are defined as before.
- The set $F_{R,S,\underline{x}}$, but the equivalence relation $\sim$ such that $\mathscr{F}_{R,S,\underline{x}} = F_{R,S,\underline{x}}/\sim$ is defined by:

$$M \sim M^{(\Gamma,\Delta)} := \Gamma^{-1}M\Delta, \text{ where } \Gamma, \Delta \in \mathrm{GL}_n(\mathbf{C}) \text{ are such that } [\Gamma, R] = [\Delta, S] = 0,$$

i.e., Γ commutes with R and Δ commutes with S.

Note that under the restricted assumptions on R, S in 12.2.1.1, the definition of $\sim$ coincides with the previous one.

THEOREM 12.3. *We obtain again a bijection* $\mathscr{E}_{R,S,\underline{x}} \simeq \mathscr{F}_{R,S,\underline{x}}$.

PROOF. See [**176**, 3.2 p 37]. □

EXERCISE 12.21. Try by yourself to adapt the proof in 12.2.1.4 to this more general setting.

12.2.2. Geometry of $F_{R,S,\underline{x}}$ and $\mathscr{F}_{R,S,\underline{x}}$. We work in the generic case only, that of 12.2.1.4 (and so we apply only the theorem that has been completely proven). We shall give some basic information, a more complete treatment may be found in **[176]**. To begin with, we shall illustrate the algebro-geometric approach in the general case (n, μ arbitrary). Later, in 12.2.4, we shall tackle the case $n = 2$, with $\mu = 1$ (this is related to q-hypergeometric equations) and then $\mu = 2$ (this is related to q-Painlevé equations). By necessity we use some elementary affine and projective algebraic geometry, for which we refer to **[173]** for instance.

12.2.2.1. *Some spaces of functions.* If $M = (m_{i,j}) \in \mathrm{Mat}_n(\mathscr{O}(\mathbf{C}^*,0))$ satisfies $\sigma_q M = RM(Sx^\mu)^{-1}$, then each coefficient $m_{i,j}$ satisfies $\sigma_q m_{i,j} = \dfrac{\rho_i}{\sigma_j} x^{-\mu} m_{i,j}$. We are thus led to introduce the following complex vector spaces:

$$\forall \rho \in \mathbf{C}^* \ , \ \forall \mu \in \mathbf{N}^* \ , \ V_{\mu,\rho} := \{f \in \mathscr{O}(\mathbf{C}^*,0) \mid \sigma_q f = \rho x^{-\mu} f\}.$$

We recall from Chapters 2, 3 and 4 that, if $\mu < 0$, the above definition would yield a trivial space; and, if $\mu = 0$, either a trivial space (if $\rho \notin q^{\mathbf{Z}}$) or a line, i.e., a one-dimensional space (if $\rho \in q^{\mathbf{Z}}$). We shall see soon that, under our hypotheses, $V_{\mu,\rho}$ has dimension μ. We now introduce (with a really tiny risk of confusion):

$$\forall i,j = 1,\ldots,n \ , \ V_{i,j} := V_{\mu,\rho_i/\sigma_j} \text{ and } V := \prod_{i,j=1,\ldots,n} V_{i,j}.$$

So $\dim_{\mathbf{C}} V_{i,j} = \mu$ for all $i,j = 1,\ldots,n$ and $\dim_C V = \mu n^2$. On the other hand, by our remark above (beginning of 12.2.1.5), we have a natural identification:

$$V = \{M \in \mathrm{Mat}_n(\mathscr{O}(\mathbf{C}^*,0)) \mid \sigma_q M = RM(Sx^\mu)^{-1}\}.$$

Before going into proofs, note that, if $\sigma_q M = RM(Sx^\mu)^{-1}$, then $\sigma_q \det M = \dfrac{\det R}{\det S} x^{-N} \det M$ (as usual, $N := n\mu$) so that:

$$\forall M \in V \ , \ \det M \in W := V_{N,(\rho_1\cdots\rho_n)/(\sigma_1\cdots\sigma_n)},$$

and $\dim_C W = N$.

PROPOSITION 12.22. (i) *The complex vector space $V_{\mu,\rho}$ has dimension μ.*
(ii) *Its nonzero element are products of Theta functions; more precisely, they have the form $C \prod_{k=1,\ldots,\mu} \theta_q(x/a_k)$, where $C \in \mathbf{C}^*$, all $a_k \in \mathbf{C}^*$ and $\prod_{k=1,\ldots,\mu} a_k = \rho$. (Clearly, any such product belongs to $V_{\mu,\rho}$.)*

PROOF. (i) Write $f \in \mathscr{O}(\mathbf{C}^*,0)$ as a convergent Laurent power series $f = \sum_{m\in\mathbf{Z}} a_m x^m$. The condition $\sigma_q f = \rho x^{-\mu} f$ is equivalent to $\forall m \in \mathbf{Z}\ , \ q^m a_m = \rho a_{m+\mu}$. Thus, fixing arbitrarily $a_0, \ldots, a_{\mu-1}$ determines f uniquely. We just need to check that the resulting series converges on $\mathbf{C}^*$. This will follow from the quick decay of a_m for $m \to \pm\infty$.

Fix some remainder $r \in \{0, \ldots, \mu - 1\}$ and consider $b_l := a_{l\mu+r}$. Then, setting $Q := q^\mu$:

$$\begin{aligned}\forall l \in \mathbf{Z}\ , \ b_{l+1} = \rho^{-1} q^r Q^l b_l \Longrightarrow \forall l \in \mathbf{Z}\ , \ b_l &= (\rho^{-1}q^r)^l Q^{l(l-1)/2} b_0 \\ &= (\rho^{-1}q^r)^l q^{\mu l(l-1)/2} a_r.\end{aligned}$$

From that, it is easy (and left to the reader) to deduce that, for some $C > 0$:

$$a_m = O\left(C^m \, |q|^{m^2/(2\mu)}\right) \text{ when } m \to \pm\infty.$$

(ii) Let $f \in V_{\mu,\rho}$, $f \neq 0$. By the lemma below, f admits at least a zero, say b. By the functional equation $\sigma_q f = \rho x^{-\mu} f$, it has a q-spiral of zeroes $[b; q]$, so it can be factored as $f = \theta_q(x/a) f_1$, $g \in \mathcal{O}(\mathbf{C}^*, 0)$, $f_1 \neq 0$, $a := -b$. Clearly f_1 is such that $\sigma_q f_1 = \rho_1 x^{-\mu_1} f_1$, where $\mu_1 := \mu - 1$ and $\rho_1 := \rho/a$. If $\mu_1 = 0$, as noted above, $a = q^l$ and $f_1 = Cx^l$ for some $l \in \mathbf{Z}$ and $C \in \mathbf{C}^*$; otherwise, $f_1 \in V_{\mu_1,\rho_1}$ and we can iterate the process. In the end we obtain $f = Cx^l \prod_{k=1,\ldots,\mu} \theta_q(x/a_k)$, where $C \in \mathbf{C}^*$, all $a_k \in \mathbf{C}^*$ and $q^l \prod_{k=1,\ldots,\mu} a_k = \rho$. Now we replace for instance a_μ by $q^l a_\mu$ and the conclusion flows.

□

LEMMA 12.23. *Let $f \in V_{\mu,\rho}$, $f \neq 0$. Then f admits at least one zero modulo $q^{\mathbf{Z}}$ (actually it admits μ such zeroes).*

PROOF. Let $g := f'/f$ the logarithmic derivative of f, a meromorphic function over $\mathbf{C}^*$ such that $qg(qx) = g(x) - \mu/x$ (as follows from logarithmic differentiation of $f(qx) = \rho x^{-\mu} f(x)$). Since g has at most denumerably many poles, we can choose $R > 0$ such that, writing $r := R\,|q| < R$, the circles $\partial\mathrm{D}(0, R)$ and $\partial\mathrm{D}(0, r)$ do not meet the polar locus of g. Moreover, g has finitely many poles in the closed annulus of radii r and R (since it is compact). We shall apply Cauchy residue theorem to the contour made of $\partial\mathrm{D}(0, R)$ and $\partial\mathrm{D}(0, r)$, respectively oriented positively and negatively; this contour is the boundary of the corresponding open annulus. The residues of g are exactly the multiplicities of the zeroes of f, so the number of zeroes of f in the open annulus, counted with their multiplicities, is:

$$\begin{aligned}
&\frac{1}{2i\pi}\left(\int_{\partial\mathrm{D}(0,R)} g(x)dx - \int_{\partial\mathrm{D}(0,r)} g(x)dx\right) \\
&\quad = \frac{1}{2i\pi}\left(\int_{\partial\mathrm{D}(0,R)} g(x)dx - \int_{\partial\mathrm{D}(0,R)} g(qx)d(qx)\right) \\
&\quad = \frac{1}{2i\pi}\int_{\partial\mathrm{D}(0,R)} \frac{\mu}{x}\,dx = \mu.
\end{aligned}$$

□

We shall need to study the subspaces of $V_{\mu,\rho}$ defined by vanishing at particular points of $\mathbf{C}^*$, so, for every $x_0 \in \mathbf{C}^*$ we consider the linear form $\mathrm{ev}_{x_0} : V_{\mu,\rho} \to \mathbf{C}$, $f \mapsto f(x_0)$. By the exercise below, its kernel $H_{x_0} := \operatorname{Ker} \mathrm{ev}_{x_0}$ is a hyperplane (although our notation for the hyperplane does not mention μ and ρ, we hope that there will be no ambiguity).

EXERCISE 12.24. Using Theta functions, show that ev_{x_0} is not trivial on $V_{\mu,\rho}$.

Of course, because of the functional equation, functions in $V_{\mu,\rho}$ vanish over q-spirals, so $x_0 \equiv x_1 \pmod{q^{\mathbf{Z}}} \Rightarrow H_{x_0} = H_{x_1}$. Apart from that, the following proposition says that in most cases those hyperplanes are "in general position":

PROPOSITION 12.25. *Let $x_1, \ldots, x_l \in \mathbf{C}^*$ pairwise nonresonant (i.e., noncongruent modulo $q^{\mathbf{Z}}$).*

(i) *If $l < \mu$, then $H_{x_1} \cap \cdots \cap H_{x_l}$ has codimension l, thus dimension $\mu - l$.*

(ii) *If $l=\mu$ and $x_1\cdots x_\mu \not\equiv \rho$, then:*

$$H_{x_1}\cap\cdots\cap H_{x_\mu} = \begin{cases}\{0\} \text{ if } x_1\cdots x_\mu \not\equiv (-1)^\mu\rho \pmod{q^{\mathbf{Z}}},\\ \mathbf{C}\theta_q(-x/x_1)\cdots\theta_q(-x/x_\mu) \text{ if } x_1\cdots x_\mu = (-1)^\mu\rho.\end{cases}$$

(iii) *More generally, if $l=\mu$ and $x_1\cdots x_\mu \equiv (-1)^\mu\rho \pmod{q^{\mathbf{Z}}}$, then $H_{x_1}\cap\cdots\cap H_{x_\mu}$ is a line (it has dimension 1).*

PROOF. To ease induction we extend the definition of $V_{\mu,\rho}$ to the limiting case $\mu=0$, for which we know that $V_{0,\rho} = \begin{cases}\{0\} \text{ if } \rho\notin q^{\mathbf{Z}},\\ \mathbf{C}x^m \text{ if } \rho = q^m, m\in\mathbf{Z}.\end{cases}$. With that convention, we start from $f\in H_{x_1}$, so that, as in the proof of the previous proposition, $f=\theta_q(-x/x_1)g$ for some $g\in\mathcal{O}(\mathbf{C}^*,0)$. By the nonresonancy assumption, g vanishes at $x_2,\dots,x_l$ and we may iterate the process: any $f\in H_{x_1}\cap\cdots\cap H_{x_l}$ may be written as $\theta_q(-x/x_1)\cdots\theta_q(-x/x_l)g$ and from the functional equation for f we find that $g\in V_{\mu-l,\rho'}$, where $\rho' := \dfrac{\rho}{(-1)^l x_1\cdots x_l}\cdot$ We deduce that, for $l\le\mu$:

$$H_{x_1}\cap\cdots\cap H_{x_l} = \theta_q(-x/x_1)\cdots\theta_q(-x/x_l)V_{\mu-l,\rho/((-1)^l x_1\cdots x_l)},$$

from which all the statements of the proposition flow. □

EXERCISE 12.26. (i) Complete the description in case (iii) of the proposition.

(ii) What if $l>\mu$?

12.2.2.2. *The determinant as a polynomial function $V\to W$.* Recall that $V_{i,j} := V_{\mu,\rho_i/\sigma_j}$ and that we identified the space of candidate monodromy matrices $\{M\in \mathrm{Mat}_n(\mathcal{O}(\mathbf{C}^*,0)) \mid \sigma_q M = RM(Sx^\mu)^{-1}\}$ ("candidate" because we did not yet take in account conditions on the determinant) with $V := \prod\limits_{i,j=1,\dots,n} V_{i,j}$, which has dimension $n^2\mu$. Also recall, from Exercise 12.18, the space:

$$W := \{f\in\mathcal{O}(\mathbf{C}^*,0)\mid \sigma_q f = \frac{\rho_1\cdots\rho_n}{\sigma_1\cdots\sigma_n}x^{-N}f\} = V_{N,(\rho_1\cdots\rho_n)/(\sigma_1\cdots\sigma_n)}.$$

As seen there, $\dim W = N := n\mu$.

LEMMA 12.27. *The map* det : $V\to W$ *is polynomial; more precisely it is n-linear as a map:*

$$\prod_{j=1}^n V_{.,j} = V_{.,1}\times\cdot\times V_{.,n}\to W, \text{ resp. as a map } \prod_{i=1}^n V_{i,.} = V_{1,.}\times\cdot\times V_{n,.}\to W,$$

where we define the spaces of columns, resp. the spaces of lines:

$$V_{.,j} := \prod_{i=1}^n V_{i,j}, \text{ resp. } V_{i,.} := \prod_{j=1}^n V_{i,j},$$

so that both $\prod V_{.,j}$ and $\prod V_{i,.}$ are naturally identified to V.

PROOF. For a permutation $\tau\in\mathfrak{S}_n$ of $\{1,\dots,n\}$ and a matrix $M:=(m_{i,j})$, write temporarily $D_\tau(M) := \prod\limits_{i=1}^n m_{i,\tau(i)}$. Obviously D_τ is n-linear from $\prod\limits_{i=1}^n V_{i,\tau(i)}$ to W and can be considered as the restriction of a n-linear map either from $\prod V_{.,j}$ or from $\prod V_{i,.}$ to W. To understand in which sense this makes it a polynomial map, choose a basis $(e_{i,j}^{(k)})_{k=1,\ldots,\mu}$ for each $V_{i,j}$: then writing each $m_{i,j}$ in coordinates as $\sum\limits_{k=1,\ldots,\mu} \lambda_{i,j}^{(k)} e_{i,j}^{(k)}$ and then expanding $\prod\limits_{i=1}^{n} m_{i,\tau(i)}$ yields a polynomial expression in the $\lambda_{i,j}^{(k)}$, with coefficients of the form $\prod\limits_{i=1}^{n} e_{i,\tau(i)}^{(k)} \in W$: this is the definition of a polynomial function with values in W.

As for the determinant itself, $\det = \sum\limits_{\tau \in \mathfrak{S}_n} \operatorname{sgn}(\tau) D_\tau$. □

Returning to our framework, let $x_1, \ldots, x_N \in \mathbf{C}^*$ be pairwise noncongruent modulo $q^{\mathbf{Z}}$ and satisfying Fuchs relation $x_1 \cdots x_N = (-1)^N \dfrac{\rho_1 \cdots \rho_N}{\sigma_1 \cdots \sigma_N}$; and set:

$$\overline{F} := \{M \in V \det M \text{ vanishes at all } x_k\} \text{ and } F := \{M \in \overline{F} \mid \det M \neq 0\}.$$

(The condition $\det M \neq 0$ means that $\det M \in W$ is not the trivial function $0 \in W$.) With the notations $\operatorname{ev}_{x_k} : W \to \mathbf{C}, f \mapsto f(x_k)$ and $H_k := \operatorname{Ker} \operatorname{ev}_{x_k}$, we see that $H_1, \ldots, H_N$ are hyperplanes in W such that:

$$\Delta := H_1 \cap \cdots \cap H_N$$

is a line (actually $\Delta = \bigcap\limits_{\substack{k=1,\ldots,N \\ k \neq l}} H_k$ for any $l \in \{1, \ldots, N\}$) and:

$$\overline{F} = \det{}^{-1}(\Delta) \subset V, \quad F = \overline{F} \setminus \det{}^{-1}(0).$$

Being the preimage of a line by a polynomial map, $\overline{F}$ is an *affine algebraic subset* of V. Similarly, $\det^{-1}(0)$ is also an affine algebraic subset of V. Therefore:

PROPOSITION 12.28. *F is a nonempty* Zariski open subset *of $\overline{F}$.*

PROOF. We just have to check that F is nonempty. Indeed, if it was, its quotient $\mathscr{F} = F/\sim$ would also be, so (by Riemann-Hilbert-Birkhoff correspondence) $\mathscr{E} = E/\sim$ would be, so E would be empty, which is obviously not true since we can produce plenty of matrices $A \in E$. □

12.2.2.3. *Application to classification.* The complete geometrical study of $F_{R,S,\underline{x}}$ and $\mathscr{F}_{R,S,\underline{x}}$ in the general case has not been realized. We shall detail in 12.2.4 the case $n = 2$, for $\mu = 1$ and for $\mu = 2$. This is however what could be hoped for:

(1) The *quasi-affine variety* F is *Zariski-dense* in $\overline{F}$, i.e., $\overline{F}$ is the smallest algebraic subset of V containing F. (This is certainly true if $\overline{F}$ is *irreducible.*) Both have dimension $n^2\mu - n\mu + 1$ (see the naive count in 12.1.1.5).
(2) The set theoretic quotient $\mathscr{F} = F/\sim$ behaves nicely in the sense of algebraic geometry, it is perhaps a *geometric quotient*, and $\mathscr{F}$ is itself an algebraic variety. Its dimension is $(n-1)(n\mu - 2)$.

Altogether this would mean that $\mathscr{F}_{R,S,\underline{x}}$ is a nice classifying space for individual q-difference equations of the class considered. However, one would wish for more: one of the motivations for the study (see 12.2.4) is to understand the behaviour of *continuous families* of such equations. Ideally, the structure of $\mathscr{F}_{R,S,\underline{x}}$ should reflect the (topological, analytical, algebraic ...) structure of such families. This is called a "moduli problem". We shall barely touch upon that in *loc. cit.*, see [CV] and [**176, 196**].

EXERCISE 12.29 (Open minded!). Suppose that $A_t(x) = A_0(t) + \cdots + A_\mu(t)x^\mu$, where each $A_i(t)$ is (say) analytic for t near $0 \in \mathbf{C}$ and, for each particular t, $A_t \in E_{R,S,\underline{x}}$. Can we define an associated family $M_t \in F_{R,S,\underline{x}}$ varying, in some sense, analytically as a function of t?

12.2.3. Reducibility via monodromy matrices.

12.2.3.1. *Preliminary facts.* Recall from Chapter 8 that a left module over a not necessarily commutative ring is simple if it is nontrivial and contains no nontrivial proper submodule. In our setting (functional equations), they are rather called *irreducible* and we know that $\mathcal{D}/\mathcal{D}P$ is irreducible if, and only if, the q-difference operator P is. On the opposite side, (K^n, Φ_A) is *reducible* if, and only if, A is (may be up to gauge equivalence) in block triangular form. We shall then say that A itself is reducible.

Another way to characterize the reducibility of A is to translate directly the existence of non trivial proper submodules of (K^n, Φ_A):

$$A \in \mathrm{GL}_n(K) \text{ is reducible} \iff \exists \begin{cases} A' \in \mathrm{GL}_{n'}(K), \\ F \in \mathrm{Mat}_{n,n'}(K) \end{cases} : \begin{cases} (\sigma_q F)A' = AF, \\ \text{rank } F = n', 0 < n' < n. \end{cases}$$

We now intend to characterize reducibility in terms of monodromy data. Although our final goal involves global monodromy data, so a study over $\mathbf{C}(x)$, we first say a word about the local data R and S. Note that the use of diagonal matrices amounts to implicitly order the exponents at 0 and at ∞, although what should really matter is the *sets* $\{\rho_1, \ldots, \rho_n\}$ and $\{\sigma_1, \ldots, \sigma_n\}$ (which are unordered objects). Actually, changing the orders amounts to using diagonal matrices $R' := \Gamma[R] = \Gamma R \Gamma^{-1}$ and $S' := \Delta[S] = \Delta S \Delta^{-1}$, where Γ and Δ are permutation matrices. We then write a diagram:

$$\begin{array}{ccccc} R & \xrightarrow{M_0} & A & \xleftarrow{M_\infty} & Sx^\mu. \\ \Gamma \downarrow & & & & \downarrow \Delta \\ R & \xrightarrow[M_0']{} & A & \xleftarrow[M_\infty']{} & Sx^\mu \end{array}$$

From this diagram we see that the construction starting from R', S' instead of R, S would yield the Birkhoff matrix $M' = \Gamma M \Delta^{-1}$. This being said, we have a simple minded characterization of local reducibility (we keep the usual notations):

LEMMA 12.30. *Every nontrivial proper submodule of* $(\mathbf{C}(\{x\})^n, \Phi_R)$ *is isomorphic to a* $(\mathbf{C}(\{x\})^{n'}, \Phi_{R'})$, *where* $0 < n' < n$ *and* R' *is extracted from* R *(i.e., it is diagonal, Sp* $R' \subset$ *Sp* R *and the order is respected). A similar statement holds for* Sx^μ *at* ∞.

PROOF. The first statement is an easy consequence of the properties of exponents such as presented in 10.1.2, along with the classification of rank 1 q-difference modules. We leave to the reader the adaptation at ∞. □

To use the lemma, we note that, if (V, Φ) is a q-difference module over $\mathbf{C}(x)$ and (V', Φ') a submodule, then it induces (by extension of scalars) a submodule over $\mathbf{C}(\{x\})$ and also one over $\mathbf{C}(\{1/x\})$.

12.2.3.2. *A criterion of reducibility.* The following theorem was proved in [**176**] through Galois theory, we propose here a more direct proof. First, an important definition:

DEFINITION 12.31. The monodromy matrix $M = (m_{i,j}) \in F_{R,S,\underline{x}}$ is *reducible* if there exist n', n'' such that $0 < n', n'' < n$, $n' + n'' = n$ and $i_1 < \cdots < i_{n'}$, $j_1 < \cdots < j_{n''}$, all in $\{1, \ldots, n\}$, such that:

$$\forall (i,j) \in \{i_1, \ldots, i_{n'}\} \times \{j_1, \ldots, j_{n''}\}\ ,\ m_{i,j} = 0.$$

In other words, up to reordering of the indices (i.e., up to permutations of lines and of columns) M can be put in block triangular form $\begin{pmatrix} M' & \star \\ 0 & M'' \end{pmatrix}$, where M' has format $n' \times n'$ and M'' has format $n'' \times n''$.

We leave it to the reader to prove the following:

EXERCISE 12.32. Reducibility is preserved by the H-action, i.e., M is reducible if, and only if, $\Gamma M \Delta^{-1}$ is ($\Gamma, \Delta \in \mathbf{D}_n(\mathbf{C})$).

It therefore makes sense to speak of a *reducible class* in $\mathscr{F}_{R,S,\underline{x}}$.

THEOREM 12.4. *Let M be associated to A by the Riemann-Hilbert-Birkhoff correspondence. Then M is reducible if, and only if, $(\mathbf{C}(x)^n, \Phi_A)$ is.*

PROOF. By the previous exercise and remark, it does not matter which of all the possible M is used here.

<u>*One way.*</u> Suppose M reducible. Up to reordering indices, we may assume that $M = \begin{pmatrix} M' & \star \\ 0 & M'' \end{pmatrix}$ (same conventions as in the definition). On the side of A, this will mean reordering exponents in R, S, which does not change (ir)reducibility. Here $M' \in \mathrm{GL}_{n'}(\mathscr{M}(\mathbf{C}^*))$, $M'' \in \mathrm{GL}_{n''}(\mathscr{M}(\mathbf{C}^*, 0))$ and $U \in \mathrm{Mat}_{n',n''}(\mathscr{M}(\mathbf{C}^*))$.

Let $R' := \mathrm{Diag}(\rho_1, \ldots, \rho_{n'})$ and $S := \mathrm{Diag}(\sigma_1, \ldots, \sigma_{n'})$ and let Γ, Δ both equal to $\begin{pmatrix} I_{n'} \\ 0_{n'',n'} \end{pmatrix} \in \mathrm{Mat}_{n,n'}(\mathbf{C})$ (which has rank n'), so that $\Gamma : R' \to R$ and $\Delta : S' \to S$ are injective morphisms: indeed, $(\sigma_q \Gamma) R' = \Gamma R' = R \Gamma$ and $(\sigma_q \Delta) S' = \Delta S' = S \Delta$. The equation $\sigma_q M = R M (S x^\mu)^{-1}$, expressed in blocks, entails $\sigma_q M' = R' M' (S' x^\mu)^{-1}$ and we have a commutative diagram (see the left one):

$$\begin{array}{ccc} R' & \xleftarrow{\ M'\ } & S'x^\mu \\ {\scriptstyle\Gamma}\downarrow & & \downarrow{\scriptstyle\Delta} \\ R & \xleftarrow[\ M\ ]{} & Sx^\mu \end{array} \qquad\qquad \begin{array}{ccccc} R' & \xrightarrow{\ M'_0\ } & A' & \xleftarrow{\ M'_\infty\ } & S'x^\mu \\ {\scriptstyle\Gamma}\downarrow & & \vdots\,{\scriptstyle F} & & \downarrow{\scriptstyle\Delta} \\ R & \xrightarrow[\ M_0\ ]{} & A & \xleftarrow[\ M_\infty\ ]{} & Sx^\mu. \end{array}$$

We apply to both horizontal lines the Riemann-Hilbert-Birkhoff correspondence and get, as in the diagram in the right, analytic factorisations of M' and M. The commutativity of the first diagram implies that $M_0 \Gamma M_0'^{-1} = M_\infty \Delta M_\infty'^{-1}$, call it F. It has rank n' and coefficients in $\mathrm{GL}_n(\mathscr{M}(\mathbf{C})) \cap \mathrm{GL}_n(\mathscr{M}(\mathbf{C}_\infty)) = \mathrm{GL}_n(\mathscr{M}(\mathbf{S})) = \mathrm{GL}_n(\mathbf{C}(x))$, so it is an injective morphism $A' \to A$ and since $0 < n' < n$, this gives the reducibility of A.

<u>*And back.*</u> We now suppose A reducible and let $F : A' \to A$ as in 12.2.3.1. Then, by extension of scalars, $(\mathbf{C}(\{x\})^{n'}, A')$ is isomorphic to a submodule of $(\mathbf{C}(\{x\})^n, A)$, thus to a submodule of $(\mathbf{C}(\{x\})^n, R)$, so, up to a reordering, we

may suppose that $A' \simeq R'$ over $\mathbf{C}(\{x\})$, with $R' := (\rho_1, \ldots, \rho_{n'})$. Similarly, at ∞, we may assume that $A' \simeq S'x^\mu$, with $S' := (\sigma_1, \ldots, \sigma_{n'})$. So we write $A' = M_0'[R'] = M_\infty'[S'x^\mu]$ with obvious conditions on M_0' and M_∞'. We then have a commutative diagram as the one at left, where we have *defined* $\Gamma := M_0^{-1} F M_0'$ and $\Delta := M_\infty^{-1} F M_\infty'$.

$$
\begin{array}{ccccc}
R' & \xrightarrow{M_0'} & A' & \xleftarrow{M_\infty'} & S'x^\mu \\
{\scriptstyle\Gamma}\,\vdots & & \downarrow{\scriptstyle F} & & \vdots\,{\scriptstyle\Delta} \\
R & \xrightarrow[M_0]{} & A & \xleftarrow[M_\infty]{} & Sx^\mu
\end{array}
\qquad\qquad
\begin{array}{ccc}
R' & \xleftarrow{M'} & S'x^\mu \\
{\scriptstyle\Gamma}\downarrow & & \downarrow{\scriptstyle\Delta} \\
R & \xleftarrow[M]{} & Sx^\mu.
\end{array}
$$

As F has rank n', so do Γ and Δ. As F has coefficients in $\mathbf{C}(x)$, the coefficients of Γ, resp. of Δ, belong to $\mathscr{M}(\mathbf{C})$, resp. to $\mathscr{M}(\mathbf{C}_\infty)$. We concentrate on $\Gamma = (\gamma_{i,j})_{\substack{1 \leq i \leq n \\ 1 \leq j \leq n'}}$: since it is a morphism $R' \to R$, it satisfies $\sigma_q \gamma_{i,j} = \dfrac{\rho_i}{\rho_j} \gamma_{i,j}$. By the nonresonancy assumption, this is only possible if $i = j$ (and then $\gamma_{i,j} \in \mathbf{C}$) or $\gamma_{i,j} = 0$. So we find that $\Gamma = \begin{pmatrix} \Gamma' \\ 0 \end{pmatrix}$ where $\Gamma' \in \mathbf{D}_{n'}(\mathbf{C})$ (because of the rank). Similarly $\Delta = \begin{pmatrix} \Delta' \\ 0 \end{pmatrix}$ where $\Delta' \in \mathbf{D}_{n'}(\mathbf{C})$.

We now modify the diagram at left in the usual way and get the one on the right. From the relation $\Gamma M' = M \Delta$, we get, by block calculation (obvious notations):

$$
\begin{pmatrix} \Gamma' \\ 0 \end{pmatrix} M' = \begin{pmatrix} M_{1,1} & M_{1,2} \\ M_{2,1} & M_{2,2} \end{pmatrix} \begin{pmatrix} \Delta' \\ 0 \end{pmatrix} \Longrightarrow \begin{pmatrix} \Gamma' M_{1,1} & \Gamma' M_{1,2} \\ 0 & 0 \end{pmatrix} = \begin{pmatrix} M_{1,1}\Delta' & 0 \\ M_{2,1}\Delta' & 0 \end{pmatrix},
$$

so $M_{2,1}\Delta' = 0 \Rightarrow M_{2,1} = 0$ and M is reducible as wanted. More precisely:

$$
M = \begin{pmatrix} \Gamma' M' \Delta'^{-1} & M_{1,2} \\ 0 & M_{2,2} \end{pmatrix},
$$

so, up to equivalence, M' is a "subobject" of M. □

COROLLARY 12.33. *If M is reducible, the subobject A' of A can be taken in the form $A_0' + \cdots + A_\mu' x^\mu$ defining $E_{R,S,\underline{x}}$.*

PROOF. Note that this is by no way guaranteed by the theorem in the form above: a priori, A' as given by the first implication has *rational* coefficients. Starting from R', $S'x^\mu$ and M' as in the proof and having found A', we can independently use Riemann-Hilbert correspondence and find $A'' = A_0'' + \cdots + A_\mu'' x^\mu$, fitting in the diagram:

$$
\begin{array}{ccccc}
R' & \xrightarrow{M_0''} & A''. & \xleftarrow{M_\infty''} & S'x^\mu \\
& {\scriptstyle M_0'}\searrow & \vdots\,{\scriptstyle G} & \swarrow{\scriptstyle M_\infty'} & \\
& & A' & &
\end{array}
$$

From $M' = M_0'^{-1} M_\infty' = M_0''^{-1} M_\infty''$, we deduce that $M_0' M_0''^{-1} = M_\infty' M_\infty''^{-1}$, which we take as definition for G. Usual arguments say that G is an isomorphism over $\mathbf{C}(x)$ and so A'' is the wanted subobject (the injective morphism $A'' \to A$ is $F \circ G$ with $F : A' \to A$ given by the theorem). □

EXERCISE 12.34. Show that $A'' \in E_{R',S',\underline{x}'}$ for some $\underline{x}'$ extracted from $\underline{x}$.

12.2.4. Example: rank 2, low degree. Here we shall take $n = 2$ (anyway, the vast majority of q-difference equations met "in nature" have rank 2) and the two most elementary cases, $\mu = 1$ and (very briefly) $\mu = 2$, respectively corresponding to the q-hypergeometric and to the so-called *Jimbo-Sakai* families. For more details; see **[176]**.

12.2.4.1. *Rewriting the q-hypergeometric equation.* In 4.4.1.5, we wrote the q-hypergeometric equation in the form $\sigma_q X = BX$ (we slightly change notation); we write $B = fA$, where:

$$f := \frac{1}{c/q - abx} \text{ and } A := \begin{pmatrix} 0 & c/q - abx \\ -(1-x) & (1+c/q) - (a+b)x \end{pmatrix}.$$

On the one hand, since $F[fA] = fF[A]$ classifying fA up to gauge transformations is the same as classifying A; on the other hand:

$$A = A_0 + A_1 x, \text{ where } A_0 := \begin{pmatrix} 0 & c/q \\ -1 & 1+c/q \end{pmatrix} \text{ and } A_1 := \begin{pmatrix} 0 & -ab \\ 1 & -(a+b) \end{pmatrix},$$

so it clearly falls under our framework with $n = 2$, $\mu = 1$.

Since $\mathrm{Sp} A_0 = \{\{1, c/q\}\}$ and $\mathrm{Sp}\ A_1 = \{\{a, b\}\}$, the conditions we imposed upon R and S in 12.2.1 here amount to: $a \not\equiv b$ and $c \not\equiv 1$. The zeroes of $\det A$ are 1 and $c/(qab)$, so the last nonresonancy condition is $ab \not\equiv c$. Last, Fuchs relation (FR) is automatically satisfied by construction.

EXERCISE 12.35. Check compatibility with the requirements in 4.4.

12.2.4.2. *The "basic hypergeometric family".* So we consider the adaptation of 12.2.1 to the case $n = 2$, $\mu = 1$. Then $N = 2$ and the dimension predicted in 12.2.1.5 for $\mathscr{F}_{R,S,\underline{x}}$ is $(n-1)(N-2) = 0$. This is interpreted as *the rigidity of basic hypergeometric equations*, namely: once fixed the local monodromy data (for us: $R, S, \underline{x}$) they remain no "continuous moduli". Such facts are well known for classical hypergeometric equations and have been studied for general basic hypergeometric equations notably by Roques and Sakai-Yamaguchi **[208, 215]**.

Matrices M in $F_{R,S,\underline{x}}$ satisfy $\sigma_q M = RM(Sx)^{-1}$, equivalently (coefficientwise) $\sigma_q m_{i,j} = \frac{\rho_i}{\sigma_j x} m_{i,j}$, $i, j = 1, 2$, so they have the form:

$$M = \begin{pmatrix} \alpha_{1,1}\theta_q(\sigma_1 x/\rho_1) & \alpha_{1,2}\theta_q(\sigma_2 x/\rho_1) \\ \alpha_{2,1}\theta_q(\sigma_1 x/\rho_2) & \alpha_{2,2}\theta_q(\sigma_2 x/\rho_2) \end{pmatrix}, \text{ where } (\alpha_{1,1}, \alpha_{1,2}, \alpha_{2,1}, \alpha_{2,2}) \in \mathbf{C}^4.$$

Comparing the zeroes of $\theta_q(\sigma_1 x/\rho_2)\theta_q(\sigma_2 x/\rho_1)$ and $\theta_q(\sigma_1 x/\rho_1)\theta_q(\sigma_2 x/\rho_2)$, one sees that the condition that $\det M$ does not vanish identically is equivalent to the condition that $(\alpha_{1,1}\alpha_{2,2}, \alpha_{1,2}\alpha_{2,1}) \neq (0, 0)$; so there is a well defined associated point $\alpha(M) := (\alpha_{1,1}\alpha_{2,2} : \alpha_{1,2}\alpha_{2,1}) \in \mathbf{P}^1(\mathbf{C})$ (the complex projective line).

Moreover, the equivalence relation $M \sim \Gamma^{-1} M \Delta$ on $\mathscr{F}_{R,S,\underline{x}}$, in coefficients: $(m_{i,j})_{i,j=1,2} \sim (\frac{\delta_j}{\gamma_i} m_{i,j})_{i,j=1,2}$, is compatible with that construction: $M \sim M' \Rightarrow \alpha(M) = \alpha(M')$, so that we actually get a well defined map $\overline{\alpha} : \mathscr{F}_{R,S,\underline{x}} \to \mathbf{P}^1(\mathbf{C})$. One can (and you should!) check that $\overline{\alpha}$ is injective when restricted to the subset $\mathscr{F}^*_{R,S,\underline{x}}$ of irreducible classes (those such all $\alpha_{i,j} \neq 0$ by 12.2.3.2). Now matrices M in $F_{R,S,\underline{x}}$ are also required to satisfy $\det\ M(x_1) = 0$ (or, equivalently, the same at x_2), which we write:

$$\alpha_{1,1}\alpha_{2,2}\theta_q(\sigma_1 x_1/\rho_1)\theta_q(\sigma_2 x_1/\rho_2) = \alpha_{1,2}\alpha_{2,1}\theta_q(\sigma_1 x_1/\rho_2)\theta_q(\sigma_2 x_1/\rho_1), \text{ i.e.:}$$

$$(\alpha_{1,1}\alpha_{2,2} : \alpha_{1,2}\alpha_{2,1}) = (\theta_q(\sigma_1 x_1/\rho_2)\theta_q(\sigma_2 x_1/\rho_1) : \theta_q(\sigma_1 x_1/\rho_1)\theta_q(\sigma_2 x_1/\rho_2)),$$

$$\text{equivalently: } \overline{\alpha}(\mathrm{cl}M) = \Phi(x_1), \text{ where } \Phi(x) := \frac{\theta_q(\sigma_1 x/\rho_2)\theta_q(\sigma_2 x/\rho_1)}{\theta_q(\sigma_1 x/\rho_1)\theta_q(\sigma_2 x/\rho_2)}.$$

Here we consider Φ with values in $\mathbf{P}^1(\mathbf{C})$ identified with $\mathbf{C} \cup \{\infty\}$ through $(u : v) \leftrightarrow u/v$. So $\mathscr{F}^*_{R,S,\underline{x}}$ contains only the preimage by $\overline{\alpha}$ of $\Phi(x_1)$.

EXERCISE 12.36. (i) Study the injectivity of $\overline{\alpha}$ on $\mathscr{F}_{R,S,\underline{x}} \setminus \mathscr{F}^*_{R,S,\underline{x}}$.
(ii) Show that $\Phi(x_1) = \Phi(x_2)$, thus getting rid of the apparent dissymmetry.

12.2.4.3. *The Jimbo-Sakai family.* The *q-Painlevé equation* (a nonlinear q-difference) will be studied in [CV], but, ever since the seminal article **[129]** by Jimbo and Sakai, it has been related to so-called "q-isomonodromy conditions" on a family of linear q-difference equations which fall under our framework with $n = 2$, $\mu = 2$. Prediction $(n-1)(n\mu - 2) = 2$ here suggests a *surface* of monodromy data. We barely touch upon its study.

Given $\rho_1, \rho_2, \sigma_1, \sigma_2, x_1, x_2, x_3, x_4 \in \mathbf{C}^*$ such that ρ_1/ρ_2, σ_1/σ_2 and all x_k/x_l, $1 \leq k < l \leq 4$ are in $\mathbf{C}^* \setminus q^{\mathbf{Z}}$ and $x_1x_2x_3x_4 = (\rho_1\rho_2)/(\sigma_1\sigma_2)$, we are interested in the space $E_{R,S,\underline{x}}$ of 2×2 matrices $A_0 + A_1 x + A_2 x^2$ such that $A_0 \sim R$, $A_2 \sim S$ and $\det A(x)$ vanishes at x_1, x_2, x_3, x_4, modulo rational equivalence. Monodromy matrices $M \in F_{R,S,\underline{x}}$ have the form $(m_{i,j})_{i,j=1,2}$, where each $m_{i,j} \in V_{i,j} := V_{2,\rho_i/\sigma_j,2}$ (see 12.2.2.1), a space of dimension 2. There are three independent vanishing conditions on $\det M$, reducing the dimension to $4 \times 2 - 3 = 5$. Then the gauge action by pairs of diagonal matrices reduces the dimension to 2, as predicted.

Sending M to $(m_{1,1}m_{2,2}, m_{1,2}m_{2,1})$ maps $\prod V_{i,j}$ into a quadric hypersurface of the eight-dimensional space $V^2_{4,(\rho_1\rho_2)/(\sigma_1\sigma_2)}$. The monodromy surface $\mathscr{F}_{R,S,\underline{x}}$ maps injectively (at least generically) into the quotient by the gauge action, which opens the possibility of an algebro-geometric study; see **[176]** by Ohyama, Ramis and Sauloy and **[132]** by Joshi and Roffelsen (also see **[196]**).

12.2.5. Riemann-Hilbert-Birkhoff with parameters. The previous application to the Jimbo-Sakai family is incomplete, since it avoids the problem of dependency on a parameter. This problem is tackled in **[196]**. Here we merely lay the technical basis for a Riemann-Hilbert-Birkhoff correspondence in the case of an analytical dependency on parameters. The approach is to some extent the same as in Subsection 11.1.5. To begin with, we consider the problem *locally with respect to* t (which varies in a polydik Δ in $\mathbf{C}^d$, possibly a $\mathring{D}(0, r) \subset \mathbf{C}$) but *globally with respect to* x (which will eventually vary in $\mathbf{C}^*$). Global issues with respect to t are tackled in **[196]**.

The situation without parameters is illustrated by the diagram:

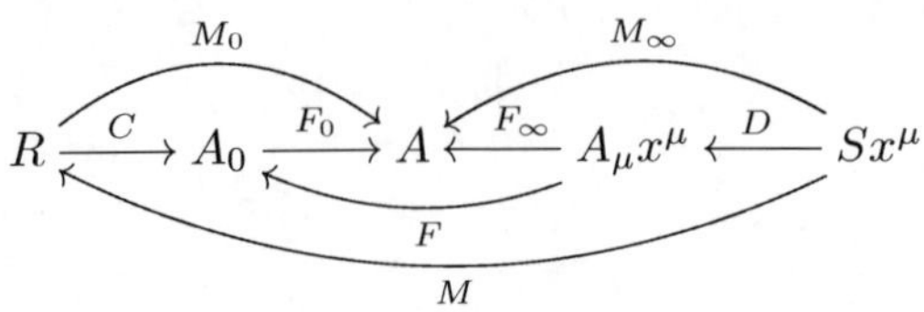

The diagram displays the standard data $C, D \in \mathrm{GL}_n(\mathbf{C})$, $F_0 \in \mathrm{GL}_n(\mathbf{C}\{x\})$ and $F_\infty \in \mathrm{GL}_n(\mathbf{C}\{1/x\})$ such that $F_0(0) = F_\infty(\infty) = I_n$, $A_0 = C[R] = CRC^{-1}$, $A_\mu = D[S] = DSD^{-1}$, $A = F_0[A_0] = M_0[R] = F_\infty[A_\mu x^\mu] = M_\infty[Sx^\mu]$, $R = M[Sx^\mu]$, $M_0 = F_0 C$, $M_\infty = F_\infty D$, $M = M_0^{-1}M_\infty$. But we also have added $F :=$

$F_0^{-1}F_\infty = CMD^{-1} \in \mathrm{Mat}_n(\mathscr{O}(\mathbf{C}^*,0))$ such that $F[A_\mu x^\mu] = A_0$. (By Proposition 12.14, matrix M is holomorphic over $\mathbf{C}^*$.)

In what follows (the parameterized situation), we shall assume that:

$$A = A_t(x) = A(t,x) = A_0(t) + \cdots + A_\mu(t)x^\mu,$$

all $A_k(t)$ being analytic over Δ and A_0, A_μ being regular (analytic with analytic inverse). Moreover we shall make the following hypotheses:

- $|||A_m(t)|||$ is bounded on Δ for $m = 0,\ldots,\mu$;
- $A_0(t)$ and $A_\mu(t)$ are nonresonant for all $t \in \Delta$.

Actually, Subsubsection 12.2.5.1 has an auxiliary technical role as we return to the local situation in x at 0. So, *stricto sensu*, the above assumptions hold from Subsubsection 12.2.5.2 on.

12.2.5.1. *Local parameterized reduction to the constant term.* First recall the (parameterless) algorithm of "Reduction of nonresonant to constant" from Subsubsection 11.1.1.4: if $A = A_0 + A_1x + \cdots \in \mathrm{GL}_n(\mathbf{C}\{x\})$ is nonresonant, i.e., $\mathrm{Sp}\, A_0 \cap \mathrm{Sp}\, q^k A_0 = \emptyset$ for $k \geq 1$, then the unique Birkhoff transformation[8] $F = F_0 + F_1x + \cdots \in \mathrm{GL}_n(\mathbf{C}\{x\})$ such that $F[A_0] = A$ is given by the initialisation $F_0 := I_n$ and the recursive relation $\forall m \geq 1$, $F_m := -\Phi^{-1}_{A_0,q^mA_0}(A_1F_{m-1} + \cdots + A_mF_0)$, where, generally speaking, $\Phi_{P,Q}(M) := PM - MQ$.

In the local (in x near 0) parameterized setting, we take $A(t,x) := \sum_{m\geq 0} A_m(t)x^m$, an invertible functional matrix analytic in $\Delta \times \overset{\circ}{\mathrm{D}}(0,R)$ for some $R > 0$. More precisely, we shall assume that all $A_m(t)$ are analytic matrix valued functions on Δ and there exist $C, D > 0$ such that:

$$\forall t \in \Delta\ ,\ \forall m \geq 0\ ,\ |||A_m(t)||| \leq CD^{-m}.$$

(So any $R < D$ is suitable.) If $A(t,x)$ is polynomial in x, this of course amounts to requiring all A_m to be bounded on Δ. We shall moreover make that $A_0(t)$ is nonresonant for all $t \in \Delta$.

LEMMA 12.37. *Then $A_0(t)$ is uniformly nonresonant on every compact subset Δ' of Δ in the sense that there exists $M > 0$ such that:*

$$\forall t \in \Delta'\ ,\ \forall k \geq 1\ ,\ \left|\left|\left|\Phi^{-1}_{A_0,q^kA_0}\right|\right|\right| \leq M.$$

PROOF. For every t, let $R(t) = \mathrm{Diag}(\rho_1(t),\ldots,\rho_n(t))$ and $C(t) \in \mathrm{GL}_n(\mathbf{C})$ such that $A_0(t) = C(t)R(t)C(t)^{-1}$. By proposition 12.40 further below (and you may check that there is here no circular argument), they can be chosen analytic; we just need their continuity. Then, by direct calculation:

$$\Phi_{A_0,q^kA_0} = L_C \circ \Phi_{R,q^kR} \circ L_C^{-1},\ \text{where}\ L_C(M) := CMC^{-1}.$$

Since $C(t)$ is continuous and invertible, one easily checks that $|||L_C|||$ and $\left|\left|\left|L_C^{-1}\right|\right|\right|$ are bounded on Δ'. it remains to show that $\left|\left|\left|\Phi^{-1}_{R,q^kR}\right|\right|\right|$ is also bounded. In the canonical basis, it is diagonal with eigenvalues the $\dfrac{1}{\rho_i(t) - q^k\rho_j(t)}$, $i,j = 1,\ldots,n$.

Letting $\alpha_{i,j}(t) := \rho_i(t)/\rho_j(t)$ and noting that all $1/\rho_i(t)$ are bounded on Δ', we are left to prove that all distances $d\left(\alpha_{i,j}(t), q^{\mathbf{N}^*}\right)$ are bounded away from 0 on Δ'. But those are continuous functions and they do not vanish by the (simple) nonresonancy assumption, so the compactness allows us to conclude. □

[8]Beware that the use of symbols F_0, F here is not exactly the same as in the diagram above.

Under the previous assumptions, there is, for every $t \in \Delta$, a unique power series $F_t(x) := F(t,x) = \sum_{m \geq 0} F_m(t)x^m \in \mathscr{O}(\Delta)[[x]]$ such that $F_0(t) = I_n$ and $F_t[A_0] = A$, i.e., $F(t,qx)A_0(t) = A(t,x)F(t,x)$: it is determined by the initialisation and the recursive relations $F_m(t) := -\Phi^{-1}_{A_0(t),q^m A_0(t)}(A_1(t)F_{m-1}(t) + \cdots + A_m(t)F_0(t))$.

THEOREM 12.5. *The series $F(t,x)$ is analytic on $\Delta \times \overset{\circ}{D}(0,R)$ for some $R > 0$. The series $F(t,x)^{-1}$ is analytic on $\Delta \times \overset{\circ}{D}(0,R_0)$ for some $R_0 > 0$.*

PROOF. We may replace Δ by a relatively compact polydisk, thus assume from start uniform boundedness and nonresonancy. From the triangular inequality and the above assumptions we draw:

$$|||F_m||| \leq MC\left(\frac{|||F_{m-1}|||}{D} + \cdots + \frac{|||F_0|||}{D^m}\right) \Longrightarrow D^m\, |||F_m||| \leq MC\sum_{i=0}^{m-1} D^i\, |||F_i|||\,.$$

Using the lemma recalled at the beginning of this subsection and inequality:

$$\forall t \in \Delta\,,\ \forall m \geq 1\,,\ |||F_m(t)||| \leq C\left(\frac{1+MC}{D}\right)^m,$$

(derived from the usual computations), we see that (i) holds with $R := \dfrac{D}{1+MC}$. Now, setting $R_0 := R/(1+E)$ and restricting to $x \in \overset{\circ}{D}(0,R_0)$, we get the following further estimate which implies statement (ii):

$$|||F - I_n||| = \left|\left|\left|\sum_{m\geq 1} F_m x^m\right|\right|\right| < E\sum_{m \geq 1}(FR_0)^m = 1.$$

□

12.2.5.2. *Connection of $A_0(t)$ with $A_\mu(t)$.* First we parameterize the reduction of A to A_0. From Subsection 11.1.5 we know that there is a $F_0(t,x)$ analytic over some $\Delta \times \overset{\circ}{D}(0,R)$, $R > 0$, such that $F_0(t,0) = I_n$ (Birkhoff transformation) for all $t \in \Delta$ and moreover:

$$F_0(t,qx)A_0(t) = A(t,x)F_0(t,x) \Longrightarrow F_0(t,x)^{-1} = A_0(t)^{-1}F_0(t,qx)^{-1}A(t,x).$$

The usual process of meromorphic continuation by expansion of the disk of definition from Subsubsection 5.3.2.2 then ensures that $F_0(t,x)^{-1}$ is analytic over $\Delta \times \mathbf{C}$.

Now consider the corresponding problem for $F_\infty(t,x)$. Write it as $G(t,y)$, $y := 1/x$. We have $G(t,0) = I_n$ and:

$$F_\infty(t,qx)A_\mu(t)x^\mu = A(t,x)F_\infty(t,x) \Longrightarrow G(t,y/q)B_0(t) = B(t,y)G(t,y),$$

where we have set $x^{-\mu}A(t,x) = B(t,y) = B_0(t) + \cdots + B_\mu(t)y^\mu$. Writing it as $G(t,y) = B(t,qy)G(t,y)B_0(t)^{-1}$, we conclude as before that $G(t,y)$ is analytic for $(t,y) \in \Delta \times \mathbf{C}$. We summarize:

PROPOSITION 12.38 (Local unfolding of F). *The (uniquely defined) connecting map $F(t,x)$ is analytic for $(t,x) \in \Delta \times \mathbf{C}^*$.*

PROOF. Indeed $F(t,x) = F_0(t,x)^{-1}G(t,y)$ □

COROLLARY 12.39 (Global unfolding of F). *The result remains if Δ is replaced (in assumptions and in conclution) by an arbitrary connected analytic manifold.*

PROOF. Indeed, F being uniquely determined by A, the local families patch. □

12.2.5.3. *Reduction by conjugacy of $A_0(t)$ to $R(t)$ and of $A_\mu(t)$ to $S(t)$.*

PROPOSITION 12.40. *Let $P(t)$, $t \in \Delta$ an analytic family of separable matrices (i.e., such that all their eigenvalues are pairwise distinct).*

(i) *Then there is an analytic family $Q(t)$, $t \in \Delta$ of diagonal matrices such that $P(t) \sim Q(t)$ (conjugacy) for all $t \in \Delta$.*

(ii) *For every $t_0 \in \Delta$ and $C_0 \in GL_n(\mathbf{C})$ such that $P(t_0) = C_0 Q(t_0) C_0^{-1}$, there exists a neighborhood $\Delta' \subset \Delta$ of t_0 and an analytic family $C(t)$, $t \in \Delta'$ such that $C(t_0) = C_0$ and $P(t) = C(t)Q(t)C(t)^{-1}$ for all $t \in \Delta'$.*

PROOF. (i) The characteristic polynomial $\chi_{P(t)}(X)$ is unitary, separable and has coefficients that are analytic on Δ. So the statement follows from Corollary A.27 at the end of Appendix A.

(ii) For each particular eigenvalue $\lambda_i(t_0)$ of $P(t_0)$ (i.e., diagonal coefficient on $Q(t_0)$), the transpose of the comatrix of $P(t) - \lambda(t) I_n$ has rank 1 (its columns generate the corresponding eigenspace). Let $X_i(t)$ a particular column not vanishing at t_0, whence in a neighborhood $\Delta'_i \subset \Delta$ of t_0. Taking $\Delta' := \bigcap \Delta'_i$ and $C(t)$ the matrix with columns the $X_i(t)$ yields the answer. (We leave it to the reader to adapt the argument to satisfy the supplementary requirement $C(t_0) = C_0$.) □

This applies of course to $A_0(t)$ and $A_\mu(t)$.

12.2.5.4. *Local analytical unfolding of M.* Putting everything together, we arrive at our main (local) result. We now assume that the analytic family:

$$A(t,x) = A_0(t) + \cdots + A_\mu(t) x^\mu \ , \ (t,x) \in \Delta \times \mathbf{C}$$

is subjected to the conditions (uniform boundedness and uniform resonancy) stated at the beginning of this subsection, and also that the $N = n\mu$ roots $\underline{x}(t)$ of $\det A(t,x)$ (seen as a polynomial in x) are nonresonant for all $t \in \Delta$.

THEOREM 12.6. *Every $t_0 \in \Delta$ admits a neighborhod $\Delta' \subset \Delta$ on which there are analytic families $R(t), S(t), \underline{x}(t), M(t,x)$ such that, for all $t \in \Delta'$:*

1. *$R(t), S(t) \in GL_n(\mathbf{C})$ and $\underline{x}(t) \in \mathbf{C}^{*N}$ are nonresonant;*
2. *$M(t,x) \in F_{R(t),S(t),\underline{x}(t)}$ is associated to $A(t,x)$ by the Riemann-Hilbert-Birkhoff correspondence.*

PROOF. Only the case of $\underline{x}(t)$ is not covered by the previous results, but it follows immediately from Corollary A.27 applied to the polynomial $\det A_t(x)$. □

12.2.5.5. *A slightly more global variant.* Here we let $A(t,x)$ be defined for $t \in X$, an open Riemann surface, and assume that $R(t)$ and $S(t)$ can be found analytic all over $t \in X$. (This of course holds true if X is simply connected but our true goal is $X := \mathbf{C}^*$!) Then, the only obstructions to a global family coming from the reductions by conjugacy, in this setting they are taken care of by the results in Appendix B.7, whence:

COROLLARY 12.41. *An analytic family $M(t,x)$ as above can be found for $t \in X$.*

12.3. Confluence again: global study

In this section (as in the previous one one), we stick to the general convention of this book: $0 < |q| < 1$. We shall here explain one of the reasons that allow us to consider Birkhoff's connection matrix $P := \left(\mathcal{X}^{(\infty)}\right)^{-1} \mathcal{X}^{(0)}$ from 12.1.1.3 as a q-analogue of monodromy for differential equations (more reasons will come in Chapter 13). This is related to degeneracy of q-difference systems to differential equations when $q \to 1$, of which we gave the local theory in 11.2.

Note that related work **[74]** due to Thomas Dreyfus aims at understanding "q-isomonodromy", while **[73]** deals with a somewhat different model.

12.3.1. General theory.

12.3.1.1. *Solutions at* ∞. Starting from system $\sigma_q X = AX$, where $A \in \mathrm{GL}_n(\mathbf{C}(x))$, the local study at infinity runs as follows. Set $y := 1/x$ and $Y(y) := X(x) = X(1/y)$. Then:

$$Y(y/q) = X(q/y) = A(1/y)X(1/y) = A(1/y)Y(y) \Longrightarrow Y(qy) = A(1/qy)^{-1}Y(y),$$

so the local study of $\sigma_q X = AX$ at ∞ boils down to the local study at 0 of q-difference equation:

$$\sigma_q Y = BY, \text{ where } B(y) := A(1/qy)^{-1}.$$

Now suppose that, as in the study of local confluence in 11.2 (of which we use the conventions and notations[9]), $q = q_0^\epsilon$ and A depends on $\epsilon > 0$ in such a way that $(A - I_n)(q-1) \to \tilde{B}$ when $q \to 1$ (i.e., when $\epsilon \to 0^+$). Then a small calculation (left to the pleasure of the reader) shows that:

$$\frac{A(1/qy)^{-1} - I_n}{q-1} \to -\tilde{B}(y).$$

On the other hand, we can operate the transformation $y := 1/x$ on the limit differential system $\delta\tilde{X} = \tilde{B}\tilde{X}$ by setting $\tilde{Y}(y) := \tilde{X}(x) = \tilde{X}(1/y)$. Then:

$$y\,d\tilde{Y}(y)/dy = y\,d\tilde{X}(1/y)/dy = -1/y\,(d\tilde{X}/dx)(1/y) = -x(d\tilde{X}/dx)(x).$$

Thus, using the symbol δ to denote $y\,d/dy$ when applied to a function of y:

$$\delta\tilde{Y} = -\tilde{B}(y)\tilde{Y}.$$

Therefore, applying the results of 11.2, we find that a canonical fundamental matricial solution $\mathcal{X}^{(\infty)}(x) := \mathcal{Y}(y)$ of the q-difference system degenerates to a canonical fundamental solution $\tilde{\mathcal{X}}^{(\infty)}(x) := \tilde{\mathcal{Y}}(y)$ of the differential system. Now, showing explicitly the dependency on ϵ:

$$\mathcal{X}_\epsilon^{(\infty)} \to \tilde{\mathcal{X}}^{(\infty)}.$$

EXERCISE 12.42. Of course, the assumptions of 11.2 of the nature of the degeneracy must be justified, or (for SDJS at ∞) added. State precisely what is needed.

Figure 12.2 displays the domains of definition of the limits $\mathcal{X}_\epsilon^{(0)}$ and $\mathcal{X}_\epsilon^{(\infty)}$. The small paths γ_1 and γ_2 joining points a and b will be explained soon.

[9]But for one discrepancy: the letter B here has a different meaning.

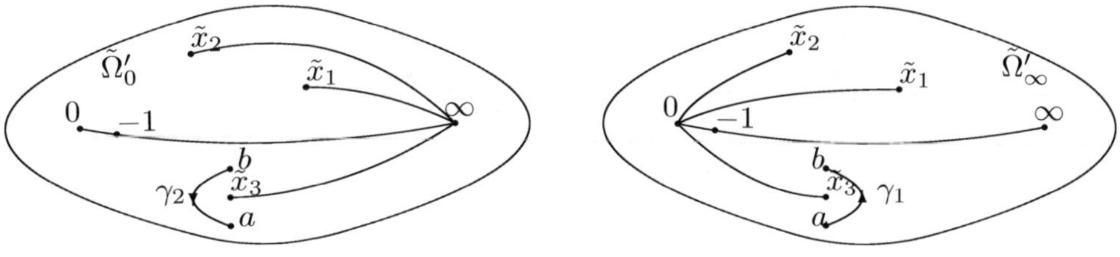

FIGURE 12.2. Domains of definition $\tilde{\Omega}'_0$ of $\tilde{\mathcal{X}}^{(0)}$ and $\tilde{\Omega}'_\infty$ of $\tilde{\mathcal{X}}^{(\infty)}$

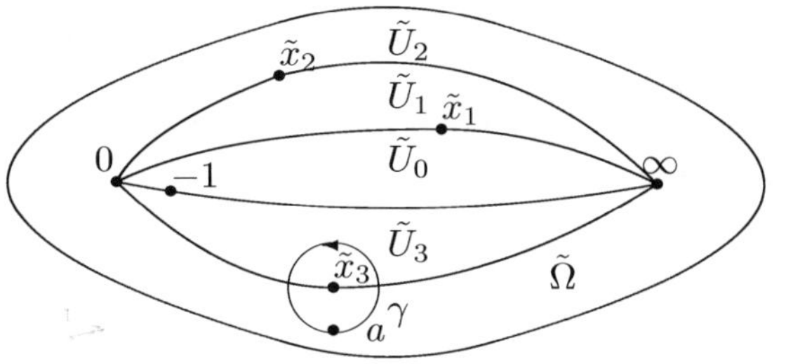

FIGURE 12.3. Domain of definition $\tilde{\Omega} := \tilde{\Omega}'_0 \cap \tilde{\Omega}'_\infty$ of $\tilde{P}$

12.3.1.2. *Connecting solutions at* 0 *and at* ∞. Assuming all the necessary hypotheses (see exercise above), we can form for each $\epsilon \approx 0$ the connection matrix $P_\epsilon := \left(\mathcal{X}_\epsilon^{(\infty)}\right)^{-1} \mathcal{X}_\epsilon^{(0)}$. Recall that the open set $\tilde{\Omega}'_0$ was defined in 11.2.4.3; there is a corresponding naturally defined open set $\tilde{\Omega}'_\infty$. On the intersection $\tilde{\Omega} := \tilde{\Omega}'_0 \cap \tilde{\Omega}'_\infty$, the matrix P_ϵ degenerates into a matrix $\tilde{P} := \left(\tilde{\mathcal{X}}^{(\infty)}\right)^{-1} \tilde{\mathcal{X}}^{(0)}$ (see Figure 12.3, in which we took $r := 3$; the loop γ based at point a will be explained soon).

The open set $\tilde{\Omega}$ is equal to $\mathbf{C}^*$ deprived of the continuous q_0-spirals $\tilde{x}_i q_0^{\mathbf{R}}$, $i = 0, \ldots, r$ (for convenience we have set $\tilde{x}_0 := -1$). it looks like a sliced watermelon. Numbering the $\tilde{x}_i$ so that a small positive circle centerd at 0 successively crosses the spirals in circular order $0, 1, \ldots, r, 0$, we call U_i the connected component bounded by $\tilde{x}_i q_0^{\mathbf{R}}$ and $\tilde{x}_{i+1} q_0^{\mathbf{R}}$ (with the obvious convention when $i = r$).

Differentiating relation $\tilde{\mathcal{X}}^{(\infty)} = \tilde{\mathcal{X}}^{(0)} \tilde{P}$ with respect to x, we deduce that:

$$\forall x \in \tilde{\Omega}\,,\ \frac{d}{dx} \tilde{P}(x) = 0.$$

Therefore the matricial function $\tilde{P}$ is constant on every connected component of $\tilde{\Omega}$. We call $P_i \in \mathrm{GL}_n(\mathbf{C})$ the value of $\tilde{P}$ on U_i.

THEOREM 12.7. *The monodromy matrix of the differential system $\delta \tilde{X} = \tilde{B} \tilde{X}$ around $\tilde{x}_i$, $i = 1, \ldots, r$, taken with respect to the fundamental matricial solution $\tilde{\mathcal{X}}^{(0)}$, is $P_i^{-1} P_{i-1}$.*

PROOF. Let $a \in \tilde{U}_{i-1}$ near $\tilde{x}_i$ (in the figure, we took $i := 3$). A small loop γ around $\tilde{x}_i$ and based at a can be taken as the composition of a path γ_1 from a to $b \in \tilde{U}_i$, while staying inside $\tilde{\Omega}'_\infty$; followed by a path γ_2 from b to a, while staying inside $\tilde{\Omega}'_0$.

To begin with, $\tilde{\mathcal{X}}^{(0)} = \tilde{\mathcal{X}}^{(\infty)} P_{i-1}$ near a. Continuation along γ_1 has a meaning for $\tilde{\mathcal{X}}^{(\infty)}$, so (along that way) $\tilde{\mathcal{X}}^{(0)}$ transforms to $\tilde{\mathcal{X}}^{(\infty)} P_{i-1}$ near b. There, $\tilde{\mathcal{X}}^{(\infty)} =$

$\tilde{\mathcal{X}}^{(0)} P_i^{-1}$, so $\tilde{\mathcal{X}}^{(\infty)} P_{i-1} = \tilde{\mathcal{X}}^{(0)} P_i^{-1} P_{i-1}$; but the right-hand side expression is well defined all along γ_2, so, in the end, $\tilde{\mathcal{X}}^{(0)}$ has been transformed into $\tilde{\mathcal{X}}^{(0)} P_i^{-1} P_{i-1}$. $\square$

12.3.2. Confluence of q-hypergeometric series. We go to some detail into an important example. However, this is not really intended to be a direct application of the general theory shown in 12.3.1, rather an illustration of the methods and phenomena involved. Indeed, since many choices (for instance: the choice of a family of q-characters $e_{q,c}$) are not uniquely determined, the precise formulation of a confluence may depend on the particular situation. Besides, the approach here is a variation of that featured in [**217**].

Note that all that follows here falls under the fuchsian case, but an interesting instance of the irregular case will be discussed in 14.2.5.

12.3.2.1. *Degeneracy of the scalar q-hypergeometric equation.* Recall from Chapter 4 that the *basic hypergeometric series* $\Phi(a, b; c; q, x) := \sum_{n \geq 0} \dfrac{(a;q)_n (b;q)_n}{(c;q)_n (q;q)_n} x^n$ satisfies the q-difference equation $(\sigma_q^2 - \lambda \sigma_q + \mu)\Phi = 0$, where (see 4.4.1):

$$\lambda := \frac{(1 + c/q) - (a+b)x}{c/q - abx} \text{ and } \mu := \frac{1-x}{c/q - abx}.$$

By 4.4.2, if $c \notin q^{\mathbf{Z}}$, there is a fundamental basis of solutions at 0:

$$\phi^{(0)}(x) := \Phi(a, b; c; q, x), \quad \gamma^{(0)}(x) := e_{q,q/c}(x)\Phi(aq/c, bq/c; q^2/c; q, x).$$

By 4.4.3, if $a/b \notin q^{\mathbf{Z}}$, there is a fundamental basis of solutions at ∞:

$$\begin{cases} \phi^{(\infty)}(y) := e_{q,a}(y)\Phi(a, qa/c; qa/b; q, (qc/ab)y), \\ \gamma^{(\infty)}(y) := e_{q,b}(y)\Phi(b, qb/c; qb/a; q, (qc/ab)y), \end{cases} \quad \text{where } y := 1/x.$$

We now set $q := q^\alpha$, $b := q^\beta$, $c := q^\gamma$ and assume that $\gamma, \alpha - \beta \notin \mathbf{Z}$ and find (from 4.4.4) that, if $q = q_0^\epsilon$ and $\epsilon \to 0^+$:

$$\begin{aligned} \phi^{(0)}(x) &\to f^{(0)}(x) := F(\alpha, \beta; \gamma; x), \\ \gamma^{(0)}(x) &\to g^{(0)}(x) := x^{1-\gamma} F(\alpha - \gamma + 1, \beta - \gamma + 1; 2 - \gamma; x), \\ \phi^{(\infty)}(y) &\to f^{(\infty)}(y) := y^\alpha F(\alpha, \alpha - \gamma + 1; \alpha - \beta + 1; y), \\ \gamma^{(\infty)}(y) &\to g^{(\infty)}(y) := y^\beta F(\beta, \beta - \gamma + 1; \beta - \alpha + 1; y), \end{aligned}$$

where $F(\alpha, \beta; \gamma; x) := \sum\limits_{n \geq 0} \dfrac{(\alpha)_n (\beta)_n}{(\gamma)_n (1)_n} x^n$ is the *classical hypergeometric series*.

The series F is solution of the differential equation $(\delta^2 - \tilde{\lambda}\delta + \tilde{\mu})F = 0$, where $\tilde{\lambda}(x) := \dfrac{(\alpha+\beta)x + (1-\gamma)}{1-x}$ and $\tilde{\mu}(x) := \dfrac{\alpha\beta x}{1-x}$. From [**125**, 2.1], [**222**, §11] or [**255**, 14.3], we draw that $(f^{(0)}(x), g^{(0)}(x))$ and $(f^{(\infty)}(y), g^{(\infty)}(y))$ introduced above are respectively the fundamental bases of solutions at 0 and at ∞ obtained by Fuchs-Frobenius method.

The degeneracy of bases of solutions of the q-difference equation to bases of solutions of the differential equation suggests that there is a degeneracy of the equations themselves. This can be verified as follows. We rewrite σ_q as $1 + (q-1)\delta_q$

and find that:

$$\begin{aligned}\sigma_q^2 - \lambda\sigma_q + \mu &= 1 + 2(q-1)\delta_q + (q-1)^2\delta_q^2 - \lambda - (q-1)\lambda\delta_q + \mu \\ &= (q-1)^2\left(\delta_q^2 - \lambda'\delta_q + \mu'\right), \\ \text{where } \lambda' &:= \frac{\lambda-2}{q-1} = \frac{(1-c/q)/(q-1) - ((a+b-2ab)(q-1))x}{c/q - abx}, \\ \mu' &:= \frac{\mu-\lambda+1}{(q-1)^2} = \frac{-((a-1)(b-1)/(q-1)^2)x}{c/q-abx}.\end{aligned}$$

If now $(a-1)/(q-1) \to \alpha$, $(b-1)/(q-1) \to \beta$, $(c-1)/(q-1) \to \gamma$, we see that $\lambda' \to \tilde{\lambda}$ and $\mu' \to \tilde{\mu}$, confirming the suggestion.

12.3.2.2. *Degeneracy of connection formulas through the scalar equation.* We are going to connect $\phi^{(0)}(x)$ to $\phi^{(\infty)}(y)$ and $\gamma^{(\infty)}(y)$, then see how the connection coefficients degenerate when $q \to 1$. The case of $\gamma^{(0)}(x)$ is similar (though maybe a bit more complicated) and left to the reader's pleasure. We start from a connection formula that may be attributed[10] to Barnes-Mellin-Watson and taken from [**93**, 4.3]. We do not attempt to prove it as it relies on techniques that have been little used in this book.

Recall the multiple parameters form of the q-Pochhammer symbol from 3.3.1: $(a_1, \ldots, a_k; q)_\infty := (a_1; q)_\infty \cdots (a_k; q)_\infty$. Then:

$$\begin{aligned}\Phi(a, b; c; q, x) = {} & \frac{(b, c/a; q)_\infty}{(c, b/a; q)_\infty} \frac{(ax, q/ax; q)_\infty}{(x, q/x; q)_\infty} \Phi(a, aq/c; aq/b; q, cq/abx) \\ & + \frac{(a, c/b; q)_\infty}{(c, a/b; q)_\infty} \frac{(bx, q/bx; q)_\infty}{(x, q/x; q)_\infty} \Phi(b, bq/c; bq/a; q, cq/abx).\end{aligned}$$

From Jacobi's Triple Product Formula 2.2.1.3: $(x, q/x; q)_\infty = \theta_q(-x)/(q; q)_\infty$. Now we appeal to Heine's q-analogue of Gauß summation formula [**93**, 1.5], which says that:

$$\frac{(c/a, c/b; q)_\infty}{(c, c/ab; q)_\infty} = \Phi(a, b; c; q, c/ab),$$

in which we replace $b \leftarrow c/b$, resp. $a \leftarrow c/a$, and combine all that to get:

$$\begin{aligned}\Phi(a, b; c; q, x) = {} & \Phi(a, c/b; c; q, b/a) \frac{\theta_q(-ax)}{\theta_q(-x)} \Phi(a, aq/c; aq/b; q, cq/abx) \\ & + \Phi(b, c/a; c; q, a/b) \frac{\theta_q(-bx)}{\theta_q(-x)} \Phi(b, bq/c; bq/a; q, cq/abx).\end{aligned}$$

Using definitions in 12.3.2.1, this yields the *connection formula*:

$$\phi^{(0)}(x) = \frac{\Phi(a, c/b; c; q, b/a)}{e_{q,a}(-x)e_{q,a}(y)} \phi^{(\infty)}(y) + \frac{\Phi(b, c/a; c; q, a/b)}{e_{q,b}(-x)e_{q,b}(y)} \gamma^{(\infty)}(y).$$

Now we set $a := q^\alpha$, etc, and have $q \to 1$. Using 12.3.2.1, along with the classical Gauss-Kummer summation formula [**125**, §2 3.7], we obtain:

$$f^{(0)}(x) = \frac{\Gamma(\gamma)\Gamma(\beta-\alpha)}{\Gamma(\gamma-\alpha)\Gamma(\beta)} \frac{1}{(-x)^\alpha (1/x)^\alpha} f^{(\infty)}(y) + \frac{\Gamma(\gamma)\Gamma(\alpha-\beta)}{\Gamma(\gamma-\beta)\Gamma(\alpha)} \frac{1}{(-x)^\beta (1/x)^\beta} f^{(\infty)}(y).$$

This is a worthy connection formula for the differential setting. Indeed, the coefficients are obviously locally constant, the respective degeneracies of $e_{q,b}(-x)$ and

[10] Even in the classical case of hypergeometric series some well known formulas receive different attributions in, for instance, [**93, 125, 255**].

$e_{q,b}(y)$ have given rise to cuts along $\pm q_0^{\mathbf{R}}$ leaving two connected components corresponding to choices of path of continuation. (Compare with the formulas in [**125**] and in [**217**].)

EXERCISE 12.43. (i) Detail the possible values.
(ii) Redo the study for $g^{(0)}$.

12.3.2.3. *Through the matricial equation.* From the scalar equation in 12.3.2.3, we deduce that $X := \begin{pmatrix} \Phi \\ \sigma_q \Phi \end{pmatrix}$ satisfies $\sigma_q X = AX$, where $A := \begin{pmatrix} 0 & 1 \\ -\mu & \lambda \end{pmatrix}$. A first consequence is that $A(0) = \begin{pmatrix} 0 & 1 \\ -q/c & 1 + q/c \end{pmatrix}$ has spectrum $\{1, q/c\}$, whence the condition $c \notin q^{\mathbf{Z}}$. Then:

$$B(y) := A(1/qy)^{-1} = \begin{pmatrix} \dfrac{(a+b) - (c+q)y}{1 - qy} & -\dfrac{ab - cy}{1 - qy} \\ 1 & 0 \end{pmatrix},$$

so $B(0)$ has spectrum $\{a, b\}$, whence the condition $a/b \notin q^{\mathbf{Z}}$.

Now it unfortunately appears that $\dfrac{A - I_n}{q - 1}$ has no particular limit when $q \to 1$. But this can be repaired as follows. Instead of vectorializing the scalar equation in the usual way, rather set $X := \begin{pmatrix} \Phi \\ \delta_q \Phi \end{pmatrix}$ (remember that $\delta_q := (\sigma_q - \mathrm{Id})/(q - 1)$). Then find the matrix B' such that $\delta_q X = B'X$. Now $A' := I_n + (q - 1)B'$ satisfies $\sigma_q X = A'X$ and degenerates nicely.

EXERCISE 12.44. Apply the above program.

CHAPTER 13

Galois Theory of Fuchsian Systems

Anyone who considers transcendental methods of producing algebraic objects is, of course, in a state of sin.
—Anonymous [1]

This chapter will require a bit more autonomy from the reader. We describe all new ideas in some detail, but go swiftly through standard technical arguments.

13.1. Introduction to the Galois theory of q-difference equations

In this section, we shall try to explain the path followed toward a Galois theory of q-difference equations, through some historical and epistomological remarks. The reason for the above quotation will then maybe appear clearly! Our justification for such a detour is that we follow a path that maybe seen as unusual in a domain that was previously thoroughly algebraized (see [**249**]); in turn, our justification for this unusual path is that it seems to have been the most fruitful so far in many (but not all[2]) aspects. As for the prerequisites in Tannakian theory, they are modest and summarized in Appendix G (also see the presentation in [**222**, Appendix D]).

13.1.1. Galois theory of ordinary differential equations.

13.1.1.1. *Early history.* Galois himself (see [**6**, **91**, **192**, **245**] left a few cryptical remarks on the application of his methods to transcendental objects. This has been analyzed by Ramis, André, Malgrange, Cartier, Umemura, and others and we will not say more.

Riemann [**203**] developed in his works on Abelian functions and on hypergeometric differential equation what can legitimately been considered as a galoisian theory of some classes of functions (either algebraic or solutions of linear complex analytic differential equations). In the case of differential equations, he extracts from solutions *intrinsic transcendental invariants* (the monodromy): first displayed in the form of encoding complex numbers (the connection coefficients) [**98**, **255**], they were later modelled as monodromy matrices and seen by Hilbert as the *monodromy representation*, a linear representation of a fundamental group π_1 [**9**, **125**, **169**, **222**]. The first form of so-called *Riemann-Hilbert correspondence* was thus the classification of linear complex analytic differential equations by their monodromy representations. In particular, the image of the representation attached to a particular equation is its *monodromy group* Mon; it has a finite presentation (i.e., a finite number of defining generators and relations) prescribed by the geometry of the singular locus (actually, the topology of its complementary subset).

[1] Actually, von Neumann said (1951): "Anyone who considers arithmetical methods of producing random digits is, of course, in a state of sin". The spirit seems, to me, to be the same.

[2] The main success of the purely algebraic approach is that it provides effective methods [**47**, **117**, **228**] ...; but those will be tackled in [CV].

A parallel but independent path was followed by Picard and Vessiot, who algebraized the theory along lines close to the formalization of the Galois theory of algebraic numbers by Dedekind (fields), Jordan (groups), etc. The Picard-Vessiot theory is founded on differential fields and their groups of automorphisms. They thus obtain a *linear algebraic group* Gal. For sources of those developments; see [**193**, **250**]; a breezy minimal introduction appears in [**222**].

The link of Picard-Vessiot theory with Riemann-Hilbert correspondence is given by the *density theorem of Schlesinger*: the monodromy group Mon embeds into the Galois group Gal and the latter is the Zariski closure of the former (i.e., the smallest algebraic subgroup containing it). The second part of the statement actually only holds for *fuchsian* equation, see further below what can be said for irregular equations.

However, there is a significant discrepancy between Riemann-Hilbert and Picard-Vessiot theories: while in the former the groups Mon are universally parameterized by the π_1 (through its linear representations), there does not seem to be such a universal object, say $\tilde{\pi}_1$, for the groups Gal. Actually there really is one but it is not apparent, it is better displayed through another approach; see 13.1.1.2.

Let us just say a word about irregular equations. In order to classify them, Birkhoff introduced the *Stokes operators*; then Malgrange and Sibuya formalized the underlying structure by cohomological methods. The corresponding Galois group, together with a density theorem, was described by Ramis by adding to the monodromy group Mon (which had been sufficient as far as regular singularities were involved) some well selected Stokes operators along with the so-called *exponential tori*[3], see the given references. The *wild monodromy group* thus obtained is Zariski-dense in Gal; and all wild monodromy groups thus obtained are images of representations of a *wild fundamental group*. On all those matters, see [**150**, **193**, **250**] and the references therein.

REMARK 13.1. When one wants to "understand" Gal (e.g. direct problem: compute it; inverse problem: determine which algebraic groups can be obtained that way; etc), many purely algebraic methods are available, with roots in Picard-Vessiot theory and differential algebra (and in the theory of linear algebraic groups as developed by Kolchin, [**134**, **138**]). Yet using the embedding Mon $\subset$ Gal (or that of the wild monodromy group) along with the density theorem, one can take advantage of the rich combinatorial structure of Mon as an image of the finitely presented group π_1. Here again (as in Weyl's quotation in Subsection 0.1.3 of the Introduction) using transcendental methods in algebraic problems turns out fruitful.

13.1.1.2. *The Tannakian way.* How can one get a universal group[4] parameterizing all Picard-Vessiot Galois groups? The Tannakian theory, which allows for it, has ancestors in the theory of representations of compact Lie groups (Tannaka-Krein in [**42**, pp 146 et sq]) and of algebraic groups ([**49**, Chapter VI] and [**239**, p 37]); there is also a source in abstract category theory, the theorem of Barr-Beck [**53**, **56**]. However its direct origin lies in the approach by Grothendieck to the theory of etale coverings and the fundamental group in the context of algebraic curves. Smooth complex algebraic curves are Riemann surfaces and algebraic topology in

[3]These are tori in the sense of the theory of linear algebraic groups, i.e., products of multiplicative groups $\mathbf{C}^*$.

[4]An ad hoc method was proposed in [**222**] but here we look for something more conceptual.

the style of Riemann directly applies to them. For algebraic geometry over arbitrary fields something else must be used.

For the following discussion, see **[72]** (also see **[49]** for an intermediate approach to the theory of the fundamental group). Note that if X is a topological space submitted to some conditions relating to connectivity, the "representations" of $\pi_1(X,x)$ (here referring to sets endowed with an action of $\pi_1(X,x)$, also called $\pi_1(X,x)$-sets) are in natural correspondence with etale coverings of X. Fixing a base point $x \in X$, for any such etale covering $p : X' \to X$, the group $\pi_1(X,x)$ acts on the fiber $p^{-1}(x)$, making it a $\pi_1(X,x)$-set. If expressed in functorial terms involving at the same time all etale coverings and all $\pi_1(X,x)$-sets, *plus an extra-structure, the fibered product* $X' \times_X X''$ *of spaces*, this setting allows conversely to recover the group $\pi_1(X,x)$ as *the group of* $\times$*-compatible automorphisms of the "fibre functor"* $(p : X' \to X) \rightsquigarrow p^{-1}(x)$. So Grothendieck took an adequate definition of etale coverings of curves (over an arbitrary fields) and defined: first a fibre functor ω on their category; second the fundamental groups as the groups $\mathrm{Aut}^{\times}(\omega)$ obtained that way. In the case of a complex curve, this yields a group $\hat{\pi}_1$ strictly bigger than π_1, but coming with a density theorem rather similar to Schlesinger's theorem[5].

Returning to differential equations, we note that they give rise to *linear* actions of π_1 on the spaces of local solutions (equivalently, on space of initial conditions), so we have *linear representations.* It was understood early **[146]** that the conditions defining elements of the Picard-Vessiot Galois groups (conservation of algebraic relations among the solutions) can be expressed in terms of tensor products. Also, clearly, the space of initial conditions at a point bears some resemblance with a fiber $p^{-1}(x)$. So one might guess that the Galois group could be recovered from the "category of equations", endowed with a tensor product $\otimes$ and a fibre functor ω, in the form of $\mathrm{Aut}^{\otimes}(\omega)$.

The formalism making possible such a theory again came from algebraic geometry, indeed, from Grothendieck. With completely different motivations (the "motives") Grothendieck's student Saavedra developed the point of view that a category endowed with a tensor product and a fibre functor, all subjected to precise conditions (Appendix G), should give rise[6] to a "proalgebraic group" $\mathrm{Aut}^{\otimes}(\omega)$ whose "rational representations" parameterize a lot of linear algebraic groups, the individual Galois groups of its objects: the image $\mathrm{Im}\, \rho_A$ of the representation ρ_A associated to the object A by the said equivalence is an algebraic group, the Galois group of A **[41, 53, 56, 213]**.

Deligne and Katz **[53, 136]** applied this approach to linear differential equations. In particular, in **[53]**, Deligne recovered (and enhanced in many ways) the bulk of Picard-Vessiot theory. The extension of such methods to irregular equations was done by Ramis **[193, 250]**. As we shall see later, this is the path we followed for the Galois theory of q-difference equations whence our insistance on that approach (see also Appendix G devoted to the basics).

13.1.1.3. *Mon and Gal in Riemann-Hilbert or in Tannakian style.* We now synthetize the relation between Mon (and Riemann-Hilbert correspondence) on one

[5]The structure of $\hat{\pi}_1$ moreover contains information on the Galois group of the involved extensions of fields of definition; see **[72]** for the deep link with classical Galois theory.

[6]True Tannakian theory actually gives rise to "group schemes" but for our purposes the associated proalgebraic groups are quite sufficient.

side, and Gal (and Tannaka duality) on the other side. We shall nickname π_1^{alg} the Tannakian Galois group $\mathrm{Aut}^{\otimes}(\omega)$ obtained from the category of fuchsian linear differential equations, with its natural tensor product and a fibre functor of the type "space of initial conditions" (at some base point x). We then have three $\otimes$-equivalences of categories:

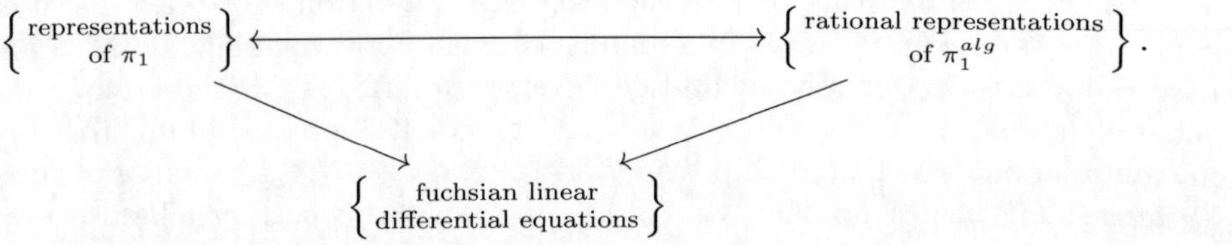

The leftmost arrow is Riemann-Hilbert correspondence and the rightmost one is Tannaka duality The upper arrow comes from the natural group embedding $\pi_1 \subset \pi_1^{alg}$ by restriction of representations. The fact that it is an equivalence is expressed by saying that π_1^{alg} is the *proalgebraic hull* of π_1, whence its notation. A consequence is that, for any representation ρ_A of π_1 restriction of the representation ρ_A^{alg} of π_1^{alg}, the group $\mathrm{Mon}(A) = \rho_A(\pi_1)$ is Zariski-dense in $\mathrm{Gal}(A) = \rho_A^{alg}(\pi_1^{alg})$. So this is a kind of universal version of Schlesinger's theorem.

EXAMPLE 13.2. Looking at complex analytic equations with a regular singularity at 0, we have $\pi_1 = \mathbf{Z}$. Then, as we shall see in 13.2:

$$\pi_1^{alg} = \mathbf{Z}^{alg} = \mathbf{C}^* \times \mathrm{Hom}_{gr}(\mathbf{C}^*, \mathbf{C}^*), \text{ with embedding } \begin{cases} \mathbf{Z} \to \mathbf{Z}^{alg}, \\ n \mapsto (n, (\gamma_n : x \mapsto x^n)). \end{cases}$$

13.1.2. First breakthroughs for q-difference equations. As has been illustrated in previous chapters (see for instance Chapter 12), analytic continuation does not provide us with such a powerful classification tool for q-difference equations as for ordinary differential equations. So, after Birkhoff, for quite some time, only purely algebraic methods were used[7].

13.1.2.1. The first significant result relating Birkhoff's connection matrix P to a group was Etingof's 1995 article **[82]**. He considered rational equations $\sigma_q X = AX$, $A \in \mathrm{GL}_n(\mathbf{C}(x))$, which he assumed to be regular at 0 and ∞, i.e., subjected to the initial conditions $A(0) = A(\infty) = I_n$. In a typical russian way, he said that the general case is similar, which, however, is hardly the case: there are in general local contributions at 0 and ∞ (see section 13.2), and the way they combine with the contribution of P may be complicated (see section 13.3). However, the progress was significant. It can be summarized as follows. In the regular case, there are canonical solutions "at 0" and "at ∞" given by infinite products:

$$\mathcal{X}^{(0)} := A^{-1}(\sigma_q A)^{-1}({\sigma_q}^2 A)^{-1} \cdots \text{ and } \mathcal{X}^{(\infty)} := ({\sigma_q}^{-1} A)({\sigma_q}^{-2} A) \cdots$$

whence a canonically defined connection matrix given by a doubly infinite product:

$$P := \left(\mathcal{X}^{(\infty)}\right)^{-1} \mathcal{X}^{(0)} = \left(\cdots ({\sigma_q}^2 A)(\sigma_q A) A ({\sigma_q}^{-1} A)({\sigma_q}^{-2} A) \cdots\right)^{-1}.$$

Let the singular locus of A:

$$\Sigma := \mathrm{Sing}(A) := \{\text{poles of } A\} \cup \{\text{poles of } A^{-1}\} = \{\text{poles of } A\} \cup \{\text{zeroes of } \det A\}.$$

[7]The analytic works by Bézivin and Ramis described in Chapter 6 predate those we are dealing with here, but they do not touch classification.

Then P is invertible, q-invariant and meromorphic over $\mathbf{C}^*$ (thus elliptic) and its singular locus is in general $\mathrm{Sing}(P) = q^{\mathbf{Z}}\Sigma$ (in particular cases, there might be cancellations of singularities). At any rate, P defines is a map from $\mathbf{C}^* \setminus q^{\mathbf{Z}}\Sigma$ to $\mathrm{GL}_n(\mathbf{C})$ and the set of "round trip values":

$$\{P(a)^{-1}P(b) \mid a, b \in \mathbf{C}^* \setminus q^{\mathbf{Z}}\Sigma\}$$

generates an algebraic subgroup of $\mathbf{C}^*$, which Etingof proves to be the Galois group of A, here understood in the sense of a q-analogue of Picard-Vessiot theory [**62**, **249**] (also [**51**] for difference algebra).

The second breakthrough was accomplished in 1996 by Baranovsky and Ginzburg (with a contribution by Kontsevitch) in [**17**]. Motivated by the theory of loop groups, they classified local fuchsian q-difference equations seen as *formal* objects with the help of *holomorphic* flat vector bundles over $\mathbf{E}_q$ (see section 13.2). In an appendix, Kontsevitch deduces from this the local Galois group of such equations[8]. Note that the results of Etingof and of Baranovsky-Ginzburg (as far as Galois theory is meant) are subsumed in [**249**].

13.1.2.2. Last, van der Put and Singer [**249**] extended both approaches (Picard-Vessiot and Tannakian) to more general settings. However, the most natural way to apply either approach to q-difference equations meets an important technical difficulty. If we use analytic functions e_c to solve $\sigma_q e_c = c e_c$ ($c \in \mathbf{C}^*$), we obtain plenty of nonconstant elliptic functions $e_c e_d / e_{cd}$ which generate a big "field of constants" (here "constants" in the sense of q-difference theory; see Chapters 2 and 3) and this should a priori yield algebraic groups over that big field, while we would wish for one over $\mathbf{C}$ (in Birkhoff's language: numerical invariants). In Picard-Vessiot theory, this big field (which is not even algebraically closed) prohibits the application of most fundamental theorems. In Tannakian theory, we get a Galois group over $\mathscr{M}(\mathbf{E}_q)$.

The problem does not occur in the regular case tackled by Etingof and is hidden in the rather roundabout path of Baranovsky and Ginzburg. Marius van der Put and Michael Singer confronted and solved it by replacing the analytic functions e_c by *symbols*[9] constrained to satisfy the relations:

$$\forall c \in \mathbf{C}^* \ , \ \sigma_q e_c = c e_c \text{ and } \forall c, d \in \mathbf{C}^* \ , \ e_c e_d = e_{cd}.$$

In Picard-Vessiot theory this leads to a secondary problem (reduced nonintegral algebras appear instead of fieds) which they overcome successfully. In Tannakian theory, the consequences are detailed in *ref. loc.* but we shall not dwell on that, since our contention is that transcendental methods lead to stronger results.

13.1.2.3. Beyond van der Put and Singer, purely algebraic approaches in Picard-Vessiot style have been fruitfully pursued by such as Di Vizio, Hardouin, ... [**62**]. A totally different path was followed by Yves André in [**4**], allowing for a unified treatment (in straight line of Birkhoff!) of the variation of Galois group of differential, difference and q-difference equations. This, among other things, provides a conceptual explanation for what we called here "confluence" in 11.2. It is well beyond the scope of this book and cannot be presented here.

---

[8] Similar results were obtained independently by the present author; see [**218**] for a comparison.

[9] It seems that van der Put and others [**247**] had already used such ideas in the context of ordinary differential equations with the goal of effective computations.

However, it is our opinion that the transcendental approach, using "true" analytic functions and relying on Riemann-Hilbert-Birkhoff correspondence, is the one that has so far led to the most powerful results (confluence, Schlesinger type density theorems, direct and inverse problems ...). This (along with this author's taste for it) motivates our choice to follow that path.

13.1.2.4. It is necessary to say a word about the Galois theory of *nonlinear* equations. As regards differential equations, it seems that only recently has a succesfull approach been found, first by Malgrange [**157, 160, 161**], using methods of differential geometry; then by Umemura [**171, 245, 246**] on a more algebraic basis. The extension of Malgrange's approach to q-difference equations is due to Anne Granier [**99**]. Umemura himself started an application of his theory to q-difference equations [**166**].

13.1.3. Contents of this chapter. Since we are entering a more sophisticated area within q-difference theory, more effort will be required from the reader. We shall try to explain clearly all relevant new ideas, but application of standard techniques will be less detailed (and sometimes left as an exercise). All the calculations in this chapter make heavy use of the abelian and tensor structure of the category of q-difference modules, as expounded in Chapters 8 and 9.

13.1.3.1. In section 13.2, we present the local theory. This means that the base field is most of the time $K := \mathbf{C}(\{x\})$, while $\hat{K} := \mathbf{C}((x))$ sometimes plays an auxiliary role[10]. In Chapter 11 (precisely, Subsubsection 11.1.3.2), we introduced the category ${\mathcal{E}_f}^{(0)}$ of analytic q-difference modules fuchsian at 0. In section 13.3, we identify ${\mathcal{E}_f}^{(0)}$ with its full subcategory of objects (K^n, Φ_A). We loosely speak of the object A (instead of (K^n, Φ_A)) and likewise of morphisms $F : A \to B$. We show that the tensor category ${\mathcal{E}_f}^{(0)}$ is equivalent to the category of representations of a proalgebraic group $G_f^{(0)}$, which we thus see as the universal Galois group for q-difference equations fuchsian at 0. We describe the Galois group $G_f^{(0)}(A)$ of a particular equation A. This rests on the equivalence of ${\mathcal{E}_f}^{(0)}$ with its full subcategory $\mathcal{P}^{(0)}$ of objects (K^n, Φ_A) with $A \in \mathrm{GL}_n(\mathbf{C})$ (which itself can be interpreted in terms of vector bundles and related to so-called "Weil's correspondence"). All this is related to the above mentioned paper of Baranovsky and Ginzburg.

13.1.3.2. To obtain something resembling the Riemann-Hilbert correspondence (in terms of representations), we find inside of $G_f^{(0)}$ a Zariski-dense subgroup, a natural candidate for the role of fundamental group. This being central for our approach of Galois theory, we first detail in 13.2.2 a baby example rooted in classical algebraic topology, that of the proalgebraic hull $\mathbf{Z}^{alg}$ of $\pi_1(\mathbf{C}^*, 0) = \mathbf{Z}$; which, by the way, appears indirectly in the subsequent arguments.

13.1.3.3. In section 13.3, we compute the global (universal, fuchsian) Galois group for the category $\mathcal{E}_f$ of rational q-difference equations $A \in \mathrm{GL}_n(\mathbf{C}(x))$ that are fuchsian at 0 and ∞. This is done, in some sense, by glueing the local Galois groups $G_f^{(0)}$ and $G_f^{(\infty)}$ (the latter defined in an obvious way) by a process akin to Van Kampen's theorem (which says how to compute $\pi_1(X \cup Y)$ from $\pi_1(X)$, $\pi_1(Y)$, $\pi_1(X \cap Y)$, and the way the latter one embeds into the two former ones). The glueing here is provided by the connection matrix (more precisely, its regular

[10] As we shall see, for *fuchsian* equations, the results are essentially the same in the formal and in the convergent case. This will be very different in the irregular case; see Chapter 15.

values). When the local groups $G_f^{(0)}$ and $G_f^{(\infty)}$ are trivial, we recover Etingof's result. In the general (fuchsian) case, there are complications and we propose two ways to overcome them. One of them involves a twisting of Birkhoff's matrix taking in account the local monodromies encoded in $G_f^{(0)}$ and $G_f^{(\infty)}$; for the other one, we replace Birkhoff's matrix by its central part freed of the local contributions, just like we used M instead of P in section 12.2 of Chapter 12. In either case, and in the same spirit as in some versions of Van Kampen's theorem, we find out that, although in the end we better like groups, the global landscape is better understood in terms of *groupoids*.

As in 13.2, we can then define a nice Zariski-dense subgroup of the global fuchsian Galois group G_f, but the result here is less precise, as we do not have here a countable (let alone finite) set of generators. Actually, the big difference with the differential case is that *we are not presently (2023) able to isolate local contributions to the monodromy* at the intermediate singularities (those in $\mathbf{C}^*$). It can be done under a very strong assumption (abelianity of G_f) and it does yield some interesting information (on the abelianized of G_f), but it gives a complete description only in rather trivial cases. However, in [CV], we shall present a significant application to q-hypergeometric series, due to Julien Roques [**207**].

13.1.3.4. In section 13.3.4, we explore the degeneracy of G_f when $q \to 1$. Of course this rests to a large extent on the confluence results from Chapters 11 and 12. Since we rely on rather computational methods, what we really describe is the confluence of the Zariski-dense subgroup found previously. A really algebraic description of the variation of G_f is to be found in the already quoted work of Yves André [**4**].

13.2. Local fuchsian Galois theory

The irregular theory will be presented in Chapter 15. The results of this section are similar to those of [**17**] but totally orthogonal to those of [**82**].

Here, the field of interest is $K := \mathbf{C}(\{x\})$, while $\hat{K} := \mathbf{C}((x))$ plays an auxiliary role, rather unimportant because, for local fuchsian equations, the formal and analytic classification are the same. In Chapter 15, they will be completely different!

In the next section 13.3, we shall return to the historically most important base field $\mathbf{C}(x)$ and then $\mathbf{C}(\{x\})$ will play an auxiliary role.

13.2.1. Categories.

13.2.1.1. The Galois group (Tannakian style) is deduced from the tensor structure of some category, and, to reduce the argument from one such category to a more manageable one, we need to prove some tensor equivalence. We shall prove that the tensor category $\mathscr{E}_f{}^{(0)}$ of fuchsian q-difference equations at 0, i.e., of q-difference modules M over the q-difference field $\mathbf{C}(\{x\})$ such that $S(M) = \{0\}$, is tensor equivalent to its full category $\mathscr{P}^{(0)}$ of objects (K^n, Φ_A) with $A \in \mathrm{GL}_n(\mathbf{C})$ and then compute the proalgebraic group $G_f^{(0)}$ such that $\mathscr{P}^{(0)} \simeq \mathcal{R}ep_{rat}(G_f^{(0)})$ (for notation $\mathcal{R}ep_{rat}$ see 0.4).

The general underlying principle (see Appendix G) is as follows. Let $\mathscr{C}$ a Tannakian category and ω a fibre functor on $\mathscr{C}$ over $\mathbf{C}$, so that $G := \mathrm{Aut}^{\otimes}(\omega)$ is a proalgebraic group and $\mathscr{C} \simeq \mathcal{R}ep_{rat}(G)$. Let $\mathscr{C}'$ a full subcategory of $\mathscr{C}$ which is stable with respect to $\otimes$ and to dualizing $M \rightsquigarrow M^{\vee}$. Remember that *full* means that $\mathrm{Hom}_{\mathscr{C}'}(A,B) = \mathrm{Hom}_{\mathscr{C}}(A,B)$ for any two objects A, B of $\mathscr{C}'$. Then, if $\mathscr{C}'$ is *essential*, meaning that every object of $\mathscr{C}$ is isomorphic to one in $\mathscr{C}'$, then $\mathscr{C}$ and $\mathscr{C}'$ are equivalent tensor categories and $\mathscr{C}' \simeq \mathcal{R}ep_{rat}(G)$. Still more precisely, the restriction ω' of ω to $\mathscr{C}'$ is a fibre functor and G can be canonically identified to $G' := \mathrm{Aut}^{\otimes}(\omega')$. We shall generally omit such very general sorites in the sequel, leaving it for the reader to fill in the details.

Here is an obvious example. In the category $\mathscr{C}$ of all q-difference modules over some q-difference field (K, σ_q), the subcategory $\mathscr{C}'$ of all (K^n, Φ_A), $A \in \mathrm{GL}_n(K)$, has the quoted properties. The equivalence is so fundamental that we shall identify the two categories without caring to mention it. The same thing is done in linear algebra when one identifies the category of finite-dimensional vector spaces over K (and linear maps) with the category of spaces K^n (and rectangular matrices). Moreover, in order to further alleviate the terminology, we shall speak of the objct $A \in \mathrm{GL}_n(K)$, instead of the object (K^n, Φ_A) (and of morphisms $A \to B$, etc).

13.2.1.2. Beyond ${\mathscr{E}_f}^{(0)}$, and just for the sake of completeness, we denote $\widehat{{\mathscr{E}_f}^{(0)}}$ the tensor category ${\mathscr{E}_f}^{(0)}$ of *formal* fuchsian q-difference equations at 0, i.e., of q-difference modules M over the q-difference field $\mathbf{C}((x))$ such that $S(M) = \{0\}$. We already met (starting from Chapter 8) the formalization functor from ${\mathscr{E}_f}^{(0)}$ to $\widehat{{\mathscr{E}_f}^{(0)}}$: this is just the base change functor from $\mathbf{C}(\{x\})$ to $\mathbf{C}((x))$ (see Section 8.3.3). In matricial description, it is simply described as $(\mathbf{C}(\{x\})^n, \Phi_A) \rightsquigarrow (\mathbf{C}((x))^n, \Phi_A)$ (the matrix stays the same), so the object A of ${\mathscr{E}_f}^{(0)}$ is sent to the object A of $\widehat{{\mathscr{E}_f}^{(0)}}$; and morphism $F : A \to B$ in ${\mathscr{E}_f}^{(0)}$ is likewise sent to morphism $F : A \to B$ in $\widehat{{\mathscr{E}_f}^{(0)}}$.

PROPOSITION 13.3. *The formalization functor* $(\mathbf{C}(\{x\})^n, \Phi_A) \rightsquigarrow (\mathbf{C}((x))^n, \Phi_A)$ *is a tensor equivalence* ${\mathscr{E}_f}^{(0)} \simeq \widehat{{\mathscr{E}_f}^{(0)}}$.

PROOF. All tools for that proof are present in Chapters 8 and 9, so we leave it for the reader to use them to build a proof. Note however that a shortcut could be to prove that both inclusions of $\mathscr{P}^{(0)}$ into ${\mathscr{E}_f}^{(0)}$ and into $\widehat{{\mathscr{E}_f}^{(0)}}$ are tensor equivalences. □

EXERCISE 13.4. Just in order to refresh your basic thechniques, show directly the following: if $A \in \mathrm{GL}_n(\mathbf{C}(\{x\}))$ and $B \in \mathrm{GL}_p(\mathbf{C}(\{x\}))$ and if $F \in \mathrm{Mat}_{p,n}(\mathbf{C}((x)))$ is a morphism $A \to B$, then $F \in \mathrm{Mat}_{p,n}(\mathbf{C}(\{x\}))$.

13.2.1.3. The method used in the proof of the following proposition applies exactly the same for the inclusion functor from $\mathscr{P}^{(0)}$ to $\widehat{{\mathscr{E}_f}^{(0)}}$, and so would provide an alternative argument for the above proposition.

PROPOSITION 13.5. *The inclusion functor* $\mathscr{P}^{(0)} \rightsquigarrow {\mathscr{E}_f}^{(0)}$ *is a tensor equivalence.*

PROOF. This relies on the general principle stated in 13.2.1.1. The only nontrivial point is that $\mathscr{P}^{(0)}$ is essential, i.e., that any fuchsian $A \in \mathrm{GL}_n(\mathbf{C}(\{x\}))$ is gauge equivalent (analytically) to some $A^{(0)} \in \mathrm{GL}_n(\mathbf{C})$; but this is the content of Chapter 11. □

To be of any practical use, this equivalence has to be completed:

LEMMA 13.6. *Let A, B objects in $\mathscr{P}^{(0)}$ and $F : A \to B$ a morphism in $\mathscr{P}^{(0)}$ or ${\mathscr{E}_f}^{(0)}$ (by definition, they are the same). Then F is a Laurent polynomial:*

$$F = \sum_{k=k_0}^{k_1} F_k x^k, \text{ all } F_k \in Mat_{p,n}(\mathbf{C}).$$

PROOF. Let $F = \sum F_k x^k$ a Laurent power series expansion. The relation $(\sigma_q F)A = BF$ is equivalent to: $q^k F_k A = BF_k$ for all $k \in \mathbf{Z}$. The equation $q^k XA = BX$ has a nontrivial solution X only if the spectra of $q^k A$ and of B intersect (see about $\Phi_{M,N}$ in Section F.4 of Appendix F), which can only happen for a finite number of values of k. □

COROLLARY 13.7. *A morphism F in $\mathscr{P}^{(0)}$ is holomorphic all over $\mathbf{C}^*$; an isomorphism is regular over $\mathbf{C}^*$ (i.e., F and F^{-1} are holomorphic there).*

Clearly the categories $\mathscr{P}^{(0)}$ and ${\mathscr{E}_f}^{(0)}$ are $\mathbf{C}$-linear abelian tensor categories. From Chapter 9, they are rigid (terminology from Appendix G). So, to prove that they are Tannakian, we just need to define a fibre functor. This will come in 13.2.3; but first, we shall tackle a baby example: it actually comes from the Galois theory of differential equations (see [**222**]), and, in 13.2.2, it will be seen to be relevant for our goals.

CAUTION 13.8. We use the subscript "f" in ${\mathscr{E}_f}^{(0)}$, $G_f^{(0)}$ ..., to mean *fuchsian*, not *formal*. In the fuchsian case, per Proposition 13.3, we need no particular denotation for "formal". In the irregular case, we have seen that "formal" is equivalent to "pure" (see 10.3), so we shall use the subscript "p".

On the other hand, in 13.2.6 we shall deal with so-called *flat* vector bundles. The French for "flat" is "plat", whence the letter $\mathscr{P}$ for the category $\mathscr{P}^{(0)}$, and, in particular places, the subscript p to denote flatness.

Last, to make the confusion more complete, "f" sometimes means "finite", as in $\mathcal{V}ect_{\mathbf{C}}^f$!

13.2.2. A baby example: $\mathcal{R}ep_{\mathbf{C}}^f(\mathbf{Z}) = \mathcal{R}ep_{rat}(\mathbf{Z}^{alg})$. (For notations $\mathcal{R}ep_{\mathbf{C}}^f$ and $\mathcal{R}ep_{rat}$; see 0.4 and Appendix G.) This section is intended as a good training in practical Tannakian calculations; moreover, it is a useful step toward the computation of $G_f^{(0)}$. We intend to find a proalgebraic group $\mathbf{Z}^{alg}$ such that the category $\mathcal{R}ep_{\mathbf{C}}^f(\mathbf{Z})$ of finite-dimensional complex representations of $\mathbf{Z}$ identifies naturally to the category $\mathcal{R}ep_{rat}(\mathbf{Z}^{alg})$ of *rational* finite-dimensional complex representations of $\mathbf{Z}^{alg}$.

13.2.2.1. Let $\mathcal{R}ep_{\mathbf{C}}^f(\mathbf{Z})$ the category of finite-dimensional complex representations of $\mathbf{Z}$. Any such object is equivalent to some morphism $\mathbf{Z} \to \mathrm{GL}(V)$, V in $\mathcal{V}ect_{\mathbf{C}}^f$, or even, more precisely, $\mathbf{Z} \to \mathrm{GL}_n(\mathbf{C})$; and the latter is in turn determined by the image of 1, say $\phi \in \mathrm{Aut}(V)$ or $A \in \mathrm{GL}_n(\mathbf{C})$. So we encode objects of $\mathcal{R}ep_{\mathbf{C}}^f(\mathbf{Z})$ either as pairs (V, ϕ) or as matrices A (not to be confused with matrices encoding q-difference equations; however, see 13.2.2.2). A morphism from (V, ϕ) to (W, ψ), resp. from $A \in \mathrm{GL}_n(\mathbf{C})$ to $B \in \mathrm{GL}_p(\mathbf{C})$, is a linear map $f : V \to W$ intertwining ϕ and ψ, i.e., such that $f \circ \phi = \psi \circ f$; resp. a matrix $F \in \mathrm{Mat}_{p,n}(\mathbf{C})$ such that $FA = BF$.

Note that $\mathcal{R} := \mathcal{R}ep^f_{\mathbf{C}}(\mathbf{Z})$ embeds into $\mathscr{P}^{(0)}$ as a subcategory by sending object $A \in \mathrm{GL}_n(\mathbf{C})$ to itself and $F \in \mathrm{Mat}_{p,n}(\mathbf{C})$ to $F \in \mathrm{Mat}_{p,n}(\mathbf{C}(\{x\}))$. The objects are the same, but there are more morphisms in $\mathscr{P}^{(0)}$; for instance, $x : 1 \to q$.

EXERCISE 13.9. There are no nontrivial morphisms $1 \to q$ in $\mathcal{R}$. Give a condition for the existence of a nontrivial morphism $A \to B$ in $\mathcal{R}$, resp. in $\mathscr{P}^{(0)}$.

Also note that exact sequences, tensor products[11], dual, etc, are the same in $\mathcal{R}$ and in $\mathscr{P}^{(0)}$, and we summarize all that by saying that $\mathcal{R} \rightsquigarrow \mathscr{P}^{(0)}$ is a faithful exact $\mathbf{C}$-linear tensor compatible functor.

Now, $\mathcal{R}$ is endowed with an obvious fibre functor $\omega : \mathcal{R} \rightsquigarrow \mathcal{V}ect^f_{\mathbf{C}}$. In intrinsic terms, it sends $\rho : \mathbf{Z} \to \mathrm{GL}(V)$ to V, and the morphism $f : \rho \to \rho'$ to the underlying linear map $f : V \to V'$. In matricial terms, it sends $A \in \mathrm{GL}_n(\mathbf{C})$ to $\mathbf{C}^n$ and $F \in \mathrm{Mat}_{p,n}(\mathbf{C})$ to the linear map $F : \mathbf{C}^n \to \mathbf{C}^p$. (Recall from Appendix G that the meaning of "fibre functor" is that it is faithful, exact, $\mathbf{C}$-linear and tensor compatible.).

EXERCISE 13.10. Check that ω is not fully faithful.

The Galois group of $\mathcal{R}$ (Tannakian style) is the proalgebraic group $G := \mathrm{Aut}^{\otimes}(\omega)$, which we presently proceed to describe in concrete matricial terms (the letter G is a temporary notation). First, an element ϕ of G (thus a tensor automorphism of ω):

- consists of a collection $\{\phi_A\}$, where, for each $A \in \mathrm{GL}_n(\mathbf{C})$ (n arbitrary), $\phi_A \in \mathrm{GL}_n(\mathbf{C})$ (in more intrinsic terms, to each $\rho : \mathbf{Z} \to \mathrm{GL}(V)$ is associated a $\phi_\rho \in \mathrm{GL}(\omega(\rho)) = \mathrm{GL}(V)$);
- is submitted to a functoriality condition: if $F : A \to B$ is a morphism, then we have a commutative square as here below (in other words, $FA = BF \Rightarrow F\phi_A = \phi_B F$);

$$\begin{array}{ccc} \mathbf{C}^n & \xrightarrow{\phi_A} & \mathbf{C}^n \\ {\scriptstyle F}\downarrow & & \downarrow{\scriptstyle F} \\ \mathbf{C}^p & \xrightarrow[\phi_B]{} & \mathbf{C}^p \end{array}$$

- and to so-called tensor compatibility: with the usual conventions and up to the identification of $\omega(A \otimes B) = \omega(A) \otimes \omega(B)$, we have $\phi_{A\otimes B} = \phi_A \otimes \phi_B$ (the former should belong to $\mathrm{GL}(\omega(A \otimes B))$, the latter to $\mathrm{GL}(\omega(A) \otimes \omega(B))$).

Then the group law on G is defined componentwise by the group laws on all the linear groups $\mathrm{GL}(\omega(A))$, resp. (in intrinsic terms) of the $\mathrm{GL}(\omega(\rho)) = \mathrm{GL}(V)$:

$$\{\phi_A\}\{\psi_A\} := \{\phi_A \circ \psi_A\}, \text{ resp. } \{\phi_\rho\}\{\psi_\rho\} := \{\phi_\rho \circ \psi_\rho\}.$$

EXERCISE 13.11. (i) Describe the neutral element of G.
(ii) Check that setting $\phi_A := A$ for all A one gets an element $\{\phi_A\}$ of G.
(iii) Using the group law of G, deduce from (ii) a canonical embedding $\mathbf{Z} \to G$.

[11] Recall how tensor products of matrices were encoded as matrices in Chapters 8 and 9.

We intend to determine the precise structure of the group G and show how $\mathcal{R}$ identifies to the category of rational representations of G. Before that, we will describe that identification in more or less tautological terms (independent of the particular structure).

First, fix an object $A_0 \in \mathrm{GL}_n(\mathbf{C})$ of $\mathcal{R}$. To every $\phi = \{\phi_A\}$ of G, associate its component $\phi_{A_0} \in \mathrm{GL}(\omega(A_0)) = \mathrm{GL}_n(\mathbf{C})$. From the description of the group law on G, it follows that the map $\phi \mapsto \phi_{A_0}$, $G \to \mathrm{GL}_n(\mathbf{C})$, is a group representation. The general theory guarantees that it is a rational representation. We denote it ρ_{A_0}. Second, let $F : A_0 \to B_0$ a morphism in $\mathcal{R}$. It follows from the functoriality condition that $F\phi_{A_0} = \phi_{B_0}F$, i.e., $F\rho_{A_0}(\phi) = \rho_{B_0}(\phi)F$, i.e., the linear map $F : \mathbf{C}^n \to \mathbf{C}^p$ intertwines the representations ρ_{A_0} and ρ_{B_0}, i.e., it is a morphism of representations of G. So we have a functor from $\mathcal{R}$ to $\mathcal{R}ep_{rat}(G)$. The general theory guarantees that this indeed yields a tensor equivalence from $\mathcal{R}$ to the category $\mathcal{R}ep_{rat}(G)$ of rational representations of G.

EXERCISE 13.12. Do the same constructions in intrinsic terms, starting from an object of $\mathcal{R}$ described as $\rho_0 : \mathbf{Z} \to \mathrm{GL}(V_0)$ instead of $A_0 \in \mathrm{GL}_n(\mathbf{C})$.

13.2.2.2. Using Exercise 13.11, we obtain an embedding from $\mathbf{Z}$ into $G = \mathrm{Aut}^{\otimes}(\omega)$; it sends $m \in \mathbf{Z}$ to the element $\phi = \{\phi_A\}$ given by $\phi_A := A^m$ for all A.

More generally, let $\lambda \in \mathbf{C}^*$ and $\gamma \in \mathrm{Hom}_{gr}(\mathbf{C}^*, \mathbf{C}^*)$. (The case above corresponds to the special choices $\lambda := n$ and $\gamma : x \mapsto x^n$.) For any $A \in \mathrm{GL}_n(\mathbf{C})$ and $B \in \mathrm{GL}_p(\mathbf{C})$, by F.4, Appendix F:

- $\gamma(A_s)$ and A_u^λ belong to $\mathbf{C}[A]$, in particular they commute to each other.
- If $F \in \mathrm{Mat}_{p,n}(\mathbf{C})$ is such that $FA = BF$, then $F\gamma(A_s) = \gamma(B_s)F$ and $FA_u^\lambda = B_u^\lambda F$.

It follows that, setting $\phi_A := \gamma(A_s)A_u^\lambda$ for all A, the functoriality condition for an element of G is satisfied. The tensor compatibility flows from the formulas:

$$(A \otimes B)_s = A_s \otimes B_s, \quad (A \otimes B)_u = A_u \otimes B_u,$$

and, for any semisimple D_1, D_2 and unipotent U_1, U_2:

$$\gamma(D_1 \otimes D_2) = \gamma(D_1) \otimes \gamma(D_2), \quad (U_1 \otimes U_2)^\lambda = U_1^\lambda \otimes U_2^\lambda.$$

So indeed $\phi \in G$.

EXERCISE 13.13. Let $\phi' \in G$ be similarly defined from $\lambda' \in \mathbf{C}^*$ and $\gamma' \in \mathrm{Hom}_{gr}(\mathbf{C}^*, \mathbf{C}^*)$. Show that $\phi'_A \circ \phi_A = (\gamma\gamma')(A_s)A_u^{\lambda+\lambda'}$. (Recall that the product in $\mathrm{Hom}_{gr}(\mathbf{C}^*, \mathbf{C}^*)$ is defined by $(\gamma\gamma')(x) = \gamma(x)\gamma'(x)$.)

From that exercice, we deduce that the map just defined $(\lambda, \gamma) \mapsto \phi$ from $\mathbf{C} \times \mathrm{Hom}_{gr}(\mathbf{C}^*, \mathbf{C}^*)$ to G is a morphism of groups. From now on, we shall use the notation $\mathbf{C}^{*\vee}$ for $\mathrm{Hom}_{gr}(\mathbf{C}^*, \mathbf{C}^*)$ (see in Appendix G, section G.1.1 the definition of the proalgebraic structure on $\Gamma^\vee := \mathrm{Hom}_{gr}(\Gamma, \mathbf{C}^*)$ for an arbitrary group Γ).

THEOREM 13.1. *Let $\mathbf{Z}^{alg} := \mathbf{C} \times \mathbf{C}^{*\vee}$, seen as a proalgebraic group. The above defined map $(\lambda, \gamma) \mapsto \phi$ from $\mathbf{Z}^{alg}$ to $Aut^{\otimes}(\omega)$ is an isomorphism of proalgebraic groups.*

PROOF. We just checked that it was a morphism. The fact that it is an isomorphism is proved in **[218, 222]** and we shall admit it here (but see the following exercise). Some details on the proalgebraic structure of $\mathbf{Z}^{alg}$ will be given soon. □

EXERCISE 13.14. Prove the injectivity.

General Tannakian theory says that $\mathcal{R}$ is $\otimes$-equivalent to the category $\mathcal{R}ep_{rat}(\mathbf{Z}^{alg})$ of rational representations of $\mathbf{Z}^{alg}$. Without going into theoretical explanations, let us try to explain what that means. We look for pairs of rational representations $\mathbf{C} \to \mathrm{GL}_n(\mathbf{C})$ and $\mathbf{C}^{*\vee} \to \mathrm{GL}_n(\mathbf{C})$ with commuting images.

Clearly the first factor $\mathbf{C}$ of $\mathbf{Z}^{alg}$ is an algebraic group. Its rational representations are well known (see **[222]**): in matricial form, they are all representations $\mathbf{C} \to \mathrm{GL}_n(\mathbf{C})$ which have the form $\lambda \mapsto U^\lambda$ for some unipotent matrix U.

As for representations of the second factor $\mathbf{C}^{*\vee}$, note first that any $\Gamma^\vee$ is the inverse limit of the groups $\Gamma_i^\vee$ where Γ_i runs among the finitely generated subgroups of Γ (this is of course because Γ is the direct limit of the Γ_i). The proalgebraic structure of $\Gamma^\vee$ is defined in such a way that a representation $\Gamma^\vee \to \mathrm{GL}_n(\mathbf{C})$ is rational whenever it factors through a rational representation of one of the algebraic groups $\Gamma_i^\vee$.

We now describe the equivalence of $\mathcal{R}$ with $\mathcal{R}ep_{rat}(\mathbf{Z}^{alg})$. One way is obvious: since $\mathbf{Z}$ embeds into $\mathbf{Z}^{alg}$, any representation (rational or not) of the latter restricts to a representation of the former. The other way is more interesting and instructive, it rests on the following fact: for any $A \in \mathrm{GL}_n(\mathbf{C})$, the only rational representation of $\mathbf{Z}^{alg}$ restricting to $m \mapsto A^m$ (for $m \in \mathbf{Z}$) is the representation $(\lambda, \gamma) \mapsto \gamma(A_s)A_u^\lambda$.

13.2.3. Returning to ${\mathcal{E}_f}^{(0)}$ and $\mathcal{P}^{(0)}$: Fibre functors.

13.2.3.1. Since $\mathcal{P}^{(0)} \hookrightarrow {\mathcal{E}_f}^{(0)}$ is a tensor equivalence of Tannakian categories, it is logically sufficient to define fibre functors on $\mathcal{P}^{(0)}$. This is indeed easier and that's what we shall do; see however in 13.2.3.5 to what extent it can be done on ${\mathcal{E}_f}^{(0)}$.

Let $a \in \mathbf{C}^*$. Selecting this particular a is analogous to the choice of a base point x_0 when defining a fundamental group $\pi_1(X, x_0)$ (we shall return to that). We define a functor:

$$\omega_a : \mathcal{P}^{(0)} \rightsquigarrow \mathcal{V}ect^f_{\mathbf{C}}, \begin{cases} (A \in \mathrm{GL}_n(\mathbf{C})) \rightsquigarrow \mathbf{C}^n, \\ (F : A \to B) \rightsquigarrow (F(a) : \mathbf{C}^n \to \mathbf{C}^p). \end{cases}$$

PROPOSITION 13.15. *The functor ω_a is faithful, exact, $\mathbf{C}$-linear and $\otimes$-compatible.*

PROOF. All properties are obvious from the definition of the abelian and tensor structure on ${\mathcal{E}_f}^{(0)}$ in Chapter 8 and 9, except for the faithfulness. We must check that, if $F : A \to B$ is a morphism such that $F(a) = 0$, then $F = 0$. The relation $(\sigma_q F)A = BF$, written in the form $F(qx) = BF(x)A^{-1}$, implies that $F(q^m a) = 0$ for all $m \in \mathbf{Z}$, so that there are infinitely many zeroes of F. Since, by Lemma 13.6, the latter is a Laurent polynomial, it must be 0. □

EXERCISE 13.16. Show that ω_a is not fully faithful. (Take $A := 1$ and $B := c \in \mathbf{C}^*$.) More generally, compute the rank of the linear map $\mathrm{Hom}_{\mathcal{P}^{(0)}}(A, B) \to \mathcal{L}(\mathbf{C}^n, \mathbf{C}^p)$ induced by ω_a.

In Tannakian theory, $\mathrm{Aut}^\otimes(\omega_a)$ is considered as the Galois group of $\mathcal{P}^{(0)}$, hence of ${\mathcal{E}_f}^{(0)}$. It depends[12] on a, and this dependency is rather innocuous. Analogously,

[12]In general Tannakian theory, the Galois group $\mathrm{Aut}^\otimes(\omega)$ depends on the choice of a particular fibre functor ω and it is the set of all those fibre functors that acts as the base of the Galois groupoid.

in algebraic topology, if the space X is arcwise connected, all $\pi_1(X, x_0)$ are isomorphic; but, if X is not simply connected, the isomorphisms are not unique in general (they are unique if the groups $\pi_1(X, x_0)$ are commutative). In our setting, the corresponding isomorphisms between the groups $\mathrm{Aut}^{\otimes}(\omega_a)$ will happen to be unique *in the local fuchsian case* (this is because all those groups are commutative). However, since this will no longer be the case for irregular q-difference equations (see Chapter 15) nor, by the way, in the global fuchsian case (see Section 13.3), it is maybe sensible to already introduce the groupoid of all $\mathrm{Iso}^{\otimes}(\omega_a, \omega_b)$.

CAUTION 13.17. Here, ω is a functor on $\mathcal{R}$ while the ω_a are functors on $\mathscr{P}^{(0)}$.

13.2.3.2. According to the general definition (see Appendix G), a "galoisian path" joining the "points" ω_a, ω_b in the base is an element $\phi \in \mathrm{Iso}^{\otimes}(\omega_a, \omega_b)$, and this, in turn, is defined by the following data and constraints:

- For each object $A \in \mathrm{GL}_n(\mathbf{C})$ of $\mathscr{P}^{(0)}$, an isomorphism:
$$\phi_A : \omega_a(A) = \mathbf{C}^n \to \omega_b(A) = \mathbf{C}^n, \text{ i.e. } \phi_A \in \mathrm{GL}_n(\mathbf{C}).$$
- Functoriality: every morphism $F : A \to B$ in $\mathscr{P}^{(0)}$ satisfies $F(b)\phi_A = \phi_B F(a)$ (both sides belong to $\mathrm{Mat}_{p,n}(\mathbf{C})$).
- $\otimes$-compatibility: up to usual identifications, $\phi_{A\otimes B} = \phi_A \otimes \phi_B$.

EXERCISE 13.18. Check that, if $\{\phi_A\} \in \mathrm{Iso}^{\otimes}(\omega_a, \omega_b)$ and $\{\psi_A\} \in \mathrm{Iso}^{\otimes}(\omega_b, \omega_c)$, then $\{\psi_A \circ \phi_A\} \in \mathrm{Iso}^{\otimes}(\omega_a, \omega_c)$.

We shall describe explicitly the elements of $\mathrm{Iso}^{\otimes}(\omega_a, \omega_b)$. First, we shall prove two lemmas.

LEMMA 13.19. *$Iso^{\otimes}(\omega_a, \omega_b)$ naturally embeds into $Aut^{\otimes}(\omega) = \mathbf{Z}^{alg}$ and those embeddings are collectively compatible with the composition laws.*

PROOF. Since all $\omega_a : \mathscr{P}^{(0)} \rightsquigarrow \mathcal{V}ect^f_{\mathbf{C}}$ restrict to $\omega : \mathcal{R} \rightsquigarrow \mathcal{V}ect^f_{\mathbf{C}}$, there is an obvious inclusion $\mathrm{Iso}^{\otimes}(\omega_a, \omega_b) \hookrightarrow \mathrm{Aut}^{\otimes}(\omega)$. Actually, the data described above are exactly the same as those described in the baby example 13.2.2, the only difference being that the functoriality constraint is now more restrictive: it applies to all morphisms F in $\mathscr{P}^{(0)}$, but, if F is a morphism in $\mathcal{R}$ (i.e., $F \in \mathrm{Mat}_{p,n}(\mathbf{C})$), then $F(a) = F(b) = F$ and the constraint here boils down to the one in 13.2.2.1. Also the composition laws described here and there (i.e., in 13.2.2) also are the same. □

EXERCISE 13.20. Describe in abstract algebraic terms the relationship between the groupoid of all $\mathrm{Iso}^{\otimes}(\omega_a, \omega_b)$ and the group $\mathrm{Aut}^{\otimes}(\omega)$.

It follows in particular from Lemma 13.19 that each Galois group $\mathrm{Aut}^{\otimes}(\omega_a)$ embeds as a subgroup of $\mathrm{Aut}^{\otimes}(\omega)$, i.e., of $\mathbf{Z}^{alg}$. This is a particular case of [**56**, Proposition 2.21 (b)], which moreover tells us that it embeds as a *closed* subgroup. This last fact will be verified directly as a consequence of the explicit description.

13.2.3.3. It follows from Lemma 13.19 and from 13.2.2 that an element $\{\phi_A\}$ of $\mathrm{Iso}^{\otimes}(\omega_a, \omega_b)$ can be identified to a particular element of $\mathrm{Aut}^{\otimes}(\omega)$, hence described as $A \rightsquigarrow \gamma(A_s)A_u^{\lambda}$ for some $(\gamma, \lambda) \in \mathbf{Z}^{alg}$, i.e., $\lambda \in \mathbf{C}$ and $\gamma \in \mathbf{C}^{*\vee}$; and that it is moreover submitted to a stronger functoriality constraint. To experiment with the latter, we test the morphism $x : 1 \to q$ in $\mathscr{P}^{(0)}$; the relation $F(b)\phi_A = \phi_B F(a)$ here writes:
$$b\gamma(1) = \gamma(q)a \Longleftrightarrow \gamma(q) = b/a.$$

We are going to prove that this is indeed the sole condition imposed by the stronger functoriality.

LEMMA 13.21. *Let* $F := \sum F_k x^k$ *a Laurent polynomial with coefficients in* $Mat_{p,n}(\mathbf{C})$. *Then* F *is a morphism* $A \to B$ *in* $\mathscr{P}^{(0)}$ *if, and only if, each* F_k *is a morphism* $q^k A \to B$ *in* $\mathscr{P}^{(0)}$ *or equivalently in* $\mathcal{R}$.

PROOF. A mere computation:

$$(\sigma_q F)A = BF \Longleftrightarrow \left(\sum F_k q^k x^k\right) A = B \sum F_k x^k \Longleftrightarrow \forall k\ ,\ F_k q^k A = BF_k.$$

□

THEOREM 13.2. *The pair* $(\gamma, \lambda) \in \mathbf{Z}^{alg}$ *defines (as above) an element of* $Iso^{\otimes}(\omega_a, \omega_b)$ *if, and only if,* $\gamma(q) = b/a$.

PROOF. We only have to prove the "if" part, and, for that, to check the stronger functoriality. So let $F : A \to B$ a morphism in $\mathscr{P}^{(0)}$. By Lemma 13.6 it is a Laurent polynomial $\sum F_k x^k$ and by the above lemma all $F_k : q^k A \to B$ are morphisms in $\mathcal{R}$, whence (by functoriality in $\mathcal{R}$) commutative diagrams:

$$\begin{array}{ccc} \mathbf{C}^n & \xrightarrow{\phi_{q^k A}} & \mathbf{C}^n. \\ {\scriptstyle F_k}\downarrow & & \downarrow{\scriptstyle F_k} \\ \mathbf{C}^p & \xrightarrow[\phi_B]{} & \mathbf{C}^p \end{array}$$

But $q^k A = (q^k) \otimes A$ (the left factor is a 1×1 matrix), so, by $\otimes$-compatibility:

$$\phi_{q^k A} = \phi_{(q^k)} \otimes \phi_A = \phi_{(q^k)} \phi_A = \gamma(q^k) \phi_A = (b/a)^k \phi_A,$$

so the commutative diagram entails $(b/a)^k F_k \phi_A = \phi_B F_k \Rightarrow b^k F_k \phi_A = \phi_B F_k a^k$. Summing for all k, we get the wanted equality $F(b)\phi_a = \phi_B F(a)$ (i.e., functoriality in $\mathscr{P}^{(0)}$). □

COROLLARY 13.22. (i) *The Galois groupoid of all* $Iso^{\otimes}(\omega_a, \omega_b)$ *is commutative and connected.*

(ii) *The Galois group* $Aut^{\otimes}(\omega_a)$ *is independent of* $a \in \mathbf{C}^*$.

(iii) *It is a closed subgroup of* $\mathbf{Z}^{alg}$ *(characterized by the condition* $\gamma(q) = 1$ *on* (γ, λ)*, see further below).*

PROOF. (i) Both properties are immediate from the theorem, connectedness requiring moreover the conclusion of question (i) in the following exercise (applied to $c := b/a$).

(ii) is an immediate consequence of the commutativity.

(iii) By definition of the proalgebraic structure, closedness is checked in the finite components, so here it flows from the conclusion of question (ii) in the following exercise.

□

EXERCISE 13.23. (i) Let $c \in \mathbf{C}^*$; then there exists $\gamma \in \mathbf{C}^{*\vee}$ such that $\gamma(q) = c$.

(ii) Let $\Gamma \subset \mathbf{C}^*$ a finitely generated subgroup and let $\alpha \in \Gamma$, $\beta \in \mathbf{C}^*$. Then $\{\gamma \in \Gamma^\vee \mid \gamma(\alpha) = \beta\}$ is closed (for the Zariski topology).

13.2.3.4. *Structure of* $G_f^{(0)}$. As follows implicitly from the remark at the very beginning of 13.2.3.1, since $\mathscr{P}^{(0)} \hookrightarrow \mathscr{E}_f{}^{(0)}$ id an equivalence of $\otimes$-categories, we can deduce the Galois group $G_f^{(0)}$ of $\mathscr{E}_f{}^{(0)}$ from the Galois groupoid of $\mathscr{P}^{(0)}$, i.e., of all $\mathrm{Iso}^{\otimes}(\omega_a, \omega_b)$ found in 13.2.3.3; so $G_f^{(0)}$ may be identified to any of the $\mathrm{Aut}^{\otimes}(\omega_a)$, and the identification will be canonical since, as we just saw, the groupoid is commutative.

COROLLARY 13.24. *As a proalgebraic group, split into its semisimple and unipotent parts:*

$$G_f^{(0)} = G_{f,s}^{(0)} \times G_{f,u}^{(0)}, \text{ where } \begin{cases} G_{f,s}^{(0)} := \{\gamma \in \mathbf{C}^{*\vee} \mid \gamma(q) = 1\}, \\ G_{f,u}^{(0)} := \mathbf{C}. \end{cases}$$

The elements of $G_{f,s}^{(0)}$ are the group morphisms $\gamma : \mathbf{C}^* \to \mathbf{C}^*$ such that $\gamma(q) = 1$, i.e., those such that $q^{\mathbf{Z}} \subset \mathrm{Ker}\ \gamma$. They can be identified to group morphisms[13] $\mathbf{E}_q = \mathbf{C}^*/q^{\mathbf{Z}} \to \Gamma$; beware here $\mathbf{E}_q$ just denotes an "abstract" group, meaning that we do not take in account other structures like topological or algebraic or Lie group. So we have a natural identification:

$$G_{f,s}^{(0)} = \mathbf{E}_q^{\vee}.$$

From 13.2.2, we can describe the representation of $G_f^{(0)}$ associated to an object $A \in \mathrm{GL}_n(\mathbf{C})$ of $\mathscr{P}^{(0)}$:

$$\rho_A : (\lambda, \gamma) \mapsto A_u^{\lambda} \gamma(A_s).$$

By general Tannakian theory:

$$G_f^{(0)} = \varprojlim \mathrm{Gal}(A), \text{ where } \mathrm{Gal}(A) = G_f^{(0)}(A) = \mathrm{Im}\ \rho_A = \{A_u^{\lambda} \gamma(A_s) \mid \lambda \in \mathbf{C}, \gamma \in \mathbf{E}_q^{\vee}\}.$$

By the remark at the beginning of 13.2.2.2, $\gamma(A_s)$ and A_u^{λ} belong to $\mathbf{C}[A]$ for all $\gamma \in \mathbf{C}^{*\vee}$ and $\lambda \in \mathbf{C}$, so that, $\mathrm{Gal}(A) \subset \mathbf{C}[A]$. In particular, it is commutative. (Of course, it also follows from the fact that $G_f^{(0)}$ is.) This implies that it splits as a product of it semisimple and unipotent part **[28]**. They are easy to describe:

- The unipotent component is $\{A_u^{\lambda} \mid \lambda \in \mathbf{C}\}$. If $A_u = I_n$, that is, if A is semisimple, that component is trivial. Otherwise, writing $A_u = I_n + N$, where N has index of nilpotency k (i.e., the dimension of $\mathbf{C}[N] = \mathbf{C}[A_u]$), it is isomorphic to $\mathbf{C}$ through the morphism:

$$\lambda \mapsto A_u^{\lambda} = I_n + \lambda N + \cdots + \binom{\lambda}{k-1} N^{k-1}.$$

- The semisimple component is $\{\gamma(A_s) \mid \gamma \in \mathbf{E}_q^{\vee}\}$. Writing as usual $A_s = P\mathrm{Diag}(c_1, \ldots, c_n)P^{-1}$, we find:

$$\{P\mathrm{Diag}(d_1, \ldots, d_n)P^{-1} \mid (d_1, \ldots, d_n) \text{ is a "}q\text{-replica" of } (c_1, \ldots, c_n)\}.$$

[13]Identifying morphisms $\mathbf{E}_q \to \mathbf{C}^*$ with morphisms $\mathbf{C}^* \to \mathbf{C}^*$ which vanish on $q^{\mathbf{Z}}$ may lead to some ambiguity, since such a γ could be applied either to a nonzero complex number or to its image in $\mathbf{E}_q$; we cross fingers in the hope that no confusion will thus arise. (Besides it is not worse than working with integers modulo m, as is often done in number theory, e.g. dealing with quadratic residues; or working with functions defined almost everywhere, as is often done in real analysis, e.g. dealing with Lebesgue integrals).

We shall now explain what are "q-replicas" and characterize them. First recall from [**222**, 14.1] that a replica[14] of $(c_1, \dots, c_n)$ is, equivalently:

- any $(\gamma(c_1), \dots, \gamma(c_n))$, where $\gamma \in \mathbf{C}^{*\vee}$;
- any $(d_1, \dots, d_n)$ such that:

$$\forall (k_1, \dots, k_n) \in \mathbf{Z}^n\,,\; c_1^{k_1} \cdots c_n^{k_n} = 1 \Longrightarrow d_1^{k_1} \cdots d_n^{k_n} = 1.$$

We call *q-replica* of $(c_1, \dots, c_n)$ any n-uple satisfying the following properties.

LEMMA 13.25. *Let $(c_1, \dots, c_n), (d_1, \dots, d_n) \in \mathbf{C}^{*n}$. The following conditions are equivalent:*

(a) *For all $(k_1, \dots, k_n) \in \mathbf{Z}^n$, if $c_1^{k_1} \cdots c_n^{k_n} \in q^{\mathbf{Z}}$, then $d_1^{k_1} \cdots d_n^{k_n} = 1$.*
(b) *There exists $\gamma \in \mathbf{E}_q^\vee$ such that $(d_1, \dots, d_n) = (\gamma(c_1), \dots, \gamma(c_n))$.*

PROOF. Indeed, both conditions are equivalent to: $(d_1, \dots, d_n, 1)$ is a replica of $(c_1, \dots, c_n, q)$. □

From *loc.cit.* it also follows that the semisimple component of $G_f^{(0)}$ is indeed (as it should!) an algebraic subgroup of $\mathrm{GL}_n(\mathbf{C})$.

EXERCISE 13.26. What are the q-replicas of $(c_1, \dots, c_n)$ if $c_1^2 = \cdots = c_n^2 = q$?

13.2.3.5. *Fibre functors on ${\mathscr{E}_f}^{(0)}$*. If A is an object of ${\mathscr{E}_f}^{(0)}$ (instead of $\mathscr{P}^{(0)}$), it is still possible to define $\omega_a(A)$ as $\mathbf{C}^n$; but, if $F : A \to B$ is a morphism in ${\mathscr{E}_f}^{(0)}$ (instead of $\mathscr{P}^{(0)}$), it may happen the radius of convergence of F is smaller than $|a|$, so that $F(a)$ is not defined. To overcome that apparent difficulty, we introduce some (slightly) more involved definitions.

For every $r > 0$, let $\mathscr{C}_r$ the full subcategory[15] of ${\mathscr{E}_f}^{(0)}$ made of objects A which are regular (i.e., holomorphic with holomorphic inverse) on the punctured disk $\dot{\mathrm{D}}(0, r)$. Clearly, for $0 < r < r'$:

$$\mathscr{P}^{(0)} \subset \mathscr{C}_{r'} \subset \mathscr{C}_r \subset {\mathscr{E}_f}^{(0)} \text{ (full embeddings)},$$

and ${\mathscr{E}_f}^{(0)} = \varinjlim \mathscr{C}_r$. The following facts are then formally clear and not to hard to rigorously state and check:

(1) If $F : A \to B$ is a morphism in $\mathscr{C}_r$ and $|a| < r$, then $F(a)$ is well defined, and contributes to a fibre functor ω_a on $\mathscr{C}_r$.
(2) Let G_r the groupoid of all $\mathrm{Iso}^{\otimes}(\omega_a, \omega_b)$ in $\mathscr{C}_r$ with $|a|, |b| < r$. Then $G_{r'}$ embeds into G_r for $r' > r$, in such a way that $G_r(a, b) = G_{r'}(a, b)$.
(3) For any choice of the a_r such that $|a_r| < r$, the corresponding Galois groups $G_r(a_r, a_r)$ form an inverse system of proalgebraic groups, the corresponding morphisms being isomorphisms; and $G_f^{(0)}$ is the inverse limit.

Practically, given an object A of ${\mathscr{E}_f}^{(0)}$, we look for an equivalent object $A^{(0)} \in \mathrm{GL}_n(\mathbf{C})$ of $\mathscr{P}^{(0)}$ and set $\mathrm{Gal}(A) := \mathrm{Gal}(A^{(0)})$. However, for the general study, we shall stick to fibre functors ω_a on $\mathscr{P}^{(0)}$.

EXERCISE 13.27. (i) State and prove rigorously all the facts above.
(ii) Describe an algorithm for the "practical" procedure when $n = 2$.

13.2.4. Representations of the local fuchsian Galois group.

---

[14] Compare the present definition of (classical) replicas to that in [**232**, 5.14].
[15] This notation will be used in this subsection only.

13.2.4.1. Fix $a \in \mathbf{C}^*$, whence a fibre functor ω_a on $\mathscr{P}^{(0)}$ and an identification of $G_f^{(0)}$ with $\mathrm{Aut}^{\otimes}(\omega_a)$. For every $\phi = \{\phi_A\} \in G_f^{(0)}$, each ϕ_A belongs to the same linear group $\mathrm{GL}_n(\mathbf{C})$ as A. Conversely, if we fix a particular object $A \in \mathrm{GL}_n(\mathbf{C})$ of $\mathscr{P}^{(0)}$, the map $\phi \mapsto \phi_A$ is a rational representation ρ_A of $G_f^{(0)}$ in $\mathrm{GL}_n(\mathbf{C})$. Under the previous identification of $G_f^{(0)}$ as a subgroup of $\mathbf{Z}^{alg}$ in 13.2.3.3:

$$\forall(\gamma,\lambda) \in \mathbf{E}_q^{\vee} \times \mathbf{C}\ ,\ \rho_A(\gamma,\lambda) = \gamma(A_s)A_u^{\lambda}.$$

(Recall that in the notation $\gamma(A_s)$, we see $\gamma \in \mathbf{E}_q^{\vee}$ as an element of $\mathbf{C}^{*\vee}$ such that $\gamma(q) = 1$.) By Tannakian theory, $A \rightsquigarrow \rho_A$ is a $\otimes$-equivalence of $\mathscr{P}^{(0)}$ with $\mathcal{R}ep_{rat}(G_f^{(0)})$. So, in principle, we should be able to recover object A (morphisms will be considered in 13.2.4.2) from ρ_A at least up to gauge equivalence in $\mathscr{P}^{(0)}$. This can be done as follows.

First, writing 1 the unit in $\mathbf{E}_q^{\vee}$ (that is the trivial morphism $x \mapsto 1$), we see that[16] $\rho_A(1,1) = A_u$, the unipotent component of A. We similarly get all $\gamma(A_s) = \rho_A(\gamma,0)$ for $\gamma \in \mathbf{E}_q^{\vee}$. They are codiagonalisable matrices of the form $P\mathrm{Diag}(\gamma(c_1),\ldots,\gamma(c_n))P^{-1}$ for some fixed $\underline{c} = (c_1,\ldots,c_n) \in \mathbf{C}^{*n}$, and we are thus reduced to recover $\underline{c}$ from its q-replicas. Obviously, the c_i can only be determined modulo $q^{\mathbf{Z}}$, but this is all right since transformations $c_i \leftarrow qc_i$ correspond to gauge transformations in $\mathscr{P}^{(0)}$.

EXERCISE 13.28. State and prove rigorously the last statement: if A_s, A'_s are semisimple and such that $\gamma(A_s) = \gamma(A'_s)$ for all $\gamma \in \mathbf{E}_q^{\vee}$, then $A_sA_u \sim A'_sA_u$ in $\mathscr{P}^{(0)}$. Do it first algebraically, then through Tannakian general nonsense.

Anticipating on 13.2.5, we introduce two particular elements of $G_f^{(0)}$. For that, we use the following decomposition of $\mathbf{C}^*$ as a topological group:

$$\mathbf{C}^* = \mathbf{U} \times q^{\mathbf{R}} = \mathbf{U} \times e^{2\mathrm{i}\pi\tau\mathbf{R}}.$$

We then define $\gamma_1 : uq^y \mapsto u$ and $\gamma_2 : uq^y \mapsto e^{2\mathrm{i}\pi y}$.

EXERCISE 13.29. (i) Give explicit formulas for the individual decomposition $x = uq^y = e^{2\mathrm{i}\pi\tau y}$, $u \in U$, $y \in \mathbf{R}$, of $x \in \mathbf{C}^*$.

(ii) Check that $\gamma_1, \gamma_2 \in \mathbf{E}_q^{\vee}$

(iii) Check that $\mathrm{Im}\ \gamma_1 = \mathrm{Im}\ \gamma_2 = \mathbf{U}$ and that $\mathrm{Ker}\gamma_1 \cap \mathrm{Ker}\gamma_2$ is trivial (if we see them as morphisms $\mathbf{E}_q \to \mathbf{C}^*$), resp. equal to $q^{\mathbf{Z}}$ (if we see them as morphisms $\mathbf{C}^* \to \mathbf{C}^*$).

Now, if $\underline{c} = (c_1,\ldots,c_n) \in \mathbf{C}^{*n}$ and we know all $(\gamma(c_1),\ldots,\gamma(c_n))$, $\gamma \in \mathbf{E}_q^{\vee}$, we can in particular set $u_i := \gamma_1(c_i)$, y_i such that $e^{2\mathrm{i}\pi y_i} = \gamma_2(c_i)$ and $c'_i := u_iq^{y_i}$ for $i = 1,\ldots,n$. The resulting $\underline{c'} = (c'_1,\ldots,c'_n) \in \mathbf{C}^{*n}$ is the one we are looking for.

13.2.4.2. Now we turn to morphisms. Let objects A, B be fixed in $\mathscr{P}^{(0)}$ and let $\Phi : \rho_A \to \rho_B$ a morphism of representations. We look for $F : A \to B$, a morphism $\mathscr{P}^{(0)}$ such that $\Phi = F(a)$. The assumption on Φ means that $\Phi\gamma(A_s)A_u^{\lambda} = \gamma(B_s)B_u^{\lambda}\Phi$ for all $\lambda \in \mathbf{C}$, $\gamma \in \mathbf{E}_q^{\vee}$. Successively taking $\lambda = 1$ and $\lambda = 0$, we draw that $\Phi A_u = B_u\Phi$ and that $\Phi\gamma(A_s) = \gamma(B_s)\Phi$ for all $\gamma \in \mathbf{E}_q^{\vee}$.

EXERCISE 13.30. Check conversely that the last equalities imply the assumption.

[16]Be aware that the first 1 in the argument belongs to $\mathbf{E}_q^{\vee}$ and is trivial, while the second 1 belongs to $\mathbf{C}$ and is not trivial.

We shall first assume that $\mathrm{Sp}\, A, \mathrm{Sp}\, B \subset \mathcal{C}_q$. Write $A_s = P\mathrm{Diag}(c_1, \ldots, c_n)P^{-1}$ and $B_s = Q\mathrm{Diag}(d_1, \ldots, d_p)Q^{-1}$, then set $\Psi := Q^{-1}\Phi P$, so that:

$$\Psi\,\mathrm{Diag}(\gamma(c_1), \ldots, \gamma(c_n)) = \mathrm{Diag}(\gamma(d_1), \ldots, \gamma(d_p))\,\Psi$$

for all $\gamma \in \mathbf{E}_q^\vee$. We shall check that this entails relation $\Psi\mathrm{Diag}(\overline{c_1}), \ldots, \overline{c_n})) = \mathrm{Diag}(\overline{d_1}), \ldots, \overline{d_p}))\Psi$; but, taking in account our assumption on the spectra, this means that $\Psi\mathrm{Diag}(c_1, \ldots, c_n) = \mathrm{Diag}(d_1, \ldots, d_p)\Psi$, i.e., $\Phi A_s = B_s\Phi$. Since $\Phi A_u = B_u\Phi$, we draw that $\Phi A = B\Phi$ (this uses the commutations) and taking F constant equal to Φ is then a valid choice.

Before turning to the general case, we must justify our contention. The given relation says that $\Psi_{i,j}\gamma(c_j) = \gamma(d_i)\Psi_{i,j}$, i.e., that, each time $\Psi_{i,j} \neq 0$, we have $\gamma(c_j) = \gamma(d_i)$ for all $\gamma \in \mathbf{E}_q^\vee$. By 13.2.4.2, we know that this means $\overline{c_j} = \overline{d_i}$. This being true each time $\Psi_{i,j} \neq 0$, we have $\Psi_{i,j}\overline{c_j} = \overline{d_i}\Psi_{i,j}$ for all i, j, which is indeed equivalent to $\Psi\mathrm{Diag}(\overline{c_1}, \ldots, \overline{c_n}) = \mathrm{Diag}(\overline{d_1}, \ldots, \overline{d_p})\Psi$.

Now we relax the assumption on the spectra. We appeal to results in 11.1. Using those results, and what we just did, we get a commutative diagram (on the left) and, applying ω_a to it, another (on the right):

$$\begin{array}{ccc} A & \xrightarrow{F} & B \\ {\scriptstyle R}\downarrow & & \downarrow{\scriptstyle S} \\ \overline{A} & \xrightarrow[\phi]{} & \overline{B} \end{array} \qquad\qquad \begin{array}{ccc} \mathbf{C}^n & \xrightarrow{F(a)} & \mathbf{C}^p \\ {\scriptstyle R(a)}\downarrow & & \downarrow{\scriptstyle S(a)} \\ \mathbf{C}^n & \xrightarrow[\phi]{} & \mathbf{C}^p. \end{array}$$

The second diagram says that $R(a)$, $S(a)$ provide isomorphims of representations $\rho_A \to \rho_{\overline{A}}$ and $\rho_B \to \rho_{\overline{B}}$, whence a conjugacy of Φ to a morphism $\rho_A \to \rho_B$ which lifts to $F : A \to B$.

13.2.4.3. Since $\mathcal{R} \to \mathscr{P}^{(0)}$ is a $\otimes$-functor which is identity on objects and an inclusion on morphisms, the group morphisms $\mathrm{Aut}^\otimes(\omega_a) \to \mathrm{Aut}^\otimes(\omega)$ induced by the restriction of ω_a to $\mathcal{R}$ *selects* among the elements of $\mathrm{Aut}^\otimes(\omega)$ those which satisfy the stronger functoriality conditions: so we actually have a group *embedding* $\mathrm{Aut}^\otimes(\omega_a) \hookrightarrow \mathrm{Aut}^\otimes(\omega)$. We know from 13.2.3.3 that it realizes the former as a closed subgroup of the latter: $G_f^{(0)} \hookrightarrow \mathbf{Z}^{alg}$.

Dually, a rational representation of $\mathbf{Z}^{alg}$ restricts to a rational representation of $G_f^{(0)}$ and we have a $\otimes$ functor $\mathcal{R}ep_{rat}(\mathbf{Z}^{alg}) \rightsquigarrow \mathcal{R}ep_{rat}(G_f^{(0)})$; by Tannakian duality, those two categories are in essence $\mathcal{R}$ and $\mathscr{P}^{(0)}$ and this functor is just the embedding $\mathcal{R} \to \mathscr{P}^{(0)}$.

EXERCISE 13.31. (i) Check this using the way we recovered A from ρ_A in 13.2.4.1.

(ii) Do the same at the level of morphisms using 13.2.4.2.

13.2.5. A Zariski dense subgroup.

13.2.5.1. We wish to do for $G_f^{(0)}$ something similar to the equivalence $\mathcal{R}ep^f_{\mathbf{C}}(\mathbf{Z}) = \mathcal{R}ep_{rat}(\mathbf{Z}^{alg})$ from 13.2.2 (and, by the way, to Schlesinger theorem mentioned in 13.1.1), i.e., find an "abstract" or "discrete"[17] Zariski-dense subgroup $\Gamma_f^{(0)}$ such that the obvious restriction functor $\mathcal{R}ep_{rat}(G_f^{(0)}) \to \mathcal{R}ep^f_{\mathbf{C}}(\Gamma_f^{(0)})$ is an equivalence.

[17]The adjective just means: without more structure than being a group.

To do that, we use the natural splitting $G_f^{(0)} = G_{f,s}^{(0)} \times G_{f,u}^{(0)}$ introduced in 13.2.3.4 and we likewise look for $\Gamma_f^{(0)}$ as $\Gamma_{f,s}^{(0)} \times \Gamma_{f,u}^{(0)}$. Either for $\mathcal{R}ep_{rat}(G_f^{(0)})$ or for $\mathcal{R}ep_{\mathbf{C}}^f(\Gamma_f^{(0)})$ we use the fact that representations of a product are pairs of commuting representations to reduce our problem to correponding problems on the two factors.

On the unipotent side, it is clear that we can take for instance $\Gamma_{f,u}^{(0)} := \mathbf{Z}$. However, this may be somewhat misleading as the particular chosen generator $1 \in \mathbf{C}$ plays no particular role: any $\mathbf{Z}u$ with $u \in \mathbf{C}^*$ would be equally suitable, so our choice really is "up to scaling".

On the semisimple side, there is unhappily no embedding of $\mathbf{Z}$ into $G_{f,s}^{(0)}$ that would serve our purpose, so we'll have to proceed differently. We shall use, as an intrinsically defined subgroup of $G_{f,s}^{(0)} = \mathbf{E}_q^\vee$; let:

$$\begin{aligned}\mathrm{Hom}_{grtop}(\mathbf{E}_q, \mathbf{C}^*) &:= \{\gamma \in \mathrm{Hom}_{gr}(\mathbf{E}_q, \mathbf{C}^*) \mid \gamma \text{ is continuous}\} \\ &= \{\gamma \in \mathrm{Hom}_{gr}(\mathbf{C}^*, \mathbf{C}^*) \mid \gamma \text{ is continuous and } \gamma(q) = 1\}.\end{aligned}$$

We shall take $\Gamma_{f,s}^{(0)} := \mathrm{Hom}_{grtop}(\mathbf{E}_q, \mathbf{C}^*)$, whence in the end our candidate q-analogue for the (fuchsian) fundamental group π_1:

$$\Gamma_f^{(0)} = \mathrm{Hom}_{grtop}(\mathbf{E}_q, \mathbf{C}^*) \times \mathbf{C}.$$

We next study the structure of $\Gamma_{f,s}^{(0)} = \mathrm{Hom}_{grtop}(\mathbf{E}_q, \mathbf{C}^*)$. Note that, $\mathbf{E}_q^\vee$ being compact, all elements of $\mathrm{Hom}_{grtop}(\mathbf{E}_q, \mathbf{C}^*)$ must have as image a compact subgroup of $\mathbf{C}^*$, whence a subgroup of $\mathbf{U}$.

EXERCISE 13.32. Prove that the only compact subgroups of $\mathbf{C}^*$ are $\mathbf{U}$ and the finite cyclic groups $\mu_n := \{e^{2ki\pi/n} \mid k = 0, \dots, n-1\}$, $n \geq 1$.

13.2.5.2. Recall from 13.2.4.1 the two morphisms from $\mathbf{C}^* = \mathbf{U} \times q^{\mathbf{R}}$ to itself given by formulas $\gamma_1 : uq^y \mapsto u$ and $\gamma_2 : uq^y \mapsto e^{2i\pi y}$. They are continuous morphisms trivial on $q^{\mathbf{Z}}$, so they are elements of $\mathrm{Hom}_{grtop}(\mathbf{E}_q, \mathbf{C}^*)$. We know their images are $\mathrm{Im}\, \gamma_1 = \mathrm{Im}\, \gamma_2 = \mathbf{U}$, in accordance with the above exercise.

THEOREM 13.3. (i) *The group $\Gamma_{f,s}^{(0)}$ is abelian free on two generators γ_1, γ_2.*

(ii) *The subgroup $\Gamma_f^{(0)}$ of $G_f^{(0)}$ is Zariski-dense and the restriction functor from $\mathcal{R}ep_{rat}(G_f^{(0)})$ to $\mathcal{R}ep_{\mathbf{C}}^f(\Gamma_f^{(0)})$ is an equivalence of $\otimes$-categories.*

PROOF. (i) We shall first describe all continuous morphisms $\mathbf{C}^* \to \mathbf{C}^*$. Since $q^{\mathbf{R}} \simeq \mathbf{R}$ as topological groups (through $q^y \leftrightarrow y$), it amounts to list all continuous morphisms $\mathbf{U} \times \mathbf{R} \to \mathbf{U} \times \mathbf{R}$. Working componentwise, we have four classes to examine:

(a) Continuous morphisms from $\mathbf{R}$ to $\mathbf{R}$: they have the form $y \mapsto ay$ for some $a \in \mathbf{R}$.
(b) Continuous morphisms from $\mathbf{R}$ to $\mathbf{U}$: they have the form $y \mapsto e^{iby}$ for some $b \in \mathbf{R}$.
(c) Continuous morphisms from $\mathbf{U}$ to $\mathbf{R}$: there is only the trivial one.
(d) Continuous morphisms from $\mathbf{U}$ to $\mathbf{U}$: they have the form $u \mapsto u^n$ for some $n \in \mathbf{Z}$.

EXERCISE 13.33. Prove those four statements.

We deduce from that list that the continuous morphisms $\mathbf{C}^* \to \mathbf{C}^*$ have the form $uq^y \mapsto vq^z$, where $v := u^n e^{iby}$ and $z := ay$. Those that send $q = 1.q^1$ to 1 correspond to $b \in 2\pi\mathbf{Z}$ and $a = 0$, i.e., they have the form $uq^y \mapsto u^n e^{2i\pi py}$ (where $b = 2\pi p$); but this is just $\gamma_1^n \gamma_2^p$.

EXERCISE 13.34. Check that n, p are unique, whence the freeness.

(ii) An element uq^y belongs to Ker $\gamma_1 \cap$ Kerγ_2 if, and only if, $u = 1$ and $e^{2i\pi y} = 1$, i.e., $y \in \mathbf{Z}$, so Ker $\gamma_1 \cap$Ker $\gamma_2 = q^{\mathbf{Z}}$ as a subgroup of $\mathbf{C}^*$ (and therefore is trivial as a subgroup of $\mathbf{E}_q$). The conclusion will be a consequence of the following lemma (density criterion in 13.2.5.3), which is slightly more general than needed, but of independent interest.

□

COROLLARY 13.35. *The Zariski-dense "fuchsian q-monodromy subgroup" Mon of Gal, here defined as Mon(A) := $\rho_A(\Gamma_f^{(0)})$, is abelian free on three generators (namely 1 in the factor $\mathbf{C}$ and γ_1, γ_2 in the factor $Hom_{grtop}(\mathbf{E}_q, \mathbf{C}^*)$).*

13.2.5.3. *A density criterion.*

LEMMA 13.36. *A subgroup Γ of the proalgebraic group $\mathbf{E}_q^\vee$ is Zariski dense if, and only if, $\bigcap_{\gamma \in \Gamma} Ker\, \gamma$ is trivial.*

PROOF. We shall use the following Tannakian criterion due to Chevalley and Kolchin **[50]** (see also Appendix G):

Let A an object of $\mathscr{P}^{(0)}$ and let L a line in $\omega_a(A) = \mathbf{C}^n$ stabilized by all $\gamma(A) \in \mathrm{GL}_n(\mathbf{C})$, $\gamma \in \Gamma$. Assume that this implies that L is stabilized by all $\gamma(A) \in \mathrm{GL}_n(\mathbf{C})$, $\gamma \in G_{f,s}^{(0)} = \mathbf{E}_q^\vee$. Then Γ is Zariski-dense in $G_{f,s}^{(0)}$.

So we must prove the above implication, i.e., we start from such a line stabilized by $\Gamma(A)$. Up to conjugacy (whence equivalence in $\mathscr{P}^{(0)}$), we may assume A to be diagonal: $A = \mathrm{Diag}(c_1, \ldots, c_n)$. To say that L is stabilized by $\Gamma(A)$ is equivalent to say that it is generated by a common eigenvector $X = {}^t(x_1, \ldots, x_n)$. The reader will check the following elementary characterization of those eigenvectors:

$$x_i x_j \neq 0 \Longrightarrow \gamma(\overline{c_i}) = \gamma(\overline{c_j}).$$

By the assumption on Γ, we can deduce from those equalities for all $\gamma \in \Gamma$:

$$x_i x_j \neq 0 \Longrightarrow \overline{c_i} = \overline{c_j}.$$

This in turn implies that $\gamma(\overline{c_i}) = \gamma(\overline{c_j})$ for all $\gamma \in \mathbf{E}_q^\vee$. □

EXERCISE 13.37. As written hereabove, the proof yields only an implication (which, by the way, is enough for our application of the criterion). Modify it to obtain logical equivalence.

COROLLARY 13.38. *The Galois group Gal(A) of a particular $A \in GL_n(\mathbf{C})$ is trivial if, and only if, $\gamma_1(A_s) = \gamma_2(A_s) = A_u = I_n$.*

EXERCISE 13.39. Which matrices in $\mathrm{GL}_n(\mathbf{C})$ possess that property?

13.2.5.4. *Miscellaneous remarks.* Although the Galois group $G_f^{(0)}$ has been realized as a subgroup of the Galois group $\mathbf{Z}^{alg}$ (thus expressing a particular relation of their categories (cf. the reference to [**56**, Proposition 2.21 (b)] in 13.2.3.2), the same cannot be said of their respective "fundamental" subgroups $\Gamma_f^{(0)}$ and $\mathbf{Z}$; the following diagram cannot be completed by an upper horizontal arrow:

$$\begin{array}{ccc} \Gamma_f^{(0)} & & \mathbf{Z} \\ \big\downarrow & & \big\downarrow \\ G_f^{(0)} & \hookrightarrow & \mathbf{Z}^{alg}. \end{array}$$

The definition of γ_1 and γ_2 is not totally intrinsic from the knowledge of q, it also depends on the choice of a particular logarithm $\log q = 2i\pi\tau$. Since τ is defined modulo $\mathbf{Z}$, let us try with $\tau' = \tau + 1$ and the corresponding alernative:

$$x = \begin{cases} ue^{2i\pi\tau y}, \\ u'e^{2i\pi\tau' y'} \end{cases} \Longrightarrow \begin{cases} \gamma_1(x) = u, \\ \gamma_1'(x) = u' \end{cases} \text{and} \begin{cases} \gamma_2(x) = e^{2i\pi y}, \\ \gamma_2'(x) = e^{2i\pi y'}. \end{cases} .$$

From the equalities $x = ue^{2i\pi\tau y} = u'e^{2i\pi(\tau+1)y'} = (u'e^{2i\pi y'})e^{2i\pi\tau y'}$ and from the uniqueness of the decomposition, we draw that $u = u'e^{2i\pi y'}$ and $y = y'$, whence:

$$\gamma_1(x) = \gamma_1'(x)\gamma_2(x), \text{ while } \gamma_2(x) = \gamma_2'(x).$$

Therefore, γ_2 is intrinsically defined, while γ_1 is defined only through a power of γ_2.

The loops γ_1 and γ_2 can be thought as generating loops on an "infinitesimal torus" $\mathbf{E}_q$ (obtained by folding $\mathbf{C}^*$ near 0 through identifications $\Gamma \equiv q\Gamma$, where Γ is for instance any circle $\partial\mathrm{D}(0, |q|^m)$. Other arguments lead to the same interpretation. See 13.3.3.5 and Figure 13.3.

In [**190**], J.-P. Ramis states that the loops γ_1 and γ_2 (denoted differently there) satisfy a "transcendental relation". A possible interpretation is the following:

$$\gamma_1\gamma_2^\tau = \mathrm{Id}_{\mathbf{C}^*}, \text{ said otherwise: } \forall x \in \mathbf{C}^*\ ,\ \gamma_1(x)\gamma_2(x)^\tau = x.$$

Indeed, letting $x = uq^y$, we see that, up to some heuristic accomodations:

$$\gamma_1(x)\gamma_2(x)^\tau = u(e^{2i\pi y})^\tau = ue^{2i\pi\tau y} = x.$$

In the theory of algebraic Lie groups, one finds that every semisimple complex algebraic group admits a so-called "compact real form", i.e., a compact real Lie subgroup that is Zariski dense; a most typical example is $\mathrm{SU}_n(\mathbf{C}) \subset \mathrm{SL}_n(\mathbf{C})$. In some sense, $\Gamma_s^{(0)}$ is a compact real form of $G_{f,s}^{(0)}$.

13.2.6. Flat vector bundles. What is explained here is partly redundant with Section 7.3 but in a form more adapted to Galois theory; the goal is theorem 13.4.

13.2.6.1. Recall from Appendix B.5.5 that so called *flat vector bundles* on a Riemann surface[18] are those for which transition matrices can be chosen constant (i.e., independent of the variable on the base); whence the following basic construction for every $A \in \mathrm{GL}_n(\mathbf{C})$. We define on $\mathbf{C}^* \times \mathbf{C}^n$ the equivalence relation $\sim_A$ as

[18]Beware that the word "flat" here comes from differential geometry and must not be confused with its use in commutative algebra for "flat modules", etc.

generated by the relations $(x, X) \sim_A (qx, AX)$, for all $x \in \mathbf{C}^*$, $X \in \mathbf{C}^n$. We write $\mathfrak{F}_A := (\mathbf{C}^* \times \mathbf{C}^n)/\sim_A$ the quotient set and get a commutative diagram:

$$\begin{array}{ccc} \mathbf{C}^* \times \mathbf{C}^n & \longrightarrow & \mathbf{C}^* \\ \big\downarrow & & \big\downarrow \pi \\ \mathfrak{F}_A & \xrightarrow{\;p\;} & \mathbf{E}_q \end{array}$$

This way, we obtain a flat vector bundle $p : \mathfrak{F}_A \to \mathbf{E}_q$ of rank n. The vector bundle $\mathbf{E}_q$ is related to the q-difference equation $\sigma_q X = AX$ through the sections of the former and the "solutions" of the latter[19]. Recall from Appendix B.5.3 the definition of the sheaf of sections:

$$\forall \text{ open subset } V \subset \mathbf{E}_q\,,\ \Gamma(V, \mathfrak{F}_A) := \{\phi \in \mathscr{O}(V, \mathbf{C}^n) \mid \pi \circ \phi = \mathrm{Id}_V\}.$$

This sheaf is canonically isomorphic to the sheaf of solutions[20], also defined over $\mathbf{E}_q$; for every open subset $V \subset \mathbf{E}_q$:

$$\mathscr{F}_A(V) := \{X \in \mathscr{O}(\pi^{-1}(V), \mathbf{C}^n) = \mathscr{O}(\pi^{-1}(V))^n \mid \sigma_q X = AX \text{ on } \pi^{-1}(V)\}.$$

We shall identify $\mathscr{F}_A(V) = \Gamma(V, \mathfrak{F}_A)$ (anyhow, the notation $\mathfrak{F}_A(V)$ is traditional for the sheaf of sections of the vector bundle $\mathfrak{F}_A$). If V is *small*, that is, if $\pi^{-1}(V)$ is a disjoint union $\bigsqcup_{k \in \mathbf{Z}} q^k U$, where each U projects bijectively to V, the map $\mathscr{F}_A(V) \to \mathscr{O}(U)^n$, sending solution X on $\pi^{-1}(V)$ to its restriction to U, is bijective (thus a linear isomorphism). Indeed, any $X_0 \in \mathscr{O}(U)^n$ uniquely extends to every $q^k U$, whence to the whole of $\pi^{-1}(V)$, thanks to the functional equation $\sigma_q X = AX$. It follows (applying the above argument to all $V' \subset V$) that the restriction of the sheaf $\mathscr{F}_A$ to V is isomorphic to $\mathscr{O}_V^n$, and therefore that $\mathscr{F}_A$ is locally free (as should be the sheaf of sections of a vector bundle).

EXERCISE 13.40. Give explicitly the formula for X on $q^k U$.

REMARK 13.41. The condition of holomorphy over $\pi^{-1}(V)$ is natural in the context of vector bundles but maybe not so much in the context of functional equations: "solutions" in the above sense could be very wild at 0, while we usually require the solutions of a q-difference equation to be meromorphic at 0. Therefore the use of the above sheaf to understand a q-difference equation will not always be straightforward.

As for *global meromorphic solutions*, we have the following:

PROPOSITION 13.42. *There is a natural isomorphism of the space* $\Gamma(\mathbf{E}_q, \mathfrak{F}_A)$ *of global holomorphic sections of* $\mathfrak{F}_A$ *with the space of all solutions of* $\sigma_q X = AX$ *which are meromorphic over* $\mathbf{C}$*:*

$$\Gamma(\mathbf{E}_q, \mathfrak{F}_A) \simeq \{X \in \mathscr{M}(\mathbf{C})^n \mid \sigma_q X = AX\}.$$

PROOF. Let X such a solution: it has to be holomorphic over $\mathbf{C}^*$; indeed, equation $X(qx) = AX(x)$ implies that any pole a in $\mathbf{C}^*$ would produce a q-spiral $[a; q]$ of poles accumulating at 0, thus a wild singularity there, contradicting the

[19]The quote marks around "solutions" will be explained in a little moment.

[20]We use the letter $\mathfrak{F}$ to denote vector bundles and the letter $\mathscr{F}$ to denote sheaves; however, since giving a vector bundle or a locally free sheaf is equivalent, we may sometimes be sloppy and elude distinction.

meromorphy assumption. As explained above, X therefore corresponds to a global section (since $\pi^{-1}(\mathbf{E}_q) = \mathbf{C}^*$).

Conversely, let X a holomorphic global section of $\mathfrak{F}_A$, i.e., for the same reasons, a solution of $\sigma_q X = AX$ which is holomorphic over $\pi^{-1}(\mathbf{E}_q) = \mathbf{C}^*$. We want to prove that it is meromorphic at 0. It is at any rate continuous in the compact annulus $|q| \leq |x| \leq 1$, so it is bounded there: $\|X(x)\| \leq M$.

When $x \to 0$, let $k \in \mathbf{Z}$ such that $|q|^{k+1} < |x| \leq |q|^k$, i.e.:

$$k := \left\lfloor \frac{\log |x|}{\log |q|} \right\rfloor \sim C \log |x| \text{ where } C := \frac{1}{\log |q|} < 0.$$

Then $x_0 := q^{-k} x$ belongs to the compact annulus, so, using $X(x) = X(q^k x_0) = A^k X(x_0)$:

$$\|X(x)\| = \left\|A^k X(x_0)\right\| \leq M \|A\|^k \approx \|A\|^{C \log |x|} = |x|^K \text{ for some real } K.$$

(We write $u \approx v$ for $\log u \sim \log v$.) We now use the usual polynomial growth criterion: if $|x^m f(x)| \to 0$ for some m, then $x^m f$ has a removable singularity at 0 [**211**, Theorem 10.20], so f has a pole there. □

13.2.6.2. Let $F : A \to B$ a morphism in $\mathscr{P}^{(0)}$. We know from the proof of proposition 13.15 that F is regular over $\mathbf{C}^*$, so any holomorphic solution X of $\sigma_q X = AX$ over some $\pi^{-1}(V)$ gives rise to a holomorphic solution $Y := FX$ of $\sigma_q Y = BY$ over $\pi^{-1}(V)$. This way, we get a morphism of sheaves, whence of vector bundles $\mathfrak{F}_A \to \mathfrak{F}_B$. In the following proof, we shall denote it $\tilde{F}$. Together those facts define a functor $A \rightsquigarrow \mathfrak{F}_A$ from $\mathscr{P}^{(0)}$ to the category $\mathbf{Fib}_p(\mathbf{E}_q)$ of flat vector bundles[21] over $\mathbf{E}_q$.

THEOREM 13.4. *The functor $A \rightsquigarrow \mathfrak{F}_A$ is a tensor equivalence of categories from $\mathscr{P}^{(0)}$ to $\mathbf{Fib}_p(\mathbf{E}_q)$.*

PROOF. We are going to sketch the main arguments of the proof, which, if detailed, is rather long; see [**216, 218**] for details. It should be more or less clear from the above and from Appendix B.5.5 that every flat vector bundle over $\mathbf{E}_q$ can be realized as a $\mathfrak{F}_A$ for some $A \in \mathrm{GL}_n(\mathbf{C})$, so our functor is essentially surjective. Also, with the notations given just before the statement of the theorem, it is obvious that $\tilde{F} = 0 \Rightarrow F = 0$, so the functor is faithful. We are therefore reduced to check that it is full and $\otimes$-compatible.

We first tackle fullness. Let A, B objects in $\mathscr{P}^{(0)}$ and $\Phi : \mathfrak{F}_A \to \mathfrak{F}_B$ a morphism. We look for a morphism $F : A \to B$ such that $\Phi = \tilde{F}$. We write $\Psi : \mathscr{F}_A \to \mathscr{F}_B$ the corresponding morphism of sheaves. For every small open subset $V \subset \mathbf{E}_q$ (meaning that the restrictions of $\mathfrak{F}_A$ and $\mathfrak{F}_B$ to V are both trivial, equivalently, that the restrictions of $\mathscr{F}_A$ and $\mathscr{F}_B$ to V are both free), the corresponding linear map $\mathscr{F}_A(V) \to \mathscr{F}_B(V)$ is given by a matrix Ψ_V which is holomorphic over $\pi^{-1}(V)$ and satisfies there $(\sigma_q \Psi_V)A = B\Psi_V$. Because Ψ is a morphism of sheaves, the Ψ_V glue into a unique matrix F holomorphic over $\mathbf{C}^*$ and such that $(\sigma_q F)A = BF$. If we prove that F is meromorphic at 0, it will be the morphism we are looking for. But F can be considered as a candidate solution for the internal Hom q-difference module $\underline{\mathrm{Hom}}(A, B)$ (defined in 9.1.1) and Proposition 13.42 entails exactly that.

[21]Here again we use the subscript "p" for "flat", which is "plat" in French.

Now we come to $\otimes$-compatibility. We shall give some more details, because none of the quoted references gives them. The first step is to notice that, up to the usual indentifications (of $\mathbf{C}^n \otimes \mathbf{C}^p$ with $\mathbf{C}^{np}$, etc):

$$(\sigma_q X = AX \text{ and } \sigma_q Y = BY) \Longrightarrow \sigma_q(X \otimes Y) = (A \otimes B)(X \otimes Y).$$

Therefore we have a natural morphism of sheaves $\mathscr{F}_A \otimes \mathscr{F}_B \to \mathscr{F}_{A \otimes B}$ (it is clear that the tensor product preserves holomorphy and is compatible with restriction to smaller open subsets). On a small open subset V, writing $U := \pi^{-1}(V)$, the above morphisms identifies to $\mathscr{O}(U)^n \otimes \mathscr{O}(U)^p \to \mathscr{O}(U)^{np}$, which is an isomorphism. Therefore $\mathscr{F}_A \otimes \mathscr{F}_B \to \mathscr{F}_{A \otimes B}$ is an isomorphism of sheaves.

It is a general fact that there is a one-to-one correspondence between morphisms (resp. isomorphisms) of vector bundles and morphisms (resp. isomorphisms) of their associated sheaves of sections, and also that the tensor product of bundles corresponds to the tensor product of sheaves. Therefore, we shall write:

$$\mathfrak{F}_A \otimes \mathfrak{F}_B = \mathfrak{F}_{A \otimes B}. \tag{13.1}$$

□

Actually, most of the above holds for arbitrary q-difference equations (fuchsian or irregular). Only *fullness* of the functor $A \rightsquigarrow \mathfrak{F}_A$ is directly related to fuchsianity.

EXAMPLE 13.43. Let $A := x$ and $B := 1$. Then $\mathfrak{F}_B$ is the trivial line bundle, while $\mathfrak{F}_A$ is a line bundle with the meromorphic section $1/\theta_q$, whence it has degree -1; but it has no holomorphic global section. The morphisms $\mathfrak{F}_A \to \mathfrak{F}_B$ are functions $f \in \mathscr{O}(\mathbf{C}^*, 0)$ such that $(\sigma_q f)x = 1f$, i.e., $\sigma_q f = x^{-1} f$. So they are the functions $c\theta_q$, $c \in \mathbf{C}$. The particular morphism $\theta_q : \mathfrak{F}_A \to \mathfrak{F}_B$ cannot be meromorphically continued to 0, so it does not come from a morphism $A \to B$.

EXERCISE 13.44. In the above example, we have admitted that the description of bundles, sections and morphisms works all the same for any matrix A which is *regular* over $\mathbf{C}^*$. Check the relevant details.

13.2.6.3. Here, we refer to 7.3. In case A is not constant, the formalism of vector bundles and sheaves may still be used as follows. We first assume the most general case (in the local analytic setting[22]): $A \in \mathrm{GL}_n(\mathbf{C}(\{x\}))$. We consider the *germ of space* $(\mathbf{C}^*, 0)$ and the trivial bundle over it $(\mathbf{C}^*, 0) \times \mathbf{C}^n$. On the latter, we define the partial equivalence relation $\sim_A$ generated by the relations $(x, X) \sim_A (qx, A(x)X)$ and set:

$$\mathfrak{F}_A := \frac{(\mathbf{C}^*, 0) \times \mathbf{C}^n}{\sim_A} \xrightarrow{p} \frac{(\mathbf{C}^*, 0)}{x \sim qx} = \mathbf{E}_q,$$

a holomorphic vector bundle of rank n. Now similar arguments as in the narrower case of $\mathscr{P}^{(0)}$ allow us to prove the following:

(1) The sheaf of sections $\mathscr{F}_A$ of the bundle $\mathfrak{F}_A$ can be identified with the following:

$$\mathscr{F}_A(V) := \{X \in \mathscr{O}\left(\pi^{-1}(V), 0\right)^n \mid \sigma_q X = AX \text{ near } 0\},$$

where $\mathscr{O}\left(\pi^{-1}(V), 0\right)$ is the algebra of germs at 0 of holomorphic functions on $\pi^{-1}(V)$.

[22]When A is *formal* but fuchsian, for instance $A \in \mathrm{GL}_n(\mathbf{C}[[x]])$, Baranovsky and Ginzburg **[17]** still are able to build an associated holomorphic vector bundle by completely different methods. We summarize their results in 13.2.6.5.

(2) The functor $A \rightsquigarrow \mathfrak{F}_A$ (or $\rightsquigarrow \mathscr{F}_A$) is well defined, it is $\otimes$-compatible, exact, $\mathbf{C}$-linear and faithful.

All that is more or less obvious; but we saw Example 13.43 showing that it is not necessarily *fully* faithful. This defect can be repaired by taking in account the slope filtrations; see Chapter 15.

REMARK 13.45. Using the formalism of [**53**], we can consider $A \rightsquigarrow \mathfrak{F}_A$ as a "fibre functor over the curve $\mathbf{E}_q$" giving rise to a *group scheme* $\underline{\mathrm{Aut}}^{\otimes}(\omega)$ defined over $\mathbf{E}_q$. Our proalgebraic group $\mathrm{Aut}^{\otimes}(\omega_a)$ is actually the fiber at $\overline{a} \in \mathbf{E}_q$ of $\underline{\mathrm{Aut}}^{\otimes}(\omega)$. However, taking $a \in \mathbf{C}^*$ (and not $\overline{a} \in \mathbf{E}_q$) as a base point endows our Galois groupoid with more transcendental structure, which is important for us.

Returning to the fuchsian case, we deduce:

COROLLARY 13.46. *The functor $A \rightsquigarrow \mathfrak{F}_A$ is a tensor equivalence of ${\mathscr{E}_f}^{(0)}$ with* $\mathbf{Fib}_p(\mathbf{E}_q)$.

EXERCISE 13.47. Prove directly that morphisms $A \to B$ in ${\mathscr{E}_f}^{(0)}$ correspond bijectively to morphisms $\mathfrak{F}_A \to \mathfrak{F}_B$ in $\mathbf{Fib}_p(\mathbf{E}_q)$.

13.2.6.4. *Relation with Weil's correspondence.* For Weil's correspondence in its original 1938 form (without the language of vector bundles), see [**254**]. For its analysis in the language of algebraic topology and of analytic and algebraic geometry, see Grothendieck's Bourbaki talk [**106**].

Let $\mathfrak{F} \to X$ a holomorphic vector bundle over a compact Riemann surface X and $\pi : \tilde{X} \to X$ the universal covering. Lifting $\mathfrak{F}$ to $\tilde{X}$ yields a trivial bundle $\tilde{\pi} : \tilde{\mathfrak{F}} \simeq \tilde{X} \times V \to \tilde{X}$ and a cartesian square (i.e., induced by a pullback $\tilde{\mathfrak{F}} := \pi^* \mathfrak{F}$):

$$\begin{array}{ccc} \tilde{\mathfrak{F}} & \longrightarrow & \tilde{X} \\ \big\downarrow & & \big\downarrow \pi \\ \mathfrak{F} & \longrightarrow & X \end{array}$$

endowed with an equivariant action of the fundamental group[23] $\Gamma := \pi_1(X) = \mathrm{Aut}(\tilde{X}/X)$. Here, *equivariant* means that $\gamma \in \Gamma$ sends any fiber $\{\tilde{x}\} \times V \subset \tilde{\mathfrak{F}}$ to $\{\gamma.\tilde{x}\} \times V$. More precisely, the action being linear:

$$\gamma.\left(\tilde{x}, v\right) = \left(\gamma.\tilde{x}, A_\gamma\left(\tilde{x}, v\right)\right)$$

for some $A_\gamma\left(\tilde{x}, -\right) \in \mathrm{GL}(V)$ depending holomorphically on γ.

EXERCISE 13.48. Translate the fact that this is a group action into a "cocycle condition" on the family of all A_γ, $\gamma \in \Gamma$.

If the vector bundle $\mathfrak{F}$ is flat, A_γ maybe taken independent of $\tilde{x}$, i.e.:

$$\gamma.\left(\tilde{x}, v\right) = \left(\gamma.\tilde{x}, A_\gamma(v)\right).$$

Then the cocycle condition of the exercise above becomes simply $A_\gamma A_{\gamma'} = A_{\gamma\gamma'}$, i.e., $\gamma \mapsto A_\gamma$ is a linear representation of Γ in V.

Weil's correspondence asserts that the procedure is bijective:

$$\left\{ \begin{array}{c} \text{flat vector bundles over } X \\ \text{up to isomorphism} \end{array} \right\} \longleftrightarrow \left\{ \begin{array}{c} \text{linear representations of } \pi_1(X) \\ \text{up to equivalence} \end{array} \right\}.$$

[23] We denote $\mathrm{Aut}(Y/X)$ the group of automorphisms of an etale covering $Y \to X$.

Here, the equivalence of two representations $\gamma \mapsto A_\gamma \in \mathrm{GL}(V)$ and $\gamma \mapsto B_\gamma \in \mathrm{GL}(W)$ has the form of a linear map $F \in \mathscr{L}(V, W)$ such that $FA_\gamma = B_\gamma F$ for all $\gamma \in \Gamma$. In case $\Gamma = \mathbf{Z}$, this boils down to a morphism in $\mathcal{R}$, as in 13.2.2. However, when extending the scope to $\mathscr{P}^{(0)}$, we found that we had to take in account the dependency $F(x)$ on a variable $x \in \mathbf{C}^*$, so we deal with a kind of equivariant form of Weil's correspondence. We shall try to understand what it means.

EXERCISE 13.49. When $X = \mathbf{E}_q$, the universal covering is $\tilde{X} = \mathbf{C}$ and the fundamental group Γ maybe identified to a lattice $\Lambda := \mathbf{Z} + \mathbf{Z}\tau$, $\Im \tau > 0$ (Appendix A). Check that a representation of Λ in $\mathbf{C}^n$ is totally encoded by two commuting matrices $A, B \in \mathrm{GL}_n(\mathbf{C})$ and show how to recover from them a flat bundle over $\mathbf{E}_q$.

LEMMA 13.50. *Any representation of* $\Lambda := \mathbf{Z} + \mathbf{Z}\tau$ *in* $\mathbf{C}^n$ *is equivalent (in the equivariant way, with* $F(z)$ *depending on* $z = \tilde{x} \in \mathbf{C}$*) to one such that* $1 \mapsto I_n$

PROOF. The statement means that we can find a holomorphic $F : \mathbf{C} \to \mathrm{GL}_n(\mathbf{C})$ such that $F(z+1)A = A'F(z)$ and $F(z+\tau)B = B'F(z)$ (this expresses the "equivariant" equivalence of two representations of Λ), moreover requiring that $A' = I_n$ (this is the special requirement of the lemma). Write A as $\exp(C)$ for some $C \in \mathrm{Mat}_n(\mathbf{C})$. Then $F(z) := \exp(-zC)$ satisfies the first equality and B' is then defined by the second one. □

EXERCISE 13.51. One may instead impose that $\mathrm{Sp}\, A', \mathrm{Sp}\, B'$ be included in the fundamental annulus $\mathcal{C}_q$ (see **[249]**).

The geometric interpretation of the lemma is that a representation of $\Lambda = \mathrm{Aut}(\mathbf{C}/\mathbf{E}_q)$ can be reduced to one of $\mathrm{Aut}\left(\frac{\mathbf{C}}{\mathbf{Z}}/\mathbf{E}_q\right) = \mathrm{Aut}(\mathbf{C}^*/\mathbf{E}_q) = q^{\mathbf{Z}}$, which will be encoded by a single matrix $A' \in \mathrm{GL}_n(\mathbf{C})$. There is another way to understand that fact.

The description of a flat bundle over X by a matrix relied on the fact that its pullback over the simply connected $\tilde{X}$ is a trivial bundle. However, in the case of $\mathbf{E}_q$, there is another covering Riemann surface $\pi : \mathbf{C}^* \to \mathbf{E}_q$; and $\mathbf{C}^*$ is not simply connected, but it is an open Riemann surface and (Subsection B.5.6 of Appendix B) every holomorphic vector bundle over it is trivial. So $\pi^*\mathfrak{F} \simeq \mathbf{C}^* \times \mathbf{C}^n$ and the equivariant action takes the form $(x, X) \mapsto (qx, AX)$: we find again our construction of $\mathfrak{F}_A$ as a quotient $(\mathbf{C}^* \times \mathbf{C}^n) / \sim_A$.

13.2.6.5. *The classification theorem of Baranovsky and Ginzburg.* What follows comes directly from **[218]**. Let G denote a complex algebraic group and write $G((z))$ for the group of $\mathbf{C}((z))$ rational points of G, where $\mathbf{C}((z))$ is the field of formal Laurent series, a kind of "loop group". Write $G[[z]]$ for its subgroup of $\mathbf{C}[[z]]$ rational points. Then define a "twisted" conjugation action of $G((z))$ on itself by putting, for $a(z), g(z) \in G((z))$:

$$^{g}a(z) = g(q.z).a(z).g(z)^{-1}.$$

This twisted conjugation action can actually be seen as a plain conjugation action in some larger group: putting $a(z) \mapsto a(t.z)$ defines a $\mathbf{C}^*$-action on the loop group $G((z))$ (the "rotation of the loop"), hence a semi-direct product, and twisted conjugacy classes are actually ordinary conjugacy classes in a Kac-Moody group.

The following is stated and proved in **[17]**:

Theorem. - *If G is connected and semi-simple, there is a natural bijection between the set of integral twisted conjugacy classes in $G((z))$ and the set of isomorphism classes of semi-stable holomorphic principal G-bundles on the elliptic curve $E = \mathbf{C}^*/q^{\mathbf{Z}}$.*

Here, integral twisted conjugacy classes are those which contain an element of $G[[z]]$: they are analogous to our fuchsian equations. One subtlety of this theorem (and the main difficulty in its proof) is that it provides an *analytic* classification of *formal* objects. The authors give a Tannakian extension of this theorem. They define a tensor category $\mathcal{M}_q$ of formal integral q-difference modules and prove:

Theorem. - *It is equivalent to the tensor category of degree zero semi-stable holomorphic vector bundles on E.*

In the electronic (IMRN) version of their paper, they moreover quote a computation by Kontsevich of the corresponding Galois group: the result is the same as our Corollary 2.22.

13.3. Global fuchsian Galois theory

13.3.1. Introduction.

13.3.1.1. We now turn to the category $\mathscr{E}_f$ of *global (or rational) fuchsian equations*, i.e., of those $A \in \mathrm{GL}_n(\mathbf{C}(x))$ which are fuchsian at 0 and at ∞. When $A \in \mathrm{GL}_n(\mathbf{C}(x))$ is fuchsian at 0 and at ∞, we can define $G_f^{(0)}(A)$, its local Galois group at 0, using the embedding $\mathbf{C}(x) \subset \mathbf{C}(\{x\})$; and similarly $G_f^{(\infty)}(A)$, using the embedding $\mathbf{C}(x) \subset \mathbf{C}(\{1/x\})$. Both groups are images of the universal Galois groups $G_f^{(0)}$ and $G_f^{(\infty)}$.

EXERCISE 13.52. Prove rigorously that $G_f^{(0)} \simeq G_f^{(\infty)}$. (You must examine the effect of the change of variable $x \mapsto 1/x$.) Is it true that $G_f^{(0)}(A) \simeq G_f^{(\infty)}(A)$?

Let us see more precisely how the category $\mathscr{E}_f$ can be looked at "locally". It comes with two natural "localisation functors", from $\mathscr{E}_f$ to ${\mathscr{E}_f}^{(0)}$ and to ${\mathscr{E}_f}^{(\infty)}$. Those functors are $\mathbf{C}$-linear, exact, faithful, $\otimes$-compatible and essentially surjective. Write temporarily $\iota^{(0)}$ and $\iota^{(\infty)}$ those functors. Using the fibre functors $\omega_a^{(0)}$ on ${\mathscr{E}_f}^{(0)}$ and $\omega_a^{(\infty)}$ on ${\mathscr{E}_f}^{(\infty)}$, we get two families of fibre functors $\omega_a^{(*0)} := \omega_a^{(0)} \circ \iota^{(0)}$ and $\omega_a^{(*\infty)} := \omega_a^{(\infty)} \circ \iota^{(\infty)}$ on $\mathscr{E}_f$. This induces morphisms of Galois groups:

$$\mathrm{Aut}^{\otimes}(\omega_a^{(0)}) \to \mathrm{Aut}^{\otimes}(\omega_a^{(*0)}) \text{ and } \mathrm{Aut}^{\otimes}(\omega_a^{(\infty)}) \to \mathrm{Aut}^{\otimes}(\omega_a^{(*\infty)}).$$

As we did when comparing the Galois groups of $\mathcal{R}$ and of $\mathscr{P}^{(0)}$, we can argue that each time the source of the arrow embeds into the target, because automorphisms are taken in a category with more morphisms, whence a selection by more functoriality constraints:

$$\mathrm{Aut}^{\otimes}(\omega_a^{(0)}) \hookrightarrow \mathrm{Aut}^{\otimes}(\omega_a^{(*0)}) \text{ and } \mathrm{Aut}^{\otimes}(\omega_a^{(\infty)}) \hookrightarrow \mathrm{Aut}^{\otimes}(\omega_a^{(*\infty)}).$$

EXERCISE 13.53. Similar statements hold at the level of the groupoids $\mathrm{Iso}^{\otimes}(\omega_a^{(0)}, \omega_b^{(0)})$, etc. Make them precise.

Of course, we hope that the local Galois groups $G_f^{(0)}$ and $G_f^{(\infty)}$ will contribute to the global Galois group G_f of the Tannakian category $\mathscr{E}_f$. To understand why we cannot just directly compose them, we take on the analogy with Van Kampen theorem in Algebraic Topology. Let (X, x_0) and (Y, y_0) two pointed spaces, both

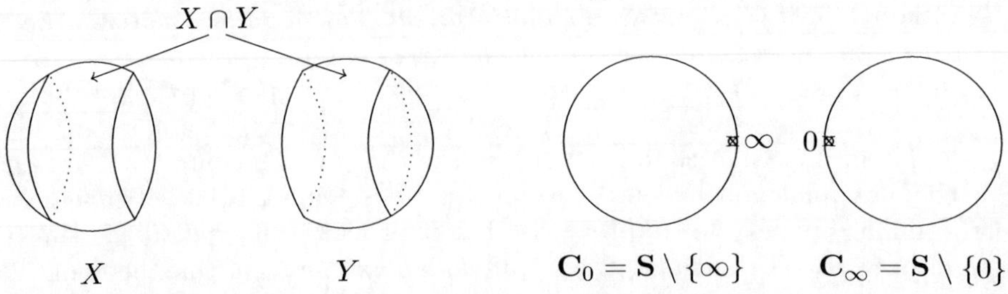

FIGURE 13.1. Van Kampen and the connection matrix

lying in some bigger space, so that $X \cup Y$ and $X \cap Y$ make sense as spaces. We assume implicitly whatever conditions are usual in the theory of the fundamental group (for instance, local arcwise connectedness) and we forget base points in the notation. Van Kampen theorem then says how to compute $\pi_1(X \cup Y)$ from $\pi_1(X)$ and $\pi_1(Y)$ *plus* the way they glue along $\pi_1(X \cap Y)$, equivalently: the way $\pi_1(X \cap Y)$ "embeds" into $\pi_1(X)$ and $\pi_1(Y)$ (quotation marks are here because those "embeddings" need not be injective, as shows the case of $X = \mathbf{C}_0 = \mathbf{C}$, $Y = \mathbf{C}_\infty$).

13.3.1.2. The analogy[24] clearly leads us to the idea of glueing $G_f^{(0)}$ and $G_f^{(\infty)}$ with the help of Birkhoff's connection matrix P. In the regular case tackled by Etingof, the local Galois groups are trivial (they correspond to $\pi_1(X)$ and $\pi_1(Y)$) and only P contributes (it corresponds to the glueing). We see more precisely that points x_0, y_0 correspond to $0, \infty$ and paths from x_0 to y_0 correspond to particular values $P(a)$.

In Tannakian terms, one would hope that the $P(a)$ can be used to define galoisian isomorphisms connecting 0 and ∞. More precisely, fixing a and setting $\phi_A := P(a)$, where P is "the" connection matrix attached to A, one would hope that the collection $\{\phi_A\}$ be an element of $\mathrm{Iso}^{\otimes}(\omega^{(0)}, \omega^{(\infty)})$ for suitable fibre functors $\omega^{(0)}, \omega^{(\infty)}$. This does work all right in the regular case (compare with **[82]** and with **[249**, 12.3.2**]**).

However, in the general fuchsian case the connection matrix P cannot be used directly because it behaves badly with respect to tensor products: if equations A_1, A_2 respectively give rise to connection matrices P_1, P_2, then $A_1 \otimes A_2$ gives rise to a connection matrix $P \neq P_1 \otimes P_2$. This can already be seen for systems of rank 1:

EXAMPLE 13.54. Let $a_1, a_2 \in \mathbf{C}(x)^*$ and let $\alpha_i := a_i(0)$, $\beta_i := a_i(\infty)$, $i = 1, 2$, all assumed to lie in $\mathbf{C}^*$ (fuchsianity). Let $m_i \in \mathbf{C}(\{x\})$ and $n_i \in \mathbf{C}(\{1/x\})$ such that $a_i = m_i[\alpha_i] = n_i[\beta_i]$, $i = 1, 2$, whence the usual gauge transformation diagrams and their tensor product:

$$\begin{cases} \alpha_1 \xrightarrow{m_1} a_1 \xleftarrow{n_1} \beta_1, \\ \alpha_2 \xrightarrow{m_2} a_2 \xleftarrow{n_2} \beta_2 \end{cases} \Longrightarrow \alpha_1\alpha_2 \xrightarrow{m_1 m_2} a_1 a_2 \xleftarrow{n_1 n_2} \beta_1\beta_2$$

The connection matrices are respectively given by:

$$\begin{cases} p_1 = (n_1 e_{\beta_1})^{-1}(m_1 e_{\alpha_1}), \\ p_2 = (n_2 e_{\beta_2})^{-1}(m_2 e_{\alpha_2}) \end{cases} \quad \text{and } p = (n_1 n_2 e_{\beta_1\beta_2})^{-1}(m_1 m_2 e_{\alpha_1\alpha_2}).$$

[24]The analogy with Van Kampen theorem is maybe enhanced by the results of Etingof and by our interpretation of the connection matrix suggested by the confluence phenomenon.

But in general $p \neq p_1 p_2$ since $e_{\alpha_1\alpha_2} \neq e_{\alpha_1}e_{\alpha_2}$ and $e_{\beta_1\beta_2} \neq e_{\beta_1}e_{\beta_2}$. Indeed, $e_{\alpha_1\alpha_2}/e_{\alpha_1}e_{\alpha_2}$ and $e_{\beta_1\beta_2}/e_{\beta_1}e_{\beta_2}$ are arbitrary elliptic functions of order 2 (see 2.3 and 2.2.2).

It appears from the above example that the difficulties with P really appear in the local data, whence in the local monodromies. The connection matrix P mixes the global monodromies with the local ones. There are two ways to overcome that difficulty[25]. The first one is to "twist" P into something different $\check{P}$ which satisfies the requirements. (By the way, the notation $\check{P}$ will be found to be slightly misleading since the corresponding twisting factors really involve local data $A^{(0)}$ and $A^{(\infty)}$, not only P.) The method works well, but, at first sight, it may look bizarre. So we shall first explain the other method (although it was invented later); it will maybe help understanding all that twisting affair.

13.3.1.3. So we choose to use, instead of P, the matrix M. (Note however that here $M := (M^{(\infty)})^{-1}M^{(0)}$ goes from 0 to ∞, as in Birkhoff, while in Section 12.2 we had $M := (M^{(0)})^{-1}M^{(\infty)}$ thus going the other way round.)

Recall from 12.1.1 and 12.2.1 that M comes from the diagram:

$$A^{(0)} \xrightarrow{M^{(0)}} A \xleftarrow{M^{(\infty)}} A^{(\infty)}$$

(meaning that $A^{(0)}, A^{(\infty)} \in \mathrm{GL}_n(\mathbf{C})$, $M^{(0)} \in \mathrm{GL}_n(\mathbf{C}(\{x\}))$, $M^{(\infty)} \in \mathrm{GL}_n(\mathbf{C}(\{1/x\}))$ and $M^{(0)}[A^{(0)}] = A = M^{(\infty)}[A^{(\infty)}]$) by inverting one arrow, thus getting:

$$A^{(0)} \xrightarrow{M} A^{(\infty)},$$

where $M := (M^{(\infty)})^{-1}M^{(0)}$. On the other side, $P = (e_{A^{(\infty)}})^{-1}Me_{A^{(0)}}$. Then:

$$\sigma_q P = P, \text{ while } \sigma_q M = A^{(\infty)}M(A^{(0)})^{-1}.$$

Using M, the tensor compatibility offers no difficulty: if $A = A_1 \otimes A_2$, then, from: $M_1^{(0)}[A_1^{(0)}] \otimes M_2^{(0)}[A_2^{(0)}] = (M_1^{(0)} \otimes M_2^{(0)})[A_1^{(0)} \otimes A_2^{(0)}]$ and similarly at ∞, we see that we can take $M^{(0)} := M_1^{(0)} \otimes M_2^{(0)}$ (and similarly at ∞!), whence:

$$M := \left(M_1^{(\infty)} \otimes M_2^{(\infty)}\right)^{-1}\left(M_1^{(0)} \otimes M_2^{(0)}\right) = M_1 \otimes M_2.$$

EXERCISE 13.55. Check that property on the rank 1 example in 13.3.1.2.

Now the problem comes from functoriality. As we shall next see, the construction of M gives rise to a functor isomorphism $A \rightsquigarrow M(a)$ from $\omega_a^{(0)}$ to $\omega_a^{(\infty)}$. So, contrary to each $P(a)$ which connects (functorially but not $\otimes$-compatibly) the same fibre functors $\omega^{(0)}$, $\omega^{(\infty)}$ (independent of a), now we have to deal with two families of fibre functors. This means that we are going to use a groupoid instead of a group. Remembering that the family of all $\omega_a^{(0)}$, $a \in \mathbf{C}^*$, makes up by itself a local groupoid with base $\mathbf{C}^*$ and sets of arrows $\mathrm{Iso}^{\otimes}(\omega_a^{(0)}, \omega_b^{(0)})$, we now picture our global groupoid as in Figure 13.2.

In each of the planes $\mathbf{C}^*$ (left at 0, right at ∞) the arrows come from the sets of arrows $\mathrm{Iso}^{\otimes}(\omega_a^{(0)}, \omega_b^{(0)})$ described in 13.2.3 and their analogs at ∞. The horizontal arrows, connecting pairs of corresponding base points from both planes, are given by the transformation $A \rightsquigarrow M(a)$ quoted above. Each such arrow belongs

[25] As already told, since we insist on using the full power of function theoretic tools, we do not follow the method of van der Put and Singer in **[249]** of replacing the "true functions" $e_{q,c}$ by *symbols* e_c constrained to satisfy $e_c e_d = e_{cd}$.

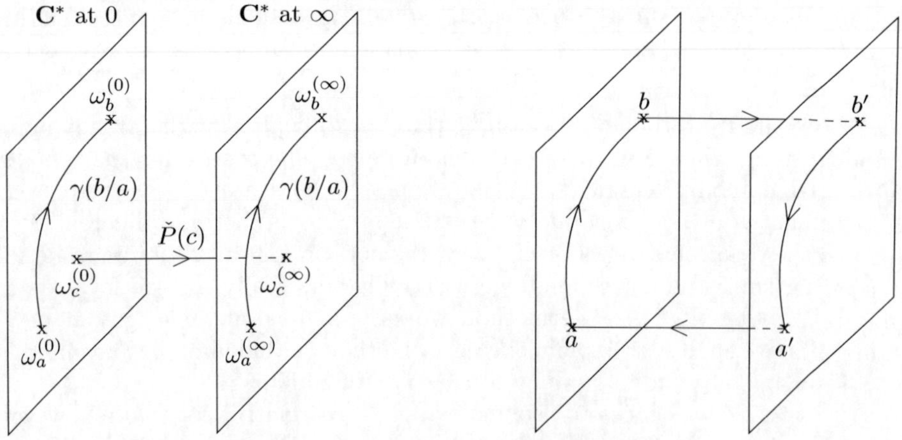

The global Galois groupoid A global loop based at a

FIGURE 13.2. The Global Galois Groupoid and a global loop

to $\mathrm{Iso}^{\otimes}(\omega_a^{(0)}, \omega_a^{(\infty)})$. From this we can describe the global Galois group G_f as "incarnated" at any particular point a, say on the 0 side, i.e., by taking $\mathrm{Aut}^{\otimes}(\omega_a^{(0)})$; but now $\omega_a^{(0)}$ is a fibre functor on the global category $\mathscr{E}_f$ and new galoisian loops may appear as in Figure 13.2 (right).

13.3.1.4. There remains one more technical difficulty in constructing the fibre functor isomorphisms $A \rightsquigarrow M(a)$: the matrix M is not uniquely determined from A, it involves a non deterministic algorithm to produce $M^{(0)}$ and $M^{(\infty)}$. We have to *choose* normal forms $A^{(0)}$ and $A^{(\infty)}$ and even then the rest of the process is not deterministic. The reader may tries to think of a process that would be deterministic (this is indeed possible) and at the same time produce functorial and $\otimes$-compatible results; the present author thinks there is no such *explicit* process[26].

In order to avoid bizarre uses of the Axiom of Choice, we choose to use as intermediate step objects bearing maximum information: these are the 4-uples $(A^{(0)}, M^{(0)}, A^{(\infty)}, M^{(\infty)})$; and to project them, one one side, to A; on the other side, to the "monodromy triples" $(A^{(0)}, A^{(\infty)}, M)$. At the level of objects, the corresponding diagram of categories then will look like this:

$$\begin{array}{ccc} & (A^{(0)}, M^{(0)}, A^{(\infty)}, M^{(\infty)}) & \\ \swarrow & & \searrow \\ A & & (A^{(0)}, A^{(\infty)}, M). \end{array}$$

Both downward arrows will be $\otimes$-equivalences, and this will allow us to use the right-hand side to determine the global Galois groupoid and groups.

EXERCISE 13.56. In anticipation of what will follows:

(i) Write down explicitly the formulas for those arrows;

[26]It seems likely that the constructions described in 13.3.2 together with a very fancy form of the Axiom of Choice (fancy because used with categories, not sets!) entail the existence of such a process; but not an *explicit* one.

(ii) they mention objects: do the same for morphisms.

13.3.2. First construction of the global Galois group. We know proceed to realize the program sketched in 13.3.1.

13.3.2.1. The category $\mathscr{E}_f$ of interest, that of global fuchsian q-difference equations, is the full subcategory of $\mathcal{D}_{K,q}$ having as objects those $A \in \mathrm{GL}_n(\mathbf{C}(x))$ that are fuchsian at 0 and ∞. Since we know in advance that it is Tannakian, all we need in order to compute the global fuchsian Galois group $G_f = \mathrm{Gal}(\mathscr{E}_f)$ is one (or many) fibre functor.

The category $\mathscr{E}_f$ naturally embeds into ${\mathscr{E}_f}^{(0)}$ through a functor $\iota^{(0)}$ sending each object $A \in \mathrm{GL}_n(\mathbf{C}(x))$, resp. each morphism $F \in \mathrm{Mat}_{n,p}(\mathbf{C}(x))$, to itself $A \in \mathrm{GL}_n(\mathbf{C}(\{x\}))$, resp. $F \in \mathrm{Mat}_{n,p}(\mathbf{C}(\{x\}))$. There is a similar functor[27] $\iota^{(\infty)}$ embedding $\mathscr{E}_f$ into ${\mathscr{E}_f}^{(\infty)}$, and both functors are $\mathbf{C}$-linear, exact, faithful and $\otimes$-compatible.

EXERCISE 13.57. Are they fully faithful? (You must provide either a proof or an explicit counter-example.)

REMARK 13.58. Since every object in ${\mathscr{E}_f}^{(0)}$ is equivalent to some $A^{(0)} \in \mathrm{GL}_n(\mathbf{C}) \subset \mathrm{GL}_n(\mathbf{C}(x))$, we see that $\iota^{(0)}$ is moreover essentially surjective (as is $\iota^{(\infty)}$). (Actually, the argument is that $\mathscr{P}^{(0)}$ is essential in ${\mathscr{E}_f}^{(0)}$ and is contained in $\mathscr{E}_f$.) So the main difference between $\mathscr{E}_f$ and ${\mathscr{E}_f}^{(0)}$ is that the latter has more morphisms. In particular it has more isomorphisms, so it confuses more objects: while morphisms in $\mathbf{C}(x)$ permit global observation, morphisms in $\mathbf{C}(\{x\})$ allow only for local observation. In the same way, in analysis, global functions give a global geometric view while germs only allow for a local view. That is the reason why ${\mathscr{E}_f}^{(0)}$ may be considered as a localisation of $\mathscr{E}_f$. On a more computational level, the excedentary morphims of ${\mathscr{E}_f}^{(0)}$ impose more functoriality conditions for the isomorphisms and automorphisms of fibre functors. We already saw something similar when going from $\mathcal{R}$ to $\mathscr{P}^{(0)}$ in 13.2.3.3.

For each fibre functor $\omega_a^{(0)}$ on ${\mathscr{E}_f}^{(0)}$, we introduce the fibre functor:

$$\omega_a^{(*0)} := \omega_a^{(0)} \circ \iota^{(0)}$$

on $\mathscr{E}_f$. The same formal arguments as for $\mathcal{R}$ et $\mathscr{P}^{(0)}$ then yield an embedding of proalgebraic groups:

$$\mathrm{Aut}^{\otimes}\left(\omega_a^{(0)}\right) \subset \mathrm{Aut}^{\otimes}\left(\omega_a^{(*0)}\right).$$

Indeed, elements of the left-hand side group are, among those collections $\{\phi_A\}$ in the right-hand side group, those which satisfy the excedentary functoriality conditions imposed by the excedentary morphisms in ${\mathscr{E}_f}^{(0)}$. So, if we realize the respective Galois groups at the point a (in the left plane as regards G_f since we use $\omega_a^{(*0)}$), we get the embedding:

$$G_f^{(0)} \subset G_f,$$

as a proalgebraic subgroup. For exactly the same reasons, we have more general embeddings:

$$\mathrm{Iso}^{\otimes}\left(\omega_a^{(0)}, \omega_b^{(0)}\right) \subset \mathrm{Iso}^{\otimes}\left(\omega_a^{(*0)}, \omega_b^{(*0)}\right),$$

[27]From now on, in this section, each construction or statement "local at 0" is tacitly understood to be completed by the corresponding construction or statement "local at ∞".

actually a morphism of groupoids, i.e., they are compatible with multiplications (compositions). However, while the sets of arrows on the left-hand side define the Galois groupoid of ${\mathcal{E}_f}^{(0)}$, those on the right-hand side define only a part of the Galois groupoid of $\mathcal{E}_f$. We are going to describe the latter:

(1) One part of it has base $\mathbf{C}^*$ "on the left" and sets of arrows the said $\mathrm{Iso}^{\otimes}\left(\omega_a^{(*0)}, \omega_b^{(*0)}\right)$. It looks a little like the Galois groupoid of ${\mathcal{E}_f}^{(0)}$, it has the same base; but its sets of arrows are bigger.
(2) A second part has base $\mathbf{C}^*$ "on the right" and sets of arrows the corresponding $\mathrm{Iso}^{\otimes}\left(\omega_a^{(*\infty)}, \omega_b^{(*\infty)}\right)$.
(3) The base is now complete, with its two copies of $\mathbf{C}^*$, but one must add the sets of arrows $\mathrm{Iso}^{\otimes}\left(\omega_a^{(*0)}, \omega_a^{(*\infty)}\right)$ connecting the left and right side.

13.3.2.2. In order to introduce the first auxiliary category, recall that our standard fundamental matricial solutions at 0 have the form $\mathcal{X}^{(0)} = M^{(0)} e_{A^{(0)}}$, $M^{(0)} \in \mathrm{GL}_n(\mathbf{C}(\{x\}))$, $A^{(0)} \in \mathrm{GL}_n(\mathbf{C})$. Since we are starting with global equations $A \in \mathrm{GL}_n(\mathbf{C}(x))$, we moreover have $M^{(0)} \in \mathrm{GL}_n(\mathcal{M}(\mathbf{C}))$ (see 7.2.3).

With the hope of not losing too much information too soon, we speak of a *solution pair* (at 0) $(A^{(0)}, M^{(0)}) \in \mathrm{GL}_n(\mathbf{C}) \times \mathrm{GL}_n(\mathcal{M}(\mathbf{C}))$. Each time such a pair will be given, we shall use the notations:

$$\mathcal{X}^{(0)} = M^{(0)} e_{A^{(0)}} \text{ and } A := (\sigma_q \mathcal{X}^{(0)})(\mathcal{X}^{(0)})^{-1} = (\sigma_q M^{(0)}) A^{(0)} (M^{(0)})^{-1}.$$

As usual, similar conventions hold at ∞.

Now suppose we are given solution pairs $(A^{(0)}, M^{(0)})$ and $(A^{(\infty)}, M^{(\infty)})$ respectively at 0 and at ∞. We say that they are *connected* if they give rise to the same matrix A:

$$\begin{aligned}
\begin{matrix}(A^{(0)}, M^{(0)}) \\ (A^{(\infty)}, M^{(\infty)})\end{matrix} \text{ are connected } &\overset{def}{\iff} (\sigma_q \mathcal{X}^{(0)})(\mathcal{X}^{(0)})^{-1} = (\sigma_q \mathcal{X}^{(\infty)})(\mathcal{X}^{(\infty)})^{-1} \\
&\iff (\sigma_q M^{(0)}) A^{(0)} (M^{(0)})^{-1} = (\sigma_q M^{(\infty)}) A^{(\infty)} (M^{(\infty)})^{-1} \\
&\iff (\sigma_q M) A^{(0)} = A^{(\infty)} M,
\end{aligned}$$

where we have set $M := (M^{(\infty)})^{-1} M^{(0)}$; as we shall do each time connected solution pairs are given. Note that in this case:

$$A = (\sigma_q M^{(0)}) A^{(0)} (M^{(0)})^{-1} = (\sigma_q M^{(\infty)}) A^{(\infty)} (M^{(\infty)})^{-1}$$

entails:

$$A \in \mathrm{GL}_n(\mathcal{M}(\mathbf{C})) \cap \mathrm{GL}_n(\mathcal{M}(\mathbf{C}_\infty)) = \mathrm{GL}_n(\mathcal{M}(\mathbf{C} \cup \mathbf{C}_\infty)) = \mathrm{GL}_n(\mathcal{M}(\mathbf{S})) = \mathrm{GL}_n(\mathbf{C}(x)).$$

So connected solution pairs give rise to rational q-difference equations. They fit in a diagram:

$$A^{(0)} \xrightarrow{M^{(0)}} A \xleftarrow{M^{(\infty)}} A^{(\infty)}.$$

Definition 13.59. A *solution quadruple* $(A^{(0)}, M^{(0)}, A^{(\infty)}, M^{(\infty)})$ is made of two connected solution pairs $(A^{(0)}, M^{(0)})$ and $(A^{(\infty)}, M^{(\infty)})$ at 0 and ∞.

Now we speak of morphisms. We wish to insert two solutions quadruples into a diagram:

$$
\begin{array}{ccccc}
A^{(0)} & \xrightarrow{M^{(0)}} & A & \xleftarrow{M^{(\infty)}} & A^{(\infty)} \\
 & & \Big\downarrow{\scriptstyle F} & & \\
B^{(0)} & \xrightarrow[N^{(0)}]{} & B & \xleftarrow[N^{(\infty)}]{} & B^{(\infty)}.
\end{array}
$$

The natural way is to define the leftmost and rightmost vertical arrows $S^{(0)} : A^{(0)} \to B^{(0)}$ and $S^{(\infty)} : A^{(\infty)} \to B^{(\infty)}$ so that the completed diagram commutes. From equality $S^{(0)} := (N^{(0)})^{-1} F M^{(0)}$, we draw that $S^{(0)}$ has coefficients in $\mathrm{GL}_n(\mathscr{M}(\mathbf{C}))$, so it is a morphism in $\mathscr{P}^{(0)}$. Also, reversing the rightmost horizontal arrows, we get the diagram:

$$
\begin{array}{ccc}
A^{(0)} & \xrightarrow{M} & A^{(\infty)} \\
{\scriptstyle S^{(0)}}\Big\downarrow & & \Big\downarrow{\scriptstyle S^{(\infty)}} \\
B^{(0)} & \xrightarrow[N]{} & B^{(\infty)}.
\end{array}
$$

Given a solution quadruple $(A^{(0)}, M^{(0)}, A^{(\infty)}, M^{(\infty)})$, we say that the morphisms $S^{(0)} : A^{(0)} \to B^{(0)}$ in $\mathscr{P}^{(0)}$ and $S^{(\infty)} : A^{(\infty)} \to B^{(\infty)}$ in $\mathscr{P}^{(\infty)}$ are *connected* if $(N^{(0)}) S^{(0)} (M^{(0)})^{-1} = (N^{(\infty)}) S^{(\infty)} (M^{(\infty)})^{-1}$ (we then write this F); equivalently, if $N S^{(0)} = S^{(\infty)} M$. The same calculations show conversely that if $S^{(0)}$ and $S^{(\infty)}$ are connected, then $F : A \to B$ is a morphism in $\mathscr{E}_f$.

DEFINITION 13.60. A *morphism* of solution quadruples:

$$(S^{(0)}, F, S^{(\infty)}) : (A^{(0)}, M^{(0)}, A^{(\infty)}, M^{(\infty)}) \to (B^{(0)}, N^{(0)}, B^{(\infty)}, N^{(\infty)})$$

is a triple consisting in connected morphisms $S^{(0)} : A^{(0)} \to B^{(0)}$ in $\mathscr{P}^{(0)}$ and $S^{(\infty)} : A^{(\infty)} \to B^{(\infty)}$ in $\mathscr{P}^{(\infty)}$ and of the morphism:

$$F := (N^{(0)}) S^{(0)} (M^{(0)})^{-1} = (N^{(\infty)}) S^{(\infty)} (M^{(\infty)})^{-1} : A \to B \text{ in } \mathscr{E}_f.$$

The corresponding completed diagram then reads:

$$
\begin{array}{ccccc}
A^{(0)} & \xrightarrow{M^{(0)}} & A & \xleftarrow{M^{(\infty)}} & A^{(\infty)}. \\
{\scriptstyle S^{(0)}}\Big\downarrow & & \Big\downarrow{\scriptstyle F} & & \Big\downarrow{\scriptstyle S^{(\infty)}} \\
B^{(0)} & \xrightarrow[N^{(0)}]{} & B & \xleftarrow[N^{(\infty)}]{} & B^{(\infty)}
\end{array}
$$

The category $\mathscr{S}_f$ of *(fuchsian) solution quadruples* has objects and morphisms as defined above. Using the corresponding properties of $\mathscr{E}_f$, $\mathscr{E}_f{}^{(0)}$, $\mathscr{E}_f{}^{(\infty)}$, we easily see that it is a $\mathbf{C}$-linear abelian category. We make it a tensor category by setting:

$$
\begin{aligned}
&(A_1^{(0)}, M_1^{(0)}, A_1^{(\infty)}, M_1^{(\infty)}) \otimes (A_2^{(0)}, M_2^{(0)}, A_2^{(\infty)}, M_2^{(\infty)}) := \\
&(A_1^{(0)} \otimes A_2^{(0)}, M_1^{(0)} \otimes M_2^{(0)}, A_1^{(\infty)} \otimes A_2^{(\infty)}, M_1^{(\infty)} \otimes M_2^{(\infty)}),
\end{aligned}
$$

and

$$(S_1^{(0)}, F_1, S_1^{(\infty)}) \otimes (S_2^{(0)}, F_2, S_2^{(\infty)}) := (S_1^{(0)} \otimes S_2^{(0)}, F_1 \otimes F_2, S_1^{(\infty)} \otimes S_2^{(\infty)}).$$

Of course, we must check that we thus obtain objects and morphisms in $\mathscr{S}_f$. The fact that the result is a $\mathbf{C}$-linear abelian tensor category is then easy.

EXERCISE 13.61. (i) Check those facts.
(ii) Describe the additional tensor structures (unit object, internal Hom, dual).

THEOREM 13.5. *The functor from* $\mathscr{S}_f$ *to* $\mathscr{E}_f$ *defined by:*

$$\begin{cases} (A^{(0)}, M^{(0)}, A^{(\infty)}, M^{(\infty)}) \rightsquigarrow A, \\ (S^{(0)}, F, S^{(\infty)}) \rightsquigarrow F, \end{cases}$$

is a tensor equivalence.

PROOF. The functor is clearly well defined and $\otimes$-compatibility follows from the computation:

$$A_1 \otimes A_2 = M_1^{(0)}[A_1^{(0)}] \otimes M_2^{(0)}[A_2^{(0)}] = (M_1^{(0)} \otimes M_2^{(0)})[A_1^{(0)} \otimes A_2^{(0)}],$$

and the corresponding computation at ∞ (the middle equality flows from usual formulas on tensor products and the fact that σ_q is a morphism of $\mathbf{C}$-algebras. Full faithfulness comes from the arguments leading to the definition of morphisms in $\mathscr{S}_f$. The only really nontrivial point is essential surjectivity, which is a consequence of the reduction (in $\mathscr{E}_f{}^{(0)}$) of a fuchsian A to a constant $A^{(0)}$ through a gauge transformation $M^{(0)}$. □

It is important to draw consequences about fibre functors. Temporarily note SE the above functor: then $\omega_a^{(*0)} \circ \mathrm{SE}$ and $\omega_a^{(*\infty)} \circ \mathrm{SE}$ are fibre functors on $\mathscr{S}_f$. For instance, $\omega_a^{(*0)} \circ \mathrm{SE}$ sends $(A^{(0)}, M^{(0)}, A^{(\infty)}, M^{(\infty)})$ to the $\mathbf{C}^n$ underlying $A^{(0)}$ and $(S^{(0)}, F, S^{(\infty)})$ to $S^{(0)}(a)$.

13.3.2.3. We now introduce the category $\mathscr{C}_f$ of *(fuchsian) connection data*, which exhibits the galoisian meaning of Birkhoff's connection matrix P (except that we replace it by M). First, as noted in 13.3.2.2, if $(A^{(0)}, M^{(0)}, A^{(\infty)}, M^{(\infty)})$ is an object of $\mathscr{S}_f$, we have, with the same current notations, $M[A^{(0)}] = A^{(\infty)}$; also, from $M^{(0)} \in \mathrm{GL}_n(\mathscr{M}(\mathbf{C}))$ and $M^{(\infty)} \in \mathrm{GL}_n(\mathscr{M}(\mathbf{C}_\infty))$, we draw that $M := (M^{(\infty)})^{-1} M^{(0)} \in \mathrm{GL}_n(\mathscr{M}(\mathbf{C}^*))$.

Next, as also noted there, a morphism $(S^{(0)}, S^{(\infty)})$ satisfies $NS^{(0)} = S^{(\infty)} M$ (notations as before), i.e., we have a commutative diagram:

$$\begin{array}{ccc} A^{(0)} & \xrightarrow{M} & A^{(\infty)}. \\ {\scriptstyle S^{(0)}}\big\downarrow & & \big\downarrow{\scriptstyle S^{(\infty)}} \\ B^{(0)} & \xrightarrow[N]{} & B^{(\infty)} \end{array}$$

Therefore, our category $\mathscr{C}_f$ has

- as objects *connection triples (or data)* $(A^{(0)}, M, A^{(\infty)})$, where $A^{(0)}, A^{(\infty)} \in \mathrm{GL}_n(\mathbf{C})$, $M \in \mathrm{GL}_n(\mathscr{M}(\mathbf{C}^*))$ and $M[A^{(0)}] = A^{(\infty)}$;
- as morphisms $(A^{(0)}, M, A^{(\infty)}) \to (B^{(0)}, N, B^{(\infty)})$, pairs $(S^{(0)}, S^{(\infty)})$ giving rise to such a commutative diagram of morphisms (vertical ones in $\mathscr{P}^{(0)}$, resp. $\mathscr{P}^{(\infty)}$; horizontal ones in $\mathrm{GL}_n(\mathscr{M}(\mathbf{C}^*))$).

The tensor structure is defined in the obvious way:

(13.2)
$$(A_1^{(0)}, M_1, A_1^{(\infty)}) \otimes (A_2^{(0)}, M_2, A_2^{(\infty)}) := (A_1^{(0)} \otimes A_2^{(0)}, M_1 \otimes M_2, A_1^{(\infty)} \otimes A_2^{(\infty)}),$$

(13.3) $$(S_1^{(0)}, S_1^{(\infty)}) \otimes (S_2^{(0)}, S_2^{(\infty)}) := (S_1^{(0)} \otimes S_2^{(0)}, S_1^{(\infty)} \otimes S_2^{(\infty)}).$$

EXERCISE 13.62. Check that we thus obtain a $\mathbf{C}$-linear abelian tensor category. Complete the description of the tensor structure.

By construction, we have a well defined functor SC from $\mathscr{S}_f$ to $\mathscr{C}_f$ sending $(A^{(0)}, M^{(0)}, A^{(\infty)}, M^{(\infty)})$ to $(A^{(0)}, M, A^{(\infty)})$ and $(S^{(0)}, F, S^{(\infty)})$ to $(S^{(0)}, S^{(\infty)})$. That functor is clearly exact, $\mathbf{C}$-linear and $\otimes$-compatible.

THEOREM 13.6. *It is an equivalence of tensor categories.*

PROOF. It is fully faithful because $(S^{(0)}, F, S^{(\infty)})$ can be uniquely recovered from $(S^{(0)}, S^{(\infty)})$ by setting $F := (N^{(0)})S^{(0)}(M^{(0)})^{-1} = (N^{(\infty)})S^{(\infty)}(M^{(\infty)})^{-1}$.

The only real difficulty is essential surjectivity, i.e., recovering $M^{(0)}$ and $M^{(\infty)})$ from M. This rests on Birkhoff's Preliminary Theorem from Chapter 12. We split M as $(M^{(\infty)})^{-1}M^{(0)}$ and find mechanically that $(A^{(0)}, M^{(0)})$ and $(A^{(\infty)}, M^{(\infty)})$ are automatically connected. □

EXERCISE 13.63. (i) Do the "mechanical" verification.
(ii) Describe the relevant fibre functors of $\mathscr{C}_f$.

13.3.2.4. From the local study, we already have at our disposal many galoisian isomorphisms $\omega_a^{(0)} \to \omega_b^{(0)}$ (and, as ususal, corresponding isomorphisms $\omega_a^{(\infty)} \to \omega_b^{(\infty)}$; namely, for each $\lambda \in \mathbf{C}$ and for each $\gamma \in \mathbf{C}^{*\vee}$ such that $\gamma(q) = b/a$):

$$(A^{(0)}, M, A^{(\infty)}) \rightsquigarrow \gamma(A_s^{(0)})(A_u^{(0)})^\lambda$$

defines an element of $\mathrm{Iso}^{\otimes}(\omega_a^{(0)}, \omega_b^{(0)})$. We must complete this by connecting 0 to ∞ with the help of the M component. The answer is natural:

$$(A^{(0)}, M, A^{(\infty)}) \rightsquigarrow M(a).$$

Clearly this is $\otimes$-compatible. Functoriality is almost tautological. Let $(S^{(0)}, S^{(\infty)})$ a morphism from $X := (A^{(0)}, M, A^{(\infty)})$ to $Y := (B^{(0)}, N, B^{(\infty)})$; then we should prove that the following diagram is commutative:

$$\begin{array}{ccc}
\omega_a^{(0)}(X) = \mathbf{C}^n & \xrightarrow{M(a)} & \omega_a^{(\infty)}(X) = \mathbf{C}^n. \\
\Big\downarrow{\scriptstyle S^{(0)}(a)} & & \Big\downarrow{\scriptstyle S^{(\infty)}(a)} \\
\omega_a^{(0)}(Y) = \mathbf{C}^p & \xrightarrow[N(a)]{} & \omega_a^{(\infty)}(Y) = \mathbf{C}^p
\end{array}$$

But this immediately follows from equality $NS^{(0)} = S^{(\infty)}M$ by the very definition of morphisms in $\mathscr{C}_f$.

However, there remains a problem with the above construction: for arbitrary $a \in \mathbf{C}^*$ and $M \in \mathrm{GL}_n(\mathscr{M}(\mathbf{C}^*))$, $M(a)$ needs not be defined. We now tackle this difficulty. Recall from the construction of M coming from $A \in \mathrm{GL}_n(\mathbf{C}(x))$ that $M^{(0)}(qx) = A(x)M^{(0)}(x)(A^{(0)})^{-1}$ and similarly at ∞, from which we deduced in Chapter 12:

$$\mathrm{Sing}(A) \subset \mathrm{Sing}(M) \subset q^{\mathbf{Z}}\mathrm{Sing}(A) \Longrightarrow q^{\mathbf{Z}}\mathrm{Sing}(M) = q^{\mathbf{Z}}\mathrm{Sing}(A),$$

where the singular locus $\mathrm{Sing}(A)$ (same for $\mathrm{Sing}(M)$) consists of poles of A and poles of A^{-1}, equivalently, from poles of A and zeroes of det A. So $M(a)$ is well defined (and invertible) for $a \in \mathbf{C}^* \setminus q^{\mathbf{Z}}\mathrm{Sing}(A)$.

We now define, for any finite subset $\Sigma \subset \mathbf{C}^*$, the full subcategory $\mathscr{E}_{f,\Sigma}$ of $\mathscr{E}_f$ whose objects A are such that $\mathrm{Sing}(A) \subset q^{\mathbf{Z}}\Sigma$; and the full subcategory $\mathscr{C}_{f,\Sigma}$ of $\mathscr{C}_f$

whose objects $(A^{(0)}, M, A^{(\infty)})$ are such that $\mathrm{Sing}(M) \subset q^{\mathbf{Z}}\Sigma$. From the relations $\mathrm{Sing}(A_1 \otimes A_2) \subset \mathrm{Sing}(A_1) \cup \mathrm{Sing}(A_2)$, etc, we deduce easily that $\mathscr{E}_{f,\Sigma}$ and $\mathscr{C}_{f,\Sigma}$ are Tannakian subcategories, and that they are $\otimes$-equivalent.

EXERCISE 13.64. Fill in necessary details in the way we fixed the problem of singularities.

The above argument about $(A^{(0)}, M, A^{(\infty)}) \rightsquigarrow M(a)$ now goes without problem in $\mathscr{C}_{f,\Sigma}$ and we do get an element Γ_a of $\mathrm{Iso}^{\otimes}(\omega_a^{(*0)}, \omega_a^{(*\infty)})$, the fibre functors $\omega_a^{(*0)}, \omega_a^{(*\infty)}$ denoting the restrictions to $\mathscr{C}_{f,\Sigma}$) of the natural fibre functors $\omega_a^{(*0)}, \omega_a^{(*\infty)}$. We consider Γ_a as defining an horizontal arrow in the pictures of the global groupoid, see the figures 13.2 (left and right).

Among the galoisian paths in our groupoid, we now count:

- all those coming from the local groupoid at 0, i.e., the elements of all the sets $\mathrm{Iso}^{\otimes}(\omega_a^{(*0)}, \omega_b^{(*0)})$ coming from our local study (and properly restricted to the relevant category);
- same thing at ∞
- all paths $\Gamma_a \in \mathrm{Iso}^{\otimes}(\omega_a^{(*0)}, \omega_a^{(*\infty)})$ just defined.

Together, they generate a subgroupoid, the Zariski closure of which is the smallest proalgebraic subgroupoid containing them all.

THEOREM 13.7 (Global fuchsian density theorem). *That groupoid is the whole groupoid of all the* $\mathrm{Iso}^{\otimes}(\omega_a^{(*x)}, \omega_b^{(*y)})$, $a, b \in \mathbf{C}^*$, $x, y \in \{0, \infty\}$.

PROOF. We use the following form of Chevalley-Kolchin criterion (Appendix G). Let $\mathcal{X} := (A^{(0)}, M, A^{(\infty)})$ an object of $\mathscr{C}_{f,\Sigma}$ and, for each basepoint $\omega_a^{(*x)}$ of the groupoid ($x \in \{0, \infty\}$), let $\mathcal{D}_a^{(*x)} \subset \omega_a^{(*x)}(\mathcal{X})$ a line. Assume that this family of lines is globally stable under our candidate Zariski generators. This means that:

(1) for all $a, b \in \mathbf{C}^*$, all arrows of $\mathrm{Iso}^{\otimes}(\omega_a^{(*0)}, \omega_b^{(*0)})$ send $\mathcal{D}_a^{(*0)}$ to $\mathcal{D}_b^{(*0)}$;
(2) same thing at ∞;
(3) for all $a \in \mathbf{C}^*$, the previously defined Γ_a sends $\mathcal{D}_a^{(*0)}$ to $\mathcal{D}_a^{(*\infty)}$.

We are going to deduce from those assumptions that our family of lines is globally stable under the whole Galois groupoid, i.e., that, for all $a, b \in \mathbf{C}^*$, $x, y \in \{0, \infty\}$, all elements of $\mathrm{Iso}^{\otimes}(\omega_a^{(*x)}, \omega_b^{(*y)})$ send $D_a^{(*x)}$ to $D_b^{(*y)}$. Chevalley-Kolchin criterion then says that the conclusion of our theorem holds.

We now proceed to prove our contention. The conditions at 0 mean that the subspaces $\mathcal{D}_a^{(*0)}$ define a subrepresentation of the representation of $G_f^{(0)} = \mathrm{Gal}(\mathscr{E}_f^{(0)})$ corresponding to $A^{(0)}$ by 13.2.3.4, so, by Tannakian duality, it corresponds to a rank 1 subobject $x^{(0)} : a^{(0)} \hookrightarrow A^{(0)}$ such that $\mathcal{D}_a^{(*0)}$ is the image of $\omega_a^{(*0)}(x^{(0)}) : \omega_a^{(*0)}(a^{(0)}) \to \omega_a^{(*0)}(A^{(0)})$, i.e., of $x^{(0)}(a) : \mathbf{C} \to \mathbf{C}^n$; we may even take $a^{(0)}$ in $\mathscr{P}^{(0)}$, i.e., $a^{(0)} \in \mathbf{C}^*$; and $x^{(0)} \in \mathrm{Mat}_{n,1}(\mathbf{C}(\{x\}))$ is such that $a^{(0)}\sigma_q x^{(0)} = A^{(0)}x^{(0)}$. By the usual argument, $x^{(0)}$ is holomorphic all over $\mathbf{C}^*$; and last, $\mathcal{D}_a^{(*0)} = \mathbf{C}x^{(0)}(a)$. Similar statement hold at ∞, which we abstain to completely write down.

The condition that Γ_a sends $\mathcal{D}_a^{(*0)}$ to $\mathcal{D}_a^{(*\infty)}$ means in turn that $M(a)x^{(0)}(a) = m(a)x^{(\infty)}(a)$ for some $m(a) \in \mathbf{C}^*$. Taken for all $a \in \mathbf{C}^* \setminus q^{\mathbf{Z}}\Sigma$, this defines a meromorphic function m on $\mathbf{C}^*$ with poles in $q^{\mathbf{Z}}\Sigma$, whence a rank 1 object $(a^{(0)}, m, a^{(\infty)})$ of $\mathscr{C}_{f,\Sigma}$.

EXERCISE 13.65. A condition *should* have been checked here. Can you state it and check it?

Taken together, the relations $\begin{cases} a^{(0)}\sigma_q x^{(0)} = A^{(0)} x^{(0)}, \\ a^{(\infty)}\sigma_q x^{(\infty)} = A^{(\infty)} x^{(\infty)}, \\ M(a)x^{(0)}(a) = m(a)x^{(\infty)}(a), \end{cases}$ mean that $\underline{x} := (a^{(0)}, m, a^{(\infty)})$ is a subobject of $\mathcal{X} := (A^{(0)}, M, A^{(\infty)})$ in $\mathscr{C}_{f,\Sigma}$, the injection morphism being $\phi := (x^{(0)}, x^{(\infty)})$. We now apply the functoriality condition for an arbitrary $h \in \mathrm{Iso}^{\otimes}(\omega_a^{(*0)}, \omega_b^{(*0)})$ and obtain a commutative diagram:

$$\begin{array}{ccc} \omega_a^{(0)}(\underline{x}) = \mathbf{C}^n & \xrightarrow{h(\underline{x})} & \omega_b^{(0)}(\underline{x}) = \mathbf{C}^n. \\ {\scriptstyle \omega_a^{(0)}(\phi)} \big\downarrow & & \big\downarrow {\scriptstyle \omega_b^{(0)}(\phi)} \\ \omega_a^{(0)}(\mathcal{X}) = \mathbf{C}^p & \xrightarrow[h(\mathcal{X})]{} & \omega_b^{(0)}(\mathcal{X}) = \mathbf{C}^p \end{array}$$

It implies in particular that $h(\mathcal{X})$ sends $D_a^{(*0)}$ to $D_b^{(*0)}$; and a similar argument holds at ∞.

EXERCISE 13.66. Sketch the corresponding argument for $h \in \mathrm{Iso}^{\otimes}(\omega_a^{(*0)}, \omega_a^{(*\infty)})$.

In conclusion, as announced, the family of all lines $D_a^{(*x)}$ is stable under the whole Galois groupoid and Chevalley-Kolchin criterion applies, yielding the desired Zariski-density statement. □

EXERCISE 13.67. Let $A \in \mathrm{GL}_n(\mathbf{C}(x))$. Give a condition for $\mathrm{Gal}(A)$ to be trivial. (Use the exercise at the end of 13.2.5.4.)

13.3.3. Second construction of the global Galois group.

13.3.3.1. Here we look for an other approach, using the connection matrix P itself, hence closer to Birkhoff's approach and its galoisian interpretation by Etingof (using values of P). The main differences with 13.3.2 are:

- We use as connection data $(A^{(0)}, P, A^{(\infty)})$ instead of $(A^{(0)}, M, A^{(\infty)})$;
- Our Galois groupoid will only have two base points, 0 and ∞, represented by fibre functors $\omega^{(0)}$ and $\omega^{(\infty)}$.

The local contributions at 0 and ∞ will not differ much, as compared to 13.3.2; but galoisian paths from 0 to ∞ will involve *twisted* values $\check{P}(a)$ of the connection matrix P, emphasizing the added colmplexity of the fuchsian case with respect to the regular case considered by Etingof.

Our first step is to define the relevant category $\mathscr{B}_f$ of *Birkhoff's connection data*; first, the objects. If $(A^{(0)}, M, A^{(\infty)})$ is an object of $\mathscr{C}_f$, the corresponding Birkhoff matrix is:

$P = e_{A^{(\infty)}}^{-1} M e_{A^{(0)}}$, which we decide to write at will as $\begin{cases} e_{A^{(\infty)}} P = M e_{A^{(0)}}, \text{ or} \\ M = e_{A^{(\infty)}} P e_{A^{(0)}}^{-1}. \end{cases}$

Under those relations, the reader is invited to check that:

$$(\sigma_q M)A^{(0)} = A^{(\infty)} M \Longleftrightarrow \sigma_q P = P.$$

So we define the objects of $\mathscr{B}_f$ as triples:

$$(A^{(0)}, P, A^{(\infty)}) \in \mathrm{GL}_n(\mathbf{C}) \times \mathrm{GL}_n(\mathscr{M}(\mathbf{E}_q)) \times \mathrm{GL}_n(\mathbf{C})$$

and conclude that they are in bijective correspondence with objects of $\mathscr{C}_f$.

Let $(S^{(0)}, S^{(\infty)}) : (A^{(0)}, M, A^{(\infty)}) \to (B^{(0)}, N, B^{(\infty)})$ a morphism in $\mathscr{C}_f$ and let $Q = e_{B^{(\infty)}}^{-1} N e_{B^{(0)}}$. We want to deduce from $(S^{(0)}, S^{(\infty)})$ a morphism $(A^{(0)}, P, A^{(\infty)}) \to (B^{(0)}, Q, B^{(\infty)})$ in $\mathscr{B}_f$. We want to translate conditions $(\sigma_q S^{(0)})A^{(0)} = B^{(0)} S^{(0)}$, $(\sigma_q S^{(\infty)})A^{(\infty)} = B^{(\infty)} S^{(\infty)}$ and $NS^{(0)} = S^{(\infty)} M$ into conditions involving P, Q in lieu of M, N. Refering to Chapter 11, in particular 11.1.2.5, we use the diagrams:

$$
\begin{array}{ccc}
\overline{A^{(0)}} & \xrightarrow{X} & A^{(0)} \\
{\scriptstyle R^{(0)}}\downarrow & & \downarrow{\scriptstyle S^{(0)}} \\
\overline{B^{(0)}} & \xrightarrow[Y]{} & B^{(0)}
\end{array}
\qquad
\begin{array}{ccc}
\overline{A^{(\infty)}} & \xrightarrow{X'} & A^{(\infty)}, \\
{\scriptstyle R^{(\infty)}}\downarrow & & \downarrow{\scriptstyle S^{(\infty)}} \\
\overline{B^{(\infty)}} & \xrightarrow[Y']{} & B^{(\infty)}
\end{array}
$$

where $R^{(0)}, R^{(\infty)} \in \mathrm{Mat}_{p,n}(\mathbf{C})$. Also recall that $S^{(0)} e_{A^{(0)}} = e_{B^{(0)}} B^{(0)}$ and similarly at ∞, so the correspondence $(R^{(0)}, R^{(\infty)}) \leftrightarrow (S^{(0)}, S^{(\infty)})$ is bijective.

EXERCISE 13.68. Check that $NS^{(0)} = S^{(\infty)} M \Leftrightarrow QR^{(0)} = R^{(\infty)} P$.

So we define morphisms $(A^{(0)}, P, A^{(\infty)}) \to (B^{(0)}, Q, B^{(\infty)})$ in $\mathscr{B}_f$ as pairs $(R^{(0)}, R^{(\infty)}) \in \mathrm{Mat}_{p,n}(\mathbf{C}) \times \mathrm{Mat}_{p,n}(\mathbf{C})$ such that:

$$
\begin{cases}
R^{(0)} \overline{A^{(0)}} = \overline{B^{(0)}} R^{(0)}, \\
R^{(\infty)} \overline{B^{(\infty)}} = \overline{A^{(\infty)}} R^{(\infty)}, \\
QR^{(0)} = R^{(\infty)} P.
\end{cases}
$$

DEFINITION 13.69. The category $\mathscr{B}_f$ has objects triples $(A^{(0)}, P, A^{(\infty)})$ and morphisms $(R^{(0)}, R^{(\infty)})$ as described above.

EXERCISE 13.70. (i) Define composition and check the relevant axioms.
(ii) Check that $\mathscr{B}_f$ is a $\mathbf{C}$-linear abelian category.
(iii) Describe the functors from $\mathscr{C}_f$ to $\mathscr{B}_f$ and back, and state precisely their properties with respect to the $\mathbf{C}$-linear abelian structures.

Note that those functors are inverse of each other, so they are *isomorphisms*, which is a (much) stronger property than equivalence (the latter invoves only the existence of a quasi-inverse; see **[154]**).

13.3.3.2. To define a tensor product on $\mathscr{B}_f$, we use the above isomorphism to transfer the one on $\mathscr{C}_f$. We use the definition given by equation (13.2), and we replace $M_i, S_i^{(0)}, S_i^{(\infty)}$, $i = 1, 2$, by their expressions in terms of $P_i, R_i^{(0)}, R_i^{(\infty)}$:

$$(A_1^{(0)}, P_1, A_1^{(\infty)}) \otimes (A_2^{(0)}, P_2, A_2^{(\infty)}) := (A_1^{(0)} \otimes A_2^{(0)}, P_1 \underline{\otimes} P_2, A_1^{(\infty)} \otimes A_2^{(\infty)}), \tag{13.4}$$

$$(R_1^{(0)}, R_1^{(\infty)}) \otimes (R_2^{(0)}, R_2^{(\infty)}) := (R_1^{(0)} \otimes R_2^{(0)}, R_1^{(\infty)} \otimes R_2^{(\infty)}). \tag{13.5}$$

There is a lot of checking to be done (left to you, patient reader!), but first we must explain $P_1 \underline{\otimes} P_2$. Note that:

$$
\begin{aligned}
P_1 \otimes P_2 &= (e_{A_1^{(\infty)}}^{-1} M_1 e_{A_1^{(0)}}) \otimes (e_{A_2^{(\infty)}}^{-1} M_2 e_{A_2^{(0)}}) \\
&= (e_{A_1^{(\infty)}} \otimes e_{A_2^{(\infty)}})^{-1} (M_1 \otimes M_2)(e_{A_1^{(0)}} \otimes e_{A_2^{(0)}}),
\end{aligned}
$$

while the formulas of 13.3.3.1 should rather attach to the tensor product $(A_1^{(0)} \otimes A_2^{(0)}, M_1 \otimes M_2, A_1^{(\infty)} \otimes A_2^{(\infty)})$ in $\mathscr{C}_f$ the connection matrix:

$$P := e_{A_1^{(\infty)} \otimes A_2^{(\infty)}}^{-1} (M_1 \otimes M_2) e_{A_1^{(0)} \otimes A_2^{(0)}}.$$

We call the latter $P_1\underline{\otimes}P_2$; it is defined by the formula:

$$P_1\underline{\otimes}P_2 := \Phi(A_1^{(\infty)}, A_2^{(\infty)})(P_1 \otimes P_2)\Phi(A_1^{(0)}, A_2^{(0)})^{-1}, \text{ where } \Phi(A_1, A_2) := (e_{A_1} \otimes e_{A_2})e_{A_1\otimes A_2}^{-1}.$$

So the notation $P_1\underline{\otimes}P_2$ is somewhat misleading, since, along with P_1 and P_2, the twisted tensor product involves the local data $A_1^{(0)}, A_2^{(0)}, A_1^{(\infty)}, A_2^{(\infty)}$.

In rank 1, we retrieve the "cocycle" of elliptic functions $\phi(c,d) := e_{q,c}e_{q,d}/e_{q,cd}$ from 2.3.1. The usual way of extending scalar functions to semi-simple matrices (F.4), Appendix F then shows that $\Phi(A_1, A_2)$ is really simply $\Phi((A_1)_s, (A_2)_s)$. Indeed, for unipotent matrices U_1, U_2 we have $e_{U_1} \otimes e_{U_2} = e_{U_1\otimes U_2}$, so $(A_1)_u, (A_2)_u$ do not contribute to the twisting. We emphasize the general formula:

$$e_{A_1} \otimes e_{A_2} = \Phi(A_1, A_2)e_{A_1\otimes A_2}.$$

EXERCISE 13.71. Justify equality $e_{U_1} \otimes e_{U_2} = e_{U_1\otimes U_2}$.

Summarizing our constructions:

THEOREM 13.8. (i) *The tensor categories $\mathscr{C}_f$ and $\mathscr{B}_f$ are isomorphic.*
(ii) *We have on $\mathscr{B}_f$ two fibre functors:*

$$\omega^{(0)} : \begin{cases} (A^{(0)}, P, A^{(\infty)}) \rightsquigarrow \mathbf{C}^n \text{ underlying } A^{(0)}, \\ (R^{(0)}, R^{(\infty)}) \rightsquigarrow R^{(0)}, \end{cases}$$

and:

$$\omega^{(\infty)} : \begin{cases} (A^{(0)}, P, A^{(\infty)}) \rightsquigarrow \mathbf{C}^n \text{ underlying } A^{(\infty)}, \\ (R^{(0}, R^{(\infty)}) \rightsquigarrow R^{(\infty)}. \end{cases}$$

PROOF. (i) is immediate from the previous discussion.
(ii) is easy and left as an exercise to the reader. □

13.3.3.3. The next step is to produce galoisian iso- and automorphisms in $\mathscr{B}_f$. We begin by local automorphisms.

LEMMA 13.72. (i) *Let $\gamma : \mathbf{C}^* \to \mathbf{C}^*$ an arbitrary map and let $\lambda \in \mathbf{C}$. Then $(A^{(0)}, P, A^{(\infty)}) \rightsquigarrow \gamma\left(A_s^{(0)}\right)(A_u^{(0)})^\lambda$ is an element of $Aut\left(\omega^{(0)}\right)$; same at ∞.*

(ii) *If moreover γ is a group morphism such that $\gamma(q) = 1$, then the above is an element of $Aut^{\otimes}\left(\omega^{(0)}\right)$; same at ∞.*

PROOF. In (i), only functoriality has to be checked, not $\otimes$-compatibility. It rests on the fact, recalled in F.4, Appendix F, that:

$$RA = BR \Longrightarrow \left(R\gamma(A_s) = \gamma(B_s)R \text{ and } RA_u^\lambda = B_u^\lambda R\right).$$

(ii) is then easy. □

EXERCISE 13.73. Where exactly is used the fact that $\gamma(q) = 1$?

Clearly, the lemma allows us to define two proalgebraic subgroups $G^{(0)} \subset \mathrm{Aut}^{\otimes}\left(\omega^{(0)}\right)$ and $G^{(\infty)} \subset \mathrm{Aut}^{\otimes}\left(\omega^{(\infty)}\right)$, both having the (by now!) well known structure $\mathbf{E}_q^\vee \times \mathbf{C}$.

Now we must connect $\omega^{(0)}$ and $\omega^{(\infty)}$, i.e., glue $G^{(0)}$ and $G^{(\infty)}$, hopefully using P. First, without caring whether $P(a)$ is defined or not (as in 13.3.2.4, this will require that we restrict $\mathscr{B}_f$) we try to ensure that $(A^{(0)}, P, A^{(\infty)}) \rightsquigarrow P(a)$ defines an element of $\mathrm{Iso}^{\otimes}(\omega_a^{(0)}, \omega_a^{(\infty)})$; we begin by functoriality, so let $(R^{(0)}, R^{(\infty)}) : (A^{(0)}, P, A^{(\infty)}) \to (B^{(0)}, Q, B^{(\infty)})$ a morphism in $\mathscr{B}_f$. From conditions $QR^{(0)} = R^{(\infty)}P$, we get that $Q(a)R^{(0)} = R^{(\infty)}P(a)$ (recall that $R^{(0)}, R^{(\infty)}$ are constant); but that is exacly the functoriality condition (you *should* draw the corresponding diagrams). So indeed we have an element of $\mathrm{Iso}(\omega_a^{(0)}, \omega_a^{(\infty)})$. However, because $P_1 \otimes P_2$ is not $P_1 \underline{\otimes} P2$, we cannot ensure that $(P_1 \underline{\otimes} P)(a) = P_1(a) \otimes P_2(a)$, so the above is not $\otimes$ compatible.

EXERCISE 13.74. Check that, restricting to regular equations, the problem disappears.

We can compensate the defect of $\otimes$-compatibility by using instead (again, without taking care of the definedness of the values at a):

$$(A^{(0)}, P, A^{(\infty)}) \rightsquigarrow (e_{A^{(\infty)}} P e_{A^{(0)}}^{-1})(a).$$

EXERCISE 13.75. Check that this now $\otimes$-compatible (this should, at this stage, require no calculation at all) but no more functorial.

After having twisted the tensor product, we shall now overcome all those miseries by twisting the connection matrix in a convenient way. First, we choose for every $a \in \mathbf{C}^*$ a group morphism $g_a : \mathbf{C}^* \to \mathbf{C}^*$ such that[28] $q \mapsto a$. We know by Exercise 13.23 (or by the following one) that it can be done. In 13.3.3.5 we shall discuss a particular choice of g_a and its consequences.

EXERCISE 13.76. Let K an algebraically closed field of characteristics 0, let $x \in K^*$ not a root of unity, and let $a \in K^*$ arbitrary. Show that there exists a group morphism $g : K^* \to K^*$ such that $g(x) = a$. (As a group, $K^* \simeq (\mathbf{Q}/\mathbf{Z}) \times V$ for some $\mathbf{Q}$-linear space V.)

Now define the *mistigri*:

$$\forall c \in \mathbf{C}^* , \; \psi_a(c) := \frac{e_{q,c}(a)}{g_a(c)} \cdot$$

EXERCISE 13.77. Check that $\psi_a(c)$ really depends on $\overline{c} \in \mathbf{E}_q$ only and that:

$$\phi_{c,d}(a) = \frac{\psi_a(c)\psi_a(d)}{\psi_a(cd)} \cdot$$

We now extend ψ_a to matrices in the usual way: if $A_s = P\mathrm{Diag}(c_1, \ldots, c_n)P^{-1}$, then $\Psi_a(A) := P\mathrm{Diag}(\psi_a(c_1), \ldots, \psi_a(c_n))P^{-1}$. From the exercise, it follows that $\Psi_a(A)$ depends on $\overline{A_s}$ only and that:

$$\Psi_a(A_1) \otimes \Psi_a(A_2) = \Psi_a(A_1 \otimes A_2)\Phi(A_1, A_2)(a).$$

We then set[29]:

$$\check{P}(a) := \left(\Psi_a(A_s^{(\infty)})\right) P(a) \left(\Psi_a(A_s^{(0)})\right)^{-1}.$$

Exacly as for the twisted tensor product $\underline{\otimes}$, the notation for the twisted connection matrix $\check{P}$ is somewhat misleading, since it does not depend on the sole P but also on the local data $A_s^{(0)}, A_s^{(\infty)}$.

[28]In [**218**, bottom p 95], the condition is improperly stated as $q \mapsto 1$.

[29]Again a slight discrepancy with [**218**], where the exponents ± 1 are misplaced.

PROPOSITION 13.78. *The transformation* $(A^{(0)}, P, A^{(\infty)}) \rightsquigarrow \check{P}(a)$ *is a tensor isomorphism* $\omega^{(0)} \to \omega^{(\infty)}$.

PROOF. This follows readily from the various steps of the construction. □

EXERCISE 13.79. Check functoriality. (The necessary property to do that has been obtained along the way.)

13.3.3.4. Now we fill some holes and then state the density result which motivates all that fuss. First we note that $P(a)$ involves $M(a)$, the singularities of which lay inside $q^{\mathbf{Z}}\mathrm{Sing}(A)$. It also involves all $e_{q,c}(a)$, c an exponent of $A^{(0)}$ or $A^{(\infty)}$. In 13.3.2.4, we used the fact that $\mathrm{Sing}(A_1 \otimes A_2) \subset \mathrm{Sing}(A_1) \cup \mathrm{Sing}(A_2)$; but exponents are multiplied (and inverted by dualisation), so $\otimes$-stability will impose on us that the set of prohibited exponents be a group. We are thus led to the following definition.

Let Σ a finite subset and C a finitely generated subgroup of $\mathbf{C}^*$. Let $\mathscr{C}_{f,\Sigma,C}$ the full category of $\mathscr{C}_f$ made up of objects $(A^{(0)}, M, A^{(\infty)})$ such that $\mathrm{Sing}(M) \subset \Sigma$ and $\mathrm{Sp}\ A^{(0)}, \mathrm{Sp}\ A^{(\infty)} \subset C$. Let $\mathscr{B}_{f,\Sigma,C}$ its image in $\mathscr{B}_f$. Both $\mathscr{C}_{f,\Sigma,C}$ and $\mathscr{B}_{f,\Sigma,C}$ are Tannakian subcategories, respectively of $\mathscr{C}_f$ and $\mathscr{B}_f$ and we have restrictions on them of our fibre functors $\omega^{(0)}, \omega^{(\infty)}$. So we can compute thet Galois group $\mathrm{Gal}(A)$ of any object of $\mathscr{E}_f$ by using a corresponding triple in some $\mathscr{B}_{f,\Sigma,C}$.

THEOREM 13.9 (Yet another variant of the global density theorem). *The local components already described, along with all values* $\check{P}(a)^{-1}\check{P}(b)$, $a, b \in \mathbf{C}^* \setminus q^{\mathbf{Z}}(\Sigma \cup C)$ *generate a Zariski-dense subgroupoid of the Galois groupoid of* A.

PROOF. It goes along the same lines as usual, details may be found in [**218**]. □

EXERCISE 13.80. How does one choose the set Σ and the group C?

COROLLARY 13.81. *For any given* A*, the following is a family of Zariski generators (or topological generators) of* $\mathrm{Gal}(A)$:

- $\gamma_1(A_s^{(0)})$, $\gamma_2(A_s^{(0)})$, $A_u^{(0)}$;
- *one conjugate of those by one value* $\check{P}(a)$;
- *all the* $\check{P}(a)^{-1}\check{P}(b)$, a, b *as above (and* a *may be a fixed one).*

13.3.3.5. As promised, we now discuss the choice of the family of morphisms g_a, $a \in \mathbf{C}^*$, and its consequences. As a preliminary remark, note that changing such a choice for another family, say g'_a, we have $g'_a = \gamma_a g_a$ for a $\gamma_a \in \mathbf{E}_q^\vee$ (the multiplication there is that in $\mathbf{C}^{*\vee}$, i.e., that such $\gamma_a g_a : x \mapsto \gamma_a(x) g_a(x)$; it is more generally that of functions with values in $\mathbf{C}^*$, as in the next equality). The twisting factor ψ_a (the "mistigri") is thereby transformed to $\psi'_a := \dfrac{1}{\gamma_a}\psi_a$ and, accordingly, the twisted connection matrix to:

$$\check{P}'(a) = \check{P}^{\gamma_a}(a) := \gamma_a\left(A_s^{(\infty)}\right)^{-1} \check{P}(a)\gamma_a\left(A_s^{(0)}\right).$$

Since $\gamma_a\left(A_s^{(0)}\right) \in G_f^{(0)}$ and $\gamma_a\left(A_s^{(\infty)}\right) \in G_f^{(\infty)}$, we see that the galoisian isomorphism $\check{P}(a) \in \mathrm{Iso}^{\otimes}(\omega_a^{(0)}, \omega_a^{(\infty)})$ is conjugated by the Galois loops in $G_f^{(0)}$ and in $G_f^{(\infty)}$ induced by $\gamma_a \in \mathbf{C}^{*\vee}$. This explains why any particular choice of the g_a will lead to the same conclusions (including the density theorem).

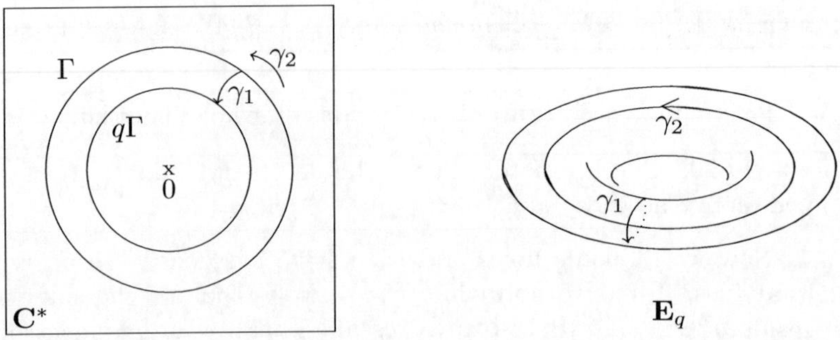

FIGURE 13.3. Generating loops on the infinitesimal torus $\mathbf{E}_q$

EXERCISE 13.82. What if we do not use the same g_a at 0 and at ∞?

From now on, we decide to use continuous morphisms $g_a \in \mathrm{Hom}_{grtop}(\mathbf{C}^*, \mathbf{C}^*)$, with the hope of getting intuitive interpretations. From 13.2.5.2, we know that:

$$g_a(uq^y) = u^n e^{2i\pi\lambda y} q^{\mu y} \text{ for some } n \in \mathbf{Z}\ ,\ \lambda, \mu \in \mathbf{R}.$$

The condition $g_a(q) = a$ reads $e^{2i\pi(\lambda+\tau\mu)} = a$. So we take $n = 0$, choose a logarithm $2i\pi\alpha$ of a and split α as $\lambda+\tau\mu$, $\lambda, \mu \in \mathbf{R}$. (After the choice of the logarithm, λ, μ are unique, since $\mathbf{C} = \mathbf{R}\oplus\mathbf{R}\tau$.) Of course, α cannot be chosen continuous in a; but g_a is continuous for each given a. The resulting morphism is given by $g_a(uq^y) = e^{2i\pi\alpha y}$.

EXERCISE 13.83. Show that $g_a(c)$ cannot be chosen continuous in a and c simultaneously.

We now find two ambiguities in our construction. First, the complex α is defined modulo $\mathbf{Z}$, so we set $\alpha' := \alpha + 1$ and thus obtain $g'_a(uq^y) = e^{2i\pi y} g_a(y)$, i.e., $g'_a = \gamma_2 g_a(y)$, so $\check{P}$ is transformed into $\check{P}^{\gamma_2}$. This being related to the indeterminacy of the logarithm, we interpret γ_2 as the fundamental loop around 0 in $\mathbf{C}^*$.

The second ambiguity (or phenomenon which can be called so) is that $\check{P}$ is not elliptic, contrary to P. Transforming a to qa transforms α to $\alpha + \tau$, so (writing $c = uq^y$):

$$\frac{g_{qa}(c)}{g_a(c)} = e^{2i\pi\tau y} = \frac{c}{\gamma_1(c)} \Longrightarrow \psi_{qa}(c) = \gamma_1(c)\psi_a(c) \text{ since } e_c(qa) = ce_c(a).$$

This time, $\check{P}$ transforms to $\check{P}^{\gamma_1}$. To interpret γ_1, which is related to the move $a \to qa$, we think as $\mathbf{E}_q$ as a folding of $\mathbf{C}^*$ along those waves, and γ_2 thus corresponds to the resulting fundamental loop on $\mathbf{E}_q$ while γ_1 corresponds to the other one, inherited from $\mathbf{C}^*$.

Recalling that we use those $\gamma \in \mathbf{E}_q^\vee$ either for the Galois group at 0 or to the one at ∞, we think of two "infinitesimal elliptic curves" $\mathbf{E}_q$, one at 0, one at ∞. On each we have two fundamental loops γ_1 and γ_2, see Figure 13.3. However, we should recall that those represent only the semi-simple part of what we would like to see as q-monodromy. There is a unipotent part, which is less intuitive. The general picture will be completed in Chapter 15 when we find that the unipotent part (related to the q-logarithm) is just the first step on an infinite scale of *Stokes operators*.

13.3.3.6. Now we try to relate the two approaches (the one in 13.3.2 and the one here in 13.3.3). First, given that $\mathscr{C}_f$ and $\mathscr{B}_f$ are isomorphic, not simply equivalent, we can identify them: we consider $(A^{(0)}, M, A^{(\infty)})$ and $(A^{(0)}, P, A^{(\infty)})$ as two descriptions (models, encodings ...) of a same object; and similarly with morphisms $(S^{(0)}, S^{(\infty)})$ and $(S^{(0)}, S^{(\infty)})$. Of course, all that assuming that the adequate relations from 13.3.3.1 are satisfied!) Actually, we consider them all as models of objects $\mathcal{X}$, $\mathcal{Y}$ and morphism $F : \mathcal{X} \to \mathcal{Y}$ in $\mathscr{E}_f$.

Exercise 13.84. Make precise the relevant restrictions which allow one to comute $M(a)$, etc.

On the category $\mathscr{C}_f \simeq \mathscr{B}_f$, we now have two groupoids of fibre functors. The first one has the base $\mathbf{C}^* \sqcup \mathbf{C}^*$ (two copies of $\mathbf{C}^*$, one "at 0" and the other "at ∞"), or equivalently the two families of fibre functors $\omega_a^{(0)}, \omega_a^{(\infty)}$, $a \in \mathbf{C}^*$; while the other has base $\{0, \infty\}$, or equvalently the two fibre functors $\omega^{(0)}, \omega^{(\infty)}$. In both cases, the sets of arrows are the $\mathrm{Iso}^{\otimes}(-,-)$. In order to have clear notations, we consider:

- for the first groupoid, the "galoisian paths" $\Gamma_a \in \mathrm{Iso}^{\otimes}(\omega_a^{(0)}, \omega_a^{(\infty)})$ defined by $\mathcal{X} \to M(a)$;
- for the second groupoid, the "twisted galoisian paths" $\check{\Gamma}_a \in \mathrm{Iso}^{\otimes}(\omega^{(0)}, \omega^{(\infty)})$ defined by $\mathcal{X} \to \check{P}(a)$.

In a (hopefully) very natural way, we shall now define other galoisian paths linking those two groupoids:

$$\Gamma_a^{(0)} \in \mathrm{Iso}^{\otimes}(\omega^{(0)}, \omega_a^{(0)}) \text{ and } \Gamma_a^{(\infty)} \in \mathrm{Iso}^{\otimes}(\omega^{(\infty)}, \omega_a^{(\infty)}),$$

in such a way that:

$$\check{\Gamma}_a = \left(\Gamma_a^{(\infty)}\right)^{-1} \circ \Gamma_a \circ \Gamma_a^{(0)}.$$

The whole picture then looks like:

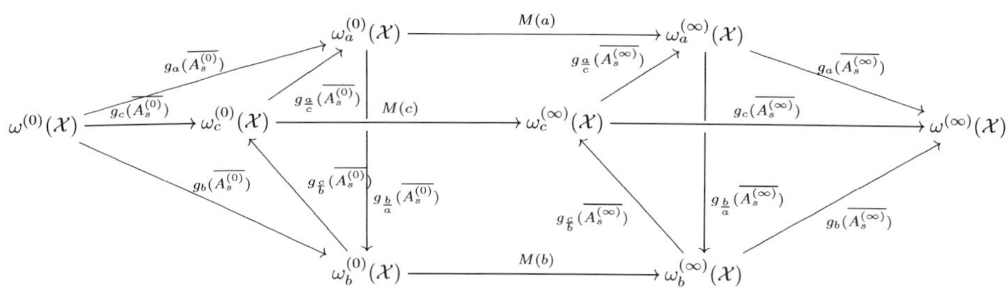

So first we recall that $P(a) = e_{A^{(\infty)}}(a)^{-1} M(a) e_{A^{(0)}}(a)$, so that:

$$\check{P}(a) = \left(\Psi_a(A_s^{(\infty)})\right) P(a) \left(\Psi_a(A_s^{(0)})\right)^{-1} = g_a(A_s^{(\infty)}) M(a) g_a(A_s^{(0)})^{-1},$$

which leads us to define $\Gamma_a^{(0)}$ as $\mathcal{X} \rightsquigarrow g_a(A_s^{(0)})^{-1}$ and $\Gamma_a^{(\infty)}$ as $\mathcal{X} \rightsquigarrow g_a(A_s^{(\infty)})^{-1}$ (check the exponents!). Obviously this enforces the stated relation between $\check{\Gamma}_a$ and Γ_a. We now need to check that $\Gamma_a^{(0)}$ and $\Gamma_a^{(\infty)}$ are indeed tensor isomorphisms $\omega^{(0)} \to \omega_a^{(0)}$ and $\omega^{(\infty)} \to \omega_a^{(\infty)}$. Since g_a is a group morphism, $\otimes$-compatibility is obvious.

So we go for functoriality. We do it at 0, it works the same at ∞. First note that $\mathcal{X} \rightsquigarrow e_{A^{(0)}}(a)$ is a functorial transformation $\omega^{(0)} \to \omega_a^{(0)}$; this follows from the diagram:

$$\begin{array}{ccc} \overline{A^{(0)}} & \xrightarrow{X} & A^{(0)} \\ {\scriptstyle R^{(0)}}\Big\downarrow & & \Big\downarrow{\scriptstyle S^{(0)}} \\ \overline{B^{(0)}} & \xrightarrow[Y]{} & B^{(0)} \end{array}$$

because it entails $S^{(0)}e_{A^{(0)}} = e_{B^{(0)}}R^{(0)} \Rightarrow S^{(0)}e_{A^{(0)}}(a) = e_{B^{(0)}}(a)R^{(0)}$.

EXERCISE 13.85. Check that this is exacly the required equality. (Write down the functoriality diagram.)

Then for any function $\psi(c)$ depending on $\overline{c}$ only, $\mathcal{X} \rightsquigarrow \psi\left(\overline{A_s^{(0)}}\right)$ is also functorial, this time $\omega^{(0)} \to \omega^{(0)}$, as flows from the already seen implication:

$$R\overline{A} = \overline{B}R \Longrightarrow R\psi\left(\overline{A_s}\right) = \psi\left(\overline{B_s}\right) R.$$

Combining those two facts, we get that $\mathcal{X} \rightsquigarrow \psi\left(\overline{A_s^{(0)}}\right)^{-1} e_{A^{(0)}}(a)$ is a functorial isomorphism $\omega^{(0)} \to \omega_a^{(0)}$; but it is precisely $\Gamma_a^{(0)}$.

13.3.4. Confluence of the global Galois group. We shall just mention *pour mémoire* two threads in the understanding of the degeneracy of the Galois group when $q \to 1$.

13.3.4.1. In [**4**], based on a theory of "noncommutative differentials" (and connections), (also see the preprint "On Galois theory of q-deformations of differential equations", URL `https://www.imj-prg.fr/preprints/333.pdf`), Yves André develops a Galois theory for difference, q-difference and differential equations at the same time. The presentation is purely algebraic.

From that, considering the setting of confluence (as we developed it in 11.2 and 12.3) as involving of families over the base $\mathbf{C}[[q-1]]$, he obtains a description of the degeneracy of the q-difference Galois group (q generic) into the differential Galois group ($q = 1$) as a *specialisation* (in the sense of algebraic geometry). In a sense, this meets the point of view of Fermat at the very beginning of the story as related in Chapter 1.

13.3.4.2. In [**218**, 4.1] a related story is told, more in the spirit of 11.2 and 12.3. The degeneracy of the local components of the Galois group is followed through its generators γ_1, γ_2 and the degeneracy of the connection component is linked to that of the connection matrix to monodromy.

The main result is that a particular Zariski-dense subgroup of the q-difference Galois group (depending on q) degenerates into a particular Zariski-dense subgroup of the differential Galois group, a kind ow twisted version of the monodromy group.

The result of [**218**, 4.1] is much less general but a bit more precise than the corresponding result of [**4**] (which, anyhow, covers much more ground).

CHAPTER 14

Irregular Equations

In the classical theory of (linear, complex, analytic) ordinary differential equations, divergent series appear as soon as the study is done near an irregular singular point. Physicists and astronomers used such series for numerical computations even before Poincaré invented asymptotic theory to give meaning to their computations, and even before Borel (and many others) proposed systematic procedures (of *summation*) to produce "true" functions from formal power series **[29, 113]**. Note that the theory was by nature local as it involved mostly series expansions.

In its computational aspects (i.e., not considering classification or Galois problems), the main modern progress of the theory was the introduction of Gevrey asymptotics, an "exact" asymptotics (see **[197]** for an explanation in our context) by Ramis. A nice survey of that story may be found in **[192]** and more systematic accounts in **[15, 120, 121, 150]**; see also **[188, 189]**. In particular, this opened the door to a deeper understanding of the Stokes phenomenon **[55]** (see a recent survey in **[27]**).

After Birkhoff **[23]** had thoroughly studied fuchsian q-difference equations (see Chapters 11, 12, and 13) and Carmichael had shown the existence of "enough" formal solutions **[44]**, the first surprise came with the discovery by Adams of the general presence of convergent solutions **[1, 2]**, even for irregular equations. This was further exploited by Birkhoff and his collaborators to tackle the study of irregular q-difference equations**[25, 26]** (also see Chapter 5).

Then, apart from breakthroughs by Bézivin and Ramis in the direction of index theory **[20, 191]** (also see Chapter 6), the revival of q-difference equation theory under the impulsion of Ramis **[190]** started bearing fruits for irregular equations only at the beginning of this millenium. Ramis and Zhang developed a q-Gevrey asymptotic theory and Zhang a "continuous" summation theory, then (with Marotte) a continuous multisummation theory, then a discrete summation theory **[163, 197, 198, 263, 265, 266]**.

We note, although that theme will not be very much pursued here (see however 14.2.5), that only more recently has the $q \to 1$ degeneracy phenomenon (confluence) been extended to irregular equations or systems (study of what happens to the q-Borel-Ramis transformation $\mathcal{B}_{q;1}$), notably by Di Vizio and Zhang **[66]**, and by Dreyfus **[73]** (although the latter work rather deals with q-difference systems, while the former deals with q-difference equations).

In this chapter, we present the summation theory of irregular scalar q-difference equations under rather restrictive assumptions. In the first section, after recalling why, thanks to Adams-Birkhoff factorisation, the heart of the study revolves around first order nonhomogeneous equations, we study the "archetypal" example of Tshakaloff series, the standard q-analogue of Euler series. This illustrates most of the themes of this chapter and quite a few of the next chapter. In the second section, we present discrete summation theory from [**265**], maybe sometimes sketchily in regards to proofs, but enough details are given to allow the reader to fill holes. In the third section, we briefly present continuous summation theory from [**263**], with bare hints of its extension in [**163**]. Indeed, although it came earlier, that theory involves in an essential way multivalued functions, while the general choice in this book has been to use uniform meromorphic functions.

We say little about the extension of Zhang's discrete summation to the more general case, as presented in [**197**, Chapter 5]; indeed, although it is presently the most powerful summation theory, it requires delicate analytic tools, it is rather complicated (a simpler approach will be detailed in the next chapter) and it provides no explicit procedure to produce sums. More about the alternative approach (the so-called "algebraic summation") will be said in the next chapter.

So the present chapter centers on the precision instruments (see the quotation of Weil at the beginning of Chapter 15) elaborated mostly by Changgui Zhang and his coauthors. We shall deal with scalar equations only: systems will be considered in the next chapter, with the help of coarser (but robust) more algebraic tools; but the main *analytic* ideas appear here.

The study being purely local[1], the field of interest is $K := \mathbf{C}(\{x\})$; its completion $\hat{K} := \mathbf{C}((x))$ plays an auxiliary (sometimes a preliminary) role.

CAUTION 14.1. In all our comparisons with the classical theory, we use power series in $\mathbf{C}[[x]]$ (asymptotic behaviour near 0), although the corresponding litterature rather deals with power series in $\mathbf{C}[[1/x]]$ (asymptotic behaviour near ∞).

14.1. Introduction through an archetypal example

In this introduction, we shall look at a very simple example (obviously related to the Tshakaloff series already met in 5.3.1) illustrating many of the themes and problems that will follow. When solving an homogeneous linear equation $Lf = 0$, where $L \in \mathcal{D}_{K,q}$, we can always reduce it to a succession of nonhomogeneous linear equations of order 1 of the particular form:

$$\sigma_q f - axf = 1, \tag{14.1}$$

for some $a \in \mathbf{C}^*$. Indeed, by Sections 5.4 and 5.5 of Chapter 5, Adams factorisation, ramification and gauge changes of unknown function of the form $f(x) = u(x)g(x)$ for some simple u will lead us either to the above equation, or to $\sigma_q f - af = 1$ or $\sigma_q f - ax^{-1}f = 1$, both of which can be solved by elementary means.

As for (14.1), it admits a unique formal, wildly divergent solution:

$$\hat{f} := 1 + ax/q + a^2x^2/q^3 + a^3x^3/q^6 + \cdots = \sum_{n\geq 0} a^n q^{-n(n+1)/2} x^n. \tag{14.2}$$

EXERCISE 14.2. (i) Redo the sequence of reductions invoked above.
(ii) In particular, solve "directly" $\sigma_q f - af = 1$ and $\sigma_q f - ax^{-1}f = 1$.
(iii) Check the statement on $\hat{f}$ and relate it to the Tshakaloff series.
(iv) Show how (14.1) could be reduced to $\sigma_q f - xf = 1$.

[1]No dealing with global problems in this chapter or the next one; for such an attempt in Galois theory, see for instance [**195**]. For q-special functions, see [CV].

The classical theory of irregular differential equations[2] then suggests that $\hat{f}$ should be considered as the asymptotic expansion of some (actually many) "true" solutions. Moreover there is an algorithmic "summation" process which allows us to derive those solutions from $\hat{f}$; and there is an abstract theory that uniquely characterizes each such solution from its asymptotic properties along with the functional equation. On our leading example, we shall see how similar ideas work for q-difference equations.

The classical theory also features something called "Stokes phenomenon": the various possibilities of summation introduce ambiguities that bear some resemblance to the multivaluedness and monodromy for fuchsian differential equations. And, like the latter, those ambiguities can be encoded in such a way to allow classification (Riemann-Hilbert style) or Galois theory. Like we did in Chapters 11 and 13, for the fuchsian case, we shall be able to extend those classical irregular ideas to q-difference equations; but it will rather be done using systems, thus in the next chapter.

14.1.1. Preliminary remarks.

14.1.1.1. For equation (14.1) to have any meaning, we need $f(qx)$ to be defined whenever $f(x)$ is, so we require that $qU \subset U$. Actually, the functional equation itself would allow us to extend any solution on U to one over $q^{-1}U$ (as we already did quite a few times, see for instance Chapter 4, e.g., 4.4.1.3), so that we might as well assume that $qU = U$; but it does not matter so much since we are mainly interested in the local behaviour near 0.

EXERCISE 14.3. Suppose we rather consider the equation $\sigma_q f - a(x)f = 1$, where $a(x) \in \mathbf{C}(x)^*$; what kind of constraints would result for U?

Then, we must decide what kind of "true solutions" we are looking for. There are three obvious possibilities: we could look for *analytic* or *continuous* or (in between) $\mathscr{C}^p$ solutions (for some $p \in [1; +\infty]$). Suppose first we look for an analytic solution f defined in some punctured neighborhood of 0 in $\mathbf{C}^*$. From the Laurent expansion $f = \sum\limits_{n \in \mathbf{Z}} a_n x^n$, substituting into equation (14.1), we draw that $a_0 = 1 + aa_{-1}$, and that $q^n a_n = aa_{n-1}$ if $n \neq 0$, whence $a_n = q^{-n(n+1)/2} a^n a_0$ if $n \geq 0$ and $a_n = q^{-n(n+1)/2} a^{n+1} a_{-1}$ if $n < 0$, so (since a_0 and a_{-1} cannot both be 0) the series necessarily diverges. Of course, it is *a fortiori* impossible to find an analytic solution on a neighborhood of 0.

14.1.1.2. Now we test the opposite choice and look for continuous solutions. Here is the recipe to build many of those. First choose any continuous function on the unit circle $\mathbf{U} = \partial \mathrm{D}(0, 1)$; extend it (uniquely) to all circles $q^n \mathbf{U}$, using equation (14.1). Then choose an arbitrary continuous extension to the fundamental annulus $\mathcal{C}_q$ (for instance using affine interpolation on segments $[x, qx]$, $x \in U$), and similarly extend it to all other annuli using the functional equation. This yields a continuous solution on $\mathbf{C}^*$.

We thus see that there are *very many*, indeed *too many* such solutions all over $\mathbf{C}$: this would yield a theory too weak for our taste; after all, ever since Riemann[3],

---

[2]In the course of our example, we shall (very briefly) summarize some of those ideas on the corresponding archetypal example; see 14.1.3.2 and 14.1.4.1.

[3]The realm of complex analytic functions is a paradise, and, to paraphrase Hilbert, *from that paradise given to us by Cauchy, Riemann, and Weierstraß we do not want to be chased.*

complex function theory was supposed to have a kind of "rigidity" not too far from that of algebraic geometry!

EXERCISE 14.4. (i) Using the kind of estimates that will be obtained in 14.1.2.1, show that the function f just found has limit 1 at 0, and therefore extends to a continuous solution on $\mathbf{C}$.

(ii) Adapt the construction above to the case of $\mathscr{C}^\infty$ solutions.

14.1.2. Some "elementary" asymptotics: behaviour of continuous solutions near 0. We call it "elementary" because we shall later (see section 14.2) refine it by controlling constants and behaviour near singularities. To begin with, in 14.1.2.1, we shall accept continuous solutions $f \in \mathscr{C}(U)$ to equation (14.1), U denoting a nonempty open subset of $\mathbf{C}$ such that $qU \subset U$. This, by the way, is the true reason why we obtain only weak results: only when we deal with analytic solutions, in the next section, shall we be able to develop a more sophisticated asymptotic theory; but we first want to find out what can be proved using weaker assumptions.

However, relying on our hope to later find analytic solutions, and taking in account the conclusions of 14.1.1.1, we assume that $U \subset \mathbf{C}^*$. Since $qU \subset U$ implies that $q^n U \subset U$ for all $n \geq 0$, it is immediate that $0 \in \overline{U}$, i.e., 0 adheres to U. It therefore makes sense to wonder about the limit of $f(x)$ at 0. If f could be continuously extended to 0, the functional equation (14.1) would imply:

$$f(0) = \lim_{\substack{x \to 0 \\ x \in U}} f(x) = 1.$$

We shall see that this is not generally the case. However we do hope that this (failing) argument can somehow lead us to some information about the asymptotic behaviour of f near 0.

14.1.2.1. In all the following, we set $r := |q| \in]0;1[$.

LEMMA 14.5. *Let $K \subset U$ a compact subset, so that each $K_n := q^n K \subset U$, $n \geq 0$, and let $\Sigma(K) := q^{\mathbf{N}} K = \bigcup_{n \geq 0} K_n$. Then any continuous solution f is bounded on $\Sigma(K)$.*

PROOF. Let $\alpha := |a|$, $d_n := \|K_n\| = \max_{x \in K_n} |x|$ and $M_n := \|f\|_{K_n}$ (for these notations see 0.4). Clearly $d_n = r^n d_0$, and equation (14.1) implies:

$$\forall x \in K_n \,,\ |f(qx)| \leq 1 + |a|\,|x|\,|f(x)| \leq 1 + \alpha d_n M_n,$$

so that $M_{n+1} \leq 1 + \alpha d_n M_n$. Also $\|f\|_{\Sigma(K)} = \sup_{n \geq 0} M_n$, so that we just have to check that the M_n are bounded above.

Let n_0 be such that $\beta := \alpha d_{n_0} < 1$ (remember that $r = |q| < 1$) and define a sequence $(u_n)_{n \geq n_0}$ by $u_{n_0} := M_{n_0}$ and $u_{n+1} := 1 + \beta u_n$. Then clearly $M_n \leq u_n$ for all $n \geq n_0$. On the other hand, the recursive relation yields:

$$\forall n \geq n_0 \,,\ u_n = \frac{1 - \beta^{n-n_0}}{1 - \beta} + \beta^{n-n_0} u_{n_0} \leq \frac{1}{1 - \beta} + u_{n_0};$$

so that the sequence (u_n) is bounded. The conclusion then flows easily. □

(According to Poincaré, the theory of analytic functions is almost totally a creation of Gauß, Cauchy, Riemann, and Weierstraß; but one can hardly say that Gauß *gave it* to us ...).

DEFINITION 14.6. We let $\mathcal{B}_c(U) := \{f \in \mathscr{C}(U) \mid f \text{ is bounded over all such } \Sigma(K)\}$.

COROLLARY 14.7. *Any continuous solution of* (14.1) *belongs to* $\mathcal{B}_c(U)$.

The sets $\Sigma(K)$ are the *strict stable subsets* of U (terminology from [**197**, Chapter 4]). We shall eventually be more interested in analytic solutions, so that we also define $\mathcal{B}(U) := \mathcal{B}_c(U) \cap \mathscr{O}(U)$. Then $\mathcal{B}_c(U)$ and $\mathcal{B}(U)$ are easily seen to be subalgebras, respectively of $\mathscr{C}(U)$ and $\mathscr{O}(U)$.

EXERCISE 14.8. Show that $\Sigma(K) \cup \{0\}$ is compact in $\mathbf{C}$ and that the image $\pi(\Sigma(K)) \subset \mathbf{E}_q$ is compact.

14.1.2.2. The formal solution (14.2) can be obtained as a fixpoint of the x-adically contracting operator:

$$\Phi : \phi \mapsto {\sigma_q}^{-1}(1 + ax\phi)$$

by the usual iterative process $\phi_{n+1} := \Phi(\phi_n)$, starting from $\phi_0 := 0$. We then find $\phi_1 = 1$, $\phi_2 = 1 + ax/q$, etc, to be the truncatures $\phi_n := T_n\hat{f}$ (see notation in 0.4), whence the congruences $\hat{f} \equiv \phi_n \pmod{x^n}$.

We wish to see the ϕ_n as "true" approximations of a "true" solution over some q-invariant open subset $U \subset \mathbf{C}^*$, and not only in the formal algebraic sense as x-adic approximations to $\hat{f}$. So we set $f = \phi_n + R_n$ and see R_n as a remainder term. Also, having in mind the most usual form of asymptotic expansions (e.g., Taylor expansions), we define S_n by $R_n = x^n S_n$. We then obviously have $R_n, S_n \in \mathscr{C}(U)$ and $R_0 = S_0 = f$; also:

$$\begin{cases} \sigma_q f = 1 + axf \\ \sigma_q \phi_n = 1 + ax\phi_{n-1} \end{cases} \Longrightarrow \sigma_q R_n = axR_{n-1} \Longrightarrow \sigma_q S_n = aq^{-n} S_{n-1}.$$

LEMMA 14.9.

$$\forall n \geq 0\,,\, S_n \in \mathcal{B}_c(U).$$

PROOF. We first prove an implication:

$$(S \in \mathscr{C}(U) \text{ and } \sigma_q S \in \mathcal{B}_c(U)) \Longrightarrow S \in \mathcal{B}_c(U).$$

Let indeed $K \subset U$ be compact. Then S is bounded on K (because it is continuous) and on $\Sigma(qK) = q\Sigma(K)$ (because $\sigma_q S$ is bounded on $\Sigma(K)$), so it is bounded on $K \cup q\Sigma(K) = \Sigma(K)$, justifying our contention.

Now $S_0 = f \in \mathcal{B}_c(U)$, so, by induction, all $S_n \in \mathcal{B}_c(U)$ (we use the obvious fact that $S \in \mathcal{B}_c(U) \Longrightarrow aq^{-n}S \in \mathcal{B}_c(U)$). □

DEFINITION 14.10. We say that a formal power series $\hat{\phi}$ is the asymptotic expansion of $\phi \in \mathscr{C}(U)$ if $\phi \in \phi_n + x^n\mathcal{B}_c(U)$ for all $n \geq 0$, the ϕ_n being the truncatures of $\hat{\phi}$. (For the uniqueness, see the exercise at the end of Subsection 14.1.2.)

Let $\mathcal{A}_c(U)$ the set of those $\phi \in \mathscr{C}(U)$ admitting an asymptotic expansion and $\mathcal{A}_{c,0}(U)$ those for which the asymptotic expansion is trivial:

$$\mathcal{A}_{c,0}(U) = \bigcap_{n \geq 0} x^n \mathcal{B}_c(U).$$

COROLLARY 14.11. (i) *Any solution of equation* (14.1) *in* $\mathscr{C}(U)$ *admits the asymptotic expansion* $\hat{f}$.

(ii) *If* f *and* g *are solutions respectively in* $\mathscr{C}(U)$ *and* $\mathscr{C}(U')$, *then the restricted difference* $f-g$ *is in* $\mathcal{A}_{c,0}(U\cap U')$

The statement (i) *does not allow* to conclude that $\lim_{x\to 0} f(x)=1$. Indeed, we shall see in 14.1.3.1 that there are analytic solutions over $\mathbf{C}^*\setminus[c;q]$ for almost $c\in\mathbf{C}^*$, that have simple poles at all points of $[c;q]$. Such a solution f being given, define a sequence (x_n) such that $|x_n-cq^n|<1/n$ and $|f(x_n)|>n$ for all $n\geq 0$ (it is possible since f is not bounded near a pole). Then $\lim x_n=0$, while $\lim|f(x_n)|=+\infty$.

One source of the problem is that an asymptotic expansion is not, by far, a converging series. The constants involved in the bounds M_n behave badly as $n\to+\infty$. From the recursive definition $\sigma_q S_n=aq^{-n}S_{n-1}$, we can only draw that $M_n=O(\alpha^n r^{-n(n-1)/2})$, which is a rather rapid explosion.

We shall really be more interested in holomorphic solutions, so we shall rather use $\mathcal{A}(U):=\mathcal{A}_c(U)\cap\mathscr{O}(U)$ and $\mathcal{A}_0(U):=\mathcal{A}_{c,0}(U)\cap\mathscr{O}(U)=\bigcap_{n\geq 0}x^n\mathcal{B}(U)$.

EXERCISE 14.12. (i) Show that, if ϕ admits an asymptotic expansion, it is unique. (If $\deg P<n$ and $P\in x^n\mathcal{B}_c(U)$, then $P=0$.)

(ii) Show that $\mathcal{A}_c(U)$ and $\mathcal{A}(U)$ are respectively subalgebras of $\mathscr{C}(U)$ and $\mathscr{O}(U)$. What can be said in this respect of $\mathcal{A}_{c,0}(U)$ and $\mathcal{A}_0(U)$?

14.1.3. Some "elementary" summation: construction of explicit analytic solutions. Recall from 2.1.1.2 in Chapter 2 that the function $\theta_q(x):=\sum_{n\in\mathbf{Z}}q^{n(n-1)/2}x^n$ is holomorphic on $\mathbf{C}^*$ and satisfies the functional equations $\theta_q(qx)=x^{-1}\theta_q(x)=\theta_q(1/x)$, whence $\theta_q(q/x)=\theta_q(x)$. Also, θ_q has simple zeroes on $[-1;q]$ and nowhere else, which can be encoded either as $\operatorname{div}_{\mathbf{C}^*}(\theta_q)=[-1;q]$, or as $\operatorname{div}_{\mathbf{E}_q}(\theta_q)=[\overline{-1}]$.

14.1.3.1. Inspired by the method of "variation of constants" for nonhomogeneous linear differential equations of first order, we begin the resolution of equation (14.1) by solving the associated homogeneous equation, i.e., we look for $f_0\neq 0$ such that $\sigma_q f_0=axf_0$. A simple natural choice (the simplest one up to a constant) is $f_0(x):=\dfrac{1}{\theta_q(ax)}\cdot$ Then we look for f as $f_0 g$; our equation becomes:

$$\sigma_q(g/\theta_q(ax))=1+axg/\theta_q(ax)\Longleftrightarrow\sigma_q g-g=\frac{1}{ax}\theta_q(ax)=\sum_{n\in\mathbf{Z}}q^{n(n+1)/2}(ax)^n.$$

It is easy to solve $\sigma_q g_1-g_1=\sum_{n\neq 0}q^{n(n+1)/2}(ax)^n$, direct identification yields the Laurent series $g_1=\sum_{n\neq 0}\dfrac{q^{n(n+1)/2}}{q^n-1}(ax)^n\in\mathscr{O}(\mathbf{C}^*)$. Then we would want to write $g:=g_0+g_1$, where $\sigma_q g_0-g_0=1$; the latter cannot be solved with $g_0\in\mathscr{O}(\mathbf{C}^*)$: by Chapter 2, its simplest solutions are q-logarithms and have a q-spiral of simple poles. So, modulo $q^{\mathbf{Z}}$, the corresponding solution $f=f_0 g$ would have either two simple poles or a double pole.

There is however a nice trick invented by Changgui Zhang[4], which we shall nickname "twisted variation of constants". Instead of $f_0(x) := \dfrac{1}{\theta_q(ax)}$, we shall rather take $f_0 := 1/\theta_{q,c}$, i.e., $f_0(x) := \dfrac{1}{\theta_q(x/c)}$, for some $c \in \mathbf{C}^*$ to be adequately chosen (twisted Theta functions $\theta_{q,c}$ were introduced in Subsection 5.5.3). So $\sigma_q f_0 = (c/x) f_0$ and $\operatorname{div}_{\mathbf{C}^*}(f_0) = -[-c;q]$, $\operatorname{div}_{\mathbf{E}_q}(f_0) = -[\overline{-c}]$. Our equation becomes:

$$\sigma_q(g/\theta_{q,c}) = 1 + axg/\theta_{q,c} \Longleftrightarrow \sigma_q g - acg = \frac{c}{x}\theta_q(x/c) = \sum_{n\in\mathbf{Z}} q^{n(n+1)/2}(x/c)^n.$$

Provided $ac \notin q^{\mathbf{Z}}$, this can be uniquely solved in $\mathscr{O}(\mathbf{C}^*)$ by setting $g := g_c$, $f := f_c$, where:

$$g_c(x) := \sum_{n\in\mathbf{Z}} \frac{q^{n(n+1)/2}}{q^n - ac}(x/c)^n \in \mathscr{O}(\mathbf{C}^*) \text{ and } f_c(x) := \frac{1}{\theta_q(x/c)} \sum_{n\in\mathbf{Z}} \frac{q^{n(n+1)/2}}{q^n - ac}(x/c)^n.$$

So we get a whole family $(f_c)_{c\in\mathbf{C}^*\setminus[a^{-1};q]}$ of solutions, each solution f_c having poles characterized by: $\operatorname{div}_{\mathbf{C}^*}(f_c) = -[-c;q]$, $\operatorname{div}_{\mathbf{E}_q}(f_c) = -[\overline{-c}]$.

EXERCISE 14.13. Check that all those poles are indeed present, i.e., there can be no cancellation with zeroes of g_c.

Note that even the weaker property $\operatorname{div}_{\mathbf{E}_q}(f) \geq -[\overline{-c}]$ uniquely characterizes the solution f_c: indeed, it implies that $g := f\theta_{q,c} \in \mathscr{O}(\mathbf{C}^*)$ and $\sigma_q g - acg = \dfrac{c}{x}\theta_{q,c}$ has a unique solution in $\mathscr{O}(\mathbf{C}^*)$, namely g_c (using the Laurent series expansion). But the characterization $\operatorname{div}_{\mathbf{E}_q}(f) \geq -[\overline{-c}]$ only depends on c modulo $q^{\mathbf{Z}}$, i.e., it really depends on $\overline{c} \in \mathbf{E}_q$. Therefore, we are entitled to write $f_{\overline{c}}$ our solutions, with the condition $\overline{c} \neq \overline{a^{-1}}$.

EXERCISE 14.14. Check by direct computation on the explicit formula above that $f_{qc} = f_c$.

14.1.3.2. Inspired by the theory of summation of divergent formal solutions of irregular linear differential equations, Changgui Zhang showed in **[265]** how one can recover $f_{\overline{c}}$ directly from $\hat{f}$, and see it as the result of *discrete summation of* $\hat{f}$ *in the "direction"* $\overline{c}$, thus denote it as $S_{\overline{c}}\hat{f}$.

We shall first illustrate here the classical summation theory (at least a small fragment of it) on an equally archetypal example, of which our $\hat{f}$ is a q-analogue, the Euler[5] series:

$$\hat{F}(x) := \sum_{n\geq 0}(-1)^n n!\, x^{n+1} = x - x^2 + 2x^3 - 6x^4 + \cdots$$

This series is a formal solution of the differential equation $x^2F' + F = x$, so we look for a "true solution"; variation of constants does it: we solve the associated

[4]It seems that this presentation is an a posteriori explanation of a different process, that was more closely based on ideas from the theory of moments, see 14.2.2.

[5]Euler wanted to give a numerical meaning to the expression $\sum_{n\geq 0}(-1)^n n! = 1-1+2-6+\cdots$ by evaluating $\hat{F}(1)$. See **[192]** for the complete history of this circle of ideas and **[150**, exa 1.1.4 p 4 and comments 1.4.3 p 30**]** for a thorough treatment of the example.

homogeneous equation $x^2F_0'+F_0=0$ by taking $F_0 := e^{1/x}$, then look for F as F_0G, where G satisfies:

$$x^2F_0G' + x^2F_0'G + F_0G = x \Longleftrightarrow G' = \frac{1}{x}e^{-1/x} \Longleftrightarrow G = \int_{x_0}^{x} \frac{1}{t}e^{-1/t}\,dt.$$

Requiring the natural condition $F(0) = 0$ (look at the equation), we have to take $x_0 := 0$ and easy changes of variables (left for the reader to find) give:

$$G(x) = \int_0^x e^{-1/t}\,\frac{dt}{t} = \int_0^{+\infty} e^{-\xi/x}\,\frac{d\xi}{1+\xi}.$$

We recognize a Laplace transform. The resulting function F is well defined and holomorphic on the half-plane $\Re(x) > 0$: it satisfies the differential equation and moreover admits $\hat{F}$ as an asymptotic expansion.

More generally, we can integrate on the half-line $\mathbf{R}_+^*d =]0, +\infty d[$ for any direction $d = e^{i\theta}$, $\theta \in \mathbf{R}$, as long as the half-line does not go through $\xi = -1$, i.e., is not $\mathbf{R}_-^*$. The resulting $F_d(x) := \displaystyle\int_0^{+\infty d} e^{-\xi/x}\,\frac{d\xi}{1+\xi}$ is holomorphic in the half-plane H_d bisected by $\mathbf{R}_+^*d$; it is a solution of the equation and admits $\hat{F}$ as an asymptotic expansion there. Moreover it can be shown that those properties uniquely characterize F_d.

EXERCISE 14.15. Show that for any two possible directions d, d', one has $F_d - F_{d'} = Ce^{1/x}$ on $H_d \cap H_{d'}$ for some $C \in \mathbf{C}$.

The consequences of the *ambiguity* created by the many possibilities of summation are very important, we shall return to them in 14.1.4. For the moment, we want to understand in what sense we have a *summation process* $S_d : \hat{F} \mapsto F_d$. This is defined in two steps:

(1) There is first a rather formal, algebraic step, the *Borel transformation.* In our case[6] we have a mapping:

$$\mathcal{B}_1 : \mathbf{C}[[x]] \to \mathbf{C}[[\xi]], \quad \sum a_n x^n \mapsto \sum a_{n+1}\frac{\xi^n}{n!}.$$

(It is customary to denote ξ the target variable and to call it "coordinate in the Borel plane".) In our example, $\mathcal{B}_1\hat{F}(\xi) = \dfrac{1}{1+\xi}$.

(2) Then we see that $F_d(x) = \mathcal{L}_d\mathcal{B}_1\hat{F}(x)$, where

$$\mathcal{L}_dG(x) := \int_0^{+\infty d} e^{-\xi/x}G(\xi)\,d\xi,$$

clearly a Laplace transform in direction d.

So the whole summation process is called a "Borel-Laplace" transformation (here "at level 1").

[6] In the general case, a notion of *level* is involved, related to the slopes of the Newton polygon; then the denominators may be powers $(n!)^s$ for some $s > 0$.

14.1.3.3. We now return to the q-difference case and to equation (14.1). To plainly see how our explicit formula for $f_{\overline{c}}$ can be made to resemble the Borel-Laplace transformation $\mathcal{L}_d\mathcal{B}_1$, we shall arrange it with the help of the identities:

$$\theta_q(qx) = \frac{\theta_q(x)}{x} \Longrightarrow \theta_q(q^n x) = \frac{\theta_q(x)}{q^{n(n-1)/2}x^n} \Longrightarrow \theta_{q,c}(q^n x) = \frac{c^n\theta_{q,c}(x)}{q^{n(n-1)/2}x^n},$$

whence (using at the second line the change $\xi = cq^{-n}$ and at the third line the identity $\theta_q(x) = \theta_q(q/x)$):

$$\begin{aligned}
f_{\overline{c}}(x) &= \frac{1}{\theta_q(x/c)} \sum_{n\in\mathbf{Z}} \frac{q^{n(n+1)/2}}{q^n - ac}(x/c)^n = \sum_{n\in\mathbf{Z}} \frac{q^{n(n+1)/2}}{(q^n - ac)\theta_{q,c}(x)}(x/c)^n \\
&= \sum_{n\in\mathbf{Z}} \frac{q^n}{(q^n - ac)\theta_{q,c}(q^n x)} = \sum_{\xi\in[c;q]} \frac{1}{(1 - a\xi)\theta_q(x/\xi)} \\
&= \sum_{\xi\in[c;q]} \frac{1}{(1 - a\xi)\theta_q(q\xi/x)} = \sum_{\xi\in[c;q]} \frac{1}{(1 - a\xi/q)\theta_q(\xi/x)} \\
&= \mathcal{L}_{q,\overline{c}}\mathcal{B}_{q,1}(x), \text{ where:}
\end{aligned}$$

(1) The *q-Borel-Ramis*[7] *transformation* $\mathcal{B}_{q,1}$ sends $\sum a_n x^n$ to $\sum a_n q^{n(n-1)/2}\xi^n$, so that:

$$\mathcal{B}_{q,1}\hat{f}(\xi) = \sum_{n\geq 0} a^n q^{-n(n+1)/2} q^{n(n-1)/2}\xi^n = \frac{1}{1 - a\xi/q}.$$

(2) The *q-Laplace-Zhang transformation in "direction" $\overline{c}$*, denoted $\mathcal{L}_{q,\overline{c}}$, sends $g(x)$ to $\sum_{\xi\in[c;q]} \frac{1}{\theta_q(\xi/x)} g(\xi)$.

So in $\mathcal{B}_{q,1}$ we use the q-analogy of $q^{n(n-1)/2}$ with $n!$; and in $\mathcal{L}_{q,\overline{c}}$ we replace the integration in direction d by a discrete integration (Jackson style!) in "direction" $\overline{c}$, and the kernel e^{-u} by the kernel $\frac{1}{\theta_q(u)}$. We shall explain in 14.1.4 why $\overline{c}$ can be understood as a direction.

EXERCISE 14.16. We could use the q-factorial $[n!]_q$ as a q-analogue instead of $q^{n(n-1)/2}$. This would imply rewriting our q-difference equation in terms of q-derivation (see no 3.1.1.4 in Chapter 3). Do it. (Also see [**265**].)

14.1.4. Some "elementary" Stokes phenomenon: ambiguity of the summation process. Let us say that here "elementary" means that we deal with a very simple situation (only one prohibited direction, whether in the differential or in the q-difference case) where the full powers of sophisticated methods (cohomological or Galois) may seem superfluous. The next chapter will hopefully confirm their usefulness, but a baby case may help. It would probably help even more to read in [**192**] the history of this circle of ideas, starting from Stokes [**242**] (and Galois!). More details can also be found in [**150**].

---

[7] We decide to call it that way, partly because it was used by Ramis in [**191**] for other reasons, see Chapter 6, and in spite of the fact that Bézivin might have used it in previous works; but mostly because Ramis guided most of the modern efforts to give q-difference theory analytical tools as poweful as those of the classical theory of differential equations.

14.1.4.1. First we return to the "classical" case (irregular differential equations) and stick to the example of the Euler series from 14.1.3.2; recall that we found that, for all directions d but one, there is a solution F_d asymptotic to the Euler series $\hat{F}$ in the half-plane H_d; and we considered it as the sum $S_d\hat{F}$ of $\hat{F}$ in direction d. It is customary to identify the direction d with the unique point of intersection $e^{\mathrm{i}\theta}$ of the half-line $\mathbf{R}_+^* d$ with the unit circle[8] $S^1 := \mathbf{U}$; the argument θ is therefore well defined modulo 2π and $\not\equiv \pi$, so we can take it in $]-\pi, +\pi[$.

For two different authorized directions d, d' the difference $F_d - F_{d'}$ is an analytic solution on the sector $H_d \cap H_{d'}$ of the homogeneous equation $x^2 f' + f = 0$, so, the sector being connected, it has the form $Ce^{1/x}$ for some $C \in \mathbf{C}$. It is proved in [**150**] that, if for instance $\theta, \theta' \in]-\pi/2, +\pi/2[$, then $C = 0$; however, if θ approaches $+\pi$, resp. $-\pi$, the corresponding solutions rather satisfy $F^+ - F^- = 2\mathrm{i}\pi e^{1/x}$: there is a *jump*.

So if we try to use the overlapping half-planes to analytically continue a solution, there will be an ambiguity, exactly as in the case of multivalued solutions near regular singular points and their monodromy (see [**169, 222**]). Numerically, that ambiguity of summation was discovered by Stokes (on the more complicated example of Airy's equation). More theoretically, Birkhoff saw that, as in the case of monodromy and Riemann-Hilbert correspondence, the ambiguity can be used to classify irregular equations. In modern terms, due to Malgrange and Sibuya, the family of all differences $F_{d,d'} := F_d - F_{d'}$ forms a *cocycle* (because $F_{d,d'} + F_{d',d''} = F_{d,d''}$) of the *sheaf* $\mathscr{F}$ *of solutions*[9] *of the associated homogeneous equation*. The sheaf $\mathscr{F}$ on the circle of directions is well defined by the following conditions. Let $I_{d,d'}$ a small open arc limited by d, d' and let $\mathfrak{s}_{d,d'} := \mathbf{R}_+^* I_{d,d'}$ the corresponding sector. Then:

$$\mathscr{F}(I_{d,d'}) := \{\text{solutions of } x^2 f' + f = 0 \text{ analytic over } \mathfrak{s}_{d,d'}\}.$$

The cocycle $(F_{d,d'})$ has a *cohomology class* in $H^1(S^1, \mathscr{F})$ and the theorem of Birkhoff-Malgrange-Sibuya[10] says in which way the space $H^1(S^1, \mathscr{F})$ can be used for classification of a particular class of irregular differential equations.

The other way to encode ambiguity is through the "jump" $2\mathrm{i}\pi e^{1/x}$ from F^+ to F^-. Exactly like the variation along a loop could be encoded as a monodromy matrix, and the latter embedded in the fuchsian Galois group, the said jump can be encoded as a *Stokes operator* and, as Ramis showed, embedded in the irregular Galois group [**55, 189, 193, 250**]. There even is a density theorem extending Schlesinger's theorem.

14.1.4.2. We now abandon the classical case to see how those ideas can be transposed to irregular q-difference equations, at least on our example. First, since summations $S_{\overline{c}}\hat{f}$ are parameterized by (authorized) elements $\overline{c} \in \mathbf{E}_q$, we should show why those can be imagined as directions. The unifying theme was explained

---

[8]The corresponding "Malgrange circle of directions" is usually denoted S^1 and rather visualized as an infinitesimal circle at 0, i.e., we imagine puncturing 0 and enlarging the (punctual) hole into an infinitesimal disk of which S^1 is the boundary. This is because we use it to encode what happens asymptotically near 0. See [**150**, II.2.1.5, p 41] and [**27**] about the so-called "real blow-up".

[9]The modern theory takes in account not only the fact that $F_{d,d'}$ satisfies the homogeneous equation, but also that it is *exponentially flat*, i.e., it has trivial asymptotic expansion..

[10]It is usually called "of Malgrange-Sibuya", but, according to Malgrange (personal communication), it comes directly from Birkhoff.

in J.L. Martins's thesis "Lemmes de Watson en dimension supérieure et applications à la sommabilité des solutions des D-modules holonomes" (to be found at URL `https://www.theses.fr/2002TOU30038`); also see [**197**, Chapter 4]. In the differential case, a semigroup $e^{]-\infty,0[}$ acts upon $\mathbf{C}^*$: it expresses the fact that we are looking at asymptotic phenomena while going toward 0, usually along half-lines. What happens "in the end" is associated to a point of the *horizon*, which is the quotient of $\mathbf{C}^*$ by the semi-group action. So a point of the horizon S^1 is associated to a direction d of approach, quite like a point on the complex projective line "at infinity" is associated to a direction in $\mathbf{C}^2$.

In the q-difference case, the semigroup is $q^{\mathbf{N}}$ and we tend to 0 along half q-spirals (they replace the half-lines); the horizon is the quotient $\mathbf{E}_q$: points on $\mathbf{E}_q$ are naturally in bijection with q-spirals. So $\overline{c} \in \mathbf{E}_q$ will be used to encode an asymptotic behaviour along $cq^{\mathbf{N}} \subset [c;q]$ (see [**197**, Chapter 4] for a more complete exposition of the relevant formalism). We saw in 14.1.3 that, except from the prohibited $\overline{a^{-1}}$, all directions $\overline{c} \in \mathbf{E}_q$ give rise to a sum $S_{\overline{c}}\hat{f}$. We shall now present q-analogues of both encodings (cocycles and jumps) displayed in 14.1.4.1.

As for the cocycle, we set $f_{\overline{c},\overline{d}} := f_{\overline{c}} - f_{\overline{d}}$, which is a holomorphic solution of $\sigma_q f = axf$ on:

$$V_{\overline{c},\overline{d}} := V_{\overline{c}} \cap V_{\overline{d}}, \text{ where } V_{\overline{c}} := \mathbf{C}^* \setminus [-c;q].$$

Actually (remembering the projection $\pi : \mathbf{C}^* \to \mathbf{E}_q$ from Appendix A, Subsection 2.1.1), $V_{\overline{c}} = \pi^{-1}(U_{\overline{c}})$, where $U_{\overline{c}} := \mathbf{E}_q \setminus \{\overline{c}\}$ and $V_{\overline{c},\overline{d}} = \pi^{-1}(U_{\overline{c},\overline{d}})$, where $U_{\overline{c},\overline{d}} := U_{\overline{c}} \cap U_{\overline{d}}$, so the family of all $f_{\overline{c},\overline{d}}$ is a cocycle for the following sheaf over $\mathbf{E}_q$:

$$\mathscr{F}(U) := \{\text{solutions of } \sigma_q f = axf \text{ analytic over } \pi^{-1}(U)\}.$$

Calling $\mathfrak{U} := (U_{\overline{c}})$ the above open covering of $\mathbf{E}_q$, we write $(f_{\overline{c},\overline{d}}) \in Z^1(\mathfrak{U}, \mathscr{F})$ and consider its cohomology class in $H^1(\mathfrak{U}, \mathscr{F})$, thus inducing a cohomology class in $H^1(\mathbf{E}_q, \mathscr{F})$: we shall see in Section 15.3 of Chapter 15 how the space $H^1(\mathbf{E}_q, \mathscr{F})$ allows us to classify irregular systems ("q-analogue of Birkhoff-Malgrange-Sibuya theorem").

Now we try to find a q-analogue for the jump. The unique prohibited "direction of summation" $\overline{a^{-1}}$ is isolated in $\mathbf{E}_q$ and does not seem to give rise to any sort of frontier crossing, as was the case for $-1 \in S^1$: our set of "directions" is now a complex space, not a real one. But, in complex analysis, there is a typical way of detecting that something special occurs at a punctual spot: taking the residue. So we recall that $\overline{a^{-1}}$, or more precisely the values $c \in [a^{-1};q]$, were prohibited because they caused vanishing of denominators: we want to see them as poles of $c \mapsto S_{\overline{c}}\hat{f}$. Since the explicit formula for the latter involves another variable x (apart the variable c presently of interest), we fix $x := x_0$ for some arbitrarily chosen[11] x_0 (there might be restrictions on the choice but it does not presently matters).

So we are led to study the poles of the meromorphic function (we use the first explicit formula; see 14.1.3.1):

$$c \mapsto S_{\overline{c}}\hat{f}(x_0) = \frac{1}{\theta_q(x_0/c)} \sum_{n \in \mathbf{Z}} \frac{q^{n(n+1)/2}}{q^n - ac} (x_0/c)^n$$

[11]We shall find when applying the process to Galois theory (in 15.4) that this amounts to choosing a fibre functor, i.e., it is similar to the choice of a base point for a fundamental group and just as inessential.

For instance at $c = a^{-1}$, only the term with index $n = 0$ is polar, and we easily find the residue:

$$\mathrm{Res}_{c=a^{-1}} S_{\overline{c}} \hat{f}(x_0) = \frac{-a^{-1}}{\theta_q(ax_0)}.$$

Note that it would be more natural to consider a meromorphic function on $\mathbf{E}_q$, namely $\overline{c} \mapsto S_{\overline{c}} \hat{f}(x_0)$, and to compute its unique residue at $\overline{c} = \overline{a^{-1}}$: we shall do so in all generality in section 15.4 of Chapter 15. Meanwhile, to get a feeling of the more general formula, we suggest the

EXERCISE 14.17. Compute $\mathrm{Res}_{c=q^n a^{-1}} S_{\overline{c}} \hat{f}(x_0)$ for all $n \in \mathbf{Z}$.

14.2. Irregular equations: discrete summation and asymptotics

This section relies essentially on **[265]**. We shall describe a discrete asymptotics and summation process, but only in the restricted case of nonhomogeneous equations $Lf = g$, where $g \in \mathbf{C}\{x\}$ and L is a q-difference operator such that $S(L) = \{-1\}$ (typical case: $g := 1$ and $L := \sigma_q - x$). Zhang extended his theory to much more general cases (arbitrary integral slopes), see e.g., [**197**, Chapter 5], but, as explained at the beginning of this chapter, we prefer to tackle said general case by other means.

The general conventions and notations are the following[12]: first, $0 < q < 1$ and we sometimes use $p := 1/q > 1$. We only deal with power series (formal or convergent) without poles, i.e., in $\mathbf{C}[[x]]$ or $\mathbf{C}\{x\}$. Let $\hat{f} \in \mathbf{C}[[x]]$ (that notation will be pervasive); its *q-Borel-Ramis transform* (justification for that terminology was given in footnote 7 in 14.1.3.3) is the map $\mathcal{B}_{q;1} : \mathbf{C}[[x]] \to \mathbf{C}[[\xi]]$ (note the use of the coordinate ξ in the "q-Borel plane") defined (here with a slight difference as regards the exponent) as:

$$\mathcal{B}_{q;1}\hat{f}(\xi) := \sum_{n \geq 0} q^{n(n-1)/2} a_n \xi^n, \text{ where } \sum_{n \geq 0} a_n x^n := \hat{f}.$$

The space of power series of *q-Gevrey order* 1 is then:

$$\mathbf{C}[[x]]_{q;1} := \mathcal{B}_{q;1}^{-1}(\mathbf{C}\{\xi\}) = \{\hat{f} \in \mathbf{C}[[x]] \mid \mathcal{B}_{q;1}\hat{f} \in \mathbf{C}\{\xi\}\}.$$

Thus $\hat{f} := \sum a_n x^n \in \mathbf{C}[[x]]_{q;1}$ if, and only if, $a_n = O\left(A^n q^{-n(n-1)/2}\right)$ for some $A > 0$. Clearly the radius of convergence of $\mathcal{B}_{q;1}\hat{f}$ is at least as big as that of $\hat{f}$; in particular, if $\hat{f} \in \mathbf{C}\{x\}$, then $\mathcal{B}_{q;1}\hat{f}$ defines an entire function, so that:

$$\mathbf{C}\{x\} \subset \mathbf{C}[[x]]_{q;1}.$$

EXERCISE 14.18. Check that $\mathcal{B}_{q;1}(x\sigma_q\hat{f}) = \xi\mathcal{B}_{q;1}\hat{f}$. (This will be useful in application to q-difference equations, see 14.2.4.)

14.2.1. Formal q-Borel-Ramis transformation.

[12]According to *loc. cit*, extension to the case of a complex q is straightforward, here and for section 14.3, but we do not attempt it.

14.2.1.1. Following **[265]**, let $\Delta := a_0 + \cdots + a_m \sigma_p^m$, all $a_i = \sum_{j \geq 0} a_{i,j} x^j \in \mathbf{C}\{x\}$ and $a_0 a_m \neq 0$; caution: σ_p was used, which, as we know, changes nothing essential. As a substitute for the Newton polygon, consider the upper convex hull of $\{(i,j) \mid a_{i,j} \neq 0\} \subset \mathbf{R}^2$ and assume that it has exactly one nonvertical side, having slope $+1$.

LEMMA 14.19. *That condition on Δ is equivalent to the following:*

$$\Delta = P(x\sigma_p) + x\tilde{\Delta}, \text{ where } P = p_0 + \cdots + p_m X^m \in \mathbf{C}[X], p_0 p_m \neq 0, \text{ and } \tilde{\Delta} \in \mathbf{C}\{x\}[x\sigma_p].$$

PROOF. Easy and left to the reader using the following exercise. □

EXERCISE 14.20. (i) Let $\Delta' := \sigma_q^m \Delta$ be written as $\Delta' = a'_0 + \cdots + a'_m \sigma_q^m$. State and prove the formulas relating the a_i to the a'_i and check that the above Newton polygon is symmetrical (with respect to a vertical axis) to the usual Newton polygon of Δ', so the nonvertical slopes are opposite.

(ii) State and prove a criterion involving Δ' similar to that of the lemma.

(iii) Relate P to a characteristic equation for Δ' (see 5.2.2.6 in Chapter 5).

(iv) On account of Chapter 5, what kind of factorisation is possible for Δ?

From now on, and for the whole of the present section 14.2, we assume that the conditions in the lemma are satisfied.

DEFINITION 14.21. (i) A *$q*$-set* is a subset of $\mathbf{C}$ of the form $\mathbf{C} \setminus \bigcup_{i=1}^{m} \alpha_i q^{-\mathbf{N}}$, where $\alpha_1, \ldots, \alpha_m \in \mathbf{C}^*$. (So it is open and connected.)

(ii) Let Ω a $q*$-set and let $\phi \in \mathscr{O}(\Omega)$. We say that ϕ has *q-exponential growth of order* 1 *(at infinity)* if:

$$\forall \epsilon > 0 \ , \ \exists C, A > 0 \ : \ \forall \xi \in \Omega \ , \ d\left(q^{\mathbf{N}}\xi, \partial\Omega\right) > \epsilon \Longrightarrow |\phi(\xi)| < C\theta_{|q|}(A\,|\xi|).$$

The use of the letter ξ suggests that those notions will be used for subsets in and functions on the q-Borel plane, typically $\phi := \mathcal{B}_{q;1}\hat{f}$ (where, by necessity, $\hat{f} \in \mathbf{C}[[x]]_{q;1}$. In particular, we define a subspace $\mathbf{H}_{q;1}(\Omega)$ as follows; if $\phi \in \mathbf{C}\{\xi\}$, then, by definition:

$$\phi \in \mathbf{H}_{q;1}(\Omega) \underset{def}{\Longleftrightarrow} \begin{array}{l} \phi \text{ admits a meromorphic continuation to } \mathbf{C}, \text{ holomorphic} \\ \text{on } \Omega, \text{ with } q\text{-exponential growth of order 1 (at infinity).} \end{array}$$

EXERCISE 14.22. (i) Check that the condition $d\left(q^{\mathbf{N}}\xi, \partial\Omega\right) > \epsilon$ means:

$$\forall n \geq 0 \ , \ \forall i = 1, \ldots, m \ , \ \left|\xi - \alpha_i q^{-n}\right| > \epsilon q^{-n}.$$

Caution: one of the implications is not tautological.

(ii) Compare the inequality involved in the q-exponential growth condition with those encountered in 6.3.1.3 of Chapter 6 and look for the mutual dominations mentioned in remark 6.57.

14.2.1.2. Now let $g \in \mathbf{C}\{x\}$ and consider the nonhomogeneous q-difference equation $\Delta\hat{f} = g$.

EXERCISE 14.23. Show that it admits a unique formal power series solution $\hat{f} \in \mathbf{C}[[x]]$.

THEOREM 14.1. *With the notations of the previous lemma, let* $\{\{\alpha_1, \ldots, \alpha_m\}\}$ *the multiset (see definition in 0.4) of roots of* P *and let* $\Omega := \mathbf{C} \setminus \bigcup_{i=1}^{m} \alpha_i q^{-\mathbf{N}}$*, thus a* $q*$*-set. Let* $\hat{f} \in \mathbf{C}[[x]]$ *the unique formal power series solution of* $\Delta \hat{f} = g$ *and let* $\phi := \mathcal{B}_{q;1}\hat{f} \in \mathbf{C}[[\xi]]$*. Then* $\phi \in \mathbf{H}_{q;1}(\Omega)$*.*

PROOF. We reorder the α_i in such a way that $i < j \Rightarrow \alpha_j/\alpha_i \notin q^{\mathbf{N}^*}$. By the first exercise in 14.2.1.1, there is a factorisation:

$\Delta = \lambda h_1(x\sigma_p - \alpha_1) \cdots h_m(x\sigma_p - \alpha_m)$, where $\lambda \in \mathbf{C}^*$ and all $h_i \in \mathbf{C}\{x\}, h_i(0) = 1$. Define $\hat{f}_m := \hat{f}$, then by descending recursion $\hat{f}_{i-1} := h_i(x\sigma_p - \alpha_i)\hat{f}_i$ (so that $\lambda \hat{f}_0 = \Delta \hat{f} = g$). Then apply iteratively the following lemma. □

LEMMA 14.24. *Let* $\hat{g} \in \mathbf{C}[[x]]$*.*
(i) *If* $h \in \mathbf{C}\{x\}$*, then* $\mathcal{B}_{q;1}\hat{g} \in \mathbf{H}_{q;1}(\Omega) \Rightarrow \mathcal{B}_{q;1}(h\hat{g}) \in \mathbf{H}_{q;1}(\Omega)$*.*
(ii) *If* $\alpha \in \mathbf{C}^* \setminus \Omega$*, then* $\mathcal{B}_{q;1}\hat{g} \in \mathbf{H}_{q;1}(\Omega) \Leftrightarrow \mathcal{B}_{q;1}((x\sigma_p - \alpha)\hat{g}) \in \mathbf{H}_{q;1}(\Omega)$*.*

PROOF. See the exercise. □

EXERCISE 14.25. Prove the lemma. For (i) use the exercise at the beginning of section 14.2. For (ii), setting $\hat{h} = \sum h_n \xi^n$ and $\gamma := \mathcal{B}_{q;1}\hat{g}$ (recall that this is an entire function), check that $\mathcal{B}_{q;1}\hat{g}(\xi) = \sum_{n \geq 0} q^{n(n-1)/2} h_n \xi^n \gamma(q^n \xi)$.

14.2.2. Moments and the q-Laplace-Zhang transformation.

14.2.2.1. *Moments.* The so-called "problem of moments" is a classical theme in functional analysis (orthogonal polynomials, continued fractions, functional equations, summation theory ...); see, e.g., [**15**, Chapters 5 and 6], [**93**, Chapter 7] and [**251**, Chapter 14]. In classical Borel summation, $\sum a_n x^n \mapsto \sum a_n \dfrac{\xi^{n+1}}{n!} = \sum a_{n-1} \dfrac{\xi^n}{\Gamma(n)}$, the identity $\Gamma(n) = \displaystyle\int_0^\infty \frac{t^n}{e^t} \frac{dt}{t}$ plays a prominent role. To extend that line of thought to the q-Borel-Ramis transformation (at level 1), we should look for some function $e_q(t)$ (in lieu of e^t) such that:

$$\forall n \geq 0 \, , \; q^{-n(n-1)/2} = \int_0^\infty \frac{t^n}{e_q(t)} \frac{dt}{t}.$$

EXERCISE 14.26. According to [**265**], the above is "almost equivalent" to condition $e_q(qt) = t^{-1} e_q(t)$. Try to give a more precise meaning and justification for that statement.

We note that there is a possible *multivalued* choice for e_q, namely:

$$\sqrt{2\pi \log(1/q)}\, e^{\left(\log(q^{-1/2}t)\right)^2/(2\log(1/q))}.$$

(You should of course check that it works!) It would lead us to a continuous summation process involving the continuous integral transformation:

$$\forall \theta \in \mathbf{R} \, , \; \mathcal{L}^{\theta}_{q;1}\phi(x) := \int_0^{\infty e^{i\theta}} \frac{\phi(\xi)}{e_q(\xi/x)} \frac{d\xi}{\xi}.$$

Although historically earlier [**263**] (probably because it bears more similarity to the classical Borel-Laplace summation process), this is not so much in the spirit of this book, where we try to stick to uniform meromorphic functions. So this approach will only be summarized in section 14.3.

14.2.2.2. *q-Laplace-Zhang transformation.* Following instead **[265]**, we rather choose:

$$e_q(t) := \log(1/q)\,\theta_q(t).$$

EXERCISE 14.27. Compute $\displaystyle\int_0^\infty \frac{t^n}{e_q(t)}\,\frac{dt}{t}$ for that choice. (Calling it c_n, get a recurrence relation; c_0 is given by Proposition 2.18 from 2.2.1.4.)

However continuous integration also frequently produces multivalued functions (partly because of Cauchy residue theorems) and Zhang was instead led to use Jackson type discrete summation, whence the introduction of the *q-Laplace-Zhang transformation with Theta kernel*:

$$\forall \lambda \in \mathbf{C}^* \;,\; \mathrm{L}_{q;1}^{[\lambda]}\,\phi(x) := \sum_{n\in\mathbf{Z}} \frac{\phi(q^n\lambda)}{\theta_q(q^n\lambda/x)} = \sum_{\xi\in[\lambda;q]} \frac{\phi(\xi)}{\theta_q(\xi/x)}.$$

Since it depends on $\overline{\lambda} \in \mathbf{E}_q$ only, we call it *q-Laplace-Zhang transformation in direction* $\overline{\lambda}$ (with Theta kernel).

EXERCISE 14.28. Check that $\mathrm{L}_{q;1}^{[\lambda]}\,\phi(x) = \dfrac{1}{\theta_q(\lambda/x)} \sum\limits_{n\in\mathbf{Z}} q^{n(n-1)/2}(\lambda/x)^n\phi(q^n\lambda)$.

EXAMPLE 14.29. Let $\phi_m(\xi) := \xi^m$, $m \in \mathbf{N}$. Then:

$$\begin{aligned}
\mathrm{L}_{q;1}^{[\lambda]}\,\phi(x) &= \frac{1}{\theta_q(\lambda/x)} \sum_{n\in\mathbf{Z}} q^{n(n-1)/2}(\lambda/x)^n(q^n\lambda)^m = \frac{\lambda^m}{\theta_q(\lambda/x)} \sum_{n\in\mathbf{Z}} q^{n(n-1)/2}(q^m\lambda/x)^n \\
&= \frac{\lambda^m}{\theta_q(\lambda/x)}\theta_q\left(q^m\lambda/x\right) = q^{-m(m-1)/2}x^m,
\end{aligned}$$

from the general identity $\theta_q(q^m y) = \dfrac{1}{q^{m(m-1)/2}y^m}$ used with $y := \lambda/x$. So we see that $\mathrm{L}_{q;1}^{[\lambda]} \circ \mathcal{B}_{q;1}(x^m) = x^m$, i.e., $\mathrm{L}_{q;1}^{[\lambda]}$ formally inverts $\mathcal{B}_{q;1}$.

14.2.2.3.

THEOREM 14.2. *Let $\hat{f} \in \mathbf{C}[[x]]_{q;1}$, so that $\phi := \mathcal{B}_{q;1}\hat{f} \in \mathbf{C}\{\xi\}$. Let $\lambda \in \mathbf{C}^*$. Assume that ϕ admits an analytic continuation to some connected open set Ω containing $[\lambda;q]$, such that:*

$$\exists C, A > 0 \;:\; \forall n \leq 0 \;,\; |\phi(\lambda q^n)| \leq CA^n q^{-n(n-1)/2}.$$

(So the growth of ϕ is at most q-exponential over the half q-spiral $\lambda q^{-\mathbf{N}}$.) Then:

(i) *Setting $f := L_{q;1}^{[\lambda]}\phi$ defines a meromorphic germ in $\mathscr{M}(\mathbf{C}^*, 0)$, say on some punctured open disk $\dot{D}(0,r)$, $r > 0$, holomorphic over $\dot{D}(0,r) \setminus [-\lambda;q]$.*

(ii) *The germ f is asymptotic to $\hat{f}$ on $\dot{D}(0,r) \setminus [-\lambda;q]$ in the following sense:*

$$\forall r_0 \in]0,r[\;,\; \exists C, A > 0 \;:\; \forall n \geq 0 \text{ and } \epsilon > 0 \text{ small enough}\;,$$

$$\left(x \in \dot{D}(0,r_0) \setminus q^{\mathbf{Z}}\mathring{D}(-\lambda,\epsilon)\right) \Longrightarrow \left(\left|f(x) - \overline{T}_n\hat{f}(x)\right| < \frac{C}{\epsilon}A^n q^{-n(n+1)/2}\,|x|^{n+1}\right).$$

(Notations for truncations in 0.4.) The premise may also be written as: $x \in \dot{D}(0,r_0) \setminus \bigcup\limits_{n\in\mathbf{Z}} \{q^n y \mid |y+\lambda| \leq \epsilon\}$.

PROOF. (Sketched) Up to obvious changes, we may assume $q \leq |\lambda| < 1$ and radius of convergence of $\phi \geq 1$. We write $\mathrm{L}_{q;1}^{[\lambda]}\phi(x)$ as $\dfrac{1}{\theta_q(\lambda/x)}(f_0(x)+f_\infty(x))$, where $f_0 := \sum\limits_{n<0} q^{n(n-1)/2}(\lambda/x)^n\phi(q^n\lambda)$ and $f_\infty := \sum\limits_{n\geq 0} q^{n(n-1)/2}(\lambda/x)^n\phi(q^n\lambda)$. From the growth assumption on $\phi(\lambda q^n)$, we draw that $f_0 \in \mathbf{C}\{x\}$. Then, since $\phi(\xi)$ is bounded on $|\xi| \leq |\lambda|$ (because $|\lambda| < 1$ and radius of convergence of $\phi \geq 1$), $f_\infty \in \mathscr{O}(\mathbf{C}_\infty)$. Together, those two facts yield conclusion (i).

We omit the proof of (ii). □

14.2.3. The discrete summation process S_λ.

14.2.3.1. *q-Gevrey asymptotics.*

DEFINITION 14.30. Let $\lambda \in \mathbf{C}^*$ and $\hat{f} \in \mathbf{C}[[x]]$. We say that $f \in \mathscr{M}(\mathbf{C}^*, 0)$ is *(q-Gevrey of order 1) asymptotic to $\hat{f}$ in direction $\overline{\lambda}$ (or $[\lambda; q]$)*, and we write $f \overset{[\lambda]}{\underset{q;1}{\sim}} \hat{f}$, if f is analytic out of $[-\lambda; q]$ and:

$$\forall r_0 \text{ small enough },\ \exists C, A > 0\ :\ \forall n \geq 0 \text{ and } \epsilon > 0 \text{ small enough },$$

$$\left(x \in \overset{\bullet}{\mathrm{D}}(0, r_0) \setminus q^{\mathbf{Z}}\overset{\circ}{\mathrm{D}}(-\lambda, \epsilon)\right) \Longrightarrow \left(\left|f(x) - \overline{T}_n\hat{f}(x)\right| < \frac{C}{\epsilon}A^n q^{-n(n+1)/2}\,|x|^{n+1}\right).$$

Clearly, for $\alpha, \beta \in \mathbf{C}$:

$$f \overset{[\lambda]}{\underset{q;1}{\sim}} \hat{f} \text{ and } g \overset{[\lambda]}{\underset{q;1}{\sim}} \hat{g} \Longrightarrow \alpha f + \beta g \overset{[\lambda]}{\underset{q;1}{\sim}} \alpha\hat{f} + \beta\hat{g}.$$

The corresponding rule for mutiplication $f \overset{[\lambda]}{\underset{q;1}{\sim}} \hat{f}$ and $g \overset{[\lambda]}{\underset{q;1}{\sim}} \hat{g} \Longrightarrow fg \overset{[\lambda]}{\underset{q;1}{\sim}} \hat{f}\hat{g}$ also holds true, thanks to identity:

$$fg - \overline{T}_n(\hat{f}\hat{g}) = \sum_{j=0}^{n}(f - \overline{T}_{n-j}\hat{f})\hat{g}_j x^j + f(g - \overline{T}_n\hat{g})$$

(of course the $\hat{g}_j$ are the coefficients of $\hat{g}$) and obvious majorations. We denote $\mathbf{A}_{q;1}^{[\lambda]}$ the set of meromorphic germs $f \in \mathscr{M}(\mathbf{C}^*, 0)$ which are analytic out of $[-\lambda; q]$ and which admit a (q-Gevrey of order 1) asymptotic expansion in direction $\overline{\lambda}$. By the above, it is a $\mathbf{C}$-algebra. Also, as we shall soon see that the asymptotic expansion, if it exists, is unique, we get a mapping $f \mapsto \hat{f}$ from $\mathbf{A}_{q;1}^{[\lambda]}$ to $\mathbf{C}[[x]]_{q;1}$, which, to a large extent, is compatible with algebraic operations (but not quite a morphism of $\mathbf{C}$-algebras since $\mathbf{A}_{q;1}^{[\lambda]}$ is not one: indeed, the factor C/ϵ in the above definition entails simple poles).

LEMMA 14.31. *Let $\lambda, \mu \in \mathbf{C}^*$ such that $\lambda \not\equiv \mu \pmod{q^{\mathbf{Z}}}$. Then:*

$$\mathbf{A}_{q;1}^{[\lambda]} \cap \mathbf{A}_{q;1}^{[\mu]} = \mathbf{C}\{x\}.$$

PROOF. The inclusion $\supset$ is evident. Let f belong to $\mathbf{A}_{q;1}^{[\lambda]} \cap \mathbf{A}_{q;1}^{[\mu]}$. Since $[-\lambda; q] \cap [-\mu; q] = \emptyset$, we see first that $f \in \mathscr{O}(\mathbf{C}^*, 0)$. Then the asymptotic condition taken with $n = 0$ implies that it is bounded near 0, whence $f \in \mathscr{O}(\mathbf{C}, 0) = \mathbf{C}\{x\}$. □

14.2.3.2. *Summation.* Recall that $\mathbf{H}_{q;1}(\Omega)$ was defined in 14.2.1.1.

DEFINITION 14.32. Let $\lambda \in \mathbf{C}^*$ and $\hat{f} \in \mathbf{C}[[x]]$. We say that $\hat{f}$ *is S_λ-summable (at level 1)*, and one writes $\hat{f} \in \mathbf{C}\{x\}_{q;1}^{[\lambda]}$, if $\mathcal{B}_{q;1}\hat{f} \in \mathbf{H}_{q;1}(\Omega)$ for some $q*$-set $\Omega \supset [\lambda; q]$. One then defines the *sum of $\hat{f}$ in direction $\overline{\lambda}$* as:

$$S_\lambda \hat{f} := \mathrm{L}_{q;1}^{[\lambda]} \circ \mathcal{B}_{q;1}\left(\hat{f}\right).$$

Clearly (but you should make sure by yourself!) it depends on $\overline{\lambda}$ only and then $S_\lambda \hat{f} \in \mathbf{A}_{q;1}^{[\lambda]}$. Note that, if $[\lambda; q]$ and $[\mu; q]$ are close enough (which, equivalently means: $\overline{\lambda}$ and $\overline{\mu}$ are close enough in $\mathbf{E}_q$; or: λ and some μq^n are close enough), and if $S_\lambda \hat{f}$ exists, then $S_\mu \hat{f}$ also exists but is in general different:

EXERCISE 14.33. If $S_\lambda \hat{f} = S_\mu \hat{f}$ and $\lambda \not\equiv \mu \pmod{q^{\mathbf{Z}}}$, then $\hat{f} \in \mathbf{C}\{x\}$.

Setting $\phi := \mathcal{B}_{q;1}\hat{f} \in \mathbf{H}_{q;1}(\Omega)$, the sum of $\hat{f}$ in direction $\overline{\lambda}$ can be written:

$$\begin{aligned} S_\lambda \hat{f} &= \frac{1}{\theta_q(\lambda/x)} \sum_{m \in \mathbf{Z}} q^{m(m-1)/2} (\lambda/x)^m \mathcal{B}_{q;1}\hat{f}(q^m\lambda) \\ &= \frac{1}{\theta_q(x/\lambda)} \sum_{m \in \mathbf{Z}} q^{m(m-1)/2} (\lambda/x)^{m-1} \mathcal{B}_{q;1}\hat{f}(q^m\lambda). \end{aligned}$$

(We used equality $\theta_q(\lambda/x) = (\lambda/x)\theta_q(x/\lambda)$.) That function is meromorphic in a small enough punctured open disk centerd at 0, with at most simple poles on $[-\lambda; q]$ and residues can be computed from the above formula:

$$\mathrm{Res}_{x=-\lambda q^n} S_\lambda \hat{f} = \frac{-\lambda q^n}{(q;q)_\infty^3} \sum_{m \in \mathbf{Z}} (-1)^m q^{m(m-1)/2} \mathcal{B}_{q;1}\hat{f}(q^{n+m}\lambda).$$

EXERCISE 14.34. Check the formula for $n = 0$ using exercise 2.16 from Chapter 2; then conclude using the fact that $S_\lambda \hat{f}$ depends on $\overline{\lambda}$ only.

By arguments similar to those used in the lemma above, one sees that $S_\lambda \hat{f}$ is the only germ asymptotic to $\hat{f}$ and having those poles and residues. The sequence of residues is thus a kind of encoding of the sum, with interesting properties; see [**197**, Chapter 5] and above all [**266**].

COROLLARY 14.35. *If $\hat{f} \in \mathbf{C}\{x\}$, then $S_\lambda \hat{f} = \hat{f}$.*

EXERCISE 14.36. Prove the corollary, first formally (see Example 14.29), then using the uniqueness just stated.

14.2.4. Summation of formal solutions of irregular q-difference equations. In what follows, we shall stick to the following notations: $\hat{f} = \sum a_n x^n \in \mathbf{C}[[x]]$, $\hat{g} = \sum b_n x^n \in \mathbf{C}[[x]]$, $\phi := \mathcal{B}_{q;1}\hat{f}$, $\gamma := \mathcal{B}_{q;1}\hat{g}$, $f := \mathrm{L}_{q;1}^{[\lambda]}\phi = S_\lambda \hat{f}$, $g := \mathrm{L}_{q;1}^{[\lambda]}\gamma = S_\lambda \hat{g}$.

EXERCISE 14.37. (i) Check that, if $\hat{g} = x\hat{f}$, then $\gamma = \xi\sigma_q\phi$ and $g = xf$.
(ii) Check that, if $\hat{g} = \sigma_q\hat{f}$, then $\gamma = \sigma_q\phi$ and $g = \sigma_q f$.

14.2.4.1. Using the exercise, we deduce formally that $S_\lambda(h\hat{f}) = hS_\lambda(\hat{f})$ whenever $h \in \mathbf{C}\{x\}$. "Formally" means that it flows easily when h is a polynomial and that we guess - or at least hope - that it can't be worse when h converges. As usual, writing the equality, we implicitly understand that if one side is defined, then the other is. A more rigorous argument would be: both sides are asymptotic to $h\hat{f}$ and have the same residues [**265**].

COROLLARY 14.38. *Let L a q-difference operator over $\mathbf{C}\{x\}$. Then each $\mathbf{C}\{x\}_{q;1}^{[\lambda]}$ is stable under L and:*

$$\forall \hat{f} \in \mathbf{C}\{x\}_{q;1}^{[\lambda]}\,,\; S_\lambda(L\hat{f}) = LS_\lambda(\hat{f}).$$

So, from any formal solution $\hat{f} \in \mathbf{C}[[x]]$ of $L\hat{f} = g$ with adequate properties, we deduce a whole family of meromorphic solutions asymptotic to $\hat{f}$. Some of the said possible "adequate properties" were described in this section.

Algebraically, the mentioned stability says that each $\mathbf{C}\{x\}_{q;1}^{[\lambda]}$ is a module over the ring $\mathcal{D}_{\mathbf{C}\{x\},q}$. However, it is not itself an algebra; indeed, as we shall now see, it is not even closed under squaring.

14.2.4.2. *A summable series with nonsummable square.*

EXAMPLE 14.39. Let $L := \sigma_q - qx$. The unique solution of $L\hat{f} = 1$ is $\hat{f} := \sum\limits_{n \geq 0} q^{-n(n-1)/2} x^n$, so here $\phi := \mathcal{B}_{q;1}\hat{f} = \dfrac{1}{1-\xi} \in \mathbf{H}_{q;1}(\Omega)$, where $\Omega := \mathbf{C} \setminus q^{-\mathbf{N}}$ and $\hat{f}$ is λ-summable for every $\lambda \in \mathbf{C}^* \setminus [1;q]$, with sum (using the formulas from 14.2.3.2):

$$S_\lambda \hat{f} = \frac{1}{\theta_q(\lambda/x)} \sum_{m \in \mathbf{Z}} (\lambda/x)^m \frac{q^{m(m-1)/2}}{1 - q^m\lambda} = \frac{1}{\theta_q(x/\lambda)} \sum_{m \in \mathbf{Z}} (\lambda/x)^{m-1} \frac{q^{m(m-1)/2}}{1 - q^m\lambda}.$$

EXERCISE 14.40. (i) Check that $\Delta := \sigma_p L$ falls under the assumptions of 14.2.1.1 What about the q-exponential growth of ϕ?

(ii) Compare $S_\lambda \hat{f}$ found here with the one in 14.1.3.

(iii) Check that the above series converges; is there normal convergence?

EXAMPLE 14.41. Let $\hat{g} := \hat{f}^2 = \sum b_n x^n = \sum q^{-n(n-1)/2} c_n x^n$, where:

$$b_n = \sum_{\substack{k+l=n \\ k,l \geq 0}} q^{-k(k-1)/2} q^{-l(l-1)/2} \Longrightarrow c_n = \sum_{\substack{k+l=n \\ k,l \geq 0}} q^{kl}.$$

Note that $1 \leq c_n \leq n+1$, so that $\gamma := \mathcal{B}_{q;1}\hat{g} = \sum c_n \xi^n$ has radius of convergence 1. We write $\gamma = \sum\limits_{k,l \geq 0} q^{kl} \xi^k \xi^l$, whence $q\xi^2 \sigma_q \gamma = \sum\limits_{k,l \geq 1} q^{kl} \xi^k \xi^l$. Decomposing $\sum\limits_{k,l \geq 0}$ into $\sum\limits_{k,l=0}$, $\sum\limits_{k \geq 1, l \geq 0}$ and $\sum\limits_{k \geq 0, l \geq 1}$, we draw:

$$\gamma = 1 + \frac{2\xi}{1-\xi} + qx^2 \sigma_q \gamma = \frac{1+\xi}{1-\xi} + q\xi^2 \sigma_q \gamma.$$

It follows that γ admits a meromorphic continuation to $\mathbf{C}$ with simple poles on $q^{-\mathbf{N}}$. On the other hand, when $|\xi| \to +\infty$, we have $\gamma(q^{-1}\xi) = O(1) + q^{-1}\xi^2\gamma(\xi)$, which easily implies q-exponential growth of order > 1, so $\gamma \notin \mathbf{H}_{q;1}(\Omega)$ and $\hat{g}$ is not summable.

The right way out of this difficulty is a theory of *multisummability*; see [**163**].

EXERCISE 14.42. (i) Iterating, prove that $\gamma = \sum_{n\geq 0} q^{n^2}\xi^{2n}\frac{1+q^n\xi}{1-q^n\xi}$.

(ii) Show that for $\xi_0 \notin [1;q]$ there is a constant C such that $\gamma(\xi_0 q^{-n}) \sim C\xi_0^{2n}q^{-n^2}$ when $n \to +\infty$.

(iii) Check that $L'\hat{g} = 1 + qx$, where $L' := (\sigma_q - qx)(\sigma_q - x^2)$. What would you guess considering the slopes of L'?

14.2.4.3. *An example of q-Stokes phenomenon.* This is about the "ambiguity of analytical interpretation" of the formal solution $\hat{f}$ according to whether one sums it along direction $\overline{\lambda} \in \mathbf{E}_q$ or along direction $\overline{\mu} \neq \overline{\lambda}$. Since both $S_\lambda \hat{f}$ and $S_\mu \hat{f}$ are solutions of the same nonhomogeneous equation $L\hat{f} = 1$, their difference is a solution of $Lf = 0$:

$$S_{\lambda,\mu}\hat{f} := S_\lambda \hat{f} - S_\mu \hat{f} \Longrightarrow LS_{\lambda,\mu}\hat{f} = 0.$$

As such, it has the form $S_{\lambda,\mu}\hat{f} = \dfrac{U_{\lambda,\mu}}{\theta_q(qx)}$ for some $U_{\lambda,\mu} \in \mathscr{M}(\mathbf{E}_q)$. On the other hand, it has simple poles over $[-\lambda;q] \cup [-\mu;q]$, so it can be written in the form $S_{\lambda,\mu}\hat{f} = \dfrac{V_{\lambda,\mu}}{\theta_q(x/\lambda)\theta_q(x/\mu)}$ for some $V_{\lambda,\mu} \in \mathscr{O}(\mathbf{C}^*,0)$ (actually, in $\mathscr{O}(\mathbf{C}^*)$). The only possibility is that:

$$S_{\lambda,\mu}\hat{f} = C_{\lambda,\mu}\frac{\theta_q(x/q\lambda\mu)}{\theta_q(x/\lambda)\theta_q(x/\mu)} \text{ for some } C_{\lambda,\mu} \in \mathbf{C}^*.$$

EXERCISE 14.43. Check that in two ways: first by determining all the possible q-elliptic $U_{\lambda,\mu}$ with those zeroes and poles; second by writing the functional equation for $V_{\lambda,\mu}$ and determining all its solutions in $\mathscr{O}(\mathbf{C}^*,0)$.

We determine $C_{\lambda,\mu}$ by identifying the residues at $x = -\lambda$ of both sides. On the left-hand side, note that $S_\mu \hat{f}$ does not contribute, so we may use the formula given in 14.2.3.2. On the right-hand side, only the factor $\theta_q(x/\lambda)$ (at the denominator) contributes.

EXERCISE 14.44. The residue of $1/\theta_q(x/\lambda)$ at $x = -\lambda$ is $\lambda/\theta_q{}'(-1)$.

So we get:

$$C_{\lambda,\mu}\frac{\theta_q(-1/q\mu)}{\theta_q(-\lambda/\mu)}\frac{\lambda}{(q;q)_\infty^3} = \frac{-\lambda}{(q;q)_\infty^3}\sum_{m\in\mathbf{Z}}(-1)^m q^{m(m-1)/2}\frac{1}{1-q^m\lambda} \Longrightarrow$$

$$C_{\lambda,\mu} = -\frac{\theta_q(-\lambda/\mu)}{\theta_q(-1/q\mu)}\sum_{m\in\mathbf{Z}}(-1)^m q^{m(m-1)/2}\frac{1}{1-q^m\lambda}.$$

REMARK 14.45. The $S_{\lambda,\mu}\hat{f}$ satisfy a "cocycle condition" $S_{\lambda,\mu} + S_{\mu,\nu} = S_{\lambda,\nu}$ which was already met in 14.1.4.2 and which will be exploited later in 15.3. It implies in particular that $S_{\lambda,\mu}\hat{f} = -S_{\mu,\lambda}\hat{f}$, whence $C_{\lambda,\mu} = -C_{\mu,\lambda}$, which is not at all obvious in the above form. Actually, it implies a nice identity.

EXERCISE 14.46. Prove the following identity:

$$\frac{\theta_q(-\lambda)}{\lambda}\sum_{m\in\mathbf{Z}}(-1)^m q^{m(m-1)/2}\frac{1}{1-q^m\lambda} = C,$$

where C does not depend on λ and can be determined by letting $\lambda \to 1$.

14.2.5. Application to some confluent basic hypergeometric series.

14.2.5.1. $q \to 1$ *degeneracy in the divergent world.* Let the q-hypergeometric series ${}_2\Phi_0(a,b;-;q,x)$ with radius of convergence 0 be defined as:

$$ {}_2\Phi_0(a,b;-;q,x) := \sum_{n\geq 0} \frac{(a;q)_n(b;q)_n}{(0;q)_n(q;q)_n} x^n = \sum_{n\geq 0} \frac{(a;q)_n(b;q)_n}{(q;q)_n} q^{-n(n-1)/2}(-1)^n x^n. $$

(See Section 4.4 of Chapter 4.) It can be considered as a q-deformation of the *confluent hypergeometric series* ${}_2F_0(\alpha,\beta;-;-z) := \sum_{n\geq 0} \frac{(\alpha)_n(\beta)_n}{n!}(-z)^n$ in the following way; letting $a_n(q) := \frac{(a;q)_n(b;q)_n}{(q;q)_n} q^{-n(n-1)/2}$ and substituting $x(q) := z/(1-q)$, and provided (as usual) that $a = q^\alpha$ and $b = q^\beta$, the series:

$$ {}_2\Phi_0(a(q),b(q);-;q,x(q)) = \sum_{n\geq 0} a_n(q)(-1)^n x(q)^n $$

tends coefficientwise to ${}_2F_0$.

CAUTION 14.47. Beware that the word "confluent" in the classical world of hypergeometric series has a meaning from which we have departed in our Chapters 11 and 12: it means that, starting from a hypergeometric function with regular singular points only and having one of those points tend to another (that is the confluence), the function degenerates into one with an irregular singular point (see in particular **[194, 255]**). We took up the analogy on the basis of the fact that, when $q \to 1$, singular points on q-spirals tend to each other making up the cuts we observed in 11.2.

In **[265]**, Zhang gives an analytic meaning to that formal degeneracy by showing how the S_λ-sum of ${}_2\Phi_0(a,b;-;q,x)$ degenerates in the Borel sum **[192]** of the divergent series ${}_2F_0$ when $q \to 1$. We shall not take up here that complete study, but merely look at the summability of ${}_2\Phi_0(a,b;-;q,x)$ in the sense of this section. Beyond **[265]**, more information can be found in **[197**, Chapter 7**]**.

14.2.5.2. *Functional equation.* Write $\hat{f} := {}_2\Phi_0(a,b;-;q,x) := \sum a_n x^n$. From equalities:

$$ \frac{a_{n+1}}{a_n} = \frac{(1-aq^n)(1-bq^n)}{1-q^{n+1}}(-q^{-n}) \Longrightarrow q^{n+1}(1-q^{n+1})a_{n+1} = -q(1-aq^n)(1-bq^n)a_n, $$

multiplying both sides by x^{n+1} and summing for $n \in \mathbf{N}$, we get the q-difference equation:

$$ \hat{f}(qx) - \hat{f}(q^2x) = -qx\left(\hat{f}(x) - (a+b)\hat{f}(qx) + ab\hat{f}(q^2x)\right) \Longrightarrow L\hat{f} = 0, $$

where $L := {\sigma_q}^2 - \frac{1-q(a+b)x}{1-qabx}\sigma_q - \frac{qx}{1-qabx}$, a q-difference operator over $\mathbf{C}\{x\}$. According to Chapter 10, the Newton polygon of L has slopes -1 and 0, the former corresponding to the convergent solution (see 14.2.5.3). The latter has characteristic equation $s^2 - s = 0$ and gives rise to the divergent solution $\hat{f}$.

EXERCISE 14.48. Chasing denominators in equation $L\hat{f} = 0$, show that it leads to a recursive equation for coefficients which can be explicitly solved and obtain conversely $\hat{f} = {}_2\Phi_0$.

EXERCISE 14.49. Following (once more) **[265]**, make $c \to \infty$ in ${}_2\Phi_1(a, b; c; q, x)$ and see how the connection formulas of 4.4 degenerate.

14.2.5.3. *A convergent solution* f_0. In its simplest guise, the q-difference equation leading to a convergent solution f_0 associated to slope -1 of $Lf_0 = 0$ is obtained by setting $g_0 := f_0\theta_q$, which should make null the following q-difference operator (the fraction in the head is meant to simplify obvious common factors):

$$L' := \frac{1}{qx}\theta_q L {\theta_q}^{-1} = x{\sigma_q}^2 - \frac{1}{q}\frac{1 - q(a+b)x}{1 - qabx}\sigma_q - \frac{1}{1 - qabx},$$

which has slopes 0 and 1, the former providing the convergent solution. The corresponding characteristic equation is $-\frac{1}{q}s - 1 = 0$, so we have an expression of the type $f_0 = {\theta_q}^{-1}e_{-q}(1 + \cdots)$ and the best choice for the character e_{-q} is clearly $e_{-q}(x) := -\theta_q(x)/\theta_q(-qx) = x\theta_q(x)/\theta_q(-x)$. So, slightly changing notations, we set $f_0(x) = xg_0(x)/\theta_q(-x)$, where g_0 satisfies:

$$\left(qx{\sigma_q}^2 + \frac{1 - q(a+b)x}{1 - qabx}\sigma_q - \frac{1}{1 - qabx}\right) g_0 = 0.$$

This one has again slopes 0 and 1, but characteristic equation $s - 1 = 0$, whence indeed a convergent solution $g_0 = 1 + \cdots \in \mathbf{C}\{x\}$, so that in the end our convergent solution for L is:

$$f_0(x) = \frac{x}{\theta_q(-x)}g_0, \quad g_0 = 1 + \cdots \in \mathbf{C}\{x\}.$$

14.2.5.4. *Factorisation.* Setting $v := \dfrac{\sigma_q f_0}{f_0}$, equation $Lf_0 = 0$ entails a factorisation $L = (\sigma_q - u)(\sigma_q - v)$, with $u, v \in \mathbf{C}(\{x\})$ and $uv = \dfrac{-qx}{1 - qabx}$. Since $\dfrac{\sigma_q f_0}{f_0} = -qx\dfrac{\sigma_q g_0}{g_0} = -qx(1 + \cdots)$, we see that $u = 1 + \cdots \in \mathbf{C}\{x\}$. Then equation $L\hat{f} = 0$ amounts to: $\sigma_q\hat{f} - v\hat{f} = g$, where g is a formal solution of $\sigma_q - u$. But since $\sigma_q\hat{f} - v\hat{f}$ has the form $1 + \cdots \in \mathbf{C}[[x]]$, the only possibility is $\sigma_q\hat{f} - v\hat{f} = g$, where $g := \dfrac{1}{u\sigma_q u{\sigma_q}^2 u \cdots} \in \mathbf{C}\{x\}$ (see Chapter 4, Section 4.1). That equation falls under the assumptions of 14.2.1.1, so that in particular $\phi := \mathcal{B}_{q;1}\hat{f} \in \mathbf{H}_{q;1}(\Omega)$. We are going to recover that fact directly.

EXERCISE 14.50. Find Δ, $\tilde{\Delta}$, P and Ω as in 14.2.1.1

14.2.5.5. *Summation.* The q-Borel-Ramis transform $\phi := \mathcal{B}_{q;1}\hat{f}$ is given by:

$$\begin{aligned}\phi(\xi) = \sum_{n\geq 0} q^{n(n-1)/2}\frac{(a;q)_n(b;q)_n}{(0;q)_n(q;q)_n}\xi^n &= \sum_{n\geq 0}\frac{(a;q)_n(b;q)_n}{(q;q)_n}(-1)^n\xi^n \\ &= {}_2\Phi_1(a, b; 0; q, -\xi).\end{aligned}$$

The coefficients $a_n := \dfrac{(a;q)_n(b;q)_n}{(q;q)_n}(-1)^n$ satisfy the recurrence relation:

$$a_{n+1}/a_n = -(1 - aq^n)(1 - bq^n)/(1 - q^{n+1}) \Longrightarrow (1 - q^{n+1})a_{n+1} = -(1 - aq^n)(1 - bq^n).$$

On the one hand, $a_{n+1}/a_n \to -1$ as $n \to +\infty$, so the radius of convergence of ϕ is 1. On the other hand, multiplying the second relation by ξ^{n+1} and summing, we get the functional equation:

$$\phi(\xi) - \phi(q\xi) = -\xi(\phi(\xi) - (a+b)\phi(q\xi) + ab\phi(q^2\xi)) \Longrightarrow$$
$$(ab\xi{\sigma_q}^2 - (1 + (a+b)\xi)\sigma_q + (1+\xi))\phi = 0.$$

The slopes are 0 and 1, the characteristic equation attached to the former is $s - 1 = 0$ and it gives rise to the convergent solution $\phi = {}_2\Phi_1(a, b; 0; q, -\xi)$. Moreover, writing:

$$\phi(\xi) = \frac{1}{1+\xi}\left((1 + (a+b)\xi)\phi(q\xi) - ab\xi\phi(q^2\xi)\right),$$

we obtain a meromorphic continuation to the whole of $\mathbf{C}$, with simple poles over $-q^{-\mathbf{N}}$. By the general theory $\phi \in \mathbf{H}_{q;1}(\Omega)$, $\Omega := \mathbf{C} \setminus (-q^{-\mathbf{N}})$, and its q-Laplace-Zhang transform $\mathrm{L}_{q;1}^{[\lambda]}\phi$ is defined for all $\lambda \notin [-1; q]$ by the formula:

$$\mathrm{L}_{q;1}^{[\lambda]}\phi(x) = \frac{1}{\theta_q(\lambda/x)} \sum_{m \in \mathbf{Z}} (\lambda/x)^m q^{m(m-1)/2}\, {}_2\Phi_1(a, b; 0; q, -q^m\lambda).$$

Using a connection formula from **[93]**, Zhang, after some algebraic manipulations, obtains:

$$S_\lambda \hat{f}(x) = \frac{(b;q)_\infty}{(b/a;q)_\infty} \frac{\theta_q(a\lambda)}{\theta_q(\lambda)} \frac{\theta_q(qax/\lambda)}{\theta_q(qx/\lambda)}\, {}_2\Phi_1(a, 0; aq/b; q, q/abx)$$
$$+ \frac{(a;q)_\infty}{(a/b;q)_\infty} \frac{\theta_q(b\lambda)}{\theta_q(\lambda)} \frac{\theta_q(qbx/\lambda)}{\theta_q(qx/\lambda)}\, {}_2\Phi_1(b, 0; bq/a; q, q/abx).$$

14.2.5.6. *Going to infinity ...* We follow a different path inspired by the global study on the Riemann sphere in section 12.1 of Chapter 12. To simplify, we write f a solution (formal or not) of q-difference equation $Lf = 0$ from 14.2.5.2. We let $y := 1/x$ and $g(y) := f(x) = f(1/y)$. Then:

$$Lf = 0 \Longleftrightarrow (1 - qab/y)g(y/q^2) - (1 - q(a+b)/y)g(y/q) - (q/y)g(y) = 0$$
$$\Longleftrightarrow g(q^2y) - ((a+b) - qy)g(qy) + (ab - qy)g(y) = 0.$$

(We chased denominators, then replaced y by q^2y and simplified by $-q$.) This is now a fuchsian equation with exponents a and b. We shall tackle only the generic case $a \not\equiv b \pmod{q^{\mathbf{Z}}}$.

EXERCISE 14.51. Redo it all when $a \equiv b \pmod{q^{\mathbf{Z}}}$.

From Chapter 11, we know that there are fundamental solutions of the form $e_a(y)u(y)$, $e_b(y)v(y)$, where $u, v \in \mathbf{C}\{y\}$, $u(0) = v(0) = 1$, and e_a, e_b are such that $e_a(qy) = ae_a(y)$, $e_b(qy) = be_b(y)$. Of course, a and b play a symmetric role, so let us look for $u(y) = \sum_{n \geq 0} u_n y^n$, $u_0 = 1$:

$$\text{(equation above)} \Longleftrightarrow a^2u(q^2y) - a((a+b) - qy)u(qy) + (ab - qy)u(y) = 0$$
$$\Longleftrightarrow \forall n \geq 1\,,\ (a^2q^{2n} - a(a+b)q^n + ab)u_n = (q - aq^n)u_{n-1}$$
$$\Longleftrightarrow \forall n \geq 1\,,\ a(b - aq^n)(1 - q^n)u_n = q(1 - aq^{n-1})u_{n-1}$$

(note how the limit case $n = 0$ is predictably tautological), which is readily solved:

$$\forall n \geq 0\,,\ u_n = (q/ab)^n \frac{(a;q)_n}{(q;q)_n(qa/b;q)_n} \Longrightarrow u(y) = {}_2\Phi_1(a, 0; qa/b; q, qy/ab).$$

In the end, we get the fundamental solutions $e_a(y)\ {}_2\Phi_1(a,0;qa/b;q,qy/ab)$ and $e_b(y)\ {}_2\Phi_1(b,0;qb/a;q,qy/ab)$.

14.2.5.7. *... and back again.* On account of 14.2.5.6, f is a linear combination over $\mathscr{M}(\mathbf{E}_q)$ of $e_{1/a}(x){}_2\Phi_1(a,0;qa/b;q,q/abx)$ and $e_{1/b}(x){}_2\Phi_1(b,0;qb/a;q,q/abx)$. In the latter functions, all poles over $\mathbf{C}^*$ come from $e_{1/a}(x)$ and $e_{1/b}(x)$, so we shall choose characters $e_{1/a}(x) := \dfrac{\theta_q(ax/\lambda)}{\theta_q(x/\lambda)}$ and $e_{1/b}(x) := \dfrac{\theta_q(bx/\lambda)}{\theta_q(x/\lambda)}$. (It does not matter whether we use x/λ or λ/x since $\theta_q(x/\lambda) = (x/\lambda)\theta_q(\lambda/x)$.) So we reach for a formula:

$$S_\lambda \hat{f}(x) = A\frac{\theta_q(ax/\lambda)}{\theta_q(x/\lambda)}\,{}_2\Phi_1(a,0;qa/b;q,q/abx) + B\frac{\theta_q(bx/\lambda)}{\theta_q(x/\lambda)}\,{}_2\Phi_1(b,0;qb/a;q,q/abx),$$

where $A, B \in \mathbf{C}$ (since they are elliptic and have no poles). This "simplette" method does not give A and B, which would amount to explicit connection formulas as those used in **[93, 265]**.

14.2.5.8. *Stokes phenomenon for* ${}_2\Phi_0(a,b;-;q,x)$. We shall sketch its description along the same lines as in 14.2.4.3. We use the nonhomogeneous equation $\sigma_q\hat{f} - v\hat{f} = g$ from 14.2.5.4. It implies that $\sigma_q S_\lambda\hat{f} - vS_\lambda\hat{f} = g$ for all allowable λ, so $\sigma_q S_{\lambda,\mu}\hat{f} = vS_{\lambda,\mu}\hat{f}$, where $S_{\lambda,\mu} := S_\lambda - S_\mu$. Since $v = \dfrac{\sigma_q f_0}{f_0}$, we deduce that $S_{\lambda,\mu}\hat{f}/f_0$ is σ_q-invariant, i.e., $S_{\lambda,\mu}\hat{f} = -f_0 E_{\lambda,\mu}$ for some $E_{\lambda,\mu} \in \mathscr{M}(\mathbf{E}_q)$ (the minus sign is meant to slightly simplify formulas to come). Using now 14.2.5.3, we have $f_0(x) = \dfrac{x}{\theta_q(-x)}g_0 = \dfrac{-1}{\theta_q(-qx)}g_0$, we conclude:

$$S_{\lambda,\mu}\hat{f} = \frac{-1}{\theta_q(-qx)}g_0 E_{\lambda,\mu}, \text{ where } g_0 = 1 + \cdots \in \mathbf{C}\{x\}.$$

On the other hand, from 14.2.5.8 we draw:

$$S_{\lambda,\mu}\hat{f} = \frac{h_{\lambda,\mu}}{\theta_q(x/\lambda)\theta_q(x/\mu)}, \text{ where } h_{\lambda,\mu} \in \mathscr{O}(\mathbf{C}^*, 0).$$

The only way to realize both conditions is:

$$S_{\lambda,\mu}\hat{f} = C_{\lambda,\mu}g_0\frac{\theta_q(-x/q\lambda\mu)}{\theta_q(x/\lambda)\theta_q(x/\mu)}, \text{ where } C_{\lambda,\mu} \in \mathbf{C}^*.$$

EXERCISE 14.52. Why is it the only way? In particular, why $C_{\lambda,\mu} \neq 0$?

14.3. A variant: continuous summation and asymptotics

14.3.1. General presentation.

14.3.1.1. In this section, we again discuss asymptotic and summation theory for divergent solutions of q-difference equations, along the lines of a more ancient work of Zhang **[263]**. The restrictive assumption (only one slope, equal to 1) still holds, although this approach also has a more general scope, as shown for instance in the article **[163]** cowritten with Fabienne Marotte. The main difference with the previous section is that here, contrary to the general conventions of this book, multivalued solutions are allowed. This presentation being informative only (the goal is to give an idea of the *flavour* of the theory) we give even less details and proofs than in section 14.2.

Although conventions $0 < |q| < 1$ and $|q| > 1$ are essentially equivalent, it maybe a dangerous game to go from one to the other; maybe even more dangerous in the analytic part of q-difference theory. So, exceptionally, in this whole section, we take up the convention $q > 1$ from the quoted papers.

14.3.1.2. Multivalued functions (like the logarithm) being involved, it will be necessary to work over $\widetilde{\mathbf{C}^*}$, the *Riemann surface of the logarithm*[13]: this is just the universal covering of $\mathbf{C}^*$, but described in a different way.

We know the universal covering is really the exponential map $\mathbf{C} \xrightarrow{\exp} \mathbf{C}^*$, $a + ib \mapsto e^a e^{ib}$ but we choose to denote $a + ib$ in the form re^{ib}, where $r := e^a \in \mathbf{R}_+^*$, and where e^{ib}, $b \in \mathbf{R}$ is not understood to be 2π-periodic in b. So $re^{ib} \neq re^{i(b+2\pi)}$ in $\widetilde{\mathbf{C}^*}$: they are supposed to live in different leaves of the covering, above the usual $re^{ib} \in \mathbf{C}^*$. In this way, $\log(re^{ib}) = \ln r + ib$ is a perfectly defined uniform analytic map from $\widetilde{\mathbf{C}^*}$ to $\mathbf{C}$ (actually, an isomorphism).

Recall that we used quite a few times the diffeomorphism $\mathbf{U} \times \mathbf{R} \simeq \mathbf{C}^*$ given by $(u, y) \mapsto uq^y$. Here we shall use the decomposition $\widetilde{\mathbf{C}^*} \simeq \mathbf{R}_+^* \times \mathbf{R}$ in the form $x = |x|\, q^{iy}$, $y := \arg_q x := \Im(\log_q x)$, where of course, in $\log_q x := \log x / \log q$ the logarithm $\log q \in \mathbf{R}_+^*$ has its usual meaning.

To a certain extent, the topology on $\widetilde{\mathbf{C}^*}$ is an extension of that of $\mathbf{C}^*$. In particular, we define in $\widetilde{\mathbf{C}^*}$ punctured disks:

$$\overset{\bullet}{\mathrm{D}}(0, r) := \{ae^{ib} \in \widetilde{\mathbf{C}^*} \mid 0 < a < r \text{ and } b \in \mathbf{R}\},$$

allowing us to define the algebra $\tilde{\mathscr{O}} := \mathscr{O}(\widetilde{\mathbf{C}^*}, 0)$ of analytic germs at 0 on $\widetilde{\mathbf{C}^*}$ (those are functions holomorphic on some $\overset{\bullet}{\mathrm{D}}(0, r)$, with the usual rules of identification); and *sectorial neighborhoods (bisected by θ)*:

$$S(\theta, \epsilon) := \{ae^{ib} \in \widetilde{\mathbf{C}^*} \mid \theta - \epsilon < b < \theta + \epsilon \text{ and } a > 0\}.$$

EXERCISE 14.53. To get accustomed to $\widetilde{\mathbf{C}^*}$, check that $q^{\log_q x} = x$ and that $\log_q(qx) = \log_q x + 1$. Then, letting $\vartheta(x) := q^{\log_q x (\log_q x - 1)/2}$, compute $\vartheta(qx)$. We shall keep the notation ϑ in the whole section.

REMARK 14.54. The underlying idea of covering maps, going back to the study of algebraic functions by Riemann [**203**], is that a multivalued function can be made uniform by considering its branches on a particular open set as as many functions on leaves "étalées" over that set U. For instance the ill defined $\log : \mathbf{C}^* \to \mathbf{C}$ (diagram on the left) can be seen as a flattening of the well defined $\log : \widetilde{\mathbf{C}^*} \to \mathbf{C}$ (diagram on the right):

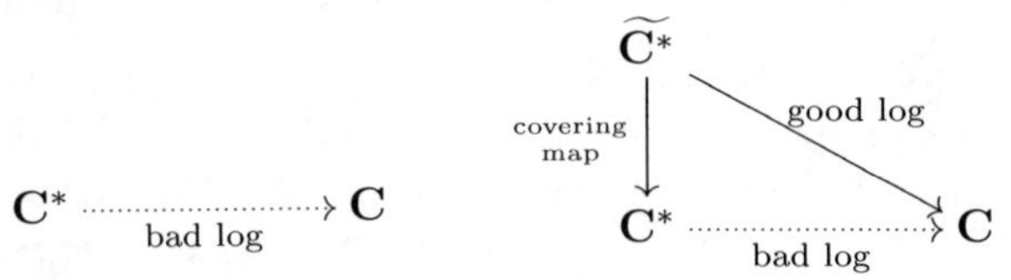

14.3.2. The basic example.

14.3.2.1. *Tshakaloff series.* When $q > 1$, the standard divergent series has the form $\hat{f}(x) := \sum_{n \geq 0} q^{n(n-1)/2} x^n$, a solution of the nonhomogeneous equation:

$$f = 1 + x\sigma_q f. \tag{14.3}$$

[13]See [**169**, Part II, §6.7] for more complete (and nice) explanations on $\widetilde{\mathbf{C}^*}$.

The associated homogeneous equation $f = x\sigma_q f$ has the obvious solution (for the notation, see the exercise at the end of Subsection 14.3.1) $q^{-\log_q x(\log_q x-1)/2} = \vartheta^{-1}(x)$, which we shall write $\vartheta'(x)$ from now on. Following standard Lagrange method ("variation of constants"), we look for a solution of the non homogeneous equation in the form $f = \vartheta' g$ and arrive at equation:

$$\sigma_q g - g = -\vartheta(x). \tag{14.4}$$

In $\widetilde{\mathbf{C}^*}$, the change of variable $x = q^t$, $t \in \mathbf{C}$, makes good sense and, setting $h(t) := g(x)$, it leads us to:

$$h(t+1) - h(t) = -q^{t(t-1)/2}. \tag{14.5}$$

This is a rather classical kind of *difference equation.*

14.3.2.2. *Use of the Fourier-Laplace transformation.* We now resort to *Fourier-Laplace transformation*[14]. A real number a being fixed, set:

$$\mathscr{F}\phi(u) := \frac{1}{2i\pi}\int_{a-\infty i}^{a+\infty i} \phi(t)e^{-ut}\,dt \text{ and } \mathscr{F}^{-1}\psi(t) := \int_{\infty-ia}^{+\infty+ia} \psi(u)e^{ut}\,du.$$

As the notation suggests, for adequate classes of functions, those transformations are inverse to each other. The following is easily checked (exercise for you!):

$$\text{If } \phi_0(t) := q^{t(t-1)/2}, \text{ then } \mathscr{F}\phi_0(u) = \frac{q^{-\frac{1}{2}\left(\frac{u}{\log q}+\frac{1}{2}\right)^2}}{\sqrt{2\pi\log q}}.$$

Now, if ϕ is analytic and has no poles in the vertical strip bounded by $a + i\mathbf{R}$ and $(a+1) + i\mathbf{R}$, twe deduce, using Cauchy residue formula:

$$\text{If } \overline{\phi}(t) := \phi(t+1), \text{ then } \mathscr{F}\overline{\phi}(u) = e^u\,\mathscr{F}\phi(u).$$

14.3.2.3. *Summation in allowable directions.* Using 14.3.2.2, we first deduce a family of solutions of (14.5):

$$\forall\theta \not\equiv 0 \pmod{2\pi}\ ,\ h_\theta(t) := \frac{1}{\sqrt{2\pi\log q}}\int_{-\infty+i\theta}^{+\infty+i\theta} \frac{q^{-\frac{1}{2}\left(\frac{u}{\log q}+\frac{1}{2}\right)^2}}{1-e^u}\,e^{ut}\,du,$$

then a family of solutions analytic over $\widetilde{\mathbf{C}^*}$ of (14.3):

$$\forall\theta \not\equiv 0 \pmod{2\pi}\ ,\ f_\theta(t) := \frac{q^{-1/8}}{\sqrt{2\pi\log q}}\int_{d_\theta} \frac{\vartheta'(x/\xi)}{1-\xi}\,\frac{d\xi}{\xi}.$$

It is customary to write d_θ the path of integration $\mathbf{R}_+^* e^{i\theta}$ oriented from 0 to $+\infty e^{i\theta}$ in $\widetilde{\mathbf{C}^*}$.

Using Cauchy residue formula, we find that, if $\theta, \theta' \in\ ]2k\pi, 2(k+1)\pi[$ for some $k \in \mathbf{Z}$, then $f_\theta = f_{\theta'}$. (Indeed, $1 - e^u$ has no zeroes in the strip bounded by $\mathbf{R} + i\theta$ and $\mathbf{R} + i\theta'$, so $h_\theta = h_{\theta'}$.) We write them all as f_k (guessing that f_1 will be rightly understood for $k = 1$, not for $\theta = 1 \ldots$).

---

[14] In real analysis, Fourier and Laplace transformations are very different, but it is not so in complex analysis (where they are also akin to Mellin transformation); see **[211, 212, 259]** and especially **[118**, Chapter 10**]**.

14.3.2.4. *Asymptotic characterisation.* Among the family of all found solutions, f_0 is characterized by the following asymptotic property:

$$\forall \eta \in]0,\pi[\ ,\ \exists K, A > 0 \text{ such that } \forall \nu \in]\eta, 2\pi - \eta[\ ,\ \forall n \in \mathbf{N}\ ,\ \forall x \in \widetilde{\mathbf{C}^*}\ ,$$
$$\left| f_0(x) - T_n \hat{f}(x) \right| < K A^n q^{\frac{1}{2}\left(n^2 + \arg_q^2(xe^{-\mathrm{i}\nu})\right)} |x|^n .$$

It is an essential fact that this property uniquely characterizes it within $\tilde{\mathcal{O}}$. For that reason, q-Gevrey asymptotics is considered as *exact asymptotics*, in contrast to Poincaré asymptotics (compare with the discussion on Gevrey asymptotics in [**192**]).

14.3.2.5. *Stokes phenomenon.* Since all f_k satisfy the same nonhomogeneous equation (14.3), their mutual differences satisfy the corresponding homogeneous equation $f = x\sigma_q f$. In 14.3.2.1 we found the particular solution ϑ', and we deduce:

$$\forall k \in \mathbf{Z}\ ,\ f_k - f_0 = C_k \vartheta'$$

for some σ_q-invariant analytic function C_k on $\widetilde{\mathbf{C}^*}$. However, C_k is not q-elliptic since it is not uniform. Using Cauchy residue formula, it can be determined exactly:

$$C_k(x) = \mathrm{i}q^{-1/8}\sqrt{\frac{2\pi}{\log q}} \times \begin{cases} \sum_{j=1}^{k} (-1)^j q^{2\pi j^2/\log^2 q} x^{2\pi \mathrm{i} j/\log q} & \text{if } k \geq 0, \\ \sum_{j=k+1}^{0} (-1)^{j+1} q^{2\pi j^2/\log^2 q} x^{2\pi \mathrm{i} j/\log q} & \text{if } k < 0. \end{cases}$$

More generally, of course, $f_k - f_l = (C_k - C_l)\vartheta'$.

EXERCISE 14.55. Measure the "defect of ellipticity" of C_k by comparing $C_k(xe^{2\mathrm{i}\pi})$ with $C_k(x)$.

14.3.3. Exact q-Gevrey asymptotics (at order 1).

14.3.3.1. Let $\theta \in \mathbf{R}$ and let $\hat{f} \in \mathbf{C}[[x]]$. We say that $f \in \tilde{\mathcal{O}}$ has *q-Gevrey asymptotic expansion $\hat{f}$ in direction θ (at order 1)*, and we write $f \underset{q;1}{\overset{\theta}{\sim}} \hat{f}$, if:

$$\exists A, K > 0\ :\ \forall n \in \mathbf{N}\ ,\ \forall x \in \widetilde{\mathbf{C}^*} \text{ small enough}\ ,$$
$$\left| f(x) - T_n \hat{f}(x) \right| < K A^n q^{\frac{1}{2}\left(n^2 + \arg_q^2(xe^{-\mathrm{i}\theta})\right)} |x|^n .$$

For instance, in 14.3.2.4, f_0 has expansion $\hat{f}$ in all directions $\nu \in]0; 2\pi[$, and, in some sense, "uniformly away from $\{0, 2\pi\}$". The q-Gevrey asymptotics as in the above definition is stronger than usual Poincaré asymptotics, so in particular the expansion $\hat{f}$, if it exists, is unique. It can be shown that then $\hat{f} \in \mathbf{C}[[x]]_{p;1}$, $p := q^{-1}$ (see definition at the beginning of 14.2; remember that there $q < 1$).

14.3.3.2. For every $\theta \in \mathbf{R}$, let $\mathcal{A}_{q;1}^{\theta}$ the space of those $f \in \tilde{\mathcal{O}}$ which admit a q-Gevrey asymptotic expansion in direction θ (at order 1). From the above uniqueness statement, we have a well defined map $J_\theta : \mathcal{A}_{q;1}^{\theta} \to \mathbf{C}[[x]]_{p;1}$, $f \mapsto \hat{f}$. Clearly both source and target are $\mathbf{C}$-linear spaces and J_θ is $\mathbf{C}$-linear. There is a q-Gevrey analogue of the classical Borel-Ritt theorem [**150, 192**]: it says that J_θ is surjective. Elements of its kernel may be characterized as those elements of $\mathcal{A}_{q;1}^{\theta}$ which have *q-exponential decay (of order 1) in direction θ*; for any germ $f \in \tilde{\mathcal{O}}$, this means:

$$\exists \mu \in \mathbf{R}\ :\ f(x) = O\left(x^\mu q^{\frac{-1}{2}\log^2(xe^{-\mathrm{i}\theta})}\right) \text{ when } |x| \to 0.$$

Having exponential decay in two directions is hardly possible:

$$\theta_1 \neq \theta_2 \Longrightarrow \operatorname{Ker} J_{\theta_1} \cap \operatorname{Ker} J_{\theta_2} = \{0\}.$$

It is this fact which will allow to characterize functions uniquely by their asymptotic behaviour, whence the statement that *q-Gevrey asymptotics is an exact asymptotic theory.*

14.3.4. q-Borel-Ramis and q-Laplace-Zhang transformations.

14.3.4.1. *q-Borel-Ramis.* Under the present convention $q > 1$, the q-Borel-Ramis transformation $\mathbf{C}[[x]] \to \mathbf{C}[[\xi]]$ takes the form

$$\mathcal{B}_{q;1} : \sum a_n x^n \mapsto \sum q^{-n(n-1)/2} \xi^n.$$

We keep the notation $\mathbf{C}[[x]]_{q;1}$ for $\mathcal{B}_{q;1}^{-1}(\mathbf{C}\{x\})$. The basic formulas now are (with obvious notations):

$$\mathcal{B}_{q;1}(x\sigma_q \hat{f}) = \xi \mathcal{B}_{q;1}\hat{f} \text{ and } \mathcal{B}_{q;1}(\hat{f}\hat{g}) = \sum a_n q^{-n(n-1)/2} \xi^n \mathcal{B}_{q;1}\hat{g}(\xi q^{-n}).$$

Let $\hat{f} \in \mathbf{C}[[x]]_{q;1}$ such that $\phi := \mathcal{B}_{q;1}\hat{f}$ has infinite radius of convergence. (This is clearly the case for instance if $\hat{f} \in \mathbf{C}\{x\}$.) Then a necessary and sufficient condition for $\hat{f}$ to have radius of convergence at least R is:

$$\forall \epsilon > 0 \;,\; \phi(\xi) = O\left(q^{\frac{1}{2}\log_q|\xi|(\log_q|\xi|+1)+(-\log_q R+\epsilon)}\right) \text{ when } |\xi| \to +\infty.$$

As a consequence, if $\hat{f} \in \mathbf{C}\{x\}$, then $\phi := \mathcal{B}_{q;1}\hat{f}$ has *q-exponential growth of order 1 and of finite type at infinity*, meaning:

$$\exists \mu \in \mathbf{R} \text{ such that } |\phi(\xi)| = O\left(|\xi|^\mu \, q^{\frac{1}{2}\log_q^2|\xi|}\right) \text{ when } |\xi| \to +\infty.$$

This makes sense if ϕ is defined on a radially unbounded subset of $\widetilde{\mathbf{C}^*}$, which will hold true in the cases of interest herebelow.

14.3.4.2. *q-Laplace-Zhang.* Now let $\phi \in \mathbf{C}\{\xi\}$ and assume that ϕ can be analytically continued over some domain $V \subset \widetilde{\mathbf{C}^*}$ containing a direction d_θ, $\theta \in \mathbf{R}$, in such a way that the extend function ϕ has q-exponential growth of order 1 and of finite type at infinity. Then the following *q-Laplace-Zhang transform* of ϕ is a well defined analytic germ in $\tilde{\mathcal{O}}$:

$$\mathcal{L}_{q;1}^\theta \phi(x) := \frac{q^{-1/8}}{\sqrt{2\pi \log q}} \int_{d_\theta} \vartheta'(x/\xi)\,\phi(\xi)\,\frac{d\xi}{\xi}.$$

As an exercise, you can prove that, if $\phi_n(\xi) := \xi^n$, $n \in \mathbf{N}$, then $\mathcal{L}_{q;1}^\theta \phi_n(x) = q^{n(n-1)/2} x^n$; therefore $\mathcal{L}_{q;1}^\theta$ formally inverts $\mathcal{B}_{q;1}$. Now let:

$$\mathbf{H}_{q;1}^\theta := \left\{ \phi \in \tilde{\mathcal{O}} \;\middle|\; \begin{array}{c} \phi \text{ can be analytically continued over some } S(\theta,\epsilon) \text{ with} \\ q\text{-exponential growth of order 1 and of finite type at infinity.} \end{array} \right\}$$

If $\phi \in \mathbf{H}_{q;1}^\theta$, then $\mathcal{L}_{q;1}^\theta \phi \in \tilde{\mathcal{O}}$; and, for θ' close enough to θ, $\phi \in \mathbf{H}_{q;1}^{\theta'}$ and $\mathcal{L}_{q;1}^{\theta'}\phi = \mathcal{L}_{q;1}^\theta \phi$. Moreover, $f := \mathcal{L}_{q;1}^\theta \phi \in \tilde{\mathcal{O}}$ has asymptotic expansion $\hat{f} := \mathcal{B}_{q;1}^{-1}\phi \in \mathbf{C}[[x]]_{q;1}$, i.e., there exists $\epsilon > 0$ and $A > 0$ such that:

$$\forall n \in \mathbf{N}, \forall \alpha \in]\theta-\epsilon, \theta+\epsilon[\;,\; \left| f(x) - T_n\hat{f}(x) \right| = O\left(A^n q^{\frac{1}{2}(n^2+\arg_q^2(xe^{-i\alpha})}\,|x|^n\right) \text{ when } |x| \to 0.$$

14.3.5. Summability and q-difference equations.

14.3.5.1. *Summability.* So we are led to say that $\hat{f} \in \mathbf{C}[[x]]$ *is Gq-summable of order* 1 *in direction* $\theta \in \mathbf{R}$ *with sum* $f \in \tilde{\mathcal{O}}$, and write $\hat{f} \in \mathbf{C}\{x\}^{\theta}_{q;1}$, if the just stated formula holds. The Gq-sum f is then unique, we write it $S^{\theta}_{q;1}\hat{f}$. For instance, if $\hat{f} \in \mathbf{C}\{x\}$, it is Gq-summable in all directions and all the Gq-sums are equal to $\hat{f}$ itself (considered as a germ).

We call *singular direction* $\theta \in \mathrm{DS}(\hat{f})$ one such that $\hat{f} \notin \mathbf{C}\{x\}^{\theta}_{q;1}$. For instance, $\mathrm{DS}(\hat{f}) = \emptyset \Leftrightarrow \hat{f} \in \mathbf{C}\{x\}$. We say that $\hat{f}$ is *Gq-summable* if $\mathrm{DS}(\hat{f})$ is discrete. In fact this is equivalent to $\mathrm{DS}(\hat{f}) \cap [0; 2\pi]$ being finite, for:

$$(\theta' = \theta + 2k\pi, k \in \mathbf{Z}) \Longrightarrow \left(\mathbf{C}\{x\}^{\theta'}_{q;1} = \mathbf{C}\{x\}^{\theta}_{q;1} \text{ and } S^{\theta'}_{q;1}\hat{f}(x) = S^{\theta}_{q;1}\hat{f}(xe^{-2i\pi})\right).$$

If $f \in \mathbf{C}\{x\}^{\theta}_{q;1}$ and θ' is close enough to θ, then $f \in \mathbf{C}\{x\}^{\theta'}_{q;1}$ and $S^{\theta'}_{q;1}\hat{f} = S^{\theta}_{q;1}\hat{f}$. However, in general $S^{\theta'}_{q;1}\hat{f} \neq S^{\theta}_{q;1}\hat{f}$ and this gives rise to a q-Stokes phenomenon encoded in the cocycle of all $S^{\theta'}_{q;1}\hat{f} - S^{\theta}_{q;1}\hat{f}$ (more precisely, in its cohomology class, as in 14.1.4.1).

The first main result of the theory is then that, if $\hat{f} \in \mathbf{C}[[x]]_{q;1}$ and we set $\phi := \mathcal{B}_{q;1}\hat{f}$, then $\hat{f}$ is Gq-summable in direction θ if, and only if, $\phi \in \mathbf{H}^{\theta}_{q;1}$ and[15] then $S^{\theta}_{q;1}\hat{f} = \mathcal{L}^{\theta}_{q;1}\phi = \mathcal{L}^{\theta}_{q;1} \circ \mathcal{B}_{q;1}\hat{f}$.

14.3.5.2. *Aplication to q-difference equations.* It rests on the following facts. First, $S^{\theta}_{q;1}$ is $\mathbf{C}$-linear. Then, if $\hat{f} \in \mathbf{C}\{x\}$ and $\hat{g} \in \mathbf{C}\{x\}^{\theta}_{q;1}$, then $\hat{f}\hat{g} \in \mathbf{C}\{x\}^{\theta}_{q;1}$ and[16] $S^{\theta}_{q;1}\left(\hat{f}\hat{g}\right) = \hat{f}S^{\theta}_{q;1}(\hat{g})$. Last, the change of variable $x \leftarrow qx$ in integral $\mathcal{L}^{\theta}_{q;1}\phi(x)$ yields equality $\mathcal{L}^{\theta}_{q;1}(\xi\phi) = x\sigma_q\mathcal{L}^{\theta}_{q;1}\phi$. Combined with $\mathcal{B}_{q;1}x\sigma_q = \xi\mathcal{B}_{q;1}$ and with the previous property, this entails that $\Delta S^{\theta}_{q;1} = S^{\theta}_{q;1}\Delta$ for every $\Delta \in \mathbf{C}\{x\}[x\sigma_q]$.

So, under conditions similar to that given in 14.2.4, formal solutions of q-difference equations of type $\Delta f = g$, $g \in \mathbf{C}\{x\}$, are Gq-summable.

However, while $\hat{f}(x) := \sum\limits_{n \geq 0} q^{n(n-1)/2}x^n$ is Gq-summable (take $\Delta := 1 - x\sigma_q$ and $g := 1$), its square is not and this requires a theory of "multisummability", see 14.3.6.2.

14.3.6. Generalisations. We shall just say a few words about adaptation to the more general case of many integral slopes.

14.3.6.1. *Slope* $\neq 1$. We return to the example used in 14.1 (slightly modified to take in account convention $q > 1$). If $\hat{f} = \sum\limits_{n \geq 0} a_n x^n$ satisfies $\hat{f} = 1 + ax\sigma_q\hat{f}$, its coefficients satisfy the recursive relation $a_n = q^{n-1}aa_{n-1}$ for $n \geq 1$. Setting $\tilde{a}_n := q^{-n(n-1)/2}a_n$, we have $\tilde{a}_n = a\tilde{a}_{n-1}$ and the corresponding q-Borel-Ramis transform satisfies $\mathcal{B}_{q;1}\hat{f} = 1 + a\xi\mathcal{B}_{q;1}\hat{f}$, which is not anymore a q-difference equation and which is immediately soved as $\mathcal{B}_{q;1}\hat{f} = \dfrac{1}{1 - a\xi}$, whence $\hat{f} = \mathcal{B}^{-1}_{q;1}\dfrac{1}{1 - a\xi}$ (assuming that we can give a meaning to $\mathcal{B}^{-1}_{q;1}$).

[15]Note that, in *loc. cit.*, Zhang writes $S^{\theta}_{q;1} = \mathcal{L}^{\theta}_{q;1} \circ S^{\theta} \circ \mathcal{B}_{q;1}$ instead of $S^{\theta}_{q;1} = \mathcal{L}^{\theta}_{q;1} \circ \mathcal{B}_{q;1}$ because he distinguishes ϕ from its analytic continuation $S^{\theta}\phi$ in direction θ.

[16]Note that, in *loc. cit.*, Zhang writes $S\left(\hat{f}\right)$ instead of $\hat{f}$ in the right-hand side member of that equality, because he distinguishes the power series $\hat{f}$ from its "natural sum", a function.

Now suppose instead that $\hat{f} = 1 + ax^k\sigma_q\hat{f}$ for some $k \geq 2$ (so the corresponding q-difference operator has the unique slope k instead of 1). The recursive relation becomes $a_n = q^{n-k}aa_{n-k}$. One can show that, up to a geometric factor $O(A^n)$, the coefficient a_n grows like $q^{sn^2/2}$, where $s := 1/k$ is the relevant *order*. This suggests using the *formal q-Borel-Ramis transformation at order s*:

$$\hat{\mathcal{B}}_{q;s} : \sum_{n\geq 0} a_n x^n \mapsto \sum_{n\geq 0} a_n q^{-sn(n-1)/2}\xi^n.$$

This is paralleled by an *analytic q-Borel-Ramis transformation at order s*, defined by:

$$\mathcal{B}_{q;s}f(\xi) := -\mathrm{i}\frac{q^{1/8k}\sqrt{k}}{\sqrt{2\pi\log q}}\int_D q^{\frac{k}{2}\log_q(x/\xi)\left(\log_q(x/\xi)-1/k\right)} f(x)\,\frac{dx}{x},$$

where the integral is taken on some contour in $\widetilde{\mathbf{C}^*}$ (see [**163**, p. 1865]). A summation operator at order s is then obtained as the composition of a q-Borel-Ramis transformation with an adequate q-Laplace-Zhang transformation (see 14.3.6.2).

EXERCISE 14.56. (i) Check the order of growth $a_n \approx q^{sn^2/2}$.

(ii) Let a sequence (t_n) in $\mathbf{C}^*$ be such that $t_{n-k} = q^{n-k}t_n$ and set $\tilde{a}_n := t_n a_n$. What recursive relation follows? What functional relation follows for $\sum \tilde{a}_n x^n$?

(iii) Using a functional relation for $T(x) := \sum t_n x^n$, suggest a possible choice for the sequence (t_n).

14.3.6.2. *More than one slope.* We shall illustrate that case on the example $\hat{F} := \hat{f}^2$, where $\hat{f} := \sum q^{n(n-1)/2}x^n$. Note that, although [**163**] offers a rather complete theory and [**197**, Chapter 7] addresses the example, few details are given there about how to practically deal with concrete cases; there is (a little) more in [**64**].

From $\hat{f} = 1 + x\sigma_q\hat{f}$, through some computations, one draws that $L\hat{F} = x + 1$, where $L := (x\sigma_q - 1)(x^2\sigma_q - 1)$ (so that L has slopes 1 and 2). We try q-Borel-Ramis transformation at order 1 and denote $\Phi := \mathcal{B}_{q;1}\hat{F}$. Setting $\hat{G} := (x^2\sigma_q - 1)\hat{F}$ and $\hat{H} := (x\sigma_q - 1)\hat{G}$ (so that $\hat{H} = x + 1$), and using formulas from the exercises below, we successively get $\mathcal{B}_{q;1}\hat{G} = (q^{-1}\xi^2{\sigma_q}^{-1} - 1)\mathcal{B}_{q;1}\hat{\mathscr{F}}$ and $\mathcal{B}_{q;1}\hat{H} = (\xi - 1)\mathcal{B}_{q;1}\hat{\mathscr{F}}$, whence (given that $\mathcal{B}_{q;1}\hat{H} = 1 + \xi$):

$$(1 - q^{-1}\xi^2{\sigma_q}^{-1})\Phi = \frac{1+\xi}{1-\xi} \Longrightarrow \Phi = \sum_{n\geq 0} q^{-n^2}\xi^{2n}\frac{1+q^{-n}\xi}{1-q^{-n}\xi}.$$

(The latter equality is obtained from the former either by iterating equality $\Phi = \dfrac{1+\xi}{1-\xi} + q^{-1}\xi^2{\sigma_q}^{-1}\Phi$, or, equivalently, by inverting the q-difference operator $1 - q^{-1}\xi^2{\sigma_q}^{-1}$, see exercise below.) One sees easily that Φ has a meromorphic continuation to $\mathbf{C}$ with simple poles on $q^{-\mathbf{N}}$. However, it has not q-exponential growth of order 1, so $\hat{F}$ is not Gq-summable at order 1. Something similar was met in 14.2.4.2.

Dealing with this case requires a theory of q-*multisummability*; see [**163**]. This involves q-Borel-Ramis transformation at order s (see 14.3.6.1) and q-Laplace-Zhang transformation at order s:

$$\mathcal{L}^{\theta}_{q;s}\phi(x) := \frac{q^{-1/8k}\sqrt{k}}{\sqrt{2\pi\log q}}\int_{d_\theta} q^{-\frac{k}{2}\log_q(x/\xi)\left(\log_q(x/\xi)-1/k\right)}\phi(\xi)\,\frac{d\xi}{\xi}.$$

The symmetry of forms between $\mathcal{B}_{q;s}$ and $\mathcal{L}^{\theta}_{q;s}$ reminds us of the inversion of the Fourier-Laplace transformation in 14.3.2.2.

Then, according to **[64]**, a sum of $\hat{F}$ is obtained by applying transformation $\mathcal{L}^{\theta}_{q;1/2} \circ \mathcal{L}^{\theta}_{q^{1/2};1} \circ \hat{B}_{q;1}$. Here, 1/2 appears more than once but in various guises. Setting $s_1 := 1/2$ and $s_2 := 1$ (the inverses of the slopes), one should read: $\mathcal{L}^{\theta}_{q;s_1} \circ \mathcal{L}^{\theta}_{q;s_2-s_1} \circ \hat{B}_{q;s_2}$.

EXERCISE 14.57. Invert the q-difference operator $1 - q^{-1}\xi^2{\sigma_q}^{-1}$.

CHAPTER 15

Irregular Systems

Le vrai analyste, pour se sentir à son aise, aime à disposer d'un arsenal d'instruments de précision, prêts à se fausser au moindre geste maladroit. Pour moi, dès qu'il s'agit d'analyse, mon instinct va à réduire cet attirail à un nombre aussi petit que possible d'outils grossiers mais robustes dont le mode d'emploi tienne en quelques règles simples; une fois ce résultat atteint, l'algébriste en moi peut reprendre le dessus.

—André Weil (1979)[1]

In this chapter again the study is local, the main fields of interest are $\mathbf{C}(\{x\})$ and $\mathbf{C}((x))$. We shall give substantial results for the case of integral slopes only; for the general case, see [**248**], [**223**] (respectively regarding formal and analytical classification) and [**224**] (regarding Galois theory); and also [CV].

Even in the restricted setting of integral slopes, we shall entirely omit the process of "algebraic summation with general polar divisors" as presented in [**197**, **219**]: indeed, although I believe that it *should* be the right framework, it has not yet met meaningful applications. Also we shall be rather brief on the general asymptotic point of view from [**197**, Chapter 4], for it gives useful information only at the price of difficult and sophisticated methods well beyond the scope of this book. All those restrictions are further commented in the course of the chapter.

15.1. The "additive" case of two slopes

We shall go into great detail on that case: first, it encompasses at least all systems of rank 2 (at least when slopes are integral), thus almost all interesting q-difference equations; second it features all the (rather simple) analytic ideas without the algebraic complications of the nonabelian case.

15.1.1. The archetypal example again: rank 2 systems. The following should be compared with the classification of rank 2 vector bundles over a general Riemann surface [**108**, §5] or over an elliptic curve [**13**, **140**, **248**].

15.1.1.1. *Setting.* We recall in a more concrete form some results from 8.3.2, 10.3.4 and 10.3.5, aiming at a (slightly) more analytical point of view.

Fix slopes $\mu > \nu$ in $\mathbf{Z}$ and exponents $a, b \in \mathbf{C}^*$ and denote $A_0 := \mathrm{Diag}(ax^\mu, bx^\nu)$. Any matrix in $\mathrm{GL}_2(\mathbf{C}((x)))$ with integral slopes is gauge equivalent to such an A_0 (and actually any matrix in $\mathrm{GL}_2(\mathbf{C}(\{x\}))$ with integral slopes is *formally* gauge equivalent to such an A_0). Let $\mathscr{E}(A_0)$ the set of all $A_u := \begin{pmatrix} ax^\mu & u \\ 0 & bx^\nu \end{pmatrix}$ with $u \in \mathbf{C}(\{x\})$. Any matrix in $\mathrm{GL}_2(\mathbf{C}(\{x\}))$ which is formally equivalent to A_0 can

[1]This is an extract of Weil's comments on his book "L'intégration dans les groupes topologiques et ses applications", see commentary [1940d] in *Œuvres scientifiques*, Springer.

be put in form A_u through an analytic gauge transformation. So we can see $\mathscr{E}(A_0)$ as the formal class of A_0 inside the category $\mathscr{E}^{(0)}$ of all q-difference modules over $\mathbf{C}(\{x\})$.

We have two distinct analytic equivalence relations on $\mathscr{E}(A_0)$. The plain analytic gauge equivalence, denoted $\simeq$, through gauge matrices $F \in \mathrm{GL}_2(\mathbf{C}(\{x\}))$; and the finer *isograded* or *isoformal*[2] analytic equivalence, denoted $\sim$. We give in parallel the two definitions (the first one being, of course, already known); let $A_u, A_v \in \mathscr{E}(A_0)$, then:

$$A_u \simeq A_v \Longleftrightarrow \exists F \in \mathrm{GL}_2(\mathbf{C}(\{x\})) \; : \; F[A_u] = A_v,$$

$$A_u \sim A_v \Longleftrightarrow \exists f \in \mathbf{C}(\{x\}) \; : \; F_f[A_u] = A_v, \text{ where } F_f := \begin{pmatrix} 1 & f \\ 0 & 1 \end{pmatrix}.$$

EXERCISE 15.1. Check that $F_f[A_u] = A_v \Leftrightarrow v - u = bx^\nu \sigma_q f - ax^\mu f$.

REMARK 15.2. It follows a priori from the functoriality of the slope filtration that, in the above definition of $\simeq$, the matrix F must be upper triangular. This can be checked directly using that $F[A_u] = A_v$ (with $F = (f_{i,j})_{i,j=1,2}$) implies $bx^\nu \sigma_q f_{2,1} - ax^\mu f_{2,1} = 0$, which in turn implies $f_{2,1} = 0$.

We defined in 8.3.2 the following sets:

$$\mathscr{F}(A_0) := \frac{\mathscr{E}(A_0)}{\sim} \text{ and } \widetilde{\mathscr{F}}(A_0) := \frac{\mathscr{E}(A_0)}{\simeq},$$

which clearly can respectively be identified with the Ext and the $\widetilde{\mathrm{Ext}}$ of $(\mathbf{C}(\{x\}), \Phi_{bx^\nu})$ by $(\mathbf{C}(\{x\}), \Phi_{ax^\mu})$. Although the latter seems a more natural object, since it is based on thenatural equivalence relation $\simeq$, our prime interest will be for the former. To understand why[3], we first note that $A_u \sim A_v$ obviously implying $A_u \simeq A_v$, there is a natural surjective map $\mathscr{F}(A_0) \to \widetilde{\mathscr{F}}(A_0)$, sending the isoformal (or isograded) analytic class of A_u to its plain analytic class.

PROPOSITION 15.3. (i) *The group $Aut(A_0) \simeq Aut(ax^\mu) \times Aut(bx^\nu)$ operates on $\mathscr{E}(A_0)$ by gauge transformations. (Obviously, $Aut(ax^\mu) \simeq Aut(bx^\nu) \simeq \mathbf{C}^*$, but we do not want to take that in account for the time being.)*

(ii) *The action is compatible with the isoformal equivalence relation $\sim$ in $\mathscr{E}(A_0)$:*

$$\forall F_0 \in Aut(A_0) \; , \; A_u \sim A_v \Longrightarrow F_0[A_u] \sim F_0[A_v].$$

(iii) $A_u \simeq A_v \Longleftrightarrow \exists F_0 \in Aut(A_0) \; : \; F_0[A_u] \sim A_v$.

PROOF. (i) The identification $\mathrm{Aut}(A_0) \simeq \mathrm{Aut}(ax^\mu) \times \mathrm{Aut}(bx^\nu)$ should be clear and is left to check for the reader's pleasure. Let $F_0 := \mathrm{Diag}(f_1, f_2) \in \mathrm{Aut}(A_0)$, i.e., $f_1[ax^\mu] = ax^\mu$ and $f_2[bx^\nu] = bx^\nu$. Then $F_0[A_u] = A_{u'}$, where $u' := u\sigma_q f/f$.

(ii) Let $F_f[A_u] = A_v$ and F_0 as in (i), so that $F_0[A_u] = A_{u'}$ and $F_0[A_v] = A_{v'}$. Then $G[A_{u'}] = A_{v'}$, where $G := F_0 F_f F_0^{-1} = F_g$, with $g := f_1 f f_2^{-1}$.

---

[2]For the moment, we rely on the general principle from section 10.3 (in Chapter 10) that "graduation is the exact algebraic counterpart of formalisation"; this will be taken up in more detail in 15.3.4. Meanwhile, we shall rather use the term "isoformal" for its closeness with the classical case of irregular ordinary differential equations.

[3]As we are going to find out, the natural relation $\simeq$ leads to badly behaved quotient spaces. Actually something similar was already observed; see example 8.76 in Subsection 8.3.2.

(iii) If $F_0[A_u] \sim A_v$, then $F_f[F_0[A_u]] = A_v$ for some $f \in \mathbf{C}(\{x\})$, whence $F[A_u] = A_v$, where $F := F_f F_0 \in \mathrm{GL}_2(\mathbf{C}(\{x\}))$, so $A_u \simeq A_v$. Conversely, if $F[A_u] = A_v$ with $F \in \mathrm{GL}_2(\mathbf{C}(\{x\}))$, by the remark above, $F = (f_{i,j})_{i,j=1,2}$ with $f_{2,1} = 0$, and we can write $F = F_f F_0$ with $F_0 := \mathrm{Diag}(f_1, f_2)$ and $f := f_{1,2}/f_{2,2}$, so $F_0[A_u] \sim A_v$.

□

COROLLARY 15.4. *The group $Aut(A_0)$ operates on the set $\mathscr{F}(A_0)$ with quotient $\widetilde{\mathscr{F}}(A_0)$.*

Actually $\mathrm{Aut}(ax^\mu) \simeq \mathrm{Aut}(bx^\nu) \simeq \mathbf{C}^*$, because for instance $f[ax^\mu] = ax^\mu$ is equivalent to $\sigma_q f = f$ which, in $\mathbf{C}(\{x\})$, implies $f \in \mathbf{C}$ (and of course we look for invertible matrices). So we shall presently describe more precisely $\mathscr{F}(A_0)$ and the group action of $\mathbf{C}^* \times \mathbf{C}^*$ on it.

15.1.1.2. *The set $\mathscr{F}(A_0)$ of isoformal classes.* Again we fix a, b, μ, ν and keep the notations A_0, A_u, also F_f from 15.1.1.1. We noted that $F_f[A_u] = A_v \Leftrightarrow v - u = bx^\nu \sigma_q f - ax^\mu f$. Using the bijection $u \mapsto A_u$ from $\mathbf{C}(\{x\})$ to $\mathscr{E}(A_0)$, we deduce an identification of $\mathscr{F}(A_0)$ with the cokernel of the map[4]:

$$\Phi_{ax^\mu, bx^\nu} : \mathbf{C}(\{x\}) \to \mathbf{C}(\{x\}),\ f \mapsto bx^\nu \sigma_q f - ax^\mu f.$$

Let $\delta := \mu - \nu \in \mathbf{N}^*$ the corresponding *q-Gevrey level.* Using for instance 6.4, together with the fact that multiplication by ax^μ (or by bx^ν) is an automorphism of $\mathbf{C}(\{x\})$, we conclude that there is a splitting:

$$\mathbf{C}(\{x\}) = \mathbf{C}[x]_{\delta-1} \oplus \mathrm{Im}\ \Phi_{ax^\mu, bx^\nu}.$$

(Recall that $\mathbf{C}[x]_d := \mathbf{C} \oplus \cdots \oplus \mathbf{C}x^d$ for any $d \in \mathbf{N}$; the first component might actually be replaced by $K_{\mu,\nu} := \mathbf{C}x^\nu \oplus \cdots \oplus \mathbf{C}x^{\mu-1} = x^\nu \mathbf{C}[x]_{\delta-1}$.) Therefore, each isoformal analytic class in $\mathscr{E}(A_0)$ contains a unique representative A_u with $u \in \mathbf{C}[x]_{\delta-1}$, alternatively in $K_{\mu,\nu}$; hence $u \mapsto \mathrm{cl}(A_u)$ (the class of A_u) is a bijection $\mathbf{C}[x]_{\delta-1} \to \mathscr{F}(A_0)$, alternatively $K_{\mu,\nu} \to \mathscr{F}(A_0)$. (The corresponding computation was done in Chapter 6 and appears again in 15.1.2.)

15.1.1.3. *Algebraic summation.* The cokernel of Φ_{ax^μ, bx^ν} can be understood as the space of *obstructions* to the resolution of $\Phi_{ax^\mu, bx^\nu}(f) = u$, $u \in \mathbf{C}(\{x\})$ given, $f \in \mathbf{C}(\{x\})$ sought for. So we proceed to study that nonhomogeneous equation:

$$bx^\nu \sigma_q f - ax^\mu f = u.$$

Transposing Lagrange method of "variation of constants" would lead us to first choose a solution to the associated homogeneous equation $bx^\nu \sigma_q f - ax^\mu f = 0$, i.e., $\sigma_q f = (a/b) x^\delta f$. We know that this has no nontrivial solution in $\mathscr{O}(\mathbf{C}^*, 0)$ and the simplest possible meromorphic solution could be $f_0 := \theta_q(x/c)^{-\delta}$, where $c^\delta = a/b$. Then setting $f = f_0 g$, we are led to equation $\sigma_q g - g = v$, where $v := u/(ax^\mu f_0) \in \mathscr{O}(\mathbf{C}^*, 0)$. Using a Laurent series expansion $v = \sum v_n x^n$, this boils down to a system of equations $(q^n - 1) g_n = v_n$, but then we have a problem at $n = 0$.

EXERCISE 15.5. (Check that $\sigma_q f = (a/b) x^\delta f$ has no nontrivial solution in $\mathscr{O}(\mathbf{C}^*, 0)$. Look for solutions in the form $\dfrac{g}{\theta_q(x/c_1) \cdots \theta_q(x/c_\delta)}, g \in \mathscr{O}(\mathbf{C}^*, 0)$.

[4]We recognize an instance of the so-called "additive q-de Rham complex" from 10.3.5.1, Chapter 10.

So, as advocated in 14.1.3 (Chapter 14), we go for *twisted Lagrange method* and set instead $f_0 := \theta_q(x/c)^{-\delta}$, where this time $c^\delta \not\equiv a/b$. The change of unknown $f = f_0 g$ now leads us to equation $(b/a)c^{-\delta}\sigma_q g - g = v$, where $v := u/(ax^\mu f_0) \in \mathcal{O}(\mathbf{C}^*, 0)$ as before. Using expansion $v = \sum v_n x^n$, we meet the system:

$$\forall n \in \mathbf{Z}\,,\ ((b/a)c^{-\delta}q^n - 1)g_n = v_n,$$

and imposing the condition:

$$\forall n \in \mathbf{Z}\,,\ ((b/a)c^{-\delta}q^n - 1) \neq 0, \text{ i.e. } c^\delta \notin [b/a; q],$$

we obtain the solutions:

$$g_c(x) := \sum_{n \in \mathbf{Z}} \frac{v_n}{(b/a)c^{-\delta}q^n - 1} x^n, \quad f_c(x) := \frac{1}{\theta_q(x/c)^\delta} g_c(x).$$

EXERCISE 15.6. Check that $g_c \in \mathcal{O}(\mathbf{C}^*, 0)$ and compare the punctured disks of convergence of u, v, g.

So f_c is meromorphic on some punctured neighborhood of 0 with poles concentrated on $[-c; q]$ and with multiplicities at most δ. We note that this polar condition entirely characterizes a solution f: indeed, it implies that $g := \theta_q(x/c)^\delta f$ is holomorphic and solution of $(b/a)c^{-\delta}\sigma_q g - g = v$, which implies (by Laurent expansion) that $g = g_c$, whence $f = f_c$. On the other hand, the polar condition depends on the class $\overline{c} \in \mathbf{E}_q$ only, so we could (and will) denote our solution $f_{\overline{c}}$. Also the condition $c^\delta \notin [b/a; q]$ then translates to:

$$\overline{c} \in \mathbf{E}_q \setminus \Sigma_{A_0}, \text{ where } \Sigma_{A_0} := \{\overline{c} \in \mathbf{E}_q \mid \overline{c}^\delta = \overline{b/a}\}.$$

EXERCISE 15.7. (i) The *prohibited locus* Σ_{A_0} has δ^2 elements.
(ii) Check that g_c does depend on c itself and compare g_c with g_{qc}.
(iii) Extend the process to solutions of the form $\dfrac{g}{\theta_q(x/c_1)\cdots\theta_q(x/c_\delta)}$.

We know from 4.2 that equation $bx^\nu\sigma_q f - ax^\mu f = u$ has a unique formal solution $\hat{f} \in \mathbf{C}((x))$ (it can for instance be obtained by iterating

$$f \mapsto {\sigma_q}^{-1}\left((bx^\nu)^{-1}(u + ax^\mu f)\right)$$

toward a fixed point). In the spirit of the summation theory of Chapter 14, we are entitled to consider $f_{\overline{c}}$ as the "sum of $\hat{f}$ in direction $\overline{c}$", and, accordingly, to write:

$$S_{\overline{c}}\hat{f} := f_{\overline{c}}.$$

Similarly, we may set:

$$\widehat{F} := F_{\hat{f}} = \begin{pmatrix} 1 & \hat{f} \\ 0 & 1 \end{pmatrix}, \text{ and then: } S_{\overline{c}}\widehat{F} := F_{f_{\overline{c}}} = \begin{pmatrix} 1 & f_{\overline{c}} \\ 0 & 1 \end{pmatrix}.$$

We shall often abreviate the latter to $F_{\overline{c}} := S_{\overline{c}}\widehat{F}$. since clearly:

$$\widehat{F}[A_0] = F_{\overline{c}}[A_0] = A_u,$$

we can consider $\widehat{F}$ and $F_{\overline{d}}$ as gauge transformations from A_0 to A_u, the first formal, the second meromorphic.

REMARK 15.8. Note however that, contrary to what was done in Chapter 14, $S_{\overline{c}}\widehat{F}$ has not been directly obtained from $\widehat{F}$: there is not really an "algebraic summation process", this is one of the weaknesses of the present approach.

EXERCISE 15.9. (i) Solve $\sigma_q X = A_0 X$ in terms of basic special functions.
(ii) Deduce analytic solutions of $\sigma_q X = A_u X$ with the help of $S_{\overline{c}}\widehat{F}$.

15.1.1.4. *Algebraic summation and classification.* Let $u_1, u_2 \in \mathbf{C}(\{x\})$ and let us use the abreviations $\widehat{F}_i, S_{\overline{c}}\widehat{F}_i$, $i = 1, 2$ for $\widehat{F_{\overline{c_i}}}, S_{\overline{c}}\widehat{F_{\overline{c_i}}}$; similarly, denote $\hat{f}_i, S_{\overline{c}}\hat{f}_i$ the $(1,2)$ component of $\widehat{F}_i, S_{\overline{c}}\widehat{F}_i$. It follows from the above calculations that:

$$\widehat{F}_{1,2} := \widehat{F}_1^{-1}\widehat{F}_2 = F_{\hat{f}_{1,2}} := \begin{pmatrix} 1 & \hat{f}_{1,2} \\ 0 & 1 \end{pmatrix}, \text{ where } \hat{f}_{1,2} := \hat{f}_1 - \hat{f}_2$$

is the unique formal isomorphism $A_{u_1} \to A_{u_2}$, and that:

$$S_{\overline{c}}\widehat{F}_{1,2} := (S_{\overline{c}}\widehat{F}_1)^{-1} S_{\overline{c}}\widehat{F}_2 = F_{S_{\overline{c}}\hat{f}_{1,2}} := \begin{pmatrix} 1 & S_{\overline{c}}\hat{f}_{1,2} \\ 0 & 1 \end{pmatrix}, \text{ where } S_{\overline{c}}\hat{f}_{1,2} := S_{\overline{c}}\hat{f}_1 - S_{\overline{c}}\hat{f}_2$$

is the unique meromorphic isomorphism $A_{u_1} \to A_{u_2}$ with at worst simple poles on $[-c; q]$. Therefore it amounts to the same to say that $A_{u_1} \sim A_{u_2}$ or that $A_0 \sim A_{u_1 - u_2}$, and maybe to look for a criterion involving the $S_{\overline{c}}\hat{f}_i$. (To completely understand that last remark, look at what follows and then return here!)

Now, if $A_0 \sim A_u$, there is an $f \in \mathbf{C}(\{x\}) \subset \mathbf{C}((x))$ such that $F_f[A_0] = A_u$ and it satisfies the properties that uniquely characterize $\widehat{F}_u$, so indeed $\widehat{F}_u = F_f$. Also, it has no poles at all near 0, so it also satisfies the properties that uniquely characterize $S_{\overline{c}}\widehat{F}_u$, so $S_{\overline{c}}\widehat{F}_u = F_f$ for all $\overline{c} \notin \Sigma_{A_0}$.

Conversely assume that $S_{\overline{c}}\widehat{F}_u = S_{\overline{d}}\widehat{F}_u$ for some $\overline{c} \neq \overline{d}$ both in $\mathbf{E}_q \setminus \Sigma_{A_0}$. The function $f_{\overline{c}} = f_{\overline{d}} =: f$ has no poles (near 0) out of $[-c; q] \cap [-d; q] = \emptyset$, so $f \in \mathscr{O}(\mathbf{C}^*, 0)$. We want to deduce more, that $f \in \mathbf{C}(\{x\})$ so that in fact $A_0 \sim A_u$, thus providing a kind of converse to the above.

We have already met this kind of argument. Translating equation $bx^\nu \sigma_q f - ax^\mu f = u$ into relations of the coefficients of the Laurent expansions entails (with obvious notations) $bq^{n-\nu} f_{n-\nu} = a f_{n-\mu}$ for all $n < n_0$ (denoting n_0 the smallest index of the terms of u), i.e., $f_{n-\delta} = (b/a) q^n f_n$ for $n < n_0 - \nu$. This yields $f_{n-k\delta} = (b/a)^k q^{-\delta k(k-1)/2} f_n$, so, if $f_n \neq 0$, wild growth of the coefficients and divergence of the Laurent series. So indeed $f \in \mathbf{C}(\{x\})$.

The conclusion is that if some $S_{\overline{c},\overline{d}}\widehat{F}_u := (S_{\overline{c}}\widehat{F}_u)^{-1} S_{\overline{d}}\widehat{F}_u$ is trivial (with $\overline{c} \neq \overline{d}$), then $A_0 \sim A_u$. The trick of comparing "sums" in two different directions fits nicely into a general scheme which goes back to Birkhoff (we already met it in 14.1.4.1, Chapter 14). Given $u \in \mathbf{C}(\{x\})$, we shall associate to A_u the family of all matrices:

$$S_{\overline{c},\overline{d}}\widehat{F}_u := (S_{\overline{c}}\widehat{F}_u)^{-1} S_{\overline{d}}\widehat{F}_u = \begin{pmatrix} 1 & S_{\overline{c},\overline{d}}\hat{f}_u \\ 0 & 1 \end{pmatrix}, \text{ where } S_{\overline{c},\overline{d}}\hat{f}_u := f_{\overline{d}} - f_{\overline{c}} = S_{\overline{c}}\hat{f}_u - S_{\overline{c}}\hat{f}_u.$$

By construction:

(1) It is a *multiplicative cocycle*: writing for short $F_{\overline{c},\overline{d}} := S_{\overline{c},\overline{d}}\widehat{F}_u$, we have for all admissible $\overline{c}, \overline{d}, \overline{e}$ (equal or not) $F_{\overline{c},\overline{d}} F_{\overline{d},\overline{e}} = F_{\overline{c},\overline{e}}$.
(2) Each $F_{\overline{c},\overline{d}}$ has the form $\begin{pmatrix} 1 & \star \\ 0 & 1 \end{pmatrix}$, with $\star$ meromorphic in a punctured neighborhood of 0 and holomorphic out of $[-c, -d; q]$.
(3) Each $F_{\overline{c},\overline{d}}$ is a meromorphic automorphism of A_0, i.e., $F_{\overline{c},\overline{d}}[A_0] = A_0$.

Due to the special form $F_{\overline{c},\overline{d}} = \begin{pmatrix} 1 & f_{\overline{c},\overline{d}} \\ 0 & 1 \end{pmatrix}$, those boil down to:

(1) We have an *additive cocycle*: $f_{\overline{c},\overline{d}} + f_{\overline{d},\overline{e}} = f_{\overline{c},\overline{e}}$.
(2) Each $f_{\overline{c},\overline{d}} \in \mathscr{M}(\mathbf{C}^*, 0)$, with at most simple poles on $[-c, -d; q]$.
(3) All $f_{\overline{c},\overline{d}}$ are solutions of $bx^\nu \sigma_q f - ax^\mu f = 0$.

In 14.1 we met (or say at least touched upon) that formalism, in the case of scalar equations. We shall here stick to the matricial point of view. However, the common point of both frameworks is that we have a covering $\mathfrak{U}$ of $\mathbf{E}_q$, here by the open sets $U_{\overline{c}} := \mathbf{E}_q \setminus \{\overline{c}\}$, $\overline{c} \in \mathbf{E}_q \setminus \Sigma_{A_0}$ (and by the way any two of those open sets are enough to cover $\mathbf{E}_q$). We shall commonly associate to it notations like $U_{\overline{c},\overline{d}} := U_{\overline{c}} \cap U_{\overline{d}}$, $V_{\overline{c}} := \pi^{-1}(U_{\overline{c}}) = \mathbf{C}^* \setminus [-c; q]$ (giving rise to an open covering $\mathfrak{V}$ of $\mathbf{C}^*$), $V_{\overline{c},\overline{d}} := V_{\overline{c}} \cap V_{\overline{d}} = \pi^{-1}(U_{\overline{c},\overline{d}})$, etc.

Then there is a sheaf Λ over $\mathbf{E}_q$ (later on it will be denoted $\Lambda_I(A_0)$, or, in a more intrinsic way, $\Lambda_I(M_0)$ if M_0 is the corresponding q-difference module) defined as:

$$\Lambda(V) := \left\{ \text{matrices } F \left\{ \begin{array}{l} \text{of the form } \begin{pmatrix} 1 & \star \\ 0 & 1 \end{pmatrix}, \\ \text{holomorphic over } \pi^{-1}(V), \\ \text{such that } F[A_0] = A_0. \end{array} \right. \right\}.$$

Since $f \mapsto F_f$ is a morphism of groups, it is a sheaf of abelian groups (in the general nonadditive case of $k > 2$ slopes, the groups will no longer be abelian). Returning to our constructions, we see that $F_{\overline{c},\overline{d}} \in \Lambda(U_{\overline{c},\overline{d}})$ so that, in the standard sheaf theoretic notation (see Appendix I):

$$(F_{\overline{c},\overline{d}})_{\overline{c},\overline{d} \in \mathbf{E}_q \setminus \Sigma_{A_0}} \in Z^1(\mathfrak{U}, \Lambda).$$

So we have a mapping $\mathscr{E}(A_0) \to Z^1(\mathfrak{U}, \Lambda)$ sending A_u to the cocycle made of the $S_{\overline{c},\overline{d}} \widehat{F}_u$, $\overline{c}, \overline{d} \in \mathbf{E}_q \setminus \Sigma_{A_0}$.

Now suppose $A_u \sim A_v$ and let $f \in \mathbf{C}(\{x\})$ such that $F_f[A_u] = A_v$. Then $F_f \widehat{F}_u$ is a formal isomorphism $A_0 \sim A_v$ in the usual form, so $F_f \widehat{F}_u = \widehat{F}_v$. For similar reasons, $F_f \, S_{\overline{c}} \widehat{F}_u = S_{\overline{c}} \widehat{F}_v$ (both are isomorphisms $A_0 \sim A_v$ satisfying the required conditions). It follows that:

$$S_{\overline{c},\overline{d}} \widehat{F}_v = (S_{\overline{c}} \widehat{F}_v)^{-1} S_{\overline{d}} \widehat{F}_v = (F_f S_{\overline{c}} \widehat{F}_u)^{-1} (F_f S_{\overline{d}} \widehat{F}_u) = (S_{\overline{c}} \widehat{F}_u)^{-1} S_{\overline{d}} \widehat{F}_u = S_{\overline{c},\overline{d}} \widehat{F}_u.$$

Therefore A_u and A_v have the same image in $Z^1(\mathfrak{U}, \Lambda)$, and we have a well defined map:

$$\mathscr{F}(A_0) \to Z^1(\mathfrak{U}, \Lambda), \quad \mathrm{cl}(A_u) \mapsto (S_{\overline{c},\overline{d}} \widehat{F}_u)_{\overline{c},\overline{d} \in \mathbf{E}_q \setminus \Sigma_{A_0}}.$$

Also note that this map arrives in a particular subspace $Z^1_{pr}(\mathfrak{U}, \Lambda)$, that of so-called *privileged* cocycles: those such that the $(\overline{c}, \overline{d})$ component has at worst simple poles over $[-c, -d; q]$.

15.1.1.5. *Classification and cohomology.* Let $(F_{\overline{c},\overline{d}}), (G_{\overline{c},\overline{d}}) \in Z^1(\mathfrak{U}, \Lambda)$ two arbitrary cocycles. They are said to be *equivalent* (Appendix I; this will later be taken in greater generality) if there exists a family of $\Phi_{\overline{c}} \in \Lambda(U_{\overline{c}})$ such that:

$$\forall \overline{c}, \overline{d} \in \mathbf{E}_q \setminus \Sigma_{A_0} \ , \ G_{\overline{c},\overline{d}} = \Phi_{\overline{c}}^{-1} F_{\overline{c},\overline{d}} \Phi_{\overline{d}}.$$

This equality is meant, of course, over $U_{\overline{c},\overline{d}}$ (so for functions restricted to $V_{\overline{c},\overline{d}}$). It is easily checked that we have indeed an equivalence relation. The quotient is called the *first cohomology set* and denoted $H^1(\mathfrak{U}, \Lambda)$. In our particular case, because Λ is a sheaf of *abelian* groups, $H^1(\mathfrak{U}, \Lambda)$ is actually a *group*.

EXERCISE 15.10. Using the map $f \mapsto F_f$, give $Z^1(\mathfrak{U},\Lambda)$ the structure of an abelian group and show that the equivalence relation above is the congruence relative to a subgroup, whence a group structure on $H^1(\mathfrak{U},\Lambda)$.

Since we intend to deal with more general cases later, we do not insist on the group structure. However, we do note that $H^1(\mathfrak{U},\Lambda)$ is by nature a *pointed set*, because it comes with a marked point, class of the trivial cocycle (corresponding to the class of A_0 in $\mathscr{F}(A_0)$) such that all $F_{\overline{c},\overline{d}} = I_2$. Composing the map $\mathscr{F}(A_0) \to Z^1(\mathfrak{U},\Lambda)$ with the quotient map $Z^1(\mathfrak{U},\Lambda) \to H^1(\mathfrak{U},\Lambda)$, we thus get a map of pointed sets $\mathscr{F}(A_0) \to H^1(\mathfrak{U},\Lambda)$.

Although we intend to prove more later, let us already see that this map is injective. So let $u, v \in \mathbf{C}(\{x\})$ such that the cocycles $(S_{\overline{c},\overline{d}}\widehat{F}_u), (S_{\overline{c},\overline{d}}\widehat{F}_v) \in Z^1(\mathfrak{U},\Lambda)$ are equivalent, i.e., there is a family of $\Phi_{\overline{c}} \in \Lambda(U_{\overline{c}})$ such that $S_{\overline{c},\overline{d}}\widehat{F}_v = \Phi_{\overline{c}}^{-1} S_{\overline{c},\overline{d}}\widehat{F}_u \Phi_{\overline{d}}$ for all $\overline{c},\overline{d} \in \mathbf{E}_q \setminus \Sigma_{A_0}$. We want to deduce from that assumption that $A_u \sim A_v$. From the obvious equivalence:

$$S_{\overline{c},\overline{d}}\widehat{F}_v = \Phi_{\overline{c}}^{-1} S_{\overline{c},\overline{d}}\widehat{F}_u \Phi_{\overline{d}} \Longleftrightarrow S_{\overline{c}}\widehat{F}_u \Phi_{\overline{c}} (S_{\overline{c}}\widehat{F}_v)^{-1} = S_{\overline{d}}\widehat{F}_u \Phi_{\overline{d}} (S_{\overline{d}}\widehat{F}_v)^{-1},$$

we draw that there is a fixed $F = S_{\overline{c}}\widehat{F}_u \Phi_{\overline{c}} (S_{\overline{c}}\widehat{F}_v)^{-1}$ for all $\overline{c} \notin \Sigma_{A_0}$. Clearly, it has no poles on $(\mathbf{C}^*, 0)$ and it has the form $\begin{pmatrix} 1 & \star \\ 0 & 1 \end{pmatrix}$ so $F = F_f$ for some $f \in \mathscr{O}(\mathbf{C}^*, 0)$. Moreover:

$$F_f[A_v] = S_{\overline{c}}\widehat{F}_u \Phi_{\overline{c}} (S_{\overline{c}}\widehat{F}_v)^{-1}[A_v] = S_{\overline{c}}\widehat{F}_u \Phi_{\overline{c}}[A_0] = S_{\overline{c}}\widehat{F}_u[A_0] = A_u,$$

i.e., $bx^\nu \sigma_q f - ax^\mu f = u - v$. By the same argument as in 15.1.1.4, since $u - v \in \mathbf{C}(\{x\})$, such a solution $f \in \mathscr{O}(\mathbf{C}^*, 0)$ must actually belong to $\mathbf{C}(\{x\})$ and therefore $A_u \sim A_v$ as wanted.

EXERCISE 15.11. (i) Complete the previous exercise by making $H^1(\mathfrak{U},\Lambda)$ into a $\mathbf{C}$-linear space.

(ii) By the above, $\mathscr{F}(A_0)$ embeds into $H^1(\mathfrak{U},\Lambda)$; but we have identified it to a cokernel, so it is a $\mathbf{C}$-linear space. Check that the structures are compatible.

15.1.2. The "additive" case: normal forms. What follows relies on the generalities on extensions, see 8.3.2 in Chapter 8.

15.1.2.1. *Some generalities.* We fix a Newton polygon, given by two slopes $\mu_1 > \mu_2$ with multiplicities $r_1, r_2 \in \mathbf{N}^*$ and we set $n := r_1 + r_2$. Slopes will be assumed integral from the end of the present Subsection 15.1.2.1 on, but, for the moment, $\mu_1, \mu_2 \in \mathbf{Q}$. We let $A_0 := \mathrm{Diag}(B_1, B_2)$ a pure module in block diagonal form, where B_i is pure isoclinic of slope μ_i, rank r_i, $i = 1, 2$ (recall that we frequently identify the q-difference module $(\mathbf{C}(\{x\})^n, \Phi_{A_0})$ with the matrix A_0).

Those data given, we define a *(unipotent) linear algebraic group*:

$$\mathcal{G}_{A_0} := \begin{pmatrix} I_{r_1} & \star \\ 0 & I_{r_2} \end{pmatrix}.$$

This means that, for any commutative ring R (in our case, always $\mathbf{C}$-algebras):

$$\mathcal{G}_{A_0}(R) := \left\{ F_f := \begin{pmatrix} I_{r_1} & f \\ f & I_{r_2} \end{pmatrix} \mid f \in \mathrm{Mat}_{r_1,r_2}(R) \right\},$$

thus an algebraic subgroup of $\mathrm{GL}_n(R)$. Note that the structure of $\mathcal{G}_{A_0}$ really depends only[5] on the Newton polygon of A_0.

Actually, the group $\mathcal{G}_{A_0}(R)$ is isomorphic with the abelian group $\mathrm{Mat}_{r_1,r_2}(R)$ through the map $f \mapsto F_f$. This is the main reason why we call this the "additive" case (compare with the general case in Section 15.2). In this section, we shall abreviate $E_{\underline{r}} := \mathrm{Mat}_{r_1,r_2}(\mathbf{C}(\{x\}))$, so $E_{\underline{r}} \simeq \mathcal{G}_{A_0}(\mathbf{C}(\{x\}))$.

CAUTION 15.12. Be aware that we used a lower case letter f to denote a $r_1 \times r_2$ matrix, for we wish to keep capital F for the bigger matrix.

Any q-difference module M endowed with a specified isomorphism $\mathrm{gr}M \simeq M_0 := (\mathbf{C}(\{x\})^n, \Phi_{A_0})$ may be put into the form $M_U := (\mathbf{C}(\{x\})^n, \Phi_{A_U})$, where, as usual, $U \in \mathrm{Mat}_{r_1,r_2}(\mathbf{C}(\{x\}))$ and $A_U := \begin{pmatrix} B_1 & U \\ 0 & B_2 \end{pmatrix}$; this incarnates the exact sequence coming from the slope filtration:

$$0 \to P_1 \to M_U \to P_2 \to 0, \text{ where } P_i := (\mathbf{C}(\{x\})^{r_i}, \Phi_{B_i}), i = 1, 2.$$

In particular, $M_0 = P_1 \oplus P_2$.

From 8.3.2 we know that the module of extensions $\mathrm{Ext}(P_2, P_1)$ can be described as a quotient set $\mathscr{F}(A_0) := \mathscr{E}(A_0)/\sim$, where:

$$\mathscr{E}(A_0) := \{A_U \mid U \in E_{\underline{r}}\} \text{ and } A_U \sim A_V \underset{def}{\Longleftrightarrow} (\exists F \in \mathcal{G}_{A_0}(\mathbf{C}(\{x\})) \ : \ F[A_U] = A_V).$$

Since $F_f[A_U] = A_V \Leftrightarrow (\sigma_q f)B_2 - B_1 f = V - U$ (easy computation by now, left as an exercise!), we are led to introduce the $\mathbf{C}$-linear map:

$$\Phi_{B_1,B_2} : E_{\underline{r}} \to E_{\underline{r}}, f \mapsto (\sigma_q f)B_2 - B_1 f,$$

and to identify the $\mathbf{C}$-linear spaces (this is justified in 8.3.2):

$$\mathrm{Ext}(P_2, P_1) = \mathscr{F}(A_0) = \mathrm{Coker}\ \Phi_{B_1,B_2}.$$

We shall sometimes write $\mathscr{E}(M_0)$, $\mathscr{F}(M_0)$ instead of $\mathscr{E}(A_0)$, $\mathscr{F}(A_0)$. Note that Φ_{B_1,B_2} is injective (since B_1, B_2 are pure isoclinic with different slopes, the only solution $f \in E_{\underline{r}}$ of $(\sigma_q f)B_2 - B_1 f = 0$ is trivial).

Before going to our main theme here in section 15.1, which is the description of $\mathscr{F}(A_0)$, $\mathscr{F}(M_0)$, we comment (as we did in 15.1.1.1 for the particular case $r_1 = r_2 = 1$) on the seemingly more natural "plain" natural equivalence $A_U \simeq A_V$ (analytic gauge equivalence) and the corresponding quotient $\widetilde{\mathscr{F}}(A_0) := \mathscr{E}(A_0)/\simeq$. It can be identified with $\widetilde{\mathrm{Ext}}(P_2, P_1)$ which we know, by 8.3.2, to be the quotient of $\mathrm{Ext}(P_2, P_1)$ by the group action of $\mathrm{Aut}(P_1) \times \mathrm{Aut}(P_2)$. An element of $\mathrm{Aut}(P_i)$, $i = 1, 2$, can be represented by an element of $\mathrm{Aut}(B_i)$, i.e., an $f_i \in \mathrm{GL}_{r_i}(Ka)$ such that $(\sigma_q f_i)B_i = B_i f_i$. So let $(f_1, f_2) \in \mathrm{Aut}(B_1) \times \mathrm{Aut}(B_2)$; we shall describe its action on $\mathrm{Ext}(P_2, P_1) = \mathscr{F}(A_0)$ through the identification with $\mathrm{Coker}\ \Phi_{B_1,B_2}$.

So let $U, V \in E_{\underline{r}}$ representing the same class in $\mathrm{Coker}\ \Phi_{B_1,B_2}$, i.e., $V - U = (\sigma_q f)B_2 - B_1 f$ for some $f \in E_{\underline{r}}$. (Note that, Φ_{B_1,B_2} being injective, this f is unique). Let $U' := (\sigma_q f_1)U f_2$, $V' := (\sigma_q f_1)V f_2$ and $g := f_1 f f_2$, all in $E_{\underline{r}}$. Then $(\sigma_q g)B_2 - B_1 g = V' - U'$ (see exercise below), so $U', V' \in E_{\underline{r}}$ represent the same class in $\mathrm{Coker}\ \Phi_{B_1,B_2}$. Denoting $\mathrm{cl}U$, etc those classes, we thus have a well defined map: $\mathrm{cl}A_U \mapsto \mathrm{cl}A_{U'}$ from $\mathrm{Coker}\ \Phi_{B_1,B_2}$ to itself: this is the action we were looking for.

[5]Actually, the structure of that group depends only on $\underline{r} := (r_1, r_2)$. However, in Section 15.2, we shall have to deal with subgroups depending on more, and we prefer a unified notation.

EXERCISE 15.13. (i) Check the computation of $(\sigma_q g)B_2 - B_1 g$.

(ii) Express the above argument in the form of a commutative diagram.

(iii) Check that this is the same action as the one described in 15.1 to define the quotient $\widetilde{\mathrm{Ext}}$ of Ext. You might have to be careful around the fact that in our definition above $\mathrm{Aut}(P_1)$ operates on the left while $\mathrm{Aut}(P_2)$ operates on the right.

From now on, we assume integrality of the slopes: $S(A_0) = \{\mu_1, \mu_2\} \subset \mathbf{Z}$. So we may (and will!) take $B_i = x^{\mu_i} A_i$, $A_i \in \mathrm{GL}_{r_i}(\mathbf{C})$, $i = 1, 2$. We define the *level* $\delta := \mu_1 - \mu_2 \in \mathbf{N}^*$. We shall use sets $I \subset \mathbf{Z}$ of representatives modulo $\delta\mathbf{Z}$ and (temporarily) call them *admissible*: for instance any interval (in $\mathbf{Z}$) of length δ is admissible, and in particular:

$$I(\mu_1, \mu_2) := \{\mu_2, \ldots, \mu_1 - 1\}$$

is admissible (it is particularly useful in relation with multiplicative and tensor properties, as we shall see later). Last, for any admissible set I, we introduce a δ-dimensional linear subspace of $\mathbf{C}(\{x\})$:

$$K_I := \sum_{d \in I} \mathbf{C} x^d.$$

Most of what follows relies on arguments similar to those of Chapters 6 (on index theory) and 8 (on extensions); however, since the calculation techniques are very fundamental and very useful, we do it all explicitly, to the risk of redundancy.

15.1.2.2. *An important lemma.* First we extend lemma 6.8 from Chapter 6.

LEMMA 15.14. *Let $c \in \mathbf{C}^*$ and let $\phi_c : \mathscr{O}(\mathbf{C}^*, 0) \to \mathscr{O}(\mathbf{C}^*, 0)$, $f \mapsto c x^{-1} \sigma_q f - f$.*

(i) *For all $g \in \mathscr{O}(\mathbf{C}^*, 0)$, there is a unique pair $(f, \alpha) \in \mathscr{O}(\mathbf{C}^*, 0) \times \mathbf{C}$ such that $g = \phi_c(f) + \alpha$. Actually, $\alpha = \mathcal{B}_{q;1} f(c)$ (see beginning of proof for the meaning of $\mathcal{B}_{q;1} f$).*

(ii) *If $g \in \mathbf{C}(\{x\})$, then $f \in \mathbf{C}(\{x\})$.*

PROOF. Recall from 6.1.3 in Chapter 6, or from 14.2 in Chapter 14, that $f = \sum f_m x^m \in \mathbf{C}[[x]]$ is sent by the q-Borel-Ramis transformation (at level 1) to $\mathcal{B}_{q;1} f(\xi) := \sum q^{m(m+1)/2} f_m \xi^m \in \mathbf{C}[[\xi]]$. So clearly, since $f_m = O(A^m)$ for some $A > 0$ implies that $q^{m(m+1)/2} f_m = O(B^m)$ for all $B > 0$:

$$f \in \mathscr{O}(\mathbf{C}^*, 0) \Rightarrow \mathcal{B}_{q;1} f \in \mathscr{O}(\mathbf{C}^*) \text{ and } f \in \mathbf{C}(\{x\}) \Rightarrow \mathcal{B}_{q;1} f \in \mathscr{M}(\mathbf{C}) \cap \mathscr{O}(\mathbf{C}^*).$$

To simplify calculations (or at least their notations), we shall set: $\tilde{f}_m := q^{m(m+1)/2} f_m$ and $\tilde{g}_m := q^{m(m+1)/2} g_m$, so that $\mathcal{B}_{q;1} f(\xi) = \sum \tilde{f}_m \xi^m$, to be abreviated into $\tilde{f}(\xi)$, and $\mathcal{B}_{q;1} g(\xi) = \sum \tilde{g}_m \xi^m$, to be abreviated into $\tilde{g}(\xi)$. By assumption, $\tilde{g} \in \mathscr{O}(\mathbf{C}^*)$ and we at least want that $\tilde{f} \in \mathscr{O}(\mathbf{C}^*)$ (we do not mean that this will imply that $f \in \mathscr{O}(\mathbf{C}^*, 0)$, this will be tackled in due time). Then (using the standard Kronecker index notation):

$$\begin{aligned} g = \phi_c(f) + \alpha &\Longleftrightarrow \forall m\,,\ c q^{m+1} f_{m+1} - f_m = g_m - \alpha \delta_{m,0} \\ &\Longleftrightarrow \forall m\,,\ c \tilde{f}_{m+1} - \tilde{f}_m = \tilde{g}_m - \alpha \delta_{m,0} \\ &\Longleftrightarrow (c/\xi - 1) \tilde{f}(\xi) = \tilde{g}(\xi) - \alpha. \end{aligned}$$

Therefore, if the equation can be solved, then, by necessity, $\mathcal{B}_{q;1} g(c) = \tilde{g}(c) = \alpha$ and $\tilde{f}$ is uniquely determined as:

$$\frac{\tilde{g}(\xi) - \alpha}{c/\xi - 1} = -\xi \, \frac{\tilde{g}(\xi) - \tilde{g}(c)}{\xi - c}.$$

This at least proves the uniqueness of (f,α). Moreover, since clearly $\widetilde{g-\alpha} = \tilde{g}-\alpha$, we can assume that $\mathcal{B}_{q;1}g(c) = 0$ and try to solve $\phi_c(f) = g$.

So we assume that $\mathcal{B}_{q;1}g(c) = 0$, set $\tilde{f}(\xi) := -\xi\tilde{g}(\xi)/(\xi - c)$ and proceed to show that $f \in \mathscr{O}(\mathbf{C}^*,0)$; and then that $f \in \mathbf{C}(\{x\})$ if $g \in \mathbf{C}(\{x\})$. We begin by the latter contention, assuming the former one (which will be proved afterward). So here we know that $f \in \mathscr{O}(\mathbf{C}^*,0)$ and that, for some m_0, $g_m = 0$ whenever $m < m_0$. From the recursive relation $cq^{m+1}f_{m+1} - f_m = g_m$ we see that $f_m = cq^{m+1}f_{m+1}$ whenever $m < m_0$, so, iterating, that either all $f_m = 0$ when $m < m_0$, or $|f_m|$ grows wildly for $m \to -\infty$, contradicting the fact that $f \in \mathscr{O}(\mathbf{C}^*,0)$. This proves what we wanted.

Now we go for the convergence of the Laurent series $\sum f_m x^m$ under the mere assumption that $g \in \mathscr{O}(\mathbf{C}^*,0)$. But first, one more reduction to get rid of the factor c: upon respectively replacing $f(x)$, $g(x)$, $\tilde{f}(\xi)$, $\tilde{g}(\xi)$, f_m, g_m, $\tilde{f}_m$, $\tilde{g}_m$, by $f(cx)$, $g(cx)$, $\tilde{f}(c\xi)$, $\tilde{g}(c\xi)$, $c^m f_m$, $c^m g_m$, $c^m \tilde{f}_m$, $c^m \tilde{g}_m$, we can assume that $c = 1$ and therefore that $\sum \tilde{g}_m = 0$ (this is the assumption $\mathcal{B}_{q;1}g(c) = \mathcal{B}_{q;1}g(1) = 0$) and that $\tilde{f}_{m+1} - \tilde{f}_m = \tilde{g}_m$ for all m (this is the definition $\tilde{f}(\xi) := -\xi\tilde{g}(\xi)/(\xi-1)$). Also, since $\tilde{f} \in \mathscr{O}(\mathbf{C}^*,0)$, we know that $\lim\limits_{m\to\pm\infty} \tilde{f}_m = 0$.

Altogether, this gives us two explicit formulas for $\tilde{f}_m$, thus for f_m:

$$\tilde{f}_m = \begin{cases} -\sum_{p\geq m} \tilde{g}_p \\ \sum_{p\leq m-1} \tilde{g}_p \end{cases} \Longrightarrow f_m = \begin{cases} -q^{-m(m+1)/2}\sum_{p\geq m} q^{p(p+1)/2} g_p \\ q^{-m(m+1)/2}\sum_{p\leq m-1} q^{p(p+1)/2} g_p \end{cases}$$

Since $g \in \mathscr{O}(\mathbf{C}^*,0)$, there exist $C, A > 0$ such that $|g_p| \leq CA^p$ for all $p \geq 0$. So, setting as usual $r := |q| \in]0;1[$, we have for $m \geq 0$:

$$\begin{aligned} |f_m| &\leq r^{-m(m+1)/2} \sum_{p\geq m} r^{p(p+1)/2} CA^p \\ &= CA^m \sum_{l\geq 0} r^{l(l+1)/2} r^{lm} A^l \quad \text{(setting } p = m+l) \\ &\leq CC'A^m, \text{ where } C' := \sum_{m\geq 0} r^{m(m+1)/2} A^m < \infty. \end{aligned}$$

So the positive part of the Laurent series f has positive radius of convergence. As for the negative part, we set $u_m := f_{-m}$, $v_m := g_{-m}$ (so that $\sum_{m\geq 0} v_m x^m$ has infinite radius of convergence) and use equality $u_m = q^{-m(m-1)/2}\sum_{p\geq m+1} q^{p(p-1)/2} v_p$; then a very similar calculation shows that $\sum_{m\geq 0} u_m x^m$ has infinite radius of convergence. □

EXERCISE 15.15. Do it.

15.1.2.3. *Building on our important lemma.* Now we reach the main application of our lemma on the road to normal forms.

PROPOSITION 15.16. *For $i = 1, 2$, let $A_i \in GL_{r_i}(\mathbf{C})$, $B_i := x^{\mu_i}A_i$ and let A_0, A_U as in 15.1.2.1. Let $I \subset \mathbf{Z}$ an admissible set and denote $E_{\underline{r},I} := Mat_{r_1,r_2}(K_I)$. Then the target space of $\Phi := \Phi_{B_1,B_2}$ splits (as a $\mathbf{C}$-linear space) as:*

$$E_{\underline{r}} = Im\ \Phi \oplus E_{\underline{r},I}.$$

EXERCISE 15.17. State and prove the same result for $\mathscr{O}(\mathbf{C}^*,0)$ in lieu of $\mathbf{C}(\{x\})$, i.e., $\mathrm{Mat}_{r_1,r_2}(\mathscr{O}(\mathbf{C}^*,0)) = \mathrm{Im}\ \Phi_{B_1,B_2} \oplus E_{\underline{r},I}$ (for the extended Φ_{B_1,B_2} acting upon $\mathrm{Mat}_{r_1,r_2}(\mathscr{O}(\mathbf{C}^*,0))$).

REMARK 15.18. For nonintegral slopes, no such simple explicit decomposition is known, preventing the definition of nice normal forms (like Birkhoff-Guenther Normal Form, which we are heading at). Virginie Bugeaud obtained partial substitutes for that; see **[43]**.

PROOF (OF THE PROPOSITION). It goes through a sequence of successive reductions, all the way to the above "important" lemma in 15.1.2.2. Note by the way that part (ii) of the lemma is a particular case of the present proposition (in the same way as part (i) is a particular case of the exercise above): can you see how? First, right compose Φ with left multiplication by $(x^{\mu_1}A_1)^{-1}$, i.e., set:

$$\Psi(f) := (x^{\mu_1}A_1)^{-1}\Phi(f) = x^{-\delta}A_1^{-1}(\sigma_q f)A_2 - f.$$

(As usual, $\delta := \mu_1 - \mu_2$.). Since left multiplication by $(x^{\mu_1}A_1)^{-1}$ is a $\mathbf{C}$-linear automorphism of $E_{\underline{r}}$ sending $E_{\underline{r},I}$ to $\mathrm{Mat}_{r_1,r_2}(K_J)$, with $J := -\mu_1 + I$ also an admissible subset, we are led to prove a similar statement for Ψ.

Identifying $\mathrm{Mat}_{r_1,r_2}(\mathbf{C}(\{x\}))$ with $\mathbf{C}(\{x\})^r$, $r := r_1r_2$, we see f as a vector in $\mathbf{C}(\{x\})^r$ and $f \mapsto A_1^{-1}fA_2$ as a linear automorphism of $\mathbf{C}(\{x\})^r$ with matrix $C \in \mathrm{GL}_r(\mathbf{C})$. So we are dealing with the map $\psi : X \mapsto x^{-\delta}CX - X$ from $\mathbf{C}(\{x\})^r$ to itself; we want to prove that its image has supplementary K_I^r in $\mathbf{C}(\{x\})^r$ (there is no harm going back to letter I instead of J since anyhow this should apply to any admissible set).

We decompose $\mathbf{C}(\{x\})$ as $\bigoplus\limits_{i\in I} x^i\mathbf{C}(\{x^\delta\})$, and accordingly $\mathbf{C}(\{x\})^r$ as $\bigoplus\limits_{i\in I} E_i$, where $E_i := x^i\mathbf{C}(\{x^\delta\})^r$, $i \in I$. Each factor E_i is stable under ψ, thus inducing ψ_i from E_i to itself. We are going to prove that $E_i = \mathrm{Im}\ \psi_i \oplus (\mathbf{C}x^i)^r$; then putting all those δ equalities together, it will follow that $\mathbf{C}(\{x\})^r = \mathrm{Im}\ \psi \oplus K_I^r$, as hoped for.

To do that we use the isomorphism $\mathbf{C}(\{x^\delta\})^r \simeq E_i$ (multiplication by x^i) to conjugate ψ_i with a map ϕ_i from $\mathbf{C}(\{x^\delta\})^r$ to itself:

$$\begin{array}{ccc} \mathbf{C}(\{x^\delta\})^r & \xrightarrow{\ \phi_i\ } & \mathbf{C}(\{x^\delta\})^r. \\ {\scriptstyle \times x^i}\downarrow & & \downarrow{\scriptstyle \times x^i} \\ E_i & \xrightarrow[\ \Psi_i\ ]{} & E_i \end{array}$$

This reduces our goal to proving that $\mathrm{Im}\ \phi_i$ has supplementary $\mathbf{C}^r$. But (see exercise between this proof and the next lemma), using the shorthands $X := x^\delta$, $Q := q^\delta$, $C_i := q^iC$ and $G(X) := g(x^\delta)$, we find that ϕ_i sends G to $X^{-1}C_i\sigma_Q G - G$. The desired decomposition $\mathbf{C}(\{x^\delta\})^r = \mathrm{Im}\ \phi_i \oplus \mathbf{C}^r$ is (in the Q, X setting) just the special case of our proposition when $\delta = 1$ and $I = \{0\}$. That case will be dealt with in the lemma herebelow. □

EXERCISE 15.19. Confirm the form given to ϕ_i by writing $f(x) = x^ig(x^\delta)$ and $\psi_i(f)$ in the form $x^ih(x^\delta)$ (so $h = \phi_i(g)$).

LEMMA 15.20. *Let $C \in GL_r(\mathbf{C})$ and let $\Psi_C : \mathbf{C}(\{x\})^r \to \mathbf{C}(\{x\})^r$, $f \mapsto x^{-1}C\sigma_q f - f$. Then $\mathbf{C}(\{x\})^r = Im\ \Psi_C \oplus \mathbf{C}^r$.*

EXERCISE 15.21. State and prove the same result for $\mathscr{O}(\mathbf{C}^*, 0)$ in lieu of $\mathbf{C}(\{x\})$ (in order to solve the corresponding exercise in 15.1.2.2).

PROOF OF THE LEMMA. If $C = PTP^{-1}$ (conjugacy in $\mathrm{GL}_n(\mathbf{C})$), then a small calculation (using the fact that $\sigma_q P = P$ and $\sigma_q T = T$) shows that $\Psi_C = P\Psi_T P^{-1}$, where $P \in \mathrm{GL}_n(\mathbf{C}) \subset \mathrm{GL}_n(\mathbf{C}(\{x\}))$ acts naturally on $\mathbf{C}(\{x\})^r$. Since it leaves $\mathbf{C}^r \subset \mathbf{C}(\{x\})^r$ stable, we are reduced to the similar question for T, i.e., we can assume that $C = (c_{i,j})$ is triangular: $c_{i,j} = 0$ whenever $1 \leq j < i \leq r$.

Referring to Lemma 15.14, we write $dec(c, g)$ for the pair (f, α) obtained there. We want a decomposition:

$$\begin{pmatrix} g_1 \\ \vdots \\ g_r \end{pmatrix} = x^{-1}C \begin{pmatrix} \sigma_q f_1 \\ \vdots \\ \sigma_q f_r \end{pmatrix} + \begin{pmatrix} \alpha_1 \\ \vdots \\ \alpha_r \end{pmatrix} \Longleftrightarrow \begin{cases} x^{-1}(c_{1,1}\sigma_q f_1 + \cdots + c_{1,n}\sigma_q f_n) + \alpha_1 = g_1, \\ \ldots, \\ x^{-1} c_{n,n}\sigma_q f_n + \alpha_n = g_n, \end{cases}$$

all $f_i \in \mathbf{C}(\{x\})$, all $\alpha_i \in \mathbf{C}$. This can be recursively solved: first $(f_r, \alpha_r) := dec(c_{r,r}, g_r)$, then, by downward induction, for $i = n-1, \ldots, 1$:

$$(f_i, \alpha_i) := dec\left(c_{i,i}, g_i - x^{-1} \sum_{j=i+1}^{r} c_{i,j}\sigma_q f_j \right).$$

□

EXERCISE 15.22. Try to build a direct proof of the lemma by using series with vectorial coefficients: write $f = \sum f_m x^m$, $g = \sum g_m x^m$ (of course here the f_m, g_m are in $\mathbf{C}^r$) and $q^{m+1} C f_{m+1} - f_m = g_m$ (plus some constant vector $\alpha \in \mathbf{C}^r$ for $m = 0$), then a criterion of the form $\mathcal{B}_{q;1} f(C) = 0$ (be careful with the order of factors there!).

15.1.2.4. *The normal forms at last.*

COROLLARY 15.23 (of proposition 15.16). *Let $A_1, A_2, \mu_1, \mu_2, A_0$ as before and I an admissible subset of $\mathbf{Z}$.*

(i) *In each class within $\mathscr{F}(A_0)$, there is a unique representative A_U such that $U \in Mat_{r_1,r_2}(K_I)$. (Remember K_I was defined at the end of Subsubsection 15.1.2.1.)*

(ii) *For each A_V in the class of A_U, there is a unique $F \in \mathcal{G}_{A_0}(\mathbf{C}(\{x\}))$ such that $F[A_U] = A_V$.*

PROOF. Only the uniqueness of F is not yet proved, but it immediately follows from the injectivity of $\Phi_{x^{\mu_1}A_1, x^{\mu_2}A_2}$ (see exercise). □

REMARK 15.24. Such a matrix A_U is automatically regular, i.e., it is holomorphic over $\mathbf{C}^*$ and meromorphic at 0 and the same holds for

$$A_U^{-1} = \begin{pmatrix} x^{-\mu_1} A_1^{-1} & U' \\ 0 & x^{-\mu_2} A_2^{-1} \end{pmatrix},$$

where $U' := -x^{-\mu_1-\mu_2} A_1^{-1} U A_2^{-1}$.

EXERCISE 15.25. Prove the injectivity of $\Phi_{x^{\mu_1}A_1, x^{\mu_2}A_2}$ operating on $\mathrm{Mat}_{r_1,r_2}(\mathscr{O}(\mathbf{C}^*, 0))$, a fortiori on $\mathrm{Mat}_{r_1,r_2}(\mathbf{C}(\{x\}))$.

COROLLARY 15.26. (i) *This yields a bijection $Mat_{r_1,r_2}(K_I) \to \mathscr{F}(A_0)$, actually an isomorphism of $\mathbf{C}$-linear spaces.*

(ii) *Each class within $\mathscr{F}(A_0)$ is a torsor under $\mathcal{G}_{A_0}(\mathbf{C}(\{x\}))$ (i.e., the left action of that group is simply transitive).*

PROOF. Only the fact that $U \mapsto \mathrm{cl}(A_U)$ is new, but it follows from the descip-tion of $\mathscr{F}(A_0)$ as Coker $\Phi_{x^{\mu_1}A_1, x^{\mu_2}A_2}$ and from the $\mathbf{C}$-*linear* splitting of $\mathrm{Mat}_{r_1,r_2}(\mathbf{C}(\{x\}))$ into its components Im $\Phi_{x^{\mu_1}A_1, x^{\mu_2}A_2}$ and $\mathrm{Mat}_{r_1,r_2}(K_I)$. □

DEFINITION 15.27. In the particular case $I = I(\mu_1, \mu_2) = \{\mu_2, \ldots, \mu_1 - 1\}$, the matrix A_U is said to be *in Birkhoff-Guenther Normal Form*, or to be *the Birkhoff-Guenther Normal Form* of all matrices in its class. Given V and the other notations as above, the unique pair (f, U) such that $F_f[A_U] = A_V$, $F_f \in \mathcal{G}_{A_0}(\mathbf{C}(\{x\}))$, A_U in Birkhoff-Guenther Normal Form will be denoted $Red(\mu_1, A_1, \mu_2, A_2, V)$.

The name "Birkhoff-Guenther Normal Form" comes from the paper **[25]** where the corresponding normal form for q-difference operators was given first, and where was strongly asserted its importance for the theory (see the quotation at the beginning of 15.2.4). Note however that we only have here the additive (two slopes) form, the complete version will come in 15.2.4.

Using the "wild part" of the important lemma in 15.1.2.2, we get:

PROPOSITION 15.28. *Let* $A_V := \begin{pmatrix} x^{\mu_1}A_1 & V \\ 0 & x^{\mu_2}A_2 \end{pmatrix}$ *where* $V \in Mat_{r_1,r_2}(\mathscr{O}(\mathbf{C}^*, 0))$, *and let I an admissible subset of* $\mathbf{Z}$. *Then there is a unique pair* (f, U), $f \in \mathscr{O}(\mathbf{C}^*, 0)$, $U \in Mat_{r_1,r_2}(K_I)$ *such that* $F_f[A_U] = A_V$.

PROOF. Left to the reader, using the relevant previous exercises. □

15.1.3. The "additive" case: algebraic summation.

15.1.3.1. *Some general notations.* Let $\mathrm{Div}(\mathbf{C}^*)$ (Appendix B) the group of *divisors* on $\mathbf{C}^*$, i.e., the free abelian group with a basis of symbols $[c]$, one for each $c \in \mathbf{C}^*$; and let similarly $\mathrm{Div}(\mathbf{E}_q)$ the group of divisors on $\mathbf{E}_q$. For every $D := \sum n_i[c_i] \in \mathrm{Div}(\mathbf{C}^*)$ (all $c_i \in \mathbf{C}^*$, all $n_i \in \mathbf{Z}$, almost all of them null) we set $\deg D := \sum n_i \in \mathbf{Z}$ and $\mathrm{ev} D := \prod c_i^{n_i} \in \mathbf{C}^*$. So we have group morphisms $\deg : \mathrm{Div}(\mathbf{C}^*) \to \mathbf{Z}$ and $\mathrm{ev} : \mathrm{Div}(\mathbf{C}^*) \to \mathbf{C}^*$. With the same notations and the same defining formulas[6], we have group morphisms $\deg : \mathrm{Div}(\mathbf{E}_q) \to \mathbf{Z}$ and $\mathrm{ev} : \mathrm{Div}(\mathbf{E}_q) \to \mathbf{E}_q$.

To every $D := \sum n_i[c_i] \in \mathrm{Div}(\mathbf{C}^*)$ we associate a function Θ_D defined by:

$$\Theta_D(x) := \prod \theta_q(x/c_i)^{n_i} \in \mathscr{M}(\mathbf{C}^*)^*.$$

The map $D \mapsto \Theta_D$ is a group morphism $\mathrm{Div}(\mathbf{C}^*) \to \mathscr{M}(\mathbf{C}^*)^*$. If D is *effective*, i.e., all $n_i \geq 0$, then $\Theta_D \in \mathscr{O}(\mathbf{C}^*)$, but the converse is not true since there may be cancellations (e.g., $D = [qc] - [c]$).

EXERCISE 15.29. Describe the kernel of $D \mapsto \Theta_D$.

In some cases, we shall exceptionally allow for divisors with infinite support (the *support* of D is $\{c_i \mid n_i \neq 0\}$): we then only require the support to be discrete; on a compact Riemann surface, like $\mathbf{E}_q$, this is of course no generalisation, but for instance we can consider the divisor of zeroes and poles of Θ_D:

$$\mathrm{div}_{\mathbf{C}^*} \Theta_D = \sum_i \sum_{m \in \mathbf{Z}} n_i[-q^m c_i].$$

Note that there might be cancellations: for instance if $D := [qc] - [c]$, then $\Theta_D = x/qc$ so $\mathrm{div}_{\mathbf{C}^*} \Theta_D = 0$. What is more important is that the zeroes and poles of Θ_D

[6]Remember that we denote multiplicatively the group $\mathbf{E}_q := \mathbf{C}^*/q^{\mathbf{Z}}$.

form, in every case, q-spirals $[-c_i; q]$ along which the orders of those zeroes and poles are constant. This means that we can associate to Θ_D a divisor on $\mathbf{E}_q$:

$$\mathrm{div}_{\mathbf{E}_q}(\Theta_D) = \sum_i n_i \overline{[-c_i]}.$$

EXERCISE 15.30. (i) Prove the following formula (which will be required later):

$$\Theta_D(qx) = \frac{\prod c_i^{n_i}}{x^{\sigma n_i}} \Theta_D(qx), \text{ i.e. } \sigma_q \Theta_D = \frac{\mathrm{ev}(D)}{x^{\deg D}} \Theta_D.$$

(ii) Let $D' := \mathrm{div}_{\mathbf{E}_q}(\theta_D)$. Compute $\deg D'$ and $\mathrm{ev}(D')$.

15.1.3.2. For the rest of this section, we return to the (by now) usual notations: $A_1, A_2, \mu_1, \mu_2, B_1, B_2, A_0, \delta, \ldots$; and we use lower case letters to denote the upper right block in for instance A_u and F_f, $u, f \in \mathrm{Mat}_{r_1,r_2}(\mathbf{C}(\{x\}))$. We know that, for any given $u, f \in \mathrm{Mat}_{r_1,r_2}(\mathbf{C}(\{x\}))$, there is a unique $\hat{f} \in \mathbf{C}((x))$ and so a unique $\widehat{F} := F_{\hat{f}} \in \mathcal{G}_{A_0}(\mathbf{C}((x)))$ such that $\widehat{F}[A_0] = A_u$. We write it $\widehat{F}_u$ (later, in a more general setting, it will sometimes be denoted $\widehat{F}_{A_u}$, see 15.2). We are now going to look for some analytic incarnation of that formal object.

EXERCISE 15.31. Recall why (and when) $\widehat{F}_u$ itself is not an analytic object.

We first need some simple analytic tools. Let $D_1, D_2 \in \mathrm{Div}(\mathbf{C}^*)$ and denote $e_i := \mathrm{ev} D_i$ and $d_i := \deg D_i$, $i = 1, 2$. By the exercise above, $(\sigma_q \Theta_{D_i})/\Theta_{D_i} = e_i/x^{d_i}$. Set $T := \mathrm{Diag}(\Theta_{D_1} I_{r_1}, \Theta_{D_2} I_{r_2})$ (thus a diagonal matrix of rank n). Then, given A_0, A_u as above, we have:

$$T[A_0] = A'_0 := \begin{pmatrix} A'_1 & 0 \\ 0 & A'_2 \end{pmatrix} \text{ and } T[A_u] = A'_v := \begin{pmatrix} A'_1 & v \\ 0 & A'_2 \end{pmatrix},$$

where $A'_i := e_i x^{\mu_i - d_i} A_i$ and $v := \dfrac{\sigma_q \Theta_{D_1}}{\Theta_{D_2}} u = (e_1/x^{d_1}) \Theta_{D_1 - D_2} u$.

We shall use the following logical equivalence:

$$F_f[A_0] = A_u \Longleftrightarrow F_g[A'_0] = A'_v, \text{ where } g := \Theta_{D_1 - D_2} f.$$

This can be checked by direct computation, or with the help of the commutative diagram, in which $G := T F_f T^{-1} = F_g$ defined as above:

$$\begin{array}{ccc} A_0 & \xrightarrow{F_f} & A_u. \\ {\scriptstyle T}\downarrow & & \downarrow{\scriptstyle T} \\ A'_0 & \xrightarrow[G]{} & A'_v \end{array}$$

We shall especially use the particular examples: $D_i := \mu_i[c]$ for some $c \in \mathbf{C}^*$ (to be discussed later). Then the above formulas hold with:

$$A'_0 = \begin{pmatrix} c^{\mu_1} A_1 & 0 \\ 0 & c^{\mu_2} A_2 \end{pmatrix}, \quad A'_v = \begin{pmatrix} c^{\mu_1} A_1 & (c/x)^{\mu_1} \theta_q(x/c)^\delta u \\ 0 & c^{\mu_2} A_2 \end{pmatrix}, \quad g = \theta_q(x/c)^\delta f.$$

15.1.3.3. *Algebraic summation with a single pole.* Two summation theories (one for scalar q-difference equations, one for systems) with general polar divisors are presented in [**197, 219**] but they do not seem to have met useful applications yet. Although it is the opinion of the present author that they *should* be the right framework to work in, they will not be presented here. Instead polar divisors supported at one point will be considered; this will also hold in the sequel, i.e., in sections 15.2 and 15.3). So we begin by fixing $c \in \mathbf{C}^*$ and, as at the end of 15.1.3.2, of which we retain the notations, we set $D_i := \mu_i[c]$, $i = 1, 2$. In the same way as $F_f[A_0] = A_u \Leftrightarrow (\sigma_q f)x^{\mu_2}A_2 - x^{\mu_1}A_1 f = u$, we see by similar calculations that $F_g[A'_0] = A'_v \Leftrightarrow (\sigma_q g)c^{\mu_2}A_2 - c^{\mu_1}A_1 g = v$. With the convention that the $r_1 \times r_2$ matrix u has its $r_1 r_2$ components in $\mathbf{C}(\{x\})$, we find that the $r_1 \times r_2$ matrix $v = (c/x)^{\mu_1}\theta_q(x/c)^\delta u$ has its $r_1 r_2$ components in $\mathscr{O}(\mathbf{C}^*, 0)$ (recall that $\delta > 0$ and that $\theta_q(x/c) \in \mathscr{O}(\mathbf{C}^*)$) so it admits a Laurent expansion $v = \sum v_m x^m$. We can look for g in the form of a Laurent expansion $\sum g_m x^m$ and draw:

$$(\sigma_q g)c^{\mu_2}A_2 - c^{\mu_1}A_1 g = v \Longleftrightarrow \forall m \in \mathbf{Z}\ ,\ q^m g_m(c^{\mu_2}A_2) - (c^{\mu_1}A_1)g_m = v_m.$$

Recall also that all the v_m, g_m are matrices in $\mathrm{Mat}_{r_1,r_2}(\mathbf{C})$. The endomorphism $X \mapsto q^m X(c^{\mu_2}A_2) - (c^{\mu_1}A_1)X$ of $\mathrm{Mat}_{r_1,r_2}(\mathbf{C})$ is known to have as eigenvalues the $\lambda_2 - \lambda_1$ such that λ_1 is an eigenvalue of $c^{\mu_1}A_1$ and λ_2 is an eigenvalue of $q^m c^{\mu_2}A_2$. So it is invertible if, and only if, the spectra of those two matrices do not meet, i.e., if:

$$\forall m \in \mathbf{Z}\ ,\ \forall \lambda_1 \in \mathrm{Sp}(A_1)\ ,\ \forall \lambda_2 \in \mathrm{Sp}(A_2)\ , q^m c^{\mu_2}\lambda_2 \neq c^{\mu_1}\lambda_1.$$

It appears that, taken as a condition on $c \in \mathbf{C}^*$, it is q-invariant, hence really a condition on $\overline{c}$. We are thus led to the following definition:

DEFINITION 15.32. We say that $\overline{c} \in \mathbf{E}_q$ is a *prohibited direction of summation* if for one of its representatives $c \in \mathbf{C}^*$ (hence for all of them) $c^{\mu_1}\mathrm{Sp}(A_1)$ meets $q^m c^{\mu_2}\mathrm{Sp}(A_2)$. Writing:

$$\mathrm{Sp}(A_2)/\mathrm{Sp}(A_1) := \{\lambda_2/\lambda_1 \mid \lambda_1 \in \mathrm{Sp}(A_1), \lambda_2 \in \mathrm{Sp}(A_2)\} \text{ and then}$$
$$\overline{\mathrm{Sp}(A_2)/\mathrm{Sp}(A_1)} := \pi\left(\mathrm{Sp}(A_2)/\mathrm{Sp}(A_1)\right) \subset \mathbf{E}_q,$$

this equivalent to: $[c^\delta; q]$ meets $\mathrm{Sp}(A_2)/\mathrm{Sp}(A_1)$, which in turn is equivalent to: $\overline{c}^\delta \in \overline{\mathrm{Sp}(A_2)/\mathrm{Sp}(A_1)}$. We denote Σ_{A_0} the set of prohibited directions:

$$\Sigma_{A_0} = \{\overline{c} \in \mathbf{E}_q \mid \overline{c}^\delta \in \overline{\mathrm{Sp}(A_2)/\mathrm{Sp}(A_1)}\}.$$

Since $\overline{\mathrm{Sp}(A_2)/\mathrm{Sp}(A_1)}$ has at most $r_1 r_2$ elements and since every element in $\mathbf{E}_q$ has δ^2 distinct δth roots, we see that Σ_{A_0} has at most $r_1 r_2 \delta^2$ elements.

So if $\overline{c} \notin \Sigma_{A_0}$, the equations above admit a unique solution g_m for every m. Admitting temporarily that the series $\sum g_m x^m$ defines a matrix with components in $\mathscr{O}(\mathbf{C}^*, 0)$, we see that $f = \dfrac{1}{\theta_q(x/c)^\delta} g$ has (near 0) at most poles on $[-c; q]$ with multiplicity δ. We shall denote that polar property like this[7]:

$$\mathrm{div}_{\mathbf{E}_q}(f) \geq -\delta[\overline{-c}].$$

[7] Although the poles of f near 0 are well defined and form a q-invariant set (near 0) if f satisfies equation $(\sigma_q f)x^{\mu_2}A_2 - x^{\mu_1}A_1 f = u$, $u \in \mathbf{C}(\{x\})$, nothing of the sort can be said about its zeroes, so the notation $\mathrm{div}_{\mathbf{E}_q}(f)$ is not really meaningful by itself, yet the relation $\mathrm{div}_{\mathbf{E}_q}(f) \geq -\delta[\overline{-c}]$ is.

Conversely, a solution f of $F_f[A_0] = A_u$ such that $\operatorname{div}_{\mathbf{E}_q}(f) \geq -\delta[\overline{-c}]$ automatically has the form $\dfrac{1}{\theta_q(x/c)^\delta} g$ where g has components in $\mathscr{O}(\mathbf{C}^*, 0)$ and satisfies $F_g[A'_0] = A'_v$, so has the Laurent expansion uniquely determined above.

PROPOSITION 15.33. *Let $\overline{c} \notin \Sigma_{A_0}$ and let $u \in \mathbf{C}(\{x\})$. Then the equation $F_f[A_0] = A_u$ admits a unique solution f meromorphic in $(\mathbf{C}^*, 0)$ and such that $\operatorname{div}_{\mathbf{E}_q}(f) \geq -\delta[\overline{-c}]$.*

PROOF. The only thing that remains to be proved is that, given $u \in \mathbf{C}(\{x\})$, the series $\sum g_m x^m$ computed as above does define a matrix with components in $\mathscr{O}(\mathbf{C}^*, 0)$. Note that its coefficients are defined by relations $g_m := L_m^{-1}(v_m)$, where L_m is the invertible endomorphism $X \mapsto q^m X(c^{\mu_2} A_2) - (c^{\mu_1} A_1)X$ of $\operatorname{Mat}_{r_1,r_2}(\mathbf{C})$ (it is invertible for all m because of the assumption $\overline{c} \notin \Sigma_{A_0}$). Using adequate matricial norms, we have $\|g_m\| \leq |||L_m^{-1}|||\, \|v_m\|$. On the other hand, for $m \to +\infty$, the operator L_m tends to left multiplication by $-(c^{\mu_1} A_1)$, so L_m^{-1} tends to left multiplication by $-(c^{\mu_1} A_1)^{-1}$, so $|||L_m^{-1}|||$ tends to $|||(c^{\mu_1} A_1)^{-1}|||$. In the same way, when $m \to -\infty$, we have $|||L_m^{-1}||| \to |||(q^m c^{\mu_2} A_2)^{-1}|||$. Combined with the majorations of $\|v_m\|$ expressing the fact that it has components in $\mathscr{O}(\mathbf{C}^*, 0)$, this yields the result. □

EXERCISE 15.34. Fill in the details of the convergence proof (with constants C, R, majorations, etc).

DEFINITION 15.35. The unique matrix F_f, f meromorphic in $(\mathbf{C}^*, 0)$, such that $F_f[A_0] = A_u$ and $\operatorname{div}_{\mathbf{E}_q}(f) \geq -\delta[\overline{-c}]$ is called *summation of $\widehat{F}_u$ in (authorized) direction $\overline{c} \notin \Sigma_{A_0}$* and it is denoted $S_{\overline{c}}\widehat{F}_u$.

EXERCISE 15.36. Check that, for a fixed $\overline{c}$, the map $u \mapsto S_{\overline{c}}\widehat{F}_u$ is $\mathbf{C}$-linear.

15.1.4. The "additive" case: Stokes operators. By analogy with the theory of differential equations [**150, 192**], also see Chapter 14, we consider as *Stokes phenomenon* the "ambiguity" arising from the variety of directions of summation.

15.1.4.1. We retain the usual notations A_i, μ_i, etc, and recall that the set $\Sigma_{A_0} \subset \mathbf{E}_q$ of prohibited directions of summation is defined in 15.1.3.3, as well as the summations $S_{\overline{c}}\widehat{F}_u$ in the various authorized directions $\overline{c} \notin \Sigma_{A_0}$.

Now fix $u \in \mathbf{C}(\{x\})$ and let $\overline{c}, \overline{d} \notin \Sigma_{A_0}$. Define the *Stokes operator*:

$$S_{\overline{c},\overline{d}}\widehat{F}_u := (S_{\overline{c}}\widehat{F}_u)^{-1} S_{\overline{d}}\widehat{F}_u = \begin{pmatrix} 1 & S_{\overline{c},\overline{d}}\hat{f}_u \\ 0 & 1 \end{pmatrix}, \text{ where } S_{\overline{c},\overline{d}}\hat{f}_u := S_{\overline{d}}\hat{f}_u - S_{\overline{c}}\hat{f}_u.$$

The matrix $f := S_{\overline{c},\overline{d}}\hat{f}_u$ (temporary short notation) has the following properties, flowing immediately from the known properties of $S_{\overline{c}}\hat{f}_u$, $S_{\overline{d}}\hat{f}_u$:

(1) $(\sigma_q f)(x^{\mu_2} A_2) - (x^{\mu_1} A_1) f = 0$,
(2) $\operatorname{div}_{\mathbf{E}_q}(f) \geq -\delta([\overline{-c}] + [\overline{-d}])$.

Similarly, the operator $F := S_{\overline{c},\overline{d}}\widehat{F}_u$ (temporary short notation) satisfies $F[A_0] = A_u$, which follows either from the first property above, or from the known properties of $S_{\overline{c}}\widehat{F}_u$, $S_{\overline{d}}\widehat{F}_u$.

Note that a priori f has components in $\mathscr{O}(\mathbf{C}^* \setminus [-c, -d; q], 0)$ (germs), but, thanks to the functional equation, it can be analytically continued to $\mathscr{O}(\mathbf{C}^* \setminus$

$[-c, -d; q]$) by the usual process of q^{-1}-dilatation of disks of definition. Then the second condition means that f is meromorphic all over $\mathbf{C}^*$ with, at worst, poles over $[-c, -d; q]$ with multiplicities $\leq \delta$.

We therefore introduce the following notations:

(1) For every effective divisor D on $\mathbf{E}_q$, let:

$$\mathscr{O}_{\mathbf{C}^*}(D) := \{\phi \in \mathscr{M}(\mathbf{C}^*) \mid \operatorname{div}_{\mathbf{E}_q}(\phi) \geq -D\}.$$

(2) For every $\delta \in \mathbf{N}$ and $c \in \mathbf{C}^*$, let $D_{\overline{c}} := \delta[\overline{-c}]$.

(3) For every $c, d \in \mathbf{C}^*$, let $E_{\overline{c},\overline{d}}(A_1, \mu_1, A_2, \mu_2)$ the following space:

$$\left\{ f \in \operatorname{Mat}_{r_1,r_2}(\mathscr{M}(\mathbf{C}^*)) \mid \begin{cases} (\sigma_q f)(x^{\mu_2} A_2) - (x^{\mu_1} A_1) f = 0, \\ f \text{ has components in } \mathscr{O}_{\mathbf{C}^*}(D_{\overline{c}} + D_{\overline{d}}). \end{cases} \right\}.$$

(You should check that the latter indeed depends on $\overline{c}, \overline{d}$ and not on $c; d$ themselves.) We shall abbreviate $E_{\overline{c},\overline{d}}(A_1, \mu_1, A_2, \mu_2)$ into $E_{\overline{c},\overline{d}}$.

PROPOSITION 15.37. *$E_{\overline{c},\overline{d}}$ is a $\mathbf{C}$-linear space of dimension $\delta r_1 r_2$.*

PROOF. It is obviously a $\mathbf{C}$-linear subspace of $\operatorname{Mat}_{r_1,r_2}(\mathscr{M}(\mathbf{C}^*))$. The mapping $f \mapsto \theta_q(x/c)^\delta \theta_q(x/d)^\delta f$ sends it isomorphically to the subspace of $\operatorname{Mat}_{r_1,r_2}(\mathscr{O}(\mathbf{C}^*))$ made up of matrices g such that $(\sigma_q g) = (cd/x)^\delta A_1 g A_2^{-1}$. Using Laurent expansions $g = \sum g_m x^m$, that equation is equivalent to the system $q^m g_m = (cd)^\delta A_1 g_{m+\delta} A_2^{-1}$. The latter admits a unique solution for every given set $g_0, \ldots, g_{\delta-1}$ (for instance) of $r_1 \times r_2$-matrices over $\mathbf{C}$, and is linearly determined by them; and the resulting g converges on $\mathbf{C}^*$. So it is a $\mathbf{C}$-linear space of dimension $\delta r_1 r_2$. □

EXERCISE 15.38. (i) Check the convergence of g.

(ii) If you know the Riemann-Roch theorem for vector bundles over a Riemann surface, use it to get an alternative proof (look at **[108]** for that theorem, or at Subsection B.6.6 of Appendix B).

15.1.4.2. So we have defined a map $A_u \mapsto S_{\overline{c},\overline{d}} \hat{f}_u$ from $\mathscr{E}(A_0)$ to $E_{\overline{c},\overline{d}}$. Seen as a map from $\operatorname{Mat}_{r_1,r_2}(\mathbf{C}(\{x\}))$ to $E_{\overline{c},\overline{d}}$, it is linear (this follows from Exercise 15.38 and from the definition of $S_{\overline{c},\overline{d}} \hat{f}_u$ as $S_{\overline{d}} \hat{f}_u - S_{\overline{c}} \hat{f}_u$).

LEMMA 15.39. *The kernel of the mapping $u \mapsto S_{\overline{c},\overline{d}} \hat{f}_u$, $Mat_{r_1,r_2}(\mathbf{C}(\{x\})) \to E_{\overline{c},\overline{d}}$, is Im $\Phi_{B_1;B_2}$ (recall our usual notation $B_i := x^{\mu_i} A_i$, $i = 1, 2$).*

PROOF. We shall use simplified notations $f_{\overline{c}} := S_{\overline{c}} \hat{f}_u$, so that $S_{\overline{c},\overline{d}} \hat{f}_u = f_{\overline{c}} - f_{\overline{d}}$. The statement of the proposision means that $f_{\overline{c}} = f_{\overline{d}} \Leftrightarrow \exists f \in \operatorname{Mat}_{r_1,r_2}(\mathbf{C}(\{x\}))$: $(\sigma_q f)(x^{\mu_2} A_2) - (x^{\mu_1} A_1) f = u$. So we first assume the right-hand side of the equivalence: $u = (\sigma_q f)(x^{\mu_2} A_2) - (x^{\mu_1} A_1)$, $f \in \operatorname{Mat}_{r_1,r_2}(\mathbf{C}(\{x\}))$. Then obviously $\operatorname{div}_{\mathbf{E}_q}(f) \geq -\delta[\overline{-c}]$, so $f = f_{\overline{c}}$; and, for the same reason, $f = f_{\overline{d}}$ and indeed $f_{\overline{c}} = f_{\overline{d}}$ as desired.

Assume conversely that $f_{\overline{c}} = f_{\overline{d}}$ and call it f. It satisfies $(\sigma_q f)(x^{\mu_2} A_2) - (x^{\mu_1} A_1) f = u$ and it has no poles (near 0) out of $[-c; q] \cap [-d; q] = \emptyset$, so f has components in $\mathscr{O}(\mathbf{C}^*, 0)$. But then, as already seen quite a few times (e.g., in the proof of the "important lemma" in 15.1.2.2), necessarily $f \in \operatorname{Mat}_{r_1,r_2}(\mathbf{C}(\{x\}))$. □

PROPOSITION 15.40. *We thus obtain a* $\mathbf{C}$*-linear isomorphism Coker* $\Phi_{B_1,B_2} \to E_{\overline{c},\overline{d}}$, *whence a bijection* $\mathscr{F}(A_0) \to E_{\overline{c},\overline{d}}$.

PROOF. By the lemma, the mapping $u \mapsto S_{\overline{c},\overline{d}}\hat{f}_u$ induces a linear injection Coker $\Phi_{B_1,B_2} \to E_{\overline{c},\overline{d}}$. But Coker Φ_{B_1,B_2} has dimension $\delta r_1 r_2$, as follows from Birkhoff-Guenther Normal Form, or actually any of the normal forms obtained in 15.1.2. □

The dimension of Coker Φ_{B_1,B_2} could also be computed with arguments from index theory (as in Chapter 6).

15.1.5. The "additive" case: Stokes cocycles. In 15.1.4, we have "measured" the ambiguity of summation by selecting two particular authorized directions $\overline{c}, \overline{d} \in \mathbf{E}_q \setminus \Sigma_{A_0}$, $\overline{c} \neq \overline{d}$. We now describe a related procedure, but one that does not require such arbitrary choices. Moreover, it leads to a *cohomological* description, thus offering a wealth of powerful tools and results.

Again we use notations A_i, μ_i, B_i, $i = 1, 2$, A_0, δ, Σ_{A_0}, ...; together with the related assumptions,and conventions. We also recall the algebraic group $\mathcal{G}_{A_0}$ and the open covering $\mathfrak{U}$ by the $U_{\overline{c}} := \mathbf{E}_q \setminus \{\overline{c}\}$, $\overline{c} \in \mathbf{E}_q \setminus \Sigma_{A_0}$. Note that any two open sets of the family suffice to cover $\mathbf{E}_q$, so it is rather redundant as a covering; but it is more symmetrical than $(U_{\overline{c}}, U_{\overline{d}})$. Although we shall not explicitly use that fact[8], it is important that $\mathbf{E}_q$ is actually a projective algebraic curve and that we cover it by Zariski open subsets (i.e., complements of finite subsets): so a more algebro-geometric approach would be possible.

15.1.5.1. *The sheaves* $\Lambda_I(A_0)$ *and* $\lambda_I(A_0)$. We shall introduce two sheaves of groups. In the present additive case, they are abelian and isomorphic (although one is written additively, the other multiplicatively) but in the general case, in section 15.2, the multiplicative one will no longer be abelian (it will however be unipotent, which is not too bad). So, A_0 being given, for every open subset $U \subset \mathbf{E}_q$, we set[9]:

$$H^0(U, \lambda_I(A_0)) := \{f \in \mathrm{Mat}_{r_1,r_2}\left(\mathscr{O}(\pi^{-1}(U))\right) \mid (\sigma_q f)(x^{\mu_2}A_2) = (x^{\mu_1}A_1)f\},$$

$$H^0(U, \Lambda_I(A_0)) := \left\{F_f = \begin{pmatrix} I_{r_1} & f \\ 0 & I_{r_2} \end{pmatrix} \mid f \in H^0(U, \lambda_I(A_0))\right\}.$$

The map $f \mapsto F_f$ is an isomorphism from the former to the later, which we express concisely by $\Lambda_I(A_0) = \begin{pmatrix} I_{r_1} & \lambda_I(A_0) \\ 0 & I_{r_2} \end{pmatrix}$. Actually, here and in the general case, $\Lambda_I(A_0)$ is a sheaf of unipotent algebraic groups and $\lambda_I(A_0)$ is the sheaf of their Lie algebras.

Moreover, according to 9.3, $\lambda_I(A_0)$ is a locally free sheaf of modules over $\mathbf{E}_q$, hence the sheaf of sections of a holomorphic vector bundle over $\mathbf{E}_q$. (According to *loc. cit*, it is the vector bundle associated to $\mathrm{End}^{>}(A_0)$.) This bundle can be geometrically described as follows:

$$\frac{\mathbf{C}^* \times \mathrm{Mat}_{r_1,r_2}(\mathbf{C})}{(x, f) \sim (qx, (x^{\mu_1}A_1)f(x^{\mu_2}A_2)^{-1})} \longrightarrow \frac{\mathbf{C}^*}{x \sim qx} = \mathbf{E}_q.$$

---

[8]We use it indirectly to describe the cohomology on the Riemann surface $\mathbf{E}_q$ through a "GAGA theorem" in Appendix H.

[9]From now on, due to the complicated names of our sheaves of main interest, we shall write $H^0(U, \mathscr{F})$ instead of $\mathscr{F}(U)$, when $\mathscr{F}$ is a sheaf and U an open set.

Recall from 9.3 how the local freeness of the sheaf and the local triviality of the bundle are realized: choose U so that $\pi^{-1}(U)$ is the disjoint union of the $q^m V_0$, $m \in \mathbf{Z}$, when V_0 maps homeomorphically to U. Then $f \in H^0(U, \lambda_I(A_0))$ is totally determined by its restriction to V_0, which can be chosen freely in $\mathrm{Mat}_{r_1,r_2}(\mathscr{O}_{\mathbf{C}^*}(V_0))$, i.e. in $\mathrm{Mat}_{r_1,r_2}(\mathscr{O}_{\mathbf{E}_q}(U))$. In other words, $\lambda_I(A_0)$ is locally isomorphic to the free sheaf (or trivial bundle) $\mathrm{Mat}_{r_1,r_2}(\mathscr{O}_{\mathbf{E}_q})$.

15.1.5.2. Returning to A_u and the associated families $(f_{\overline{c},\overline{d}})$ and $(F_{\overline{c},\overline{d}})$, we see that:

- $\forall \overline{c}, \overline{d} \in \mathbf{E}_q \setminus \Sigma_{A_0}$, $f_{\overline{c},\overline{d}} \in H^0(U_{\overline{c},\overline{d}}, \lambda_I(A_0))$ and $F_{\overline{c},\overline{d}} \in H^0(U_{\overline{c},\overline{d}}, \Lambda_I(A_0))$ (recall that $U_{\overline{c},\overline{d}} := U_{\overline{c}} \cap U_{\overline{d}}$).
- $\forall \overline{c}, \overline{d}, \overline{e} \in \mathbf{E}_q \setminus \Sigma_{A_0}$, $f_{\overline{c},\overline{d}} + f_{\overline{d},\overline{e}} = f_{\overline{c},\overline{e}}$ and $F_{\overline{c},\overline{d}} F_{\overline{d},\overline{e}} = F_{\overline{c},\overline{e}}$.

So we have defined cocycles of those sheaves relative to the covering $\mathfrak{U}$:

$$(f_{\overline{c},\overline{d}}) \in Z^1(\mathfrak{U}, \lambda_I(A_0)) \text{ and } (F_{\overline{c},\overline{d}}) \in Z^1(\mathfrak{U}, \Lambda_I(A_0)).$$

Those cocycles are very particular, in that, for instance, $f_{\overline{c},\overline{d}}$ is not merely holomorphic over $\pi^{-1}(U_{\overline{c},\overline{d}})$: it is meromorphic over $\mathbf{C}^*$ and satisfies the polar condition $\mathrm{div}_{\mathbf{E}_q}(f) \geq -\delta([-c] + [-d])$ (so this is also a condition on the upper right block of $F_{\overline{c},\overline{d}}$). We say they are *privileged cocycles* and write:

$$(f_{\overline{c},\overline{d}}) \in Z^1_{pr}(\mathfrak{U}, \lambda_I(A_0)) \text{ and } (F_{\overline{c},\overline{d}}) \in Z^1_{pr}(\mathfrak{U}, \Lambda_I(A_0)),$$

where:

$$Z^1_{pr}(\mathfrak{U}, \lambda_I(A_0)) := \{(f_{\overline{c},\overline{d}}) \in Z^1(\mathfrak{U}, \lambda_I(A_0)) \mid \forall \overline{c}, \overline{d} \in \mathbf{E}_q \setminus \Sigma_{A_0} \text{ , } f_{\overline{c},\overline{d}} \in E_{\overline{c},\overline{d}}\}$$

(spaces $E_{\overline{c},\overline{d}}$ were defined in 15.1.4.1), and a similar definition for $Z^1_{pr}(\mathfrak{U}, \Lambda_I(A_0))$ involving the upper right blocks.

EXERCISE 15.41. Check that $Z^1_{pr}(\mathfrak{U}, \lambda_I(A_0))$ is a $\mathbf{C}$-linear subspace of $Z^1(\mathfrak{U}, \lambda_I(A_0))$ and that $Z^1_{pr}(\mathfrak{U}, \Lambda_I(A_0))$ is a subgroup of $Z^1(\mathfrak{U}, \Lambda_I(A_0))$.

LEMMA 15.42. *The map $cl(A_u) \mapsto (f_{\overline{c},\overline{d}})$ is a linear isomorphism $\mathscr{F}(A_0) \to Z^1_{pr}(\mathfrak{U}, \lambda_I(A_0))$.*

PROOF. This is immediate from Proposition 15.40. The linear structure on $\mathscr{F}(A_0)$ is either that induced by Coker Φ_{B_1,B_2} or that of any of the $E_{\overline{c},\overline{d}}$. □

The canonical equivalence relation for cocycles in $Z^1(\mathfrak{U}, \lambda_I(A_0))$ (leading to the definition of the cohomology space) is the congruence modulo the subspace of special cocycles, the coboundaries, i.e., those families $(f_{\overline{c},\overline{d}})$ such that there exists a family $(\phi_{\overline{c}})$, each $f_{\overline{c}} \in H^0(U_{\overline{c}}, \lambda_I(A_0))$ (for all $\overline{c} \notin \Sigma_{A_0}$) such that $f_{\overline{c},\overline{d}} = \phi_{\overline{d}} - \phi_{\overline{c}}$ (for all $\overline{c}, \overline{d} \notin \Sigma_{A_0}$). The space of coboundaries is denoted $B^1(\mathfrak{U}, \lambda_I(A_0))$ and the corresponding cohomology space is:

$$H^1(\mathfrak{U}, \lambda_I(A_0)) := Z^1(\mathfrak{U}, \lambda_I(A_0))/B^1(\mathfrak{U}, \lambda_I(A_0)).$$

The cohomology "space" (the quotation marks will be explained soon) for the sheaf of multiplicative groups $\Lambda_I(A_0)$ is defined differently, although here, in the additive case, it will turn out to be obviously the same. We say that two cocycles $(F_{\overline{c},\overline{d}})$ and $(G_{\overline{c},\overline{d}})$ in $Z^1(\mathfrak{U}, \Lambda_I(A_0))$ are equivalent if there is a family $(\Phi_{\overline{c}})$, each $\Phi_{\overline{c}} \in H^0(U_{\overline{c}}, \Lambda_I(A_0))$ (for all $\overline{c} \notin \Sigma_{A_0}$) such that:

$$\forall \overline{c}, \overline{d} \in \mathbf{E}_q \setminus \Sigma_{A_0} \text{ , } G_{\overline{c},\overline{d}} = (\Phi_{\overline{c}})^{-1} F_{\overline{c},\overline{d}} \Phi_{\overline{d}}.$$

EXERCISE 15.43. Check directly that this is an equivalence relation. Then check that it corresponds to the one on $Z^1(\mathfrak{U}, \lambda_I(A_0))$ through the isomorphism $f \mapsto F_f$.

In the general (nonadditive) case, the resulting quotient set:

$$H^1(\mathfrak{U}, \Lambda_I(A_0)) := Z^1(\mathfrak{U}, \Lambda_I(A_0))/(\text{above equivalence relation})$$

has no natural structure of group or linear space[10]. It just has a distinguished element, the class of the trivial cocycle (all $F_{\overline{c},\overline{d}} = I_n$), so it is a pointed set. But here, in the additive case, it may be identified with the $\mathbf{C}$-linear space $H^1(\mathfrak{U}, \lambda_I(A_0))$.

At any rate, composing the maps from $\mathscr{F}(A_0)$ to $Z^1(\mathfrak{U}, \lambda_I(A_0))$ and to $Z^1(\mathfrak{U}, \Lambda_I(A_0))$ with the projection maps from cocycles to cohomology, we obtain two maps from $\mathscr{F}(A_0)$ to $H^1(\mathfrak{U}, \lambda_I(A_0))$ and to $H^1(\mathfrak{U}, \Lambda_I(A_0))$, the first one sending $\mathrm{cl}A_u$ to the cohomology class of the cocycle $(S_{\overline{c},\overline{d}}\hat{f}_u)$. From now on in the present section on the additive case, we shall concentrate on the sheaf $\lambda_I(A_0))$ since anyhow everything immediately transposes to its multiplicative avatar.

PROPOSITION 15.44. *The map $Z^1_{pr}(\mathfrak{U}, \lambda_I(A_0)) \to H^1(\mathfrak{U}, \lambda_I(A_0))$ is injective.*

PROOF. We have a morphism of groups (indeed of linear spaces), so let $(f_{\overline{c},\overline{d}})$ a privileged cocycle in the kernel: by definition, this means that there is a family $(\phi_{\overline{c}})$ such that each $\phi_{\overline{c}} \in H^0(U_{\overline{c}}, \lambda_I(A_0))$ and each $f_{\overline{c},\overline{d}} = \phi_{\overline{d}} - \phi_{\overline{c}}$. Set for notational simplicity $L(f) := (\sigma_q f)(x^{\mu_2} A_2) - (x^{\mu_1} A_1) f$. By definition, all $L(\phi_{\overline{c}}) = 0$.

On the other hand, from the previous lemma, there is an $u \in \mathbf{C}(\{x\})$ such that, denoting $f_{\overline{c}} := S_{\overline{c}}\hat{f}_u$, we have $f_{\overline{c},\overline{d}} = f_{\overline{d}} - f_{\overline{c}}$ for all $\overline{c}, \overline{d}$. From $f_{\overline{c},\overline{d}} = \phi_{\overline{d}} - \phi_{\overline{c}}$ we thus draw that all $f_{\overline{c}} - \phi_{\overline{c}}$ are equal, so have no poles in $(\mathbf{C}^*, 0)$ i.e., they belong to $\mathscr{O}(\mathbf{C}^*, 0)$. Call f their common value. We have $L(f) = L(f_{\overline{c}} - \phi_{\overline{c}}) = L(f_{\overline{c}}) = u$ by construction. But we know that $L(f) \in \mathbf{C}(\{x\})$ for $f \in \mathscr{O}(\mathbf{C}^*, 0)$ actually implies $f \in \mathbf{C}(\{x\})$. But then all $f_{\overline{c}} = S_{\overline{c}}\hat{f}_u = f$, so $(f_{\overline{c},\overline{d}})$ is trivial. □

Actually, we shall next see that the above map is bijective.

15.1.5.3. We use here general facts from Section 10.3 and Appendix I.

For every open covering $\mathfrak{V} := (V_\alpha)$ of $\mathbf{E}_q$, we can define (exactly as above for $\mathfrak{U}$) a space $Z^1(\mathfrak{V}, \lambda_I(A_0))$ of cocycles, a subspace $B^1(\mathfrak{V}, \lambda_I(A_0))$ of coboundaries and a quotient space $H^1(\mathfrak{V}, \lambda_I(A_0))$ of cohomology classes. Then, for every refinement $\mathfrak{W} := (W_\beta)$ of $\mathfrak{V}$ (i.e., an open covering such that each W_β is included in some specified $V_{s(\beta)}$) there is a natural linear map $Z^1(\mathfrak{V}, \lambda_I(A_0)) \to Z^1(\mathfrak{W}, \lambda_I(A_0))$ sending $B^1(\mathfrak{V}, \lambda_I(A_0))$ into $B^1(\mathfrak{W}, \lambda_I(A_0))$ thus inducing $H^1(\mathfrak{V}, \lambda_I(A_0)) \to H^1(\mathfrak{W}, \lambda_I(A_0))$. The open coverings form a right directed system and the direct limit is the cohomology space:

$$H^1(\mathbf{E}_q, \lambda_I(A_0)) := \lim_{\longrightarrow \mathfrak{V}} H^1(\mathfrak{V}, \lambda_I(A_0)).$$

What was just done for $\mathbf{E}_q$ and its open coverings can be done for every Riemann surface, in particular for every open subset U of $\mathbf{E}_q$; for such an open subset, we shall thus write $H^1(U, \lambda_I(A_0))$, tacitly understanding that $\lambda_I(A_0)$ denotes here the restriction to U of the said sheaf (that restriction is obviously also locally free). From the "general facts" mentioned above:

[10]We shall however prove later, in Section 15.3, that it has a kind of geometrical structure and deserves to be called a space; but this does not come from general cohomological nonsense.

(1) Each proper open subset of $\mathbf{E}_q$ being an open Riemann surface,

$$H^1(U, \lambda_I(A_0))$$

is trivial.

(2) The covering $\mathfrak{U}$ being acyclic by the first point, the natural map

$$H^1(\mathfrak{U}, \lambda_I(A_0)) \to H^1(\mathbf{E}_q, \lambda_I(A_0))$$

is an isomorphism.

Therefore, by Proposition 15.44, we have an injective linear map $\mathscr{F}(A_0) \to H^1(\mathbf{E}_q, \lambda_I(A_0))$.

THEOREM 15.1. *That map is an isomorphism* $\mathscr{F}(A_0) \simeq H^1(\mathbf{E}_q, \lambda_I(A_0))$.

PROOF. It is $\mathbf{C}$-linear, injective, and the spaces have the same dimension $r_1 r_2(\mu_1 - \mu_2)$ by the following lemma, applied to the q-difference module $M := \mathrm{Hom}(M_2, M_1)$, where $M_i := (K a^{r_i}, \Phi_{B_i})$ is pure isoclinic of slope μ_i, rank r_i, $i = 1, 2$, so that M is pure isoclinic of slope $\mu_1 - \mu_2$, rank $r_1 r_2$ (and by 15.1.4 as regards $\dim_{\mathbf{C}} \mathscr{F}(A_0)$). □

LEMMA 15.45. *Let M a pure isoclinic q-difference module of rank r and slope $\mu \neq 0$. Let $\mathscr{F}_M$ the asociated sheaf of solutions. Then* $\dim_{\mathbf{C}} H^1(\mathbf{E}_q, \mathscr{F}_M) = -r\mu$.

PROOF. We do the proof for an integral slope and take $M := (\mathbf{C}(\{x\})^r, \Phi_{x^\mu A})$. We know that $\dim_{\mathbf{C}} H^0(\mathbf{E}_q, \mathscr{F}_M) = 0$ because the equation $\sigma_q X = x^\mu A X$, $A \in \mathrm{GL}_r(\mathbf{C})$ has no nontrivial solution over $\mathbf{C}(\{x\})$ (by the usual power series expansion argument). So for such modules $-\dim_{\mathbf{C}} H^1$ is the Euler characteristic $\chi(\mathbf{E}_q, \mathscr{F}_M)$. But the latter is additive with respect to exact sequences. So we use the triangularisation of A in $\mathrm{GL}_r(\mathbf{C})$ to reduce the proof to the case that $r = 1$, $A = a \in \mathbf{C}^*$. Then the bundle $\mathscr{F}_M$ has the meromorphic section $\theta_q(x/b)^{-\mu}$, where b is any μth root of a. The degree of the bundle is the degree of the divisor of that section. The divisor is $\mu[\overline{-b}]$ (since θ_q has a simple zero and no pole), it has degree μ, which is also the Euler characteristic of the bundle. □

EXERCISE 15.46. Try to generalize the conclusion of the lemma to the case of an arbitrary slope.

15.2. The "general" case of k slopes: normal forms, basic tools

In sections 15.2, 15.3 and 15.4, we shall almost always assume the slopes to be integral. Most results come from [**197, 219**]. For information on the more complicated case of arbitrary slopes, the reader should look at [**248**] (for the formal study) and [**223, 224**] (for the analytic study); and also to [CV].

15.2.1. Conventions, notations, basics.

We group here notations and facts of general use and of a preliminary or purely algebraic nature.

15.2.1.1. We let all along $\mu_1 > \cdots > \mu_k$ the slopes (a priori in $\mathbf{Q}$) and $r_1, \ldots, r_k \in \mathbf{N}^*$ their multiplicities, whence a total rank $n := r_1 + \cdots + r_k$. Indices i, j are meant to run from 1 to k, most of the time with $1 \leq i < j \leq k$. We let P_i be fixed pure isoclinic modules of rank r_i, slope μ_i; we usually take them in the form $(\mathbf{C}(\{x\})^{r_i}, \Phi_{B_i})$. When the slopes are integral, we moreover let $B_i := x^{\mu_i} A_i$, $A_i \in \mathrm{GL}_{r_i}(\mathbf{C})$. We take up our (by now) standard notations and generalize them:

$$A_0 := \begin{pmatrix} B_1 & \dots & \dots & \dots & \dots \\ \dots & \dots & \dots & 0 & \dots \\ 0 & \dots & \dots & \dots & \dots \\ \dots & 0 & \dots & \dots & \dots \\ 0 & \dots & 0 & \dots & B_k \end{pmatrix} \text{ and } A_U := \begin{pmatrix} B_1 & \dots & \dots & \dots & \dots \\ \dots & \dots & \dots & U_{i,j} & \dots \\ 0 & \dots & \dots & \dots & \dots \\ \dots & 0 & \dots & \dots & \dots \\ 0 & \dots & 0 & \dots & B_k \end{pmatrix},$$

where $U \in \prod_{1 \le i < j \le k} \mathrm{Mat}_{r_i,r_j}(\mathscr{O}(\mathbf{C}^*, 0))$ (actually, most of the time

$$\prod \mathrm{Mat}_{r_i,r_j}(\mathbf{C}(\{x\}))).$$

We introduce a *unipotent algebraic group*:

$$\mathcal{G}_{A_0} := \begin{pmatrix} I_{r_1} & \dots & \dots & \dots & \dots \\ \dots & \dots & \dots & \star & \dots \\ 0 & \dots & \dots & \dots & \dots \\ \dots & 0 & \dots & \dots & \dots \\ 0 & \dots & 0 & \dots & I_{r_k} \end{pmatrix}.$$

This means that, for any commutative $\mathbf{C}$-algebra R, the group $\mathcal{G}_{A_0}(R)$ of *R-valued points of $\mathcal{G}_{A_0}$* is the subgroup of $\mathrm{GL}_n(R)$ consisting of matrices:

$$F_f := \begin{pmatrix} I_{r_1} & \dots & \dots & \dots & \dots \\ \dots & \dots & \dots & f_{i,j} & \dots \\ 0 & \dots & \dots & \dots & \dots \\ \dots & 0 & \dots & \dots & \dots \\ 0 & \dots & 0 & \dots & I_{r_k} \end{pmatrix},$$

where, $f := (f_{i,j})$ runs in $\prod_{1 \le i < j \le k} \mathrm{Mat}_{r_i,r_j}(R)$. The fact that it is defined inside Mat_n by polynomial equations on the coefficients is obvious. To see that it is a group, we use the fact that $F_f F_g = F_h$ and $F_f^{-1} = F_{f'}$, where:

(15.1)

$$h_{i,j} := f_{i,j} + \sum_{i<l<j} f_{i,l} g_{l,j} + g_{i,j}, \quad f'_{i,j} := -f_{i,j} + \sum_{i<l<j} f_{i,l} f_{l,j} - \sum_{i<l<m<j} f_{i,l} f_{l,m} f_{m,j} + \cdots$$

(For the inverse, write $F_f = I_n + N$ and compute it as $F_f^{-1} := I_n - N + N^2 - \cdots$.) We note for future use the following consequence. If the R_d, $d \in \mathbf{N}^*$, are subgroups of R such that $R_d R_e \subset R_{d+e}$, then

$$\{F_f \in \mathcal{G}_{A_0}(R) \mid \forall i,j \ , \ f_{i,j} \in R_{\mu_i - \mu_j}\}$$

is a subgroup of $\mathcal{G}_{A_0}(R)$. We shall apply that in particular with $R := \mathscr{M}(\mathbf{C}^*, 0)$ and, for some $a \in \mathbf{C}^*$ (fixed), R_d is the set of functions with at most poles of order d on $[a; q]$ (near 0).

15.2.1.2. The algebraic group $\mathcal{G}_{A_0}$ being unipotent, we have use for its Lie algebra (see Section F.4 in Appendix F):

$$\mathfrak{g}_{A_0} := \mathrm{Lie}(\mathcal{G}_{A_0}) = \begin{pmatrix} 0_{r_1} & \dots & \dots & \dots & \dots \\ \dots & \dots & \dots & \star & \dots \\ 0 & \dots & \dots & \dots & \dots \\ \dots & 0 & \dots & \dots & \dots \\ 0 & \dots & 0 & \dots & 0_{r_k} \end{pmatrix},$$

meaning that for every commutative $\mathbf{C}$-algebra R, the Lie algebra $\mathfrak{g}_{A_0}(R)$ is the space of matrices:

$$N_g = \begin{pmatrix} 0_{r_1} & \dots & \dots & \dots & \dots \\ \dots & \dots & \dots & g_{i,j} & \dots \\ 0 & \dots & \dots & \dots & \dots \\ \dots & 0 & \dots & \dots & \dots \\ 0 & \dots & 0 & \dots & 0_{r_k} \end{pmatrix},$$

where $N_g := F_g - I_n$, all $g_{i,j} \in \mathrm{Mat}_{r_i,r_j}(R)$.

EXERCISE 15.47. (i) Give formulas similar to those of (15.1) to relate f and g when $F_f = \exp(N_g)$ and $N_g = \log(F_f)$. (We want formulas going both ways.)
(ii) What can be said about the precise value of nilpotency index of N_g?

Since $R = \mathscr{M}(\mathbf{C}^*, 0)$, as a $\mathbf{C}$-algebra, contains $\mathbf{Q}$, we have a group isomorphism $\exp : \mathfrak{g}_{A_0}(R) \to \mathcal{G}_{A_0}(R)$ (also a bijection $N \mapsto I_n + N$). We write $\log : \mathcal{G}_{A_0}(R) \to \mathfrak{g}_{A_0}(R)$ the inverse of exp. The formulas (15.1) at the end of 15.2.1.1 entail similar consequences: the subset of $\mathfrak{g}_{A_0}(R)$ of matrices such that all $g_{i,j} \in R_{\mu_i - \mu_j}$ form a sub Lie algebra and it is mapped bijectively to the analogous subgroup of $\mathcal{G}_{A_0}(R)$ either by exp or by $N \mapsto I_n + N$.

15.2.1.3. *Filtration and graduation by upper diagonals.* We shall just state the following facts, the verification of which is easy using the formulas (15.1).

EXERCISE 15.48. Check them all (define morphisms when appropriate).

Let $\delta \in \mathbf{N}^*$ and set, on the one hand:

$$\mathcal{G}_{A_0}^{\geq \delta}(R) := \{F_f \in \mathcal{G}_{A_0}(R) \mid f_{i,j} = 0 \text{ whenever } 0 < \mu_i - \mu_j < \delta\}$$

and:

$$\mathcal{G}_{A_0}^{> \delta}(R) := \{F_f \in \mathcal{G}_{A_0}(R) \mid f_{i,j} = 0 \text{ whenever } 0 < \mu_i - \mu_j \leq \delta\},$$

so that $\mathcal{G}_{A_0}^{>\delta}(R) = \bigcup_{\delta' > \delta} \mathcal{G}_{A_0}^{\geq \delta'}(\mathrm{R})$; and, on the other hand a graduation of $\mathfrak{g}_{A_0}$:

$$\mathfrak{g}_{A_0}^{(\delta)}(R) := \{N_g \in \mathfrak{g}_{A_0}(R) \mid g_{i,j} = 0 \text{ whenever } \mu_i - \mu_j \neq \delta\}$$

and:

$$\mathfrak{g}_{A_0}^{\geq \delta}(R) := \{N_g \in \mathfrak{g}_{A_0}(R) \mid g_{i,j} = 0 \text{ whenever } \mu_i - \mu_j < \delta\},$$

so that $\mathfrak{g}_{A_0}^{\geq \delta}(R) = \bigoplus_{\delta' \geq \delta} \mathfrak{g}_{A_0}^{(\delta')}(R)$.

The $\mathcal{G}_{A_0}^{\geq \delta}$ form a decreasing sequence of normal subgroups of $\mathcal{G}_{A_0}$ and we have two natural exact sequences associated to those subgroups:

$$1 \to \mathcal{G}_{A_0}^{>\delta} \to \mathcal{G}_{A_0}^{\geq \delta} \to \mathfrak{g}_{A_0}^{(\delta)} \to 0,$$

where $F_f \in \mathcal{G}_{A_0}^{\geq \delta}(R)$ is sent to $N_g \in \mathfrak{g}_{A_0}^{(\delta)}(R)$, where $g_{i,j} := f_{i,j}$ whenever $\mu_i - \mu_j = \delta$; and:

$$0 \to \mathfrak{g}_{A_0}^{(\delta)} \to \mathcal{G}_{A_0}/\mathcal{G}_{A_0}^{>\delta} \to \mathcal{G}_{A_0}/\mathcal{G}_{A_0}^{\geq \delta} \to 1,$$

where $N_g \in \mathfrak{g}_{A_0}^{(\delta)}(R)$ is sent to the class of either $I_n + N_g$ or $\exp N_g$. Indeed,

$$\exp N_g/(I_n + N_g) = I_n + N_g^2/2 + \cdots \in \mathcal{G}_{A_0}^{>\delta}(R).$$

There is a further important property of the second exact sequence: it is a *central extension*, i.e., $\mathfrak{g}_{A_0}^{(\delta)}(R)$ embeds into the center of $\mathcal{G}_{A_0}(R)/\mathcal{G}_{A_0}^{>\delta}(R)$.

15.2.2. Formal facts.

15.2.2.1. As already noted in 10.3, the general theory of the slope filtration and associated graduation entails that, for every A_U, A_V in $\mathscr{E}(A_0)$, i.e., for every $(U_{i,j}), (V_{i,j})$ in $\prod_{1 \leq i<j \leq k} \mathrm{Mat}_{r_i,r_j}(\mathbf{C}(\{x\}))$, there is a unique $F_f \in \mathcal{G}_{A_0}(\mathbf{C}((x)))$ such that $F_f[A_U] = A_V$. This can be checked by direct computation; $F_f[A_U] = A_V \Leftrightarrow (\sigma_q F_f)A_U = A_V F$, which, in turn, is equivalent to:

$$\forall i,j \text{ s.t. } 1 \leq i<j \leq k, U_{i,j} + \sum_{i<l<j} (\sigma_q f_{i,l})U_{l,j} + (\sigma_q f_{i,j})B_j = B_i f_{i,j} + \sum_{i<l<j} V_{i,l} f_{l,j} + V_{i,j},$$

then to:

$$\forall i,j \text{ s.t. } 1 \leq i<j \leq k\,,\ \Phi_{B_i,B_j}(f_{i,j}) = V_{i,j} - U_{i,j} + \sum_{i<l<j} V_{i,l} f_{l,j} - \sum_{i<l<j} (\sigma_q f_{i,l})U_{l,j}.$$

Recall from 10.3.5 that Φ_{B_i,B_j} is a bijection from $\mathrm{Mat}_{r_i,r_j}(\mathbf{C}((x)))$ to itself. For $j = i+1$, both sums $\sum_{i<l<j}$ vanish and we solve directly $f_{i,j} := \Phi^{-1}_{B_i,B_j}(V_{i,j} - U_{i,j})$. Then we solve iteratively $f_{i,j} := \Phi^{-1}_{B_i,B_j}(\text{right-hand side})$ for increasing values of $j-i$. Since $l-i, j-l < j-i$, the induction is well founded.

Let us denote $\widehat{F}_{U,V}$ for the F_f defined above and in particular $\widehat{F}_U := \widehat{F}_{0,U}$. From the composition $A_U \overset{\widehat{F}_{U,V}}{\to} A_V \overset{\widehat{F}_{V,W}}{\to} A_W$, we immediately deduce that $\widehat{F}_{U,W} = \widehat{F}_{V,W}\widehat{F}_{U,V}$, and, as a consequence, the particular case $\widehat{F}_{U,V} = \widehat{F}_V(\widehat{F}_U)^{-1}$.

One application (not the main one) of $\widehat{F}_U$ is the production of *formal fundamental matricial solutions* for the systems $\sigma_q X = A_U X$ more or less in the style of the Founding Fathers (who, of course, rather did that for scalar equations). For instance, solving $\sigma_q \mathcal{Y}_i = B_i \mathcal{Y}_i$, $i = 1, \ldots, k$, can be done by taking $\mathcal{Y}_i := \theta_q^{-\mu_i} e_{A_i}$; then $\sigma_q \mathcal{X}_O = A_0 \mathcal{X}_0$ by taking $\mathcal{X}_0 := \mathrm{Diag}(\mathcal{Y}_1, \ldots, \mathcal{Y}_k)$; last $\sigma_q \mathcal{X} = A_U \mathcal{X}$ by taking $\mathcal{X} := \widehat{F}_U \mathcal{X}_0$.

15.2.2.2. *Order of growth of the coefficients of* $\widehat{F}_u$. The q-Gevrey orders of growth we are going to compute could probably be obtained using the results of Chapter 6 and maybe even in a sharper form (with, however, a need for adaptation, since the series considered there are entire while here they are meromorphic at 0). However, we shall sketch a direct self-contained approach, leaving to the reader the pleasure of providing the details.

First, we introduce the notion of *family of q-Borel-Ramis coefficients of q-Gevrey order* $\delta \geq 1$. This is a bilateral sequence $(t_n)_{n\in\mathbf{Z}} \in \mathbf{C}^{*\mathbf{Z}}$ such that:

$$\forall n \in \mathbf{Z};\ t_{n+\delta} = q^{n+\delta} t_n.$$

Such a family is given by setting $t_n := q^{n(n+\delta)/2\delta}$; another one by taking for the t_n the coefficients of the Laurent expansion of $(x^{-1}\theta_q)^\delta$ (provided they are all $\neq 0$; a discussion of this problem appears in **[225]**, also see **[224]**). One may also use the coefficients of $\theta_{q'}$, where $q'^\delta = q$. At any rate, for any to such families (t_n), (t'_n), the sequence (t'_n/t_n) is δ-periodic, so $t'_n = O(t_n)$ and $t_n = O(t'_n)$ when $n \to \pm\infty$.

We select a particular family and, if we want to stress the relation to δ, we denote it $(t_n^{(\delta)})$. by analogy with notations from 6.2.3, setting $s := 1/\delta$, we say that:

$$f := \sum f_n x^n \in \mathbf{C}((x))_{q,s} \text{ if } f_n = O(R^n/t_n) \text{ for some } R > 0,$$

$$g := \sum g_n x^n \in \mathbf{C}((x))_{q,(s)} \text{ if } g_n = O(R^n/t_n) \text{ for every } R > 0.$$

Clearly, if $s' < s$, i.e., $\delta' > \delta$, then $\mathbf{C}((x))_{q,s} \subset \mathbf{C}((x))_{q,(s')}$.

LEMMA 15.49. *Let $c \in \mathbf{C}^*$, $\delta \in \mathbf{N}^*$. If $cx^{-\delta}\sigma_q f - f = g$ and $g \in \mathbf{C}((x))_{q,(s)}$, then $f \in \mathbf{C}((x))_{q,s}$.*

PROOF. We write $\tilde{f}_n := t_n f_n$, $\tilde{g}_n := t_n g_n$, $\tilde{f}(\xi) := \sum \tilde{f}_n \xi^n$ and $\tilde{g}(\xi) := \sum \tilde{g}_n \xi^n$. Then the functional equation translates into:

$$\forall n \in \mathbf{Z}\ ,\ cq^{n+\delta} f_{n+\delta} - f_n = g_n \Longleftrightarrow \forall n \in \mathbf{Z}\ ,\ c\tilde{f}_{n+\delta} - \tilde{f}_n$$
$$= \tilde{g}_n \Longleftrightarrow (cx^{-\delta} - 1)\tilde{f}(\xi) = \tilde{g}(\xi).$$

Assuming $g \in \mathbf{C}((x))_{q,(s)}$, we see that $\tilde{g}$ is entire and the obvious following necessary condition makes sense:

$$\forall d \in \mathbf{C}^*\ ,\ d^\delta = c \Longrightarrow \tilde{g}(d) = 0.$$

If that condition is realized, the same kind of estimate as in 15.1.2 entails that, setting $\tilde{f}(\xi) := \tilde{g}(\xi)/((cx^{-\delta} - 1))$, expanding it into a Laurent series $\sum \tilde{f}_n \xi^n \in \mathbf{C}(\{\xi\}) \cap \mathscr{O}(\mathbf{C}^*)$ then taking $f_n := \tilde{f}_n / t_n$, we draw that $f \in \mathbf{C}((x))_{q,(s)}$.

If the vanishing condition does not hold from start, we can enforce it by substracting from $\tilde{g}$, hence from g, an interpolation polynomial of degree $\leq \delta - 1$. So, invoking $\mathbf{C}$-linearity and the obvious inclusions of $\mathbf{C}$-linear spaces $\mathbf{C}[x] \subset \mathbf{C}((x))_{q,(s)} \subset \mathbf{C}((x))_{q,s}$, we just have to deal with the case where $g = x^m$ for some $m \in \{0, \dots, \delta - 1\}$. But then it follows from explicit calculation. □

COROLLARY 15.50. *The same inference is valid for vectors $f, g \in \mathbf{C}((x))^r$ and c replaced by $C \in GL_r(\mathbf{C})$.*

PROOF. Up to replacing C by PCP^{-1}, $P \in \mathrm{GL}_r(\mathbf{C})$, and f, g by Pf, Pg, we may assume that C is upper triangular. Then the standard recursive resolution of triangular systems reduces the problem to the previous lemma. □

PROPOSITION 15.51. *Let $\widehat{F}_u = F_f$ (notations from 15.2.2.1) and, for $1 \leq i < j \leq k$, set $\delta_{i,j} := \mu_i - \mu_j$. Then all coefficients of $f_{i,j}$ are $O(R^n/t_n^{(\delta_{i,j})})$ for some $R > 0$.*

PROOF. For $\delta' < \delta$, we have the implication:

$$a_n = O(R^n/t_n^{(\delta')}) \text{ for some } R > 0 \Longrightarrow a_n = O(R^n/t_n^{(\delta)}) \text{ for all } R > 0.$$

Using that fact and the recursive relations $f_{i,j} := \Phi_{B_i,B_j}^{-1}(\text{right-hand side})$ (as in 15.2.2.1), which can take the form of the corollary (replacing $r_i \times r_j$ matrices by $r_i r_j$-vectors), the statement is reduced to the corollary. □

15.2.3. Analytic isoformal classification.

15.2.3.1. We shall first formulate an abstract, intrinsic description of the problem[11]. Then in 15.2.4, we shall translate it into a matricial form fit for computation.

Let M a q-difference module over $\mathbf{C}(\{x\})$ and let $M_0 := \mathrm{gr}\ M$ its graded module with respect to the slope filtration. We know from 10.3 that the formal class of M is the same as the formal class of M_0 (this was confirmed by direct computation in 15.2.2.1). Also, since formal morphisms between pure modules are analytic by 11.1.2, giving M_0 up to analytic or formal isomorphism amounts to the same (so M_0 is, up to analytic isomorphism, the unique pure q-difference module in his formal class). Last, any q-difference module N is formally equivalent to M (and M_0) if, and only if, $\mathrm{gr}\mathbf{N} \simeq M_0$ (analytic isomorphism). So from now on we fix the pure q-difference module M_0 and see it as incarnating a formal class.

We shall denote $\mathscr{E}(M_0)$ the class[12] of all pairs (M, u), where M is a q-difference module over $\mathbf{C}(\{x\})$ and $u : \mathrm{gr}\ M \to M_0$ is an analytic isomorphism. The q-difference modules appearing in such pairs (M, u) (within $\mathscr{E}(M_0)$) are exactly those in the formal class M_0, but they somehow come with a certificate.

Instead of the plain analytic equivalence $M \simeq M'$, we define *isoformal analytic equivalence* in M_0 as follows:

$$(M, u) \sim (M', u') \underset{def}{\Longleftrightarrow} \exists f : M \to M'(\text{ analytic isomorphism}) \quad : u = u' \circ \mathrm{gr} f.$$

We then define the *space of isoformal analytic classes* as:

$$\mathscr{F}(M_0) := \frac{\mathscr{E}(M_0)}{\sim}.$$

The reader should notice the similarity with the definition of $\widetilde{\mathrm{Ext}}$ from that of Ext in Chapter 8. The idea of this "rigidification" comes straight from [**14**].

EXERCISE 15.52. (i) Check that f is automatically an isomorphism,
(ii) that $\sim$ is indeed an equivalence relation,
(iii) and, if you care for that kind of things, that $\mathscr{F}(M_0)$ is a set. (The reason to call it a *space* will appear later, see 15.2.3.3.)

15.2.3.2. Here is an alternative description of $\mathscr{F}(M_0)$, which will later allow for some kind of induction on the number of slopes. Call P_i, $i = 1, \ldots, k$, the pure isoclinic components of M_0, so that $M_0 = P_1 \oplus \cdots \oplus P_k$, $\mathrm{rank}(P_i) = r_i$ and $S(P_i) = \{\mu_i\}$ (where, as usual, $\mu_1 > \cdots > \mu_k$). Let first $\mathscr{E}(P_1, \ldots, P_k)$ the class of all pairs $(\underline{M}, \underline{u})$, where $\underline{M} := (0 \subset M_1 \subset \cdots \subset M_k)$ is a tower of q-difference modules over $\mathbf{C}(\{x\})$ and $\underline{u} := (u_1, \ldots, u_k)$, subject to relations:

$$\forall i = 2, \ldots, k\ ,\ u_i : M_i/M_{i-1} \to P_i \text{ is an analytic isomorphism, as is } u_1 : M_1 \to P_1.$$

This implies of course that $\underline{M}$ is the slope filtration of $M := M_k$ and that $P_1 \oplus \cdots \oplus P_k$ is the associated graded module. We introduce an equivalence relation $\sim$ on $\mathscr{E}(P_1, \ldots, P_k)$ as follows: $(\underline{M}, \underline{u}) \sim (\underline{M}', \underline{u}')$ if, and only if, there is a morphism $f : M := M_k \to M' := M'_k$ such that $f(M_i) \subset M'_i$, i $= 1, \ldots, k$, and the induced morphisms $\overline{f}_i : M_i/M_{i-1} \to M'_i/M'_{i-1}$, $i \geq 2$ and $\overline{f}_1 : M_1 \to M'_1$ are such that

[11] What we call here "isoformal class", should rather be called "isograded class" as will soon be clear. While we already stated heuristically that it amounts to the same, the precise equivalence of the two formulations will be detailed in 15.3.4.

[12] Defined this way, it is not a set. It could be made into one for instance by restricting (as is obviously possible) to q-difference modules having as underlying set the underlying set of M_0. But, above all, it does not really matter for what will be done here.

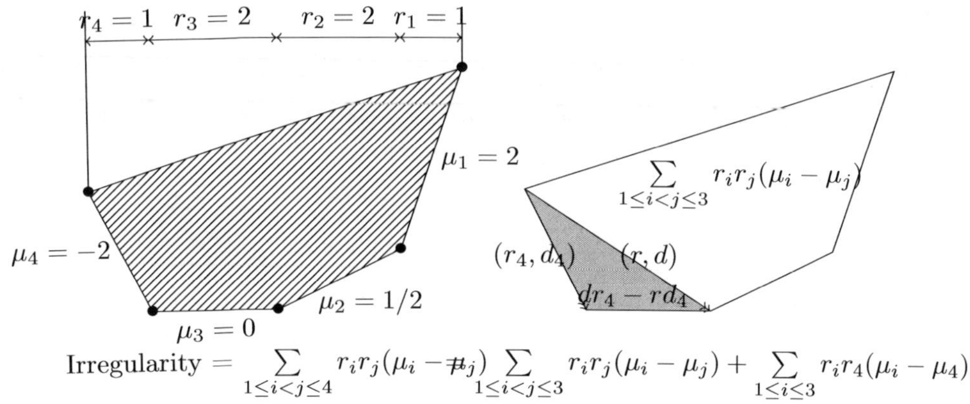

FIGURE 15.1. Reading Irregularity on the Newton polygon

$u_i = u_i' \circ \overline{f}_i$. You can check that it is indeed an equivalence relation (symmetry comes from the remark that f and the $\overline{f}_i$ are automatically isomorphisms). Then we set:

$$\mathscr{F}(P_1, \ldots, P_k) := \frac{\mathscr{E}(P_1, \ldots, P_k)}{\sim}.$$

That our previous $\mathscr{E}(M_0)$ and $\mathscr{F}(M_0)$ can be described as $\mathscr{E}(P_1, \ldots, P_k)$ and $\mathscr{F}(P_1, \ldots, P_k)$ is an easy consequence of the functoriality of slope filtration (10.2.2; recall that it was even *strict*: $f^{-1}(M_i') = M_i$). However the definition just above keeps a meaning for arbitrary $P_1, \ldots, P_k$.

15.2.3.3. *$\mathscr{F}(P_1, \ldots, P_k)$ as an affine space.* We give now a hint of why we could consider $\mathscr{F}(P_1, \ldots, P_k)$ as a *space*, indeed an affine space of dimension

$$\sum_{1\le i<j\le k} r_ir_j(\mu_i - \mu_j)$$

(see Figure 15.1). Note that this dimension is $\mathrm{Irr}(\underline{\mathrm{End}}(M_0)^{>0})$ according to the definition of $\mathrm{Irr}(M)$ in Chapter 6; since it says how big is the space of isoformal analytic classes within a formal class, it is the true measure of the defect of being pure and would deserve to be called irregularity of M_0.

EXERCISE 15.53. Check the geometric interpretation of Figure 15.1 by decomposing:

$$\sum_{1\le i<j\le k} r_ir_j(\mu_i - \mu_j) = \sum_{1\le i<j\le k-1} r_ir_j(\mu_i - \mu_j) + \sum_{1\le i\le k-1} r_ir_k(\mu_i - \mu_k)$$

and noting that the second term in the right-hand in turn decomposes as:

$$\sum_{1\le i\le k-1} (d_ir_k - r_id_k) = dr_k - rd_k, \text{ where } (r, d) := (r_1, d_1) + \cdots + (r_{k-1}, d_{k-1}),$$

so that it is the area of a triangle to be seen on the second figure (gray filling).

This is done in algebro-geometric style in **[197, 223]**, but we now propose a more down to earth, if incomplete, presentation (there will be one more after 15.2.4, and again one more in section 15.3).

Let $(\underline{M}, \underline{u})$ in $\mathscr{E}(P_1, \ldots, P_k)$, with the same notations as in 15.2.3.2. Extract from them $\underline{M}' := (0 \subset M_1 \subset \cdots \subset M_{k-1})$ and $\underline{u}' := (u_1, \ldots, u_{k-1})$, so that $(\underline{M}', \underline{u}')$ is in $\mathscr{E}(P_1, \ldots, P_{k-1})$. The mapping from $\mathscr{E}(P_1, \ldots, P_k)$ to $\mathscr{E}(P_1, \ldots, P_{k-1})$ which sends $(\underline{M}, \underline{u})$ to $(\underline{M}', \underline{u}')$ is clearly compatible with the equivalence relations $\sim$ and therefore induces a mapping:

$$\mathscr{F}(P_1, \ldots, P_k) \to \mathscr{F}(P_1, \ldots, P_{k-1}), \quad \mathrm{cl}(\underline{M}, \underline{u}) \mapsto \mathrm{cl}(\underline{M}', \underline{u}').$$

Now, we not only show that this map is onto, but describe the preimages. So let an element in the target be the class $\mathrm{cl}(\underline{M}', \underline{u}')$ for some data $(\underline{M}', \underline{u}')$ in $\mathscr{E}(P_1, \ldots, P_{k-1})$. The class of $(\underline{M}, \underline{u})$ in $\mathscr{E}(P_1, \ldots, P_k)$ has image the class of $(\underline{M}', \underline{u}')$ if, and only if, it can be realized as $((0 \subset M_1 \subset \cdots \subset M_k), (u_1, \ldots, u_k))$ (see exercise) where[13] $M_k \supset M_{k-1}$ and $u_k : M_k/M_{k-1} \to P_k$ is an isomorphism. Now checking the isomorphy condition, we deduce (see exercise) that the preimage of $\mathrm{cl}(\underline{M}', \underline{u}')$ in $\mathscr{F}(P_1, \ldots, P_k)$ is in natural bijection with $\mathrm{Ext}(P_k, M') = \mathrm{Ext}(P_k, M_{k-1})$.

EXERCISE 15.54. The above deduction is more or less obvious, but not so easy to prove rigorously. Do it.

The following step is to note that, whatever the $\underline{M}'$ and M' involved, the **C**-linear space $\mathrm{Ext}(P_k, M')$ is isomorphic to $\bigoplus_{1 \leq i \leq k-1} \mathrm{Ext}(P_k, P_i)$. Indeed, each exact sequence $0 \to M_{i-1} \to M_i \to P_i \to 0$ gives rise to an exact sequence:

$$\mathrm{Hom}(P_k, P_i) \to \mathrm{Ext}(P_k, M_{i-1}) \to \mathrm{Ext}(P_k, M_i) \to \mathrm{Ext}(P_k, P_i) \to 0,$$

as follows from general homological properties (8.3, 9.2, and **[32]**). Since $\mathrm{Hom}(P_k, P_i)$ is trivial (two pure isoclinic modules with unequal slopes) and since exact sequences of vector spaces split, $\mathrm{Ext}(P_k, M_i) \simeq \mathrm{Ext}(P_k, M_{i-1}) \oplus \mathrm{Ext}(P_k, P_i)$ and our contention follows by induction.

For $k = 2$, we know beforehand (see 15.1.2.1) that $\mathscr{F}(P_1, P_2)$ can be identified with the finite-dimensional **C**-linear space $\mathrm{Ext}(P_2, P_1)$ (recall that it has dimension $r_1 r_2(\mu_1 - \mu_2)$). Assume by induction that $\mathscr{F}(P_1, \ldots, P_{k-1})$ can be given a structure of an affine space with underlying vector space:

$$\bigoplus_{1 \leq i < j \leq k-1} \mathrm{Ext}(P_j, P_i).$$

Then using the above remarks on preimages, we can choose a bijection of

$$\mathscr{F}(P_1, \ldots, P_k)$$

with $\mathscr{F}(P_1, \ldots, P_{k-1}) \oplus \bigoplus_{1 \leq i \leq k-1} \mathrm{Ext}(P_k, P_i)$ and then use the affine structure on $\mathscr{F}(P_1, \ldots, P_{k-1})$ to obtain one on $\mathscr{F}(P_1, \ldots, P_k)$ with unserlying space:

$$\bigoplus_{1 \leq i < j \leq k} \mathrm{Ext}(P_j, P_i).$$

Clearly the affine space thus obtained has dimension:

$$\sum_{1 \leq i < j \leq k} r_i r_j (\mu_i - \mu_j).$$

[13]Here and after we leve to you the adaptation to the extreme case $k = 1$, where, of course, M_{k-1} must be replaced by 0 (since M_0 has a different meaning).

A major flaw of the above argument is that it rests on many arbitrary choices of bijections, and does not prove at all that the affine structure is canonical, as it actually is. (This is proved in [**223**], correcting an error in [**197**].)

15.2.4. The Birkhoff-Guenther Normal Form. In [**25**], Birkhoff remarked that, up to then, "the theory of q-difference equations [had] lagged noticeably behind the sister theories of linear difference and q-difference equations" and held the opinion that "the use of the canonical system [...was] destined to carry the theory of q-difference equations to a comparable degree of completeness"; he then developed a considerable program, which unhappily he never pursued. The "canonical system" introduced by himself and Guenther is, in essence, what we call the Birkhoff-Guenther Normal Form and we will now study it.

15.2.4.1. To give a matricial formulation to the problem of analytic isoformal classification introduced in 15.2.3, we choose to take M_0 in the standard form $(\mathbf{C}(\{x\})^n, \Phi_{A_0})$, with the usual notations $A_0 := \text{Diag}(B_1, \ldots, B_k)$, $B_i := x^{\mu_i} A_i$, so that the pure isoclinic components of M_0 are the $P_i := (\mathbf{C}(\{x\})^{r_i}, \Phi_{B_i})$; and then to restrict to the pairs (M, u) in $\mathscr{E}(M_0)$ of the form $M_U := (\mathbf{C}(\{x\})^n, \Phi_{A_U})$, the isomorphism $u : \text{gr } M \to M_0$ then being canonically determined by the fact that the block diagonal of A_U is A_0. So we set:

$$\mathscr{E}(A_0) := \{A_U \mid U \in \prod_{1 \le i < j \le k} \text{Mat}_{r_i, r_j}(\mathbf{C}(\{x\}))\}.$$

For every $l = 1, \ldots, k$, we shall denote $A_U^{\le l}$ the upper left square submatrix of A_U of size $r_1 + \cdots + r_l$:

$$A_U^{\le l} := \begin{pmatrix} B_1 & \ldots & \ldots & \ldots & \ldots \\ \ldots & \ldots & \ldots & U_{i,j} & \ldots \\ 0 & \ldots & \ldots & \ldots & \ldots \\ \ldots & 0 & \ldots & \ldots & \ldots \\ 0 & \ldots & 0 & \ldots & B_l \end{pmatrix}, \text{ where } (U_{i,j}) \in \prod_{1 \le i < j \le l} \text{Mat}_{r_i, r_j}(\mathbf{C}(\{x\})).$$

So, denoting $M_l := (\mathbf{C}(\{x\})^{r_1 + \cdots + r_l}, \Phi_{A_U^{\le l}})$, the slope filtration of M_U is $M_1 \subset \cdots \subset M_k = M_U$ (we omit the term of index 0 for obvious reasons). With the help of those notations, we see that $\mathscr{E}(A_0)$ is indeed a model for $\mathscr{E}(M_0)$. (Note however that, contrary to $\mathscr{E}(M_0)$, $\mathscr{E}(A_0)$ is a set all right!)

In this model, a morphism $F : A_U \to A_V$ automatically respects the slope filtration (because of the functoriality of the latter) so it has the form:

$$F := \begin{pmatrix} f_1 & \ldots & \ldots & \ldots & \ldots \\ \ldots & \ldots & \ldots & f_{i,j} & \ldots \\ 0 & \ldots & \ldots & \ldots & \ldots \\ \ldots & 0 & \ldots & \ldots & \ldots \\ 0 & \ldots & 0 & \ldots & f_k \end{pmatrix}$$

for some $(f_i) \in \prod_{1 \le i \le k} \text{GL}_{r_i}(\mathbf{C}(\{x\}))$ and $(f_{i,j}) \in \prod_{1 \le i < j \le k} \text{Mat}_{r_i, r_j}(\mathbf{C}(\{x\}))$. Each f_i is then an automorphism of P_i, so that $\text{Diag}(f_1, \ldots, f_k)$ is an automorphism of M_0. Isoformal (or isograded) equivalence $\sim$ in $\mathscr{E}(M_0)$ means, in the present model, that $\mathrm{Diag}(f_1, \dots, f_k)$ should be the identity, i.e., that:

$$F = F_f := \begin{pmatrix} I_{r_1} & \dots & \dots & \dots & \dots \\ \dots & \dots & \dots & f_{i,j} & \dots \\ 0 & \dots & \dots & \dots & \dots \\ \dots & 0 & \dots & \dots & \dots \\ 0 & \dots & 0 & \dots & I_{r_k} \end{pmatrix} \in \mathcal{G}_{A_0}(\mathbf{C}(\{x\}))$$

for some $(f_{i,j}) \in \prod_{1 \leq i < j \leq k} \mathrm{Mat}_{r_i, r_j}(\mathbf{C}(\{x\}))$. So we are led to define the relation $\sim$ in $\mathscr{E}(A_0)$ by:

$$A_U \sim A_V \underset{def}{\Longleftrightarrow} \exists F \in \mathcal{G}_{A_0}(\mathbf{C}(\{x\})) \ : \ F[A_U] = A_V.$$

Then, setting:

$$\mathscr{F}(A_0) := \frac{\mathscr{E}(A_0)}{\sim},$$

we can identify $\mathscr{F}(M_0)$ and $\mathscr{F}(A_0)$. Concretely, the bijection $\mathscr{F}(A_0) \to \mathscr{F}(M_0)$ is given as $\mathrm{cl}(A_U) \mapsto \mathrm{cl}(M_U)$.

15.2.4.2. For each $A_U \in \mathscr{E}(A_0)$, there is a unique $\widehat{F}_U \in \mathcal{G}_{A_0}(\mathbf{C}((x)))$ such that $\widehat{F}_U(A_0) = A_U$. Yet, it is not true that conversely any $\widehat{F} \in \mathcal{G}_{A_0}(\mathbf{C}((x)))$ gives rise to a $A_U := \widehat{F}(A_0)$ in $\mathrm{GL}_n(\mathbf{C}(\{x\}))$, the transform may be divergent. So we denote:

$$\mathcal{G}^{A_0}(\mathbf{C}((x))) := \{\widehat{F} \in \mathcal{G}_{A_0}(\mathbf{C}((x))) \mid \widehat{F}(A_0) \in \mathrm{GL}_n(\mathbf{C}(\{x\}))\}$$

the set of admissible $\widehat{F}$. So we have a bijection $\mathscr{E}(A_U) \leftrightarrow \mathcal{G}^{A_0}(\mathbf{C}((x)))$, where A_U corresponds to $\widehat{F}_U$.

We know that $A_U \sim A_V$ is equivalent to $F[A_U] = A_V$ for some $F \in \mathcal{G}_{A_0}(\mathbf{C}(\{x\}))$, i.e., that $(F\widehat{F}_U)[A_0] = \widehat{F}_V[A_0]$; since clearly $F\widehat{F}_U \in \mathcal{G}_{A_0}(\mathbf{C}((x)))$ and invoking the uniqueness, we conclude that $F[A_U] = A_V \Leftrightarrow F\widehat{F}_U = \widehat{F}_V$, and, in the end:

$$A_U \sim A_V \Longleftrightarrow \exists F \in \mathcal{G}_{A_0}(\mathbf{C}(\{x\})) \ : \ F\widehat{F}_U = \widehat{F}_V \Longleftrightarrow \widehat{F}_V(\widehat{F}_U)^{-1} \in \mathcal{G}_{A_0}(\mathbf{C}(\{x\})).$$

This is translated in the following way: the group $\mathcal{G}_{A_0}(\mathbf{C}(\{x\}))$ operates on the left on the set $\mathcal{G}^{A_0}(\mathbf{C}((x)))$ by left multiplication: $(F, \widehat{F}) \mapsto F\widehat{F}$; and the quotient set is naturally identified with $\mathscr{F}(A_0)$:

$$\mathscr{F}(A_0) = \frac{\mathcal{G}^{A_0}(\mathbf{C}((x)))}{\mathcal{G}_{A_0}(\mathbf{C}(\{x\}))}.$$

This quotient by a left group action is sometimes written $\mathcal{G}_{A_0}(\mathbf{C}(\{x\}))\backslash\mathcal{G}^{A_0}(\mathbf{C}((x)))$.

15.2.4.3. *The Birkhoff-Guenther Normal Form.* Choose for each i, j such that $1 \leq i < j \leq k$ a set $I_{i,j} \subset \mathbf{Z}$ of representatives for the congruence modulo $\delta_{i,j} := \mu_i - \mu_j \in \mathbf{N}^*$; then denote by abreviation (recalling the notation K_I from 15.1.2.1):

$$K_{i,j} := K_{I_{i,j}} = \bigoplus_{d \in I_{i,j}} \mathbf{C}x^d \subset \mathbf{C}(\{x\}).$$

The following process yields, for every $A_V \in \mathscr{E}(A_0)$, the unique $A_U \in \mathscr{E}(A_0)$ such that $A_U \sim A_V$ and $U \in \prod_{1 \leq i < j \leq k} \mathrm{Mat}_{r_i, r_j}(K_{i,j})$; and, at the same time, the unique $F_f \in \mathcal{G}_{A_0}(\mathbf{C}(\{x\}))$ such that $F_f[A_U] = A_V$.

Remember that the latter condition is equivalent to:

$$\forall 1 \leq i < j \leq k \ , \ U_{i,j} + \sum_{i<l<j} (\sigma_q f_{i,l}) U_{l,j} + (\sigma_q f_{i,j}) B_j = B_i f_{i,j} + \sum_{i<l<j} V_{i,l} f_{l,j} + V_{i,j}.$$

Considering it as a system of equations in the unknowns U, f, we solve it recursively along increasing values of $j - i$, in the form:

$$\Phi_{B_i,B_j}(U_{i,j}) + f_{i,j} = V_{i,j} + \sum_{i<l<j} V_{i,l} f_{l,j} - \sum_{i<l<j} (\sigma_q f_{i,l}) U_{l,j}.$$

This admits a unique solution $(U_{i,j}, f_{i,j}) \in \mathrm{Mat}_{r_i,r_j}(K_{i,j}) \times \mathrm{Mat}_{r_i,r_j}(\mathbf{C}(\{x\}))$ by the additive case, see 15.1.2.

THEOREM 15.2 (Birkhoff-Guenther Normal Form). *For every $A_V \in \mathscr{E}(A_0)$, there is a unique $A_U \in \mathscr{E}(A_0)$ and a unique $F \in \mathcal{G}_{A_0}(\mathbf{C}(\{x\}))$ such that $U \in \prod\limits_{1 \le i < j \le k} Mat_{r_i,r_j}(K_{i,j})$ and $F[A_U] = A_V$.*

PROOF. This is indeed the solution corresponding to the particular choices:

$$I_{i,j} := I(\mu_i, \mu_j), \text{ where } I(a, b) := \{b, \ldots, a-1\}.$$

(Assuming of course that $a, b \in \mathbf{Z}$, $a > b$; see Definition 15.27.) □

That *Birkhoff-Guenther Normal Form* was first obtained by Birkhoff and Guenther for scalar q-difference equations in **[25]**, also see **[163]**.

EXERCISE 15.55. Show that A_U is regular on $\mathbf{C}^*$, i.e., $A_U \in \mathrm{GL}_n(\mathscr{O}(\mathbf{C}^*))$.

15.2.4.4. *$\mathscr{F}(A_0)$ as an affine space.* From the theorem we deduce that the mapping $U \mapsto \mathrm{cl}(A_U)$ from $\prod\limits_{1 \le i < j \le k} \mathrm{Mat}_{r_i,r_j}(K_{i,j})$ to $\mathscr{F}(A_0)$ is a bijection. Now, $\dim_{\mathbf{C}} K_I = \mathrm{card}\, I$, so $\dim_{\mathbf{C}} \mathrm{Mat}_{r_i,r_j}(K_{i,j}) = r_i r_j (\mu_i - \mu_j)$, whence:

$$\dim_{\mathbf{C}} \prod_{1 \le i < j \le k} \mathrm{Mat}_{r_i,r_j}(K_{i,j}) = N := \sum_{1 \le i < j \le k} r_i r_j (\mu_i - \mu_j),$$

so we can see (by transport of structure) $\mathscr{F}(A_0)$ as an affine space with underlying linear space $\prod\limits_{1 \le i < j \le k} \mathrm{Mat}_{r_i,r_j}(K_{i,j})$, thus with dimension N over $\mathbf{C}$.

This is consistent with the discussion in 15.2.3.3 and suggests an identification of each factor $\mathrm{Mat}_{r_i,r_j}(K_{i,j})$ with $\mathrm{Ext}(P_j, P_i)$ there. However we have here at best a heuristic indication. A more rigorous discussion appears in **[223]**.

15.2.5. Algebraic summation.

15.2.5.1. *General divisors.* There are two versions of this in **[197]**. We use the one closer to that of **[219]**.

Let A_0 and A_U have the usual meaning (see 15.2.1.1); in particular, the coefficients of U are in $\mathbf{C}(\{x\})$. We are interested in matrices of functions $F = F_f \in \mathcal{G}_{A_0}(\mathscr{M}(\mathbf{C}^*, 0))$ such that $F[A_0] = A_U$. We have been able to produce such matrices in the additive case in 15.1.3. In the general case, we found there is a unique formal one $\widehat{F}_U \in \mathcal{G}_{A_0}(\mathbf{C}((x)))$. Looking for f with coefficients in $\mathscr{M}(\mathbf{C}^*, 0)$, we consider a change of unknowns encoded in the following commutative diagram (similar in form to the one in 15.1.3.2, where some of the notations used here come from):

$$\begin{array}{ccc} A_0 & \xrightarrow{F_f} & A_U. \\ {\scriptstyle T}\big\downarrow & & \big\downarrow{\scriptstyle T} \\ A'_0 & \xrightarrow[G]{} & A'_V \end{array}$$

We take for T a block diagonal (actually scalar) matrix made of theta functions: $T := \mathrm{Diag}(\Theta_{D_1} I_{r_1}, \ldots, \Theta_{D_k} I_{r_k})$. It follows readily that $G := TFT^{-1} = F_g$, where $g_{i,j} := \Theta_{D_i - D_j} f_{i,j} \in \mathrm{Mat}_{r_i,r_j}(\mathscr{O}(\mathbf{C}^*, 0))$ and that:

$$A'_V := T[A_U] = \begin{pmatrix} B'_1 & \ldots & \ldots & \ldots & \ldots \\ \ldots & \ldots & \ldots & V_{i,j} & \ldots \\ 0 & \ldots & \ldots & \ldots & \ldots \\ \ldots & 0 & \ldots & \ldots & \ldots \\ 0 & \ldots & 0 & \ldots & B'_k \end{pmatrix},$$

where $B'_i := \dfrac{\sigma_q \Theta_{D_i}}{\Theta_{D_i}} B_i = \dfrac{\mathrm{ev} D_i}{x^{\deg D_i}} B_i$ and $V_{i,j} := \dfrac{\sigma_q \Theta_{D_i}}{\Theta_{D_j}} U_{i,j} = \dfrac{\mathrm{ev} D_i}{x^{\deg D_i}} \Theta_{D_i - D_j} U_{i,j}$. We want to solve the second line of the diagram for G, from which we shall deduce $F := T^{-1} G T$, i.e., $f_{i,j} := \Theta_{D_j - D_i} g_{i,j}$. For that we try to make the equation $G[A'_0] = A'_V$ (with unknown G) look somehow like a fuchsian equation, i.e., make the corresponding system with unknown the $g_{i,j}$:

$$\forall 1 \leq i < j \leq k \, , \, (\sigma_q g_{i,j}) B'_j - B'_i g_{i,j} = \sum_{i<l<j} V_{i,l} g_{l,j} + V_{i,j}$$

look like a fuchsian system. Something we can do in that direction is to have the matrices B'_i constant matrices (i.e., in $\mathrm{GL}_{r_i}(\mathbf{C})$) and the matrices $V_{i,j}$ holomorphic in a punctured neighborhood of 0, so that they have Laurent expansions (but we will not be able to prevent them to be wild at 0). There will also appear soon a third condition, a kind of nonresonancy.

The first condition $B'_i \in \mathrm{GL}_{r_i}(\mathbf{C})$ is easily obtained by requiring that $\deg D_i = \mu_i$, so that $B'_i = \mathrm{ev}(D_i) A_i$ (recall that $B_i = x^{\mu_i} A_i$). The second condition automatically flows if we require that $\mathcal{D}_j \leq D_i$ whenever $i < j$, because $\mathcal{D}_i - D_j$ is then an effective divisor and $\Theta_{D_i - D_j} \in \mathscr{O}(\mathbf{C}^*)$, so, since $U_{i,j}$ has coefficieny in $\mathbf{C}(\{x\})$, coefficients of $V_{i,j}$ lie in $\mathscr{O}(\mathbf{C}^*, 0)$.

The system above, which has the form $\Phi_{B'_i, B'_j}(g_{i,j})$ = some right-hand side depending on the $g_{l,j}$ is recursively solved with increasing $i - j$ and thus boils down to a sequence of equations $\Phi_{c_i A_i, c_j A_j}(u) = v$ (we have denoted $c_i := \mathrm{ev}(D_i)$ where we inductively assume $v \in \mathrm{Mat}_{r_i,r_j}(\mathscr{O}(\mathbf{C}^*, 0))$ and wish to prove that $u \in \mathrm{Mat}_{r_i,r_j}(\mathscr{O}(\mathbf{C}^*, 0))$. Using Laurent expansions $u = \sum u_m x^m$, $v = \sum v_m x^m$, each such equation is equivalent to a system:

$$\forall m \in \mathbf{Z} \, , \, u_m(q^m c_j A_j) - (c_i A_i) u_m = v_m.$$

So if we can enforce that $q^m c_j A_j$ and $c_i A_i$ have disjoint spectra for all m, all those equations can be uniquely solved and moreover we know that convergence of v entails convergence of u (same calculation as in 15.1.3.3). Therefore the third condition alluded to take the form:

$$\forall m \in \mathbf{Z} \, , \, q^m c_j \mathrm{Sp}(A_j) \cap c_i \mathrm{Sp}(A_i) = \emptyset, \text{ i.e. } c_j / c_i \notin q^{\mathbf{Z}} \mathrm{Sp}(A_i) / \mathrm{Sp}(A_j),$$

where here $S/T := \{s/t \mid s \in S, t \in T\}$ and the condition should be met for all $1 \leq i < j \leq k$.

15.2.5.2. *Algebraic summation with a single pole.* Although the present author thinks (as already asserted more than once!) that general summation divisors are the "right" framework for classification and Galois theory, no significant result has been obtained so far in that direction, so we shall (regretfully) specialize the study to families (D_i) supported at a single point.

So we fix $c \in \mathbf{C}^*$ and set $D_i := \mu_i[c]$. Clearly that choice satisfies the first two requirements above: deg $D_i = \mu_i$ and $D_j \leq D_i$ whenever $i < j$. Practically, this gives us $\Theta_{D_i} = \theta_q(x/c)^{\mu_i} \in \mathscr{M}(\mathbf{C}^*)^*$ and $\Theta_{D_i - D_j} = \theta_q(x/c)^{\delta_{i,j}} \in \mathscr{O}(\mathbf{C}^*)^*$. As for the third condition (at the very end of 15.2.5.1), we see that, since here we have $c_i := \text{ev}D_i = c^{\mu_i}$, it translates to $c^{\mu_j - \mu_i} \not\in q^{\mathbf{Z}}\text{Sp}(A_i)/\text{Sp}(A_j)$ for all $1 \leq i < j \leq k$. Noting that it really is a condition on $\overline{c} \in \mathbf{E}_q$ rather than on $c \in \mathbf{C}^*$, we are led to:

DEFINITION 15.56 (Prohibited directions of summation). The set of *prohibited directions of summation* (for A_0) is the following subset of $\mathbf{E}_q$:

$$\begin{aligned}\Sigma_{A_0} &:= \{\overline{c} \in \mathbf{E}_q \mid c^{\mu_j - \mu_i} \in q^{\mathbf{Z}}\text{Sp}(A_i)/\text{Sp}(A_j) \text{ for some } 1 \leq i < j \leq k\} \\ &= \{\overline{c} \in \mathbf{E}_q \mid \overline{c} \in \bigcup_{1 \leq i < j \leq k} \pi\left(\text{Sp}(A_j)/\text{Sp}(A_i)\right)^{1/(\mu_i - \mu_j)}\},\end{aligned}$$

where we denote here $S^{1/d}$ the set of dth roots of elements of $S \subset \mathbf{E}_q$.

Since an element of $\mathbf{E}_q$ has d^2 dth roots, $\text{card}S^{1/d} = d^2\text{card}S$; and since $\text{Sp}(A_j)/\text{Sp}(A_i)$ has at most $r_i r_j$ elements, we conclude:

$$\text{card}\Sigma_{A_0} \leq \sum_{1 \leq i < j \leq k} r_i r_j (\mu_i - \mu_j)^2,$$

the equality holding generically.

THEOREM 15.3 (Algebraic summation with a single pole in an authorized direction of summation). *Matrices A_0, A_U being given as above, let $\overline{c} \in \mathbf{E}_q \setminus \Sigma_{A_0}$ an authorized direction of summation. Then there is a unique $F_f \in \mathcal{G}_{A_0}(\mathscr{M}(\mathbf{C}^*, 0))$ such that $F_f[A_0] = A_U$ and $\text{div}_{\mathbf{E}_q} f_{i,j} \geq -(\mu_i - \mu_j)\boxed{-c}$ for all $1 \leq i < j \leq k$.*

PROOF. Recall that the above contition on $f_{i,j}$ means that all its components have (in some punctured neighborhood of 0) at worst poles on $[-c; q]$ with multiplicity $\mu_i - \mu_j$. So that condition is actually equivalent to:

$$\theta_q(x/c)^{\mu_i - \mu_j} f_{i,j} \in \text{Mat}_{r_i, r_j}(\mathscr{O}(\mathbf{C}^*, 0)).$$

Then we use the procedure described in 15.2.5.1 with our present choice $D_i := \mu_i[c]$, i.e., $T := \text{Diag}(\theta_q(x/c)^{\mu_1} I_{r_1}, \ldots, \theta_q(x/c)^{\mu_k} I_{r_k})$ (a block diagonal description of a scalar matrix) and so we set accordingly $g_{i,j} := \theta_q(x/c)^{\mu_i - \mu_j} f_{i,j}$ and $V_{i,j} := (c/c)^{\mu_i} \theta_q(x/c)^{\mu_i - \mu_j} U_{i,j}$. Hence the component of $U_{i,j}$ belongs to $\mathscr{O}(\mathbf{C}^*, 0)$ and that we look for the $g_{i,j}$ there too, satisfying $F_g[A_0'] = A_V'$. The block diagonal of A_0' and A_V' is $\text{Diag}(B_1', \ldots, B_k')$ where $B_i' := c^{\mu_i} A_i$, whence the system:

$$\forall 1 \leq i < j \leq k \,,\; (\sigma_q g_{i,j})(c^{\mu_j} A_j) - (c^{\mu_i} A_i) g_{i,j} = \sum_{i < l < j} V_{i,l} g_{l,j} + V_{i,j}.$$

We recursively solve the system for increasing values of $i - j$. Each step has the form $(\sigma_q g)Q - Pg = h$ (unknown g, recursively given data h) and Laurent expansions of g and h in turn reduce such an equation to a system of independent equations:

$$\forall m \in \mathbf{Z} \,,\; \Phi_{P, q^m Q}(g_m) := g_m(q^m Q) - P g_m = h_m,$$

where $\text{Sp}(P) \cap q^{\mathbf{Z}}\text{Sp}(Q) = \emptyset$ precisely because $\overline{c} \not\in \Sigma_{A_0}$, so that each $\Phi_{P, q^m Q}$ is an automorphism of the corresponding space of complex matrices. So we set $g_m := \Phi_{P, q^m Q}^{-1}(h_m)$ for all m and the asymptotic study of $\left|\left|\left|\Phi_{P, q^m Q}^{-1}\right|\right|\right|$ allows to conclude to the convergence (again, same calculation as in 15.1.3.3). $\square$

Note that the condition defining F_f in the theorem and its corollary rest on $\overline{c} \in \mathbf{E}_q$ and not on $c \in \mathbf{C}^*$ itself, whence the following:

DEFINITION 15.57 (Sum in authorized directions). The unique F_f defined in the theorem is called the *sum of $\widehat{F}_U$ in (authorized) direction $\overline{c}$* and denoted $S_{\overline{c}}\widehat{F}_U$.

15.2.5.3. *Summation and classification.* Recalling the equations (15.1) at the end of 15.2.1.2 and the consequences we drew of them, one easily proves the following:

EXERCISE 15.58. The direction $\overline{c} \in \mathbf{E}_q$ being fixed, the set:

$$\{F_f \in \mathcal{G}_{A_0}(\mathscr{M}(\mathbf{C}^*,0)) \mid \forall 1 \leq i < j \leq k \,,\ \operatorname{div}_{\mathbf{E}_q} f_{i,j} \geq -(\mu_i - \mu_j)[-c]\}.$$

is a subgroup of $\mathcal{G}_{A_0}(\mathscr{M}(\mathbf{C}^*,0))$. (Use the R_d trick from the end of 15.2.1.1.)

COROLLARY 15.59. *Matrices A_U, A_V being given as usual, along with $\overline{c} \in \mathbf{E}_q \setminus \Sigma_{A_0}$, there is a unique $F_f \in \mathcal{G}_{A_0}(\mathscr{M}(\mathbf{C}^*,0))$ such that $F_f[A_U] = A_V$ and $\operatorname{div}_{\mathbf{E}_q} f_{i,j} \geq -(\mu_i - \mu_j)[-c]$ for all $1 \leq i < j \leq k$.*

PROOF. Indeed, it can only be $(S_{\overline{c}}\widehat{F}_V)(S_{\overline{c}}\widehat{F}_U)^{-1}$. □

The unique F_f defined in the corollary is called the *sum of $\widehat{F}_{U,V}$ in (authorized) direction $\overline{c}$* and denoted $S_{\overline{c}}\widehat{F}_{U,V} := (S_{\overline{c}}\widehat{F}_V)(S_{\overline{c}}\widehat{F}_U)^{-1}$.

COROLLARY 15.60. $A_U \sim A_V \Longleftrightarrow S_{\overline{c}}\widehat{F}_{U,V} \in \mathcal{G}_{A_0}(\mathbf{C}(\{x\}))$.

PROOF. If $F \in \mathcal{G}_{A_0}(\mathbf{C}(\{x\}))$ (which is contained in the subgroup of the exercice) is such that $F[A_U] = A_V$, then $F(S_{\overline{c}}\widehat{F}_U) = S_{\overline{c}}\widehat{F}_V$ by the usual uniqueness argument, so $S_{\overline{c}}\widehat{F}_{U,V} = F \in \mathcal{G}_{A_0}(\mathbf{C}(\{x\}))$. The converse implication is obvious. □

EXERCISE 15.61. (i) Deduce from this corollary a new description of $\mathscr{F}(A_0)$. (Note however that, in section 15.3, the classification will rather rest on variation of the direction of summation.)

(ii) Prove this variant of the corollary: if $S_{\overline{c}}\widehat{F}_{U,V} \in \mathcal{G}_{A_0}(\mathscr{O}(\mathbf{C}^*,0))$, then $A_U \sim A_V$.

REMARK 15.62. It is an important requirement in the classical theory of irregular differential equations [**15, 120, 150, 192**] (more generally, of divergent series) that the summation process be compatible with algebraic operations; linearity is usually easy, but not so for multiplicativity. In the case of q-difference equations, the latter requirement causes essential difficulties [**163, 197**]. The above property, which could be formulated as $S_{\overline{c}}\widehat{F}_{U,W} = S_{\overline{c}}\widehat{F}_{U,V}\, S_{\overline{c}}\widehat{F}_{V,W}$, is a very weak substitute - yet essential for Galois theory.

15.2.5.4. *Summation and resolution.* An immediate application of the algebraic summation process is the construction of explicit meromorphic solutions for the system $\sigma_q X = A_U X$. First solve $\sigma_q X = A_0 X$ with Theta functions as in 15.2.2.1, yielding a fundamental matricial solution $\mathcal{X}_0$; then each $\mathcal{X}_{\overline{c}} := (S_{\overline{c}}\widehat{F}_U)\mathcal{X}_0$ ($\overline{c}$ an authorized direction) is a fundamental matricial solution of $\sigma_q X = A_U X$ over $\mathscr{M}(\mathbf{C}^*,0)$.

Although the theorem of Praagman (see 7.3) guarantees in principle the existence of such a fundamental matricial solution over $\mathscr{M}(\mathbf{C}^*,0)$, the above process is explicit, though restricted to systems with integral slopes.

15.3. The Stokes phenomenon and the classification of irregular systems

Again we assume the slopes to be integral; for the general case, which is more complicated, see **[223]** and [CV]. We take up notations $S_{\overline{c}}\widehat{F}_U$ and $S_{\overline{c}}\widehat{F}_{U,V}$ from section 15.2. As explained in 15.1.4 (also see 14.2.4.3 in Chapter 14), we interpret here the Stokes phenomenon, more to the point its q-analogue, as *the ambiguity of summation in various directions.* Thus, letting $\overline{c}, \overline{d} \in \mathbf{E}_q \setminus \Sigma_{A_0}$, we set:

$$S_{\overline{c},\overline{d}}\widehat{F}_U := (S_{\overline{c}}\widehat{F}_U)^{-1}(S_{\overline{d}}\widehat{F}_U) \in \mathcal{G}_{A_0}(\mathscr{M}(\mathbf{C}^*, 0)),$$

thus, as a compositum $A_0 \xrightarrow{S_{\overline{d}}\widehat{F}_U} A_U \xrightarrow{(S_{\overline{c}}\widehat{F}_U)^{-1}} A_0$, a meromorphic automorphism of A_0.

15.3.1. Stokes operators and Stokes cocycles.

15.3.1.1. *Stokes operators.* To begin with, we select two particular authorized directions of summation $\overline{c}, \overline{d} \in \mathbf{E}_q \setminus \Sigma_{A_0}$, $\overline{c} \neq \overline{d}$. Using the known properties of $S_{\overline{c}}\widehat{F}_U, S_{\overline{c}}\widehat{F}_U$ from 15.2.5 follow those:

- Let $(f_{i,j})$, $1 \leq i \leq j \leq k$, the (i,j) block of $S_{\overline{c},\overline{d}}\widehat{F}_U \in \mathcal{G}_{A_0}(\mathscr{M}(\mathbf{C}^*, 0))$; then $\operatorname{div}_{\mathbf{E}_q}(f_{i,j}) \geq -(\mu_i - \mu_j)([\overline{-c}] + [\overline{-d}])$. This is a consequence of equations (15.1) at the end of 15.2.1.2 and the corresponding relations for the (i,j) blocks of $S_{\overline{c}}\widehat{F}_U, S_{\overline{d}}\widehat{F}_U$.
- As already noted with the help of a diagram, $S_{\overline{c},\overline{d}}\widehat{F}_U$ is a meromorphic automorphism of A_0:

$$(S_{\overline{c},\overline{d}}\widehat{F}_U)[A_0] = (S_{\overline{c}}\widehat{F}_U)^{-1}(S_{\overline{d}}\widehat{F}_U[A_0]) = (S_{\overline{c}}\widehat{F}_U)^{-1}[A_U] = A_0.$$

We denote $E_{\overline{c},\overline{d}}$ the subset of $\mathcal{G}_{A_0}(\mathscr{M}(\mathbf{C}^*, 0))$ made of those F_f such that each (i,j) block satisfies $\operatorname{div}_{\mathbf{E}_q}(f_{i,j}) \geq -(\mu_i - \mu_j)([\overline{-c}] + [\overline{-d}])$ and such that $F_f[A_0] = A_0$.

PROPOSITION 15.63. *The map $A_U \mapsto S_{\overline{c},\overline{d}}\widehat{F}_U$ from $\mathscr{E}(A_0)$ to $E_{\overline{c},\overline{d}}$ induces an injection $clA_U \mapsto S_{\overline{c},\overline{d}}\widehat{F}_U$ from $\mathscr{F}(A_0)$ to $E_{\overline{c},\overline{d}}$.*

PROOF. We are to show that $A_U \sim A_V \Leftrightarrow S_{\overline{c},\overline{d}}\widehat{F}_U = S_{\overline{c},\overline{d}}\widehat{F}_V$. So first assume that $A_U \sim A_V$, and let $F \in \mathcal{G}_{A_0}(\mathbf{C}(\{x\}))$ realizing that; then, by the usual uniqueness argument $F(S_{\overline{c}}\widehat{F}_U) = S_{\overline{c}}\widehat{F}_V$ and $F(S_{\overline{d}}\widehat{F}_U) = S_{\overline{d}}\widehat{F}_V$, whence:

$$S_{\overline{c},\overline{d}}\widehat{F}_V = (S_{\overline{c}}\widehat{F}_V)^{-1}(S_{\overline{d}}\widehat{F}_V) = (FS_{\overline{c}}\widehat{F}_U)^{-1}(FS_{\overline{d}}\widehat{F}_U) = (S_{\overline{c}}\widehat{F}_U)^{-1}(S_{\overline{d}}\widehat{F}_U) = S_{\overline{c},\overline{d}}\widehat{F}_U.$$

Assume conversely that $S_{\overline{c},\overline{d}}\widehat{F}_U = S_{\overline{c},\overline{d}}\widehat{F}_V$, i.e., $(S_{\overline{c}}\widehat{F}_U)^{-1}(S_{\overline{d}}\widehat{F}_U) = (S_{\overline{c}}\widehat{F}_V)^{-1}(S_{\overline{d}}\widehat{F}_V)$, from which one easily draws that $(S_{\overline{c}}\widehat{F}_V)(S_{\overline{c}}\widehat{F}_U)^{-1} = (S_{\overline{d}}\widehat{F}_V)(S_{\overline{d}}\widehat{F}_U)^{-1}$, call it F. Immediate consequences are: $F \in \mathcal{G}_{A_0}(\mathscr{M}(\mathbf{C}^*, 0))$; $F[A_U] = A_V$; and the poles of F near 0 lay inside $[-c; q] \cap [-d; q] = \emptyset$ (since $\overline{c} \neq \overline{d}$), so that actually $F \in \mathcal{G}_{A_0}(\mathscr{O}(\mathbf{C}^*, 0))$. From (ii) in the following lemma, we draw that $F \in \mathcal{G}_{A_0}(\mathbf{C}(\{x\}))$, whence the conclusion. □

LEMMA 15.64. (i) *Let B_1, B_2 with integral slopes $\mu_1 > \mu_2$ and let $Y := \Phi_{B_1,B_2}(X)$. If X has components in $\mathscr{O}(\mathbf{C}^*, 0)$ and Y has components in $\mathbf{C}(\{x\})$, then X has components in $\mathbf{C}(\{x\})$.*

(ii) *If $F \in \mathcal{G}_{A_0}(\mathscr{O}(\mathbf{C}^*, 0))$ and $F[A_U] = A_V$ (where A_U, A_V have components in $\mathbf{C}(\{x\})$), then $F \in \mathcal{G}_{A_0}(\mathbf{C}(\{x\}))$.*

PROOF. (i) Using Laurent expansions $X = \sum X_m x^m$, $Y = \sum Y_m x^m$, the equation $Y := \Phi_{B_1,B_2}(X)$ becomes a system of equations $q^{m-\mu_2} X_{m-\mu_2} A_2 - A_1 X_{m-\mu_1} = Y_m$, $m \in \mathbf{Z}$. If m_0 is such that $Y_m = 0$ for $m < m_0$, we have (setting $\delta := \mu_1 - \mu_2 \in \mathbf{N}^*$ as usual) $q^m X_m = A_1 X_{m-\delta} A_2^{-1}$ for $m < m_0 - \mu_2$. If those X_m were nonzero, this would imply wild growth of $\|X_m\|$ as $m \to -\infty$, contradicting the fact that X has components in $\mathscr{O}(\mathbf{C}^*, 0)$.

(ii) We write the relation $F[A_U] = A_V$ in the usual recursive form:

$$\Phi_{B_i,B_j}(f_{i,j}) = V_{i,j} - U_{i,j} + \text{ terms involving pairs of indices } (i,l),(l,j), i < l < j.$$

Then induction with increasing values of $j - i$ allows us to apply (i).

□

EXERCISE 15.65. (i) Extend the lemma to the case that the slopes are not supposed integral.

(ii) Deduce from the lemma a description of $\mathscr{F}(A_0)$ looking like $\dfrac{\mathcal{G}^{A_0}(\mathbf{C}((x)))}{\mathcal{G}_{A_0}(\mathbf{C}(\{x\}))}$ (see 15.2.4.2) but with $\mathcal{G}_{A_0}(\mathscr{O}(\mathbf{C}^*, 0))$ at the denominator.

EXERCISE 15.66. Using 15.1.2.4, 15.2.4.3 and, by anticipation, 15.3.1.3, show that $\mathscr{F}(A_0) \to E_{\overline{c},\overline{d}}$ is bijective.

15.3.1.2. *Stokes cocycles.* We now aim at a cohomological formulation[14] of the Stokes phenomenon, generalizing that in 15.1.5 for the additive case of two slopes. For that, we introduce the following sheaf of nonabelian groups[15] on $\mathbf{E}_q$:

$$\Lambda_I(A_0) : U \mapsto \left\{F \in \mathcal{G}_{A_0}(\mathscr{O}(\pi^{-1}(U))) \mid F[A_0] = A_0\right\}.$$

Note that since $F[A_0] = A_0$ translates to $f_{i,j} = B_i^{-1}(\sigma_q f_{i,j}) B_j$, any germ of solution at $(\pi^{-1}(U), 0)$ can be continued (by successive dilatations of the disk of definition) to the whole of $\pi^{-1}(U)$, whence the condition $F \in \mathcal{G}_{A_0}(\mathscr{O}(\pi^{-1}(U)))$. The reader will check that we indeed thus defined a sheaf of nonabelian groups.

Now recall the covering $\mathfrak{U}$ of $\mathbf{E}_q$ by the open subsets $U_{\overline{c}} := \mathbf{E}_q \setminus \{\overline{c}\}$, $\overline{c} \in \mathbf{E}_q \setminus \Sigma_{A_0}$. We denote $U_{\overline{c},\overline{d}} := U_{\overline{c}} \cap U_{\overline{d}}$, $U_{\overline{c},\overline{d},\overline{e}} := U_{\overline{c}} \cap U_{\overline{d}} \cap U_{\overline{e}}$, etc. So, from 15.3.1.1:

$$\forall \overline{c}, \overline{d} \in \mathbf{E}_q \setminus \Sigma_{A_0}\ ,\ S_{\overline{c},\overline{d}} \widehat{F}_U \in H^0(U_{\overline{c},\overline{d}}, \Lambda_I(A_0)).$$

Since obviously $(S_{\overline{c},\overline{d}} \widehat{F}_U)(S_{\overline{d},\overline{e}} \widehat{F}_U) = S_{\overline{c},\overline{e}} \widehat{F}_U$ on $U_{\overline{c},\overline{d},\overline{e}}$, we have a cocycle:

$$(S_{\overline{c},\overline{d}} \widehat{F}_U)_{\overline{c},\overline{d} \in \mathbf{E}_q \setminus \Sigma_{A_0}} \in Z^1(\mathfrak{U}, \Lambda_I(A_0)).$$

Taking in account the properties listed in 15.3.1.1, we introduce a definition[16].

DEFINITION 15.67. A *privileged cocycle* is an element

$$(F_{\overline{c},\overline{d}})_{\overline{c},\overline{d} \in \mathbf{E}_q \setminus \Sigma_{A_0}} \in Z^1(\mathfrak{U}, \Lambda_I(A_0))$$

such that $F_{\overline{c},\overline{d}} \in E_{\overline{c},\overline{d}}$ whenever $\overline{c}, \overline{d} \in \mathbf{E}_q \setminus \Sigma_{A_0}$, $\overline{c} \neq \overline{d}$. The set of privileged cocycles is denoted $Z^1_{pr}(\mathfrak{U}, \Lambda_I(A_0))$.

The next proposition can be seen as a q-analogue of [**150**, 3.5.11].

---

[14]The basic formalism of nonabelian cocycles and cohomology classes is explained in Appendix I.

[15]A different (but equivalent) definition of that sheaf, closer to the classical case (using asymptotic theory) will be given in 15.3.3.

[16]To be compared with the definition of Stokes cocycles in [**148–150**].

PROPOSITION 15.68. *We have a well defined injective map*

$$clA_U \mapsto (S_{\overline{c},\overline{d}}\widehat{F}_U)_{\overline{c},\overline{d}\in\mathbf{E}_q\setminus\Sigma_{A_0}}$$

from $\mathscr{F}(A_0)$ *to* $Z^1_{pr}(\mathfrak{U},\Lambda_I(A_0))$.

PROOF. Immediate from 15.3.1.1. □

In the additive case, we were able to prove bijectivity using that source and target are complex linear spaces of the same finite dimension and that the map itself is **C**-linear. Here, neither source nor target has a structure of linear space, not even of group, essentially because the sheaf $\Lambda_I(A_0)$ is nonabelian. However, it is unipotent and can be "peeled off" (we Frenchmen speak of "dévissage", see footnote in 10.1.1.2) using a filtration by normal subsheaves with abelian quotients (thus a sheaf version of what we saw in 15.2.1.3), actually locally free sheaves, hence vector bundles. This is the program for 15.3.2.

15.3.1.3. *Stokes cohomology classes.* However, we can still improve our results without resorting to dévissage.

PROPOSITION 15.69. *The natural map* $Z^1_{pr}(\mathfrak{U},\Lambda_I(A_0)) \to H^1(\mathfrak{U},\Lambda_I(A_0))$ *sending a cocycle to its cohomology class (relative to covering* $\mathfrak{U}$*) is injective.*

PROOF. Let $(F_{\overline{c},\overline{d}}),(G_{\overline{c},\overline{d}})$ two privileged cocycles having the same class in $H^1(\mathfrak{U},\Lambda_I(A_0))$. That means that there is a family of $\Phi_{\overline{c}}\in H^0(U_{\overline{c}},\Lambda_I(A_0))$ such that:

$$\forall\overline{c},\overline{d}\in\mathbf{E}_q\setminus\Sigma_{A_0}\ ,\ G_{\overline{c},\overline{d}} = \Phi_{\overline{c}}^{-1}F_{\overline{c},\overline{d}}\Phi_{\overline{d}}.$$

We write this as $\Phi_{\overline{c}} = F_{\overline{c},\overline{d}}\Phi_{\overline{d}}{G_{\overline{c},\overline{d}}}^{-1}$, which shows that $\Phi_{\overline{c}}$ has at worst poles on $[-c;q]$, with multiplicities $\mu_i-\mu_j$ on block (i,j) (this uses the fact that the three factors are such and, once more, the exercise at the beginning of 15.2.5.3). Since $\Phi_{\overline{c}}[A_0]=A_0$, the uniqueness property implies that $\Phi_{\overline{c}} = S_{\overline{c}}\widehat{F}_0 = I_n$; same for $\Phi_{\overline{d}}$ and so $F_{\overline{c},\overline{d}} = G_{\overline{c},\overline{d}}$. □

COROLLARY 15.70. *We have injective maps:*

$$\mathscr{F}(A_0)\to Z^1_{pr}(\mathfrak{U},\Lambda_I(A_0))\to H^1(\mathfrak{U},\Lambda_I(A_0))\to H^1(\mathbf{E}_q,\Lambda_I(A_0)).$$

PROOF. Only the injectivity of the last arrow still requires justification. But this is a general fact; see Appendix I or [**89**, Lemma 3.1] together with the fact that injectivity is preserved by direct limits. □

Here is a last improvement which can be obtained by elementary means.

THEOREM 15.4. *The map* $\mathscr{F}(A_0)\to H^1(\mathfrak{U},\Lambda_I(A_0))$ *is bijective. As a consequence, the first two maps in the corollary above are bijective.*

PROOF. Let $(F_{\overline{c},\overline{d}})\in Z^1(\mathfrak{U},\Lambda_I(A_0))$. We want to prove that it is equivalent to the cocycle $(S_{\overline{c},\overline{d}}\widehat{F}_U)$ for some $A_U\in\mathscr{E}(A_0)$. We denote $V_{\overline{c}} := \pi^{-1}(U_{\overline{c}}) = \mathbf{C}^*\setminus[-c;q]$ and $\mathfrak{V} := (V_{\overline{c}})_{\overline{c}\in\mathbf{E}_q\setminus\Sigma_{A_0}}$, thus an open covering of $\mathbf{C}^*$. The $(F_{\overline{c},\overline{d}})$ can also be considered as a cocycle of the sheaf $\mathcal{G}_{A_0}(\mathscr{O}_{\mathbf{C}^*})$ relative to the covering $\mathfrak{V}$. As such, they are the transition cocycle of a holomorphic vector bundle over $\mathbf{C}^*$ with structure group $\mathcal{G}_{A_0}$, which is connected. By Appendix I or [**205**, Theorem 1.0], such a bundle can be trivialized as a $\mathcal{G}_{A_0}$-bundle, which means that there is a family of $F_{\overline{c}}\in\mathcal{G}_{A_0}(\mathscr{O}(V_{\overline{c}}))$, $\overline{c}\in\mathbf{E}_q\setminus\Sigma_{A_0}$, such that $F_{\overline{c},\overline{d}} = {F_{\overline{c}}}^{-1}F_{\overline{d}}$. Since $F_{\overline{c},\overline{d}}[A_0]=A_0$, this implies that all $F_{\overline{c}}[A_0]$ are equal; of course, they have the form A_V but here $V = (V_{i,j}) \in \prod \mathrm{Mat}_{r_i,r_j}(\mathscr{O}(\mathbf{C}^*,0))$ (no poles since they would lie in the intersection of disjoint q-spirals). However, by 15.2.4, such a "wild" A_V also has a Birkhoff-Guenther Normal Form A_U, i.e., $A_V = F[A_U]$ for some $F \in \mathcal{G}_{A_0}(\mathscr{O}(\mathbf{C}^*,0))$. Now $F^{-1}F_{\overline{c}}[A_0] = F^{-1}[A_V] = A_U$, and by the usual uniqueness argument, $F^{-1}F_{\overline{c}} = S_{\overline{c}}\widehat{F}_U$, from which $F_{\overline{c},\overline{d}} = S_{\overline{c},\overline{d}}\widehat{F}_U$ follows immediately, as wanted. □

Our main goal is now to show that $H^1(\mathfrak{U}, \Lambda_I(A_0)) \to H^1(\mathbf{E}_q, \Lambda_I(A_0))$ is bijective. This will be done in 15.3.2.

15.3.2. The q-analogue of Birkhoff-Malgrange-Sibuya Classification Theorem. There are actually two theorems deserving that name[17]. Here we deal with the more down-to-earth of them; a fancier one will be discussed in 15.3.3. The former involves solutions of functional equations, while the latter tackles more general classes of functions. Their classical analogues are respectively called *classification theorem* and *isomorphism theorem* in [**150**, 3.5.2].

15.3.2.1. *Statement of the classification theorem.*

THEOREM 15.5. *The map* $\mathscr{F}(A_0) \to H^1(\mathbf{E}_q, \Lambda_I(A_0))$ *is bijective.*

PROOF. The mentioned map is of course the compositum:

$$\mathscr{F}(A_0) \to H^1(\mathfrak{U}, \Lambda_I(A_0)) \to H^1(\mathbf{E}_q, \Lambda_I(A_0)),$$

so, relying on 15.3.1.3, we just have to prove that the second map is bijective. This rests on a *nonabelian version of Leray's theorem of acyclic coverings*: let Λ a sheaf of groups on $\mathbf{E}_q$; if an open covering $\mathfrak{V} = (V_\alpha)$ of $\mathbf{E}_q$ is such that all $H^1(V_\alpha, \Lambda)$ are trivial (i.e., are singletons), then the natural direct limit map $H^1(\mathfrak{V}, \Lambda_I(A_0)) \to H^1(\mathbf{E}_q, \Lambda_I(A_0))$ is bijective (Appendix I or [**14**, 1.2.4, p. 113]. So we are led to prove that all $H^1(U_{\overline{c}}, \Lambda_I(A_0))$ are trivial. We shall next (in 15.3.2.2 and 15.3.2.3) prove it for all proper open subsets of $\mathbf{E}_q$; but first: □

COROLLARY 15.71. *Each of the maps:*

$$\mathscr{F}(A_0) \to Z^1_{pr}(\mathfrak{U}, \Lambda_I(A_0)) \to H^1(\mathfrak{U}, \Lambda_I(A_0)) \to H^1(\mathbf{E}_q, \Lambda_I(A_0))$$

is bijective.

EXERCISE 15.72. Check that for $\overline{c} \neq \overline{d}$ the natural map $Z^1_{pr}(\mathfrak{U}, \Lambda_I(A_0)) \to E_{\overline{c},\overline{d}}$ is bijective. Interpret the resulting bijections $E_{\overline{c},\overline{d}} \to E_{\overline{c},\overline{e}}$.

15.3.2.2. Recall from the end of 15.2.1 the exact sequences:

$$1 \to \mathcal{G}_{A_0}^{>\delta} \to \mathcal{G}_{A_0}^{\geq\delta} \to \mathfrak{g}_{A_0}^{(\delta)} \to 0,$$

$$0 \to \mathfrak{g}_{A_0}^{(\delta)} \to \mathcal{G}_{A_0}/\mathcal{G}_{A_0}^{>\delta} \to \mathcal{G}_{A_0}/\mathcal{G}_{A_0}^{\geq\delta} \to 1,$$

[17]In the classical theory of complex differential equations [**150**, 3.5] (see also [**250**]), they are usually called Malgrange-Sibuya theorems, but, according to Malgrange (private communication), the main idea comes from Birkhoff. Strangely enough, [**250**] attributes the Isomorphism Theorem to Birkhoff-Sibuya.

the second one being a central extension. We correspondingly define three sheaves $\Lambda_I^{\geq\delta}(A_0)$, $\Lambda_I^{>\delta}(A_0)$ and $\lambda_I^{(\delta)}(A_0)$, hereafter abreviated as $\Lambda^{\geq\delta}$, $\Lambda^{>\delta}$ and $\lambda^{(\delta)}$ (and $\Lambda_I(A_0)$, $\lambda_I(A_0)$ will likewise be abreviated Λ, λ):

$$\Lambda^{\geq\delta}(U) := \Lambda(U) \cap \mathcal{G}_{A_0}^{\geq\delta}(\mathscr{M}(\mathbf{C}^*,0)),$$
$$\Lambda^{>\delta}(U) := \Lambda(U) \cap \mathcal{G}_{A_0}^{>\delta}(\mathscr{M}(\mathbf{C}^*,0)),$$
$$\lambda^{(\delta)}(U) := \lambda(U) \cap \mathfrak{g}_{A_0}^{(\delta)}(\mathscr{M}(\mathbf{C}^*,0)).$$

Correspondingly, we have two exact sequences of sheaves:

$$1 \to \Lambda^{>\delta} \to \Lambda^{\geq\delta} \to \lambda^{(\delta)} \to 0,$$
$$0 \to \lambda^{(\delta)} \to \Lambda/\Lambda^{>\delta} \to \Lambda/\Lambda^{\geq\delta} \to 1.$$

Last, note that, the slopes being supposed integral, $\Lambda^{>\delta} = \Lambda^{\geq\delta+1}$.

Now we apply Appendix I or [**89**, Theorem 1.2] (together with the obvious fact that, if there is an exact sequence of pointed sets $\{.\} \to E \to \{.\}$, then $E = \{.\}$) to the first exact sequence, getting an exact sequence of pointed sets:

$$H^1(V,\Lambda^{\geq\delta+1}) \to H^1(V,\Lambda^{\geq\delta}) \to H^1(V,\lambda^{(\delta)}).$$

If $\delta \geq \mu_1 - \mu_k$, we have $\Lambda^{\geq\delta+1} = 0$ and $H^1(V,\Lambda^{\geq\delta+1})$ is trivial. So if we prove that $H^1(V,\lambda^{(\delta)})$ is trivial, by descending induction, all $H^1(V,\Lambda^{\geq\delta})$ will be trivial and (taking $\delta := 0$) so will be $H^1(V,\Lambda)$. We now proceed to prove that this indeed holds for all propers open subsets $V \subset \mathbf{E}_q$.

15.3.2.3. To understand better the sheaves $\lambda := \lambda_I(A_0)$ and all the $\lambda^{(\delta)} := \lambda_I^{(\delta)}(A_0)$, $\delta \geq 0$, we begin with a short intrinsic description. The module $M_0 := \left(\mathbf{C}(\{x\})^n, \Phi_{A_0}\right)$ splits into its pure components as a direct sum $M_0 = P_1 \oplus \cdots \oplus P_k$, where $P_i := \left(\mathbf{C}(\{x\})^{r_i}, \Phi_{B_i}\right)$, $i = 1,\ldots,k$. This causes a splitting of the module "internal End":

$$\underline{\mathrm{End}}(M_0) := \underline{\mathrm{Hom}}(M_0,M_0) = \bigoplus_{i,j=1,\ldots,k} \underline{\mathrm{Hom}}(P_j,P_i).$$

The q-difference module $\underline{\mathrm{End}}(M_0)$ can be described as the vector space

$$\mathrm{Mat}_n(\mathbf{C}(\{x\}))$$

endowed with the structural σ_q-linear automorphism $F \mapsto A_0^{-1}(\sigma_q F)A_0$.

The q-difference module $\underline{\mathrm{Hom}}(P_j,P_i)$ can be described as the $\mathbf{C}(\{x\})$-vector space $\mathrm{Mat}_{r_i,r_j}(\mathbf{C}(\{x\}))$ endowed with the structural σ_q-linear automorphism $f \mapsto B_i^{-1}(\sigma_q f)B_j$. So its solutions in any q-difference ring R are the $f \in \mathrm{Mat}_{r_i,r_j}(R)$ such that $(\sigma_q f)B_j = B_i f$, i.e., the morphisms $P_j \to P_i$ over R. The associated sheaf $\lambda^{(i,j)}$ of solutions of $\underline{\mathrm{Hom}}(P_j,P_i)$ is locally free of rank $r_i r_j$, thus a holomorphic vector bundle. The sheaf of solutions of $\underline{\mathrm{End}}(M_0)$ is $\bigoplus_{i,j=1,\ldots,k} \lambda^{(i,j)}$.

Each q-difference module $\underline{\mathrm{Hom}}(P_j,P_i)$ is pure isoclinic of slope $\mu_i - \mu_j$. So the positive component is:

$$\underline{\mathrm{End}}^{>0}(M_0) = \bigoplus_{1\leq i<j\leq k} \underline{\mathrm{Hom}}(P_j,P_i).$$

Its associated sheaf of solutions is:

$$\lambda = \lambda_I(A_0) = \bigoplus_{\delta\geq 1} \lambda^{(\delta)} \text{ and each } \lambda^{(\delta)} = \bigoplus_{1\leq i<j\leq k} \lambda^{(i,j)}.$$

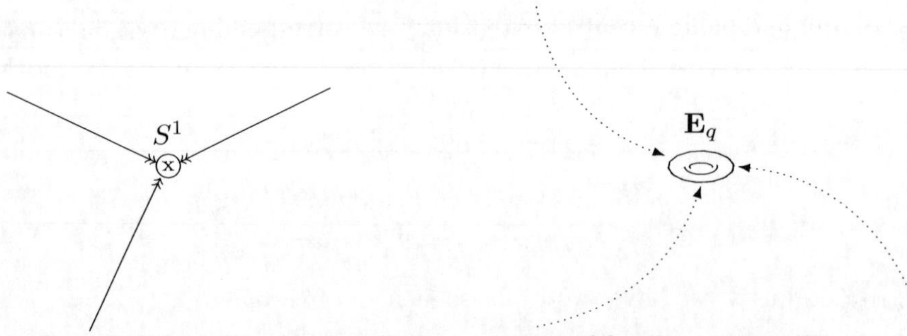

Classical horizon at $0 \in \mathbf{C}^*$ with quotient the infinitesimal circle of directions

q-spiral horizon at $0 \in \mathbf{C}^*$ with quotient the infinitesimal torus $\mathbf{E}_q$

FIGURE 15.2. The two horizons corresponding to semi-groups $e^{]-\infty,0]}$ and $q^{\mathbf{N}}$

Returning to our pending proof, we see that $\lambda = \lambda_I(A_0)$ is a locally free sheaf over $\mathbf{E}_q$, and its restriction to the proper subset $V \subset \mathbf{E}_q$ is a holomorphic vector bundle over an open Riemann surface. By Appendix I or by combining

(1) the proposition in [**100**, Chapter 2, §5.3], which guarantees that $\lambda_I(A_0)$ is coherent;
(2) [**230**, Théorème 1], which ensures us that we can legitimately use the cohomology over the projective algebraic curve $\mathbf{E}_q$;
(3) and [**114**, Theorem 3.7 of III.3], which entails triviality,

we conclude that $H^1(V, \lambda_I(A_0))$ is trivial. This completes the proof of the theorem.

15.3.3. Relation to asymptotic theory: the q-analogue of Birkhoff-Malgrange-Sibuya Isomorphism Theorem. This section is intended to sketch the q-analogue of Birkhoff-Malgrange-Sibuya Isomorphism Theorem (versus the Classification Theorem proved in 15.3.2).

15.3.3.1. The classical Birkhoff-Malgrange-Sibuya theorems are stated in terms of sheaves defined by asymptotic conditions, and so is the q-analogue of the Isomorphism Theorem stated and proved in [**197**, Chapter 4]. However, that part of the theory is much difficult and it does not seem possible to simplify it by the algebraic approach through filtration and graduation. The q-asymptotic theory involved has been developed by Ramis and Zhang in the last two decades; it is summarized in [**197**]. The formulation of classical asymptotics (plain or Gevrey) in terms of a semigroup of motions toward a "horizon" is due to Jose-Luis Martins; see 14.1.4.

In the classical case, the semigroup is continuous: $\Sigma := e^{]-\infty,0]}$, it operates on $\mathbf{C}^*$ (or on the germ $(\mathbf{C}^*, 0)$) and the *horizon* is the quotient $\mathbf{C}^*/\Sigma$ (or $(\mathbf{C}^*, 0)/\Sigma$), which can naturally be identified with the Malgrange circle of directions S^1 (think of an infinitesimal circle with center 0). In our setting, the semigroup will be $\Sigma := q^{\mathbf{N}}$, also operating on $\mathbf{C}^*$ or on $(\mathbf{C}^*, 0)$; and therefore, the horizon will be $\mathbf{E}_q$; see Figure 15.2.

EXERCISE 15.73. It is easy to attach to each half line in the left figure a point on the circle. How do you attach to a half q-spiral in the right figure a point on the torus? How does varying the spiral affect the point?

15.3.3.2. As explained above, we have the semigroup $\Sigma := q^{\mathbf{N}}$ acting by multiplication on $\mathbf{C}^*$, or, more frequently, on the germ of space $(\mathbf{C}^*, 0)$. For every subset $A \subset \mathbf{C}^*$, we set:

$$\Sigma(A) := q^{\mathbf{N}} A = \{q^m a \mid m \in \mathbf{N}, a \in A\}.$$

We call *Σ-invariant* a (generally open) subset $V \subset \mathbf{C}^*$ such that $\Sigma(V) \subset V$, equivalently $qV \subset V$. We then define the *horizon* of V as $V_\infty := \pi(V) \subset \mathbf{E}_q$ (recall that $\pi : \mathbf{C}^* \to \mathbf{E}_q$ is the natural projection). All open subsets of $\mathbf{E}_q$ are horizons of open subsets of $\mathbf{C}^*$ but, of course, V_∞ being given, V is not unique. $V \subset \mathbf{C}^*$ being such a Σ-invariant subset, we call *stable strict subset* $K \subset V$ a subset of the form $K = \Sigma(K')$ for some compact $K' \subset V$. One can then show that $K \cup \{0\} \subset \mathbf{C}$ is compact, as is $\pi(K) \subset \mathbf{E}_q$.

Let $V \subset \mathbf{C}^*$ a Σ-invariant subset and let $f \in \mathscr{O}(V)$. We say that *f is q-asymptotic to $\hat{f} \in \mathbf{C}[[x]]$*, abbreviated indifferently as $f \underset{V}{\sim} \hat{f}$ or as $f \underset{U}{\sim} \hat{f}$ (where U denotes the horizon of V), if[18], with usual notations $\hat{f} = \sum a_n x^n$ and $S_m \hat{f} := a_0 + \cdots + a_m x^m$:

$$\forall n \geq 0\ ,\ \forall K \subset V \text{ stable strict}\ ,\ x^{-n}(f - S_{n-1}\hat{f}) \text{ is bounded over } K.$$

If f admits such a q-asymptotic expansion, it is unique (we assume of course that V is nonvoid) and we then set $f^{(n)}(0) := n! a_n$. We say that f is *(infinitely) flat on V* if it is q-asymptotic to 0.

We shall now define sheaves on $\mathbf{E}_q$ to express those conditions. Following **[197]** and shunning no ambiguity, we use overloaded notations. So let first $V \subset \mathbf{C}^*$ a Σ-invariant subset and set:

$$\mathcal{A}(V) := \{f \in \mathscr{O}(V) \mid f \text{ admits an asymptotic expansion}\},$$
$$\mathcal{A}_0(V) := \{f \in \mathscr{O}(V) \mid f \text{ admits a trivial asymptotic expansion}\},$$
$$\mathcal{B}(V) := \{f \in \mathscr{O}(V) \mid f \text{ is bounded on every strict stable subset of } V\}.$$

EXERCISE 15.74. Define $\mathcal{A}$ in terms of $\mathcal{B}$. Check that $\mathcal{A}(V)$ is an algebra and $\mathcal{A}_0(V)$ an ideal.

Now for every open subset $U \subset \mathbf{E}_q$, we define:

$$\mathcal{A}(U) := \varinjlim_{V_\infty = U} \mathcal{A}(V), \quad \mathcal{A}_0(U) := \varinjlim_{V_\infty = U} \mathcal{A}_0(V), \quad \mathcal{B}(U) := \varinjlim_{V_\infty = U} \mathcal{B}(V).$$

THEOREM 15.6 (Weak q-analogue of Borel-Ritt). **[197**, Theorem 4.1.4**]** *The natural map from $\mathcal{A}$ to the constant sheaf $\mathbf{C}[[x]]$ over $\mathbf{E}_q$ is surjective (as a morphism of sheaves).*

This means that, $\hat{f} \in \mathbf{C}[[x]]$ being given, $\mathbf{E}_q$ can be covered by open subsets U_i such that there are $f_i \in \mathcal{A}(U_i)$ satisfying $f_i \underset{U_i}{\sim} \hat{f}$.

15.3.3.3. If we apply (componentwise) the above theorem to $\widehat{F} \in \mathrm{GL}_n(\mathbf{C}[[x]])$, we obtain an open covering (U_i) of $\mathbf{E}_q$ and invertible matrices F_i with components in $\mathcal{A}(V_i)$ (where Vi has horizon U_i) such that $F_i \underset{V_i}{\sim} \widehat{F}$, so $F_{i,j} := F_i^{-1} F_j \underset{V_{i,j}}{\sim} I_n$ (as

[18]Note that, as in Chapter 14, and according to what seems the taste of analysts, we rather use here $\mathbf{C}[[x]]$ than $\mathbf{C}((x))$.

usual $V_{i,j} := V_i \cap V_j$). we now introduce a new sheaf Λ_I on $\mathbf{E}_q$ (its precise relation with the sheaf $\Lambda_I(A_0)$ will be explored in 15.3.4):

$$\Lambda_I(U) := \{F \in \mathrm{GL}_n(\mathcal{A}(U)) \mid F \underset{U}{\sim} I_n\} = I_n + \mathrm{Mat}_n(\mathcal{A}_0(U)).$$

With in view the restriction to $\mathcal{G}_{A_0}$, we then define, for every algebraic subgroup G of GL_n:

$$\Lambda_I^G(U) := \Lambda_I(U) \cap G(\mathcal{A}(U)) = \{F \in G(\mathcal{A}(U)) \mid F \underset{U}{\sim} I_n\} = I_n + \mathrm{Mat}_n(\mathcal{A}_0(U)).$$

So the $F_{i,j}$ obtained above by application of weak q-Borel-Ritt form a cocycle for Λ_I relative to the covering (U_i). It is not hard to check (say it is an exercise for you) that, if $F \in \mathrm{GL}_n(\mathbf{C}\{x\})$, then $FF_i \underset{V_i}{\sim} F\widehat{F}$, so that the cocycle associated with $F\widehat{F}$ is the same. Actually, to be perfectly rigorous, one must note that the covering provided by weak q-Borel-Ritt is not unique but any two such coverings admit a common refinement and a well defined cohomology class in $H^1(\mathbf{E}_q, \Lambda_I)$.

THEOREM 15.7 (q-analogue of Birkhoff-Malgrange-Sibuya Isomorphism Theorem). [**197**, Theorem 4.3.11] *We thus obtain a bijection:*

$$\frac{GL_n(\mathbf{C}[[x]])}{GL_n(\mathbf{C}\{x\})} \to H^1(\mathbf{E}_q, \Lambda_I).$$

More generally, for every algebraic subgroup G of GL_n,we obtain a bijection:

$$\frac{G(\mathbf{C}[[x]])}{G(\mathbf{C}\{x\})} \to H^1(\mathbf{E}_q, \Lambda_I^G).$$

Since source and target are, in both cases, pointed sets, those bijections are understood to respect the marked point (can you see what class is represented by that point?) The proof of this theorem, due to Ramis, is considerably more sophisticated than the general level of this book.

15.3.3.4. The application of q-Birkhoff-Malgrange-Sibuya Isomorphism Theorem to classification rests on the following refinement of weak q-Borel-Ritt:

THEOREM 15.8 (Existence of q-asymptotic solutions). [**197**, Theorem 4.2.1] *Let $A \in GL_n(\mathbf{C}(\{x\}))$ and $\hat{X} \in \mathbf{C}[[x]]^n$ a solution of $\sigma_q \hat{X} = A\hat{X}$. Let $V_0 \in \mathbf{C}^*$ an open subset which projects homeomorphically to $U := \pi(V_0) \subset \mathbf{E}_q$, and $V := \Sigma(V_0)$, so that $V_\infty = U$. Then there exists $X \in \mathcal{A}(U)^n$ such that $\sigma_q X = AX$ and $X \underset{U}{\sim} \hat{X}$.*

Thus every formal solution is the asymptotic expansion of a "true" solutions, actually many of them. We shall apply the theorem to equations of the form $(\sigma_q F)A_0 = A_U F$ (with unknown F): the unique formal solution $\widehat{F_U} \in \mathcal{G}_{A_0}(\mathbf{C}((x)))$ lies, under proper assumptions, in $\mathcal{G}_{A_0}(\mathbf{C}[[x]])$, so it is the expansion of many "true solutions" $F_i : A_0 \to A_U$ on $V_i \subset \mathbf{C}^*$ such that the U_i cover $\mathbf{E}_q$. The $F_{i,j}$ are automorphisms of A_0 in Λ_I, which gives a meaning (if not a complete proof ...) to the following:

THEOREM 15.9 (q-analogue of Birkhoff-Malgrange-Sibuya Classification Theorem). [**197**, Theorem 4.4.1] *We thus obtain a bijection from $\mathscr{F}(A_0) = \dfrac{GL_n^{A_0}(\mathbf{C}[[x]])}{GL_n(\mathbf{C}\{x\})}$ to $H^1(\mathbf{E}_q, \Lambda_I(A_0))$.*

Of course $\mathrm{GL}_n^{A_0}(\mathbf{C}[[x]])$ consists of those $\widehat{F} \in \mathrm{GL}_n(Rf)$ such that $\widehat{F}[A_0] \in \mathrm{GL}_n(Ra)$; as explained at the end of 15.2.4.3, the use of $\mathbf{C}[[x]]$, $\mathbf{C}\{x\}$ instead of $\mathbf{C}((x))$, $\mathbf{C}(\{x\})$ is not significant (this will be further detailed in 15.3.4). However, we abuse here the notation $\Lambda_I(A_0)$ by giving it a sense related to asymptotics, not a priori the same as in 15.3.2.1. We will now justify this abuse.

LEMMA 15.75. *Let X be a solution of $\sigma_q X = x^\delta C X$, $C \in GL_n(\mathbf{C})$, $X \in \mathscr{O}(\pi^{-1}(U))^n$.*

(i) *If $\delta > 0$, then $X \in \mathcal{A}_0(U)^n$.*
(ii) *If $\delta \leq 0$ and $X \in \mathcal{A}_0(U)^n$, then $X = 0$.*

PROOF. (i) If $K' \subset U$ is compact and $K = \Sigma(K')$, let $M := \max_{K'} \|X\|$. Then, on $q^m K'$, using $X(q^m x) = q^{\delta m(m-1)/2} x^{\delta m} C^m X(x)$, we have $\|X\| \leq M D^{\delta m} |q|^{\delta m^2/2}$ for some fixed $D > 0$. Flatness follows immediately.

(ii) Similar arguments when $\delta \leq 0$ show that if $X(x_0) \neq 0$ for some x_0, then X cannot be flat.

□

Now if $F[A_0] = F$ the (i,j) blocks satisfy $\sigma_q f_{i,j} = x^{\mu_i - \mu_j} A_i f_{i,j} A_j^{-1}$. If $F \sim I_n$, the lemma implies that F is in $\mathcal{G}_{A_0}$. Conversely, if F is in $\mathcal{G}_{A_0}$, by the lemma again, $F \sim I_n$. So our two definitions of $\Lambda_I(A_0)$ agree.

We should add that the asymptotic solutions provided by the theorem at the beginning of 15.3.3.4 can as well be taken to be the $S_{\overline{c}} \widehat{F}_U$, so that the maps used in both versions of the Classification Theorem are indeed the same.

15.3.4. *Bona fide* analytic isoformal classification. Based on the results of Chapter 10, we have taken for granted in 15.2.3 that isoformal and isograded classification were exactly the same thing. This was, at best, a heuristic argument and it simplified (or at least it algebraized) what should have been a more faithful adaptation of the approach in **[14]**. It may sometimes be useful to check in more detail what our intuition tells us, and that we shall do here in a very down-to-earth way.

15.3.4.1. Recall first from 8.3.3, Chapter 8, the formalization functor which sends a q-difference module M over $\mathbf{C}(\{x\})$ to the q-difference module $\hat{M} := \mathbf{C}((x)) \otimes_{\mathbf{C}(\{x\})} M$ over $\mathbf{C}((x))$; and the morphim $f : M \to N$ to the morphism $\mathrm{Id}_{\mathbf{C}((x))} \otimes f : \hat{M} \to \hat{N}$. As noted there, the formalization of $M := (\mathbf{C}(\{x\})^n, \Phi_A)$ is $\hat{M} := (\mathbf{C}((x))^n, \Phi_A)$ and the formalization of $F : (\mathbf{C}(\{x\})^n, \Phi_A) \to (\mathbf{C}(\{x\})^m, \Phi_B)$, $F \in \mathrm{Mat}_{m,n}(\mathbf{C}(\{x\}))$ is $\widehat{F} := F \in \mathrm{Mat}_{m,n}(\mathbf{C}((x)))$ itself. It follows that $f : M \to N$ is an isomorphism if, and only if, $\hat{f} : \hat{M} \to \hat{N}$ is.

So we fix a q-difference module M_0 over $\mathbf{C}(\{x\})$ (later to be taken pure) and define the class $\widehat{\mathscr{E}}(M_0)$ (as usual it is not a set, but nevertheless ...) of all pairs (M, u), where M is a q-difference module over $\mathbf{C}(\{x\})$ and $u : \hat{M} \to \hat{M}_0$ is an isomorphism over $\mathbf{C}((x))$; and then, an equivalence relation $\sim$ in $\widehat{\mathscr{E}}(M_0)$:

$$(M, u) \sim (M', u') \underset{def}{\iff} \exists f : M \to M' \ : \ u = u' \circ \hat{f}.$$

Thus $\hat{f}$, whence f has to be an isomorphism. We shall say that f *enforces* the equivalence $\sim$. It is clear that $\sim$ is indeed an equivalence relation and we set:

$$\widehat{\mathscr{F}}(M_0) := \frac{\widehat{\mathscr{E}}(M_0)}{\sim}.$$

This can rightly be called the set (see below) of *isoformal analytic classes.* Clearly, $\widehat{\mathscr{F}}(M_0)$ projects surjectively to the set of analytic isomorphism classes of q-difference modules formally equivalent to M (this is indeed a set: take all of them with underlying set M_0). By the matricial description that is coming now, $\widehat{\mathscr{F}}(M_0)$ itself is a set. Before giving that description, we make two preliminary remarks.

1- Restriction to the case of a pure M_0: Let $\phi : \hat{M}_0 \to \hat{N}_0$ a formal isomorphism. Then $(M, u) \mapsto (M, \phi \circ u)$ is a bijection from $\widehat{\mathscr{E}}(M_0)$ to $\widehat{\mathscr{E}}(N_0)$, and it is compatible with the equivalence relation $\sim$, because if $f : M \to M'$ enforces that $(M, u) \sim (M', u')$, it also enforces that $(M, \phi \circ u) \sim (M', \phi \circ u')$. So, in order to construct $\widehat{\mathscr{F}}(M_0)$, we can choose any particular q-difference module in the class $\widehat{\mathscr{E}}(M_0)$. We shall therefore take for M_0 a pure q-difference module $(\mathbf{C}(\{x\})^n, \Phi_{A_0})$, with the usual notations and conventions $A_0 := \mathrm{Diag}(B_1, \ldots, B_k)$, etc.

2- Restriction to a subclass of $\widehat{\mathscr{E}}(M_0)$: Let (M, u) in $\widehat{\mathscr{E}}(M_0)$ and let $f : M \to M'$ an analytic isomorphism. Then $(M, u) \sim (M', u')$, where $u' := u \circ \hat{f}^{-1}$ (this is enforced by f itself). So we can restrict $\widehat{\mathscr{E}}(M_0)$ to any subclass containing enough representatives M for analytic equivalence. We shall of course use the subclass of q-difference modules $(\mathbf{C}(\{x\})^n, \Phi_A)$.

This leads us to define (with the above possible choices of A_0):

$$\widehat{\mathscr{E}}(A_0) := \{(A, \widehat{F}) \mid \widehat{F} \in \mathrm{GL}_n(\mathbf{C}((x))) \text{ and } \widehat{F}[A_0] = A\},$$

and the equivalence relation:

$$(A, \widehat{F}) \sim (B, \widehat{G}) \underset{def}{\iff} \exists \Phi \in \mathrm{GL}_n(\mathbf{C}(\{x\})) \ : \ \widehat{G} = \Phi \widehat{F}.$$

(Note that $\widehat{F}[A_0] = A$, $\widehat{G}[A_0] = B$ and $\widehat{G} = \Phi \widehat{F}$ together imply $\Phi[A] = B$.) By the first remark above, we can decide that $M_0 := (\mathbf{C}(\{x\})^n, \Phi_{A_0})$; then, by the second remark , the natural inclusion from $\widehat{\mathscr{E}}(A_0)$ into $\widehat{\mathscr{E}}(M_0)$ sending $(A, \widehat{F})$ to $((\mathbf{C}(\{x\})^n, \Phi_A), \widehat{F})$ induces a bijection from $\widehat{\mathscr{F}}(A_0) := \dfrac{\widehat{\mathscr{E}}(A_0)}{\sim}$ to $\widehat{\mathscr{F}}(M_0)$. We immediately deduce the following identification:

$$\widehat{\mathscr{F}}(A_0) \simeq \frac{\mathrm{GL}_n{}^{A_0}(\mathbf{C}((x)))}{\mathrm{GL}_n(\mathbf{C}(\{x\}))}.$$

15.3.4.2. Letting $A_0 := \mathrm{Diag}(B_1, \ldots, B_k)$, $B_i := x^{\mu_i} A_i$, $\mu_1 > \cdots > \mu_k$ in $\mathbf{Z}$ and $A_i \in \mathrm{GL}_{r_i}(\mathbf{C})$ as always, we do not restrict generality (by the first remark in 15.3.4.1). By the second remark, we can restrict $\widehat{\mathscr{E}}(A_0)$ to pairs $(A_U, \widehat{F})$, $U \in \prod\limits_{1 \leq i < j \leq k} \mathrm{Mat}_{r_i, r_j}(\mathbf{C}(\{x\}))$; and even, thanks to Birkhoff-Guenther Normal Form in 15.2.4.3, to the case that each $U_{i,j} = x^{\mu_j} V_{i,j}$, $V_{i,j} \in \mathrm{Mat}_{r_i, r_j}(\mathbf{C} + \cdots + \mathbf{C} x^{\mu_i - \mu_j - 1})$.

LEMMA 15.76. *With the above assumptions, let $\widehat{F} \in GL_n(\mathbf{C}((x)))$ such that $\widehat{F}[A_0] = A_U$. Then $\widehat{F} = F_f \mathit{Diag}(F_1, \ldots, F_k)$, where each $F_i \in GL_{r_i}(\mathbf{C}(\{x\}))$ is such that $F_i[A_i] = A_i$, $i = 1, \ldots, k$, and where $F_f \in \mathcal{G}_{A_0}(\mathbf{C}[[x]])$ is such that $F_f[A_0] = A_U$.*

PROOF. We write $F_{i,j}$ the (i,j) block of $\widehat{F}$ (it has format $r_i \times r_j$). Equation $\widehat{F}[A_0] = A_U$ then writes:

$$\forall i,j = 1,\ldots,k\ ,\ (\sigma_q F_{i,j})B_j - B_i F_{i,j} = \sum_{i<l\leq k} U_{i,l}F_{l,j}.$$

For $i = k$, the right-hand side is 0 and either $j < k$, in which case, since B_k, B_j are pure of different slopes, the only possible solution $F_{k,j}$ with components in $\mathbf{C}((x))$ is 0; or $j = k$ and $F_k := F_{k,k}$ is a formal automorphism of a pure B_k, whence an analytic automorphism (actually, a Laurent polynomial).

Then we go bottom up in the matrix (decreasing values of i). At each step, by induction, the right-hand side is 0 for $j \leq i$ and the same argument implies that $F_{i,j} = 0$ for $j < i$ and that $F_i := F_{i,i}$ is a formal automorphism of a pure B_i, whence an analytic automorphism. In the end, it is clear that $\widehat{F}\mathrm{Diag}(F_1,\ldots,F_k)^{-1}$ has the form $F_f \in \mathcal{G}_{A_0}(\mathbf{C}((x)))$ and that $F_f[A_0] = A_U$.

Now we write the last relation in the usual form of a systems of equations for blocks:

$$\forall 1 \leq i < j \leq k\ ,\ (\sigma_q f_{i,j})B_j - B_i f_{i,j} = \sum_{i<l<j} U_{i,l}f_{l,j} + U_{i,j},$$

equivalently:

$$\forall 1 \leq i < j \leq k\ ,\ (\sigma_q f_{i,j})A_j - x^{\mu_i-\mu_j}A_i f_{i,j} = \sum_{i<l<j} x^{\mu_l-\mu_j}V_{i,l}f_{l,j} + V_{i,j}.$$

This last equation has the form $\sigma_q u - x^\delta Cu = v$, $\delta \geq 1$, $C \in \mathrm{GL}_r(\mathbf{C})$ and the resolution by iteration of $u \mapsto {\sigma_q}^{-1}(v + x^\delta Cu)$ shows that, if $v \in \mathbf{C}[[x]]^r$, then the unique formal solution $u \in \mathbf{C}((x))^r$ is in $\mathbf{C}[[x]]^r$ too. So inductively all $f_{i,j}$ have components in $\mathbf{C}[[x]]$. □

Therefore, in the terms of the second remark in 15.3.4.1, we can restrict to the subclass of $\widehat{\mathscr{E}}(A_0)$ involving pairs (A_U, F_f) instead of $(A_U, \widehat{F})$. But then, by the usual uniqueness argument, F_f is necessarily $\widehat{F}_U$. Moreover, any $F : A_U \to A_V$ ensuring that $(A_U, \widehat{F}_U) \sim (A_V, \widehat{F}_V)$ has to be in $\mathcal{G}_{A_0}$. In other terms, we have recovered our previous definition of $\mathscr{F}(A_0)$ as the quotient of the set $\mathscr{E}(A_0)$ by the relation $\sim$ defined over it. More precisely: mapping A_U to $(A_U, \widehat{F}_U)$ induces a bijection of $\mathscr{F}(A_0)$ with $\widehat{\mathscr{F}}(A_0)$.

EXERCISE 15.77. Use the arguments and calculations above to check that the following identifications are valid; $\mathscr{F}(A_0) = \widehat{\mathscr{F}}(A_0)$ and both equal:

$$\frac{\mathrm{GL}_n^{A_0}(\mathbf{C}((x)))}{\mathrm{GL}_n(\mathbf{C}(\{x\}))} = \frac{\mathrm{GL}_n^{A_0}(\mathbf{C}[[x]])}{\mathrm{GL}_n(\mathbf{C}\{x\})} = \frac{\mathcal{G}_{A_0}^{A_0}(\mathbf{C}((x)))}{\mathcal{G}_{A_0}(\mathbf{C}(\{x\}))} = \frac{\mathcal{G}_{A_0}^{A_0}(\mathbf{C}[[x]])}{\mathcal{G}_{A_0}(\mathbf{C}\{x\})}.$$

15.4. The Galois group of irregular systems: a breezy account

Even restricting, as we will, to q-difference modules of integral slopes, the theory is rather intricate. So we shall content us with a dogmatic exposition[19] , including constructions and statements but no proof: see **[195]** for them. And, for the general case of arbitrary slopes, see **[224]** and [CV]. The vocabulary and basics of Tannakian theory in the form we need are summarized in Appendix G; see also a short overview in **[222]**.

[19]We also totally omit the resolution of the *inverse problem* as given in **[195]**.

15.4.1. General structure.

15.4.1.1. *The categories and functors at stake.* We shall denote for short $\mathscr{C}$ the category of q-difference modules over $\mathbf{C}(\{x\})$ such that $S(M) \subset \mathbf{Z}$; and $\mathscr{C}_p$ its full subcategory with objects the pure modules. In comparison with the classical theory, and in accordance to our general strategy of using the graduation functor in lieu of the formalization functor (as in 15.3.4), $\mathscr{C}_p$ plays here the role of the formal category (compare with $\mathscr{C}^\wedge{}_0$ *vs* $\mathscr{C}_0$ in **[14]** or with $\mathrm{Diff}_{\hat{K}}$ *vs* Diff_K in **[250]**). Both $\mathscr{C}$ and $\mathscr{C}_p$ are $\mathbf{C}$-linear abelian categories.

The category $\mathscr{C}$ is equipped with "tensor operations": a tensor product $\otimes$, internal Homs $\underline{\mathrm{Hom}}$, a dual $M^\vee$ for each q-difference module M, and a unit object $\underline{1}$; and $\mathscr{C}_p$ is stable under all those operations. There are adjunction and rigidity formulas, most of which have been stated and proved in section 9.1, to be completed by Chapter 10 as regards integral slopes and pure modules.

There is an inclusion functor ι from $\mathscr{C}_p$ to $\mathscr{C}$ and a graduation functor gr going back, which is a retraction of the former, i.e., $\mathrm{gr} \circ \iota$ is the identity functor of $\mathscr{C}_p$. Both functors are exact, $\mathbf{C}$-linear, faithful and $\otimes$-compatible.

We shall *identify* those categories with their full subcategories having as objects the models $(\mathbf{C}(\{x\})^n, \Phi_A)$, $A \in \mathrm{GL}_n(\mathbf{C}(\{x\}))$. Usually, we shall speak of an object $A \in \mathrm{GL}_n(\mathbf{C}(\{x\}))$ (in lieu of $(\mathbf{C}(\{x\})^n, \Phi_A)$) and then a morphism from A to $B \in \mathrm{GL}_p(\mathbf{C}(\{x\}))$ will be understood to be a matrix $F \in \mathrm{Mat}_{p,n}(\mathbf{C}(\{x\}))$ such that $(\sigma_q F)A = BF$. Since any pure object is isomorphic to one of the form A_0 and any object to one of the form A_U (same standard conventions of notation as in 15.3), most constructions will be described in those terms. And we shall often take A_U in Birkhoff-Guenther Normal Form.

15.4.1.2. *Fibre functors and (universal) Galois groups.* We are going to define *fibre functors* on $\mathscr{C}$ and $\mathscr{C}_p$, that is, exact $\mathbf{C}$-linear faithful $\otimes$-compatible functors to the category $\mathcal{V}ect^f_{\mathbf{C}}$ of finite-dimensional complex vector spaces. For that we first fix a particular element $a \in \mathbf{C}^*$; its role is unessential (in some sense, like fixing a particular base point to define a fundamental group).

On $\mathscr{C}_p$ we take the fibre functor $\hat{\omega}$ which sends A_0 to the underlying space $\mathbf{C}^n$; and $F_0 : A_0 \to B_0$ to $F_0(a) \in \mathrm{Mat}_{p,n}(\mathbf{C})$, seen as a morphism $\mathbf{C}^n \to \mathbf{C}^p$. Note that F_0 is a Laurent polynomial, thus automatically holomorphic over $\mathbf{C}^*$, so $F_0(a)$ is well defined. One can prove that $\hat{\omega}$ is indeed a fibre functor on $\mathscr{C}_p$. Then we set $\omega := \hat{\omega} \circ \mathrm{gr}$, which is a fibre functor on $\mathscr{C}$. Moreover, it restricts to $\hat{\omega}$ on $\mathscr{C}_p$ since $\omega \circ \iota = \hat{\omega} \circ \mathrm{gr} \circ \iota = \hat{\omega}$.

Let $G := \mathrm{Aut}^{\otimes}(\omega)$ the group of $\otimes$-automorphisms of the functor ω and let $G_p := \mathrm{Aut}^{\otimes}(\hat{\omega})$ (the group law will be described in a moment). These are actually *proalgebraic groups* over $\mathbf{C}$, but we do not take in account that fact for the moment. An element ϕ of G has the form of a "collection" (ϕ_A) indexed by the objects A of $\mathscr{C}$, where each ϕ_A is a $\mathbf{C}$-linear automorphism of $\omega(A) = \mathbf{C}^n$, and ϕ is subjected to some collective constraints (functoriality and tensor compatibility) which we do not need to make explicit (compare with the description of the fuchsian Galois group in Subsection 13.2.2). The internal law that makes G a group defines the product of collections (ϕ_A) and (ϕ'_A) as the collection $(\phi_A \phi'_A)$. An element ψ of G_p is similarly defined with respect to $\mathscr{C}_p$ as a collection (ψ_{A_0}).

So let $\phi \in G$. By restriction, it defines an element of G_p, the collection $(\psi_{A_0}) := (\phi_{\iota(A_0)})$ where A_0 runs among the objects of $\mathscr{C}_p$. This way we obtain a map $\iota^* : G \to G_p$, which happens to be a morphism of groups (as tautologically checked if one takes in account the definitions). It is actually, in the terminology of proalgebraic groups, a *rational* morphism of proalgebraic groups.

Conversely let ψ an element of G_p. Setting $\phi_A := \psi_{\mathrm{gr}A}$ for every object A of $\mathscr{C}$, we define an element $\mathrm{gr}^*\psi := \phi$ of G; and the map gr^* from G_p to G is again a rational morphism of proalgebraic groups. Moreover, it is immediately checked that $\iota^*(\mathrm{gr}^*(\psi)) = \psi$; indeed, its A_0 component is:

$$(\iota^*(\mathrm{gr}^*(\psi)))_{A_0} = (\mathrm{gr}^*(\psi))_{\iota(A_0)} = \psi_{\mathrm{gr}(\iota(A_0))} = \psi_{A_0}.$$

So $\iota^* \circ \mathrm{gr}^* = \mathrm{Id}_{G_p}$. Therefore, denoting $\mathfrak{St} := \mathrm{Ker}\ \iota^*$, we have a semi-direct product:

$$G = \mathfrak{St} \ltimes G_p.$$

Note that the *Stokes group* $\mathfrak{St}$ is, by construction, a proalgebraic subgroup of G; but so is also G_p thanks to the embedding gr^*. The structure of G is entirely determined by the structures of G_p and of $\mathfrak{St}$, together with the conjugacy action on G_p on the normal subgroup $\mathfrak{St}$.

15.4.1.3. *Objects and representations.* Fix some matrix $A = A_U$, seen as an object of $\mathscr{C}$ and let $V := \omega(A)$. For each element ϕ of G, denote $\rho_A(\phi) := \phi_A \in \mathrm{GL}(V)$. Then ρ_A is a group morphism from G to $\mathrm{GL}(V)$ (this is immediate from the definition of the product on G), thus a finite dimensional complex linear representation of G. (It is moreover a rational representation.) In exactly the same way, we associate to every object A_0 of $\mathscr{C}_p$ a representation $\hat{\rho}_{A_0}$ of G_p. We might actually use the same denotation ρ_{A_0} because, A_0 seen as an object of $\mathscr{C}$, defines a representation of G which restricts to $\hat{\rho}_{A_0}$ on G_p. More formally:

$$\rho_{\iota(A_0)} \circ \mathrm{gr}^* = \hat{\rho}_{A_0},$$

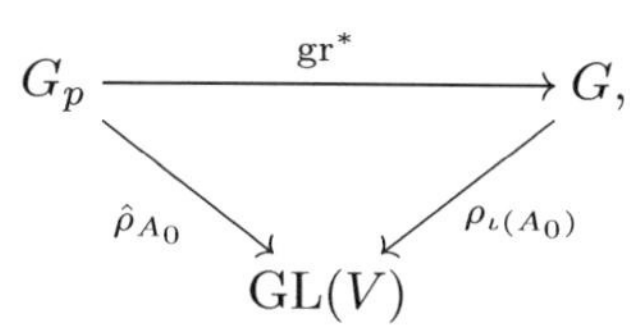

The fundamental theorem of Tannakian theory (Appendix G) then states that the correspondences:

$$A \longleftrightarrow \rho_A, \quad A_0 \longleftrightarrow \hat{\rho}_{A_0},$$

are equivalences of $\otimes$-categories: the first one between $\mathscr{C}$ and the category $\mathcal{R}ep_{rat}(G)$ of (rational) finite-dimensional complex linear representations of G; the second one, between $\mathscr{C}_p$ and $\mathcal{R}ep_{rat}(G_p)$. Practically, this means that, given such a representation ρ of G, resp. $\hat{\rho}$ of G_p, we can recover an object A of $\mathscr{C}$, resp. A_0 of $\mathscr{C}_p$ such that $\rho = \rho_A$, resp. $\hat{\rho} = \hat{\rho}_{A_0}$.

When we describe more precisely the groups G and G_p, we shall make explicit how to to recover object from representation. Meanwhile, we can at least take advantage of the semi-direct product to say a little more. Write $s \mapsto s^g$ the conjugacy action of $g \in G_p$ on $\mathfrak{St}$, i.e., s^g is just $g^{-1}sg$ computed in G. Then a representation ρ of G in some space V induces a representation ρ' of $\mathfrak{St}$ and a representation $\hat{\rho}$ of G_p and:

$$\rho'(s^g) = \rho(g^{-1}sg) = \rho(g)^{-1}\rho(s)\rho(g) = \rho'(s)^{\hat{\rho}(g)},$$

where the last term reflects the conjugacy action inside $\mathrm{GL}(V)$. Conversely, any pair of representations ρ' of $\mathfrak{St}$ and $\hat{\rho}$ of G_p such that $\rho'(s^g) = \rho'(s)^{\hat{\rho}(g)}$ for all s, g defines a representation of G.

The Galois group of an object. The image $G(A) := \text{Im } \rho_A \subset \text{GL}(\omega(A)) = \text{GL}_n(\mathbf{C})$ is an algebraic subgroup, the Galois group of A. If $A = A_0$ is pure, it is naturally identified $G_p(A_0) := \text{Im } \hat{\rho}_{A_0} \subset \text{GL}(\hat{\omega}(A_0)) = \text{GL}_n(\mathbf{C})$, so there is no ambiguity speaking of the Galois group of A_0 as an object of $\mathscr{C}$ or of $\mathscr{C}_p$.

The family of all groups $G(A)$ (resp. of all groups $G_p(A_0)$) can be made into an inverse system and the universal group is the inverse limit:

$$G = \lim_{\longleftarrow A} G(A), \quad G_p = \lim_{\longleftarrow A_0} G_p(A_0), \quad \mathfrak{St} = \lim_{\longleftarrow A} \mathfrak{St}(A).$$

(The first two equalities are general facts, the third is a consequence here.)

15.4.1.4. *The formal (or pure) Galois group G_p.* It can be shown that (equality meaning here canonical isomorphism):

$$G_p = G_{p,u} \times G_{p,s} \text{ where } \begin{cases} G_{p,u} = \mathbf{C} \text{ (unipotent component)}, \\ G_{p,s} = \mathbf{E}_q^\vee \times \mathbf{C}^* \text{ (semisimple component)}. \end{cases}$$

The components $\mathbf{C}$ and $\mathbf{E}_q^\vee := \text{Hom}_{gr}(\mathbf{E}_q, \mathbf{C}^*)$ have already been encountered in Chapter 13. The component $\mathbf{C}^*$ is related to the slopes, as we will soon see; it is the form $G_p =$ (something) $\times \mathbf{C}^*$ of the pure group that changes most when admitting arbitrary slopes, as shown by van der Put and Reversat in **[248]**; and then the action of the pure group on the Stokes group also becomes more complicated **[224]**.

In our case, let $A_0 := \text{Diag}(x^{\mu_1} A_1, \ldots, x^{\mu_k} A_k)$ and let $\phi := (\lambda, \gamma, t) \in G_p = \mathbf{C} \times \mathbf{E}_q^\vee \times \mathbf{C}^*$. Then the following relation (in which the first equality is by definition, see 15.4.1.3) expresses completely the duality of G_p and $\mathscr{C}_p$):

$$\phi_{A_0} = \rho_{A_0}(\phi) = \text{Diag}\left(\gamma(A_{1,s}) A_{1,u}^\lambda t^{\mu_1}, \ldots, \gamma(A_{k,s}) A_{k,u}^\lambda t^{\mu_k}\right).$$

15.4.2. The Stokes group $\mathfrak{St}$ and the adjoint action of G_p.

15.4.2.1. *The Stokes group and its Lie algebra.* Elements of $\mathfrak{St}$ are those who have trivial effect on A_0, thus those who "see" only the off-diagonal parts of $A = A_U$. It can be proved that $\mathfrak{St}(A) \subset \mathcal{G}_{A_0}(\mathbf{C})$. So in particular each $\mathfrak{St}(A)$ is a unipotent algebraic group and $\mathfrak{St}$ itself is a prounipotent algebraic group.

The study of unipotent algebraic groups is easier through their Lie algebras. We set:

$$\mathfrak{st}(A) := \text{Lie}(\mathfrak{St}(A)) \subset \mathfrak{g}_{A_0}(\mathbf{C}),$$

thus a nilpotent Lie algebra. Then $\mathfrak{st} := \text{Lie}(\mathfrak{St})$ is the inverse limit of all $\mathfrak{st}(A)$ and is a pronilpotent Lie algebra. It can be described intrinsically as the algebra of Lie-like tensor endomorphisms of ω. Such a thing can be described as a collection (D_A), each D_A an endomorphism of $\omega(A)$ and a $\otimes$-compatibility condition which makes it look like a kind of derivation. The group $\mathfrak{St}$ can be recovered as $\exp(\mathfrak{st})$, with multiplication defined by the Baker-Campbell-Hausdorff formula.

It is convenient to attach to $\mathfrak{St}$ the unipotent part of G_p, and to set:

$$\widetilde{\mathfrak{St}} := \mathfrak{St} \times \mathbf{C}, \quad \widetilde{\mathfrak{st}} := \text{Lie}(\widetilde{\mathfrak{St}}) = \mathfrak{st} \oplus \mathbf{C}\tau,$$

with the convention that $[\mathfrak{st}, \mathbf{C}\tau] = 0$ (this is because $\mathbf{C} \subset G_p$ commutes with $\mathfrak{St}$). The element τ is any generator, for instance the one defined by $\tau_A := \log A_u$.

15.4.2.2. *The conjugacy action of G_p and the adjoint action of $G_{p,s}$.* The conjugacy action of G_p on $\mathfrak{St}$ is best defined in terms of $G_p(A)$ acting on $\mathfrak{St}(A)$. Since the unipotent part $G_{p,u} = \mathbf{C}$ of G_p commutes with $\mathfrak{St}$, the corresponding conjugacy action on $\mathfrak{St}$ is trivial.

With the ususal notations for A_U and A_0, an element of $G_{p,s}(A)$ is a block diagonal $\text{Diag}(C_1, \ldots, C_k) \in \text{GL}_n(\mathbf{C})$ (each $C_i = t^{\mu_i}\gamma(A_{i,s})$) and an element of $\mathfrak{St}(A)$ is a $F_f \in \mathcal{G}_{A_0}(\mathbf{C})$ (so here the blocks $f_{i,j}$ are complex matrices). Then the conjugacy action sends each $f_{i,j}$ to $C_i^{-1} f_{i,j} C_j$.

The adjoint action of $G_p(A)$ on $\mathfrak{st}(A)$ is then given as follows: the unipotent part $G_{p,u}(A)$ acts trivially; the element $\text{Diag}(C_1, \ldots, C_k) \in G_{p,s}(A) \subset \text{GL}_n(\mathbf{C})$ acts on an element $N_g \in \mathfrak{st}(A) \subset \mathfrak{g}_{A_0}(\mathbf{C})$ (notation from 15.2.1.2) by sending each $g_{i,j}$ to $C_i^{-1} g_{i,j} C_j$.

If one wants instead to describe G through $G_{p,s}$ and $\widetilde{\mathfrak{St}}$, and the conjugacy action of $G_{p,s}$ on $\widetilde{\mathfrak{St}}$ through its action on $\widetilde{\mathfrak{st}} = \mathfrak{st} \oplus \mathbf{C}\mathcal{T}$, one combines the above description of the action of $G_{p,s}$ on $\mathfrak{st}$ with the condition that the action on $\mathbf{C}\mathcal{T}$ is trivial.

15.4.2.3. *The spectral decomposition.* The semi-simple proalgebraic group $G_{p,s} = \mathbf{C}^* \times \mathbf{E}_q^\vee$ is commutative and totally determined by its group of characters:

$$\mathbf{X}(G_{p,s}) = \mathbf{X}(\mathbf{C}^*) \times \mathbf{X}(\mathbf{E}_q^\vee) = \mathbf{Z} \times \mathbf{E}_q.$$

Its linear rational action on the space $\widetilde{\mathfrak{st}}$ then gives rise to a decomposition into eigenspaces:

$$\widetilde{\mathfrak{st}} = \bigoplus_{\substack{\delta \in \mathbf{Z} \\ \overline{c} \in \mathbf{E}_q}} \widetilde{\mathfrak{st}}^{(\delta,\overline{c})}, \text{ where } \widetilde{\mathfrak{st}}^{(\delta,\overline{c})} := \{X \in \widetilde{\mathfrak{st}} \mid \forall (t,\gamma) \in G_{p,s} \ , \ (t,\gamma).X = t^\delta \gamma(\overline{c}) X\}.$$

Actually, because of the properties of the slope filtration, all components with $\delta < 0$ are trivial and the only component with $\delta = 0$ is $\widetilde{\mathfrak{st}}^{(\delta,\overline{1})} = \mathbf{C}\mathcal{T}$. More will be said about the other components in 15.4.3.

15.4.3. Explicit galoisian Stokes operators.

15.4.3.1. *q-analogues of alien derivations.* Alien derivations were invented by Jean Ecalle in order to understand the "resurgence" properties of irregular differential equations; see **[150, 250]**. The following q-analogue is due to Ramis; see **[195]**.

Fix $A = A_U$ and A_0 with the usual conventions. Recall a, the so-to-say base-point which served to define fibre functors $\hat{\omega}$ and ω, thus also the universal Galois groups G and G_p and their particular images $G(A)$, $G_p(A_0)$ and $\mathfrak{St}(A)$, the kernel of $G(A) \to G_p(A_0)$.

Then it can be proved that $S_{\overline{c}}\widehat{F}_U(a) \in \mathfrak{St}(A)$ for every authorized direction $\overline{c}$. Since those are unipotent matrices, their logarithms are well determined and $\log S_{\overline{c}}\widehat{F}_U(a) \in \mathfrak{st}(A)$. Last, the mapping $\overline{c} \mapsto \log S_{\overline{c}}\widehat{F}_U(a)$ is a meromorphic mapping from $\mathbf{E}_q$ to $\mathfrak{st}(A)$, with simple poles at the points of Σ_{A_0}. Since residues can be expressed as integrals, they stay inside the linear space $\mathfrak{st}(A)$ and one can set:

$$\Delta_{\overline{d}}(A) := \text{Res}_{\overline{c}=\overline{d}} \log S_{\overline{c}}\widehat{F}_U(a) \in \mathfrak{st}(A),$$

which is therefore 0 if $\overline{d} \notin \Sigma_{A_0}$. One can prove that, for every $\overline{d} \in \mathbf{E}_q$, there is a "universal" $\Delta_{\overline{d}} \in \mathfrak{st}$ which specializes into all the $\Delta_{\overline{d}}(A) \in \mathfrak{st}(A)$.

The spectral decomposition of 15.4.2.3 applies to each such $\Delta_{\overline{d}}$:

$$\Delta_{\overline{d}} = \sum_{\substack{\delta \geq 1 \\ \overline{c} \in \mathbf{E}_q}} \Delta_{\overline{d}}^{(\delta,\overline{c})}, \text{ where each } \Delta_{\overline{d}}^{(\delta,\overline{c})} \in \mathfrak{st}^{(\delta,\overline{c})}.$$

From the explicit formulas for $S_{\overline{c}}\widehat{F}_U$ in the case of two slopes, one finds that $\Delta_{\overline{d}}^{(\delta,\overline{c})}$ is nonzero only if $\overline{d}^{\delta} = \overline{c}^{-1}$. So, since an element of $\mathbf{E}_q$ has δ^2 such roots, we have defined δ^2 particular elements of $\mathfrak{st}^{(\delta,\overline{c})}$.

15.4.3.2. *The density theorem and the wild fundamental group.* By classification arguments (resting on the results of 15.3), one proves that the above family of δ^2 particular elements of $\mathfrak{st}^{(\delta,\overline{c})}$ generates the quotient $\mathbf{C}$-linear space $\dfrac{\mathfrak{st}^{(\delta,\overline{c})}}{[\mathfrak{st}^{(\delta,\overline{c})}, \mathfrak{st}^{(\delta,\overline{c})}]}$, which besides has exactly dimension δ. So we can extract from it a family $\Delta_i^{(\delta,\overline{c})}$, $i = 1, \ldots, \delta$, having as image a basis of said quotient space. Unhappily, no explicit way to define such a family is presently known[20].

At any rate, returning to $\widetilde{\mathfrak{st}}$, we can put together the element τ (which generates $\widetilde{\mathfrak{st}}^{(0,\overline{1})}$) and all families of $\Delta_i^{(\delta,\overline{c})}$. The adjoint action of $G_{p,s}$ upon that big (infinite) family is explicit (they are all eigen elements). A density theorem says that $G_{p,s}$ and the one parameter subgroups of $\mathfrak{St}$ generated by those elements together generate topologically G.

In order to express that statement in a form close to Riemann-Hilbert correspondence, one introduces the free graded Lie algebra L generated by the special elements above and an hybrid object christened "q-analogue of the wild fundamental group" (an invention of J.-P. Ramis, see **[55, 164]** and the comments at the beginning of **[195]**). It is denoted $L \ltimes G_{p,s}$ and consists in the data of a Lie algebra L and a semisimple group G_p together with an action of the latter on the former; and representations of that object are simply compatible pairs of representations (in the same spirit as 15.4.1.3).

Then the final form of the theorem is that there is a natural bijective correspondence between representations of the q-wild fundamental group above and rational representations of G; and therefore (by Tannakian duality) with (isomorphism classes of) objects of $\mathscr{C}$.

[20]An explicit procedure is known for generic values of q; see **[224]**.

APPENDIX A

Some Classical Special Functions

This appendix and the next one are more detailed than the following ones, because the theory of special functions is not any more so much taught in France—and, I think, in Western countries. So I always took advantage of my courses to insist on it more than strictly necessary. Also, for lack of convenient references, I have added a section on "continuity of roots" from the analytical point of view.

The most famous special functions are the Gamma and Zeta functions, both invented by Euler. However, they do not play such a central role in the theory of q-difference equations; but they still are an important tool, so we will content ourselves here with a brief summary in the last section of this appendix. We shall add an application of the Gamma function to the real Theta function, because it is used in the functional equation of the Zeta function, but above all, because it will help us to prove the *modular functional equation for the complex Theta function.*

Other classical special functions commonly appearing in books are those which satisfy a linear complex analytic ordinary differential equation, usually one with polynomial coefficients (and often of order 2). The most prominent one are the hypergeometric functions (also invented by Euler!) along with some of their variants or degeneracies, like the Bessel functions for instance. For us, they rather come as an illustration of degeneracy (when $q \to 1$) of their q-analogs, the so-called basic hypergeometric functions. The latter are introduced in 1.6 and studied more closely in 4.4 (see also Appendix C and [CV]).

In the present appendix we shall concentrate on elliptic and modular functions (first three sections) then give a brief exposition of the necessary facts on Gamma and Zeta functions. Indeed, elliptic functions naturally appear as the constants of q-difference theory; more generally, solutions of q-difference equations can be seen as sections of vector bundles over complex tori (elliptic curves), so we find ourselves in the universe of elliptic functions. Modular functions spontaneously arise in the world of q-series; the contributions of Ramanujan for instance often involve them. They also occur here because degeneracy of solutions of q-difference equations to solutions of ordinary differential equations are best proved from the asymptotic properties of Theta functions when their modular parameter degenerates, and in turn those asymptotic properties flow from the modular functional equation of Theta functions.

REMARK A.1. Theta functions, invented by Gauß and Jacobi[1], are an essential tool for solving q-difference equations if one wants to avoid the difficulties caused by multivalued solutions (which is our slogan). We present in the main text, section 2.1 the part of the theory which is relevant to us, aiming at the "multiplicative" model, the one in which the parameter is $q := e^{2i\pi\tau}$ and the variable is $x := e^{2i\pi z}$.

[1]There also is a real form of Theta functions, which Fourier introduced in his theory of heat.

They also are a way to generate elliptic functions, and the best way for us since it gives control on zeroes and poles, which in turn allows for a geometric view[2].

Special functions are intimately related to Riemann surfaces, which are often the only framework to truly understand them. So Appendix A will be followed by an appendix on Riemann surfaces, with special emphasis on the complex tori, which, in their $\mathbf{E}_q := \mathbf{C}^*/q^{\mathbf{Z}}$ form, are central to the domain of q-difference equations. This is to be completed by a section on vector bundles over Riemann surfaces; but we already give here a heuristic hint on line bundles while studying Theta functions and q-characters: indeed, they are best understood as sections of line bundles over $\mathbf{C}^*/q^{\mathbf{Z}}$.

The most comprehensive source on much of what follows is **[255]**.

A.1. Elliptic functions and $\mathbf{E}_\Lambda = \mathbf{C}/\Lambda$

For the general content of this section, see **[144, 255]**; for the beginning (lattices), see **[38, 97]**.

A.1.1. Lattices. Every discrete subgroup Λ of $\mathbf{R}^n$ is free abelian with a basis made of $\mathbf{R}$-linearly independent vectors: $\Lambda = \mathbf{Z}e_1 \oplus \cdots \oplus \mathbf{Z}e_k$, where the free family $(e_1, \ldots, e_k)$ can be completed into a basis $(e_1, \ldots, e_n)$ of $\mathbf{R}^n$, so that:

$$\mathbf{R}^n/\Lambda \simeq (\mathbf{R}/\mathbf{Z})^k \times \mathbf{R}^{n-k}$$

for a unique *rank* $r \in \mathbf{N}$.

The isomorphism above is one of topological groups, so Λ is cocompact in $\mathbf{R}^n$ (i.e., $\mathbf{R}^n/\Lambda$ is compact) if, and only if, $k = n$, i.e., Λ is generated by a basis $(e_1, \ldots, e_n)$ of $\mathbf{R}^n$. In that case it is called a *lattice* of $\mathbf{R}^n$ and:

$$\mathbf{R}^n/\Lambda \simeq (\mathbf{R}/\mathbf{Z})^n \simeq (S^1)^n \simeq \mathbf{U}^n,$$

where S^1 denotes the circle either in its incarnation $\mathbf{R}/\mathbf{Z}$ or in its incarnation $\mathbf{U}$ (the unit circle $|z| = 1$ in $\mathbf{C}$). The topological group $(S^1)^n$ is a n-dimensional real torus, an important kind of compact abelian group.

In particular, the discrete subgroups of $\mathbf{C}$ are either trivial, or monogeneous, or of the form $\mathbf{Z}u \oplus \mathbf{Z}v$, where u and v are $\mathbf{R}$-linearly independent, i.e., $u, v \neq 0$ and $v/u \notin \mathbf{R}$. In the last case, they are lattices of $\mathbf{C}$ and the quotient $\mathbf{C}/\Lambda$ is a two-dimensional torus. Since we can interchange u and v and since $\Im(u/v)$, $\Im(v/u)$ have opposite signs, we may, and will, assume that $\tau := v/u \in \mathcal{H}$ (Poincaré's half plane), i.e., $\Im\, \tau > 0$. Then:

$$\Lambda = u\Lambda_\tau, \text{ where } \Lambda_\tau := \mathbf{Z} \oplus \mathbf{Z}\tau.$$

So every lattice in $\mathbf{C}$ is similar (in the geometric sense of similarity in $\mathbf{C}$: multiplication by a nonzero complex number) to one of the form Λ_τ, $\tau \in \mathcal{H}$. We shall discuss in A.3 the freedom of choice of τ and the classification of complex lattices up to similarity.

Let $a \in \mathbf{C}$ and let (u, v) a $\mathbf{Z}$-basis of the lattice $\Lambda \subset \mathbf{C}$. The associated *fundamental parallelogram*:

$$P := a + [0; 1]u + [0; 1]v = \{a + su + tv \mid 0 \leq s, t \leq 1\}$$

[2]As Ramis said **[190]**: *Je suis convaincu que, tout comme pour les fonctions spéciales solutions d'équations différentielles, les formules intéressantes dérivent de considérations "géométriques" simples.*

has the following properties:

(1) $\forall z \in \mathbf{C}\ ,\ \exists z_0 \in P\ :\ z \equiv z_0 \pmod{\Lambda}$ (i.e., $z - z_0 \in \Lambda$).
(2) Let $z \neq z' \in P$ such that $z \equiv z' \pmod{\Lambda}$. Then $z, z' \in \partial P$.

The boundary ∂P of P is

$$\{a + su + tv \mid (s \in [0;1] \text{ and } t \in \{0,1\}) \text{ or } (s \in \{0,1\} \text{ and } t \in [0;1])\}.$$

In other words, the translates $\lambda + P$, $\lambda \in \Lambda$, of P by the lattice cover $\mathbf{C}$:

$$\mathbf{C} = \Lambda + P = \{\lambda + z_0 \mid \lambda \in \Lambda, z_0 \in P\} = \bigcup_{\lambda \in \Lambda} (\lambda + P),$$

and their interiors $\lambda + \overset{\circ}{P}$, $\lambda \in \Lambda$, are pairwise disjoint:

$$\forall \lambda, \lambda' \in \Lambda \text{ such that } \lambda \neq \lambda'\ ,\ (\lambda + \overset{\circ}{P}) \cap (\lambda' + \overset{\circ}{P}) = \emptyset.$$

Any subset of $\mathbf{C}$ having these two properties is called a *fundamental domain* for the group action of Λ on $\mathbf{C}$ (here by translations). It is essential for our purposes that fundamental parallelograms are compact and connected.

EXERCISE A.2. Exhibit a noncompact, nonconnected fundamental domain.

A.1.2. Elliptic functions. From now on, we shall write $\mathbf{E}_\Lambda := \mathbf{C}/\Lambda$. Let f a nonconstant meromorphic functions on $\mathbf{C}$. The set $\Lambda := \{\lambda \in \mathbf{C} \mid f(z+\lambda) = f(z)\}$ of its periods is a discrete subgroup of $\mathbf{C}$ (by the principle of analytic continuation), so it is either trivial, or monogeneous, or a lattice. Any meromorphic function (constant or not) such that the set of its periods contains the lattice Λ (so this also includes the constants) is said to be Λ-*elliptic* or for short *elliptic* if no ambiguity results. If $\Lambda = \mathbf{Z}u + \mathbf{Z}v$, this is equivalent to $f(z+u) = f(z+v) = f(z)$.

The set of elliptic functions is a subfield of $\mathscr{M}(\mathbf{C})$ denoted $\mathscr{M}(\mathbf{E}_\Lambda)$; the notation will be explained in A.1.5. It is the fixed subfield of $\mathscr{M}(\mathbf{C})$ under the action of the group Λ by automorphisms $f(z) \mapsto f(z+\lambda)$, $\lambda \in \Lambda$, so we write: $\mathscr{M}(\mathbf{E}_\Lambda) = \mathscr{M}(\mathbf{C})^\Lambda$.

Note that every continuous Λ-periodic function on $\mathbf{C}$ is bounded on a fundamental parallelogram (since they are compact), whence everywhere (by Λ-periodicity), so every holomorphic elliptic function must be constant by Liouville's theorem: nonconstant elliptic functions have poles (and zeroes: just invert them). This will soon made more precise.

There are two main ways of constructing explicit elliptic functions. One, using Weierstraß $\mathfrak{P}$ function, will be detailed in A.2; the other, more in the spirit of Jacobi, is an application of Theta functions, see section 2.1 of the main text, but we shall presently give a foretaste.

So let $\tau \in \mathcal{H}$, denote $q := e^{2i\pi\tau}$ (so that $0 < |q| < 1$) and define the Theta function (so are they called):

$$T(z) := \sum_{n \in \mathbf{Z}} q^{n(n-1)/2} e^{2i\pi nz}.$$

The quadratic exponent for q guarantees normal convergence over any compact subset of $\mathbf{C}$, so $T(z)$ is an entire function. Clearly $T(z+1) = T(z)$ (since z appears only through $e^{2i\pi nz}$). Now:

$$T(z+\tau) = \sum_{n \in \mathbf{Z}} q^{n(n-1)/2} e^{2i\pi n(z+\tau)} = \sum_{n \in \mathbf{Z}} q^{n(n-1)/2} q^n e^{2i\pi nz} = \sum_{n \in \mathbf{Z}} q^{n(n+1)/2} e^{2i\pi nz},$$

which is equal to $e^{-2i\pi z}T(z)$ by shifting index $n \leftarrow n-1$, and we get the functional equation:

$$T(z+\tau) = e^{-2i\pi z}T(z).$$

Also, by standard theorems on trigonometric series, T is not the zero function. As a consequence, let $a_1, \dots, a_k, b_1, \dots, b_k \in \mathbf{C}$ such that $a_1 + \cdots + a_k = b_1 + \cdots + b_k$. Then $f(z) := \dfrac{T(z-a_1)\cdots T(z-a_k)}{T(z-b_1)\cdots T(z-b_k)}$ is meromorphic (because the denominator does not vanish identically), satisfies $f(z+1) = f(z)$ (obviously) and $T(z+\tau) = T(z)$ (a factor $e^{2i\pi(\sum a_i - \sum b_i)}$ appears, but it is equal to 1), so it is Λ-elliptic. We shall see later that, for a generic choice of the a_i, b_i, it is not constant. So we see that, while every meromorphic function on $\mathbf{C}$ is a quotient of holomorphic functions [**211**], this does not hold true if we restrict to elliptic functions.

EXERCISE A.3. Prove that $T(1/2) = 0$ (use manipulation of indices involving $n+1$ and $-n$). Deduce some candidate poles and zeroes for f above. Compare with Abel-Jacobi theorem in A.1.3.

We just produced elliptic functions for a lattice of the form Λ_τ. For a general lattice, we use the following fact: if $\Lambda' = w\Lambda$, $w \in \mathbf{C}^*$, (i.e., the lattices are similar), then the automorphism $f(z) \mapsto f(z/w)$ of $\mathscr{M}(\mathbf{C})$ induces an isomorphism $\mathscr{M}(\mathbf{E}_\Lambda) \simeq \mathscr{M}(\mathbf{E}_{\Lambda'})$. So can apply the same process, but with a change of variable.

A.1.3. Divisors of elliptic functions. Let $f \in \mathscr{M}(\mathbf{E}_\Lambda)$, $f \neq 0$. Then its set of zeroes $f^{-1}(0)$ and its set of poles $f^{-1}(\infty)$ are discrete, thus denumerable. Therefore it is possible to choose a fundamental parallelogram P such that its boundary ∂P meets neither $f^{-1}(0)$ nor $f^{-1}(\infty)$. (Argument: let Λ have $\mathbf{Z}$-basis (u, v) and let X, resp. Y, the set of all abcissas, resp. ordinates, of all zeroes and poles in the $\mathbf{R}$-basis (u, v): they are denumerable; then take $x \notin X$, $y \notin Y$ and $a := xu + uv$: the corresponding fundamental parallelogram fits.) Then the sets $P \cap f^{-1}(0)$ and $P \cap f^{-1}(\infty)$ are finite (by compactness of P) and contained in $\overset{\circ}{P}$ (by the choice we just made). Now call $a_1, \dots, a_k$, resp. $b_1, \dots, b_l$, the zeroes, resp. the poles, of f in $\overset{\circ}{P}$. Call $m_1, \dots, m_k$, resp. $n_1, \dots, n_l$, their multiplicities. The Λ-periodicity then implies that the zeroes of f are exactly the $a_i + \lambda$, $i = 1, \dots, k$, $\lambda \in \Lambda$, and each $a_i + \lambda$ has multiplicity m_i; and its poles are exactly the $b_j + \lambda$, $j = 1, \dots, l$, $\lambda \in \Lambda$, and each $b_j + \lambda$ has multiplicity n_j.

Slightly anticipating things to come next, we write $\overline{a}$ the class in $\mathbf{E}_\Lambda$ of any $a \in \mathbf{C}$ and encode the above situation by a formal sum, the *divisor (of zeroes and poles)* of f (on $\mathbf{E}_\Lambda$):

$$\operatorname{div}_{\mathbf{E}_\Lambda}(f) := \sum_{i=1}^{k} m_i[\overline{a_i}] - \sum_{j=1}^{l} n_j[\overline{b_j}].$$

The brackets $[-]$ are the trick to make it formal: $[\alpha]$ is just a symbol attached to $\alpha \in \mathbf{E}_\Lambda$. The general formalism on divisors is explained in Appendix B. By a suitable application of Cauchy's residue theorem [**144**], one proves:

$$m_1 + \cdots + m_k = n_1 + \cdots + n_l,$$
$$m_1a_1 + \cdots + m_ka_k \equiv n_1b_1 + \cdots + n_lb_l \pmod{\Lambda}.$$

Note that the first equation can be written $\sum m_i - \sum n_j = 0$ and the second $\sum m_i\overline{a_i} - \sum n_j\overline{b_j} = 0$ (computation in the abelian group $\mathbf{E}_\Lambda$). The *degree* $\deg D$ of

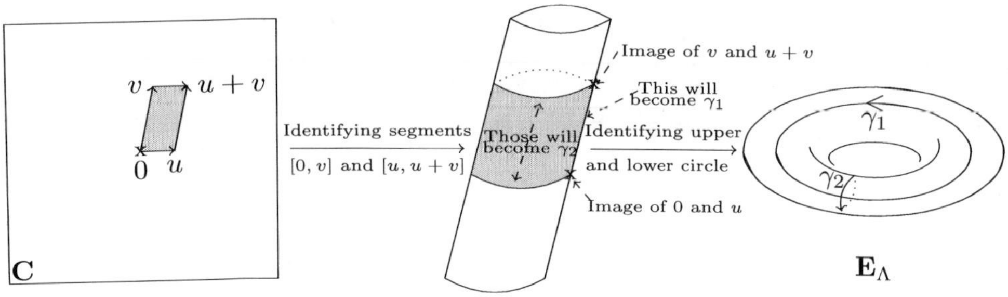

FIGURE A.1. Folding of the fundamental parallelogram into a torus

a general divisor $D := \sum p_i[\alpha_i]$ (all $p_i \in \mathbf{Z}$, all $\alpha_i \in \mathbf{E}_\Lambda$) is defined as $\sum p_i$, so the first property above states that $\deg \operatorname{div}_{\mathbf{E}_\Lambda}(f) = 0 \in \mathbf{Z}$. The *evaluation* $\operatorname{ev}_{\mathbf{E}_\Lambda}(D)$ of D is the value, computed in the group $\mathbf{E}_\Lambda$, of $\sum p_i\alpha_i$, so the second property states that $\operatorname{ev}_{\mathbf{E}_\Lambda}(\operatorname{div}_{\mathbf{E}_\Lambda}(f)) = 0 \in \mathbf{E}_\Lambda$.

DEFINITION A.4. The *order* of a nonconstant elliptic function is $m_1+\cdots+m_k = n_1 + \cdots + n_l$, i.e., it is its total number of zeroes (counted with multiplicities) or, equivalently, its total number of poles (counted with multiplicities). The order of a nonzero constant elliptic function is 0.

Note that the order cannot be 1: indeed, we would have a divisor $[\overline{a}] - [\overline{b}]$ with evaluation $\overline{a} - \overline{b} = 0$ in Λ, so the function would have neither zeroes nor poles and its order would actually be 0.

Conversely, if the a_i, m_i and the b_j, n_j are given satisfying such relations, then there exists an elliptic function having (in P) exactly those zeroes and poles, with those multiplicities. This is proved in an effective way as an application of Theta functions in Chapter 2 of the main text. Altogether, this gives:

THEOREM A.1 (Abel-Jacobi). *The divisor $D := \sum p_i[\alpha_i]$ on $\mathbf{E}_\Lambda$ is the divisor of a nonzero Λ-elliptic function if, and only if, $\deg D = 0 \in \mathbf{Z}$ and $ev_{\mathbf{E}_\Lambda}(D) = 0 \in \mathbf{E}_\Lambda$.*

A.1.4. Topology of $\mathbf{E}_\Lambda$. To better express the consequences of Λ-periodicity, it is convenient to endow the topological group $\mathbf{E}_\Lambda$ (which we already know to be compact, connected and abelian) with more structure. First we note that the projection map $p : \mathbf{C} \to \mathbf{E}_\Lambda$, $a \mapsto \overline{a} := a \pmod{\Lambda}$ is at the same time a group morphism and an etale covering (see explanations in the General Notations, Subsection 0.4.3); and second, that $\mathbf{E}_\Lambda$ is not simply connected, its fundamental group is $\pi_1(\mathbf{E}_\Lambda) = \mathbf{Z}^2$. The fact that p is a group morphism is simple and obvious, we shall now attempt ot visualize the statements on "etale covering" and on the fundamental group.

Let P a fundamental parallelogram. We already know that the restriction $p : P \to \mathbf{E}_\Lambda$ is onto and that distinct points have the same image only if they lie on the boundary (we merely translated the two properties of P stated in A.1.1). So we can imagine $\mathbf{E}_\Lambda$ as P folded in such a way that points having the same image by p are identified. First we fold to identify $A := a + [0;1]u$ with $A' := a + [0;1]u + v$ (each z is identified with $z + v$); second we fold to identify $B := a + [0;1]v$ with $B' := a + u + [0;1]u$ (each z is identified with $z + u$). That way, we obtain first a finite cylinder, second a torus. The images of A (or A') and B (or B') form two loops based at $\overline{a} := p(a)$. The homotopy classes of those two loops commute and freely generate $\pi_1(\mathbf{E}_\Lambda, a)$ (see Figure A.1).

The etale covering thing means that $\mathbf{E}_\Lambda$ can be covered with *trivializing open subsets*: these are open subsets V such that $U := p^{-1}(V)$ is a disjoint union of open subsets U_λ of $\mathbf{C}$ such that each restriction $p : U_\lambda \to V$ is a homeomorphism. Actually, fixing one of them, say U_0, we obtain them all by translations $U_\lambda = U_0 + \lambda$. We also call them small open subsets of $\mathbf{E}_\Lambda$. For instance if U_0 is contained in the interior $\overset{\circ}{P}$ then $V := p(U_0)$ is trivializing (same thing it is contained in any translate $b + \overset{\circ}{P}$, or, indeed, the interior of any fundamental parallelogram) (see Figure A.2).

The (quotient) topology of $\mathbf{E}_\Lambda$ can be either characterized by the fact that $V \subset \mathbf{E}_\Lambda$ is open if, and only if, $p^{-1}(V) \subset \mathbf{C}$ is open; or by the fact that a map $f : V \to Y$ from an open set $V \subset \mathbf{E}_\Lambda$ to a topological space Y is continuous if, and only if, $f \circ p : p^{-1}(V) \to Y$ is continuous. We favor the second criterion, which leads us to sheaves of functions, which will be the easiest why for us to endow $\mathbf{E}_\Lambda$ with an analytic structure. So we write $\mathscr{C}_{\mathbf{C}}$, $\mathscr{C}_{\mathbf{E}_\Lambda}$ the sheaves of complex continuous functions respectively on $\mathbf{C}$ and on $\mathbf{E}_\Lambda$. For an open set $V \subset \mathbf{E}_\Lambda$, any $f \in \mathscr{C}_{\mathbf{E}_\Lambda}(V)$ (i.e., any continuous function $V \to \mathbf{C}$) gives rise to $f \circ p \in \mathscr{C}_{\mathbf{C}}(p^{-1}(V))$ (i.e., a continuous function $p^{-1}(V) \to \mathbf{C}$) and we thus define an injective morphism of $\mathbf{C}$-algebras: $\mathscr{C}_{\mathbf{E}_\Lambda}(V) \to \mathscr{C}_{\mathbf{C}}(p^{-1}(V))$. Now the sheaf (on $\mathbf{E}_\Lambda$) $V \mapsto \mathscr{C}_{\mathbf{C}}(p^{-1}(V))$ is the so-called *direct image sheaf* $p_* \mathscr{C}_{\mathbf{C}}$, so we obtain (up to identification) an inclusion:

$$\mathscr{C}_{\mathbf{E}_\Lambda} \subset p_* \mathscr{C}_{\mathbf{C}}.$$

We can easily characterize the exact image of $\mathscr{C}_{\mathbf{E}_\Lambda}$ in $p_* \mathscr{C}_{\mathbf{C}}$: a continuous function $g : p^{-1}(V) \to \mathbf{C}$ factors as $g = f \circ p$ if, and only if, it is Λ-periodic, i.e., $f(z) = f(z+\lambda)$ for all $\lambda \in \Lambda$ (and the condition makes sense because f is defined on $p^{-1}(V)$, which is invariant by Λ-translations). So we have Λ act on each $\mathscr{C}_{\mathbf{C}}(p^{-1}(V))$ by translations of the variable: $f(z) \mapsto f(z+\lambda)$ and write $\mathscr{C}_{\mathbf{C}}(p^{-1}(V))^\Lambda$ the subalgebra fixed by this action. Then $V \mapsto \mathscr{C}_{\mathbf{C}}(p^{-1}(V))^\Lambda$ is a subsheaf of $p_* \mathscr{C}_{\mathbf{C}}$, denoted $p_* \mathscr{C}_{\mathbf{C}}^\Lambda$,

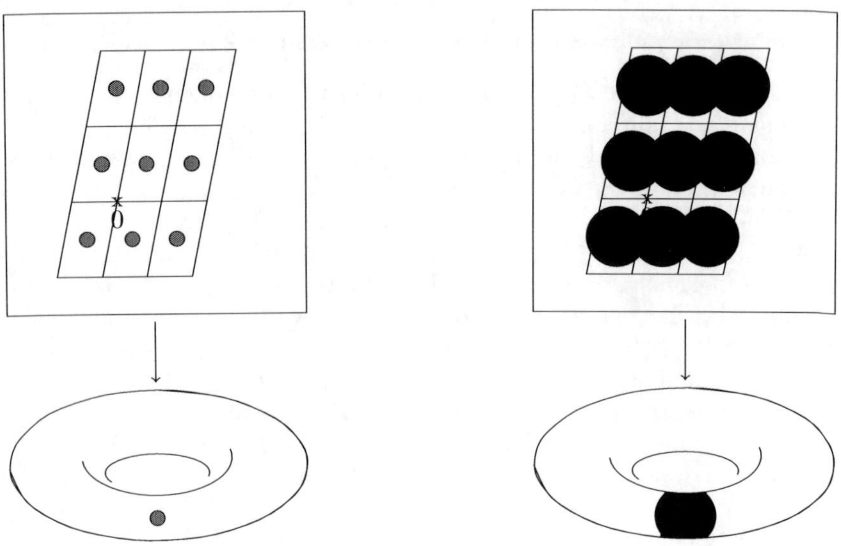

FIGURE A.2. Trivializing *vs* nontrivializing open sets in $\mathbf{E}_\Lambda$

and we eventually get (up to identification[3]):

$$\mathscr{C}_{\mathbf{E}_\Lambda} = (p_*\mathscr{C}_{\mathbf{C}})^\Lambda.$$

A.1.5. The Riemann surface $\mathbf{E}_\Lambda$. Now an essential feature of our topological space $\mathbf{E}_\Lambda$ is that it is a Riemann surface. In Appendix B, we recall the basics of the theory. Here we introduce it in the most convenient way (for us), by defining the sheaf $\mathscr{O}_{\mathbf{E}_\Lambda}$ of complex-valued holomorphic functions on $\mathbf{E}_\Lambda$, starting from the sheaf $\mathscr{O}_{\mathbf{C}}$ of holomorphic functions on $\mathbf{C}$; for every open subset $V \subset \mathbf{E}_\Lambda$:

$$\mathscr{O}_{\mathbf{E}_\Lambda}(V) := \{f \in \mathscr{C}_{\mathbf{E}_\Lambda}(V) \mid f \circ p \in \mathscr{O}_{\mathbf{C}}(p^{-1}(V))\}.$$

Obviously, $\mathscr{O}_{\mathbf{E}_\Lambda}$ is a subsheaf of $\mathscr{C}_{\mathbf{E}_\Lambda}$; but also, from the same arguments as above, it can be identified to a subsheaf of $p_*\mathscr{O}_{\mathbf{C}}$, consisting of Λ-periodic holomorphic functions:

$$\mathscr{O}_{\mathbf{E}_\Lambda} = (p_*\mathscr{O}_{\mathbf{C}})^\Lambda.$$

The interest of those sheaves is that, although there are no nonconstant global holomorphic functions on $\mathbf{E}_\Lambda$, locally there are many. In the same way, we can define the sheaf of meromorphic functions on $\mathbf{E}_\Lambda$ and deduce:

$$\mathscr{M}_{\mathbf{E}_\Lambda} = (p_*\mathscr{M}_{\mathbf{C}})^\Lambda.$$

Therefore, global meromorphic functions on $\mathbf{E}_\Lambda$ are exactly the Λ-elliptic functions on $\mathbf{C}$: we recognize the previously seen identification $\mathscr{M}(\mathbf{E}_\Lambda) = \mathscr{M}_{\mathbf{E}_\Lambda}(\mathbf{E}_\Lambda)$ as the field of elliptic functions.

Now let $a \in \mathbf{C}$, $\overline{a} := p(a) \in \mathbf{E}_\Lambda$ and $V \subset \mathbf{E}_\Lambda$ an open neighborhood of $\overline{a}$. A function $f \in \mathscr{M}_{\mathbf{E}_\Lambda}(V)$ has a zero, resp. a pole at $\overline{a}$ if, and only if, the corresponding function $f \circ p \in \mathscr{M}_{\mathbf{C}}(p^{-1}(V))$ has a zero, resp. a pole at a, and the multiplicity is the same. The order $v_{\overline{a}}(f)$ of f at $\overline{a}$, which is the multiplicity of the zero, or the opposite of the multiplicity of the pole, or 0, satisfies: $v_{\overline{a}}(f) = v_a(f \circ p)$. If $f \in \mathscr{M}(\mathbf{E}_\Lambda)$ is nonzero, it has a finite number of zeroes and poles and the divisor is well defined as the formal sum (with finite support):

$$\mathrm{div}_{\mathbf{E}_\Lambda}(f) := \sum_{\alpha \in \mathbf{E}_\Lambda} v_\alpha(f)[\alpha].$$

We recognize the divisor $\mathrm{div}_{\mathbf{E}_\Lambda}(f \circ p)$ previously introduced (with a slightly different notation since there f was defined on $\mathbf{C}$).

We define $\mathrm{Div}(\mathbf{E}_\Lambda)$, the group of divisors of $\mathbf{E}_\Lambda$, as the free abelian group with basis the symbols $[\alpha]$, $\alpha \in \mathbf{E}_\Lambda$. We have a degree morphism sending $D := \sum n_\alpha[\alpha]$ to $\deg D := \sum n_\alpha \in \mathbf{Z}$. Its kernel is denoted $\mathrm{Div}^0(\mathbf{E}_\Lambda)$. We also have an evaluation morphism sending D to $\mathrm{ev}_{\mathbf{E}_\Lambda} D := \sum n_\alpha \alpha \in \mathbf{E}_\Lambda$. Its restriction to $\mathrm{Div}^0(\mathbf{E}_\Lambda)$ is onto, since $[\alpha] - [0] \mapsto \alpha$ (we write $0 \in \mathbf{E}_\Lambda$ instead of $\overline{0}$). We know, from Abel-Jacobi theorem, that the principal divisors on $\mathbf{E}_\Lambda$, i.e., those of the form $\mathrm{div}_{\mathbf{E}_\Lambda}(f)$ for some nonzero $f \in \mathscr{M}(\mathbf{E}_\Lambda)$, are exactly the divisors with zero degree and trivial evaluation. The kernel of $f \mapsto \mathrm{div}_{\mathbf{E}_\Lambda}(f)$ consists of nonzero functions with neither

[3]This identification will allow us to consider an elliptic function either as a Λ-periodic meromorphic function on $\mathbf{C}$; or as a function on $\mathbf{C}/\Lambda$. A more elementary similar situation is when we consider a trigonometric function like sin either as defined on $\mathbf{C}$ and 2π-periodic; or as defined over $\mathbf{C}/2\pi\mathbf{Z}$ seen as the trigonometric circle.

poles nor zeroes, i.e., constant functions: it is $\mathbf{C}^* \subset \mathscr{M}(\mathbf{E}_\Lambda)$. Altogether we obtain the *Abel-Jacobi exact sequence*:

$$1 \longrightarrow \mathbf{C}^* \longrightarrow \mathscr{M}(\mathbf{E}_\Lambda)^* \xrightarrow{\mathrm{div}_{\mathbf{E}_\Lambda}} \mathrm{Div}^0(\mathbf{E}_\Lambda) \xrightarrow{\mathrm{ev}_{\mathbf{E}_\Lambda}} \mathbf{E}_\Lambda \longrightarrow 0. \tag{A.1}$$

We denote the trivial group 1 on the left because the groups $\mathbf{C}^*$ and $\mathscr{M}(\mathbf{E}_\Lambda)^*$ are denoted multiplicatively; and 0 on the right because the groups $\mathrm{Div}^0(\mathbf{E}_\Lambda)$ and $\mathbf{E}_\Lambda$ are denoted additively.

The subgroup $P(\mathbf{E}_\Lambda) \subset \mathrm{Div}^0(\mathbf{E}_\Lambda)$ of principal divisors is (by definition) the image of $\mathrm{div}_{\mathbf{E}_\Lambda}$ and the corresponding quotient (thus the cokernel of $\mathrm{div}_{\mathbf{E}_\Lambda}$) is the *Picard group* $\mathrm{Pic}(\mathbf{E}_\Lambda) := \mathrm{Div}^0(\mathbf{E}_\Lambda)/P(\mathbf{E}_\Lambda)$. It follows from Abel-Jacobi theorem that it is isomorphic to $\mathbf{E}_\Lambda$:

$$\mathrm{Pic}(\mathbf{E}_\Lambda) \simeq \mathbf{E}_\Lambda.$$

A.2. Elliptic curves

Compact Riemann surfaces are, in some sense, "the same thing as" smooth complex projective algebraic curves **[168]**, but, in the case of complex tori $\mathbf{E}_\Lambda = \mathbf{C}/\Lambda$, that correspondence can be made quite explicit: they are "the same thing as" nonsingular projective cubic curves. This algebraic character is very important because it defines a bridge between complex analysis and algebraic geometry, and this has been used in quite a few questions about q-difference equations. We shall have an indirect use for that in Appendix H.

We make a preliminary remark before going into computations (see Appendix B for precisions). Any meromorphic function f can be seen as holomorphic with values in the Riemann sphere $\mathbf{S} := \mathbf{C} \cup \{\infty\}$: the extended function sends poles at ∞. So in particular every nonconstant elliptic function can be thought of as a nonconstant *morphism (of Riemann surfaces)* $\mathbf{E}_\Lambda \to \mathbf{S}$. Both source and target being compact, it is a *ramified covering* of degree the *order* of the elliptic function, in the sense of the theory of Riemann surfaces; see Appendix B.

This equals in particular the number of zeroes (counted with multiplicities) or the number of poles (counted with multiplicities) (these numbers are equal since the divisor has degree 0); so this definition of the order is compatible with the definition given in A.1.3, but the order is also the number of antecedents of any point of $\mathbf{S}$ (for branch points, which are in finite number, multiplicity has to be taken in account).

A.2.1. Weierstraß' $\mathfrak{P}$-function.

We begin by the construction of a Λ-elliptic function, Weierstraß' $\mathfrak{P}$-function. So let Λ a lattice in $\mathbf{C}$. We define:

$$\mathfrak{P}_\Lambda(z) := \frac{1}{z^2} + \sum_\Lambda{}' \left(\frac{1}{(z-\lambda)^2} - \frac{1}{\lambda^2} \right).$$

For short, we shall just write $\mathfrak{P}$ instead of $\mathfrak{P}_\Lambda$. The notation $\sum'$ means that the summation is extended to $\lambda \in \Lambda$ *excluding* 0. From the easy estimation $\dfrac{1}{(z-\lambda)^2} - \dfrac{1}{\lambda^2} = O\left(\dfrac{1}{\lambda^3}\right)$, one draws that the series above converges normally on every compact subset of $\mathbf{C} \setminus \Lambda$, whence defines a holomorphic function there. Moreover, for every particular $\lambda \in \Lambda$, the function $\mathfrak{P}(z) - \dfrac{1}{(z-\lambda)^2}$ is holomorphic at λ, so $\mathfrak{P}$ is meromorphic over $\mathbf{C}$ with a double pole at each $\lambda \in \Lambda$ and no other pole. Then termwise derivation is allowed (see why?):

$$\mathfrak{P}'(z) = -2 \sum_{\Lambda} \frac{1}{(z-\lambda)^3}.$$

This one is immediately seen to be Λ-periodic with a triple pole at each $\lambda \in \Lambda$ and no other pole. This implies that, for every fixed $\lambda \in \Lambda$, the function $\mathfrak{P}(z+\lambda) - \mathfrak{P}(z)$ is constant; evaluating it at $-\lambda/2$ and taking in account the (obvious) parity of $\mathfrak{P}$, we conclude that it is 0, i.e., $\mathfrak{P}$ is Λ-periodic, so in the end $\mathfrak{P}, \mathfrak{P}' \in \mathscr{M}(\mathbf{E}_\Lambda)$; moreover $\mathfrak{P}$ is even and $\mathfrak{P}'$ is odd.

EXERCISE A.5. The argument fails if $\lambda/2 \in \Lambda$: fix it.

Using the preliminary remark just before A.2.1, we see that $\mathfrak{P}$ has order 2 and $\mathfrak{P}'$ has order 3. Let (u, v) be a $\mathbf{Z}$-basis of Λ. Then $u/2$, $v/2$ and $(u+v)/2$ do not belong to Λ, and using jointly periodicity and oddity, we see that (modulo Λ) they are the three zeroes of $\mathfrak{P}'$. As for $\mathfrak{P}$, we see that each preimage $\mathfrak{P}^{-1}(b)$, $b \in \mathbf{S}$, has the form $\{-a, a\}$ modulo Λ, and that the ramification points (i.e., those such that $-a \equiv a$) are $a = 0$ (corresponding to $b = \infty$) and $u/2$, $v/2$, $(u+v)/2$: we write e_1, e_2, e_3 for the images of those three ones; they are pairwise distinct.

THEOREM A.2. (i) *There exist complex numbers g_2, g_3 (depending on Λ) such that* ${\mathfrak{P}'}^2 = 4\mathfrak{P}^3 - g_2\mathfrak{P} - g_3$.
(ii) $\mathscr{M}(\mathbf{E}_\Lambda) = \mathbf{C}(\mathfrak{P}, \mathfrak{P}')$.

Actually, defining the *Eisenstein series*:

$$G_n(\Lambda) := {\sum_{\Lambda}}' \lambda^{-2n}, \text{which converges absolutely for } n \geq 2,$$

and setting $g_2 := 60G_2$, $g_3 := 140G_3$, a (somewhat grubby) computation of the expansion of ${\mathfrak{P}'}^2 - (4\mathfrak{P}^3 - g_2\mathfrak{P} - g_3)$ at $z = 0$ shows that it has no pole (so is constant since it is elliptic) and vanishes at 0, thus everywhere. For the proof of (i), see for instance [**144**].

This algebraic differential equation together with the known zeroes of $\mathfrak{P}$ shows that:

$$4x^3 - g_2X - g_3 = 4(X - e_1)(X - e_2)(X - e_3).$$

Since the e_i are pairwise distinct, the discriminant of that polynomial is not zero:

$$\Delta := g_2^3 - 27g_3^2 \neq 0.$$

Note that g_2, g_3 and Δ are actually functions of Λ, whence of u, v.

A.2.2. Uniformisation of $\mathbf{E}_\Lambda$. The image of the holomorphic map $z \mapsto (\mathfrak{P}(z), \mathfrak{P}'(z))$ from $\mathbf{C} \setminus \Lambda$ to $\mathbf{C}^2$ is included in the cubic algebraic curve $\Gamma : y^2 = 4x^3 - g_2x - g_3$ by the theorem. That curve is nonsingular since $\Delta := g_2^3 - 27g_3^2 \neq 0$ ([**144, 167, 168**]). Because of Λ-periodicity, the map above can be factored through a holomorphic mapping $\overline{z} \mapsto (\mathfrak{P}(z), \mathfrak{P}'(z))$, $\mathbf{E}_\Lambda \setminus \{0\} \to \mathbf{C}^2$ with image included in Γ.

When $z \to 0$, $\mathfrak{P}(z)$ and $\mathfrak{P}'(z)$ respectively go to infinity like $1/z^2$ and $-2/z^3$, i.e., nearly along an asymptotic direction $y^2 \sim 4x^3$. It therefore seems fit to introduce the projective curve:

$$\overline{\Gamma} := \{(X : Y : T) \in \mathbf{P}^2(\mathbf{C}) \mid Y^2T = 4X^3 - g_2XT^2 - g_3T^3\}.$$

Recall that a point in $\mathbf{P}^2(\mathbf{C})$ has projective coordinates $(X:Y:T)$ in $\mathbf{C}^3 \setminus \{(0,0,0\}$ defined up to a nonzero scalar factor: $(X:Y:T) = (cX:cY:cT)$ for $c \in \mathbf{C}^*$. The affine part "at finite distance" of $\overline{\Gamma}$ is where $T \neq 0$, setting $x := X/T$, $y := Y/T$ we see it is Γ. At infinity, we have $T = 0$ and there is only the point $X = 0$, so:

$$\overline{\Gamma} = \Gamma \cup \{(0:1:0)\}.$$

Going back to our estimation of $(\wp(z), \wp'(z))$ when $z \to 0$, it suggests that 0 should be sent to $(0:1:0) \in \overline{\Gamma}$. This can be made rigorous and one can prove:

THEOREM A.3. *The map $\overline{z} \mapsto (\wp(z), \wp'(z))$ from $\mathbf{E}_\Lambda$ to Γ admits a unique continuous extension $\mathbf{E}_\Lambda \to \overline{\Gamma}$ which is an isomorphism of Riemann surfaces.*

It is implicitly understood here that $\overline{\Gamma}$ is a noncomplex singular projective algebraic curve and that, as such, it can be given an analytic structure making it a compact Riemann surface [**168**].

A.2.3. Rational and meromorphic functions on $\mathbf{E}_\Lambda$. The analytic theory of $\mathbf{E}_\Lambda$ and the algebraic theory of $\overline{\Gamma}$ are related by the following facts:

(1) The meromorphic functions on $\mathbf{E}_\Lambda$ are the rational functions on $\overline{\Gamma}$. This is actually equivalent to the fact that $\mathscr{M}(\mathbf{E}_\Lambda) = \mathbf{C}(\wp, \wp')$. More precisely, the field of rational functions on $\overline{\Gamma}$ is $\mathbf{C}(\overline{\Gamma}) := \mathbf{C}(x,y)$, the quotient field of $\mathbf{C}[x,y]/(y^2 - (4x^3 - g_2x - g_3))$, and the isomorphism $\mathbf{C}(\overline{\Gamma}) \simeq \mathscr{M}(\mathbf{E}_\Lambda)$ sends x to $\wp$ and y to $\wp'$ (or rather their factorisations through $\mathbf{E}_\Lambda$).
(2) If $\Lambda' = w\Lambda$, the coefficients $g_2' := g_2(\Lambda')$, $g_3' := g_3(\Lambda')$, are related to $g_2 := g_2(\Lambda)$, $g_3 := g_3(\Lambda)$, by $g_2' = w^{-4}g_2$, $g_3' = w^{-6}g_3$, and there is an isomorphism $\Gamma \to \Gamma'$, $(x,y) \mapsto (w^{-2}x, w^{-3}y)$ of the corresponding algebraic curves which extends to their projective completions as $\overline{\Gamma} \to \overline{\Gamma'}$, $(X:Y:T) \mapsto (wX:Y:w^3T)$.
(3) Every nonsingular cubic projective curve over $\mathbf{C}$ can be put in the form $Y^2T = 4X^3 + aXT^2 + bT^3$ and the latter comes from a complex torus, i.e., there is a lattice Λ such that $a = -g_2(\Lambda)$, $b = -g_3(\Lambda)$.
(4) Two such nonsingular cubic projective curves are isomorphic (as curves) if, and only if, the corresponding complex tori are isomorphic (as Riemann surfaces); and this holds if, and only if, the corresponding lattices are similar.

This is part of a more general "AMAZING SYNTHESIS" ([**172**]), but we shall not delve into that.

A.3. Modular functions

Modular functions play an important role in the works of Jacobi and Ramanujan and thus in the domain of q-series. Most of what we present is detailed in [**231**].

A.3.1. Classification of complex tori. First we describe all holomorphic functions $f : \mathbf{E}_\Lambda \to \mathbf{E}_{\Lambda'}$, Λ, Λ' two lattices in $\mathbf{C}$. Such a function necessarily lifts to a continuous function $F : \mathbf{C} \to \mathbf{C}$, i.e., it fits into a commutative diagram:

$$\begin{array}{ccc} \mathbf{C} & \xrightarrow{F} & \mathbf{C} \\ {\scriptstyle p}\downarrow & & \downarrow{\scriptstyle p'} \\ \mathbf{C}/\Lambda & \xrightarrow{f} & \mathbf{C}/\Lambda' \end{array}$$

Indeed $f \circ p$ is continuous from the simply connected $\mathbf{C}$ to $\mathbf{E}_{\Lambda'} = \mathbf{C}/\Lambda'$, so it can be factored through the universal covering $p' : \mathbf{C} \to \mathbf{E}_{\Lambda'}$. Now, the a priori continuous function F is holomorphic because this is a local property and p, p' are local homeomorphisms (in down to earth terms: holomorphy can be checked on small trivializing open sets). Since F goes to the quotient as $f : \mathbf{E}_\Lambda \to \mathbf{E}_{\Lambda'}$, it must satisfy the following consistency property:

$$\forall z, z' \in \mathbf{C}\,,\ z \equiv z' \pmod{\Lambda} \Longrightarrow F(z) \equiv F(z') \pmod{\Lambda'}.$$

So for any fixed $\lambda \in \Lambda$, the map $z \mapsto F(z+\lambda) - F(z)$ sends $\mathbf{C}$ into Λ'. A continuous map sending a connected set into a discrete set has to be constant, so this one has zero-derivative: $F'(z+\lambda) - F'(z)$, so F' is Λ-periodic. Since it is holomorphic, it must be constant: $F' = a$, $a \in \mathbf{C}$, so $F(z) = az + b$. And the consistency condition above says that $a\Lambda \subset \Lambda'$. Conversely, such a map obviously induces some holomorphic $f : \mathbf{E}_\Lambda \to \mathbf{E}_{\Lambda'}$.

PROPOSITION A.6. *The holomorphic maps $\mathbf{E}_\Lambda \to \mathbf{E}_{\Lambda'}$ are the maps induced by functions $z \mapsto az + b$, $a, b \in \mathbf{C}$, $a\Lambda \subset \Lambda'$.*

We leave it to the reader to deduce:

COROLLARY A.7. *The complex tori $\mathbf{E}_\Lambda$, $\mathbf{E}_{\Lambda'}$, are isomorphic (as Riemann surfaces) if, and only if, the lattices Λ, Λ' are similar, i.e., if $\exists a \in \mathbf{C}^* \ :\ \Lambda' = a\Lambda$.*

EXERCISE A.8. Find all automorphisms of $\mathbf{E}_\Lambda$. (If stuck, read in **[144]** about "complex multiplication".)

A.3.2. Classification of the lattices of C. Classification of complex tori up to isomorphism boils down to classification of lattices up to similarity, which we now tackle. We saw in A.1.1 that every lattice in $\mathbf{C}$ is similar to a Λ_τ, $\tau \in \mathcal{H}$, so we are reduced to determine when Λ_τ and $\Lambda_{\tau'}$ are similar. So suppose that $\Lambda_{\tau'} = w\Lambda_\tau$ for some $w \in \mathbf{C}^*$. This means that $(w, w\tau)$ is a basis for $\Lambda_{\tau'}$. But all bases of a given free abelian group are images of each other by invertible integral matrices, so there are $a, b, c, d \in \mathbf{Z}$ such that $ad - bc = \pm 1$ and:

$$(\tau', 1) = (w\tau, w)\begin{pmatrix} a & c \\ b & d \end{pmatrix} \Longrightarrow \begin{cases} \tau' = w(a\tau + b), \\ 1 = w(c\tau + d) \end{cases} \Longrightarrow \tau' = \frac{a\tau + b}{c\tau + d}.$$

Conversely, if $\tau' = \dfrac{a\tau + b}{c\tau + d}$, then, taking $w := \dfrac{1}{c\tau + d}$, you can (and should!) prove the same way that $\Lambda_{\tau'} = w\Lambda_\tau$. Now, if τ' is given by the formula above:

$$\Im\tau' = \frac{1}{2\mathrm{i}}\left(\frac{a\tau + b}{c\tau + d} - \frac{a\overline{\tau} + b}{c\overline{\tau} + d}\right) = \frac{ad - bc}{|c\tau + d|^2}\Im\tau,$$

so $ad - bc$ must be positive: $ad - bc = 1$. So we introduce the action of the group:

$$\mathrm{SL}_2(\mathbf{Z}) := \left\{\begin{pmatrix} a & b \\ c & d \end{pmatrix} \mid a, b, c, d \in \mathbf{Z}, ad - bc = 1\right\}$$

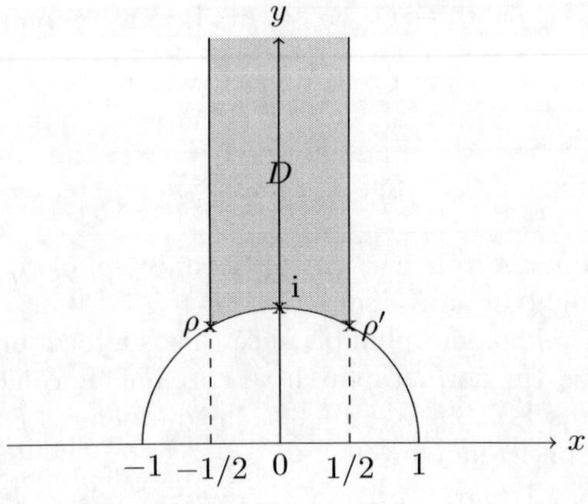

FIGURE A.3. The fundamental domain for the action of G on $\mathcal{H}$

on Poincaré's half plane by *homographies* (also called in English linear or Mobius transformations):

$$\forall g := \begin{pmatrix} a & b \\ c & d \end{pmatrix} \in \mathrm{SL}_2(\mathbf{Z}) \ , \ \forall \tau \in \mathcal{H} \ , \ g.\tau := \frac{a\tau + b}{c\tau + d}.$$

One can (meaning: you should) readily check that it is a group action (axioms: $I_2.\tau = \tau$ and $g'.(g.\tau) = (g'g).\tau$). We conclude the discussion:

THEOREM A.4. *The map which to $\tau \in \mathcal{H}$ associates the similarity class of Λ_τ induces a bijection from the orbit space $\mathcal{H}/SL_2(\mathbf{Z})$ onto the set of similarity classes of lattices of $\mathbf{C}$.*

Note that, by A.3.1, it also induces a bijection with the set of isomorphism classes of complex tori. Now $X := \mathcal{H}/\mathrm{SL}_2(\mathbf{Z})$ is the quotient of a Riemann surface by the proper action of a discrete group, and the general theory (into which we shall not go) allows us to endow it with an analytic structure making it a Riemann surface in such a way that the natural projection map $\mathcal{H} \to X$ is a morphism of Riemann surfaces (actually a ramified covering of infinite degree). We shall see (in two ways) that X can be identified with $\mathbf{C}$.

A.3.3. The modular group $\mathbf{PSL}_2(\mathbf{Z})$. We now make more precise the content of the theorem. The kernel of the action of $\mathrm{SL}_2(\mathbf{Z})$ on $\mathcal{H}$ (i.e., the subgroup of those g acting trivially) is $\{\pm I_2\}$ (actually its center), so the group of interest really is the *modular group*:

$$G := \mathbf{PSL}_2(\mathbf{Z}) := \mathrm{SL}_2(\mathbf{Z})/\{\pm I_2\}.$$

We define two particular homographies on $\mathcal{H}$:

$$S : z \mapsto \pm \begin{pmatrix} 0 & -1 \\ 1 & 0 \end{pmatrix} .z = \frac{-1}{z} \text{ and } T : z \mapsto \pm \begin{pmatrix} 1 & 1 \\ 0 & 1 \end{pmatrix} .z = z + 1.$$

We also define a particular subset of $\mathcal{H}$ (see Figure A.3):

$$D := \{z \in \mathcal{H} \mid |z| \geq 1 \text{ and } -1/2 \leq \Re z \leq 1/2\}.$$

Note the special positions in D of $\mathrm{i} = S.\mathrm{i}$, $\rho := e^{2\mathrm{i}\pi/3}$ and $\rho' := e^{\mathrm{i}\pi/3} = S.\rho = T.\rho$.

PROPOSITION A.9. (i) *Every orbit $G.z$ under G meets D.*

(ii) *If $z \in \overset{\circ}{D}$, then $G.z \cap D = \{z\}$.*

(iii) *If $z \in D \setminus \{i, \rho, \rho'\}$, then the only possibilities that $g.z \in D$ with $g \neq Id$ are: z lies on the lower arc of circle and $g = S$; or z is on one of the vertical half lines and $g = T^{\pm 1}$ (anyway, $z \in \partial D$).*

(iv) *The only points in D with nontrivial stabilizer are i (stabilizer $\{Id, S\}$), ρ (stabilizer $\{Id, (ST), (ST)^2\}$) and ρ' (stabilizer $\{Id, (TS), (TS)^2\}$).*

(v) *S and T together generate G, with only relations $S^2 = (ST)^3 = Id$.*

It follows that the restriction of $\mathcal{H} \to X$ to $D \to X$ is surjective, while $\overset{\circ}{D} \to X$ is injective. The projection map $D \to X$ identifies the left vertical half-line with the right one and the left part of the arc of circle with the right part, so the result is a cylinder with bottom filled, i.e., topologically it is **C**. We shall later compactify it by adding a point at infinity at the top of the cylinder, thus getting (topologically) the Riemann sphere **S**.

EXERCISE A.10. Make a drawing to illustrate how those identifications produce **C**; and then, adding the point $+\infty i$, produces **S**.

A.3.4. The quotient of $\mathcal{H}$ by $\mathbf{PSL}_2(\mathbf{Z})$. The identification of X as a Riemann surface is best done by using functions on X, or, equivalently, G-invariant functions on $\mathcal{H}$. Exactly as in the use of Theta functions to produce elliptic functions (as we did in A.1.2; also see the Chapter 2), we shall first produce functions on $\mathcal{H}$ that are "nicely" transformed under the action of G. The handiest ones are the Eisenstein functions.

Recall from A.2.1 the Eisenstein series; Λ is a lattice in **C** and $k \geq 2$:

$$G_k(\Lambda) := \sum_{\Lambda}{}' \lambda^{-2k} := \sum_{\lambda \in \Lambda \setminus \{0\}} \lambda^{-2k}.$$

It is immediate that $G_k(w\Lambda) = w^{-2k} G_k(\Lambda)$. Now, if $\tau \in \mathcal{H}$ and $g \in G$ comes from matrix $\pm \begin{pmatrix} a & b \\ c & d \end{pmatrix}$, setting $\tau' := g.\tau$ and $w := 1/(c\tau + d)$:

$$\Lambda_{\tau'} = w\Lambda_\tau \Longrightarrow G_k(\Lambda_{\tau'}) = w^{-2k} G_k(\Lambda_\tau),$$

i.e., if we define (with a slightly abusive overload of notation) the *Eisenstein function*:

$$G_k(\tau) := G_k(\Lambda_\tau) = \sum_{\substack{m,n \in \mathbf{Z} \\ (m,n) \neq (0,0)}} \frac{1}{(m\tau + n)^{2k}},$$

we have, again for $g := \pm \begin{pmatrix} a & b \\ c & d \end{pmatrix}$:

$$G_k(g.\tau) = (c\tau + d)^{2k} G_k(\tau).$$

With the same extension of notations: $g_2(\tau) := g_2(\Lambda_\tau)$, $g_3(\tau) := g_3(\Lambda_\tau)$, $\Delta(\tau) := \Delta(\Lambda_\tau)$, it follows immediately that:

$$g_2(g.\tau) = (c\tau + d)^4 g_2(\tau), \quad g_3(g.\tau) = (c\tau + d)^6 g_2(\tau), \quad \Delta(g.\tau) = (c\tau + d)^{12} \Delta(\tau),$$

so, defining $J(\Lambda) := \dfrac{g_2^3(\Lambda)}{\Delta(\Lambda)}$ $J(\tau) := J(\Lambda_\tau)$, we conclude:

$$\forall g \in G\,,\ J(g.\tau) = J(\tau).$$

All this is formal. On the analytic side, it can be shown that the series $G_k(\tau)$ converges normally on compact subsets of $\mathcal{H}$, so the Eisenstein functions $G_k(\tau)$, and the discriminant function $\Delta(\tau)$ are holomorphic. Since, by A.2.1, Δ never vanishes, we conclude that J is holomorphic and G-invariant on $\mathcal{H}$. Therefore, it defines a holomorphic function on X; with a slight abuse we still call it $J : X \to \mathbf{C}$.

THEOREM A.5. *$J : X \to \mathbf{C}$ is an isomorphism of Riemann surfaces.*

DEFINITION A.11. A meromorphic function on $\mathcal{H}$ satisfying (with the above notations) $f(g.z) = (c\tau + d)^{2k} f(z)$ is said to be *weakly modular of weight*[4] $2k$.

So G_k has weight $2k$, Δ has weight 12 and J has weight 0. Note that:

$$\frac{d(gz)}{dz} = \frac{d}{dz}\frac{az+b}{cz+d} = \frac{1}{(cz+d)^2},$$

so the condition $f(g.z) = (c\tau + d)^{2k} f(z)$ is equivalent to: the differential form $f(z)(dz)^k$ is G-invariant. In this form, it is easy to infer that it is sufficient to check the condition on generators. In particular, using S and T, it is equivalent to:

$$f(z+1) = f(z) \text{ and } f(-1/z) = z^{2k} f(z).$$

The first relation will enable us to study what goes on "at infinity". Indeed, it implies that f can be written as a function of $q := e^{2i\pi\tau}$ (isn't it a nice world!). Now note that "infinity" on D means way up the stripe, where $\Im\tau \to +\infty$, i.e., $q \to 0$. So we study the behaviour at infinity of a weakly modular function $f(\tau)$ by writing it as $F(q)$ and studying $F(q)$ near $q = 0$. We shall do that in A.3.5. But we shall anticipate a bit by completing the above statement $X \simeq \mathbf{C}$.

It will follow from the computations (to come) that J, as a function of q, has the expansion $\dfrac{1}{q}+$ something in $\mathbf{C}\{q\}$. This means that we can make the compactification $\hat{X} := X \cup \{\infty\}$ a Riemann surface with uniformizing parameter q at ∞, and also extend J as a function $\hat{X} \to \mathbf{S}$ by sending ∞ to ∞. We then conclude:

THEOREM A.6. *The extended $J : \hat{X} \to \mathbf{S}$ is an isomorphism of Riemann surfaces.*

DEFINITION A.12. (i) A *modular function of weight* $2k$ is a weakly modular function of weight $2k$ on $\mathcal{H}$ which is moreover meromorphic at infinity (i.e., if $F(q)$ is meromorphic at $q = 0$).

(ii) We say that it is a *modular form* if it is moreover holomorphic on $\mathcal{H}$ and has weight 0.

(iii) A modular form is *parabolic* if $F(0) = 0$.

It will follow from A.3.5 that all weakly modular functions mentioned above are modular functions. Also, J is a modular form (not a parabolic one though); but we shall rather use the more standard modular function $j := 1728J$. From the expansion of J mentioned above, one draws (invoking Appendix B):

COROLLARY A.13. *The field $\mathscr{M}(\hat{X})$ of modular functions of weight* 0 *is* $\mathbf{C}(j)$.

[4]The whole terminology on weights varies in the literature, as well as the indexing system for Eisenstein functions; we adopt the conventions from **[231]**.

A.3.5. Fourier expansions. For 1-periodic functions f on $\mathcal{H}$, we shall wite $f(\tau) = F(q)$ where $q := e^{2i\pi\tau}$. We start by expansions of Eisenstein functions. We use the well-known[5] formula:

$$\cos(zt) = \frac{\sin \pi z}{\pi z} + \frac{\sin \pi z}{\pi} \sum_{n\geq 1} (-1)^n \left(\frac{1}{z+n} + \frac{1}{z-n} \right) \cos nt. \tag{A.2}$$

EXERCISE A.14. Derive the Fourier expansion of $t \mapsto \cos(zt)$ on $[-\pi, \pi]$ and deduce from it the formula:

$$\pi z \cot(\pi z) = 1 - \sum_{n\geq 1} \zeta(2n) z^{2n}.$$

Taking $t := \pi$ and $z := \tau$ in equation (A.2), we deduce:

$$\pi \cot(\pi\tau) = \frac{1}{\tau} + \sum_{n\geq 1} \left(\frac{1}{\tau+n} + \frac{1}{\tau-n} \right).$$

On the other hand, $\cot(\pi\tau) = i\dfrac{e^{i\pi\tau} + e^{-i\pi\tau}}{e^{i\pi\tau} - e^{-i\pi\tau}} = i\dfrac{q+1}{q-1}$, so we have:

$$\frac{1}{\tau} + \sum_{n\geq 1} \left(\frac{1}{\tau+n} + \frac{1}{\tau-n} \right) = i\pi \frac{q+1}{q-1} = i\pi - 2i\pi \sum_{n\geq 0} q^n.$$

We are going to differentiate many times that equality with respect to τ. Note that $\dfrac{dq^n}{d\tau} = \dfrac{d}{d\tau} e^{2ni\pi\tau} = 2ni\pi q^n$, so, iterating, $\left(\dfrac{d}{d\tau}\right)^k q^n = (2ni\pi)^k q^n$. So, differentiating $(k-1)$ times the above equality, we find, for $k \geq 2$:

$$(-1)^k (k-1)! \sum_{m\in\mathbf{Z}} \frac{1}{(\tau+m)^k} = (2i\pi)^k \sum_{n\geq 1} n^{k-1} q^n.$$

Denote temporarily $f_k(\tau) := \sum\limits_{m\in\mathbf{Z}} \dfrac{1}{(\tau+m)^k}\cdot$ It appears in Eisenstein function as follows:

$$\begin{aligned} G_k(\tau) &= \sum_{\substack{m,n\in\mathbf{Z} \\ (m,n)\neq(0,0)}} \frac{1}{(m\tau+n)^{2k}} = \sum_{\substack{n\in\mathbf{Z} \\ n\neq 0}} \frac{1}{n^{2k}} + \sum_{\substack{m,n\in\mathbf{Z} \\ m\neq 0}} \frac{1}{(m\tau+n)^{2k}} \\ &= 2\zeta(2k) + 2\sum_{n\geq 1} f_{2k}(n\tau) \\ &= 2\zeta(2k) + 2\sum_{n\geq 1} \frac{(-2i\pi)^{2k}}{(2k-1)!} \sum_{p\geq 1} p^{2k-1} q^{np}. \end{aligned}$$

But $\sum\limits_{n\geq 1} \sum\limits_{p\geq 1} p^{2k-1} q^{np} = \sum\limits_{m\geq 1} \sigma_{2k-1}(m) q^m$, where $\sigma_l(m) := \sum\limits_{d|m} d^l$. In the end, we obtain:

$$G_k(\tau) = 2\zeta(2k) + 2\frac{(2i\pi)^{2k}}{(2k-1)!} \sum_{m\geq 1} \sigma_{2k-1}(m) q^m.$$

We may already conclude that G_k is holomorphic at infinity, so it is a modular form of weight $2k$.

---

[5]It can be obtained by classical Fourier expansion of the left-hand side *qua* function of t; see further below A.4.1.

Now we introduce the Bernoulli numbers B_k by the expansion:

$$\frac{z}{e^z - 1} = 1 - z/2 + \sum_{k \geq 1} (-1)^{k+1} B_k z^{2k}/(2k)!.$$

A recursive formula can be deduced from the identification of coefficients on both sides of the equality $(e^z - 1)(1 - z/2 + \sum (-1)^{k+1} B_{2k} z^{2k}/(2k)!) = z$ and one computes $B_1 = 1/6$, $B_2 = 1/30$, $B_3 = 1/42$, etc. (All B_k are rational.) From the above expansion of $\pi \cot(\pi\tau)$ as:

$$\mathrm{i}\pi(1 + 2\frac{1}{e^{2\mathrm{i}\pi\tau} - 1}) = \frac{1}{\tau} + \sum_{n \geq 1} \left(\frac{1}{\tau + n} + \frac{1}{\tau - n} \right),$$

one can deduce that $\zeta(2k) = \dfrac{2^{2k-1}}{(2k)!} B_k \pi^{2k}$, for instance $\zeta(2) = \pi^2/6$, $\zeta(4) = \pi^4/90$, $\zeta(6) = \pi^6/945$, etc. This allows us a slight change in the expression of Eisenstein series; setting:

$$E_k(\tau) := 1 + \gamma_k \sum_{n \geq 1} \sigma_{2k-1}(n) q^n, \text{ where } \gamma_k := \frac{(-1)^k 4k}{B_k},$$

we can write $G_k = 2\zeta(2k) E_k$. Note that $E_k(\infty) = 1$ (take $q := 0$). We are now in position to compute precise behaviours at infinity; first, $G_2(\infty) = 2\zeta(4) = \pi^4/45$, $G_3(\infty) = 2\zeta(6) = 2\pi^6/945$; and then:

$$\begin{aligned}
\Delta = g_2^3 - 27 g_3^2 = (60G_2)^3 - 27(140G_3)^2 &= \frac{64}{27}\pi^{12}(E_2^3 - E_3^2) \\
&= \frac{64}{27}\pi^{12}\left((1 + 240(q + \cdots)^3 - (1 - 504(q + \cdots)^2\right) \\
&= \frac{64}{27}\pi^{12}(1728q + \cdots),
\end{aligned}$$

so $\Delta = (2\pi)^{12}(q + \cdots)$. We immediately mention a consequence that will be used in Subsection 2.2.4 of the main text as an application of Theta functions:

PROPOSITION A.15. *Let f a parabolic modular form of weight* 12. *Then it is equal to $C\Delta$ for some $C \in \mathbf{C}$. In other words, the space of parabolic forms of weight* 12 *has dimension* 1.

PROOF. The function f/Δ is holomorphic and G-invariant on $\mathcal{H}$ and, because of the q-expansion of Δ we just found, it is holomorphic at infinity; so it defines a holomorphic function on the compact Riemann surface $\hat{X}$, so it is constant. □

We return to J and j. From the above:

$$g_2^3 = (60G_2)^3 = \frac{64}{27}\pi^{12} E_2^3 = \frac{64}{27}\pi^{12}(1 + \cdots),$$

so by the expansion of Δ:

$$J = \frac{\frac{64}{27}\pi^{12}(1 + \cdots)}{(2\pi)^{12}(q + \cdots)} = \frac{1}{1728}\left(\frac{1}{q} + \text{ positive powers}\right) \Longrightarrow j = \frac{1}{q} + \text{ positive powers},$$

as announced (and used) in A.3.4. Serre **[231]** gives the more complete expansion $j = \dfrac{1}{q} + 744 \sum_{n \geq 1} c_n q^n$, where the coefficients $c_n \in \mathbf{Z}$ "enjoy remarkable divisibility properties".

A.3.6. Number theoretic applications. From the above expansions, we see that $\Delta = (2\pi)^{12} \sum_{n\geq 1} \tau(n)q^n$, where $\tau(1) = 1$ and all $\tau(n) \in \mathbf{Z}$: this is *Ramanujan's Tau function.* We show in Subsection 2.2.4 as a consequence of the proposition above and of our study of Theta functions that actually:

$$\sum_{n\geq 1} \tau(n)q^n = q(q;q)_\infty^{24} = q \prod_{n\geq 1} (1-q^n)^{24}.$$

The function τ also enjoys remarkable arithmetical properties, some proved by Ramanujan, others conjectured by him and proved later. One of those conjectures was that $|\tau(p)| < 2p^{11/2}$ for every prime p. It was proved by Deligne as a consequence of his proof of Weil's conjecture, which brought him fame and a Fields medal. The proof of the weaker estimate $\tau(n) = O(n^6)$ is easier and given in [**231**] as an application of the theory of modular forms. (Actually Deligne attended the course of which the book was a transcription.)

We give an example of such arithmetical properties based on the general theory. From the general formula $E_k(\tau) := 1+\gamma_k \sum_{n\geq 1} \sigma_{2k-1}(n)q^n$, we deduce that $E_6(\tau) - E_3^2(\tau) = (\gamma_6 - 2\gamma_3)q + \cdots$ is a parabolic form of weight 12, whence, by the same dimension argument (i.e., the proposition):

$$E_6(\tau) - E_3^2(\tau) = (\gamma_6 - 2\gamma_3) \sum_{n\geq 1} \tau(n)q^n.$$

we know that $\gamma_3 = -504$ and, using the value $B_6 = \dfrac{691}{2730}$ (given by Serre [**231**, VII 4.1 p 148] or computable from the recursive formula alluded to above), we see that $\gamma_6 = \dfrac{24 \times 2730}{691} = \dfrac{2^4.3^2.5.7.13}{691}$ (an irreducible fraction since 691 is prime), so:

$$(\gamma_6 - 2\gamma_3)\tau(n) = \gamma_6\sigma_{11}(n) - 2\gamma_3\sigma_5(n) - \gamma_3^2 \sum_{m=1}^{n-1} \sigma_5(m)\sigma_5(n-m).$$

Since γ_3 and all $\sigma_k(n)$ are integers, we deduce that so is $\gamma_6(\tau(n) - \sigma_{11}(n))$. We deduce one of Ramanujan's famous congruences:

$$\forall n \geq 1\ ,\ \tau(n) \equiv \sigma_{11}(n) \pmod{691}.$$

We give another example of arithmetical property deducible from the general theory. Serre gives the dimensions of the spaces of modular forms. For weights 8 and 10, the dimension is 1. This immediately implies the identities $E_2^2 = E_4$ and $E_2E_3 = E_5$ (indeed the implied constant factor is found to be 1 by taking values at $q = 0$). To apply the first identity, we use the fact that $\gamma_2 = 240$ and (Serre giving the value $B_4 = 1/30$) that $\gamma_4 = 480$. Then, expanding the identity:

$$\forall n \geq 1\ ,\ \sigma_7(n) = \sigma_3(n) + 120 \sum_{m=1}^{n-1} \sigma_3(m)\sigma_3(n-m).$$

EXERCISE A.16. Given that $B_5 = 5/66$, find γ_5 and expand formula $E_2E_3 = E_5$ to get an identity involving $\sigma_9(n)$.

A.3.7. A nonmodular Eisenstein series.

A.3.7.1. First, a slightly different form to the already defined Eisenstein series:

LEMMA A.17. *For all $l \geq 0$, the double series* $\sum_{m,n\geq 1} m^l q^{mn}$ *converges absolutely and has the two evaluations:*

$$\sum_{m,n\geq 1} m^l q^{mn} = \sum_{m\geq 1} \frac{m^l q^m}{1-q^m} = \sum_{n\geq 1} \sigma_l(n) q^n.$$

PROOF. Let $r := |q| < 1$. From the easy majoration $\sum_{mn=N} \left|m^l q^{mn}\right| \leq N^{l+1} r^N$, absolute convergence flows at once. Then writing the sum as $\sum_{m\geq 1}\sum_{n\geq 1}$ yields the first evaluation. The second one comes by grouping sums for $mn = N$. □

Series of the form $\Lambda_l(q) := \sum_{m\geq 1} \frac{m^l q^m}{1-q^m}$ are called *Lambert series* and Ramanujan often uses them to denote Eisenstein series and other modular q-series. Note that they appear in A.3.5 with $l = 2k-1$ and $k \geq 2$, so $l \geq 3$:

$$G_k(\tau) = 2\zeta(2k) E_k(\tau) = 2\zeta(2k)\left(1 + \gamma_k \Lambda_{2k-1}(q)\right).$$

However the lemma shows that for $k = l = 1$ the Lambert series still makes sense.

A.3.7.2. If we try to define $G_1(\tau)$ as the sum of the $(m+n\tau)^{-2}$ (avoiding $m = n = 0$), we are immediately stuck on the fact that the double series does not converge absolutely, so it is at least necessary (and maybe not sufficient) to fix an order of summation. (Only for absolutely convergent series are we certain of commutative convergence.) We shall define:

$$G_1(\tau) := \sum_{n,m}{}' (m+n\tau)^{-2},$$

where m, n vary in $\mathbf{Z}$, the summation symbol $\sum_{n,m}$ stands for $\sum_n \sum_m$ and the prime symbol means that we avoid $m = n = 0$. The serious reader will carefully check step by step that the following calculation makes sense and is right.

PROPOSITION A.18.

$$G_1(\tau) = \pi^2/3 - 8\pi^2 \Lambda_1(q).$$

PROOF. We begin by splitting the sum according to whether $n = 0$ or not:

$$\begin{aligned}
\sum_{n,m}{}' (m+n\tau)^{-2} &= \sum_{m\neq 0} m^{-2} + \sum_{n\neq 0}\sum_{m\in\mathbf{Z}} (m+n\tau)^{-2} \\
&= 2\zeta(2) + \sum_{n\geq 1}\sum_{m\in\mathbf{Z}} (m+n\tau)^{-2} + \sum_{n\leq -1}\sum_{m\in\mathbf{Z}} (m+n\tau)^{-2} \\
&= 2\zeta(2) + \sum_{n\geq 1}\sum_{m\in\mathbf{Z}} (m+n\tau)^{-2} + \sum_{n\geq 1}\sum_{m\in\mathbf{Z}} (m-n\tau)^{-2} \\
&= 2\zeta(2) + \sum_{n\geq 1}\sum_{m\in\mathbf{Z}} (m+n\tau)^{-2} + \sum_{n\geq 1}\sum_{m\in\mathbf{Z}} (-m+n\tau)^{-2} \\
&= 2\zeta(2) + \sum_{n\geq 1}\sum_{m\in\mathbf{Z}} (m+n\tau)^{-2} + \sum_{n\geq 1}\sum_{m\in\mathbf{Z}} (m+n\tau)^{-2} \\
&= 2\zeta(2) + 2\sum_{n\geq 1} f(n\tau),
\end{aligned}$$

where we temporarily set:

$$f(\tau) := \sum_{m\in\mathbf{Z}} \left((m+\tau)^{-2} + (m-\tau)^{-2}\right) = \sum_{m\in\mathbf{Z}} \left((m+\tau)^{-2} + (\tau-m)^{-2}\right)$$
$$= -4\pi^2 \sum_{m\geq 1} mq^m$$

(by the beginning of A.3.5) and accordingly $f(n\tau) = -4\pi^2 \sum_{m\geq 1} mq^{mn}$. So in the end:

$$G_1(\tau) = 2\zeta(2) + \sum_{n\geq 1} \left(-8\pi^2 \sum_{m\geq 1} mq^{mn}\right),$$

as stated. □

A.3.7.3. When we prove the transformation formula for the Eisenstein series under the substitution $\tau \leftarrow \tau' := -1/\tau$, the crux of the argument is the following:

$$\sum_{m,n\in\mathbf{Z}} (m+n\tau')^{-2k} = \sum_{m,n\in\mathbf{Z}} (m-n/\tau)^{-2k} = \tau^{2k} \sum_{m,n\in\mathbf{Z}} (-n+m\tau)^{-2k}$$
$$= \sum_{m,n\in\mathbf{Z}} (m+n\tau)^{-2k}.$$

The last equality is justified because $(m,n) \mapsto (-n,m)$ is a permutation of $\mathbf{Z}^2 \setminus \{(0,0)\}$, so only the order of the terms is changed. But this is only allowed for absolutely convergent series, so it is not a correct deduction when $k=1$. And indeed:

PROPOSITION A.19. *$G_1(-1/\tau) = \tau^2 G_1(\tau) - 2i\pi\tau$.*

PROOF. There are many more or less complicated proofs. The most elementary seems the following one, taken from **[231]**. We shall write $\sum_{n,m}'$ with the same meaning as before, i.e. $\sum_n \sum_m$ while avoiding $m=n=0$. So what we can do is the following:

$$G_1(-1/\tau) = \sum_{n,m}{}'(m-n/\tau)^{-2} = \tau^2 \sum_{n,m}{}'(-n+m\tau)^{-2} = \tau^2 \sum_{n,m}{}'(n+m\tau)^{-2} = \tau^2 G(\tau),$$

where $G(\tau) := \sum_{n,m}{}'(n+m\tau)^{-2} = \sum_{m,n}{}'(m+n\tau)^{-2}$. (The reader can check that changing $-n$ to n was innocuous.) In order to overcome the curse of noncommutative convergence, we introduce two other double series:

$$H_1(\tau) := \sum_{n,m}{}'' \frac{1}{(m+n\tau)(m-1+n\tau)} \text{ and } H(\tau) := \sum_{m,n}{}'' \frac{1}{(m+n\tau)(m-1+n\tau)}.$$

where $\sum_{n,m}''$ stands for $\sum_n \sum_m$ while avoiding $m=n=0$ and also $m=1$, $n=0$. There are two combined reasons to do that:

(1) $\dfrac{1}{(m+n\tau)(m-1+n\tau)} - \dfrac{1}{(m+n\tau)^2} = \dfrac{1}{(m+n\tau)^2(m-1+n\tau)}$, the general term of an absolutely convergent double series, and (taking in account

that G_1 and G do have a term with $m = 1$, $n = 0$, which is missing in H_1 and H):

$$H_1 - G_1 = \sum_{n,m}'' \left(\frac{1}{(m + n\tau)(m - 1 + n\tau)} - \frac{1}{(m + n\tau)^2} \right) - 1,$$

$$H - G = \sum_{m,n}'' \left(\frac{1}{(m + n\tau)(m - 1 + n\tau)} - \frac{1}{(m + n\tau)^2} \right) - 1,$$

so $H-G+1$ and H_1-G_1+1 are commutatively convergent and $H-G+1 = H_1 - G_1 + 1$, i.e., $G_1 - G = H_1 - H$;

(2) $\dfrac{1}{(m + n\tau)(m - 1 + n\tau)} = \dfrac{1}{m - 1 + n\tau} - \dfrac{1}{m + n\tau}$, so H and H_1 are in some sense telescopic and can be explicitly evaluated.

The evaluation of H is more involved than that of H_1. We refer to the lemma below, stating that $H_1 = 2$ and $H = 2 - 2i\pi/\tau$. Therefore $H_1 - H = 2i\pi/\tau$, i.e., $G_1 - G = 2i\pi/\tau$, which is equivalent to the desired equality. □

LEMMA A.20. (i) $H_1 = 2$.
(ii) $H = 2 - 2i\pi/\tau$.

PROOF. (i) We split $\sum_n$ in two parts according to whether $n = 0$ or not:

$$H_1 = \sum_{m\neq 0,1} \frac{1}{m(m-1)} + \sum_{n\neq 0} \sum_m \left(\frac{1}{m - 1 + n\tau} - \frac{1}{m + n\tau} \right).$$

The first sum is equal to:

$$\sum_{m\geq 2} \left(\frac{1}{m-1} - \frac{1}{m} \right) + \sum_{m\leq -1} \left(\frac{1}{m-1} - \frac{1}{m} \right) = 1 - (-1) = 2.$$

The internal sum in the second term is:

$$\sum_m \left(\frac{1}{m - 1 + n\tau} - \frac{1}{m + n\tau} \right) = 0,$$

because it has the form $\sum_m (u_m - u_{m-1})$ with $u_m \to 0$ when $m \to \pm\infty$. So in the end we find $H_1 = 2 + \sum_n 0 = 2$.

(ii) Recall from A.3.5 the formula:

$$\pi \cot(\pi\tau) = \frac{1}{\tau} + \sum_{n\geq 1} \left(\frac{1}{\tau + n} + \frac{1}{\tau - n} \right).$$

We split $\sum_n$ in three parts according to whether $m = 0$, $m = 1$ or else:

$$H = \sum_{n\neq 0} \left(\frac{1}{-1 + n\tau} - \frac{1}{n\tau} \right) + \sum_{n\neq 0} \left(\frac{1}{n\tau} - \frac{1}{1 + n\tau} \right) + \sum_{m\neq 0,1} \Phi(m-1, m),$$

where we introduce temporarily $\Phi(a, b) := \sum_{n\in\mathbf{Z}} \left(\dfrac{1}{a + n\tau} - \dfrac{1}{b + n\tau} \right)$.

The first two sums composing H combine into one (we write $\tau' := -1/\tau$):

$$\sum_{n\neq 0}\left(\frac{1}{-1+n\tau}-\frac{1}{n\tau}\right)+\sum_{n\neq 0}\left(\frac{1}{n\tau}-\frac{1}{1+n\tau}\right)=\sum_{n\neq 0}\left(\frac{1}{-1+n\tau}-\frac{1}{1+n\tau}\right)$$
$$=-\sum_{n\neq 0}\left(\frac{1}{1-n\tau}+\frac{1}{1+n\tau}\right)=-2\sum_{n\geq 1}\left(\frac{1}{1-n\tau}+\frac{1}{1+n\tau}\right)$$
$$=-2\sum_{n\geq 1}\left(\frac{1}{1+n/\tau'}+\frac{1}{1-n/\tau'}\right)=-2\tau'\sum_{n\geq 1}\left(\frac{1}{\tau'+n}+\frac{1}{\tau'-n}\right)$$
$$=-2\tau'(\pi\cot\pi\tau'-1/\tau')=2-2\pi\tau'\cot\pi\tau'.$$

Going for the third term in H, we first evaluate generally:

$$\begin{aligned}\Phi(a,b)&=\sum_{n\in\mathbf{Z}}\left(\frac{1}{a+n\tau}-\frac{1}{b+n\tau}\right)\\&=\frac{1}{a}-\frac{1}{b}+\sum_{n\geq 1}\left(\frac{1}{a+n\tau}-\frac{1}{b+n\tau}\right)+\sum_{n\leq -1}\left(\frac{1}{a+n\tau}-\frac{1}{b+n\tau}\right)\\&=\frac{1}{a}-\frac{1}{b}+\sum_{n\geq 1}\left(\frac{1}{a+n\tau}-\frac{1}{b+n\tau}\right)+\sum_{n\geq 1}\left(\frac{1}{a-n\tau}-\frac{1}{b-n\tau}\right)\\&=\frac{1}{a}-\frac{1}{b}+\sum_{n\geq 1}\left(\frac{1}{a+n\tau}+\frac{1}{a-n\tau}-\frac{1}{b+n\tau}-\frac{1}{b-n\tau}\right)\\&=\frac{1}{a}-\frac{1}{b}+\sum_{n\geq 1}\left(\frac{1}{a+n\tau}+\frac{1}{a-n\tau}\right)-\sum_{n\geq 1}\left(\frac{1}{b+n\tau}+\frac{1}{b-n\tau}\right)\\&=\Psi(a)-\Psi(b),\end{aligned}$$

where we introduced:

$$\Psi(a)=\frac{1}{a}+\sum_{n\geq 1}\left(\frac{1}{a+n\tau}+\frac{1}{a-n\tau}\right)=\frac{1}{\tau}\left(\frac{\tau}{a}+\sum_{n\geq 1}\left(\frac{1}{a/\tau+n}+\frac{1}{a/\tau-n}\right)\right),$$

i.e., $\Psi(a)=\dfrac{\pi}{\tau}\cot\pi a/\tau=\pi\tau'\cot a\pi\tau'$. Therefore, setting $u:=e^{2\mathrm{i}\pi\tau'}$ and, for short, $u^a:=e^{2\mathrm{i}a\pi\tau'}$, etc, and noting that $\cot t=\mathrm{i}\dfrac{e^{2\mathrm{i}\pi t}+1}{e^{2\mathrm{i}\pi t}-1}$:

$$\Phi(a,b)=\mathrm{i}\pi\tau'\left(\frac{u^a+1}{u^a-1}-\frac{u^b+1}{u^b-1}\right).$$

Now here comes the third term in H:

$$\sum_{m\neq 0,1}\Phi(m-1,m)=\mathrm{i}\pi\tau'\left[\sum_{m\geq 2}\left(\frac{u^{m-1}+1}{u^{m-1}-1}-\frac{u^m+1}{u^m-1}\right)+\sum_{m\leq -1}\left(\frac{u^{m-1}+1}{u^{m-1}-1}-\frac{u^m+1}{u^m-1}\right)\right]$$

We note that $\Im\tau'>0$, so $|u|<1$, so we rearrange the sums as:

$$\sum_{m\geq 2}\left(\frac{u^{m-1}+1}{u^{m-1}-1}-\frac{u^m+1}{u^m-1}\right)=\sum_{m\geq 2}\left(\frac{2u^{m-1}}{u^{m-1}-1}-\frac{2u^m}{u^m-1}\right)=\frac{2u}{u-1},$$

and

$$\sum_{m\leq -1}\left(\frac{u^{m-1}+1}{u^{m-1}-1}-\frac{u^m+1}{u^m-1}\right)=\sum_{m\leq -1}\left(\frac{2}{u^{m-1}-1}-\frac{2}{u^m-1}\right)=\frac{-2}{u^{-1}-1}=\frac{2u}{u-1},$$

so the third term in H boils down to:

$$\begin{aligned}\mathrm{i}\pi\tau'\left(\frac{2u}{u-1}+\frac{2u}{u-1}\right)=4\mathrm{i}\pi\tau'\frac{u}{u-1}=2\pi\tau'\frac{e^{\mathrm{i}\pi\tau'}}{\sin\pi\tau'}&=2\pi\tau'(\cot\pi\tau'+\mathrm{i})\\&=2\pi\tau'\cot\pi\tau'+2\mathrm{i}\pi\tau'.\end{aligned}$$

We conclude this tiring calculation:

$$H=2-2\pi\tau'\cot\pi\tau'+2\pi\tau'\cot\pi\tau'+2\mathrm{i}\pi\tau'=2+2\mathrm{i}\pi\tau'=2-2\mathrm{i}\pi/\tau,$$

as desired.

□

EXERCISE A.21. Evaluate the sum $\sum_{\substack{m,n\in\mathbf{Z}\\ m\pm n\neq 0}}\frac{1}{m^2-n^2}$ in two ways: $\sum_m\sum_n$, then $\sum_n\sum_m$. (The results should differ.)

A.4. The Gamma and the Zeta function

A.4.1. the Gamma function. The Gamma function was invented by Euler to "interpolate" factorial numbers. It is defined by:

$$\Gamma(x):=\int_0^\infty t^{x-1}e^{-t}\,dt=\int_0^\infty t^x e^{-t}\,\frac{dt}{t}.$$

The second form is meant to emphasize that it is a kind of Mellin transform, so the domain of integration really is really $\mathbf{R}_+^*$ rather than $\mathbf{R}_+$. The integral is well defined for a complex x such that $\Re(x)>0$, and it defines an analytic function of x. Integration by parts immediately gives the functional equation:

$$\forall x\ ,\ \Gamma(x+1)=x\Gamma(x).$$

Since obviously $\Gamma(1)=1$, we get that $\Gamma(n)=(n-1)!$ for $n\in\mathbf{N}^*$. We shall see soon that $\Gamma(1/2)=\sqrt{\pi}$, so that:

$$\forall n\in\mathbf{N}\ ,\ \Gamma(n+1/2)=\frac{(2n)!\sqrt{\pi}}{2^{2n}n!}.$$

Using again the functional equation in the form $\Gamma(x)=\dfrac{1}{x}\Gamma(x+1)$, we see that Γ can be iteratively continued to the half planes $\Re(x)>-n$, $n\in\mathbf{N}$, and, in the limit, that it admits a meromorphic continuation to the whole of $\mathbf{C}$ with a simple pole at each $-n$, $n\in\mathbf{N}$, with residue $\dfrac{(-1)^n}{n!}$.

Before stating the next theorem, we recall that the Fourier series expansion of the 2π-periodic function defined by $t\mapsto\cos zt$ on $[-\pi;\pi]$ yields the formula:

$$\cos(zt)=\frac{\sin\pi z}{\pi z}+\frac{\sin\pi z}{\pi}\sum_{n\geq 1}(-1)^n\left(\frac{1}{z+n}+\frac{1}{z-n}\right)\cos nt.$$

Setting $t:=0$ and $z:=x$, we deduce that:

$$\frac{\pi}{\sin\pi x}=\frac{1}{x}+\sum_{n\geq 1}(-1)^n\left(\frac{1}{x+n}+\frac{1}{x-n}\right).$$

THEOREM A.7 (Euler's reflection formula).

$$\Gamma(x)\Gamma(1-x) = \frac{\pi}{\sin \pi x}.$$

PROOF. We sketch it for $0 < x < 1$, the general case follows by analytic continuation. We have:

$$\Gamma(x)\Gamma(1-x) = \int_0^\infty \int_0^\infty t^{x-1} u^{-x} e^{-t-u}\, dt\, du,$$

which we transform to polar coordinates: $t = r\cos\theta$, $u = r\sin\theta$, so:

$$\begin{aligned}\Gamma(x)\Gamma(1-x) &= \int_0^{\pi/2} \left(\int_0^\infty r^{-1} e^{-r(\cos\theta+\sin\theta)}\, r\, dr \right) (\cos\theta)^{x-1} (\sin\theta)^{-x}\, d\theta \\ &= \int_0^{\pi/2} \frac{(\cos\theta)^{x-1}(\sin\theta)^{-x}}{\cos\theta + \sin\theta}\, d\theta = \int_0^\infty \frac{v^{-x}}{1+v}\, dv.\end{aligned}$$

(The last equality follows from substitution $v := \tan\theta$.) We split the last integral as $\int_0^1 + \int_1^\infty$, then change v to $1/v$ in the second term:

$$\Gamma(x)\Gamma(1-x) = \int_0^1 \frac{v^{-x} + v^{x-1}}{1+v}\, dv = \sum_{m\geq 0} (-1)^m \int_0^1 (v^{m-x} + v^{m+x-1})\, dv,$$

by power series expansion of $1/(1+v)$. So we get the value:

$$\begin{aligned}\Gamma(x)\Gamma(1-x) &= \sum_{m\geq 0} (-1)^m \left(\frac{1}{m+1-x} + \frac{1}{m+x} \right) \\ &= \sum_{m\geq 0} (-1)^m \frac{1}{m+1-x} + \sum_{m\geq 0} (-1)^m \frac{1}{m+x} \\ &= \sum_{n\geq 1} (-1)^n \frac{1}{x-n} + \sum_{m\geq 0} (-1)^m \frac{1}{m+x} \\ &= \frac{1}{x} + \sum_{n\geq 1} (-1)^n \left(\frac{1}{x+n} + \frac{1}{x-n} \right) = \frac{\pi}{\sin \pi x}.\end{aligned}$$

□

COROLLARY A.22. (i) $\Gamma(1/2) = \sqrt{\pi}$.
(ii) $\int_{-\infty}^\infty e^{-\pi x^2}\, dx = 1$.

PROOF. (i) From the reflection formula, $\Gamma(1/2)^2 = \pi$ and clearly $\Gamma(1/2) > 0$.

(ii) The integral equals $2\int_0^\infty e^{-\pi x^2}\, dx$ and the change of variable $x = \sqrt{t/\pi}$ gives:

$$\int_0^\infty e^{-\pi x^2}\, dx = \int_0^\infty e^{-t} \frac{dt}{2\sqrt{\pi}\sqrt{t}} = \frac{1}{2\sqrt{\pi}} \Gamma(1/2) = 1/2.$$

□

The second relation above is used in the proof of the modular functional relation for the real Theta function.

A.4.2. The Zeta function. The real version of the Zeta function was invented by Euler, but the complex version seems to be due to Riemann (although Dirichlet used something similar to prove his theorem on primes in arithmetic progressions). The function:

$$\zeta(s) := \sum_{n\geq 1} n^{-s} := \sum_{n\geq 1} e^{-s\log n}$$

is analytic over the half-plane $\Re(s) > 1$. From the fundamental theorem of arithmetic, one deduces the *eulerian product expansion*:

$$\zeta(s) = \prod_{p\in P} (1-p^{-s})^{-1}, \text{ where } P := \{2,3,5,\ldots\} \text{ (the prime numbers)}.$$

It can be quickly justified as follows: each factor of the right-hand side has the form $1 + p^{-s} + p^{-2s} + \cdots$ and their product is the sum of all terms $(p_1^{r_1}\cdots p_k^{r_k})^{-s}$ for all finite sequences of prime numbers and exponents, and this is exactly $\sum n^{-s}$ by the fundamental theorem of arithmetic. The argument has the same combinatorial flavour as the q-expansion of $\mathcal{P}(q)$ (the generating function of the numbers of partitions) in 1.2.1 and the analogy even goes further, since in both cases the interplay between the generating series and its product expansion yields, through analysis of singularities, asymptotic information (on the distribution of prime numbers in the case of Zeta). That seems a good reason to devote here some place to $\zeta(s)$. We refer to **[112]** for elementary examples of such arguments. To study the behaviour of ζ in the neighborhood of $s = 1$, we note that, since $\int_0^{+\infty} t^{-s}\,dt = \dfrac{1}{s-1}$:

$$\zeta(s) - \frac{1}{s-1} = \sum_{n\geq 1} \int_0^{+\infty} (n^{-s} - t^{-s})\,dt.$$

LEMMA A.23. *If* $0 < a < b$ *and* $\sigma := \Re(s) > 0$, *then:*

$$\left|b^{-s} - a^{-s}\right| \leq \frac{(b-a)\,|s|}{a^{\sigma+1^3}}.$$

PROOF.

$$b^{-s} - a^{-s} = \int_a^b -st^{-s-1}\,dt \Longrightarrow \left|b^{-s} - a^{-s}\right| \leq (b-a) \max_{t\in[a;b]} \left|-st^{-s-1}\right|.$$

□

PROPOSITION A.24. *The Zeta function has a meromorphic continuation to the half-plane* $\Re(s) > 0$ *with a unique pole at* $s = 1$*; the pole is simple with residue* 1.

PROOF. It suffices to prove that $s \mapsto \sum_{n\geq 1} \int_0^{+\infty} (n^{-s} - t^{-s})\,dt$ is analytic for $\Re(s) > 0$, which readily follows from the lemma. □

EXERCISE A.25. (i) Prove the equality $(1 - 2^{1-s})\zeta(s) = -\sum_{n\geq 1} (-1)^n n^{-s}$. (Split the right-hand side into two sums, one with even, the other with odd n.)

(ii) Deduce from it another proof of the proposition.

COROLLARY A.26. (i) $\zeta(s) = \dfrac{1}{s-1} + O(1)$ *as* $s \to 1$.

(ii) $\sum_{p\in P} \dfrac{1}{p^s} = \log \dfrac{1}{s-1} + O(1)$ *as* $s \to 1$.

PROOF. (i) is immediate from the proposition.

(ii) Combine (i) with the estimation deduced from the eulerian expansion:

$$\log \zeta(s) = \sum_{k \geq 1} \sum_{p \in P} \frac{1}{kp^{ks}} = \sum \frac{1}{p^s} + O(1),$$

because the terms with $k \geq 2$ add up to a normally convergent series. □

Using the second fact above, mathematicians studying analytic number theory defined the "density" of a set of primes $P' \subset P$ as $\lim_{s \to 1} \frac{\sum_{p \in P'} \frac{1}{p^s}}{\log \frac{1}{s-1}}$. For instance, from a theorem of Dirichlet, if $a, b \in \mathbf{N}^*$ are relatively prime, the density of $P_{a,b} := \{p \in P \mid p \equiv b \pmod{a}\}$ is $\frac{1}{\phi(a)}$ (where ϕ is Euler's totient function). This is an example of what we meant by asymptotical information flowing from the eulerian expansion and analysis of singularity. Of course the true power of Riemann's theory lies much deeper **[80]**.

THEOREM A.8 (Riemann). *The Zeta function has a meromorphic continuation to the whole complex plane with a unique pole at $s = 1$; the pole is simple with residue* 1. *That continuation satisfies the functional equation $\Lambda(s) = \Lambda(1-s)$, where $\Lambda(s) := \pi^{-s/2}\Gamma(s/2)\zeta(s)$.*

PROOF. If satisfied in the stripe $0 < \Re(s) < 1$, the functional equation (which is a priori meaningful there) will allow us to *define* said continuation, so we proceed to prove it. We use function ϑ from 2.1.2. Let:

$$A(s) := \int_1^{+\infty} t^{s/2}(\vartheta(t) - 1)\,\frac{dt}{t} \text{ and } B(s) := \int_0^1 t^{s/2}(\vartheta(t) - \sqrt{1/t})\,\frac{dt}{t}.$$

Those are holomorphic functions. Using equation $\vartheta(t) = \sqrt{1/t}\,\vartheta(1/t)$ from 2.1.2.2 and a change of variable $t \mapsto 1/t$, we get $A(s) = B(1-s)$. Let $\Phi(s) := A(s) + B(s)$, so $\Phi(s) = \Phi(1-s)$. Also:

$$\Phi(s) = \frac{2}{s} + \frac{2}{1-s} + 2\sum_{n \geq 1} \int_0^{+\infty} t^{s/2} e^{-\pi n^2 t}\,\frac{dt}{t} = \frac{2}{s} + \frac{2}{1-s} + 2\sum_{n \geq 1} n^{-s}\pi^{-s/2}\Gamma(s/2),$$

i.e., $\Phi(s) = \frac{2}{s} + \frac{2}{1-s} + 2\Lambda(s)$ and the conclusion is immediate. □

Using what we found about the poles of Γ in A.4.1, we can deduce that ζ has simple zeroes at all negative even numbers $-2n$, $n \in \mathbf{N}^*$: they are nicknamed the "trivial zeroes". Riemann's Hypothesis is the conjecture that all nontrivial zeroes lie on the line $\Re(s) = 1/2$.

A.5. Analyticity of roots

A.5.1. Preliminary computations. Let $\underline{x} := (x_1, \ldots, x_n) \in \mathbf{C}^n$ and denote $\underline{s} := (s_1, \ldots, s_n) \in \mathbf{C}^n$ the n-uple of the associated *elementary symmetric polynomials* $s_k := \sum_{1 \leq i_1 < \cdots < i_k \leq n} x_{i_1} \cdots x_{i_k}$, so that for instance $s_1 := x_1 + \cdots + x_n$ and $s_n := x_1 \cdots x_n$. The associated polynomial $P_{\underline{s}}(X) := X^n - s_1 X^{n-1} + \cdots + (-1)^n s_n$ factors as $P_{\underline{s}}(X) = (X - x_1) \cdots (X - x_n)$. Some of the computations that follow may be performed either directly on the s_k or on the polynomial $P_{\underline{s}}(X)$. First of all:

$$\forall j = 1, \ldots, n \,,\ \sum_{i=1}^{n} (-1)^{i-1} \frac{\partial s_i}{\partial x_j} X^{n-i} = (X - x_1) \cdots \widehat{(X - x_j)} \cdots (X - x_n),$$

whence, using the usual convention on "hatted" missing terms or factors:

$$\forall j, k = 1, \ldots, n \,,\ \sum_{i=1}^{n} \frac{\partial s_i}{\partial x_j} (-x_k)^{n-i} = (x_1 - x_k) \cdots \widehat{(x_j - x_k)} \cdots (x_n - x_k)$$

In matricial terms, this relates a jacobian and a Vandermonde matrix:

$$\begin{pmatrix} \partial s_1/\partial x_1 & \ldots & \partial s_n/\partial x_1 \\ \vdots & \ddots & \vdots \\ \partial s_1/\partial x_n & \ldots & \partial s_n/\partial x_n \end{pmatrix} \begin{pmatrix} (-x_1)^{n-1} & \ldots & (-x_n)^{n-1} \\ \vdots & \ddots & \vdots \\ (-x_1)^{n-n} & \ldots & (-x_n)^{n-n} \end{pmatrix} = \begin{pmatrix} \delta_1(\underline{x}) & \ldots & 0 \\ \vdots & \ddots & \vdots \\ 0 & \ldots & \delta_n(\underline{x}) \end{pmatrix}.$$

where we have set $\delta_i(\underline{x}) := (x_1 - x_i) \cdots \widehat{(x_i - x_i)} \cdots (x_n - x_i)$, $i = 1, \ldots, n$. We then deduce from classical determinant formulas that the jacobian determinant $J(\underline{s}/\underline{x}) := \det(\partial s_i/\partial x_j)_{i,j=1,\ldots,n}$ is equal to $\delta(\underline{x}) := \prod_{1 \leq i < j \leq n} (x_i - x_j)$.

Note that $\delta(\underline{x})^2$ is a symmetric polynomial in the x_i, and therefore equal to a polynomial $\Delta(\underline{s}) \in \mathbf{C}[s_1, \ldots, s_n]$, the so-called *discriminant* of $P_{\underline{s}}$.

A.5.2. An important etale covering. Geometrically, we thus obtain a holomorphic map $f : \mathbf{C}^n \to \mathbf{C}^n$, $\underline{x} \mapsto \underline{s}$. We emphasize the special loci $Y := \delta^{-1}(0)$ (at the source) and $Z := \Delta^{-1}(0)$ (at the target): obviously $Y = f^{-1}(Z)$.

THEOREM A.9. (i) *The map f is surjective, proper and has finite fibres.*
(ii) *Its restriction/corestriction $\mathbf{C}^n \setminus Y \to \mathbf{C}^n \setminus Z$ is an etale covering of degree $n!$.*

PROOF. (i) Surjectivity comes from the fact that $\mathbf{C}$ is algebraically closed. Each fibre consists in the permutations of a particular n-uple of roots, so it has at most $n!$ elements. Properness come from the well known bounds on roots, for instance $|x_i| \leq \max(1, \sum |s_j|)$ e.g., [**258**, §3.3].

(ii) Since $Y = f^{-1}(Z)$, the map is still surjective. From the above computation of its jacobian determinant, it is (by the Inverse Function Theorem) a local homeomorphism and all its fibres have exactly $n!$ elements. Therefore it is an etale covering by [**95**, Proposition 2.5] or [**181**, Chapter VII §2] or Exercise 1 of [**67**, XVI.12].

□

COROLLARY A.27. *Let $F_t(X) := X^n + a_1(t)X^{n-1} + \cdots + a_n(t) \in \mathscr{O}(\Delta)[X]$ a parameterized unitary polynomial with analytic coefficients in a polydisk $\Delta \subset \mathbf{C}^d$. Assume that F_t is separable (i.e., has n pairwise distinct roots) for all $t \in \Delta$. Then there exist n analytic functions $x_i(t)$ on Δ such that:*

$$\forall t \in \Delta \,,\ F_t(X) = (X - x_1(t)) \cdots (X - x_n(t)).$$

PROOF. Use statement (ii) from the theorem and the simple connectedness of Δ. The latter is crucial, as shown by the example of $F_t(X) := X^2 - t$, $t \in \mathbf{C}^*$. □

APPENDIX B

Riemann Surfaces and Vector Bundles

General references for this appendix are **[88, 168]**. (For the Riemann sphere **S** and in part for elliptic functions, also see **[3, 45, 211]**.) All the contents are standard, except for some direct applications to the example of $\mathbf{E}_q := \mathbf{C}^*/q^{\mathbf{Z}}$. However, since there are essentially no proofs, the order is not always the most logical one; it rather aims at pedagogy.

B.1. General definitions

B.1.1. Riemann surfaces. A *topological surface* (or *topological variety of dimension* 2) is a connected Hausdorff topological space X endowed with a family $(\phi_i : U_i \to V_i)_i$ of homeomorphisms, with sources an open covering $(U_i)_i$ of X and targets V_i being open subsets of $\mathbf{R}^2$. Every such homeomorphism $\phi : U \to V$ is called a *(topological) chart* and a family of charts whose sources cover X is called a *(topological) atlas.* In this book, we shall use $\mathbf{C}$ instead of $\mathbf{R}^2$ (with the usual identification) and therefore *complex* charts or atlases. If $x \in X$ and ϕ is a chart in a neighborhood of x, we consider $\phi(x) \in \mathbf{C}$ as the *coordinate* of x (in that chart). If there are two charts $\phi : U \to V$, $\phi' : U' \to V'$ near x, the restriction map $\phi' \circ \phi^{-1}$ going from $\phi(U \cap U') \subset V$ to $\phi'(U \cap U') \subset V'$ is therefore a *change of coordinates* map.

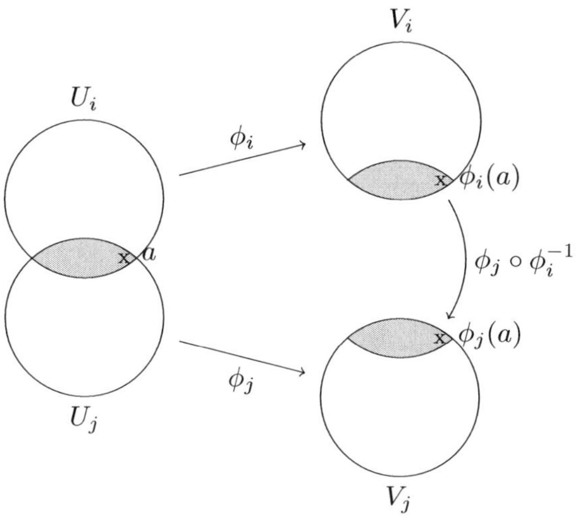

$a \in U_i \cap U_j$ has coordinates $\phi_i(a) \in V_i$ and $\phi_j(a) \in V_j$

FIGURE B.1. Change of charts

Two topological charts $\phi : U \to V$, $\phi' : U' \to V'$ on the topological surface X are said *(analytically) compatible* if either $U \cap U' = \emptyset$ or the corresponding changes of coordinates from $\phi(U \cap U') \subset V$ to $\phi'(U \cap U') \subset V'$ and conversely are both holomorphic (hence biholomorphic). An *analytical atlas* is a topological atlas whose charts are pairwise analytically compatible. Two analytical atlases $(\phi_i : U_i \to V_i)_i$ and $(\phi'_j : U'_j \to V'_j)_j$ are *compatible* if each ϕ_i is compatible with each ϕ'_j; equivalently, joining the two atlases yields an atlas. Any atlas on X endows it with an *analytical structure* and, by definition, equivalent atlases define the same analytical structure. Therefore there is a biggest atlas defining that analytical structure (the union of them all). Whenever an analytical structure has been selected, we shall call chart any member of that maximal atlas.

A *Riemann surface* is a topological surface equipped with an analytical structure. Note that any connected open subset of X automatically inherits a structure of Riemann surface.

REMARK B.1. Replacing everywhere "holomorphic" by "$\mathscr{C}^\infty$-differentiable", one defines $\mathscr{C}^\infty$ varieties, the object of differential geometry. Since holomorphic maps are $\mathscr{C}^\infty$-differentiable, a Riemann surface is a $\mathscr{C}^\infty$ variety and the methods from differential geometry can be applied to it.

B.1.2. The sheaf $\mathscr{O}_X$. A map $f : U \to \mathbf{C}$, U an open subset of X is *holomorphic* if, for some analytical atlas $(\phi_i : U_i \to V_i)_i$ (and therefore for all) all maps:

$$f \circ \phi_i^{-1} : \phi_i(U_i \cap U) = V_i \cap \phi_i(U) \to \mathbf{C}$$

are holomorphic. The above is "f read off in chart ϕ_i" or "$f(x)$ expressed in coordinates $z_i = \phi_i(x)$". The *sheaf $\mathscr{O}_X$ of holomorphic functions on X* is defined by:

$$\forall U \subset X \text{ (open subset)}\ ,\ \mathscr{O}_X(U) := \{f : U \to \mathbf{C} \mid f \text{ is holomorphic on } U\}.$$

It is a sheaf of $\mathbf{C}$-algebras ($\mathbf{C}$-linear combinations and products of holomorphic maps are holomorphic, as well as constant functions). Recall that $\mathscr{O}_X$ being a sheaf means that there are restriction morphisms $\mathscr{O}_X(U) \to \mathscr{O}_X(U')$ whenever $U' \subset U$ (with obvious transitivity properties); and that, if $(U_i)_i$ is a covering of U, then the $f_i \in \mathscr{O}_X(U_i)$ glue into a (unique) $f \in \mathscr{O}_X(U)$ if, and only if, the restrictions of $f_i \in \mathscr{O}_X(U_i)$ and $f_j \in \mathscr{O}_X(U_j)$ to a section of $\mathscr{O}_X(U_i \cap U_j)$ always coincide.

The analytic structure on X can be completely defined by giving the sheaf $\mathscr{O}_X$ as a subsheaf of the sheaf $\mathscr{C}_X$ of all complex valued continuous functions of X. (Of course not every subsheaf of $\mathscr{C}_X$ can be used for that; necessary and sufficient conditions are stated in the litterature.)

B.1.3. Examples. We describe them through atlases and through their sheaves of holomorphic functions. As a preliminary remark, note that, if $Y \subset X$ is a connected open subset seen as a Riemann surface, the sheaf $\mathscr{O}_Y$ is defined as $\mathscr{O}_Y(U) = \mathscr{O}_X(U)$ for every open subset $U \subset Y$.

B.1.3.1. *Connected open subsets of* $\mathbf{C}$. Clearly, for every open subset U of $\mathbf{C}$, one has $\mathscr{O}_\mathbf{C}(U) = \mathscr{O}(U)$, the $\mathbf{C}$-algebra of holomorphic functions $\mathbf{U} \to \mathbf{C}$. Also, by the above remark, every connected open subset Y of $\mathbf{C}$ is a Riemann surface and $\mathscr{O}_Y(U) = \mathscr{O}(U)$ for every open subset U of Y. Two prominent examples are the Poincaré half-plane $\mathcal{H}$ and the unit disk $\mathcal{D}$:

$$\mathcal{H} := \{z \in \mathbf{C} \mid \Im\, z > 0\} \text{ and } \mathcal{D} := \{z \in \mathbf{C} \mid |z| < 1\}.$$

The map $z \mapsto \dfrac{z-\mathrm{i}}{z+\mathrm{i}}$ is a biholomorphism $\mathcal{H} \to \mathcal{D}$ so (anticipating a definition to come) *they are isomorphic Riemann surfaces.* But none of them is isomorphic to $\mathbf{C}$, since the identity $z \mapsto z$ is a bounded holomorphic function on $\mathcal{D}$, while there are no non constant bounded holomorphic functions on $\mathbf{C}$.

B.1.3.2. *The Riemann sphere* $\mathbf{S}$. The Riemann sphere $\mathbf{S} := \mathbf{C} \cup \{\infty\}$ is a compact Riemann surface. The simplest atlas on it consists on the canonical chart $\mathrm{Id} : \mathbf{C} \to \mathbf{C}$; and on the "symmetric" (in some sense) chart with support $\mathbf{C}_\infty := \mathbf{S} \setminus \{0\} = \mathbf{C}^* \cup \{\infty\}$ and coordinate function $\phi : \mathbf{C}_\infty \to \mathbf{C}$, $w \mapsto 1/w$ (of course $\infty \mapsto 0$). So, to test if a function f on an open set $U \subset \mathbf{S}$ is holomorphic, test holomorphy the usual ways at points of $U \cap \mathbf{C}$; and, if $\infty \in U$, set $g(w) := f(1/w)$ and test if g is holomorphic at $w = 0$. This defines completely the sheaf $\mathcal{O}_{\mathbf{S}}$. Note that, by Liouville theorem, $\mathcal{O}_{\mathbf{S}}(\mathbf{S}) = \mathbf{C}$ (since continuous complex functions on compact spaces are bounded and bounded holomorphic functions are constant).

B.1.3.3. *Complex tori* $\mathbf{E}_\Lambda := \mathbf{C}/\Lambda$. This example is standard and thoroughly discussed in textbooks, so we only sketch it. Let Λ a lattice in $\mathbf{C}$ and let $p : \mathbf{C} \to \mathbf{E}_\Lambda := \mathbf{C}/\Lambda$ the canonical projection; it is an etale covering (see General Notations, 0.4.3): every sufficiently small open subset $U \subset \mathbf{E}_\Lambda$ is such that $p^{-1}(U)$ is a disjoint image $\bigsqcup_{\lambda \in \Lambda} (V + \lambda)$, where V is some open subset of $\mathbf{C}$ and each $V + \lambda$ projects homeomorphically to U. Then any of the inverse maps $U \to V + \lambda$ is an analytic chart. The change of coordinates maps all are translations $z \mapsto z + \lambda$ for some $\lambda \in \Lambda$ (and so they are biholomorphisms as they should). The sheaf of holomorphic functions is defined as follows:

$$\mathcal{O}_{\mathbf{E}_\Lambda}(U) = \{f : V \to \mathbf{C} \mid f \circ p \in \mathcal{O}_{\mathbf{C}}(p^{-1}(U))\}.$$

Functions of the form $f \circ p$ on $p^{-1}(U)$ are exactly the Λ-periodic functions. Moreover, since p is surjective, $f \mapsto f \circ p$ is injective. Therefore, the map $f \mapsto f \circ p$ identifies $\mathcal{O}_{\mathbf{E}_\Lambda}(V)$ to the subalgebra of $\mathcal{O}_{\mathbf{C}}(p^{-1}(V))$ made of Λ-periodic functions:

$$\mathcal{O}_{\mathbf{E}_\Lambda}(U) = \left(\mathcal{O}_{\mathbf{C}}(p^{-1}(U))\right)^\Lambda.$$

Since $U \mapsto \mathcal{O}_{\mathbf{C}}(p^{-1}(U)$ is also a sheaf on $\mathbf{E}_\Lambda$, the so-called *direct image sheaf* $p_*\mathcal{O}_{\mathbf{C}}$, one also writes:

$$\mathcal{O}_{\mathbf{E}_\Lambda} = (p_*\mathcal{O}_{\mathbf{C}}))^\Lambda.$$

Again $\mathbf{E}_\Lambda$ being compact, $\mathcal{O}_{\mathbf{E}_\Lambda}(\mathbf{E}_\Lambda) = \mathbf{C}$.

B.1.3.4. *Complex tori* $\mathbf{E}_q := \mathbf{C}^*/q^{\mathbf{Z}}$. Let $q \in \mathbf{C}^*$ such that $|q| < 1$. Let $\mathbf{E}_q := \mathbf{C}^*/q^{\mathbf{Z}}$ and $\pi : \mathbf{C}^* \to \mathbf{E}_q$ the canonical projection $a \mapsto \overline{a} := a \pmod{q^{\mathbf{Z}}}$. We endow $\mathbf{E}_q$ with the quotient topology, i.e., $U \subset \mathbf{E}_q$ is open if, and only if, $\pi^{-1}(U) \subset \mathbf{C}^*$ is. Then π is an etale covering; we do not prove it as it will directly flow from the coming constructions (however, a direct, down to earth proof is detailed in 7.3.1.2). First, we select two particular annuli of interest in $\mathbf{C}^*$; we denote $r := |q|$:

$$V_1 := \mathring{\mathcal{C}}(r, 1) \text{ and } V_2 := \mathring{\mathcal{C}}(r^{1/2}, r^{-1/2}).$$

(Recall from General Notations 0.4 that $\mathring{\mathcal{C}}(R', R) := \{x \in \mathbf{C}^* \mid R' < |x| < R\}$.) Note that for any $k, l \in \mathbf{Z}$, $k \neq l$, one has $q^k V_i \cap q^l V_i = \emptyset$ for $i = 1, 2$. We write

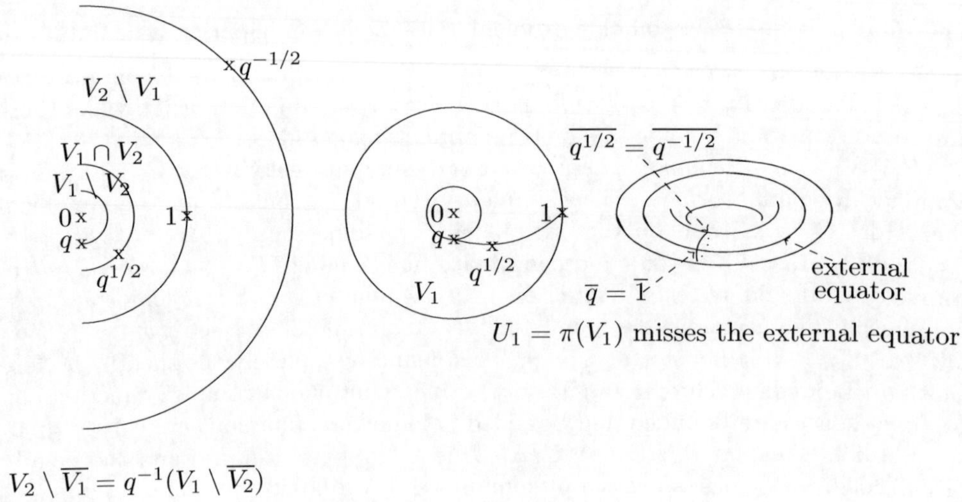

FIGURE B.2. Charts on $\mathbf{E}_q$ (here $q := -i/4$)

$U_1 := \pi(V_1)$ and $U_2 := \pi(V_2)$, so that (see the leftmost part of Figure B.2):

$$\begin{cases} U_1 = \{\overline{x} \mid |x| \notin r^{\mathbf{Z}}\}, \\ U_2 = \{\overline{x} \mid |x| \notin r^{\mathbf{Z}+1/2}\}, \end{cases} \quad \text{and} \quad \begin{cases} \pi^{-1}(U_1) = \bigsqcup_{k \in \mathbf{Z}} q^k V_1, \\ \pi^{-1}(U_2) = \bigsqcup_{k \in \mathbf{Z}} q^k V_2 \end{cases} \quad \text{(two disjoint unions).}$$

We see that U_1, resp. U_2, resp. $U_1 \cap U_2$, is $\mathbf{E}_q$ deprived of its "external equator", resp. of its "internal equator", resp. of both equators (see Figure B.2). As a consequence, $U_1 \cap U_2$ has two connected components the southern one $U_S := \pi(V_1 \cap V_2)$ and the northern one $U_N := \pi(V_1 \setminus V_2) = \pi(V_2 \setminus V_1)$ (in all cases, these are homeomorphic images); more precisely:

$$U_S = \{\overline{x} \mid |x| \in \bigsqcup_{m \in \mathbf{Z}}]r^{m+1/2}, r^m[\} \text{ and } U_N = \{\overline{x} \mid |x| \in \bigsqcup_{m \in \mathbf{Z}}]r^m, r^{m-1/2}[\}$$

Let $\phi_i : U_i \to V_i$, $i = 1, 2$, the two charts (locally inverting π). The change of charts on $U_1 \cap U_2$ is $\phi_2 \circ \phi_1^{-1}$, with source:

$$\phi_1(U_1 \cap U_2) = \phi_1(U_S \sqcup U_N) = \phi_1(U_S) \sqcup \phi_1(U_N) = (V_1 \cap V_2) \sqcup (V_1 \setminus \overline{V_2})$$

and target:

$$\phi_2(U_1 \cap U_2) = \phi_2(U_S \sqcup U_N) = \phi_2(U_S) \sqcup \phi_2(U_N) = (V_2 \cap V_1) \sqcup (V_2 \setminus \overline{V_1}).$$

On $V_1 \cap V_2$, it is the identity; on $V_1 \setminus \overline{V_2}$ it is the map $x \mapsto x/q$.

As for the sheaf $\mathscr{O}_{\mathbf{E}_q}$, arguments similar to the previous ones show that:

$$\mathscr{O}_{\mathbf{E}_q}(U) = \{f : U \to \mathbf{C} \mid f \circ \pi \in \mathscr{O}_{\mathbf{C}^*}(\pi^{-1}(U))\},$$

and, since functions of the form $f \circ \pi$ on $\pi^{-1}(U)$ are exactly the functions g such that $g(qx) = g(x)$, and since $f \mapsto f \circ \pi$ realizes an identification of $\mathscr{O}_{\mathbf{E}_q}(U)$ with the corresponding sub-algebra of $\mathscr{O}_{\mathbf{C}^*}(\pi^{-1}(U)) = \pi_* \mathscr{O}_{\mathbf{C}^*}(U)$, one can write:

$$\mathscr{O}_{\mathbf{E}_q} = (\pi_* \mathscr{O}_{\mathbf{C}^*})^{\sigma_q}.$$

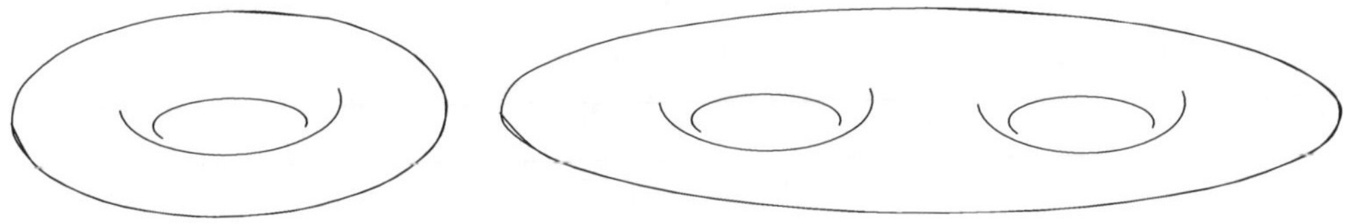

$\mathbf{E}_q$ has genus $g = 1$ This one has genus $g = 2$

FIGURE B.3. Genus of $\mathbf{E}_q$

Here we denote σ_q the operator sending $g(x)$ to $g(qx)$. It is an automorphism of any of the subalgebras $A := \mathscr{O}_{\mathbf{C}^*}(\pi^{-1}(U))$ and A^{σ_q} denotes the subalgebra of invariant elements. Again, $\mathbf{E}_q$ being compact, the only global holomorphic functions are the constant: $\mathscr{O}_{\mathbf{E}_q}(\mathbf{E}_q) = \mathbf{C}$.

B.1.4. Compact and open Riemann surfaces. Since the most important examples for our study are $\mathbf{S}$ and $\mathbf{E}_q$, we shall insist on compact Riemann surfaces. Any compact oriented topological surface is homeomorphic to a "generalized" torus, i.e., a sphere with g "handles" for some $g \geq 0$ called the *genus*: $\mathbf{S}$ has genus 0 and $\mathbf{E}_q$ has genus 1 (beyond that see **[88, 168]**); see Figure B.3. Conversely, any Riemann surface of genus 0, resp. of genus 1, is isomorphic (this will be properly defined soon) to $\mathbf{S}$, resp. to some $\mathbf{E}_\Lambda$, equivalently to some $\mathbf{E}_q$. Note that by A.3, there is an infinite family of pairwise nonisomorphic complex tori (parameterized by $\mathbf{C}$ through their modular invariant j).

A noncompact Riemann surface is said to be *open* (this is not to be confused with the notion of open subset in topology). For instance $\mathbf{C}$, $\mathbf{C}^*$, $\mathcal{H}$ and $\mathcal{D}$ are open Riemann surfaces, as well as $X \setminus S$ for any finite nonempty subset S of a compact Riemann surface X.

B.1.5. Universal covering. According to the Riemann Mapping Theorem, there are, up to isomorphism, three simply connected Riemann surfaces: $\mathbf{S}$, $\mathbf{C}$ and $\mathcal{H}$ (or $\mathcal{D}$)[1]. We already saw that $\mathbf{C}$ and $\mathcal{H}$ are not isomorphic (argument relying on the existence of nonconstant bounded holomorphic functions), although they are homeomorphic (and even $\mathscr{C}^\infty$-diffeomorphic). But $\mathbf{S}$ is compact, thus homeomorphic neither to $\mathbf{C}$ nor to $\mathcal{H}$.

Any connected etale covering $p : Y \to X$ of a Riemann surface X can be made into one in such a way that p is a morphism (and that, uniquely): one "lifts" the charts on X to charts on Y. This applies in particular to the universal covering $\tilde{X} \to X$. By the remark above, there are three possibilities:

(1) $\tilde{X} = \mathbf{S}$: then necessarily $X = \mathbf{S}$ and p is the identity.
(2) $\tilde{X} = \mathbf{C}$: then necessarily $X = \mathbf{E}_\Lambda$ for some lattice and p is the natural projection.
(3) $\tilde{X} = \mathcal{H}$: then there is a discrete group $G \subset \mathbf{PSL}_2(\mathbf{R})$ (acting by homographies) such that $X = \mathcal{H}/G$ and p is the natural projection. This case does not appear in the book.

[1]The "model" $\mathcal{D}$ is used more frequently in the general litterature, but $\mathcal{H}$ is closer to the spirit of this book, because of the use of modular functions.

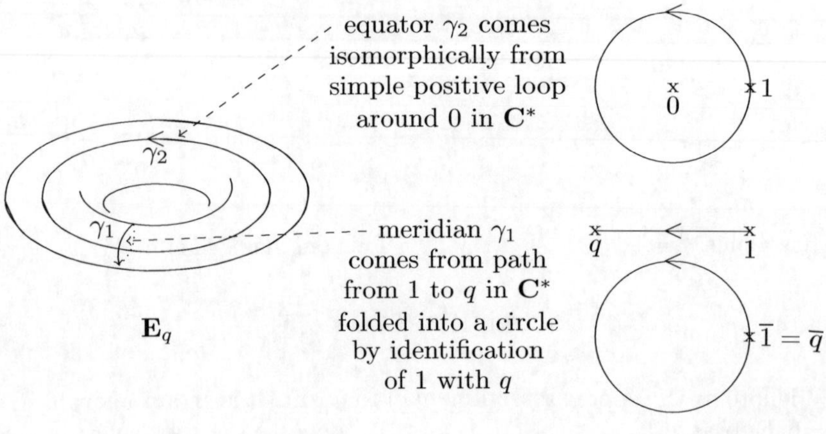

FIGURE B.4. Fundamental loops of $\mathbf{E}_q$ and where they come from

Our main cases of interest are of course $\mathbf{E}_\Lambda$ and $\mathbf{E}_q$. In the first case, if $\Lambda = \mathbf{Z}u + \mathbf{Z}v$, it can be identified with the fundamental group: indeed, $[0;1]u$ folds into a loop and $[0;1]v$ into another and those are free commuting generators of $\pi_1(\mathbf{E}_\Lambda, \overline{0})$ (see Figure A.1 in A.1.4). In the case of $\mathbf{E}_q$, this can be visualized as in Figure B.4.

B.2. Functions

B.2.1. Holomorphic germs. A complex valued function f defined in an open neighborhood U of $a \in X$ is *holomorphic* at a if it is when read off in a chart, i.e., for some chart ϕ defined at a the function $f \circ \phi^{-1}$ is holomorphic at $\phi(a) \in \mathbf{C}$. It follows that f is holomorphic on U (in the sense of A.1.5) if, and only if, it is holomorphic at all points of U. If f does not vanish in a neighborhood of a, the order of $f \circ \phi^{-1}$ at $\phi(a)$ is well defined and does not depend on the chart: we call it the *order* of f at a and write it $\nu_a(f)$.

Let f_1, f_2 defined in open neighborhoods U_1, U_2 of a and write temporarily $f_1 \sim_a f_2$ if, and only if, there is an open neighborhood $V \subset U_1 \cap U_2$ of a where the restrictions coincide: $(f_1)_{|V} = (f_2)_{|V}$. This is an equivalence relation and the equivalence class is called the *germ* of f_1, f_2 at a; and f_1, f_2 are *representatives* of this germ. If one (hence all) representative is holomorphic, the germ is holomorphic. All representatives of a holomorphic germ f at a have the same order at a, called the *order* of the germ at a and denoted $\nu_a(f)$. The set $\mathscr{O}_{X,a}$ of holomorphic germs at a is a local $\mathbf{C}$-algebra, called the *stalk* of the sheaf $\mathscr{O}_X$ at a. Its maximal ideal $\mathfrak{m}_{X,a}$ consists in germs vanishing at a.

A chart ϕ defined at a is *centered* at a if $\phi(a) = 0$. The germ of a centered chart is called a *uniformizing parameter* at a. A germ is a uniformizing parameter if, and only if, its order is 1. For every uniformizing parameter $u \in \mathscr{O}_{X,a}$, one has: $\mathscr{O}_{X,a} = \mathbf{C}\{u\}$ and $\mathfrak{m}_{X,a} = (u)$. It follows that $\mathscr{O}_{X,a}$ is a discrete valuation ring with valuation ν_a: every nonzero germ f can be uniquely written $f = \epsilon u^n$, where $\epsilon \in \mathscr{O}_{X,a}^*$ (a *regular* germ, i.e., defined at a and nonzero there) and $n = \nu_a(f)$.

B.2.2. Standard function theory on X. From the equality $f = \epsilon u^n$, it follows that the zeroes of a nontrivial germ are isolated, so the *principle of analytic continuation* holds on X: if $f \in \mathscr{O}_X(U)$ has a nonisolated zero and if U is connected, then $f = 0$. As a consequence, for every nonempty open subset V of U, the restriction map $\mathscr{O}_X(U) \to \mathscr{O}_X(V)$ is injective (still assuming U connected).

Another classical theorem on complex functions holds: if a germ has a local maximum or a nonzero local minimum (in modulus), it is a constant germ. By analytic continuation, all its representatives on connected open sets are constant. As a corollary, if X is a compact Riemann surface, then $\mathscr{O}_X(X) = \mathbf{C}$.

THEOREM B.1. *Let $a \in U$, an open domain in X, let u a uniformizing parameter at a and let $f \in \mathscr{O}_X(U \setminus \{a\})$. Then:*

(i) *It f is bounded on $U \setminus \{a\}$, it admits a (unique) holomorphic continuation to U.*

(ii) *If $\lim_a |f| = +\infty$, there is a unique $n_0 \in \mathbf{N}^*$ such that $u^{n_0} f$ is bounded on $U \setminus \{a\}$. The integer n_0 does not depend on the choice of u.*

In the second case, f is said to be *meromorphic at a* with a *pole of order n_0*; and one defines the *order of f at a* as $\nu_a(f) = -n_0$. Those definitions hold for functions or germs alike. The set of meromorphic germs at a is the quotient field $\mathscr{M}_{X,a}$ of $\mathscr{O}_{X,a}$, it is equal to $\mathbf{C}(\{u\})$ for any uniformizing parameter u at a. In the missing case of the theorem (when neither (i) nor (ii) holds), f is said to have an *essential singularity* at a. The behaviour of $|f|$ is then very wild (think of $e^{1/x}$ when x approaches 0 in $\mathbf{C}^*$).

B.2.3. The field $\mathscr{M}(X)$. Let $U \subset X$ a nonempty open subset and $S \subset U$ a discrete subset. A function holomorphic on $U \setminus S$ and having at worst poles at S is said to be *meromorphic* on U. The set $\mathscr{M}_X(U)$ of functions meromorphic on U is a $\mathbf{C}$-algebra containing $\mathscr{O}_X(U)$ (actually a field if U is connected) and $U \mapsto \mathscr{M}_X(U)$ is a sheaf. We write $\mathscr{M}(X)$ for $\mathscr{M}_X(X)$, the field of (global) meromorphic functions on X. If X is an open Riemann surface, then $\mathscr{M}(X)$ is the quotient field of $\mathscr{O}(X) := \mathscr{O}_X(X)$ **[200]**. If X is a compact Riemann surface, this is never true, as $\mathscr{O}(X) = \mathbf{C}$ (B.2.2) while $\mathscr{M}(X)$ is then a transcendental extension as we will see. Here are more explicit examples.

Example 1. Let $f := P/Q \in \mathbf{C}(x)$, P, Q coprime polynomials, a nonzero rational function. It is clearly meromorphic on $\mathbf{C}$. It is also meromorphic at ∞ with order $\deg Q - \deg P$ (just use the uniformizing parameter $w := 1/x$ at ∞ to get a local expansion). There is therefore a natural embedding $\mathbf{C}(x) \subset \mathscr{M}(\mathbf{S})$. It is an equality: $\mathscr{M}(\mathbf{S}) = \mathbf{C}(x)$; this can be proved using Liouville's theorems.

Example 2. Let $X := \mathbf{E}_\Lambda$. Then $\wp_\Lambda, \wp'_\Lambda$ generate $\mathscr{M}(X)$ (A.2). Actually, $\mathscr{M}(X) = \mathbf{C}(\wp_\Lambda, \wp'_\Lambda)$ is a quadratic extension of its subfield $\mathbf{C}(\wp_\Lambda)$ which is itself isomorphic to $\mathbf{C}(x)$. Hence the transcendance degree of $\mathscr{M}(X)$ over $\mathbf{C}$ is 1.

Example 3. As a consequence, if $X := \mathbf{E}_q$, the transcendance degree of $\mathscr{M}(X)$ over $\mathbf{C}$ is again 1.

B.2.4. Projective algebraic curves as Riemann surfaces. We only describe here the case of *plane* algebraic curves, which is the most important for us. Let $F(X, Y, T)$ a nonconstant homogeneous irreducible polynomial over $\mathbf{C}$ and assume that its partial derivatives F'_X, F'_Y, F'_T have no other common zero that (maybe) $(0, 0, 0)$. Then the projective nonsingular plane algebraic curve:

$$\begin{aligned}\Gamma &:= \{[x : y : t] \in \mathbf{P}^2(\mathbf{C}) \mid F(x, y, t) = 0\} \\ &= \frac{\{(x, y, t) \in \mathbf{C}^3 \setminus \{(0, 0, 0)\} \mid F(x, y, t) = 0\}}{(x, y, t) \sim (\lambda x, \lambda y, \lambda t),\ \lambda \in \mathbf{C}^*}\end{aligned}$$

is a closed subset of the complex projective plane

$$\mathbf{P}^2(\mathbf{C}) := \frac{\mathbf{C}^3 \setminus \{(0,0,0)\}}{\text{homothety action of } \mathbf{C}^*}.$$

It can be made into a compact Riemann surface as follows.

The open subset $t \neq 0$ of $\mathbf{P}^2(\mathbf{C})$ is identified to $\mathbf{C}^2$ by $[x : y : t] \mapsto (x/t, y/t)$. The corresponding subset of Γ has equation $P(x,y) := F(x,y,1) = 0$, it is an affine nonsingular plane algebraic curve. At every point, either $P'_x \neq 0$, then y can serve as a chart (actually a uniformizing parameter); or $P'_y \neq 0$ and x can be used. This all follows from the complex implicit function theorem. Similar statements hold on the subsets $x \neq 0$, $y \neq 0$ of $\mathbf{P}^2(\mathbf{C})$.

THEOREM B.2. (i) *This defines a compact Riemann surface X and $\mathscr{M}(X)$ is (for instance) the quotient field of* $\mathbf{C}[X,Y]/(P)$.

(ii) *Up to isomorphism (this word is defined in* B.3*) every compact Riemann surface can be obtained that way.*

B.2.5. Divisors. Let X a compact Riemann surface and $f \in \mathscr{M}(X)$, $f \neq 0$. Then f has a finite number of zeroes $a_1, \ldots, a_k$, of orders $n_1, \ldots, n_k$; and a finite number of poles $b_1, \ldots, b_l$, of orders $p_1, \ldots, p_l$. Equivalently, the points where $\nu_a(f) > 0$ are the a_i and $n_i = \nu_{a_i}(f)$; and the points where $\nu_b(f) < 0$ are the b_j and $p_j = -\nu_{b_j}(f)$. We define the *divisor* of f as:

$$\operatorname{div}_X(f) := \sum_{a \in X} \nu_a(f)[a] = \sum_{i=1}^{k} n_i[a_i] - \sum_{j=1}^{l} p_j[b_j].$$

Then $\operatorname{div}_X(fg) = \operatorname{div}_X(f) + \operatorname{div}_X(g)$ and, X being compact, $\operatorname{div}_X(f) = 0 \Leftrightarrow f \in \mathbf{C}^*$, so div_X is a group morphism $\mathscr{M}(X)^* \to \operatorname{Div}(X)$ with kernel $\mathbf{C}^*$. It will follow from the "theorem of residues" in B.4 that $\deg \operatorname{div}_X(f) = 0$, so the image of div_X lies in the kernel $\operatorname{Div}^0(X)$ of the surjective group morphism $\deg : \operatorname{Div}(X) \to \mathbf{Z}$.

The group of *principal divisors* is $P(X) := \operatorname{Im} \operatorname{div}_X \subset \operatorname{Div}^0(X)$. The *Picard group* of X is the group of *divisor classes* $\operatorname{Pic}(X) := \operatorname{Div}(X)/P(X)$; the subgroup of classes of degree 0 is denoted $\operatorname{Pic}^0(X) := \operatorname{Div}(X)^0/P(X)$, whence an exact sequence:

$$\{1\} \to \mathbf{C}^* \to \mathscr{M}(X)^* \to \operatorname{Div}^0(X) \to \operatorname{Pic}^0(X) \to \{0\}.$$

(At left, the trivial group is denoted $\{1\}$ because $\mathbf{C}^*$ is written multiplicatively; at right it is denoted $\{0\}$ because $\operatorname{Div}(X)$, thus $\operatorname{Div}^0(X)$, $\operatorname{Pic}(X)$ and $\operatorname{Pic}^0(X)$, are written additively.)

B.2.6. The theorem of Abel-Jacobi. On the Riemann sphere $\mathbf{S}$, all divisors of degree 0 are principal. Indeed, any such divisor can be written $D := \sum n_i[a_i] - N[\infty]$ where all $a_i \in \mathbf{C}$ and $N := \sum n_i$. Then one checks that $f := \prod (x - a_i)^{n_i}$ has divisor $\operatorname{div}_{\mathbf{S}}(f) = D$. Therefore $\operatorname{Pic}^0(\mathbf{S}) = \{0\}$.

On a complex torus $E := \mathbf{E}_\Lambda$ or $E := \mathbf{E}_q$, we have an evaluation morphism $\operatorname{Div}(E) \to E$. For $\mathbf{E}_\Lambda$ it is written additively: $\sum n_i[a_i] \mapsto \sum n_i a_i$; for $\mathbf{E}_q$ it should be written multicatively: $\sum n_i[a_i] \mapsto \prod a_i^{n_i}$, but this can lead to awkward formulas and we shall sometimes break this rule. At any rate, Abel-Jacobi theorem says that degree zero divisors with trivial evaluation are principal, so $\operatorname{Pic}^0(E) \simeq E$. The inverse map from E to $\operatorname{Pic}^0(E)$ can be realized as follows: write a_0 the neutral element of the group E (so it is $\overline{0}$ in the additive notation for $E := \mathbf{E}_\Lambda$ and $\overline{1}$ in the multiplicative notation for $E := \mathbf{E}_q$); then the inverse map is $a \mapsto$ class of $[a]-[a_0]$. Accordingly, the above exact sequence can be modified:

$$\{1\} \to \mathbf{C}^* \to \mathscr{M}(E)^* \to \mathrm{Div}^0(E) \to E \to \{0\}.$$

(Here we assumed that E is written additively.)

B.3. Morphisms

B.3.1. Morphisms and functions. Let X, X' two Riemann surfaces. A map $f : X \to X'$ is said to be *holomorphic* or *a morphism* if it is when read off in charts, i.e., for any charts $\phi : U \to V$ for X and $\phi' : U' \to V'$ for X' such that $f(U) \subset U'$, the map $\phi' \circ f \circ \phi^{-1} : V \to V'$ is holomorphic:

$$\begin{array}{ccccccc} & & U & \xrightarrow{\quad f \quad} & U' & & \\ & & \Big\downarrow{\scriptstyle \phi} & & \Big\downarrow{\scriptstyle \phi'} & & \\ \mathbf{C} & \hookleftarrow & V & \xrightarrow[\text{holomorphic}]{\text{this must be}} & V' & \hookrightarrow & \mathbf{C}. \end{array}$$

Clearly identities and compositions of morphisms are morphisms, so we obtain a *category of Riemann surfaces.* In that category, bijective morphisms are isomorphisms, i.e., their inverse maps are morphisms. Morphisms from X to $\mathbf{C}$ are the same thing as holomorphic functions; morphisms from X to $\mathbf{S}$ are meromorphic functions (the poles map to ∞), except for the constant function ∞.

If $f : X \to X'$ is a morphism and if g is a holomorphic function on $U' \subset X'$, then $g \circ f$ is a holomorphic function on $f^{-1}(U') \subset X$, so we obtain a morphism of $\mathbf{C}$-algebras $\mathscr{O}_{X'}(U') \to \mathscr{O}_X(f^{-1}(U')) = f_*\mathscr{O}_X(U')$. Altogether, this yields a morphism of sheaves of $\mathbf{C}$-algebras $\mathscr{O}_{X'} \to f_*\mathscr{O}_X$.

B.3.2. Local description of morphisms. Nonconstant morphisms can be completely described locally through charts. So let $f : X \to X'$ a nonconstant morphism and let $a \in X$, $a' := f(a) \in X'$. Then one can choose charts $\phi : U \to V$, $\phi' : U' \to V'$ respectively centered at a and a' such that $\phi' \circ f \circ \phi^{-1} : V \to V'$ (i.e., the lower horizontal arrow in the previous diagram) is the map $x \mapsto x^k$ for some $k \in \mathbf{N}^*$. The integer k depends on f and a only, it is called the *multiplicity of f at a*; we write it $\mu_a(f)$. For instance, if f is a meromorphic function and if a is a zero or a pole, then $e_a(f)$ is its order: $\mu_a(f) = |\nu_a(f)|$.

Topologically, $k := \mu_a(f)$ can be described as an *index of ramification*: there is a simply connected open neighborhood U of a (so, topologically, a disk) such that, $U' := f(U)$ is also topologically a disk and the restriction of f to $U \setminus \{a\} \to U' \setminus \{a'\}$ is an etale covering of degree k: it is only "ramified" at a. (More on this in B.3.3.)

Algebraically $k := \mu_a(f)$ can be described as follows. The map $g \mapsto g \circ f$ from $\mathscr{O}_{X'}(U')$ to $\mathscr{O}_X(f^{-1}(U'))$ goes to a map of germs $\mathscr{O}_{X',a'} \to \mathscr{O}_{X,a}$. Choosing uniformizing parameters, this can be identified with $\mathbf{C}\{u'\} \to \mathbf{C}\{u\}$. For an adequate choice, this is the unique morphism such that $u' \mapsto u^k$.

B.3.3. Nonconstant morphisms of compact Riemann surfaces. The global description of nonconstant morphisms is nice too. First note that it follows from the local description that such a morphism is an open map.

THEOREM B.3. *Let $f : X \to X'$ a nonconstant morphism of Riemann surfaces and assume X is compact.*

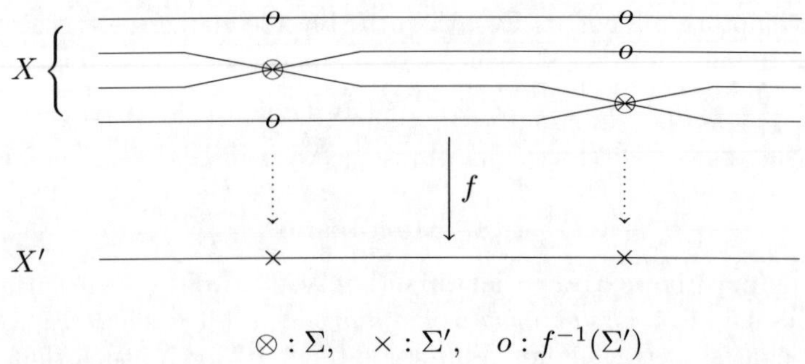

FIGURE B.5. Ramified coverings of compact Riemann surfaces

(i) *Then* X' *is compact too and* f *is onto.*

(ii) *There is an integer* $n \geq 1$ *such that, for all* $a' \in X'$, *the preimage* $f^{-1}(a')$ *has* n *elements "counted with multiplicity":* $\sum_{f(a)=a'} \mu_a(f) = n$.

(iii) *Let* $\Sigma := \{a \in X \mid \mu_a(f) \geq 2\}$ *the* ramification locus *of* f *and* $\Sigma' := f(\Sigma)$ *its* branch locus. *Then* Σ, Σ' *and* $f^{-1}(\Sigma')$ *are finite sets; and* $f : X \setminus f^{-1}(\Sigma') \to X' \setminus \Sigma'$ *is an etale covering of degree* n.

We then say that f is a *ramified covering of degree* n. We define the *ramification divisor of* f on X, with support Σ, and its *branch divisor* on X', with support Σ', as:

$$R_f := \sum_{a\in X} (\mu_a(f)-1)[a] \text{ and } B_f := \sum_{a'\in X'} \sum_{f(a)=a'} (\mu_a(f)-1)[a'] = \sum_{a\in X} (\mu_a(f)-1)[f(a)].$$

They are *effective divisors* (all coefficients are ≥ 0).

Let $f : X \to X'$ a nonconstant morphism of compact Riemann surfaces. For every meromorphic function g on X', the function $g \circ f$ is meromorphic on X. We thus obtain a ring morphism $g \mapsto g \circ f$, $\mathscr{M}(X') \to \mathscr{M}(X)$ between fields, thus a field extension $\mathscr{M}(X)/\mathscr{M}(X')$.

COROLLARY B.2. *The degree of the extension is* $[\mathscr{M}(X) : \mathscr{M}(X')] = n$, *the degree of the ramified covering* f.

In particular, if $f \in \mathscr{M}(X) \setminus \mathbf{C}$ is a nonconstant meromorphic function seen as a morphism $X \to \mathbf{S}$, the image of $\mathscr{M}(\mathbf{S}) = \mathbf{C}(x)$ in $\mathscr{M}(X)$ is $\mathbf{C}(f)$ and $[\mathscr{M}(X) : \mathbf{C}(f)] = n$, the number of zeroes or the number of poles of f, counted with multiplicities. This is also the number of solutions (counted with multiplicities) of $f(a) = b$ for any $b \in \mathbf{C}$.

EXAMPLE B.3. Let Λ a lattice with basis u, v. The Weierstrass function $\mathfrak{P}_\Lambda$ is a ramified covering $\mathbf{E}_\Lambda \to \mathbf{S}$ of degree 2 and $[\mathscr{M}(\mathbf{E}_\Lambda) : \mathbf{C}(\mathfrak{P}_\Lambda)] = 2$. The ramification locus is $\{\overline{u}, \overline{v}, \overline{\frac{u+v}{2}}, \infty\}$ (see A.2).

B.3.4. The Riemann-Hurwitz formula.

THEOREM B.4. *Let* X, *resp.* X', *a compact Riemann surface of genus* g, *resp.* g', *and let* $f : X \to X'$ *a nonconstant morphism of degree* n. *Then:*

$$2g - 2 = (\deg\ f)(2g' - 2) + \deg\ R_f = n(2g' - 2) + \sum_{a\in X} (\mu_a(f) - 1).$$

The proof actually uses differential forms (see B.4). Here we draw consequences. We use the fact that the genus is always ≥ 0 (it is 0 for $\mathbf{S}$, 1 for a complex torus, ≥ 2 otherwise); and that deg R_f is always ≥ 0 (it is 0 for an etale covering, ≥ 1 otherwise).

Case $g = 0$. If $g = 0$, then $g' = 0$. So, if $X = \mathbf{S}$, then $X' = \mathbf{S}$, and deg $R_f = 2n-2$; so, if f is not ramified (thus an etale covering), it is an isomorphism[2] (which is only natural since $\mathbf{S}$ is simply connected). Note that conversely $g' = 0$ does not imply $g = 0$: so-called "hyperelliptic curves" fall under that case [**88**, **168**].

Case $g = 1, g' = 0$. If $g = 1$, then $g' = 0$ or 1. The former case ($g = 1$, $g' = 0$) is provided by any nonconstant elliptic function (see the definition of their order in A.1.3 and A.2). Since a complex torus is not isomorphic to $\mathbf{S}$, the minimal possible value of n is 2; and then deg $R_f = (2.1 - 2) - 2(2.0 - 2) = 4$. This is realized for instance by $\mathfrak{P}_\Lambda$.

Case $g = g' = 1$. In the later case $g = g' = 1$, there is no ramification: any nonconstant morphism between complex tori is an etale covering, it is called an *isogeny*. Every such isogeny can be described (up to isomorphism) in the form of the natural surjection $\mathbf{E}_\Lambda \to \mathbf{E}_{\Lambda'}$, where $\Lambda \subset \Lambda'$.

B.4. Differential forms (an informal presentation)

B.4.1. Holomorphic differential forms on C. Let $V \subset X$ an open subset. A *(holomorphic) differential form on* V is an expression $f(x)\,dx$ where $f \in \mathscr{O}(V)$. The space of differential forms on V is denoted $\Omega^1(V)$; it is obviously a free $\mathscr{O}(V)$-module of rank 1. For any $g \in \mathscr{O}(V)$, we define $dg := g'(x)\,dx$. Such a form is said to be *exact*: hence a form $f(x)\,dx$ is exact if, and only if, f has a primitive on V. For instance $\dfrac{1}{x}\,dx \in \Omega^1(\mathbf{C}^*)$ is not exact; but, if V is simply connected, all forms on V are exact.

Let $\phi : V_1 \to V_2$ a holomorphic function and let $\omega := f(x)\,dx \in \Omega^1(V_2)$. We define its *pull back by* ϕ as $\phi^*\omega := (f \circ \phi)(x)\phi'(x)\,dx \in \Omega^1(V_1)$. This defines a $\mathbf{C}$-linear map $\phi^* : \Omega^1(V_2) \to \Omega^1(V_1)$. If $\psi : V_2 \to V_3$ is another holomorphic function, then $(\psi \circ \phi)^* = \phi^* \circ \psi^*$ (this just translates the chain rule for derivatives of composed functions).

The particular case where $V_1 \subset V_2$ and ϕ is the inclusion defines the *restriction* of $\omega \in \Omega^1(V_2)$ to $\omega_{|V_1} \in \Omega^1(V_1)$, but we may improperly (exactly as we do for functions) keep the notation ω for the restricted form.

B.4.2. Holomorphic differential forms on Riemann surfaces. Let X a Riemann surface and $\phi : U \to V$, $\phi' : U' \to V'$, two overlapping charts (i.e., $U \cap U' \neq \emptyset$; see Figure B.1). We say that the differential forms $\omega \in \Omega^1(V)$ and $\omega' \in \Omega^1(V')$ are *compatible* (understand: with respect to those two charts) if, writing $\psi := \phi' \circ \phi^{-1} : \phi(U \cap U') \to \phi'(U \cap U')$, one has $\psi^*(\omega') = \omega$ (of course, those are the adequate restrictions ...). We then consider ω and ω' to define the same differential form on $U \cap U'$, but read off in different charts (i.e., written in different coordinates).

A *(holomorphic) differential form* on X is given by a form $\omega_\phi \in \Omega^1(V)$ for each chart $\phi : U \to V$, those being pairwise compatible. It is enough to give the forms ω_ϕ for some particular atlas (understand: with the corresponding compatibility

[2] This can be used to prove Lüroth's theorem that every subfield of $\mathbf{C}(x)$ other than $\mathbf{C}$ is isomorphic to $\mathbf{C}(x)$.

conditions). Note that, if X is an open subset of $\mathbf{C}$, this definition is consistent with the one in B.4.1.

For every open subset U of X, considered itself as a Riemann surface, we thus define the space $\Omega^1_X(U)$ of differential forms. If U is the source of a chart, $\Omega^1_X(U)$ is therefore a free $\mathscr{O}_X(U)$-module of rank 1. Globally, Ω_X is a sheaf of $\mathscr{O}_X$-modules and it is locally free of rank 1.

B.4.3. Examples. We illustrate on our basic cases of interest the fact that, on a compact Riemann surface X of genus g, *the space* $\Omega^1(X) := \Omega^1_X(X)$ *of global holomorphic differential forms has dimension* g (over $\mathbf{C}$).

Example 1. There are no nontrivial holomorphic forms on $\mathbf{S}$ (genus 0: the space $\Omega^1(\mathbf{S})$ should be trivial). Indeed such a form should be $f(x)\,dx$ on $\mathbf{C}$, with $f \in \mathscr{O}(\mathbf{C})$; and $g(1/x)\,d(1/x)$ on $\mathbf{C}_\infty$, with $g \in \mathscr{O}(\mathbf{C})$ also. This imposes $f(x) = -g(1/x)/x^2$ on $\mathbf{C}^*$. From $g(1/x) = -x^2 f(x)$, one draws that g is an entire function vanishing at infinity, hence, from Liouville's theorems, $f = g = 0$.

Example 2. A holomorphic form on $\mathbf{E}_\Lambda$ should have the form $f(z)\,dz$, where f is an entire elliptic function, Thus a constant. So $\Omega^1(\mathbf{E}_\Lambda) = \mathbf{C}\,dz$.

Example 3. Now consider the commutative diagram:

$$\begin{array}{ccc} \mathbf{C} & \xrightarrow{\underline{e}} & \mathbf{C}^* \\ {\scriptstyle p}\downarrow & & \downarrow{\scriptstyle \pi} \\ \mathbf{E}_\Lambda & \xrightarrow{\sim} & \mathbf{E}_q \end{array},$$

where, as usual, $\Lambda := \Lambda_\tau$, $q := e^{2i\pi\tau}$ and $\underline{e}(z) := e^{2i\pi z}$. From $x := \underline{e}(z) = e^{2i\pi z}$, we deduce that $dz = \dfrac{1}{2i\pi}\dfrac{dx}{x}$, so $\Omega^1(\mathbf{E}_q) = \mathbf{C}\,\dfrac{dx}{x}\cdot$ Note that the change of charts $\psi :$ $x \mapsto x/q$ found in B.1.3.4 satisfies $\psi^*\left(\dfrac{dx}{x}\right) = \dfrac{d(x/q)}{(x/q)} = \dfrac{dx}{x}$, so the compatibility with respect to changes of charts is satisfied.

B.4.4. Meromorphic differential forms and the canonical divisor class. A *meromorphic differential form* on the Riemann surface X is defined exactly as a holomorphic one, except that in its coordinate avatars $f(x)\,dx$, meromorphic functions f are allowed. The big difference is that any two nontrivial meromorphic differential forms ω, ω' are proportional: there is a unique $f \in \mathscr{M}(X)^*$ such that $\omega' = f\omega$.

An example of nontrivial meromorphic differential form on $\mathbf{S}$ is the one that reads dx in the chart $\mathrm{Id} : \mathbf{C} \to \mathbf{C}$ and $-dw/w^2$ in the chart $w \mapsto w^{-1}$ with support $\mathbf{C}_\infty$. The meromorphic differential forms on $\mathbf{E}_\Lambda$, resp. on $\mathbf{E}_q$, are the $f(z)\,dz$, $f \in \mathscr{M}(\mathbf{E}_\Lambda)$, resp. the $f(x)\,dx/x$, $f \in \mathscr{M}(\mathbf{E}_q)$.

Let $a \in X$ and u a uniformizing parameter at a. Any meromorphic differential form ω read off in a neighborhood of a has the form $f(u)\,du$, $f = \sum f_n u^n \in \mathbf{C}(\{u\})$.

LEMMA B.4. *The integer* $\min\{n \in \mathbf{Z} \mid f_n \neq 0\}$ *and the complex* f_{-1} *depend on* ω *and* a *only, not on the particular choice of a uniformizing parameter.*

We call $\nu_a(\omega) := n \in \mathbf{Z}$ the *order of* ω *at* a and $\mathrm{Res}_a(\omega) := f_{-1} \in \mathbf{C}$ the *residue of* ω *at* a.

THEOREM B.5 (Residue theorem). *If X is a compact Riemann surface and ω a nontrivial meromorphic differential form on X, the residues of ω are trivial except at a finite number of points and* $\sum_{a \in X} \mathrm{Res}_a(\omega) = 0$.

Applying this to $\omega := df/f$, $f \in \mathscr{M}(X)^*$, one deduces that $\sum_{a \in X} \nu_a(f) = 0$ (so $\mathrm{div}_X(f)$ has degree 0, as stated in B.2.5).

Now we define the divisor of a nonzero meromorphic differential form in the obvious way:

$$\mathrm{div}_X(\omega) := \sum_{a \in X} \nu_a(\omega)[a].$$

Clearly $\nu_a(f\omega) = \nu_a(f) + \nu_a(\omega)$, so $\mathrm{div}_X(f\omega) = \mathrm{div}_X(f) + \mathrm{div}_X(\omega)$, and the divisor class of $\mathrm{div}_X(\omega)$ in $\mathrm{Pic}(X)$ is therefore the same for all ω: this is the *canonical class.* As an element of $\mathrm{Pic}(X)$, it has a well defined degree.

THEOREM B.6. *The degree of the canonical class is $2g - 2$.*

For instance, the differential form on $\mathbf{S}$ defined by dx on $\mathbf{C}$ and $-dw/w^2$ on $\mathbf{C}_\infty$ has divisor $-2[\infty]$, hence degree -2 as it should since here $g = 0$. Likewise, the divisor of the form dz on $\mathbf{E}_\Lambda$, resp. dx/x on $\mathbf{E}_q$ is 0 in both cases, it has degree $0 = 2g - 2$ since here $g = 1$.

B.5. General vector bundles

Since in this book we only deal with holomorphic vector bundles, we shall often call them simply vector bundles or even bundles.

B.5.1. General definitions. Let X a Riemann surface and $p : \mathfrak{F} \to X$ a map, a priori arbitrary. For every $a \in X$, we say that the elements of the *fiber* $\mathfrak{F}_a := p^{-1}(a)$ are "above a". We call *(vector) bundle chart* for p, or, more improperly, on $\mathfrak{F}$, a pair (U, ψ) where U is an open subset of X and $\psi : p^{-1}(U) \to U \times \mathbf{C}^n$ a bijection sending, for every $a \in U$, the elements above a in $p^{-1}(U)$ to those above a in $U \times \mathbf{C}^n$, thus realizing a bijection $\mathfrak{F}_a \to \{a\} \times \mathbf{C}^n$; equivalently, the following diagram commutes:

$$\begin{array}{ccc} p^{-1}(U) & \xrightarrow{\ \psi\ } & U \times \mathbf{C}^n. \\ & {\scriptstyle p}\searrow \quad \swarrow{\scriptstyle pr_1} & \\ & U & \end{array}$$

So the bundle chart ψ gives to each element of $\mathfrak{F}$ above $a \in U$ coordinates[3] $(a, v) \in \mathbf{U} \times \mathbf{C}^n$. We say (this is temporary terminology) that n is the rank of the chart.

If $U' \subset U$, a bundle chart (U, ψ) clearly restricts to a bundle chart (U', ψ') of the same rank. Thus, if $\psi_1 : p^{-1}(U_1) \to U_1 \times \mathbf{C}^n$ and $\psi_2 : p^{-1}(U_2) \to U_2 \times \mathbf{C}^n$ are two charts of the same rank over overlapping open sets U_1, U_2, we get a bijection

[3]This is but a partial coordinatization, since we would in practice use a Riemann surface analytic chart ϕ at a to give it a coordinate $\phi(a) \in \mathbf{C}$.

$\psi_2 \circ \psi_1^{-1} : (U_1 \cap U_2) \times \mathbf{C}^n \to (U_1\ U_2) \times \mathbf{C}^n$ and actually a commutative diagram:

$$\begin{array}{ccc} (U_1 \cap U_2) \times \mathbf{C}^n & \xrightarrow{\psi_2 \circ \psi_1^{-1}} & (U_1 \cap U_2) \times \mathbf{C}^n \\ & {\scriptstyle pr_1} \searrow \quad \swarrow {\scriptstyle pr_1} & \\ & U_1 \cap U_2 & \end{array}$$

The "change of coordinates" in the upper horizontal arrow has the form $(a, v) \mapsto (a, \Psi(a, v))$ for some map $\Psi : (U_1 \cap U_2) \times \mathbf{C}^n \to \mathbf{C}^n$.

REMARK B.5. In Appendix I, the maps $\psi_2 \circ \psi_1^{-1}$ will be collectively considered as forming a *cocycle*, thus leading to the (nonabelian) cohomological classification of vector bundles.

DEFINITION B.6. (i) The bundles charts ψ_1, ψ_2 are *(analytically) compatible* if either U_1, U_2 are disjoint, or, in the case they overlap, if the change of coordinates has the form $\Psi(a, v) := g(a)(v)$, where the *transition function* g is holomorphic from $U_1 \cap U_2$ to $\mathrm{GL}_n(\mathbf{C})$ (i.e., the n^2 coefficients of $g(a)$ are holomorphic functions of $a \in U_1 \cap U_2$).

(ii) A *(vector) bundle atlas* is a family of pairwise compatible bundle charts ψ_i over open sets U_i which cover X. Two atlases are *compatible* if their union is an atlas (equivalently, their charts are all mutually compatible).

(iii) A *(holomorphic) (vector) bundle* is a map $p : \mathfrak{F} \to X$ (or improperly $\mathfrak{F}$ itself) together with an equivalence class of atlases (equivalently, with one atlas in the class; equivalently, with the maximal atlas in the class, i.e., the union of them all).

(iv) The *rank* of a vector bundle is the common rank of all its charts (they are equal by connexity of X).

(v) An open subset of X is *trivializing* (for the bundle) if there is a bundle chart over it. If X itself is trivializing, $\mathfrak{F}$ is a *trivial bundle.* It is then isomorphic (this word will be defined soon!) to the *trivial bundle* $X \times \mathbf{C}^n \xrightarrow{pr_1} X$.

It follows that each fiber $\mathfrak{F}_a$, through its bijections with $\{a\} \times \mathbf{C}^n$, inherits a unique structure of linear space of dimension n: a holomorphic vector bundle can be seen as a holomorphic family of linear spaces parameterized by X.

B.5.2. Morphisms. Let $p : \mathfrak{F} \to X$ and $p' : \mathfrak{F}' \to X$ two bundles of ranks n, n' over X and let $f : \mathfrak{F} \to \mathfrak{F}'$ a map commuting to the projections: $p' \circ f = p$; equivalently, p sends each fiber $\mathfrak{F}_a$, $a \in X$, to the fiber $\mathfrak{F}'_a$. Let $U \subset X$ an open subset trivializing for both $\mathfrak{F}$ and $\mathfrak{F}'$ (such open sets clearly cover X) and let ψ, ψ' corresponding charts over U. Then $\psi' \circ f \circ \psi^{-1} : U \times \mathbf{C}^n \to U \times \mathbf{C}^{n'}$ has the form $(a, v) \mapsto (a, F(a, v))$, for some map $F : U \times \mathbf{C}^n \to \mathbf{C}^{n'}$. We say that f is a *morphism of vector bundles* if every such map has the form $F(a, v) := L(a)(v)$ where $L : U \to \mathscr{L}(\mathbf{C}^n, \mathbf{C}^{n'}) = \mathrm{Mat}_{n',n}(\mathbf{C})$ is holomorphic.

Clearly identities and compositions of morphisms are morphisms, so we obtain a *category of vector bundles over X*. In that category, bijective morphisms are isomorphisms, i.e., their inverse maps are morphisms. Isomorphic bundles have the same rank, and trivial bundles of the same rank are isomorphic.

A morphism $\mathfrak{F} \to \mathfrak{F}'$ induces for each $a \in X$ a linear map $L_a : \mathfrak{F}_a \to \mathfrak{F}'_a$. The function $a \mapsto \mathrm{rank}(L_a)$ is lower semi-continuous (as is the case for any continuous family of matrices), i.e., the condition $\operatorname{rank}(L_a) \geq n_0$ defines an open subset of X. Let r the maximum value of $\operatorname{rank}(L_a)$ (so $r \leq n, n'$). Then the subset of X defined by $\operatorname{rank}(L_a) < r$ is finite, as the zero locus of some nontrivial holomorphic functions (extracted determinants).

B.5.3. Sheaf of sections. A *holomorphic section* of a bundle $\mathfrak{F}$ over an open subset $U \subset X$ is a holomorphic map $s : U \to \mathfrak{F}$ such that $p \circ s = \mathrm{Id}_U$; equivalently $s(a) \in \mathfrak{F}_a$ for all $a \in U$. Here, holomorphy should be read in bundle charts: for each such chart ϕ defined at a, the map $\phi \circ s$ from a neighborhood $V \subset U$ of a to $V \times \mathbf{C}^n$ has the form $x \mapsto (x, v(x))$, and it is required that $v : V \to \mathbf{C}^n$ should have holomorphic components.

Holomorphic sections can be added and multiplied by holomorphic functions, so the set $\mathscr{F}(U)$ of holomorphic sections over U is a $\mathscr{O}_X(U)$-module. There are various notations for this module, related to various approaches (the last one is related to sheaf cohomology; see Appendix H):

$$\Gamma(X, \mathfrak{F}) = \mathscr{F}(U) = \Gamma(X, \mathscr{F}) = H^0(X, \mathscr{F}).$$

Moreover, $U \mapsto \mathscr{F}(U)$ is a sheaf of $\mathscr{O}_X$-modules. We hope that the similarity of symbols $\mathfrak{F}$ (for the bundle) and $\mathscr{F}$ (for the sheaf) will cause no ambiguity, but it is an empirical fact that even giving them the same symbol is usually innocuous.

CAUTION B.7. Note however that a possible confusion however could arise from the notation $\mathscr{F}_a$ for sheaves (stalk: it consists of germs and carries *local* information around a) vs $\mathfrak{F}_a$ for bundles (fiber: it carries *punctual* information at a). The confusion is even worse in French, where the words "stalk" and "fiber" have the same translation "fibre".

EXERCISE B.8. Check that for a locally free sheaf of rank 1 a stalk looks like $\mathbf{C}\{x\}$, while for a line bundle a fiber looks like $\mathbf{C}$.

PROPOSITION B.9. *The sheaf $\mathscr{F}$ of sections of a rank n bundle $\mathfrak{F}$ is locally free of rank n.*

Practically, that means that, for every trivializing open set U for $\mathfrak{F}$, letting ψ a bundle chart over U, a section s over U satisfies $\psi(s(a)) = (a, g(a))$ for some holomorphic $g : U \to \mathbf{C}^n$, i.e., $g \in \mathscr{O}_X(U)^n$. This defines an isomorphism $\mathscr{F}(U) \simeq \mathscr{O}_X^n(U)$ and this is compatible with restrictions to $U' \subset U$, so that we have actually an isomorphism of sheaves on U: $\mathscr{F}_{|U} \simeq \mathscr{O}_U^n$, i.e., the restricted sheaf $\mathscr{F}_{|U}$ is free. Another way to see it is to say that there are n sections $s_1, \dots, s_n \in \mathscr{F}(U)$ such that each $\mathscr{F}(U')$, $U' \subset U$, is free over $\mathscr{O}_X(U')$ with basis $(s_1)_{|U'}, \dots, (s_n)_{|U'}$.

Holomorphic sections of $\mathfrak{F}$ on X can be described in coordinates (i.e., in bundle charts) as follows. Let $(U_i, \psi_i)_i$ a bundle atlas with transition functions[4] $g_{i,j} : U_i \cap U_j \to \mathrm{GL}_n(\mathbf{C})$: recall that $\psi_j \circ \psi_i^{-1} : (U_i \cap U_j) \times \mathbf{C}^n \to (U_i \cap U_j) \times \mathbf{C}^n$ sends (x, v) to $(x, g_{i,j}(x)(v))$. Then $\psi_i \circ s$ sends x to $(x, s_i(x))$, where $s_i : U_i \to \mathbf{C}^n$ is holomorphic, and similarly for $\psi_j \circ s$ and some $s_j : U_j \to \mathbf{C}^n$; moreover, $s_j(x) = g_{i,j}(x)s_i(x)$ on $U_i \cap U_j$. Conversely, every family of holomorphic maps $s_i : U_i \to \mathbf{C}^n$ satisfying the compatibility conditions $s_j(x) = g_{i,j}(x)s_i(x)$ defines a unique section s.

The above description involves global sections $s \in \mathscr{F}(X)$. For a section $s \in \mathscr{F}(U)$, the description is the same except that we use an atlas on U (i.e., charts (U_i, ψ_i) such that the U_i cover U).

[4]The $g_{i,j}$ form a *cocycle*: the restrictions to $x \in U_i \cap U_j \cap U_k$ satisfy $g_{i,k}(x) = g_{j,k}(x)g_{i,j}(x)$; but we shall deal with that only in the case of line bundles; see Appendix I.

B.5.4. Equivalence with locally free sheaves. Let $f : \mathfrak{F} \to \mathfrak{F}'$ a morphism of vector bundles over X. For every section $s \in \mathscr{F}(U)$ of $\mathfrak{F}$ over $U \subset X$, the compositum is a section $f \circ s \in \mathscr{F}'(U)$ of $\mathfrak{F}'$ over U. This defines a morphism of sheaves $\mathscr{F} \to \mathscr{F}'$.

THEOREM B.7. *We thus obtain an equivalence of the category of vector bundles over X with the category of locally free sheaves of $\mathscr{O}_X$-modules.*

We shall not describe the general way to recover a vector bundle (a geometric object) from its sheaf of sections (an algebraico-topological object) but we shall detail it in the important case (for us) $X = \mathbf{E}_q$ when dealing with sheaves of solutions of q-difference equations (4.5 and, in a more general form, 7.3) and also for line bundles in B.6.

B.5.5. Some vector bundles over $\mathbf{E}_q := \mathbf{C}^*/q^{\mathbf{Z}}$. We now describe a way[5] to define vector bundles over $\mathbf{E}_q := \mathbf{C}^*/q^{\mathbf{Z}}$, which will be useful in the study of q-difference equations. This is related to a construction that is sketched (in a particular case) in 2.2.3 and (in a slightly more general case) in 4.5. We use charts on the Riemann surface X as they were described in B.1.3.4 (see in particular Figure B.2): recall that, denoting $U_i := \pi(V_i)$ for $i = 1, 2$, where V_1 and V_2 respectively are the annuli $|q| < |x| < 1$ and $|q|^{1/2} < |x| < |q|^{-1/2}$ in $\mathbf{C}^*$, those charts were the maps $\phi_i : U_i \to V_i$ inverting the homeomorphisms $\pi : V_i \xrightarrow{\sim} U_i$. The change of charts $\phi_2 \circ \phi_1^{-1} : \phi_1(U_1 \cap U_2) \to \phi_2(U_1 \cap U_2)$ is identity on $V_1 \cap V_2$; on $V_1 \setminus \overline{V_2}$ it is the map $x \mapsto x/q$.

We consider a holomorphic map $A : \mathbf{C}^* \to \mathrm{GL}_n(\mathbf{C})$ and the corresponding q-difference equation[6], which we write here $v(qy) = A(y)v(y)$ (keeping the letter x for points on $\mathbf{E}_q$). We associate to that equation the $q^{\mathbf{Z}}$ action on the trivial bundle $\mathbf{C}^* \times \mathbf{C}^n$ generated by the elementary action (that of the generator q) $(y, V) \mapsto q.(y, v) := (qy, A(y)v)$. It is easy to check that:

$$q^k.(y,v) = (q^k y, A_k(y)v), \text{ where } A_k(y) := \begin{cases} A(q^{k-1}y) \cdots A(y) \text{ if } k \geq 0, \\ \left(A(q^{-1}y) \cdots A(q^k y)\right)^{-1} \text{ if } k \leq 0. \end{cases}$$

Note that, by the usual conventions on empty products, both cases give $A_0(y) = I_n$.

EXERCISE B.10. Write the action as $q^k.(y,v) := (q^k y, A_k(y)v)$ with A_k a priori unknown. Check that the group action axioms imply that $A_0(y) = I_n$ and $A_{k+l}(y) = A_l(q^k y)A_k(y)$ and deduce the value of each A_k from $A := A_1$.

This group action on $\mathbf{C}^* \times \mathbf{C}^n$ is compatible with the obvious action of $q^{\mathbf{Z}}$ on $\mathbf{C}^*$, i.e., $pr_1(q^k.(y,v)) = q^k.pr_1(y,v)$, thus making it an *equivariant bundle.* The group action also defines an equivalence relation $\sim_A$ on $\mathbf{C}^* \times \mathbf{C}^n$, actually the equivalence relation generated by $(y,v) \sim_A (qy, A(y)v)$. As a consequence, the projection map

---

[5] Actually this is a very particular case of a process explained in **[106]**.

[6] In real life, A may have singularities on $\mathbf{C}^*$, e.g., $A \in \mathrm{GL}_n(\mathbf{C}(x))$ and it will be required to be meromorphic at 0; see 7.3 for the way to deal with that case.

goes to the quotient and we have a commutative diagram:

$$\begin{array}{ccc} \mathbf{C}^* \times \mathbf{C}^n & \xrightarrow{\pi_A} & \dfrac{\mathbf{C}^* \times \mathbf{C}^n}{\sim_A} =: \mathfrak{F}_A. \\ {\scriptstyle pr_1}\Big\downarrow & & \Big\downarrow{\scriptstyle p_A} \\ \mathbf{C}^* & \xrightarrow{\quad\pi\quad} & \mathbf{C}^*/q^{\mathbf{Z}} =: \mathbf{E}_q \end{array}$$

The right vertical arrow $p_A : \mathfrak{F}_A \to \mathbf{E}_q$ is a holomorphic vector bundle of rank n. This follows from the "generalities" in **[106]** but we shall give a down-to-earth description with bundle charts. The supports of our bundle charts will be the open subsets $U_i' := p_A^{-1}(U_i)$, $i = 1, 2$, of $\mathfrak{F}_A$. We note that:

$$\pi_A^{-1}(U_i') = \pi_A^{-1}(p_A^{-1}(U_i)) = (p_A \circ \pi_A)^{-1}(U_i) = (\pi \circ pr_1)^{-1}(U_i) = pr_1^{-1}(\pi^{-1}(U_i)) =$$

$$pr_1^{-1}\left(\bigsqcup_{k\in\mathbf{Z}} q^k V_i\right) = \left(\bigsqcup_{k\in\mathbf{Z}} q^k V_i\right) \times \mathbf{C}^n = \bigsqcup_{k\in\mathbf{Z}} (q^k V_i \times \mathbf{C}^n).$$

Every element of this open subset of $\mathfrak{F}_A$ is equivalent (modulo $\sim_A$) to a unique representant in $V_i \times \mathbf{C}^n$, so we deduce that the restriction $\pi_A : V_i \times \mathbf{C}^n \to U_i'$ is a homeomorphism for $i = 1, 2$. On the other hand, we also have a homeomorphism $\pi \times \mathrm{Id} : V_i \times \mathbf{C}^n \to U_i \times \mathbf{C}^n$, so, we can combine them into a homeomorphism:

$$\underbrace{U_i' \xleftarrow{\ \pi_A\ } V_i \times \mathbf{C}^n \xrightarrow{\pi\times\mathrm{Id}} U_i \times \mathbf{C}^n}_{\psi_i}.$$

The $\psi_i := (\pi \times \mathrm{Id}) \circ \pi_A^{-1} : U_i' \simeq U_i \times \mathbf{C}^n$, $i = 1, 2$, are our bundle charts; clearly, their supports cover $\mathfrak{F}_A$ (since U_1, U_2 cover $\mathbf{E}_q$) but we must also check their compatibility. So let $x \in U_1 \cap U_2$. There is a unique $y_1 \in V_1$, resp. $y_2 \in V_2$, such that $\pi(y_1) = \pi(y_2) = x$. An element z of $\mathfrak{F}_A$ above x, i.e., an element of $p_A^{-1}(x) \in p_A^{-1}(U_1 \cap U_2) = U_1' \cap U_2'$ is the image by π_A of a unique $(y_1, v_1) \in V_1 \times \mathbf{C}^n$, resp. of a unique $(y_2, v_2) \in V_2 \times \mathbf{C}^n$; and the images of those by $\pi \times \mathrm{Id}$ are $\psi_1(z) = (x, v_1)$ and $\psi_2(z) = (x, v_2)$. Those are the "coordinates" of z in the two bundle charts and the corresponding change of coordinates is $(x, v_1) \mapsto (x, v_2) = (x, g_{1,2}(x)(v_1))$. So we must understand what is the transformation $g_{1,2}(x) : v_1 \mapsto v_2$. For that we note that $(y_1, v_1) \sim_A (y_2, v_2)$ in $\mathbf{C}^* \times \mathbf{C}^n$ (since they both lie above z in $\mathfrak{F}_A$). So there are two possible cases:

- If $y_1 \in V_1 \setminus V_2$, i.e., if $x \in U_N := \pi(V_1 \setminus V_2)$, then $y_2 = y_1$ and $v_2 = v_1$;
- if $y_1 \in V_1 \cap V_2$, i.e., if $x \in U_S := \pi(V_1 \cap V_2)$, then $y_2 = y_1/q$, so $v_2 = A(y_1)v_1$.

So $g_{1,2}(x)$ is the constant function $x \mapsto I_n$ on U_N and the function $x \mapsto A(y_1)$ on U_S (where y_1 depends holomorphically on x): this has the right form.

We now describe the associated sheaf $\mathscr{F}_A$. First let $U \subset \mathbf{E}_q$ a small open set, i.e., one such that $\pi^{-1}(U) = \bigsqcup q^k V$ for some V such that $\pi : V \simeq U$. The same argument as that used for U_1 and U_2 above yields a homeomorphism $p_A^{-1}(U) \simeq V \times \mathbf{C}^n$ and also an identification of $V \times \mathbf{C}^n$ with the quotient of $\left(\bigsqcup q^k V\right) \times \mathbf{C}^n$ by relations $(y, v) \sim_A (q^k y, A_k(y)v)$. Under those identifications, a holomorphic section $s : U \to p_A^{-1}(U)$ is the same thing as a holomorphic function $v : \pi^{-1}(U) \to \mathbf{C}^n$ compatible with $\sim_A$, i.e., such that $v(q^k y) = A_k(y)v(y)$ for all y and k. By construction, this in turn is equivalent to the unique condition $v(qy) = A(y)v(y)$ for all $y \in \pi^{-1}(U)$. So we conclude that:

$$\mathscr{F}_A(U) = \{v \in \mathscr{O}_{\mathbf{C}^*}(\pi^{-1}(U)) \mid v(qy) = A(y)v(y)\}.$$

Over an arbitrary open subset of $\mathbf{E}_q$, a section would be obtained by glueing sections on small open subsets. We summarize (remembering that $\mathscr{O}_{\mathbf{E}_q}(U)$ is naturally identified with the $\mathbf{C}$-algebra $\mathscr{O}_{\mathbf{C}^*}(\pi^{-1}(U))^{\sigma_q}$ of q-invariant holomorphic functions over $\pi^{-1}(U)$):

PROPOSITION B.11. *For every open subset $U \subset \mathbf{E}_q$, the $\mathscr{O}_{\mathbf{E}_q}(U)$-module of sections $\mathscr{F}_A(U)$ is naturally identified with the $\mathscr{O}_{\mathbf{C}^*}(\pi^{-1}(U))^{\sigma_q}$-module of holomorphic solutions of $\sigma_q v = Av$ over $\pi^{-1}(U) \subset \mathbf{C}^*$.*

REMARK B.12. As already noted, the construction described here is a special case of [**106**]. In [**233**, p. 39], Seshadri tackles the following situation: $\tilde{X} \to X$ is an etale covering of Riemann surfaces, $\tilde{X}$ being simply connected; then holomorphic vector bundles over X can be obtained starting from trivial bundles $\tilde{X} \times V$ over $\tilde{X}$ endowed with an equivariant action of the fundamental group $\pi_1(X)$. Seshadri deals in particular with the case of *flat* vector bundles (see Subsubsection 13.2.6.4 of Chapter 13); for them the equivariant action is encoded in a linear representation ρ of $\pi_1(X)$ on the typical fiber V, yielding a bundle V_ρ on X. Actually, the same constructions work if one only assumes that all holomorphic vector bundles over $\tilde{X}$ are trivial (e.g. if $\tilde{X}$ is an open Riemann surface), up to replacing $\pi_1(X)$ by the automorphism group $\mathrm{Aut}(\tilde{X}/X)$ of the covering. Our construction here is the particular case where $X = \mathbf{E}_q$, $\tilde{X} = \mathbf{C}^*$ and $\mathrm{Aut}(\tilde{X}/X) = \mathbf{Z}$ (realized as $q^{\mathbf{Z}}$).

B.5.6. Holomorphic/Meromorphic triviality. The following results are nontrivial[7] and are usually proved using heavy functional analysis. Standard references for statement (i) are [**205**, Theorem 1.0], [**88**, Chapter 3, §30, Theorem 30.4], [**107**, Chapter V, §4, Theorem 3]. Also see [**169**, Theorem 4.29, p. 99] where a direct proof is sketched in the case that X is an open subset of $\mathbf{C}$. Statement (ii) will be proved in Appendix I.

THEOREM B.8. *Let $\mathfrak{F}$ a holomorphic vector bundle on a Riemann surface X.*

(i) *If X is open, $\mathfrak{F}$ is trivial. (See an application in Section* B.7.*)*

(ii) *If X is compact, $\mathfrak{F}$ is* meromorphically trivial.

We first comment (i) before explaining (ii). Equivalently, (i) says that every locally free sheaf on X is free. Still equivalently, it means that there are n (the rank of $\mathfrak{F}$) global holomorphic sections $s_1, \ldots, s_n \in \mathscr{F}(X)$ such that, for every open sets $U \subset X$, the $\mathscr{O}_X(U)$-module $\mathscr{F}(U)$ is free with basis $(s_1)_{|U}, \ldots, (s_n)_{|U}$. However, while local trivialisations (i.e., on small open subsets) automatically come along each time we define a vector bundle, it is not always explicitly feasible to find a global trivialisation. We shall find that in some important cases (see Chapter 15) we are able to do so on $\mathbf{E}_q \setminus \Sigma$ where $\Sigma \subset \mathbf{E}_q$ is a nonempty finite set.

The statement (ii) means that there are global *meromorphic* sections $s_1, \ldots, s_n \in \mathscr{F}(X)$ such that, for every connected open subset $U \subset X$, every $s \in \mathscr{F}(U)$ can be uniquely written as a linear combination $\sum_{i=1}^{n} f_i (s_i)_{|U}$, where the $f_i \in \mathscr{M}_X(U)$. We

[7]We omit facts about *flat* vector bundles and Weil's correpondence; see Chapter 13, in particular Subsubsection 13.2.6.4 about them.

shall further detail that fact in the case of line bundles in B.6. However, the most important consequence is a theorem due to Praagman stating that *rational q-difference equations have fundamental meromorphic solutions*, see 7.3.

B.5.7. Pullbacks. Let $f : X' \to X$ a morphism of Riemann surfaces and let $p : \mathfrak{F} \to X$ a holomorphic vector bundle. The *pullback of* $\mathfrak{F}$ *by* f is defined as the *fibered product* $\mathfrak{F}' := \{(x', z) \in X' \times \mathfrak{F} \mid f(x') = p(z)\}$ together with the map $p' : (x', z) \mapsto x$. Then $p' : \mathfrak{F}' \to X'$ is a holomorphic vector bundle and, along with the map $F : \mathfrak{F}' \to \mathfrak{F}$, $(x', z) \mapsto z$, it gives rise to a *cartesian square*, the commutative diagram:

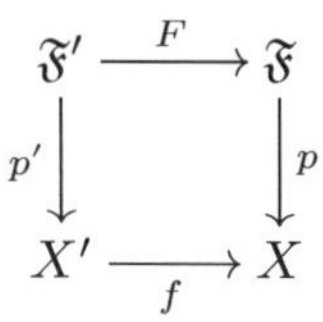

If $\mathscr{F} = X \times \mathbf{C}^n$ and $p = pr_1$, then $\mathfrak{F}' = \{(x', f(x'), v) \mid x' \in X', v \in \mathbf{C}^n\}$ can be identified with $X' \times \mathbf{C}^n$ and p' can be identified with pr_1. It follows that, if $\mathfrak{F}$ is trivial, so is $\mathfrak{F}'$ (and the map F is easily explicited). More generally, if $U \subset X$ is an open subset and $\psi : p^{-1}(U) \to U \times \mathbf{C}^n$ is a bundle chart for $\mathfrak{F}$, let us set $U' := f^{-1}(U)$. Then ${p'}^{-1}(U') = (f \circ p')^{-1}(U) = F^{-1}(p^{-1}(U))$; we shall presently define a bundle chart for $\mathfrak{F}'$ with source ${p'}^{-1}(U')$. Let $z' := (x', z) \in {p'}^{-1}(U')$, so that $f(x') = p(z) =: x$, and $p'(z') = x'$, $F(z') = z \in p^{-1}(U)$. By construction of the chart ψ, there is a $v \in \mathbf{C}^n$ such that $\psi(z) = (x, v)$. We set $\psi'(z') := (x', v)$. This looks complicated, but maybe the following commutative "prism" may help:

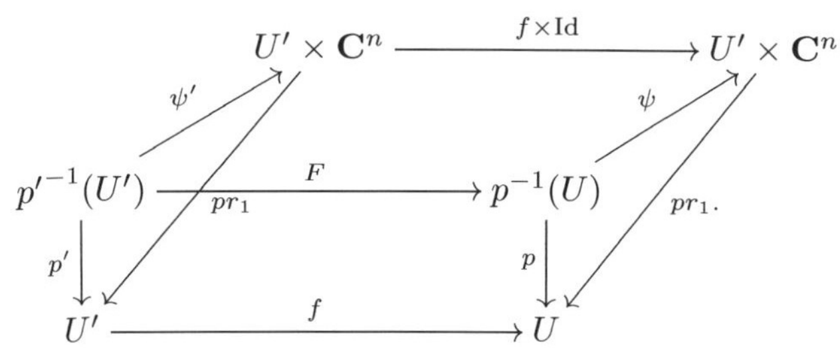

We write $f^*\mathfrak{F} := \mathfrak{F}'$ this pullback. The associated sheaf is the *inverse image* $f^*\mathscr{F}$ of $\mathscr{F}$ by f. It is not very easy to describe $f^*\mathscr{F}(U')$, even when $U' = p^{-1}(U)$, but there is a nice formula about stalks, which are the spaces of germs of sections:

$$\forall a' \in X', (f^*\mathscr{F})_{a'} = \mathscr{F}_{f(a')}.$$

This equality (rather a canonical isomorphism by the way) can be understood as follows. An element of $\mathscr{F}_{f(a')}$ is the germ of a section $\mathfrak{F}$ near $a := f(a')$. So it can be considered as a section s on an open neighborhood U of a. For every $x' \in U' := f^{-1}(U)$, letting $x := f(x')$, we see that $z := s(x)$ has $p(z) = x = f(x')$, so $z' := (x', z)$ is an element of $\mathfrak{F}'$ above x'. So the map $s' : x' \mapsto (x', s(f(x))$ is a section of $\mathfrak{F}'$ over U'. Our equality above means that we can identify the germ of s (an element of $\mathscr{F}_a$) to the germ of s' (an element of $(f^*\mathscr{F})_{a'}$).

B.6. Line bundles

First some preliminary standard notational conventions[8]. Let $\mathfrak{U} := (U_i)_i$ an open covering of the Riemann surface X. We shall write $U_{i,j} := U_i \cap U_j$, $U_{i,j,k} := U_i \cap U_j \cap U_k$, etc. A *refinement* of $\mathfrak{U}$ is a covering $\mathfrak{V} := (V_j)_j$ together with a map s on indices such that $\forall j$, $V_j \subset U_{s(j)}$. Any two covering $(U_i)_i$, $(U'_{i'})_{i'}$ have common refinements, for instance $(U_i \cap U'_{i'})_{i,i'}$. Also a bundle atlas $(\psi_i : p^{-1}(U_i) \to U_i \times \mathbf{C})_i$ will be said to be "supported" by the open covering $\mathfrak{U} := (U_i)_i$ (this is not a standard terminology, it is for local use only).

B.6.1. Line bundles and cocycles. A *(holomorphic) line bundle* is a holomorphic vector bundle of rank 1. As we saw in B.5.3, a locally free sheaf of rank 1 is associated to a line bundle; this will be called an *invertible sheaf.* (This is related to the fact that, under tensor product, they give rise to the group of isomorphy classes of locally free sheaves of rank 1, which is actually the Picard group, see theorem B.10 further below.)

An atlas on a line bundle supported by the open covering $\mathfrak{U} := (U_i)_i$ gives rise to holomorphic transition functions $g_{i,j} : U_{i,j} \to \mathrm{GL}_1(\mathbf{C}) = \mathbf{C}^*$ defined (where they make sense) by the formulas $\psi_j \circ \psi_i^{-1}(x, v) = (x, g_{i,j}(x)v)$ (B.5.1). The $g_{i,j}$ satisfy the cocycle condition (B.5.3):

$$\forall x \in U_{i,j,k} \ , \ g_{i,k}(x) = g_{i,j}(x) g_{j,k}(x).$$

We shall denote $C^1(\mathfrak{U}, \mathscr{O}_X^*)$ the group $\prod_{i,j} \mathscr{O}_X(U_{i,j})^*$ of all families of holomorphic functions $g_{i,j} : U_{i,j} \to \mathbf{C}^*$ and $Z^1(\mathfrak{U}, \mathscr{O}_X^*)$ the subgroup of those satisfying the above cocycle condition. So to every line bundle $\mathfrak{F}$, and to every bundle atlas, we associate a cocycle $(g_{i,j})_{i,j} \in Z^1(\mathfrak{U}, \mathscr{O}_X^*)$ where $\mathfrak{U}$ is the covering of X supporting the atlas.

If $\mathfrak{V}$ is a refinement of $\mathfrak{U}$, restrictions from $U_{s(i)}$ to their subsets V_i allow one to define group morphisms $C^1(\mathfrak{U}, \mathscr{O}_X^*) \to C^1(\mathfrak{V}, \mathscr{O}_X^*)$, which in turn induce group morphisms $Z^1(\mathfrak{U}, \mathscr{O}_X^*) \to Z^1(\mathfrak{V}, \mathscr{O}_X^*)$. They can be interpreted as follows: the bundle atlas $(\psi_i)_i$ supported by $\mathfrak{U}$ induces by restrictions a bundle atlas supported by $\mathfrak{V}$ and the cocycle associated to the latter is the image of the cocycle $(g_{i,j})_{i,j} \in Z^1(\mathfrak{U}, \mathscr{O}_X^*)$ by the map $Z^1(\mathfrak{U}, \mathscr{O}_X^*) \to Z^1(\mathfrak{V}, \mathscr{O}_X^*)$.

Now suppose we have two line bundles $\mathfrak{F}$ and $\mathfrak{F}'$. One can find for them bundle atlases supported by a common covering $\mathfrak{U}$. Let $(g_{i,j})_{i,j}, (g'_{i,j})_{i,j} \in Z^1(\mathfrak{U}, \mathscr{O}_X^*)$ their associated cocycles. Then one can prove (it is not very difficult):

$$\text{(B.1)} \qquad \mathfrak{F} \simeq \mathfrak{F}' \iff \left(\exists (f_i)_i \in \prod_i \mathscr{O}_X(U_i)^* \ : \ \forall i, j \ , \ \frac{g'_{i,j}}{g_{i,j}} = \frac{f_j}{f_i} \right).$$

We are thus led to introduce the map $\prod_i \mathscr{O}_X(U_i)^* \to C^1(\mathfrak{U}, \mathscr{O}_X^*)$ which sends the family $(f_i)_i$ to the family $\left(\dfrac{f_j}{f_i}\right)_{i,j}$. Its image is a subgroup $B^1(\mathfrak{U}, \mathscr{O}_X^*)$ of $Z^1(\mathfrak{U}, \mathscr{O}_X^*)$ and the relation on the right-hand side of the logical equivalence (B.1) is the congruence relation modulo this subgroup. So we define the *Čech cohomology group* associated to the covering $\mathfrak{U}$:

$$\check{H}(\mathfrak{U}, \mathscr{O}_X^*) := Z^1(\mathfrak{U}, \mathscr{O}_X^*) / B^1(\mathfrak{U}, \mathscr{O}_X^*).$$

---

[8]When dealing with families $(U_i)_{i \in I}$, we shall usually omit the mention of the set of indices and instead write $(U_i)_i$ for short.

This will be generalized in appendices H and I.

If $\mathfrak{V}$ is a refinement of $\mathfrak{U}$, restrictions induce group morphisms $Z^1(\mathfrak{U}, \mathcal{O}_X^*) \to Z^1(\mathfrak{V}, \mathcal{O}_X^*)$ and $B^1(\mathfrak{U}, \mathcal{O}_X^*) \to B^1(\mathfrak{V}, \mathcal{O}_X^*)$, whence $\check{H}^1(\mathfrak{U}, \mathcal{O}_X^*) \to \check{H}^1(\mathfrak{V}, \mathcal{O}_X^*)$. Because of what we said on refinements of bundle atlases, we are then led to identify an element of $\check{H}^1(\mathfrak{U}, \mathcal{O}_X^*)$ with its image in $\check{H}^1(\mathfrak{V}, \mathcal{O}_X^*)$. The classes obtained by those identifications make up a the direct limit (or inductive limit) group:

$$\check{H}^1(X, \mathcal{O}_X^*) := \lim_{\to} \check{H}^1(\mathfrak{U}, \mathcal{O}_X^*).$$

THEOREM B.9. *Associating to a line bundle the class in $\check{H}^1(X, \mathcal{O}_X^*)$ of any of its cocycles of transition functions (for any of its bundle atlases) yields a bijection:*

$$\{\textit{isomorphism classes of line bundles on } X\} \longleftrightarrow \check{H}^1(X, \mathcal{O}_X^*).$$

For instance the class consisting of all the trivial line bundles (those isomorphic to $pr_1 : X \times \mathbf{C} \to X$) corresponds to the neutral element of the group $\check{H}^1(X, \mathcal{O}_X^*)$.

The analogue construction for vector bundles of ranks $n \geq 2$ gives rise to "nonabelian cohomology"; see **[89, 108]** and Appendix I; in that case the groups $\check{H}^1(X, \mathcal{O}_X^*)$ are replaced by pointed sets $\check{H}^1(X, \mathrm{GL}_n(\mathcal{O}_X))$.

B.6.2. Sections. We now describe sections of a line bundle $p : \mathfrak{F} \to X$ with bundle atlas $(\psi_i : p^{-1}(U_i) \to U_i \times \mathbf{C})_i$ supported by the covering $\mathfrak{U}$ and transition maps $\psi_j \circ \psi_i^{-1} : (x, v) \mapsto (x, g_{i,j}(x)v)$, $g_{i,j} \in \mathcal{O}_X^*(U_{i,j})$. According to the general case (Subsection B.5.3), a global section $s : X \to \mathfrak{F}$ is determined by a family $(s_i \in \mathcal{O}_X(U_i))_i$ such that $s_j(x) = g_{i,j}(x)s_i(x)$ on $U_{i,j}$ for all i, j. The section s then satisfies $\psi_i(s(x)) = (x, s_i(x))$ for all i. More generally, a section $s : X \to \mathfrak{F}$ is determined by a family $(s_i \in \mathcal{O}_X(U \cap U_i))_i$ satisfying similar compatibility conditions on the $U \cap U_{i,j}$. Meromorphic sections are described the same way, with $\mathcal{M}_X$ instead of $\mathcal{O}_X$. A consequence of B.5.6 is:

COROLLARY B.13. *Let $\mathfrak{F}$ a holomorphic line bundle on a Riemann surface X.*

(i) *If X is open, $\mathfrak{F}$ is trivial, i.e., it admits a nowhere vanishing holomorphic section. Any two such sections s, s' are such that $s' = fs$ for some $f \in \mathcal{O}(X)^*$.*

(ii) *If X is compact, $\mathfrak{F}$ is* meromorphically trivial, *i.e., it admits a global nontrivial meromorphic section. Any two such sections s, s' are such that $s' = fs$ for some $f \in \mathcal{M}(X)^*$.*

B.6.3. Divisors of sections. An important observation is then the following:

$$\forall a \in U_{i,j} \ , \ \nu_a(s_i) = \nu_a(s_j).$$

This comes from the fact that $g_{i,j}(a) \in \mathbf{C}^*$. It stays valid if $s_i = s_j = 0$ (with the usual convention that the order is then $+\infty$).

DEFINITION B.14. The *order* $\nu_a(s)$ of a nonzero section s of $\mathfrak{F}$ on $U \subset X$ at a point $a \in U$ is the order $\nu_a(s_i)$ for any U_i such that $a \in U_i$. If $\nu_a(s) > 0$, we say that a is a *zero of order* $\nu_a(s)$. If $\nu_a(s) < 0$, we say that a is a *pole of order* $-\nu_a(s)$.

These definitions do not depend on the bundle atlas. If $u \in \mathcal{O}_{X,a}$ is a uniformizing parameter, any nonzero section germ $s \in \mathcal{F}_a$ (that is a section in an unspecified neighborhood of a) can be uniquely written $s = u^k s_0$, where $k := \nu_a(s)$ and $\nu_a(s_0) = 0$.

PROPOSITION B.15. *The line bundle $\mathfrak{F}$ is trivial if, and only if, there is a global holomorphic section that vanishes nowhere, i.e., $\forall a \in X \ , \ \nu_a(s) = 0$.*

If U is connected and $s \in \mathscr{F}(U)$ is not trivial, the points $a \in U$ where $\nu_a(s) \neq 0$ are isolated. We can therefore define the divisor of a nonzero global meromorphic section on a compact Riemann surface:

$$\operatorname{div}_X(s) := \sum_{a \in X} \nu_a(s)[a].$$

Holomorphic sections are precisely meromorphic sections with no poles, i.e., those such that $\operatorname{div}_X(s) \geq 0$. If $f \in \mathscr{M}(X)$, $f \neq 0$, one sees immediately that $\operatorname{div}_X(fs) = \operatorname{div}_X(f) + \operatorname{div}_X(s)$. So, by Corollary B.13, one sees that to any line bundle $\mathfrak{F}$ is associated a well defined divisor class (the class of any $\operatorname{div}_X(s)$, s a nonzero global meromorphic section), thus an element of $\operatorname{Pic}(X)$.

THEOREM B.10. (i) *We thus obtain a one-to-one correspondence:*

$$\{\textit{isomorphism classes of line bundles on } X\} \longleftrightarrow \textit{Pic}(X).$$

(ii) *The resulting bijection* $\textit{Pic}(X) \leftrightarrow \check{H}^1(X, \mathscr{O}_X^*)$ *is a group isomorphism (see the first paragraph of* B.6.1*).*

Another consequence of equality $\operatorname{div}_X(fs) = \operatorname{div}_X(f) + \operatorname{div}_X(s)$ is that $\operatorname{div}_X(fs)$ and $\operatorname{div}_X(s)$ have the same degree, which therefore depends only on the line bundle: it is called the *degree* $\deg \mathfrak{F}$ of the line bundle $\mathfrak{F}$. Examples will flow from the proposition to come herebelow.

B.6.4. The sheaf $\mathscr{O}_X(D)$. From statement (i) in the previous theorem, we should be able to associate to every divisor a line bundle; we shall explicitly describe the corresponding invertible sheaf. So let X a compact Riemann surface and let $D := \sum_{i=1}^{k} n_i[a_i] = \sum_{a \in X} D(a)[a]$ a divisor on X: so, every $a \in X$, we write $D(a) := 0$ if $a \notin \{a_1, \ldots, a_k\}$ (the support of D); and $D(a_i) := n_i$. We now define for every open subset $U \subset X$:

$$\mathscr{O}_X(D)(U) := \{f \in \mathscr{M}_X(U) \mid \forall a \in U\ ,\ \nu_a(f) \geq -D(a)\},$$

or, more informally $\mathscr{O}_X(D)(U) := \{f \in \mathscr{M}_X(U) \mid \operatorname{div}_U(f) \geq -D_{|U}\}$.

PROPOSITION B.16. (i) *We thus define an invertible sheaf on X. It admits a nontrivial global meromorphic section s such that $\operatorname{div}_X(s) = D$.*

(ii) *The sheaves $\mathscr{O}_X(D)$ and $\mathscr{O}_X(D')$ are isomorphic if, and only if, D and D' have the same class in Pic(X). The resulting induced map is the right-to-left map in statement (i) of the previous theorem.*

EXAMPLE B.17. We saw in B.4.4 that, on a compact Riemann surface X of genus g, there are nonzero global meromorphic differential forms and that the divisors of any two of them are equivalent, thus defining the canonical divisor class $K \in \operatorname{Pic}(X)$, which has degree $2 - 2g$. The descriptions of the corresponding line bundle and invertible sheaf (more rigorously: of their isomorphism classes) are given in [**88**, **168**]. The important case for us is $X := \mathbf{E}_q$. We saw that the canonical class is then trivial, so the canonical bundle is trivial too: it is (isomorphic to) $pr_1 : \mathbf{E}_q \times \mathbf{C} \to \mathbf{E}_q$; and the canonical sheaf is free: it is (isomorphic to) $\mathscr{O}_{\mathbf{E}_q}$. (This example will be referred to in B.6.6.)

As a consequence of the above proposition, the degree of $\mathscr{O}_X(D)$ (as defined at the end of B.6.3) is equal to the degree of D:

$$\deg \mathscr{O}_X(D) = \deg D.$$

EXERCISE B.18. Under the notations of B.5.5, let $u \in \mathbf{C}\{x\}^*$, and $\mathfrak{F}_u := \mathfrak{F}_A$, where $A := u$ considered as a 1×1 invertible matrix. Compute $\deg \mathfrak{F}_u$.

B.6.5. Some line bundles over $\mathbf{E}_q := \mathbf{C}^*/q^{\mathbf{Z}}$. We now return to B.5.5 but for $n = 1$. So let $a \in \mathcal{O}_X^*(\mathbf{C}^*)$; the corresponding q-difference equation $v(qy) = a(y)v(y)$ gives rise to a line bundle $\mathfrak{F}_a := \dfrac{\mathbf{C}^* \times \mathbf{C}^n}{\sim_a}$ over $\mathbf{E}_q$ fitting in the commutative diagram:

$$\begin{array}{ccc} \mathbf{C}^* \times \mathbf{C} & \xrightarrow{\ \pi_a\ } & \mathfrak{F}_a \\ {\scriptstyle pr_1}\downarrow & & \downarrow{\scriptstyle p_a} \\ \mathbf{C}^* & \xrightarrow{\ \pi\ } & \mathbf{E}_q \end{array},$$

where the relation $\sim_a$ is generated by $(y, v) \sim_a (qy, a(y)v)$. From the description by transition functions in B.6.1, one draws that $\mathfrak{F}_a \simeq \mathfrak{F}_b$ if, and only if, there exists $f \in \mathcal{O}_X^*(\mathbf{C}^*)$ such that $b(y)/a(y) = f(qy)/f(y)$.

The associated sheaf is given by $\mathscr{F}_a(U) := \{v \in \mathcal{O}_{\mathbf{C}}(\pi^{-1}(U)) \mid v(qy) = a(y)v(y)\}$. If $a(y) = cy^m$, there is a nonzero global meromorphic solution $f_0 := {\theta_q}^{m+1}(y)/\theta_q(cy)$; and all solutions have the form $f_0 g$, where g is q-invariant. The divisor of f_0 has degree m and evaluation $\overline{(-1)^m c} \in \mathbf{E}_q$. Since a divisor class on $\mathbf{E}_q$ is entirely determined by degree and evaluation (B.2.6), we see that every line bundle on $\mathbf{E}_q$ can be obtained this way; and, as a corollary, that every $a \in \mathcal{O}_X^*(\mathbf{C}^*)$ can be written $cy^m f(qy)/f(y)$ for some $f \in \mathcal{O}_X^*(\mathbf{C}^*)$.

B.6.6. Riemann-Roch theorem. Let X a compact Riemann surface, $\mathfrak{F}$ a line bundle and $\mathscr{F}$ the sheaf of its sections. In this section, we shall use the notation $H^0(X, \mathscr{F})$ for $\mathscr{F}(U)$.

THEOREM B.11. *The $\mathbf{C}$-linear space $H^0(X, \mathscr{F})$ is finite-dimensional.*

We write $h^0(\mathscr{F}) := \dim_{\mathbf{C}} H^0(X, \mathscr{F})$. If D is a divisor, we write $h^0(D) := \dim_{\mathbf{C}} H^0(X, \mathcal{O}_X(D))$. Of course, after B.6.4, this number depends only on the class of D in $\mathrm{Pic}(X)$. So, writing K the canonical class (see the example at the end of B.6.4) the slightly abusive notation $h^0(K - D)$ has a nonambiguous meaning. We write[9] it $h^1(D)$.

THEOREM B.12 (Riemann-Roch). $h^0(D) - h^1(D) = \deg D + 1 - g$.

Note that the involution $D \mapsto K - D$ leaves the formula invariant since $\deg K = 2 - 2g$. Here are some consequences of the theorem. The second one is related to the following obvious fact: if $\mathcal{O}_X(D)$ has a nontrivial global holomorphic section, then $\deg D \geq 0$, with equality only if D is a principal divisor.

(1) Every line bundle (or invertible sheaf) has a nonzero global meromorphic section.
(2) If $\deg D > 2g - 2$, then $h^0(D) = \deg D + 1 - g$.

And here is the application to $\mathbf{E}_q$ of the theorem and its consequences:

(1) $h^0(D) - h^0(-D) = \deg D$.
(2) If $\deg D > 0$, then $h^0(D) = \deg D$.

[9] The notation refers to the first cohomology space $H^1(X, \mathscr{F})$; see H and is consistent with *Serre's duality theorem*: $H^1(X, \mathcal{O}_X(K - D))$ and $H^0(X, \mathcal{O}_X(D))$ are dual to each other.

The latter can be directly proved as follows. Let $D := \sum n_i[\overline{a_i}]$, the $a_i \in \mathbf{C}^*$ being noncongruent modulo $q^{\mathbf{Z}}$, the $n_i \neq 0$ and $N := \sum n_i > 0$. Let $g \in \mathscr{M}(\mathbf{E}_q)^*$ be such that $\mathrm{div}_{\mathbf{E}_q}(g) \geq -D$, i.e., a q-elliptic function such that $h := g \prod \theta_q(-x/a_i)^{n_i}$ is holomorphic over $\mathbf{C}^*$. We thus define an isomorphism between $H^0(X, \mathscr{O}_X(D))$ and the space of those $h \in \mathscr{O}(\mathbf{C}^*)$ such that $h(qy) = ay^{-N}h(y)$, where $a \in \mathbf{C}^*$ (exercise: compute it!). Using power series expansion $h = \sum h_n y^n$, we see that the condition amounts to $q^n h_n = a h_{n+N}$. This recursion has N free degrees of freedom, for instance $h_0, \ldots, h_{N-1}$; and any choice of initial conditions yields a convergent power series, because h_{iN+j}/h_j decays like $q^{Ni^2/2}$.

B.7. Analyticity of diagonalisation

We propose here an application of the holomorphic triviality theorem of Subsection B.5.6 to a basic problem in Linear Algebra. This is useful in the kind of problems studied in Subsection 12.2.5, also see **[196]**. So we let X be an open Riemann surface, or, more generally, any complex manifold such that every holomorphic vector bundle on X is trivial: for instance, a so-called simply connected "Stein variety"; this includes the case of polydisks.

B.7.1. Analyticity of eigenvectors. Let $A(t)$, $t \in X$, an analytically parameterized matrix (with values in $\mathrm{Mat}_n(\mathbf{C})$). Suppose that there is a complex valued analytic function $\rho(t)$, $t \in X$, such that $\rho(t)$ is a simple eigenvalue of $A(t)$ for every $t \in X$. Note that the simplicity entails that the corresponding eigenspace $E_{\rho(t)} := \mathrm{Ker}(A(t) - \rho(t)I_n)$ has dimension 1.

PROPOSITION B.19. *Then there is an analytically parameterized vector $U(t)$, $t \in X$, (with values in $\mathbf{C}^n$) such that $U(t)$ vanishes nowhere and:*

$$\forall t \in X\ ,\ A(t)U(t) = \rho(t)U(t).$$

PROOF. Define the total space $\mathfrak{F} := \{(t, U) \in X \times \mathbf{C}^n \mid A(t)U = \rho(t)U\}$ and the projection $p : \mathfrak{F} \to X$, $(t, U) \mapsto t$. Each fiber $\mathfrak{F}_{t_0} := p^{-1}(t_0)$ has the form $\{t_0\} \times E_{\rho(t_0)}$, so it identifies to a line. Moreover a particular generator $U(t_0)$ can be chosen as any of the nonzero columns of the transpose of the comatrix of $A(t_0) - \rho(t_0)I_n$, thus analytically extended to a generator in a neighborhood of t_0. Therefore $\mathfrak{F}$ is a holomorphic line bundle over X, so it is trivial and any nonvanishing global section is an adequate solution $U(t)$. □

B.7.2. Analyticity of conjugacy matrix. Now keeping the analytically parameterized matrix $A(t)$ we suppose that there is an analytically parameterized separable diagonal matrix $R(t)$ such that $A(t) \sim R(t)$ (conjugacy) for every $t \in X$. (If we simply assume that $A(t)$ is separable for every t, such a $R(t)$ exists *locally*, see Subsubsection 12.2.5.2).

PROPOSITION B.20. *Then there is an analytically parameterized invertible matrix $C(t) \in GL_n(\mathbf{C})$, $t \in X$, such that $A(t) = C(t)R(t)C(t)^{-1}$ for every $t \in X$.*

PROOF. Indeed we apply the previous proposition to each of the diagonal coefficients $\rho_i(t)$ of $R(t)$ and form the matrix $C(t)$ with the resulting columns $U_i(t)$. □

The existence a priori of $R(t)$ all over X is a relevant precondition, as shows the example of $A(t) := \begin{pmatrix} 0 & 1 \\ t & 0 \end{pmatrix}$, which admits no nonvanishing eigenvector $U(t)$ analytic all over $\mathbf{C}^*$.

APPENDIX C

Classical Hypergeometric Functions

We deal only with the most standard case, that of functions $F := {}_2F_1$ and restrict to the fuchsian ones (and even generally assume nonresonancy). The general references are [**125**, Chapter 2] [**255**, Chapter XIV] and [**222**, Chapter 11] (also see [**93**]).

C.1. The series and the equation

Recall the *Pochhammer symbols*, defined for $\alpha \in \mathbf{C}$ and $n \in \mathbf{N}$:

$$(\alpha)_n := \alpha(\alpha+1)\cdots(\alpha+n-1) = \frac{\Gamma(\alpha+n)}{\Gamma(\alpha)}.$$

For instance $(\alpha)_0 = 1$ and $(1)_n = n!$.

C.1.1. The hypergeometric series. It is defined for $\alpha, \beta, \gamma \in \mathbf{C}$ as:

$$F(\alpha,\beta,\gamma;x) := \sum_{n\geq 0} \frac{(\alpha)_n(\beta)_n}{n!(\gamma)_n} x^n.$$

We impose a priori the following restrictions: $\gamma \notin -\mathbf{N}$ (lest terms be undefined); $\alpha, \beta \notin -\mathbf{N}$ (to have true power series, not polynomials); $\alpha-\gamma, \beta-\gamma, \notin \mathbf{Z}$ (to avoid simplifications of the fractions).

For reasons to appear later in the study (nonresonancy), we moreover impose the following restrictions: $\alpha - \beta, \gamma, \gamma - \alpha - \beta \notin \mathbf{Z}$.

The series F has radius of convergence 1, so it defines a holomorphic functions in $\overset{\circ}{\mathrm{D}}(0,1)$. As we will see, this admits analytic continuations in almost all of $\mathbf{S}$.

C.1.2. The hypergeometric differential equation. The series F satisfies the following (analytic linear) differential equation:

$$x(1-x)F''(x) + (\gamma - (\alpha+\beta+1)x)F'(x) - \alpha\beta F(x) = 0.$$

This shows that all points $\neq 0, 1$ are ordinary, so the function F on $\overset{\circ}{\mathrm{D}}(0,1)$ admits a unique analytic continuation to any simply connected subset of $\mathbf{C} \setminus \{0,1\}$. it is customary to take domains of the form $\mathbf{S} \setminus L$ where L is any line in $\mathbf{S}$ joining $0, 1, \infty$. The real line is a possible choice, but so are continuous spirals:

$$q_0^{\overline{\mathbf{R}}} = q_0^{\mathbf{R}} \cup \{0, \infty\}, \text{ where we denote } \overline{\mathbf{R}} := \mathbf{R} \cup \{+\infty, -\infty\}.$$

If we write the above equation using Euler's differential operator $\delta := xd/dx$:

$$(1-x)\delta^2 F + ((\gamma-1) - (\alpha+\beta)x)\delta F - \alpha\beta x F = 0,$$

we find that 0 is a regular singular point. The indicial equation is $s^2 + (\gamma-1)s = 0$, so it has exponents 0 and $1-\gamma$. The solution corresponding to exponent 0 is of course $f^{(0)}(x) := F(\alpha, \beta; \gamma; x)$. The solution corresponding to exponent $1 - \gamma$ is found to be:

$$g^{(0)}(x) := x^{1-\gamma} F(\alpha - \gamma + 1, \beta - \gamma + 1; 2 - \gamma; x).$$

of course we have to choose some determination of $x^{1-\gamma}$, which is possible after the choice of a cut. We then have a basis $\mathcal{B}_0 := (f^{(0)}, g^{(0)})$ of solutions at 0.

A similar study at infinity[1] shows that ∞ is a regular singular point, with exponents α, β (or $1/\alpha, 1/\beta$ depending whether we refer to variable x or to local variable $y := 1/x$), so there is a basis $\mathcal{B}_\infty := (f^{(\infty)}, g^{(\infty)})$ of solutions at ∞, where:

$$f^{(\infty)}(y) := y^\alpha F(\alpha, \alpha-\gamma+1; \alpha-\beta+1; y) \text{ and } g^{(\infty)}(y) := y^\beta F(\beta, \beta-\gamma+1; \beta-\alpha+1; y).$$

As a consequence, the local monodromies at 0, resp. at ∞, relative to the above bases have matrices:

$$\begin{pmatrix} 1 & 0 \\ 0 & e^{-2i\pi\gamma} \end{pmatrix}, \text{ resp. } \begin{pmatrix} e^{2i\pi\alpha} & 0 \\ 0 & e^{2i\pi\beta} \end{pmatrix}.$$

C.1.3. Special values. Although those cases were prohibited, we mention the obvious equalities: $F(1, 1, 2, x) = -\log(1 - x)/x$ and $F(\alpha, \beta, \beta; x) = (1 - x)^{-\alpha}$. More relevant are, for instance:

$$F(1/2, 1/2, 3/2, x^2) = \frac{\arcsin x}{x} \text{ and } F(1/2, 1, 3/2, -x^2) = \frac{\arctan x}{x}.$$

The above involve special values of the parameter. As regards special values of the variable, the most important is Gauß' summation formula:

$$F(\alpha, \beta, \gamma; 1) = \frac{\Gamma(\gamma)\Gamma(\gamma - \alpha - \beta)}{\Gamma(\gamma - \alpha)\Gamma(\gamma - \beta)} \text{ assuming } \Re(\gamma - \alpha - \beta) > 0.$$

C.2. Integral and connection formulas

C.2.1. Barnes's integral representation. It reads as follows (see figure):

$$F(\alpha, \beta, \gamma; x) = \frac{1}{2i\pi} \frac{\Gamma(\gamma)}{\Gamma(\alpha)\Gamma(\beta)} \int_C \frac{\Gamma(\alpha + s)\Gamma(\beta + s)\Gamma(-s)}{\Gamma(\gamma + s)} (-x)^s \, ds.$$

C.2.2. Barnes's connection formula. Using the above integral representation, one proves first the following connection formula:

$$\begin{aligned} F(\alpha, \beta, \gamma; x) = {} & \frac{\Gamma(\gamma)\Gamma(\beta - \alpha)}{\Gamma(\beta)\Gamma(\gamma - \alpha)} (-x)^{-\alpha} F(\alpha, \alpha - \gamma + 1; \alpha - \beta + 1; 1/x) \\ & + \frac{\Gamma(\gamma)\Gamma(\alpha - \beta)}{\Gamma(\alpha)\Gamma(\gamma - \beta)} (-x)^{-\beta} F(\beta, \beta - \gamma + 1; \beta - \alpha + 1; 1/x). \end{aligned}$$

Then one deduces the following one:

$$\mathcal{B}_0 = \mathcal{B}_\infty \begin{pmatrix} e^{-i\pi\alpha} \dfrac{\Gamma(\gamma)\Gamma(\beta - \alpha)}{\Gamma(\beta)\Gamma(\gamma - \alpha)} & e^{-i\pi\alpha'} \dfrac{\Gamma(\gamma')\Gamma(\beta' - \alpha')}{\Gamma(\beta')\Gamma(\gamma' - \alpha')} \\ e^{-i\pi\beta} \dfrac{\Gamma(\gamma)\Gamma(\alpha - \beta)}{\Gamma(\alpha)\Gamma(\gamma - \beta)} & e^{-i\pi\beta'} \dfrac{\Gamma(\gamma')\Gamma(\alpha' - \beta')}{\Gamma(\alpha')\Gamma(\gamma' - \beta')} \end{pmatrix},$$

where one sets $\alpha' := \alpha - \gamma + 1$, $\beta' := \beta - \gamma + 1$ and $\gamma' := 2 - \gamma$.

[1]The regular singular point 1 is worth a local study too, see the quoted references; but we do not need it here (although it plays a role in **[217]**.)

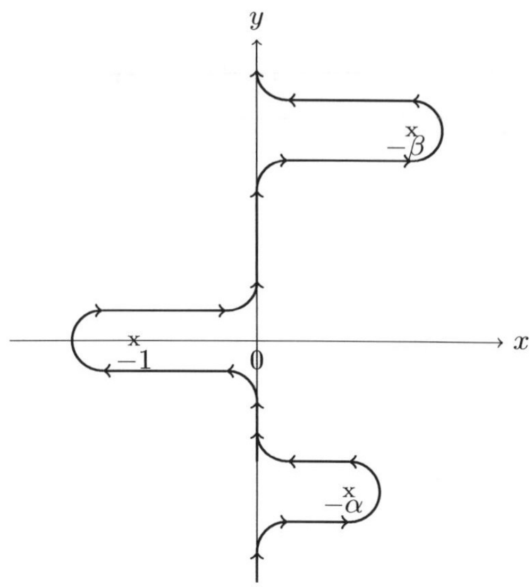

FIGURE C.1. The contour for Barnes's integral representation

APPENDIX D

Basic Index Theory

General references for this appendix: [**226**, **234**] and especially [**188**]; see also [**103**, **141**, **142**].

D.1. Indices in linear algebra

All spaces are over some fixed (commutative) field K.

D.1.1. Exact sequences and dimensions. Recall that, if $0 \to E' \xrightarrow{f} E \xrightarrow{g} E'' \to 0$ is an exact sequence (i.e., f is injective, g is surjective and $\text{Im } f = \text{Ker } g$), then, up to identifications, $E'' = E/E'$ and we have:

$$\dim E < \infty \iff (\dim E' < \infty \text{ and } \dim E'' < \infty).$$

Moreover, if those dimensions are finite, then $\dim E = \dim E' + \dim E''$.

Lemma D.1. (i) *If $E' \xrightarrow{f} E \xrightarrow{g} E''$ is exact (i.e., $Imf = Ker\ g$), then:*

$$(\dim E' < \infty \text{ and } \dim E'' < \infty) \Rightarrow \dim E < \infty.$$

(ii) *If the sequence $0 \xrightarrow{u_0} E_1 \xrightarrow{u_1} E_2 \xrightarrow{u_2} \cdots \xrightarrow{u_{m-1}} E_m \xrightarrow{u_m} 0$ of finite-dimensional spaces is exact, then $\sum_{i=1}^{m} (-1)^i \dim E_i = 0$.*

Proof. (i) There is an exact sequence $0 \to \text{Ker } g \to E \to \text{Im } g \to 0$, whence $\dim E < \infty$ if $\dim \text{Ker } g, \dim \text{Im } g < \infty$. Suppose E', E'' are finite dimensional. Since $\text{Im } g \subset E''$, $\text{Im } g$ is finite-dimensional; and since $f : E' \to \text{Im } f = \text{Ker } g$ is onto, $\text{Ker } g$ is finite-dimensional.

(ii) Let $E'_i := \text{Ker } u_i$ and $E''_i := \text{Im } u_i$ for $i = 0, \dots, m$, whence the exact sequences $0 \to E'_i \to E_i \to E''_i \to 0$. From the exactness of the long sequence, we draw that $E'_i = E''_{i-1}$ for $i = 1, \dots, m$, so $\dim E_i = \dim E''_{i-1} + \dim E''_i$. Taking in account cancellations and the extremities of the long exact sequence, we then compute:

$$\sum_{i=1}^{m} (-1)^i \dim E_i = \sum_{i=1}^{m} (-1)^i (\dim E''_{i-1} + \dim E''_i) = -\dim E''_0 + (-1)^m \dim E''_m = 0.$$

□

D.1.2. The snake lemma. This is about a commutative diagram of exact sequences:

$$\begin{array}{ccccccccc} 0 & \longrightarrow & E' & \longrightarrow & E & \longrightarrow & E'' & \longrightarrow & 0 \\ & & \downarrow{\scriptstyle u'} & & \downarrow{\scriptstyle u} & & \downarrow{\scriptstyle u''} & & \\ 0 & \longrightarrow & F' & \longrightarrow & F & \longrightarrow & F'' & \longrightarrow & 0 \end{array}$$

The snake lemma says that there is an exact sequence:

$$0 \to \text{Ker } u' \to \text{Ker } u \to \text{Ker } u'' \to \text{Coker } u' \to \text{Coker } u \to \text{Coker } u'' \to 0.$$

The left part with the kernels is immediate to construct, the right part with the cokernels is not too difficult. The "connection arrow" Ker $u'' \to$ Coker u' is more tricky (more fun also). Complete construction and proof can be found in [**37**, Chapter 1, §1, no 2], except from injectivity at left (which is immediate since it is induced by restriction of the injective map $E' \to E''$); and surjectivity at right, see the exercise.

EXERCISE D.2. From the commutative diagram

$$\begin{array}{ccc} E_1 & \longrightarrow & E_2 \\ \Big\downarrow{\scriptstyle u_1} & & \Big\downarrow{\scriptstyle u_2} \\ F_1 & \longrightarrow & F_2 \end{array}$$

deduce a map Coker $u_1 \to$ Coker u_2 and show that it is onto if $F_1 \to F_2$ is.

D.1.3. Index of a linear map. A linear map between linear spaces $E \xrightarrow{u} F$ is said to *have an index* if its kernel and cokernel are finite-dimensional. The *index* of u then is:

$$\chi(u; E, F) := \dim \text{Ker } u - \dim \text{Coker } u.$$

If $E = F$ we write $\chi(u, E)$. We may also abreviate into $\chi(u)$ if no ambiguity arises.

EXERCISE D.3. If E and F are finite-dimensional, all maps have the same index $\chi(u; E, F) = \dim E - \dim F$.

REMARK D.4. The index is a particular case of a notion from homological algebra, the *Euler characteristics of a complex* (see Appendix E). Let a **Z**-*graded complex*:

$$\cdots \to C_{n+1} \xrightarrow{d_{n+1}} C_n \xrightarrow{d_n} C_{n-1} \to \cdots$$

The word "complex" means that $d_n \circ d_{n+1} = 0$ for all n. Then one can define the *homology spaces* $H_n :=$ Ker $d_n/$Im d_{n+1}. If they are all finite-dimensional and almost all trivial, $\sum(-1)^n \dim H_n$ is well defined: this is the Euler characteristics.

If $d_0 := u$, $C_0 := E$, $C_{-1} := F$ and all other C_n are trivial, the Euler characteristics is the index $\chi(u; E, F)$. Many properties of the index we state and use are particular cases of properties of the Euler characteristics.

PROPOSITION D.5 (additivity of the index). (i) *Let the same diagram as in the snake lemma. If u' and u'' have an index, then so does u and then $\chi(u; E, F) = \chi(u'; E', F') + \chi(u''; E'', F'')$.*

(ii) *Let $f : E \to F$ and $g : F \to G$ have an index. Then so does $g \circ f$ and $\chi(g \circ f; E, G) = \chi(f; E, F) + \chi(g; F, G)$.*

PROOF. (i) From the snake diagram and part (i) of lemma D.1, we see that Ker u and Coker u are finite-dimensional. Then from part (ii) of the lemma we get:

$$-\dim \text{Ker } u' + \dim \text{Ker } u - \dim \text{Ker } u'' + \dim \text{Coker } u' - \dim \text{Coker } u + \dim \text{Coker } u'' = 0,$$

which immediately gives the formula.

(ii) We have two exact sequences:

$$0 \to \text{Ker } f \to \text{Ker } g \circ f \xrightarrow{f} \text{Ker } g \to \frac{\text{Ker } g}{\text{Ker } g \cap \text{Im } f} \to 0$$

(note indeed that $f\,(\text{Ker } g \circ f) = \text{Ker } g \cap \text{Im } f$); and:

$$0 \to \frac{\text{Ker } g + \text{Im } f}{\text{Im } f} \to \frac{F}{\text{Im } f} \xrightarrow{g} \frac{G}{\text{Im } g \circ f} \to \frac{G}{\text{Im } g} \to 0$$

Then the vanishing of the alternating sums of dimensions along with the isomorphism theorem $\dfrac{X+Y}{X} \simeq \dfrac{Y}{Y \cap X}$ yield the formula.

□

EXERCISE D.6. Check the meaning and validity of the second exact sequence.

EXAMPLES D.7. (1) Let L the q-difference operator x (multiplication by x) in E, where E is any of $\mathbf{C}[x]$, $\mathbf{C}[[x]]$, $\mathbf{C}\{x\}$, $\mathscr{O}(\mathbf{C})$ or $C(R)$ (recall that $C(R)$ is defined in the main text, 6.2.1.2). Then Ker $L = \{0\}$ and Im L has codimension 1 ($\mathbf{C}$ is a supplementary space: $E = \text{Im } L \oplus \mathbf{C}$), so $\chi(L, E) = -1$.

(2) Let L the q-difference operator σ_q in E as above, except that E is not $C(R)$. Then L is an automorphism and $\chi(L, E) = 0$.

(3) Let L the q-difference operator σ_q in $C(R)$. It is injective but its image is $C(R')$ where $R' := |q|^{-1} R > R$. The image has not finite codimension (for instance the functions $1/(x - R'')$ with $R < R'' < R'$ are elements of $C(R)$ and they are linearly independent modulo $C(R') \subset C(R)$) so L acting on $C(R)$ has no index.

EXERCISE D.8. (i) Compute $\chi(L, \mathbf{C}[x])$ where L is the multiplication by some polynomial P.

(ii) Prove the linear independence stated in the third example.

D.2. Indices in functional analysis

D.2.1. Banach spaces. Let $u : E \to F$ a continuous linear map of Banach spaces. So E, F are complete normed spaces and u is linear and *bounded*, i.e.:

$$\|u\| := \sup_{x \neq 0} \frac{\|u(x)\|}{\|x\|} < \infty.$$

EXERCISE D.9. We defined $C(R)$ along with its norm $\|-\|_R$ in 6.2.1.2. Then, if $R' > R > 0$, the natural restriction map $C(R') \to C(R)$ has norm 1.

DEFINITION D.10. The linear map $u : E \to F$ (Banach spaces) is said to be *compact* if the image by u of the unit ball in E is relatively compact in F (i.e., its closure is compact). (The said image is therefore bounded, so u is automatically continuous.)

PROPOSITION D.11 (Sequential criterion of compactness). *A continuous operator $T : X \to Y$ between two Banach spacesis compact if, and only if, every sequence (x_n) in X admits a subsequence (x_{n_k}) such that the image sequence $(T(x_{n_k}))$ converges in Y.*

PROOF. Rudin, Functional analysis, no 4.16 □

COROLLARY D.12. *Let $R' > R > 0$. Then the restriction map $C(R') \to C(R)$ is compact.*

PROOF. Continuity is obvious. The sequential criterion above can be applied as an immediate consequence of Rudin, Real and Complex Analysis, ch 14, th 14.6, here with $\Omega := \mathring{\mathrm{D}}(0,R') \supset \overline{\mathrm{D}}(0,R)$: any bounded sequence (f_n) in $C(R')$ restricts to a bounded sequence in $\mathscr{O}(\Omega)$ and the latter admits a subsequence such that the restrictions to $\overline{\mathrm{D}}(0,R)$ converge uniformly, that is for the norm on $C(R)$. □

DEFINITION D.13. The continuous linear map $u : E \to F$ (Banach spaces) is said to be *Fredholm* if it has an index.

Some books require u to have a closed image, but by [**234**, lemma 8.1], this actually is a consequence of having an index.

THEOREM D.1. *Let $u : E \to F$ a Fredholm map and $v : E \to F$ a compact map of Banach spaces. Then $u+v$ is Fredholm and $\chi(u;E,F) = \chi(u+v;E,F)$.*

PROOF. [**234**, Proposition 8.3] □

Note however that u and $u+v$ may have kernels, resp. cokernels, of different dimensions: only the index is guaranteed to be the same, and this is where lies the full power of the notion!

EXAMPLE D.14. Since σ_q acting on $C(R)$ is compact, $\sigma_q - 1$ has the same index as -1, that is 0. However $\mathrm{Ker}(\sigma_q - 1, C(R)) = \mathbf{C}$ and the image is the subspace of functions vanishing at 0 so the cokernel also has dimension 1.

EXERCISE D.15. Prove the last two statements.

More generally, Shubin proves that the set of Fredholm operators is open in the normed space of continuous linear operators, and that the index is a locally constant (i.e., continuous) map.

D.2.2. Beyond Banach spaces: direct (or inductive) limits. Clearly $\mathbf{C}\{x\} = \bigcup_{R>0} C(R)$ and $\mathscr{O}(\mathbf{C}) = \bigcap_{R>0} C(R)$. These are particular cases of *direct*, resp. *inverse* limits. We shall not go into the general formalism, see for instance [**226**], but we shall precise what it means here. We shall mention FN and DFN spaces without defining them: see the references.

We start with direct (or inductive) limits. Let (R_n) a strictly decreasing sequence with limit 0. Clearly $\mathbf{C}\{x\} = \bigcup C(R_n)$. The *direct limit topology* on $\mathbf{C}\{x\}$ is defined by the condition that $F : \mathbf{C}\{x\} \to Y$ (target: an arbitrary topological space) is continuous if, and only if, all maps $F_n : C(R_n) \to Y$ (induced by restriction) are continuous. Equivalently: a subset of $\mathbf{C}\{x\}$ is open, resp. closed, if, and only if, all its intersections with the $C(R_n)$ are. It is a fact that this definition does not depend on a particular choice of the sequence (R_n). The topological vector space $\mathbf{C}\{x\}$ thus obtained is said to be DFN (for "dual of Frechet nuclear", cf. why "dual" in the section D.3 on dualities).

A purely formal consequence of this definition is that, if a linear continuous map L_R from $C(R)$ to itself is given for every $R > 0$ in such a way that L_R and $L_{R'}$ are compatible when $0 < R < R'$ (i.e., L_R restrictd to $C(R')$ is $L_{R'}$), then they patch together uniquely into a linear continuous map L from $\mathbf{C}\{x\}$ to itself. Using [**188**, Lemme 1.3.8 (i) p 20] one then gets:

LEMMA D.16. *With the above notation, suppose that each L_R is Fredholm and that for R small enough $\chi(L_R, C(R))$ does not depend on R. Then the natural maps Ker $L_R \to$ Ker L and Coker $L_R \to$ Coker L are isomorphisms for R small enough, L is Fredholm and $\chi(L, \mathbf{C}\{x\})$ is the common value of the $\chi(L_R, C(R))$.*

PROOF. In order to apply the lemma from **[188]**, we have to check that, for $0 < R < R'$, the restriction map $C(R') \to C(R)$ is injective and has dense image. The former flows from the uniqueness of analytic continuation; the latter of Stone-Weierstrass theorem since $\mathbf{C}[x]$ embeds densely in both $C(R)$ and $C(R')$. □

D.2.3. Beyond Banach spaces: inverse (or projective) limits. Let (R_n) a strictly increasing sequence with limit $+\infty$. Clearly $\mathscr{O}(\mathbf{C}) = \bigcap C(R_n)$. The *inverse limit topology* on $\mathscr{O}(\mathbf{C})$ is defined by the condition that $F : X \to \mathscr{O}(\mathbf{C})$ (source: an arbitrary topological space) is continuous if, and only if, all maps $F_n : X \to C(R_n)$ (induced by restriction) are continuous; equivalently, embed $\mathscr{O}(\mathbf{C})$ in the product $\prod C(R_n)$ and take the induced topology. Again, this topology does not depend on a particular choice of the sequence (R_n). The topological vector space $\mathscr{O}(\mathbf{C})$ thus obtained is said to be FN (for "Frechet nuclear").

A formal consequence of this definition is that, if a linear continuous map L_R from $C(R)$ to itself is given for every $R > 0$ in such a way that L_R and $L_{R'}$ are compatible when $0 < R < R'$ (i.e., L_R restrictd to $C(R')$ is $L_{R'}$), then they define uniquely into a linear continuous map L from $\mathscr{O}(\mathbf{C})$ to itself. Using **[188**, lemme 1.3.8 (ii) p 20**]** one then gets in the same way as before:

LEMMA D.17. *With the above notation, suppose that each L_R is Fredholm and that for R big enough $\chi(L_R, C(R))$ does not depend on R. Then the natural maps Ker $L \to$ Ker L_R and Coker $L \to$ Coker L_R are isomorphisms for R big enough, L is Fredholm and $\chi(L, \mathscr{O}(\mathbf{C}))$ is the common value of the $\chi(L_R, C(R))$.*

Note that in **[188]** the setting is more general and conditions for (ii) are slightly more restrictive than for (i); in our setting, there is no such difference.

D.2.4. One more lemma [188, p 5]. In the commutative diagram of topological vector spaces and continuous linear maps:

$$\begin{array}{ccc} E_1 & \xrightarrow{u_1} & F_1 \\ {\scriptstyle v}\downarrow & & \downarrow{\scriptstyle w} \\ E_2 & \xrightarrow{u_2} & F_2 \end{array}$$

suppose that:

(1) both E_1 and F_1 are Banach spaces, or both are Frechet, or both are DFS;
(2) both E_2 and F_2 are Banach spaces, or both are Frechet, or both are DFS;
(3) u_1 and u_2 are Fredholm;
(4) v and w are injective and w has a dense image.

Then the induced map $\overline{u}_2 : E_2/E_1 \to F_2/F_1$ is surjective and Fredhol, and:

$$\chi\left(\overline{u}_2\right) = \dim \operatorname{Ker} \overline{u}_2 = \chi(u_2) - \chi(u_1).$$

We do not define "Frechet" or "DFS" (for "dual of Frechet-Schwartz") but note that the lemma can be applied for $\mathscr{O}(\mathbf{C})$, $\mathbf{C}[[x]]$ and $\mathbf{C}[[x]]_{q,(s)}$, which are FN, which implies Frechet; and to $\mathbf{C}\{x\}$, $\mathbf{C}[x]$ and $\mathbf{C}[[x]]_{q,s}$, which are FN, which implies DFS.

D.3. Dualities

The main references here are [**105**, **188**, **191**].

D.3.1. A DFN space. Let $E := \mathcal{O}(\mathbf{C})$, endowed with its inverse limit topology[1], making it a FN space; and $F := \mathbf{C}\{x\}$, endowed with its direct limit topology, making it (for reasons to be seen soon) a DFN space. There is a well defined bilinear map $E \times F \to \mathbf{C}$, denoted $(f, g) \mapsto \langle f, g \rangle$ and given by the formula:

$$\langle \sum a_n x^n, \sum b_n x^n \rangle := \sum a_n b_n.$$

Classically, this induces a map $f \mapsto (g \mapsto \langle f, g \rangle)$ from E to the dual of F; and, actually, to the continuous dual F' of F. Symmetrically we get a map $g \mapsto (f \mapsto \langle f, g \rangle)$ from F to E'. Moreover, it can be proved that, for some adequate definition of the topology on the continuous dual, those maps are isomorphisms of topological vector spaces $E \simeq F'$ and $F \simeq E'$. Therefore, each of $\mathbf{C}\{x\}$ and $\mathcal{O}(\mathbf{C})$ can be identified with the continuous dual of the other (and so, if $\mathcal{O}(\mathbf{C})$ is FN, $\mathbf{C}\{x\}$ has to be DFN).

D.3.2. A DFN space and a FN space. If we endow $\mathbf{C}[[x]] \simeq \mathbf{C}^{\mathbf{N}}$ with the product topology (a particular case of inverse limit topology), we make it a FN space. If we endow $\mathbf{C}[x] = \bigcup(\mathbf{C} \oplus \cdots \oplus \mathbf{C}x^N)$ with the direct limit topology, we make it a DFN. Actually the same formula as above defines a linear pairing $\mathbf{C}[[x]] \times \mathbf{C}[x] \to \mathbf{C}$ and allows us to identify each of $\mathbf{C}[[x]]$, $\mathbf{C}[x]$ with the continuous dual of the other.

As for $\mathbf{C}[[x]]_{q,s}$ and $\mathbf{C}[[x]]_{q,(s)}$, they can be topologized by transport of structure from $\mathbf{C}\{x\}$ and $\mathcal{O}(\mathbf{C})$ through the q-Borel-Ramis $\mathcal{B}_{q;s}$, respectively making them into a DFN and a FN.

D.3.3. Indices. Let $u : V \to W$ a continuous linear map of topological vector spaces. For any $\lambda \in W'$, the compositum $\lambda \circ u \in V'$. This defines the *transpose map* $u' : W' \to V'$. It is continuous and linear. Moreover $u \mapsto u'$ is linear and reverses composition: $(v \circ u)' = u' \circ v'$. In particular, u is an isomorphism if, and only if, u' is.

LEMMA D.18. *If u has an index, so has u' and $\chi(u'; W', V') = -\chi(u; V, W)$.*

After [**188**, p 4, th 0.10 (iii)], a sufficient condition for the lemma to hold is that V and W are both Banach or both FS or both DFS. In particular, if u has source and target V, a FN or DFN space, then $\chi(u', V') = -\chi(u, V)$.

In the case of the duality maps seen above, the transpose of $u : E \to E$ can be identified with the *adjoint map* $v : F \to F$, characterized by the formula:

$$\langle u(x), y \rangle = \langle x, v(y) \rangle.$$

So we would have $\chi(u, E) = -\chi(v, F)$.

---

[1] Instead of $\mathcal{O}(\mathbf{C})$ and x, Ramis deals with the space $\mathcal{O}(\mathbf{S} \setminus \{0\})$ coming with coordinate x^{-1} and similarly for various spaces. The formulation here is equivalent and suffices for our needs.

APPENDIX E

Cochain Complexes

General references for this appendix are: [**32, 46, 94, 153**].

Complexes come in two guises: chain complexes, which arise naturally in topology and provide an algebraic frame for homology; and their dual counterparts, cochain complexes, which are the natural frame for cohomology (in homological algebra). Here, we describe the latter only.

E.1. Cochain complexes and cohomology

E.1.1. Cochain complexes. The following definitions and theorems are stated for linear spaces over a fixed commutative field K. However, much remains valid for modules over a commutative ring. (Whenever dimension is involved, modules should be supposed of finite rank.)

DEFINITION E.1. A *(cochain) complex* $(C^\bullet, d^\bullet)$ consists of a $\mathbf{Z}$-sequence $(C^i)_{i\in\mathbf{Z}}$ of K-vector spaces and a $\mathbf{Z}$-sequence $(d^i : C^i \to C^{i+1})_{i\in\mathbf{Z}}$ of K-linear maps such that:

$$\forall i \in \mathbf{Z}\ ,\ d^{i+1} \circ d^i = 0. \tag{E.1}$$

The elements of C^i are called *cochains of degree* i. The elements of $Z^i := \operatorname{Ker} d^i$ are called *cocycles of degree* i. The elements of $B^i := \operatorname{Im} d^{i-1}$ are called *coboundaries of degree* i.

By assumption (E.1), we have $B^i \subset Z^i \subset C^i$. The quotient space $H^i := Z^i/B^i$ is the i*th cohomology space* of the complex, its elements are called *cohomology classes of degree* i.

We denote $H^\bullet$ the $\mathbf{Z}$-sequence of all spaces H^i. It is endowed with no structural map, the H^i are somehow "independent" of each other (contrary to the C^i which are "linked" by the differential). We may at will identify $H^\bullet$ with a graded vector space:

$$H^\bullet = \bigoplus_{i\in\mathbf{Z}} H^i.$$

When a bounded sequence indexed by $i \geq i_0$ or $i \leq i_1$, etc, is used, it is tacitly completed by null spaces and null maps at left and/or at right.

If another complex $(C'^\bullet, d'^\bullet)$ appears, notations can be disambiguated either by writing $Z^i(C^\bullet, d^\bullet)$, $Z^i(C'^\bullet, d'^\bullet)$, $B^i(C^\bullet, d^\bullet)$, $B^i(C'^\bullet, d'^\bullet)$, $H^i(C^\bullet, d^\bullet)$, $H^i(C'^\bullet, d'^\bullet)$, or even, more lightly, $Z^i(C^\bullet)$, $Z^i(C'^\bullet)$, $B^i(C^\bullet)$, $B^i(C'^\bullet)$, $H^i(C^\bullet)$, $H^i(C'^\bullet)$, etc; or else by writing such as Z^i, Z'^i, B^i, B'^i, H^i, H'^i.

REMARK E.2. An equivalent but more formal definition of a complex says that $C^\bullet$ is a $\mathbf{Z}$-graded space $\bigoplus_{i\in\mathbf{Z}} C^i$, endowed with a *differential* $d : C^\bullet \to C^\bullet$ (this means that $d \circ d = 0$) of degree $+1$ (this means that $d(C^i) \subset C^{i+1}$ for all i).

E.1.2. Euler-Poincaré characteristics of a complex. It is defined when all cohomology spaces H^i are finite-dimensional and almost all of them (that is, all but a finite number) are null.

DEFINITION E.3. The *Euler* or *Euler-Poincaré characteristics* of a complex $(C^\bullet, d^\bullet)$ is the rational integer:

$$\chi(C^\bullet, d^\bullet) := \sum_{i \in \mathbf{Z}} (-1)^i \dim_K H^i(C^\bullet, d^\bullet).$$

EXAMPLE E.4. This is the application appearing in 7.1.5.4. To a linear map $f : E \to F$ between two vector spaces, we associate a complex $(C^\bullet, d^\bullet)$ defined by: $C^0 := E, C^1 := F, d^0 := f$, all other C^i, d^i are null. We write it as:

$$\cdots \to 0 \to \underset{0}{E} \xrightarrow{f} \underset{1}{F} \to 0 \to \cdots$$

Then we find that $Z^0 = E, B^0 = 0, H^0 = \operatorname{Ker} f, Z^1 = F, B^1 = \operatorname{Im} f, H^1 = \operatorname{Coker} f$ while all other Z^i, B^i, H^i are null. Its Euler-Poincaré characteristics is therefore the index (D.3.3) of f:

$$\chi(C^\bullet, d^\bullet) = \chi(f; E, F).$$

E.2. Morphisms of complexes, functoriality

E.2.1. Morphisms of complexes. A *morphism of complexes* $\phi^\bullet : (C^\bullet, d^\bullet) \to (C'^\bullet, d'^\bullet)$ is a $\mathbf{Z}$-sequence of linear maps $\phi^i : C^i \to C'^i$ such that:

(E.2) $\quad \forall i \in \mathbf{Z}\ ,\ d'^i \circ \phi^i = \phi^{i+1} \circ d^i$, i.e., the following square is commutative:

$$\begin{array}{ccc} C^i & \xrightarrow{d^i} & C^{i+1}. \\ {\scriptstyle \phi^i}\downarrow & & \downarrow{\scriptstyle \phi^{i+1}} \\ C'^i & \xrightarrow[d'^i]{} & C'^{i+1} \end{array}$$

The following facts are easily checked:

- There are identity morphisms $\operatorname{Id}_{C^\bullet} : (C^\bullet, d^\bullet) \to (C^\bullet, d^\bullet)$ (a sequence of identity maps Id_{C^i}) and morphisms $\phi^\bullet : (C^\bullet, d^\bullet) \to (C'^\bullet, d'^\bullet)$, $\phi'^\bullet : (C'^\bullet, d'^\bullet) \to (C''^\bullet, d''^\bullet)$ can be composed componentwise into a morphism $\phi'^\bullet \circ \phi^\bullet : (C^\bullet, d^\bullet) \to (C''^\bullet, d''^\bullet)$. Usual rules (neutrality, associativity) hold: in other words, cochain complexes form a category.
- Morphisms from $(C^\bullet, d^\bullet)$ to $(C'^\bullet, d'^\bullet)$ form a vector space: if $\phi^\bullet$ and $\psi^\bullet$ are such morphisms and $\lambda, \mu \in K$, then the $\mathbf{Z}$-sequence of all $\lambda\phi^i + \mu\psi^i$ is a morphism $\lambda\phi^\bullet + \mu\psi^\bullet$ from from $(C^\bullet, d^\bullet)$ to $(C'^\bullet, d'^\bullet)$. The composition is bilinear. In other words, we have a K-linear additive category.
- It is actually an abelian category, that is, in essence, exact sequences (in particular kernels and cokernels) behave "the right way", i.e., they obey the same rules as exact sequences of modules over a ring.

E.2.2. Functoriality. Let $\phi : C^\bullet \to C'^\bullet$ a morphism of complexes (we turn to light notation, with differentials $d^\bullet, d'^\bullet$ implicit). It is easily checked, using (E.1), that $\phi^i(B^i) \subset B'^i$ and $\phi^i(Z^i) \subset Z'^i$, so that ϕ^i induces a K-linear map in cohomology $\overline{\phi}^i : H^i \to H'^i$, more formally denoted as:

$$H^i(\phi) : H^i(C^\bullet) \to H^i(C'^\bullet).$$

The **Z**-sequence of all those maps can be written:

$$H^\bullet(\phi) : H^\bullet(C^\bullet) \to H^\bullet(C'^\bullet).$$

The whole process is functorial: $H^\bullet(\mathrm{Id}_{C^\bullet})$ is the identity of $H^\bullet(C^\bullet)$; and if we compose ϕ above with $\phi' : C'^\bullet \to C''^\bullet$, we find that:

$$\forall i \in \mathbf{Z}\ ,\ H^i(\phi' \circ \phi) = H^i(\phi') \circ H^i(\phi), \text{ which we summarize as } H^\bullet(\phi' \circ \phi) = H^\bullet(\phi') \circ H^\bullet(\phi).$$

Moreover each mapping $\phi \mapsto H^i(\phi)$ is K-linear:

$$H^i(\lambda\phi + \mu\psi) = \lambda H^i(\phi) + \mu H^i(\psi).$$

In the end we conclude that $H^\bullet$ is a K-linear functor from the category of cochain complexes to the category of **Z**-graded vector spaces.

E.2.3. The cohomology long exact sequence. We keep light notation with differentials implicit.

THEOREM E.1. *Let* $0 \to C'^\bullet \to C^\bullet \to C''^\bullet \to 0$ *an exact sequence of complexes, i.e., the morphisms* $\phi : C'^\bullet \to C^\bullet$ *and* $\psi : C^\bullet \to C''^\bullet$ *are such that each sequence* $0 \to C'^i \to C^i \to C''^i \to 0$ *is exact. Then the morphisms* $H^i(C'^\bullet) \to H^i(C^\bullet) \to H^i(C''^\bullet)$ *resulting from functoriality (*E.2.2*) can be completed by a* **Z***-sequence of* connecting homomorphisms $H^i(C''^\bullet) \to H^{i+1}(C'^\bullet)$ *in such a way that the following long sequence (in obvious abridged notation) is exact:*

$$\cdots \to H'^i \to H^i \to H''^i \to H'^{i+1} \to H^{i+1} \to H''^{i+1} \to \cdots$$

The following uses computations of dimensions with exact sequences (see Appendix D).

COROLLARY E.5. *In the above case,* $\chi(C^\bullet) = \chi(C'^\bullet) + \chi(C''^\bullet)$ *(if they have meanings).*

EXAMPLE E.6. Pursuing our example from E.1.2, suppose we have a commutative diagram of exact sequences:

$$\begin{array}{ccccccccc} 0 & \longrightarrow & E' & \longrightarrow & E & \longrightarrow & E'' & \longrightarrow & 0. \\ & & \downarrow{\scriptstyle f'} & & \downarrow{\scriptstyle f} & & \downarrow{\scriptstyle f''} & & \\ 0 & \longrightarrow & F' & \longrightarrow & F & \longrightarrow & F'' & \longrightarrow & 0 \end{array}$$

The associated complexes $C^\bullet := \cdots \to 0 \to \underset{0}{E} \xrightarrow{f} \underset{1}{F} \to 0 \to \cdots$ and likewise for $C'^\bullet$ and $C''^\bullet$ form an exact sequence $0 \to C'^\bullet \to C^\bullet \to C''^\bullet \to 0$. Then:

(1) The corresponding long exact sequence of cohomology from the theorem yields the snake lemma from D.1.2, i.e., an exact sequence:

$$0 \to \mathrm{Ker}\ f' \to \mathrm{Ker}\ f \to \mathrm{Ker}\ f'' \to \mathrm{Coker}\ f' \to \mathrm{Coker}\ f \to \mathrm{Coker}\ f'' \to 0.$$

(2) The additivity of Euler-Poincaré characteristics from the corollary yields the additivity of the index with respect to exact sequences from D.1.3:

$$\chi(f; E, F) = \chi(f'; E', F') + \chi(f''; E'', F'').$$

EXERCISE E.7. Prove those two statements.

E.3. Homotopy

E.3.1. Homotopy of morphisms. Let $C^\bullet, C'^\bullet$ two complexes and suppose we are given a $\mathbf{Z}$-sequence of linear maps $h^i : C^i \to C'^{i-1}$, subject to no condition. Define:

$$\phi^i := h^i \circ d^i + d'^{i-1} \circ h^{i-1} : C^i \to C'^i$$

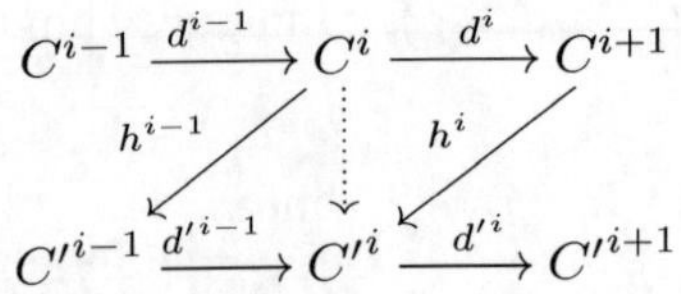

Then:

$$d'^i \circ \phi^i = d'^i \circ h^i \circ d^i + d'^i \circ d'^{i-1} \circ h^{i-1} = d'^i \circ h^i \circ d^i,$$

$$\phi^{i+1} \circ d^i = h^{i+1} \circ d^{i+1} \circ d^i + d'^i \circ h^i \circ d^i = d'^i \circ h^i \circ d^i,$$

since $d'^i \circ d'^{i-1} = 0$ and $d^{i+1} \circ d^i = 0$. Therefore $d'^i \circ \phi^i = \phi^{i+1} \circ d^i$ for all i, i.e., the $\mathbf{Z}$-sequence of all ϕ^i is a morphism of complexes $\phi : C^\bullet \to C'^\bullet$. Such particular morphisms are said to be *null homotopic*. Note that null homotopic morphisms form a subspace of the space of all morphisms $C^\bullet \to C'^\bullet$. Indeed, if $\phi_1^i := h_1^i \circ d^i + d'^{i-1} \circ h_1^{i-1}$ and $\phi_2^i := h_2^i \circ d^i + d'^{i-1} \circ h_2^{i-1}$, then $\phi := \lambda_1\phi_1 + \lambda_2\phi_2$ is given by $\phi^i := h^i \circ d^i + d'^{i-1} \circ h^{i-1}$, where $h^i := \lambda_1 h_1^i + \lambda_2 h_2^i$.

LEMMA E.8. *For a null homotopic morphism, the associated map in cohomology is trivial.*

PROOF. With the above notation, if $x \in Z^i$:

$$\phi^i(x) = h^i(d^i(x)) + d'^{i-1}(h^{i-1}(x)) = d'^{i-1}(h^{i-1}(x)) \in B'^i.$$

Now, since $\phi^i(Z^i) \subset B'^i$, the induced map $H^i = Z^i/B^i \to B'^i = Z'^i/B'^i$ is trivial. $\square$

E.3.2. Homotopy and cohomology. We say that morphisms $\phi, \psi : C^\bullet \to C'^\bullet$ are *homotopic* if $\psi - \phi$ is null homotopic. Then:

$$H^i(\phi^i) - H^i(\psi^i) = H^i(\phi^i - \psi^i) = 0$$

by the lemma, i.e., *homotopic morphisms induce the same maps on cohomology.*

EXAMPLE E.9. Let a diagram:

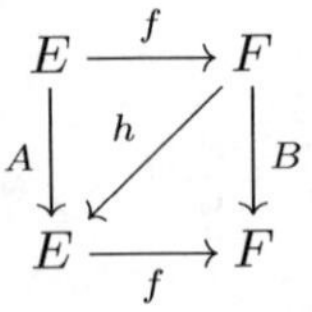

in which we suppose $A = h \circ f$ and $B = f \circ h$. Then the induced maps $\overline{A} : \operatorname{Ker} f \to \operatorname{Ker} f$ and $\overline{B} : \operatorname{Coker} f \to \operatorname{Coker} f$ are trivial. Indeed h (completed by zero maps) induces a null homotopic endomorphism of the associated complex. If $u : E \to E$ and $v : F \to F$ are such that $u - \mathrm{Id}_E = h \circ f$ and $v - \mathrm{Id}_F = f \circ h$, then the induced maps $\overline{u} : \operatorname{Ker} f \to \operatorname{Ker} f$ and $\overline{v} : \operatorname{Coker} f \to \operatorname{Coker} f$ are identities. This is in essence the argument used in 7.1.5.4 (with XX', UU' in place of u, v).

APPENDIX F

Base Change and Tensor Products (and some more facts from linear algebra)

F.1. Base change for vector spaces

The general references here are **[37, 127, 145]**.

Let K'/K a field extension (recall that all our fields are commutative). If V' is a K'-vector space, it can be seen as a K-vector space V by *restriction of scalars*; if necessary (that is, if we wish to distinguish those two avatars of the same group V'), we shall write $V'_{(K)} := V$. If (e'_i) is a basis of V' over K and if (u_j) is a basis of K' over K, then $(u_j v_i)$ is a basis of $V'_{(K)}$ over K. In particular:

$$\dim_K V'_{(K)} = [K' : K] \dim_{K'} V'.$$

Conversely, if we start from a a K-vector space V, there is in general no natural way to consider it as a K'-vector space, i.e., to have K' operate on it by left multiplication. Instead, we associate canonically to V a K'-vector space V' which contains V. This can be done using tensor products (see F.2) but we find it more convenient to introduce it axiomatically here. So we admit:

PROPOSITION F.1. *We can associate to every K-vector space V a K'-vector space V' denoted $K' \otimes_K V$ such that:*

(i) *V is a sub K-vector space of $V'_{(K)}$;*

(ii) *Every K-basis (e_i) of V is a K'-basis of V'.*

□

Practically, every element of V' can be written as a linear combination $\sum \lambda_i v_i$, all $\lambda_i \in K'$, all $v_i \in V$; if the v_i are K-free, this decomposition is unique.

EXAMPLE F.2. $K' \otimes_K K^n = K'^n$. More generally, $K' \otimes_K \mathrm{Mat}_{n,p}(K) = \mathrm{Mat}_{n,p}(K')$. Similarly, $K' \otimes K[X] = K'[X]$.

CAUTION F.3. Although $\mathbf{R} \otimes_{\mathbf{Q}} \mathbf{Q}^n$ may be identified with $\mathbf{R}^n$ for every $n \in \mathbf{N}$, and $\mathbf{R} \otimes_{\mathbf{Q}} \mathbf{Q}^{(\mathbf{N})}$ may likewise be identified with $\mathbf{R}^{(\mathbf{N})}$ (space of sequences with finite support), $\mathbf{R} \otimes_{\mathbf{Q}} \mathbf{Q}^{\mathbf{N}}$ is identified with the $\mathbf{R}$-subspace of $\mathbf{R}^{\mathbf{N}}$ made of sequences which are finite linear combinations of rational sequences. In particular the real sequence $(\sqrt{n})_{n \in \mathbf{N}}$ belongs to $\mathbf{R}^{\mathbf{N}}$ but not to $\mathbf{R} \otimes_{\mathbf{Q}} \mathbf{Q}^{\mathbf{N}}$. Similarly, while $\mathbf{R} \otimes \mathbf{Q}[X] = \mathbf{R}[X]$ (as we just saw), $\mathbf{R} \otimes \mathbf{Q}[[X]]$ is a strict subspace of $\mathbf{R}[[X]]$.

Note that the operations of restriction and extension of scalars are not inverse to each other. For instance, if V' a n-dimensional $\mathbf{C}$-vector space, then $V'_{(\mathbf{R})}$ has dimension $2n$ over $\mathbf{R}$, so $\mathbf{C} \otimes_{\mathbf{R}} V'_{(\mathbf{R})}$ has dimension $2n$ over $\mathbf{C}$, so it is not isomorphic to V'. Similarly, if V is a n-dimensional $\mathbf{R}$-vector space, then $\mathbf{C} \otimes_{\mathbf{R}} V$ is a n-dimensional $\mathbf{C}$-vector space, so $(\mathbf{C} \otimes_{\mathbf{R}} V)_{(\mathbf{R})}$ has dimension $2n$ over $\mathbf{R}$, so it is not isomorphic to V. (For a general K-extension K', replace 2 by $[K' : K]$.) The precise relationship between extension and restriction of scalars is the *adjunction formula*, involving a space V over K and a space W' over K':

$$\mathscr{L}_K(V, W'_{(K)}) = \mathscr{L}_{K'}(K' \otimes_K V, W').$$

Equality here means that there is a canonical isomorphism between those two K-linear spaces. Practically, every K-linear map $f : V \to W'$ (rigorously the target should be written $W'_{(K)}$) extends uniquely to a K'-linear map $f : K' \otimes_K V \to W'$ in the following way: if (e_i) is a K-basis of V, thus a K'-basis of $K' \otimes_K V$, then we set $f'(\sum \lambda'_i e_i) := \sum \lambda'_i \otimes f(e_i)$.

Just from the axiomatic description in the proposition, we easily draw the following facts (beyond the above adjunction formula):

(1) Every K-linear map $f : V \to W$ extends uniquely to a K'-linear map $K' \otimes_K V \to K' \otimes_K W$ sending $\sum \lambda_i v_i$ to $\sum \lambda_i f(v_i)$ (same notations as before). In this way, we get a covariant functor from the category of K-spaces to the category of K'-spaces. That functor is additive: $K' \otimes_K (V_1 \oplus V_2) = (K' \otimes_K V_1) \oplus (K' \otimes_K V_2)$. It is also exact; in particular, if $V_1 \subset V$ (subspace), then $K' \otimes_K V_1 \subset K' \otimes_K V$ and $K' \otimes_K (V/V_1) = (K' \otimes_K V)/(K' \otimes_K V_1)$.

(2) There is an injective K'-linear map:

$$K' \otimes_K \mathscr{L}_K(V, W) \to \mathscr{L}_K(K' \otimes_K V, K' \otimes_K W).$$

If $[K' : K] < \infty$ or $\dim_K V < \infty$, then it is an isomorphism. (Using the previous exercise, we see that, taking $K := \mathbf{Q}$, $K' := \mathbf{R}$, $V := \mathbf{Q}^{(\mathbf{N})}$ and $W := \mathbf{Q}$, the map would not be onto.)

(3) Let K''/K' and K'/K successive extensions. Then:

$$K'' \otimes_{K'} (K' \otimes_K V) = K'' \otimes_K V.$$

If, as in point (1) above, $V_1 \subset V$ (subspace), the subspace $V'_1 := K' \otimes_K V_1$ of $V' := K' \otimes_K V$ is generated by the subset $V_1 \subset V'$, so we sometimes write $V'_1 = K'V_1$. It also follows from point (3) that the functor $V \rightsquigarrow K' \otimes_K V$ is faithful, i.e., the map $\mathscr{L}_K(V, W) \to \mathscr{L}_K(K' \otimes_K V, K' \otimes_K W)$ is injective. And from that (applied to the identity map) follows in turn that $K' \otimes_K V$ is trivial only if V is.

F.2. Tensor products of vector spaces

To any two K-vector spaces V, W is associated, by a construction that we will not recall, their *tensor product*, a K-vector space $V \otimes_K W$ together with a bilinear map:

$$V \times W \to V \otimes_K W, (x, y) \mapsto x \otimes y.$$

They together satisfy a "universal property" which completely, and, in some sense, uniquely characterizes them: *Any bilinear map $V \times W \to U$ can uniquely be written $(x, y) \mapsto f(x \otimes y)$ for some linear map $f : V \otimes W \to U$.*

There are two useful ways to express that universal property:

(1) Note that, for every linear map $f : V \otimes W \to U$, the map $\phi_f : (x, y) \mapsto f(x \otimes y)$ (the notation ϕ_f is provisional) from $V \times W$ to U is bilinear. So we have a map: $f \mapsto \Phi_f, \mathscr{L}_K(V \otimes W, U) \Rightarrow \mathrm{Bil}_K(V, W; U)$. The universal property says that the map $f \mapsto \Phi_f$ is a bijection (thus an isomorphism, since it is clearly K-linear).

(2) There is a commutative diagram:

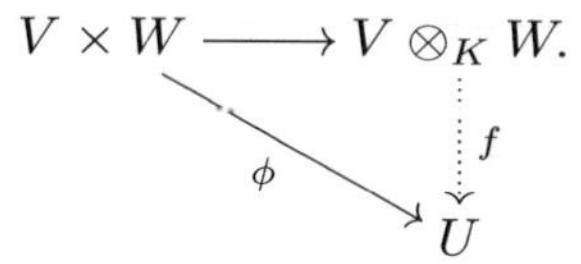

The horizontal map is the structural map $V \times W \to V \otimes_K W$ (the one that defines the tensor product). The diagram says that for any bilinear ϕ there is a unique linear f. The arrow for f is dotted because it does not belong to data, it is a result.

The bilinearity of $V \times W \to V \otimes_K W$ is expressed by the following rules of computation on *pure tensors* (those are the elements of the form $x \otimes y$). Let $x, x' \in V$, $y, y' \in W$ and $\lambda \in K$; then:

$$\lambda x \otimes y = (\lambda x) \otimes y = x \otimes (\lambda y), \quad (x + x') \otimes y = x \otimes y + x' \otimes y, \quad x \otimes (y + y') = x \otimes y + x \otimes y'.$$

EXAMPLE F.4. (i) The map $x \mapsto 1 \otimes x$ is an isomorphism from V to $K \otimes_K V$. We shall identify them. Likewise, we shall identify $V \otimes_K K$ and V.

(ii) More generally one can identify $K^n \otimes V$ and $V \oplus \cdots \oplus V$ (n terms). Writing $(e_i)_{i=1,\dots,n}$ the canonical basis of K^n (so each e_i has ith component 1 and all other components 0), the map from $K^n \otimes V$ to $V \oplus \cdots \oplus V$ sends $e_i \otimes v$ to $0 \oplus \cdots \oplus v \oplus \cdots \oplus 0$ (v is the ith term).

EXERCISE F.5. Describe the map from $V \oplus \cdots \oplus V$ to $K^n \otimes V$.

Now let $f : V \to V'$ and $g : W \to W'$ two linear maps. Then the map $V \times W \to V' \otimes_K W'$, $(x, y) \mapsto f(x) \otimes g(y)$ is bilinear, so, by the universal property, there is a unique linear map $V \otimes_K W \to V' \otimes_K W'$ factoring it. We call it $f \otimes g$ and see that iot is entirely characterized by the relation:

$$\forall x \in V\ ,\ \forall y \in W\ ,\ (f \otimes g)(x \otimes y) = f(x) \otimes g(y).$$

Then one easily checks that the tensor product is a *bifunctor*, covariant in each argument and in some sense bilinear:

- Fixing V, we see that $W \rightsquigarrow V \otimes_K W$, $g \rightsquigarrow \mathrm{Id}_V \otimes g$ is a covariant K-linear functor from K-vector spaces to K-vector spaces;
- fixing W, we see that $V \rightsquigarrow V \otimes_K W$, $f \rightsquigarrow f \otimes \mathrm{Id}_W$ is a covariant K-linear functor from K-vector spaces to K-vector spaces.

As for the *structure* of $V \otimes_K W$, one can prove the following.

PROPOSITION F.6. *If $(e_i)_{i \in I}$ is a basis of V and $(f_j)_{j \in J}$ is a basis of W, then $(e_i \otimes f_j)_{(i,j) \in I \times J}$ is a basis of $V \otimes_K W$.*

Consequences then flow:

COROLLARY F.7. *If V and W are finite-dimensional, so is $V \otimes_K W$ and:*

$$\dim_K V \otimes_K W = (\dim_K V)(\dim_K W).$$

COROLLARY F.8. *If $(x_i)_{i \in I}$ is a free family in V and $(y_j)_{j \in J}$ a free family in W, then $(x_i \otimes y_j)_{(i,j) \in I \times J}$ is a free family in $V \otimes_K W$. As a consequence:*

$$x \otimes y = 0 \Longleftrightarrow x = 0 \text{ or } y = 0; \quad V \otimes_K W = \{0\} \Longleftrightarrow V = \{0\} \text{ or } W = \{0\}.$$

COROLLARY F.9. *If $(x_i)_{i\in I}$ is a generating family in V and $(y_j)_{j\in J}$ a generating family in W, then $(x_i \otimes y_j)_{(i,j)\in I\times J}$ is a generating family in $V\otimes_K W$.*

In particular, the family of all pure tensors $x\otimes y$ is generating in $V\otimes_K W$.

Note that generally all tensors are not pure.

EXERCISE F.10. (i) If $\dim_K V = 1$ or $\dim_K W = 1$, then show that all tensors are pure.

(ii) If $\dim_K V \geq 2$ and $\dim_K W \geq 2$, show that it is false.

(For both statements, use the corollaries above.)

F.3. Matricial view

Let us now take $V := K^n$, with canonical basis $(e_1,\dots,e_n)$ and $W := K^p$, with canonical basis $(f_1,\dots,f_p)$. Then $(e_i\otimes f_j)_{\substack{i=1,\dots,n\\ j=1,\dots,p}}$ is a basis of $K^n\otimes_K K^p$. In order to define an isomorphism $K^n \otimes_K K^p \xrightarrow{\sim} K^{np}$, we have to put a total order on that basis and write it $(g_1,\dots,g_{np})$. This amounts to choose a bijection $\tau : \{1,\dots,n\}\times\{1,\dots,p\}\to\{1,\dots,np\}$ and then to set $g_{\tau(i,j)} := e_i\otimes f_j$. There are two natural ways to do this. The first is as follows: $(1,1)\leftrightarrow 1$, $(2,1)\leftrightarrow 2$, ..., $(n,1)\leftrightarrow n$, $(1,2)\leftrightarrow n+1$, etc. Graphically, it amounts to visit the (i,j) array by going down column after column. The corresponding map $\{1,\dots,n\}\times\{1,\dots,p\}$ is $(i,j)\mapsto i+n(j-1)$. The inverse bijection is $k\mapsto(1+i',1+j')$ where one has performed the euclidean division $k-1 = j'n+i'$.

EXERCISE F.11. Describe similarly the second natural bijection.

Let us denote $\tau_{n,p}$ the above bijection. If we have three natural integers m,n,p, we find that there are two bijections from $\{1,\dots,m\}\times\{1,\dots,n\}\times\{1,\dots,p\}$ to $\{1,\dots,mnp\}$ depending on the order of association $(E\times F)\times G$ or $E\times(F\times G)$. One bijection is $(i,j,k)\mapsto \tau_{mn,p}(\tau_{m,n}(i,j),k)$; the other is $(i,j,k)\mapsto\tau_{m,np}(i,\tau_{n,p}(j,k))$. The divine surprise is that they coincide, so that defining consistently the isomorphisms $K^n\otimes_K K^p\xrightarrow{\sim}K^{np}$ with $\tau_{n,p}$ yields well defined isomorphisms $K^m\otimes_K K^n\otimes_K K^p\xrightarrow{\sim}K^{mnp}$. From now on, we use such identifications without further notice.

EXERCISE F.12. Discuss identification of $V\otimes_K W$ with $W\otimes_K V$ and then the same in terms of K^n, K^p.

Now let linear maps $f : K^n\to K^{n'}$ and $g : K^p\to K^{p'}$ be respectively encoded in matrices $A\in \mathrm{Mat}_{n',n}(K)$ and $B\in\mathrm{Mat}_{p',p}(K)$. Let (e_i), $(e'_{i'}$, (f_j), $(f'_{j'})$ the respective canonical bases of K^n, $K^{n'}$, K^p and $K^{p'}$. Then, writing $A=(a_{i',i})$ and $B=(b_{j',j})$, we have:

$$\forall i\in\{1,\dots,n\}\,,\; f(e_i) = \sum_{i'=1}^{n'} a_{i',i}e'_{i'}\; \forall j\in\{1,\dots,p\}\,,\; g(f_j) = \sum_{j'=1}^{p'} b_{j',j}f'_{j'},$$

which together imply:

$$(f\otimes g)(e_i\otimes f_j) = \sum_{i'=1,j'=1}^{i'=n',j'=p'} a_{i',i}b_{j',j}e'_{i'}\otimes f'_{j'},$$

so, in some sense, the "matrix" of $f\otimes g$ relative to the bases $(e_i\otimes f_j)_{\substack{i=1,\dots,n\\ j=1,\dots,p}}$ of $K^n\otimes_K K^p$ and $(e'_{i'}\otimes f'_{j'})_{\substack{i'=1,\dots,n'\\ j'=1,\dots,p'}}$ of $K^{n'}\otimes_K K^{p'}$ is $(c_{(i',j'),(i,j)})_{\substack{(i',j')\in\{1,\dots,n'\}\times\{1,\dots,p'\}\\ (i,j)\in\{1,\dots,n\}\times\{1,\dots,p\}}}$, where $c_{(i',j'),(i,j)} := a_{i',i}b_{j',j}$.

Such a "matrix" (actually a *tensor*) with its bi-indexation by pairs of pairs $((i',j'),(i,j))$ maybe inconvenient to deal with, but we can resort to the above identifications of $K^n \otimes_K K^p$ with K^{np} and $K^{n'} \otimes K^{p'}$ with $K^{n'p'}$. In that model, $f \otimes g : K^n \otimes_K K^p \to K^{n'} \otimes_K K^{p'}$ corresponds to a well defined linear map $K^{np} \to K^{n'p'}$, itself encoded by a matrix in $\mathrm{Mat}_{n'p',np}(K)$. We call that matrix $A \otimes B$. It can be written $(c_{k',k})_{\substack{k'=1,\ldots,n' \\ k=1,\ldots,n}}$, where[1] $c_{\tau_{n',p'}(i',j'),\tau_{n,p}(i,j)} := a_{i',i}b_{j',j}$.

Practically, using the chosen identifications, you can check that $A \otimes B$ has the following block structure:

$$A \otimes B = \begin{pmatrix} Ab_{1,1} & \ldots & Ab_{1,p} \\ \vdots & \ddots & \vdots \\ Ab_{p',1} & \ldots & Ab_{p',p} \end{pmatrix}$$

The other identification would have given $\begin{pmatrix} a_{1,1}B & \ldots & a_{1,n}B \\ \vdots & \ddots & \vdots \\ a_{n',1}B & \ldots & a_{n',n}B \end{pmatrix}$.

EXERCISE F.13. Describe $\begin{pmatrix} 1 & 0 \\ 0 & 1 \end{pmatrix} \otimes \begin{pmatrix} 1 & 1 \\ 1 & 1 \end{pmatrix}$ and $\begin{pmatrix} 1 & 1 \\ 1 & 1 \end{pmatrix} \otimes \begin{pmatrix} 1 & 0 \\ 0 & 1 \end{pmatrix}$ under both conventions.

F.4. Some miscellaneous useful facts from linear algebra

F.4.1. Some spectra. Let $A \in \mathrm{Mat}_m(\mathbf{C})$ and $B \in \mathrm{Mat}_n(\mathbf{C})$. Then the map $M \mapsto AM - MB$ is an endomorphism $\Phi_{A,B}$ of the linear space $\mathrm{Mat}_{m,n}(\mathbf{C})$. We denote $\{\{\lambda_1, \ldots, \lambda_m\}\}$ and $\{\{\mu_1, \ldots, \mu_n\}\}$ the respective spectra of A and B seen as *multisets* (this means that values may appear more than one time). In other words, their respective characteristic polynomials are:

$$\chi_A(T) := \det(TI_m - A) = \prod_{i=1}^{m}(T - \lambda_i) \text{ and } \chi_B(T) := \det(TI_n - B) = \prod_{j=1}^{n}(T - \mu_j).$$

We are going to prove that the spectrum of $\Phi_{A,B}$ is:

$$\mathrm{Sp}\ \Phi_{A,B} = \{\{\lambda_i - \mu_j \mid 1 \le i \le m, 1 \le j \le n\}\} = \{\{\lambda_1 - \mu_1, \ldots, \lambda_m - \mu_n\}\}.$$

First assume that A and B are upper triangular, with respective diagonals $(\lambda_1, \ldots, \lambda_m)$ and $(\mu_1, \ldots, \mu_n)$, and call $(E_{i,j})_{\substack{1 \le i \le m \\ 1 \le j \le n}}$ the canonical basis of $\mathrm{Mat}_{m,n}(\mathbf{C})$. Then ("l.c." stands for "linear combination"):

$$\Phi_{A,B}(E_{i,j}) = (\lambda_i - \mu_j)E_{i,j} + \text{ l.c. of the } E_{l,j} \text{ s.t. } l < i \text{ and the } E_{i,m} \text{ s.t. } m > j.$$

Therefore, if the $E_{i,j}$ are totally ordered in a way compatible with those relations (exercise: describe the relevant total order), then $\Phi_{A,B}$ is represented by an upper triangular matrix with diagonal made up of all the $\lambda_i - \mu_j$, which proves our contention in that case. The general case boils down to the previous one thanks to the following equality:

$$\Phi_{PAP^{-1},QBQ^{-1}} = \Psi_{P,Q} \circ \Phi_{A,B} \circ \Psi_{P,Q}^{-1}, \text{ where } \begin{cases} P \in \mathrm{GL}_m(\mathbf{C}), Q \in \mathrm{GL}_n(\mathbf{C}), \\ \Psi_{P,Q} : M \in \mathrm{Mat}_{m,n}(\mathbf{C}) \mapsto PMQ^{-1}. \end{cases}$$

[1] This flow of stacked bi-indices is temporary and just meant to illustrate the possibility of using $A \otimes B$ as a *bona fide* matrix. In practice we will never have to do it in such explicit terms.

F.4.2. Around multiplicative Dunford decomposition. The (usual, or additive) Dunford decomposition of a matrix $A \in \mathrm{Mat}_n(\mathbf{C})$ is $A = A_s + A_n$, where A_s is semi-simple (i.e., diagonalisable), A_n is nilpotent and $A_s A_n = A_n A_s$. It is unique and satisfies the following properties: $A_s, A_n \in \mathbf{C}[A]$; and A_s has the same spectrum as A (as multisets).

The multiplicative Dunford decomposition of a matrix $A \in \mathrm{GL}_n(\mathbf{C})$ is easily deduced: $A = A_s A_u$, where A_s is semi-simple, A_u is unipotent and $A_s A_u = A_u A_s$. It is unique, the A_s component is the same as in the additive Dunford decomposition, $A_u = I_n + A_s^{-1} A_n$ and $A_u \in \mathbf{C}[A]$. Beware the index n in A_n means "nilpotent" while in I_n it denotes the size of the identiy matrix; sorry for that!. (Multiplicative Dunford decomposition is also called Jordan decomposition in the Anglo-Saxon world, but we prefer to keep that terminology for linear algebraic groups.)

F.4.3. Around functions of semi-simple matrices. Let $A = PDP^{-1}$, where $D = \mathrm{Diag}(c_1, \ldots, c_n)$ (a diagonal matrix). Let f a function with values in some $\mathbf{C}$-algebra $\mathcal{A}$, defined on a subset of $\mathbf{C}$ containing the spectrum of A. Then we *define*:

$$f(A) := P\mathrm{Diag}(f(c_1), \ldots, f(c_n))P^{-1}.$$

Indeed, this matrix is independent of the particular choice of a diagonalisation of A, which can be seen (for instance) as follows: if F is any polynomial interpolating f on the set $\{c_1, \ldots, c_n\}$, then:

$$f(A) = P\mathrm{Diag}(f(c_1), \ldots, f(c_n))P^{-1} = P\mathrm{Diag}(F(c_1), \ldots, F(c_n))P^{-1} = F(A).$$

(This argument[2] was kindly provided by Charles Torossian.) Since $f(A) \in \mathbf{C}[A]$, it commutes with all elements of $\mathbf{C}[A]$.

F.4.4. Around functions of general matrices. Let $A := A_s A_u$ a multiplicative Dunford decomposition in $\mathrm{GL}_n(\mathbf{C})$. If the definition domain of f contains $\mathrm{Sp}\, A$, we sometimes set $f(A) := f(A_s)A_u = A_u f(A_s)$ (commutation is consequence of $f(A_s), A_u \in \mathbf{C}[A]$). Then one easily proves (using polynomial interpolation) that:

(1) $AB = BA \Longrightarrow f(A)g(B) = g(B)f(A)$ whenever defined;
(2) $RA = BR \Longrightarrow Rf(A_s) = f(B_s)R$ and $RA_u^\lambda = B_u^\lambda R$.

Complex powers of a unipotent matrix $U = I_n + N$, N nilpotent, are obtained from generalized binomial formula:

$$U^\lambda = I_n + \lambda N + \binom{\lambda}{2} N^2 + \cdots \text{(actually a finite sum)}.$$

F.4.5. Lie algebras of unipotent linear algebraic groups. Let G an algebraic subgroup of $\mathrm{GL}_n(\mathbf{C})$ (this means that it is the zero set of a family of polynomials in the coefficients of the matrices). Assume that all elements of G are unipotent. Then:

$$\mathfrak{g} := \mathrm{Lie}(G) = \{M \in \mathrm{Mat}_n(\mathbf{C}) \mid \exp(M) \in G\}$$

is a Lie subalgebra of $\mathrm{Mat}_n(\mathbf{C})$, i.e., it is a linear subspace stable under the Lie bracket $[M, N] := MN - NM$. The exponential map is bijective from $\mathfrak{g}$ to G with inverse the logarithm map $U := I_n + N \mapsto N - N^2/2 + \cdots$.

[2]Another proof, of a more combinatorial nature, is given in **[222]**; both are very special instances of functional calculus.

APPENDIX G

Tannaka Duality (without schemes)

We present a very simplified version of the theory of Tannakian categories, replacing group schemes by proalgebraic groups and restricting to **C**-linear categories (so all our algebraic and proalgebraic groups are over **C**). General references are [**53, 56, 213, 250**]; also see [**72**] for the underlying historical motivation.

CAUTION G.1. All representations considered are meant in *finite-dimensional* **C**-*linear spaces*. This is reflected in the notation $\mathcal{R}ep^f_{\mathbf{C}}(\Gamma)$ for the category of plain representations of an abstract group Γ (i.e., without supplementary structure); but the category $\mathcal{R}ep_{rat}(G)$ of rational representations of a proalgebraic goup G obeys the same principle, although the notation does not reflect it.

G.1. The weak form of Tannaka duality

G.1.1. Proalgebraic groups and rational representations.

G.1.1.1. A linear algebraic group is a subgroup of some $\mathrm{GL}_n(\mathbf{C})$ (or equivalently some $\mathrm{GL}(V)$, V a finite-dimensional **C**-linear space) that can be defined as the zero set of a family of polynomial equations. A morphism (or rational map) of linear groups is a morphism of groups that can be expressed (in coordinates in some $\mathrm{GL}_n(\mathbf{C})$) by polynomial maps. Note that in this context a polynomial map $\mathrm{GL}_n(\mathbf{C}) \to \mathbf{C}$ is understood as a map $A := (a_{i,j}) \mapsto F(A)/(\det A)^m$ where F is a polynomial in the coefficients $a_{i,j}$ and $m \in \mathbf{N}$.

A rational representation of a linear algebraic group is a rational map $G \to \mathrm{GL}_n(\mathbf{C})$ or $G \to \mathrm{GL}(V)$. In the latter form, it can be seen as a linear representation $G \times V \to G$, with the usual axioms and the constraint that $(A, v) \mapsto Av$ be polynomial in the coefficients of A and the coordinates of v (negative powers of $\det A$ being authorized). Note that in the above definitions, the case of $\mathrm{GL}(V)$ reduces to that of $\mathrm{GL}_n(\mathbf{C})$ modulo the choice of a basis and the definitions do not depend on the particular choice.

G.1.1.2. A proalgebraic group is a group G endowed with a compatible family of group morphisms $f_i : G \to G_i$, where the $G_i, f_i^j : G_j \to G_i$ form an inverse system of linear algebraic groups and rational maps; compatibility means that $f_i = f_i^j \circ f_j$ for all $j > i$. The family of all (f_i) may be thought as a kind of atlas (as in B.1.1) and different families may define the same structure of proalgebraic group on G: this can be best understood through the definition of morphisms and isomorphisms herebelow, we leave it to the reader to make the condition explicit.

If H is an algebraic group and G a proalgebraic group, as above, a group morphism $H \to G$ is said to be rational (or a morphism of proalgebraic groups) if all compositions $H \to G_i$ are. Then if H is a proalgebraic group with structure defined by a family of $H \to H_k$, a group morphism $H \to G$ is said to be rational if it factors through some $H_k \to G$ which is rational. In other words, if $G = \varprojlim G_i$

and $H = \varprojlim H_k$ (in the sense that those inverse limits of algebraic groups give G and H their structures of proalgebraic groups), then:

$$\mathrm{Hom}(H,G) = \varinjlim_{H_k} \varprojlim_{G_i} \mathrm{Hom}(H_k, G_i).$$

Proalgebraic groups with rational maps form a category.

G.1.1.3. Let Γ an arbitrary "discrete" or "abstract" group, i.e., a mere group with no additional structure. Let $\Gamma^\vee := \mathrm{Hom}_{gr}(\Gamma, \mathbf{C}^*)$ the group of all group morphisms from Γ to $\mathbf{C}^*$ (structure induced by multiplication in the target: fg is defined as $x \mapsto f(x)g(x)$). From the fact that $\Gamma = \bigcup \Gamma_i$ is the union (and therefore the direct limit) of its finitely generated subgroups, one deduces that:

$$\Gamma^\vee = \varprojlim \Gamma_i^\vee.$$

On the other hand, if Γ_0 is a finitely generated group and Γ_0^{ab} its abelianized (hence a finitely generated abelian group), then $\Gamma_0^{ab} \simeq \mathbf{Z}^r \times T$, where T is a product of cyclic groups C_j. Now:

$$\Gamma_0^\vee = (\Gamma_0^{ab})^\vee \simeq (\mathbf{Z}^r)^\vee \times T^\vee = \mathbf{C}^{*r} \times \prod C_j^\vee,$$

and each $C_j^\vee$ is isomorphic to a cyclic group of roots of unity in $\mathbf{C}^*$. This makes $\Gamma_0^\vee$ an algebraic group and one can show that its algebraic group structure is intrinsic.

Combining with the inverse limit above, one deduces that $\Gamma^\vee$ has a natural structure of proalgebraic group.

G.1.1.4. A rational representation of a proalgebraic group G is a rational map $G \to \mathrm{GL}_n(\mathbf{C})$ or $G \to \mathrm{GL}(V)$ which is at the same time a group morphism; we then say that the representation is in V. The group G being fixed, a morphism of representations in V and in W is a linear map $u : V \to W$ intertwining the representations, i.e., such that $u(g.x) = g.u(x)$ for all $g \in G$, $x \in V$.

The rational representations of the proalgebraic group G form a $\mathbf{C}$-linear abelian category $\mathcal{Rep}_{rat}(G)$. It comes with "tensor constructions":

(1) There is a unit object, which is the space $V := \mathbf{C}$ together with the trivial representation (all $g \in G$ are mapped to the identity of $\mathbf{C}$); alternatively, it is the trivial morphism $G \to \mathrm{GL}(\mathbf{C}) = \mathbf{C}^*$ (all $g \in G$ are mapped to 1).
(2) Any representation of G in V admits a dual, the contragredient representation of G in the dual $V^\vee$ (if $\rho : G \to \mathrm{GL}(V)$, then $\rho^\vee : G \to \mathrm{GL}(V^\vee)$ sends $g \in G$ to the contragredient ${}^t\rho(g)^{-1}$).
(3) There is a tensor product of representations of G in V, W. It is defined by the formulas $g.(x \otimes y) := (g.x) \otimes (g.y)$.
(4) There is an internal Hom. With the above notations, it is a representation of G in $\mathscr{L}_{\mathbf{C}}(V,W)$ such that, if $g \in G$ and $u \in \mathscr{L}_{\mathbf{C}}(V,W)$, then $g.u \in \mathscr{L}_{\mathbf{C}}(V,W)$ is the map $x \mapsto g.u(g^{-1}x)$.

All basic properties seen in Appendix F (neutral element, associativity, adjunctions ...) remain valid here, thus making $\mathcal{Rep}_{rat}(G)$ a *tensor category*.

G.1.1.5. The category $\mathcal{Rep}_{rat}(G)$ also comes equipped with the forgetful functor ω to the category $\mathcal{Vect}_{\mathbf{C}}^f$ of finite-dimensional $\mathbf{C}$-linear spaces, sending a representation of G in V to its underlying space V (and a morphism of such representations to the underlying linear map $u \in \mathscr{L}_{\mathbf{C}}(V,W)$). This functor is faithful, exact, $\mathbf{C}$-linear and $\otimes$-compatible: it is a *fibre functor*.

G.1.2. Automorphisms of the fibre functor on $\mathcal{Rep}_{rat}(G)$.

G.1.2.1. Fix $g \in G$. For every rational representation $\rho : G \to \mathrm{GL}(V)$, the linear map $x \mapsto g.x$ is an automorphism of $V = \omega(\rho)$. If $\rho' : G \to \mathrm{GL}(V')$ is another rational representation, and $u \in \mathscr{L}_{\mathbf{C}}(V, V')$ underlies a morphism $\rho \to \rho'$, we have a commutative diagram:

$$\begin{array}{ccc} V & \xrightarrow{\rho(g)} & V \\ {\scriptstyle u}\downarrow & & \downarrow{\scriptstyle u.} \\ V' & \xrightarrow{\rho'(g)} & V' \end{array}$$

Therefore we have an automorphism (an invertible natural transformation) of the functor ω. It is, in a sense easy to make explicit (do it!), $\otimes$-compatible. So it is a tensor automorphism, an element of $\mathrm{Aut}^{\otimes}(\omega)$. Denote it (temporarily) ϕ_g. Then there is an obvious group structure on $\mathrm{Aut}^{\otimes}(\omega)$ and the map $g \mapsto \phi_g$ is (this is equally obvious) a group morphism.

G.1.2.2. If $\phi \in \mathrm{Aut}^{\otimes}(\omega)$ is any tensor automorphism, it induces, for every representation $\rho : G \to \mathrm{GL}(V)$, an element of $\mathrm{Aut}(\omega(\rho)) = \mathrm{GL}(V)$. So the group $\mathrm{Aut}^{\otimes}(\omega)$ comes equipped with a family of group morphisme to linear groups $\mathrm{GL}(V)$. It can be proved that the image subgroups of the $\mathrm{GL}(V)$ form an inverse system of algebraic groups and that the above family of maps define a structure of proalgebraic group on $\mathrm{Aut}^{\otimes}(\omega)$. Moreover, the group morphism $g \mapsto \phi_g$ from G to $\mathrm{Aut}^{\otimes}(\omega)$ is rational.

The weak form (or let's say the easy part) of Tannaka duality says that the structure of proalgebraic group on G can be recovered from the tensor category $\mathcal{R}ep_{rat}(G)$ of its rational representations, together with its fibre functor ω:

THEOREM G.1. *The map $G \to Aut^{\otimes}(\omega)$ is an isomorphism of proalgebraic groups.*

The strong form (or let's say the harder part) of Tannaka duality says that conversely any tensor category equipped with a fibre functor ω and subject to some constraints (see G.2) is isomorphic to $\mathcal{R}ep_{rat}(G)$, where $G := \mathrm{Aut}^{\otimes}(\omega)$.

G.2. The strong form of Tannaka duality

G.2.1. Tannakian categories.

G.2.1.1. To begin with, let $\mathscr{C}$ a $\mathbf{C}$-linear abelian category. We suppose that it is endowed with a tensor product and tensor constructions (unit object, dual, internal Hom) subject to the same constraints as already quoted. This is called a rigid tensor category. The rigidity esscentially refers to the adjunction properties involving the dual.

Note that algebraic properties such as $\underline{\mathbf{1}} \otimes X = X$ or $X \otimes (Y \otimes Z) = (X \otimes Y) \otimes Z$ are really more subtle to properly state than it would seem. Indeed, plain equalities of objects seldom have a meaning in general categories. What is implied is (for instance) a collection of isomorphisms $\underline{\mathbf{1}} \otimes X \simeq X$ that are collectively subject to some functorial and compatibility conditions. To get the flavour of those, see the associativity constraint in [**56**, §1, p. 104]. Alternatively, try yourself on the following exercise.

EXERCISE G.2. With the standard (i.e., Bourbaki ...) definition of ordered pairs $(a, b) := \{\{a\}, \{a, b\}\}$, it is not true that $(a, (b, c)) = ((a, b), c)$, so it is not true either (sets X, Y, Z being given) that $X \times (Y \times Z) = (X \times Y) \times Z$. What is true is that there are canonical bijections $\phi_{X,Y,Z} : X \times (Y \times Z) \to (X \times Y) \times Z$.

(i) State and check the functoriality of the collection of all $\phi_{X,Y,Z}$.

(ii) How many cartesian products can be formed with sets X, Y, Z, T in that order? Write down isomorphisms among them made of compositions of bijections such as $\phi_{X,Y,Z}$.

(iii) Form a diagram with all those entities and check commutativity.

G.2.1.2. Now we define a fibre functor on $\mathscr{C}$ over $\mathbf{C}$ as a $\mathbf{C}$-linear exact faithful $\otimes$-compatible functor $\mathscr{C} \rightsquigarrow \mathcal{V}ect^f_{\mathbf{C}}$. Here $\otimes$-compatibility can be abusively expressed as an equality $\omega(X \otimes Y) = \omega(X) \otimes \omega(Y)$ but, in the same spirit as above, it should rather be stated as the existence of an isomorphism of the corresponding bifunctors, moreover subject to some compatibility conditions (see for instance Definition 1.8 in [**56**, §1, p. 113]).

A (neutral) Tannakian category (over $\mathbf{C}$) is a rigid tensor category equipped with a fibre functor over $\mathbf{C}$. We shall omit the word "neutral" (it refers to the fact that the target of ω is the category of spaces over the base field $\mathbf{C}$).

G.2.2. Automorphisms of the fibre functor.

G.2.2.1. Let $(\mathscr{C}, \omega)$ a Tannakian category. An element of $G := \mathrm{Aut}^{\otimes}(\omega)$ can be described as a family of linear automorphisms $\phi_X : \omega(X) \to \omega(X)$ subject to conditions:

(1) Functoriality: if $f : X \to Y$ is a morphism in $\mathscr{C}$, then $\phi_Y \circ \omega(f) = \omega(f) \circ \phi_X$.
(2) Tensor compatibility: up to identifications, $\phi_{X \otimes Y} = \phi_X \otimes \phi_Y$.

If (ϕ_X), (ψ_X) are two such collections, the collection of all $\psi_X \circ \phi_X : \omega(X) \to \omega(X)$ satisfies the same properties and this allows to define a group structure on G.

There are natural mappings $(\phi_X) \mapsto \phi_{X_0}$ from G to the linear groups $\mathrm{Aut}(\omega(X_0)) = \mathrm{GL}(V_0)$ (where $V_0 := \omega(X_0)$) and together those endow G with a structure of proalgebraic group.

G.2.2.2. Let X_0 a fixed object in $\mathscr{C}$ and let $V_0 := \omega(X_0)$. To every $g \in G$ consisting in a collection (ϕ_X) as above, we associate the element $\phi_{X_0} \in \mathrm{GL}(V_0)$. This defines a representation $G \to \mathrm{GL}(V_0)$. That representation is actually rational. If $X_0 \to X_1$ is a morphism in $\mathscr{C}$, the linear map $\omega(X_0) \to \omega(X_1)$ coming from functoriality underlies a morphism of the corresponding representations. In this way we get a functor $\mathscr{C} \rightsquigarrow \mathcal{R}ep_{rat}(G)$. This functor can be seen to be $\mathbf{C}$-linear, exact and $\otimes$-compatible. The strong form (or let's say the harder part) of Tannaka duality says that $\mathscr{C}$ can thereby be identified with the category of rational representations of $\mathrm{Aut}^{\otimes}(\omega)$:

THEOREM G.2. *The functor $\mathscr{C} \rightsquigarrow \mathcal{R}ep_{rat}(G)$ is an equivalence of tensor categories.*

G.2.2.3. The above description relies on the choice of a particular fibre functor. Things can be made more intrinsic in terms of groupoids[1]. Generally speaking, a groupoid[2] consists in a base B, a group $G(a)$ for every $a \in B$ and a set of

[1] And even more intrinsic in terms of "gerbes", but we don't use them: see [**53**, §3.2] or [**41**] for that point of view.

[2] It is customary to define a groupoid as a small category (this means that the class of all its objects is a set) all the arrows of which are invertible; you can check that the following description is equivalent.

isomorphisms $G(a,b)$ from $G(a)$ to $G(b)$ for every $a, b \in B$. This comes with the folloxing axioms: each $G(a,a)$ contains the identity of $G(a)$; composition induces maps $G(a,b) \times G(b,c) \to G(a,c)$; and inversion induces maps $G(a,b) \to G(b,a)$. The groupoid is connected if all $G(a,b)$ are nonempty. A proalgebraic groupoid is one such that all $G(a)$ are proalgebraic groups and all elements of all $G(a,b)$ are rational. A representation of G involves a family of vector spaces $V(a)$ indexed by B and, for every $a, b \in B$, a map (rational if G is taken proalgebraic) from $G(a,b)$ to $\mathscr{L}(V(a), V(b))$, with easy to state compatibilities with compositions.

EXERCISE G.3. Suppose G is a connected proalgebraic groupoid on base B and let $a \in B$. Any rational representation of G induces a representation of $G(a)$. Compare $\mathcal{R}ep_{rat}(G)$ to $\mathcal{R}ep_{rat}(G(a))$.

In the case of a Tannakian category, one can take for B any set of fibre functors (or even the collection of all fibre functors, but this, of course, is certainly not a set) and for $G(\omega, \omega')$ the set $\mathrm{Iso}^{\otimes}(\omega, \omega')$ of all tensor isomorphisms. A general theorem of Deligne says that it is necessarily connected. So, in some sense (see the exercise), the dependence on ω of the group $\mathrm{Aut}^{\otimes}(\omega)$ is unessential. Examples of such bases and groupoids appear in Chapters 13 and 15.

G.2.3. Tannakian Galois groups and groupoids.

G.2.3.1. The group $\mathrm{Aut}^{\otimes}(\omega)$ is called the *(Tannakian) Galois group of* $\mathscr{C}$ and denoted $\mathrm{Gal}(\mathscr{C}, \omega)$. Due to the nonessentiality of the choice of ω mentioned above, we shall rather write it $\mathrm{Gal}(\mathscr{C})$; (and maybe we can even sometimes use the same denotation for any naturally defined Galois groupoid).

If $\mathscr{C}$ and $\mathscr{C}'$ are two Tannakian categories with say fibre functors ω and ω', any tensor compatible functor $F : \mathscr{C} \rightsquigarrow \mathscr{C}'$ induces a map from $\mathrm{Aut}^{\otimes}(\omega')$ to $\mathrm{Aut}^{\otimes}(\omega)$ (to an element of the former described as a collection $(\phi'_{X'})$, setting $\phi_X := \phi'_{F(X)}$, we associate the collection (ϕ_X) describing an element of the latter; actually we have to require that $\omega' \circ F = \omega$, so it is not any F that will do it, but it will be all right in our examples). This yields a rational morphism of proalgebraic groups $\mathrm{Gal}(\mathscr{C}') \to \mathrm{Gal}(\mathscr{C})$. In **[56]** it is shown how properties of the functor are related to properties of the morphism.

G.2.3.2. Let X an object of $\mathscr{C}$. The Tannakian category generated by X, denoted $< X >$, is the full subcategory having as objects all subquotients of "polynomials in $X, X^\vee$, i.e., arbitrary direct sums of tensor products-powers $X^{\otimes m} \otimes (X^\vee)^{\otimes n}$. One can prove that $\mathrm{Gal}(< X >)$ is an algebraic group. More precisely, the natural map from $\mathrm{Gal}(< X >)$ to $\mathrm{GL}(\omega(X))$ is an isomorphism of $\mathrm{Gal}(< X >)$ with an algebraic subgroup of $\mathrm{GL}(\omega(X))$. It is denoted $\mathrm{Gal}(X)$ and seen as the Galois group of X.

The inclusion functor of $< X >$ in $\mathscr{C}$ induces a rational morphism $\mathrm{Gal}(\mathscr{C}) \to \mathrm{Gal}(X)$. That morphism is surjective. If $X' = X \oplus Y$, then $< X >$ injects into $< X' >$ and the corresponding rational morphism $\mathrm{Gal}(X') \to \mathrm{Gal}(X)$ is also surjective. In this way we obtain an inverse system of algebraic groups $\mathrm{Gal}(X)$ with surjective rational morphisms, and then the family of morphisms $\mathrm{Gal}(\mathscr{C}) \to \mathrm{Gal}(X)$ realizes $\mathrm{Gal}(\mathscr{C})$ as the inverse limit of that system:

$$\mathrm{Gal}(\mathscr{C}) = \varprojlim \mathrm{Gal}(X).$$

This induces the proalgebraic structure on $\mathrm{Gal}(\mathscr{C})$.

G.2.3.3. The groupoid of all $\mathrm{Iso}^{\otimes}(\omega, \omega')$ is sometimes called the "(Tannakian) fundamental groupoid of $\mathscr{C}$", and accordingly denoted $\Pi_1(\mathscr{C})$. This is related to a topological language introduced (I believe) by Katz: we consider the collection of all fibre functors as a space (even if it is not a set); and any element of $\mathrm{Iso}^{\otimes}(\omega, \omega')$, resp. of $\mathrm{Aut}^{\otimes}(\omega)$, as a path from ω to ω', resp. as a loop based at ω.

However, it is my feeling (and motivation for much of the Galois theory presented in this book and in [**222**]) that the term "fundamental group" should be reserved to some strictly smaller subgroup or subgroupoid, with probably no algebro-geometric structure of its own, but with some property of density (see Chapters 13 and 15).

APPENDIX H

Čech Cohomology of Abelian Sheaves

In this appendix, we provide the bare necessary information needed to justify some assertions in the text about the cohomology of coherent sheaves on a Riemann surface. We do not give the definition of a coherent sheaf but state enough basic facts (with precise references[1]) to guarantee that the chain of arguments is complete.

The approach to cohomology is (as in all our book!) down-to-earth, in the style of **[229]** (Čech cohomology), which is our main reference (however, you may look at [**150**, §2.2] for a comparable presentation). Moreover, since our main goal is to understand sheaves over the Riemann surfaces $\mathbf{C}^*$, $\mathbf{S}$ or $\mathbf{E}_q$, statements are sometimes less general than would be possible.

Recall that, if $\mathscr{F}$ is a sheaf over a topological space X, one denotes $\Gamma(U, \mathscr{F}) := \mathscr{F}(U)$ the set (group, module ...) of sections of $\mathscr{F}$ over the open subset U of X.

H.1. Coherent sheaves

We let X a topological space, $\mathscr{O}_X$ a sheaf of commutative rings over X and consider only sheaves of $\mathscr{O}_X$-modules. So all words such as free, locally free, coherent ..., refer to that structure; in particular, if X is a Riemann surface and $\mathscr{O}_X$ its sheaf of holomorphic functions (see H.1.2).

H.1.1. Generalities on coherent sheaves.

For the definition of a coherent sheaf, see [**229**, I §2.12 p 208], where it is said that this is a local property.

Let $0 \to \mathscr{F} \to \mathcal{G} \to \mathcal{H} \to 0$ an exact sequence of sheaves. Then, if two of them are coherent, so is the third [**229**, I §2.12 p 208]. In particular, a finite direct sum of coherent sheaves is coherent [**229**, I §2.12 p 209].

As a consequence, of the above facts, if $\mathscr{O}_X$ is coherent, then every locally free sheaf of finite rank is coherent.

H.1.2. The case of a Riemann surface.

THEOREM H.1. *Let X a Riemann surface. Then the sheaf of its analytic functions is coherent.*

This is a particular case of Oka's theorem, which holds for arbitrary complex spaces [**100**, Chapter 2 §5.3].

COROLLARY H.1. *Every locally free sheaf of finite rank over a Riemann surface, in particular the sheaf of sections of every holomorphic vector bundle, is coherent.*

[1] Unhappily I was not able to find a single source sufficient for all needs, so I had to take in some places facts about analytic sheaves, in some other places facts about algebraic sheaves, and, of course, quote [**230**] to ensure the interoperability of both lines of thought.

H.2. Čech cohomology

We follow [**229**, Chapter I §3] (also see Appendix E).

H.2.1. The groups $H^p(\mathfrak{U}, \mathscr{F})$. Let X a topological space and $\mathscr{O}_X$, $\mathscr{F}$ as in H.1.1. To every open covering $\mathfrak{U} := (U_i)_{i\in I}$ of X and every $p \geq 0$, we associate the module of p-cochains:

$$C^p(\mathfrak{U}, \mathscr{F}) := \prod_{\underline{i}\in I^{p+1}} \Gamma(U_{\underline{i}}, \mathscr{F}), \text{ where } U_{\underline{i}} := U_{i_0} \cap \cdots \cap U_{i_p} \text{ whenever } \underline{i} = (i_0, \ldots, i_p).$$

In other words, a p-cochain is a family $(f_{i_0,\ldots,i_p})$ indexed by I^{p+1} and such that every $f_{i_0,\ldots,i_p} \in \Gamma(U_{i_0} \cap \cdots \cap U_{i_p}, \mathscr{F})$.

For every $r = 0, \ldots, p+1$, we let $\rho_r : I^{p+2} \to I^{p+1}$ the map sending $(i_0, \ldots, i_{p+1})$ to $(i_0, \ldots, \hat{i_r}, \ldots, i_{p+1})$ (i.e., we omit component i_r). Then we define $d^p : C^p(\mathfrak{U}, \mathscr{F}) \to C^{p+1}(\mathfrak{U}, \mathscr{F})$ as the mapping[2]:

$$(f_{\underline{i}})_{\underline{i}\in I^{p+1}} \mapsto (g_{\underline{j}})_{\underline{j}\in I^{p+2}}, \text{ where } \forall \underline{j} \in I^{p+2} \ , \ g_{\underline{j}} := \sum_{r=0}^{p+1} (-1)^r f_{\rho_r(\underline{j})}.$$

This makes the family of all $C^p(\mathfrak{U}, \mathscr{F})$ a cochain complex, i.e., $d^p \circ d^{p-1} = 0$ for all $p \geq 1$. So we have:

$$B^p(\mathfrak{U}, \mathscr{F}) := \text{Im } d^{p-1} \subset Z^p(\mathfrak{U}, \mathscr{F}) := \text{Ker } d^p \subset C^p(\mathfrak{U}, \mathscr{F}),$$

and we can define the pth cohomology group:

$$H^p(\mathfrak{U}, \mathscr{F}) := Z^p(\mathfrak{U}, \mathscr{F})/B^p(\mathfrak{U}, \mathscr{F}).$$

It is easily checked that $H^0(\mathfrak{U}, \mathscr{F}) = \Gamma(X, \mathscr{F})$ [**229**, I §3.20 p 213]. In general, elements of $Z^p(\mathfrak{U}, \mathscr{F})$, resp. of $B^p(\mathfrak{U}, \mathscr{F})$ are called *p-cocycles*, resp. *p-coboundaries* and the image of a p-cocycle in $H^p(\mathfrak{U}, \mathscr{F})$ is its *cohomology class.*

REMARK H.2. A 1-cocycle is a family $(f_{i,j})$ such that $f_{i,j} + f_{j,k} = f_{i,k}$. But the three terms belong to to different groups. To make sense of that, it is implicitly understood that all three sections are restricted to $U_i \cap U_j \cap U_k$. Such precautions will remain implicit in all similar cases.

H.2.2. The groups $H^p(X, \mathscr{F})$. We say that covering $\mathfrak{U} := (U_i)_{i\in I}$ is finer than covering $\mathfrak{V} := (V_j)_{j\in J}$ if there is a map $\tau : I \to J$ such that $U_i \subset V_{\tau(i)}$ for all i. Then, for every $p \geq 0$, there is a naturally defined map $C^p(\mathfrak{V}, \mathscr{F}) \to C^p(\mathfrak{U}, \mathscr{F})$ (we let you write it down ...) and it sends $Z^p(\mathfrak{V}, \mathscr{F})$ into $Z^p(\mathfrak{U}, \mathscr{F})$ and $B^p(\mathfrak{V}, \mathscr{F})$ into $B^p(\mathfrak{U}, \mathscr{F})$, so it induces $H^p(\mathfrak{V}, \mathscr{F}) \to H^p(\mathfrak{U}, \mathscr{F})$. Moreover, the latter map does not depend on τ [**229**, I §3.21 p 214]. This way, we get a direct system of groups $H^p(\mathfrak{U}, \mathscr{F})$ (indexed by coverings) [**229**, I §3.22 p 215] and define the cohomology groups:

$$H^q(X, \mathscr{F}) := \varinjlim_{\mathfrak{U}} H^p(\mathfrak{U}, \mathscr{F}).$$

It immediately follows that $H^0(X, \mathscr{F}) = \Gamma(X, \mathscr{F})$.

---

[2]See the remark further below regarding the meaning of the right-hand side of $g_{\underline{j}}$.

H.2.3. Functoriality. The space X being fixed, any morphism of sheaves $\mathscr{F} \to \mathcal{G}$ induces maps $C^p(\mathfrak{U},\mathscr{F}) \to C^p(\mathfrak{U},\mathcal{G})$ sending $Z^p(\mathfrak{U},\mathscr{F})$ into $Z^p(\mathfrak{U},\mathcal{G})$ and $B^p(\mathfrak{U},\mathscr{F})$ into $B^p(\mathfrak{U},\mathcal{G})$, thus inducing $H^p(\mathfrak{U},\mathscr{F}) \to H^p(\mathfrak{U},\mathcal{G})$. In this way, we get covariant functors $\mathscr{F} \rightsquigarrow H^p(\mathfrak{U},\mathscr{F})$ and (going to the direct limit) $\mathscr{F} \rightsquigarrow H^p(X,\mathscr{F})$.

The functors $\mathscr{F} \rightsquigarrow H^p(\mathfrak{U},\mathscr{F})$ and $\mathscr{F} \rightsquigarrow H^p(X,\mathscr{F})$ are additive [**229**, I §3.23 p 216].

H.2.4. Algebraic and analytic coherent sheaves. The following are the main results of [**230**].

Recall that any smooth algebraic curve X over $\mathbf{C}$ can be considered[3] as a Riemann surface X^h. Affine curves give rise to open Riemann surfaces, projective curves to compact Riemann surfaces. Moreover every compact Riemann surface is of the form X^h for some smooth projective algebraic curve X (essentially unique).

This extends to sheaves: any coherent sheaf $\mathscr{F}$ of $\mathscr{O}_X$ modules over X ("coherent algebraic sheaf") gives rise to a coherent sheaf $\mathscr{F}^h$ of $\mathscr{O}_{X^h}$ modules over X^h ("coherent analytic sheaf"). The correspondence is functorial: a morphism $\mathscr{F} \to \mathcal{G}$ gives rise to a morphism $\mathscr{F}^h \to \mathcal{G}^h$. The functoriality extends to the cohomology groups in form of naturally defined maps $H^p(X,\mathscr{F}) \to H^p(X^h,\mathscr{F}^h)$. Those groups are actually $\mathbf{C}$-linear spaces and the maps are $\mathbf{C}$-linear. Then [**230**, §3 no 12 pp 19–20]:

THEOREM H.2. (i) *If $\mathscr{F}$ is a coherent algebraic sheaf over X, the maps $H^p(X,\mathscr{F}) \to H^p(X^h,\mathscr{F}^h)$ are isomorphisms.*

(ii) *If $\mathscr{F}$, $\mathcal{G}$ are coherent algebraic sheaves over X, the map sending morphisms $\mathscr{F} \to \mathcal{G}$ to morphisms $\mathscr{F}^h \to \mathcal{G}^h$ is bijective.*

(iii) *Every coherent analytic sheaf over X^h has the form $\mathscr{F}^h$ for a coherent algebraic sheaf $\mathscr{F}$ which is essentially unique.*

So we see that coherent sheaves over $\mathbf{S}$ and $\mathbf{E}_q$ and their cohomology spaces are "essentially the same thing" as the corresponding sheaves over the projective line $\mathbf{P}^1(\mathbf{C})$, resp. some smooth cubic projective curve. Thanks to H.1, this will in particular apply to all locally free sheaves of finite rank and to sheaves of sections of holomorphic vector bundles.

H.3. Cohomology of algebraic coherent sheaves

We consider here sheaves over algebraic varieties over $\mathbf{C}$. However, thanks to H.2.4, all statements about projective resp. affine varieties may be applied to $\mathbf{E}_q$ resp. $\mathbf{E}_q$ deprived of any nonempty finite subset. All statements come from [**229**].

H.3.1. Acyclic coverings. If X is an affine variety, $\mathscr{F}$ a coherent sheaf over X and $\mathfrak{U}$ a finite open covering by affine open subsets, then $H^n(\mathfrak{U},\mathscr{F}) = 0$ for all $n > 0$ [**229**, II §3.47 p 239]. Since X is quasi-compact, it follows that $H^n(X,\mathscr{F}) = 0$ for all $n > 0$.

If now X is an arbitrary algebraic variety and $\mathscr{F}$, $\mathfrak{U}$ are as above, the naturally defined map $H^n(\mathfrak{U},\mathscr{F}) \to H^n(X,\mathscr{F})$ is bijective[4] for all $n \geq 0$ [**229**, II §3.47 p 239].

[3]The superscript h (for holomorphic) in X^h will be omitted as soon as the theorem is stated, since it will then be clear that the algebraic and analytic setting are essentially equivalent.

[4]This is actually a particular case of a theorem of Leray on acyclic coverings, which will be stated in a more general form where we need it, in Appendix I.

H.3.2. The long exact sequence. Let X an arbitrary algebraic variety and $0 \to \mathscr{F} \to \mathcal{G} \to \mathcal{H} \to 0$ an exact sequence of coherent sheaves. Then the morphisms in cohomology flowing from functoriality may be completed by connecting morphisms $H^p(X, \mathcal{H}) \to H^{p+1}(X, \mathscr{F})$ (the connecting morphisms also behave functorially) in such a way that we obtain a long exact cohomology sequence [**229**, II §3.47 p 240]:

$$\begin{aligned} 0 \to H^0(X, \mathscr{F}) \to H^0(X, \mathcal{G}) \to H^0(X, \mathcal{H}) \to \\ \to H^1(X, \mathscr{F}) \to H^1(X, \mathcal{G}) \to H^1(X, \mathcal{H}) \to \\ \to H^2(X, \mathscr{F}) \to H^2(X, \mathcal{G}) \to H^2(X, \mathcal{H}) \to \cdots \end{aligned}$$

H.3.3. Vanishing and finite-dimensional cohomologies. If X is moreover projective, then all $H^p(X, \mathscr{F})$ are finite-dimensional over $\mathbf{C}$ and vanish for $p > \dim X$ [**229**, III §3.66 p 259]. In particular, for a compact Riemann surface, the long exact cohomology sequence takes the form:

$$\begin{aligned} 0 \to H^0(X, \mathscr{F}) \to H^0(X, \mathcal{G}) \to H^0(X, \mathcal{H}) \to \\ \to H^1(X, \mathscr{F}) \to H^1(X, \mathcal{G}) \to H^1(X, \mathcal{H}) \to 0 \end{aligned}$$

As a consequence, the Euler characteristics

$$\chi(X, \mathscr{F}) := \dim_{\mathbf{C}} H^0(X, \mathscr{F}) - \dim_{\mathbf{C}} H^1(X, \mathscr{F})$$

is additive with respect to exact sequences of sheaves.

H.4. Some special sheaves

Here X is a compact Riemann surface and the reference is [**88**, Chapter 2 §16]. We denote $h^i(X, \mathscr{F}) := \dim_{\mathbf{C}} H^i(X, \mathscr{F})$, $i = 0, 1$. So $\chi = h^0 - h^1$.

H.4.1. Skyscraper sheaves. Let $a \in X$. Denote $\mathbf{C}_X(a)$ the sheaf such that $U \mapsto \mathbf{C}$ if $a \in U$ and $U \mapsto \{0\}$ otherwise. Then $h^0(X, \mathbf{C}_X(a)) = 1$ and $h^1(X, \mathbf{C}_X(a)) = 0$.

It follows that, if a coherent sheaf $\mathscr{F}$ has finite support, its stalks have finite dimension over $\mathbf{C}$ and $h^0(X, \mathscr{F}) = \sum_{x \in X} \dim_{\mathbf{C}} \mathscr{F}_x$ and $h^1(X, \mathscr{F}) = 0$.

This holds, in particular, for every coherent torsion sheaf over a Riemann surface.

Although standard terminology restricts the word *skyscraper* to a sheaf like $\mathbf{C}_X(a)$, in 7.3.4 we abusively extended it to arbitrary coherent torsion sheaves with the idea that they may have quite a few skyscrapers (so Manhattan is a skyscraper island).

H.4.2. Sheaves $\mathscr{O}_X(D)$. They were defined in B.6.4 (see Appendix B; beware they are denoted $\mathscr{O}_D$ in [**88**]). If $g := h^1(\mathscr{O}_X) = h^0(\Omega^1_X)$ is the genus of X:

$$\chi(\mathscr{O}_X(D)) = \deg D - g + 1.$$

Moreover, if $\deg D < 0$ then $h^0(\mathscr{O}_X(D)) = 0$, while if $\deg D > 2g - 2$ then[5] $h^1(\mathscr{O}_X(D)) = 0$.

[5]This fact uses Serre Duality theorem [**88**, Chapter 2 §17], as well as the equality $h^1(\mathscr{O}_X) = h^0(\Omega^1_X)$.

APPENDIX I

Čech Cohomology of Nonabelian Sheaves

In this appendix, we provide information needed to justify some facts used in Chapter 15 about the cohomology of nonabelian sheaves on a Riemann surface. We approach "nonabelian cohomology" in a very down-to-earth way, in the style of our main reference, [**89**] (but you may also look at [**150**, §2.3]).

REMARK I.1. We shall deal with pointed sets (X, x), $x \in X$, (indeed, the cohomology sets) and all maps $(X, x) \to (Y, y)$ will be understood to be such that $x \mapsto y$, even when we abusively write X, Y, $X \to Y$ instead of (X, x), (Y, y), $(X, x) \to (Y, y)$.

I.1. Čech cohomology for sheaves of nonabelian groups

To begin with, the general setting here is this: X is a topological space and $\mathscr{F}$ a sheaf of (possibly nonabelian) groups on X. We shall denote e the unit elements of all the groups $\mathscr{F}(U)$. All this section comes from [**89**, Chapter I §3], except for one result coming from [**14**].

I.1.1. Coverings, cochains, cohomology pointed sets. Let $\mathfrak{U} := (U_i)_{i \in I}$ an open covering of X. If $\underline{i} := (i_0, \ldots, i_p) \in I^{p+1}$, we set $U_{\underline{i}} := U_{i_0} \cap \cdots \cap U_{i_p}$. The group of p-cochains of $\mathscr{F}$ with respect to the covering $\mathfrak{U}$ is:

$$C^p(\mathfrak{U}, \mathscr{F}) := \prod_{\underline{i} \in I^{p+1}} \Gamma(U_{\underline{i}}, \mathscr{F}).$$

The unit element of that group is the trivial p-cochain, all the components of which are $f_{\underline{i}} = e \in \Gamma(U_{\underline{i}}, \mathscr{F})$.

We shall only consider 0-cochains, i.e., families $(f_i) \in \prod \Gamma(U_i, \mathscr{F})$ and 1-cochains, i.e., families $(f_{i,j}) \in \prod \Gamma(U_i \cap U_j, \mathscr{F})$.

To every 0-cochain (f_i) we can associate the 1-cochain $(f_{i,j})$ defined by $f_{i,j} := f_i^{-1} f_j$ (f_i, f_j implicitly restricted to $U_i \cap U_j$). It is trivial if, and only if, there is a global section $f \in \Gamma(X, \mathscr{F})$ such that all f_i are its restrictions: $f_i = f_{|U_i}$. Although the map $C^0(\mathfrak{U}, \mathscr{F}) \to C^1(\mathfrak{U}, \mathscr{F})$ thus defined is not a group morphism, we can define its kernel $H^0(\mathfrak{U}, \mathscr{F})$ as the preimage of the trivial 1-cochain. As we just saw, it is in natural bijection with $H^0(X, \mathscr{F})$.

We say that a 1-cochain $(f_{i,j})$ is a 1-cocycle if:

$$\forall i, j, k \in I \ , \ f_{i,j} f_{j,k} = f_{i,k} \text{ in restriction to } U_i \cap U_j \cap U_k.$$

For example the 1-cochain associated juste above to a 0 cochain is a 1-cocycle.

The set $Z^1(\mathfrak{U}, \mathscr{F})$ is not a subgroup of $C^1(\mathfrak{U}, \mathscr{F})$ (except of course if $\mathscr{F}$ is abelian), but it contains the trivial 1-cochain and we can see it as a pointed set. Note that every 1-cocycle $(f_{i,j})$ satisfies $f_{i,i} = e$ and $f_{i,j} = f_{j,i}^{-1}$ for all i, j.

We say that the 1-cochains $(f_{i,j})$, $(f'_{i,j})$ are equivalent if there is a 0-cochain (g_i) such that:

$$\forall i,j \in I\ ,\ f'_{i,j} = f_i^{-1} f_{i,j} f_j \text{ in restriction to } U_i \cap U_j.$$

Actually this equivalence relation comes from a group action of $C^0(\mathfrak{U}, \mathscr{F})$ on $C^1(\mathfrak{U}, \mathscr{F})$. The subset $Z^1(\mathfrak{U}, \mathscr{F})$ is stable under this group action. The resulting quotient set of $Z^1(\mathfrak{U}, \mathscr{F})$ is the cohomology set $H^1(\mathfrak{U}, \mathscr{F})$. It is a pointed set with distinguished element the trivial cohomology class (the class of the trivial cocycle).

I.1.2. The direct system of cohomology sets; good coverings. Let $\mathfrak{U}$, $\mathfrak{V}$ two open coverings and suppose that $\mathfrak{U}$ is finer than $\mathfrak{V}$. Then, as in Appendix H, there is a natural well defined mapping $H^1(\mathfrak{V}, \mathscr{F}) \to H^1(\mathfrak{U}, \mathscr{F})$. This is actually a mapping of pointed sets since it sends the trivial class to the trivial class. In this way, we obain a direct system of cohomology pointed sets and we define the cohomology pointed set of X as the direct limit:

$$H^1(X, \mathscr{F}) := \lim_{\longrightarrow \mathfrak{U}} H^p(\mathfrak{U}, \mathscr{F}).$$

According to [**89**, Chapter I §3 Lemma 3.1], all maps $H^1(\mathfrak{V}, \mathscr{F}) \to H^1(\mathfrak{U}, \mathscr{F})$ are injective. As a consequence of the general properties of direct limits, the resulting maps $H^1(\mathfrak{U}, \mathscr{F}) \to H^1(X, \mathscr{F})$ are also injective.

We shall say that $\mathfrak{U}$ is a good covering if the map $H^1(\mathfrak{U}, \mathscr{F}) \to H^1(X, \mathscr{F})$ is bijective. We say that it is acyclic if all $H^1(U_i, \mathscr{F})$ are trivial. As a consequence of [**14**, II §1.2 cor 1.2.4] (which actually is a little stronger), every acyclic covering is good. This is a version of *Leray's theorem of acyclic coverings.*

I.1.3. Functoriality. Any morphism $\mathscr{F} \to \mathcal{G}$ of sheaves of groups induces an obvious mapping $C^1(\mathfrak{U}, \mathscr{F}) \to C^1(\mathfrak{U}, \mathcal{G})$, which sends $Z^1(\mathfrak{U}, \mathscr{F})$ into $Z^1(\mathfrak{U}, \mathcal{G})$ and respects the equivalence relation, so it induces in turn $H^1(\mathfrak{U}, \mathscr{F}) \to H^1(\mathfrak{U}, \mathcal{G})$, a mapping of pointed sets. Then for every fixed covering $\mathfrak{U}$, we get a covariant functor $\mathscr{F} \rightsquigarrow H^1(\mathfrak{U}, \mathscr{F})$; and, going to the limit, a covariant functor $\mathscr{F} \rightsquigarrow H^1(X, \mathscr{F})$.

A special case occurs if the sheaf of groups $\mathcal{G}$ acts on the sheaf of groups $\mathscr{F}$, i.e., each group $\mathcal{G}(U)$, $U \subset X$ an open subset, acts on the group $\mathscr{F}(U)$ (by morphisms) and the actions are compatible with the restriction maps of the sheaves. Then functoriality entails that the group $H^0(X, \mathcal{G})$ acts on the pointed set $H^1(X, \mathscr{F})$.

I.2. Cohomology exact sequences

All this section comes from [**89**, Chapter I §5–8]. We shall say that a sequence of pointed sets $\cdots \to A \to B \to C \to \cdots$ is exact at B if the image of A is equal to the preimage of the distinguished element of C. For a sequence of groups, this coincides with the usual definition. In mixed cases (A is a group acting on B with quotient C) we shall give precisions where needed.

I.2.1. We shall say that $e \to \mathscr{F} \to \mathcal{G} \to \mathcal{H} \to e$ is an exact sequence of sheaves of groups if $\mathscr{F}$ and $\mathcal{G}$ are sheaves of (possibly nonabelian) groups, $\mathscr{F}$ is a (possibly nondistinguished) sheaf of subgroups of $\mathcal{G}$ and $\mathcal{H}$ is the corresponding sheaf of homogeneous spaces. Then by [**89**, I §5 theorem 1.1], the functoriality maps embed into an exact sequence of pointed sets:

$$e \to H^0(X, \mathscr{F}) \to H^0(X, \mathcal{G}) \to H^0(X, \mathcal{H}) \to H^1(X, \mathscr{F}) \to H^1(X, \mathcal{G}).$$

Exactness implies that $H^0(X,\mathscr{F}) \to H^0(X,\mathcal{G})$ is injective. We have the complementary informations that:

- two elements of $H^0(X,\mathcal{G})$ have the same image in $H^1(X,\mathcal{H})$ if, and only if, they lie in the same coset of the subgroup $H^0(X,\mathscr{F})$;
- the group $H^0(X,\mathcal{G})$ acts on $H^0(X,\mathcal{H})$ and two elements of $H^0(X,\mathcal{H})$ have the same image in $H^1(X,\mathscr{F})$ if, and only if, they lie in the same orbit.

I.2.2. We now suppose that $\mathscr{F}$ is distinguished in $\mathcal{G}$, so that $e \to \mathscr{F} \to \mathcal{G} \to \mathcal{H} \to e$ is an exact sequence of sheaves of groups in the usual sense. Then by [**89**, I §6 theorem 1.2], the above cohomology exact sequence can be completed into an exact sequence of pointed sets:

$$e \to H^0(X,\mathscr{F}) \to H^0(X,\mathcal{G}) \to H^0(X,\mathcal{H}) \to H^1(X,\mathscr{F}) \to H^1(X,\mathcal{G}) \to H^1(X,\mathcal{H}).$$

We have the complementary information that the group $H^0(X,\mathcal{H})$ acts on the set $H^1(X,\mathscr{F})$ and that two elements of $H^1(X,\mathscr{F})$ have the same image in $H^1(X,\mathcal{G})$ if, and only if, they lie in the same orbit.

I.2.3. We now moreover suppose that $\mathscr{F}$ is central in $\mathcal{G}$, so in particular it is abelian and $H^1(X,\mathscr{F})$, $H^2(X,\mathscr{F})$, are sheaves of abelian groups. Then by [**89**, I §8 theorem 1.4], and assuming moreover than X is paracompact (see further below), the above cohomology exact sequence can be completed into an exact sequence of pointed sets:

$$e \to H^0(X,\mathscr{F}) \to H^0(X,\mathcal{G}) \to H^0(X,\mathcal{H}) \to$$
$$\to H^1(X,\mathscr{F}) \to H^1(X,\mathcal{G}) \to H^1(X,\mathcal{H}) \to H^2(X,\mathscr{F}).$$

We have the complementary informations that:

- Two elements of $H^1(X,\mathscr{F})$ have the same image in $H^1(X,\mathcal{G})$ if, and only if, their difference is in the image of $H^0(X,\mathcal{H})$.
- The group $H^1(X,\mathscr{F})$ acts on $H^1(X,\mathcal{G})$ and two elements of $H^1(X,\mathcal{G})$ have the same image in $H^1(X,\mathcal{H})$ if, and only if, they lie in the same orbit.

A space is paracompact if every open covering has an open refinement that is locally finite. This is in particular true for a Riemann surface X. Moreover, in that case, since we shall apply the theorem to coherent sheaves of $\mathscr{O}_X$-modules, $H^2(X,\mathscr{F})$ will be trivial (see Appendix H).

I.3. Nonabelian cohomology and vector bundles

I.3.1. Cocycles associated to a holomorphic vector bundle. Let $\mathfrak{F}$ a vector bundle over the Riemann surface X and let $\mathfrak{U} = (U_i)$ a covering by trivializing open sets. Combining trivialisation diagrams:

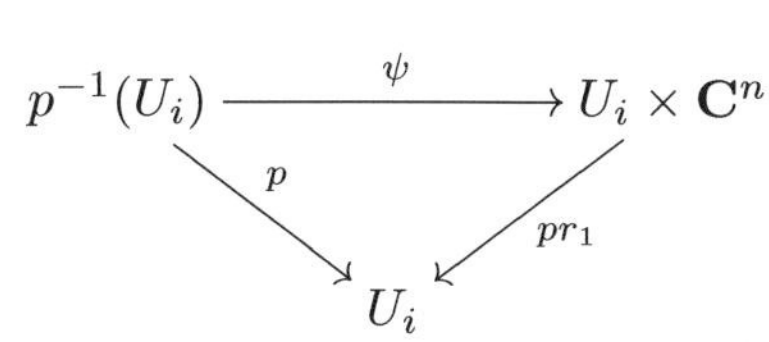

restricted to $U_i \cap U_j$, we obtain chart changes diagrams:

$$\begin{array}{ccc} (U_i \cap U_j) \times \mathbf{C}^n & \xrightarrow{\psi_j \circ \psi_i^{-1}} & (U_i \cap U_j) \times \mathbf{C}^n. \\ & {\scriptstyle pr_1} \searrow \quad \swarrow {\scriptstyle pr_1} & \\ & U_i \cap U_j & \end{array}$$

The "change of coordinates" $\psi_{j,i} := \psi_j \circ \psi_i^{-1}$ in the upper horizontal arrow has the form $(x, v) \mapsto (x, f_{j,i}(x)(v))$ for some map $f_{j,i} : U_i \cap U_j \to \mathrm{GL}_n(\mathbf{C})$. We may see $f_{i,j}$ as an element of $\mathrm{GL}_n(\mathscr{O}_X(U_i \cap U_j))$, i.e., of a section over the open set $U_i \cap U_j$ of the sheaf of groups $\mathrm{GL}_n(\mathscr{O}_X)$ such that $U \mapsto \mathrm{GL}_n(\mathscr{O}_X(U))$.

From the obvious relation $\psi_{k,j} \circ \psi_{j,i} = \psi_{k,i}$ (on $U_i \cap U_j \cap U_k$) we draw that $f_{k,j} f_{j,i} = f_{k,i}$, i.e., the family of all $f_{j,i}$ is a cocycle:

$$(f_{j,i}) \in Z^1(\mathfrak{U}, \mathrm{GL}_n(\mathscr{O}_X)).$$

Given a cocycle $(f_{j,i}) \in Z^1(\mathfrak{U}, \mathrm{GL}_n(\mathscr{O}_X))$ associated in this way to a vector bundle $\mathfrak{F}$, sections can be described as follows (we do it for global sections only, the generalisation is straightforward). A section $s \in \mathscr{F}(X)$ defines compatible maps $s_i : U_i \to p^{-1}(U_i)$, whence mappings $\psi_i \circ s_i : U_i \to U_i \times \mathbf{C}^n$, which have the form $x \mapsto (x, f_i(x))$ where $f_i \in \mathscr{O}_X(U_i)^n$. Then, since $s_i := s_{|U_i}$ and $s_j := s_{|U_i}$ coincide on $U_i \cap U_j$, we have there $\psi_{j,i} \circ \psi_i \circ s_i = \psi_j \circ s_j$, which implies that $f_{j,i}(x).f_i(x) = f_j(x)$ (the left-hand side is a matrix in $\mathrm{GL}_n(\mathbf{C})$ applied to a vector in $\mathbf{C}^n$, the right-hand side is a vector in $\mathbf{C}^n$).

So a global section of $\mathfrak{F}$ can be identified with a family of $f_i \in \mathscr{O}_X(U_i)^n$ such that $f_{j,i}.f_i = f_j$ on $U_i \cap U_j$ for all i, j (the left-hand side is a matrix in $\mathrm{GL}_n(\mathscr{O}_X(U_i \cap U_j))$ applied to a vector in $\mathscr{O}_X(U_i \cap U_j)^n$, the right-hand side is a vector in $\mathscr{O}_X(U_i \cap U_j)^n$).

If we consider the f_i as column vectors acted upon (on the left) by invertible matrices $f_{j,i}$, we may form matrices $M_i := [f_i f_i' \ldots]$ satisfying similar relations $f_{j,i}.M_i = M_j$ and decide to call them "matricial sections" (in the same way we spoke of "matricial solutions"). If all M_i are invertible (thus square) we have a fundamental matricial global section. The existence of such a thing is equivalent to the triviality of the bundle.

I.3.2. Classification of vector bundles by cohomology classes. It is not hard to see that isomorphic vector bundles $\mathfrak{F}$, $\mathfrak{F}'$ have the same trivializing covers and give rise to equivalent cocycles.

Conversely, there is in [**88**, Chapter 3 §29 theorem 29.7] a construction of a vector bundle from a cocycle; and it is proved[1] in [**108**, §1 (b) Theorem 1.11] that we obtain in this way a bijective correspondence:

$$\left\{ \begin{array}{c} \text{isomorphism classes} \\ \text{of vector bundles} \end{array} \right\} \longleftrightarrow H^1(X, \mathrm{GL}_n(\mathscr{O}_X)).$$

The same construction entails that a vector bundle is trivial if, and only if, the associated cohomology class is trivial; equivalently, the associated cocycle $(f_{i,j})$ on some covering $\mathfrak{U}$ is a coboundary, i.e., has the form $f_{i,j} = f_i f_j^{-1}$ for some family (f_i).

---

[1] Beware however of the terminology in Gunning: on the left-hand side, he actually considers locally free sheaves of finite rank (but we know that this is the same as vector bundles); on the right-hand side, he calls vector bundle what is here (and usually!) called cocycles.

If X is an open (i.e., noncompact) Riemann surface, it follows from the triviality of holomorphic vector bundles (see B.5.6) that $H^1(X, \mathrm{GL}_n(\mathscr{O}_X))$ is trivial (i.e., reduced to the trivial class) [**88**, Chapter 3 §30 Corollary 30.5].

The case of a compact Riemann surface will be tackled next.

I.3.3. Meromorphic triviality of vector bundles on a compact Riemann surface. In 7.3, we use theorem B.8 from Appendix B to prove Praagman's theorem on the existence of fundamental matricial meromorphic solution for a q-difference system. The part on the meromorphic triviality of holomorphic vector bundles over a compact Riemann surface seems to be "folklore" but it is not easy to find a complete explicit proof. A proof should be reconstituable either based a remark in [**108**, §2 (c) p 43] (replacing Theorem 7 with Theorem 10 in page 63 of §4 (a)); or based on the sketch of proof given in [**249**, 12.3.1], which is itself based on the original proof of Praagman. So I shall propose a direct proof.

EXERCISE I.2. The case $n = 1$ says that $H^1(X, \mathscr{M}_X^*)$ is trivial. It is generally presented in textbooks (see a reference further below). Interpret it in terms of meromorphic functions. Find associated classical problems in complex analysis, for instance when $X = \mathbf{C}$.

In principle, we are supposed to prove that $H^1(X, \mathrm{GL}_n(\mathscr{M}_X))$ is trivial, where $\mathscr{M}_X$ denotes the sheaf of meromorphic functions on X. However, according to [**108**, §4 (a) corollary of Theorem 10], if X is a compact Riemann surface and $\mathfrak{F}$ a holomorphic vector bundle on X, the associated cocycle may be chosen in upper triangular form. So, if we denote T_n^* the group of invertible upper triangular matrices, $\mathfrak{F}$ gives rise to a class in $H^1(X, \mathrm{T}_n^*(\mathscr{O}_X))$. We are therefore reduced to prove that the cohomology pointed set $H^1(X, \mathrm{T}_n^*(\mathscr{M}_X))$ is trivial.

We do that using a "dévissage" argument rather similar to the one used in 15.3.2. We use the fact that, if $e \to \mathscr{F} \to \mathcal{G} \to \mathcal{H} \to e$ is an exact sequence of sheaves of groups in the usual sense and if $H^1(X, \mathscr{F})$ and $H^1(X, \mathcal{H})$ are both trivial, then so is $H^1(X, \mathcal{G})$; this follows from the exact cohomology sequence in I.2 and the fact that, if there is an exact sequence of pointed sets $\{.\} \to E \to \{.\}$, then $E = \{.\}$.

Now the sheaf $\mathrm{T}_n^*(\mathscr{M}_X)$ takes place inside an exact sequence:

$$e \to \mathrm{U}_n^*(\mathscr{M}_X) \to \mathrm{T}_n^*(\mathscr{M}_X) \to \mathrm{D}_n^*(\mathscr{M}_X) \to e.$$

The group D_n^* is that of diagonal invertible matrices, the map $\mathrm{T}_n^* \to \mathrm{D}_n^*$ projects a matrix to its diagonal part and the kernel U_n^* consists in unipotent triangular matrices. The sheaf $\mathrm{D}_n^*(\mathscr{M}_X)$ is of course isomorphic to the product of n copies of $\mathscr{M}_X^*$ and, by the previous remark, we are reduced to prove the triviality of $H^1(X, \mathrm{U}_n^*(\mathscr{M}_X))$ and that of $H^1(X, \mathscr{M}_X^*)$.

The latter says in essence that every line bundle over X is meromorphically trivial, which (by the argument given in [**108**] to prove the above mentioned theorem 10) is a consequence of [**88**, Chapter 3 §29 Corollary 29.17].

As for $\mathrm{U}_n^*(\mathscr{M}_X)$, it admits a sequence of distinguished subsheaves of groups defined by:

$$\mathrm{U}_n^*(k) := \{\text{unipotent triangular matrices } (a_{i,j}) \text{ s.t. } 0 < j - i < k \Longrightarrow a_{i,j} = 0\}.$$

Sending a matrix to its overdiagonal $j - i = k$ projects $\mathrm{U}_n^*(k)(\mathscr{M}_X)$ onto $\mathscr{M}_X^{n-k}$ with kernel $\mathrm{U}_n^*(k+1)(\mathscr{M}_X)$, so, since $\mathrm{U}_n^*(n+1)(\mathscr{M}_X)$ is obviously trivial and

$\mathrm{U}_n^*(1) = \mathrm{U}_n^*$, descending induction reduces the fact that $H^1(X, \mathscr{M}_X)$ is trivial. But this is the content of [**88**, Chapter 2 §17 Corollary 17.17] (it is proved using the fact that, if $\deg D > 2g - 2$, then $H^1(X, \mathscr{O}_X(D)) = 0$ is trivial; see Appendix H; and also the fact that $\mathscr{M}_X$ is the direct limit of all $\mathscr{O}_X(D)$); also see [**101**, Chapter VII, §5].

Bibliography

[1] C. Raymond Adams, *On the linear ordinary q-difference equation*, Ann. of Math. (2) **30** (1928/29), no. 1-4, 195–205, DOI 10.2307/1968274. MR1502876

[2] C. R. Adams, *Linear q-difference equations*, Bull. Amer. Math. Soc. **37** (1931), no. 6, 361–400, DOI 10.1090/S0002-9904-1931-05162-4. MR1562160

[3] Lars V. Ahlfors, *Complex analysis: An introduction to the theory of analytic functions of one complex variable*, 3rd ed., International Series in Pure and Applied Mathematics, McGraw-Hill Book Co., New York, 1978. MR510197

[4] Yves André, *Différentielles non commutatives et théorie de Galois différentielle ou aux différences* (French, with English and French summaries), Ann. Sci. École Norm. Sup. (4) **34** (2001), no. 5, 685–739, DOI 10.1016/S0012-9593(01)01074-6. MR1862024

[5] Yves André, *Slope filtrations*, Confluentes Math. **1** (2009), no. 1, 1–85, DOI 10.1142/S179374420900002X. MR2571693

[6] Yves André, *Idées galoisiennes* (French), Histoire de mathématiques, Ed. Éc. Polytech., Palaiseau, 2012, pp. 1–16. MR2905515

[7] George E. Andrews, *q-series: their development and application in analysis, number theory, combinatorics, physics, and computer algebra*, CBMS Regional Conference Series in Mathematics, vol. 66, Published for the Conference Board of the Mathematical Sciences, Washington, DC; by the American Mathematical Society, Providence, RI, 1986. MR858826

[8] George E. Andrews, *The theory of partitions*, Camb. Math. Libr., Cambridge: Cambridge University Press, 1998 (English).

[9] D. V. Anosov and A. A. Bolibruch, *The Riemann-Hilbert problem*, Aspects of Mathematics, E22, Friedr. Vieweg & Sohn, Braunschweig, 1994, DOI 10.1007/978-3-322-92909-9. MR1276272

[10] K. Aomoto, *A note on holonomic q-difference systems*, Algebraic analysis, Vol. I, Academic Press, Boston, MA, 1988, pp. 25–28. MR992444

[11] Kazuhiko Aomoto, *q-analogue of de Rham cohomology associated with Jackson integrals. I*, Proc. Japan Acad. Ser. A Math. Sci. **66** (1990), no. 7, 161–164. MR1078398

[12] Tom M. Apostol, *Introduction to analytic number theory*, Undergraduate Texts in Mathematics, Springer-Verlag, New York-Heidelberg, 1976. MR434929

[13] M. F. Atiyah, *Vector bundles over an elliptic curve*, Proc. London Math. Soc. (3) **7** (1957), 414–452, DOI 10.1112/plms/s3-7.1.414. MR131423

[14] D. G. Babbitt and V. S. Varadarajan, *Local moduli for meromorphic differential equations* (English, with French summary), Astérisque **169-170** (1989), 217. MR1014083

[15] Werner Balser, *Formal power series and linear systems of meromorphic ordinary differential equations*, Universitext, Springer-Verlag, New York, 2000. MR1722871

[16] Constantin Bănică and Octavian Stănăşilă, *Algebraic methods in the global theory of complex spaces*, Translated from the Romanian, Editura Academiei, Bucharest; John Wiley & Sons, London-New York-Sydney, 1976. MR0463470

[17] Vladimir Baranovsky and Victor Ginzburg, *Conjugacy classes in loop groups and G-bundles on elliptic curves*, Internat. Math. Res. Notices **15** (1996), 733–751, DOI 10.1155/S1073792896000463. MR1413870

[18] Bruce C. Berndt, *What is a q-series?*, Ramanujan rediscovered, Ramanujan Math. Soc. Lect. Notes Ser., vol. 14, Ramanujan Math. Soc., Mysore, 2010, pp. 31–51. MR2856954

[19] Bruce C. Berndt and Robert A. Rankin (eds.), *Ramanujan: essays and surveys*, History of Mathematics, vol. 22, American Mathematical Society, Providence, RI; London Mathematical Society, London, 2001, DOI 10.1090/hmath/022. MR1862732

[20] Jean-Paul Bézivin, *Sur les équations fonctionnelles aux q-différences* (French, with English summary), Aequationes Math. **43** (1992), no. 2-3, 159–176, DOI 10.1007/BF01835698. MR1158724

[21] B. J. Birch, *Some calculations of modular relations*, Modular functions of one variable, I (Proc. Internat. Summer School, Univ. Antwerp, Antwerp, 1972), Lecture Notes in Math., Vol. 320, Springer, Berlin-New York, 1973, pp. 175–186. MR332658

[22] Garrett Birkhoff and Gian-Carlo Rota, *Ordinary differential equations*, 3rd ed., John Wiley & Sons, New York-Chichester-Brisbane, 1978. MR507190

[23] George D. Birkhoff, *The generalized riemann problem for linear differential equations and the allied problems for linear difference and q-difference equations*, Proc. Amer. Acad. **49** (1913), 521–568.

[24] George D. Birkhoff, *Fifty years of American Mathematics*, Science, **88** (1938). 461–467, DOI 10.1126/science. 88.2290.461.

[25] George D. Birkhoff and Paul E. Guenther, *Note on a canonical form for the linear q-difference system*, Proc. Nat. Acad. Sci. U.S.A. **27** (1941), 218–222, DOI 10.1073/pnas.27.4.218. MR4047

[26] George D. Birkhoff and W. J. Trjitzinsky, *Analytic theory of singular difference equations*, Acta Math. **60** (1933), no. 1, 1–89, DOI 10.1007/BF02398269. MR1555364

[27] P. P. Boalch, *Topology of the Stokes phenomenon*, Integrability, quantization, and geometry. I. Integrable systems, Proc. Sympos. Pure Math., vol. 103, Amer. Math. Soc., Providence, RI, 2021, pp. 55–100, DOI 10.1090/pspum/103.1/01832. MR4285677

[28] Armand Borel, *Linear algebraic groups*, 2nd ed., Graduate Texts in Mathematics, vol. 126, Springer-Verlag, New York, 1991, DOI 10.1007/978-1-4612-0941-6. MR1102012

[29] Emile Borel, *Leçons sur les séries divergentes. 2. éd. revue et entièrement remaniée avec le concours de G. Bouligand.*, Collection de monographies sur la théorie des fonctions. Paris: Gauthier-Villars. iv, 260 pp. (1928)., 1928.

[30] Nicolas Bourbaki, *Éléments de mathématique. VI. Première partie: Les structures fondamentales de l'analyse. Livre II: Algèbre. Chapitre II: Algèbre linéaire* (French), Actualités Scientifiques et Industrielles [Current Scientific and Industrial Topics], No. 1032, Hermann & Cie, Paris, 1947. MR23803

[31] N. Bourbaki, *Éléments de mathématique. XII. Ière Partie: Les structures fondamentales de l'analyse. Livre IV: Fonctions d'une variable réelle. Théorie élémentaire. Chap. IV: Équations différentielles. Chap. V: Étude locale des fonctions. Chap. VI: Développements tayloriens généralisés, formule sommatoire d'Euler-MacLaurin. Chap. VII: La fonction gamma*, Actualités Scientifiques et Industrielles. 1132. Paris: Hermann & Cie., 1951.

[32] N. Bourbaki, *Éléments de mathématique. Algèbre. Chapitre 10: Algèbre homologique.*, Paris: Masson. VII, 1980.

[33] Nicolas Bourbaki, *Éléments de mathématique* (French), Masson, Paris, 1981. Algèbre. Chapitres 4 à 7. [Algebra. Chapters 4–7]. MR643362

[34] Nicolas Bourbaki, *Éléments de mathématique. Algèbre commutative. Chapitres 1 à 4.*, reprint of the 1985 original ed., Berlin: Springer, 2006 (French).

[35] Nicolas Bourbaki, *Éléments de mathématique. Algèbre commutative. Chapitres 5 à 7.*, reprint of the 1985 original ed., Berlin: Springer, 2006 (French).

[36] Nicolas Bourbaki, *Éléments de mathématique. Algèbre commutative. Chapitres 8 et 9*, reprint of the 1983 original ed., Berlin: Springer, 2006 (French).

[37] Nicolas Bourbaki, *Éléments de mathématique. Algèbre. Chapitre 9* (French), Springer-Verlag, Berlin, 2007. Reprint of the 1959 original. MR2325344

[38] Nicolas Bourbaki, *Éléments de mathématique. Topologie générale. Chapitres 5 à 10* (French), Hermann, Paris, 1974. MR3822133

[39] Nicolas Bourbaki, *Éléments de mathématique* (French), Masson, Paris, 1985. Algèbre commutative. Chapitres 1 à 4. [Commutative algebra. Chapters 1–4]; Reprint. MR782296

[40] Nicolas Bourbaki, *Elements of Mathematics. Algebra. Chapter 8*, English translation of [3027127]; Translated by Reinie Erné, Springer, Cham, 2022, DOI 10.1007/978-3-031-19293-7. MR4696567

[41] Lawrence Breen, *Tannakian categories*, Motives (Seattle, WA, 1991), Proc. Sympos. Pure Math., vol. 55, Amer. Math. Soc., Providence, RI, 1994, pp. 337–376, DOI 10.1090/pspum/055.1/1265536. MR1265536

[42] Theodor Bröcker and Tammo tom Dieck, *Representations of compact Lie groups*, Translated from the German manuscript; Corrected reprint of the 1985 translation, Graduate Texts in Mathematics, vol. 98, Springer-Verlag, New York, 1995. MR1410059

[43] Virginie Bugeaud, *Local Galois group of irregular q-difference equations* (English, with English and French summaries), Ann. Inst. Fourier (Grenoble) **68** (2018), no. 3, 901–964, DOI 10.5802/aif.3181. MR3805765

[44] R. D. Carmichael, *The General Theory of Linear q-Difference Equations*, Amer. J. Math. **34** (1912), no. 2, 147–168, DOI 10.2307/2369887. MR1506145

[45] Henri Cartan, *Elementary theory of analytic functions of one or several complex variables*, Éditions Scientifiques Hermann, Paris; Addison-Wesley Publishing Co., Inc., Reading, Mass.-Palo Alto, Calif.-London, 1963. MR154968

[46] Henri Cartan and Samuel Eilenberg, *Homological algebra*, Princeton University Press, Princeton, NJ, 1956. MR77480

[47] Shao Shi Chen, Ru Yong Feng, Guo Feng Fu, and Jin Kang, *Multiplicative decompositions of multivariate q-hypergeometric terms* (Chinese, with English and Chinese summaries), J. Systems Sci. Math. Sci. **32** (2012), no. 8, 1019–1032. MR3036885

[48] Claude Chevalley, *Introduction to the Theory of Algebraic Functions of One Variable*, Mathematical Surveys, No. VI, American Mathematical Society, New York, 1951. MR42164

[49] Claude Chevalley, *Theory of Lie groups. I*, Fifteenth printing; Princeton Landmarks in Mathematics, Princeton Mathematical Series, vol. 8, Princeton University Press, Princeton, NJ, 1999. MR1736269

[50] C. Chevalley and E. Kolchin, *Two proofs of a theorem on algebraic groups*, Proc. Amer. Math. Soc. **2** (1951), 126–134, DOI 10.2307/2032633. MR40305

[51] Richard M. Cohn, *Difference algebra*, Interscience Publishers John Wiley & Sons, New York-London-Sydney, 1965. MR205987

[52] P. Deligne, *Le groupe fondamental de la droite projective moins trois points* (French), Galois groups over **Q** (Berkeley, CA, 1987), Math. Sci. Res. Inst. Publ., vol. 16, Springer, New York, 1989, pp. 79–297, DOI 10.1007/978-1-4613-9649-9_3. MR1012168

[53] P. Deligne, *Catégories tannakiennes* (French), The Grothendieck Festschrift, Vol. II, Progr. Math., vol. 87, Birkhäuser Boston, Boston, MA, 1990, pp. 111–195. MR1106898

[54] Pierre Deligne, *Théorie de Hodge. II* (French), Inst. Hautes Études Sci. Publ. Math. **40** (1971), 5–57. MR498551

[55] Pierre Deligne, Bernard Malgrange, and Jean-Pierre Ramis, *Singularités irrégulières: Correspondance et documents. [Correspondence and documents]* (French), Documents Mathématiques (Paris) [Mathematical Documents (Paris)], vol. 5, Société Mathématique de France, Paris, 2007. MR2387754

[56] Pierre Deligne and J.S. Milne, *Tannakian categories.*, Hodge cycles, motives, and Shimura varieties, Lect. Notes Math. 900, 101-228 (1982)., 1982.

[57] Jean-Pierre Demailly, *Analyse numérique et équations différentielles* (French), 4th ed., Grenoble Sciences, EDP Sciences, Les Ulis, 2016. MR3752612

[58] Lucia Di Vizio, *Arithmetic theory of q-difference equations: the q-analogue of Grothendieck-Katz's conjecture on p-curvatures*, Invent. Math. **150** (2002), no. 3, 517–578, DOI 10.1007/s00222-002-0241-z. MR1946552

[59] Lucia Di Vizio, *Introduction to p-adic q-difference equations (weak Frobenius structure and transfer theorems)*, Geometric aspects of Dwork theory. Vol. I, II, Walter de Gruyter, Berlin, 2004, pp. 615–675. MR2099081

[60] Lucia Di Vizio, *Local analytic classification of q-difference equations with* $|q| = 1$, J. Noncommut. Geom. **3** (2009), no. 1, 125–149, DOI 10.4171/JNCG/33. MR2457039

[61] Lucia Di Vizio, *Difference Galois theory for the "applied" mathematician*, Arithmetic and geometry over local fields—VIASM 2018, Lecture Notes in Math., vol. 2275, Springer, Cham, 2021, pp. 29–59. MR4238486

[62] Lucia Di Vizio and Charlotte Hardouin, *Galois theories of q-difference equations: comparison theorems*, Confluentes Math. **12** (2020), no. 2, 11–35. MR4236700

[63] Lucia Di Vizio, Charlotte Hardouin, and Michael Wibmer, *Difference Galois theory of linear differential equations*, Adv. Math. **260** (2014), 1–58, DOI 10.1016/j.aim.2014.04.005. MR3209348

[64] L. Di Vizio, J.-P. Ramis, J. Sauloy, and C. Zhang, *Équations aux q-différences* (French), Gaz. Math. **96** (2003), 20–49. MR1988639

[65] Lucia Di Vizio and Jacques Sauloy, *Outils pour la classification locale des équations aux q-différences linéaires complexes* (French, with English and French summaries), Arithmetic and Galois theories of differential equations, Sémin. Congr., vol. 23, Soc. Math. France, Paris, 2011, pp. 169–222. MR3076082

[66] Lucia Di Vizio and Changgui Zhang, *On q-summation and confluence* (English, with English and French summaries), Ann. Inst. Fourier (Grenoble) **59** (2009), no. 1, 347–392, DOI 10.5802/aif.2433. MR2514868

[67] J. Dieudonné, *Éléments d'analyse. Tome III: Chapitres XVI et XVII* (French), Cahiers Scientifiques [Scientific Reports], Fasc. XXXIII, Gauthier-Villars Éditeur, Paris, 1970. MR270377

[68] Jean Dieudonné, *History of functional analysis. 1st reprint*, North-Holland Math. Stud., vol. 49, Elsevier, Amsterdam, 1983.

[69] Jean Dieudonné, *Abrégé d'histoire des mathématiques 1700–1900. Nouv. version. (A short history of mathematics 1700–1900)*, Paris: Hermann. VIII, 517 p.; FF 210.00 (1986)., 1986.

[70] Jean Dieudonné, *Calcul infinitésimal. 2ième éd., rev. et corr. (Nouv. tirage). (Infinitesimal calculus. Reprint of the 2nd rev. and corr. ed.)*, Collection Méthodes. Paris: Hermann, 480 p. (1986)., 1986.

[71] John D. Dollard and Charles N. Friedman, *Product integration with applications to differential equations. Foreword by Felix E. Browder. Appendix by P. R. Masani*, reprint of the 1979 hardback ed. ed., Encycl. Math. Appl., vol. 10, Cambridge: Cambridge University Press, 2011 (English).

[72] Régine Douady and Adrien Douady, *Algèbre et théories galoisiennes. 2* (French), CEDIC, Paris, 1979. Théories galoisiennes. [Galois theories]. MR595328

[73] Thomas Dreyfus, *Confluence of meromorphic solutions of q-difference equations* (English, with English and French summaries), Ann. Inst. Fourier (Grenoble) **65** (2015), no. 2, 431–507, DOI 10.5802/aif.2937. MR3449160

[74] Thomas Dreyfus, *Isomonodromic deformation of q-difference equations and confluence*, Proc. Amer. Math. Soc. **145** (2017), no. 3, 1109–1120, DOI 10.1090/proc/13173. MR3589311

[75] Anne Duval, *Lemmes de Hensel et factorisation formelle pour les opérateurs aux différences* (French), Funkcial. Ekvac. **26** (1983), no. 3, 349–368. MR748022

[76] Anne Duval, *Séries de q-factorielles, opérateurs aux q-différences et confluence* (French, with English and French summaries), Ann. Fac. Sci. Toulouse Math. (6) **12** (2003), no. 3, 335–374. MR2030091

[77] Anne Duval, *Confluence q-différence vers différence pour un système fuchsien* (French, with English summary), Pacific J. Math. **217** (2004), no. 2, 221–245, DOI 10.2140/pjm.2004.217.221. MR2109932

[78] Anne Duval, *Une remarque sur les "logarithmes" associés à certains caractères* (French, with English summary), Aequationes Math. **68** (2004), no. 1-2, 88–97, DOI 10.1007/s00010-003-2721-7. MR2167011

[79] Anne Duval and Julien Roques, *Familles fuchsiennes d'équations aux (q-)différences et confluence* (French, with English and French summaries), Bull. Soc. Math. France **136** (2008), no. 1, 67–96, DOI 10.24033/bsmf.2549. MR2415336

[80] H. M. Edwards, *Riemann's zeta function*, Dover Publications, Inc., Mineola, NY, 2001. Reprint of the 1974 original [Academic Press, New York; MR0466039 (57 #5922)]. MR1854455

[81] Thomas Ernst, *A comprehensive treatment of q-calculus*, Birkhäuser/Springer Basel AG, Basel, 2012, DOI 10.1007/978-3-0348-0431-8. MR2976799

[82] Pavel I. Etingof, *Galois groups and connection matrices of q-difference equations*, Electron. Res. Announc. Amer. Math. Soc. **1** (1995), no. 1, 1–9, DOI 10.1090/S1079-6762-95-01001-8. MR1336694

[83] Ulrich Felgner, *The term 'adaption' (παρισότης, adaequatio) in the work of Diophant and Fermat*, (German) Sudhoffs Arch. **100** (2016), no. 1, 83–109.

[84] Nathan J. Fine, *Basic hypergeometric series and applications*, Mathematical Surveys and Monographs, vol. 27, American Mathematical Society, Providence, RI, 1988, DOI 10.1090/surv/027. MR956465

[85] Gerd Fischer, *Complex analytic geometry*, Lecture Notes in Mathematics, Vol. 538, Springer-Verlag, Berlin-New York, 1976. MR430286

[86] Philippe Flajolet and Robert Sedgewick, *Analytic combinatorics*, Cambridge University Press, Cambridge, 2009, DOI 10.1017/CBO9780511801655. MR2483235

[87] Monique Fleinert-Jensen, *Théorèmes d'indices précisés et convergences des solutions pour une équation linéaire aux q-différences* (French, with English and French summaries), C. R. Acad. Sci. Paris Sér. I Math. **321** (1995), no. 4, 425–428. MR1351090

[88] Otto Forster, *Lectures on Riemann surfaces*, Translated from the German by Bruce Gilligan, Graduate Texts in Mathematics, vol. 81, Springer-Verlag, New York-Berlin, 1981. MR648106

[89] Jean Frenkel, *Cohomologie non abélienne et espaces fibrés* (French), Bull. Soc. Math. France **85** (1957), 135–220. MR98200

[90] William Fulton, *Algebraic topology: A first course*, Graduate Texts in Mathematics, vol. 153, Springer-Verlag, New York, 1995, DOI 10.1007/978-1-4612-4180-5. MR1343250

[91] Évariste Galois, *Œuvres mathématiques* (French), Éditions Jacques Gabay, Sceaux, 1989. Publiées en 1846 dans le Journal de Liouville, suivies d'une étude par Sophus Lie, "Influence de Galois sur le développement des mathématiques" (1895). [Published in 1846 in the *Journal de Liouville*, followed by a study by Sophus Lie, "Influence of Galois on the development of mathematics" (1895)]; With a foreword by J. Liouville. MR1188873

[92] Stavros Garoufalidis and Thang T. Q. Lê, *A survey of q-holonomic functions*, Enseign. Math. **62** (2016), no. 3-4, 501–525, DOI 10.4171/LEM/62-3/4-7. MR3692896

[93] George Gasper and Mizan Rahman, *Basic hypergeometric series*, 2nd ed., With a foreword by Richard Askey, Encyclopedia of Mathematics and its Applications, vol. 96, Cambridge University Press, Cambridge, 2004, DOI 10.1017/CBO9780511526251. MR2128719

[94] Sergei I. Gelfand and Yuri I. Manin, *Methods of homological algebra*, 2nd ed., Springer Monographs in Mathematics, Springer-Verlag, Berlin, 2003, DOI 10.1007/978-3-662-12492-5. MR1950475

[95] Claude Godbillon, *Éléments de topologie algébrique* (French), Hermann, Paris, 1971. MR301725

[96] Roger Godement, *Topologie algébrique et théorie des faisceaux* (French), Publications de l'Institut de Mathématique de l'Université de Strasbourg, XIII, Hermann, Paris, 1973. Troisième édition revue et corrigée; Actualités Scientifiques et Industrielles [Current Scientific and Industrial Topics], No. 1252. MR345092

[97] Roger Godement, *Introduction to the theory of Lie groups*, Translated from the 2004 French edition [MR2047118] by Urmie Ray, Universitext, Springer, Cham, 2017, DOI 10.1007/978-3-319-54375-8. MR3616786

[98] E. Goursat, *Propriétés générales de l'équation d'Euler et de Gauss. (Leçons sur les séries hypergéométriques et sur quelques fonctions qui s'y rattachent. I.)*, Actual. sci. industr. 333, 1936.

[99] Anne Granier, *A Galois D-groupoid for q-difference equations* (English, with English and French summaries), Ann. Inst. Fourier (Grenoble) **61** (2011), no. 4, 1493–1516 (2012), DOI 10.5802/aif.2648. MR2951501

[100] Hans Grauert and Reinhold Remmert, *Coherent analytic sheaves*, Grundlehren der mathematischen Wissenschaften [Fundamental Principles of Mathematical Sciences], vol. 265, Springer-Verlag, Berlin, 1984, DOI 10.1007/978-3-642-69582-7. MR755331

[101] Hans Grauert and Reinhold Remmert, *Theory of Stein spaces*, Translated from the German by Alan Huckleberry, Grundlehren der Mathematischen Wissenschaften, vol. 236, Springer-Verlag, Berlin-New York, 1979. MR580152

[102] Marvin J. Greenberg and John R. Harper, *Algebraic topology: A first course*, Mathematics Lecture Note Series, vol. 58, Benjamin/Cummings Publishing Co., Inc., Advanced Book Program, Reading, MA, 1981. MR643101

[103] A. Grothendieck, *Espaces vectoriels topologiques* (French), Universidade de São Paulo, Instituto de Matemática Pura e Aplicada, São Paulo, 1954. MR77884

[104] Alexander Grothendieck, *Sur quelques points d'algèbre homologique* (French), Tohoku Math. J. (2) **9** (1957), 119–221, DOI 10.2748/tmj/1178244839. MR102537

[105] Alexandre Grothendieck, *Sur certains espaces de fonctions holomorphes. II* (French), J. Reine Angew. Math. **192** (1953), 77–95, DOI 10.1515/crll.1953.192.77. MR62335

[106] Alexandre Grothendieck, *On Weil's memoir: Generalization of abelian functions*, Sémin. Bourbaki 9 (1956/57), Exp. No. 141, 1959.

[107] J. Guenot and R. Narasimhan, *Introduction à la théorie des surfaces de Riemann*, Enseign. Math. (2) **21** (1975), no. 2-4, 123–328. MR399449

[108] R. C. Gunning, *Lectures on vector bundles over Riemann surfaces*, University of Tokyo Press, Tokyo; Princeton University Press, Princeton, NJ, 1967. MR230326

[109] G. H. Hardy, *Ramanujan: twelve lectures on subjects suggested by his life and work*, Chelsea Publishing Co., New York, 1959. MR106147

[110] G. H. Hardy and J. E. Littlewood, *Notes on the theory of series. XXIV. A curious power-series*, Proc. Cambridge Philos. Soc. **42** (1946), 85–90, DOI 10.1017/s0305004100022805. MR15529

[111] G. H. Hardy and S. Ramanujan, *Asymptotic Formulae in Combinatory Analysis*, Proc. London Math. Soc. (2) **17** (1918), 75–115, DOI 10.1112/plms/s2-17.1.75. MR1575586

[112] G. H. Hardy and E. M. Wright, *An introduction to the theory of numbers*, 6th ed., Revised by D. R. Heath-Brown and J. H. Silverman; With a foreword by Andrew Wiles, Oxford University Press, Oxford, 2008. MR2445243

[113] G.H. Hardy, *Divergent series*, 2nd (textually unaltered) ed., New York, NY: Chelsea, 1991.

[114] Robin Hartshorne, *Algebraic geometry*, Graduate Texts in Mathematics, No. 52, Springer-Verlag, New York-Heidelberg, 1977. MR463157

[115] Allen Hatcher, *Algebraic topology*, Cambridge University Press, Cambridge, 2002. MR1867354

[116] E. Heine, *Über die Reihe* $1 + \frac{(q^\alpha-1)(q^\beta-1)}{(q-1)(q^\gamma-1)}x + \frac{(q^\alpha-1)(q^{\alpha+1}-1)(q^\beta-1)(q^{\beta+1}-1)}{(q-1)(q^2-1)(q^\gamma-1)(q^{\gamma+1}-1)}x^2 + \dots$, J. Reine Angew. Math. **32** (1846), 210–212 (German).

[117] Peter A. Hendriks, *An algorithm for computing a standard form for second-order linear q-difference equations: Algorithms for algebra (Eindhoven, 1996)*, J. Pure Appl. Algebra **117/118** (1997), 331–352, DOI 10.1016/S0022-4049(97)00017-0. MR1457845

[118] Peter Henrici, *Applied and computational complex analysis. Vol. 2: Special functions—integral transforms—asymptotics—continued fractions; Reprint of the 1977 original; A Wiley-Interscience Publication*, Wiley Classics Library, John Wiley & Sons, Inc., New York, 1991. MR1164865

[119] Lars Hörmander, *An introduction to complex analysis in several variables*, 3rd ed., North-Holland Mathematical Library, vol. 7, North-Holland Publishing Co., Amsterdam, 1990. MR1045639

[120] Po-Fang Hsieh and Yasutaka Sibuya, *Basic theory of ordinary differential equations*, Universitext, Springer-Verlag, New York, 1999, DOI 10.1007/978-1-4612-1506-6. MR1697415

[121] Yulij Ilyashenko and Sergei Yakovenko, *Lectures on analytic differential equations*, Graduate Studies in Mathematics, vol. 86, American Mathematical Society, Providence, RI, 2008, DOI 10.1090/gsm/086. MR2363178

[122] Geertrui K. Immink, *Asymptotics of analytic difference equations*, Lecture Notes in Mathematics, vol. 1085, Springer-Verlag, Berlin, 1984, DOI 10.1007/BFb0100174. MR765699

[123] E. L. Ince, *Ordinary Differential Equations*, Dover Publications, New York, 1944. MR10757

[124] Katsunori Iwasaki, *A modular group action on cubic surfaces and the monodromy of the Painlevé VI equation*, Proc. Japan Acad. Ser. A Math. Sci. **78** (2002), no. 7, 131–135. MR1930217

[125] Katsunori Iwasaki, Hironobu Kimura, Shun Shimomura, and Masaaki Yoshida, *From Gauss to Painlevé: A modern theory of special functions*, Aspects of Mathematics, E16, Friedr. Vieweg & Sohn, Braunschweig, 1991, DOI 10.1007/978-3-322-90163-7. MR1118604

[126] F. H. Jackson, *q-Difference Equations*, Amer. J. Math. **32** (1910), no. 4, 305–314, DOI 10.2307/2370183. MR1506108

[127] Nathan Jacobson, *Lectures in abstract algebra: Volume II: Linear algebra; Reprint of the 1953 edition [Van Nostrand, Toronto, Ont.]*, Graduate Texts in Mathematics, No. 31, Springer-Verlag, New York-Berlin, 1975. MR369381

[128] Nathan Jacobson, *Basic algebra I. 2nd ed*, New York: W. H. Freeman and Company. XVIII, 1985.

[129] Michio Jimbo and Hidetaka Sakai, *A q-analog of the sixth Painlevé equation*, Lett. Math. Phys. **38** (1996), no. 2, 145–154, DOI 10.1007/BF00398316. MR1403067

[130] Warren P. Johnson, *An introduction to q-analysis*, American Mathematical Society, Providence, RI, 2020. MR4249564

[131] Nalini Joshi, *Discrete Painlevé equations*, Published for the Conference Board of the Mathematical Sciences, CBMS Regional Conference Series in Mathematics, vol. 131, American Mathematical Society, Providence, RI, 2019, DOI 10.1090/cbms/131. MR3931704

[132] Nalini Joshi and Pieter Roffelsen, *On the Riemann-Hilbert problem for a q-difference Painlevé equation*, Comm. Math. Phys. **384** (2021), no. 1, 549–585, DOI 10.1007/s00220-021-04024-y. MR4252884

[133] Victor Kac and Pokman Cheung, *Quantum calculus*, Universitext, Springer-Verlag, New York, 2002, DOI 10.1007/978-1-4613-0071-7. MR1865777

[134] Irving Kaplansky, *An introduction to differential algebra*, 2nd ed., Publications de l'Institut de Mathématique de l'Université de Nancago, No. V, Hermann, Paris, 1976. Actualités Scientifiques et Industrielles [Current Scientific and Industrial Topics], No. 1251. MR460303

[135] Christian Kassel, *Quantum groups*, Graduate Texts in Mathematics, vol. 155, Springer-Verlag, New York, 1995, DOI 10.1007/978-1-4612-0783-2. MR1321145

[136] Nicholas M. Katz, *On the calculation of some differential Galois groups*, Invent. Math. **87** (1987), no. 1, 13–61, DOI 10.1007/BF01389152. MR862711

[137] Donald E. Knuth, *The art of computer programming. Vol. 1: Fundamental algorithms*, Third edition [of MR0286317], Addison-Wesley, Reading, MA, 1997. MR3077152

[138] Ellis Kolchin, *Selected works of Ellis Kolchin with commentary*, Commentaries by Armand Borel, Michael F. Singer, Bruno Poizat, Alexandru Buium and Phyllis J. Cassidy; Edited and with a preface by Hyman Bass, Buium and Cassidy, American Mathematical Society, Providence, RI, 1999. MR1677530

[139] Jacob Korevaar, *Tauberian theory: A century of developments*, Grundlehren der mathematischen Wissenschaften [Fundamental Principles of Mathematical Sciences], vol. 329, Springer-Verlag, Berlin, 2004, DOI 10.1007/978-3-662-10225-1. MR2073637

[140] Jean-Louis Koszul, *Fibrés vectoriels sur les courbes elliptiques* (French), Secrétariat mathématique, Paris, 1958. Séminaire Bourbaki; 10e année: 1957/1958; Textes des conférences; Exposés 152 á 168; 2e éd. corrigée, Exposé 154, 10. MR123571

[141] Gottfried Köthe, *Topological vector spaces. I*, Translated from the German by D. J. H. Garling, Die Grundlehren der mathematischen Wissenschaften, Band 159, Springer-Verlag New York, Inc., New York, 1969. MR248498

[142] Gottfried Köthe, *Topological vector spaces. II*, Grundlehren der Mathematischen Wissenschaften, vol. 237, Springer-Verlag, New York-Berlin, 1979. MR551623

[143] Marek Kuczma, *Functional equations in a single variable*, Monografie Matematyczne [Mathematical Monographs], Tom 46, Państwowe Wydawnictwo Naukowe, Warsaw, 1968. MR228862

[144] Serge Lang, *Elliptic functions*, 2nd ed., With an appendix by J. Tate, Graduate Texts in Mathematics, vol. 112, Springer-Verlag, New York, 1987, DOI 10.1007/978-1-4612-4752-4. MR890960

[145] Serge Lang, *Algebra*, 3rd ed., Graduate Texts in Mathematics, vol. 211, Springer-Verlag, New York, 2002, DOI 10.1007/978-1-4613-0041-0. MR1878556

[146] A. H. M. Levelt, *Differential Galois theory and tensor products*, Indag. Math. (N.S.) **1** (1990), no. 4, 439–449, DOI 10.1016/0019-3577(90)90012-C. MR1106091

[147] Alexander Levin, *Difference algebra*, Algebra and Applications, vol. 8, Springer, New York, 2008, DOI 10.1007/978-1-4020-6947-5. MR2428839

[148] M. Loday-Richaud, *Stokes cocycle and differential Galois groups*, J. Math. Sci., New York **124** (2003), no. 5, 5262–5274 (English).

[149] Michèle Loday-Richaud, *Stokes phenomenon, multisummability and differential Galois groups* (English, with English and French summaries), Ann. Inst. Fourier (Grenoble) **44** (1994), no. 3, 849–906, DOI 10.5802/aif.1419. MR1303886

[150] M. Loday-Richaud, *Divergent series, summability and resurgence II. Simple and multiple summability*, Lect. Notes Math., vol. 2154, Cham: Springer, 2016 (English).

[151] M. Loday-Richaud and G. Pourcin, *On index theorems for linear ordinary differential operators* (English, with English and French summaries), Ann. Inst. Fourier (Grenoble) **47** (1997), no. 5, 1379–1424, DOI 10.5802/aif.1604. MR1600379

[152] D. S. Lubinsky, *On q-exponential functions for $|q| = 1$*, Canad. Math. Bull. **41** (1998), no. 1, 86–97, DOI 10.4153/CMB-1998-014-x. MR1618963

[153] Saunders Mac Lane, *Homology*, Reprint of the 1975 edition, Classics in Mathematics, Springer-Verlag, Berlin, 1995. MR1344215

[154] Saunders Mac Lane, *Categories for the working mathematician*, 2nd ed., Graduate Texts in Mathematics, vol. 5, Springer-Verlag, New York, 1998. MR1712872

[155] Wilhelm Magnus, *Rings of Fricke characters and automorphism groups of free groups*, Math. Z. **170** (1980), no. 1, 91–103, DOI 10.1007/BF01214715. MR558891

[156] Wilhelm Magnus, Fritz Oberhettinger, and Raj Pal Soni, *Formulas and theorems for the special functions of mathematical physics*, Third enlarged edition, Die Grundlehren der mathematischen Wissenschaften, Band 52, Springer-Verlag New York, Inc., New York, 1966. MR232968

[157] B. Malgrange, *On nonlinear differential Galois theory*, Dedicated to the memory of Jacques-Louis Lions, Chinese Ann. Math. Ser. B **23** (2002), no. 2, 219–226, DOI 10.1142/S0252959902000213. MR1924138

[158] Bernard Malgrange, *Sur les points singuliers des équations différentielles* (French), Enseign. Math. (2) **20** (1974), 147–176. MR368074

[159] Bernard Malgrange, *Sur la réduction formelle des équations différentielles à singularités irrégulières.*, (1979).

[160] Bernard Malgrange, *The Galois groupoid of a foliation*, Essays on geometry and related topics. Mémoires dédiés à André Haefliger. Vol. 2, Genève: L'Enseignement Mathématique, 2001, pp. 465–501 (French).

[161] Bernard Malgrange, *An analogue of the Galois correspondence for foliations*, The mathematical legacy of Leon Ehrenpreis, Berlin: Springer, 2012, pp. 223–232 (English).

[162] Yuri I. Manin, *Theta functions, quantum tori and Heisenberg groups*, Lett. Math. Phys. **56** (2001), no. 3, 295–320, DOI 10.1023/A:1017909525397. EuroConférence Moshé Flato 2000, Part III (Dijon). MR1855265

[163] F. Marotte and C. Zhang, *Multisommabilité des séries entières solutions formelles d'une équation aux q-différences linéaire analytique* (French), Ann. Inst. Fourier (Grenoble) **50** (2000), no. 6, 1859–1890 (2001), DOI 10.5802/aif.1809. MR1817385

[164] Jean Martinet and Jean-Pierre Ramis, *Elementary acceleration and multisummability. I*, Ann. Inst. H. Poincaré Phys. Théor. **54** (1991), no. 4, 331–401. MR1128863

[165] William S. Massey, *A basic course in algebraic topology*, Graduate Texts in Mathematics, vol. 127, Springer-Verlag, New York, 1991. MR1095046

[166] Akira Masuoka, Katsunori Saito, and Hiroshi Umemura, *Toward quantization of Galois theory* (English, with English and French summaries), Ann. Fac. Sci. Toulouse Math. (6) **29** (2020), no. 5, 1319–1431, DOI 10.5802/afst.1663. MR4245863

[167] Henry McKean and Victor Moll, *Elliptic curves: Function theory, geometry, arithmetic*, Cambridge University Press, Cambridge, 1997, DOI 10.1017/CBO9781139174879. MR1471703

[168] Rick Miranda, *Algebraic curves and Riemann surfaces*, Graduate Studies in Mathematics, vol. 5, American Mathematical Society, Providence, RI, 1995, DOI 10.1090/gsm/005. MR1326604

[169] Claude Mitschi and David Sauzin, *Divergent series, summability and resurgence. I: Monodromy and resurgence*, With a foreword by Jean-Pierre Ramis and a preface by Éric Delabaere, Michèle Loday-Richaud, Claude Mitschi and David Sauzin, Lecture Notes in Mathematics, vol. 2153, Springer, [Cham], 2016, DOI 10.1007/978-3-319-28736-2. MR3526111

[170] Rached Mneimné and Frédéric Testard, *Introduction à la théorie des groupes de Lie classiques* (French), Collection Méthodes. [Methods Collection], Hermann, Paris, 1986. MR903847

[171] Shuji Morikawa and Hiroshi Umemura, *On a general difference Galois theory. II* (English, with English and French summaries), Ann. Inst. Fourier (Grenoble) **59** (2009), no. 7, 2733–2771, DOI 10.5802/aif.2506. MR2649332

[172] David Mumford, *Curves and their Jacobians*, University of Michigan Press, Ann Arbor, MI, 1975. MR419430

[173] David Mumford, *Algebraic Geometry. I: Complex projective varieties.*, reprint of the corr. 2nd print. 1976 ed., Class. Math., Berlin: Springer-Verlag, 1995.

[174] David Mumford, *Tata lectures on theta. I*, With the collaboration of C. Musili, M. Nori, E. Previato and M. Stillman; Reprint of the 1983 edition, Modern Birkhäuser Classics, Birkhäuser Boston, Inc., Boston, MA, 2007, DOI 10.1007/978-0-8176-4578-6. MR2352717

[175] Kumiko Nishioka and Seiji Nishioka, *Algebraic theory of difference equations and Mahler functions*, Aequationes Math. **84** (2012), no. 3, 245–259, DOI 10.1007/s00010-012-0132-3. MR2996417

[176] Yousuke Ohyama, Jean-Pierre Ramis, and Jacques Sauloy, *The space of monodromy data for the Jimbo-Sakai family of q-difference equations* (English, with English and French summaries), Ann. Fac. Sci. Toulouse Math. (6) **29** (2020), no. 5, 1119–1250, DOI 10.5802/afst.1659. MR4245859

[177] M. A. Olshanetsky and V.-B. K. Rogov, *Liouville quantum mechanics on a lattice from geometry of quantum Lorentz group*, J. Phys. A **27** (1994), no. 13, 4669–4683. MR1294969

[178] Ken Ono, *The web of modularity: arithmetic of the coefficients of modular forms and q-series*, CBMS Regional Conference Series in Mathematics, vol. 102, Published for the Conference Board of the Mathematical Sciences, Washington, DC; by the American Mathematical Society, Providence, RI, 2004. MR2020489

[179] Oystein Ore, *Theory of non-commutative polynomials*, Ann. of Math. (2) **34** (1933), no. 3, 480–508, DOI 10.2307/1968173. MR1503119

[180] Alexey Ovchinnikov and Michael Wibmer, *σ-Galois theory of linear difference equations*, Int. Math. Res. Not. IMRN **12** (2015), 3962–4018, DOI 10.1093/imrn/rnu060. MR3356746

[181] Frédéric Pham, *Géométrie et calcul différentiel sur les variétés* (French), InterEditions, Paris, 1992. Cours, études et exercices pour la maîtrise de mathématiques. [Course, studies and exercises for the Masters in mathematics]. MR1195791

[182] C. Praagman, *The formal classification of linear difference operators*, Nederl. Akad. Wetensch. Indag. Math. **45** (1983), no. 2, 249–261. MR705431

[183] Hans Rademacher, *Topics in analytic number theory*, Edited by E. Grosswald, J. Lehner and M. Newman, Die Grundlehren der mathematischen Wissenschaften, Band 169, Springer-Verlag, New York-Heidelberg, 1973. MR364103

[184] A. Ramani, B. Grammaticos, and V. Papageorgiou, *Singularity confinement*, Symmetries and integrability of difference equations (Estérel, PQ, 1994), CRM Proc. Lecture Notes, vol. 9, Amer. Math. Soc., Providence, RI, 1996, pp. 303–318, DOI 10.1090/crmp/009/28. MR1416848

[185] Srinivasa Ramanujan, *Notebooks. Vols. 1, 2*, Tata Institute of Fundamental Research, Bombay, 1957. MR99904

[186] Srinivasa Ramanujan, *The Lost Notebook and other unpublished papers. With an introduction by George E. Andrews*, New Delhi: Narosa Publishing House; Berlin (FRG): Springer-Verlag, 1988 (English).

[187] Srinivasa Ramanujan, *Collected papers of Srinivasa Ramanujan*, Edited by G. H. Hardy, P. V. Seshu Aiyar and B. M. Wilson; Third printing of the 1927 original; With a new preface and commentary by Bruce C. Berndt, AMS Chelsea Publishing, Providence, RI, 2000. MR2280843

[188] Jean-Pierre Ramis, *Théorèmes d'indices Gevrey pour les équations différentielles ordinaires* (French, with English summary), Mem. Amer. Math. Soc. **48** (1984), no. 296, viii+95, DOI 10.1090/memo/0296. MR733946

[189] Jean-Pierre Ramis, *Phénomène de Stokes et filtration Gevrey sur le groupe de Picard-Vessiot* (French, with English summary), C. R. Acad. Sci. Paris Sér. I Math. **301** (1985), no. 5, 165–167. MR801953

[190] Jean-Pierre Ramis, *Fonctions Theta et équations aux q-différences.*, UNPUBLISHED, see `http://www.cantoperdic.fr/LivresRamis/RamisTheta.zip`, 1990.

[191] Jean-Pierre Ramis, *About the growth of entire functions solutions of linear algebraic q-difference equations* (English, with English and French summaries), Ann. Fac. Sci. Toulouse Math. (6) **1** (1992), no. 1, 53–94. MR1191729

[192] Jean-Pierre Ramis, *Séries divergentes et théories asymptotiques* (French), Bull. Soc. Math. France **121** (1993), no. Panoramas et Synthèses, suppl., 74. MR1272100

[193] Jean-Pierre Ramis and Jose Cano, *Théorie de Galois différentielle, multisommabilité et phénomènes de Stokes.*, UNPUBLISHED, see `http://www.math.univ-toulouse.fr/~ramis/Cano-Ramis-Galois`, 2009.

[194] J.-P. Ramis and J. Martinet, *Théorie de Galois différentielle et resommation* (French), Computer algebra and differential equations, Comput. Math. Appl., Academic Press, London, 1990, pp. 117–214. MR1038060

[195] Jean-Pierre Ramis and Jacques Sauloy, *The q-analogue of the wild fundamental group and the inverse problem of the Galois theory of q-difference equations* (English, with English and French summaries), Ann. Sci. Éc. Norm. Supér. (4) **48** (2015), no. 1, 171–226, DOI 10.24033/asens.2241. MR3335841

[196] Jean-Pierre Ramis and Jacques Sauloy, *Geometry of the space of monodromy data.*, (2024).

[197] Jean-Pierre Ramis, Jacques Sauloy, and Changgui Zhang, *Local analytic classification of q-difference equations* (English, with French summary), Astérisque **355** (2013), vi+151. MR3185985

[198] Jean-Pierre Ramis and Changgui Zhang, *Développement asymptotique q-Gevrey et fonction thêta de Jacobi* (French, with English and French summaries), C. R. Math. Acad. Sci. Paris **335** (2002), no. 11, 899–902, DOI 10.1016/S1631-073X(02)02586-4. MR1952546

[199] Michael Reed and Barry Simon, *Methods of modern mathematical physics. I. Functional analysis*, Academic Press, New York-London, 1972. MR493419

[200] Éric Reyssat, *Quelques aspects des surfaces de Riemann* (French), Progress in Mathematics, vol. 77, Birkhäuser Boston, Inc., Boston, MA, 1989. MR1034955

[201] Robert C. Rhoades, *On Ramanujan's definition of mock theta function*, Proc. Natl. Acad. Sci. USA **110** (2013), no. 19, 7592–7594, DOI 10.1073/pnas.1301046110. MR3067051

[202] Robert C. Rhoades, *A unified approach to partial and mock theta functions*, Math. Res. Lett. **25** (2018), no. 2, 659–675, DOI 10.4310/MRL.2018.v25.n2.a16. MR3826840

[203] Bernhard Riemann, *The collected works of Bernhard Riemann. The complete German texts. With an English introduction by Hans Lewy*, reprint of the 1953 Dover republication of the 2nd edition of Bernhard Riemann's 'Gesammelte Mathematische Werke' published in 1892 and the Supplement (Nachträge) published 1902 ed., Mineola, NY: Dover Publications, 2017 (German; English).

[204] P. Robba, *Lemmes de Hensel pour les opérateurs différentiels. Application à la réduction formelle des équations différentielles* (French), Enseign. Math. (2) **26** (1980), no. 3-4, 279–311 (1981). MR610528

[205] Helmut Röhrl, *Holomorphic fiber bundles over Riemann surfaces*, Bull. Amer. Math. Soc. **68** (1962), 125–160, DOI 10.1090/S0002-9904-1962-10715-0. MR146854

[206] Julien Roques, *Classification rationnelle et confluence des systèmes aux différences singuliers réguliers* (French, with English and French summaries), Ann. Inst. Fourier (Grenoble) **56** (2006), no. 6, 1663–1699, DOI 10.5802/aif.2224. MR2282672

[207] Julien Roques, *Generalized basic hypergeometric equations*, Invent. Math. **184** (2011), no. 3, 499–528, DOI 10.1007/s00222-010-0294-3. MR2800693

[208] Julien Roques, *Birkhoff matrices, residues and rigidity for q-difference equations*, J. Reine Angew. Math. **706** (2015), 215–244, DOI 10.1515/crelle-2013-0065. MR3393368

[209] Julien Roques and Jacques Sauloy, *Euler characteristics and q-difference equations*, Ann. Sc. Norm. Super. Pisa Cl. Sci. (5) **19** (2019), no. 1, 129–154. MR3923842

[210] Peter Roquette, *Analytic theory of elliptic functions over local fields*, Hamburger Mathematische Einzelschriften (N.F.), Heft 1, Vandenhoeck & Ruprecht, Göttingen, 1970. MR260753

[211] Walter Rudin, *Real and complex analysis*, 3rd ed., McGraw-Hill Book Co., New York, 1987. MR924157

[212] Walter Rudin, *Functional analysis*, 2nd ed., International Series in Pure and Applied Mathematics, McGraw-Hill, Inc., New York, 1991. MR1157815

[213] Neantro Saavedra Rivano, *Catégories Tannakiennes* (French), Lecture Notes in Mathematics, Vol. 265, Springer-Verlag, Berlin-New York, 1972. MR338002

[214] Claude Sabbah, *Systèmes holonomes d'équations aux q-différences* (French, with French summary), D-modules and microlocal geometry (Lisbon, 1990), de Gruyter, Berlin, 1993, pp. 125–147. MR1206016

[215] Hidetaka Sakai and Masashi Yamaguchi, *Spectral types of linear q-difference equations and q-analog of middle convolution*, Int. Math. Res. Not. IMRN **7** (2017), 1975–2013, DOI 10.1093/imrn/rnw089. MR3658190

[216] Jacques Sauloy, *Équations aux q-différences et fibrés vectoriels holomorphes sur la courbe elliptique* $\mathbf{C}^*/q^{\mathbf{Z}}$ (French, with English and French summaries), Astérisque **323** (2009), 397–429. MR2647980

[217] Jacques Sauloy, *Systèmes aux q-différences singuliers réguliers: classification, matrice de connexion et monodromie* (French, with English and French summaries), Ann. Inst. Fourier (Grenoble) **50** (2000), no. 4, 1021–1071, DOI 10.5802/aif.1784. MR1799737

[218] Jacques Sauloy, *Galois theory of Fuchsian q-difference equations* (English, with English and French summaries), Ann. Sci. École Norm. Sup. (4) **36** (2003), no. 6, 925–968 (2004), DOI 10.1016/j.ansens.2002.10.001. MR2032530

[219] Jacques Sauloy, *Algebraic construction of the Stokes sheaf for irregular linear q-difference equations*, Analyse complexe, systèmes dynamiques, sommabilité des séries divergentes et théories galoisiennes. I., Astérisque **296** (2004), 227–251,

[220] Jacques Sauloy, *La filtration canonique par les pentes d'un module aux q-différences et le gradué associé* (French, with English and French summaries), Ann. Inst. Fourier (Grenoble) **54** (2004), no. 1, 181–210, DOI 10.5802/aif.2015. MR2069126

[221] Jacques Sauloy, *équations aux q-différences linéaires: factorisation, résolution et théorèmes d'indices*, Revista del Seminario Iberoamericano de Matematicas (2010).

[222] Jacques Sauloy, *Differential Galois theory through Riemann-Hilbert correspondence: An elementary introduction*, With a foreword by Jean-Pierre Ramis, Graduate Studies in Mathematics, vol. 177, American Mathematical Society, Providence, RI, 2016, DOI 10.1090/gsm/177. MR3585800

[223] Jacques Sauloy, "Classification analytique locale des équations aux q-différences de pentes arbitraires", **33** (2024), fascicule n° 2, *Annales de la Faculté de Sciences de Toulouse*, DOI 10.5802/afst.

[224] Jacques Sauloy, *Théorie analytique locale des équations aux q-différences de pentes arbitraires.*, UNPUBLISHED, see `https://arxiv.org/abs/2006.03237`, 2023.

[225] Jacques Sauloy and Changgui Zhang, *On the vanishing of coefficients of the powers of a theta function*, Ramanujan J. **57** (2022), no. 3, 1125–1167, DOI 10.1007/s11139-020-00366-8. MR4382183

[226] Helmut H. Schaefer, *Topological vector spaces*, Graduate Texts in Mathematics, Vol. 3, Springer-Verlag, New York-Berlin, 1971. Third printing corrected. MR342978

[227] Martin Schechter, *Principles of functional analysis*, 2nd ed., Graduate Studies in Mathematics, vol. 36, American Mathematical Society, Providence, RI, 2002, DOI 10.1090/gsm/036. MR1861991

[228] Carsten Schneider, *Symbolic summation in difference rings and applications*, Proceedings of the 2016 ACM International Symposium on Symbolic and Algebraic Computation, ACM, New York, 2016, pp. 9–12, DOI 10.1145/2930889.2930945. MR3565689

[229] Jean-Pierre Serre, *Faisceaux algébriques cohérents* (French), Ann. of Math. (2) **61** (1955), 197–278, DOI 10.2307/1969915. MR68874

[230] Jean-Pierre Serre, *Algebraic geometry and analytic geometry*, Ann. Inst. Fourier **6** (1956), 1–42 (French).

[231] Jean-Pierre Serre, *A course in arithmetic*, Translated from the French, Graduate Texts in Mathematics, No. 7, Springer-Verlag, New York-Heidelberg, 1973. MR344216

[232] Jean-Pierre Serre, *Lie algebras and Lie groups*, 2nd ed., 1964 lectures given at Harvard University, Lecture Notes in Mathematics, vol. 1500, Springer-Verlag, Berlin, 1992. MR1176100

[233] C. S. Seshadri, *Fibrés vectoriels sur les courbes algébriques* (French), Notes written by J.-M. Drezet from a course at the École Normale Supérieure, June 1980, Astérisque, vol. 96, Société Mathématique de France, Paris, 1982. MR699278

[234] M. A. Shubin, *Pseudodifferential operators and spectral theory*, 2nd ed., Translated from the 1978 Russian original by Stig I. Andersson, Springer-Verlag, Berlin, 2001, DOI 10.1007/978-3-642-56579-3. MR1852334

[235] L. J. Slater, *Confluent hypergeometric functions*, Cambridge University Press, New York, 1960. MR107026

[236] L. J. Slater, *Generalized hypergeometric functions*, Cambridge: At the University Press. XIII, 1966.

[237] Yan Soibelman, *Quantum tori, mirror symmetry and deformation theory*, Lett. Math. Phys. **56** (2001), no. 2, 99–125, DOI 10.1023/A:1010984821575. EuroConférence Moshé Flato 2000, Part II (Dijon). MR1854130

[238] Edwin H. Spanier, *Algebraic topology*, Springer-Verlag, New York, [1995?]. Corrected reprint of the 1966 original. MR1325242

[239] T. A. Springer, *Linear algebraic groups*, 2nd ed., Modern Birkhäuser Classics, Birkhäuser Boston, Inc., Boston, MA, 2009. MR2458469

[240] G. W. Starcher, *A solution of a simple functional equation as a basis for readily obtaining certain fundamental formulas in the theory of elliptic functions*, Bull. Amer. Math. Soc. **36** (1930), no. 8, 577–581, DOI 10.1090/S0002-9904-1930-05003-X. MR1561997

[241] Henning Stichtenoth, *Algebraic function fields and codes*, 2nd ed., Graduate Texts in Mathematics, vol. 254, Springer-Verlag, Berlin, 2009. MR2464941

[242] G. G. Stokes, *On the discontinuity of arbitrary constants that appear as multipliers of semi-convergent series*, A letter to the editor, Acta Math. **26** (1902), no. 1, 393–397, DOI 10.1007/BF02415503. MR1554970

[243] V. Tarasov and A. Varchenko, *Geometry of q-hypergeometric functions, quantum affine algebras and elliptic quantum groups* (English, with English and French summaries), Astérisque **246** (1997), vi+135. MR1646561

[244] E. C. Titchmarsh, *The theory of functions*, 2nd ed., Oxford University Press, Oxford, 1939. MR3728294

[245] Hiroshi Umemura, *Invitation to Galois theory*, Differential equations and quantum groups, IRMA Lect. Math. Theor. Phys., vol. 9, Eur. Math. Soc., Zürich, 2007, pp. 269–289. MR2322334

[246] Hiroshi Umemura, *Sur l'équivalence des théories de Galois différentielles générales* (French, with English and French summaries), C. R. Math. Acad. Sci. Paris **346** (2008), no. 21-22, 1155–1158, DOI 10.1016/j.crma.2008.09.025. MR2464256

[247] Marius van der Put, *Symbolic analysis of differential equations*, Some tapas of computer algebra, Algorithms Comput. Math., vol. 4, Springer, Berlin, 1999, pp. 208–236, DOI 10.1007/978-3-662-03891-8_9. MR1679926

[248] Marius van der Put and Marc Reversat, *Galois theory of q-difference equations* (English, with English and French summaries), Ann. Fac. Sci. Toulouse Math. (6) **16** (2007), no. 3, 665–718. MR2379057

[249] Marius van der Put and Michael F. Singer, *Galois theory of difference equations*, Lecture Notes in Mathematics, vol. 1666, Springer-Verlag, Berlin, 1997, DOI 10.1007/BFb0096118. MR1480919

[250] Marius van der Put and Michael F. Singer, *Galois theory of linear differential equations*, Grundlehren der mathematischen Wissenschaften [Fundamental Principles of Mathematical Sciences], vol. 328, Springer-Verlag, Berlin, 2003, DOI 10.1007/978-3-642-55750-7. MR1960772

[251] H. S. Wall, *Analytic Theory of Continued Fractions*, D. Van Nostrand Co., Inc., New York, 1948. MR25596

[252] Wolfgang Wasow, *Asymptotic expansions for ordinary differential equations*, Reprint of the 1965 edition, Robert E. Krieger Publishing Co., Huntington, NY, 1976. MR460820

[253] G. N. Watson, *The Final Problem : An Account of the Mock Theta Functions*, J. London Math. Soc. **11** (1936), no. 1, 55–80, DOI 10.1112/jlms/s1-11.1.55. MR1573993

[254] A. Weil, *Généralisation des fonctions abéliennes.*, J. Math. Pures Appl. (9) **17** (1938), 47–87 (French).

[255] E. T. Whittaker and G. N. Watson, *A course of modern analysis*, Cambridge Mathematical Library, Cambridge University Press, Cambridge, 1996. An introduction to the general theory of infinite processes and of analytic functions; with an account of the principal transcendental functions; Reprint of the fourth (1927) edition, DOI 10.1017/CBO9780511608759. MR1424469

[256] Michael Wibmer, *Geometric Difference Galois Theory*, ProQuest LLC, Ann Arbor, MI, 2010. Thesis (Ph.D.)–Ruprecht-Karls-Universitaet Heidelberg (Germany). MR3828665

[257] Herbert S. Wilf, *generatingfunctionology*, 3rd ed., A K Peters, Ltd., Wellesley, MA, 2006. MR2172781

[258] Herbert S. Wilf, *Mathematics for the physical sciences*, reprint of the 1962 original ed., Mineola, NY: Dover Publications, 2006 (English).

[259] Kōsaku Yosida, *Functional analysis*, Classics in Mathematics, Springer-Verlag, Berlin, 1995. Reprint of the sixth (1980) edition, DOI 10.1007/978-3-642-61859-8. MR1336382

[260] D. Zagier, *On an approximate identity of Ramanujan*, Proc. Indian Acad. Sci. Math. Sci. **97** (1987), no. 1-3, 313–324 (1988), DOI 10.1007/BF02837833. MR983624

[261] Don Zagier, *The dilogarithm function*, Frontiers in number theory, physics, and geometry II. On conformal field theories, discrete groups and renormalization. Papers from the meeting, Les Houches, France, March 9–21, 2003, Berlin: Springer, 2007, 3–65 (English).

[262] Don Zagier, *Ramanujan's mock theta functions and their applications (after Zwegers and Ono-Bringmann)*, Séminaire Bourbaki. Volume 2007/2008. Exposés 982–996, Paris: Société Mathématique de France (SMF), 2009, 143–164, ex (English).

[263] Changgui Zhang, *Développements asymptotiques q-Gevrey et séries Gq-sommables* (French, with English and French summaries), Ann. Inst. Fourier (Grenoble) **49** (1999), no. 1, vi–vii, x, 227–261, DOI 10.5802/aif.1672. MR1688144

[264] Changgui Zhang, *Sur la fonction q-gamma de Jackson* (French, with English summary), Aequationes Math. **62** (2001), no. 1-2, 60–78, DOI 10.1007/PL00000144. MR1849140

[265] Changgui Zhang, *A discrete summation for linear q-difference equations with analytic coefficients: General theory and examples*, Differential equations and the Stokes phenomenon. Proceedings of the conference, Groningen, Netherlands, May 28–30, 2001, Singapore: World Scientific, 2002, 309–329 (French).

[266] Changgui Zhang, *Asymptotic and meromorphic solutions of q-difference equations*, Théories asymptotiques et équations de Painlevé, Paris: Société Mathématique de France, 2006, 341–356 (French).

[267] Changgui Zhang, *On the mock-theta behavior of Appell-Lerch series* (English, with English and French summaries), C. R. Math. Acad. Sci. Paris **353** (2015), no. 12, 1067–1073, DOI 10.1016/j.crma.2015.09.028. MR3427910

[268] Changgui Zhang, *Mock theta-functions of the third order of Ramanujan in terms of Appell-Lerch series*, Frontiers in orthogonal polynomials and q-series. Papers based on the international conference, University of Central Florida, Orlando, FL, USA, May 10–12, 2015. Dedicated to Professor Mourad Ismail on his 70th birthday, Hackensack, NJ: World Scientific, 2018, 523–544 (English).

[269] Changgui Zhang, *Analytical study of the pantograph equation using Jacobi theta functions*, J. Approx. Theory **296** (2023), 21 (English), Id/No 105974.

Index of terms

Index of notations

Index of names

Selected Published Titles in This Series

287 **Jacques Sauloy,** Basic Modern Theory of Linear Complex Analytic q-Difference Equations, 2024

282 **Michael Kapovich and Pranab Sardar,** Trees of Hyperbolic Spaces, 2024

281 **Horst R. Thieme,** Discrete-Time Dynamics of Structured Populations and Homogeneous Order-Preserving Operators, 2024

280 **Tadashi Ochiai,** Iwasawa Theory and Its Perspective, Volume 2, 2024

279 **Ellen Elizabeth Eischen, Wee Teck Gan, Aaron Pollack, and Zhiwei Yun,** Automorphic Forms Beyond GL_2, 2024

278 **Delaram Kahrobaei, Ramón Flores, Marialaura Noce, Maggie E. Habeeb, and Christopher Battarbee,** Applications of Group Theory in Cryptography, 2024

277 **George Metcalfe, Francesco Paoli, and Constantine Tsinakis,** Residuated Structures in Algebra and Logic, 2023

276 **Balázs Bárány, Károly Simon, and Boris Solomyak,** Self-similar and Self-affine Sets and Measures, 2023

275 **Alekos Vidras and Alain Yger,** Multidimensional Residue Theory and Applications, 2023

274 **Willi Freeden and M. Zuhair Nashed,** Recovery Methodologies: Regularization and Sampling, 2023

273 **Gennadiy Feldman,** Characterization of Probability Distributions on Locally Compact Abelian Groups, 2023

272 **Tadashi Ochiai,** Iwasawa Theory and Its Perspective, Volume 1, 2023

271 **Hideto Asashiba,** Categories and Representation Theory, 2022

270 **Leslie Hogben, Jephian C.-H. Lin, and Bryan L. Shader,** Inverse Problems and Zero Forcing for Graphs, 2022

269 **Ralph S. Freese, Ralph N. McKenzie, George F. McNulty, and Walter F. Taylor,** Algebras, Lattices, Varieties, 2022

268 **Ralph S. Freese, Ralph N. McKenzie, George F. McNulty, and Walter F. Taylor,** Algebras, Lattices, Varieties, 2022

267 **Dragana S. Cvetković Ilić,** Completion Problems on Operator Matrices, 2022

266 **Kate Juschenko,** Amenability of Discrete Groups by Examples, 2022

265 **Nabil H. Mustafa,** Sampling in Combinatorial and Geometric Set Systems, 2022

264 **J. Scott Carter and Seiichi Kamada,** Diagrammatic Algebra, 2021

263 **Vugar E. Ismailov,** Ridge Functions and Applications in Neural Networks, 2021

262 **Ragnar-Olaf Buchweitz,** Maximal Cohen–Macaulay Modules and Tate Cohomology, 2021

261 **Shiri Artstein-Avidan, Apostolos Giannopoulos, and Vitali D. Milman,** Asymptotic Geometric Analysis, Part II, 2021

260 **Lindsay N. Childs, Cornelius Greither, Kevin P. Keating, Alan Koch, Timothy Kohl, Paul J. Truman, and Robert G. Underwood,** Hopf Algebras and Galois Module Theory, 2021

259 **William Heinzer, Christel Rotthaus, and Sylvia Wiegand,** Integral Domains Inside Noetherian Power Series Rings, 2021

258 **Pramod N. Achar,** Perverse Sheaves and Applications to Representation Theory, 2021

257 **Juha Kinnunen, Juha Lehrbäck, and Antti Vähäkangas,** Maximal Function Methods for Sobolev Spaces, 2021

256 **Michio Jimbo, Tetsuji Miwa, and Fedor Smirnov,** Local Operators in Integrable Models I, 2021

For a complete list of titles in this series, visit the
AMS Bookstore at **www.ams.org/bookstore/survseries/**.